World Soybean Research Conference III:

Proceedings

edited by
Richard Shibles

Routledge
Taylor & Francis Group

NEW YORK AND LONDON

T0141176

First published 1985 by Westview Press, Inc.

Published 2021 by Routledge
605 Third Avenue, New York, NY 10017
2 Park Square, Milton Park, Abingdon, Oxon OX14 4RN

Routledge is an imprint of the Taylor & Francis Group, an informa business

Copyright © 1985 by Taylor & Francis

Library of Congress Catalog Card Number: 85-51426
ISBN: 0-8133-0091-6

ISBN13: 978-0-3672-1385-5 (hbk)
ISBN13: 978-0-3672-1666-5 (pbk)

World Soybean Research Conference III: Proceedings

ABOUT THE BOOK AND EDITOR

WORLD SOYBEAN RESEARCH CONFERENCE III: PROCEEDINGS

edited by Richard Shibles

This volume consists of full length manuscripts of 159 of the 165 invited papers presented at World Soybean Research Conference III that was held in the Scheman Continuing Education Building at Iowa State University August 12-17, 1984. The authors, widely recognized as world authorities in their fields, represent all aspects of soybean research activity: breeding and genetics, crop and soil management, economics, entomology, food science, international programs, nematology, pathology, physiology, plant nutrition, rhizobiology, utilization, and weed science. This proceedings, which contains more than 1200 pages of information including many tables and figures, represents the most extensive compilation of soybean research results since the previous proceedings were published in 1980. It should be of value to research scientists, students and administrators alike.

Richard Shibles is a professor and crop physiologist in the Department of Agronomy at Iowa State University, Ames.

CONTENTS

ECONOMICS

Plenary Paper

Supply and Demand Prospects

Government Policy Impacts

Microeconomics of Production and Marketing

UTILIZATION

BREEDING AND GENETICS

PATHOLOGY

Diseases of Seeds

Nematology

Fungus Diseases

ENTOMOLOGY

PHYSIOLOGY

RHIZOBIUM

SOIL AND CROP MANAGEMENT

Soil-Crop Interactions

WEED SCIENCE

DIMENSIONS OF WORLD PRODUCTION, UTILIZATION, AND RESEARCH

Plenary Paper

World Soil Erosion: Problems and Solutions

Seed Programs in the Tropics

Regional Production and Utilization

International Programs and Networks

FOREWORD

The soybean is a major component of the diet of millions of people throughout the world. The ability of the international community of soybean specialists to enhance the supply and usefulness of this important crop will be a vital factor in the future health and welfare of our global society. Our effectiveness as scientists in meeting this challenge will be enhanced by opportunities to exchange the results of our research and to discuss new possibilities for investigation. World Soybean Research Conference III (WSRC III) provided the opportunity for such communication.

WSRC III was the culmination of three years of planning by numerous individuals. Iowa State University was asked to serve as host of the conference by Robert Judd, a member of the Executive Committee of WSRC II and a long-time leader in the support of soybean research in the United States. Thanks to the dedicated effort of persons at Iowa State University and throughout the world, an extensive program was developed. Generous donations from public and private institutions made it possible to ensure the participation of international scientists and to minimize the cost of the conference for everyone.

The 1,050 persons from 66 countries who attended WSRC III approved the orderly continuation of future conferences by adoption of a formal constitution. Because the constitution represents an historic document, it has been included in its entirety as a part of this proceedings. A major provision of the constitution is the establishment of a Continuing Committee that will select the host country for WSRC IV to be held in 1989.

Recognition was given at WSRC III to three persons who have played key roles in the initiation and continuation of the conferences. Robert Howell was thanked for providing the leadership that led to the First World Soybean Research Conference in 1975, which was held at the University of Illinois, Champaign-Urbana. Billy Caldwell was recognized for serving as the chair of WSRC II, held in 1979 at North Carolina State University, Raleigh. Robert Judd was honored for his major contribution in providing for the continuation of the conferences.

This proceedings was made possible by the dedicated work of Richard M. Shibles, who has given so generously of his time to edit for publication the many invited papers that were presented at WSRC III. We are indebted to him for this exceptional service to the international community of soybean scientists.

Walter R. Fehr
Conference Chair

PREFACE

The cultivated soybean, *Glycine max* (L.) Merr., is one of the most extensively researched plant species extant, principally because of its economic significance as a food and feed crop, but also because it is widely used as a model system in basic research. Because of international interest in the soybean and its products—soybeans, soyoil, and soymeal—the world soybean research conferences were conceived and held at 5-year intervals as an opportunity for scientists from many disciplines to exchange information and ideas. *World Soybean Research Conference-III: Proceedings*, contains 159 of the 165 invited papers presented at the third conference.

The organizers of WSRC-III designed it as a series of mini-symposia focused upon specific research topics within the several disciplines. This proceedings keeps rather faithfully to the conference organization. It consists of ten parts (disciplines), each partitioned into several sections representing the various symposia. With few exceptions SI units have been used throughout. A few manuscripts dealing with U.S. domestic policy and grain trade employ imperial units. And, both the authors and I opted for the kilogram or metric ton (t), instead of the megagram (Mg), for the unit of mass.

I thank Walter Fehr, conference chairman, for his support and guidance throughout this project. His leadership and encouragement were invaluable. I am deeply grateful to Carolyn Taylor, who typed the manuscript with an IBM Selectric Composer typewriter. Her task was an arduous and frustrating one, because of the numerous fonts, type sizes and symbols the composer allows. Nevertheless, she produced virtually flawless copy. I am indebted to Elizabeth Todd and Ida Morgan for careful proofing of the manuscript, and to Karen Osborn for handling all correspondence with the authors.

Lastly, I thank my graduate students and my family for their patience and understanding during this long, time-consuming task.

1 June 1985　　　　　　　　　　　　　　　　　　　　　　　Richard Shibles
Ames, Iowa　　　　　　　　　　　　　　　　　　　　　　　Editor

ECONOMICS

Plenary Paper

WORLD SUPPLY, DEMAND AND COMPETITION FOR SOYBEANS AND PRODUCTS IN THE EIGHTIES AND NINETIES

S. Mielke

Whatever forecasting is—a science, an art or a sort of handicraft—it is as necessary as it is challenging. Necessary, because people in general and investors in particular have to make decisions, the results of which will be seen only in the future. Challenging, because it represents one of the most difficult tasks which, just because of that, stimulates both interest and effort.

All this applies to both the producers and the consumers of forecasts. Of course, the boundaries between the two groups that are fluent: a producer of forecasts is one of his own best consumers. On the other hand, there is hardly anyone who consumes a forecast as it is. He will rather try to gather various opinions about the future and then produce his own forecasts. After all, he alone has to bear the risks involved. It is in this sense that I would like to present to you my findings about the future of soybeans.

One thing is easy to forecast: the soybean economy will continue to be embedded in the political, economic and technological conditions of this world in general and in the markets of the competing oilseeds, oils, fats and meals in particular. Having stated that, I can quickly add that forecasting the future of soybeans certainly cannot be an exact science as men, politics and the unpredictability of technological development, as well as of the climate, will be involved. Yet, I feel that useful conclusions can be drawn from the analysis of certain general and specific factors that have shaped and will continue to shape the development of the soybean economy.

THE GENERAL CONDITIONS

Politics

One can hardly formulate projections without a number of assumptions in this field. My first assumption for the eighties and nineties is that there will be no major war or change of political system. I need not depict how devastating and disrupting a major war would be.

Of utmost importance also is our assumption that the present political systems; i.e., the communist/socialist countries on the one hand and the countries with free market economies on the other hand, will be roughly maintained. In the former group we see not only the USSR, East Europe and China, but also many developing countries with more or less democratic political systems but with a high degree of central planning and relatively low economic incentives. The failures of the centrally planned countries—low agricultural profitability, insufficient utilization of resources and lower yields—have been one of the major reasons for the sharper than domestically required increase in soybean production of countries like the U.S., Canada, Brazil, Argentina and Malaysia.

But my assumption that the present political systems will be maintained applies also to countries that have no important direct influence on our fields but which may have a great indirect influence. I mean, for instance, the major mineral oil producing countries such as Saudi Arabia and others where a coup d'etat could result in a new mineral oil price explosion with all its negative effects on the world economy.

As regards the EEC, all the data used in this analysis for the past and the future refer to the ten countries presently belonging to the Community, although some of them joined the EEC only a few years ago and two more (Spain and Portugal) may join it in the second half of the eighties.

My third assumption in the field of politics is that there will be no significant increase in government intervention. With its many forms of protective measures and subsidy incentives, it has played an enormous role in our fields in the past. I have little hope that it will be reduced materially during the next two decades, but on the other hand, I am assuming—perhaps somewhat optimistically—that it will not be materially increased either.

General Economy

The dependence of the per capita disappearance of oils, fats, livestock products and meals on the general economy is not as clear, or at least not as uniform, as is generally assumed. It is relatively high in the developed, industrialized countries and rather low in the developing countries. In the latter group, the availability of domestic crops and/or foreign exchange and credit usually is at least as important as, and sometimes more important than, the income.

In China, for instance, the per capita disappearance of oils and fats increased only 2.6% during the five years ended 1977, although the income per capita increased about 4%. This was due above all to the low domestic production of oilseeds and the inability or unwillingness of the government to import more oils and fats. As against this, the per capita disappearance of oils and fats increased by around 12% annually during the five years ended 1982, although the income per capita rose only by an average of 5%. This, in turn, was made possible by the exceptional increase in oilseed crops, resulting mainly from the sharp expansion of acreage at the expense of grains, but also from higher yields.

On the other hand, the development of the economy will not be unimportant. In general, I assume slower growth rates for the next two decades than for the past two. For the industrialized countries I assume the real rate of growth to average around 3% annually up to about 1998 (when the present downswing of the long economic cycle should end) and of around 4 to 5% thereafter. This compares with over 5% during 1958 to 1973 and 2.2/ during 1974 to 1982. This group includes North America, West Europe and Japan. For the developing countries as a whole, I assume an average annual growth rate of 5% from now up to 1998 and of 6% thereafter. This compares with 5% during 1974 to 1982 and 5.7% during 1958 to 1973.

Population

The prospects of population growth have always fascinated forecasters. And, the combination of population and agricultural production forecasts has sometimes scared mankind. Malthus' theory has been one of the most famous examples. Around 1800 he predicted that population would increase at a geometric rate while subsistence would accumulate arithmetically, so that population would tend to outstrip the food supplies.

In fact, the rate of population growth did increase dramatically in the 18th, 19th and most of the present century. From only about 0.1% annually up to around 1700, it rose to 2.1% at the end of the sixties. Up to that time and well into the seventies many population scientists and economists expected the rate of growth to continue to rise or at least stay at or near the record level of 2.1%. In the meantime it has become clear, however, that this rate represented the climax of the population explosion that lasted around 2½ centuries. Since 1973 the rate has been declining. It may be down to 1.4% by the year 2000 (Figure 1).

The year 1973 was a milestone in a double or even triple sense. First, it marked the end of the long economic upswing (which had begun in 1948) and the beginning of the downswing. According to the long-cycle theory, this downswing should last until 1998. Secondly, it was the year of the first mineral oil price explosion. It marked the beginning of "The Limits To Growth" proclaimed by the Club of Rome. This scared men throughout the world and certainly has had, and will continue to have, its impact on birthrates. Thirdly, the micro-electronic revolution began in the early seventies. It has been a major, but by no means the only, factor for the tremendous increase in unemployment which we have seen in the past ten years and which is likely to be with us for many years to come.

Of course, the fact that the *rate* of growth is declining does not mean that the same is true for absolute growth, too. But *absolute* growth is slowing substantially and will probably cease altogether around the turn of the century (Figure 2). Certainly the growth of *total world population* will remain substantial. I expect it to rise from 4.42 million in 1980 to 6.07 million in the year 2000.

It is interesting to see that population growth is slowing more and more in the major developed countries. Their share therefore is likely to decline from 22% in 1980 to 18% by the end of this century. Even the growth of Chinese and Indian population is falling below the world growth rate, but that of the other countries (almost all of them developing) is likely to remain substantial (Figure 3).

Technology

Considering the limited availability of land, technological progress has been and will continue to be a major precondition for increasing agricultural production through

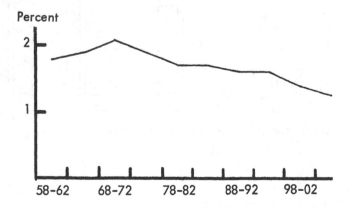

Figure 1. Annual rate of growth of world population, 5-year averages.

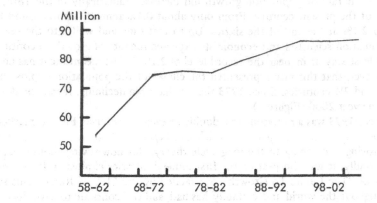

Figure 2. Absolute annual growth of world population, 5-year averages.

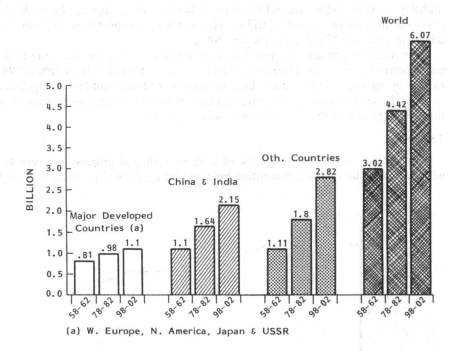

Figure 3. World population, 5-year averages.

higher yields. During the twenty years ending 1982 the world's total agriculturally used land (arable land plus land under permanent crops and pasture) increased only by 0.7% to 4.6 Gha. Among the major producing countries, above average increases took place (and are still possible in the future) only in Brazil, Argentina, Canada, Malaysia, Indonesia and to a moderate extent also in China and India. But, these increases have been to a considerable extent offset by more or less sharp declines in Europe, the U.S. and Japan.

Higher production will thus have to come more and more from increased yields. In the past several decades substantial improvements have been achieved through conventional breeding, i.e. mainly crossing and hybridization. This potential now seems to be nearing exhaustion. Future yield increases will have to come from biotechnology in general and genetic engineering in particular.

In the latter field, tissue culture will probably be the most important development, at least for the next ten years, as its commercial use has already started in the Malaysian oil palm industry. It means taking tissues from exceptionally high yielding individual plants and culturing them in test tubes through various growth media, inducing the formation of plantlets. Each plantlet is genetically identical to the original tissue of the plant or tree from which the tissue was obtained.

The oil palm industry thus appears to have an edge over oilseeds including soybeans, at present, and probably for the next ten years or so. But, since the beginning of this decade increasing efforts have been undertaken to improve the yields and reduce production costs through genetic engineering of oilseeds, too. According to D. D. Adams (pers. comm.), this includes the use of soil and leaf bacteria which improve plant growth and health and provide frost protection, but it also includes the developing of new crop plant cultivars with resistance to herbicides (which can result in higher yields and lower growing costs) and tolerance to stresses such as drought, chill, excess sunlight and male sterility (which make possible the production of higher yielding cultivars). This is being done mainly by means of tissue culture, molecular biology, biochemistry, and plant genetics. Also, improved nutrient uptake by plants (thus reduced fertilizer costs) and better pest control are goals of genetic engineering.

While companies commercially engaged in genetic engineering (such as that of Mr. Adams) are optimistic, probably overoptimistic about the time when commercial products become available for use, unbiased experts are more conservative. They do not expect commercial results to become available before the second half of the nineties. In fact, Adams also admits that only in the areas of frost protection and growth-inducing bacteria are commercial products either here today or just around the corner. Of course, even in this field many problems remain to be solved. For instance, how can one protect the soybean crop from frost damage and at the same time avoid protecting also weeds and even perhaps insects from similar damage? The ecological problems will be enormous.

When thinking about the future yield potential for the soybean and competing crops, including oil palm, we also have to take into consideration other constraints and even negative factors. The first, certainly, is that Mother Nature cannot be easily fooled. She has a certain biological equilibrium and capacity, which can be improved only gradually. The Malaysian oil palm industry provides the latest example: After introducing the weevil, the crude palm oil yield per hectare jumped by 11% above the previous three-year average in 1982. But, the next year the trees reacted by aborting female inflorescences and produing more male flowers. As a result, in 1983 the crude palm oil yield per hectare dropped by 16% below the previous three-year average (and as much as 22% below the previous year).

An important negative factor is to be seen in yield losses through soil erosion. The sharply increased use of mineral fertilizer during the past several decades has depleted many soils of organic matter. Studies over many years have pegged yield reductions for U.S. soybeans through erosion at 8% to 44% nationally and an average 12% in the Midwest where topsoils are deeper and subsoils more desirable, according to S. Marking (*Soybean Digest*, 1984), a USDA soil scientist.

Technological breakthroughs are difficult, if not impossible, to forecast. My feeling is that the increase in the three-year moving-average yield of soybeans will lag behind that of oil palm at least until the mid-nineties. But, a beginning has been made and a breakthrough could well come some time in the mid or late nineties.

THE MARKETS FOR OILS AND FATS AND THE PROSPECTIVE SHARE OF SOYBEAN OIL

Any long-run projection must start with the demand side, as over a period of two decades production will be largely determined by it. In the past two decades the demand for meals increased at a considerably higher rate than that for oils and fats. Therefore, the supply of seed oils, especially from soybeans, was primarily a function of the demand for meal. In the eighties and nineties, as against this, I expect demand for oils to be relatively stronger than for meals. This will be due mainly to the exceptionally sharp increase in palm oil and palmkernel oil production and the relatively lower oil prices resulting therefrom. The lower prices will make the oils increasingly attractive in non-food uses such as for feed, fuel and other uses. But, also the food use of oils and fats will continue to show a good increase. On the other hand, the demand for meals will be affected not only by the relatively higher meal prices, but also by the increasing competition from high-lysine corn and the faltering increase in the consumption of meat and other livestock products.

In the twenty years ending 1980, per capita disappearance of oils and fats increased by an average 0.16 kg or 1.5% annually—from 9.5 kg in 1960 to 12.7 twenty years later. The world average hides wide extremes, the 1980 data ranging from around 35 kg in the EEC and the U.S. to less than 5 in China. The world total is of course depressed by the fact that the population of the countries with below average per capita usage is far greater than that of countries with a higher per capita usage.

This will remain so in the eighties and nineties. In addition, per capita GNP is likely to grow more slowly than in the past two decades, while foreign exchange and credit availabilities will be more restricted in many developing countries. In some of them, particularly in China, the increase in domestic oilseed supplies is likely to be slower than in the six years ending 1983 as grains and some technical crops are given a higher priority. This also will limit the growth of per capita usage of oils and fats. Finally, in the industrialized countries, the slower growth of real GNP, together with the fact that present per capita disappearance already is near the saturation level, will probably slow the increase in per capita disappearance.

On the other hand, some factors will favor a continued increase in per capita disappearance; namely, (1) the expected relatively low prices for oils and fats, and (2) the already mentioned rising use of oils and fats in non-food usages, especially feeds. As a result, I expect world per capita disappearance of oils and fats to be around 15.6 kg in the year 2000 (more precisely, as all my predictions refer to five-year periods, on the average of the five years ending 2002). Most of the countries whose per capita disappearance was below the world average in 1980 will remain below it (Figure 4).

World per capita disappearance of oils and fats thus is likely to rise an average 1% annually in the twenty years ending 2000, and thus slightly less than in the past two

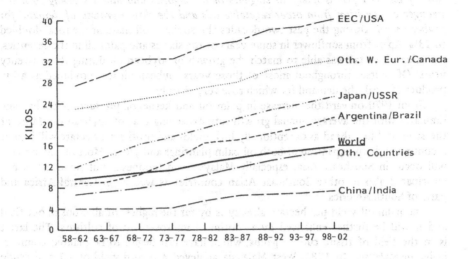

Figure 4. Per capita disappearance of 17 selected oils and fats, 5-year averages.

decades. The slower population growth will further reduce the growth rate of *total disappearance of oils and fats*. I expect it to be around 95 Mt (million metric tons) by the year 2000. This implies an average annual growth of almost 2 Mt during the eighties and nineties. This is sizeably below the 2.4 Mt achieved on the average of the five years ending 1982, but it is sharply above the 1.1 Mt registered during the twenty years ending 1977.

The major industrial countries' share of total world disappearance is likely to decline to 38% by the year 2000 compared with 47% in 1980. This will be due mostly to the slower growth of population and real GNP and to the fact that per capita disappearance is nearer satiation than in the other, mostly developing, countries (Figure 5).

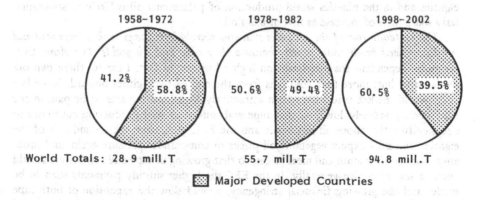

Figure 5. Total world disappearance of oils and fats and the share of the major developed countries, 5-year averages (T = metric tons).

After substantial growth in the past, the soybean oil share of total world production of oils and fats is likely to stagnate in the eighties and nineties owing to much stronger competition from other vegetable oils and the slower growth of demand for soybean meal. During the past two decades the soybean oil share of the total doubled to 23%. Apart from sunflower in some years of the sixties and palm oil in the seventies, no other oil or fat was able to match the growth of soybean oil during those twenty years. Of course, throughout most of those years soybean oil was produced as a by-product of meal, the demand for which rose very sharply.

From 1970 onward the increase in palm oil and kernel oil production has become dramatic with the average annual growth rate exceeding that of soybean oil. More of the same still lies ahead as exceptionally high yields and profits per hectare will induce a continued sharp expansion of both oil palm plantings and yields. Most of the increase will occur in Southeast Asia, especially Malaysia and Indonesia, but some increase is anticipated also in other Southeast Asian countries as well as in Central Africa and parts of South America.

The palm oil yield per hectare already is by far the highest of all crops in our field and it will be further enhanced by a continued improvement of cultivars. The latest is in the field of tissue culture palms, which have just begun to be planted commercially in Malaysia. In 1982, West Malaysia achieved a record yield of 4.1 t of crude palm oil plus 0.35 t of palmkernel oil and 0.47 t of meal per hectare. This compares with record U.S. soybean yields of 0.4 t of oil and 1.7 t of meal (1982) and EEC rape yields of 1.06 t of oil and 1.45 t of meal (1982) per hectare.

The high and probably further rising profitability will be a strong motor for a continued sharp expansion of oil palm acreage and the usage of cloning and, later on, possibly recombinant DNA or RNA techniques in the propagation of planting material. The latter techniques, when applied to single cell organisms, mean the transfer of single genes that allow the biosynthesis of important proteins such as insulin and interferon, improving, among other characteristics, also yield.

The rate of growth in world palmkernel output was even sharper in 1982 than for crude palm oil as the introduction of the pollinating weevil in Malaysia raised the palm-kernel/palm oil ratio by almost 40%. By now the weevil has been introduced throughout Southeast Asia, sharply hiking the palmkernel yield there, too. In the remaining eighties and in the nineties world production of palmkernel oil is likely to show similarly sharp rates of increase as that of palm oil.

The meteoric rise of the oil palm industry should hurt the growth of rapeseed and sunflower seed production, mostly because they have high oil yields. Therefore, they are more dependent than soybeans on high oil prices. In fact, I expect these two oil-seeds to be hurt more than soybeans from the end of the eighties onward. Since last year, world market prices have been attractively high (and the aids to be paid in the EEC correspondingly low), so that rape and sunflower seed production continues to rise significantly, above all in Canada and the EEC. But, from the second half of the eighties onward I expect vegetable oil prices to come under pressure again, and probably more so than grain and meal prices, so that growing of rape and sunflowers should become less attractive generally. In the EEC the higher subsidy payments then to be made, and the growing financial stringency, should slow the expansion of both rape and sunflower seed production. Actually, the support prices have already been reduced by 1% for the 84-85 season, for the first time in the history of the EEC.

The production shares of all other oils and fats are expected to continue to decline. These include, first, the oils and fats derived as by-products such as lard, tallow and greases, cotton oil, fish oil, and butter (although the latter is not to the same extent a by-product as the others). All these by-product oils and fats have one major factor in common: The increase in consumption, and thus in the production of the main products, is likely to be slower in the eighties and nineties than that of all oils and fats, taken as a group. In the case of lard and tallow a further slight decline of the fat yields per ton slaughtered will be an additional factor. In the case of fish oil the limited fish resources, with their restriction either by quotas or by nature, should continue to limit production growth.

Also the production shares of the other tree crop oils—coconut, olive and tung—are expected to continue to decline. This is due chiefly to the relatively low yields and profitability, the high labor costs and problems, as well as the lack of capital and the economic conditions necessary to replace the existing trees quickly enough by higher-yielding and lower-cost cultivars.

Also for the remaining seed oils—groundnut, sesame, linseed and castor—I expect production shares to continue to decline. Generally speaking, these crops are difficult to grow, not very profitable, mostly too high-priced, are subject to keen competition from other (especially subsistence) crops in the developing countries, and they are problematic as far as meal quality is concerned.

Thus, the market shares of 12 out of the 17 oils and fats included in my analysis are expected to decline in the eighties and nineties, which is actually not a big change from the past two decades. The change will be in the growth leadership of the remaining five oils—soybean, sunflower, rapeseed, palm and palmkernel. Almost all the growth will be on account of palm oil and kernel oil, a little in sunflower and rape and *almost none in soybean* (Figure 6).

Research into the recombinant DNA/RNA techniques will probably produce earlier results in annual crops. Unfortunately, not only soybeans and other oilseeds but also corn and other grains will benefit from it, possibly earlier than oilseeds. The aim is to produce maize with higher protein and oil contents, which will mean another strong competitor for soybean oil and meal.

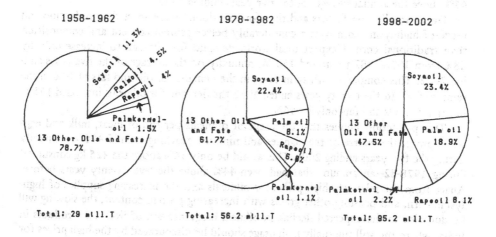

Figure 6. Shares of total production for the major oils and fats, 5-year averages (T = metric tons).

THE MARKET FOR OILMEALS AND THE PROSPECTIVE SHARE OF SOYBEAN MEAL

The meteoric rise of oil palm production and its share of the total oils and fats market will have significant effects on the oilmeal markets. The characteristics of the oil palm complex are diametrically opposed to those of the soybean complex. As soybeans contain 80% meal and only 18% oil, the meal/oil ratio is 4.4:1. As against this, 100 kg of fresh palm fruit bunches contain only 3% meal and 22.8% oil (thereof 2.3 palmkernel oil). The meal/oil ratio thus is only 0.13:1.

Therefore, a complete rethinking will be necessary. The fact that, contrary to the past, much of the growth of vegetable oil production will be on account of the oil palm complex will tend to produce a relative surplus of oil and a relative deficit of meal. This should result in a predominant tendency towards relatively higher prices for oilmeals during the second half of the eighties and especially the nineties when tissue culture palm oil starts to be produced in more significant amounts. This could well mean that the soybean meal share of the combined product value could rise to or above 75% in the nineties.

As a result of this and other factors the increase in demand for oilmeals is likely to slow. Owing partly to rising oilmeal prices and partly to slower GNP growth, financial and foreign exchange stringencies, and the nearly satiated demand in Europe and North America, the increase in per capita consumption of meat, milk and eggs is likely to slow in the course of the eighties and nineties. The increase in per capita production (which is in the long run more or less equivalent to consumption), reached a preliminary high of around 48 kg in 1980 (milk on a dry basis), after increasing around 1.1% in the sixties and around 0.8% in the seventies. Since 1981 it has declined slightly and is unlikely to begin to recover before 1985. In the eighties and nineties the increase in per capita production of meat, milk and eggs may average only 0.3% annually.

The expected slowing of the increase in per capita demand as well as in population will result in a considerably slower increase in total meat, milk and egg production during the eighties and nineties. I expect the five-yearly rates of increase to diminish to 10 to 9% compared with 18 to 13% during the past two decades. Nevertheless, production is likely to rise substantially to an average 306 Mt annually in 1998-2002. This is 45% above the annual average for the five years ending 1982.

Owing to all these factors and the aforementioned increase in the production and usage of high-lysine corn with a considerably better protein content and composition than traditional corn, I expect total world demand for oilmeals to increase only by 68% from 1978-1982 to around 150 Mt annually on the average of the five years ending 2002. This compares with an increase in the demand for oils and fats of 70% in the same period. In the twenty years ended 1982 the demand for oilmeals increased 155%, but that for oils and fats only by 96%.

This prediction implies that the usage of oilmeals per ton of meat, milk and eggs will slow in the course of the eighties and nineties, reaching around 490 kg annually during the five years ending 2002. This would be only 16% above the 425 kg consumed during 1978-82—an amount that had been 44% above the level twenty years before. Apart from the relatively high prices for oilmeals and the increasing supplies of high-lysine corn, and possibly other grains with increasing protein content, the slowing will be due also to the expected decline in oilmeal usage per ton of livestock products in India, where the still unusually high usage should be discouraged by the high prices for oilmeals.

Soybean meal will of course remain the leader of the protein complex and, contrary to oil, its market share will even rise, though more slowly. On the average of the

five years ending 2002, I expect it to reach around 66% compared with 61% two decades earlier and 41% four decades earlier. The only other meals that I expect to be able
to extend their market shares will be rape and palmkernel meals. Of course, the share
of the latter remains tiny and after all it is, with its low protein content, not an immediate competitor of soybean meal. The market share of the other seven meals taken
as a group will continue to decline significantly (Figure 7).

THE CONSEQUENCES FOR SOYBEANS

From all the foregoing, I must draw the conclusion that the growth of the world
soybean industry is very likely to be slowed in the eighties and nineties by sharply increasing competition from palm oil and the slower increase in meal demand. It will
be only thanks to the "mealseed" character of soybeans that there will still be considerable absolute growth left. Of course, the average growth rates of the past two
decades will not be attained again. *On the average of the ten crop years ending 81/82
world soybean production increased 7.5% annually. For the following two decades I
expect an average annual rate of growth of only 3%.* The growth rate would thus be
more than halved.

However, the absolute growth should remain substantial. I expect world soybean
production to reach an average 144 Mt during the five crop years ending 2001-02.
This would be almost 62 Mt above the average production in the five years ending
81-82. It translates into an average annual increase of a little over 3 Mt compared with
an average 2.8 Mt during the twenty years ended 81-82.

Figure 8 tells the story of the development of world soybean production and the
shares of the four leading countries. It shows (as five-year averages) that the U.S. share
continued to increase sharply up to the beginning of the seventies when it reached a
record 72%. This was mainly at the expense of China whose share declined dramatically. Since the first half of the seventies, however, the share of Brazil and Argentina
increased sharply, which was at the expense of both China and the U.S.

Chinese soybean production has suffered from the limited land availability and the
inability to increase yields along with the world trend. In fact, in the recent past the
yields have stagnated at the rather low level of slightly over 1 t per hectare. As against
this, ample new land is available in South America where acreage has been expanding
sharply from the early seventies. In addition, yields have been relatively high from the

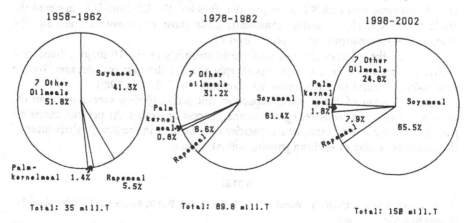

Figure 7. Shares of total production for the major oilmeals, 5-year averages (T = metric tons).

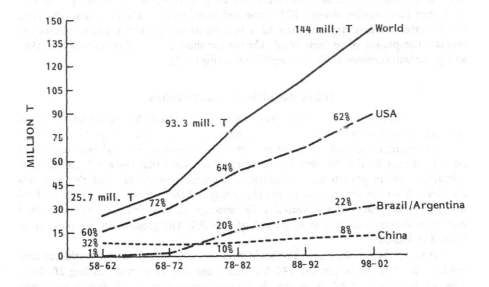

Figure 8. Soybean production of the major countries and the world, 5-year averages (T = metric tons).

beginning of the expansion and have been further improved, especially in Argentina.

Most of the increase in soybean production expected for the eighties and nineties is likely to come from higher yields in the U.S. and from acreage expansion and yield increase in South America. The final development of yields will depend to a considerable extent on when and to what extent biotechnologists and plant breeders will be successful in the research activities I mentioned in the section on technology.

After more than quadrupling from 17 in 1958 to 75 Mt in 1982, the increase in world soybean crushings will slow substantially in the eighties and nineties. By the year 2000 I expect them to reach around 125 Mt. This represents an average annual increase of only 3% compared with 8% in the past two decades. The EEC and U.S. shares of the world total are likely to decline somewhat, while those of Argentina, India and the group of "other countries" are expected to rise.

However, the soybean share of total world crushings of the 10 major oilseeds will probably continue to rise, though less sharply than in the past two decades. By the year 2000 I expect it to reach almost 59% compared with 56% in 1980.

The expansion of the crushing capacities will probably slow even more than the crushings themselves, considering the existing excess capacities. At present excess capacities are excessive in total, new capacities may be built in the North where most of the future expansion of soybean growing will take place.

NOTES

S. Mielke, Editor-in-Chief, Oil World, ISTA Mielke GmbH, P.O.B. 90 08 03, 2100 Hamburg 90, Federal Republic of Germany.

Supply and Demand Prospects

SUPPLY, USE AND PRICE PROSPECTS FOR SOYBEANS AND OTHER OILSEEDS

Jim L. Matthews

OVERVIEW OF WORLD AND U.S. OILSEEDS PROSPECTS

World Oilseed Production

World production of the eight major oilseeds is forecast to reach a record 185 million tons in 1984-85, exceeding the previous high of 180 million tons two years ago and some 12% above last year when U.S. crops were seriously impacted by drought. Soybeans are expected to account for most of this year's increase in oilseed production rising to a forecasted 93 million tons. However, soybeans may not exceed the record 94 million tons in 1979-80, while many other oilseeds may establish new record highs (Figure 1). Overall, world oilseed growth has slowed perceptibly in the past five

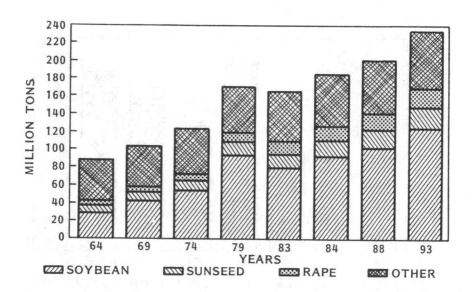

Figure 1. World production of oilseeds by type.

15

years, largely due to a slowdown in world economic growth and the need for many countries to make substantial financial adjustments. Some modest acceleration in world oilseed production is seen in the next five-year period. By 1993 world oilseed output could approximate 240 million metric tons. Soybeans will likely account for a little more than half of this total.

U.S. Oilseed Production

In the United States oilseed production is forecast to show a sharp rebound in 1984-85, reaching 63 million tons or 30% above 1983-84's drought-reduced levels (Figure 2). A forecast recovery in soybeans to 55 million tons compared to only a 43 million ton crop accounts for most of the rise. U.S. output of oilseeds is not expected to exceed 1979-80 levels until late in the 1980s or early 1990s when stronger economic growth around the world is expected to boost protein meal demand. U.S. oilseed production could approximate 80 million tons a decade from now, with soybeans accounting for about 88% of the total, about the same share as in 1983-84. Sunflower seeds could constitute a slightly larger share of U.S. output, while the share held by cottonseed and peanuts may decline fractionally.

Rest of World Supply-Use Balances

Though oilseed supplies in the rest of the world are likely to continue to set new records in the decade ahead, growth is likely to remain quite modest while demand and use expand more rapidly (Figure 3). Soybeans will be the major beneficiary of this likely trend with net demand by the rest of the world rising to a projected 34 million tons for soybeans compared to the 22.5 million tons forecast for 1984-85. For the

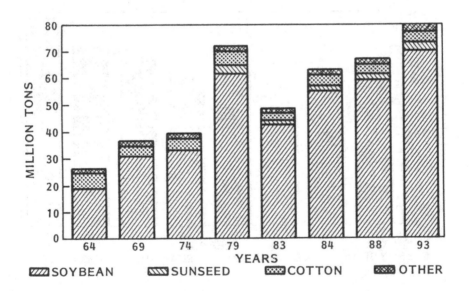

Figure 2. U.S. production of oilseeds by type.

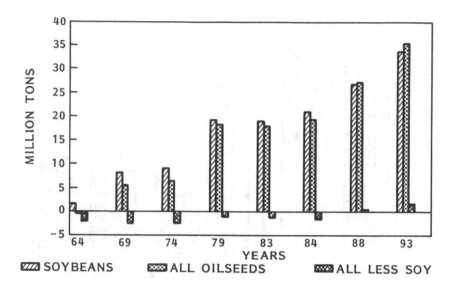

Figure 3. Use deficits by oilseed type, rest-of-world.

other oilseeds rest-of-the-world supply-use deficits will likely remain quite small showing only slight gains in the decade ahead. Consequently, U.S. exports of sunflower seed
and peanuts may show only very slight gains from recent levels.

BASIC ASSUMPTION

The forecasts and projection in this report are based on a number of key assumptions, perceptions and interpretations of background trends that strongly impact the
U.S. and world oilseeds economies. In brief, they are as follows.

Macro Economic Trends

- Economic growth, measured in terms of real gross domestic product (GDP), will
  average around 3% over the next 5-year period and 3.5% in the 1989 to 1993
  period for the developed economies (Figure 4), compared to slightly less than
  2% during 1979 to 1983. For the developing countries, real growth is assumed to
  recover to a 4% annual average rate in 1984 to 1988 and to near 5% over the
  1980 to 1993 period, compared to about 2.5% in the 1979-83 period.
- The U.S. dollar will remain relatively strong showing only a modest decline from
  1984 levels.
- World population growth will average about 1.7% annually or nearly 80 million
  persons per year. Rapid population growth of 2.3% will continue for developing
  countries as a group.
- Most developing countries will ultimately achieve more manageable debt loads
  and the world financial system will be able efficiently and equitably to accommodate needs of individual countries for economic growth and development
  purposes. However, the transition period for many countries is assumed to

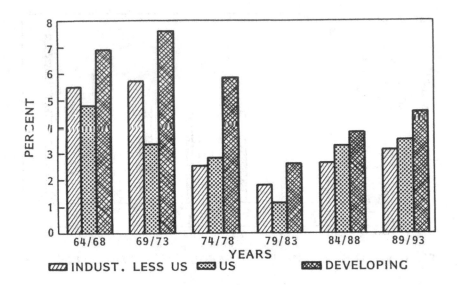

Figure 4. Real gross domestic product growth rates for major world regions.

persist for awhile, keeping demand for U.S. exports somewhat below what might be expected on the basis of real GDP growth rates.
- Inflation in the U.S. and the rest of the world is assumed to continue at much more modest rates than in the 1970s, with U.S. inflation averaging 4 to 7% over much of the period. In the U.S., this assumes a better balance between fiscal and monetary policies.

Commodity Specific Considerations

- Price and income support policies for most major producing and consuming countries will be tilted significantly toward a free-market bias. In the U.S., price supports for soybeans are set based on market clearing prices in the world economy. Commodities that compete with soybeans are assumed to adopt price and income support measures that are more in keeping with those for soybeans. In the EEC, supports for most commodities are assumed to drop on an inflation-adjusted basis.
- Measures to restrict trade in oilseeds are assumed to be only marginally effective and transitory in nature and trade will continue to have strong free market traits.
- Supply projections for all oilseeds basically assume that historical yield trends over the past 10 to 15 years will largely continue.

KEY COUNTRY MARKET ASSESSMENTS

United States Soybeans

A key factor in U.S. soybean prospects is the slowdown in growth for domestic crush in response to slowed product demand and use, particularly protein meal.

Following a record crush of 30.6 million tons in 1979-80, U.S. soybean crush activity has averaged around 27.3 million tons over the past 5-year period, reaching a recent low this past year partly because of a short U.S. crop (Figure 5).

For 1984-85, U.S. crush may again fall considerably short of the 1979-80 highs, reaching perhaps only 26.9 million tons. Crush may not exceed the 1979 high on a sustainable basis until late in this decade, because demand growth for protein meal in the domestic and export markets may continue to grow at a rate much below that experienced in the 1970s. A probable acceleration in economic activity and livestock feeding, particularly in foreign markets, should help boost U.S. crushing rates in the 1988 to 1993 period because of stronger meal export demand. Crush levels will most likely reach 34 to 35 million tons annually in the United States. But, growth in crush for the next decade may be only half that experienced in the 1960s and early 1970s.

Stronger demand growth in markets abroad for soybean products than in the United States is expected to boost U.S. soybean export growth more rapidly than growth for domestic crush. By the early 1990s, U.S. soybean exports may actually exceed domestic crush. In addition to stronger relative foreign growth encouraging soybean exports, emphasis by many countries on policies to encourage local process-ing of oilseeds will also encourage a stronger rise in U.S. soybean exports relative to meal exports. In Argentina, a sharp expansion in crushing capacity has occurred this year with a sharp shift toward more soybean meal exports rather than soybeans. The principal consequence is to boost U.S. soybean exports to a forecasted 22.4 million tons in 1984-85, up 9% from this year, as other foreign crushers turn to the U.S. for soybeans. But, a secondary consequence will be to curb U.S. meal exports.

The larger concern for the U.S. soybean industry is the slowdown in some tradi-tional, developed-country markets, with most future growth likely to be concentrated

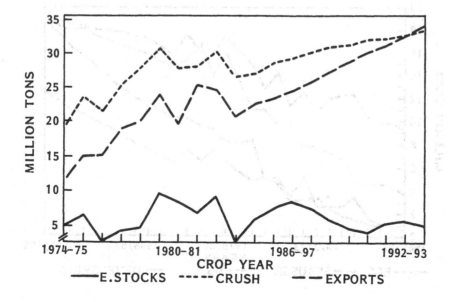

Figure 5. U.S. soybean supply and use data.

in developing, importing countries and perhaps some centrally planned economies. More rapid growth in internal consumption by some countries that are key producers of soybeans and other competing oilseeds and porducts is also an important development shaping the future for U.S. soybeans and other oilseeds.

European Economic Community and Other Developed Countries

In recent years the EEC has imported between one-third and nearly one-half of all soybeans moving in world trade. During the 1960s and early 1970s the EEC's imports grew at an annual rate of 15% (Figure 6), though the rate of feeding per animal unit was probably near a peak in the late 1970s. The recent rise in the U.S. dollar and a slowing in the European economic growth rates have been major contributors to the EEC's decline in usage rates in recent years. In the next decade, protein meal use may grow only 1 to 2% in the Community as feeding rates remain near current levels while livestock numbers expand slowly. EEC trade in protein meal with the Centrally Planned Economies (CPE) may provide the biggest boost for their soybean imports. Growth in other developed countries, mainly other West Europe countries and Japan, is also projected to slow in the coming 10 years, but not as much as in the EEC.

Centrally Planned Economies

East Europe and the USSR have constituted an important source of growth in world oilseed trade. Though they import mostly meal, principally from Brazil, Argentina and the EEC, they indirectly support U.S. oilseed export prospects. U.S. soybean exports to Europe have been increasingly influenced by Soviet meal buying out of European origins. In 1982-83 the USSR's soybean meal imports more than doubled, setting up expectations that the USSR would move quickly to more efficient feed

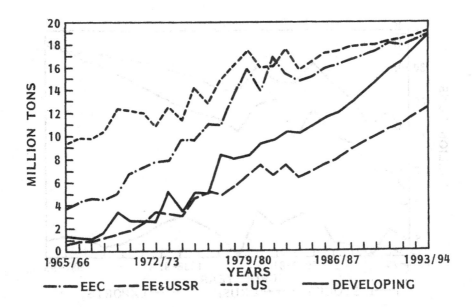

Figure 6. Soybean meal consumption by major world regions.

rations because they may be at only 60 to 65% of their protein feeding potential. Nevertheless, Soviet meal import buying and use was off sharply in 1983-84, and the forecast for 1984-85 is for very little increase. Over the decade ahead, expansion in meal use by both the USSR and East European countries is expected to show modest to strong growth with each area probably reaching consumption levels between 6 to 7 million metric tons.

Developing Exporter-Producer Countries

Growth in competing exportable supplies of soybeans and products will continue to be centered in Brazil and Argentina, with annual percentage growth in output expected to exceed the United States, though financial difficulties in both countries will hinder output expansion particularly in the next five years. In Brazil soybean production may reach or exceed 22 million metric tons by 1993 compared to this year's 15-million-ton crop. Most of this will be crushed internally, with meal production up about 50% from current levels by 1983-84 (Figure 7). However, rapid growth in domestic meal use is likely to slow future exportable supplies. Argentina, with a much smaller population base is expected to have much more rapid growth in exportable supplies of soybean meal, growing at close to 10% annually compared to only 5% for Brazil.

Developing Importer Countries

The area most likely to support a major share of future consumption growth for protein meal and soybeans is a group of developing countries located in East-Southeast Asia and the Middle East. Many of these countries are experiencing rapid economic growth and have emerging livestock-feed economies. Most are expected to show growth

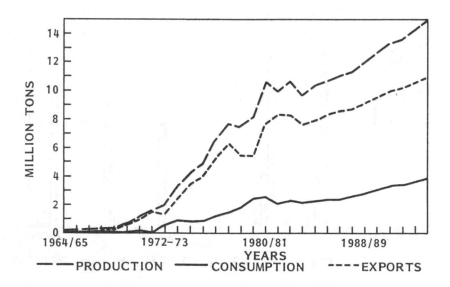

Figure 7. Supply and use balances in soybean meal for Brazil.

in protein meal use during the next 10 years near the rate of the late 1970s.

Implications for U.S. Soybean and Meal Exports

Despite the likelihood of strong growth in soybean meal use in many developing countries, as well as for some Centrally Planned Economies in the next decade, U.S. soybean exports are likely to grow slower than during the 1960s and 1970s. Combined with slow domestic growth and trend yield assumptions for production, soybean area expansion is likely to rise by about 4.5 million hectares, putting little or no strain on U.S. land resource availabilities. Growth in U.S. soybean meal exports is likely to be quite modest and may not exceed levels reached in 1979-80. Consequently, U.S. crush growth may only average around 2% annually.

Some Implications for Vegetable Oils

A probable slowing in U.S. and world meal consumption and oilseed growth rates in the decade ahead point to marginally tighter supply-use balances for vegetable oils in spite of continued good gains in world palm oil output. Per capita consumption of the 10 major vegetable oils outside the United States may rise to only 9.6 kg per person in the next 10 years compared to 8.2 currently and 6.4 a decade earlier (Figure 8). For major population based countries like India, per capita use is likely to show very marginal gains from the current 5.9 kg. Compared to the United States per capita use in the rest of the world is only about one third its potential. Vegetable oil supply-use balances could be somewhat tighter in the next few years, particularly if developing countries expand their economic growth rates. In the United States, soybean oil exports as well as inventories of soybean oil may remain substantially below levels reached in the late 1970s (Figure 9).

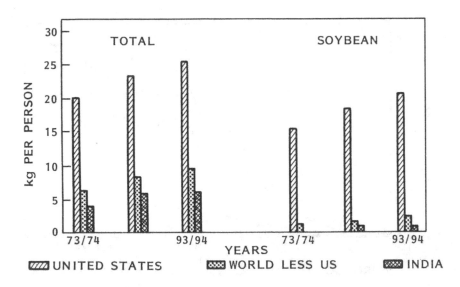

Figure 8. Per capita vegetable oil use for major world regions.

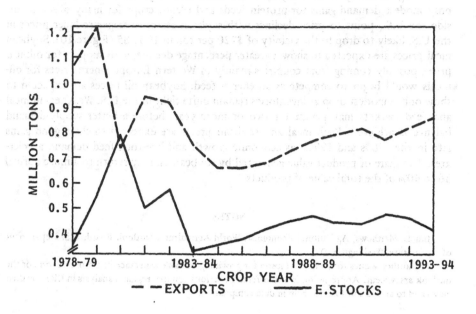

Figure 9. Soybean oil supply and use data for the U.S.

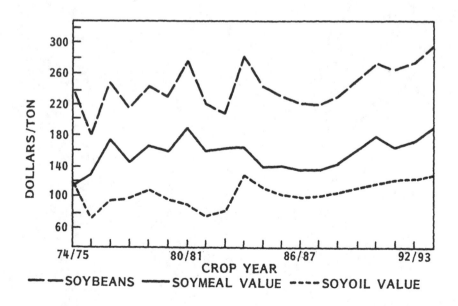

Figure 10. U.S. prices and value of soybeans and products.

PRICE PROSPECTS FOR SOYBEANS AND PRODUCTS

A large recovery in U.S. soybean and other oilseed production in 1984-85, with only modest demand gains for protein feeds and record crops for many oilseeds outside the U.S., point to price declines with soybean season-average-producer prices in the U.S. likely to drop to the vicinity of $220 per ton in 1984-85 (Figure 10). Soybean meal prices are expected to show a greater percentage decline, with support for oilmeal prices possibly coming from countries mainly in Western Europe where prices for oil-meals would begin to compete as an energy feed. Soybean oil prices are expected to show only a modest drop as inventories remain quite tight in the U.S. Weakness in meal and feed markets may persist for two or more years before a better supply-demand balance is achieved. Both meal and soybean prices are expected to show good gains late in the 1980s and 1990s, as economic growth and livestock feed demand accelerates. The share of product value generated by soybean oil may return to a more normal 36 to 40% of the total value of products.

NOTES

Jim L. Matthews, Agricultural Economist, World Agricultural Outlook Board, U.S. Department of Agriculture, Washington, D.C.

The author wishes to thank Jan Lipson for providing valuable assistance in the preparation of the outlook assessment. Appreciation is also offered to various country-regional analysts in ERS for their views and to analysts in FAS for help in data compilation.

PROJECT 2002: THE LONG-RANGE FUTURE FOR SOYBEANS FROM A U.S. PERSPECTIVE

Dennis B. Sharpe

The U.S. soybean producer has enjoyed tremendous success in the past 20 years. Plantings have more than doubled and yields per hectare have increased by more than 33% since 1963. That phenomenal expansion made the U.S. soybean the world's premier oilseed. The soybean's top ranking is the result of its role as the most prolific source of vegetable protein for animal feed. Soybeans yield high protein meal, 80% by weight, when processed, much higher than competitive oilseeds (Table 1).

The U.S. share of the world soybean market continues to be the lion's share, but intense competition has eroded the U.S. share, from 71% in 1965-66 to 63% in 1981-82. In the past ten years, the U.S. piece of the export pie was also cut from 92% to 70%.

The euphoria that accompanied the U.S. soybean success story of the past two decades has been replaced by concern over lagging exports and competitive threats to the world market share. U.S. soybean farmers are looking closely at their future. Project 2002 was conceived to help guide the U.S. soybean farmer, and others in the industry into the 21st century.

THE FUTURE BELONGS TO THOSE WHO PLAN FOR IT

Project 2002 is based on the philosophy that the future is inherently unpredictable. It is possible, however, through study and analysis to identify major issues that are likely to help shape the future, and to construct some probable alternative futures. The future that actually materializes will depend on many factors, economic, political and demographic, but one of the most important determinants of the future is *human*

Table 1. Average yield by weight of processed oilseeds.

	Protein meal	Oil
	— % —	
Soybean	80	17
Sunflower	68	31
Rapeseed	58	40
Peanut	58	42
Palm Kernel	52	46
Copra	35	64

choice. Through long-range planning and action-oriented policies, individuals and groups can shape their futures. That is the essence of Project 2002, a strategic planning process to help shape the future for the benefit of the U.S. soybean farmer. Specifically, the goals of Project 2002 were as follows:

- Provide a forum for industry-wide discussion on long-term issues that could affect soybeans.
- Define alternative futures and make a best estimate for U.S. soybean plantings, yields and production in five-year intervals to the year 2002.
- Develop strategic guidelines to help soybean farmers and others in the industry shape their futures.

A combination of techniques was utilized to achieve these goals, ranging from econometric analysis to surveys of expert opinion. Indeed, since there are no hard facts about the future (statistics are available only on the past), Project 2002 relied heavily on the reasoned opinion of literally hundreds of experts. The first phase of Project 2002 involved recruitment of a Blue Ribbon panel of industry thought-leaders.

As part of their commitment to Project 2002, each Blue Ribbon Panel member agreed to a one-hour, spontaneous interview with Project Manager, D. B. Sharpe. A specially designed questionnaire developed by the Institute for The Future, Menlo Park, California, was utilized in the interviews. The purpose was to identify the most critical long-term issues facing the U.S. soybean industry.

A Project 2002 advisory panel further clarified the issues, and helped guide the study process. Its 21 members spanned the soybean industry from farm field to export elevator and from land grant university to international bank.

In the final analysis, Project 2002 was narrowed to the following areas:

- World Economic Growth
- Energy Prices
- Trade Prospects
- Soy Foods Potential
- Soil Conservation Policy
- Soybean Yield Prospects
- Biotechnology
- Assessment of Major Competitors

Data generated from analyses of these areas were then used to make alternative projections for world demand for soybeans and for U.S. production and planting requirements.

SYNOPSIS OF RESULTS AND RECOMMENDATIONS

The following briefly summarizes the "most likely" future facing the U.S. soybean farmer and concludes with some strategic recommendations.

Slower Economic Growth

The economic scenario most favored by the experts surveyed was that the current world economic malaise will be only gradually resolved. World economic growth will average 3% over the next 20 years, significantly lower than the post-World-War-II boom of 1950-1972. Major elements of this scenario are only moderately higher energy costs and slower expansion in world trade relative to a given increase in economic growth. Protectionism will intensify as world trade expands more slowly, heightening competition.

Significant Slow Down in Protein Demand

Given the high correlation between economic growth rates and protein consumption, world demand for vegetable protein meal is projected to increase at only a 4% compound annual rate of growth. This compares to an historical growth rate of about

5% during the past 20 years.

Soybean meal's share of total oilseed protein consumption was calculated on historical trend and analysis of growth rates by individual countries. A significant slowdown in soybean meal consumption growth rates was projected. The most likely rate is 4.6%, well below the 7.5% of the past two decades.

Demand for soy-foods could triple to 860 million kg by the year 2002, but this is expected to remain a highly specialized market, comprising only about 1% of total usage. The fastest growing segment will be soy-isolate, where production is projected to grow at a 7% compound annual rate, compared to 4.5% for soy-flour and 5.4% for all products combined. Soy-protein used for pet foods will grow significantly, but wide variation in estimates of current usage prevents accurate projections.

U.S. Share of Production May Slip Further

The U.S. share of world soybean production under the most likely scenario would trend lower, from 65% in the 1983-1987 period to 61% in the 1998-2002 period. The U.S. producer would lose his share faster if economic growth is stronger than expected, since non-U.S. production of soybeans and other oilseeds would likely be spurred by higher prices to fill demand. Conversely, the slower the growth of world economies, the larger the share of the soybean market U.S. producers would probably be able to hold.

Competition More Intense

The most likely scenario implies the struggle for market share will heat up. Primary competition for U.S. soybeans, soybean meal and soybean oil will continue to be the soybeans that are produced in other countries, principally Argentina and Brazil. They will gain larger shares of both product markets in the long run.

Competition will also increase from other oil crops, especially rapeseed, sunflower and palm. Palm oil, when the new, cloned, higher-yielding cultivars come on-stream in the 1990s, will be an even more formidable competitor than it has been in the past.

Yield Projected Above Trend

A survey of over 60 plant breeders, agronomists, molecular biologists and seed managers indicates that average U.S. soybean yields could rise to nearly 3.1 t/ha by 2002. This is significantly above the trend yield projection (increasing annually at 20 to 27 kg/ha). Under optimistic circumstances they could exceed 3.7 t/ha. Under their pessimistic scenario, the experts projected a yield of only 2.5 t/ha by 2002.

Main contributors to higher yield will be:
· Narrow row spacing
· Improved cultivars
· Better herbicides
· New plant growth regulators
· Grower education

Genetic potential currently exists for soybean yields in excess of 6.7 t/ha, as proven in experimental plots. As the survey of experts indicated, higher on-farm yields are currently possible through wider adoption of existing technology. The rate of adoption depends on many factors, one of the most important being the profitability or payback to the farmer from raising more soybeans per hectare.

Nearly 40 Million Hectares Needed

Assuming a median yield of 2.96 t/ha, U.S. farmers could be called on to plant an average of 40 million ha during the 1998-2002 period. The resulting 280 million t (from 38.5 million harvested ha) would be the U.S. contribution to a projected 470 million t world demand that year. This compares to the U.S. highest production to date of 155 million t in 1979.

There is ample U.S. cropland available to meet expanded demand. Currently, some 140 million ha are planted to major crops in the U.S. According to the U.S. Soil Conservation Service's National Resource Inventories, there is another 80 million ha that could be readily brought into production with adequate economic incentive. Furthermore, improved drainage and other improvements on the existing cropland base could substantially increase its productivity. For example, in 1980 the U.S. Department of Agriculture determined that crop yields on 14 million ha of wetlands could be increased 50% with improved drainage.

Erosion Policy Crucial

Achieving these levels of yield and production will require sound policies to deal with soil erosion. As part of Project 2002, an intensive two-round survey of experts was conducted to assess the feasibility and potential effectiveness of actual and proposed erosion policies. Especially favored were:

- low interest (4% to 5%) loans to farmers to finance soil conserving capital investments
- allowing land devoted to soil conservation to remain in the base acreage calculation for federal price supports
- additional investment tax credits for purchases of conservation tillage equipment and other capital items
- redirecting Federal outlays to areas to most severe erosion. About 9% of the land area accounts for 70% of annual soil loss

The Biotechnology Wild Card

The wild card in charting the soybean's future, as identified by the Blue Ribbon Panel, is biogenetics or cell-level manipulation of the plant's traits. Although a technology in its infancy, biogenetics offers a wide array of possible benefits and challenges. The most intensive and promising work being done on soybeans would enhance yield through improved insect and disease resistance. Results could be in farmers' hands within the next five years. Quality factors, such as protein content and oil quality, also being addressed at the molecular level, are expected to emerge from a marriage of cellular and classical plant breeding techniques.

However, genetic engineers are simultaneously trying to come up with a high-protein, high-oil corn. Palm oil yields are already showing increases due to strains cloned from tissue culture. The prospect thus becomes one of increased opportunity through directed research, but also increased competition from biotechnology related developments in other crops.

STRATEGIC RECOMMENDATIONS

The projections and findings of Project 2002 are not ends in themselves. They are to provide guidelines in strategy formation. The project's ultimate goal is to stimulate thought and action to insure the continued profitability of soybeans.

Stay the Course on Market-Oriented Government
Policy for Soybeans

During the next several years the U.S. farmer should not plan for high prices. The most likely scenario is sluggish trade, increasing protectionism, lower commodity prices and abundant supplies. These developments could encourage factions within and outside the agricultural industry to seek to impose trade restrictions or sanctions, planting controls, soybean reserves, or even guaranteed prices.

Such policies should be resisted in the appropriate international or domestic arena with vigor. Aberrations of weather or other unexpected events like the 1983 drought may disrupt the balance between supply and demand in any given season. The U.S. farmer should strive to be positioned to take advantage of sudden short-term market opportunities, but more importantly, he should work for and support a strategy that will maintain or increase his share of the world market, long-term.

Devote Increased Effort to Emerging Markets

As people around the world improve their incomes, they strive to improve their diets with higher levels of animal protein. This is especially true in Asia where animal protein consumption levels are meager. Efforts must be intensified to market soybeans and soybean products in Asia as well as in the emerging livestock economies of Eastern Europe and the Middle East.

Increased buying interest in centrally planned economies, which favor long-term trade agreements and barter, will necessitate re-thinking traditional ways of doing business. Capital investment by industrialized countries in developing nations is another strategy to eventually increase the ability of Third World countries to afford quality protein.

Increase Marketing Effort and Technical Support in
Established Markets

Among established customers such as the European Community countries and Japan, new uses for known products, research into customer preferences and needs, quality control, and continued consumer education and technical support for burgeoning crushing industries would top the list of near-term strategies. Joint ventures with U.S. processors and distributors of soybean products, such as the soybean oil work already underway in West Germany and Japan, should be duplicated elsewhere.

Find New Uses for Soybean Oil

In addition to maintaining efforts to increase sales of soybean oil for cooking purposes, new uses should be pursued. Possible applications include use of soybean oil as a carrier for pesticides, as a dust suppressant in grain elevators and as a fuel oil.

Re-think Approach to Soy as Human Food

A fresh approach to marketing soy-protein is needed. The health and nutritional attributes should be emphasized to change its image as a cheap substitute for meat and cheese.

Redouble Research and Education Effort to
Gain Competitive Edge

The bottom line in any marketing effort is that the least-cost producer of the best product gains market share, long-term. It's to the U.S. farmer's advantage to be the world's most efficient soybean producer. It allows successful competition in world markets, and discourages production of "swing crops" like sunflower-seed and rapeseed.

In the last decade of this century, projected demand for soybeans will require increased plantings even if today's most optimistic yield projections are achieved.

A strong commitment to research, including basic, long-term projects, must be made. The goal should be to boost yields above the experts' most optimistic expectations. This will assure that the U.S. soybean producer will remain the most reliable supplier of the lowest cost product.

U.S. Soybean Farmers Should Become Proactive on Soil Erosion Issue

Efficient agronomic and management practices will be another major component of productivity, and soil conservation will be a major focus of future farm management. Land-use planning and zoning systems are already in place in several states. The danger exists that a passive farm constituency will end up the victim of urban-directed zoning unless initiative is taken. On the local, state and federal policy levels, producers must begin to shape regulations rather than wait to react to unfavorable legislation later. Preferred policies include those mentioned earlier (see Erosion Policy Crucial).

Form a Biotechnology Task Force on Soybeans

Biogenetics holds great promise for the soybean if its technology can be harnessed. However, currently there are several bottlenecks to its use:

1. Lack of basic information on soybean genetics.
2. Lack of a suitable vector (agent to carry genetic material into the host cell).
3. Inability to regenerate soybean plants from a single cell.

The U.S. soybean industry needs to re-commit itself to basic physiology research to assure that biogenetics can be utilized to enhance yields, improve pest resistance and improve the quality of the protein and oil.

A biogenetics task force should be established to continuously monitor biogenetics

Table 2.   Project 2002: Projected U.S. planting requirements for soybeans based upon economic alternatives and yield assumptions.

Economic Alternative[b]		Optimistic	Yield Assumption[a] Pessimistic	Most Likely
			— million ha —	
1983-1987	I	21.6	24.0	23.1
	II	23.4	26.0	24.9
	III	22.6	25.2	24.1
1988-1992	I	21.4	26.4	24.0
	II	28.2	34.9	31.7
	III	23.5	29.1	26.5
1993-1997	I	24.3	31.4	27.7
	II	35.8	46.2	40.9
	III	28.4	36.7	32.4
1998-2002	I	27.2	37.6	32.2
	II	43.3	59.9	51.2
	III	33.3	46.0	39.4

[a]Assumes harvested hectarage equals 98% of planted hectares. Yield assumptions based on median yields from 2002 survey for each five-year interval.

[b]I - Stagnation, no economic growth; II - Return to high growth rate, similar to post-World War II; III - Slow economic recovery, moderate growth.

activity on behalf of soybeans or in competition with soybeans. This task force should be composed of soybean farmers, processors and agribusiness representatives.

ALTERNATIVE DEMAND, PRODUCTION AND PLANTING REQUIREMENTS

Pessimistic and optimistic scenarios also were constructed based on survey results and expert opinion. Table 2 summarizes projected alternative U.S. soybean planting requirements for various economic situations and assumptions about yields.

NOTES

Dennis B. Sharpe, Director of Corporate Relations, American Soybean Association, 777 Craig Road, St. Louis, Missouri 63141.

Government Policy Impacts

TRADE POLICY ISSUES IN THE SOYBEAN SECTOR

James P. Houck

International trade is a vital part of the soybean and soybean product economy. Because soybean producers, processors and product users depend heavily upon international markets for sales and supplies, trade policy developments around the world take on crucial importance. This paper takes an overall look at the international soybean complex with particular emphasis on the special way trade policies of both importer and exporter nations affect this unique industry.

THE INTERNATIONAL MARKET

Anyone who knows anything about the world soybean industry is aware that this commodity and its related products are locked into heavy reliance on international trade. In fact, the rapid development of this industry since World War II, and surely since 1960, occurred because of growing world demand in countries where soybean production was not and is not economically efficient.

Approximately 31% of all soybeans produced in recent years move directly into world trade as beans.[1] This compares with the 13% of total grain production that is traded internationally. Within this total, wheat is 20%, coarse grain is 12%, and rice is 4%. Trade is also crucial for the major soybean products, with 37% of all soybean meal produced and 25% of all soybean oil produced moving into trade. Because approximately 30% of the world's production of meal and oil are based on imported soybeans, the overall reliance on international trade is 50% for the meal equivalent of world soybean output and 35% for the corresponding oil equivalent.

Since such a large share of the world's soybean output crosses international boundaries at least once and sometimes two, three, or more times on its way to final consumption, trade policy for soybeans, other oilseeds, protein meals, crude and processed edible oils, and final soy products becomes relevant in any discussion of current and future market developments.

TRADE POLICY: WHAT IS IT?

It is not easy to define the general term, "trade policy," in a precise or concise way. Overall, a nation's trade policy can be viewed as its collective attitude about the role and value of foreign trade in its international behavior. The specific elements of trade policy are deliberate decisions and actions that affect the way in which trade is conducted with regard to specific products, countries of origin or destination, and methods of exchange.

These deliberate decisions or actions can include tariffs, import quotas, domestic content regulations, packing and labeling requirements, sanitary restrictions, variable import levies, export controls, export subsidies, and so on. This web of specific government decisions is the core of a nation's trade policy in relation to other nations. A major goal of direct intervention via trade policy is to protect producers, processors, or sometimes consumers from the full force of international competition.

A nation's trade policy also can involve other, specific actions that encourage or discourage foreign trade by affecting the legal, financial, and institutional environment within which foreign transactions are made. The elements of each nation's trade policy interact directly or indirectly with those of other nations in virtually all economic transactions across international borders. Hence, trade policies form the economic buffer between one national economy and another.

A country's trade policy for agricultural products, in both general and specific dimensions, is mainly an extension of that nation's domestic food and agricultural policy, modified perhaps by balance of payments and exchange rate considerations. Agricultural trade policy extends and protects basic decisions and programs pursued in behalf of a nation's farmers, consumers, and related sectors.

Naturally, trade policy schemes can apply to both import and export transactions. Consequently, in focusing our attention on a specific commodity or product group, like soybeans and their products, one is interested in policy actions, taken by both importers and exporters of beans, meal, and oil, that affect trade volumes and values. Because all three major products in the soybean complex are storable and transportable, one also is interested in both import and export policies of some individual nations—those which import whole soybeans, for example, and export soybean meal and/or oil. Trade policies in such cases often reflect decisions to protect and promote domestic processing and handling industries.

SPECIAL CHARACTERISTICS OF THE SOYBEAN COMPLEX

The international soybean market is unique in many ways. Many of these special characteristics have implications for trade-policy-related trade issues. The following list contains relatively concise statements of several special features that impinge on trade and trade policy, some of which have been mentioned already.

1. The international market for soybeans and products is not "thin" in relation to production. Depending on the measure and the product, between 25% and 50% of the volume produced enters into international trade.
2. Production of soybeans is highly concentrated, with 92% of world output coming from four nations. Consumption of soy products is quite widely dispersed around the globe.
3. The major soybean products, meal and oil, are joint products of the crushing industry. Each contributes importantly to the total value of whole beans. In 1982-83, the meal and oil components provided, respectively, 64% and 36% of the average value of processed soybeans in the United States.
4. The processing of soybeans into meal and oil is a highly mechanized, but relatively simple, industrial process which, in principle, can be located virtually anywhere in the world. However, efficient operation of individual, modern plants requires sizable volumes of raw soybeans on a regular basis.
5. The basic joint products, meal and oil, are intermediate goods, not final products. They must be further processed, substantially in the case of crude soybean oil.

6. Because of the valuable joint products involved, soybeans straddle the strong economics of the international feed/livestock economy and the weaker economics of the edible fats and oils complex. Both sets of forces have powerful effects on the production and trade of soybeans.

7. Compared with other, major agricultural commodities, soybeans and their products are relative newcomers to world trade. The soybean commodities were virtually unknown in international commerce before World War II.

The net effect of all these general characteristics has been to foster a relatively free and open international market for soybeans and soybean meal, especially on the import side. Markets for soybean oil have been more encumbered by trade restraints, especially among importers who produce substitute oils—especially products which contain more highly refined soybean oil. In addition, various export interventions are common but not overwhelming.

THE FREE TRADE FOOTHOLD

A crucial milestone in the relatively free path that soybeans and soybean meal have followed in world trade was the 1961 zero tariff binding agreed to by the European Economic Community (EEC) (Jabara, 1981). This trade concession was achieved in the Dillon Round of multi-nation negotiations held under the auspices of the General Agreement on Tariffs and Trade (GATT). This binding appears to have set the tone for numerous similar agreements among soybean trading nations, including a zero tariff binding on soybeans and soybean meal offered by Japan, a major importer, in the 1973-79 Tokyo Round of GATT negotiations (Houck, 1979). Overall, these agreements by importers of soybeans and meal reflect the desire of importing nations for ready access (1) to raw materials for domestic crushing industries and (2) to low-cost protein material for livestock nutrition.

Since about 1947, various trade negotiating rounds in GATT have secured tariff reductions and bindings on many agricultural and industrial products including several in the fats and oils sector. These agreements are available to all contracting parties (members) of GATT. While trade policy interventions in soybeans and related sectors are by no means rare, the GATT agreements and concessions have kept international trade barriers, especially tariff and quotas, lower than they otherwise would be.

TRADE POLICY AND PROTECTION ISSUES

The various trade-encouraging agreements achieved over the years in GATT have not precluded the existence and evolution of protectionist measures in the soybean sector. We will discuss the general character and broad economic implications of these measures in this section. Specific examination of various schemes and policies conducted by major nations in the international soybean complex will appear in other papers of this section.

Figure 1 is a simplified schematic diagram illustrating the points at which trade policies alter and help shape the nature of international trade in soybeans and soybean products. The block on the left represents all soybean producing nations that export beans and, at least occasionally, meal and oil. This group contains the United States, Brazil, Argentina, Paraguay, and a few others. The block on the right represents all nations that import whole beans, process them, and then consume or re-export at least some of the meal and oil outturn. Nations of the EEC, Spain, Portugal, and some

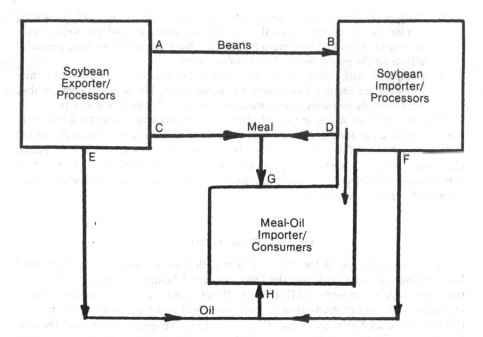

Figure 1. Trade policy points in the soybean complex.

others are in this group, as well as several nations that import beans for crushing but do not typically re-export—Japan, for example. The block in the lower middle contains all nations that import meal or oil.

The importer/processor and importer/consumer blocks are connected in the diagram to suggest that some nations import and process soybeans for re-export and also for domestic consumption along with product imports. The arrows outside the blocks indicate international trade flows, and the capital letters identify points at which trade policy schemes impinge upon exports and imports.

The next sections of this paper discuss the main classes of trade policy issues—from the standpoint of exporters or importers, producers or consumers, growers or processors, etc. In the real world, trade policies, like agricultural policies or general economic policies, usually are in continual flux. They often conflict with each other, within individual nations, from season to season, and from commodity to commodity. For example, trade policy intervention in one year to help farmers weather low prices may be abandoned the next year in favor of a program to protect oil and meal users from rapidly rising prices. Seldom are situations as neat or clear-cut as either Figure 1 or the following discussions might suggest.

Producer Protection: Import Nations

A traditional objective of trade policy is to insulate domestic producers of a commodity from competition with low-priced imports of that commodity or a close substitute. In the soybean complex, protection via tariffs, quotas, and other measures may be applied at points B, G, and H of Figure 1 in order to insulate domestic soybean growers, producers of other protein feeds, and producers of edible fats and oils, including butter. As mentioned, trade barriers against beans and meal, points B and G respectively, are not pervasive among major importers. However, almost every nation that produces edible fats and oils in any significant amount relative to domestic demand has a protective network of trade restrictions in place to favor locally produced commodities. Although not typically pointed specifically at soybean oil, these restrictions always impinge on soybean oil trade, point H. This is because soybean oil is second only to palm oil as the most important edible oil in international commerce. The basic issues of conflict here arise with exporters of the protected products who see actual and potential markets dwindle in favor of domestic producers of similar items.

Consumer Protection: Export Nations

In order to defend domestic price ceilings for soybean meal and oil, some exporting nations apply export quotas and taxes at points like A, C, D, E, and F of Figure 1. Such measures discourage exports, bottle up supplies inside the nation, and drive domestic prices down. Occasionally export sales of soybeans and products are suspended completely. Brazil, Argentina, and even the United States have resorted to export controls to hold domestic soybean product prices and availabilities in line.

Export taxes are used by some nations to generate government revenue. Although financial gain for the treasury may be the primary motivation for some export taxation, a concomitant effect is to press domestic prices of the taxed item downward.

When nations with large export market shares control or heavily tax outgoing shipments, world prices of the affected products tend to increase and overall availabilities tend to fall. Although competing exporters see such maneuvers as advantageous for them, importers who depend on international supplies may be vigorously opposed.

Producer Protection: Export Nations

Just as exports may be restricted at points A, C, D, E, and F in Figure 1, they also may be deliberately encouraged or expanded in order to stimulate sales and incomes of domestic soybean producers. Direct intervention, like export subsidies and rebates, favorable exchange rate treatment, concessional food aid sales, and special tax adjustments, may be applied. In addition, indirect export incentives may be available. These include credit subsidies and guarantees, export promotion programs, and preferential trading agreements with favored trading partners. Programs to expand exports and market shares tend to create clashes between competing producer nations, each viewing the other's expansionary policies as "unfair."

Processor Protection: Importers and Exporters

One of the most interesting aspects of the international soybean complex is the position of oilseed processors in the trade policy arena. Because all three major raw and semi-processed products (beans, meal, and oil) are readily transportable in bulk and relatively easily stored for moderate periods, processing can occur virtually anywhere in the channel between growers and users. In addition, the industrial and chemical processes are not very complex. Hence, there is a perceptible drive among nations to create at least a modest cloak of protection for domestic crushing industries and to secure the processing value added for local labor and capital. Protection can be achieved if the soybean volume moving through domestic crushers is increased and/or if the net returns per unit handled are enhanced.

Processor protection can and has occurred in soybean growing and exporting nations and in bean importing nations. In soybean producing and exporting nations, such protection can be established by systematic adjustments in export taxes or incentives so as to discourage bean exporters but encourage meal and oil product exports—points A, C, and E on Figure 1. In this way, the volume of whole soybeans moving through the domestic crushing industry is increased, and possibly, so are returns per unit to the industry.

Numerous countries process oilseed crushing facilities but do not grow sufficient soybeans domestically to keep them in efficient operation. They rely on foreign soybeans. These importer/processors may or may not re-export soybean meal and oil, depending upon internal requirements. In either case, the desire to protect a domestic processing industry can lead to special trade policies. For instance, operating at points B, G and H, nations can discourage imports of processed meal and/or oil and simultaneously encourage bean imports. Actually, to make such a program successful only one of the products imported (soybean oil, usually) needs to be discouraged in order to maintain domestic crushing throughput and returns. Nations that re-export meal and oil may protect domestic crushing by offering special export incentives and possibly subsidies or tax rebates on outward shipments of the processed products at points D and F.

Exchange Rates

In recent years, currency exchange rates for major trading nations have become much more flexible and volatile from day-to-day and month-to-month (Longmire and Morey, 1983). The level and trend of exchange rates are not usually viewed as specific commodity trade policy decisions; and usually they are not. An exception is when trading nations stipulate either favorable or unfavorable currency exchange rates for commercial sales or purchases of designated products within a national, multiple exchange rate regime.

Exchange rate changes, whether managed by central authority or allowed to occur freely in the market, affect international transactions all across the board. When a country's exchange rates increase generally, exports tend to fall as prices of those goods rise in terms of other currencies. Imports tend to increase as foreign goods prices fall inside the nation in question. Of course, these tendencies are reversed as exchange rates fall. Thus, to the extent that exchange rates reflect either a nation's general economic policies or overt intervention in international currency markets, they become trade policy in a very broad sense. Exchange rate movements are so pervasive that they easily can swamp and nullify the effect of specific trade policy interventions in behalf of individual commodities.

The complex economics of exchange rates are beyond the scope of this paper. However, one can assert that, if a nation deliberately maintains the value of its currency higher than it otherwise would be, the currency is "overvalued." A deliberate policy in the other direction would result in the currency being "undervalued." An overvalued currency, because it stifles exports and encourages imports, is protective to its domestic consumers generally, but detrimental to domestic trade-related sectors. An undervalued currency, because it promotes exports and suppresses imports, is protective to a nation's trade-related domestic sectors generally, but onerous to consumers.

MAJOR DEVELOPMENTS IN 1984 AND BEYOND

Other papers in this section will probe the specific policy developments among the main nations in the international soybean complex. Consequently, the purpose of what follows here is to enumerate and highlight, without much elaboration, what this author believes are some of the most important trade policy issues and controversies for 1984 and beyond.

European Economic Community Policies

The EEC is seriously considering a consumption tax on food fats and oils, except butter. This tax would be applied to both domestic and imported products, including the oil content of imported oilseeds such as soybeans. The central goal is to spur the consumption of surplus butter and to reduce the direct budget costs of the EEC's olive oil support program. Because the incidence of this proposed tax will fall upon products that are imported in substantial amounts, its potential enactment has become a very significant trade policy issue to nations exporting both oils and oilseeds to the EEC.

A second major question with the EEC involves the duty-free status of imported soybeans and soybean meal into the Community. This long-time GATT agreement represents a sizable gap in the internal consistency of the EEC's Common Agricultural Policy (CAP). Relatively free access of inexpensive high protein feed material has caused serious distortions in price relationships and utilization within the feed/live-stock economy. Although the strong U.S. dollar has increased bean and meal prices recently within the EEC, various maneuvers to close this expensive loophole in the CAP via trade restrictions of one kind or another can be expected in 1984 and beyond. To say that these attempts will be controversial on the international scene is to understate the matter considerably.

United States Policies

The extremely strong value of the U.S. dollar has undermined the economic recovery of U.S. agriculture, including soybean growers and processors, from its serious recession in the post-1981 period. High real interest rates, relatively low inflation, and political stability make the U.S. dollar a preferred currency in which to hold international assets. Hence, the strong demand for dollars. The trade deficit resulting in part from the very strong dollar is and will continue to be a source of pressure on U.S. policy makers as they struggle with huge federal budget deficits, surging imports, and sluggish employment growth.

The 1985 Farm Bill likely will contain price support and other provisions directly affecting soybeans and other export products. Because the United States still produces 60-65% of the world's soybeans, these decisions will undoubtedly set the tone for future growth and change in the sector, especially its crucial international dimensions.

Export Policies: Brazil and Argentina

Though relatively less important than those of the United States, the mix of price support, trade policy, and exchange rate decisions by South American soybean producers also will be a source of concern and debate in coming months and years. As they have done in the past, these nations will be trying to balance numerous conflicting objectives with trade policy decisions—price ceilings for domestic soybean products, strong producer prices for soybeans, increased foreign exchange earnings from soybeans and soybean products, and protection for the volume and value-added in their internal crushing industry.

CONCLUSION

International trade is crucial to the present and future economic health of the world soybean sector. Between 35% and 50% of the value of sales of soybeans and their products is generated by international commerce. Soybeans and products may cross international boundaries two, three, or more times on their way from growers to end product users. Consequently, trade policy decisions taken by both importing and exporting nations are and will be very important to farmers, processors, and product users. Most deliberate trade policy interventions in the soybean sector are designed to protect or shield some favored group from the full force of international competition. They may be soybean farmers, oilseed processors, producers of substitute fats, oils, and protein feeds, or even a nation's users of edible oils or protein feeds. Major issues and points of controversy in 1984 and beyond will focus on topics such as (1) how the EEC adjusts its agricultural policy and excise tax policies to affect imports of soybeans and products, (2) how the 1985 U.S. Farm Bill evolves, especially in its treatment of soybean price and income support, and how the exchange value of the U.S. dollar *vis á vis* other currencies behaves, and (3) what direction the export policies of the major soybean exporters in South America take in coming months and years.

NOTES

James P. Houck, Professor, Department of Agricultural and Applied Economics, University of Minnesota, St. Paul, Minn. 55108 (USA).

[1]Unless otherwise stated, all specific data in this paper are drawn from official statistics published by the Food and Agriculture Organization of the United Nations and the U.S. Department of Agriculture.

REFERENCES

Houck, J. P. 1979. MTN studies (2): The Tokyo/Geneva Round: Its relation to U.S. agriculture. Senate Committee on Finance, 96th Cong., 1st sess. Committee Print 96-12.

Jabara, C. L. 1981. Trade restrictions in international grain and oilseed markets. Foreign Agric. Econ. Rep. No. 162. U.S. Department of Agriculture, Washington.

Longmire, J. and A. Morey. 1983. Strong dollar dampens demand for U.S. farm exports. Foreign Agric. Econ. Rep. No. 193. U.S. Department of Agriculture, Washington.

United Nations, Food and Agriculture Organization, Intergovernmental Group on Oilseeds, Oils, and Fats, Committee on Commodity Problems. 1981-1984. Various statistical summaries and agenda documents.

U.S. Department of Agriculture, Foreign Agricultural Service. Oilseeds and oilseed products. Foreign Agricultural Circular, various issues. U.S. Government Printing Office, Washington.

IMPACT OF GOVERNMENT PROGRAMS ON SOYBEAN PRODUCTION, TRADE AND POLICIES

Abner W. Womack, Stanley R. Johnson, and Robert Young II

The U.S. agricultural industry is preparing for the implementation of a "New Farm Bill" in 1985. Several modifications of the 1981 Farm Bill now governing the agricultural industry are under consideration. These range from no government intervention (free market) to mandatory supply control programs requiring substantial government involvement. One particular design currently under consideration, and identified with the movement toward a more free market for agricultural commodities, is the farm program as it has been operated for soybeans under the 1981 Farm Bill. Unlike feed grains, cotton, rice and wheat, soybeans are not presently protected directly by a target price, base acreage or a three-year, farmer-held reserve. Soybean farmers do not have the option of removing a percent of base acreage for higher target supports, nor storing beans for longer than permitted under the regular 9-month, non-recourse loan of $5.02 per bushel. In fact, the 9-month loan at the $5.02 rate is the only direct government option open to soybean producers. Also, the loan rate is adjusted to respond to the market price.

A minimum loan of $5.02 prevails. However, if 75% of the five-year season average farm price, with high-low price years removed, is greater than $5.02, then the higher price becomes the new loan rate. The result of this program option under the 1981 Farm Bill has been to hold production and stocks at levels more consistent with prevailing market prices.

This more simplified commodity program with minimal government intervention is being considered as a model for other major crops. In addition to the prospect of lower government involvement, supporters of this program design contend that the experience with the soybean industry has demonstrated that this free-market approach will, over time, adjust production to market conditions, provide "reasonable" returns to producers and maintain adequate supplies for effective competition in domestic and export markets.

Soybeans, meal and oil are substitutes for other major agricultural commodities. Thus, the demand for soybeans, meal and oil are dependent upon prices of these other commodities. In addition, farmers make production decisions on a land base where soybeans compete with other crops that have more direct government price supports. The question of how the "positive" experience with the soybean program has been influenced by less free-market-oriented government programs for the major competitive crops on both the demand and supply sides of the market is the focus of this paper.

It will be shown that the indirect influence of government programs from competing commodities, particularly corn, has rendered the conclusions for the more free market to be overly positive with respect to soybean price maintenance. Why? Acreage control programs for feed grains in general also remove soybean acreage. On the demand side, the cross substitution of soybeans with feed grains in the domestic and foreign markets results in strong indirect influences from the feed grain program on the demand side. These policy instruments for feed grains have included the regulation of farmer-held reserves, Commodity Credit Corporation reserves and policy determined exports such as PL-480 shipments and longer term agreements with centrally planned economies. To evaluate these influences and the consequences of more free-market, "rolling"-loan-rate programs, econometric models developed at the University of Missouri for the U.S. corn and soybean industry are utilized.

ECONOMETRIC MODELS

Schematic descriptions of the econometric models for corn and soybeans are given in Figures 1 and 2. Models are annual and estimated using data from 1961 through 1981. The supply side of each industry is represented by planted area equations that include government program parameters and market prices as explanatory variables. The government program variables reflect loan rates, target prices and land diversion payments. For soybeans, the planted area is estimated to respond to direct influences from soybean market prices and loan rates as well as indirect effects from corn target prices, paid diversions and market prices. Specifically, this estimated relationship shows how the paid diversion parameter for the corn industry reduces soybean planted area. Yields in both models are exogenously determined.

The most dramatic example of the influence across commodity markets of the supply control variables was reflected in the Reduced Acreage-Paid Diversion/Payment-in-Kind Program for corn in 1983. Producers had the option of participating in land

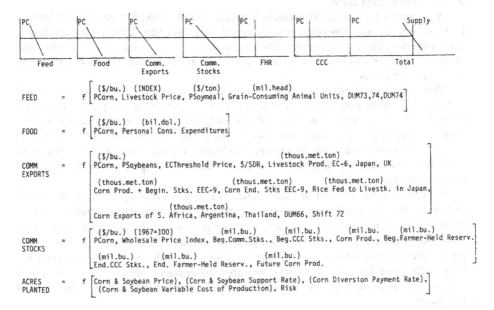

Figure 1. Corn model (see text for details).

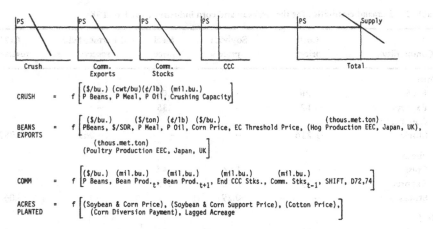

$$CRUSH = f\left[\begin{array}{l}(\$/bu.)\ (cwt/bu)(\cent/lb)\ (mil.bu.)\\ P\ Beans,\ P\ Meal,\ P\ Oil,\ Crushing\ Capacity\end{array}\right]$$

$$BEANS\ EXPORTS = f\left[\begin{array}{l}(\$/bu.)\quad (\$/ton)\quad (\cent/lb)\ (\$/bu.)\qquad\qquad (thous.met.ton)\\ PBeans,\ \$/SDR,\ P\ Meal,\ P\ Oil,\ Corn\ Price,\ EC\ Threshold\ Price,\ (Hog\ Production\ EEC,\ Japan,\ UK),\\ (thous.met.ton)\\ (Poultry\ Production\ EEC,\ Japan,\ UK)\end{array}\right]$$

$$COMM = f\left[\begin{array}{l}(\$/bu.)\ (mil.bu.)\quad (mil.bu.)\quad (mil.bu.)\quad (mil.bu.)\\ P\ Beans,\ Bean\ Prod._t,\ Bean\ Prod._{t+1},\ End\ CCC\ Stks.,\ Comm.\ Stks_{t-1},\ SHIFT,\ D72,74\end{array}\right]$$

$$ACRES\ PLANTED = f\left[\begin{array}{l}(Soybean\ \&\ Corn\ Price),\ (Soybean\ \&\ Corn\ Support\ Price),\ (Cotton\ Price),\\ (Corn\ Diversion\ Payment),\ Lagged\ Acreage\end{array}\right]$$

Figure 2. Soybean model (see text for details).

retirement options, ranging from 20 to 100% of base area. Farmers controlling approximately 67% of the USDA corn base complied, resulting in a corn planted area reduction from the previous year of 21.6 million acres (8.7 million ha). Approximately 60.2 million acres (24.4 million ha) of corn were planted for grain harvest in 1983. Interestingly, soybean planted area was also significantly reduced from 71.5 million acres in 1982 to 63.5 million acres in 1983 (29 and 25.8 million ha), an 11% reduction. Although no specific program was applied to achieve this reduction for soybeans, this shift illustrates that soybean planted area is strongly impacted by land (and supply) control decisions for feed grains.

An additional cross-commodity influence is from wheat, especially with double-cropping of soybeans and competition for the land base in cotton regions. For these reasons, the free-market, rolling-loan rate for soybeans has been distorted by programs with stronger government intervention for other commodities, both on the demand and supply sides of the markets.

These cross effects for production are summarized in Table 1 for soybean and corn planted areas. Elasticities are evaluated at simple mean values for the corresponding price and quantity variables.

A similar situation prevails on the demand side. Corn substitutes directly with soybean meal in the domestic livestock feed market and vice versa. This cross effect, shown in Table 2, is substantially stronger in the export than in the domestic market. Soybeans, meal and oil trade without restrictions. However, corn is subjected to strong levies for entry into the European Community. The result is an internal threshold price that approximately doubles the U.S. farm price of corn to European livestock

Table 1. Planted area elasticities for corn and soybeans, 1961 to 1981.

Commodity	Corn effective loan rate	Soybean effective loan rate	Corn effective diversion rate	Corn lagged market price	Soybean lagged market price
Corn	.08	-.10	-.08	.13	-.08
Soybeans	-.09	.09	-.03	-.446	+.58

Table 2. Demand elasticities for the soybean and corn industries, 1961 to 1981.

Commodity	Corn price	Soybean price	Meal price	Farmer-held reserve	CCC reserve
Corn					
Feed	-.26	–	.07	–	–
Exports	-.17	–	.35	–	–
Commercial					
Stocks	-1.216	–	–	-.061	-.083
Soybeans					
Crush	–	-1.06	.48	–	–
Exports	.29	-1.52	.75	–	–
Stocks	–	-0.29	–	–	-.04
Meal					
Feed	.14	–	-.23	–	–
Exports	.25	–	-.27	–	–

producers. As a result, responses to corn price changes are generally amplified in the European market. In contrast, the European livestock industry reacts to unconstrained soybean and meal prices. Therefore, in excess supply years with down-side price pressure, beans and bean products are relatively cheaper than corn with a corresponding expansion in export. In higher price years for beans, the economic advantage shifts to corn in the European market. For these reasons, a fairly strong cross substitution effect is estimated in the aggregate export market.

Finally, government program parameters also directly impact the feed grain industry and thus the soybean price. Commercial stock levels of feed grains are estimated to react directly to changes in farmer-held and CCC level of stocks. These estimated effects are shown in Table 2. Increasing reserve levels in these categories affect forward price expectations of commercial holders and result in corresponding down-side pressure on current demand for both corn and soybeans.

IMPACT MULTIPLIERS

The simultaneous impacts captured by the full econometric model across the feed grain-soybean complex can be reflected locally by impact multipliers. These impact multipliers that are derived from a linear approximation of the full modeling system are provided in Table 3 for soybeans. Although exports are internally generated in the fully linked corn and soybean models, proxies for these impacts are contained in the set of multipliers that includes policy exports to centrally planned economies.

It is apparent from these multipliers that both industries exhibit fairly strong cross substitution on the export side. Additionally, a dual impact is incorporated in the production changes. Our earlier discussion pointed out the planted area influence from competitive commodity program parameters. Resulting changes in supply directly influence bean prices. For example, with soybeans, season average farm price is estimated to decline approximately 15 cents per bushel for each 100 million bushel increase in corn production (Table 3). Also, a 100 million bushel increase in soybean production has an estimated direct impact of a 68 cents per bushel reduction in soybean price. The first round indirect impact on the soybean price from the supply side is the adjustment in planted area for both soybeans and corn due to the corn program design. The second

Table 3. Soybean impact multipliers.

Variable/Year	Change	1983-84	1984-85
	million bu	$	$
Corn			
Supply	100	(0.15)	(0.15)
Exports	100	0.14	0.14
Gasohol	100	0.14	0.14
FMHR stocks	100	0.15	0.15
CCC stocks	100	0.15	0.15
Future supply	100	(0.02)	(0.02)
Soybeans			
Supply	100	(0.68)	(0.68)
Exports	100	0.81	0.81
CCC stocks	100	0.68	0.68
Future supply	100	(0.10)	(0.06)
Livestock	%		
Numbers	10	1.00	1.00
Prices	10	0.13	0.13
WPI	10	0.20	0.20

round effect is associated with price pressure generated by the production impact.

Increases in farmer-held and CCC corn reserves, raise corn prices and strengthen soybean prices. Policy exports of corn have similar indirect influences on soybean prices. For example, a 100 million bushel increase in exports to the Soviet Union is estimated to increase corn prices by about 17 cents per bushel and soybean prices by approximately 14 cents per bushel (Table 3).

Thus, commodity programs aimed at balancing supply and demand for feed grains around artificially high or low loan rates have spill-over effects on soybean prices. Given that corn prices have been supported through the accumulation of CCC and farmer-held reserve stocks, it is apparent that soybean prices have benefited indirectly (Table 3) from these policy actions. This will be the case in excess supply years if the corn loan rate is supported by reserve accumulation, or in short crop years if these same reserves were allowed back on the market with upward price movements restricted by release of government controlled corn stocks. In short, floor and ceiling prices set for feed grains have served as artificial floors and ceilings for soybeans and soybean products. Moreover, these bounds have been at levels sufficient to make the soybean loan rate a redundant policy parameter. Thus, while it is certain that part of the price performance in the soybean industry is related to free-market forces, it is also clear that soybean prices have been strongly influenced by government program parameters for other commodities, specifically for feed grains and wheat.

IMPLICATIONS FOR THE 1984-85 CROP YEAR

To further illustrate the point of cross-program design, we will discuss results of several options evaluated for the 1984-85 corn program in the fall of 1983. These include:

- 0 percent reduction in planted area, or no program

- 10 percent reduction in planted area
- 10 percent reduction in planted area, plus 10 percent paid diversion
- 10 percent reduction in planted area, plus 5 percent paid diversion
- 10 percent reduction in planted area, plus 10 percent Payment-in-Kind program

All of these options were evaluated with a loan rate of $2.55, a target of $3.03, paid diversion price of $1.50 per bushel on base area and a 75% PIK payback on program certified yields. The soybean loan rate was set at $5.02 with no other options for soybeans.

A first step in the analysis focused on program participation and the corresponding planted area of corn and soybeans. In the second phase of the analysis, forward projections were made to 1984-85 for both the soybean and corn prices. These results are given in Tables 4 and 5. Corn prices varied from a low of $2.68 under the "0 Percent

Table 4. Projected prices per bushel for corn, 1984-85, for possible government program options. (See text for explanation of options.)

Variable/Year	82/83	83/84	0% RAP	10% RAP	10% RAP 5% PD	15% RAP 5% PD	10% RAP 10% PD	10% RAP 0% PD 10% PIK	15% RAP
Acreage, millions									
Planted	81.9	60.2	84.5	84.2	82.6	82.5	81.3	80.7	84.0
Harvested	73.2	51.2	74.4	74.1	72.7	72.6	71.5	71.0	73.9
Yield, bu/A	114.8	80.5	106.0	106.0	106.0	106.0	106.0	106.0	106.0
Supply, mil bu									
Beginning stocks	2,312	3,252	686	686	686	686	686	686	686
Production	8,403	4,121	7,881	7,850	7,705	7,696	7,584	7,528	7,836
Imports	1	1	1	1	1	1	1	1	1
Total supply	10,717	7,374	8,568	8,537	8,392	8,383	8,271	8,215	8,523
Domestic demand, mil bu									
Feed	4,700	3,875	4,231	4,225	4,205	4,205	4,173	4,160	4,215
Food	743	765	815	815	813	813	810	810	814
Gasohol	155	130	175	175	175	175	175	175	175
Seed	17	18	22	22	22	22	22	22	22
Total domestic	5,615	4,788	5,243	5,237	5,215	5,215	5,180	5,167	5,226
Export demand, mil bu									
USSR and PRC	212	200	189	189	189	189	189	189	189
PL480 and AID	70	25	25	25	25	25	25	25	25
Commercial	1,568	1,675	1,980	1,976	1,952	1,952	1,922	1,915	1,970
Total export	1,850	1,900	2,194	2,190	2,166	2,166	2,136	2,129	2,184
Total demand	7,465	6,688	7,437	7,427	7,381	7,381	7,316	7,296	7,410
Ending stocks, mil bu	3,252	686	1,131	1,110	1,011	1,002	955	919	1,113
Farmer-held reserves	1,600	25	150	150	50	50	50	50	150
CCC owned	1,225	75	150	150	150	150	150	150	150
"Free" stocks	427	586	831	810	811	802	755	719	813
Farm price	$2.65	$3.54	$2.68	$2.72	$2.75	$2.75	$2.97	$3.02	$2.73
Chicago price	$2.87	$3.80	$2.90	$2.93	$2.97	$2.97	$3.20	$3.25	$2.95
Loan rate	$2.55	$2.65				$2.55			
Target	$2.70	$2.86				$3.03			
Reserve entry	$2.90	$2.65				$2.55			
Reserve release	$3.25	$3.25				$3.32			

Table 5. Projected prices per bushel for soybeans, 1984-85, for possible government program options. (See text for explanation of options.)

Variable/Year	82/83	83/84	0% RAP	10% RAP	10% RAP 5% PD	15% RAP 5% PD	10% RAP 10% PD	10% RAP 0% PD 10% PIK	15% RAP
Acreage, millions									
Planted	71.5	63.3	70.8	71.2	70.7	71.0	70.3	70.1	71.4
Harvested	69.8	61.4	69.3	69.7	69.2	69.5	68.8	68.6	69.9
Yield, bu/A	31.9	24.7	29.5	29.5	29.5	29.5	29.5	29.5	29.5
Supply, mil bu									
Beginning stocks	266	387	149	149	149	149	149	149	149
Production	2,229	1,537	2,043	2,056	2,042	2,049	2,028	2,023	2,062
Total supply	2,495	1,923	2,193	2,205	2,191	2,198	2,178	2,173	2,212
Domestic demand, mil bu									
Crush	1,108	975	1,010	1,013	1,003	1,005	992	990	1,015
Seed, etc.	95	89	90	90	90	90	90	90	90
Total domestic	1,203	1,064	1,100	1,103	1,093	1,095	1,082	1,080	1,105
Export demand, mil bu									
USSR and PRC	7	15	20	20	20	20	20	20	20
Commercial	898	695	835	844	842	844	839	837	846
Total export	905	710	855	864	862	864	859	857	866
Total demand	2,108	1,774	1,955	1,967	1,955	1,959	1,941	1,937	1,971
Ending stocks, mil bu	387	149	238	238	236	239	237	236	241
CCC owned	25	0	0	0	0	0	0	0	0
"Free" stocks	362	149	238	238	236	239	237	236	241
Farm price	$5.65	$8.21	$6.61	$6.64	$6.75	$6.73	$6.95	$7.04	$6.62
Chicago price	$6.01	$8.46	$6.77	$6.80	$6.91	$6.90	$7.13	$7.23	$6.78
Loan rate	$5.02	$5.02	$5.02	$5.02	$5.02	$5.02	$5.02	$5.02	$5.02
Bean/corn price	$2.13	$2.32	$2.47	$2.44	$2.45	$2.44	$2.34	$2.33	$2.43
Decatur prices									
Meal 44%	$187.44	$231.59	$191.55	$192.42	$194.88	$194.54	$199.81	$202.15	$192.03
Oil, crude	20.5	30.6	24.3	24.4	24.9	24.8	25.7	26.1	24.4

Reduced Acreage Program" to about $3.02 under the "10 Percent RAP—10 Percent Payment-in-Kind, Program". Similarly, soybean prices are estimated to range from a low of $6.61 per bushel under the 0 percent strategy to a high of $7.04 per bushel under the PIK option.

CONCLUSIONS

When commodity program designs are considered for the 1985 Farm Bill, the "free-market concept" of the soybean industry is likely to be a model for a number of proposed changes. Our analysis indicates that the soybean market has been and will continue to be highly influenced and supported by other commodity programs, especially those for feed grains, which have involved a greater degree of government intervention. For this reason, the soybean market has not really experienced the effects of a hands-off farm program design, as many of the advocates of the rolling-loan-rates claim. Extreme care should be taken in characterizing a farm program design as "free-market oriented" based on the recent experience with soybeans. Cross substitution on both the supply and demand sides has, to an extent, precluded free-market effects on soybean prices. It is quite conceivable that experience with full free market conditions would reverse the sentiment for rolling loan rates, arguing for stronger controls in soybean market support. Factors that could generate even more pressure for a more restrictive soybean program are breakthroughs in soybean yields relative to feed grains or pressure from the feed grain industry through a corn genetic breakthrough, specifically higher protein content.

It is always easy to embrace "new wonder drugs" and "cure-all solutions". However, it is often the case that downstream implications of these panaceas are wrought with a different set of problems. Embracing the soybean program design as a "cure-all," because of the perceived free-market orientation may be such a situation. The fact is, we have not seen soybeans in a free-market environment yet!

NOTES

Abner W. Womack, Stanley R. Johnson, and Robert Young II, Associate Professor of Agricultural Economics, Professor of Economics and Agricultural Economics, and Research Associate, respectively, Department of Economics, University of Missouri, Columbia, Missouri.

THE SOUTH AMERICAN SOYBEAN INDUSTRY:
POLICY IMPACTS AND ISSUES

Gary W. Williams and Robert L. Thompson

A dramatic growth in the production and export of soybeans and products by South American countries—primarily Brazil, Argentina, and Paraguay—has greatly eroded U.S. dominance of the world market in recent years. From 2% in the mid-1960s, the collective soybean output of these three South American countries has grown to nearly 25% of world production. In 1980 they accounted for 20% of world soybean exports, 47% of world soybean meal exports, and 38% of world soybean oil exports.

To better understand the impact and context of current issues in the South American soybean industry, one must first consider the economic and policy forces behind its spectacular growth over the last decade. Then, one can focus more clearly on the impact of these policies on the world market and forces currently determining the future course of the industry.[1]

POLICY CONTEXT OF THE SOUTH AMERICAN PHENOMENON

Despite many similarities, the relationship of government policies to the development of the soybean markets of each of the three principal soybean prdoucers is unique.

Brazil

The preconditions for the marked increase in Brazilian soybean production were established during the 1960s as Brazil's development strategy shifted from import substitution-industrialization to export promotion. Brazil allowed a large devaluation of the historically overvalued cruzeiro, followed by frequent minidevaluations to maintain parity with trading partners (Leff, 1967). Along with fiscal incentives and a general easing from postwar controls, this export strategy created the Brazilian miracle of high economic growth during 1968 to 1973 (Schuh, 1977).

The cruzeiro devaluations eliminated the export taxation implicit in the overvalued exchange rate, allowing the domestic price of soybeans to rise and encouraging many farmers in Southern Brazil to experiment with producing soybeans (Schuh, 1977; Zockun, 1978). At the same time, agricultural credit was expanded. An aggressive wheat expansion program supported wheat prices and permitted farmers, especially in the fertile southern region, to acquire both machinery and production inputs on very favorable terms. Soybeans were initially double-cropped with wheat, since the wheat program largely eliminated the risk of experimenting with soybeans and also provided most of the needed inputs except seed (Savasini and Zockun, 1977). Several new soybean cultivars, better adapted to Brazilian conditions than those initially introduced from the U.S., were released during that period. Also, the perennial deficit in the production of edible oils from traditional sources (peanuts and cottonseed) assured a ready domestic market for soybeans.

The expansionary effects of these early events were reinforced by events in the 1970s, including the high world market prices of 1972 to 1974 and the U.S. soybean embargo of 1973, which sent importers to Brazil in search of an alternative source of soybeans. The net effect of these developments was that no other crops or beef cattle could compete with soybeans in Southern Brazil on a profit per hectare basis. As a consequence, production accelerated rapidly, but declined in recent years largely as a result of unfavorable weather.

Most of the growth in Brazilian soybean production has come from area expansion rather than yield improvements. Nevertheless, production is concentrated in Southern Brazil where it got its start. From Southern Brazil, soybean area expanded northward into the Central West. In the southernmost part of this region, the soils are very fertile, similar to those in certain areas of Southern Brazil. The most recent expansion, however, has been in the cerrado—an area of heavily weathered soils, with low pH and a great capacity to tie up phosphate in an unavailable form, and covered with scrub vegetation. Much of this land had previously been used only for cattle grazing.

Two economic factors have slowed the expansion into the cerrado area. First, the cost of production is much higher than in the South because of the lower inherent fertility of the soil. Second, there is little infrastructure, and the distances are much greater to the ports and to the internal markets, which are also concentrated in the South. This lowers the at-farm price of the product and raises the at-farm price of inputs in this area.

In the South there has been a marked reduction in soybean area in recent years. Both technical and economic factors account for this decline. Technical factors include increasing problems of disease, insects and nematodes, and severe erosion problems. These technical factors in the recent reductions in soybean area have been reinforced by weather problems and several economic forces that have reduced the margin of profitability of soybeans relative to other crops. These forces include the weakening of the world price of soybeans, an increasingly overvalued cruzeiro leading to trade and foreign exchange controls reminiscent of the early 1960s, and a relative increase in production subsidies for crops in competition with soybeans.

The growth in soybean production was accompanied by an even more rapid expansion of the crushing industry (Savasini and Zockun, 1977). This growth was so rapid that national crush capacity passed the level of production in 1977 and reached 27 million t in 1982, more than double the size of the crop in that year (U.S. Dept. of Agriculture). Consequently, the Brazilian Government generally restricts exports to avoid large financial losses by firms with substantial excess capacity. Under this policy, soybean exports declined from a peak of 3.5 million t in 1975 to 740,000 t in 1982. The Brazilian Government has also attempted to stimulate usage of excess capacity through a drawback scheme that allows the importation of soybeans tax-free, if the processed products are exported rather than sold domestically. Much of Paraguay's soybean crop is imported by Brazil each year under this scheme. The drawback policy was suspended in 1982 as a result of severe Brazilian foreign exchange problems, but was reintroduced in mid-1983 (U.S. Dept. of Agriculture).

As production grew, the Brazilian Government began to stimulate growth in crush capacity in order to capture the value added from processing rather than exporting soybeans. Until 1981 the general policy was to maintain domestic price ceilings on soybean meal and oil and establish export quotas to prevent their prices from moving above the ceilings. At the same time, however, exports of soybeans were taxed at a much higher rate than either soybean meal or oil and export licenses were required to export beans. This substantially raised Brazilian crushing margins. This, together with other fiscal incentives, attracted major multinational soybean crushers, as well as a number of Brazilian firms and cooperatives, to expand capacity rapidly. Domestic price ceilings were removed in 1981, and the export quota system was abolished the following year (U.S. Dept. of Agriculture). The complicated system of export taxes, subsidies, and other programs, however, was left in place (see Stancill, 1981; Thompson, 1979; and Williams, 1981 for more detail). However, soybean exports are still restricted periodically to insure adequate domestic supplies.

As the Brazilian production and crush of soybeans grew, so did the production of soybean meal and oil. Meanwhile, an increased rate of urbanization and a growth in per capita income in Brazil shifted domestic demand for edible fats and oils from lard toward vegetable oils. Since traditionally consumed vegetable oils (peanut and cottonseed oils) were declining in availability, soybean oil thus found ready market acceptance. Until 1975, exports of soybean oil were not permitted unless the domestic market was deemed by policymakers to be adequately supplied at the officially set ceiling price. Domestic soybean oil consumption thus grew rapidly. Increased domestic supplies and eased export controls allowed soybean oil exports to grow rapidly after 1975, reaching over 36% of the volume of world soybean oil exports in 1980-81.

Domestic consumption of soybean meal initially experienced slower growth than soybean oil. As a result, a substantial portion (about 95%) of annual meal production was exported. Nevertheless, a rapid growth in per capita income in the 1970s led to a marked increase in the Brazilian demand for livestock products and consequently for high protein feed supplements (Zockun, 1978). The recent recession, which has slowed income growth in Brazil, and the removal of export controls led to a 24.5% decline in soybean meal consumption in 1981. Soybean meal exports have grown rapidly, reaching nearly 40% of world exports in 1981, making Brazil the world's leading exporter of soybean meal.

Argentina

Soybeans were introduced into Argentina in the 1950s, but rapid area expansion did not begin until after accelerated growth began in Brazil. The development of short-cycle wheats and their adoption in the mid-1970s by Argentine farmers evidently precipitated the rapid growth in Argentine soybean production (Reca, Secretary of Agriculture, Government of Argentina, Buenos Aires, pers. comm.). These wheat cultivars made it possible to plant wheat in July and harvest in early December. Soybeans could then be planted in mid- to late-December and harvested in early April, leaving about 2½ months for the land to lie fallow and accumulate moisture during the autumn rains of May and June before planting wheat again in July. Double-cropped wheat and soybeans soon became highly profitable and year-to-year adjustments between corn and the wheat/soybean combination, in response to changing relative prices, is currently the common practice (U.S. Department of Agriculture).

Argentina's soybean production is concentrated in the humid Pampa areas of northern Buenos Aires province and southern Santa Fe province. This level, rain-fed, fertile area is Argentina's principal grain belt, where most of its wheat and corn are produced. Most farmers in that region rotate cropland and improved pasture. Nevertheless, from the early 1960s to the mid-1970s, crop area increased at the expense of pasture at the rate of 1.4% per year. As a consequence, beef production was pushed farther west in the Pampas where rainfall is lower (Penna, Director, Economics Dept., Instituto Nacional de Technologia Agropecuária, Buenos Aires, personal communication).

Rapid soybean growth in production occurred in Argentina despite, rather than as a result of, government policy (Recca op. cit.). The Peronists returned to power in Argentina in 1973 and raised export taxes 40 to 50%. The Argentine Grain Board was given monopoly power (from 1973 to 1975) in the exportation of crops, destroying private grain trade and hampering the flow of crops to the export market. Despite these obstacles, rapid soybean area growth was underway by the time the Peronists left office in 1976. Currently, a price support program for wheat encourages wheat production and thereby stimulates the growth of soybean area.

The most important tool the government utilizes to influence soybean production is export policy. Historically, the government has operated a complicated system of export taxes and rebates. The levels of these taxes and rebates are frequently changed in response to domestic and world market conditions, conditions in the general Argentine economy, and devaluations in the Argentine peso. In 1982 the Argentine Government set differential export taxes on soybeans at 25% and at 10% on meal and oil (U.S. Department of Agriculture). This move was an effort to promote exports of soybean products instead of soybeans by taxing exports of soybean products at a lower rate than soybeans. Overvaluation of the peso, which may be substantial given the high rate of internal inflation in Argentina, imposes an additional although implicit tax on all exports.

In recent years Argentine policymakers have attempted to stimulate domestic crushing of soybeans to capture the value-added through a system of export tax and rebate policies (U.S. Department of Agriculture). Argentine crush capacity has increased annually by 6 or 7 million t in recent years, while the proportion of soybean production crushed in Argentina has also risen. Accordingly, the proportion exported has dropped over the same period.

Consumption of soybean oil in Argentina has grown slowly relative to the growth of production, mainly because of acceptance problems by Argentine consumers. Consumer acceptance may be on the rise, however, because over half of the soybean oil produced in 1981 was consumed domestically compared with 15 to 25% in the mid-1970s. Nevertheless, this trend in consumer attitude may be more apparent than real. Argentine policies seem to favor the export of higher priced sunflower and peanut oils to bolster export revenue and service the large Argentine external debt, forcing consumers to turn to soybean oil to satisfy their demand for vegetable oils (U.S. Department of Agriculture). Consumption of soybean meal in Argentina has grown even more slowly than that of soybean oil at about half the annual rate of increase in soybean meal production. Exports, on the other hand, increased by over 650% between 1975 and 1982.

The per capita beef consumption in Argentina is among the highest in the world. As a result, the pork and poultry industries and the demand for mixed feed have grown slowly. Furthermore, consumer purchasing power has declined in recent years and the beef cattle cycle is in a retention phase, combining to reduce Argentine beef consumption markedly. Government policies, however, have stimulated an increase in beef exports, and two beefless days a week are required. The net result of these events has been a restrained increase in consumer purchases of poultry and a corresponding increase in the production of mixed feeds and the use of soybean meal.

Paraguay

The expansion of soybean area in Paraguay has been largely an extension of the growth in soybean area in Southern Brazil. Paraguay's soybean region is in the east, just across from the border from Brazil's main soybean growing area. The soils, the climate, and the topography are all similar to those in that part of Brazil. Because Paraguay is landlocked, most of the soybeans produced in Paraguay must be transported across Brazil for purchase by Brazilian crushers or to Brazilian ports for export. Soybean yields in Paraguay are similar to those in Brazil, in both level and variability.

A number of the soybean farmers in Paraguay are Brazilians who migrated to eastern Paraguay as the rapid expansion in Brazilian soybean area took place. One incentive for this migration was the differences in government policies. Unlike Brazil, Paraguay imposes no export taxes on soybeans. Producers in Paraguay receive the world market price in the port of Paranagua, Brazil, less the cost of transporting the beans to that port. Brazilian crushers must pay a comparable price for Paraguayan soybeans. Brazilian policies, however, have generally depressed prices to Brazilian farmers well below world price levels. The Paraguay Government, however, does not provide production incentives, such as subsidized credit, as Brazil does.

Soybean production in Paraguay reached 600,000 t in 1982 from only 18,000 in 1965. Of production in 1982, over 85% was exported. Until recently, Paraguayan soybeans have suffered a quality discount because the country lacked adequate drying, cleaning, and storage facilities. In addition, Paraguay has only limited crushing capacity. While capacity has been increasing in recent years, only about 40,000 t have been crushed each year, because it is generally more profitable to export the soybeans.

INTERNATIONAL IMPACTS OF SOUTH AMERICAN POLICY

Brazilian policies are commonly believed to have resulted in larger soybean meal and oil exports and smaller soybean exports than otherwise would have occurred. U.S. soybean product exports are believed to have been adversely affected as a consequence. In this view, Brazilian market intervention has tended to raise the profitability and volume of the soybean crush in Brazil and to lower its soybean exports while raising meal and oil output and exports. The logical conclusion would then be that the Brazilian policies have been effective in attaining the objective of higher soybean product exports. However, this conclusion is based mainly on observation of the short-run effects of the policies while ignoring the restrictive effect over time of artificially low soybean prices on Brazilian soybean supply. According to a recent study by Williams and Thompson (1984), when the policy-induced reduction in the domestic availability of soybeans over time is taken into account, the conclusion concerning the effectiveness of Brazilian policies is less obvious. This is because the reduced soybean supply tended to restrict soybean crush, leading to smaller domestic supplies and disappearance, and to smaller exports of soybean meal and oil from Brazil than otherwise.

The study indicates that the restriction of Brazilian soybean production over time as a result of Brazilian policies was sufficient to more than offset the stimulative effects of the policies on Brazilian soybean product supplies and exports. In other words, while in any given year the policies had the effect of stimulating Brazilian production and exports of soybean products, the increase was from a lower level than otherwise would have been the case in the absence of the restriction of soybean production. Moreover, the stimulative effects of the policies on domestic and export supplies of soybean products were insufficient to overcome the negative effects of the reduction in soybean supplies. The net effect was lower domestic and export supplies of soybean products from Brazil than otherwise might have occurred, thwarting attainment of the policy objectives.

The Brazilian policies in effect forced a divergence of world market from Brazilian price levels. While the policies lowered the bean price and increased meal and oil prices in Brazil, they had the opposite effect on world prices. Thus, the effect of the policies on the U.S. and importing countries' soybean markets was much the opposite of the effect on the Brazilian market. A higher world and consequently higher U.S. soybean price led to an increase in U.S. soybean production over time, increasing the availability of supplies for both domestic crushing and export. Further, while lower U.S. wholesale prices led to an increase in the quantities of soybean meal and oil consumed in each period, the relatively larger increase in their production allowed an increase in U.S. exports of soybean meal and oil. Consequently, by intervening in its markets, Brazil made possible larger U.S. production, domestic use, and exports of soybeans, meal, and oil than might otherwise have occurred. The combination of higher soybean prices and increased production of soybeans also led to higher gross cash receipts to U.S. farmers.

The study further suggests that Brazilian policies reduced world soybean trade by about 3% during the late 1970s. On the other hand, as a result of lower world oil and meal prices, total oil and meal shipments were higher (2 to 3% and 6%, respectively). The higher world oil trade occurred despite lower soybean oil exports by the European Community, resulting from higher oil consumption and lower soybean imports, crushing, and oil production in those countries.

ISSUES FOR THE FUTURE

Just as many different economic and policy forces have led to the emergence and growth of the South American soybean industry, divergent factors, some of which are already having an impact, will help decide the future course of the industry.

Brazil

Although further expansion of soybean acreage in Brazil will likely occur, it will be limited in the short run and occur mostly in the Central West. The expansion in this area is likely to be slower and more difficult than was the case in the southern states largely because of soil problems. A much heavier application of fertilizer and lime is required to achieve yields in the Central West comparable to those in the South. Thus, production costs will be significantly higher. Government willingness to subsidize these costs and the farm level price of soybeans will largely determine the rate at which expansion occurs in that region. However, given the Brazilian Government's efforts to reduce budget deficits by reducing production subsidies, this is not likely to occur to any extent in the near future. In addition, the isolation of Central West states from export ports and crushing facilities, and the lack of necessary infrastructure, likely will mean that soybean production will develop less rapidly in this region than occurred elsewhere in Brazil. Again, major government investments in infrastructure are not likely to be forthcoming in the near future.

Yield maintenance problems related to erosion, weeds, insects and disease in Brazil will continue to increase. Though the soybean research base in Brazil is rather thin, measures are rapidly being taken to correct this deficiency. A National Center for Soybean Research, established in 1976, is working to solve these problems as well as to breed cultivars better adapted for the low latitudes farther north. This government-funded research institution has matured sufficiently to produce a flow of yield-increasing technological improvements.

Technological improvements that have enhanced corn yields in Brazil and larger relative production subsidies for corn also will restrain future growth in soybean area, particularly in Southern Brazil. Further expansion of Brazilian crush capacity will be determined largely by the profitability of crushing relative to exporting, as influenced by government policies restricting soybean exports and stimulating exports of soybean products and by the availability of soybeans for crushing.

Capacity is not likely to expand much beyond the current level within the next few years, however, because (1) only 50 to 55% of capacity is currently utilized, (2) the availability of soybeans domestically and through imports is expected to grow slowly and (3) the cost of crushing operations in Brazil will continue to be high by international standards.

Growth in the domestic consumption of soybean oil and meal will be determined mainly by population and income growth. The Brazilian government's effort to reduce the inflationary strain of the public-sector deficit restricts the near-term prospects of a strong recovery in the Brazilian economy and, hence, in the growth rate of domestic soybean oil and meal consumption. However, spurred both by recovering world demand and government export policies, Brazilian exports of both soybean oil and meal are likely to increase, but at slower rates than those achieved in the late 1970s. The need to generate export revenue to service Brazil's huge foreign debt will likely put upward pressure on exports of both, particularly since the soybean complex has been one of the great Brazilian export successes of the last decade.

Argentina

Soybean production in Argentina will continue to be limited to the humid Pampas. Since soybean-wheat double cropping has expanded onto most of the land for which it is suited, future expansion in soybean production will be as a single crop in competition with other crops. The very low rate at which current inputs are used in Argentine soybean production will also tend to retrain the growth in yields. No major reversal in agricultural policy is foreseen that would increase the profitability of Argentine agriculture and, in turn, the intensity of input use.

Crush capacity has increased more than the volume of soybeans crushed in Argentina in recent years, resulting in some excess capacity. Consequently, exports of Argentine soybeans are unlikely to grow significantly over the next decade as the government restricts exports of raw soybeans and promotes the export of the value-added products through export policies. The keys to future growth of domestic consumption and exports of soybeans and soybean products in Argentina include conditions in the Argentine economy, changing consumer attitudes toward soybean oil, poultry and pork, and government policies. The economy is moving towards a gradual recovery, but external liquidity problems continue to restrain economic growth, restraining growth in the domestic consumption of both soybean oil and meal. As in Brazil, this should continue to provide policymakers with the incentive to stimulate exports of soybeans and, particularly, soybean meal and oil.

Paraguay

There is potential for a substantial increase in soybean production in Paraguay. This cannot occur, however, until the required infrastructure is in place. The single-most effective impact of government policy on the Paraguayan soybean industry could be intensive investment in transportation, marketing, and communication systems and other aspects of infrastructure, which could catalyze a significant increase in soybean production while both satisfying domestic demand and helping to reduce foreign exchange constraints. This does not seem imminent, and soybean production in Paraguay is expected to continue its slow recent rate of growth. No known expansion is occurring in Paraguay's crushing industry. Consequently, most of any increase in production will likely be exported.

NOTES

Gary W. Williams, Assistant Professor, Department of Economics, Iowa State University, Ames, IA 50011.

Robert L. Thompson, Senior Staff Economist, Council of Economic Advisors, Executive Office Building, Washington, D.C. (On leave from Purdue University.)

[1]The opinions expressed here are those of the authors and should not be construed as being those of the U.S. Government.

REFERENCES

Leff, N. H. 1967. Export stagnation and autarkic development in Brazil. Quart. J. Econ. 80:286-301.

Savasini, J. A. and M. H. Zockun. 1977. Diagnostico do sector soja: producao e commercializacao. Fundacao Instituto de Pesquisas Economicas, Sao Paulo.

Schuh, G. E. 1977. A politica cambial e o desenvolvimento da agricultural no Brasil. Revista de Economia Rural 15:3-24.

Stancill, M. 1981. Brazil: agricultural and trade policies. FAS M-205, Foreign Agriculture Service, U.S. Department of Agriculture, Washington, D.C.

Thompson, R. L. 1979. The Brazilian soybean situation and its impact on the world oils market. J. Am. Oil Chem. Soc. 56:391-398.

United States Department of Agriculture. Agricultural attaché reports and Foreign agriculture circulars—FOP series. Foreign Agriculture Service, Washington, D.C.

Williams, G. W. 1981. The U.S. and world oilseeds and derivatives markets: economic structure and policy interventions. Ph.D. Dissertation, Purdue University, W. Lafayette, Indiana.

Williams, G. W. and R. L. Thompson. 1984. The Brazilian soybean industry: economic structure and policy interventions. Foreign agriculture economic report no. 200. U.S. Department of Agriculture, Washington, D.C.

Zockun, M. H. 1978. A expansao da soja no Brasil: alguns aspectos da producao e consumo, Vols. 1-2. Fundacao Instituto de Pesquisas Economicas, Sao Paulo.

EUROPEAN POLICY IMPACTS ON THE SOYBEAN SECTOR

Wipada S. Huyser and William H. Meyers

Soybeans and soybean meal are among the leading agricultural products of the United States, earning over $7 billion in export revenue in 1980. Soybeans are the world's most important oilseed in terms of utilization and trade volumes. However, there are only three countries exporting substantial amounts of soybeans and soybean meal. The U.S. is the traditional exporter. Brazil and Argentina are the second and third largest net exporters, having entered the world market in the 1970s. The major importers include the European Community (EC), Japan, Eastern European Countries (EE), Spain and the Soviet Union (USSR), with the EC having the largest import share—around 40% of trading volume.

Recently, there has been an increasing concern about possible soybean and soybean meal trade barriers from the EC. Spain has been in a process of joining the EC and presumably will follow general Common Agricultural Policy (CAP) decisions. Another concern stems from the effect on soybean and soybean meal trade of the recent large appreciation of the U.S. dollar vis-a-vis other major currencies. The strong increase in exports of these products during the 1970s is partly explained by the depreciating U.S. dollar. The appreciation of the U.S. dollar and the world recession in the early 1980s has hurt U.S. agricultural product exports. Recent world economic recovery should again enhance the import ability of U.S. trade partners, although a continued strong dollar could inhibit U.S. export growth.

The objective of this paper is to evaluate the impact of hypothetical trade barrier policies in the EC and Spain, and the effect of a hypothetical U.S. dollar depreciation upon the soybean and soybean meal trade market. These policy impact analyses are based upon the regional trade model developed by Huyser (1983). Some adjustments have been made to the model to allow forecasting and analysis of these policies. The nonlinear econometric model contains ten country regions, where seven regions (U.S., Brazil, Argentina, EC, Spain, Japan, and EE) are endogenous and have their domestic market behavior estimated based on 1965 to 1980 data. Two regions (USSR and PRC) are exogenous, and the rest of the world is the residual market.

DESCRIPTION OF THE MODEL

This is a dynamic non-spatial equilibrium model estimated by two-stage least squares and principal component procedures. The basic elements of a non-spatial equilibrium supply and demand model are illustrated in Figure 1. The summation of net demands of importers (EDT) less the net supplies of other exporters (ESO) is the net excess demand facing the U.S. market (EDN). The necessary components of this

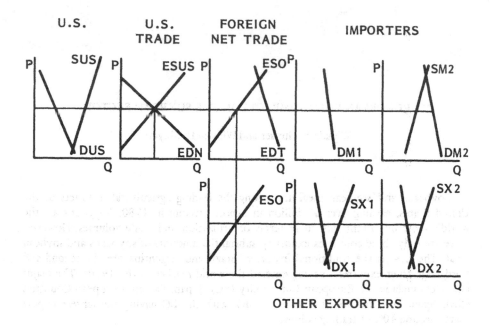

Figure 1. Illustration of Regional Supply and Demand Model. See text for details.

model are detailed in the equations below.

$$EDT = \Sigma\, DM_i - \Sigma\, SM_i = \Sigma\, f_i(P_i,X_i) - \Sigma\, h_i(P_i,Z_i) \qquad i = 1,...,n \text{ Importers} \qquad (1)$$

$$ESO = \Sigma\, SX_j - \Sigma\, DX_j = \Sigma\, f_j(P_j,X_j) - \Sigma\, h_j(P_j,Z_j) \qquad j = 1,...,m \text{ Exporters} \qquad (2)$$

$$ESUS = h_u(P_u,X_u) - f_u(P_u,Z_u) \qquad\qquad\qquad \text{United States} \qquad (3)$$

$$EDT = ESUS + ESO \qquad\qquad\qquad\qquad \text{Market Equilibrium} \qquad (4)$$

$$P_i = P_u e_i + M_i \qquad\qquad\qquad\qquad\qquad\qquad i = 1,...,n \qquad (5)$$

$$P_j = P_u e_j + M_j \qquad\qquad\qquad\qquad\qquad\qquad j = 1,...,m \qquad (6)$$

where e = exchange rate, M = trade margin (transport cost, tariff, or subsidy), P = domestic price, X = vector of demand shifters and Z = vector of supply shifters.

The model used in this study is more complex than the simplified version above. First, the soybean sector includes three distinct but closely related markets for soybeans and its two products, soybean meal and soybean oil. Second, the domestic demand in the U.S. is disaggregated into crush and inventory demand for soybeans, and consumption and inventory demand for soybean oil. To keep the model relatively small, oil markets were not endogenized outside the U.S., but soybean meal markets were. The conceptual model is described in equations (7) to (22) below.

Domestic Market of Exporter i or Importer j

Soybean production (SP_i or SP_j) $\qquad\qquad\qquad\qquad\qquad\qquad\qquad$ (7)

Soybean crush demand (SCR_i or SCR_j)   (8)

Soybean ending stock demand ($ESTK_i$ or $ESTK_j$)   (9)

Soybean meal production (SM_i or SM_j)   (10)

Soybean meal demand (MD_i or MD_j)   (11)

Soybean meal ending stock demand ($EMSTK_i$ or $EMSTK_j$)   (12)

Soybean Price Linkages to U.S. Price (PB_{us})

$$PB_i = e_i PB_{us} + MB_i \qquad (13)$$

$$PB_j = e_j PB_{us} + MM_j \qquad (14)$$

Soybean Meal Price Linkages to U.S. Price (PM_{us})

$$PM_i = e_i PB_{us} + MM_i \qquad (15)$$

$$PM_j = e_j PM_{us} + MM_j \qquad (16)$$

Excess Demands of Importer i

$$EDB_i = SCR_i + ESTK_i - LAG(ESTK_i) - SP_i \qquad (17)$$

$$EDM_i = MD_i + EMSTK_i - LAG(EMSTK_i) - SM_i \qquad (18)$$

Excess Supplies of Exporter j

$$ESB_j = SP_j - SCR_j - ESTK_j + LAG(ESTK_j) \qquad (19)$$

$$ESM_j = SM_j - MD_j - EMSTK_j - LAG(EMSTK_j) \qquad (20)$$

Market Equilibrium Identity

$$\sum_j EDB_j = \sum_j ESB_j \qquad (21)$$

$$\sum_j EDM_i = \sum_j ESM_j \qquad (22)$$

The price elasticities of the behavioral equations are summarized in Table 1. It should be noted that crush demand is not as price elastic as it may appear in the U.S. and some other countries. Since the value of meal and oil is endogenous and normally moves in the same direction as soybean prices, they have offsetting effects. In all cases a simultaneous rise of 1% in the price of soybeans and value of meal and oil would result in a relatively small change in crush demand. Soybean meal demand is the basic source of demand for soybeans, and it is price inelastic in every case.

ANALYTICAL METHOD

The dynamic nature of this model enables policy simulation into the future for given exogenous variables and controlled policy variables. Forecasts of exogenous variables are necessary in order to evaluate policy impacts over the period of 1980 to 1989. The comparison of the dynamic simulation results, with and without a given

Table 1. Price elasticities of supply and demand at 1980 values.

	Soybean price	Soybean meal price	Soybean oil price	Value of meal and oil	Corn price
U.S.					
Production	.43				-.34
Soybean crush	-1.96			1.73	
Soybean stocks	-.92				
Soybean meal demand		-.24			.20
Soybean oil demand			-.36		
Soybean oil stocks			-.26		
Brazil					
Production	.04				
Soybean crush	-.18			.29	
Soybean meal demand		-.16			-.11
Argentina					
Production	.06				
Soybean crush	-2.47			2.86	
Soybean meal demand		-.04			
EC					
Soybean crush	-.56			.64	
Soybean meal demand		-.22			.25
Spain					
Soybean crush	-2.80			3.15	
Soybean meal demand		-.41			.73
Japan					
Soybean crush	-.22			.14	
Soybean meal demand		-.08			
Eastern Europe					
Soybean crush	-1.77			1.33	

event or policy, measures the expected impact of such an event or policy for the period 1980 to 1989.

The policy scenarios of interest include (1) 20% reduction of import corn threshold price in the EC, (2) 20% import tariff on soybean and soybean meal in the EC, (3) 20% reduction of the corn price in Spain, and (4) 10% depreciation of the U.S. dollar relative to other currencies. Each policy is assumed to be introduced in 1980. The dynamic simulation, with a proper adjustment for the policy of interest, is run from 1980 to 1989 and compared to the base model results. The average impacts of the policy changes over this period on key endogenous variables are reported in Table 2.

HYPOTHESIS 1: A 20 PERCENT REDUCTION OF THE CORN
THRESHOLD PRICE IN THE EC

In order to protect domestic price support programs, the EC sets a threshold price for all grain imports. To evaluate how this price intervention has benefited soybean exporters the corn threshold price is lowered hypothetically by 20% from 1980 to 1989. This simulation studies the effects of such a policy change.

Table 2. Percent change in endogenous variables caused by hypothetical policy changes.

	Reduce EC corn price by 20%	EC tariff on soybeans and meal of 20%	Reduce Spain corn price by 20%	10% U.S. $ depreciation
U.S.				
Soybean price	-0.72	-0.84	-0.60	1.08
Soybean meal price	-1.94	-1.13	-1.59	-0.61
Soybean meal demand	0.72	0.42	0.59	0.23
Soybean exports	-2.17	-2.91	-1.68	4.69
Soybean meal exports	-2.17	-0.73	-2.21	-3.95
Export value	-3.19	-3.30	-2.60	3.85
Brazil				
Export value	-1.53	-1.11	-1.31	-2.35
Argentina				
Export value	-0.82	-1.06	-0.65	-1.30
EC				
Soybean price	-0.67	19.12	-0.54	-7.86
Soybean meal price	-1.68	18.83	-1.39	-8.95
Soybean meal demand	-4.31	-4.32	0.31	2.05
Soybean imports	-0.78	-5.08	-0.59	1.02
Soybean meal imports	-14.14	-1.93	2.88	4.90
Spain				
Soybean price	-0.61	-0.67	-0.56	-8.46
Soybean meal price	-1.63	-0.91	-1.42	-9.77
Soybean meal demand	0.94	0.52	-18.42	5.65
Soybean imports	-12.89	-0.71	-10.13	16.28
Soybean meal imports	25.61	2.72	-32.67	-13.48
Japan				
Soybean price	-0.61	-0.67	-0.53	-7.64
Soybean meal price	-1.29	-0.55	-0.80	-5.31
Soybean meal demand	0.08	0.05	0.07	0.47
Soybean imports	0.13	0.34	0.12	3.59
Soybean meal imports	-0.87	-4.33	-0.84	-46.42
Total Trade				
Meal equivalents	-1.63	-1.69	-1.27	1.39

Corn is a livestock feed substitute for soybean meal in the EC with a cross-price elasticity estimate (for 1980) of 0.25. A 20% reduction of the corn threshold price in the EC would encourage corn use in the EC and reduce soybean meal consumption. Soybean meal demand in the EC wuld fall an average of 4.3% per year. This reduction is mostly in the form of soybean meal imports which would be reduced by 14.1% per year. Reduction of soybean meal demand in the EC would lead to domestic soybean and soybean meal price decreases averaging 0.7 and 1.7%, respectively. The relative price change would reduce the crushing margin and, thus, soybean crush demand or soybean import in the EC an average of 0.8% per year.

The reduction of soybean and soybean meal import demand from the EC would decrease these products' prices elsewhere. The U.S. soybean and soybean meal average prices are predicted to decline by 0.7 and 1.9%, respectively. A greater reduction of EC soybean meal imports than of soybean imports would cause the soybean meal price to drop more than the soybean price in all regions. Spain would substitute soybean meal imports for meal crushed from bean imports. The price decrease of soybean meal also would increase Spanish soybean meal consumption by nearly 1%. Japan also would increase meal consumption slightly, but should import more beans and less meal.

The reduction of soybean meal demand in the EC would result in an average reduction of soybean and meal exports from the U.S. of over 2% per year. The value of U.S. export earnings from soybeans and meal would be expected to decrease by an average of 3.2% or 351 million dollars per year. Export values also would decrease in Brazil and Argentina an average of 1.5 and 0.8%, respectively.

A 20% reduction of the corn threshold price in the EC would be expected to reduce total world trade of soybeans and soybean meal by 1.63% or 703,000 t in soybean meal equivalent per year. As the major exporter, the impact on U.S. export would be the largest both in absolute and relative terms.

HYPOTHESIS 2: A 20 PERCENT IMPORT TARIFF ON SOYBEAN AND SOYBEAN MEAL IMPORTS IN THE EC

The EC is the world's largest soybean and soybean meal importer. Recently, there has been concern that the EC might impose an import tariff on soybeans and meal to reduce the latter's competitiveness with internally produced feedstuffs. This simulation assumes a 20% import tariff on soybeans and soybean meal, starting in 1980 in order to evaluate the impact of such a policy over the period 1980 to 1989.

A 20% import tariff on soybeans and soybean meal in the EC would increase domestic prices by almost the percentage devaluation, around 19%. The EC meal demand would be rather price inelastic, with an estimate of 1980 price elasticity of -0.22. Therefore, there would be a much smaller decrease in soybean meal demand—about 4.3%. The EC soybean price should increase slightly more (19.1%) than the soybean meal price (18.8%), which would reduce the domestic crushing margin. The soybean crush demand and soybean imports would decline by about 5% per year. Soybean meal imports would decline by 2% per year.

Leftward shifts of excess demand from the EC for soybeans and soybean meal are predicted to decrease world prices of both commodities. There would be small price decreases for soybeans and meal, with a larger price decrease for meal than for soybeans. The largest percentage soybean meal price reduction would be in the U.S., but it is predicted to be only 1.13%. Soybean meal price decreases would encourage a little more soybean meal demand in all countries other than the EC.

The soybean price in the U.S. would decrease only by $2.57 per t or 0.84% per year. This would lead to a decrease in soybean production of 909,000 t per year. The increase of domestic soybean meal demand, because of the price decrease, would offset the export decrease. Also, more soybean crush demand is predicted, averaging 0.07% per year. Soybean meal export should decrease only by 0.73% per year, compared with 2.9% annually for soybeans. The value of export earnings from soybean and soybean meal would decline, on average, $363 million per year. This is a 3.3% reduction in total value of soybean and soybean meal exports.

Brazil and Argentina would experience the same changes, but of smaller magnitude.

Both would face lower export prices and lower export sales. Consequently, production is expected to decline slightly in both countries. The net decrease of trade is predicted to be only 15,600 t in soybean meal equivalent or 0.5% for Brazil and 3,600 t or 0.1% for Argentina. So, the percentage decline in export value for these two countries would be about one-third that of the U.S.

The effects of the 20% import tariff on soybean and soybean meal imports would occur mainly in the EC. There would be a net average decrease in EC imports of 4.3% in meal equivalent per year, and most of the price impact would occur there. Soybean meal demand increases in other countries would be less than the reduction of meal demand in the EC; thus there would be a decrease of total world soybean crushing demand. We predict there would be decreases in world soybean and soybean meal trade of about 730,000 t meal equivalence or 1.7% of the normal trade volume. The U.S. would have the highest trade loss both in terms of percentage decrease and in quantity.

HYPOTHESIS 3: A 20 PERCENT REDUCTION OF THE CORN THRESHOLD PRICE IN SPAIN

Corn is a livestock feed substitute in Spain with a cross-price elasticity estimate (for 1980) of 0.73. A 20% reduction in corn price in Spain is predicted to lead to an average soybean meal demand reduction of 18.4% per year. There should be a 1.4% meal price decrease due to less demand in Spain, as well as a soybean price decrease of 0.6%. The 32.7% decline in soybean meal imports would make up about two-thirds of the reduction in meal demand. The rest would be brought about by a 10% decline in soybean crush and imports.

Substitution of corn for soybean meal consumption in Spain would lead to meal price decreases everywhere else, from 0.8% in Japan to 1.6% in the U.S. The soybean price decrease should be smaller than the meal price decrease, an average of around 0.6% per year. These price decreases would lead to a reduction of 642,000 t in soybean production in the U.S. There also would be decreases of U.S. soybean and soybean meal exports of 1.68% and 2.21%, respectively.

The total decrease in U.S. export earnings from soybeans and soybean meal is expected on the average, to be $286 million per year, or 2.61% of normal earnings. Export values in Brazil and Argentina would decline by 1.3 and 0.6%, respectively. There would be an increase in soybean meal demand in the EC, averaging 0.31% per year because of the slight price decrease. The EC soybean price would decrease less than the meal price, resulting in a reduction of soybean crushing margin, and thus soybean crush and import demand of 0.6% per year. The EC would increase their meal imports by 2.9% to substitute for the domestic crush of imported soybeans.

The 20% reduction in corn price in Spain affects Spanish soybean and soybean meal markets the most. Spain would be expected to reduce their imports of soybeans and meal an average 647,000 t per year in meal equivalence, or an 18% reduction. This would be somewhat offset by increased consumption in other countries, but a net reduction in world trade of 1.27% would occur.

HYPOTHESIS 4: A 10 PERCENT PER YEAR DEPRECIATION OF THE U.S. DOLLAR

In recent years, the U.S. dollar has appreciated relative to currencies of most of its trading partners. This has been a factor in the decline of U.S. agricultural exports. In order to evaluate its importance to the soybean sector, we consider the impact of a 10% depreciation of the U.S. dollar relative to other currencies from 1980 to 1989.

U.S. depreciation would make all U.S. exports less expensive in terms of other currencies. U.S. export supplies of soybeans and soybean meal would increase. The ten-year average percentage price decline of soybeans in the other countries would range between 7.6 (Japan) and 8.5% (Spain) per year. The only exception would be Brazilian soybeans, where the average real price decrease would be 12.3% per year, due to the government-imposed inflexibility in domestic meal price. The average percentage price decrease of meal in the other countries would range from 5.3% in Japan to 9.8% in Spain. These price decreases should encourage import demand, but the import increases also would depend upon each country's price elasticity of demand.

In the U.S., soybean exports are predicted to increase by 4.7% per year. The increase of soybean export demand, in turn, would push the U.S. soybean price up by 1.1% and soybean production would rise by 1.9% per year.

The U.S. depreciation nevertheless would not increase the U.S. soybean meal price. In fact, U.S. soybean meal export would decrease nearly 4% per year. This is because Spain, Japan and Eastern Europe would increase their own soybean crush demand and replace their soybean meal imports with domestic production of soybean meal. Overall, the depreciation of the U.S. dollar would increase the U.S. export of soybean and soybean meal by 785,000 t in soybean meal equivalent, or $423 million per year. The dollar depreciation would hurt the other exporters, Brazil and Argentina. Their export values are predicted to decrease by 2.3 and 1.3%, respectively.

The 10% depreciation of the U.S. dollar should alter the soybean and soybean meal trading pattern. Overall, there would be less meal and more soybeans traded, and total world trade in soybean meal equivalent would rise by 1.4% per year. The U.S. would face a relatively inelastic foreign demand for soybeans and soybean meal, as evidenced by the much larger price effects that are expected to occur in foreign markets. The U.S. price effect would be the smallest resulting from the U.S. dollar depreciation. The benefit to the U.S. of such a development is estimated to be a 4% increase in annual export earnings. Another measure of benefit is an increase of about $570 million in value of production annually.

SUMMARY

A 20% reduction of corn threshold price in the EC would encourage corn utilization and reduce soybean and soybean meal demand. Most of the reduction in imports would be from soybean meal imports. This would result in a larger decline of meal price than of soybean price in world markets. Soybean and soybean meal export earnings would decrease by about 3% in the U.S., 1.5% in Brazil and less than 1% in Argentina. The impacts of a reduction in Spain's corn price are similarly distributed, but smaller in magnitude. These results indicate that the soybean market has benefited substantially from the high corn price policies of Western Europe.

The price impact of a 20% import tariff on soybean and soybean meal by the EC would be mostly in the country imposing the tariff. The EC would reduce soybean meal imports more than soybean imports, so meal price would decrease more than the soybean price in world markets. The U.S. would lose more than 3% in export value, compared with about 1% for Brazil and Argentina. A similar impact analysis for Spain generated similar results.

The price impacts of a U.S. dollar depreciation would be much larger in other countries than in the U.S. This implies that the U.S. is facing a world excess demand more inelastic than the U.S. export supply. The EC would increase both soybean and soybean meal imports. Japan, Spain and the EE would increase their soybean imports and substitute domestically produced soybean meal for soybean meal imports. The depreciation would increase U.S. export earnings from soybeans and meal by nearly 4%, while the export earnings of Brazil and Argentina would decrease by over 2% and 1%, respectively.

NOTES

Wipada S. Huyser, Economist, International Monetary Fund, Washington, D.C.
William H. Meyers, Associate Professor, Department of Economics, Iowa State University. The views expressed herein are those of the authors, not of the organizations in which they work.
This research was supported by the Iowa Agric. and Home Econ. Exp. Stn. and the USDA under Cooperative Agreement No. 58-3A96-2-00495 while Ms. Huyser was at Iowa State University.

REFERENCES

Huyser, Wipada S. 1983. A Regional Analysis of Trade Policies Affecting the Soybean and Soymeal Market. Ph.D. Dissertation, Iowa State University Library, Ames.

Microeconomics of Production and Marketing

SOYBEANS AND CROPPING PATTERNS IN CHINA

P. H. Calkins and J. C. Ma

As China's population has grown, soybean has been a major source of protein and vegetable oil for the populace. Soybean cultivation can also extend soil fertility, reduce insect damage to other crops, spread out the annual use patterns of China's abundant labor, and provide by-products for animal feeding. For these reaasons, Chinese farmers developed a rich repertory of cropping patterns that included soybeans.

Traditionally, China was self-sufficient in soybean production, and exported large amounts from the northeastern provinces (Shen, 1951). Today, however, as the result of the pro-grain agricultural and pro-natal demographic policies of the 1949 to 1979 period, China has become a net importer with an annual production of 8.4 million tons, far less than the historical high of 11 million tons in 1938 (Ma and Zhang, 1983). China is, therefore, interested in improving the productive technology and the supply response by domestic producers to better meet its needs for this important crop. Recent changes in agricultural policy allow not only continued technological research, but the use of price policies to stimulate production during the period of the Agricultural Responsibility System. Farmer choice is particularly interesting in the case of soybeans, which can fit into a wide range of cropping patterns. The shifting importance of soybeans, therefore, will help dictate changes in the importance of various cropping systems. Because the new policies also promote regional specialization for trade along the lines of comparative advantaged, it is important to study the regions in which China's farmers are likely to specialize in soybean production.

SOYBEAN PRODUCTION REGIONS AND ASSOCIATE CROPPING PATTERNS

As of 1981, eight major provinces (Heilongjiang, Jilin, Liaoning, Hebei, Henan, Shandong, Jiangsu, and Anhui) accounted for 76.5% of the total area and 80.3% of the total production of soybeans. Figure 1 shows that these provinces lie along the northern and central portions of China's eastern seaboard. The agroclimate of these provinces determines three phenomena:

1. The cultivars of soybean that can be grown in each province (Wang, 1983).
2. The limits to where multiple-cropping systems can be profitably practiced.
3. The relative importance of soybean as a function of these cropping patterns.

These agroclimatic considerations allow one to divide China into five key producing regions for soybeans, the general characteristics of which are given in Table 1. It is evident that the eight key producing provinces lie in Regions I and II, where growing seasons are the shortest and the potential for multiple crops per year the lowest. Thus, the major soybean cultivation is associated with a low multiple cropping index.

Figure 1. Distribution of soybean cropping patterns in five regions of China, 1978-1983. Numbers are codes for cropping practices (see Table 2 for definitions).

Table 1. Characteristics of five major soybean producing regions in China (Source: Shandong Agricultural College, 1980).

	Annual precipitation	Average temperature	Average illumination	Frost-free period	Crops/year	Soybean as percent of Area	Soybean as percent of Output
	mm	C	h/d	d		%	%
REGION I:							
Spring—Sown in the North	500-700	-.06 to about 12	)15	120-160	1	37	42
REGION II:							
Summer—Sown along Yellow River and Huai River Basins	500-600	10 to 15	12.5-14.5	200	1-1.5	32	30
REGION III:							
Summer—Sown along Yangtzee River Basins	800-1000	15 to 18.6	(14	)200	2	23	20
REGION IV:							
Autumn—Sown on the south of Yellow River	1,000 -1,600	19 to 22	(14	300	2-3	5	5
REGION V:							
In the South	1,000 -1,600	20 to 25.5	(13.5	)300	2-4	3	3

However, soybeans are grown throughout China and are found in an astonishing array of cropping patterns. Table 2 lists a sampling of some of the more important of these patterns, while Figure 1 indicates by number the location where some were reported. These data suggest four conclusions:

1. Although the number of crops per pattern does not vary significantly by region, the length in years trends downward from Region I to Region V. Thus, the multiple cropping index is highest in the south and lowest in the north.
2. Soybean is mainly combined with wheat, sorghum and corn in Regions I-IV and rice in Region V.
3. Soybean is often intercropped with corn, particularly in Regions I-III.
4. The seasons in which soybean may be grown are more varied in Regions IV and V, allowing the farmer greater flexibility for including soybean in his cropping systems. Paradoxically, soybean is most important where there is the least flexibility.

A study of cropping patterns before 1949 (Buck, 1937) revealed another fact: the length and complexity of soybean cropping patterns have decreased under the technical and economic environment of the past 35 years. For example, Table 2 lists one pattern for Jiangsu province that includes three crops over two years. Buck listed two key patterns for Jiangsu: an eight-crop pattern over three years and a nine-crop pattern over four years.

These cropping patterns represent the fruit of centuries of experimentation by Chinese farmers. By contrast, the Qi Min Yao Shu, written when fewer crops had been introduced to China from abroad, could only say that the most favorable preceding crop for soybean was spiked millet, which in turn should not follow soybean (Shih, 1974).

MOTIVES FOR INCLUDING SOYBEAN IN CROPPING PATTERNS

Table 3 lists the motives for including soybean in cropping patterns for selected areas of China. In general, the ranking of reasons is consistent across the country. However, four facts are noteworthy:

1. Before 1979 many farmers would have listed meeting state quotas as the primary reason for growing soybeans. Today, there is a wide variety of motives, among which quota fulfillment ranks low.
2. In the first three patterns in general, and Liaoning in particular, soybean is not a "filler" crop planted primarily to fully utilize the land. This is in contrast to the south, where multiple cropping is more developed.
3. In Taiwan province, where labor is a scarcer resource, the relatively low labor-intensity of soybeans is a major reason for inclusion in cropping patterns.
4. Farmers do not generally raise soybeans for home consumption, preferring instead to purchase tofu from neighboring specialized households. This contrasts with evidence from Buck (1937), which showed that half of the farmers increased soybean area because it was "desirable as food."

The farmer responses in Table 3 are also noteworthy for the two possible motives they leave out: impact on pest control, and relative yield variability. In a separate study (Hu et al., 1983), it was found that soybeans may be included in rotations to reduce disease and insects in both itself and other crops. Table 4 gives information on the absolute and relative yield variability of soybean over the period 1965 to 1983.

Table 2. Key soybean cropping patterns in five production regions in China 1978 to 1983 (Sources: Survey questionnaires; Calkins and Huang, 1978).

Rotation No.	Province	Source year	Cropping pattern[a]	Length of pattern (yrs)	MCI[b]	Month soybean Planted	Harvested
Region 1							
1	NE	81	S-C(SO)-M	3	100	April	Oct.
2	NE	81	S-SO-C	3	100	April	Oct.
3	NE	81	S-SW(FLAX)-C	3	100	n.a.	n.a.
4	NE	81	S-SW-SW	5	100	May	Sept.
5	NE	81	S-M-C-WW	4	100	n.a.	n.a.
6	NE	81	C-S/C-C(SO)	3	133	April	Oct.
7	NE	82	C-S/WW-M	3	133	May	Sept.
8	NE	79	WW-C-S-WW-WW-S	6	100	n.a.	n.a.
9	Gansu	82	WW-S-WW	2	150	June	Oct.
10	Gansu	82	SW-S/WW	2	150	n.a.	n.a.
11	Gansu	82	SW/S-SW/C	2	200	n.a.	n.a.
12	Ningxia	82	SW/S(C/S)	2	200	n.a.	n.a.
13	Ningxia	82	SW-S-R	2	150	n.a.	n.a.
14	Qinghai	82	SW/S-SW/BB	2	200	n.a.	n.a.
15	Xinjiang	82	WW-BEETS-C/S	3	133	n.a.	n.a.
16	Xinjiang	82	C/S-SW-(CO)	2	150	n.a.	n.a.
17	Xinjiang	82	WW-S-CO(C)	3	100	n.a.	n.a.
18	Shaanxi	82	C/M/S	1	300	n.a.	n.a.
Region 2							
19	Shaanxi	79	WW-S-C	2	150	June	Oct.
20	Gansu	79	WW-S-SW (SO)	2	150	May	Sept.
21	Shaanxi	83	WW-S/C-WW-S/C	2	300	June	Oct.
22	Shaanxi	79	WW-S/C-CO	2	100	June	Oct.
23	Jiangsu	83	WWS-F-CO (Misc)	2	100	n.a.	n.a.
24		79	SW-M-S	3	100	June	Oct.
25			CO(3-5)-C-W-S	6-8	100	n.a.	n.a.
26		83	RA-S-W	3	100	May	Sept.
27	Henan	83	W-S-W	3	100	June	Sept.
Region 3							
28		83	SW-C/S-SW	2	200	June	Oct.
29		83	SW-C/S-CO	2	200	June	Oct.
30			WW-S-F-SO	n.a.	n.a.	n.a.	n.a.
31		79	CO-W-S-W-CO-GM	3	200	n.a.	n.a.
32		79	WW-S/C-WW-R (Misc)	2	200	n.a.	n.a.
33	Sichuan	82	BB-S	1	200	April	July
34	Sichuan	83	W-S-RA	3	100	June	Nov.
35	Sichuan	82	S-PEA	2	67	April	July
36	Sichuan	82	HUDOU-S	1	67	April	July
37	Sichuan	82	SP-S/C-W	3	133	June	Nov.

[a]B = Barley, BB = Broadbeans, BE = Beans, C = Corn, CO = Cotton, CP = Cowpeas, F = Fallow, FP = Field Peas, GB = Green Beans, GM = Green Manure, M = Millet, Misc = Misc, O = Oats, Op = Opium, P = Peanuts, R = Early or Late Rice, RA = Rapeseed, RAD = Radish, S = Soybeans, SE = Sesame, SG = Small Grain, SO = Sorghum, SP = Sweet Potatoes, SU = Sugar Cane, SW = Spring Wheat, T = Tobacco, TA = Redtares, TU = Turnips, WW = Winter Wheat. HUDOU is a native bean.

[b]Multiple Cropping Index, defined as crops planted and harvested per hectare per 12-month period.

Table 2. Continued.

Rotation No.	Province	Source year	Cropping pattern[a]	Length of pattern (yrs)	MCI[b]	Month soybean Planted	Month soybean Harvested
Region 4							
38		83	RA-S-W	2	300	May	Oct.
39		83	WW-S-R	2	300	March	July
40		83	WW/S-R	2	300	n.a.	n.a.
41		83	B-S-R	2	300	March	July
43		83	B/S-R	2	300	n.a.	n.a.
44		83	WW(B)-S/C-W/B	2	400	June	Nov.
Region 5							
45	Guangdong	82	R-SU/S-R	2	400	Feb.	June
46	Guangdong	82	R-S/R	2	400	July	Nov.
47	Taiwan	78	R-R-S	2	300	Sept.	Jan.
48	Taiwan	78	Misc-S	2	200	Oct.	Jan.
49	Taiwan	78	S-R-Misc	2	300	Jan.	May
50	Taiwan	78	Misc-S-Misc	2	300	June	Nov.

The data indicate that:

1. For all locations except Sichuan, soybean had lower yield variability (as meas-ured by the coefficient of variation) than did all grains, including soybeans. Thus, soybeans may play a key role in stabilizing yields and returns to Chinese farmers.
2. From Liaoning to Sichuan, the yield of all grains tends to become less variable.
3. From Liaoning to Sichuan, the importance of soybean as a percentage of tilled area, and of total grain output, also declines. This suggests that in the south, soybeans are less necessary to maintain stable yields and returns.

TRENDS IN SOYBEAN YIELD AND AREA PLANTED

The choice of cropping patterns and the motives that farmers have for choosing among them are only part of the story of soybeans in recent China. Table 5 shows that 20th-century soybean production in China can be divided into four periods, based upon technological improvements in soybean yield and politico-economic inducements to expand or contract the area planted to soybeans:

1. The pre-1949 base period. China exhibited unimpeded specialization for trade both domestically and internationally. Thus, soybean was grown primarily in the northeast, where soybean's comparative advantage was greatest. Because of a lack of modern research, soybean yields per ha were modest, but the large area planted contributed to large, total output, reaching 11.3 million tons in 1936.
2. 1949-1959; post-war recovery and growth. During the early period of collec-tivization, soybean breeding through scientific research was begun. Further-more, the recovery of China's agricultural economy as a whole meant that farmers could expand production from 5.1 million tons in 1949 to up to 11.5 million tons. This period saw relatively free crop choice.
3. 1960-1978; grain first policy, with rice as the key link. Here there was active

Table 3. Ranking[a] importance of motives for including soybeans in cropping patterns for selected areas of China, 1976 to 1983 (Sources: Survey questionnaires, 1984; Calkins and Huang, 1978).

	Baoguing, Heilongjiang	Changtu, Liaoning	Gangu, Gansu	Wugong, Shaanxi	Yongcheng, Henan	Liangzhong, Sichuan	Jiangbei, Sichuan	Taiwan	Avg. rank excluding Taiwan	Avg. rank including Taiwan
Region:	I	I	II	II	II	III	III	V		
Pattern:	W-S/C-W	C-S-SO	W-S-W	W-S/C-W	W-S-W	SP-S/C-W	HUDOU-S	All		
Fully use land	2	6	3	1	1	1	1	1	2.1	2.0
Improve fertility	1	1	1	2	5	2	5	3	2.4	2.5
Increase income	3	5	2	4	2	6	4	5	3.7	3.9
Save labor	6	3	6	5	3	3	6	2	4.5	4.3
Lower production costs	7	4	4	3	4	5	7	4	4.9	5.8
Meet State quota	5	7	5	6	6	4	2	n.a.	5.0	
Self-consumption	4	2	7	7	7	7	3	n.a.	5.2	

[a] 1 = most important; 7 = least important.

Table 4. Area planted, yield level and yield variability for soybean and all grains in selected areas of China, 1965-1983 (Source: Survey questionnaires, 1984).

	China	Changtu, Liaoning	Wupong, Shaanxi	Jiangbei, Sichuan	Liangzhong, Sichuan
Region:		I	II	III	III
Rotation:	All	C-S-SO	W-S/C-W	HUDOU-S	SP-S-C/W
Soy Area as % Tilled Area					
Range	5.5-8.8	16.4-24.1	2.7-9.5	1.3-3.6	0.4-2.8
Mean	5.84	21.6	6	2.2	1.4
C.V.	0.39	0.11	0.45	0.36	0.50
Soybean Output as % of Total Grain Output					
Range	2.2-5.0	1.6-19.1	0.7-1.4	0.4-3.2	0.001-0.3
Mean	3.0	5.5	1.0	1.9	0.12
C.V.	0.26	0.81	0.25	0.47	0.58
Yield of All Grains (kg/ha)					
Range	1173-3117	1046-6768	1500-6825	3527-5674	4289-7438
Mean	2238	3736	4590	4466	5570
C.V.	0.23	0.56	0.44	0.17	0.15
Soybean Yield (kg/ha)					
Range	685-1163	575-1257	638-1388	1077-4687	302-551
Mean	995	760	788	3208	426
C.V.	0.15	0.39	0.29	0.33	0.23

Table 5. Production costs for soybeans in selected areas of China, 1983 (Source: Survey questionnaires; Calkins and Huang, 1978).

	Changtu, Liaoning	Yongcheng, Henan	Wugong, Shaanxi	Wugong, Shaanxi	Wei, Gansu	Liangzhong, Sichuan	Ba, Sichuan	Chongqing, Sichuan	Pingtung, Taiwan
Region:	I	II	II	II	II	III	III	III	V
Rotation:	C-S/C-CO	W-S-W	W-S/C-W	W-S/C-W	W-S-C	SP-S/C-W	W-S-SP	B-S	R-R-S
Cost Item (U.S.$/ha)									
Seed	40.74	22.74	15.79	19.74	47.37	18.47	25.82	54.47	49
Fertilizer–P	29.61	39.47	33.16	27.63	22.11		13.97	2.37	
Fertilizer–N			40.26	37.11	44.21	82.89			71
Farm Manure	58.89	90.79	63.16	55.26	59.21		18.95	44.21	56
Pesticides	4.74	1.58	9.47	7.11			9.47		
Draft Power	50.13	39.47	6.32	7.89					
Labor	48.55	31.58	104.21	142.11	177.63	110.53	157.89	213.16	345
Other	48.55		35.53						3
Total Costs	275.45	225.63	307.89	296.84	350.53	211.89	249.32	314.21	524
Yield (kg/ha)	923.1	750.0	750.0	750.0	787.5	412.5	855.0	1146.0	2158
Cost (U.S.$/t)	298	301	410	396	445	514	292	274	243
Rank of Input Use									
Labor (days/ha)	8	9	6	4	2	5	3	1	7
Nonlabor (U.S.$/ha)	1	3	2	6	5	7	9	8	4
Labor: Capital ratio	0.21		0.51	0.92	1.03	1.09	1.73	2.11	1.93
Provincial average Labor: Capital ratio	0.54	n.a.	0.89	0.89	n.a.	2.38	2.38	2.38	1.88
Net Returns (U.S.$/ha)	363.33	211.11	211.11	222.16	194.42	73.56	342.34	478.82	341

discrimination against soybeans in the push for maximizing total grain tonnage per ha. To discourage soybean production, three measures were taken. First, the relative producer price of soybeans flagged as planners increased only the prices of other grains. Second, research, if not suspended entirely, as during the peak of the Cultural Revolution (1966 to 1969), was severely restricted. Thus, only token technological development was possible. Third, the cropping patterns of Table 2 were forced out by cropping systems with relatively greater grain intensity.

4. 1979-present: the Responsibility System and other reforms. Chinese farmers now have much more freedom in deciding what and how to produce, can fulfill their quotas with any type of grain, and enjoy a soybean:rice price ratio of 2.89 in 1981, up from 1.51 in 1976 (Pang and DeBoer, 1983). Thus, farmers have increased their soybean planted area faster than for any other crop except cotton. Even so, total production volume has only regained its 1952 level, while the population has increased 60%.

Two policies initiated by the Chinese government in the countryside during Period 3 should theoretically have helped to mitigate these trends. The first was the effort to increase the multiple cropping index (MCI). As a result, the MCI has increased from 121.1 in 1949 to 151. As cropping intensity increased, other things equal, soybean area should also have risen. But, other things were not equal during the pro-grain campaign. Moreover, technological improvements enjoyed by maize and rice were not shared by soybean. This has meant that within a constant set of possible cropping patterns, those including soybean have not been favored. The second policy was the "Eight-Point Charter" for agriculture, promulgated in 1956. This included emphasis on improving the soil, irrigation, fertility, plant protection, cropping intensity, cultivars, implements, and management. Again in theory, soybean should have been favored (to improve soil fertility, and for plant protection). Specifically, by-products such as meal could be used for fertilizer and livestock feed, while we have pointed out the role of soybeans in reducing diseases and insect pests. But, the "best practices" were dictated from the top, where central planners considered soybeans to be of secondary importance.

The significant shift of the policies of Period 4 was to give farmers the price incentive to grow more soybeans, to provide them with higher-yielding cultivars to further enhance profitability, and to reduce the quantitative controls on output mix that characterized the previous period. Before 1979, production levels had been set by strict target, with prices used primarily to adjust incomes. After 1978, however, prices and profit maximization played an increasing role. Furthermore, respect for local conditions has led to rationalization of cropping patterns on state farms (Pang and DeBoer, 1983) as well as collective farms. Rather than concentrating on a production function, as during the Eight-Character Charter, planners began to recognize implicitly the importance of a supply function that included the relative values of inputs, output, and alternative crops.

PROFITABILITY AND FACTOR USE BY REGION IN SOYBEAN PRODUCTION

Under the liberal economic policies of the post-1978 period, a comparison of current profitability and factor-use demands of soybean production in the various regions will help to predict in which location and in which cropping patterns soybean enjoys a comparative advantage. Table 5 points to the following facts:

1. Although the northeastern provinces, including Liaoning, are most noted for soybean production, they do not necessarily enjoy the highest yield. Recent

Chinese production data by province confirm this picture, showing Shandong, Hebei, and Ningxia all with yields exceeding 1.5 t/ha (Anonymous, 1983).

2. Nevertheless, Liaoning enjoys competitive net returns per ha when compared with the other regions of China.

3. The most successful cropping patterns have widely varying labor/capital ratios, suggesting that the new economic policies have allowed local conditions to influence the rationality of production technology. Thus, Liaoning, with relatively large amounts of land per agricultural worker, can use a production technology with only one-tenth the labor-intensity of patterns in Sichuan, where land per worker is among the lowest in China. This wide range in the labor intensity of soybean production is supported not only by recent Chinese data (Anonymous, 1983), but also data from the 1930s (Buck, 1937), when technology reflected local resource availability.

In addition to factor ratios, climatic conditions help to determine the yield and profitability of soybeans in different parts of China. We, therefore, estimated a production function based on data from 22 provinces (Anonymous, 1983), to determine the relative importance of the following agroclimatic variables: annual average temperature, annual precipitation, relative humidity, soil temperature, and frost-free days. Because of the high correlation between pairs of these agroclimatic variables, we eliminated annual precipitation and estimated the following equation:

$$Y = -2859.58 - 176.15 \, L + 124.56 \, L^2 + 0.03 \, L^3 - 24.70 \, K + 206.39 \, K^2$$
$$ (-3.84) \quad (-3.49) \qquad (3.65) \qquad (2.80) \quad (-2.42) \qquad (2.51)$$

$$- 13.34 \, AT + 3.11 \, RH + 0.05 \, ST + 0.86 \, FF$$
$$(-3.27) \qquad (1.94) \qquad (2.07) \qquad (2.23)$$

$$R^2 = 0.76 \qquad\qquad \bar{R}^2 = 0.58$$

where L = the number of worker-days per ha of soybeans, K = the value in U.S.$ of nonlabor production costs per ha, AT = average annual temperature (Celsius), RH = average annual relative humidity (percent), ST = soil temperature (Celsius), and FF = frost-free days.

The multiple regression results show that yield is a curvilinear function of labor and capital, that temperature increases have a negative impact on soybean yield, and that all other environmental factors increase soybean yield. All variables were significant at the 10% level or better, as indicated by the t-values in parentheses. These results help to explain why variations in factor proportions and agroclimate lead to such great regional variability of soybean yields in China.

PROSPECTS FOR THE FUTURE

Price policies have been shown to be an effective stimulus to soybean output. If soybean prices continue to increase relative to other crops, output may continue to grow. The recent decision to price soybean in relation to its nutritional advantages by giving it a grain-equivalent rating of two in quota fulfillment (Wiens, 1982) augurs well for continued strong price stimulus. China is currently emphasizing rural industrialization, 15-year tenure in production contracts, investments in marketing and processing infrastructure, and the shifting of agricultural labor to rural industries. All these policies could increase the attractiveness to farmers of growing soybeans, which are relatively labor-extensive, improve the soil, and may be processed in crushing plants. In this regard, the continued growth of China's population and per capita income will increase the demand for the soybean oil and oilmeal which come from such plants.

China is also moving to correct the previous underemphasis upon genetic and production research in soybeans. The crossing of low-yielding Chinese cultivars with wide genetic base and high-yielding American cultivars with narrow genetic base gives hope for higher-yielding cultivars adapted to each of China's five producing regions.

If the above trends continue, efforts to achieve maximum genetic yield potential within a rational economic environment should allow soybean to play its fullest role in China's cropping patterns, domestic consumption, and even exports.

NOTES

P. H. Calkins, Associate Professor, and J. C. Ma, Graduate Student, Department of Economics, Iowa State University.

The authors wish to thank soybean research scientists throughout China who provided them with micro-level data on soybean production.

REFERENCES

Anonymous. 1983. Handbook of agricutural technology economics. Agricultural Publishing House of China, Beijing.

Buck, J. L. 1937. Land utilization in China. University of Chicago Press, Chicago.

Calkins, P. H. and K. R. Huang. 1978. Soybean Production in Taiwan, a Farm Survey. AVRDC Tech. Bull. No. 11 (78-89). Shanhua, Taiwan.

Hu Ji-Cheng, Guo Shou-guei and Yu Zi-lin. 1983. Major diseases and pests of soybeans in China. p. 52-55. In Soybean research in China and USA, Intsoy Series Number 25. University of Illinois, Urbana.

Ma Rhu-hwa and Zhang Kan. 1983. Historical development of soybean production in China. p. 16-18. In Soybean Research in China and the United States, Intsoy Series Number 25. University of Illinois, Urbana.

Pang Chung-min and A. John DeBoer. 1983. Management decentralization on China's state farms. Am. J. of Agric. Econ. 65:657-666.

Shandong Agricultural College. 1980. Crop cultivation. Agricultural Publishing House of China, Beijing.

Shen, T. H. 1951. Agricultural resources of China. Cornell University Press. Ithaca, New York.

Shih Sheng-han. 1974. A preliminary survey of the book Ch'i Min Yao Shu: An agricultural encyclopedia of the 6th century. Science Press, Peking.

Wang Jin-Ling. 1983. Ecological distribution of soybean cultivars in China. p. 26-31. In Soybean Research in China and USA, Intsoy Series Number 25. University of Illinois, Urbana.

Wiens, T. 1982. Technological change. In R. Barker and R. Sinha (eds.) The Chinese agricultural economy. Westview Press, Boulder, Colorado.

DISTRIBUTIVE IMPLICATIONS OF THE GROWTH IN SOYBEAN PRODUCTION IN BRAZIL

Fernando Homem de Melo

This paper will focus on the distributive implications of technological innovations, and in particular concentrate on their effects on low income food consumers in Brazil. In the words of Singer and Ansari, "Even the ultimate objective of development is a great deal more than a mere increase in per capita income; questions relating to the use and distribution of this income are as important dimensions of development policies as its increase" (Singer and Ansari, 1978, p. 42). My main reference point will be an economy characterized as "semi-open", in the context of Myint's conception, in which "...a large part of the domestic economy must remain insulated from the impact of foreign trade and comparative costs..." (Myint, 1975, p. 332). In such a case, I intend to show that a pattern of technological change concentrated on exportables can impair the growth of domestic food crops, alter relative prices (domestic-exportables) and bring negative effects (in terms of real income) for low income consumers. My central reference will be the large expansion of the soybean crop in Brazil, resulting from a process of technological change, and the consequent distribution effects during 1967 to 1979, as well as the most recent pattern of soybean growth in Brazil's Center-West region.

THE TWO PHASES OF SOYBEAN GROWTH IN BRAZIL

The soybean, in Brazilian agriculture, represents the most recent and important example of expansion in acreage during a short time interval, from the early sixties to the early eighties. Two main phases characterized such an expansion. First, from the early sixties to 1979-80, representing large increases in cultivated hectarage in Southern Brazil, primarily in the states of Rio Grande do Sul and Paraná. Second, from the late seventies to the present, when the Center-West region, including the states of Mato Grosso do Sul, Mato Grosso, Goiás and Minas Gerais, showed significant increases in cultivated hectarage of soybean.

In 1960, the total soybean area in Brazil was only 177 kha, highly concentrated in the state of Rio Grande do Sul with 159 kha (Paraná had only five kha). By 1972, these figures were, respectively, 2,191, 1,460 and 453 kha. By 1980, the totals were 8,774, 3,988 and 2,411 kha. From then to 1983 total soybean hectarage in Brazil and in the states of Rio Grande do Sul and Paraná declined, with a first recovery occurring in 1984 (9,466 kha in Brazil as a whole).

On the other hand, total soybean hectarage in Brazil's Center-West region in 1977 was only 580 kha, while by 1984 it had reached 2,601 kha. The old state of Mato

Grosso (now divided into Mato Grosso do Sul and Mato Grosso) is the largest producer in the region with 1,696 kha in 1984 and an annual rate of growth in cultivated hectarage of 19.1% during 1977 to 1984. In Goiás, total soybean hectarage in 1984 was 572 kha and there was an annual rate of growth of 28.5% during 1977 to 1984, while for the state of Minas Gerais these figures were 333 kha and a growth rate of 17.6% during 1977 to 1984.

In 1984, soybean was the second crop in Brazil in terms of cultivated hectarage, the first one being corn (soybean, 9,466 and corn, 12,402 kha) and, is very likely to be the first one in value of exports (close to US$ 3.0 billion and 12% of total Brazilian foreign exchange revenue). As a result, it is a very important crop for producers' income and for the country as a whole, much more so nowadays with the difficult macroeconomic situation caused by external debt. However, such great importance only raises analytical interest in learning about the economic factors behind the rapid expansion in area and production of the soybean in Brazil, as well as its distributive implications.

Three factors can be mentioned to explain soybean growth in Brazil: (a) the adoption of technological innovations by producers, with significant effects on yields; (b) the introduction, in 1968, of the system of exchange minidevaluations at short time intervals, as compared to the previous one of a fixed exchange rate over long periods with high inflation; (c) a favorable period of international prices for soybeans, mainly during the first half of the seventies. In this paper, however, only the first factor will be analyzed in detail, for the following reasons. First, factors (b) and (c) have already received attention from studies about the degree of openness of Brazilian agriculture[1] to the external market (Mendonça de Barros, 1979). Second, the increases in international prices of soybean started most obviously in 1972 when, in real terms they were 11.3% higher than the average for 1967 to 1969, and they reached maximum levels in 1973 and 1974 (Homem de Melo, 1983b, p. 19). In 1972, as already mentioned, the total soybean area already was at 2,192 kha. Certainly, this favorable period of international prices made a positive contribution to the growth of soybean hectarage. However, this was not the principal factor behind the commencement of soybean expansion in Brazil, since during the sixties international prices remained practically constant in nominal dollars and the area increase was about 2 Mha. In addition, it should be noted that several commodities had price increases during certain years in the seventies, but none of them experienced an expansion even comparable to that of the soybean.

In line with my intended emphasis on technological change, Table 1 shows a summary of agronomic research for the soybean in Brazil, in terms of new cultivars, the time of their introduction and their impact on average yields. Two of these cultivars, Santa Rosa and Hardee, were very important for crop expansion during the late sixties and early seventies. The first one originated from Campinas Agronomic Institute, São Paulo, beginning with the introduction of North American cultivars and, later on, the development of L-326 in 1958. In the mid-sixties it became commercially available in the state of Rio Grande do Sul under the name of Santa Rosa. Hardee, also of North American origin, was studied and adapted at Campinas after 1965. The information in Table 1 reveals the importance of international knowledge transfer for the process of technological change in Brazilian soybeans, mainly by obviating the need to repeat research completed elsewhere and so leading, as in the case discussed by Guttman (1978), to a decline in research costs. Also of note is that from the 48 cultivars recommended for planting in 1980, 26 had originated in national research programs and 22

Table 1. Time of introduction and adoption of new soybean cultivars in Brazil and effects on yields (from Kaster and Bonato, 1980).

Period	Average yield Brazil	New cultivars
	kg/ha	
1960	--	Amarela comum, Abura, Pelicano, Mogiana
1960-68	1,060	Hill, Hood, Majos, Bienville, Hampton
1969-74	1,394	Bragg, Davis, Hardee, Santa Rosa, Delta, Campos Gerais, IAC-2, Viçosa, Mineira
1975-80	1,541	IAS-4, IAS-5, Planalto, Prata, Pérola, BR-1, Paraná, Bossier, Santana, Sao Luiz, IAC-4, UFV-1
1980	1,740	BR-2, BR-3, BR-4, Ivaí, Vila Rica, União, Cobb, Lancer, CO-136, IAC-5, IAC-6, IAC-7, UFV-2, UFV-3, Cristalina, Dokko

came from the United States, half of which were in the form of breeding lines.

Several other agronomic aspects were emphasized over the years by the research centers (Kaster and Bonato, 1980): selection of *Rhizobium* strains; direct planting; control of weeds, diseases and pests; plant density and planting time. The indications are that in the late seventies, soybean research was one of the most developed in the country. For instance, in recent years these research centers have been involved in developing production systems for other regions, besides Southern Brazil, like the East and Center-West regions. In addition, research is aimed at developing technology specific for soybean production in regions with latitutdes below 15° S. The prospects for obtaining cultivars specifically adapted to lower latitudes, as well for knowledge about crop management, are excellent, and new in the world (trans. from Kaster and Bonato, 1980).

The technological evidence so far presented corresponds mostly to land-saving innovations, the so-called biochemicals (Hayami and Ruttan, 1971). These innovations arise through selection and cultivar improvement, including a greater response to fertilizer application. In addition, they result in larger yields, cause practically no change in the final product and reduce production costs (Kuznets, 1972). In Table 2 I present the annual average rates of growth of yields for soybean and several other crops in the states of Rio Grande do Sul, Paraná, Mato Grosso, Goiás and Minas Gerais for the period 1965 to 1981. As mentioned above these are the most important states involved in the two phases of soybean expansion in Brazil.

From the data presented in Table 2 a reasonable coincidence is noted among states for crops with higher rates of growth in yields: soybean, cotton, sugarcane, corn, tobacco, coffee and orange. With respect specifically to the soybean case, it had, over 1965 to 1981 (1975 to 1981 in Goiás), the highest growth in yields in the states of Paraná, Goiás and Minas Gerais, the second highest in Mato Grosso, and the third highest in Rio Grande do Sul. I believe this is extremely good performance, since Table 2 considers the most important crops in Brazilian agriculture. In addition, Table 2 still shows an agreement among states for those crops with constant or declining yields: cassava, edible beans, rice (excepting the irrigated crop in Rio Grande do Sul) and

Table 2. Annual average rates of growth of yields in Brazilian agriculture, Center-South states, 1965 to 1981 (three year moving average)

Rio Grande do Sul		Paraná		Mato Grosso[a]		Goiás		Minas Gerais	
Tobacco	2.63	Soybean	3.28	Cotton	3.70	Soybean[b]	9.32	Soybean	5.95
Orange	1.96	Cotton	2.58	Soybean	2.03	Cotton	4.86	Coffee	5.33
Soybean	1.86	Corn	1.90	Corn	1.62	Sugarcane	1.80	Cotton	3.64
Rice	1.21	Tobacco	1.00	Peanuts	0.00	Corn	1.66	Tobacco	3.46
Corn	1.17	Peanuts	0.00	Rice	0.00	Cassava	-1.74	Corn	1.67
Peanut	0.00	Sugarcane	0.00	Sugarcane	0.00	Rice	-2.40	Sugarcane	1.22
Cassava	-0.51	Rice	0.00	Cassava	0.00	Edible beans	-8.10	Rice	0.00
Edible beans	-2.90	Cassava	-1.16	Edible beans	-3.91			Cassava	-0.55
		Orange	-1.79					Orange	-0.58
		Edible beans	-2.08					Edible beans	-1.10
		Coffee	-3.12						

Source: Basic data of annual average yields, FIBGE-Fundação Instituto Brasileiro de Geografia e Estatística.

[a]Corresponds to the present states of Mato Grosso do Sul and Mato Grosso.

[b]Goiás, 1975 to 1981.

peanuts (excepting São Paulo). It is important to note that, among those of the first group, the only domestic food crop is corn, while among those of the second group, that is, constant or declining yields, three (cassava, rice and edible beans) out of four are domestic food crops, this concept being discussed in the next section.

The results presented so far lend support to my argument that one of the main reasons for soybean growth in Brazil is technological innovation and reduction in production costs. This seems true for the states of Rio Grande do Sul and Paraná, the most important ones in phase one of that expansion, as well as for the states of Mato Grosso, Goiás and Minas Gerais, forming the Center-West region and characterizing phase two of that expansion.

EFFECTS ON COMPOSITION OF AGRICULTURAL PRODUCT

In this section I intend to show that extraordinary growth (phase one) resulted (together with a few other exportables) in significant changes in the composition of agricultural output, leading to an inadequate rate of output growth of domestic food crops. I also hope to show that the same pattern of growth, highly concentrated in soybean, may be going on currently in the states of the Center-West region during phase two of that expansion.

Table 3 shows the annual rates of growth of production during 1960 to 1969, 1967 to 1976 and 1970 to 1979 for 14 crops, exportables and domestics. After a relatively uniform and reasonable performance for most crops during the sixties, in the seventies the country had a substantial deterioration in the performance of domestic food crops and a great expansion of certain exported ones, a process clearly led by soybeans. As can be seen, the worst cases were cassava and edible beans, with large declines, while the production levels of rice and corn stagnated during the seventies. At the same time the population was growing at an annual rate of 2.47%. If the first five domestic-food commodities of Table 3 are aggregated in terms of per capita caloric/protein availability, the result is an annual rate of decline during 1967 to 1979 of -1.34% and -1.31%, respectively. The availability of rice, corn and edible beans was only very slightly increased by

Table 3. Annual rates of growth in Brazilian production of 14 commodities.

	1960-69	1967-76	1970-79
		– % –	
1. Domestic:			
Rice	3.20	2.47	1.46[a]
Edible beans	5.37	-1.93	-1.90
Cassava	6.05	-1.86	-2.09
Corn	4.74	3.55[a]	1.75[a]
Potatoes	4.34	1.34	3.73
Onions	3.87	4.77	9.27
2. Exportables			
Soybean	16.31	35.03	22.47
Oranges	6.01	12.73	12.57
Sugarcane	3.63	5.10	6.30
Tobacco	5.30	–	6.16
Cocoa	2.55	–	3.73
Coffee	-7.10	-6.34[a]	-1.54[a]
Peanuts	5.89	-6.80[a]	-12.06
Cotton	1.51[a]	-1.99	-4.41

Source: Production data from FIBGE-Fundação Instituto Brasileiro de Geografia e Estatística,
      Homem de Melo (1983b).
[a]Not significantly different from zero at the 5% level.

imports over the period. These five domestic-food crops, in addition to cotton and pasture land, were the agricultural activities most affected by the substantial expansion of soybean in southern Brazil (Zockun, 1980). The substitution effect of technological change, in terms of crop mix, was not limited to domestic crops. Cotton, an exported crop, was also negatively affected.

For some time Brazilian agriculture has had two subsectors, exportables and domestic crops, the first open and the second closed to international transactions (Homem de Melo, 1983a). The first group has included soybean, orange, sugar, tobacco, cocoa, coffee, peanuts and cotton, while the second, with some variations over time, has consisted of, among other crops, rice, edible beans, cassava, corn, potatoes and onions, most of which are important foods for low-income families. It should be noted that at least some of the domestic crops are potential exportables, since there exists either a well-developed international market (corn, rice) or a developing one (cassava). However, in varying degrees, over time, internal cost conditions have prevented a favorable competitive position internationally (domestic prices were above international ones) and, as a result, such crops have not been exported regularly or on a significant basis.[2]

Now the distributive implications for families of consumers can be analyzed, with a process of technological change biased, in a certain time period, towards one or more of the exportable crops, basically soybeans in this case. If the case of land-saving bio-chemicals (Table 1) is considered, the individual marginal cost curves and the market supply curve of soybean would shift to the right. With a perfectly elastic export demand (in the sixties and early seventies), the cultivated hectarage of soybean should have been increased (Castro, 1974), with these effects occurring with a constant product price. This represents the specific case where all direct benefits from technological change are appropriated by domestic soybean producers, including increases in land prices[3], mainly considering the location-specific nature of such innovations. In such a case, the process of hectarage growth would tend to be directed towards soybean (or other favored crops), in addition to causing changes in the crop mix in regions already under cultivation.

In Brazil, during phases one and two of soybean expansion, both effects appear to have happened; that is, a direct substitution of soybean for other crops, as well as a process of hectarage growth highly specific for the soybean. During the sixties and the seventies, soybean hectarage increased at an annual rate of 13.6% and 17.0%, respectively, in the state of Rio Grande do Sul, while in Paraná such rates were 40.9% and 25.7%. In the same decades, the annual growth rates for total hectarage in Rio Grande do Sul were 4.0% and 5.5%, whereas in Paraná they were 4.4% in both decades.[4]

The important point to emphasize, as a result of the unbalanced pattern of technological innovations, mostly favoring soybean and a few other exportables (as well as a favorable exchange rate policy and higher international prices), is that growth rates of output of domestic food crops were negatively affected, and their internal prices were kept above international ones (Homem de Melo, 1983a). After examining the behavior of 13 food items in São Paulo during 1967 to 1979, it was noticed that those with the largest increases were cassava, edible beans, beef, pork and corn, three of them being domestic vegetable foods.

If the so-called domestic crops include important foods, in terms of budget shares of low-income families, as is the case in Brazil, the real price increases following the change in growth rates would be like a regressive tax. As a result, the unbalanced nature of the process of technological change among crops with different market characteristics (exportables and domestics), could bring a worsening of income distribution (from the expenditure side). For that scenario, it is necessary that no compensatory imports made, even when possible, following market conditions (rice and corn)[5], or alternatively, that the international market not be a supplier able to complement domestic production (edible beans and cassava). In addition, an intermediate case should be mentioned: internal prices of domestic crops stay above the export prices but below the import ones.[6]

The strong differences in consumption structures over income (expenditure) classes and regions in Brazil (Homem de Melo, 1983b), as well as the poor performance of food production (Table 3) and availabilities, are good reasons for expecting an uneven impact in terms of relative prices and real incomes for Brazilian families. This would occur through changes in market prices and consequent income effects. In an attempt to verify this effect, I estimated (Homem de Melo, 1983b) the increase of the food price index by income classes (based on weights for 1974-1975) for São Paulo, Rio de Janeiro, as well as South and Northeast regions. These indices were computed by taking the shares (weights) of each product in total food expenditures for the two states and two regions of Brazil and the observed prices in São Paulo (Cost of Living Index).

The results show that the most serious case was in the Northeast region of Brazil. When I compared the lowest and highest income classes in terms of annual rates of growth of nominal food prices, I noted that during 1967 to 1979 they were 28.6% and 26.2%, respectively. Alternatively, the cumulated increase for the lowest income class was 32.9% higher than for the highest class. For São Paulo, Rio de Janeiro and the Southern region these greater cumulated increases for the lowest income class were 10.0%, 12.7% and 8.7%, respectively.

Two main reasons may be advanced to explain such regional differences in food price increases: (a) the greater importance of cassava and edible beans for lower income families in the Northeast as compared to other regions (26.7% as compared to 2.4% among the income extremes in that region, versus 14.2% as compared to 1.4% in the South). These two commodities were the ones with greater increases in retail prices during 1967 to 1979; (b) the relatively small importance of wheat in the consumption habits of lower income families in the Northeast (4.2% as compared to 10.0% among the extremes in the Northeast, and 8.9% as compared to 7.1% in the South). It should be recalled that, beginning in 1972 the Brazilian government subsidized wheat prices to all consumers, which, in the Northeast, had a regressive effect.

Now, with such unfavorable results in terms of domestic food production, availabilities and relative prices, it is even more important to examine what is happening during phase two of soybean expansion in Brazil. To simplify, I will analyze only the changes in crop mix occurring in the state of Mato Grosso (currently Mato Grosso do Sul and Mato Grosso), the most relevant in soybean acreage in the Center-West region of the country. In Table 4 I present, for nine crops in this state, the annual growth rates of cultivated hectarage during 1977 to 1984 and the totals including and excluding wheat (a winter crop).

From the data of Table 4 it can be noted that three out of eight crops (excepting wheat, a winter crop) had acreage growth during 1977 to 1984: sugarcane (29.1%), mostly due to the alcohol program; soybean, 19.1%; and corn, 8.1% per year. In its absolute dimension, soybean is the most expressive case, with an increase of 1,284 kha, rising from a share in total hectarage of 16.4% in 1977 to 52.5% in 1984 (Total 2). On the other hand, four out of eight crops had declines in hectarage: peanuts, rice, cassava and cotton, and one crop, edible beans, had no significant change. The most relevant case is rice, a key domestic food commodity with a reduction of 629 kha between 1977 and 1984 (-8.0% per year), and its area share declining from 61.6% to 28.4% in the same period. Of all domestic food crops listed in Table 3, only corn had a positive

Table 4. Crop hectarage and annual growth rates in the states of Mato Grosso do Sul and Mato Grosso, 1977 to 1984.

Year	Cotton	Peanuts	Rice	Sugarcane	Edible beans	Cassava	Corn	Soybean	Wheat	Total 1[a]	Total 2[a]
1977	68	29	1.547	11	116	61	247	412	36	2.548	2.513
1978	46	21	1.526	14	113	58	179	500	40	2.497	2.458
1979	51	20	1.326	18	76	56	175	599	104	2.506	2.403
1980	49	26	1.397	20	147	39	192	877	122	2.946	2.824
1981	51	12	1.275	33	115	40	242	896	81	2.825	2.745
1982	46	8	1.110	47	149	39	313	1.037	164	2.991	2.827
1983	45	6	945	58	146	42	324	1.227	114	2.907	2.793
1984	41	2	918	81	127	38	324	1.696	–	–	3.228
Annual rate, %	-4.67	-33.93	-7.95	29.12	5.92[b,c]	-7.13	8.15	19.12	21.81[c]	3.14[c]	3.06[d]

Source: Basic data from FIBGE-Fundação Instituto Brasileiro de Geografia e Estatística.
[a]Total 1 includes wheat and Total 2 excludes wheat.
[b]Not different from zero at the 10% significance level.
[c]1977 to 1983.
[d]Excludes edible beans.

hectarage growth rate. Recalling the information about growth rates for yields in several states including Mato Grosso (Table 2), the evidence of changes in crop mix in this state seems to be in accordance (the exception being cotton) with expectation; that is, changes in composition are affected by an unbalanced pattern of technological innovatons among crops, with soybean leading the process. This is an indication that phase two of soybean expansion in Brazil is repeating the negative effects observed in phase one, in terms of growth rates in area and output of domestic food crops.

CONCLUSIONS

The evidence presented leads me to conclude that lower income families have suffered the most from the behavior of food prices during 1967 to 1979. Phase two of soybean expansion may be repeating phase one in terms of its negative effects on production of domestic food crops. To the extent that real income distribution is taken into account when deciding on public investment in agricultural research, my results indicate the need to increase such investments for domestic food crops. In the beginning, the benefits will appear as lower food prices for consumers, since internal prices are above international ones. Later on, society as a whole should also gain since, as a result of declines in costs and prices, some of the crops would become exportable and provide foreign exchange (rice, corn). For crops without a well-developed international market (edible beans, cassava), most of the benefits would go to consumers, mainly to low-income ones. In both cases the country would have a good result in terms of income distribution.

NOTES

Fernando Homem de Melo, Professor Titular, Department of Economics, Universidade de São Paulo, Caixa Postal 11.474, 01000 - São Paulo-SP, Brasil.

[1]In value terms the share of agricultural exports over total production increased from 11.1% in 1967, before the change in the exchange rate system and the rise in international prices, to 22.5% in 1976 (see Mendonça Barros, 1979, p. 23).

[2]Corn, among the domestic crops, was the only one showing some exports, although quite irregularly, during 1967 to 1979. Among those thirteen years, in four exports were less than 1.2%; in two, 3.4% and 5.1%; and in seven, between 6.8% and 10.3% of domestic production.

[3]This point is theoretically discussed by Schuh (1976) in the context of United States agriculture, concluding that "it is not likely that a consumer-dominated body politic will be willing to support domestic R&D as they have in the past". This may be one of the reasons for the success of soybean research in Brazil; that is, large gains to producers.

[4]The annual rates of growth in hectarage for domestic food crops in the seventies were: (a) Rio Grande do Sul: corn, zero; rice (irrigated), 3.8%; edible beans, -3.6%; cassava, -3.8%; (b) Paraná: corn, 1.4%; rice, zero; edible beans, zero; cassava, -6.9% (Homem de Melo, 1983a).

[5]Only in 1978, 1979 and 1980 were food imports of some significance, basically for corn and rice and mostly due to serious crop failures in 1978 and 1979.

[6]See the discussion about import and export points in Hinshaw (1975).

REFERENCES

Castro, J. P. R. 1974. An economic model for establishing priorities for agricultural research and a test for the Brazilian economy. Ph.D. Dissertation. Purdue University Library, West Lafayette, Indiana.

Guttman, J. M. 1978. Interest groups and the demand for agricultural research. J. Political Econ. 86: 467-484.

Hayami, Y. and V. W. Ruttan. 1971. Agricultural development: an international perspective. The Johns Hopkins Press, Baltimore.

Hinshaw, R. 1975. Non-traded goods and the balance of payments: further reflections. J. Econ. Lit. 13:475-478.

Homem de Melo, F. 1983a. Commercial policy, technology and food prices in Brazil. Quart. Rev. Econ. Bus. 23:58-78.

Homem de Melo, F. 1983b. O problema alimentar no Brasil: a importância dos desequilíbrios tecnológicos. Editora Paz e Terra, São Paulo.

Kaster, M. and E. R. Bonato. 1980. Contribuição das ciências agrárias para o desenvolvimento: a pesquisa em soja. Revista de Economia Rural 84:405-434.

Kuznets, S. 1972. Innovations and adjustments in economic growth. The Swedish J. Econ. 74:431-451.

Mendonça de Barros, J. R. 1979. Política e desenvolvimento agrícola no Brasil. p. 9-35. *In* A. Veiga (ed.) Ensaios sobre política agrícola brasileira. Secretaria da Agricultura, São Paulo, Brazil.

Myint, H. 1975. Agriculture and economic development in the open economy. p. 327-354. *In* L. G. Reynolds (ed.) Agriculture in development theory. Yale University Press, New Haven.

Schuh, G. E. 1976. The new macroeconomics of agriculture. Am. J. Agric. Econ. 58:795-801.

Singer, H. and J. Ansari. 1978. Rich and poor countries. George Allen and Unwin, London, 1980.

Zockun, M. H. G. 1980. A expansão da soja no Brasil: alguns aspectos da produção. Instituto de Pesquisas Econômicas, Universidade de São Paulo. Ensaios Econômicos 4, São Paulo.

RISK MANAGEMENT IN U.S. SOYBEAN PRODUCTION AND MARKETING

R. N. Wisner and R. W. Jolly

Since their introduction into midwest agriculture, soybeans have come to play an integral role in farmers' risk management strategies. As we discuss later in this paper, the soybean enterprise both necessitates and contributes to the management and control of farm production and price risk.

When a farmer makes a production or marketing decision, many factors that will influence the outcome are unknown. Examples of key variables include rainfall, hail, pest outbreaks, cultivar performance, machinery performance, prices and government programs. One can characterize each of these unknown variables with a probability density function. Since the variables are not independent, one should probably think in terms of a joint density function.

Within this conceptual framework, we define risk and risk management a bit more completely. (For a more detailed discussion, see Anderson et al., 1977; Jolly, 1983.) Risk originates in two basic ways. First, because yields and prices follow a distribution, the farmer cannot guarantee an outcome. Second, the parameters of the distribution function are unknown. This latter case would also include variables omitted from consideration in making the decision. In essence, we are equating risk with variability and estimation error.

Risk creates two types of potential losses in agriculture. A producer can experience an adverse outcome such as a low yield and/or low price. Or, a particular type of decision may remove the opportunity for a high yield or other favorable outcome. Either way, the farmer may experience a loss that he or she would prefer to avoid if possible.

The impact of risk on the farm business is pervasive. In traditional economic theory, risk creates a source of friction that reduces input use and output below their optimal riskless levels. This friction can retard firm growth, lead to excessive diversification or selection of lower-income enterprises. In more catastrophic situations, risk can lead to loss of the farm business and human stress and suffering.

For these reasons, risk must be managed. Or we would prefer to say, risk must be incorporated more completely into the decision making and management processes of the firm. Farmers can manage risk in two major ways: by controlling risk exposure and by reducing the potential negative impact of risk on the farm business.

Risk exposure can be controlled in several ways including enterprise combinations, irrigation, diversification, machinery selection for timely field operations and marketing strategies. Risk impacts can be controlled by creating a business structure that can withstand adverse impacts or exploit favorable events. For example, a farmer might use credit reserves, crop insurance, or choose a resilient financial structure to improve his or her ability to withstand adverse production or marketing outcomes. Frequently, such management strategies are long run in nature. The resulting characteristic of the farm business is often referred to as its risk-bearing ability.

Risk-bearing ability is distinguished from attitude toward risk (or risk attitude) in the following way. The former refers to the ability of the business to bear risk. Risk attitude reflects the willingness or emotional ability of the farmer to live with risk—in essence, to live with the unknown. While the two concepts are different, the risk-bearing ability of the business will, in time, partially reflect the risk attitudes of the owner and other participants in the management of the farm.

RISK MANAGEMENT IN SOYBEAN PRODUCTION

Risk management is an important element in soybean production. The major risk control strategies in production are those directed toward losses from pests and weather. In many cases, a given strategy will influence both sources of production risk.

One example of a production risk-management strategy is the use of herbicides, which have greatly increased the producer's ability to control grass and other weed population in soybeans. Another risk-control strategy for the midwest is drainage. Many poorly drained northern soils are vulnerable to excessive moisture. Drainage investments can reduce yield variability and increase modal yields. Both results are desirable from a risk management perspective. The underlying economic question is whether or not the improvement in soybean incomes and reduced variability is sufficient to offset the investment costs of a drainage system.

Soybeans also contribute to a farmer's risk-management strategy through enterprise diversification. The corn-soybean rotation is a mainstay in cornbelt cash crop production. Its prominence and success are due in part to its value in risk management.

Diversification in its most elementary form involves linking two or more enterprises that exhibit negative correlation in returns. The result is a more stable income for the combination than for any individual enterprise. In some cases the expected income for the diversified business is less than some of the individual enterprises. However, trading less income for greater stability is frequently preferred by farmers attempting to manage risk.

Soybeans and corn have a number of favorable characteristics for diversification. They are biologically distinct. Consequently, they respond to weather risks in different ways. Although yields of corn and soybeans will tend to move in the same direction in response to weather, they frequently exhibit differential yield responses. In other words, in a drought, both corn and soybean yields may decline, but corn declines relatively more than soybeans. This creates an advantage to a diversified crop rotation.

Cash production costs for corn following beans are generally less than for continuous corn. Also long-term yields frequently are higher for corn grown in rotation. Both of these income improving factors contribute to better risk management. Corn and soybeans are also sold in different markets. This too creates advantages for diversifying.

MARKETING AND RISK MANAGEMENT

In the two decades before the early 1970s, the largest source of risk facing U.S. grain and soybean producers usually was weather-related, with sharply reduced yields possible due to drought, hail, floods or other natural disasters. Grain and oilseed prices were stabilized to a considerable extent by government price support programs and heavy reliance on domestic markets. Then in the early 1970s, the risk of adverse price changes increased dramatically as the U.S. dollar was devalued, major trading partners moved to a system of floating currency exchange rates, the Soviet Union and China became large grain and oilseed importers and demand grew rapidly in other nations (Wisner and Chase, 1983). This combination of events generated grain and oilseed markets which are highly sensitive to changing U.S. and world crop prospects, interest rates, currency values and international political developments. For a mid-sized, midwest, cash-grain farm, price volatility in the last few years has generated potential variations in annual income of $50,000 or more—depending solely on how and when the corn and soybean crops were marketed. And, for the majority of U.S. grain producers, price volatility has become a much greater source of financial risk than yield variability.

Financial risk-bearing ability varies greatly from one farm to another, depending on the total amount of outstanding debt, debt-to-equity ratios, repayment periods for the outstanding debt and interest rates being paid. In general, farmers with little or no debt have strong financial risk-bearing ability, whereas those who are heavily indebted for land, machinery or other items have very little capacity to bear financial risks. One indication of variations in soybean growers' abilities to bear market risks is shown in Table 1. These data show estimated total 1984 cash-flow and machinery replacement costs per hectare, per bushel and per ton for producing soybeans in Iowa under four different financial structures. The "recent buyer" column reflects a situation where approximately one-half of the land was purchased on an 11% land contract at peak land values of the early 1980s. The other half of the land is owned debt-free. With a normal soybean yield of 2.55 t/ha, this operator requires $364/t ($9.89/bu) to cover on-going cash outlays, depreciation and expenses for supporting his/her family. Break-even prices at this level would be 53% above average market prices for the 1981-82 through 1983-84 marketing years. If yields can be increased to 3.09/ha through excellent production management, the cash-flow, break-even price drops to about $300/t

Table 1. Example cash flow requirements for soybeans following corn—one hectare.

	Example 1 Recent buyer	Example 2 Full owner	Example 3 Cash tenant	Example 4 Share tenant (50% of crop)
	— U.S. dollars —			
1. Seed, fertilizer, pesticides, lime, crop insurance, miscellaneous, and interest on operating capital	172.90	172.90	172.90	91.39
2. Machinery operating and drying	41.99	41.99	41.99	34.58
3. Machinery ownership and replacement	79.04	59.28	79.04	79.04
4. Land ownership				
. Taxes	39.52	39.52	0	0
. Financing	553.28	0	0	0
. Cash rent	0	0	291.46	0
5. Family living plus income taxes	41.99	98.80	41.99	41.99
6. Total per hectare	928.72	412.49	627.38	247.00
7. Total per metric ton @ 2.55 t/ha	364.20	161.76	246.03	193.72
8. Total per metric ton @ 3.09 t/ha	300.56	133.49	203.04	159.88
9. Total per bushel @ 38 bu/acre	9.89	4.39	6.68	5.26
10. Total per bushel @ 46 bu/acre	8.17	3.80	5.52	4.35

($8.17/bu)—26% above the most recent three-year, Iowa average soybean price. In either case, this producer's ability to bear market risks is very limited. If pricing opportunities are available at or above these break-even levels, he or she should very seriously consider marketing a sizable part of the soybean crop in order to insure financial survival. While pricing opportunities have reached this level for brief periods of time, there is a very low probability that marketing-year average prices would be at or above these levels and a high probability they would be substantially lower.

For the other three example producers, risk-bearing ability is much greater. The "full-owner" example (with no land or machinery debt) can cover cash-flow and machinery depreciation expenses by marketing soybeans for as little as $162/t ($4.39/bu) if necessary. With excellent management and soybean yields at 3.09 t/ha, his/her break-even price drops as low as $140/t ($3.80/bu). Because of low cash-flow costs, soybean growers in this situation can devote very little attention to marketing and still survive financially. But for those in the "recent buyer" situation, careful attention to both marketing and production management will be absolutely essential for business survival. Cash and crop-share renters typically fall between these two extremes.

U.S. soybean growers have several tools available for managing marketing risks. For example, forward pricing tools such as hedging in the futures market or contracting at local elevators are procedures that allow the pricing decision to be separated from the time of harvest. These tools allow a producer to stretch out the time period in which he/she searches for profitable prices—to as much as 15 months before harvest.

When a crop or some portion of it is priced well in advance of harvest, production risk obviously becomes important in the decision-making process. Producers in this case face potential problems if more of the crop is sold than is eventually produced. The decision process is especially difficult if pre-harvest pricing opportunities equal or exceed minimum cash-flow requirements and are well above the average price of recent years. In this situation, the producer needs to ask himself which is the greater risk, the risk of sharply below normal yields, or the risk of a sharp decline in prices to levels that would endanger financial survival. In these circumstances, it also is important to consider tools available for controlling production risks, such as hail insurance and all-risk crop insurance. Some people view marketing part of the crop before harvest as a form of speculation. However in most cases it is probably less speculative than buying land on a 30-year variable interest rate mortgage and assuming yields and season average prices will automatically be sufficient to meet the resulting cash-flow requirements. In many parts of the U.S. corn-soybean belt, recent history indicates production risks are less than market risks and/or can be controlled sufficiently to price a substantial amount of production without major risk exposure.

Table 2 shows a 14-year corn and soybean yield history for two counties in Iowa, to illustrate geographic variations in production risk. Since we have not attempted to remove the upward trend in yields, these data may slightly exaggerate county yield variability. However, they reveal two important patterns: (1) corn yield variability in Iowa has been greater than in soybeans and (2) year-to-year yield variability in northeast Iowa (Blackhawk County) is much less than in west central Iowa (Harrison County). There are at least two marketing implications from these data. First, due to relatively lower yield variability, midwest soybean growers can safely market a higher percentage of their expected soybean production before harvest than would be the case with corn. Secondly, because of large variations in production risk from one area to another and in financial risk-bearing ability from one farm to another, there is no one best marketing plan that will fit all procedures. A successful marketing strategy must be tailored to the individual farm situation. Specific objectives of a marketing

Table 2.   Crop yield history for two Iowa counties.

Year	Blackhawk County		Harrison County	
	Corn yield	Soybean yield	Corn yield	Soybean yield
	— t/ha —			
1983	6.96	2.57	5.46	2.11
1982	8.15	2.56	6.88	2.51
1981	8.44	2.80	5.90	2.63
1980	7.39	2.81	5.02	2.21
1979	8.64	2.59	7.45	2.41
1978	7.43	2.55	7.23	2.60
1977	6.89	2.39	4.61	2.55
1976	5.78	2.06	4.75	1.81
1975	6.00	2.28	5.28	2.37
1974	6.08	2.14	3.29	1.87
1973	6.60	2.29	6.53	2.18
1972	7.21	2.42	7.16	2.35
1971	6.33	2.05	6.02	1.97
1970	5.98	2.18	4.58	1.81
Average	6.99	2.41	5.73	2.24
Extreme high yields, % above average	+23.5%	+16.8%	+30.0%	+17.4%
Extreme low yields, % below average	-17.4%	-14.8%	-42.6%	-19.2%
% of time yields more than 17% below average	7.1%	0	28.6%	14.3%

plan will vary depending on risk preferences of the operator. For many financially leveraged producers, business survival will be a major objective. For those in the debt-free category, other important objectives influencing marketing might include growth of the farm business earning a specific rate of return on investment or generation of retirement income.

Table 3 shows an example of how yield variability of one specific farm differs from its respective county average yields. The example farm is Iowa State University Foundation's Hancock Farm, located in Hancock County, Iowa, in the north central part of the state. It is a 100 ha unit in the Clarion-Webster Nicollet Soil Association area with predominant soils being silty clay loam. These data indicate yield variability on individual farms may be greater or less than that of averages for the county in which it is located, depending on soil types and a number of other factors. To develop an effective marketing strategy, grain producers ideally should have a record of average yields per hectare or per acre for their farm for the past several years, similar to the one shown in Table 3. It also should be obvious that production practices which reduce year-to-year yield variability on the individual farm will facilitate more effective marketing and financial risk management.

Table 4 shows an example farm situation illustrating the impact of alternative marketing and production management strategies on the net cash-flow position of a moderately leveraged midwest soybean farm in the last four years. The hedging strategy indicated here would involve sale of futures contracts before harvest to price

Table 3. Corn and soybean yield variability for an example farm in Hancock County in north central Iowa, 1970-1983 (L. D. Trede, pers. commun.).

	Corn		Soybean	
	Hancock Farm	Hancock County	Hancock Farm	Hancock County
	— t/ha —			
1983	7.26	6.25	2.44	2.39
1982	7.95	7.68	2.90	2.51
1981	9.65	9.00	3.04	2.48
1980	9.37	8.01	2.82	2.68
1979	8.91	8.21	2.49	2.34
1978	8.22	7.93	2.69	2.45
1977	8.47	7.08	2.62	2.52
1976	7.84	6.20	2.69	2.29
1975	6.34	5.96	2.89	2.33
1974	6.59	5.73	2.42	2.16
1973	6.77	6.73	2.49	2.23
1972	7.46	7.40	2.55	2.55
1971	7.21	6.84	2.42	2.22
1970	6.90	6.25	2.22	2.18
Average	7.78	7.09	2.62	2.38
Extreme high yields, % above average	24.0%	26.9%	16.0%	12.6%
Extreme low yields, % below average	18.5%	19.2%	15.3%	9.2%
% of time yields more than 17% below average	7.1%	7.1%	0	0

Table 4. Change in net total cash position for a moderately leveraged 200 hectare Iowa soybean farm with alternative marketing strategies.[a]

Crop	Harvest price	Hedge price 75% of crop[b]	Net hedge return		Net return, harvest sale	
			Average yields	Top yields	Average yields	Top yields
			— U.S. dollars —			
1981	5.92/bu.	7.70/bu.	-14,155	+14,865	-38,950	-15,150
1982	4.98	6.55	-35,000	-10,377	-57,950	-37,460
1983	8.26	7.85	-902	+30,908	+4,940	+37,980
1984	6.00?	8.75	+1,188	+33,438	-38,000	-14,000
Cumulative total[c]			-48,887	+68,834	-129,960	-28,630

[a]Cash-flow cost with average yield of 2.55 t/ha is $294.40/t ($8.00/bu); with yield of 3.09 t/ha it is $242/t ($6.60/bu).

[b]Other 25% of production sold at harvest.

[c]Excludes interest charge on negative cash-flow.

three-fourths of expected production for fall delivery, with the other one-fourth of the crop sold at harvest. These marketing strategies are contrasted with sale of the entire crop at harvest, under "average yield" and "top yield" situations. For this type of farm, the results in Table 4 show clearly that careful attention to both marketing and production management are essential for economic survival.

Much more could be said about risk-management strategies in soybean production and marketing. We could discuss seasonal probability distributions of prices, risk levels associated with storing soybeans for various lengths of time, timing of preharvest and postharvest marketings, linking a marketing plan to the amount and timing of cash-flow needs, alternative strategies involving use of the futures market and the potential role of the forthcoming commodity options markets in risk management and market-ing (Wisner, 1981; 1982). For farmers with limited risk-bearing ability, these all are potentially important subjects in the development of an overall risk-management strategy. However, space limitations prevent us from discussing their potential roles in more detail.

CONCLUSIONS

In the economic environment of the present and foreseeable future, moderately and heavily indebted U.S. soybean growers will need to give special attention to risk management and risk-bearing ability in order to insure financial survival. Risk manage-ment involves (1) recognition that almost all facets of agriculture involve uncertain outcomes, (2) ordering production and marketing strategies according to a realistic assessment of the degree of risk involved and (3) selecting strategies that match the operator's financial risk-bearing ability and risk attitude.

NOTES

R. N. Wisner, Professor, and R. W. Jolly, Associate Professor, Department of Economics, Iowa State University, Ames, IA.

REFERENCES

Anderson, J. R., J. L. Dillon and B. Hardaker. 1977. Agricultural decision analysis. Iowa State Uni-versity Press. Ames, Iowa.

Jolly, R. W. 1983. Risk management in agricultural production. Am. J. Agric. Econ. 65:1107-1113.

Wisner, R. N. 1981. Grain marketing tools and techniques for managing risk. Department of Eco-nomics marketing mimeo, M-1224. Iowa State University. Ames, Iowa.

Wisner, R. N. 1982. Advanced hedging techniques. Department of Economics marketing mimeo, M-1233. Iowa State University. Ames, Iowa.

Wisner, R. N. and C. Chase. 1983. World food trade and U.S. agriculture, 1960-82, third annual edi-tion. World Food Institute, Iowa State University. Ames, Iowa.

RISK MANAGEMENT IN U.S. SOYBEAN PRODUCTION AND MARKETING IN THE SOUTHEAST

T. E. Nichols, Jr.

Several previous studies have analyzed hedging strategies as risk management tools (Purcell et al., 1972; Purcell and Richardson, 1977; Kane et al., 1984; Leuthold, 1975; Miyat and McLemore, 1982). Results from these studies indicate that the ability of a soybean grower to increase net return or minimize losses through hedging depends on where and how much he hedges. A number of strategies can be used to reduce risk and increase net returns. However, these same strategies may reduce the potential for profit in some years.

This study analyzed those hedging strategies that appeared to offer southeastern growers their greatest potential to maximize net returns or minimize price risk during the period 1970-83. One alternative evaluated was agricultural options, even though they are not expected to be available until October 1984.

The data were divided into two representative groups to determine how variations in pricing and marketing strategies might affect the producer's net returns in "short" and "normal" crop years. Yields are below the trend line in a "short" year and above it in a "normal" year. Short crop years were 1974-75, 1976-77, 1977-78 and 1980-81. All other years from 1970-83 were "normal."

Producer strategies were divided between preharvest and postharvest pricing decisions. Emphasis was on comparing the results of various forward pricing strategies with traditional cash market sales.

CRITERIA USED IN COMPARING PRICING STRATEGIES

Producers use pricing strategies to accomplish one of three goals: (1) to obtain a higher mean net return compared to that in the cash market, (2) to obtain a lower variance of return (less risk) compared to that in the cash market and (3) to obtain both benefits. The risk or variability associated with each level of return is especially important to young farmers and other growers who are highly leveraged.

Pricing strategies that reduce the variance of returns and have similar or higher levels of mean net returns per bushel when compared with the unhedged position, will be considered "superior strategies." Strategies that have both higher mean net returns and variances or lower mean net returns and variances when compared to the unhedged operation will be classified as "indeterminate strategies." Strategies that yield lower net returns with more risk than the cash market will be designated "inferior strategies."

HEDGING PROCEDURES

The 1970-83 crop years for North Carolina were examined under the following assumptions: (1) Breakeven prices were calculated by adding all production expenses except a return to management and risk (Nichols, 1982). (2) Futures prices were adjusted by the Wilson, North Carolina basis to reflect local cash prices in central North Carolina markets. (3) The November futures contract was used for preharvest strategies, and the March and July futures contracts were used for postharvest strategies. (4) The brokerage fee charged was $50 per roundturn for futures and options, and the initial margin on futures contracts was $1,000 per contract regardless of price. (5) The interest foregone on soybeans in storage was at the annual rate charged by the Production Credit Association, and the interest on money used in the margin account or premium on put options was at the annual prime rate during 1970-83. (6) Physical storage costs were calculated at 1½ cents per bushel per month for the number of months stored after November 1. (7) Premiums for put options were based on 8% of the closing futures price on the date puts were purchased. When cash market prices for soybeans declined, puts were sold (offset) one month prior to the delivery date of the unhedged futures contracts. (8) Price-later contracts were based on a 15 cents minimum deferment charge plus 1½ cents per bushel per month service charge. (9) To account for the loss in value of the dollar caused by inflation, incremental dollar returns to management and risk were adjusted by the index of prices paid by farmers during the 1970-83 period.

PREHARVEST HEDGING STRATEGIES

Strategy 0: No hedge. The traditional cash market in which the farmer speculates that prices will be profitable at harvest serves as the benchmark against which other strategies are compared. Net returns are calculated by subtracting breakeven price per bushel from cash price in central North Carolina markets on November 1. Production unit is assumed to be 10,000 bushels.

Strategy 1: Hedge 50% when Localized Futures (LF) are more than the Breakeven Price (BE). In late fall or early winter, the farmer calculates his estimated breakeven price for next year's crop. Also, he determines the expected local price for the new crop at harvest by adjusting the November futures price by the local basis (difference between cash and futures prices). He hedges one 5,000-bushel contract, or 50% of the expected crop, any time the localized futures price equals or is greater than the breakeven price. He remains with the futures contract throughout the production period and offsets it at harvest. Net return from this strategy is the difference between results from hedging and the breakeven price for the 5,000 bushels hedged plus the difference between the cash market price and the breakeven price for the 5,000 bushels not hedged.

Strategy 2: Hedge 50% when LF is greater than BE plus 25 cents. Same strategy as No. 1 except that returns to management and a risk of 25 cents per bushel are added to the breakeven price.

Strategy 3: Hedge 50% when LF is greater than BE plus 50 cents. Same strategy as No. 1 except that returns to management and a risk of 50 cents per bushel are added to the breakeven price.

Strategy 4: Hedge 50% when LF is greater than BE plus 75 cents. Same strategy as No. 1 except that returns to management and a risk of 75 cents per bushel are added to the breakeven price.

Strategy 5: Hedge 50% in March if LF is greater than BE. Same strategy as No. 1 except that the farmer waits until mid-March to hedge. At that time, he hedges 5,000 bushels in the November contract without comparing localized futures with breakeven price.

Strategy 6: Hedge 25% in March and 25% in June. Same strategy as No. 1 except that the farmer hedges 2,500 bushels of his crop in mid-March and another 2,500 bushels in mid-June regardless of localized futures price and breakeven price.

Strategy 7: Hedge 25% in March, 12.5% in June and 12.5% in July. Typically, a period of dry weather each summer threatens yields and causes futures prices to rally. The farmer takes advantage of the price improvement that might come from this weather market.

Strategy 8: Sell 50% when Cash Forward Contract Price (CP) is greater than BE. Instead of hedging in the futures market, a farmer sells his expected crop (5,000 bushels) to a local buyer for a guaranteed (fixed) price whenever that price is equal to or greater than the breakeven price.

Strategies 9, 10, 11: Same as Strategy 8 except that the breakeven price is increased by 25, 50 and 75 cents per bushel, respectively.

Strategies 12, 13, 14: Same as Strategy 8 except that the farmer waits until mid-March to sell 50% of his crop with Strategy 12; he sells 25% in mid-March and 25% in mid-June with Strategy 13, and he sells 25% in mid-March and 12.5% in June and July, respectively, with Strategy 14.

Strategy 15: The farmer buys a put option for 100% of crop in March at a strike price equal to the closing price of the November futures price. If the cash market price at harvest is higher than the strike price, the producer lets the put option expire and sells the crop at a higher price. If, however, the market price declines after the put option is purchased, the farmer sells the option before it expires.

Strategies 16, 17: Same as Strategy 15 except that the farmer buys puts for 50% of the crop in mid-March, and 50% in mid-June with Strategy 16. He splits purchases into 50% March, 25% June and 25% July with Strategy 17.

POSTHARVEST HEDGING STRATEGIES

Strategy 0: No hedge. The farmer stores his soybeans on the farm unhedged at harvest (November 1) and sells 100% of the crop in July. Returns from this strategy are the difference between the cash market price in mid-July and the cost of storage.

Strategy 1: Store on farm and hedge 100% in March futures. The hedge is placed in March futures on November 1 when beans are stored, and is lifted on March 15 when beans are sold on the cash market. All stored beans are hedged. Net returns are calculated as the difference between returns from hedging and the costs of storage.

Strategy 2: Store on farm and hedge 50% in March futures and 50% in July futures. Same as Strategy 1 except that sales are split into two time periods.

Strategies 3, 4: Same as Strategies 1 and 2 except that the beans are stored in a commercial elevator instead of on-farm.

Strategy 5: Sell 100% in March futures on basis contract. The farmer stores his beans in a commercial elevator at harvest and enters into a contract with the elevator to price beans at a 9 cents discount to March futures. Net returns to storage are the difference between the contract price (March futures—9 cents basis) and the cost of storage (1½ cents per bushel per month).

Strategy 6: Sell 50% in March futures and 50% in July futures on basis contract. Same as Strategy 5 except that bean sales are split into two time periods to take advantage of any seasonal price appreciation due to storage, and July beans are sold at a 12 cents discount to July futures.

Strategy 7: Sell 100% in March futures on deferred price contract. Differs from Strategy 5 in that the farmer pays a minimum 15 cents deferment charge plus a 1½ cents per bushel service charge from January 1 until the beans are sold. The farmer receives the cash market price on the date of sale less deferment and service charges.

Strategy 8: Sell 50% in March futures and 50% in July futures on a deferred price contract. Same as Strategy 7 except that the crop is sold in two different time periods.

Strategy 9: Buy put options and sell 100% in March futures. The farmer uses the options market to protect against an adverse price movement during storage. He buys put options at harvest equal to the March futures price (strike price). He offsets the options in March if cash market prices have declined between harvest and the time of sale. He lets the put option expire if cash market prices have risen above the strike price.

Strategy 10: Buy put options and sell 50% in March futures and 50% in July futures. Same as Strategy 9 except that the crop is sold in two time periods.

Strategy 11: Nonrecourse loan. This strategy was not applicable, because the cash market price for soybeans was consistently higher than the government price support during the 13-year study period.

RESULTS OF PRICING STRATEGIES

Results of the 17 preharvest hedging strategies and 11 postharvest hedging strategies and their comparison to the cash market strategy are shown in Table 1.

Preharvest Hedges

All Years. The mean and variance of net returns for preharvest hedges are shown in Figure 1a and Table 1. The results show that the mean net return was higher in the cash market (number 0) than for strategies 5-7 in the futures market and 12-14 in the forward cash market. Selectively hedging 50% of the crop whenever the localized futures price equalled or exceeded the breakeven price plus 25, 50, and 75 cents per bushel, respectively, significantly improved mean net returns and the variance of net returns for strategies 2, 3 and 4. Similarly, selectively booking the same proportion of the crop in the same months yielded the highest mean net returns and lowest variance of all strategies. Buying a November put for 100% of the crop in March and offsetting it at harvest yielded a higher mean net return and lower variance than did the other two strategies using the options market.

Normal Crop Years. Mean and variance of net returns for normal crop years are shown in Figure 1b and Table 1. The results are similar to those obtained for all years. Eight of the 17 marketing strategies were superior to the cash market, and 9 fell into the indeterminate category. All of the selective hedging and forward contracting strategies were superior to the cash market. Determining breakeven price plus a return to management, and entering the market only when localized futures exceeded breakeven prices, yielded about twice the net returns of hedging or contracting a fixed portion of the crop in the spring and summer without regard for production costs. Buying a put option in mid-March to cover 100% of the crop at a strike price equivalent to the November futures price on that date and offsetting it at harvest yielded about 25 cents per bushel less profit than did the cash market strategy. Buying puts in mid-March for 50% of the crop and waiting until mid-June and/or mid-July to price the other 50% of the crop reduced the variance slightly, but net returns were also lower.

Table 1. Net returns from pre- and postharvest marketing strategies for North Carolina soybean producers, 1970-1983.

Strategy	All years				Normal crop years[a]				Short crop years[b]			
	Mean	Var.	Max.	Min.	Mean	Var.	Max.	Min.	Mean	Var.	Max.	Min.
	— $/bu —				— $/bu —				— $/bu —			
Preharvest Marketing Strategies												
Cash												
0	1.17	2.29	3.88	-2.09	1.27	3.02	3.88	-2.09	0.93	1.01	1.19	0.24
Futures Market												
1	1.29	0.65	2.77	0.15	1.30	0.98	2.77	0.15	1.31	0.09	1.83	1.14
2	1.52	0.44	2.77	0.86	1.66	0.73	2.77	0.86	1.31	0.09	1.83	1.14
3	1.52	0.44	2.77	0.86	1.66	0.73	2.77	0.86	1.31	0.09	1.83	1.14
4	1.64	0.45	2.77	1.14	2.13	0.54	2.77	1.44	1.31	0.09	1.83	1.14
5	0.98	1.35	2.86	-1.63	0.86	1.82	2.86	-1.63	1.25	0.42	2.00	0.58
6	1.01	1.32	2.62	-1.67	0.95	1.78	2.61	-1.67	1.16	0.52	1.99	0.38
7	1.06	1.41	2.84	-1.68	1.02	1.94	2.84	-1.68	1.17	0.46	1.99	0.47
Forward Cash												
8	1.30	0.64	2.89	0.13	1.24	0.96	2.89	0.13	1.40	0.09	1.78	1.09
9	1.61	0.40	2.89	0.96	1.78	0.65	2.89	0.96	1.40	0.09	1.78	1.09
10	1.61	0.40	2.89	0.96	1.78	0.65	2.89	0.96	1.40	0.09	1.78	1.09
11	1.75	0.42	2.89	1.09	2.21	0.57	2.89	1.40	1.40	0.09	1.78	1.90
12	0.96	1.41	2.98	-1.68	0.81	1.90	2.98	-1.68	1.30	0.35	1.95	0.63
13	0.98	1.41	2.73	-1.73	0.88	1.91	2.73	-1.73	1.20	0.45	1.94	0.43
14	1.02	1.49	2.96	-1.74	0.94	2.07	2.96	-1.74	1.20	0.39	1.94	0.51
Options												
15	1.11	1.62	3.37	-1.68	1.02	2.23	3.37	-1.68	1.31	0.46	1.88	0.46
16	1.07	1.78	3.41	-1.77	1.04	2.40	3.45	-1.77	1.13	0.69	1.89	0.10
17	1.03	1.65	3.31	-1.79	0.98	2.26	3.31	-1.79	1.14	0.53	1.89	0.26
Postharvest Marketing Strategies												
Cash												
0	0.16	2.47	2.48	-3.09	0.79	1.14	2.48	-0.66	-1.26	3.00	0.84	-3.09
Storage/Hedge												
1	0.14	0.05	0.51	-0.26	0.10	0.05	0.43	-0.26	0.22	0.05	0.51	0.08
2	-0.03	0.14	0.37	-1.01	-0.10	0.18	0.37	-1.01	0.06	0.05	0.34	-0.11
3	0.12	0.05	0.49	-0.28	0.08	0.05	0.41	-0.28	0.20	0.05	0.49	0.06
4	-0.06	0.14	0.34	-1.04	0.13	0.18	0.34	-1.04	0.03	0.05	0.31	-0.14
Basis Contract												
5	0.12	2.09	2.96	-2.70	0.37	1.16	2.96	-0.51	-0.46	4.61	1.76	-2.70
6	0.22	2.42	3.52	-2.87	0.70	1.46	3.52	-0.68	-0.86	3.53	0.92	-2.87
Deferred Price												
7	0.02	1.91	2.58	-2.83	0.25	0.92	2.58	-0.50	-0.51	4.64	1.68	-2.83
8	0.01	1.91	2.45	-3.03	0.45	0.86	2.45	-0.65	-0.96	3.55	0.80	-3.03
Storage/Put												
9	-0.25	0.16	0.57	-0.84	-0.32	0.14	0.50	-0.84	-0.10	0.21	0.57	-0.46
10	-0.20	0.32	0.94	-1.29	-0.18	0.38	0.94	-1.30	-0.25	0.24	0.47	-0.64
Nonrecourse Loan												
11	Not Applicable				Not Applicable				Not Applicable			

[a]Normal crop years are 1970-71, 1971-72, 1972-73, 1973-74, 1975-76, 1979-80, 1981-82 and 1982-83.
[b]Short crop years are 1974-75, 1976-77, 1977-78, and 1980-81.

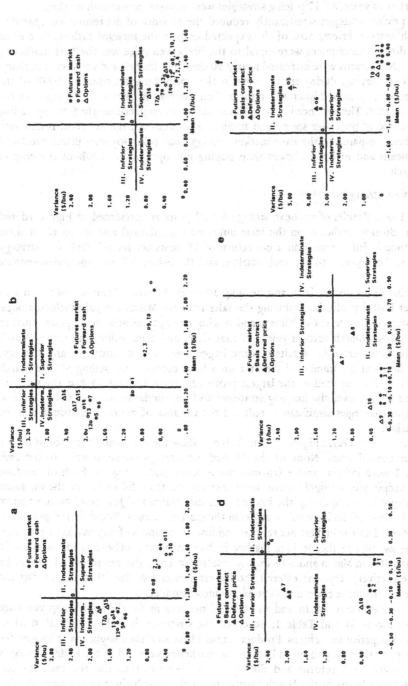

Figure 1. Mean and variance of net returns from selected strategies for North Carolina soybean producers. Panels a, b and c are preharvest strategies for all years, normal years and short crop years, respectively. (Numbers refer to items in Table 1.) Panels d, e and f are postharvest strategies for all years, normal years and short crop years, respectively. (Numbers refer to items in Table 1.)

Short Crop Years. Figure 1c and Table 1 show the mean and variance of net returns for short crop years. All 17 pricing strategies were superior to the cash market.

All pricing strategies significantly reduced the variance of net returns compared to the cash market. Pricing 50% of the expected crop in the forward cash market when the localized futures prices were equal to the breakeven price was the best marketing strategy. Net returns were increased nearly 50 cents per bushel, and variance was sharply reduced. Buying, during spring, puts in the options market equal to 100% of the expected crop yielded higher net returns than the cash market and was equal to the futures market. The variance in the options market was reduced nearly 50%. Spreading the purchase of puts over spring and summer also improved the net returns and variance when compared to the cash market strategy, but rising soybean prices produced lower means and higher variances than buying put options for 100% of the crop in mid-March.

Postharvest Strategies

All Years. Results of soybean storage for all years are presented in Figure 1d and Table 1. Storing soybeans on the farm unpriced (speculating) and selling all of them at one time in July resulted in a net return of 16 cents per bushel. Only one strategy proved to be superior to the cash market, and the others fell into the indeterminate category.

Storing beans on the farm and hedging 100% in the March futures resulted in a 14 cents per bushel profit, but splitting the sales between March and July yielded a negative return. Storing beans at harvest and buying out options at a strike price equal to the underlying March futures price on that date, and then selling the put in March when the beans are sold, resulted in the largest negative net returns of any strategy. Storing beans at a commercial elevator on a basis contract and selling 50% in March and 50% in July resulted in the largest profit (22 cents per bushel), but the variance increased sharply over the hedging strategies and was similar to the cash market. All of the hedging strategies significantly reduced the variance of net returns observed in the cash market.

Normal Crop Years. Figure 1e and Table 1 show the means and variance of net returns for normal years. None of the 10 hedging strategies were superior to the cash market; 2 were inferior and 8 fell into the indeterminate category. Each of the four storage hedges yielded significantly lower net returns than did cash, but the variances were sharply lower. Splitting the hedged sales into March and July resulted in negative returns because of the increased interest in the margin account. The deferred price contracts resulted in the highest net returns and lowest variances of any marketing alternative. Buying puts in the options market in November and offsetting all of them in March and 50% in March and 50% in July resulted in a negative net return of 32 and 18 cents, respectively. Options offered the lowest variance of the 10 strategies, but the negative returns make them unacceptable to most producers.

Short Crop Years. Mean and variance of net returns for the short crop years are shown in Figure 1f and Table 1. Seven of the strategies, including the cash market, resulted in negative net returns. Producers could have sold their soybeans at harvest for a higher price than they received later in the marketing year. All of the hedge strategies yielded positive net returns and sharply lower variances than did the cash market. Hedging stored beans in the March futures resulted in the highest net returns. Buying puts in the options market resulted in the smallest negative net returns of the strategies examined. The nonrecourse loan strategy was not applicable in any of the years analyzed because market prices were higher than the loan rate.

SUMMARY AND CONCLUSIONS

The results for all years show that a higher mean net return and a lower variance of net returns could have been obtained by selecting certain pricing strategies using cash forward contracts and the futures market instead of speculating in the cash market. Forward pricing 50% of expected production when the contract price or localized futures price equalled the breakeven price was superior to the cash market in almost every observation in both short and normal crop years. Of all strategies tested, greater use of futures and forward cash contracts appears to show the most promise for increasing net returns. In general, neither the cash forward market nor the futures market strategies was clearly superior. Differences in mean net returns were due primarily to the costs of hedging in the futures market.

Strategies that used put options performed quite well in reducing the variance of mean net return when compared with the cash market. For example, buying a put option in mid-March at a strike price equal to the closing price of the November futures price in mid-March cut the variance in half and yielded net returns similar to those obtained using the futures market strategies in short crop years. However, in normal crop years, the put options strategies appeared to increase the variance and yielded lower net returns than did most of the futures market strategies.

Results from the marketing strategies that involve storage and pricing decisions indicate significant tradeoff between net returns and risk. In normal crop years, North Carolina producers obtained the highest net returns from storing soybeans unpriced and selling them in July. However, the risk was much greater. A deferred price contract reduced the risk in storage, but it also reduced net returns compared to those obtained in the cash market. Selling at harvest instead of storing for later sale increased the net return for nearly all marketing alternatives in short crop years. Hedging beans in the March futures at the time they are stored appears to be the optimal strategy to follow in short crop years if beans are to be held for later sale.

The findings of this study provide strong support for the use of futures markets, cash forward contracts and options as tools to transfer price risks for soybean producers in North Carolina. It also showed that certain selective hedging strategies can be useful management techniques in helping soybean producers reduce the variability of returns and increase the level of these returns. The hedging strategies that a producer would follow during a "short" crop year appear to be different from those used during a "normal" crop year. The choice of strategies selected by a soybean farmer in a particular year, therefore, will depend on his desire for higher returns and his aversion to risk.

NOTES

T. Everett Nichols Jr., Extension Agronomist, Box 8109, North Carolina State University, Raleigh, NC 27695-8109.

REFERENCES

Kane, M. A., J. G. Beierlein and J. W. Dunn. 1984. Economic evaluation of alternative hedging strategies for Pennsylvania corn. Pennsylvania Agric. Exp. Stn. Bull. 852.

Leuthold, R. M. 1975. Actual and potential use of the livestock futures market by Illinois producers. Illinois Agric. Exp. Stn. AERR 141.

Miyat, C. D. and D. L. McLemore. 1982. An evaluation of hedging strategies for backgrounding feeder cattle in Tennessee. Tennessee Agric. Exp. Stn. Res. Bull. 607.

Nichols, T. E. 1982. Marketing of agricultural products—the producer's marketing plan. p. 2-3. *In* Tar Heel Economist (April). N. C. State Univ. Library, Raleigh.

Purcell, W. D., T. M. Hague and D. Holland. 1972. Economic evaluation of alternative strategies for the cattle feeder. Oklahoma Agric. Exp. Stn. Res. Bull. 702.

Purcell, W. D. and T. W. Richardson. 1977. Quantitative models to predict quarterly average cash corn prices and related hedging strategies. Oklahoma Agric. Exp. Stn. Res. Bull. 731.

Emerging Issues in Processing, Marketing and Distribution

TECHNOLOGICAL ISSUES IN THE PROCESSING INDUSTRIES

John E. Heilman

Two years ago I was asked to give an assessment of the oilseed crushing industry at the World Conference on Oilseed and Edible Oil Processing. At that time I noted that the industry was in a period of overcapacity that would probably continue for two or three years. I predicted that little capacity would be added by the industry, which would concentrate on improving efficiency.

I dearly wish I could now apologize for being so wrong in my predictions, but unfortunately I cannot. The oilseed processing industry remains in an overcapacity situation, which started in 1981. This overcapacity is fairly uniform worldwide, and while processors like to think they have seen the worst margins they will see for awhile, the industry is not now profitable. There is yet to occur the combination of a large supply of soybeans, coupled with strong demand for edible oils and protein feedstuffs, that is required to produce margins that allow reasonable return on investment to owners. As a result, the soybean processing industry continues to limp along at very low capacity, with many plants, especially those in the Midwest, shut-down on a semi-permanent basis.

One change is that the weak edible oil demand, when compared to its supply, has reversed itself. Two years ago the price of soybean oil was 37 cents per kilo; today it is 30 cents. Unfortunately, low 1983 U.S. soybean plantings, caused indirectly by the PIK program, and compounded by low yields caused by last year's drought, produced a small crop. As a result, the improvement in oil prices was matched by increased soybean prices and little improvement in soybean margins and profitability occurred. There is nothing on the horizon to make the processors optimistic about the next few years. They believe it will take some time for rising product demand to overcome the world's oilseed crushing capacity.

TECHNOLOGICAL ADVANCES

How will this continuing period of poor margins affect the soybean processing industry? I stick by what I said two years ago. I do not think there will be very much increase in capacity, but investment in efficiency will occur. Four areas come particularly to mind.

Dehulling Systems

Simpler, more energy-efficient dehulling systems will become common in the soybean industry. Fluidized-bed systems that allow the processor to process soybeans at moistures above 12%, sometimes approaching 15%, are in operation in the U.S.

Two plants have been in operation for one year or more and three more in the U.S. and the United Kingdom are scheduled to start up this year. They eliminate drying to low moistures and tempering to produce beans that are of a uniform moisture and able to be dehulled. The higher temperature, short time, and more uniform heating results in a surface heated, but still crackable, bean that is relatively easier to dehull. Figure 1 compares a conventional dehulling system with the new fluidized-bed system. In the conventional system beans are dried to about 10% moisture then cooled and tempered

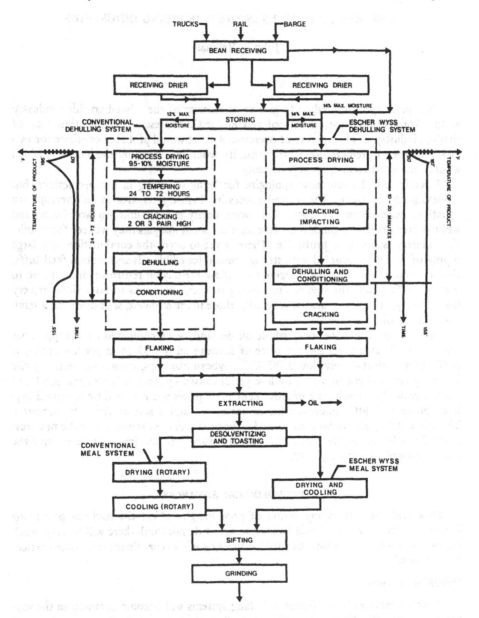

Figure 1. Comparison of conventional dehulling system with new fluidized-bed system. (Courtesy of Escher-Wyss, G.m.B.H., Ravensburg, DFR.)

to allow for moisture uniformity. These tempered beans are then cracked and dehulled cold, before being reheated to provide a softened bean meat for flaking.

In the fluidized-bed system beans are rapidly heated to dry them enough for cracking without allowing this heat to soften the bean. Recycling of the air in the dryer provides a high degree of energy efficiency. The beans are then immediately cracked to halves in a single pair of cracking chills.

These halves are fed through an "impactor", which is a specially adapted hammermill that provides additional energy to knock the hull loose from the bean. Typical aspirators provide for hull removal. Fluidized-bed conditioning and flaking complete the preparation process. Florin and Barlesch (1983) discuss this technology in more detail.

In new plants, the fluidized-bed system is attractive both from a capital cost and an energy cost standpoint. In existing plants, the capital cost saving is less, because grain drying and tempering already exist, but the energy savings still usually make the payout attractive.

Another new technology uses microwave heating under vacuum to accomplish the same results as the fluidized-bed system. It has the added advantage of heating from the inside out and, for this reason, ought to improve the loosening of the hull. This system has recently completed pilot testing at one of Continental Grains' plants. Because energy is 100% utilized, consumption is very low. Presently, utility-purchased power costs about offset these savings. The cost of initial units seems to be high. The benefits of cogeneration on electric costs and volume production of the driers could very well make the process economically attractive.

Meal Processing

A second area of improvement has been the use of the Shumacher type Desolventizer-Toaster (DT) and Drier-Cooler (DC), including the improved hollowstaybolt DT tray (Figure 1). Depending on size and plant capacity, these units are installed separately as DT and DC or together as a DTDC. In either case the more effective mixing of the Shumacher design, when compared with the classical Central Soya type DT, has resulted in significantly lower solvent losses and steam consumption in the DT. The DC, because of better mixing, heat recovery and capture of the frictional heat in the fans, has improved the energy efficiency of meal drying and cooling when compared to the separate rotary driers and coolers that were commonly used in this process.

Use of Coal as Fuel

Soybean plants are returning to coal as a fuel to generate steam. With petroleum-based fuels and natural gas costing from $3.50 up to $7.00 per million BTU, the attractiveness of coal at $2.00-$2.50 is obvious.

The emergence of the "acid rain" issue and the lower cost of high-sulfur coal have promoted the use of fluidized-bed boilers, which can use lime to absorb sulfur oxides in the bed. These types of coal boilers are already in use in the industry.

Cogeneration, the simultaneous generation of power and steam, has been slowly adopted in our facilities. I think there are two reasons for this. Such plants are very expensive to build, but perhaps more importantly, the steam demand of a soybean plant only provides enough steam to generate about a third of the electrical demand.

A better option may be to install steam turbine drives on some high horsepower drives, like the DT, to replace electrical energy with coal-generated steam.

Micro-processor Controls

A fourth technological change that is occurring in the soybean industry is the use of computers and other electronic gear in the plant. The advantage of programmable controllers, video-graphic control panels, and computer analyzed and recorded data probably won't result in a large reduction in operators More likely, we will see plants run more efficiently with fewer supervisors and clerks.

Witte and Pass (1984) recently published a very excellent and thorough review of the state of Computer Controls in Soybean Mills. They believe that specific areas of automation that will likely use computer controls are:

1. Automation of individual control loops in the continuous part of the process.
2. The use of programmable controllers.
3. The use of micro- or mini-computers for computer control of a specific area of the process.
4. Integration of several micro- or mini-computers into an overall, distributed process-control and production-control system.

These applications will show improvements in safety, product quality, and energy conservation.

NOTES

John E. Heilman, Vice President, Continental Grain Company, Processing Division, 277 Park Avenue, New York, NY 10017.

REFERENCES

Florin, G. and H. R. Bartesch. 1983. Processing of oilseeds using fluidbed technology. J. Am. Oil Chem. Soc. 60:193.

Witte, N. and D. Pass. 1984. Experience with computer controls in soybean mills. Oil Mill Gaz. 89: 32-36.

IMPORTER AND EXPORTER PROCESSING AND THE
BEAN-PRODUCT TRADE MIX

Gary W. Williams

A surge in demand after World War II for a wide variety of commodities that require either protein meals or vegetable oils spurred rapid growth in world production, processing, and trade of soybeans and soybean products.[1] The expansion in soybean processing, which is the connecting link between the bean and product markets, has occurred in both exporting and importing countries and, in general, has been accompanied by an equally rapid increase in processing capacity and efficiency. What determines the annual level of soybean processing in a country? What determines whether the processing occurs in the exporting or importing country? This paper evaluates these questions.

SOYBEAN PROCESSING AND WORLD SOYBEAN AND PRODUCTS TRADE

Crush Industries in Major Trading Countries

The U.S. soybean processing industry of the 1950s consisted of about 200 small, inefficient processing mills, mostly of the mechanical press type, used to process a wide variety of oilseeds (Williams, 1981). Over the last three decades, the industry trend has been toward fewer, larger processing facilities that are also more highly specialized. Virtually all U.S. plants now utilize the solvent extraction technique. In Brazil, the rapid growth in soybean production since the early 1970s was accompanied by such a rapid expansion and modernization of the crushing industry that national crush capacity passed the level of production in 1977 and reached 27 million tons in 1982, more than double the size of the crop that year. The Brazilian government has provided a number of incentives to expand and utilize capacity through a variety of domestic and trade policies (Williams and Thompson, 1984).

Soybean crush and crush capacity has also expanded in importing countries. Many European oil mills are tooled up to handle a wide variety of domestic and imported oilseeds. In contrast to the U.S. and Brazil, processors in Europe generally sacrifice efficiency for flexibility of operation, allowing them to maximize profits by taking advantage of changing market prices of alternative imported and domestic oilseeds and products. A rapid increase in the European demand for protein meals, however, has led to a trend toward the construction of plants intended mainly for the solvent extraction of soybeans. The Japanese crushing industry of the early 1970s was characterized by large, efficient soybean plants and small, scattered and inefficient rapeseed plants (Sabatini, 1975). In the wake of the U.S. soybean export embargo of 1973, the

Japanese processing industry has become somewhat less specialized in an effort to improve flexibility of operation to changing market conditions.

The Bean-Product Composition of World Trade

Nearly two-thirds of the world's soybeans do not enter world trade, but rather are utilized in the countries where they are produced (Table 1). Over 90% of the soybeans utilized in the United States and South American countries are processed into meal and oil, much of which is exported. In contrast, only 40% of the soybeans produced in the People's Republic of China (PRC) are crushed, with the remainder fed directly to livestock and used for food purposes.

The United States exports over 40% of the soybeans, nearly 30% of the soybean meal, and about 20% of the soybean oil it produces. South American countries, on the other hand, export only 22% of their soybeans, while exporting 80% of the meal and 40% of the oil they produce. Consequently, although the United States accounts for over 80% of world soybean exports, South American countries account for 40% of world soybean meal exports and 30% of world soybean oil exports, more of either than are exported by the United States (Table 1). The PRC is currently a small net importer of soybeans despite being the world's third largest soybean producer.

The world demand for soybeans and soybean meal is derived largely from the demand for protein feed supplements by the compound feed industries in the developed and centrally planned economic regions. The U.S. and other soybean-producing regions

Table 1. Average share of world soybean and product utilization and trade by country, 1979-80-1981-82. (Source: USDA Foreign Agriculture Circulars, FOP Series.)

	Soybeans			Soybean oil			Soybean meal		
	Utiliza-tion[a]	Ex-ports[b]	Im-ports[b]	Utiliza-tion[c]	Ex-ports[b]	Im-ports[b]	Utiliza-tion[c]	Ex-ports[b]	Im-ports[b]
					– % –				
Major producers									
United States	36.2	83.1	-	33.2	28.2	-	28.5	33.9	-
Brazil	15.9	4.6	3.2	11.1	26.2	1.4	3.1	38.4	-
Argentina	1.9	8.3	-	0.5	2.9	-	0.4	2.4	-
Paraguay	-	2.1	-	-	-	-	-	-	-
PRC	10.2	-	2.2	3.9	-	-	4.9	-	-
Major importers									
E.C. (10)	13.0	0.8[d]	42.3	11.6	26.7[e]	14.8[e]	26.3	20.6	53.5
Spain	3.5	-	10.9	2.3	12.2	-	4.1	-	0.6
E. Europe	1.5	-	2.4	3.3	-	5.3	7.9	-	19.7
USSR	2.1	-	4.8	2.9	-	3.3	4.3	-	5.3
Japan	5.3	-	15.5	5.2	-	-	5.1	-	1.1
Taiwan	1.3	-	3.8	1.2	-	-	1.3	-	-
Others[f]	9.1	1.1	14.9	24.8	3.8	75.2	14.1	4.7	19.8

[a] Includes crush and estimated non-crush use and excludes stocks.
[b] Gross exports and imports.
[c] Excludes stocks.
[d] Primarily includes trans-shipments.
[e] Includes intra-EC trade (EC net exports are about 17% of world net exports).
[f] Mainly less developed, non-producing countries in Latin America, Africa, and Asia.

utilize just over one-third of the world's soybean meal (Table 1). Most of the remainder is imported as either soybean meal or unprocessed soybeans by the European Community (EC), Spain, Eastern Europe, the USSR, and Japan. Most of these countries import and process soybeans to meet the largest share of their domestic soybean meal requirements (Table 2). The major exception is Eastern Europe which has imported soybean meal to fill about 80% of its soybean meal requirements in recent years.

The demand for soybean oil in these importing regions has tended to grow more slowly than that for soybean meal, at approximately the rate of population growth. Consequently, soybean oil production, as a by-product of the processing of soybeans, which are imported primarily for their protein content, has filled the majority of the demand for soybean oil in these importing regions (Table 2). In fact, the EC has become a major net exporter of its surplus oil in competition with the United States and South American exporters. In contrast, less developed countries account for a large share of the world consumption and over 75% of the world imports of soybean oil (Table 1). Reflecting the relatively lower efficiency and capacity of their soybean processing industries, these countries import about two-thirds of their soybean oil requirements as oil rather than as unprocessed soybeans.

THE DETERMINANTS OF SOYBEAN CRUSH

The processing of soybeans during a given time period is affected by at least three major forces: 1) the relative profitability of crushing soybeans as measured by crush margins, 2) the level of installed crush capacity, and 3) the availability of soybean supplies. An oilseed crush margin is the difference between the value of the oil and meal in a unit of the oilseed and the cost of that oilseed unit to the processor. The

Table 2. Average percentage of importing countries' soybean product requirements met through the importation of processed and unprocessed soybeans, 1979-80 to 1981-82. (Source: As for Table 1.)

	Net imports as:	
	Unprocessed soybeans	Processed soybeans
	%	%
Soybean meal		
EC	57.9	42.1
Spain	95.5	4.5
Japan	93.0	7.0
USSR	50.7	49.3
E. Europe	19.9	80.1
Other	55.0	45.0
Soybean oil		
EC	129.3	0.0
Spain	492.1	0.0
Japan	96.5	3.5
USSR	73.8	26.2
E. Europe	54.5	45.5
Other	34.0	66.0

soybean crush margin, therefore, is the "price" to which processors respond in determining the rate of operation in the short run and the level of investment in crushing facilities over the long run. A country's installed crush capacity acts as a constraint on the annual level of soybean processing. The available soybean supply may also constrain the crush volume. A supply constraint would not necessarily affect crush through squeezing the profit margin because processors, faced with excess capacity, could minimize losses in the short run by crushing to the extent of available supplies. Thus, crush volume is likely unresponsive to changes in profitability in the short run.

The short-run behavior of soybean crush in a given country can be represented by the following general equation:

$$C = f(MS/MO, K, A) \tag{1}$$

where A is the availability of soybeans for crushing; C is soybean crush volume; K is installed crush capacity; MS is the soybean crush margin; and MO is an alternative oilseed crush margin.

Table 3 presents the results of estimating soybean crush demand equations according to the above specification for the two major exporting countries—the U.S. and Brazil—and two major importing countries—the EC(9) and Japan—for 1960 to 1982.[2] Variations in the levels of crush from year to year in all four regions are explained mainly by changes in the levels of each country's installed crush capacity and the availability of soybeans for crushing as measured by soybean supplies. The soybean crush margin in most of these countries has remained high enough on average over time that movements in the margin or relative margins seem to have had little influence on the volume of soybeans crushed. Indeed, the elasticity of crush with respect to the soybean crush margin is extremely low in all four regions (Table 3).

Over the long run, however, sustained profitability of crushing activities leads to an increased level of investment in crush capacity. Nevertheless, little investment will occur if adequate supplies of soybeans are generally unavailable. This suggests the following general behavioral relationship to explain changes in the level of crush capacity in a given country over time:

$$K = f(L(MS), L(A)) \tag{2}$$

where all variables are as defined for equation (1) and L is a distributed lag operator.

Using a second degree polynomial distributed lag specification, crush capacity relations following equation (2) were estimated for the United States and Brazil (Table 4). For both countries, the level of installed crush capacity responds to past movements in the profitability of crushing and the level of production as a proxy for soybean supply availability. A shorter response time was postulated for Brazil than for the United States because soybean crushing was highly profitable in Brazil during the 1970s, leading to intense efforts to expand soybean capacity. With the current overcapacity problem, lower government incentives to expand capacity, and more narrow crush margins, the response time of Brazilian investment in crush capacity to movements in the crush margin and supplies will likely begin to lengthen substantially.

DETERMINANTS OF THE BEAN-PRODUCT TRADE MIX

The particular bean-product composition of world trade is largely a by-product of the many interrelated forces that determine the geographic location and levels of

Table 3. Soybean crush equations for selected countries.[a]

(1) CSBUS[b] $= 132.8900 + 15.5140$ MSBUS $+ 0.1115$ CAPUS $+ 0.3442$ (ISBUS(-1) + SSBUS)
 $\quad\quad\quad\quad\quad$ (16.6829) $\quad$ (9.9003) $\quad\quad\quad$ (0.0661) $\quad\quad\quad$ (0.0380)
 $\quad\quad\quad\quad\quad\quad\quad\quad\quad\quad\quad$ [0.01] $\quad\quad\quad\quad\quad$ [0.14] $\quad\quad\quad\quad$ [0.65]
 $\quad\quad\quad\quad\quad$ $R^2 = .989$ $\quad$ $\bar{R}^2 = .988$ $\quad\quad$ DW = 1.65

(2) CSBBZ $= -173.1060 + 186.3430$ (MSBBZ/MPNBZ) $+ 0.5524$ CAPBZ $+ 0.3195$ ISBBZ(-1)
 $\quad\quad\quad\quad\quad$ (95.2504) $\quad$ (92.4168) $\quad\quad\quad\quad\quad\quad$ (0.0359) $\quad\quad\quad$ (0.2725)
 $\quad\quad\quad\quad\quad\quad\quad\quad\quad$ [0.01] $\quad\quad\quad\quad\quad\quad\quad\quad$ [0.76] $\quad\quad\quad\quad$ [0.02]

 $\quad\quad\quad\quad$ $+ 0.1857$ SSBBZ
 $\quad\quad\quad\quad$ (0.0395)
 $\quad\quad\quad\quad$ [0.25]
 $\quad\quad\quad\quad$ $R^2 = .998$ $\quad$ $\bar{R}^2 = .997$ $\quad\quad\quad$ DW = 2.79

(3) CSBEC $= 2327.2500 + 5.1037$ (MSBEC/WPIEC) $- 3.1314$ (MRSEC/WPIEC)
 $\quad\quad\quad\quad\quad$ (282.0510) (3.6570) $\quad\quad\quad\quad\quad$ (1.2818)
 $\quad\quad\quad\quad\quad\quad\quad\quad\quad$ [0.02] $\quad\quad\quad\quad\quad\quad$ [-0.02]

 $\quad\quad\quad\quad$ $- 10.3099$ (MPNEC/WPIEC) $- 5.1793$ (MCPEC/WPIEC) $+ 427.4000$ CAPEC
 $\quad\quad\quad\quad$ (2.4301) $\quad\quad\quad\quad\quad\quad$ (0.9849) $\quad\quad\quad\quad\quad$ (12.9620)
 $\quad\quad\quad\quad$ [-0.11] $\quad\quad\quad\quad\quad\quad$ [-0.11] $\quad\quad\quad\quad\quad\quad$ [0.81]

 $\quad\quad\quad\quad$ $+ 539.7930$ DIEC $+ 843.0250$ D2EC
 $\quad\quad\quad\quad$ (115.4930) $\quad\quad\quad$ (123.2700)
 $\quad\quad\quad\quad$ $R^2 = .995$ $\quad$ $\bar{R}^2 = .993$ $\quad\quad\quad$ DW = 1.66

(4) CSBJP $= -6805.6200 + 42.3486$ (MSBJP/MRSJP) $+ 10.0436$ (MSBJP/MCSJP)
 $\quad\quad\quad\quad\quad$ (391.6820) (21.1086) $\quad\quad\quad\quad\quad\quad$ (5.6400)
 $\quad\quad\quad\quad\quad\quad\quad\quad\quad$ [0.03] $\quad\quad\quad\quad\quad\quad\quad$ [0.01]

 $\quad\quad\quad\quad$ $+ 1.3919$ CAPJP $+ 245.0110$ EMBGO
 $\quad\quad\quad\quad$ (0.0608) $\quad\quad\quad\quad$ (131.4280)
 $\quad\quad\quad\quad$ [4.05]
 $\quad\quad\quad\quad$ $R^2 = 0.97$ $\quad$ $\bar{R}^2 = 0.96$ $\quad\quad\quad$ DW = 0.97

[a]Numbers in parentheses are standard errors. Numbers in brackets are elasticities. R^2 is the coefficient of multiple determination while $\bar{R}^2$ is the R^2 corrected for degrees of freedom. DW is the Durbin-Watson statistic for autocorrelation. Data used ranged from 1960 to 1982. (-1) indicates a one-year lag.

[b]Variable definitions: CAPBZ = soybean crush capacity in Brazil (1000 t) year beginning March; CAPEC = time trend representing EC crush capacity; CAPJP = total oilseed crush capacity in Japan (1000 t) calendar year; CAPUS = soybean crush capacity in the U.S. (mil. bu) year ending August; CSBBZ = soybean crush in Brazil (1000 t) year beginning March; CSBEC = soybean crush in the EC (1000 t) calendar year; CSBJP = soybean crush in Japan (1000 t) calendar year; CSBUS = soybean crush in the U.S. (mil. bu) year ending August; DIBZ = dummy variable, 1978 = -1, 1980 = 1, all other years = 0; D2BZ = dummy variable, 1974 = 1, 1981 = -1, all other years = 0; D1EC = dummy variable, 1979 = 1, all other years = 0; D2EC = dummy variable, 1973 = 1, all other years = 0; EMBGO = dummy variable representing U.S. soybean export embargo in 1973; ISBBZ = ending inventories of soybeans (1000 t) February 28; ISBUS = ending inventories of soybeans in U.S. (mil. bu) August 31; MCPEC = copra crush margin in the EC (DM/t) calendar year; MCSJP = cottonseed crush margin in Japan (yen/t) calendar year; MPNBZ = peanut crush margin in Brazil (cruzeiros/t) calendar year; MPNEC = peanut crush margin in the EC (DM/t) calendar year; MRSEC = rapeseed crush margin in the EC (DM/t) calendar year; MRSJP = rapeseed crush margin in Japan (yen/t) calendar year; MSBBZ = soybean crush margin in Brazil (cruzeiros/t) calendar year; MSBEC = soybean crush margin in the EC (DM/t) calendar year; MSBJP = soybean crush margin in Japan (yen/t) calendar year; MSBUS = soybean crush margin in the U.S. ($/bu) year ending August; SSBBZ = soybean production in Brazil (1000 t) year beginning March; SSBUZ = soybean production in the U.S. (mil. bu) year ending August; WPIEC = wholesale price index in the EC (1970 = 100).

Table 4. Soybean crush capacity equations for the U.S. and Brazil.[a]

(1) $CAPUS$ = 158.0830 + 5.0221 MSBUS(-3) + 10.2945 MSBUS(-4) + 15.8172 MSBUS(-5) + 21.5901 MSBUS(-6) + 27.6134 MSBUS(-7)
  (20.0478) (4.8362)   (7.7306)   (8.9171)   (9.0084)   (9.5543)
            [.003]   [0.01]   [0.01]   [0.01]   [0.02]

  + 33.8869 MSBUS(-8) + 0.09242 SSBUS(-3) + 0.15185 SSBUS(-4) + 0.1783 SSBUS(-5) + 0.1717 SSBUS(-6)
    (12.8311)   (0.0310)   (0.0427)   (0.0353)   (0.0096)
    [0.02]   [0.10]   [0.15]   [0.17]   [0.16]

  + 0.1322 SSBUS(-7) + 0.0596 SSBUS(-8)
    (0.0384)   (0.1035)
    [0.11]   [0.05]

  R^2 = .992 $\bar{R}^2$ = .991   DW = 1.04

(2) $CAPBZ$ = 129.3960 + 1.0313 MSBBZ(-1) + 1.3751 MSBBZ(-2) + 1.0313 MSBBZ(-3) + 0.2657 SBBZ(-1) + 0.3543 SSBBZ(-2)
  (57.0510) (0.0360)   (0.0480)   (0.0360)   (0.006)   (0.008)
            [0.14]   [0.13]   [0.071]   [0.23]   [0.25]

  + 0.2657 SSBBZ(-3) + 1640.2500 DIBZ + 570.6140 D2BZ
    (0.006)   (134.4860)   (105.1880)
    [0.17]

  R^2 = .999 $\bar{R}^2$ = .999   DW = 2.71

[a]Numbers in parentheses are standard errors. Numbers in brackets are elasticities. Variable definitions are as given in Table 3.

soybean crushing in the world. Primarily these include: (1) the relative costs of trans-porting soybeans in the raw and in the processed forms; (2) the profitability of soy-bean crushing and the availability of supplies; (3) the relative efficiencies of the soy-bean processing industries among countries; (4) the relative strengths of the demands for processed soybeans between producing and importing countries; (5) the relative levels of product demand and crush capacity in a country; (6) the relative sizes of the meal and oil markets in producing and importing countries; and (7) government inter-vention in soybean and product markets.

Processing facilities generally tend to locate close to potential markets for the processed products, both within and among countries. The cost of transporting soy-beans from the producing region to the processing plant is generally lower than the cost of transprting equivalent amounts of soybean products from the plant to their respective markets (Houck et al., 1972). Because it is also normally cheaper to trans-port oil than an equivalent amount of meal, largely because of the relative difficulty in handling and shipping meal, processing plants tend to locate close to livestock feed industries and markets. Rapid economic growth after World War II led to the develop-ment of intensive livestock industries, a switch to compound feeds, and as a conse-quence, an increase in soybean crushing capacity in developed non-soybean-producing countries. This restricted world trade of soybean products compared to that of soy-beans. As economic development proceeds in Taiwan, South Korea, and other newly industrialized countries, the result will be more pressure to shift the mix of world trade toward soybeans.

An increase in the value of soybean meal and oil on the world market, relative to the cost of soybeans, provides crushers in importing countries the opportunity to capture the value added in the production of soybean products, prompting a relative increase in soybean trade to that of soybean products. Also, changes in the values of alternative meals and oils and of competing oilseeds can lead to changes in the level and type of oilseed processed and a corresponding shift in the bean-product mix of trade. Crush margins in importing countries, particularly Europe, have been squeezed recently by the high value of the U.S. dollar. An increase in the value of the dollar in-creases both the value of the products as well as the cost of the soybeans to importing countries. Nevertheless, the cost of a unit of soybeans increases by more than the value of the product in that unit because of losses in processing. The result is, again, a shift of the trade mix toward soybeans.

A difference in the efficiency with which importing and exporting countries process soybeans also affects the bean-product composition of trade. The U.S. process-ing industry is relatively more efficient in terms of output per unit of input and unit costs than competing industries in importing and other exporting countries (Mogush, 1979). This provides the U.S. processing industry with an economic advantage in the export of soybean products over competitors, and a price incentive to non-soybean-producing countries to import their meal and oil requirements in the processed rather than in the raw form. Nevertheless, an increase in the processing efficiency of import-ing countries is slowly eroding the U.S. comparative advantage in the production and export of soybean products. At the same time, the modernization and rapid growth of the Brazilian soybean industry has reduced the U.S. share of world soybean product exports.

A strong demand for soybean products in importing countries relative to that in exporting countries also tends to shift the world trade mix toward soybeans. The growth in soybean meal demand by importing countries has been much more rapid than that in the U.S. since World War II. For example, the EC use of soybean meal was identical to that in the United States in 1982, compared to being only 10 to 15% of use in the United States in the 1950s (Anon., 1982). The result has been a tendency for the processing industries in importing countries to grow more rapidly than in soybean-producing countries and a corresponding increase in the soybean share of world trade.

Soybean crush capacity constraints shift the mix of world trade toward soybeans or products depending on whether the constraint occurs in exporting or importing countries. A constraint in exporting countries shifts the mix toward soybeans, whereas a constraint in importing countries shifts the mix toward soybean products. In most soybean-producing countries, processing capacity has expanded rapidly enough not only to fill domestic requirements but also to provide an increasing volume of soybean products to the world market. In contrast, the growth in demand for soybean meal in developed and centrally planned countries, such as Western Europe and the Soviet bloc, and for soybean oil, primarily in developing countries, has outstripped the growth of their processing capacity (Williams, 1984). Consequently, many countries import processed soybeans to fill a significant portion of their domestic soybean meal and oil requirements. Processing capacity growth in other countries, notably Japan, has kept pace with that of product demand so that domestic meal and oil requirements are met almost entirely through the importation and processing of soybeans.

An imbalance in the domestic demands for soybean meal and oil in an importing or exporting country can alter the trade mix by forcing a restriction in the processing of soybeans or other measures to avoid a surplus of one or the other product in the market. This is particularly a problem in the EC where a relatively strong meal demand has resulted in a surplus of soybean oil. The EC is currently considering a tax on vegetable oils to reduce the surplus, stimulate butter demand, and help reduce the large surpluses of dairy products under the Common Agricultural Policy (CAP) (Williams, 1984).

Government policies exert a major influence on the location of processing world-wide and the bean-product mix of trade. Brazil's complicated system of financial and export policies stimulated the rapid increase in domestic crush capacity in the 1970s and continue to provide incentives to crush rather than export soybeans (Williams and Thompson, 1984). The EC CAP supports cereal prices, making soybean meal more price competitive in feed formulation. The zero binding on EC import tariffs for soybeans and soybean meal negotiated by the United States during the "Dillion Round" of the Multilateral Trade Negotiations also has helped keep soybean meal prices low relative to cereal grain substitutes and enhanced the attractiveness of locating and expanding processing facilities in the EC. The EC also supports the price of domestically produced rapeseed, which essentially amounts to a subsidy for crushing rapeseed. Many similar policies by exporting and importing countries to protect their domestic oilseed producers and processors effectively reduce the soybean share of world soybean and soybean product trade.

SUMMARY AND CONCLUSIONS

Soybeans are unique among the major grains traded in world markets in that the demand for soybeans is derived primarily from the demand for its joint products—vegetable oil and protein meal. Although these joint products are related in production, their market demands are independent and influenced by many different forces. Soybeans are processed in both producing and non-producing countries so that both soybeans and soybean products are traded in world markets. Whether the market demands for soybean products in non-producing countries is satisfied through the importation and processing of soybeans or through direct importation of the soybean products is determined by the interaction of a multitude of diverse forces. The net effect of these forces is largely an incentive to shift processing over time from producing to importing countries. Nevertheless, the relative efficiency of the U.S. processing industry, together with protectionistic government policies, will likely continue to provide producing countries with a competitive edge.

NOTES

Gary W. Williams, Assistant Professor, Department of Economics, Iowa State University, Ames, Iowa, 50011.

[1] See Williams (1984) for an analysis of the economic and non-economic forces behind the growth of the world soybean market.

[2] These equations are part of the World Soybean Model first constructed by the author while working for the International Economics Division of the U.S. Department of Agriculture (Williams, 1983) and later revised by the author while a senior economist at Chase Econometrics.

REFERENCES

Anonymous. 1982. Oil World, the past 25 years and the prospects for the next 25 in the markets for oilseeds, oil, fats and meals. Oil World, Hamburg.

Houck, J. P., M. E. Ryan, and A. V. Subotnik. 1972. Soybeans and their products: markets, models, and policy. University of Minnesota Press, Minneapolis.

Mogush, J. J. 1979. Soybean processing industry—U.S. and foreign. p. 889-896. *In* F. T. Corbin (ed.) World soybean research conference II: Proceedings. Westview Press, Boulder, Colorado.

Sabatini, O. 1975. Canada's export market development for agricultural products. FAER-107. Economic Research Service. U.S. Department of Agriculture, Washington.

Williams, G. W. 1981. The U.S. and world oilseeds and derivatives markets: economic structure and policy interventions. Ph.D. Dissertation. Purdue University, West Lafayette, Indiana.

Williams, G. W. 1983. A world soybean model. I.E.D. Staff Report, Economic Research Service, U.S. Department of Agriculture, Washington.

Williams, G. W. 1984. Development and future direction of the world soybean market. Quart. J. Int. Agric. (in press).

Williams, G. W. and R. L. Thompson. 1984. The Brazilian soybean industry: economic structure and policy interventions. FAER-200. Economic Research Service. U.S. Department of Agriculture, Washington.

THE IMPACTS OF TRANSPORTATION RATES ON WORLD SOYBEAN TRADE COMPETITION

Tenpao Lee, C. Phillip Baumel and Robert W. Acton

World soybean exports are dominated by the United States and Brazil. The United States shipped almost two-thirds of world soybean exports in the late seventies and 58% in the early eighties. Brazilian soybean exports fluctuated between 15 and 24% of world soybean trade during the same time period.

Economic competition in the world soybean market depends on production costs, product quality and availability, marketing costs and government trade policies. A large share of marketing costs are transportation related, because of the long distances between production regions in exporting countries to importing ports. Table 1 presents the distances from production regions in central Iowa of the U.S. and Cruz Alta of Brazil to the importing ports of Rotterdam, The Netherlands and Yokohama, Japan. The distance from central Iowa to Japan via West Coast ports is less than half the distance from Brazil to Japan and about 20% less via New Orleans than from Brazil. The distances to Rotterdam are about the same from central Iowa via New Orleans and from Brazil. Central Iowa is near the western fringe of the major soybean production area of the United States. Thus, the shipping distances for a large share of the U.S. soybean production via New Orleans are less than those shown in Table 1.

Barge and ocean freight rates are the major determinants of U.S. soybean transportation costs, since most U.S. export soybeans are carried by barge. Almost all international trade is carried by ocean vessels. Ocean freight rates have declined dramatically

Table 1. Distances between major soybean production regions and importing ports in km (Defense Mapping Agency, 1980).

Modes	Central Iowa, USA to Yokohama, Japan via		Central Iowa, USA to Rotterdam, Netherlands via New Orleans	Cruz Alta, Brazil to Yokohama, Japan via Rio Grande	Cruz Alta, Brazil to Rotterdam, Netherlands via Rio Grande
	West Coast	New Orleans			
Rail	3,218	2,108	323[a]	2,108	323[a]	490[a]	490[a]
Barge	--	--	2,204	--	2,204	--	--
Ocean	7,865	16,866	16,866	9,067	9,067	23,379	11,026
Total	11,083	18,974	19,393	11,175	11,594	23,869	11,516

[a]Truck distances to barge loading elevators are approximately equal to the rail distances.

116

in the past three years. From the U.S. Gulf to Rotterdam they were approximately $23.00 per long ton in January, 1981, but declined to about $6.40 per long ton in August, 1982. Barge rates on the Mississippi River were over 300% of tariff in early 1980, and fell to as low as 100% in 1982. The major reason for these declines is the recent large increase in numbers of barge and ocean vessels at the same time that the volume of grain and other world trade declined. Rail and truck rates in the U.S. have also declined in the past three years, but less sharply than barge and ocean rates. Both truck and rail rates in Brazil may change due to the devaluated cruzeiro and unstable economic conditions.

The purpose of this paper is to compare the current transportation costs of U.S. and Brazilian soybean shipments to Japan and EEC countries and to examine the impacts of changing barge and ocean rates on the relative competitive positions of these two soybean exporting countries.

METHOD OF ANALYSIS

Recent freight rates of all modes were used to estimate total soybean transportation costs from U.S. and Brazil origins to Yokohama, Japan and to Rotterdam, The Netherlands. Secondly, barge and ocean freight rates were assumed to double to estimate the impact of higher freight rates on total soybean transport costs from Brazil and several U.S. soybean origins.

Since distances between origins and exporting port areas vary widely in the United States, the selection of origins significantly affects the results. As major soybean producing areas of the United States, several Iowa and Illinois origins were selected as the basis for analysis. Figure 1 shows the grain routings assumed in the U.S. and Brazil.

U.S. Inland Transportation Rates

The transport rates used in this analysis were the actual rates in effect during April-May, 1984. Truck rates were obtained from grain and marketing cooperatives in Iowa and Illinois. The rail rates are the X-084 rate levels in effect during April, 1984, for unit-train size shipments. The rail rate from Paxton, Illinois to New Orleans is for shipper-owned or leased cars; an additional 3 cents per bushel was added to the rate for the cost of leased cars.

Barge rates were obtained from the Merchants Exchange of St. Louis daily barge auction. Barge rates are quoted as a percent of benchmark rates. To determine actual rate, the benchmark rate was multiplied by the quoted percent. Barge rates used in this analysis are 125% of the benchmark rate.

Brazil Inland Transportation Rates

Brazil inland transportation rates were obtained from the USDA attaché in Rio de Janeiro. The selected origin of Brazilian soybeans is Cruz Alta located in the heart of the soybean producing area in the state of Rio Grande do Sul. The transport cost of soybeans from Cruz Alta to the port of Rio Grande was 1,200 cruzeiros per 60 kilo bag by truck and 1,180 cruzeiros by rail at the end of April, 1984. The exchange ratio was 1,428 cruzeiros per U.S. dollar at that time.

Ocean Freight Rates

Ocean freight rates were estimated based on an accounting model shown in the Appendix. This model is based on daily ocean vessel charter rates, steaming days and fuel costs. This model was selected because of its convenience in estimating per ton

118

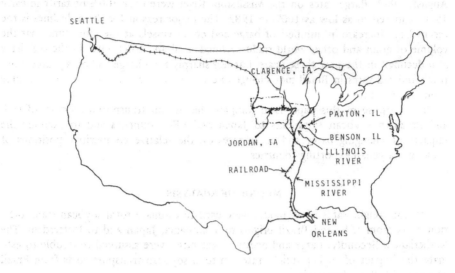

Figure 1. Typical U.S. and Brazilian inland routings for soybean exports.

rates as ocean vessel daily charter rates change. The model generates per ton ocean rates that are very close to April-May 1984 per ton ocean freight rates.

THE RESULTS

Current Total Transportation Costs

Table 2 presents the estimated total transportation costs for U.S.-Japan, U.S.-EEC, Brazil-Japan and Brazil-EEC shipments via alternative modes and routes based on April-May, 1984 rates. The following conclusions can be drawn.

1. The most cost efficient modal combination to ship soybeans from central Iowa to either Japan or Rotterdam is via the unit-train-barge combination to New Orleans, and ocean vessel to Japan or Rotterdam.
2. The unit-train rate direct to New Orleans is $6.23/t more costly than the unit-train-barge combination. The unit-train to the West Coast is $7/t higher than the unit-train direct to New Orleans. The higher unit-train cost to the Pacific Northwest (PNW) just offsets the ocean freight savings from the PNW to Japan compared to New Orleans to Japan.
3. The 490 km rail shipments from Cruz Alta to Rio Grande, Brazil is $1.20/t less costly than the 2,527 km unit-train-barge movement from central Iowa shipment, but is more costly than truck-barge 2,146 km shipments out of central Illinois.
4. Total transportation costs of the U.S.-Japan soybean movement in 30,000 dwt ocean vessels from central Iowa via any modal combination are $0.98 to $7.55/t

Table 2. Estimated transportation costs for soybean shipments from selected U.S. and Brazil origins to Yokohama, Japan and Rotterdam, The Netherlands in 30,000 dwt ocean vessels ($/t).

Mode	Via PNW[a], USA	Via New Orleans, USA					Via Rio Grande
	Jordan, IA	Jordan, IA	Jordan, IA	Clarence, IA	Benson, IL	Paxton, IL	Cruz Alta, Brazil
To Yokohama							
Unit-train	$28.20	$21.20	$5.80	--	--	$11.57	$13.77[b]
Truck	--	--	--	$4.04	$3.31	--	--
Barge[c]	--	--	9.17	9.17	8.46	--	--
Ocean	8.08	14.74	14.74	14.74	14.74	14.74	23.49
Total	$36.28	$35.94	$29.71	$27.95	$26.51	$26.31	$37.26
To Rotterdam							
Unit-train		$21.20	$5.80	--	--	$11.57	$13.77[b]
Truck		--	--	$4.04	$3.31	--	--
Barge[c]		--	9.17	9.17	8.46	--	--
Ocean		9.43	9.43	9.43	9.43	9.43	15.58
Total		$30.63	$24.40	$22.64	$21.20	$21.00	$29.35

[a]Pacific Northwest.

[b]Rail rate is 24 cents cheaper than truck rate.

[c]Including $1.84/t transfer costs from rail or truck to barge.

cheaper than the Brazil-Japan soybean movements.

5. Soybean shipments by unit-train-barge-ocean from central Iowa to Rotterdam in 30,000 dwt vessels have a cost advantage of $4.95/t over Brazil-Rotterdam shipments. The unit-train shipments from central Iowa direct to New Orleans and ocean to Rotterdam at current rate levels are $1.28 higher than the total costs of Brazil-Rotterdam soybean shipments.

6. For soybeans originating in eastern Iowa and in central Illinois, the U.S. has a $6.71 to $8.35/t transportation cost advantage to Rotterdam and a $9.31 to $10.95/t transportation cost advantage to Japan over Brazil shipments.

Costs Based on Increased Barge and Ocean Rates

Most transportation analysts believe that current barge and ocean rates are near variable cost levels. Therefore, any significant barge or ocean rate change is likely to be upward. Table 3 shows the impact of a 100% increase in barge rates and ocean daily values. The assumptions made for sensitivity analysis are:

1. Barge rates increase from 125 to 250% of tariff.
2. Ocean vessel daily values increase from $4,000 to $8,000.
3. All other costs remain unchanged.

The following conclusions can be drawn from Table 3.

1. A 100% increase in ocean vessel daily values would result in ocean freight rates increasing $6.78/t from Brazil to Japan, $5.72 from New Orleans to Japan and $3.35 from the U.S. PNW to Japan. Thus, higher ocean freight rates improve the competitive position of U.S. soybean exports to Japan.
2. If U.S. barge rates double in combination with doubling the daily ocean vessel

Table 3. Estimated total transportation costs, assuming a 100% increase of barge and ocean freight rates, for soybean shipments from selected U.S. and Brazil origins to Yokohama, Japan and Rotterdam, The Netherlands in 30,000 dwt ocean vessels ($/t).

Mode	Via PNW, USA	Via New Orleans, USA					Via Rio Grande
	Jordan, IA	Jordan, IA	Jordan, IA	Clarence, IA	Benson, IL	Paxton, IL	Cruz Alta, Brazil
To Yokohama							
Rail	$28.20	$21.20	$5.80	--	--	$11.57	$13.77[a]
Truck	--	--	--	$4.04	$3.31	--	--
Barge[b]	--	--	16.50	16.50	15.08	--	--
Ocean	11.33	20.46	20.46	20.46	20.46	20.46	30.27
Total	$39.63	$41.66	$42.76	$41.00	$38.85	$32.03	$44.04
To Rotterdam							
Rail		$21.20	$5.80	--	--	$11.57	$13.77[a]
Truck		--	--	$4.04	$3.31	--	--
Barge[b]		--	16.50	16.50	15.08	--	--
Ocean		12.85	12.85	12.85	12.85	12.85	19.67
Total		$34.05	$35.15	$33.39	$31.24	$24.42	$33.44

[a]Rail rate is 24 cents cheaper than truck rate.
[b]Including $1.84/t transfer costs from rail or truck to barge.

values, total transport costs increase more from New Orleans than from PNW or Brazil ports. In this event, total U.S. soybean transport costs are still lower than Brazil shipments to Japan. The cost advantage of unit-train-barge shipments to New Orleans declines, whereas the cost advantage of unit-trains direct to the PNW and New Orleans increases over Brazil.

3. The net effect of a 100% increase in U.S. barge rates and ocean vessel daily values is that U.S. soybeans retain their absolute transport cost advantages, but the net advantage of barge transported soybeans declines relatively.

SUMMARY

Table 4 presents a summary of transportation costs. Based on April-May 1984 freight rates, transport costs from Iowa and Illinois producing areas to Japan are lower than from Brazil to Japan. The U.S. transport cost advantage is greatest for soybeans moving to New Orleans by barge. Soybean transport costs from the Iowa and Illinois origins are also lower to Rotterdam than from Brazil except for central Iowa soybeans moving directly to New Orleans by unit-trains. The major reasons for the U.S. transport cost advantages are the shorter ocean distances to these importing areas and the depressed barge rates on the Mississippi River system.

With a 100% increase in barge rates and daily ocean vessel charter rates soybeans originating in Iowa and Illinois would retain the transport cost advantage to Japan over Brazil soybeans, but the absolute level of cost advantage would shrink for soybeans shipped by barge. Higher barge and ocean vessel daily values would essentially eliminate the transport cost advantage of Iowa soybeans over Brazil soybeans shipped to the EEC and would sharply reduce the cost advantage of Illinois soybeans shipped by barge. This analysis assumes no change in rail or truck rates. However, neither U.S. rail nor truck rates have fallen as sharply as barge or ocean rates in recent years, so it would seem reasonable to expect that they would not increase as rapidly as barge and ocean rates. Moreover, if rail and truck rates increase, it also seems reasonable to expect both U.S. and Brazil truck and rail rates to increase.

Table 4. Estimated total transportation cost for soybeans from selected U.S. and Brazil origins to Japan and The Netherlands in 30,000 dwt vessels for current rates and for a 100% increase in barge and ocean rates ($/t).

Origin	Inland transport mode	Export port	To Japan			To Netherlands		
			Early 1984 rates	100% barge and ocean rate increase	Change	Early 1984 rates	100% barge and ocean rate increase	Change
Jordan, IA	Rail	PNW	$36.28	$39.63	$3.35			
Jordan, IA	Rail	New Orleans	35.94	41.66	5.72	$30.63	$34.05	$3.42
Jordan, IA	Rail-barge	New Orleans	29.71	42.76	13.05	24.40	35.15	10.75
Clarence, IA	Truck-barge	New Orleans	27.95	41.00	13.05	22.64	33.39	11.31
Benson, IL	Truck-barge	New Orleans	26.51	38.85	12.34	21.60	31.24	10.04
Paxton, IL	Rail	New Orleans	26.31	32.03	5.72	21.00	24.42	3.42
Cruz Alta, Brazil	Truck	Rio Grande	37.26	44.04	6.78	29.35	33.44	4.09

IMPLICATIONS FOR U.S. SOYBEAN EXPORTS

The major implication of this analysis for U.S. soybean exports is that the U.S. has a real transportation advantage to the major import markets. And, this advantage remains even if barge and ocean freight rates double. The absolute rate advantage for Illinois soybeans at April-May 1984 rail, barge and ocean rates is about $11.00/t or about $0.30/bu for Japan and $8.00/t or about $0.22/bu for The Netherlands. If the barge and ocean rates double, the real advantage for barge shipments from Illinois drops to $0.14/bu for Japan and $0.06/bu for The Netherlands

A second implication is that part of the real transportation advantage for U.S. exports comes from the lack of a low-cost inland transportation system in Brazil. Currently, nearly all of the soybeans in Brazil move relatively short distances to port by truck at $0.50 to $0.75/bu, primarily because soybean production regions have no navigable rivers or efficient railroad systems. Should Brazil ever improve and expand its rail system, the transportation advantage U.S. soybean exports enjoy could be reduced. The likelihood of Brazil expanding its railroad system in the short run is quite small because of its large international debt ($94 billion). However, extending railroads to the soybean producing areas is feasible, since soybean production is concentrated in several well-defined areas. Extending rail facilities into the newer developing soybean production areas has been promoted by several groups.

The impact of an improved rail system in Brazil would change over time. In the short run, a more efficient transportation system would reduce the cost of bringing soybeans to market, and result in higher local farm prices and/or lower costs for exporters. The allocation of these transportation savings would depend largely on the competition among grain handlers within Brazil. The greater the competition, the greater the share allocated to farmers in the form of higher prices, and the less available for exporters to pass on to compete with U.S. products.

In the long run, the transportation savings would likely be capitalized into land values. This is happening in Brazil today. Land near the ports is more expensive than land in the interior. Part of the difference, of course, is the relative development, but soybean transportation cost is a major factor in land prices.

Transportation costs are only one element of competition in the international market. Foreign exchange rates, product quality, basic production costs, product availability, and government trade policies all combine to determine real competitive advantage in the international market.

NOTES

Tenpao Lee, Assistant Professor, Southern University, New Orleans, Louisiana; C. Phillip Baumel, Charles F. Curtiss Distinguished Professor of Agriculture, Iowa State University, Ames, Iowa; and Robert W. Acton, Senior Economist, American Soybean Association, St. Louis, Missouri.

REFERENCES

Defense Mapping Agency, Hydrographic Center. 1976. Distances Between Ports, 1976, Pub. 151, U.S. Government Printing Office, Washington.

APPENDIX

Ocean freight rates in this analysis are based on daily ocean charter market rates and the following assumptions:

1. The assumed market value for a 30,000 dwt vessel is $4,000 per day. The daily market ship values include the ship, crews, and stores. Additional costs include fuel, tolls and port charges.
2. Bunker fuel consumption is 35 tons/day.
3. Generator diesel fuel consumption is 1.5 tons/day.
4. Fuel is priced at $160/ton for bunker fuel and $260/ton for generator fuel.
5. Voyage speeds are 14.5 knots per hour, where one knot equals 1.15 miles.
6. There is a 2-day delay at the Panama Canal and the Canal toll is $1.83 per Panama Canal ton, which is equal to 100 cubic feet of revenue producing cargo space.
7. U.S.-Japan shipping patterns are:
   a. a grain vessel is loaded from PNW ports to Japan and returns under ballast to the PNW.
   b. a grain vessel moves under ballast from Rotterdam to New Orleans and is loaded from New Orleans to Japan.
8. The U.S.-EEC shipping pattern is a ship loaded from New Orleans to Rotterdam and returning under ballast from Rotterdam to New Orleans.
9. The Brazil to Japan and Rotterdam shipping patterns are ships moving under ballast from Gibraltar to Rio Grande where they are loaded for shipment to Japan and Rotterdam.
10. Ballast fuel consumption is 90% of fully loaded fuel consumption.

The accounting model used to estimate total ocean freight rates is as follows:

Total ocean freight costs = (ship costs) + (fuel costs) + (Panama Canal tolls)

Ship costs = (daily market value) x (ship steaming days + 2 days for each Panama Canal transit)

$$\text{Ship steaming days} = \frac{\text{distance in nautical miles}}{(14.5 \text{ knots}) \times (24 \text{ hours})}$$

Fuel costs = (bunker fuel consumption in tons per day) x (ship steaming days) x ($160) + (generator fuel consumption in tons per day) x (ship steaming days + 2 days for each Panama Canal transit) x ($260)

$$\text{Panama Canal tolls} = (\frac{\text{cargo earning capacity in cubic feet}}{100}) \times \$1.83$$

$$\text{Ocean vessel cost per metric ton} = (\frac{\text{total ship cost}}{\text{grain capacity in cubic feet}}) \times \text{stowage factor} \div 1.016$$

All costs were adjusted to dollars per metric ton.

DEMAND FOR U.S. AND BRAZILIAN SOYBEANS AND SOYBEAN MEAL IN SELECTED EUROPEAN COMMUNITY COUNTRIES

K. D. Sisson and S. C. Schmidt

The major objective of this study was to identify and estimate the effects of economic factors that influence the demand for U.S. and Brazilian soybeans and soybean meal in The Netherlands, Germany, Belgium-Luxembourg (Bel-Lux) and Italy. The study also attempts to discern whether (a) the demand for U.S. soybeans and soybean meal is affected by the same factors as those determining Brazilian soybean and soybean meal exports, and (b) whether importers differentiate between sources of supply in response to changes in the trade-affecting factors.

To meet these objectives an econometric model was constructed consisting of eight, country, import-demand relationships. The behavioral relations are based on soybean meal equivalents (SME)[1] and were estimated over the period 1965 through 1979, using ordinary least squares procedures. The import demand for the four countries is specified to be dependent on (a) relative export prices, (b) export prices of soybeans and soybean meals relative to European corn prices, (c) European domestic corn prices, (d) soybean oil import price, (e) consumption of competing meals, and (f) livestock units. In the U.S. functions, the relative soybean price is the price of soybeans at U.S. ports; in the Brazilian functions, the price used is the price of soybean meal at Brazilian ports. All of the equations are linear in logarithms, and hence, the elasticity estimates can be read directly from the coefficient values.

ESTIMATION RESULTS

The variables have about the same explanatory power for both U.S. and Brazilian export-demand equations (Tables 1-4). Some 61 to 97% of the variation in the dependent variables is explained by the independent variables when looking at the highest R^2 in each set of equations. Whereas some of the estimated coefficients (elasticities) of the explanatory variables accord well with previous estimates by other authors, several are not consistent with expectations.

Price of U.S. Soybeans Relative to Brazilian Soybean Meal (PSM)[2]

The coefficients have the expected negative signs in all but the Bel-Lux model of the U.S. equations. The most significant elasticities range from very inelastic (-0.21) in The Netherlands equation to slightly elastic (-1.21) in the Italian equation (Table 5). By contrast, the estimates of coefficients for this variable in the Brazilian equations have consistently incorrect positive signs. Since none of the estimated coefficients meet

Table 1. Soybean-meal-equivalent import-demand equations for The Netherlands (Sisson, 1982).

Inter.	Soybn/Soyml Price Ratio PSM	Soybn/Corn Price Ratio PSCNU	Corn Price PCN	Soybean Oil Price PO	Competing Meal Qty. NCMC	Livestock Unit Nos. QLUN	Comp. Meal/ Livestock Ratio MLN	R^2	DW
				U.S. Exports to The Netherlands = TQMENU					
-14.45 (-7.30)**	-0.21 (-0.72)		0.75 (2.46)**	0.42 (1.86)*	0.83 (2.70)**	2.95 (12.85)**		.97	2.929 (i)
0.60 (0.30)	-0.19 (-0.24)		1.07 (1.28)	-0.79 (-1.70)			-1.43 (-3.85)**	.72	1.643 (i)
-11.62 (-4.67)**		-0.51 (-2.12)*		0.56 (2.47)**	0.77 (1.97)*	2.64 (9.97)**		.95	1.919 (n)
1.95 (2.02)*		-1.12 (-2.71)**		-0.29 (-0.91)			-1.31 (-5.93)**	.78	1.220 (i)
				Brazilian Exports to The Netherlands = TQMENB					
-13.86 (-1.38)	0.85 (0.56)		-0.46 (-0.29)	-0.17 (-0.15)	-2.12 (-1.36)	5.58 (4.78)**		.92	1.649 (i)
-0.08 (-0.02)	0.83 (0.52)		-0.17 (-0.10)	-1.27 (-1.41)			-4.19 (-5.77)**	.90	2.169 (n)
-18.06 (-2.01)*		0.96 (1.49)		-0.22 (-0.26)	-1.50 (-1.13)	5.95 (6.14)**		.93	2.017 (n)
0.27 (0.14)		0.54 (0.77)		-1.53 (-2.35)**			-4.12 (-8.90)**	.90	2.293 (n)

Correlation Coefficient Matrix

	TQMENU	TQMENB	PSM	PSCNU	PMS	PMCNB	PCN	PO	QLUN	NCMC	MLN
TQMEGU	1.00	.83	-.24	.12	.24	.03	-.41	-.30	.92	-.31	-.74
TQMECB		1.00	-.08	.35	.08	.35	-.69	-.22	.94	-.64	-.92
PSM			1.00	-.08	-1.00	-.50	-.26	.40	-.09	-.23	-.06
PSCGU				1.00	-.08	.82	-.51	.35	.14	-.63	-.41
PMS					1.00	.50	.26	-.40	.09	.23	.06
PMCGB						1.00	-.29	-.08	.17	-.41	-.32
PCG							1.00	.26	-.69	.52	.70
PO								1.00	-.32	-.39	-.00
GCMC									-.52	1.00	.84
QLUG									1.00	-.39	-.90
MLG											1.00

Table 2. Soybean-meal-equivalent import-demand equations for West Germany (Sisson, 1982).

Inter.	Soybn/Soyml Price Ratio PSM	Soybn/Corn Price Ratio PSCGU	Corn Price PCG	Soybean Oil Price PO	Competing Meal Qty. GCMC	Livestock Unit Nos. QLUG	Comp. Meal/Livestock Ratio MLG	R^2	DW
U.S. Exports to West Germany = TQMEGU									
-8.78 (-1.92)*	-0.05 (-0.12)				-0.69 (-1.61)	3.38 (3.10)**		.32	1.798 (i)
4.21 (3.77)**	-0.68 (-1.73)		-0.87 (-2.00)*				-0.47 (-0.85)	.66	1.952 (n)
-11.69 (-4.75)**		-0.49 (-2.25)**		-0.29 (-2.40)**	-1.33 (-4.19)**	4.47 (7.51)**		.88	2.108 (n)
2.33 (2.63)**		-0.24 (-0.56)		-0.35 (-1.45)			-1.55 (-2.47)**	.48	0.662 (i)
Brazilian Exports to West Germany = TQMEGB									
-11.72 (-0.61)	0.55 (0.36)		-2.24 (-1.29)	0.10 (0.08)	-1.53 (-0.85)	5.38 (1.16)		.71	1.533 (i)
6.82 (1.91)*	1.46 (1.16)		-3.25 (-2.33)**	0.70 (0.68)			-1.22 (-0.69)	.68	1.588 (i)
-31.30 (-2.23)*		-0.51 (-0.67)		-1.18 (-1.71)	-3.73 (-2.75)**	10.89 (3.49)**		.67	1.752 (n)
0.43 (0.18)		0.11 (0.14)		-1.19 (-1.46)			-4.24 (-2.69)**	.50	1.190 (i)

Correlation Coefficient Matrix

	TQMEGU	TQMEGB	PSM	PSCGU	PMS	PMCGB	PCG	PO	GCMC	QLUG	MLG
TQMEGU	1.00										
TQMEGB	.88	1.00									
PSM	-.18	-.07	1.00								
PSCGU	.41	.45	-.10	1.00							
PMS	.18	.06	-1.00	.10	1.00						
PMCGB	.42	.40	-.60	.85	.60	1.00					
PCG	-.70	-.76	-.06	-.52	.06	-.39	1.00				
PO	-.08	-.10	.40	.23	-.40	-.03	.38	1.00			
GCMC	-.32	-.41	-.38	-.66	.38	-.33	.35	.30	1.00		
QLUG	.80	.63	-.29	.41	.29	.48	-.48	.05	.02	1.00	
MLG	-.61	-.62	-.24	-.77	.24	-.50	.55	-.34	.92	-.38	1.00

Table 3. Soybean-meal-equivalent import-demand equations for Bel-Lux (Sisson, 1982).

Inter.	Soybn/Soyml Price Ratio PSM	Soybn/Corn Price Ratio PSCBU	Corn Price PCB	Soybean Oil Price PO	Competing Meal Qty. BCMC	Livestock Unit Nos. QLUB	Comp. Meal/ Livestock Ratio MLB	R^2	DW
				U.S. Exports to Bel-Lux = TQMEBU					
-5.73 (-1.24)	0.57 (0.78)		-0.10 (-0.15)	-0.54 (-1.26)	-0.81 (-1.61)	3.00 (2.50)**		.60	2.728 (i)
1.68 (0.89)	0.03 (0.04)		0.13 (0.17)	-0.38 (-0.84)			-1.09 (-2.10)*	.47	1.729 (n)
-3.53 (-0.86)		-0.42 (-0.98)		-0.33 (-1.03)	-1.16 (-3.25)**	2.48 (2.35)**		.61	2.669 (i)
1.29 (1.30)		-0.59 (-1.44)		-0.22 (-0.70)			-1.27 (-3.63)**	.55	2.064 (n)
				Brazilian Exports to Bel-Lux = TQMEBB					
-16.29 (-1.21)	2.81 (1.33)		-1.42 (-0.72)	1.44 (1.16)	-3.01 (-2.05)*	6.25 (1.79)		.69	2.397 (i)
-5.33 (-1.07)	3.61 (1.90)*		-1.08 (-0.56)	1.68 (1.39)			-3.43 (-2.50)**	.66	2.105 (n)
-23.46 (-1.84)*		0.65 (0.68)		0.35 (0.37)	-3.13 (-3.15)**	7.97 (2.40)**		.63	2.103 (n)
-5.13 (-1.75)		0.63 (0.63)		0.63 (0.63)			-3.29 (-3.17)**	.55	1.433 (i)

Correlation Coefficient Matrix

	TQMEBU	TQMEBB	PSM	PSCBU	PMS	PMCBB	PCB	PO	BCMC	QLUB	MLB
TQMEBU	1.00										
TQMEBB	.65	1.00									
PSM	.14	-.01	1.00								
PSCBU	.04	.35	.04	1.00							
PMS	-.14	.01	-1.00	-.04	1.00						
PMCBB	-.04	.30	-.51	.84	.51	1.00					
PCB	-.49	-.43	-.22	-.53	.22	-.33	1.00				
PO	-.11	.24	.40	.30	-.40	.04	.30	1.00			
BCMC	-.54	-.60	-.42	-.55	.42	-.24	.70	-.10	1.00		
QLUB	.36	.40	-.28	-.16	.28	.02	.16	.21	.15	1.00	
MLB	-.65	-.72	-.34	-.51	.34	-.25	.65	-.17	.96	-.14	1.00

Table 4. Soybean-meal-equivalent import demand equations for Italy (Sisson, 1982).

Inter.	Soybn/Soyml Price Ratio PSM	Soybn/Corn Price Ratio PSCIU	Corn Price PCI	Soybean Oil Price PO	Competing Meal Qty. ICMC	Livestock Unit Nos. QLUI	Comp. Meal/ Livestock Ratio MLI	R²	DW
				U.S. Exports to Italy = TQMEIU					
-28.20 (-3.29)**	-0.64 (-0.96)		-0.32 (-0.67)	0.39 (1.08)	-0.27 (-0.57)	7.05 (4.02)**		.78	1.537 (i)
4.35 (2.12)*	-1.21 (-1.22)		-1.05 (-1.55)	0.01 (0.01)			-0.55 (-0.78)	.43	0.626 (i)
-32.00 (-4.62)**		0.37 (1.30)		0.23 (0.73)	-0.06 (-0.19)	7.69 (5.50)**		.79	1.829 (n)
3.38 (2.50)**		0.33 (0.63)		-0.64 (-1.32)			-0.65 (-1.27)	.23	0.449 (r)
				Brazilian Exports to Italy = TQMEIB					
-72.34 (-1.96)*	0.88 (0.31)		-2.04 (-0.98)	1.01 (0.65)	0.14 (0.07)	17.00 (2.25)**		.61	1.565 (i)
9.86 (1.45)	2.34 (0.71)		-3.88 (-1.72)	0.03 (0.02)			-0.58 (-0.25)	.39	0.914 (i)
-89.76 (-2.93)**		0.96 (0.92)		0.28 (0.20)	-0.62 (-0.48)	20.73 (3.35)**		.60	1.706 (n)
3.50 (0.85)		1.15 (0.83)		-2.03 (-1.32)			-2.16 (-1.36)	.22	0.697 (i)

Correlation Coefficient Matrix

	TQMEIU	TQMEIB	PSM	PSCIU	PMS	PMCIB	PCI	PO	ICMC	QLUI	MLI
TQMEIU	1.00										
TQMEIB	.86	1.00									
PSM	-.06	-.02	1.00								
PSCIU	.22	.20	.09	1.00							
PMS	.06	.02	-1.00	-.09	1.00						
PMCIB	.22	.19	-.47	.84	.47	1.00					
PCI	-.58	-.59	-.25	-.58	.25	-.38	1.00				
PO	-.19	-.23	.40	.27	-.40	.03	.29	1.00			
ICMC	-.13	-.09	-.72	-.38	.72	.05	.38	-.47	1.00		
QLUI	.84	.74	-.10	-.05	.10	.01	-.58	-.44	.05	1.00	
MLI	-.31	-.24	-.69	-.36	.69	.05	.49	-.38	.98	-.15	1.00

Table 5.   Price elasticities for imports of U.S. and Brazilian soybean meal equivalents.[a]

Import demand relationships	U.S. soybean/ Brazilian meal price ratio (PSM)	Brazilian meal/U.S. soybean price ratio (PMS)	U.S. soybean/ domestic corn price ratio (PSC.U)	Brazilian meal/domestic corn price ratio (PMC.B)	Domestic corn price (PC.)	Soybean oil import price (PO)
Imports from U.S.						
Netherlands	-.21		-1.12*[b]		.75*	-.79
West Germany	-.68		-.49*		-.87+	-.29*
Bel-Lux	.57		-.59		-.10	-.54
Italy	-1.21		.37		-1.05	-.64
Imports from Brazil						
Netherlands		.85		.96	-.46	-1.53*
West Germany		1.46		-.51	-3.25*	-1.18
Bel-Lux		3.61+		.65	-1.42	1.68
Italy		2.34		.96	-3.88	-2.03

[a]The most significant price elasticity with the correct sign according to theory was selected from each set of equations. In some cases, none of the elasticities had the correct sign; therefore, the most significant estimate is the one shown here. U.S. soybean price is f.o.b., U.S. ports. Brazilian soybean meal price is f.o.b., Brazilian ports. Corn price is each country's domestic wholesale corn price. Soybean oil price is crude, U.S. origin, Rotterdam.

[b]+ and * indicate statistically significant differences at the 10 and 5% levels, respectively.

the statistical test of significance at the 5% level, it may be inferred that relative export prices had a rather minor influence on importers' decisions as to the source of their supplies.

Price of U.S. Soybeans (PSC.U) or Brazilian Soybean Meal (PMC.B) Relative to Domestic European Corn Prices

U.S. elasticities that are significant range from -0.49 (Germany) to -1.12 (The Netherlands). The coefficient in the Bel-Lux equation, while carrying the proper sign (-0.59), is not significant, nor is the coefficient in the Italian equation (0.37).

Excepting the elasticity coefficient pertaining to Brazilian exports to Germany (-0.51), all other Brazilian elasticity coefficients for this variable carry positive signs. The unlikely implication is that Brazilian soybean meal exports tend to benefit from a relative rise in soybean meal prices.

European Domestic Corn Prices (PC.)

The magnitude of corn price elasticities with respect to imports from the U.S. varies between markets, but in the majority of cases the coefficients display negative signs. The elasticities lie in the range -0.10 (Bel-Lux) to -1.05 (Italy). Only the coefficients in the Dutch equation (0.75 and 1.07) have unanticipated positive signs, probably because of multico-linearity. Among the estimated coefficients, most reliance can be placed on the one relating to German import demand (-0.87). All of the corn price elasticities relating to imports from Brazil are negative and have a rather wide range, -0.17 to -3.88. As in the case of the negative estimates from the U.S. equations, only coefficients pertaining to Brazilian exports to Germany have acceptable levels of significance.

Soybean Oil Import Price (PO)

Soybean oil price elasticity estimates are mixed and vary by countries. The majority of the coefficients for imports from the U.S. have the correct negative sign and are inelastic, lying in the range -0.29 to -0.79 (Table 5). However, only the coefficient pertaining to U.S. exports to Germany, -0.29, is significant. All but one of the U.S. soybean oil price elasticities have incorrect positive signs in relation to exports to Italy.

Only Brazilian exports to The Netherlands have shown a significant response to soybean oil prices. They have an elasticity of -1.53. While Brazilian exports to Germany and Italy were also found to be elastic with respect to soybean oil import prices, these coefficients do not have reliable statistical properties.

Consumption of Competing Meals (CMC)

Consumption of competing meals had a significant negative effect on U.S. exports to Germany and to Bel-Lux and a negative though not significant effect on exports to Italy. The former coefficients were -1.33 and -1.16, respectively. Positively signed coefficients, 0.83 and 0.77, were found in exports to The Netherlands. Negative coefficients for this variable were estimated for all but one Brazilian export equation. They range from slightly inelastic (-0.62) to highly elastic (-3.73).

Livestock Units (QLU.)

Livestock units were the single most important factors determining the volume and patterns of imports into the four EC countries considered. The coefficients had the expected positive sign in all the equations, both U.S. and Brazilian, and were highly significant in all of the U.S. equations and in all but two of the Brazilian equations. The most significant U.S. elasticities range from 2.95 to 7.69, while the range for Brazilian elasticities is from 5.95 to 20.73.

Consumption of Competing Meals Per Livestock Unit (ML.)

The estimates of coefficients for per capita consumption of competing meals also turned out well. The coefficients displayed the expected negative sign in all U.S. and Brazilian export equations. The significant U.S. elasticities range from -1.09 to -1.55, and those for Brazil range from -3.29 to -4.24.

IMPLICATIONS

In comparing coefficients for the explanatory variables used in U.S. and Brazilian export demand equations, it appears that the existing livestock population and consumption, or availability, of competing protein meals, were the primary determinants of imports from both countries. Judging from the order of magnitude of elasticity coefficients, Brazilian exports benefited to a greater extent from the growth of the four EC countries livestock population than did U.S. exports. Changes in the consumption of competing protein meal also had a greater proportionate effect on Brazilian exports than on U.S. exports. Similar export response patterns are discernible from the per capita consumption elasticity coefficients.

SME exports from both the U.S. and Brazil were inversely related to changes in corn prices in the four EC countries, indicating linkages between the meal and feedgrain markets. The negative relationship suggests a complementary role for corn in the feeds of EC countries, as it provides the energy and soybean meal the protein for the feed ration. The size of the elasticity coefficients implies that Brazilian exports tended to be more responsive to changes in corn prices than U.S. exports, which is in line with the dominance of soybean meal in Brazilian shipments.

Findings by Paarlberg and Thompson also confirm a complementary relationship between corn and soybean meal or high protein meals in feed rations. However, Paarlberg and Thompson also admit that their "...results do not preclude that substitution (of soybean meal for corn exists), but suggest that the complementary effect dominates."[3] In a recent study Williams (1983) has shown that high protein meal consumption in the EC was inversely related to changes in the import price of corn as well as to its own price.

The coefficients of the ratio of soybean export price to that of the domestic corn price variable used in this analysis do not allow the drawing of firm conclusions on the nature of relationships between soybean meal and domestic corn in feed use. The variable tends to exhibit rather low negative elasticities with respect to U.S. exports to The Netherlands, Germany and Bel-Lux. This is consistent with the findings of an FAS study (FOP 10-83) using the soybean meal corn price ratio expressed in ECUs as one of the factors influencing EC soybean meal use (USDA, 1983).

It must be pointed out, however, that besides relative soybean meal and corn prices, several other factors also affect EC oilseed meal use. First, soybean meal use varies by livestock rations. Its use in dairy feeds, for example, has continued to decrease, and by 1983 the soybean meal content of German and Dutch dairy rations was down at 2 and 5%, respectively. Reports (USDA, 1984a,b) indicate that corn gluten feed as a source of protein competes against protein meals rather than against grains in German and Dutch dairy rations, and has contributed to reduced use of soybean meals. Soybean meal is still an important source of protein in hog and poultry rations.

Overall, it may be concluded that, as long as the price of soybean meal and corn stay close to each other, soybean is substituted for corn as an energy source. High prices for corn and other grains, however, also increase the profitability of grain by-product feeds including corn gluten feed. Use of corn gluten feed, in turn, has resulted in some displacement of soybean meal in feed rations. On the other hand, use of tapioca pellets has expanded EC meal import requirements. These two effects probably canceled each other.

U.S. soybean oil import prices had a rather small effect on the volume of U.S. and Brazilian SME exports. In part, this may be because, among the four EC countries considered, only Italy is a net importer of soybean oil, while the other three countries are net exporters. Exportable soybean oil surpluses in these countries are generated by the crushing of imported soybeans to reduce heavy dependence upon the world market for protein feeds.

It should be noted that while the demand for all oils in the EC is quite inelastic, ranging between -0.15 (Griffith and Meilke, 1982) and -0.56 (Williams, 1983) with respect to price, the demand for soybean oil with respect to its own price is -1.06 (Griffith and Meilke, 1982). The income elasticity of demand for all oils was found to be 0.60 (Griffith and Meilke, 1982) and 1.12 (Williams, 1983).

Our choice to limit the scope of this study to aggregated soybean meal equivalent trade caused some interpretational difficulties with respect to importers' trade behavior. This is because the level of aggregation does not allow for interaction among the three interrelated markets, notably soybeans, soybean meal and soybean oil. Each of these has multiple market outlets and is in competition with many other commodities. Another factor that inhibits interpretation of results is Brazil's various policies for the promotion of exports of value-added products (meal and oil) rather than soybeans. The frequency of change in these policies, and the lack of information, has made it impossible to include them in the model as explanatory variables.

The difficulties encountered in explaining interdependencies in the oilseed and product markets, and related trade behavior, point up the need for the further extension of the model, including multi-product specification.

NOTES

K, D. Sisson, Foreign Agricultural Service, and J. G. Schmidt, Trade Policy Branch of IED, U.S. Department of Agriculture, Washington, D.C. 20250.

[1]Conversion rate of soybeans to soybean meal is 80.5%.

[2]In the Brazilian equation the ratio is reversed with the Brazilian export price as the numerator.

[3]Paarlberg, Phillip L. and Robert L. Thompson, "Joint Products and the Impact of a Tariff: The Case of Soybean Meal in West Germany", contributed paper presented at the 1978 Annual Meeting of the American Agricultural Economics Association at Virginia Polytechnic Institute and State University, Department of Agricultural Economics, Purdue University, West Lafayette, Indiana, 1978.

REFERENCES

Griffith, G. R. and K. D. Meilke. 1982. A structural econometric model of the world markets for rapeseed, soybeans, and their products. Ontario Agricultural College, University of Guelph, AEEE/82/5, Guelph.

Sisson, K. D. 1982. A comparative analysis of the foreign demand for U.S. and Brazilian soybeans and soybean meal. Masters Thesis. University of Illinois Library. Urbana-Champaign.

USDA. 1983. Oilseeds and products, FOP 10-83. FAS, Foreign Agriculture Circular. U.S. Department of Agriculture, Washington.

USDA. 1984a. Nongrain feed ingredients, FAS, Attache Report GE-4105. U.S. Department of Agriculture, Washington.

USDA. 1984b. Feed utilization in the Netherlands under varying price assumptions, FAS, Attache Report NL-4033. U.S. Department of Agriculture, Washington.

Williams, G. W. 1983. A world soybean model. *In* F. W. Williams and R. L. Thompson. Supplement to: The Brazilian soybean industry: economic structure and policy intervention. Economic Research Service, U.S. Department of Agriculture, Washington.

UTILIZATION

Plenary Paper

EXPANDING OPPORTUNITIES FOR UTILIZATION OF SOYBEAN OIL AND PROTEIN

A. R. Baldwin and R. W. Fulmer

For nearly 500 years, the soybean has been cultivated for human consumption. Even today, 95% of soybean oil produced is consumed as food. It is only within the last 40 to 50 years that the effectiveness of feeding soy proteins to animals was demonstrated, and the confinement feeding of large numbers of animals in one location was adopted. Today, 97% of the soybean meal produced is consumed in commercial animal feeds.

Population projections estimate that the earth will have 6.5 billion inhabitants by the year 2000. These population numbers alone assure dramatically increased fat and oil consumption to meet demands for both calories and nutrition. While experts are generally confident that genetics and farming practices can combine to produce soybeans to meet expected needs, there remains the persistent question as to whether the economic reality of being able to purchase the production (and thereby pay the farmer to produce) is achievable. Again, considering only the projected change in total population, it becomes evident that a dramatic increase in soybean meal utilization will occur through consumption of meat and eggs.

Both of the aforementioned are well established uses. For many years, researchers have been adding to our knowledge about the characteristics of soybean products and the effects of processing variables. As knowledge is increased, potential uses for soybean products are added. However, most of the new knowledge leads individually to small increases in use.

The objective of this paper is to highlight some opportunities (and thereby perhaps to imply research needs) that may result in significantly increased consumption of soybeans and soybean products. Some of these uses are nearly at hand; some are remote. Some are needed for the well-being and nutrition of major populations in the world. Some are feasible under today's world economic situations, and some must wait for changes in prices or availability in order to substitute for and displace higher priced or scarcer materials.

POTENTIAL PROTEIN USES

Products for Rumen Feeds

High-quality proteins such as soybean meal are susceptible to rumen-bacterial deamination before they reach the absorption sites in the lower digestive tract of ruminants. Thus, feeding costly proteins to ruminant animals results in degradation to ammonia, necessitating energy and the nutritional inputs to re-synthesize amino acids for subsequent formation of body protein tissue.

Several active research programs are focused on "protected protein" or "rumen by-pass protein". Several approaches to reducing rumen degradation are being tried: partial denaturation by heat processing (Stern, 1981), cross-linking protein with organic and inorganic chemicals (Moller, 1983) to reduce solubility and make unavailable certain enzyme sites (examples would be aldehydes, tannic acids and lignin products), and encapsulation with materials not readily soluble under the pH and temperature conditions of the rumen or resistant to the fermentation processes in the rumen.

None of these treatments should be so severe, however, as to destroy or greatly reduce the post-rumen availability to the animal. Since very little high quality protein currently is being fed to ruminant animals, this achievement would represent a sizable new potential market for "protected soy proteins".

Protein Meals in Animal Feeds

Another major opportunity for increased consumption of protein meals exists in countries determined to expand significantly their poultry and swine production. Notable among these countries are the USSR and China, where government policy decisions cover broad population bases. The transfer of technology is important to capture this large potential market. Through cooperative-sponsored feeding studies that show the efficiency and cost-benefit ratios of using protein meals, this market should be more available. Already the American Soybean Association, National Soybean Processors Association, and others have sent marketing and technical specialists to instruct customers on intensive animal feeding and the resultant benefits. In invited seminars with the USSR, plans were discussed for a joint demonstration project to gain first-hand experience and practical performance data in Russian feeding trials. Could this unexploited market double the need for vegetable protein meals? Probably, yes.

Other Animal and Pet Foods

Keeping in mind that humans, generally, judge the suitability of feed ingredients for pet foods by some kind of anthropomorphic extension of themselves, areas requiring further development include improvements in flavor and palatability of soybean meal products for pet foods. Some economic method is needed to prevent or remove the characteristic bitter, beany, astringent flavor component, perhaps by early enzyme deactivation.

The reduction or elimination of flatus factors in soy products used for dog foods is, in practice, achievable by wet processing to dissolve away the poorly-digested complex carbohydrates that are ultimately converted by intestinal microorganisms to gaseous products. Process steps and decreased yields increase the final cost of these improvements.

Expanded usage of soy proteins in the form of flours in specialty feeds, such as calf and sow milk replacers and pig starters, should result from improved functional properties (protein dispersibility, suspendability and ease of reconstitution into milk-like fluids).

Human Foods/Ingredients

Earlier we referred to the fact that 97% of soy protein consumed is now used in animal feeds. Obviously, improvements in flavor, flatus and functional properties would also permit more extensive use in human foods. Whether these improvements result from genetic developments, processing modifications or further product modifications is not important. Just a doubling of the current usage in human foods would increase soybean meal disappearance by 3%. More acceptable food ingredients might raise this to 20%.

One attractive alternative may resolve two current problems. The process to produce soy-concentrate with less flavor and flatulence involves a water wash at the protein isoelectric pH to remove soluble sugars and salts. If milk or dilute whey from the cheese industry were used as the make-up liquid in an ultrafiltration process to recover the milk protein while washing the soy protein, then both processes would benefit. Protein would be recovered from whey, and the soy protein would be improved as a combined protein with predictably improved properties.

Use of Soy Flour in Breads and Bread-like, Dough-Based Products

A slight and economically practical change in the recipe for one of the world's most basic foods, bread, could make a major contribution to improving worldwide nutrition. Replacing 5% of the wheat flour in commercially baked bread with soybean flour would increase the protein content in the bread by 12 to 15%. This is due to the 44 to 52% protein level in defatted soy flour vs. 10 to 13% protein content of wheat flour. While this change would generally add less than 0.3 cents to the cost of a one-pound loaf of bread, this protein increase could make the difference between good nutrition and malnutrition to many millions of people throughout all countries of the world.

By current estimates, 36% of the people of the world rely on bread for more than 50% of their daily calories. Fortifying commercially baked bread or blending vegetable protein into retail or subsidized wheat flour programs would best reach those most protein deficient. The basic research of how to formulate this is nearly complete. Simple incorporation of full-fat soy-flour, or of defatted soy-flour into many baked-goods formulations is very successful in the 5 to 8% range. Higher levels contribute off flavors.

Research is needed in product formulation, nutritional contributions, methods of blending, economics of the additions and in marketing to promote and sell the more wholesome bread and flour. This may require a change in usual bread standards and strong promotion of the concept. Blended flour for nutritional reasons could form a basis for concessional sales and donated foodstuffs. There is no doubt that significant consumption in the U.S. would signal acceptability and encourage broadened use elsewhere.

POTENTIAL SOYBEAN OIL INDUSTRIAL USES

About 95% of the soybean oil produced finds its way into food applications. There are several major industrial uses for the oil and/or the fatty acids derived from soybeans. These fatty acids may arise from acidulation of the soap stock that is removed in the process that separates the natural, free fatty acids from the crude extracted oil. Other fatty acids are produced by intentional hydrolysis of the triglyceride oil.

Present industrial applications include use of the basic oil or fatty acids in drying-oil applications, mainly alkyd resins (about 35 kt/year domestically), industrial fatty acids, ester and amine chemicals, surfactants, plasticizer esters and in epoxidized soybean oil for polyvinylchloride plasticization (currently about 55 kt/year).

Dust Controller

Recently the Federal Food and Drug Administration approved the addition of up to 0.02% of white mineral oil as a dust suppressant on commodity grains to be used either as feed or food (FDA, 1983). Application was approved for wheat, corn, soybeans, barley, sorghum, oats, rye and rice. The USDA has warned that any changes in odor or quality may cause downgrading.

Research needs to confirm that soybean oil could also be effective at application rates of 500 g/t of grain. Application rates, multiple applications (as grain moves through extensive re-handling), development of rancid oil odors, long-term effects and other ramifications need to be studied. If this were to become an acceptable application at 500 g/t of grain, 90 kt a year might be used in handling grains throughout the storage and marketing system. White mineral oils containing emulsifier and viscosity depressants for cold weather use are more costly at about 90 cents/kilo than current vegetable oils.

Kerosene and Diesel Fuel Additive Replacement

Significant work has already occurred in testing crude, degummed and Refined-Bleached-Deodorized vegetable oils as major full replacements, especially for diesel use (USDA, 1983; Harwood, 1984). Many stationary and propulsion engines have been run with fuels and blends ranging from 0 to 100% vegetable oil. Short-term performance appears acceptable; high flash-points indicate some drawbacks long-term and in cold weather starting. Therefore, as a "neat" fuel, methyl esters of soybean fatty acids are attractive. Ester interchange to form the methyl or ethyl ester is a well-known technology and is easily accomplished; by-product glycerine credits may cover a good portion of the conversion costs. Since diesel fuel has the present price advantage, this technology is not likely to result in significant increased uses of soybean oil until petroleum prices escalate.

As a kerosene replacement in space heaters, the same general comments apply. In 1982 to 1983 an estimated 3,800 ML of kerosene was burned for producing heat in about 10 million space heaters. Sulfur dioxide (SO_2) emissions and flammability are major concerns regarding kerosene. To eliminate odor and SO_2 problems, kerosene for space heaters requires extra refining and deodorization, making it more expensive.

Soybean oil blended with kerosene would burn in heaters now on the market. A redesigned heater, with oil preheating and a different wick system, could burn soybean oil directly. A thorough study of this potential application should be made to develop the correct heater design and determine the oil product and the pricing structure at retail levels. A 10% replacement of the estimated 1982 to 1983 kerosene usage would represent 380 ML.

Carrier in Pesticide Applications

Work dating back over ten years confirms that vegetable oil, with appropriate emulsifiers, will function in place of petroleum distillates in spray application of pesticides (Erickson, 1983). Better emulsion and droplet dispersion were shown; less organic material was required in that 1 L of oil replaced 4 L of petroleum distillate; heavier loading of active ingredient was accomplished; less wind drift during aerial application was noted; and plant phytotoxicity did not appear to be a problem.

Research programs are underway to confirm these observations and to determine whether the pesticide may be longer lasting and whether soybean oil itself should be beneficial to pest control. Since this application is still in development, usage estimates are difficult. Indeed suitability and relative price will determine whether this may reach a 4 to 8 ML annual usage rate.

A Chemical Building Block

Edible oils frequently can be blended and substituted on a price/availability basis. The large industrial fatty acid business, however, is largely product performance dependent. Price plays an important role, of course, but even highly saturated tallow fatty acids are hydrogenated to achieve the lowest unsaturation for performance or stability. Soybean fatty acids face stiff price competition from tallow oil fatty acids as an ingredient in coatings resins. Any process that could reduce the cost of a fatty acid might cause the industrial chemical industry to use fatty acids and derivatives in large-volume products. Such a process improvement could involve supported or immobilized fat-splitting enzymes.

Soybean oil and soapstock are available as a non-fossil, renewable starting material for industrial chemicals which, by development of suitable technologies, could replace depletable petro-chemicals. Research efforts needed for the development of such technologies include solventless oil products for industrial coatings, additives for resins and plastics, thermoplastic engineering plastics, and urethane foam.

To enhance the rate of oxidative drying or reaction with other chemicals, isomerization of the unconjugated double bonds to produce a conjugated configuration is desirable. Better methods to achieve this are needed; biotechnology may lead to a commercial enzymatic process.

We find two divergent rationales with regard to research on industrial oil products. One rationale holds that per capita consumption of edible oils is constant and that only a population increase can escalate edible oil utilization. Therefore, research on industrial oil products holds a real hope for increasing soybean oil consumption. The other view holds that if research increases industrial oil use; e.g., by say 10%, only a modest amount of product is involved whereas, if a 10% increase in edible oil resulted, a sizable amount of product is involved. The relative merits of the above conflicting research justifications may be overriden, however, by a worsening energy situation in which soybean oil, as an annually renewable resource, may become an attractive source of industrial raw materials.

FABRICATED FOOD OPPORTUNITIES

The edible uses of soybean oil are well known. It is a source of concentrated low cost calories; it is a carrier of the fat soluble vitamins; it is an excellent cooking and frying medium; it formulates well into table spreads and margarine, salad dressings and many foods; and with reasonable care in processing it has appropriate edible and savory attributes.

The blandness of quality oil permits formulation into many food products. Large potential markets exist where calorie requirements fail to be met with existing foodstuffs. Part of this failure is certainly due to economic factors—the ability to pay for food. However, product acceptance also relates to familiar, staple foods and the ethnic heritage of the consumer. In a poor economy, people eat to survive, and where income permits people will eat what tastes good. Therefore, beyond just being available, products containing either soybean oil or soybean proteins need to be attractive, tasty, stable in quality and nutritious. The soybean thus can contribute oil as a caloric energy source in foods and beverages. The soybean flour protein fraction could provide basic protein nutrition.

One such product with good combined potential would be a milk-like product composed of non-fat, dry milk-solids for protein, emulsifiable oil for calories, water and minor additives; that is, a completely soybean-based beverage. Milk, since it is available everywhere, gives us a product model. Attempts to make a full-fat soybean milk-like drink have yielded only marginal satisfaction. Improved flavor and stability are necessary. Perhaps, due to enzymatic activity throughout processing, the protein fraction imparts a disagreeable sensory impact. This must be balanced against a higher cost soybean concentrate (70% protein) or soybean isolate (90% protein), both of which avoid much of the off-flavor complaint. Isoelecuic insolubility of soy-protein in carbonated beverages requires very finely ground flour (600-800 mesh) or use of thickeners to keep a stable suspension, again at higher cost.

Optimistically, one must believe that research will eventually lead to an acceptable protein beverage as well as to an oil-containing high-caloric protein beverage. In certain regions of the world, a technical success in developing these products will not necessarily guarantee a large, new consumption of soybean products. Economic studies are also required.

VEGETABLE OILS VS. ANIMAL FATS

Soybean oil has further opportunity as a replacement for cholesterol containing fats in meats and milks. As medical science struggles with proof of cholesterol-induced health issues, food scientists are expanding use of vegetable oils in structured foods. Current public emphasis in this country has focused on "lite" or low-calorie products, and some food processors have turned to incorporating water to introduce bulk without calories.

Processing techniques are feasible for removing animal or butter fat from normal products and replacing part or all of it with a vegetable oil. Food technologists could work with restructured meat products based on low-fat, ground or flaked meat to achieve products similar to those consumed today.

For this use to increase dramatically, soybean oil consumption will require either a high price differential favorable to the recovered animal fat or a major consumer attitude change due to overwhelming scientific proof of the health hazards of cholesterol.

SUMMARY

We have focused on several opportunities that could lead to significant increases in consumption of soybean oil or soybean protein factions. However, success of these other opportunities depends on a combination of technology, economics of production or processing, money availability, nutrition and health, governmental policies, marketing research and overall consumer acceptance.

NOTES

A. R. Baldwin and R. W. Fulmer, Cargill, Inc., 4854 Thomas Ave. South, Minneapolis, MN 55410.

REFERENCES

Erickson, D. 1983. ASA. J. Am. Oil Chem. Soc. 60:512.
FDA. 1983. Federal Register 48: (Dec. 15) 55727.

Harwood, H. J. 1984. Oilchemicals as a fuel: mechanical and economic feasibility. J. Am. Oil Chem. Soc. 61:315-324.

Moller, J. 1983. Treating feeds with formaldehyde to protect protein. Feedstuffs. May 30, p. 12f.

Stern, M. D. 1981. Effect of heat on protein utilization by ruminants. Feedstuffs. Nov. 9, p. 24ff.

USDA. 1983. Vegetable oil as diesel fuel: Seminar III. ARS/USDA Publication ARM-NC-28; Library of Congress, ISSN 0193-3787.

Protein Chemistry

STRUCTURAL CHARACTERISTICS OF SOYBEAN GLYCININ AND β-CONGLYCININ

Patricia A. Murphy

Glycinin and β-conglycinin in soybeans comprise 65 to 80% of the protein fraction or 25 to 35% of the seed weight (Murphy and Resurreccion, 1984; Medeiros, 1982). There are several excellent reviews detailing the genetics and developmental sequence of glycinin and β-conglycinin in maturing seeds (Pernollet and Mosse, 1983; Brown et al., 1982; Larkins, 1981). The two proteins belong to the two general classes of legume proteins, and have no native biological activity other than as amino acid stores for the germinating seed. Glycinin is one of the legumins, which are characterized by molecular weights of 300-400 k daltons and sedimentation coefficients of 11S ± 1S. β-Conglycinin is a vicilin. Vicilins are usually smaller, 150-250 k daltons, glycosylated, and have sedimentation coefficients of 7S ± 0.5S. The structural differences of these two groups of proteins contribute to the variation in functional properties of legume protein foods, such as solubility, coagulation temperature, sulfur and nitrogen content, and flavor retention. In addition, the amounts of the two proteins vary with environment, genetics and subunit composition (Hughes and Murphy, 1983; Murphy and Resurreccion, 1984). These variations also probably affect food functionality.

In soybeans, the two major proteins, glycinin and β-conglycinin, are frequently described by their respective sedimentation values, 11S and 7S (Table 1). Unfortunately, these values are used as misnomers at times. The 7S fraction of soy protein contains, in addition to β-conglycinin, lectin, lipoxygenase and β-amylase. In addition, in the appropriate buffer, β-conglycinin will dimerize to a pseudo 11S type. Glycinin will dissociate under the proper conditions to a pseudo 7S protein. The purification of the

Table 1. Major soybean seed proteins.

Ultracentrifuge fraction	Protein	Mol. weight
		k daltons
2S	Bowman-Birk trypsin inhibitor	6-10
	Kunitz trypsin inhibitor	20-25
7S	β-amylase	156
	Lectin (hemagglutinin)	104
	Lipoxygenase	102-108
	β-conglycinin	156
11S	Glycinin	350-360

two proteins is difficult to accomplish in simple steps. The fractionation method of Thanh and Shibasaki (1976a) for glycinin and β-conglycinin is an excellent first step and is used by most research groups. It is, however, a first step (Figure 1). Without further purification, the fractions are quite impure. Unfortunately, there are many reports in the literature claiming to use purified soy proteins for functionality studies when this is just not the case. Therefore, considerable confusion can occur in interpreting data for these two proteins. Advances in protein separation methodology, in vitro translation assays, and gene clonings have improved our understanding of the structure and complexity of these two proteins.

β-CONGLYCININ

Information on the structural properties of β-conglycinin is not very complete, as with most of the legume vicilins. A major reason for this is the difficulty in purifying these proteins to homogeneity. β-Conglycinin occurs as a trimer and/or hexamer in solution and probably in both forms in the seed. Its subunit structure is composed of three of four different peptides (Table 2, Figure 1). The protein is a glycoprotein. Affinity for concanavalin A has improved the isolation success for β-conglycinin. Glycosylation also gives a higher than actual molecular weight on sodium dodecyl-sulfate (SDS)-polyacrylamide gels due to the associated water (Figure 1). There is microheterogeneity within each of the subunits, depending on the genotype of the soybean examined (Yamanchi et al., 1976; Thanh and Shibasaki, 1976b; Lei et al., 1983). Expression of the β-conglycinin subunit genes varies with cultivar. For example, Raiden soybeans contain only traces of the γ peptide (Thanh and Shibasaki, 1977). A high-protein genotype, CX635-1-1-1, contains more γ subunit than β (Medieros, 1982). We have found Vinton soybeans to contain only traces of the γ peptide.

The amino acid composition varies to a minor extent when different cultivars are examined. Until recently, β-conglycinin was reported to contain no cysteine. More precise analysis with S-alkylated protein, or performic acid oxidation of cysteine, has revealed that the α subunit contains 1 or 2 moles of cysteine (Medieros, 1982; P. A. Murphy, unpublished data) and 1 mole of cysteine in the α' subunit. The β subunit displays the most variability in amino acid content (Table 3). This is supported by the multiple isoelectric forms of β observed by Lei et al. (1983).

There is only one published report on primary sequence analysis of the β-conglycinin subunits (Medieros, 1982). The α and α' peptides have homologous NH_2-terminals, although only the first 10 residues have been sequenced. The β peptide is significantly different from the α and α' subunits at the N-terminal (Medeiros, 1982) and in the mRNA sequence coding for α/α' versus β (Shuler et al., 1982b). Multiple isoelectric forms of α and α' evidently result from families of α and α' genes. Thus, the

Table 2. β-Conglycinin composition.

Subunit	Mol. weight	Carbohydrate (moles per subunit)
	k daltons	
α	57	4 glucosamine, 12 mannose, 2 moieties
α	57	4 glucosamine, 12 mannose, 2 moieties
β	42	2 glucosamine, 6 mannose, 1 moiety
γ	42	6%

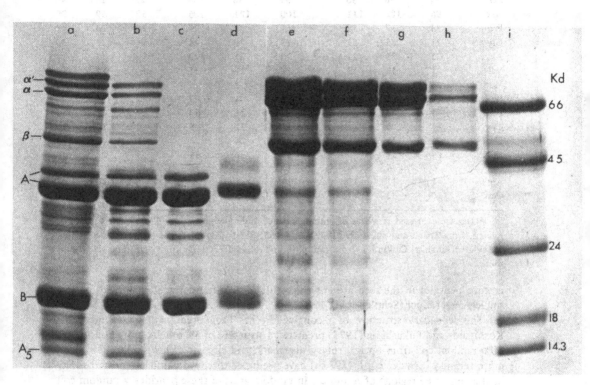

...n and β-conglycinin purification on a 10 to 15% NaDodSO$_4$-polyacrylamide gradient gel. Lanes contain the following: a) whole soy ...ectrically precipitated glycinin; c) column-purified glycinin, rocket electrophoresis standard; d) immunologically pure glycinin used ...oduction; e) isoelectrically precipitated β-conglycinin after Sepharose 6B-CL chromatography; f) e after second chromatography wi... ...B-CL; g) f after 2 times through concanavalin-A-Sepharose and 2 more times through Sepharose 6B-CL, rocket electrophoresis s... ...unologically pure β-conglycinin used for antisera production; i) molecular weight marker proteins—albumin, bovine, 66k; albumin, e... ...ogen, 24k; β-lactogobulin, 18k; and lysozyme, 14.3k (Sigma). (Bands are discussed in the text.)

Table 3.   Amino acid composition of α, α' and β subunits of β-conglycinin in different genotypes.[a]

	α			α			β		
	A	B	C	A	B	C	A	B	C
Asx	55	66	61	52	61	53	46	54	78
Thr	10	12	10	11	14	9	9	11	14
Ser	31	36	36	31	36	34	25	29	46
Glx	96	122	111	100	121	106	59	69	99
Pro	33	37	35	31	33	29	18	18	29
Gly	21	25	23	23	29	24	16	20	30
Ala	21	23	23	20	21	20	19	22	34
Val	21	18	13	23	18	11	19	16	17
Meth	2	2	1	2	2	2	0	0	0
Ile	27	23	16	23	19	12	22	17	18
Leu	41	43	39	35	36	30	37	37	55
Tyr	11	8	11	11	12	9	9	10	15
Phe	24	26	24	24	24	22	23	23	36
His	29	31	29	31	36	31	19	22	30
Lys	6	6	6	17	19	16	7	7	12
Arg	41	42	41	34	36	31	25	27	45
Cys	0	2	1	0	1	1	0	0	0

[a]Data are expressed as moles of amino acid per mole subunit; α, α' = 57,000; β = 42,000. A, Raiden, from Thanh and Shibasaki (1977); B, CX635-1-1-1, from Medeiros (1982), cysteine from S-alkylated subunits; C, Vinton, from (Murphy, unpublished), cysteine from performic acid oxidation.

mature peptides in the seed are not the result of post-translational modification of a single gene product (Schuler et al., 1982a).

The secondary structure of β-conglycinin has been predicted by two methods. Koshiyama and Fukushima (1973) predicted a structure of 5% α-helix, 35% β-sheet and 60% random coil from optical rotary dispersion and circular dichroism studies. Much more recently, Schuler et al. (1982b) have predicted three antiparallel β-pleated sheets in the conserved regions of α and α'. In random areas of these peptides, a random coil structure exists. The conformation of the β-peptide is under study. The data accumulated to date suggest that the structure of these peptides is highly conserved for the purpose of packaging the proteins in the maturing seed.

There is almost no information on the tertiary and quaternary structure of β-conglycinin. In fact, only one vicilin, canavalin from castor beans (MacPherson, 1980), has been subjected to X-ray diffraction analysis for a 3-dimensional (3-D) structure prediction. The predicted structure of canavilin is useful in theorizing the probable 3-D configuration of β-conglycinin. Seven isomers of the protein have been reported (Thanh and Shibasaki, 1976a; Sykes and Gayler, 1981; Yamauchi et al., 1981). The isomers α'β$_2$, αβ$_2$, αα'β, α$_2$β, α$_2$α', α$_3$ and β$_3$. No other combinations have been isolated in nature, although upon reconstitution of isolated subunits, the α'$_3$ isomer has been identified (Thanh and Shibasaki, 1978). When the α-subunit is present, no α' trimer forms. A β trimer could not be reconstituted by Thanh and Shibasaki (1978), but the β$_3$ was isolated in the glycinin fraction recently by two groups (Gayler and Sykes, 1981; Yamuchi et al., 1981). The trimers have molecular weights of 125 k daltons (β$_3$) to 171 k daltons (α$_3$ and α$_2$α').

In the developing seed, the α and α' subunits appear about 15 days after flowering

(DAF). The β subunit does not appear until 22 DAF (Gayler and Sykes, 1981). At maturity, the seed seems to contain the subunits in the proportions reported by Thanh and Shibasaki (1977). Therefore, rearrangement must occur in the protein bodies between newly synthesized β subunit and the α and α' subunits already present in trimers and hexamers.

The α subunit is synthesized as a higher molecular weight precursor than mature α. This is cleaved to a precursor with the same apparent molecular weight as mature α (Sengupta et al., 1981; Beachy et al., 1980). There is co- and post-translational glycosylation followed by proteolysis and deglycosylation of α before the mature glycosylated subunit appears. The α' subunit probably underoes a similar biosynthetic pathway (Sengupta et al., 1981). Details of the β-subunit biosynthesis are unknown.

GLYCININ

In contrast to the minimal information on β-conglycinin, glycinin has been purified and extensively studied. It is a dodecamer with equal numbers of subunits, with acidic and basic isoelectric points associated in a nonrandom manner through disulfide bridges. Six acidic subunits have been identified in CX635-1-1-1 (Moreira et al., 1979). In Raiden, one of these subunits is missing (A_4) and a seventh acidic subunit has been reported (A_6) (Staswick and Nielsen, 1983). Mori et al. (1981) have identified up to 7 acidic subunits and up to 8 basic subunits in 18 genotypes. Five basic subunits have been identified in CX635-1-1-1 (Moreira et al., 1979).

Most of the primary structure information has been derived from Raiden (Staswick and Nielsen, 1983) or CX635-1-1-1 (Moreira et al., 1979). The acidic peptides have molecular weights of 42, 37 and 10 k daltons, and the basic peptides are all 10 k daltons (Figure 1). There is considerable sequence homology at the N-terminal in the subunits (Table 4). The A_{1a} and A_{1b} subunits are the same except for one amino acid change at the beginning of the N-terminal (Ereken-Tumer et al., 1982a). These two peptides share antigenic determinants, suggesting conserved sequences (Moreira et al., 1981). A_2, A_4 and A_6 (Raiden) share some homology with A_{1a} and A_{1b}. A_3 and A_5 share similar sequences but are quite different from the other acidic peptides. The five basic peptides share considerable sequence homology with each other but are dissimilar

Table 4. Primary sequence of glycinin subunits.[a]

Subunit		5	10	15	20
A_{1a}	F S S R E	Q P Q Q N	E C Q I Q	K L N A L	K P D N
A_{1b}	F S F R E	Q P Q Q N	E C Q I Q		
A_2	L R E Q A	Q Q N E C	Q I Q K L	N A L K P	D N R I
A_3	I T S S K	F N E C Q	L N N L N	A L Q P D	H R V Q
A_4	R R G S R	S Q K Q Q	L Q D S H	Q K L R H	F N E G
A_5	L S S S K	L N E C Q	L N N L N	A L	
A_6[b]	G S R S Q	- Q Q L Q	D S H Q L		
B_{1a}	G I D E T	I C T M R	L R Q N I	G Q N S S	P D I Y
B_{1b}	G I D E T	I C T M R	L R H N I	G Q T S S	P D I Y
B_2	G I D E T	I C T M R	L R H N I	G Q T S S	P D I F
B_2	G V E E N	I C T L K	L H E N I	A R P S W	A D F Y
B_4	G V E E N	I C T M K	L H E N I	A R P S W	A D F Y

[a]From Ereken-Tumer et al. (1982a) except where noted, CX635-1-1-1.

[b]From Staswick and Nielsen (1983), Raiden.

to the acidic peptides (Table 4). The amino acid composition of the acidic and basic peptides can be grouped into sulfur-rich and sulfur-poor peptides. A_{1a}, A_{1b}, and A_2 and A_3 contain significant amounts of the sulfur amino acids, whereas A_4 and A_5 contain almost none. The basic subunits follow a similar pattern, with B_{1a}, B_{1b} and B_2 having higher concentrations of sulfur amino acids, whereas B_3 and B_4 have almost none (Moreira et al., 1979).

The secondary structure of glycinin has been predicted by optical rotary dispersion method to be, on one extreme, 100% β structure (Catsimpoolas et al., 1973) to 60% random coil and 35% β-structure (Koshiyama and Fukushima, 1973). Recent computer predictions based on the partial sequence data suggested that the acidic subunits are composed of β and β-turns. The basic subunits are also composed of β-turns, with an α-helix tail of varying lengths at the N-terminal end (Pernollet and Mosse, 1983).

Nothing is known concerning the tertiary structure of glycinin. The quaternary structure of glycinin has been predicted from the X-ray scattering and electron photomicrograph data of Badley et al. (1975). The double hexamers are composed of acidic and basic subunit pairs attached through disulfide bonds. The rest of the associations probably are electrostatic and hydrophobic. The nonrandom association of acidic and basic peptides has been reported by Staswick et al. (1981) and are supported by biosynthesis data by Ereken-Tumer et al. (1982a,b) and Barton et al. (1982). Table 5 presents the data for CX635-1-1-1 with total sulfur amino acid contents and predicted molecular weights. These dimers are synthesized as single, high-molecular-weight precursors (Barton et al., 1982; Ereken-Tumer et al., 1982b). The peptides are not cleaved into the acidic and basic subunits until assembly of the "half" glycinin has occurred (Barton et al., 1982). Further details on glycinin assembly are unknown. The mechanism for incorporation of multiple acidic-basic pairs into the glycinin holoprotein is unknown.

BIOCHEMISTRY IN SOY-FOODS

How does this basic biochemical information connect with food usage of these proteins? Initially, it gives us a plausible explanation for the very different kinds of properties reported for soy protein, given similar treatment conditions. Because there is microheterogeneity in both glycinin and β-conglycinin, one would anticipate that these isomers could react differently. Nakamura et al. (1984) have demonstrated these differences for gelling times and hardness of gels of glycinin with different amounts of the acidic peptides. Genotypes containing A_4 gelled faster than those without it.

Table 5. Glycinin acidic-basic subunit complexes from CX635-1-1-1 (Staswick et al., 1981).

Acidic	Basic	AB-Complex	Total sulfur amino acids	Molecular weight
				k dalton
A_{1a}	B_2	$A_{1a}B_2$	14	57
A_{1b}	B_{1b}	$A_{1b}B_{1b}$	12	57
A_2	B_{1a}	A_2B_{1a}	14	57
A_3	B_4	A_3B_4	9	62
A_5	B_3	A_5B_3	1	30
A_4	–	unknown	2	37

Hardness of glycinin gels was directly related to the amount of the A_3 subunit in the genotype.

The total amounts of glycinin and β-conglycinin, and the ratio of the two, can vary among genotypes, and within genotypes from different growing environments (Hughes and Murphy, 1983; Murphy and Resurreccion, 1984; Medeiros, 1982). Preliminary data suggest that glycinin contents correlate significantly with several textural characteristics of tofu (Wilson and Murphy, unpublished data; Saio et al., 1969). Subunit composition also probably affects curd formation.

Off-flavor binding has been reported to occur with β-conglycinin, but only minimally with glycinin isolated from soy flour in the presence of a reducing agent (Damodaran and Kinsella, 1981). The behavior of off-flavor ligands in nonreducing conditions needs to be assessed.

Heat treatment of glycinin and β-conglycinin with and without reductant, and from different sources (soy flour, Raiden and Vinton, respectively), gave very different results (Damodaran and Kinsella, 1982; Yamagishi et al., 1983). The effect of salt species greatly affects heat denaturation of mixtures of these two proteins, and their 3-D conformation changes with heat treatment without precipitation (Guzman and Murphy, unpublished). The external sulfhydral content has been related to turbidity of heated glycinin solutions (Nakamura et al., 1984), but turbidity is not related to structural changes in the proteins over different ionic strengths (Guzman and Murphy, unpublished).

CONCLUSIONS

Isolated glycinin and β-conglycinin are complex systems, given their multiple peptides and microheterogeneity within peptide groups. The complexity of these proteins has been described for only a few cultivars. In addition, most work describing the biochemical basis for soy protein functionality has been conducted with only a few cultivars or with unidentified soy protein sources. Additional work is required to adequately examine the effect of soy storage protein heterogeneity on the functionality of these proteins in food systems.

NOTES

Patricia A. Murphy, Associate Professor, Department of Food Technology, Iowa State University, Ames, Iowa 50011.

Journal Paper No. J-11549 of the Iowa Agric. and Home Econ. Exp. Stn., Ames, IA. Projects 2433 and 2164, the latter being a contributing project to the North Central Regional Project NC-136.

REFERENCES

Badley, R. A., D. Atkinson, H. Hauser. D. Oldani, J. P. Green and J. M. Stubbs. 1975. The structural, physical and chemical properties of soy bean protein glycinin. Biochim. Biphys. Acta 412:214-228.

Barton, K. A., J. F. Thompson, J. T. Madison, R. Rosenthal, N. P. Jarvis and R. N. Beachy. 1982. The biosynthesis and processing of high molecular weight precursors of soybean glycinin subunits. J. Biol. Chem. 257:6089-6095.

Beachy, R. N., K. A. Barton, J. F. Thompson and J. T. Madison. 1980. In vitro synthesis of the α and α' subunits of 7S storage proteins (conglycinin) of soybean seeds. Plant Physiol. 65:990-994.

Brown, J. W. S., D. R. Ersland and T. C. Hall. 1982. Molecular aspects of storage protein synthesis during seed development. p. 3-42. In A. A. Khan (ed.) The physiology and biochemistry of seed development, dormancy and germination. Elsevier Biomedical Press, New York.

150

Catsimpoolas, N., J. Wang and T. Berg. 1973. Spectroscopic studies on the confirmation of native and denatured glycinin. Int. J. Protein Res. 3:177-184.

Damodaran, S. and J. E. Kinsella. 1981. Interaction of carbonyls with soy protein:conformational effects. J. Agric. Food Chem. 29:1253-1257.

Damodaran, S. and J. E. Kinsella. 1982. Effect of conglycinin on the thermal aggregation of glycinin. J. Agric. Food Chem. 30:812-817.

Ereken-Turner, N., J. D. Richter and N. C. Nielsen. 1982a. Structural characterization of glycinin precursors. J. Biol. Chem. 157:1016-4018.

Ereken-Turner, N., V. H. Thanh and N. C. Nielsen. 1982b. Purification and characterization of mRNA from soybean seeds. Identification of glycinin and β-conglycinin precursors. J. Biol. Chem. 230: 8756-8760.

Gayler, K. R. and G. E. Sykes. 1981. β-Conglycinins in developing soybean seeds. Plant Physiol. 67: 958-961.

Hughes, S. A. and P. A. Murphy. 1983. Varietal influence on the quantity of glycinin in soybeans. J. Agric. Food Chem. 31:376-379.

Koshiyama, I. and D. Fukushima. 1973. Comparison of confirmation of 7S and 11S soybean globulins by optical rotary dispersion and circular dichroism studies. Cereal Chem. 50:114-121.

Larkins, B. A. 1981. Seed storage proteins:characterization and biosynthesis. p. 449-489. In P. K. Stumpf and E. E. Conn (eds.) The biochemistry of plants, Vol. 6. Academic Press, New York.

Lei, M.-G., D. Tyrell, R. Bassette and G. R. Reeck. 1983. Two-dimensional electrophoretic analysis of soy proteins. J. Agric. Food Chem. 31:963-968.

MacPherson, A. 1980. The three-dimensional structure of canavalin at 3.0 A resolution by X-ray diffraction analysis. J. Biol. Chem. 255:10472-10480.

Medeiros, J. S. 1982. Characterization of the subunits of β-conglycinin, and the application of the enzyme-linked immunoabsorbant assay (ELISA) to the determination of the contents of β-conglycinin and glycinin in soybean (Glycine max (L.) Merr.) seeds. Ph.D. dissertation, Purdue University, West Lafayette, IN.

Moreira, M. A., M. A. Hermodson, B. A. Larkins and N. C. Nielsen. 1979. Partial characterization of the acidic and basic polypeptides of glycinin. J. Biol. Chem. 254:9921-9926.

Moreira, M. A., W. C. Mahoney, B. A. Larkins and N. C. Nielsen. 1981. Comparison of the antigenic properties of glycinin polypeptide. Arch. Biochem. Biophys. 210:643-646.

Mori, T., S. Utsumi, H. Inaba, K. Kitamura and K. Harada. 1981. Differences in subunit composition of glycinin among soybean cultivars. J. Agric. Food Chem. 29:20-23.

Murphy, P. A. and A. P. Resurreccion. 1984. Varietal and environmental differences in soybean glycinin and β-conglycinin. J. Agric. Food Chem. 32:911-915.

Nakamura, T., S. Utsumi, K. Kitamura, K. Harada and T. Mori. 1984. Cultivar differences in gelling characteristics of soybean glycinin. J. Agric. Food Chem. 32:647-651.

Pernollet, J.-C. and J. Mosse. 1983. Structure and localization of legume and cereal seed storage proteins. p. 155-191. In J. Deussant, J. Mosse and J. Vaugn (eds.) Seed proteins. Academic Press, New York.

Saio, K., M. Kamiya and T. Watanabe. 1969. Food processing characteristics of soybean 11S and 7S proteins. Part I. Effect of difference of protein components among soybean varieties on formation of tofu-gel. Agric. Biol. Chem. 33:1301-1308.

Schuler, M. A., E. S. Schmitt and R. N. Beachy. 1982a. Closely related families of genes code for the α and α' subunits of soybean 7S storage protein complex. Nucl. Acids Res. 10:8225-8244.

Schuler, M. A., B. F. Ladin, J. C. Pollaco, G. Freyer and R. N. Beachy. 1982b. Structural sequences are conserved in the genes coding for α, α' and β-subunits of soybean 7S seed storage protein. Nucl. Acids Res. 10:8245-8261.

Sengupta, C., V. Deluca, D. S. Bailey and D. P. S. Verma. 1981. Past-translational processing of 7S and 11S components of soybean storage proteins. Plant. Mol. Biol. 1:19-34.

Staswick, P. E., M. A. Hermodson and N. C. Nielsen. 1981. Identification of the acidic and basic subunit complexes of glycinin. J. Biol. Chem. 256:8752-8755.

Staswick, P. E. and N. C. Nielsen. 1983. Characterization of a soybean cultivar lacking certain glycinin subunits. Arch. Biochem. Biophys. 223:1-8.

Sykes, G. E. and K. R. Gayler. 1981. Detection and characterization of a new β-conglycinin from soybean seeds. Arch. Biochem. Biophys. 210:525-530.

Thanh, V. H. nd K. Shibasaki. 1976a. Major proteins of soybean seeds. A straightforward fractionation and their characterization. J. Agric. Food Chem. 24:1117-1121.

Thanh, V. H. and K. Shibasaki. 1976b. Heterogeneity of beta-conglycinin. Biochim. Biophys. Acta 439:326-338.

Thanh, V. H. and K. Shibasaki. 1977. Beta-conglycinin from soybean proteins. Isolation and immunological and physiochemical properties of the monomeric forms. Biochim. Biophys. Acta 490: 370-384.

Thanh, V. H. and K. Shibasaki. 1978. Major proteins of soybean seeds. Reconstitution of β-conglycinin from its subunits. J. Agric. Food Chem. 26:695-698.

Yamagishi, T., A. Miyakawa, N. Noda and F. Yamauchi. 1983. Isolation and electrophoretic analysis of heat-induced products of mixed soybean 7S and 11S globulins. Agric. Biol. Chem. 47:1229-1237.

Yamauchi, F., M. Sato, W. Sato, Y. Kamata and K. Shibasaki. 1981. Isolation and identification of a new type of β-conglycinin in soybean globulins. Agric. Biol. Chem. 45:2863-2868.

Yamauchi, F., V. H. Thanh, M. Kawase and K. Shibasaki. 1976. Separation of the glycopepides from soybean 7S protein: their amino acid sequence. Agric. Biol. Chem. 40:691-696.

FUNCTIONAL CRITERIA FOR EXPANDING UTILIZATION OF SOY PROTEINS IN FOODS

J. E. Kinsella

Incentives for increasing production of soybeans ultimately depend on expanding their utilization, especially in high-value products, such as foods. A major challenge to food scientists is to transform protein-rich plant products; e.g., soybeans (42% protein), into a wide range of food products that are universally desirable and nutritious.

Nutritional value, safety, acceptability (organoleptic and functional) and cost are critical criteria in determining consumer acceptance of new foods or food ingredients for developed and developing economies. More research emphasis on the basis of these is essential for expanding food uses of soy proteins.

If soy proteins were the sole source of food protein for human growth, they would require supplementation with methionine (Wilcke et al., 1979). The nutritional value of soy proteins is limited by antitryptic factors, and by their inherent quaternary and tertiary structure. Trypsin inhibitors (TI) are heat labile and are destroyed by moist heat, spray drying or thermoplastic extrusion (Liener, 1981). Undenatured soybean protein is poorly digestible, and this is a problem in human diets, especially with soy-based infant formula and weaning foods. Dissociation of the dimeric glycinin sub-units significantly improves in vitro digestibility by 20 to 30%. Reduction of the disulfide bonds enhances digestibility approximately two-fold (Kinsella and Rothenbuhler, unpublished). Reduction of the disulfide bonds of the Kunitz and Bowman-Burk TI with heating decreases TI activity by 90% (Friedman, 1982).

FUNCTIONAL PROPERTIES

Functional properties of proteins are those physicochemical properties that govern the behavior of proteins in food systems during their preparation, processing, storage and during consumption; i.e., those properties which, in the aggregate, affect the organoleptic and quality attributes of foods.

Different foods and food applications require quite different functional properties (Table 1). For example, in a beverage, solubility, storage stability, flavor compatibility, controlled viscosity and turbidity may be required at different pH values. In meats, a range of functional properties that change in a desirable manner with processing and cooking are required; e.g., in sausage products, emulsion stabilization during cooking and subsequent gelation, good adhesive properties and water holding are important. In cheese, the protein should coagulate and align with casein and synerese to form a firm curd. Soy protein, because of its compact, stable, globular structure, does not readily incorporate into a cheese curd matrix, and because of its swelling capacity, it

Table 1. Functional requirements of proteins added to foods.

Food	Example	Functional requirements[a]
Beverage	Breakfast drink	Solubility; colloidal stability; flavor release
	Soft drink pH 3.0	Acid solubility; stability; flavor
Gel system	Yogurt	Viscosity; limited protein; protein interaction; water holding; smooth texture
	Meat/sausage	Strong protein; protein association; emulsion stability; gelation; water holding; moisture release.
	Cheese	Strong protein; protein association; matrix formation; friable to elastic texture; meltability.
Emulsion	Spreads	Interfacial spreading; film formation; cohesiveness; water retention
Foam	Meringue	Viscoelastic film; formation; rapid air entrapment; film strength; stability to heat
	Whipped topping	Surface activity; film formation; whipping with fat
	Cake batters	Surface activity for (a) foaming; (b) emulsion stabilization; heat setting; moisture retention

[a]Additionally, protein should lack color and off-flavors. It should be dispersible and not bind flavors and pigments, nor impair the functional behavior of other components.

does not release water significantly. Meltability and elasticity are important functional properties that are not met by native soy proteins, especially 11S which has a very high transition temperature, especially in the presence of salts (Kinsella et al., 1984).

The consumer is conditioned to like proteins as they function in traditional foods (meat, cheese), and it may not be possible to simulate these at reasonable cost. Hence, efforts to make new products that the consumer will accept, to direct the biological producer (plant) to make a product that behaves more like the traditional protein, or to discover how it may be modified acceptably to perform as required in specific food systems, are warranted. However, before one can productively or profitably work with the geneticist or food protein chemist the structure and physical properties of ingredient proteins and factors affecting these must be studied systematically. When the physicochemical basis of each specific functional property and the relationship(s) between protein structure and functional behavior is understood, the food scientist can instruct the 'genetic engineer' about the particular protein structure to design that will be optimal for a specific application.

USES OF SOY PROTEINS AS FUNCTIONAL INGREDIENTS

Approximately 3% of the soy crop is used in foods in the U.S.A. Soy flours are used in small amounts in bakery products as a source of lipoxygenase for bleaching and to improve handling of dough and quality attributes of bread. Soy concentrates (70% protein) are the most common functional ingredient, because of their functionality and cost, being used in comminuted meats, cereals and bakery products. Concentrates are used in sausages for water-holding, emulsifying and gelling properties. The use of textured concentrates in formed and comminuted meat products as extenders and adhesive structuring agents (in meatballs, retorted products, meatloaves, and chili products) is slowly expanding. Soy isolates (90% protein) are used in comminuted meats (hamburgers), in gelled meat products, sausages, meatloaves (to improve slicing and water holding), in dairy foods; e.g., in yogurt for thickening, water-holding and gelling properties; and for infusion into ham, pork and turkey rolls to improve juiciness and reduce cookout. Soy isolates are used in infant formulae as a substitute for milk proteins, where dispersibility, stability to pasteurization and digestibility are desirable.

The utilization of soy proteins is lower than the optimistic predictions of the 1960s. This is because of variability in functionality, occasional off-flavor problems, limited solubility and thermal stability in certain products. Thus, more information concerning structure-function relationships and factors affecting these with regard to particular applications should help solve some of these limitations.

FACTORS AFFECTING FUNCTIONAL PROPERTIES

Several factors influence the behavior of proteins in food systems; e.g., the composition, size, shape, conformation, and secondary interaction of the proteins themselves. Additionally, environmental factors, methods and conditions of isolation and processing, and modification by physicochemical or enzymatic processes also modify protein behavior.

Composition and Structure of Soy Proteins

Protein accounts for over 50, 70 and 90% of soy flour, concentrate and isolate, respectively. The globulins, conglycinin (7S) and glycinin (11S), account for approximately 80% of the proteins (Table 2).

The 7S is a quarternary trimeric glycoprotein (molecular weight 141 to 171 k daltons), composed of three subunits associated via hydrophobic interactions. The 7S possesses approximately two intramolecular disulfide bonds and contains 5% carbohydrate. Thus, it may possess amphiphillic properties for good surface activity and flavor binding.

Glycinin (11S) consists of two, six-hydrophobically associated pairs of disulfide-linked acidic and basic subunits. It has a low net charge around pH 6.0 and limited solubility. Extensive amidation of the aspartic and glutamic residues accounts for the alkaline insolubility of the basic subunits. The acidic subunits have an average of one to three disulfide links per mole, and are more heat stable than the basic subunits, which have one to two cysteine residues. Glycinin contains two, free thiol groups, which apparently are located on the surface and can engage in thiol-disulfide interchange. For example, in gelation the formation of new intermolecular disulfide links enhances strength, but they also may cause polymerization and reduction in extractability and solubility. Marked heterogeneity exists in the content and composition of the acidic subunits of soy proteins. This may be exploited in selecting soybean cultivars for different applications. Cultivars with a high concentration of acidic subunit III form firmer gels (Utsumi et al., 1983).

Protein Stability

Because the expressed functional properties of food proteins are related to structure and conformation, knowledge of factors responsible for the native conformation is important in predicting and/or explaining the behavior of soy proteins in food systems. Hence, secondary forces involved in imparting conformational stability play an important role in determining functional properties of soy proteins. The oligomeric state of both the 11S and 7S proteins reflect both intermolecular and intramolecular disulfide bonds, electrostatic interactions, hydrogen bonding and hydrophobic

Table 2. Amino acid compositions of soy globulins.

	11S	Acidic	Basic	7S	α1	α	β
ILE	4.24	3.95	4.72	6.40	4.88	5.76	6.23
TYR	2.81	1.70	2.76	3.60	2.34	2.35	2.55
PHE	3.85	3.13	4.17	7.40	5.10	5.12	6.52
PRO	6.85	7.51	5.76	4.30	6.58	7.04	5.10
LEU	7.05	5.36	10.15	10.30	7.43	8.74	10.48
VAL	4.83	4.00	7.61	5.10	4.88	4.45	5.38
LYS	4.44	4.96	3.59	7.00	7.21	6.18	5.38
MET	0.98	0.94	0.89	0.30	0.42	0.43	0.00
CYS	1.44	0.93	0.69	0.00	0.00	0.00	0.00
ALA	5.16	3.52	7.59	3.70	4.25	4.48	5.38
ARG	5.81	6.49	5.99	8.80	7.21	8.74	7.08
THR	3.91	3.64	4.09	2.80	2.33	2.13	2.54
GLY	7.50	7.99	7.23	2.90	4.88	4.48	4.53
SER	6.66	6.01	7.15	6.80	6.58	6.61	7.08
HIS	1.89	2.25	1.96	1.70	3.61	1.28	1.98
AS$_X$	11.88	12.42	12.72	14.10	11.04	11.73	13.03
GL$_X$	19.97	25.23	12.93	20.50	21.23	20.47	16.71
NH$_3$	17.39	ND	ND	1.70	ND	ND	ND
MW	360,000	37,000	20,000	175,000	58,000	57,000	42,000
H	953	827	1,044	1,092	1,048	988	1,091
pI		4.65-5.4	8-8.5	5.3	5.2	4.9	5.7-6
CHO	0	0	0	5%	5%	5%	3.5%

interactions between the subunit components. Electrostatic interactions determine the association/dissociation of both 7S and 11S as a function of ionic strength (Damodaran and Kinsella, 1982). The solubility of the dissociated basic subunit of 11S depends upon the electrostatic interaction with negatively charged proteins; e.g., with 7S, at pH 7.0 (Damodaran and Kinsella, 1982). In extruded soy products, hydrophobic interactions and hydrogen bonding are the major forces involved in texturization at 150 C (Hager, 1984), and in soy protein gels hydrogen bonding and disulfide bonds are the main stabilizing forces (Utsumi and Kinsella, unpublished).

Disulfide linkages affect physical properties of soy proteins. The numerous intra-peptide disulfide links, particularly in acidic subunits, may impair or reduce their functionality in several respects, i.e., reduced surface activity and thermal unfolding may reduce their full functionality in meats and also impair their digestibility and bioavailability (Kinsella et al., 1984).

Source

The cultivar used as the source of protein may affect the composition or relative concentration of different components and, thereby, functional properties. The amount of acidic subunit III directly affects the strength of gels (Utsumi et al., 1983).

Storage conditions affect the extractability, composition and functional properties of extracted soy proteins. Saio et al. (1982) showed that there was a progressive decrease in the solubility of proteins during storage of soy meal. Thiol reagents increased extractability of protein, indicating that disulfide-based polymerization occurred during storage. This observation may account for some of the reported variability in the relative concentrations of globulin components and the variability in functional properties of different preparations, and it clearly underlines the need for researchers in this field to be careful in using soy proteins for comparative studies, or in making extrapolations from limited observations on soy protein preparations of unknown history.

The methods used for isolation, separation and refining of proteins from soybeans dramatically affect their functional properties. For example, heat treatments required for desolventization and for inactivation of trypsin inhibitors concomitantly causes denaturation and extensive protein-protein interaction, resulting in loss in solubility and inferior functional properties. This is, perhaps, the major problem in the processing and preparation of standard functional ingredients consistently from soy protein. Research is needed to develop alternative methods for separating soy protein from soybean oil (aqueous separation) for desolventization and inactivation of trypsin inhibitors.

Loss in solubility due to protein-protein aggregation may be induced by drying procedures, by acid precipitation or by excessive shear. The addition of reducing agents may decrease polymerization. To minimize the disulfide polymerization reaction during drying, it may be expedient to oxidize the two free thiol groups of the 11S fraction before drying (Fukushima, 1980). Cryoprecipitation also results in gradual loss of solubility of soy proteins, particularly the 11S component. This may be a problem when these proteins are used in frozen foods; i.e., frozen entrees, hamburgers, etc., because following thawing the precipitated soy protein loses its water-holding capacity and becomes gritty in the mouth.

The foregoing summarizes some of the numerous factors that can affect solubility and functional behavior of soy proteins, and demonstrates the challenges involved in producing soy proteins with consistent properties. The marked disparities in information on functional properties may be ascribed in part to some of the factors cited above and discussed in other reviews (Kinsella, 1976). The physical role of isolated soy proteins in performing specific functional properties; e.g., hydration, swelling, solubility, viscosity, gelation, emulsification, flavor-binding, etc. have been reviewed in detail (Kinsella, 1979; Kinsella et al., 1984).

RESEARCH NEEDS

Research on functional properties (nomenclature, methodology, thermodynamics) needs to be critically reviewed and expanded with more rigorous standards. Food scientists must develop the fundamental scientific data, elucidate the physical-chemical basis of each functional property, clarify the relationships between structure and function of proteins, and indicate processing or modifying treatments that improve specific properties. With such data, the food technologist should be able to convert available proteins to a range of desirable products acceptable to a wide range of consumers.

The acceptability of soy protein products in foods is the ultimate determinant of their marketability in the food industry. Yet, it is rather ironic that *none* of the research objectives regarding expanding the utilization of soy protein in foods; i.e., off-flavor control and elucidation of physical-chemical properties, received a high priority rating by the Soybean Advisory Institute (Anon., 1984). Because utilization is ultimately the driving force that justifies expanding production of soy protein, it seems an oversight not to have placed high priority on fundamental research concerning the physical-chemical properties of soy proteins. This information is needed for new product development and new strategies for the utilization of soy proteins.

NOTES

J. E. Kinsella, Institute of Food Science, Cornell University, Ithaca, NY 14853.
Supported in part by a grant from American Soybean Association.

REFERENCES

Anonymous. 1984. U.S. soybean production and utilization research. Report to Senate Committee. Soybean Research Advisory Institute.

Damodaran, S. and J. E. Kinsella. 1982. Protein solubility. p. 327-357. *In* J. P. Cherry (ed.) Food protein deterioration mechanisms and functionality, ACS Series 206. Am. Chem. Soc., New York.

Friedman, M. 1982. Disulfide modification of TI. p. 359. *In* J. P. Cherry (ed.) Food protein deterioration, ACS Symp. Series 206. Am. Chem. Soc., Washington.

Fukushima, D. 1980. Soy protein deterioration. p. 359. *In* R. Feeney (ed.) Chemical deterioration or proteins. Am. Chem. Soc., Washington, D.C.

Hager, D. F. 1984. Effects of extrusion on soy solubility. J. Agric. Food Chem. 32:295.

Kinsella, J. E. 1976. Functional properties of food proteins. Crit. Rev. Food Sci. and Nutr. 7:219.

Kinsella, J. E. 1979. Functional properties of soy proteins. J. Am. Oil Chem. Soc. 56:242.

Kinsella, J. E., B. German and S. Damodaran. 1984. Physicochemical and functional properties of soy proteins. *In* (in press) A. Altschul and H. Wilcke (eds.) New protein foods. Acad. Press, New York.

Liener, I. 1981. Factors affecting nutritional quality of soya products. J. Am. Oil Chem. Soc. 58:406.

Saio, K., K. Kobayakawa and M. Kito. 1982. Effects of storage conditions on soy protein solubility. Cer. Chem. 59:408.

Utsumi, S., T. Nakamura and T. Mori. 1983. Role of subunits in properties of heat induced gels from 11S globulins. J. Agric. Food Chem. 31:503.

Wilcke, H., D. Hopkins and D. Waggle. 1979. Soy protein and human nutrition. Academic Press, New York.

FLAVOR BINDING AND THE REMOVAL OF FLAVORS FROM SOYBEAN PROTEIN

Lester A. Wilson

The flavor of soybeans is believed by many to be one of the major limiting factors to their increased utilization in human foods (Kalbrener et al., 1971; Hammonds and Call, 1972; Rackis et al., 1979). The flavor problem associated with soybeans is not new. As early as 1924, Berczeller, a Hungarian chemist, stated that soybeans, in their natural state, were "evil tasting" (Wolf, 1975). The undesirable flavors of soybeans have also been characterized as "green," "beany," "painty," "grassy," and "bitter" (Wolf and Cowan, 1975; Rackis et al., 1979). These flavors can be developed by lipoxygenase activity or oxidative rancidity in the raw, crushed soybean or soybean flours. They are believed to interact with soy protein and are often released when flours, concentrates, and isolates are used as ingredients, extenders, or fabricated foods (Arai et al., 1970; Twigg et al., 1977; Ashraf and Snyder, 1981). This off-flavor release hinders consumer acceptability of soybean products and the marketability of soybean foods.

Research into the off-flavor problem has centered around the following major areas: (1) isolation and identification of off-flavors from lipoxygenase activity (Fujimaki et al., 1965; Arai et al., 1970; Goosens, 1975; Sessa, 1979; Hsieh et al., 1982), (2) empirical methods to improve the flavor (Tables 1 and 2), (3) binding or adsorption of flavor compounds onto soy protein (Beyeler and Solms, 1974; Franzen and Kinsella, 1974; Gremli, 1974; Aspelund and Wilson, 1980; Crowther et al., 1980; Damodaran and Kinsella, 1981a,b; Thissen, 1982; Aspelund and Wilson, 1983), and (4) enzymatic improvements of the flavor of soy protein by use of proteases (Arai et al., 1970) or aldehyde and alcohol dehydrogenases (Chiba et al., 1979a,b; Takahashi et al., 1979).

A number of empirical methods to prevent, remove, or mask off-flavors in soy

Table 1. Methods to inactivate or inhibit lipoxygenase activity (modified from Kinsella and Damodaran, 1980).

Addition of antioxidants
Blanching
Dry heat or moist heat
Grinding at acidic pH
Grinding with aqueous ethanol plus heat
Grinding with H_2O_2 plus calcium chloride
Grinding in hot water (100 C)
Grinding with solvent azeotropes
Inhibition by acetylenic compounds

Table 2. Methods used to minimize off-flavors in soy protein.

> Azeotrope extraction
> Enzymatic treatment
> Masking
> Single-solvent extraction
> Steam treatment
> Vacuum treatment

protein have been used (Tables 1 and 2). Kinsella and Damodaran (1980) have presented an excellent review of this literature. Although these methods have improved soy protein utilization, they have not solved the soy protein flavor problem.

That certain flavor compounds interact with soy protein is agreed upon by most investigators, but the mode of this interaction is not. The researchers in this field have taken two different approaches to studying flavor binding to soy protein: gas/solid interactions and aqueous interactions.

GAS/SOLID FLAVOR INTERACTIONS

Aspelund and Wilson (1980, 1983) used a dynamic flavor-binding system to determine the thermodynamics of adsorption for a homologous series of alcohols, aldehydes, ketones, alkanes, and methyl esters onto soy protein isolate. They found (Table 3) that the alcohols were more strongly bound than the carbonyls, which all bind more strongly than the alkanes. The free energy of adsorption (ΔG) was negative for all the statistically significant compounds, decreased with increasing temperature, and increased with increasing chain length. They found that the binding reaction was driven by the enthalpy of adsorption in gaseous systems. Their data emphasize the importance of the ligand's functional group in the binding of flavors to soy protein. It was hypothesized that the binding of flavor compounds to soy protein isolate was a combination of nonspecific Van der Waals forces and hydrogen bonding. Crowther et al. (1980) determined the heats of adsorption and the adsorption coefficients for the binding of alcohols, aldehydes, and ketones onto soy protein isolate previously subjected to treatments involving different levels of temperature, moisture, and shear. The enthalpy (ΔH) and adsorption coefficient (K_{eq}) increased with chain length (Van der Waals interactions), and alcohols had larger ΔHs and K_{eq}s than the carbonyl compounds. Inasmuch as ΔH values were not affected by processing, the changes in K_{eq} were postulated to be the result of changes in the number of binding sites. Crowther et

Table 3. Statistically significant[a] thermodynamic quantities determined for dry soy isolate at 90 C (Aspelund and Wilson, 1983).

Compounds	-ΔH	-ΔG	-ΔS
	kJ/mol	kJ/mol	J/mol K
2-hexanone	25.3	3.59	59.75
1-hexanal	37.2	3.46	92.93
1-hexanol	58.2	10.35	131.60
Methyl pentanoate	28.1	3.83	66.83

[a]99% confidence limit.

al. (1980) postulated that their data were consistent with chain unfolding of globular proteins, which exposes the more nonpolar regions of the molecule.

AQUEOUS FLAVOR INTERACTIONS

Using headspace analysis, Gremli (1974) reported that aldehydes and ketones showed significant interaction with soy protein in an aqueous system. Unsaturated aldehydes were retained to a greater extent than the corresponding saturated ones. Increasing the chain length of the aldehyde also increased retention. In contrast, alcohols did not interact with the protein. Gremli (1974) concluded that there were both irreversible and reversible reactions of the flavor compounds with soy protein. Franzen and Kinsella (1974) also found a decrease in the headspace concentration of flavors when proteins were added to an aqueous model system. Although these studies demonstrated qualitative interactions between flavors and soy protein, they did not determine the quantitative interactions or mode(s) of binding.

A number of investigators (Beyeler and Solms, 1974; Kinsella and Damodaran, 1980; Damodaran and Kinsella, 1981a,b; Thissen, 1982) have tried to elucidate the nature of the interaction of flavors with proteins in aqueous model systems. However, these studies conflict regarding the nature and magnitude of the flavor-protein binding (Table 4). Arai et al. (1970) used a gel filtration technique to study the binding process of n-hexanol and hexanal with native, denatured, and enzymatically hydrolyzed soy protein and determined that binding constants (K_{eq}) for n-hexanol and hexanal were 80.3/M and 173.4/M, respectively (Table 4). The interactions increased with the degree of denaturation of the protein and decreased with hydrolysis of the protein. Using the data of Arai et al. (1970), Kinsella and Damodaran (1980) estimated that the total number of binding sites for partially denatured soy protein with a molecular weight of 100,000 would be about one.

Beyeler and Solms (1974), using an aqueous equilibrium dialysis technique, found a linear relationship between soy-bound ligand and the free ligand concentration that

Table 4. Flavor-binding constants for soy protein in aqueous systems (modified from Kinsella and Damodaran, 1980).

Ligand	Soy protein	K_{eq} [a]	ΔG	Reference
		1/M	kJ/mol	
1-hexanal	Native	173.4	-12.60	Arai et al., 1970
1-hexanol	Native	80.3	-10.72	Arai et al., 1970
2-butanone	Native	5174	-	Beyeler and Solms, 1974
2-heptanone	Native	110	-11.65	Kinsella and Damodaran, 1980; Damodaran and Kinsella, 1981a
2-octanone	Native	310	-14.22	Kinsella and Damodaran, 1980; Damodaran and Kinsella, 1981a
2-nonanone	Native	930	-16.95	Kinsella and Damodaran, 1980; Damodaran and Kinsella, 1981a
2-nonanone	Part. denatured	1240	-17.66	Kinsella and Damodaran, 1980; Damodaran and Kinsella, 1981a
5-nonanone	Native	541	-15.61	Kinsella and Damodaran, 1980; Damodaran and Kinsella, 1981a
Nonanal	Native	1094	-17.35	Kinsella and Damodaran, 1980; Damodaran and Kinsella, 1981a

[a] Equilibrium binding constant.

was independent of temperature and pH. The binding constants were found to decrease in the following order: aldehydes, ketones, and alcohols. Carboxylic acids did not bind to the soy protein. These results have been criticized by Kinsella and Damodaran (1980) because of the model used to determine the binding constants and possible membrane binding of the ligand. For example, the binding constant for 2-butanone at pH 7.0, 20 C, was found to be 5174/M (Table 4). Assuming a 50,000-molecular-weight soy protein, there would be 1200 binding sites, which is more than two moles of 2-butanone bound to each amino acid residue, which is unlikely (Kinsella and Damodaran, 1980).

Kinsella and Damodaran (1980) and Damodaran and Kinsella (1981a) also used an aqueous equilibrium dialysis method to quantitatively elucidate the thermodynamics of the binding of ketones (2-heptanone, 2-octanone 2-nonanone, and 5-nonanone) and nonanal with soy protein (Table 4). They determined that soy protein contains four binding sites for all the compounds studied. Binding constants (K_{eq}) increased and the free energy (ΔG) became more negative with increasing chain length. The position of the function group also influenced binding. Binding was independent of temperature from 25 to 45 C. However, markedly different values were obtained at 5 C; the binding sites decreased from four to two, and binding constants increased by more than twofold. These changes in binding site and affinity were attributed to possible changes in the tertiary and quaternary structures of the proteins at low temperatures. The nature of binding was attributed to hydrophobic interactions.

Damodaran and Kinsella (1981b) compared the binding of 2-nonanone to whole soy protein and the 7S and 11S fractions of soy protein. They determined that the 7S fraction was the major protein fraction involved with ligand binding at the concentrations used in their study. The 11S fraction had a near-zero binding affinity for 2-nonanone.

Thissen (1982) used headspace analysis to determine flavor binding in aqueous solutions of soy protein isolate. She also evaluated the various mathematical binding models (Table 5) that had been used to determine binding constants and the free energy of adsorption. Hexane did not bind with the soy protein under the conditions used in this study. This would imply that the functional group on the corresponding carbonyls is involved in the interaction between these compounds and soy protein. The binding curves for the ketones, aldehydes, and methyl ester were complex, suggesting unequal, positive cooperativity among the binding sites. This hypothesis was confirmed by using standard graphical methods (Klotz, Scatchard and Hill plots) (Price and Dwek, 1979; Van Holde, 1971; Cantor and Schimmel, 1980; Thissen, 1982) to determine the mode of binding. A modified Klotz plot procedure for positively cooperative systems was also used for all the compounds in this study. Using this procedure, all the compounds have at least three different sets of binding sites on the soy protein, with the exception of 5-nonanone, which had two different sets of binding sites.

For comparative purposes, the data for 2-hexanone at 25 C (Thissen, 1982) was used for each of the binding models used by other researchers in this field, and the results of these calculations are given in Table 5. The comparison of the parameters obtained indicates that values vary considerably (K_{eq} from 33.33 to 1,470.80) although their trends are similar. Inasmuch as the ΔG values are directly determined from K, the magnitude of these values also varies greatly between methods. The binding constant at 35 C (308 K) is approximately twice that obtained at 25 C (298 K). All the ΔGs are negative, indicating that the binding occurs spontaneously. In all instances, the $T\Delta S$ term was found to be larger than ΔH, indicating that the entropy is the driving force in this system. From these observations, we concluded that (1) the binding of

162

Table 5. Thermodynamic parameters for the binding of 2-hexanone (25 C) calculated by different methods (Thissen, 1982).

Parameter	$\bar{V}/(N-\bar{V})(C)$[a]	$\bar{V}/C$[b]	$N/(S-N)(C)$[c]	$1/\bar{V}$ vs. $\dfrac{1}{[L]^n}$[d]
K_{298}[e]	33.33	1470.80	146	289.94
K_{308}[e]	69.69	2578.52	329	589.31
ΔG_{298}[f]	-8.72	-18.1	-12.4	-14.1
ΔG_{308}[f]	-10.9	-20.1	-14.8	-16.4
ΔH[a]	56.36	42.86	62.10	54.18
$T\Delta S_{298}$[g]	65.07	60.96	74.46	67.24
$T\Delta S_{308}$[g]	67.25	62.98	76.93	70.52
ΔS[b]	218	204	250	225
ΔS[b]	218	204	250	229

[a]Where $\bar{V}$ equals average number of moles bound/mole protein, N equals maximum number of moles bound/mole protein, C equals molar concentration of nonbound ligand at equilibrium (Damodaran and Kinsella, 1981).

[b]Where $\bar{V}$ equals average number of moles ligand bound/mole protein, C equals molar concentration at equilibrium (used by Beyeler and Solms, 1974).

[c]Where C equals concentration of total ligand, N equals amount of ligand bound when initial concentration equals C, S equals amount bound when ligand concentration has reached a maximum (used by Arai et al., 1970).

[d]Where $\bar{V}$ equals average number of moles bound/mole protein, L is the concentration of total ligand, n equals lowest power to obtain a straight line that equals the number of different binding constants (Segal, 1975).

[e]Where K_{298} and K_{303} are the equilibrium binding constants (in 1/M) at 298 and 303 K.

[f]kJ/mol.

[g]J/mol K.

ligand(s) to soy protein may disrupt the tertiary and quaternary structure, which in turn, exposes additional binding sites, (2) the functional group of the flavor compound is important in the initial binding process, and (3) the published research, to date, where equal, noncooperative, reversible binding has been assumed and traditional binding models have been used, is based upon false assumptions and is, therefore, subject to considerable question.

Thus far, the lack of a suitable binding model has delayed the accurate determination of parameters that characterize flavor binding to soy protein. The stepwise equilibrium model (Fletcher et al., 1970) designed for drug protein interactions shows some promise as a method to handle positive cooperativity between binding sites. Application of this modified method to the data of hexanal at 25 C indicates three different binding constants (Thissen, 1982).

All the researchers (to date) have used equilibrium studies to measure reversible flavor binding and ignore irreversible flavor binding. Irreversible flavor binding is important, because of its influence on desirable flavor losses and potential for off-flavor release, and it must be quantified in future studies. Likewise, although protein-flavor binding has been emphasized in this paper, it must be remembered that foods are multicomponent systems containing other fractions (carbohydrate, lipids, phenols) that can interact with flavors (Solms et al., 1981). The understanding of these interactions is essential in the development of methods or processes to remove or mini-mize off-flavors and retain desired flavors in soy products.

CONCLUSION

The interactions of flavors with soy protein is very complex. There are a number of problems associated with comparing literature data describing the binding of flavors to soy protein. In aqueous systems, (1) the pH and presence of reducing agents in the buffer systems influence binding; (2) solvent extraction procedures may alter the pro-tein binding or form emulsions that increase the analytical error; (3) membranes in dialysis systems may plug or bind ligand, giving a false equilibrium and incorrect bind-ing constants; (4) the proteins used may be pure (and peptide microheterogenity may play a role) or a mixture with differing degrees of native or denatured structure; (5) the ranges of ligand concentration used in these studies are dissimilar due to saturation of headspaces, low air/water partition coefficients, solubilities, or the ligand precipitating the proteins; (6) the binding model assumptions of equal, noncooperative, reversible binding, are in question; (7) lack of statistical analysis clouds the significance of the data; and (8) use of atypical molecular weights for soy protein (50 to 100 k daltons instead of 150-400 k daltons) makes the comparison of literature data all but impossi-ble. Likewise, the misuse of the Scatchard binding model in other fields has brought considerable attention to the proper use of this model (Norby et al., 1980; Klotz, 1982; Scheinberg, 1982; Feldman, 1983; Siiteri, 1984). Thus, only by using a con-sistent approach can we better understand the nature of these flavor interactions with soy protein. This basic knowledge can then be applied to food systems (1) to eliminate off-flavors and odors, (2) to retain desirable flavors, (3) to improve process control, and (4) to increase consumer acceptability of soy foods with the concurrent increase in marketability and utilization of soy protein in human foods.

NOTES

Lester A. Wilson, Professor, Department of Food Technology, Iowa State University, Ames, IA USA.

Journal Paper No. J-11611 of the Iowa Agric. and Home Econ. Exp. Stn., Ames, Iowa. Project No. 2164, a contributing project to the North Central Regional Research Project NC-136.

REFERENCES

Arai, S., M. Noguchi, M. Yamashita, Y. Kata, and M. Fujimaki. 1970. Studies on flavor components in soybeans. Agric. Biol. Chem. 34:1569-1573.

Ashraf, H. L., and H. E. Snyder. 1981. Influence of ethanolic soaking of soybeans on flavor and lipoxygenase activity of soymilk. J. Food Sci. 46:1201-1204.

Aspelund, T. G. and L. A. Wilson. 1980. Adsorption of off-flavor compounds on soy protein. p. 32. In F. T. Corbin (ed.) World Soybean Research Conference II: abstracts. Westview Press, Boulder, Colorado.

Aspelund, T. G. and L. A. Wilson. 1983. Adsorption of off-flavor compounds onto soy protein: A thermodynamic study. J. Agric. Food Chem. 31:539-545.

Beyeler, M. and J. Solms. 1974. Interaction of flavor model compounds with soy protein and bovine serum albumin. Lebensm.-Wiss. Technol. 7:217-219.

Cantor, C. R. and P. R. Schimmel. 1980. Biophysical Chemistry Part III: The Behavior of Biological Macromolecules. W. H. Freeman and Company, San Francisco.

Chiba, H., N. Takahashi, N. Kitabatake and R. Sasaki. 1979a. Enzymatic improvement of food flavor. III. Oxidation of the soybean protein-bound aldehyde by aldehyde dehydrogenase. Agric. Biol. Chem. 43:1891-1897.

Chiba, H., N. Takahashi and R. Sasaki. 1979b. Enzymatic improvement of food flavor. II. Removal of beany flavor from soybean products by aldehyde dehydrogenase. Agric. Biol. Chem. 43:1883-1889.

Crowther, A., L. A. Wilson and C. E. Glatz. 1980. Effects of processing on adsorption of off-flavors onto soy protein. J. Food Process Eng. 4:99-115.

Damodaran, S. and J. E. Kinsella. 1981a. Interaction of carbonyls with soy protein: Conformational effects I. Agric. Food Chem. 29:1253-1257.

Damodaran, S. and J. E. Kinsella. 1981b. Interaction of carbonyls with soy protein: Thermodynamic effects. J. Agric. Food Chem. 29:1249-1253.

Feldman, H. A. 1983. Statistical limits in Scatchard analysis. J. Biol. Chem. 250:12865-12867.

Fletcher, J. E., A. A. Spector and J. D. Ashbrook. 1970. Analysis of macromolecule-ligand and binding by determination of stepwise equilibrium constants. Biochemistry 9:4580-4587.

Franzen, K. L. and J. E. Kinsella. 1974. Parameters affecting the binding of volatile flavor compounds in model food systems. I. Proteins. J. Agric. Food Chem. 22:675-678.

Fujimaki, M., S. Arai, N. Kirigaya and Y. Sakurai. 1965. Studies on flavor components in soybean. Part I. Aliphatic carbonyl compounds. Agric. Biol. Chem. 29:855-863.

Goosens, A. E. 1975. Protein flavour problems. Food Process. Ind. 44:29-30.

Gremli, H. A. 1974. Interaction of flavor compounds with soy protein. J. Am. Oil Chem. Soc. 95A-97A.

Hammonds, T. M. and D. L. Call. 1972. Protein use patterns. Chem. Technol. 2:156-162.

Hsieh, O. A. L., A. S. Huang and S. S. Chang. 1982. Isolation and identification of objectionable volatile flavor compounds in defatted soybean flour. J. Food Chem. 47:16-18.

Kalbrener, J. E., A. C. Eldridge, H. A. Moser and W. J. Wolf. 1971. Sensory evaluation of commercial soy flours, concentrates and isolates. Cereal Chem. 48:595-600.

Kinsella, J. E. and S. Damodaran. 1980. Flavor problems in soy proteins; origin, nature, control and binding phenomenon. p. 95-131. In G. Charalambous (ed.) The analysis and control of less desirable flavors in foods and beverages. Academic Press, Inc., New York.

Klotz, I. M. 1982. Numbers of receptor sites from Scatchard graphs: Facts and fantasies. Science 217:1247-1249.

Norby, J. G., P. Ottolenghi and J. Jensen. 1980. Scatchard Plot: Common misinterpretation of binding experiments. Anal. Biochem. 102:318-320.

Price, N. C. and R. A. Dwek. 1979. Binding of ligands to macromolecules. p. 41. In N. C. Price and R. A. Dwek (eds.) Principles and problems in physical chemistry for biochemists. Clarendon Press, Oxford.

Rackis, J. J., D. J. Sessa and D. H. Honig. 1979. Flavor problems of vegetable food proteins. J. Am. Oil Chem. Soc. 56:262-271.

Scheinberg, I. H. 1982. Scatchard Plots. Science 215:312-313.

Segal, I. H. 1975. Lineweaver-burk plot for allostoric enzymes. p. 365-371. In I. H. Segal (ed.) Enzyme Kinetics. John Wiley and Sons, New York.

Sessa, D. J. 1979. Biochemical aspects of lipid-derived flavors in legumes. J. Agric. Food Chem. 27:234-239.

Siiteri, P. 1984. Receptor binding sites. Science 223:191-193.

Solms, J., B. M. King and R. Wyler. 1981. Interactions of flavor compounds with food components. p. 7-18. In G. Charalambous and G. E. Inglett (eds.) The Quality of Foods and Beverages. Academic Press, New York.

Takahashi, N., R. Sasaki and H. Chiba. 1979. Enzymatic improvement of food flavor. IV. Oxidation of aldehydes in soybean extracts by aldehyde oxidase. Agric. Biol. Chem. 43:2557-2561.

Thissen, J. A. 1982. Interaction of off-flavor compounds with soy protein isolate in aqueous systems: Effects of chain length, functional group and temperature. M.S. Thesis. Iowa State University Library, Ames.

Twigg, G., A. W. Kotula and E. P. Young. 1977. Consumer acceptance of beef patties containing soy protein. J. Anim. Sci. 44:218-223.

Van Holde, K. E. 1971. Highly cooperative binding. p. 63. In K. E. Van Holde (ed.) Physical Biochemistry. Prentice-Hall, Inc., Englewood Cliffs, New Jersey.

Wolf, W. J. 1975. Lipoxygenase and flavor of soybean products. J. Agric. Food Chem. 23:126-141.
Wolf, W. J. and J. D. Cowan. 1975. Soybeans as a food source. CRC Press, Cleveland, Ohio.

CHEMICAL, FUNCTIONAL AND NUTRITIONAL PROPERTIES
OF PHYTATE-REDUCED SOY PROTEINS

C. V. Morr

The importance of phytic acid and its salts in soybeans has been the subject of numerous publications dating back to the late 1930s. Interest in this topic has been revitalized the last few years, and it has been reviewed recently (Cheryan, 1980; Maga, 1982).

This paper summarizes our research to remove phytate from soy protein extracts and to compare the physico-chemical, functional and nutritional properties of control and phytate-reduced soy proteins.

PHYTATE REMOVAL FROM SOY PROTEINS

Phytate is co-extracted with and complexes with proteins and minerals via ionic bonding during alkaline extraction and acid precipitation of soy protein isolates. Several laboratory and pilot scale processes have been proposed for removing phytate from soy protein extracts. These processes generally involve dialysis, ultrafiltration, extraction at pH 11 to 12 and selective cation precipitation of phytate. Although each of these processes can be optimized by appropriately controlling pH, ionic strength, Ca ion concentrations and addition of Ca ion complexing chemicals, each has inherent problems that limits its potential for industrial application.

Since Ca ions are required to form and stabilize the protein-phytate complex in soy protein isolates (de Rham and Jost, 1979), we devised a two-step ion exchange process to sequentially remove Ca and phytate ions from soy protein extracts at pH 8-9 (Brooks and Morr, 1982). This combination cation/anion exchange process consistently removes over 95% of the Ca, Mg and phytate from aqueous soy extracts.

The ability of added Ca, Ba and Zn ions to complex selectively with and precipitate phytate ions from soy protein extracts was investigated with the aim of producing a phytate-reduced soy protein isolate (Brooks and Morr, 1984a). Two general approaches were employed. The first was to add up to 5% (w/v) of divalent cation salts to aqueous soy protein extracts to precipitate the phytate-divalent cation complex. The second approach was to use up to 5% (w/v) solutions of the divalent cation salts as protein extractants. Although both approaches resulted in partial reduction of phytate content of soy protein isolates, yields were consistently low; e.g., 10 to 85% of those obtained by conventional isolate production. Similar results were obtained in those experiments where the proteins were extracted with divalent cation salt solutions.

PHYTATE CONTENT OF SOY PROTEIN COMPONENTS

The strong tendency of phytate to complex with soy proteins during their isolation raises the possibility that it may interfere with subsequent efforts to fractionate and characterize their major components. This possibility has been virtually ignored in all soy protein fractionation work to date.

The major soy protein components; e.g., 7S, 11S, soy whey and the soy whey precipitate fraction, were prepared from aqueous extracts of laboratory defatted Bragg soybeans and commercially defatted soy flakes by a modified Thanh and Shibasaki (1976) procedure. Each of the fractionated components was characterized with respect to protein, total phosphorus (P) and phytate content (Brooks and Morr, 1982).

Results in Table 1 demonstrated that P and phytate were effectively removed from 11S soy protein, but were not as effectively removed from 7S soy protein and the soy whey fractions (Brooks and Morr, 1984b). Most of the P and phytate were recovered in the soy-whey precipitate fraction. This confirms that the 11S soy protein nomenclature is not influenced by phytate ions. However, the 7S soy protein component needs further attention because of the possibility that phytate may interfere with efforts to fractionate and characterize it accurately.

Solubility of Phytate-Reduced Soy Protein Isolate

Compositional data for control, phytate-reduced and commercial soy protein isolates are provided in Table 2 (Chen, 1983). Phytate-reduced and commercial soy isolates contained 77 and 47% less phytate, respectively, than control isolate. Phytate-reduced soy protein also contained about 55% less ash than either of the other two isolates. Freeze-dried phytate-reduced and control isolates exhibited equivalent or higher solubility values than spray-dried commercial isolate at all three pH values (Table 2). Phytate-reduced soy protein isolate exhibited highest solubility at pH 3 (84%) and lowest solubility at pH 6 (12%). Although phytate may exert only a minor effect upon the extractibility of soy proteins from defatted soy flour (de Rham and Jost, 1979), it has a major effect upon their solubility in freeze-dried isolates where the phytate has complexed with the protein. The extensive removal of Ca and Mg ions by a combination cation/anion exchange process (Brooks and Morr, 1982) may also be a contributing factor in altering the pH-solubility of phytate-reduced soy protein isolate.

Table 1. Chemical composition of freeze-dried soy protein components (Brooks and Morr, 1984b).

Protein component	Apparent mean protein content	Mean total P content[a]	Mean phytate content[a]
		– % –	
Whole soy extract	78.95	0.60	1.41
7S component	112.60	0.68	1.40
11S component	100.80	0.08	0.07
Soy whey	16.88	0.25	0.28
Soy whey precipitate	4.49	15.20	45.37

[a]Data from three replicate trials.

Table 2. Composition and solubility of soy protein isolates used in foaming studies.

	Protein isolate		
	Control	Phytate-reduced[a]	Commercial[b]
		– % –	
Composition[c]			
Moisture	4.57	4.22	5.05
Ash	3.62	1.65	3.74
Protein[d]	74.70	82.20	77.70
Phytate	1.90	0.43	1.00
Solubility[e]			
pH 3	61.3 ± 0.26[f]	83.7 ± 2.25[f]	8.9 ± 0.96
pH 6	22.3 ± 0.28[f]	11.6 ± 0.80[f]	12.1 ± 0.45
pH 9	70.2 ± 4.27[f]	70.3 ± 9.25[f]	37.9 ± 4.45

[a]Brooks and Morr (1982).
[b]Supro 620, Ralston Purina Co., St. Louis.
[c]Mean values for three trials.
[d]Micro-kjeldahl N x 5.7.
[e]Mean ± SD. Procedure from NE-123 regional research project.
[f]Freeze-dried protein products.

Foaming Properties of Phytate-reduced Soy Proteins

Foaming studies were conducted on the three soy protein isolates (Chen, 1983) at two protein concentrations, 5 and 10% w/v; three solution preheat conditions of 25, 60 and 80 C for 30 min; and three pH conditions of 3, 6 and 9 (Table 3). At pH 3 and 10% protein concentration, both control and phytate-reduced soy protein isolate solutions formed gels during preheating at 60 and 80 C and subsequently formed low-expansion, highly stable foams. Averages of the three or four highest foam expansion values for each isolate were 956% (control), 1336% (phytate-reduced) and 320% (commercial). All three isolates had completely stable foams under these test conditions.

It was concluded that, in terms of foam expansion and stability, phytate-reduced soy protein isolate functions better than either of the other two soy protein isolates at pH 3. Phytate-reduced isolate provided optimal foaming properties at pH 3, 5% protein concentration, and with a 25 C preheat treatment. Control isolate functioned best at pH 9, irrespective of protein concentration and preheating conditions.

Properties of Phytate-reduced and Control Soy Proteins

Properties of control and phytate-reduced (Brooks and Morr, 1982) soy protein extracts were investigated by chromatography on Sepharose 6B gel columns. One aliquot of each extract was freeze-dried, dispersed in 100 mL of Briggs and Wolf (1957) phosphate buffer (pH 7.6, ionic strength 0.5 M, 0.01 M β-mercaptoethanol, 0.5 g/L sodium azide) and centrifuged at 13,000 x g. One mL of the resulting supernatant was chromatographed on a 60 x 1.5 cm Sepharose 6B gel column with a 40 mL void volume. Protein elution was monitored in the effluent at 280 nm, and results are given in Figure 1a.

A second aliquot of the control and phytate-reduced soy protein extracts was dialyzed against Briggs and Wolf (1957) buffer containing 0.005 M dithiothreitol (DTT), and chromatographed on a 63 x 2.6 cm Sepharose 6B gel column with a void

Table 3. Foaming properties of soy protein isolates.[a]

Protein concentration	30 min preheat temperature	pH	Soluble protein[b]	Maximum foam expansion[b]	Foam stability[c] Time of first drop	Drainage
%, w/v	C		%	%	min	%, w/v
CONTROL ISOLATE[d]						
5	25	3	70.8	753	6.2	47
5	25	6	34.8	645	1.1	60
5	25	9	100.0	863	9.4	39
5	60	3	79.2	726	9.1	37
5	60	6	24.0	948	1.6	63
5	60	9	100.0	833	3.7	64
5	80	3	75.5	755	10.4	33
5	80	6	8.8	897	2.4	55
5	80	9	84.0	802	4.4	63
10	25	3	65.0	763	18.5	8.6
10	25	6	51.1	666	35.5	0
10	25	9	100.0	749	22.8	3.5
10	60	3	f	475	60.0	0
10	60	6	33.7	1108	6.8	41
10	60	9	100.0	645	14.8	16
10	80	3	f	497	60.0	0
10	80	6	10.8	790	6.1	32
10	80	9	96.1	469	60.0	0
PHYTATE-REDUCED ISOLATE[d]						
5	25	3	100.0	1539	22.6	14
5	25	6	21.0	511	25.0	0
5	25	9	95.1	735	14.7	16
5	60	3	100.0	1304	23.6	11
5	60	6	4.5	541	0.7	70
5	60	9	89.5	676	3.4	63
5	80	3	93.1	741	9.8	43
5	80	6	13.6	673	2.6	60
5	80	9	64.4	611	3.5	56
10	25	3	100.0	1164	32.9	0
10	25	6	8.2	676	4.0	33
10	25	9	89.0	564	58.9	0
10	60	3	f	442	60.0	0
10	60	6	16.4	643	4.5	34
10	60	9	96.8	541	22.1	2.2
10	80	3	f	285	60.0	0
10	80	6	3.2	859	6.3	34
10	80	9	56.8	465	60.0	0
COMMERCIAL ISOLATE[e]						
5	25	3	55.9	277	13.8	6.5
5	25	6	25.2	62.5	0.02	91
5	25	9	91.3	74.3	0.02	90
5	60	3	42.7	141	7.6	1.3
5	60	6	34.2	57.0	0.02	91
5	60	9	87.2	332	0.02	91
5	80	3	33.4	250	2.0	47
5	80	6	38.1	34.8	1.8	32
5	80	9	95.4	320	0.02	94
10	25	3	11.6	36.5	60.0	0
10	25	6	25.3	136	60.0	0
10	25	9	92.4	283	60.0	0
10	60	3	36.5	36.1	60.0	0
10	60	6	30.9	44.2	27.2	0
10	60	9	75.5	315	60.0	0
10	80	3	39.4	102	60.0	0
10	80	6	27.4	31.2	8.5	84
10	80	9	63.3	315	60.0	0

[a]Average for two or three replicate trials.
[b]Procedure from NE-123 regional research project.
[c]Foam stability determined as time of first drop and % of total liquid drainage from foam placed in a 7.5 cm inner diameter pyrex funnel for 30 min following maximum foam expansion.
[d]Brooks and Morr (1982).
[e]Supro 620, Ralston Purina Company, St. Louis.
[f]Protein solution formed a gel during preheat treatment.

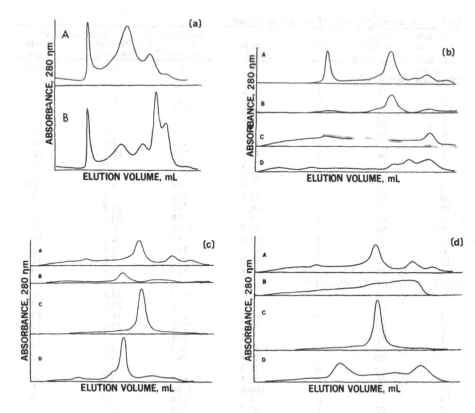

Figure 1. (a) Sepharose 6B gel column chromatograms of soy protein: A. phytate-reduced soy extract; B. Control soy extract. (b) Sepharose 6B gel column chromatograms of soy proteins: A. Control soy protein extract; B. control acid-precipitated soy proteins; C. phytate-reduced, acid-precipitated soy proteins; and D. acid soy whey proteins. (c) Sepharose 6B gel chromatograms of control soy protein extract components: A. Briggs and Wolf (1957) 7S soy protein; B. Thanh et al. (1975) 7S soy protein; C. Briggs and Wolf (1957) 11S soy protein; and D. Thanh et al. (1975) 11S soy protein. (d) Sepharose 6B gel chromatograms of control and phytate-reduced soy protein components prepared by the procedure of Briggs and Wolf (1957): A. control 7S soy protein; B. phytate-reduced 7S soy protein; C. control 11S soy protein; and D. phytate-reduced 11S soy protein. See text for details of fractionation and chromatographic procedures.

volume of 115 mL, using the same buffer as eluant (Figure 1b).

A third aliquot of each extract was adjusted to pH 4.5 with HCl and centrifuged to recover the acid-precipitated soy protein fraction. This protein fraction was dispersed in and dialyzed against Briggs and Wolf (1957) phosphate buffer containing DTT, and chromatographed on the 115 mL void volume Sepharose 6B gel column (Figure 1b).

The extracts also were fractionated into their 7S and 11S components, by the procedures of Briggs and Wolf (1957) and Thanh et al. (1975), dispersed in and dialyzed against Briggs and Wolf (1957) phosphate buffer, and chromatographed on the 115 mL void volume Sepharose 6B gel column (Figures 1c and d).

Results demonstrate that the combination cation/anion exchange phytate removal treatment (Brooks and Morr, 1982) produced a major shift in the molecular weight distribution of the proteins in soy extract. Although the range of molecular weights is similar for the proteins in both extracts, there was a shift from smaller to larger molecular weight protein components in the phytate-reduced extract. Comparison of Sepharose 6B gel chromatograms for control and phytate-reduced, acid-precipitated soy proteins and their 7S and 11S soy protein components (Figures 1b,d and Table 4) confirms that phytate removal from soy proteins results in major alterations of their molecular weights. Both 7S and 11S soy protein components prepared from phytate-reduced soy protein extracts were considerably more polydispersed with a broader molecular weight distribution than control soy extract counterparts.

EFFECT OF PHYTATE ON SOY PROTEIN DIGESTIBILITY

Soy proteins are generally recognized as being somewhat resistant to proteolytic digestion and in vitro enzymatic modification, due mainly to their compact conformational state, which renders their peptide bonds inaccessible to the enzymes (Abdul-Kadir, 1980). The protein-Ca-phytate complex may also tend to inhibit the soy protein's digestibility, by contributing to a more compact conformational state.

We presently are comparing in-vitro digestibility characteristics of control and phytate-reduced soy protein isolates, by use of a pH stat and several biochemical assay methods, and several enzymes with pH optima of 2 to 8. Preliminary results indicate that phytate-reduced soy protein isolate is slightly more susceptible to in-vitro digestibility than control isolate, especially at acid pH values.

EFFECT OF PHYTATE ON IRON BIOAVAILABILITY

There is considerable controversy concerning the role of soy proteins and phytate in lowering the bioavailability of Zn, Fe and other trace minerals (Maga, 1982;

Table 4. Summary of Sepharose 6B chromatographic data for soy protein extracts and their components.

Protein fractions and components	Sepharose 6B fraction number			
	1	2	3	4
	− Elution volume, mL[a] −			
CONTROL				
Soy extract 1	110	200	265	-
Soy extract 2	110	210	265	310
Acid-precipitated proteins	110	190-205	255-280	-
Acid whey	-	-	240-265	300
Briggs & Wolf 7S 1	-	210	275	-
Briggs & Wolf 7S 2	-	220	280	-
Briggs & Wolf 11S 1	-	215	-	-
Briggs & Wolf 11S 2	-	210	-	-
Thanh et al. 7S	-	200	250-300	-
Thanh et al. 11S	110	210	270-280	-
PHYTATE-REDUCED				
Acid-precipitated proteins	135	205	305	-
Briggs & Wolf 7S	-	-	300-310	-
Briggs & Wolf 11S	-	205	305	-

[a]Volume corresponding to maximum peak height.

Table 5. Fe bioavailability of soy-protein-containing diets determined by chick hemoglobin repletion (C. J.Rodriguez, pers. commun.).

Composition/treatment of diets	Fe content	Hemoglobin content[a]	Hematocrit value[a]
	mg/kg	g/dL	%
Control soy isolate without added Fe	41	5.14	29.6
Control soy isolate with added Fe[b]	40	5.34	31.7
Heated control soy isolate without added Fe[c]	41	6.67	34.6
Heated control soy isolate with added Fe[bc]	58	7.48	35.7
Phytate-reduced soy isolate without added Fe	40	5.42	30.6
Phytate-reduced soy isolate with added Fe[b]	52	6.75	34.8
Heated phytate-reduced soy isolate without added Fe[c]	36	7.16	35.4
Heated phytate-reduced soy isolate with added Fe[bc]	51	7.53	35.4

[a] Average for 12 chicks 10 d after diet began.
[b] 20 mg added Fe/kg diet.
[c] Heated 120 C for 20 min.

Cheryan, 1980; Abdul-Kadir, 1980). C. J. Rodriguez (pers. commun.) studied the effect of phytate upon bioavailability of Fe from soy-protein-based diets, by determining hemoglobin repletion rates in two-week-old Leghorn cockerels. This animal was specifically chosen because its digestive system reportedly contains low phytase activity, and thus would provide minimal phytate degradation.

Commercial soy protein isolate (Supro 620, Ralston Purina Company) and phytate-reduced soy protein isolate (Brooks and Morr, 1982) were used as protein sources. The experiment was designed as a factorial with two levels for each factor; e.g., phytate content, heat treatment and iron content. Ninety-six freshly hatched Leghorn cockerel chicks, fed an Fe-depletion diet for two weeks, were divided into eight groups of 12 chicks each, such that each group had an equal mean hemoglobin content. Each group was fed one of eight Fe-repletion diets for 10 days (Table 5). Four of the diets were heated in an autoclave at 120 C for 20 min to inactivate trypsin inhibitors. Four of the diets received 20 mg of Fe (60.8 mg of $FeSO_4 \cdot H_2O$)/kg of diet.

Those diets that contained added Fe resulted in greater hemoglobin repletion values than diets without added Fe (Table 5). Heated soy-isolate-containing diets resulted in greater hemoglobin repletion values than did unheated diets. Phytate-reduced soy-protein-containing diets, both with and without added Fe, also resulted in greater hemoglobin repletion values than control soy-protein-containing diets. Although generally similar trends in blood hemoglobin and hematocrit repletion were observed during the study, hemoglobin values provided a better indication of Fe repletion than hematocrit values. Partitioning of primary effects by Analysis of Variance procedures indicated that 68 to 75% of the hemoglobin repletion was due to heating, 10 to 15% was related to phytate content, and 13 to 14% was accounted for by difference in Fe content of the diets.

NOTES

C. V. Morr, Department of Food Science, Clemson University, Clemson, SC 29631.

REFERENCES

Abdul-Kadir, R. B. 1980. The effect of phytate content on the nutritional quality of soy and wheat bran proteins. Ph.D. Dissertation. University of Nebraska Library. Lincoln, NE.

Briggs, D. R. and W. J. Wolf. 1957. Studies on the cold-insoluble fraction of the water-extractable soybean proteins. I. Polymerization of the 11S component through reactions of sulfhydryl groups to form disulfide bonds. Arch. Biochem. Biophys. 72:127-144.

Brooks, J. R. and C. V. Morr. 1982. Phytate removal from soy protein isolates using ion exchange processing treatments. J. Food Sci. 47:1280-1282.

Brooks, J. R. and C. V. Morr. 1984a. Phytate removal from soybean proteins. J. Am. Oil Chem. Soc. 61: (in press).

Brooks, J. R. and C. V. Morr. 1984b. Phytate content of soybean protein components. J. Agric. Food Chem. 32: (in press).

Chen, B. H-Y. 1983. Effect of several processing pretreatments upon foaming properties of soy protein isolates. M.S. Thesis. Winthrop College Library. Rock Hill, SC.

Cheryan, M. 1980. Phytic acid interactions in food systems. CRC Critical Reviews in Food Science and Nutrition 13:297-335.

de Rham, O. and T. Jost. 1979. Phytate-protein interactions in soybean extracts and low-phytate soy protein products. J. Food Sci. 44:596-600.

Maga, J. A. 1982. Phytate: its chemistry, occurrence, food interactions, nutritional significance, and methods of analysis. J. Agric. Food Chem. 30:1-9.

Thanh, V. H., K. Okubo and K. Shibasaki. 1975. Isolation and characterization of the multiple 7S globulins of soybean proteins. Plant Physiol. 56:19-22.

Thanh, V. H. and K. Shibasaki. 1976. Major proteins of soybean seeds. A straightforward fractionation and their characterization. J. Agric. Food Chem. 24:1117-1121.

Utilization of Soybean Protein

NONFOOD USES OF SOY PROTEIN PRODUCTS

Dale W. Johnson

The nonfood uses of soy protein products consist of their utilization in industrial products and processes and in animal feed. The information in this report deals with U.S. production only.

Soy protein products are categorized as soybean meal, soy flour and grits, isolated soy protein, soy protein concentrates and hydrolized soy proteins. Each of these general categories represents a number of different products, with differences depending on particle size, degree of heat treatment at some phase in the processing, and other conditions of processing during their manufacture. The products are intended to have functional characteristics that satisfy the properties needed to obtain the desired results for each application.

SOYBEAN PROCESSING AND PRODUCTS

Figure 1 illustrates the various soy protein products. As shown, whole soybeans are processed to destroy enzyme inhibitions and to improve the nutritional value of the protein for utilization of the full-fat soybeans, usually cracked or ground, in animal feed. The use of this type of product in animal feed is insignificant today, however.

At one time the whole soybeans were processed by screw pressing (involving heat) to remove as much oil as possible, and the partially deoiled product was used in animal feed. This was the procedure used before the development of solvent processing. There may still be two or three screw presses being used in the U.S., but the volume of beans processed in this manner is almost nil. In the past, cracked, dehulled beans were processed to make a full fat soy flour, which was used mainly for human food, although some was used in calf milk replacers. Dehulled beans were also processed through screw presses to produce a low-fat soybean product, most of which was used for human food consumption, though some was used in pet foods and animal feed. These products are no longer produced.

The flakes from solvent extraction are further processed to produce a variety of products with different physical and functional characteristics as shown in Figure 1.

PRODUCTS AND USES

The major nonfood use for soy protein products is soybean meal, which is used in animal feed. However, in certain specialty animal feed products, soy flour and grits, textured soy flour, and soy protein concentrates are used in substantial quantities. There is a considerable amount of soy protein used in a variety of applications

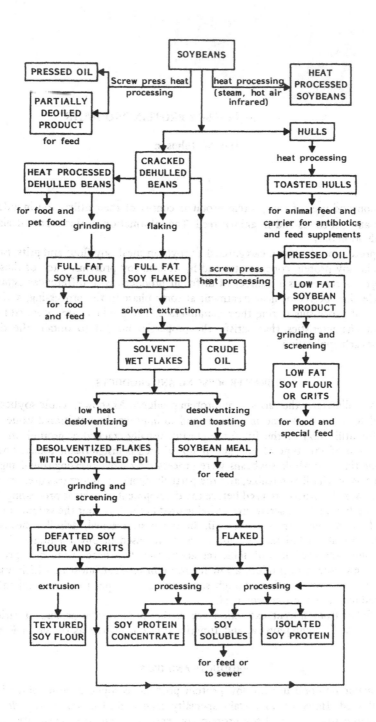

Figure 1. Flow sheet illustrating the processing of soybeans to produce soy protein products.

other than animal feed and for edible purposes, but there has been a considerable change in certain uses over the past 20 or 30 years.

Since there is no requirement by government agencies for processors to report production or sales of the specialty soy protein products, it is virtually impossible to obtain completely accurate information on the quantities produced. It is known, with reasonable accuracy, that around 20 years ago (in the early 1960s) the industrial use of soy flour in the U.S. was in the range of 50 to 55 thousand tons per year in plywood. It is estimated that the current use is in the range of 7,000 tons annually.

Industrial isolated soy protein was being produced at the rate of 18 to 20 thousand tons per year in the 1960s and increased to about 27 thousand tons per year in the early 1970s. Since that time, the largest producer of industrial soy proteins shut down its operation, so at present, there is only one producer of industrial isolated soy protein. It is estimated that the current production of industrial isolated soy protein is in the range of 9,000 tons annually.

There is a substantial production of edible isolated soy protein, but the processing to produce edible protein products is quite different from that to produce the industrial product. Edible isolated soy proteins do not have the functional and physical characteristics desired in industrial soy proteins.

It is estimated that the production of textured soy flour products could be in the range of 80 to 100 thousand tons per year. It is known that a substantial amount of the textured soy flour products are used in canned pet food, primarily dog food, but there are no accurate figures available as to the amounts going into pet foods.

It is estimated that, in the U.S., the 1984 pet food sales (including dog and cat food) will be about 4.5 million tons (Zaworski, 1983). If one assumes that the average protein content for the mixture of canned, semi-moist and dry products is in the range of 15% to 20%, the total protein consumption by these animals would be in the range of 680 thousand to 900 thousand tons of protein. If one further assumes that 60% of the protein in these products is supplied by soybeans, and calculates the quantities on the basis of 50% protein soybean meal, this would mean a usage of around 0.8 to 1.0 million tons of 50% soybean meal.

While there is a large amount of meal used in pet foods in the U.S., there are a number of pet food products, besides textured soy flour, that also utilize soy flour and grits, and lesser amounts of isolated soy protein and soy protein concentrate. In milk replacer products for feeding calves and other newborn animals, soy protein concentrates and soy flour are used in significant amounts. Very little soy protein isolate is used in calf milk replacer.

Soy Flour

Many of the major industrial uses for soy flour have been decreasing over the years, not because of price, but because other chemical polymers, which are more effective than the soy protein for particular applications, have been developed. Table 1 lists many of the nonfood applications that use or have used defatted soy flour. Some uses are for relatively small quantities of soy flour. While the listing is for defatted soy flour, there is some overlapping usage with industrial isolated soy protein, which is more useful in certain of these applications. The listing of these products is not in order of the quantities used. The use of the products in various applications depends on the particular functional characteristics the soy flour possesses to give the desired result in the finished product. In Tables 1 and 2, a code is used to indicate the particular function or functions desired for each application. In pet foods, and possibly in calf milk replacers, palatability of the ingredients is exceedingly important. Therefore,

178

Table 1.   Commercial applications for defatted soy flour.

Application	Property of soy flour of interest for the application[a]			
Pet foods (dog, cat, fish, bird)	N	E	S	
Calf and other milk replacers	N	E	S	
Feed pelleting	A			
Fermentation	N			
Bee feed	N			
Tanning	N	E		
Joint cements	A			
Wallboard	A			
Mortar cement	A			
Asphalt	E	V		
Sprays-insecticide, pesticide	A	F	E	S
Pan grease	F	E	S	
Wall paper coating	A	F	E	S
Plywood glues	A			
Putties	A	V	W	
Metal polishing	Absorbent and abrasive			

   [a]A = adhesive, E = emulsifying, F = film forming, N = nutrient, S = stabilizing, W = water holding, V = viscosity control.

Table 2.   Commercial applications for industrial isolated soy protein.

Application	Property of the soy protein of interest for the application[a]				
Paper coating	A	D	F	S	V
Joint cement	A				
Detergent products	D	DE	E	F	
Water base paints	A	D	E	F	S	V
Fire fighting foam	FO				
Wall paper coating	A	F	E	S	
Shoe polish	D	DE	F	S	
Fiber board	A				
Textiles	Sizing				
Paper board laminating	A				
Foamed concrete	FO				
Putties	A	V	W		
Printing inks	A	D	S		
Spun fibers					

   [a]A = adhesive, D = dispersant, DE = detergency, E = emulsifying, F = film forming, FO = foaming, S = stabilizing, V = viscosity control, W = water holding.

the products must be processed to give a flavor characteristic that is acceptable to the animals.

   In fermentation, a significant amount of soy flour is used as one of the ingredients in the media for growing the organisms that produce the desired products. The major use of soy products is in the "mycin" type fermentation, but they are also used in certain other fermentations. Depending on the manufacturer, meal, flour or grits may be used. It is estimated that the use in fermentation is several thousand tons per year. At one time there was some soy flour used in producing beer, where the soy served as a nutrient for better yeast growth. It never was an extensive use, and it is not known to the author whether some brewers may still be using it.

Defatted flour is used for making a diet to feed bees for honey production. Generally, a low-fat expeller soy flour or grit has been used, as well as full-fat soy flour. Soy flour has been used in the tanning industry, most probably as a nutrient for organisms responsible for digesting some of the non-collagen proteins remaining on the hides.

In joint cements, wall board and mortar cements, soy flour products have been used strictly for their adhesive property, usually in an alkaline medium, along with other additives. At one time soy flour was used as an ingredient in asphalt for roofing and for road surfacing. It apparently served as an emulsifier and for viscosity control.

Soy flour is an ingredient in insecticide, pesticide and herbicide products for spraying crops. It is used to help form uniform films on the surface of leaves and stems, and as an adhesive, so the sprayed materials will stick to the surfaces.

Soy flour is used in pan grease formulations along with edible oil, lecithin and other emulsifiers to be brushed or sprayed on the surface of the pans used for baking. It prevents the baked dough from sticking to the pan, hence it is easier to remove the finished loaves.

Some soy flour is used in blends of materials for coating wall paper. In the plywood field, soy flour is only used for internal plywood, since adhesives produced with soy flour are not water resistant. New polymers have been developed that do a better job and are lower in cost than are adhesives produced using soy flour; hence the drop in usage mentioned earlier.

Soy flour has been used as an ingredient in certain types of putties for its adhesive, water-holding and dough-forming properties. Soy flour has been an ingredient in products used in tumbling of metal castings in barrel tumblers to remove burrs and to polish the castings.

ISOLATED SOY PROTEIN

Industrial isolated soy proteins differ from edible isolated soy proteins in that industrial soy protein undergoes hydrolysis during processing. The degree of hydrolysis will determine the characteristics of the industrial soy protein. Usually, the criterion for determining the degree of hydrolysis is viscosity. Products are generally made as low, medium and high viscosity products.

Table 2 lists a variety of applications for which industrial soy protein has been used and gives the particular properties that are of interest for the different applications. The major use of industrial soy protein is in paper coatings (Bain et al., 1961). This application probably accounts for more than 90% of its use. The proteins are dispersed using a variety of alkaline agents, with the addition of clays, colors, antifoam agents, preservatives, etc. There are a number of procedures for applying the dispersion mixtures to the paper. Enamel coated papers with soy protein have high gloss, and the process results in a high quality product for printing of labels and magazine paper for advertisements using colored inks. The protein serves as the adhesive for bonding the pigment for coating the paper surface. The solids content of the dispersions being applied to the paper may be as much as 60%.

Joint cements are used in the construction field where plaster board or sheetrock is sealed at the joint using cement and tape, following which it is sanded, so that when the wall is painted there is no indication that a joint exists.

There is probably very little industrial isolated protein being used today in detergents, but it may be of interest to note that the first commercial production of industrial isolated soy protein, more than 40 years ago, was for a wall washing detergent. The product is still on the market, but other compositions are now used rather than soy protein.

At one time there was a considerable amount of soy protein used in water-based paints, but this use has also disappeared due to the developments of more useful polymers. During World War II most of the industrial isolated protein being produced was used for making fire fighting foam for the Armed Services.

At one time a very high percentage of wall paper was coated using isolated soy protein, but again, other polymers have replaced soy protein to a considerable extent. A small volume of isolated soy protein has also been used in shoe polish. Considerable amounts have been used in fiber board and in lamination of paperboard.

Hydrolized soy protein has been used in foamed concrete. When properly used the expansion ratio is 20 to 1, and the product density is about 480 kg/m^3. The foamed material is not used for structural purposes, but is used as roofing and insulation material. Some isolated soy has been used for sizing in textiles and in printing ink.

Attempts have been made to spin fibers. Whereas fibers can be made, those produced have no wet strength, so it has never worked out commercially.

Isolated soy proteins have been reacted to produce acylated derivatives that have superior adhesive properties and improved fluidity of colored coatings. These products were available commercially at one time. Soy proteins have been modified using other chemicals such as acrylonitrile, imidazole, alkylene oxides, epichlorohydrins, sodium chloroacetate and other reagents (Bain et al., 1961).

Soy Protein Concentrate

The only significant nonfood use of soy protein concentrates is in milk replacer products for calves and other young animals. It is estimated that as much as 40% of the soy protein concentrate produced goes for this application. Total production of soy protein concentrate is estimated to be in the range of 40,000 to 50,000 tons per year. The soy protein concentrate produced by the acid-water leach process is not generally acceptable as an ingredient in calf milk replacers. The type of product that is used is that produced by an aqueous alcohol extraction process.

By-Product

A by-product from the production of isolated soy protein is the spent insoluble material remaining after dissolving the protein and other soluble materials from the flakes or flour. This spent material is dried, and because of its ability to absorb substantial quantities of water, it has been used as a carrier for antibiotics, vitamins, minerals, etc. used as additives in animal feed.

ESTIMATED PRODUCTION

In the crop year 1981 to 1982, U.S. soybean meal produced, calculated as 44% protein meal, was approximately 23 million tons (American Soybean Association, 1982). If this is converted to the basis of 50% protein meal, this would be a production of about 20 million tons of 50% protein meal. Of this quantity, the domestic disappearance in the U.S. was around 14 million tons. The remainder, about 6 million tons, went for export.

Nonfood use, other than that used in major animal feed production, including pet foods, milk replacer products, fermentation and other miscellaneous uses discussed above, is estimated to be the equivalent of 0.9 to 1.1 million tons per year of 50% protein meal, or 6 to 8% of the total soybean meal produced. The equivalent of 50% protein meal used for soy protein products in food applications could be in the range of 0.4 million tons per year, or about 3% of the total soybean meal produced. Combined nonfood and food use of soybean meal is about 10% of the total meal produced, with 90% being used in animal feed.

CONCLUSION

The nonfood usage of soy protein will continue. Some uses may grow, others will decrease and occasional new uses will be found. Looking to the future, it may be possible to carry out reactions to chemically modify soy protein to develop products that may have desirable and improved characteristics for some nonfood applications, but it will be necessary to determine the economics of doing this in comparison to polymers that now exist or may be developed in the future.

NOTES

Dale W. Johnson, Food Ingredients (Minnesota), Inc., 2121 Toledo Ave. No., Golden Valley, Minnesota 55422.

REFERENCES

American Soybean Association. 1982. Soya Bluebook. p. 183.
Bain, W. M., S. J. Circle and R. A. Olson. 1961. Isolated soy proteins for paper coating. TAPPI Monograph Series No. 22.
Zaworski, Frank. 1983. Study predicts doubling of pet food sales by 1992. Pet Food Industry 25(3): 12-14.

SOY PROTEIN PRODUCTS IN PROCESSED MEAT AND DAIRY FOODS

L. Steven Young

Soy protein technology now allows extensive diversification of protein product applications. Currently, manufactured products have more functionality, better performance, more consistency and better flavor profiles than traditional soy flours and grits. Among these products are improved textured soy flours and concentrates, functional and non-functional soy protein concentrates, and highly soluble, highly functional soy protein isolates. Although continually being developed and evaluated, these commercially textured, concentrated and isolated proteins are finding new, significant applications particularly in processed meat and dairy foods.

SOY PROTEIN PRODUCTS

Given specific goals or guidelines, soy protein products offer uniqueness, versatility and superior functional characteristics for use in processed meats. Applicable meat systems include coarse ground, emulsion, and whole muscle products. Textured vegetable proteins, soy protein concentrates, and properly selected soy protein isolates may be used to help reduce ingredient cost, maintain quality and create new product opportunities. Exact formulations normally are determined by cost, nutrition, quality, process and marketing requirements. Specifically, soy proteins:

1. Increase Cooking Yields - Actual yields depend on application and the specific protein applied.
2. Maintain Protein Content in Finished Meat Products - Careful formulation can insure adequate maintenance of protein levels to satisfy local regulations or nutrient density requirements.
3. Allow Use of Existing Plant Facilities - There is no need for special process equipment or facilities.
4. Help Utilize Available Supplies of Lean Skeletal Meat As Well As Meat By-products - More finished product can be made to satisfy local market demands and regulations, economic considerations, and nutritional needs.
5. Help Reduce Ingredient Cost - Savings are more than significant.
6. Allow Precise Control of Finished Product Quality.

Textured Vegetable Proteins

Textured vegetable proteins, both textured flours (50% protein) and concentrates (70% protein dry basis), are widely used in a variety of food systems and are utilized throughout the processed meat industry to provide meat-like structure and help reduce

182

costs. Available in a variety of sizes and shapes, colored or uncolored, flavored or un-flavored, fortified or unfortified, textured soy proteins, when properly selected, can resemble any meat ingredient. Beef, pork, fish and poultry applications are possible. However, proper protein selection and hydration are critical to achieving superior finished product quality.

Soy Protein Concentrate

Both non-functional and highly functional soy protein concentrates (70% protein dry basis) are commercially available for use in meat products. Normally, a highly functional (i.e., high solubility, good emulsification, superior dispersibility) product with no bitter or harsh flavors is desirable. When used to replace lean meat, hydrated concentrate can be used at levels up to 6 or 7% (non-specific emulsion meats) in the finished product. Higher replacement levels or formulas with specific cost/nutrition requirements may use concentrate with a judicious amount of textured protein. Excellent yields, cost savings, texture, flavor, and nutrient profiles are possible. No single rule applies in all cases.

Soy Protein Isolates

Soy protein isolates (90% protein dry basis) are highly dispersible, highly soluble, highly functional soy products. Designed to replace a portion of salt soluble meat proteins, isolates bind fat and water, stabilize emulsions and help insure maintenance of structure in finished cooked meats. Proper selection is required to match specific functionality with cost, nutrition, process and marketing needs. Isolates that give good emulsification and high viscosities, and/or are stable to high salt concentrations found in processing brines, are desirable. Again, no one isolate may be applicable to all meat products processed at a single plant location.

Processed Meat Applications

Patty Products. Patties (burger-type products) offer an ideal opportunity to use soy proteins. Care is required to properly select the soy proteins that will most resemble the appropriate meat (beef, pork, chicken, fish) needed. Hydrated, textured soy proteins (hydration ratios of soy protein products for an 18% protein content include: one part textured soy flour + 1.8 parts water; one part soy protein concentrate + 2.0 parts water; one part soy protein isolate + 4.0 parts water) add cooked meat texture at 20 to 30% meat product replacement (Table 1). Higher replacement levels can be obtained by using textured products in combination with soy protein concentrate or isolate.

Emulsion Products. Soy proteins provide water and fat binding as well as texture to emulsion-type products (e.g., bologna or weiners), thus reducing dependence on lean sketetal meat. Only slight modifications differentiate finished products, and proper formula balancing is required (Table 2). Soy proteins alone or in combination offer significant opportunities to reduce cost while maintaining quality. Again, appropriate hydration ratios are required. Guidelines for beef and pork emulsion products are also applicable to other meats (e.g., chicken, turkey).

Coarse Ground Meat Products. Coarse ground sausage formulas are easily adapted to allow use of various soy protein products (Table 3). Products made by the grinder-mixer method normally require textured soy protein for structure and soy protein concentrate or soy protein isolate for functionality, while significantly reducing ingredient costs and maintaining quality.

Table 1. Ingredients used for beef patty products using soy protein products.

BEEF PATTY MIX (26.3% non-meat)	%
Beef Plate (50% fat)	41.0
Lean Beef (10% fat)	32.8
Water	16.4
Textured Soy Flour	8.2
Salt	0.8
Beef Flavor(s)	0.8
	100.0%
BEEF PATTY MIX (30% non-meat)	**%**
Beef Trim (30% fat)	70.0
Water	20.0
Textured Soy Flour, caramel color, unflavored	8.5
Soy Protein Concentrate	1.5
	100.0%

Table 2. Ingredients used for emulsion products using soy proteins.

BEEF—using soy concentrate (35.3% non-meat)	%
Lean Beef	26.4
Plates	15.8
Jowls	12.9
Bull Meat	9.6
Water	29.7
Soy Protein Concentrate	3.4
Salt	1.6
ADM Clintose® Dextrose	0.4
Cure	0.2
Seasonings	As Desired
	100.0%
PORK—using soy isolate (39% non-meat)	**%**
Regular Pork Trim (60% fat)	35.0
Lean Pork Trim (20% fat)	26.0
Water	30.0
Soy Protein Isolate	5.0
Spice Blend	3.9
Cure (6.25% Sodium Nitrite)	0.1
	100.0%
BEEF—using soy isolate and textured flour (34.4% non-meat)	**%**
Bull Meat (8% fat)	34.8
Fat Beef Trim	30.8
Water	24.5
Textured Soy Flour	5.2
Salt	2.1
Soy Protein Isolate	1.3
42% High Fructose Corn Sweetener	0.8
Cure	0.2
Lemon Juice	0.1
Garlic Powder	0.1
Liquid Smoke	0.1
	100.0%

Table 3.   Ingredients used for coarse ground meat products using soy proteins.

POLISH SAUSAGE—using textured soy protein and soy protein isolate (32.3% non-meat)	%
Regular Pork (60% fat)	28.1
Lean Pork (20% fat)	21.1
Lean Chuck (10% fat)	17.6
Water	21.1
Textured Soy Flour—uncolored, unflavored	6.7
Soy Protein Isolate	1.8
Seasoning Blend	1.1
Salt	1.6
Dextrose	0.7
Cure	0.2
	100.0%

CHICKEN ROLL—using textured soy protein and soy protein isolate (38.9% non-meat)	%
Chicken Breast Meat	41.8
Chicken Skins	19.3
Water/Ice (75/25)	25.3
Textured Soy Flour—uncolored, unflavored	7.9
Soy Protein Isolate	3.8
Salt	0.3
Sodium Tripolyphosphate	0.3
MSG	0.1
Onion Powder	0.1
White Pepper	0.1
	100.0%

Combination Whole-Muscle Product. The unique functionalities of soy protein isolates are specifically designed for increasing yields and reducing costs in injected, whole-muscle meat products. Flours, grits, concentrates and textured products do not have proper solubility and dispersibility for use in whole-muscle systems. When properly formulated and selected, injection brines with soy protein isolates (Table 4) yield high-quality finished meat products. There is no compromise of flavor, texture, mouthfeel or finished product performance. In order to satisfy U.S.D.A. regulations, care should be taken during brine preparation, so that the finished product exceeds 17% protein. The percentage of protein to be used is dependent on the percentage above green weight to be pumped. Yield increases up to 35% are possible, resulting in more than just significant savings.

Although applicable in other processed meats (i.e., absorbed/restructured applications), soy protein isolates offer exciting new opportunities for whole-muscle meat products such as ham, corned beef, roast beef, turkey ham, chicken breast, B-B-Q, etc.

Restructured/Absorbed Meat Products. When injection of whole-muscle meat is not possible (i.e., when meat pieces are too small), simple absorption of soy protein isolates yield a very acceptable finished product. Again, flours, grits, concentrates and textured products do not have adequate solubility for use. Soy protein isolates are especially adaptable for meat absorption techniques. This can be done by absorbing soy containing brine (Table 4) into the muscle using a meat mixer, paddle mixer, ribbon blender or similar piece of equipment. Finished products resemble whole muscle when stuffed into casings, cooked and sliced.

Specialty Meat Products. New products are continually being developed. When properly selected, soy protein isolates, soy protein concentrates, and textured soy

Table 4.   Ingredients for whole-muscle meat brine formulas.

COMBINATION (BRINE INJECTED) HAM PRODUCT (70% Meat)	%
Water	82.2
Soy Protein Isolate	7.5
Salt	7.0
Dextrose	2.0
Tripolyphosphate	1.2
Cure	0.1
	100.0%

RESTRUCTURED/ABSORBED CORNED BEEF (70% Beef)	%
Water	81.7
Soy Protein Isolate	7.5
Salt	7.0
Dextrose	2.0
Tripolyphosphate	1.2
Garlic Powder	0.4
Sodium Erythorbate	0.1
Cure	0.1
	100.0%

proteins offer significant opportunities to develop products with or without meat (Table 5). Any of these new ideas can be formulated to satisfy a variety of cost, nutrition, process and marketing guidelines. For example, injection or absorption of soy protein isolates can offer significant reduction in fat, calories, cholesterol and cost.

Table 5.   Ingredients of two specialty meat products using soy proteins.

MEATLESS SAUSAGE/BOLOGNA[a]	%
Water	60.2
Soybean Oil	15.8
Soy Protein Isolate	12.2
Dextrose	11.8
Color/Flavor	As Desired
	100.0%

MEATLESS CHICKEN[b]	%
Water	55.53
Textured Vegetable Protein	16.50
Vegetable Oil	13.00
Soy Protein Isolate	8.00
Egg White Solids	4.00
Chicken Flavor	2.50
Salt	0.25
White Pepper	0.12
Onion Powder	0.10
	100.00%

[a]Mix, stuff in casing, water-cook and cool.
[b]Form into desired shape; vegetable skin may be added.

Vegetable Protein Entrees—Low Fat, No Cholesterol, Reduced Calorie

When properly formulated, a wide variety of meatless, vegetable protein entrees can be developed. Proper selection of textured soy products is required to insure adequate simulation of meat color and texture. Normally, unflavored products are used, since spices, seasonings and special flavorings can be added as dry flavoring blends or during preparation. Possible are several, meatless, dry mix, no cholesterol, low fat, reduced-calorie entrees: pepper steak, taco mix, sloppy joe mix, chili mix, chicken almondine, chicken curry, chicken creole, sweet and sour pork, beef stroganoff, hearty stew.

Bacon-flavored textured soy flours can be used as condiments to lend spice to salads, sour dressings, salad dressings and other applications. A variety of colored/flavored, bits/chips are available for each desired purpose.

PROCESSED DAIRY FOODS

Special care is required when selecting soy protein products for dairy food applications. Soy flour and grits, textured products, and virtually all soy protein concentrates have limited use, due to flavor, solubility, mouth-feel, and other undesirable characteristics. If properly selected, isolated soy proteins have been shown to be excellent sources of good quality, functional proteins well adapted to conventional dairy product formulation and processing. Critical consideration of plant processes, available ingredients, legal guidelines, marketing goals, nutritional composition, economics and good judgment can allow formulation of acceptable fluid, frozen, cultured and non-traditional dairy and non-dairy foods. The use of soy protein allows:

1. Better Utilization of Existing Plant Equipment and Raw Milk Supplies - This can be particularly useful where raw milk or whole bean process soy milk supplies are limited.
2. Use of Existing Plant Facilities - No need for special process equipment or process considerations.
3. Maintenance of Product Quality - No need to compromise quality in formulating products with good market acceptance.
4. Maintenance or Improvement of Nutrient Profiles - This is possible while reducing ingredient costs via total control of product composition. Products can be designed for specific nutrient or marketing needs.
5. Reduced Ingredient Costs - There is total control of selected protein source, fat/oils, sweeteners, and other ingredients. Partial or total replacement of milk solids is possible. Special ingredient considerations may be required.
6. New Product Opportunities - With care, numerous fluid, frozen, cultured, dry or non-traditional products can be developed.

Soy protein isolates can be used to develop highly acceptable milk-like fluid and frozen products. Careful selection of ingredients can yield significant new applications and reduced ingredient costs. Products can be processed through conventional dairy plant equipment and are compatible with all currently available packaging systems. When properly formulated, few special process considerations are required.

Beverages

Whole Milk-Soy Beverage Blends. Also applicable to whole-bean process soy milk-soy beverage blends, these concepts help to reduce cost, increase plant efficiencies, maintain quality and nutrient profiles, and spare existing raw milk or soymilk supplies (Table 6). With proper application, it is possible to produce more milk drink equivalents and more cream/butter equivalents from a given milk supply. The same basic concept can be applied to help increase traditional soymilk manufacturing efficiencies and reduce cost without changing qualitative attributes important for market success.

Soy Protein Fortified Milk. Designed for improved nutrient composition, reduced ingredient costs are possible. High protein, low fat, even reduced-calorie, fortified milk can be formulated (Table 6).

Soy Beverages. Unflavored, flavored, concentrate and high protein products have been developed (Table 6). Significant process efficiencies can be achieved compared to traditional soymilk manufacture. Again, specific formulas will be determined by local market demands and available ingredients.

Coffee Whitener/Liqueur Cream Base. Liquid or dry; casein, caseinate, cream, milk or milk-solid replacements are used to reduce cost and ease raw material handling. Proper formula balancing is required to insure whitener/liqueur stability to heat, acid, pH and mineral hardness. In addition, liqueur formulas must be able to withstand the stress of alcohol concentrations up to 20%.

All Vegetable/High Protein Frozen Desserts

Partial or total replacement of standard frozen dessert constituents (e.g., milk-solids-not-fat, butterfat) is possible (Table 7) resulting in more than just significant savings. Replacement of milk-solids-not-fat can be made on a protein-for-protein basis, to protein fortify, or based on ingredient functionality. A low viscosity, non-gelling, soy protein isolate with good flavor profile is most readily adaptable. Proper formula balancing is required to yield mixes that perform normally in the plant and through distribution.

Soy protein isolates can be used effectively to produce non-traditional tofu and

Table 6. Basic formulas for soy beverages.

	Whole Milk	Soy Beverage		Soy Fortified Milk	50/50 Soy/Milk Blend
		Unflavored	Flavored		
		— Percent Dry Weight Basis —			
Milk Solids-Not Fat	9.00	-	-	9.00	4.50
Butterfat	3.25	-	-	-	1.63
Vegetable Fat	-	2.00	2.00	2.00	1.63
Soy Protein Isolate	-	3.62	3.62	3.62	1.81
42% HFCS[a]	-	-	8.00	8.00	-
42 D.E.[b] Corn Syrup	-	6.00	-	-	3.00
Sodium Citrate	-	0.20	0.20	-	-
Salt	-	0.05	0.05	-	-
Stabilizer/Emulsifier	-	0.10	0.10	-	-
Total Solids	12.25	11.97	13.97	33.62	12.57
Protein	3.24	3.26	3.26	6.49	3.26
Fat	3.25	2.00	2.00	2.00	3.25

[a]High fructose corn sweetener.
[b]Dextrose equivalent.

Table 7. Basic formulas for frozen dessert.

| | U.S. Standard Ice Cream | All Vegetable Mixes | | Soy Fortified Ice Milk-Type |
		4% Fat	2% Fat	
	— Percent Dry Weight Basis —			
Milk-Solids-Not-Fat	10.00	-	-	10.00
Butterfat	10.00	-	-	-
Vegetable Fat	-	4.00	2.00	4.00
Soy Protein Isolate	-	2.00	2.00	3.60
Sucrose	12.00	-	-	-
55% HFCS[a]	-	12.00	14.00	10.00
36 D.E.[b] Corn Syrup	5.00	13.00	14.00	10.00
Stabilizer/Emulsifier	0.30	0.45	0.45	0.30
Total solids, %	37.30	31.45	32.45	37.90

[a,b]See Table 6.

yogurt-like products. Careful formulation can result in high quality, low cost finished foods that can be manufactured in conventional process systems. Improved process efficiency, extended shelf-life, and better control of composition and quality are features of interest. Using properly selected soy protein isolate, vegetable oil, sweetener and emulsifier/stabilizer systems, several products (firm to soft texture) are easily made (Table 8). Similarly, non-cultured products and high protein prototypes can also be achieved. The characteristics of the target product can be achieved via formula balancing without additional equipment or special process procedures. Special consideration of packaging systems available at the dairy will often determine concepts immediately possible.

Imitation Cheese and Cheese-like Products

Soy protein isolate can be used successfully in formulation of imitation Mozzarella and other imitation processed cheese substitutes (Table 9). Soy protein functionality

Table 8. Yogurt/tofu-like products.

	"Yogurt" Soft Curd	"Tofu" Firm Curd	"Tofu" From Dry Mix	Chocolate Dessert
	— Percent Dry Weight Basis —			
Soy Protein Isolate	3.50	8.00	8.20	8.00
Vegetable Fat	3.40	4.20	4.30	3.00
55% HFCS[a]	5.50	2.50	-	10.00
Corn Starch	-	-	12.00	-
Cocoa	-	-	-	8.00
Emulsifier	0.25	-	1.00	-
Salt	1.00	1.00	1.00	-
Stabilizer	1.00	1.00	-	0.80
Total solids, %	14.65	16.70	26.50	29.80

[a]See Table 6.

Table 9. Cheese type products.

	Cheese Substitutes		Cheese-Like Spreads	
	Mozzarella	American	Cheddar	Cream Cheese
	— Percent Dry Weight Basis —			
Soy Protein Isolate	7 04	11.50	12.50	8.30
Casein/Caseinate	21.11	11.50	-	-
Vegetable Fat	22.50	22.50	24.00	13.80
Salt	1.90	1.00	1.90	0.90
Emulsifier/Stabilizer	0.21	0.21	0.80	0.90
Phosphates/Citrate	2.45	2.10	-	-
Preservative	0.20	0.20	0.20	0.20
Flavor	0.20	2.06	3.10	7.60
Water	44.39	48.83	57.40	66.00
Color	-	0.10	0.10	-

allows its use as a caseinate replacement in cheese product formulations. In conjunction with vegetable oils, various emulsifying salts and flavors can be formulated to produce cheese-like products resembling Mozzarella cheese substitute and processed cheese foods or spreads.

NOTES

L. Steven Young, Archer-Daniels-Midland Company, Food Research Division, Chicago, Illinois 60639.

FEASIBILITY OF A SHELF-STABLE SOYBEAN BEVERAGE

L. S. Wei, G. E. Urbanski, M. P. Steinberg and A. I. Nelson

In recent years, soybean foods, and particularly soybean beverage, have received much attention in the Western World because they are economical sources of high quality protein (Smith and Circle, 1978). Soybean beverages produced by the Illinois process are of special interest, since they are free of painty off-flavors and contain all the protein present in the raw soybeans (Nelson et al., 1976). These beverages also have excellent suspension stability without added stabilizers (Priepke et al., 1980) and are free of chalky mouth-feel if processed properly (Kuntz et al., 1978).

The pasteurized Illinois soybean beverage shows a refrigerated shelf life of over one month (Nelson et al., 1976). However, in order for such a beverage to be distributed in areas where refrigeration is not available a "shelf-stable" product is needed. For instance, spray-dried soy milk is such a product. Although it results in lower transportation costs than the pasteurized product, reconstitution of the spray-dried powders is often poor, and lipid oxidation rates are high (Aminlari et al., 1977). Commercially sterile soybean beverage would show a long shelf life without refrigeration; however, processing and shipping costs are high and the thermal processing may impart a cooked flavor (Luttrell et al., 1981).

Intermediate moisture foods represent lower transportation costs, convenience, and stability from microbial spoilage without refrigeration. They contain sufficient water for immediate consumption or ready reconstitution, but depend on a lowered water activity (a_w) of the system, to about 0.85, for preservation. Two approaches are available for decreasing a_w; one is evaporation of some of the water to increase concentration of solids, and the other is to add solids.

Both Lo et al. (1968) and Tan (1958) attempted to concentrate filtered soy beverages by conventional evaporation methods. However, the increase in viscosity was so great that above 15% solids the beverage would no longer flow in the evaporator. This was referred to as the "viscosity barrier". Additionally, sucrose, which is commonly used to preserve intermediate moisture foods, caused the viscosity barrier to be encountered at soy-beverage solid contents below 15% (Lo et al., 1968). They concluded that a concentrated soy-beverage containing enough sucrose to preserve the final product could not be produced with a reasonably acceptable viscosity.

Since the Illinois process starts with blanched soybeans containing about 40% solids, it would save much energy to use this material to formulate a high-solids, intermediate-moisture product, instead of preparing the usual liquid base at about 10% solids and attempting to evaporate. Work with whole-soybean concentrates (Wei, unpublished) showed that addition of sucrose to these highly viscous slurries greatly reduced their viscosity. Therefore, the objective of this study was to develop shelf-stable soybean beverage concentrate by formulation of blanched soybean cotyledons and sucrose. Converting the Illinois process to a technique that allows preparation of a soymilk of intermediate moisture content required careful evaluation of the effects of sugar addition on initial milling with a Reitz or Fitz mill, as well as high pressure homogenization of the final product.

MILLING EFFICIENCY

Milling of blanched soybean cotyledons was accomplished with both a Reitz mill and Fitz mill. Milling efficiency was determined by first reconstituting the milled samples in water to 6% cotyledon solids. The reconstituted sample was stored undisturbed for 24 h at 1 C. After this, two evaluations were made. The first was to visually determine the height of a line of demarcation between an upper phase containing less suspended matter and a lower phase containing more. The ratio of this height to the total height of the reconstituted sample was calculated and expressed as separation index (Priepke et al., 1980). The second evaluation was to withdraw a 15 mL portion with a pipet from the center of the upper one-third and the center of the lower one-third. Each 15 mL aliquot was analyzed for solids content and the ratio of solids top to bottom was calculated. A 30 min water blanch of dry cotyledons raised moisture content to about 61% (39% solids). Therefore, the solid contents of blanched cotyledons were adjusted above 39% by decreasing the blanching time. Solid contents were adjusted below 39% by adding water to the blanched cotyledons during milling.

For samples milled at solids contents greater than 11.8%, milling efficiency, as measured by demarcation heights, was significantly lower than samples milled at 11.8% solids. However, the difference in efficiency between solids contents of 39.0 and 62.7% was not significant. For milling efficiency, as measured by top to bottom solids ratios, it appears that milling at solids contents above 24.3% significantly decreased milling efficiency, as reflected by the lower ratios. The top to bottom solids ratios are not significantly affected by milling at solids contents above 39%. However, for concentrated soybean beverages, where a high solids content is desired, milling at solids content as high as 24.3% is recommended. Regardless of the solids content at milling, the reconstituted material had demarcation heights below 0.6 and top to bottom solids ratios below 0.35. These values are far below the values of 0.95 to 1.0 obtained for homogenized soy beverages (Priepke et al., 1980).

The effect of screen size on milling efficiency with both hammer mills was studied with solids content held at 40% by water blanching for 30 min before milling. There was a significant decrease in demarcation height ratio as the screen openings of the Reitz mill increased. This effect was also evident with the measurement of solids top to bottom ratio. The same decrease in demarcation heights and solids top to bottom ratios was noted in the Fitz mill. The larger the openings of the hammer mill screen, the poorer the milling efficiency. The type of hammer mill did not significantly affect milling efficiency.

Cotyledons were water-blanched for various lengths of time in order to vary tenderness, which was related to milling efficiency. Solids content was adjusted by soaking blanched cotyledons in water at room temperature. Tenderness of cotyledons was measured with a Lee Kramer shear press and expressed in shear press values (lbs. force/100 g sample). All cotyledons were milled in a Reitz mill with the 0.48 cm (3/16-inch) screen. Figure 1 illustrates the relationship between cotyledon shear press value and separation index. The comparison was also made between tenderness value and the solids top to bottom ratio. Tenderness significantly affected both measures of milling

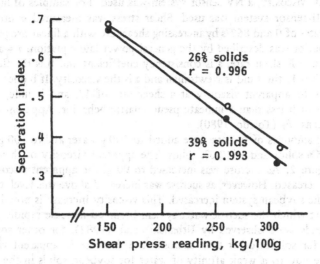

Figure 1. Effect of cotyledon tenderness on separation index for cotyledons milled at 39 or 26% solids and hydrated to 6% solids.

efficiency. There was a linear decrease in both measures of milling efficiency as the shear press values increased (decreasing tenderness). The separation index line of the 26% solids was above that of the line for the 39% solids samples in Figure 1. A similar relationship was also found for the effect of tenderness on the top to bottom solid ratio. This shows the same effect of solids content on milling efficiency previously discussed. These results are in agreement with work by Kuntz et al. (1978). They showed that the greater tenderization achieved in blanching, the greater the particle disruption in homogenization. Our results indicate that the tenderization obtained in blanching also aids the particle disruption in milling.

SORPTION ISOTHERMS OF BLANCHED COTYLEDONS

Blanched cotyledons at known moisture contents were equilibrated against several standard saturated salt solutions at known a_w in a proximity equilibration cell as described by Lang et al. (1981). The weight gain or loss of the sample upon equilibration was used to determine the equilibrium moisture content at that a_w.

The water-binding characteristics of blanched cotyledons at various a_w was very low and followed normal parabolic patterns of sorption isotherms. The isotherms for adsorption and desorption showed a slight hysteresis effect. This implies that the equilibrium moisture content of milled, whole cotyledons at a given a_w will depend upon whether water has been desorbed by drying. The isotherms also showed that the cotyledon equilibrium moisture content is only 22% at the target a_w of 0.85. The product would not flow at such a low moisture content.

EFFECT OF SUCROSE ADDITION

Viscosity

A Haake Rotovisco Model RV-3 rotational viscometer equipped with a DMK 500/50 dual head was used for measuring shear rate-shear stress relationships. For samples of low viscosity, a NV sensor system was used. For samples of higher viscosities, the SV-II sensor system was used. Shear stress was measured continuously between shear rates of 0 and 862/s by increasing shear rate with a linear program.

Flow behavior was described by the general power law equation: $\tau = a\gamma^2$, where τ is shear stress, γ is shear rate, a is consistency coefficient and b is the flow behavior index. When b = 1, the fluid is Newtonian and a is the viscosity. If b does not equal 1, a corresponds to apparent viscosity at a shear rate of 1/s and is independent of b values. Values of b less than 1 indicate pseudoplastic behavior. Apparent viscosity (μ) was calculated as τ/γ (Toledo, 1980).

Increasing amounts of sucrose were added to 100 g water and to 100 g cotyledons slurry at 14.5% solids and at 12.0% solids. The apparent viscosity of these samples is shown in Figure 2. As sucrose was increased to 90 g, the apparent viscosity of both suspensions decreased. However, as sucrose was increased above this level, the apparent viscosity of the soybean system increased. This viscosity increase is noted at the same sucrose level at which the sucrose-water system began to increase rapidly in viscosity.

These trends were observed by Urbanski et al. (1982), for other soybean components and for solutes other than sucrose. The decrease in apparent viscosity was thought to be due to a weak affinity of water for soybean solids in the presence of sucrose. These results of lowered viscosity with sucrose addition confirm the aforementioned observations (Wei, unpublished). This implies that soy slurries at higher solids content, which are too viscous for homogenization, could be sufficiently reduced in viscosity by sucrose addition. This viscosity reduction was sufficient to enable homogenization to be used to further reduce particle size to produce beverages of acceptable quality.

Milling Efficiency. In processing, it is possible to add sucrose to cotyledons before milling. We thought that addition of sugar before milling might improve milling efficiency. Therefore, the effect on milling efficiency of sucrose addition before and after milling, as measured by separation index, was determined. Table 1 shows that the

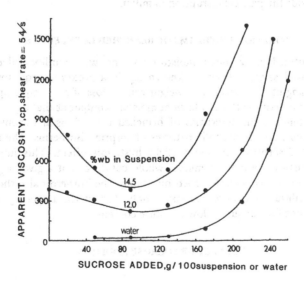

Figure 2. Effect on apparent viscosity of sucrose addition to two cotyledon suspensions and to water.

Table 1. Effect of adding sucrose before or after milling on milling efficiency of blanched cotyledons. All samples milled in a Reitz mill with a 0.48 cm screen.

	Separation index as a fraction of total height[a]		
	Sucrose not added	Sucrose added before milling	Sucrose added after milling[b]
Milled at 40% solids[c]			
Single milling	0.53	0.96	0.72
Double milling	0.69	0.98	0.88
Solids content at milling[c]			
24.3	0.51		0.72
39.0	0.51		0.71
51.8	0.50		0.69
62.7	0.47		0.61

[a]Values within a row differ significantly (P < 0.01).
[b]Sucrose added to give a 60% solution, wet basis, in the aqueous phase.
[c]Wet basis.

addition of sucrose before or after milling greatly enhanced milling efficiency as compared to the non-sugar control. This was the case regardless of single or double milling. The highest efficiency was obtained when sucrose was added to cotyledons before milling. To further check the effect of sucrose addition on milling efficiency, sucrose was added to milled soybean cotyledons of various solid contents between 24.3 and 62.7%. The separation index of these was compared to milled cotyledons without added sucrose (Table 1). These results showed that sucrose addition increased milling efficiency regardless of the soybean solids content.

Water Activity. To test the effectiveness of sucrose in lowering a_w of milled cotyledon suspensions, various amounts of sucrose were added to milled cotyledon slurries at 14.5% solids. The a_w was then determined by the methods described by McCune et al. (1981). The results showed that upon sucrose addition, a_w decreased linearly (r=0.992) from 1.0 to 0.86. This indicates that sucrose is effective in lowering water activity of milled cotyledon slurries. Addition of sucrose in levels greater than 170 g per 100 g suspension would probably not cause a substantial further reduction in a_w since sucrose binds little water at a_w below 0.86 (Lang et al., 1981).

HOMOGENIZATION PRESSURE

The effect of varying pressure of the double homogenization on particle size and flow characteristics was investigated. Homogenization at higher pressures was effective in reducing the hydrated particle size of the beverage. However, this did not significantly alter the flow characteristics of the concentrate. The concentrate was moderately pseudoplastic and had flow behavior best described by the equation $\tau = 20\gamma^{0.70}$.

The suspension stability of the beverage concentrate (14.5% solids) reconstituted to 6% soybean solids is shown in Table 2. Both homogenization pressures resulted in excellent suspension stability. Priepke et al. (1980) stated that, for an Illinois soybean beverage, the minimum top to bottom protein ratio for a beverage to be considered acceptable is 0.90, while 0.96 was obtained here. They found that such values could be obtained as long as the sum of the two homogenization pressures was at least 24.2 MPa. Thus, a double homogenization at a pressure of 24.2 MPa, as was done here,

Table 2. Effect of homogenization pressure on suspension stability after 5 days quiescent storage of soybean beverage concentrate hydrated to 6% soybean solids.

Homogenization pressure	Protein top to bottom ratio[a]	Separation Index as a fraction of total height[a]
24.2 MPa	0.96[a]	1.0[a]
41.4 MPa	0.98[a]	1.0[a]

[a]Not significantly different.

was more than adequate to produce a beverage concentrate with acceptable suspension stability.

SUGAR BLEND

Urbanski et al. (1982) found that a mixture of sugars, 70 sucrose: 30 dextrose, lowered viscosity of soy protein to a greater extent than either dextrose or sucrose alone. Interestingly, this blend of sugar ingredients shows higher solubility than either of its components individually. Addition of this blend could occur after the first milling in Step 5 (Table 3), but milling efficiency was not increased by exercising such an option.

Based on the data presented and the rheological studies by Urbanski et al. (1982), a flow sheet for production of a shelf-stable soybean beverage concentrate by formulation was established (Table 3). In our studies, a gravity feed homogenizer was used and slurries with apparent viscosity greater than 1000 cp. at a shear rate 54/s did not readily pass through the homogenizer.

SENSORY CHARACTERISTICS

The sensory characteristics of the shelf-stable soy beverage concentrate produced as shown in Table 3 were compared. Multiple comparison tests were used for organoleptic evaluations. A known reference or standard, labeled R, was presented to the panelists with several coded samples. The panelists were asked to compare each coded sample with the reference sample for flavor, color, and mouth-feel. The reference sample was sweetened, condensed cow's milk obtained from a local supermarket. Hedonic scores

Table 3. Stepwise procedure for production of a shelf-stable soybean beverage concentrate.

Procedure	Conditions
1. Blanch cotyledons	In 0.025% Na Bicarbonate for 30 minutes, 5 kg solution to 1 kg cotyledons.
2. Drain	
3. Add water	Weight calculated to give desired soy solids in final product.
4. Add sugar	A blend of 3 parts dextrose to 7 parts sucrose at a level of 63% of the aqueous phase.
5. Mill twice	Reitz Mill twice, first through 0.48 cm (3/16 inch) screen, then through 0.08 cm (1/32 inch) screen.
6. Heat	To 88 C.
7. Homogenize	Twice, maintaining temperature at 88 C and pressure at 24.2 MPa.
8. Cool and package	

Table 4. Effect of homogenization pressure on organoleptic characteristics of soybean beverage concentrate.

Product	Homogenization pressure	Multiple comparison scores [a]		
		Flavor	Color	Mouth-feel
	MPa			
Soy concentrate	24.2	3.8	3.7	4.3
Soy concentrate	41.4	3.9	4.1	3.9
Reference sweetened, condensed cow's milk	-	4.1	3.7	4.0

[a]Means not significantly different.

were assigned to the ratings with no difference equaling four, extremely better than R equaling seven and extremely worse than R equaling one. All samples were reconstituted to 3% protein. Homogenization pressure did not significantly affect flavor, color or mouth-feel (Table 4). Therefore, it can be concluded that increasing homogenization pressure above 24.2 MPa had no beneficial effect. More importantly, there was no significant difference between either of the reconstituted soybean beverages and sweetened, condensed cow's milk in flavor, color or mouth-feel.

STORAGE STABILITY

Both of the soybean beverage concentrates, as well as the sweetened, condensed cow's milk, showed water activity of 0.86, and therefore, may be expected to be free of microbial spoilage for an extended storage time at room temperature. Preliminary results showed no visual evidence of microbial spoilage or physical change after 2 months of storage at room temperature.

NOTES

L. S. Wei, M. P. Steinberg and A. I. Nelson, INTSOY and Department of Food Science, University of Illinois, Urbana, Illinois 61801. G. E. Urbanski, Express Foods Co. Inc., Louisville, Kentucky 40222.

REFERENCES

Aminlari, M., L. K. Ferrier and A. I. Nelson. 1977. Protein dispersibility of spray-dried whole soybean milk base: Effect of processing variables. J. Food Sci. 42:985-988.

Kuntz, D. A., A. I. Nelson, M. P. Steinberg and L. S. Wei. 1978. Control of chalkiness in soymilk. J. Food Sci. 43:1279-1283.

Lang, K. W., T. D. McCune and M. P. Steinberg. 1981. A proximity equilibration cell for rapid determination of sorption isotherms. J. Food Sci. 46:936-938.

Lang, K. W. and M. P. Steinberg. 1981. Predicting water activity from 0.30 to 0.95 of a multicomponent food formulation. J. Food Sci. 46:670-672.

Lo, W. Y., K. H. Steinkraus and D. B. Hand. 1968. Concentration of soymilk. Food Technol. 22: 1028-1030.

Luttrell, W. R., L. S. Wei, A. I. Nelson and M. P. Steinberg. 1981. Cooked flavor in sterile Illinois soybean beverage. J. Food Sci. 46:373-376.

McCune, T. D., K. W. Lang and M. P. Steinberg. 1981. Water activity determination with the proximity equilibration cell. J. Food Sci. 46:1978-1979.

Nelson, A. I., M. P. Steinberg and L. S. Wei. 1976. Illinois process for preparation of soymilk. J. Food Sci. 41:57-61.

Priepke, P. E., L. S. Wei, A. I. Nelson and M. P. Steinberg. 1980. Suspension stability of Illinois soybean beverage. J. Food Sci. 45:342-345.

Smith, A. K. and J. J. Circle. 1978. Soybean: chemistry and technology. AVI Publishing Company, Westport, CT.

Tan, D. H 1958. Technology of soymilk and some derivatives. Thesis. University of Wageningen, Netherlands.

Toledo, R. T. 1980. Flow of Fluids. p. 152-196. In Fundamentals of Food Processing Engineering. AVI Publishing Company, Westport, CT.

Urbanski, G. E., L. S. Wei, M. P. Steinberg and A. I. Nelson. 1982. Effect of solutes on rheological characteristics of soy flour and its components. J. Food Sci. 47:792-795.

ECONOMICAL PROCESSED BLENDS OF SOYBEAN AND CEREAL AS HUMAN FOOD

Judson M. Harper

PROCESSED CEREAL/SOY FOODS

Centrally processed nutritious soy-based foods have made major contributions to reducing malnutrition. Their advantages are: 1) precooked blends are energy efficient; 2) processing inactivates antigrowth factors present in soy; 3) blends of cereals and soy have digestible high quality protein; 4) calorie and nutritive density are increased; 5) fortification with a broad range of vitamins and minerals is possible; 6) suitable packaging protects these products and increases shelf life; 7) formulation and product quality are consistent; and, 8) time-saving to the homemaker results.

In many less developed countries (LDCs), two major sources of imported high-protein weaning foods have been commercial foods produced in the West and Title II blended foods, supplied by AID through the PL480 Food for Peace Program. These Title II blended foods include corn/soy blend (CSB), corn/soy/milk (CSM), instant corn/soy/milk (ICSM) and wheat/soy blend (WSB).

For a number of economic reasons, it is desirable to explore ways to augment the supply of weaning foods through local manufacture within LDCs, using appropriate central processing technologies. Such arrangements 1) provide a market for locally grown agricultural commodities; 2) create employment by developing small-scale industry; 3) reduce transportation costs; 4) shorten storage time; 5) allow formulation and processing of products to meet local tastes; and, 6) minimize packaging costs. These advantages motivated extensive work in appropriate central processing technologies for cereals and soy.

EXTRUSION PROCESSING

There are many alternatives for the central processing of soy-based foods. Of all the processes available, extrusion has been found to be versatile, cost effective, adaptable to LDCs, and energy efficient.

An extruder is a type of food processing equipment that can combine cereals and soy while cooking them at high temperatures (150-180 C) for short periods of time (60-120 s) at moistures of 15 to 25%. The extruder shown in Figure 1 consists of a rotating screw which forces the product through a discharge die. The screw is turned by a large electric motor whose energy is converted to heat by friction to cook the product.

The extrusion process uses whole, raw cereal grains and soy which are stored in bulk or bags, with provision for protection from insects and rodents. Cleaning consists

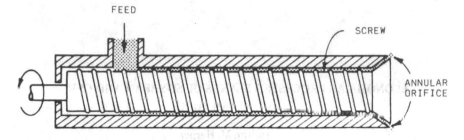

Figure 1. Cross-sectional schematic of the Brady extruder.

of size grading (scalping) and removal of stones (destoning) by gravity separation. A simple dehuller to loosen and aspirate (winnow) the hulls reduces fiber content. Cereals are mixed with soy in an approximate 70/30 weight ratio, coarsely ground, and extruded. Excess moisture in the product is removed in a cooler/dryer using air to bring the product within safe storage limits (<14% moisture). Once cooled, the product can be coarse-ground, mixed with a vitamin/mineral (V/M) fortification mix, and packaged in either large bags or individual packages. The resulting pre- or partially-cooked flour is normally prepared as a gruel by addition of potable water and may involve a short heating step.

The high capital costs, large throughput requirements, and need for a significant degree of technical skill inhibit the use of large food extruders used in the USA. However, several simple extruders, originally designed for cooking soybeans on the farm, have been found to be attractive alternatives for LDC use. These low-cost extrusion cookers (LECs) have been evaluated for the production of blended and other nutritious foods made in central processing facilities.

CAPABILITIES AND LIMITATIONS OF LECs

Characteristics of Low-cost Extrusion Cookers

In order for LEC systems to assist effectively in solving food and nutrition problems in LDCs, they must have the following characteristics.

1. Low-cost: Processing equipment must cost <$300,000 per t/ha capacity, meaning the extruder must cost <$50,000 per t/h capacity.
2. Moderate Production Rate: Production rates of 250-1,000 kg/h are suitable for most installations.
3. Simple Operation: The system must require little sophisticated technical capability to operate and maintain.
4. Minimal Auxiliary Equipment: The system should not require auxiliary boilers, dryers or pre- or post-processing/conditioning equipment that increase costs and can lead to product contamination.
5. Versatile: The process should accept a wide variety of cereals in combination with soybeans.
6. Maintenance: The system should be easily maintained and utilize locally available parts and equipment to the greatest possible extent.
7. Sanitary: Equipment must be cleanable and capable of producing human food products.

Several types of extrusion systems have been extensively evaluated and found to have these characteristics. These are all extruders, processing food ingredients at (18% moisture so that little or no product drying is required. Such extruders require high electrical energy inputs (ca. 0.10 kW-h/kg) and supply heat by friction, have greater maintenance costs, and limited product capabilities. These apparent limitations are off-set by low capital costs and their suitability for operation in LDCs. LDCs' operational capabilities and limitations vary considerably, as detailed by Harper and Jansen (1981).

The Brady (M and N Distributions, 23535 Telo Avenue, Torrance, CA 90505, USA) is well suited to small-scale operations, producing (1,000 kg/h of ground pre-cooked composite flour for gruels, weaning foods, beverages and soup bases, etc. The Brady is the lowest cost LEC available, but cannot produce shaped products.

The Insta-Pro (Insta-Pro International, 10301 Dennis Drive, Des Moines, IA 50322, USA) extruders can produce relatively dense, simple-shaped products from mixtures of cereal and oilseed. The Insta-Pro extruder comes in two sizes, which produce 250 or 600 kg/h. The Insta-Pro requires ingredients with higher fat and/or water content than other LECs. Both the Insta-Pro and Brady can effectively heat-treat whole, raw soy-beans, which can be ground into full-fat soy flour (FFSF).

The Anderson (Anderson-International Corp., 6200 Harvard Avenue, Cleveland, OH 44105, USA) extruder has the broadest range of product capabilities and is de-signed for continuous duty. It can manufacture expanded/shaped products suitable for snacks and/or ready-to-eat cereals. The cost of an Anderson exceeds both the Brady and the Insta-Pro, but it becomes cost competitive at production rates (2 t/h, where part of the energy is added as steam injected into the barrel.

COST ASSOCIATED WITH LEC PLANTS

Elements of a LEC Plant

The central processing system for nutritious foods is illustrated in Figure 2, show-ing the sequential operations of a LEC plant. Modifications to the basic system can be made for producing breakfast cereals, snacks, instant soup and beverage bases, and whole, cooked soybean flours.

Plant Costs

The LEC plant's cost is highly dependent upon the type and quantity of product to be made, the physical setting for the process and the extent of infrastructure neces-sary for the plant. Table 1 lists the capital costs necessary to construct a plant on an existing site devoid of the necessary infrastructure. Costs can often be substantially reduced by associating the LEC with a pre-existing grain mill. Building costs typically exceed equipment and installation costs. Land and site preparation (e.g., roads, rail, water, sewer, electricity) can also be expensive and substantially increase the cost of the entire installation.

Processing equipment for a 0.5 t/h LEC plant will cost about $300,000. When all related costs are considered, the completed plant may cost four times this much. Working capital is also necessary for plant start-up and to fill original product orders until sales provide for ongoing operating costs.

Manufacturing Costs

Representative manufacturing costs for an extruded food product are summarized in Table 2. LEC processing costs include raw ingredients, packaging, labor, utilities,

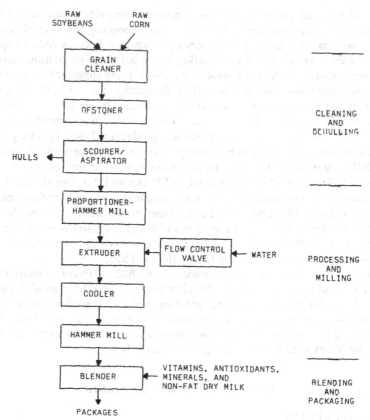

Figure 2. Diagram of sequential processing operations of a LEC processing plant (Harper and Jansen, 1981).

Table 1. Capital cost requirements for a 0.5 t/h LEC weaning food plant.

1.	Land (2 ha)		$40,000
2.	Site preparation		25,000
	Electrical		20,000
	Water/sewer		15,000
	Roads		450,000
3.	Building (450 m^2)		105,000
4.	Bulk grain storage/handling		
5.	Equipment		
	Cleaning/dehulling module	$51,000	
	Processing module	94,700	
	Blending module	62,600	
	Packaging module (manual)	41,000	
	Ancillary equipment	51,000	
	Subtotal—equipment	$300,300	
	Spare parts at 10%	30,000	
	Crate, insurance, freight at 20%	60,000	
	Contingency at 10%	30,000	
	Total—equipment		$420,300
6.	Engineering, management training		120,000
	Total cost including equipment and installation		$1,195,300

Table 2. Summary of typical manufacturing costs for a packaged, fortified blended food.

Cost category	Daily requirements (A)	Unit cost (B)	Total daily cost (C = A x B)	Specific cost, (D = C ÷ 12)
		$	$	$/t
Materials				
Cereal component	9.74 t	220/t	2,143	179
Oilseed component	4.18 t	440/t	1,839	153
Vitamins	12.01 kg	14/kg	169	14
Minerals	312.00 kg	11/kg	312	26
Packaging				
Poly-liners	48,000	5/1,000	240	20
Master bags	480	100/1,000	5	5
Operation expense				
Salaries and wages	--	--	--	75
Depreciation	--	--	--	24
Electricity	7,200 kW	0.06/kW	432	36
Other utilities	--	20/day	20	2
Maintenance	--	12/day	12	1
Spare parts	--	60/day	60	5
Transportation				
Freight	1,200 t-km	0.10/t-km	120	10
Shipping	12 t	2/t	24	2
Miscellaneous	--	360/day	360	30
				$582/t or $0.58/kg

Overview of Production Requirements:

Plant output: 500 kg/h
Operation: 3 8-h shifts/day
5 days/week
Formula: Cereal/oilseed (70/30)

Packaging: 250 g
Poly bags,
100/master bag
Packing method: manual 2 line

Losses: Processing 15%
Packaging 1%
Packaging materials 0%

depreciation and distribution costs. For a product consisting of 70% grain/30% soy placed in 250 g polyethylene consumer-type bags, the manufacturing cost is about $0.58/kg. One-half this cost is for raw ingredients. These costs are only about one-half to one-third those of similar commercial products.

CEREAL/SOY BLENDS

Centrally processed blends of cereals and soy, through the well-known principle of protein complementation as reported by Bressani (1977), offer the most practical route to low-cost weaning foods.

Specifications

The long-established specifications for Title II blended foods (USDA, 1978) have been used in the LEC program. The option is always available at any site to change these specifications if necessary.

204

Energy. The food energy supplied by a weaning food will depend on the quantity consumed and the energy per unit weight or volume; i.e., the energy density of the porridge or gruel as actally consumed. One way to increase the energy density of a weaning food is to increase the proportion of soybeans in the blend. Soy has the advantage of being both a good and economical source of energy and protein. It is desirable for the blended food to have a fat content of at least 6%, providing an energy of 100 kcal/100 g

The calorie density of extruded cereal/soy blends was reported by Jansen et al. (1981). Extruding was found to increase calorie density of gruels, since some mechanical damage to the starch occurred during processing, making it less capable of producing gruels of high viscosity.

Protein. The minimum protein level recommended is 16.7%, with a quality comparable to casein. A blended food consumed at a level of 300 kcal/day would raise the estimated net protein value of a poor quality cereal diet consumed by a 1 to 3 year-old child from 6 to 7% to 8 to 9% of calories.

Dietary Fiber. Based on digestibility studies of various blends, with infants and preschool children, a maximum crude fiber level of 2.0% is recommended for CSB or other cereal/legume blends made by the LEC process (Jansen, 1980). A general recommendation is always to dehull the soybean. Use of whole corn or rice would appear to be acceptable, but sorghums and millets should be dehulled to remove tannins that reduce digestibility.

Vitamins and Minerals. It is very easy to add micronutrients at a relatively low cost to blended foods that are centrally processed. Therefore, the general LEC program philosophy is to add complete V/M premixes to the blended food to make reasonably sure that the micronutrient needs of the targeted consumers will be met.

Protein Quality Evaluation

LEC-produced weaning foods were fed to weanling rats at 10% protein in the diet for 4 weeks for protein quality determination. Comparison was made with ANRC reference casein (assigned a PER of 2.50). Table 3 shows PERs for CSBs made from

Table 3. Effect of Brady Crop Cooker processing variables on protein quality of corn/soy blends.

Conditions	Temperature[a]	Weight gain	PER[b]	Corrected PER[c]
	C	g		
Dehulled Soy/Whole Corn				
Cooked together	154	65.7±1.7[d]	3.62±0.07	2.50±0.05
Cooked separately	138/154	66.7±2.6	3.81±0.10	2.63±0.07
Dehulled Soy/Degerminated Corn				
Cooked together	154	55.1±3.3	3.25±0.13	2.24±0.09
Cooked separately	138/154	53.7±3.6	3.73±0.13	2.57±0.09
Whole Soy/Degerminated Corn				
Cooked together	154	60.2±2.4	3.46±0.05	2.39±0.03
Cooked separately	138/154	55.4±2.3	3.61±0.09	2.49±0.06
CSB-Krause Milling		57.6±2.9	3.65±0.08	2.52±0.06
ANRC casein		48.6±2.6	3.61±0.12	2.50±0.08

[a]When cooked separately, first temperature listed is for soy.
[b]PER = weight gain (g)/protein consumed (g).
[c]Corrected PER = PER x 2.5/PER for casein.
[d]Mean ± standard error (n = 10).

dehulled or whole soybeans, combined with whole or degerminated corn and extruded singly or together. All blends gave PERs not significantly different from casein. This study showed that extruding soybean (high in lysine) in combinations with corn (high in carbohydrate) did not result in loss of lysine through the Maillard reaction.

Metabolic Studies in Human Infants and Preschool Children

Nitrogen balance studies in infants and preschool children have been carried out on extruded CSBs and extruded sorghum. Previously malnourished but recovered infants of 5 to 20 months of age were fed extruded CSB with varying fiber levels, so that protein was 8.4 to 8.7% of calories. Table 4 shows that both absorption and retention of nitrogen were significantly lower in the higher fiber blends made from whole corn and dehulled or from whole soy than for the lower fiber blend made from degermed corn and dehulled soy. There was no significant difference in absorption or retention of nitrogen between the two higher fiber blends. Retention of nitrogen from the two higher fiber blends was comparable to Title II CSM (Jansen, 1980), however. To reduce costs and simplify processing, whole corn in combination with dehulled soy can safely be used in making suitable weaning foods in a LEC processing plant.

APPLICATIONS IN DEVELOPING COUNTRIES

Sri Lanka

In 1976, the first LEC plant was built in Sri Lanka to process a variety of cereal/legume blends to extend the quantity of imported Title II blended food available there. Distribution of the product, "Thriposha," was through maternal health centers by CARE for the Ministry of Health of the government of Sri Lanka (GOSL). The original Thriposha consisted of WSB and extruded, locally grown sorghum and soybeans blended in the ratio of 80/20. Under this arrangement, 500,000 children or mothers were reached annually by the Thriposha program.

Relocated closer to the capital in 1980, the new plant has an expanded capacity to meet production commitments using locally grown inputs. This plant includes two electrically powered Brady extruders. During the first year of operation, it processed 2,300 t of CSB for blending with 5,400 t of imported ICSM. Currently, the plant operates 22 h/day, 6 days/week, and is producing 4,500 t annually of product from local ingredients for $0.56/kg. When mixed with ICSM in a 60/40 blend, the resulting Thriposha is sufficient to provide a 50 g/day ration to over 650,000 pregnant and lactating mothers and weaning aged chilren. Besides providing a high quality food supplement for individuals nutritionally at risk, the plant employs 150 people and creates a market for locally grown soybeans, whose production is being encouraged as part of a national agricultural plan.

Table 4. Effect of dehulling on nitrogen utilization by infants (Graham et al., 1979).

Corn/soy blend (70/30)	N absorption (% of intake)	N retention (% of intake)
Degermed corn/dehulled soy	81.8±4.8[a]	35.3±6.6
Whole corn/dehulled soy	74.7±7.8	25.3±7.7
Whole corn/whole soy	77.1±7.6	27.9±4.8

[a]Mean ± standard deviation.

The impact of the Sri Lankan Thriposha program, as part of an integrated health care program which includes immunizations, primary health care and education, has been significant. Drake et al. (1982) found that the improved nutritional status of children in the program was a function of age, nutritional status at the time entering the program, and length of participation. The increase in nutritional status was significant and surpassed that seen in other rigorously analyzed food supplementation programs.

In addition to the GOSL-supported free distribution program for Thriposha, the product has been tested in the marketplace (Nichols, 1983). Levor Brothers (Ceylon) Ltd. conducted consumer acceptance tests, found the concept viable, and carried the product through the test market stage. Based on Thriposha's wholesale and retail sales, it was found to be a commercial success. Distributors report good consumer satisfaction and increasing sales. The product is sold in a 300 g box with a polyethylene liner. At a sales price of $0.96/kg, projected sales are in the range of 1,300 t annually.

Costa Rica

In 1976, the Government of Costa Rica, CARE-Costa Rica and AID cooperated to construct a LEC plant which went into production in 1979. CARE and AID shared the cost of approximately $550,000 for a new building and equipment. The processing system was designed by engineers at Colorado State University and includes a Brady extruder for making precooked food blends and FFSFs. The plant has a single processing line capable of processing 450 kg/h of low-cost food products for use in nutrition programs.

Isralow (1983) indicates that the presence of the plant has helped stimulate soybean production in Costa Rica by creating a market for the newly introduced crop. This successful joint effort illustrates the capability of LEC projects for stimulating development of the necessary support infrastructure and significantly expanding the economic impact of the processing plant.

The initial Costa Rican products were formulated around CSB for use in maternal health care centers and in school feeding programs. To expand the use of soy fortified products and widen the potential market, the plant pioneered the development of a number of other products, which have achieved acceptance in the government school feeding programs. "Frescorchata" is a drink product made from extruded, dehulled soybean and corn blended with milk powder, sugar, cinnamon, and vanilla. "Vitaleche," also a powder, is a drink base made from milk powder, fortified with extrusion-cooked dehulled soybeans and blended with sugar and spices. "Masarina" is an instant tortilla flour.

Other Applications

Tanzania, Guyana, Mexico, Ecuador, and Thailand are other sites where LECs process foods for human use. The widespread geographical dispersion of production plants and research activities suggests that the LEC program is steadily moving to an expanded application phase.

NOTES

Judson M. Harper, Department of Agricultural and Chemical Engineering, Colorado State University, Fort Collins, Colorado 80523.

This work was made possible by Cooperative Agreement, 58-319R-4-29, between the Research Institute of Colorado and the U.S. Department of Agriculture, Office of International Cooperation and Development, with funding provided by the Agency for International Development, Office of Nutrition.

REFERENCES

Bressani, R. 1977. Protein supplementation and complementation. p. 204-232. *In* C. E. Bodwell (ed.) Evaluation of proteins for humans. AVI Publishing Co., Westport, CT.

Drake, W. D., J. N. Gunning, A. Horwitz, R. I. Miller, H. L. Rice and G. Thenabadu. 1982. Nutrition Programs in Sri Lanka using U.S. Food AID. USAID report No. AID/SOD/PDC/0262-I-05-1010-00.

Graham, G. G. and W. C. MacLean, Jr. 1979. Digestibility and utilization of extrusion-cooked corn-soy blends. Report submitted to Office of Nutrition, AID, Washington, D.C.

Harper, J. M. and G. R. Jansen. 1981. Nutritious foods produced by low-cost technology. LEC Report 10, Dept. of Agricultural and Chemical Engineering, Colorado State Univ., Fort Collins.

Isralow, S. 1983. Beyond better nutrition—a close-up of potential multiplier effects of food aid, combined with other development resources, in Costa Rica. Horizons 2(9):22-27.

Jansen, G. R. 1980. A consideration of allowable fiber levels in weaning foods. UNU Food and Nutrition Bull. 2(4):38-47.

Jansen, G. R., L. O'Deen, R. E. Tribelhorn and J. M. Harper. 1981. The calorie densities of gruels made from extruded corn-soy blends. UNU Food and Nutrition Bull. 3(1):39-44.

Nichols, J. P. 1983. A review of the commercial marketing program for Thriposha. Texas Agricultural Market Research and Development Center, Texas A and M University, College Station, TX.

USDA. 1978. Purchase of corn/soy blend for use in export programs. Corn/Soy Announcement CSB-2. USDA-ASCS, Shawnee Mission, KS.

Utilization of Soybean Products in Animal Feeds

IMPROVING SOYBEAN PROTEIN UTILIZATION BY RUMINANTS

Terry Klopfenstein

Soybean meal is the most commonly used protein supplement for ruminants, especially in the U.S. In fact, it is the standard to which other protein supplements are compared. In the past 10 to 15 years, it has become apparent that soybean meal is not utilized as efficiently by ruminants as is theoretically possible. Some other protein sources are utilized more efficiently than soybean meal, and have taken an increased part of the market. It is, therefore, very important to investigate means of improving soybean protein utilization. I would like to discuss four aspects of this situation: (1) ruminant protein metabolism; (2) methods of increasing bypass of soybean protein; (3) methodology for comparing protein sources; and (4) importance of amino acid composition of bypass protein.

RUMINANT PROTEIN METABOLISM

Dietary protein entering the rumen is either degraded by the rumen microbes to peptides, amino acids, volatile fatty acids and ammonia, and/or it can bypass rumen degradation to be digested and absorbed as amino acids and peptides from the intestinal tract. Rumen microbes synthesize amino acids and proteins from the products of protein degradation. The ammonia can be supplied less expensively as urea than as degradable protein.

There is evidence that some rumen degradable protein is needed to obtain maximum microbial protein synthesis. This would be especially true when the animals are fed high fiber diets. The rumen-degradable protein requirement may be for specific, branched-chain, volatile fatty acids, amino acids or peptides. The amount of rumen degradable protein required has not been established, but is likely between 25 and 30% of the total crude protein fed (Klopfenstein, 1984; Maeng et al., 1976).

The ruminant then has two sources of protein for metabolic functions; bypass protein and microbial protein. The bypass protein comes from two sources: the basal feed ingredients, such as grain, hay and silage; and from the supplemental protein souce, such as soybean meal. In some cases the microbial protein, plus the bypass protein from the basal feeds, will meet the animal's protein requirements. This would be true of fattening beef cattle and dry cows. Dairy cows during lactation and growing calves need additional bypass protein from the supplemental protein source.

The value of a supplemental protein source is dependent upon the portion of protein that bypasses rumen digestion. Rumen degradable protein is often supplied by the basal diet, or can be supplied inexpensively by by-products. The bypass value is affected by the nature of the protein, processing conditions involved in the preparation of the protein source, and animal and ration factors, such as level of intake, rumen pH and rate of particle passage from the rumen. Some proteins such as milk protein are soluble in this native state, and are completely degraded in the rumen. Conversely, corn protein is highly degraded before processing. Heat applied in the desolventizing/ toasting part of the oil extraction process reduces somewhat the solubility of soybean protein. However, commercially prepared soybean meal appears to be only 25 to 30% bypassed. Some dried animal proteins appear to be 70-80% bypassed. This indicates the amount of improvement that is possible with soybean meal.

TREATMENT OF SOYBEAN PROTEIN

There are two means of treating soybean protein that have been investigated intensively. The first is the use of formaldehyde to cross link the soybean protein making it unavailable for degradation in the rumen. The low pH in the abomasum, plus pepsin digestion, breaks the cross linking allowing digestion to occur in the intestine. This technique is not especially new. Reports over the last 5 years indicate good response (in-vitro response) to formaldehyde treatment (Table 1), although the response in animal performance has been inconsistent. This variation in animal response will be discussed later. Formaldehyde treatment is being commercially used in Europe but has not been cleared by the Food and Drug Administration for use in the U.S. Formaldehyde is on the suspected carcinogen list, and its use will likely never be permitted.

Heating to reduce ruminal degradation of soybean protein has been widely tested. The effect of heat on the protein is not well understood. Responses measured by laboratory methods have generally been good. However, animal production response has been inconsistent. Response was good in two well designed studies (Kung and Huber, 1983; Thomas et al., 1979). However, we have been unable to obtain a significant response in three experiments using extruded soybeans or soybean meal heated longer in the desolventizer toaster (Table 2). The degree of heating may be of major importance. It is my opinion that simple heating in conventional processing equipment is probably not practical. Extruded soybeans or meal may not be sufficiently improved to justify the cost.

The simplistic approach to heating in the past is likely not sufficient. Many conditions affect protein functionality and the response of protein to heating. Considerably more research needs to be conducted to study means of heating soybean protein to obtain reduced rumen availability. Basic research on protein structure, etc. is needed to determine what has been changed. Certainly, overheating, which could destroy amino acids or make them unavailable, must be considered.

Some other methods of increasing bypass of soybean protein include alcohol and acetic acid treatments and complexing with bentonite (Table 1). These treatments that alter protein structure have potential to improve bypass of soybean protein. Certainly, these and other innovative approaches to enhancing bypass of soybean protein are needed.

METHODOLOGY FOR COMPARING PROTEIN SOURCES

I previously indicated that animal production response to treatment of soybean protein was highly variable. This is due primarily to variation, or deficiencies, in experimental design. The experiments are summarized in Table 1. In most of the experiments, there was no demonstration that protein was the first limiting nutrient. In many, a response to protein was shown, but proteins were not compared below the

Table 1. Summary of responses to treatment of soybeans or soybean meal (SBM).

Reference	Treatment	In-vitro response	Animal species	Response	Protein limiting[a]
Sudweeks et al. (1978)	roasted beans	–	beef	none	no
Thomas et al. (1979)	heated SBM	high	beef	high	yes
Ahrar and Schingoethe (1979)	extruded SBM	moderate	dairy	small	no
Block et al. (1981)	extruded beans	moderate	dairy	none	no
Mielke and Schingoethe (1981)	extruded beans	moderate	dairy	none	no
Netemeyer et al. (1982)	heated SBM	moderate	dairy	small	no
Plegge et al. (1982)	roasted SBM	high	beef	none	no[b]
Thomas et al. (1979)	formaldehyde SBM	high	beef	high	yes
Spears et al. (1980)	formaldehyde SBM	high	beef	high	yes
Folman et al. (1981)	formaldehyde SBM	–	dairy	none	no
Stanton et al. (1983)	formaldehyde SBM	high	beef	variable	variable[c]
Crooker et al. (1983)	formaldehyde SBM	high	dairy	none	no
Thomas et al. (1979)	wood molasses SBM	high	beef	high	yes
Van der Aar et al. (1983a)	alcohol SBM	moderate	beef	none	no
Van der Aar et al. (1983b)	alcohol SBM	moderate	–	–	–
Vicini et al. (1983)	acetic acid SBM	moderate	–	–	–
Lundquist et al. (1982)	formaldehyde SBM	–	dairy	small	no
Crawford et al. (1983)	formaldehyde SBM	moderate	dairy	none	no[e]
Ruegsegger et al. (1983)	heated soybeans	–	dairy	small	no[e]
Madsen (1982)	formaldehyde SBM	high	dairy	none	no
Oldham et al. (1981)	formaldehyde SBM	–	dairy	none	no
Palmer et al. (1983)	extruded SBM/soybeans	–	beef	none	no
Grummer and Clark (1982)	heated SBM	moderate	dairy	none	no[e]
Satter and Stehr (1984)	extruded soybean/SBM	–	dairy	small	no[e]
Kung and Huber (1983)	heated SBM	high	dairy	some	yes
Rock et al. (1983)	bentonite SBM	high	beef	small	yes
Pena and Satter (1984)	heated SBM	–	dairy	some	no
Pena and Satter (1984)	heated soybeans	–	dairy	some	no
Hawkins et al. (1984)	heated soybeans	–	dairy	some	no
Hawkins et al. (1984)	heated/bentonite soybeans	–	dairy	some	no

[a]Was protein deficiency demonstrated in the experiment?

[b]Protein was limiting in 1 of 4 trials, and ammonia may have been limiting in low protein controls.

[c]Protein limiting in some trials, rumen ammonia or rumen degradable protein may have been limiting in all trials.

[d]Level of protein may have limited response, and rumen ammonia may have been limiting.

[e]Rumen ammonia may have been limiting.

protein requirement. In other studies, protein was treated to reduce rumen degradation and no urea was added to supply the needed rumen ammonia. Satter and Stehr (1984) mention this and Kung and Huber (1983) obtained a response to ammonia-treated corn silage.

The important points to consider in experimental design are: (1) a response to protein must be demonstrated; (2) proteins (or treatments of proteins) *must be compared below* the protein requirement; and (3) rumen ammonia must *not* be limiting.

To illustrate methodology considerations, the procedure used at Nebraska is described. Young, growing calves are fed, individually, a high-forage, low-protein diet,

Table 2. Calf performance on soybean meal, extruded soybeans and heated soybean meal (Cleale et al., 1985).

Item	Experiment 1					Experiment 2				Experiment 3		
	Urea	Soybean meal	Ext[a] Soy 1	Ext[a] Soy 2	Urea	Soybean meal	Ext[a] Soy 1	Ext[a] Soy 2	Urea	Urea	Soybean meal	Heated soybean meal
Average gain, kg/d	0.48^b	0.72^c	0.73^c	0.77^c	0.64^b	0.72^c	0.74^c	0.70^c	0.91^b		0.99^c	0.99^c
Dry matter intake, kg/d	5.3	5.5	5.4	5.3	5.6	5.7	5.7	5.7	7.0		7.0	7.0
Gain/feed	0.89^b	0.131^c	0.135^c	0.150^c	0.114^b	0.126^c	0.128^c	0.123^c	0.130^b		0.139^c	0.141^c
Protein efficiency ratio^d	–	1.564	1.589	1.907	–	0.524	0.626	0.478	–		–	–

^{a} Extruded soybeans at 185 C (1) and 175 C (2).

^{b,c} Means within an experiment and in the same row with different superscripts are statistically different (P <.05).

^{d} Represents the slope of the line obtained by regressing gain (kg/d) above urea control-fed cattle on protein intake (kg/d) above urea control-fed cattle.

generally 2/3 corn silage and 1/3 corn cobs. The control animals are supplemented with urea. Test proteins are fed at increasing levels replacing the urea. This is called the slope-ratio technique, and it has been used in monogastric studies of amino acid, mineral or vitamin availabilities. It is similar to a dose-response curve in a drug study.

The increase in live-weight gain from increasing levels of protein is a direct measure of the value of the protein. At stome point, the protein requirement of the animals is met and no further response to protein is obtained. It is absolutely essential that proteins be compared below the animal's requirement. Otherwise protein is not the first limiting nutrient, and valid comparisons cannot be made.

The amount of gain obtained per unit of test protein fed is the protein efficiency value. This value is equivalent to the slope of the regression of gain on protein intake (Figure 1). Generally, the greater the protein bypass, the higher the protein efficiency value. Blood meal, a high bypass protein compared to soybean meal, meets the animal's protein requirement (maximum gain) with about 40% as much supplemental protein (Figure 1). The protein efficiency (slope of the regression line) was over 2.5 times as great for blood meal compared to soybean meal. This indicates that soybean meal could be improved at least 2.5 times by reducing rumen protein degradation. Data from intestinally fistulated cattle and laboratory estimates support the bypass values for soybean meal and blood meal.

Likely, the production response to added protein is curvilinear rather than linear as shown in Figure 1. Satter (1983) graphically showed (Figure 2) that milk production response in the dairy cow is curvilinear. He shows that a response of 4.1 kg of milk is obtained by raising the crude protein from 12 to 15%, but only 0.45 kg was obtained by increasing protein from 15 to 16.5%. Often the response due to treated or high bypass protein is too small to detect. It should be noted that, if rumen ammonia is limiting at 12% crude protein, additional protein would add both ammonia and bypass protein, which would cause the response to appear curvilinear on the lower part of the curve because, in fact, it represents two responses combined. It should be obvious that proteins must be compared on the more linear part of the response curve.

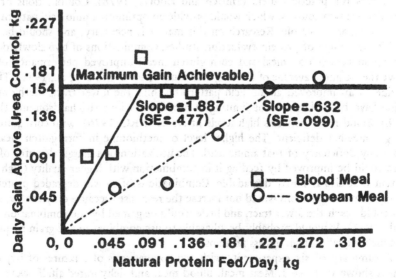

Figure 1. Natural protein fed per day vs. daily gain above that of urea control, using the nonlinear program (Stock et al., 1983).

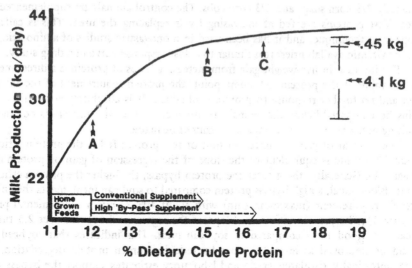

Figure 2. Milk response to added soybean meal (Satter, 1983).

This type of experiment is difficult to conduct but necessary if one is to draw logical conclusions. Most of the experiments reported in Table 1 do not meet these design criteria, so that a conclusion relative to treatments cannot be drawn. The problem is more with experimental design than with variation in treatment. Admittedly, these experiments are difficult to run, especially with dairy cows, but if they are not done correctly, they are of no value.

AMINO ACID COMPOSITION OF BYPASS PROTEINS

The amino acid profile of bypass protein is an important criterion in the evaluation of sources for practical diets (Sniffen and Hoover, 1978). Combinations of slowly degraded protein sources, which would provide an optimum amino acid pattern for the lower tract, are desirable. Research on this matter is necessary, and should be considered in the design of protein evaluation studies. Combinations of two slowly degraded protein sources (blood meal and corn gluten meal) improved performance of steers above the weighted average of the two individual sources (Stock et al., 1981). This may be due to an improved amino acid pattern reaching the lower tract. Methionine and lysine have been shown to be limiting in ruminant rations (Richardson and Hatfield, 1978). Blood meal contains high levels of lysine (NRC, 1976), with corn gluten meal being somewhat deficient. The higher level of methionine in corn gluten meal could offset any deficiency of that amino acid. The utilization of a marginal source of amino acids could be improved by feeding it in combination with higher quality (high lysine) protein that also is slowly degradable. Combinations of slowly degraded proteins and rapidly degraded proteins would not increse the response, because only the bypass protein could reach the lower tract, and little would be gained by the combination (Waller et al., 1981). It would probably be advisable to use combinations of grain by-products with higher quality, high bypass proteins.

A summary of six experiments with ten comparisons of mixtures of bypass proteins is shown in Table 3. Meat meal, blood meal and dehydrated alfalfa were sources of lysine, and corn gluten meal was the source of sulfur amino acids. The average

Table 3. Summary of gain responses to mixtures of bypass proteins.[a]

Protein sources	PE response[b]
	%
Corn gluten meal, Dehydrated alfalfa	37
Corn gluten meal, Dehydrated alfalfa	6
Corn gluten meal, Blood meal	48
Corn gluten meal, Meat meal	121
1/3 Corn gluten meal, 2/3 Blood meal	31
2/3 Corn gluten meal, 1/3 Blood meal	27
1/3 Corn gluten meal, 2/3 Meat and bone meal	0
1/3 Corn gluten meal, 2/3 Meat and bone meal	28
Corn gluten meal, Blood meal	19
Corn gluten meal, Meat and bone meal	-15
Average	30

[a]Aines et al. (1985).

[b]Protein efficiency for the mixture divided by the weighted means of protein efficiency for the appropriate protein sources fed alone.

response of the combinations was 30.8% greater than that obtained with the sources fed alone. The calculated flows of metabolizable amino acids for two of the experiments indicate that lysine (as a percentage of metabolizable protein) needed to be above 7.0% and methionine above 2.1%. This is in agreement with the values of Burroughs et al. (1974) (requirement of 7.2 and 2.1%, respectively).

Soybean protein is a high quality protein, an excellent source of lysine and is very palatable compared to many bypass protein sources. If bypass of the soybean proteins can be improved, it offers great potential for use in balancing ruminant rations for metabolizable amino acids.

NOTES

Terry Klopfenstein, Professor, Animal Science Department, University of Nebraska, Lincoln, NE 68583-0908.

REFERENCES

Ahrar, M. and D. J. Schingoethe. 1979. Heat-treated soybean meal as a protein supplement for lactating cows. J. Dairy Sci. 62:932-940.

Aines, G., B. Brown, S. Wimer and T. Klopfenstein. 1985. Bypass protein quality and evaluation techniques. Nebraska Beef Cattle Report, MP 48. University of Nebraska, Lincoln.

Block, E., L. D. Muller, L. C. Griel, Jr. and D. L. Garwood. 1981. Brown midrib-3 corn silage and heat extruded soybeans for early lactating dairy cows. J. Dairy Sci. 64:1813-1825.

Burroughs, W., A. Trenkle and R. Vetter. 1974. A system of protein evaluation for cattle and sheep involving metabolizable protein (amino acids) and urea fermentation potential of feedstuffs. Vet. Med/Small Anim. Clin. 69:713-720.

Cleale, R., T. Klopfenstein, J. Merrill, M. Nelson and W. Stroup. 1985. Heated soybean products for growing beef cattle. Nebraska Beef Cattle Report, MP48. University of Nebraska, Lincoln.

Crawford, R. J., Jr., W. H. Hoover and J. E. Brooke. 1983. Effects of particle size and formaldehyde treatment of soybean meal on milk production and composition in dairy cows. (Abstr.) J. Dairy Sci. 66(Suppl. 1):148.

216

Crooker, B. A., J. H. Clark and R. D. Shanks. 1983. Effects of formaldehyde treated soybean meal on milk yield, milk composition and nutrient digestibility in the dairy cow. J. Dairy Sci. 66:492-504.

Folman, Y., H. Neumark, M. Kaim and W. Kaufmann. 1981. Performance, rumen and blood metabolites in high-yielding cows fed varying protein percents and protected soybean. J. Dairy Sci. 64: 759-768.

Grummer, R. R. and J. H. Clark. 1982. Effect of nitrogen solubility on lactational performance and protein and dry matter degradation in situ. J. Dairy Sci. 65:1432-1444.

Hawkins, E. E. 1984. Optimizing soybean utilization in dairy rations. J. Dairy Sci. (Abstr.) 67(Suppl. 1):125.

Klopfenstein, T. 1984. Applications of protected protein formulations. In National Feed Ingredients Assoc. nutrition institute proceedings: Amino acids. NFIA, Des Moines, Iowa.

Kung, L., Jr. and J. T. Huber. 1983. Performance of high producing cows in early lactation fed protein of varying amounts, sources and degradability. J. Dairy Sci. 66:227-234.

Lundquist, R., D. E. Otterby and J. G. Linn. 1982. Influence of formaldehyde-treated soybean meal on milk production. J. Dairy Sci. (Abstr.) 65(Suppl. 1):137.

Madsen, J. 1982. The effect of formaldehyde-treated protein and urea on milk yield and composition in dairy cows. Acta Agric. Scand. 32:398-394.

Maeng, W. J., C. J. Van Nevel, R. L. Baldwin and J. G. Morris. 1976. Rumen microbial growth rates and yields: Effect of amino acids and protein. J. Dairy Sci. 59:68-79.

Mielke, C. D. and D. J. Schingoethe. 1981. Heat-treated soybeans for lactating cows. J. Dairy Sci. 64:1579-1585.

NRC. 1976. Nutrient Requirements of Domestic Animals, No. 4. Nutrient Requirements of Beef Cattle. Fifth, Revised Ed. National Academy of Sciences—National Research Council, Washington, DC.

Netemeyer, D. T., L. J. Bush, J. W. Ward and S. A. Jafri. 1982. Effect of heating soybean meal for dairy cows. J. Dairy Sci. 65:235-241.

Oldham, J. D., R. J. Fulford and D. J. Napper. 1981. Source and level of supplemental N for cows in early lactation. Proc. Nutr. Soc. 40:30A.

Palmer, L. F., L. B. Embry, J. G. Nothnagel, R. M. Luther and M. J. Goetz. 1983. Sources of supplemental protein with corn silage for growing cattle. p. 40-48. In Proc. S. Dak. State Univ. Cattle Feeders Day, 83-9, Univ. South Dakota.

Pena, F. and L. D. Satter. 1984. Effect of feeding heated soybean meal and roasted soybeans on milk production in Holstein cows. J. Dairy Sci. (Abstr.) 67(Suppl. 1):123.

Plegge, S. D., L. L. Berger and G. C. Fahey, Jr. 1982. Effect of roasting on utilization of soybean meal by ruminants. J. Anim. Sci. 55:395-401.

Richardson, C. R. and E. E. Hatfield. 1978. The limiting amino acids in growing cattle. J. Anim. Sci. 46:740-745.

Rock, D. W., T. J. Klopfenstein, J. K. Ward, R. A. Britton and M. L. McDonnell. 1983. Evaluation of slowly degraded proteins: Dehydrated alfalfa and corn gluten meal. J. Anim. Sci. 56:476-482.

Ruegsegger, G. J., L. H. Schultz and D. Sommer. 1983. Response of dairy cows in early lactation to the feeding of heat treated whole soybeans. J. Dairy Sci. (Abstr.) 66(Suppl. 1):168.

Satter, L. D. 1983. Protected protein and NPN. Animal Nutrition and Health 38(6):14-18.

Satter, L. D. and D. B. Stehr. 1984. Feeding resistant protein to dairy cows. Proceedings Distillers Feed Conference. Distillers Feed Research Council, Cincinnati, Ohio.

Sniffen, C. J. and W. H. Hoover. 1978. Amino acid profile of dietary bypass protein and its importance to ruminants. Proc. 33rd Distillers Feed Conf. Distillers Feed Research Council, Cincinnati, Ohio.

Spears, J. W., E. E. Hatfield and J. H. Clark. 1980. Influence of formaldehyde treatment of soybean meal on performance of growing steers and protein availability in the chick. J. Anim. Sci. 50: 750-755.

Stanton, T. L., F. N. Owens and K. S. Lusby. 1983. Formaldehyde-treated soybean meal for ruminants grazing winter range grass. J. Anim. Sci. 56:6-14.

Stock, R., N. Merchen, T. Klopfenstein and M. Poos. 1981. Feeding value of slowly degraded proteins. J. Anim. Sci. 53:1109.

Stock, R., T. Klopfenstein, D. Brink, S. Lowry, D. Rock and S. Abrams. 1983. Impact of weighing procedures and variation in protein degradation rate on measured performance of growing lambs and cattle. J. Anim. Sci. 57:1276-1285.

Sudweeks, E. M., L. O. Ely, L. R. Sisk and M. E. McCullough. 1978. Effect of roasting sorghum and soybeans on gains and digestibility. J. Anim. Sci. 46:867-872.

Thomas, E., A. Trenkle and W. Burroughs. 1979. Evaluation of protective agents applied to soybean meal and fed to cattle. 2. Feedlot trials. J. Anim. Sci. 49:1346-1356.

Van der Aar, P. J., G. L. Lynch, L. L. Berger, G. L. Fahey, Jr. and N. R. Merchen. 1983a. Alcohol treatment of soybean meal. p. 39. *In* Illinois Beef Cattle Report. University of Illinois, Urbana.

Van der Aar, P. J., L. L. Berger, C. G. Fahey, Jr. and S. C. Loerch. 1983b. Effects of alcohol treatments on utilization of soybean meal by lambs and chicks. J. Anim. Sci. 57:511-518.

Vicini, J. L., J. H. Clark and B. A. Crooker. 1983. Effectiveness of acetic acid and formaldehyde for preventing protein degradation in the rumen. J. Dairy Sci. 66:350-354.

Waller, J., T. Klopfenstein and M. Poos. 1981. Distillers' feeds as protein sources for growing ruminants. J. Anim. Sci. 51:1154-1167.

DIGESTIBILITY OF THE CARBOHYDRATE FRACTION OF SOYBEAN MEAL BY POULTRY

L. M. Potter and M. Potchanakorn

About one-third of the soybean meal produced in 1982 was consumed by poultry in the United States. About one-third of soybean meal is carbohydrate, the digestibility of which is the concern in this paper. For a component that makes up as much as 8% of poultry diets, it is surprising how little information is available on the digestibility of carbohydrate in soybean meal. Relatively few experiments have been conducted to determine its value for poultry. However, calculations, based on available information on the metabolizable energy and digestibility of other nutrient components in soybean meal, allows an indirect estimate of the digestibility of the carbohydrate.

COMPOSITION OF SOYBEANS AND SOYBEAN MEAL

On a dry matter basis, soybeans contain about 20% ether extract and 41% protein (Table 1). Upon extraction of most of the oil, a meal is produced containing about 49% protein, 8% crude fiber, and 36% nitrogen-free extract on a dry matter basis. Further removal of the hulls results in a product containing about 54% protein, 4% crude fiber, and 34% nitrogen-free extract.

Honig and Rackis (1979) have summarized the composition of the carbohydrate in soybean meal (Table 2). The carbohydrate portion, which includes the crude fiber and nitrogen-free extract fractions, is composed of various oligosaccharides and polysaccharides. The starch content of soybeans is very small, about 0.5% (Wilson et al.,

Table 1. Proximate composition of soybeans, soybean meal, and dehulled soybean meal on a dry matter basis (NRC, 1977).

Component	Soybeans	Soybean meal	Dehulled soybean meal
		– % –	
Protein	41.1	49.4	53.9
Ether extract	20.0	0.9	1.1
Crude fiber	6.1	8.2	4.3
Ash	5.4	5.9	6.5
Nitrogen-free extract	27.4	35.6	34.2

Table 2.   Carbohydrate content of dehulled soybean meal (Honig and Rackis, 1979).

Constituent	% of meal
Oligosaccharide content, total	15
Sucrose	6-8
Stachyose	4-5
Raffinose	1-2
Verbascose	trace
Polysaccharide content, total	15-18
Acidic polysaccharides	8-10
Arabinogalactan	5
Cellulosic material	1-2
Starch	0.5

1978), and thus most of the polysaccharides are acidic polysaccharides, arabinogalactan, and cellulosic material, which are essentially non-digestible.

The more digestible carbohydrate components of soybean meal are found in the oligosaccharide fraction. On a dry matter basis, soybeans contain 4.5 to 5.5% sucrose, 2.8 to 4.9% stachyose, and 0.4 to 1.4% raffinose, and soybean meal contains 7.4 to 9.9% sucrose, 4.7 to 4.8% stachyose, and 1.0 to 1.1% raffinose (Delente and Ladenburg, 1972; Hymowitz et al., 1972; Tanaka et al., 1975; Black and Bagley, 1978; Openshaw and Hadley, 1978). These values complement those presented in Table 2. Sucrose is highly digested by poultry; however, data on the degree of digestibility of the other oligosaccharides are lacking.

GROSS AND METABOLIZABLE ENERGY OF SOYBEAN MEAL

Gross energy of feed ingredients can be estimated based on their proximate analysis. In general, the gross energy contents of protein, ether extract, and carbohydrate of feed ingredients are 23.68, 38.91, and 17.57 kJ/g, respectively. Based on these values, the gross energy contents of soybean meal and dehulled soybean meal, on a dry matter basis, are about 19.74 and 19.96 kJ/g, respectively, whereas that of ground yellow corn is 18.80 kJ/g (Table 3).

Experiments conducted to determine the metabolizable energy of feed ingredients (Hill et al., 1960; Hill and Renner, 1960; Potter and Matterson, 1960; Sibbald and Slinger, 1962) revealed that soybean meal contains about 10.48 kJ/g, dehulled soybean meal about 11.34 kJ/g, and ground yellow corn about 16.12 kJ/g on a dry matter basis. Therefore, soybean meal and dehulled soybean meal contain about 5 and 6% more gross energy but 54 and 42% less metabolizable energy, respectively, as compared to ground yellow corn. This indicates that some components of the two soybean products are poorly digested and metabolized.

Fraps (1944) conducted extensive studies to determine the digestion coefficients of the various proximate components in feedstuffs used in poultry diets. From 117 samples of ground yellow corn, he reported that its protein was 86.1 ± 11.6% (mean and standard deviation) digested, its ether extract 89.5 ± 12.2% digested, and its nitrogen-free extract 94.1 ± 3.7% digested (Table 4). Using these digestion coefficients and the metabolizable energy values of 16.74, 37.66, and 16.74 kJ/g for the digested and metabolized protein, ether extract, and nitrogen-free extract, respectively, the total metabolizable energy of ground yellow corn is calculated to be 15.84

Table 3. Calculated gross energy content of soybean meal, dehulled soybean meal, and ground yellow corn, dry matter basis.

Ingredient and nutrient	Gross energy in nutrient	Nutrient content	Gross energy
	kJ/g	%	kJ/g
Soybean meal			
Protein	23.68	49.4	11.70
Ether extract	38.91	0.9	0.35
Crude fiber	17.57	8.2	1.44
Nitrogen-free extract	17.57	35.6	6.25
Ash	-	5.9	-
Total		100.0	19.74
Dehulled soybean meal			
Protein	23.68	53.9	12.76
Ether extract	38.91	1.1	0.43
Crude fiber	17.57	4.3	0.76
Nitrogen-free extract	17.57	34.2	6.01
Ash	-	6.5	-
Total		100.0	19.96
Ground yellow corn			
Protein	23.68	9.9	2.34
Ether extract	38.91	4.3	1.67
Crude fiber	17.57	2.5	0.44
Nitrogen-free extract	17.57	81.7	14.35
Ash	-	1.6	-
Total		100.0	18.80

kJ/g dry matter (Table 4). This value is not significantly different from 16.12 kJ/g which was reported by the National Research Council (1977), based on the research of Hill and Renner (1960), Potter and Matterson (1960), and Sibbald and Slinger (1962).

Fraps (1944) also reported comparable digestion coefficients for the average of only 6 samples of solvent extracted soybean meal. The digestion coefficients were 75.2 ± 5.1% for protein, 37.7 ± 30.6% for ether extract, and 30.0 ± 11.4% for nitrogen-free extract. The application of these digestion coefficients to the calculation of metabolizable energy of soybean meal and dehulled soybean meal (Table 4) results in values that are 22 and 24% below those expected, based on the metabolizable energy values reported by the National Research Council (1977). Therefore, a discrepancy exists, which indicates that the digestion coefficient of either 75.2% for the protein or 30.0% for the nitrogen-free extract obtained by Fraps is too low for present-day soybean meal.

DIGESTIBILITY OF SOYBEAN MEAL

Amino Acids

Soybean meal is added to poultry diets primarily for its protein content. During the past few years, several experiments have been conducted to determine the

Table 4. Calculated metabolizable energy content of soybean meal, dehulled soybean meal, and ground yellow corn on a dry matter basis.

Ingredient and nutrient	Metabolizable energy in digestible nutrient	Nutrient content	Digestibility of nutrient[a]	ME
	kJ/g	%	%	kJ/g
Soybean meal				
Protein	16.74	49.4	75.2	6.22
Ether extract	37.66	0.9	37.7	0.13
Crude fiber	16.74	8.2	0.1	0.00
Nitrogen-free extract	16.74	35.6	30.0	1.79
Ash	-	5.9	-	-
Total		100.0		8.14
NRC (1977)				10.48
Dehulled soybean meal				
Protein	16.74	53.9	75.2	6.79
Ether extract	37.66	1.1	37.7	0.16
Crude fiber	16.74	4.3	0.1	0.00
Nitrogen-free extract	16.74	34.2	30.0	1.72
Ash	-	6.5	-	-
Total		100.0		8.67
NRC (1977)				11.34
Ground yellow corn				
Protein	16.74	9.9	86.1	1.43
Ether extract	37.66	4.3	89.5	1.45
Crude fiber	16.74	2.5	21.6	0.09
Nitrogen-free extract	16.74	81.7	94.1	12.87
Ash	-	1.6	-	-
Total		100.0		15.84
NRC (1977)				16.12

[a]Fraps (1944).

digestibility or availability of the amino acids in dehulled soybean meal. Feed and excreta samples collected in experiments, in which soybean meal was the sole source of protein, were analyzed for amino acid content. Upon relating the quantity of excreta derived from a unit of feed consumed in these experiments, 87 to 99% of the quantities of amino acids in soybean meal were found to be digested and metabolized (Ivy et al., 1971; Flipot et al., 1971; Nwokolo et al., 1976; Likuski and Dorrell, 1978; El Boushy and Roodbeen, 1980; Pierson et al., 1980; Parsons et al., 1981). Therefore, the value of 75.2% digestibility for protein in soybean meal, as obtained by Fraps (1944), does not seem applicable to properly processed, solvent-extracted soybean meal.

Carbohydrates

Very few studies have been directed toward determining the digestibility of the carbohydrate fraction. Person et al. (1980) investigated the digestibility of the various components of the proximate analysis in dehulled soybean meal consumed by adult turkeys. They reported that the carbohydrate in dehulled soybean meal was only 4% digested. However, a correction needs to be made in this value, due to the presence of uric acid and other nitrogen end-products of metabolism in the excreta, a factor not previously considered in their calculations. In the calculations presented in Table 5, the excreta was estimated to contain 37.62% uric acid and other nitrogen-containing, metabolic end-products, which were previously considered as parts of the nitrogen-free extract. By making this correction, the nitrogen-free extract in the dehulled soybean meal is calculated to be about 61% digested. Applying the values of 87 and 61% digestibility to the protein and nitrogen-free extract, respectively, of soybean meal or dehulled soybean meal, the calculated metabolizable energy values of these respective ingredients of 10.71 and 11.46 kJ/g agree very well with those listed in the National Research Council (1977) publication.

The bioavailability of carbohydates in soybean meal has been determined by a slope-ratio analysis using rats and chicks fed a carbohydrate-free diet supplemented with glucose as a standard. In these experiments, the carbohydrates in soybean meal were reported to be 40% available in chicks (Lodhi et al., 1969) and 35% in rats (Karimzadegan et al., 1979).

Most of the studies concerning the digestibility of oligosaccharides in legume seeds have been conducted with human subjects, and have concerned flatulence. Cristofaro et al. (1974) and Rackis (1976) have reviewed the flatulence problem associated with oligosaccharides in soybeans.

CONCLUSION

The importance of the carbohydrate fraction in soybean meal has not been fully realized, since major emphasis has been placed on the protein fraction. Soybean meal

Table 5. Digestibility of the components of the proximate analysis in dehulled soybean meal (DSBM) on a dry matter basis (Pierson et al., 1980).

Component	DSBM	Excreta	Non-digested	Digested	Digestibility
			– % –		
Protein	54.00[a]	15.03[b]	7.02[b]	46.98[b]	87
Ether extract	4.19[c]	1.00	0.47	3.72	89
Crude fiber	5.57	10.31	4.81	0.76	14
Ash	5.27	9.94	4.64	0.63	12
Nitrogen-free extract	30.97	63.72	29.76	1.21	4
Total	100.00	100.00	46.70	53.30	
Uric acid	–	37.62[d]	(17.57)	--	
Corrected NFE	--	26.10	12.19	18.78	61
Corrected total			29.13	70.87	

[a]Crude protein in DSBM = 6.25 x %N.

[b]Excreta contains 14.94% nitrogen by Kjeldahl analysis; protein in excreta was calculated assuming 87% digestibility as indicated by the average digestibility of the amino acids.

[c]Seems high, but results were confirmed upon reanalysis.

[d]Calculated by subtracting 2.40% nitrogen (16% of 15.03% protein) from 14.94% nitrogen in excreta. Thus, the remaining nitrogen was assumed to be metabolic nitrogen end-products equivalent to 37.62% uric acid in the excreta.

contains more than 30% carbohydrate and contributes about 8% carbohydrate to poultry diets, but no experiments have been conducted to determine the digestibility of this portion of the diet by poultry. The digestibility of the protein and amino acids of soybean meal is 87% or more; the digestibility of the carbohydrates (nitrogen-free extract and crude fiber) in soybean meal is calculated to be not more than 60% by poultry. Research conducted to determine the digestibility of the oligosaccharides could provide useful leads for better utilization of soybeans and their products by poultry. Obviously, the carbohydrate portion of soybean meal needs to be extensively researched and its digestibility evaluated directly to determine whether the value of soybean meal as a major ingredient in poultry diets can be improved through increased digestion of its carbohydrate fraction.

NOTES

L. M. Potter and M. Potchanakorn, Department of Poultry Science, Virginia Polytechnic Institute and State University, Blacksburg, Virginia 24061.

REFERENCES

Black, L. T. and E. B. Bagley. 1978. Determination of oligosaccharides in soybeans by high pressure liquid chromatography using an internal standard. J. Am. Oil Chem. Soc. 55:228-232.

Cristofaro, E., F. Mottu and J. J. Wuhrmann. 1974. Involvement of the raffinose family of oligosaccharides in flatulence. p. 313-336. *In* H. L. Sipple and K. W. McNutt (eds.) Sugars in Nutrition. Academic Press, New York.

Delente, J. and K. Ladenburg. 1972. Quantitative determination of the oligosaccharides in defatted soybean meal by gas-liquid chromatography. J. Food Sci. 37:372-374.

El Boushy, A. R. and A. E. Roodbeen. 1980. Amino acid availability in Lavera yeast compared with soybean and herring meal. Poultry Sci. 59:115-118.

Flipot, P., R. J. Belzile and G. J. Brisson. 1971. Availability of the amino acids in casein, fish meal, soya protein and zein as measured in the chicken. Can. J. Anim. Sci. 51:801-802.

Fraps, G. S. 1944. Digestibility of feeds and human foods by chickens. Texas Agric. Exp. Stn. Bull. 663.

Hill, F. W., D. L. Anderson, R. Renner and L. B. Carew, Jr. 1960. Studies of the metabolizable energy of grains and grain products for chickens. Poultry Sci. 39:573-579.

Hill, F. W. and R. Renner. 1960. The metabolizable energy of soybean oil meals, soybean millfeeds and soybean hulls for the growing chick. Poultry Sci. 39:579-583.

Honig, D. H. and J. J. Rackis. 1979. Determination of the total pepsin-pancreatin indigestible content (dietary fiber) of soybean products, wheat bran, and corn bran. J. Agric. Food Chem. 27:1262-1266.

Hymowitz, T., F. I. Collins, J. Panczner and W. M. Walker. 1972. Relationship between the content of oil, protein, and sugar in soybean seed. Agron. J. 64:613-616.

Ivy, C. A., D. B. Bragg and E. L. Stephenson. 1971. The availability of amino acids from soybean meal for the growing chick. Poultry Sci. 50:408-410.

Karimzadegan, E., A. J. Clifford and F. W. Hill. 1979. A rat bioassay for measuring the comparative availability of carbohydates and its application to legume foods, pure carbohydrates and polyols. J. Nutr. 109:2247-2259.

Likuski, H. J. A. and H. G. Dorrell. 1978. A bioassay for rapid determination of amino acid availability values. Poultry Sci. 57:1658-1660.

Lodhi, G. N., R. Renner and D. R. Clandinin. 1969. Available carbohydrate in rapeseed meal and soybean meal as determined by a chemical method and a chick bioassay. J. Nutr. 99:413-418.

National Research Council. 1977. Nutrient requirements of poultry. No. 1. *In* Nutrient Requirements of Domestic Animals, 7th ed. National Academy of Science, Washington, D.C.

Nwokolo, E. N., D. B. Bragg and W. D. Kitts. 1976. The availability of amino acids from palm kernel, soybean, cottonseed and rapeseed meal for the growing chick. Poultry Sci. 55:2300-2304.

Openshaw, S. J. and H. H. Hadley. 1978. Maternal effects on sugar content in soybean seeds. Crop Sci. 18:581-584.

Parsons, C. M., L. M. Potter and R. D. Brown, Jr. 1981. True metabolizable energy and amino acid digestibility of dehulled soybean meal. Poultry Sci. 60:2687-2696.

Pierson, E. E. M., L. M. Potter and R. D. Brown, Jr. 1980. Amino acid digestibility of dehulled soybean meal by adult turkeys. Poultry Sci. 59:845-848.

Potter, L. M. and L. D. Matterson. 1960. Metabolizable energy of feed ingredients for chickens. Conn. Agric. Exp. Stn. Program Rep 9

Rackis, J. J. 1976. Flatulence problems associated with soy products. p. 892-903 In L. D. Hill (ed.) World Soybean Research. The Interstate Printers & Publishers, Inc., Danville, Illinois.

Sibbald, I. R. and S. J. Slinger. 1962. The metabolizable energy of materials fed to growing chicks. Poultry Sci. 41:1612-1613.

Tanaka, M., D. Thananunkul, T. C. Lee and C. O. Chichester. 1975. A simplified method for the quantitative determination of sucrose, raffinose and stachyose in legume seeds. J. Food Sci. 40: 1087-1088.

Wilson, L. A., V. A. Birmingham, D. P. Moon and H. E. Snyder. 1978. Isolation and characterization of starch from mature soybeans. Cereal Chem. 55:661-670.

UTILIZATION OF ENERGY FROM SOYBEAN PRODUCTS BY YOUNG PIGS

R. C. Ewan

The major products of soybean processing are soybean oil, hulls and meal. Whole soybeans contain about 36% protein and 18% oil. Efficient solvent extraction processes remove all but about 0.5 to 0.8% of the oil. Because the hulls are removed initially, the residue (dehulled soybean meal) contains about 48% protein on an air-dry basis. Addition of the hulls to the meal reduces the level of protein to about 44%. Both 48 and 44% protein meals provide high-quality protein for pigs that is easily digested. The amino acid balance of the protein present in soybeans complements the protein in cereal grains well and allows formulation of diets that are utilized efficiently by pigs.

Although soybean meals are used in swine diets to provide a supply of amino acids, they also contribute to the energy content of the diet. The utilization of energy or energy value of soybean products has been determined for dehulled soybean meal, but little information is available for other products of soybean processing.

DEFINITION OF TERMS

The following terms are used to describe the utilization of dietary energy by pigs.

Gross Energy (GE) is the energy released upon total combustion of the feed in a bomb calorimeter. It provides an indication of the total energy in the feed, but does not provide any information about the utilization of the energy by the animal.

Digestible Energy (DE) is the energy removed from a feed during passage through the intestinal tract. It is obtained by subtracting fecal energy from the GE intake. The ratio of DE/GE is the digestibility of energy.

Metabolizable Energy (ME) is the energy available to meet the needs of the animal for maintenance, growth, activity and temperature regulation. It is obtained by subtracting the energy excreted in urine from digestible energy. The ratio of ME/DE reflects the efficiency of utilization of DE.

Net Energy (NE) is the energy used for maintenance and deposited as tissue during productive functions. The ratio of NE/ME reflects the efficiency of utilization of ME for maintenance and production.

ENERGY VALUE OF SOYBEAN PRODUCTS

A series of experiments has been conducted to evaluate the utilization of energy from soybean products by young pigs. The products evaluated were soybean oil

(Phillips and Ewan, 1977), 44% soybean meal (unpublished data of Ewan), 48% soybean meal (unpublished data of Ewan, and of Long and Ewan), soybean hulls (Bailey and Ewan, 1984) and whole, extruded soybeans (unpublished data of Long and Ewan). The proximate analysis of the ingredients is presented in Table 1. For all experiments, pigs were weaned at 3 weeks of age and allowed a 7-day adjustment period. At the end of the adjustment period, pigs were killed for an estimate of initial body composition. The remaining pigs were used in a 28-day metabolism study with complete collection of urine and feces. Pigs were housed in metabolism cages at a temperature of about 26 C. At the end of the metabolism study, all remaining pigs were killed for determination of body composition. From the results of the metabolism studies, digestible and metabolizable energy values were determined. From the differences between initial and final body composition, energy gain was determined and, with an estimate of the maintenance requirement, net energy was determined.

In experiments with soybean oil and hulls, a basal diet was formulated to provide the daily requirements for all nutrients except energy, when fed at 3% of body weight daily. The basal diet was fed to all pigs at 3% of body weight (one treatment). In addition, the ingredient was fed at two levels for two other treatments. In experiments with soybean meals and whole, extruded soybeans, a basal diet was formulated with dehulled soybean meal to provide 26% crude protein. Two other treatments were the substitution of the ingredient for dehulled meal to provide 50 to 100% of the protein from the ingredient. All diets were fed at 5% of body weight daily.

The results of these studies are presented in Table 2 and expressed as GE in kJ/g of dry matter and as the efficiency of utilization of the energy components in percentages. The energy of dehulled soybean meal was digested efficiently. The digested energy was highly available for meeting body functions of the pig. About 50% of the ME was associated with the gain in tissue and the maintenance requirement. Similar

Table 1. Proximate analysis of soybean products.

Soybean product	Dry matter	Protein	Fat	Ash
	%		— % Dry matter —	
Meal, dehulled	90.4	54.6	1.1	7.4
Hulls	91.4	11.8	0.4	5.0
Meal, with hulls	90.4	52.3	0.8	7.0
Oil	100.0	1.4	91.0	0.3
Whole, extruded	93.4	40.4	17.7	5.4

Table 2. Energy content and utilization of soybean products (dry matter basis).

Soybean product	GE	DE/GE	ME/DE	NE/ME
	kJ/g		— % —	
Meal, dehulled	19.86	83.4	95.8	50.2
Hulls	17.22	74.2	94.2	20.2
Meal, with hulls	19.73	77.0	95.6	51.5
Oil	39.39	80.5	96.3	75.4
Whole, extruded	23.37	80.2	97.3	73.2

values have been reported for dehulled soybean meal by other investigators and are reported in Table 3. The GE of the product used in our studies was similar to the values reported by others. The digestibility of energy (DE/GE) of dehulled soybean meal found in our studies was lower than the average of values reported by other investigators. The utilization of digestible energy was greater, however, than the average of values reported from other studies. The efficiency of utilization of ME for growth was 50.2% and is lower than the value of 62% reported by Just et al. (1978). No other comparable values have been reported.

The energy from soybean hulls was poorly digested (74.2%), but the digested energy was used for ME that was similar to the other soybean products (Table 2). The only comparable results indicated a lower digestibility of 50.4% and a similar conversion of DE to ME of 93.5% (NRC, 1982). The conversion of ME to NE was very low for soybean hulls (20.0%). This result is similar to those obtained with other fiber-containing feeds (Stanley and Ewan, 1982). Other comparable data for soybean hulls have not been reported.

The inclusion of soybean hulls in soybean meal resulted in lower digestibility of energy (77%) than dehulled soybean meal (Table 2). The digested energy was used for ME and NE with efficiencies similar to that of dehulled soybean meal. Thus, the hulls depressed energy digestibility but did not affect the utilization of digested energy. Results of other investigators indicate a similar GE but higher digestible energy than found in our studies (Table 4). There is no obvious explanation for this discrepancy. Conversion of DE to ME was similar to the results of other studies.

Soybean oil, while not normally included in swine diets, also was studied. The GE of soybean oil was 39.4 kJ/g, typical of fat. The digestibility of energy from soybean oil was slightly lower than the digestibility of dehulled soybean meal. The efficiency of utilization of DE was high and similar to other soybean products. The ME from soybean meal was used very efficiently for tissue synthesis. Additional dietary fat is primarily deposited directly in adipose tissue, resulting in an increase in energy gain, and requires little energy for metabolism. This probably explains the high efficiency of conversion of ME to tissue gain.

Whole soybeans, as noted earlier, contain growth inhibitors that are destroyed by

Table 3. Reported values for dehulled soybean meal (dry matter basis).

Reference	GE	DE/GE	ME/DE
	kJ/g	– % –	
Diggs et al., 1965	19.53	94.1	88.5
Young and Forshaw, 1969	19.53	85.2	88.8
May and Bell, 1971	19.63	79.7	95.4
Saben et al., 1971a	20.15	87.5	93.1
Saben et al., 1971b	19.94	97.1	95.9
Yeh et al., 1971	--	70.0[a]	95.9
Young, 1972	--	84.7[a]	89.3
Morgan et al., 1975	19.78	89.8	92.5
Young et al., 1977	20.07	88.4	93.7
Just et al., 1978	19.53	83.5	91.9
NRC, 1982	--	93.8[a]	80.0
Average	19.77	86.4	91.3

[a]Estimated by using the average GE value.

Table 4. Reported energy utilization of soybean meal with hulls (dry matter basis).

Reference	GE	DE/GE	ME/DE
	kJ/g	– % –	
Tollett, 1961	18.70	85.6	97.8
Young et al., 1977	19.67	89.6	92.8
Saben et al., 1971b	20.36	89.9	95.2
NRC, 1982	--	81.8[a]	97.8
Average	19.58	86.7	95.9

[a]Estimated by using the average GE value.

heat. Several processes have been proposed for heating whole soybeans to allow feeding of soybeans. Included in our studies was a whole soybean meal that had been processed by extrusion to destroy the growth inhibitors. The GE of the whole, extruded soybeans was high, reflecting the oil content of the soybean. The digestibility of the energy of whole, extruded soybeans was greater than soybean oil and dehulled soybean meal (85.8%). The DE was efficiently converted to ME. As with soybean oil, the ME from whole soybeans was efficiently deposited as energy gain. The efficient utilization of ME from whole soybeans is also related to the deposition of the fat directly into adipose tissue, with a reduction in the energy required for the metabolism of ingested nutrients. Values reported by NRC (1982) indicate a higher digestibility of energy (88.8%) than observed in our studies, but poorer conversion of DE to ME of 87.3%. Other comparable values have not been reported.

Energy from dehulled soybean meal was efficiently utilized by the young pig. Energy from soybean hulls was not as well digested and was poorly utilized for tissue synthesis. Energy from soybean meal containing hulls was less digestible than energy from dehulled soybean meal, but the available energy was used as efficiently for tissue synthesis as dehulled soybean meal. About 80% of the energy from soybean oil was digested and was used efficiently for tissue synthesis. The digestibility of energy from whole, extruded soybeans was similar to the digestibility of the energy from soybean oil and was used efficiently for growth. The results of these experiments are similar to the average of the available values reported for growing swine.

NOTES

Richard C. Ewan, Professor, Department of Animal Science, Iowa State University, Ames, IA 50011.

Journal paper no. J-11590 of the Iowa Agric. and Home Econ. Exp. Stn., Ames, Iowa. Project No. 2374.

REFERENCES

Bailey, E. M. and R. C. Ewan. 1984. Evaluation of soybean hulls as an energy source for weanling pigs. (Abstrc.) J. Anim. Sci. 59 (Suppl. 1):258.

Diggs, B. G., D. E. Becker, A. H. Jensen and H. W. Norton. 1965. Energy value of various feeds for the young pig. J. Anim. Sci. 24:555-558.

Just, A., H. Jorgensen and J. Fernandez. 1978. The digestibility, ME and NE content of individual feedstuffs for pigs. Nat. Inst. of Anim. Sci., Rolighedsvej, Copenhagen.

May, R. W. and J. M. Bell. 1971. Digestible and metabolizable energy values of some feeds for the growing pig. Can. J. Anim. Sci. 51:271-278.

Morgan, D. J., D. J. A. Cole and D. Lewis. 1975. Energy values in pig nutrition. I. The relationship between digestible energy, metabolizable energy and total digestible nutrient values of a range of feedstuffs. J. Agric. Sci. 84:7-17.

NRC. 1982. Nutritional Data for United States and Canadian Feeds. United States-Canadian Tables of Feed Composition. Third Revised Ed. National Academy of Sciences-National Research Council, Washington, D.C.

Phillips, B. C. and R. C. Ewan. 1977. Utilization of energy of milo and soybean oil by young pigs. J. Anim. Sci. 44:900-977.

Saben, H. S., J. P. Bowland and R. T. Hardin. 1971a. Digestible and metabolizable energy values for rapeseed meals and for soybean meal fed to growing pigs. Can. J. Anim. Sci. 51:419-425.

Saben, H. S., J. P. Bowland and R. T. Hardin. 1971b. Effect of method of determination on digestible energy and nitrogen and on metabolizable energy values of rapeseed meal and soybean meal fed to growing pigs. Can. J. Anim. Sci. 51:427-431.

Stanley, D. L. and R. C. Ewan. 1982. Utilization of energy of hominy feed and alfalfa meal by young pigs. J. Anim. Sci. 54:1175-1180.

Tollett, J. T. 1961. The available energy content of feedstuffs for swine. Ph.D. Thesis. University of Illinois, Urbana.

Yeh, T., S. Lay and S. Chen. 1971. Determination of digestible and metabolizable energy of various feedstuffs. Ann. Res. Rep., Taiwan Sugar Corp.

Young, L. G. 1972. Energy value of some feeds for swine. p. 62-72. *In* Proc. 8th Ann. Univ. of Guelph Nutr. Conf. for Feed Manufacturers. Univ. of Guelph, Guelph, Ontario, Canada.

Young, L. G., G. C. Ashton and G. C. Smith. 1977. Estimating the energy value of some feeds for pigs using regression equations. J. Anim. Sci. 44:765-771.

Young, L. G. and R. P. Forshaw. 1969. Energy values of corn, barley and soybean meal for swine. (Abstr.) J. Anim. Sci. 29:1509.

Soybean Oil Utilization

NEW NON-FOOD USES OF SOYBEAN OIL

L. G. Beauregard and D. R. Erickson

Over the past three decades, there has been an ever-increasing demand for soybeans in the U.S., but this is due more to the demand for soybean meal than for soybean oil. As a result, with few exceptions over this entire period there is a history of excess of soybean oil on the market. This tends to depress soybean and soybean oil prices. To improve the market position for soybean oil, new uses must be found. Current usage in foods is at a level of some 4.5 billion kg per year, which represents over 80% of all vegetable oils and oil products currently consumed in foods. This already being the major share of the market, one cannot realistically expect an increase in current sales of soybean oil for food use, except perhaps for the small increment that might be expected from normal population increase. Our most promising approach is to seek new uses.

SOYBEAN OIL IN PESTICIDE FORMULATIONS

Pesticide applications represent the largest potential market, and thus are getting the most research attention at this time. Last February the American Soybean Association sponsored a workshop entitled "Ag Chem Uses of Soybean Oil". At this workshop 33 research papers were presented. The compilation of research data was very impressive and gives a fair cross-section of the research activities currently in progress on this new use.

Pesticide Carriers or Solvents

Most pesticides are applied as liquid sprays. As pure or technical grade products, they are usually solid or very viscous liquids. They cannot usually be handled in this manner by pesticide applicators, and so must be diluted to a standardized concentration. This initial dilution involves dissolving or dispersing the pure or technical grade material in a suitable solvent or carrier.

The solvent or carrier must fulfill several requirements. Firstly, the pesticide must be soluble or dispersible in it. Secondly, the carrier must usually be phytobland, or non-phytotoxic, causing no unwanted damage to plants or crops on which the pesticide is sprayed. Thirdly, it must be competitively priced.

Except for the few cases where pesticides are soluble in water, the most extensively used carriers have been special phytobland solvents derived from petroleum. These include xylene, toluene, and special grades of mineral oil. Petroleum oil carriers are expensive, however, so in order to keep the cost of the spray formulation as low as possible, the manufacturer uses only as much of these oils as needed to sell his products in a form that is readily usable by the applicator.

In most instances, soybean oil could replace solvents derived from petroleum. It is frequently less expensive, and it is readily available throughout the country. Several pesticide manufacturers are currently evaluating formulations using soybean oil as a carrier, and there is no doubt that in the near future there will be a substantial increase in the use of soybean oil for this purpose.

Spray Adjuvants

Another way that soybean oil can be used in pesticide formulations is as a spray adjuvant. Various mixtures of oils and surfactants are often added to water-base solutions, particularly in post-emergence herbicide applications. These materials, called adjuvants, are used to alter the physical characteristics of tank mixes to make them act in a particular way with respect to the target.

Many herbicide manufacturers recommend the use of an oil adjuvant, commonly referred to as a "crop oil," with their products. Traditionally, such crop oil adjuvants have been mixtures of phytobland mineral oil and a surfactant, but much recent research has indicated that soybean oil can be just as effective as mineral oil for this particular application. No concrete statistics are available as to how much soybean oil is currently used in this way, but it is already several million kilograms, and increasing markedly every year.

UVL Pesticide Application Systems

Ultra-low volume (ULV) application systems were first commercialized in the U.S. about 15 years ago. With UVL the total volume is 2 to 4 L/ha. As the volumes are small, pesticides that are soluble in oil can be diluted to usage strength with soybean oil, and many pesticides have been shown to be more effective when used this way; i.e., without the use of additional water.

In contrast to conventional high-volume systems, where the farmer almost drenches his field with his spray mixture, small volumes applied in ULV systems greatly diminish chances for adequate coverage, unless size of the droplets is very small.

Advantages of using soybean oil in ULV systems include:
1. Most pesticides are sufficiently soluble in soybean oil to permit direct usage without the need for more expensive surfactants or solubilizers.
2. Because soybean oil (unlike water or mineral oil) does not evaporate, the droplets remain the same size between the time they leave the spray nozzles and when they reach their intended target. This is an important characteristic both for controlling wind drift and application rate.
3. Unlike in the case of mineral oils, there have been no reports of phytotoxicity to plants due to use of soybean oil.
4. Soybean oil is more polar than mineral oil, which is non-polar. This difference generally makes soybean oil a better solvent for pesticides than mineral oil, and makes it easier to achieve the relatively high concentration required for ULV systems.

Market Potential for Soybean Oil with Pesticides

In the United States, over 140 million hectares are planted in major crops each year. Based on data obtained from various sources, and considering multiple applications of different pesticides on each hectare, it has been calculated that the pesticide formulations used require an average of about 55 kg of oil per hectare.

Most formulations now using petroleum-derived oils of various types could use soybean oil, and with proper least-cost formulation approaches, soybean oil could be price-competitive. An estimate of the current U.S. market potential for soybean oil in pesticides at around 450 million kilograms per year would be considered conservative. This represents about 10% of the soybean oil now being consumed for foods, and certainly is an attractive potential new market.

SOYBEAN OIL AS A DUST SUPPRESSANT IN GRAIN ELEVATORS

Grain dust is *dangerous* and it is *expensive*. In the U.S., between 1964 and 1973 there were 2900 grain elevator fires and 64 explosions—all attributable to grain dust. The situation got even worse after that time when grain elevators were forced to seal up their operations in response to pressure to clean up the environment.

On the economic side, grain dust is expensive because it is essentially a part of the grain flying away, and this represents a loss in weight. For each handling, and under normal conditions, it has been reported that 0.14% of the weight of the grain is lost as dust. This means that, even with a minimum handling, an average of 2.8 kg/t of grain is lost. Anything that would reduce this loss, therefore, could result in significant savings for elevator operators. In fact, calculations already made, based on work done to date, indicate that the savings resulting from the reduced weight loss will more than pay for the oil treatment.

Historical

The use of oil to reduce the dustiness of grain during handling is not a new idea. In 1952, Moen and Dalquist patented the use of an oil emulsion to reduce the dustiness in grain[1]. Even much earlier, manufacturers of dry mixes for human food and animal feeds noted that, when fats and oils were used in these mixtures, dust generated by the mixing and handling operations was greatly reduced. It was not until the end of the decade of the '70s, however, that the concept of using vegetable oils for suppressing dust in grain elevators began to emerge.

In the late '70s, work was begun at the U.S. Grain Marketing Research Laboratory in Kansas by Lai et al. (1981) on the use of oils. Laboratory tests were also initiated in Canada by Hsieh et al. (1982), who used rapeseed oil. These Lai et al. studies initially included a wide variety of materials, such as mineral oil, water, vegetable oils, lecithin, in a variety of combinations. Water was shown to be quite effective, but its effect was very short-lived, and since it evaporates, the grains must be sprayed each time they are moved. Mineral oil had a better residual effect, but it too lost its effectiveness with time, so that after a time, grains had to be sprayed again in order to maintain a reduced level of dustiness. Of the various treatments, only soybean oil and lecithin, and combinations of these two, showed long range residual dust suppressing effects. This is understandable since they do not evaporate.

The work of Hsieh et al. in Canada was more limited in that only rapeseed oil was used, but that work tended to confirm the results of Lai et al. In addition, they showed that the treatments resulted in no subsequent adverse effects on wheat products. Flours used in baking tests turned out to be comparable to the controls.

Industrial Elevator Tests

Following the aforementioned laboratory and pilot plant tests in Kansas, Lai et al. (1982), along with the National Grain and Feed Association, initiated a series of industrial trials at a grain elevator in Fostoria, Ohio. These tests were designed to study the effectiveness of water, mineral oil, and deodorized soybean oil on large scale handling operations of corn, wheat, and soybeans.

In these tests, all additives were sprayed on the grain as it was transferred from one storage bin to another. The most convenient set-up was found to be a system of two sprays continuously applying the dust suppressant on both the top and the underside of the grain stream at the first belt transfer point after the grain left the storage bin. On days when the weather was extremely cold, it was necessary to heat the additives to achieve efficient spraying.

These tests corroborated quite well the prior smaller scale experiments. Dust could be suppressed by adding between 0.17 to 0.3% water to the grain stream, but its effectiveness was only temporary. Mineral oil applied at rates of 0.02 to 0.08% was very effective, but its ability to control dust was gradually lost, so that after three months the dust levels were once again similar to those of untreated grains. Soybean oil was applied at levels of 0.03 to 0.10%. At the 0.03% rate, soybean oil reduced the dust concentration during handling by more than 90%, and this suppressing effect was retained for the entire period studied—approximately one year.

The grains treated in these industrial tests were subsequently utilized by the Mennel Milling Company and A. E. Staley at their plants also located in Fostoria, Ohio. They were monitored throughout the industrial utilization processes, and in no instances were any differences found between these and untreated grains.

Current Studies

Since experience with soybean oil treatment of grains for dust suppression is still limited, a considerable amount of testing is still needed before the widespread adoption of this technique can take place. ASA and groups originally involved in the Ohio tests are nonetheless very enthusiastic, and are currently involved in a series of further studies aimed at answering some of the still-unanswered questions. The information now being obtained includes:

1. Determining the optimum level (based on economics, grain quality, and minimum dustiness) of oil and/or lecithin that should be added during elevator operations, relating this to measurable characteristics (dustiness, percentage breakage, etc.).
2. Determining the threshold limits of odor detection for soybean oil, lecithin, and combinations thereof, on wheat and corn.
3. Evaluating the cost, uniformity, handling properties, etc. of different methods of adding several levels of soybean oil, lecithin, and combinations thereof, to grain.
4. Measuring the effect of soybean oil and/or lecithin on grain grade, odor, germination, and physical and rheological characteristics of treated grain.
5. Assessing the end-product quality of oil- and/or lecithin-treated samples of corn, soybeans and wheat.
6. Assessing the effect of oil treatment on machinery (belts, conveyors, and bag filters), and on the collection of dust on walls and bins, during commercial operation for up to a year on a continuous basis.
7. Determining the explosiveness of dust collected during the handling of grain treated with oil and/or lecithin.
8. Evaluating the effect of oil and/or lecithin treatments on the susceptibility of grain to attack by insects and molds.
9. Developing methods to detect whether grain has been treated with oil and/or lecithin.

SOYBEAN OIL IN CHEMICAL PRODUCTS MANUFACTURING AND
AS AN ALTERNATE FUEL

With current concern over depletion of non-renewable energy and raw material sources, a lot of interest has been shown in recent years in the use of soybean oil as an alternate raw material for the manufacture of a wide variety of chemical products, and as an alternative fuel to diesel oil. In the early 1960s natural fatty chemicals held a large share of the market for fatty chemicals used in the detergent, toiletries, and cosmetic industries, but a large part of this market was lost when ethylene chemistry enabled companies to turn out fatty alcohols suitable for making synthetic surfactants that were taking over a major part of the half-billion dollar detergent market. Even when prices of hydrocarbons soared in the '70s, synthetic materials maintained their edge over natural products, since their prices were still lower than fatty materials of vegetable and animal origin.

The synthetic vs. natural picture changed drastically during the big round of petroleum price hikes between 1978 and 1980. At the same time, the supplies of natural products shot upward, and while the flow of petroleum products can be shut off at the whim of foreign governments, that of natural fatty materials must remain relatively constant. The consensus is that there will be plenty of natural fats and oils around for a long time, and this is strongly reinforced by the consistent upward trends in worldwide production of soybean and palm oil. These positive factors have sparked a round of expansions by makers of fatty chemicals, particularly fatty alcohols, which make up the bulk of the detergent industry.

With regard to the use of soybean or other vegetable oils as alternate fuels for diesel engines, research thus far indicates that pure soybean oil can cause damage to motors when used for extended periods. Blends of soybean oil with diesel oil can be used, with effects on the motor in inverse proportion to the proportion of soybean oil in the mixture. Small proportions of soybean oil (10 to 20%) in diesel fuels apparently cause no harmful effects on the engines over the short term.

Overall, such industrial applications for soybean oil are primarily governed by economic factors. At times, and in areas where petroleum products are plentiful and cheaper than soybean or other vegetable oils, the thrust to use these natural materials will not be strong. This position can be the reverse in areas where petroleum is scarce, and vegetable oils are plentiful. Brazil, for example, has a lot of soybean oil, and a lot of ethanol from its recently developed fermentation industries. It wouldn't be surprising, therefore, if Brazil and other petroleum-deficient countries were soon to use vegetable oils as fuels and as feedstocks for industrial chemicals.

CONCLUSION

These are areas of extensive activity for the American Soybean Association, and all progress made in opening such new markets will contribute significantly towards reducing what, until now, has been an almost constant surplus of soybean oil on the U.S. market.

NOTES

L. G. Beauregard and D. R. Erickson, American Soybean Association, 777 Craig Road, St. Louis, MO 63141.

[1]R. Moen and M. S. Dalquist, 'Grain handling', US Pat. 2,585,026 (1952).

REFERENCES

Hsieh, F., J. K. Daun and K. H. Tipples. 1982. J. Am. Oil Chem. Soc. 59:11-15.

Lai, F. S., C. R. Martin and B. S. Miller. 1982. Examining the use of additives to control grain dust. A report to National Grain and Feed Association, Washington, D.C.

Lai, F. S., Y. Pomeranz, B. S. Miller, C. R. Martin, D. R. Aldis and C. S. Chang. 1981. Adv. Cereal Sci. Tech. 4:237-337.

SOYBEAN OIL IN FOODS

Savinay Patel

Every single day, Americans consume roughly 11 million kilos of soybean oil. An additional 3.6 million kilos of other fats and oils are also consumed daily. At this rate of usage, 225 million people use roughly 450 million kilos of fats and oils a month, or roughly 5.5 billion kilos a year. These other fats and oils include: cottonseed, corn, peanut, and sunflower oils; imported palm and coconut oils; and animal fats such as butter, lard and tallow.

Soybean oil has been called a "king without a crown" and a "hero without medals". Like other vegetable oils, soybean oil processes well and can be hydrogenated to meet the exacting demands of customer specifications for color, flavor, melting characteristics, saturated polyunsaturated ratios, and product performance criteria. The success of soybean oil resulted primarily from its being available in large quantities at a competitive price. Over a period of time, as the demand for fats and oils grew at a rapid rate, soybean oil responded to the needs at a competitive price, while other vegetable fats and oils did not.

SOYBEAN OIL PROCESSING

There are about 84 soybean processing plants, mainly located in the soybean-growing regions of the country. Many of these plants are highly capital-intensive operations, with tremendous storage capacities and crush capacities. According to the SRI International Report (1982), both crushing and refining operations in the U.S. have over-capacities. The hydrogenation capacity is estimated at about 4 billion kilos per year.

To extract the oil, soybeans are cracked, dehulled, and pressed into flakes which are extracted with hexane. During these steps and subsequent refining, the beans and oil are protected from excessive heat, trace metals, and exposure to oxygen and light, to maintain high oil quality. After removing the hexane, the crude oil is treated with alkali. During this step, free fatty acids, some colored material, metallic pro-oxidants, and residual phosphatides are removed.

Next, the oil is treated with bleaching earth at high temperatures under vacuum. This process removes pigments such as carotenoids, xanthophylls, and chlorophyll. It also removes traces of moisture, soaps, phosphatides, and possibly some trace metals.

By reaction with hydrogen, a series of hydrogenated "basestocks" are made and they are judiciously selected and blended to make a variety of margarine oils and shortenings. The hydrogenation reaction is controlled, directed, and made selective or nonselective by manipulating temperature, hydrogen gas pressure, catalyst concentration, and agitation rates. To manufacture salad and cooking oils, soybean oil is slightly hydrogenated and "winterized".

The oil is deodorized by a high temperature, high vacuum, steam distillation process. Traces of citric acid are added to the oil to chelate trace metals such as iron and copper, to minimize their pro-oxidant activity. To minimize further contamination with trace metals and oxidation of the oil, all subsequent pipings, equipment, and storage tanks are usually of stainless steel construction, and the oil is sparged with nitrogen.

SOYBEAN OIL USES AND TRENDS IN FOODS

The 5.5 billion kilo U.S. fats and oils market is divided into three main market segments: bulk ingredients, food service and retail products, as shown in Figure 1.

Ingredient Market

This is a commodity market where purchase decisions are influenced by an oil's availability, price, and quality, and the service supplied by the manufacturer. Bulk shipments are made in 18,000-kg trucks, or by 70,000-kg jumbo railcars. The demand depends upon the sales volumes enjoyed by numerous products using soybean oil. Overall, the fats and oils needs in this segment are growing. The chief uses in this category are for salad dressing and mayonnaise; margarines; baked goods; the cookies and crackers industry; snack foods; breaded/fried/frozen foods such as vegetables, fish, poultry, and meats; and the confectionary industry.

Food Service Market

This segment includes fast food outlets, restaurants, hospitals, schools, cafeterias, the airline industry, etc. It is a growing market that reflects the changing food habits and lifestyle of America. This market is served by a well-organized distributor network and is quite competitive. The fats and oils are used in packaged forms such as cubes and plastic or metal containers.

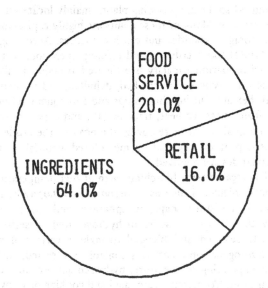

Figure 1. Fats and oils market segments—U.S.A.

Retail Market

This market caters to household needs through grocery stores and supermarket chains. The fats and oils products sold here are individually packaged, smaller-sized units of margarines, salad dressings, salad and cooking oils, and shortenings. Profit margins are higher here, and the market is dominated by branded and private label products. This market is growing very slowly.

UTILIZATION TRENDS

The consumption pattern of soybean oil use in the salad and cooking oil category is rising dramatically (Figure 2) as a result of changing dietary patterns and consumer preferences. This pattern will continue, indicating further growth in this category. Soybean oil use in shortening is rising also, but at a slower rate. The rising trend in margarines and spreads seems to have peaked out. Is this a saturated market or the influence of temporary dairy surplus and give-away programs? It is hard to tell at this time. Overall, soybean oil's share of the total fats and oils used in 1982 was 74% (Figure 3). The remaining 26% is shared by: cottonseed, corn, and peanut oil combined, 11%; animal fats, 8%; and imported coconut and palm oils, 6%. In shortenings, the interchangeability of fats and oils makes for direct competition among different types. Soybean oil usage dominates specific product categories: margarines, 85%; salad and cooking oils, 80%; and shortenings, 67% (Hazera, 1983).

ASSESSMENT AND CHALLENGES

Dietary and health trends to reduce calories, animal fats, fats and oils intake, cholesterol, and saturated fats are taking shape. The U.S. population growth rate is

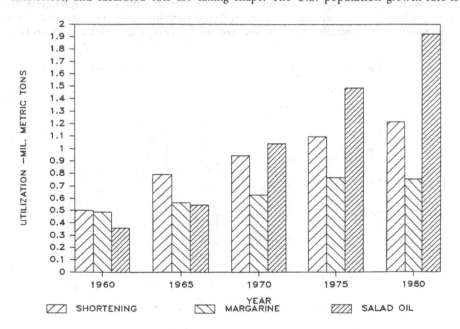

Figure 2. U.S. soybean oil utilization (American Soybean Association, 1983).

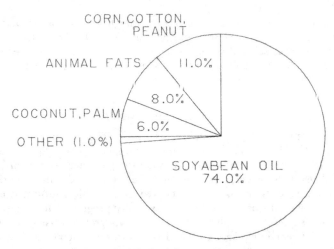

Figure 3. Percent share of U.S. fats and oils in foods in 1982 (Hazera, 1983).

stabilizing below 1% annually. At the current soybean oil usage rate, a 1% population growth rate is estimated to increase soybean oil demand roughly by 40 to 45 million kilos a year. Quick gains for soybean oil may come only at the expense of tallow, lard and imported, expensive coconut oil.

The competition from sunflower oil in the U.S. market will increase with time, as sunflower oil becomes more available and is promoted. On the international scale, soybean oil will compete against the increased availability of palm oil from Malaysia and rapeseed/Canola oil from Canada. The stiffer competition will be from Brazil and Argentina for a direct share of the soybean oil markets. Figure 4 shows the role of

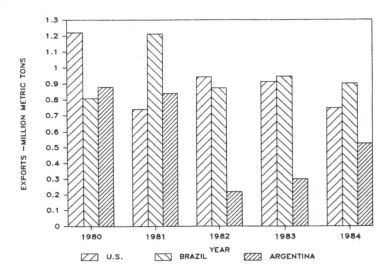

Figure 4. Soybean oil exports by key exporting nations (USDA, 1984).

Brazil and Argentina in exporting soybean oil. Brazil exports as much as the United States, and Argentina is competing strongly in the export market. Overall, the demand for soybean oil in the world market is increasing.

Increased demands for liquid soybean oil as salad and cooking oil will continue to face flavor and oxidative stability problems, due to the 7 to 10% linolenic acid it contains. The industry's expedient, and very effective, approach to this problem has been to "react-it-out" by hydrogenation. Hydrogenation, in turn, produces positional and configurational isomers in the processed oils. The metabolic fate of these isomers is still not clearly understood and requires some objective research to clarify their implications for consumers' health. Conventional breeding programs have not been sufficiently effective in reducing linolenic acid in soybean oil. Newer approaches and answers may have to come from the modern tools of biotechnology. Interesterification processes can provide alternatives to hydrogenation if positional isomers prove to be a problem.

To maintain the cost-effective edge of soybean oil compared to other oils, new industrial applications of soybean oil may have to be developed. More energy-efficient processes may have to be developed and creatively applied to soybean oil processing.

In the United States, we have the technology and capacity to produce and process large, incremental amounts of soybeans and oils; more than we need or can use domestically. It will become imperative that we explore markets for "soy-oil-intermediates" (hydrogenated basestocks) overseas, shifting our focus from large scale export of soybeans and crude oils alone to more value-added opportunities.

NOTES

Savinay Patel, Central Soya Company, Inc., Fort Wayne, Indiana.

REFERENCES

A.S.A. 1983. Soya Bluebook. American Soybean Association, St. Louis.
Hazera, J. 1983. Changing uses of fats and oils in the United States 1971-1982. p. 14-16. *In* Oil Crops. Outlook and Situation Report. USDA—Economic Research Service, Washington, D.C.
SRI International. 1982. CEH Marketing Research Report, Fats and Oils Industry Overview. October, 1982. SRI International, Inc. Menlo Park, California.
USDA. 1984. Oilseeds and Products. U.S. Department of Agriculture, Foreign Agric. Service. Washington, D.C. June 1984.

NUTRITIONAL CONSIDERATIONS IN SOYBEAN OIL USAGE

E. A. Emken

Dietary fats have a number of nutritional, physiological, and psychological roles. Nutritionally and physiologically, fats supply about 42% of the total calories consumed in the U.S. They provide the body with essential fatty acids, act as carriers for the fat-soluble vitamins (A, D, E, K), provide the body with a convenient source of stored or reserved energy, serve as one of the major components of cell membrane structures, and are precursors to the potent biologically active prostaglandins. Psychologically, fats stimulate appetite and release of digestive enzymes by providing pleasant flavors, odors, and texture to foods, and they enhance the feeling of satisfaction or satiety.

CONSUMPTION OF SOYBEAN OIL

For 1980, the disappearance of visible animal and vegetable oils and fats in the U.S. was estimated at 5.8 billion kg. Based on 1980 USDA statistical data, Rizek et al. (1983) estimated average total fat consumption at 13.5 billion kg or 169 g per person per day for 1980. However, actual consumption is considered to be about 23% (40 g) less due to waste and/or losses. Of the 169 g of total edible fat utilized, 72 g (42.6%) was visible fat, of which about 13.3% is estimated to be supplied by unhydrogenated soybean oil (SBO) and 39.6% by hydrogenated soybean oil (HSBO) plus small amounts of other hydrogenated vegetable oils from corn, peanut, cotton, and sunflower (Table 1). From the combined data for SBO and HSBO, it is apparent that soybeans are one of the most important single sources of dietary fat in the U.S., and they provide roughly 20% of the world production of edible oils.

Table 1. Dietary sources of energy for the U.S. population (Rizek et al., 1983).

	Total Calories	Visible fat		Non-visible fat		Total fat	
	%	g/person d	%	g/person d	%	g/person d	%
Butter and animal	24.5	21.2	29.4	77.4	79.8	98.6	58.3
Soybean oil	3.3	9.6	13.3	3.7	3.8	13.3	7.9
Hydrogenated oils	9.8	28.5	39.6	11.0	11.8	39.5	23.4
Vegetable oils	4.4	12.7	17.7	4.9	5.1	17.6	10.4
Total fat	42.0	72.0	42.6	97.0	57.4	169.0	100.0
Total carbohydrate	46.0						
Total protein	12.0						

Because hydrogenated soybean oil alone contributes more than 23% of the total fat intake and about 10% of the total caloric intake, questions have been raised concerning its nutritional value. However, these concerns do not involve the actual hydrogenation reaction that converts alkene bonds in unsaturated fatty acids to saturated or alkane bonds. The questions are instead based on the isomerization reaction, which is a secondary or side reaction that occurs during hydrogenation and produces the new fatty acid structures, or isomers, that are not present in unhydrogenated SBO. As a result of isomerization, a variety of mono- and polyunsaturated isomers are formed which consist of both *cis* and *trans* positional isomers. A typical HSBO sample contains the 7 through 14 positional *cis* and *trans* 18:1 isomers and a number of *c,c*-18:2, *c,t*-18:2, and *t,c*-18:2 positional isomers.

SOURCES AND LEVELS OF ISOMERS IN DIETARY FATS

In the U.S., hydrogenated vegetable fats are the main dietary source of isomeric fats, although ruminant and milk fats also contain isomeric fats. Microbial hydrogenation in the intestinal tract of nonruminants is apparently the source of isomeric fats identified in tissues of mice (Biesel et al., 1983) and kangaroos (Fogerty and Ford, 1983).

A wide variety of food products are reported to contain varying levels of hydrogenated vegetable oils. The *trans* isomer contents for various foods reported in a number of references are summarized in Table 2. These data illustrate the wide distribution and levels of isomeric fats in common food products. This knowledge of *trans* content of foods can also be used to estimate individual *cis* and *trans* positional isomer content, because the distribution of these isomers is related to the total *trans* percentage.

The total *trans* and *cis* positional isomer consumption data in Table 3 have been estimated from typical positional isomer distributions in commercial samples, and from data in Table 1 after assuming a 23% loss due to waste. This value of ca. 7 g per person per day of total 18:1 isomers is consistent with estimations of 3 to 9 g of *t*-18:1 others have reported based on actual intake (Akesson et al., 1981) and disappearance data (Fuchs and Kuivinen, 1980; Heckers et al., 1979). Adipose tissue reflects the long-term fat composition of the diet, and its *trans* content can be used also to estimate total *trans* fatty acid intake (Beynen et al., 1980; Heckers et al., 1979).

Ohlrogge (1983) has summarized data for the *trans* fatty acid content of human adipose tissue. Values reported by several groups range from 1.0 to 11.6%. These

Table 2. The *trans* and total fat content of food products containing HSBO.

Food item	Approximate fat content	*trans* content (range) of fat
	%	%
Salad and cooking oils	100	7.0- 9.4
Margarines	80	10.0-29.0
Shortenings	100	8.7-35.4
Breads/rolls	–	1.8-23.6
Cookies	15	3.6-34.2
French fries	18	3.3-35.1
Pastries	17	3.0-23.4
Animal/dairy fat	100	0.3- 6.6

Table 3.   Estimated daily consumption of specific positional octadecenoic acid isomers[a].

	Positional octadecenoic acid isomer, g							
	7	8	9	10	11	12	13	14	Total
c-18:1	0.12	0.18	4.94	0.29	0.43	0.54	0.12	0.02	6.7
t-18:1	0.21	0.50	1.20	1.18	0.97	0.63	0.37	0.22	5.3

Total isomer content minus $9c$-18:1 = 7.0 g

[a]Based on estimated content of c-18:1 and t-18:1 in hydrogenated soybean oil of 25% and 20%, respectively, and on an estimated actual daily per capita consumption of 30.4 g hydrogenated vegetable oil.

values are consistent with estimations of dietary *trans* fatty acid intake. The distribution pattern of positional isomers in adipose tissue (see Figure 1) also resembles very closely the pattern in hydrogenated oils rather than that of butter fat. Correlation coefficients indicate that hydrogenated vegetable oil is the source of approximately 95% of the isomers, and the remaining 5% is from butter plus other ruminant fats. Thus, these data establish reasonably well the level of consumption in human diets and the source of isomeric fats in adipose tissue.

Human adipose tissue data provide indirect evidence that absorption does not selectively discriminate against any specific fatty acid isomer. Analysis of the chylomicron lipids from humans fed deuterium isotope-labeled fatty acid isomers provides a direct measurement of absorption for several 18:1 isomers present in HSBO. Chylomicron triglyceride (TG) data indicate absorption of the 18:1 isomers in hydrogenated oils is essentially identical to oleic acid (Emken, 1983; Emken et al., 1984). These data agree well with results from animal experiments which used both labeled and nonlabeled isomeric fats.

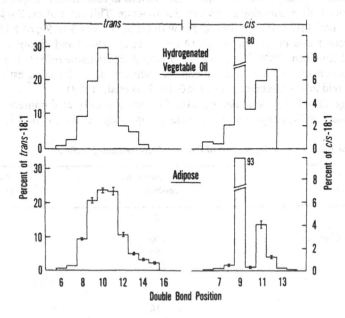

Figure 1.   Average double-bond distribution in *trans* and *cis* octadecenoate fraction of human adipose tissue and hydrogenated soybean oil.

COMPOSITION AND TURNOVER OF FATTY ACID ISOMERS

The *trans* fatty acid content of rat tissue lipids is a function of the percentage *trans*, total weight percentage of fat in the diet and the length of time the diet is fed. In rats, the *trans* fatty acid content of tissue lipids was higher than dietary levels. Specifically, *trans* levels in adipose, heart and liver lipids were 17.9 to 26.3%, 16.8%, and 7.3 to 7.5%, respectively, when diets containing 7.5 to 7.7%, 9.6%, and 2.3 to 3.2% *trans* fatty acids were fed for 3 to 6 months (Egwin and Kummerow, 1972; Bonaga et al., 1980; Hoy and Holmer, 1979; Moore et al., 1980). In comparison, the average American diet contains about 3.1% *trans*, and human adipose, heart, and liver tissue contains 3.4%, 2.8%, and 1.6% *trans*-18:1 (Ohlrogge, 1983).

The use of deuterium labeled fatty acid isomers has provided information for the distribution of several 18:1 isomers into human plasma lipid classes. Selectivity values for 18:1 isomers, relative to 9c-18:1, are summarized in Figure 2 for various human plasma lipid classes. They indicate that 18:1 isomers are weakly excluded relative to oleic acid for most lipid classes, but strong exclusion is noted for the cholesteryl ester (CE) fraction (Emken, 1983; Emken et al., 1984).

The selectivity values for those isomers in Figure 2 were also obtained for the 1- and 2-acyl position of phosphatidylcholine (PC) and indicate preferential exclusion of these 18:1 isomers from the 2-acyl position, with the exception of 12c-18:1. For the 1-acyl PC position, selective incorporation of all the isomers occurs. These data indicate the 12c-18:1 isomer is unique in terms of its recognition of phosphatidylcholine acyltransferase. Also, lecthin:cholesteryl acyltransferase is very sensitive to double-bond position and configuration, as indicated by the cholesteryl ester data. Thus, human data are qualitatively in agreement with data from a large number of animal and in-vitro studies. One exception, however, is the difference between human (Emken et al., 1984) and rat data reported by Wood (1979) for the 10c- and 10t-18:1 isomers.

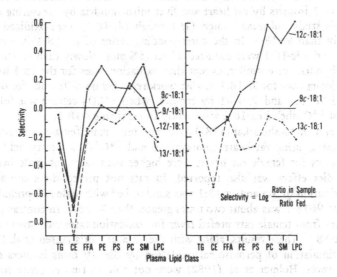

Figure 2. Selectivity values for incorporation of deuterium labeled *cis* and *trans* positional octa-decenoic acid isomers by human plasma lipid classes. Selectivity values are based on the log of the isomer/9c-18:1 ratio divided by the ratio in the fed mixture.

TURNOVER

Turnover or disappearance of *trans* isomers has been reported for various rat tissue lipids. When rats, which had accumulated *trans* acid contents of 5.9% and 15% in heart and liver TG and PL, were switched to a 0.4% *trans* diet, the levels of both 18:1 and 18:2 *trans* isomers returned to baseline values in 8 to 12 weeks (Moore et al., 1980). Lawson et al. (1982) reported the *t*-18:1 content of sciatic nerve and brain PL from second generation rats transferred from high to low *trans* acid diets declined from ca. 1.2% to ca. 0.6% over a 10-week period, whereas % *trans* in liver and heart PL declined from 10 to 15% to ca. 2% during a 6-week period. Kaufman et al.(1961) observed similar times for the *trans* content of human milk samples to reach base line values after removal of *trans* isomers from the diet.

The composition of human tissue lipids, compared to hydrogenated soybean oil, indicates that turnover rates of individual fatty acid isomers are similar and that accumulation in lipid classes from various organ tissues does not occur (Ohlrogge, 1983). These studies, in combination with relative turnover data based on deuterium-labeled fat, provide solid evidence that removal or depletion of isomeric fats from tissues generally proceeds without difficulty, with nerve and brain being a possible exception.

OXIDATION OF FATTY ACID ISOMERS

The majority of the fat consumed is oxidized to provide energy for cell functions. Muscle tissue, in general, obtains a large portion of its energy requirements from fats. Heart muscle, in particular, obtains 50% or more of its energy from dietary fat (Crass, 1972), thus it is important that fatty acid substrates be readily oxidized by the heart. Lawson and Holman (1981) have determined the relative oxidation of *cis* and *trans* positional 18:1 isomers by rat heart and liver mitochondria by measuring oxygen uptake. In this study, all *trans* isomers (7*t* through 14*t*-18:1) were oxidized 20 to 35% more slowly than 9*c*-18:1. In the corresponding series of *cis* 18:1 isomers, the 8*c*-, 10*c*-, 12*c*-, and 14*c*-18:1 were oxidized 15 to 35% more slowly than 9*c*-18:1. Data for liver mitochondria were similar, except that oxidation rates for the 18:1 isomers were 5 to 15% slower than for 9*c*-18:1. Earlier work reported no difference for oxidation of ^{14}C-labeled 9*t*-18:1 and 9*c*-18:1 by rat liver when $^{14}CO_2$ release was followed, but oxidation of ^{14}C labeled *t,t*-18:2 was slightly faster than *c,c*-18:2.

Menon and Dhopeshwarkar (1983) have reported a male/female difference for total oxidation rates, using rat heart homogenates and ^{14}C-labeled fatty acid substrates. Oxidation rates for female rat hearts were higher than rates for male hearts, and a significant diet effect was also reported. In rats not prefed diets containing *trans* isomers, oxidation of 9*c*- and 9*t*-18:1 was similar, but with male rats prefed *trans* fats, oxidation of 9*t*-18:1 was about two times greater than 9*c*-18:1. In contrast, with heart homogenates from female rats prefed *trans* fat, oxidation rates were lower for 9*t*-18:1 than for 9*c*-18:1. These results, along with the data of Christiansen et al. (1979), suggest that stimulation of peroxisomal oxidation by dietary *trans* isomers may be involved. However, Holmer et al. (1982) were not able to find evidence for increased peroxisomal oxidation in studies with male rats fed hydrogenated vegetable and marine oils. Comparison of oxidation rates for 9*c*- and 9*t*-18:1-^{14}C by male and female human and rat heart homogenates indicates a significant species difference (P<0.05) for the 9*c*/9*t*-18:1 ratios. The 9*c*/9*t*-18:1 ratios for male and female rat hearts were 1.88 and 1.71, respectively, and 1.08 for both male and female human hearts. Thus, the human heart data, in contrast to the rat heart data, indicate human heart muscle would not be predicted to have difficulty in utilizing HSBO for its energy requirements.

DESATURATION/ELONGATION/RETROCONVERSION OF ISOMERIC FATS

Fatty acid isomers in HSBO that are not oxidized for energy needs are either deposited in adipose tissue and cell membrane structures or converted to other structures via three primary reactions—desaturation, elongation, or retroconversion. These reactions involved in the interconversion of isomeric fats have received considerable attention and have been previously reviewed in detail (Emken, 1984; Holman et al., 1983).

Ample evidence is available to suggest that after positional 18:1 isomers are retroconverted (chain shortened) to 14:1 and 16:1 fatty acids, further retroconversion is slower than when 9c-18:1 is the substrate. However, rates appear to differ, and depend on the configuration and position of the double bond. The result is that 16:1 levels are often higher in tissues after HSBO or individual 18:1 isomers are fed, and the 16:1 isomers present reflect the 18:1 isomer composition of the diet.

Desaturation and elongation of 18:1 HSBO isomers to 18:2 and 20:1 isomers have been reported, but the reaction rates are relatively slow and the products are unlikely to have a significant physiological impact (Holman et al., 1983). For example, rat studies indicate that relatively large amounts of t,t-18:2 (2.5 to 5% by weight of diet) must be fed in order to induce observable physiological changes in tissue lipids and cell function (Kinsella et al., 1981).

Isomeric fats also appear to influence the desaturase and elongase enzyme reactions by depressing the conversion of 18:2 to 20:4 by at least two possible mechanisms. The net result is a lower concentration of 20:4 in tissue lipids, and is likely to be the basis for the reported increased 18:2 requirements and lowered 20:4 levels when hydrogenated oils are included in rat diets.

BIOLOGICAL AND PHYSIOLOGICAL EFFECTS OF ISOMERIC FATS

Hydrogenated oils and individual fatty acid isomers have been utilized in both in-vitro and in-vivo studies to determine the influence of isomeric fatty acids on a wide variety of physiological and biological parameters. Early studies with HSBO involved long-term feeding studies with mice, rats, and single-cell organisms to determine the effect of isomeric fats on growth, longevity, reproduction, organ size, and cell division. Except for single-cell mutants, which were not capable of fatty acid synthesis, no differences between control and hydrogenated fat diets were observed.

The influence of HSBO and other hydrogenated vegetable oil diets on human serum cholesterol, triglyceride, and phospholipid levels have been reported for several different groups of subjects. In these studies, varying levels of total fat and percentage trans were fed. These data have been reviewed in detail by Applewhite (1981) and Emken (1979). The general conclusion is that hydrogenated fats have little or no impact on serum lipid levels, provided sufficient linoleic acid is present in the diet.

Related to these human studies are animal experiments that followed the development of arterial lesions and cholesterol levels in rabbits, monkeys, and swine fed hydrogenated vegetable oils. Kritchevsky (1983) and Applewhite (1981) have reviewed these studies and conclude that no statistically significant increases in arterial lesions occur when hydrogenated fat diets are compared to diets containing unhydrogenated fats, provided adequate levels of linoleic acid are present. Comparison of recent experiments with swine to earlier results appears to support this concept (Royce et al.,

1984).

Thomas et al. (1981) have compiled data for the *trans* fatty acid content of adipose tissue from 231 British subjects who died of coronary heart disease (CHD), and from controls who died of other causes. The *trans* content of adipose tissue was reported as not significantly correlating with CHD. An additional point is that CHD in the U.S. has decreased by about 20% since 1968, but consumption of HSBO has not decreased during this period.

BIOLOGICAL ROLE OF LINOLENIC ACID

Linolenic acid ($9c,12c,15c$-18:3) has a double bond located between the third and fourth carbons from the methyl end of the acyl chain. Fats with this structural configuration are classified as omega-3 (ω3) or n-3 fatty acids. The omega-6 fatty acids are not interconverted to omega-3 fatty acids. Thus, the essential fatty acid, linoleic acid ($9c,12c$-18:2), is not a precursor for the long chain 20:5 ω3, 22:5 ω3, and 22:6 ω3 fatty acids present in animal and human tissue.

Linolenic acid, which must be obtained from dietary sources, is the accepted precursor for these ω3 fats, but explicit biological roles or functions have not been clearly demonstrated for mammals. Relatively high levels of a linolenic acid elongation-desaturation product (22:6 ω3) are present in phospholipid fractions from rat brain (31%), heart muscle (18%), and retina (23%) (Tinoco et al., 1979). The 22:6 ω3 percentage in human brain increases with age to about 34% in 80 year-old subjects, but the biological function of 22:6 ω3 and optimum tissue levels are unknown. In one study, rats raised on an 18:3-deficient diet are reported to have more difficulty learning a simple "Y" maze pattern (Lamptey and Walker, 1976).

Two additional reports, which present evidence to support a requirement for dietary 18:3, involved a child who was maintained by total intravenous infusion and primates fed a synthetic diet containing no 18:3. In both instances, symptoms developed (skin lesions, loss of hair, and scaly skin) that resembled symptoms of essential fatty acid (18:2) deficiency. These symptoms essentially disappeared when a source of 18:3 was added to the diet. Although these data are intriguing from a medical and biochemical viewpoint, normal diets contain more than adequate amounts of 18:3 necessary to prevent development of acute symptoms of this nature.

Also of potential importance to soy oil utilization are numerous recent reports showing that 18:3-derived 20 and 22 carbon omega-3 fatty acid in fish oils increases blood clotting times, decreases severity of laboratory-induced strokes, reduces serum lipid levels, lowers blood pressure, alters immune response, and influences prostaglandin metabolism. In addition, reduced incidence of CHD has been correlated with dietary intake of omega-3 fatty acids and higher levels of omega-3 fatty acids in tissue lipids. However, since optimum levels for dietary 18:3 have not been established for humans, much more information is necessary before dietary fats containing linolenic acid (i.e., soybean oil) can be recommended as nutritionally more desirable than non-linolenic-acid-containing fats and oils.

NOTES

E. A. Emken, Northern Regional Research Center, Agricultural Research Service, U.S. Department of Agriculture, Peoria, Illinois 61604.

REFERENCES

Akesson, B., M. Johansson, M. Svensson and M. D. Ockerman. 1981. Content of *trans*-octadecenoic acid in vegetarian and normal diets in Sweden, analyzed by the duplicate portion technique. Am. J. Clin. Nutr. 34:2517-2520.

Applewhite, T. H. 1981. Nutritional effects of hydrogenated soya oil. J. Am. Oil Chem. Soc. 58:260-269.

Beynen, A. C., R. J. J. Hermus and J. G. Hautvast. 1980. A mathematical relationship between the fatty acid composition of the diet and that of the adipose tissue in man. Am. J. Clin. Nutr. 33:81-85.

Biesel, E. T., L. A. Pallansch, M. Keeney and J. Sampugna. 1983. *Trans* and other fatty acids in mice fed diets lacking these components. American Chemical Society Abstr.: 123.

Bonaga, G., M. G. Trizzino, M. A. Pasquariello and P. L. Biagi. 1980. Nutritional aspects of *trans* fatty acids. Note I. Their accumulation in tissue lipids of rats fed with normolipidic diets containing margarine. Biochem. Exp. Biol. 16:51-54.

Christiansen, R. Z., E. N. Christiansen and Jon Bremer. 1979. The stimulation of erucate metabolism in isolated rat hepatocytes by rapeseed oil and hydrogenated marine oil-containing diets. Biochim. Biphys. Acta 573:417-429.

Crass, M. F., III. 1972. Exogenous substrate effects on endogenous lipid metabolism in the working rat heart. Biochim. Biophys. Acta 280:71-81.

Egwin, P. O. and F. A. Kummerow. 1972. Incorporation and distribution of dietary elaidate in the major lipid classes of rat heart and plasma lipoproteins. J. Nutr. 102:783-792.

Emken, E. A. 1979. Utilization and effects of isomeric fatty acids in humans. p. 99-129. *In* E. A. Emken and H. J. Dutton (eds.) Geometrical and positional fatty acid isomers. American Oil Chemists' Society, Champaign, Illinois.

Emken, E. A. 1983. Human studies with deuterium labeled octadecenoic acid isomers. p. 302-319. *In* E. G. Perkins and W. J. Visek (eds.) Dietary fats and health. American Oil Chemists' Society, Champaign, Illinois.

Emken, E. A. 1984. Nutrition and biochemistry of *trans* and positional fatty acid isomers in hydrogenated oils. Ann. Rev. Nutrition. 4:337-376.

Emken, E. A., W. K. Rohwedder, R. O. Adlof, W. J. Dertarlais and R. M. Gulley. 1984. *In vivo* distribution of *trans* and *cis*-10-octadecenoic acid isomers in human plasma lipids. (Abstract) J. Am. Oil Chem. Soc. 61:678.

Fogerty, A. C. and G. L. Ford. 1983. *trans* Fatty acids in kangaroo lipids. Nutr. Rep. Int. 28:375-380.

Fuchs, G. and J. Kuivinen. 1980. The presence of *trans* fatty acids in edible fats and oils. Var Föda 32(Suppl. 1):31-37.

Heckers, Von H., F. W. Melcher and K. Dittmar. 1979. Daily consumption of *trans*-isomeric fatty acids—A calculation based on composition of commercial fats and of various human depot fats. Fette Seifen Anstrichm. 81:217-226.

Holman, R. T., M. M. Mahfouz, L. D. Lawson and E. G. Hill. 1983. Metabolic effects of isomeric octadecenoic acids. p. 320-340. *In* E. G. Perkins and W. J. Visek (eds.) Dietary fats and health. American Oil Chemists' Society, Champaign, Illinois.

Holmer, G., C. E. Hoy and D. Kirstein. 1982. Influence of partially hydrogenated vegetables and marine oils on lipid metabolism in rat liver and heart. Lipids 17:585-593.

Hoy, C. E. and G. Holmer. 1979. Incorporation of *cis*- and *trans*-octadecenoic acids into the membranes of rat liver mitochondria. Lipids 14:727-733.

Kaufman, H. P., F. Volhert and G. Mankel. 1961. Use of the IR-spectroscopy in the field of fats V: Investigation of milk-fats for *trans*-unsaturated fatty acids. Fette Seifen Anstrichm. 63:261.

Kinsella, J. E., G. Bruckner, J. Mai and J. Shimp. 1981. Metabolism of *trans* fatty acids with emphasis on the effects of *trans,trans*-octadecadienoate on lipid composition, essential fatty acid, and prostaglandins: An overview. Am. J. Clin. Nutr. 34:2307-2318.

Kritchevsky, D. 1983. Influence of *trans* unsaturated fat on experimental atherosclerosis. p. 403-413. *In* E. G. Perkins and W. J. Visek (eds.) Dietary fats and health. American Oil Chemists' Society, Champaign, Illinois.

Lamptey, M. S. and B. L. Walker. 1976. A possible role for dietary linolenic acid in the development of the young rat. J. Nutr. 106:86-93.

Lawson, L. D., E. G. Hill and R. T. Holman. 1982. Accumulation and depletion of *trans* octadecenoic acid in rat peripheral nerve phospholipid. Biochim. Biophys. Acta 712:117-122.

Lawson, L. D. and R. T. Holman. 1981. Beta-oxidation of the geometric and positional isomers of octadecenoic acid by rat heart and liver mitochondria. Biochim. Biophys. Acta 665:60-65.

Menon, N. and G. Dhopeshwarkar. 1983. Differences in the fatty acid profile and beta-oxidation by heart homogenates of rats fed *cis* and *trans* octadecenoic acids. Biochim. Biophys. Acta 751:14-20.

Moore, C. E., R. B. Alfin-Slater and L. Aftergood. 1980. Incorporation and disappearance of *trans* fatty acids in rat tissues. Am. J. Clin. Nutr. 33:2318-2323.

Ohlrogge, J. B. 1983. Distribution in human tissues of fatty acid isomers from hydrogenated oils. p. 359-374. *In* E. G. Perkins and W. J. Visek (eds.) Dietary fats and health. American Oil Chemists' Society, Champaign, Illinois.

Rizek, R. L., S. O. Welsh, R. M. Marston and E. M. Jackson. 1983. Levels and sources of fat in the U.S. food supply and in diets of individuals. p. 13-43. *In* E. G. Perkins and W. J. Visek (eds.) Dietary fats and health. American Oil Chemists' Society, Champaign, Illinois.

Royce, S. M., R. P. Holmes, T. Takagi and F. A. Kummerow. 1984. The influence of dietary isomeric and saturated fatty acids on atherosclerosis and eicosanoid synthesis in swine. Am. J. Clin. Nutr. 39:215-222.

Thomas, L. H., P. R. Jones, J. A. Winter and H. Smith. 1981. Hydrogenated oils and fats: The presence of chemically-modified fatty acids in human adipose tissue. Am. J. Clin. Nutr. 34:877-886.

Tinoco, J., R. Babcock, I. Hincenbergs, B. Medwadowski, P. Miljanich and M. A. Williams. 1979. Linolenic acid deficiency. Lipids 14:166-173.

Wood, R. 1979. Distribution of isomeric fatty acids in tissue lipids from rats fed hydrogenated safflower oil. *In* E. A. Emken and H. J. Dutton (eds.) Geometrical and positional fatty acid isomers. American Oil Chemists' Society, Champaign, Illinois.

STABILITY OF SOYBEAN OIL TO OXIDATION

Earl G. Hammond

Oxidation of fats and oils by atmospheric oxygen is one of the classic problems of food technology. This reaction often limits the shelf life of foods and determines the kinds of uses for which a particular fat or oil is suitable. For many years oil technologists have noted that, compared with many other vegetable oils, soybean oil is relatively unstable and develops objectionable flavors at lower degrees of oxidation (Smouse, 1979; Frankel, 1980, 1982).

MEASURES OF OIL OXIDATION

The ultimate objection to oxidized soybean oil is to its flavor, and measurement of this defect must rest on sensory tests. The most widely used sensory test for oils is based on blandness (Mounts and Warner, 1980). An oil that has no taste at all is regarded as ideal, and the degree of deterioration is based on flavor development. Usually such judgments are made on an arbitrary 10-point scale. Only the bland end of the scale is fixed, and the intensity represented by other points on the scale is arrived at objectively by the judges. There is considerable variation from judge to judge, but this can be reduced by training and development of consensus by a panel. Such scales are quite helpful in making comparisons among samples and in distinguishing deteriorated oils from fresh oils. There also have been attempts to produce scales based on the recognition of particular defects and their intensity (Waltking, 1982). Unfortunately, there is no agreed-upon method of producing many of the flavor defects for training purposes, so this scale is useful only to those that have access to experienced panels of judges.

Seemingly, the reliability of the 10-point flavor-intensity scale could be increased and comparisons among laboratories enhanced if additional points could be fixed on the scale. Standards have seldom been used because, when oils are tasted directly, they tend to coat the mouths of the judges, and this greatly limits the number of samples that can be tasted at any one time. The inclusion of standards uses up this valuable sample capacity. Stone and Hammond (1983) found that this difficulty could be overcome by making emulsions of oils, stabilized with gum acacia, in synthetic tap water. The emulsified samples were readily rinsed from the mouth, and this allowed more samples to be observed at one time. This technique also facilitated comparison with standards.

Under the best of conditions, however, sensory tests are cumbersome, inaccurate, and require large amounts of sample. Many attempts have been made to estimate the oxidative deterioration of fats and oils by objective tests. These have been based on many properties: oxygen uptake, peroxide formation, conjugation of double bonds, and production of secondary products, especially various carbonyls. But these tests have invariably been found poor indicators of flavor deterioration (Fioriti et al., 1974; Gray, 1978). An obvious solution to this problem would be to identify the actual compounds responsible for oxidized flavors and to measure these, but the mixture of flavor compounds produced in oxidation is complex and their concentration in the oil is low. Moreover, these flavor compounds usually are in the presence of unstable peroxide intermediates, and it has been difficult to isolate the flavors without generating unknown additional amounts of flavor compounds from the peroxides.

An approach that has become popular in recent years is to distill the volatiles from an oil at high temperatures and to measure the amounts of the volatiles by gas chromatography (Dupuy et al., 1977; Williams and Applewhite, 1977; Jackson and Giacherio, 1977; Min, 1981; Waltking and Zmachinski, 1977; Gensic et al., 1984). Obviously, during such a test hydroperoxides in the oil are decomposed and flavor volatiles are formed that were not present in the unheated oil, but the total amount of volatiles produced correlates well with sensory tests under suitably restricted circumstances. Attempts to achieve better correlations by using particular peaks in the gas chromatogram rather than total volatiles have resulted in little improvement, possibly because of the limited accuracy of the sensory tests on which the correlations rest. Possibly, if hydroperoxides in the sample were reduced before the volatiles were distilled from the oils, the correlations would be less restricted in their scope.

Many of the important flavor compounds in oxidized fats and oils are carbonyl compounds, so many attempts have been made to measure these especially with 2,4-dinitrophenylhydrazine (Gaddis et al., 1959; Haverkamp-Bergeman and deJong, 1959; Swartz et al., 1963; Horikx, 1964; Pradel and Adda, 1980; Reindl and Hans-Jurgen, 1982). Often these methods have lacked the sensitivity needed to measure the intensely flavored carbonyl compounds, and the generation of carbonyl from hydroperoxides during the formation of the dinitrophenylhydrazones has been a concern. Recently, White and Hammond (1983) proposed a method for measuring the carbonyls in oil samples as the 2,4,6-trichlorophenylhydrazones. These derivatives are volatile enough to be determined by gas chromatography, and the method is sensitive enough to detect many of the carbonyls in the threshold range. Reduction of the peroxides before formation of the hydrazones did not change the results significantly, so formation of artifacts from hydroperoxides does not seem to be a problem. The primary limitation of the technique was caused by interference from traces of carbonyl in the solvents.

Once one has been able to measure the carbonyls produced in oxidized soybean oil, one needs to know what weight each of these compounds carries in the intensity of the oxidized flavor. By using an extension of the emulsion method of Stone and Hammond (1983), Dixon and Hammond (1984) were able to determine the flavor intensity of many of the carbonyls important in fat oxidation. This was done by comparing the flavor intensity of solutions of the carbonyls, dissolved in mineral oil and emulsified in water, with those of a series of standard dilutions of 2-heptanone. Studies of binary mixtures have shown that the sensory intensity of compounds in mixtures is not necessarily a simple additive function of their intensity as individual components. However, when complex mixtures of the carbonyls were made in the concentrations found in oxidized soybean oil, the flavor intensity could be predicted with reasonable approximation from the intensities of the individual components.

There is much yet to do before food technologists can predict the flavor intensity of a soybean oil sample by an analysis of the carbonyl compounds present, but strides toward that goal are being made.

FACTORS CONTRIBUTING TO SOYBEAN OIL INSTABILITY

One of the reasons that soybean oil is unstable to oxidation seems to be its content of linolenic acid. Usually, soybean oil contains about 7-9% linolenic acid. Oil technologists have demonstrated that, if linolenic acid is incorporated into oils that do not contain it, these oils develop the flavor defects characteristic of soybean oil. Reduction of the amount of linolenic acid in soybean oil by fractionation, dilution, or hydrogenation improves stability (Smouse, 1979). Other oils rich in linolenic acid, such as linseed and rapeseed oils, also are relatively unstable. Thus, there is a great deal of evidence that linolenic acid is one of the chief causes of soybean oil instability. This has led to extensive efforts to reduce the linolenic acid content of soybean oil by plant breeding (Hammond et al., 1972; Hammond and Fehr, 1975, 1983, 1984; Wilcox et al., 1984). These efforts have been partly successful, and strains of soybeans are available that contain as little as 3% linolenic acid. Comparisons of oils from these strains with those of normal soybean oil have shown that the low-linolenic acid strains are more stable (Stone and Hammond, 1983).

It is less clear why linolenic acid should have such a dramatic effect on the flavor of soybean oil. One factor is that linolenic acid oxidizes faster than the other fatty acids in soybean oil. Wong and Hammond (1977) and Fatemi and Hammond (1980) developed methods to analyze the hydroperoxides formed in oxidizing mixtures of oleic, linoleic, and linolenic esters, and they showed that the relative rates were 1:10.3:21.6, respectively. Thus, in soybean oil that is 25% oleic, 55% linoleic, and 7% linolenic acid, one would expect the relative amount of oxidation attributable to each fatty acid to be in the ratio 1:22.7:6.0. If the only effect were the relative rates of oxidation, linoleic acid should be four times as potent in developing off-flavors as linolenic acid because of the greater amount of linoleic acid in the oil.

Some have speculated that the deleterious effect of linolenic acid on oil flavor is because the products of linolenic acid oxidation have more potent flavors than those from linoleic acid (Frankel, 1980, 1982). This is possible, and there is evidence that carbonyls with isolated double bonds, such as might be expected from linolenic acid, have lower flavor thresholds than carbonyls without isolated double bonds (Keppler, 1977). Table 1 shows the carbonyl analysis of oxidized soybean oil by White and Hammond (1983) converted into relative flavor potency by the data of Dixon and Hammond (1984). The flavor threshold of 2-heptanone is about 2.5 ng/g, so at 3 days, all the individual compounds are below threshold, and most of them are at 5 days. But the intensities were approximately additive, so that, at both 3 and 5 days, the mixtures had flavor intensities well above threshold. A surprising amount of the carbonyls and flavor seem to arise from oleic acid, especially at 3 days. This may result from a greater stability of the carbonyls produced from oleic acid to further oxidation (Michalski and Hammond, 1972) or decomposition of products left from the deodorization of the oil. The flavor potency of the products from linolenic acids exceeds that from linoleic acid, but not by a large amount. The yield of carbonyl from linolenic acid seems greater than one would expect on the basis of its percentage in the oil and its rate of oxidation, especially at 8 days. Possibly, linolenic acid yields a greater amount of scission products or more stable scission products than linoleic acid.

Recent studies have revealed the formation of unconjugated and cyclic peroxides and expoxides from the oxidation of linoleic and linolenic acids (Morita, 1981;

Table 1. The concentration of carbonyls found in soybean oil oxidized for various times at 55 C and the concentration of 2-heptanone, having flavor intensity equivalent to each carbonyl when tasted in aqueous emulsion.

Carbonyl compound	Fatty acid of origin[a]	3 days[b] Conc. in oil	2-hepta-none	5 days[c] Conc. in oil	2-hepta-none	8 days[d] Conc. in oil	2-hepta-none
		µg/g	ng/g	µg/g	ng/g	µg/g	ng/g
Octanal	18:1	0.94	120	2.0	170	2	170
Nonanal		–	–	6.4	380	10	440
Decanal		0.39	170	1.1	220	1.6	240
2-Nonenal		0.39	32	1.1	57	1.6	70
2-Decenal		–	–	–	–	0.42	18
Total		1.7	322	10.6	827	15.6	938
Pentanal	18:2	–	–	0.16	2	3.5	17
Hexanal		2.4	61	35	350	45	410
2-Octanal		–	–	6.4	210	0.73	58
2,4-Decadienal		0.14	29	0.48	49	10	270
Total		2.5	90	42.0	611	59.2	755
2-Hexenal	18:3	1.2	48	6.9	160	9.9	200
2-Heptenal		0.61	67	3.5	190	5.5	290
2,4-Hexadienal		–	4	1.1	9	1.3	10
2,4-Heptadienal		–	–	0.9	130	1.6	170
2,4-Octadienal		–	–	0.8	64	1.1	74
2,4-Nonadienal		–	–	2.9	480	6.9	630
Total		1.8	119	16.1	1033	26.3	1374

[a]Assignments made on the basis of data by Frankel (1982).
[b]Peroxide value 1.5.
[c]Peroxide value 9.3.
[d]Peroxide value 16.3.

Schieberle and Grosch, 1981a,b; Frankel, 1982; Neff et al., 1982, 1983; Frankel et al., 1983; Halsbek and Grosch, 1983; Gardner and Selke, 1984; Peers et al., 1984), and the volatile decomposition products of some of these oxidation products have been investigated. Most of these products have been isolated from highly oxidized material, and the studies on scission products have been done at the temperatures of gas chromatographic injection ports to facilitate the identification of the products. It is not clear to what extent these products are produced in soybean oil oxidized under typical conditions. Studies are needed under more realistic conditions.

Soybean oil contains the natural antioxidant tocopherol, and it is generally agreed that there is sufficient tocopherol in unhydrogenated soybean oil that the addition of phenolic antioxidants is superfluous (List and Erickson, 1980). This is true even though some of the tocopherol may be lost in refining, particularly in deodorization. But soybean oil is improved in stability by the addition of metal chelating substances, and the addition of citric acid and citrates to soybean oil is routine. At the same time, care should be taken in processing to avoid contact with metals. The role of other factors in soybean oil stability is less well established. Some have suggested that incomplete removal of phosphatides during refining is responsible for instability, but there also is evidence that soybean oil stability is decreased by the removal of phosphatides (Smouse, 1979).

The initiation of oxidation in fats and oils in many instances seems caused by activation of ordinary triplet oxygen to singlet oxygen. This special high-energy form of oxygen usually is produced by the absorption of light energy by pigments and the transfer of this energy to oxygen, but singlet oxygen also can be produced by enzymatic and chemical reactions (Korycka-dahl and Richardson, 1980). There is no doubt that this reaction is important when soybean oil is exposed to light in clear glass bottles in the presence of oxygen (Frankel, 1982, 1980; List and Erickson, 1980). But it is not clear how general a role this reaction plays in initiating soybean oil oxidation in other conditions. It is not clear what pigments may be involved, although it is well known that chlorophyll can catalyze the formation of singlet oxygen. One of the problems in investigating the effect of singlet oxygen and pigment is that there is no easy way to measure the formation of singlet oxygen. Its presence can best be noted by the formation of different oxidation products from those formed by ordinary oxygen.

It is well known that, when an oil is once oxidized that, although its flavor may be restored by deodorization, the original stability is not recovered (Smouse, 1979; Frankel, 1982; Morita, 1981). This seems to be caused by oxidation products that cannot be removed from the oil by deodorization and that catalyze oxidation.

Soybeans contain the enzyme lipoxygenase, which rapidly catalyzes the oxidation of linoleic and linolenic acids and their esters with molecular oxygen (Eskin et al., 1977). This enzyme becomes active especially when a bean is damaged. Although the activity of lipoxygenase is much reduced at low moisture conditions, activity has been detected under quite dry conditions (Brockmann and Acker, 1977). There is an opportunity for this enzyme to act during the tempering, crushing, and flaking processes in preparation for solvent extraction. Once these oxidation products are formed, the future stability of the oil presumably is compromised. Soybeans contain four lipoxygenases each under separate genetic control (Eskin et al., 1977; Hildebrand and Hymowitz, 1981; Yabuuchi et al., 1982). Soybeans without two of these lipoxygenases have been produced. Some believe that lipoxygenase may be essential to the survival of the plant. Lipoxygenase may be destroyed by cooking wet beans or steeping in aqueous alcohol (Eldridge et al., 1977; Rice et al., 1981) before size reduction is initiated, but this process seems impractical for commercial processing. On a laboratory scale, it is claimed that cooking the beans results in an oil of improved stability.

The rate of oxidation of a fat or oil seems to depend also on the arrangement of its fatty acids on glycerol, the so-called glyceride structure of the fat or oil (Lau and Hammond, 1982). Like most vegetable oils, in soybean oil, the saturated fatty acids, stearic and palmitic, are located on the outer sn-1 and sn-3 positions of the glycerol, and the linoleic acid is concentrated on the sn-2 position (Fatemi and Hammond, 1977a,b; Pan and Hammond, 1983). The linolenic acid is fairly equally distributed across the three positions, but there is more on the sn-1 position than the other two positions. Oleic acid, a monounsaturated fatty acid, fills in the other spaces. There tends to be more oleic on the sn-3 position than on the other two positions. When oils with this sort of structure are randomized (that is, when they are treated with a catalyst so that their fatty acids may migrate to form a random arrangement of triglycerides) the stability decreases. Some have suggested that this change is brought about by an effect on the antioxidants (Park et al., 1983), but this does not seem to account for the observations that have been made. On the other hand, it is not clear why the glyceride structure should have this effect. Raghuveer and Hammond (1969) suggested that it was caused by the effect that the glyceride structure had on the portability of two fatty acids chains coming in contact, but Lau and Hammond (1982) were not able to verify this theory. One must proceed with caution in this research, because the rate of oxidation is so easily influenced by traces of materials. However, the effect of glyceride structure has been confirmed by several observers (Hoffmann et al., 1983; Catalino et al., 1975). According to Lau and Hammond (1982) the natural glyceride structure of oils is about three or four times more stable than the corresponding randomized mixture. Since we have no theory to account for the effect of glyceride structure, we have no way of knowing to what extent the natural glyceride structure of soybean oil is optimal for flavor stability. Oil technologists have limited ability to control the glyceride structure of oils, so mixtures with similar fatty acid compositions but different glyceride arrangements are not readily available to make stability comparisons. Pan and Hammond (1983) screened a number of soybean cultivars as well as *Glycine soja* introductions for variants in glyceride structure. They found a rather narrow range of variation. Attempts to optimize stability in this way will have to await development of a better theory and techniques to manipulate glyceride structure.

NOTES

Earl G. Hammond, Professor, Department of Food Technology, Iowa State University, Ames, Iowa 50011.

Journal Paper No. J-11609 of the Iowa Agriculture and Home Economics Experiment Station, Ames. Project 2493.

REFERENCES

Brockmann, R. and L. Acker. 1977. Lipoxygenase activity and water activity in systems of low water content. Ann. Technol. Agric. 26:167-174.

Catalino, M., M. De Felice and V. Sciancalipore. 1975. Autoxidation of monounsaturated triglycerides. Influence of fatty acid position. Ind. Aliment. 14:89-92.

Dixon, M. D. and E. G. Hammond. 1984. The flavor intensity of some carbonyl compounds important in oxidized fats. J. Am. Oil Chem. Soc. 61 (in press).

Dupuy, H. P., E. T. Rayner, J. I. Wadsworth and M. G. Legendre. 1977. Analysis of vegetable oils for flavor quality by direct gas chromatography. J. Am. Oil Chem. Soc. 54:445-449.

Eldridge, A. C., K. Warner and W. J. Wolf. 1977. Alcohol treatment of soybeans and soybean protein products. Cereal Chem. 54:1229-1237.

Eskin, N. A. M., S. Grossman and A. Pinsky. 1977. Biochemistry of lipoxygenase in relation to food quality. Crit. Rev. Food Sci. Nutrition 9:1-40.

Fatemi, S. H. and E. G. Hammond. 1977a. Glyceride structure variation in soybean varieties: I. Sterospecific analysis. Lipids 12:1032-1036.

Fatemi, S. H. and E. G. Hammond. 1977b. Glyceride structure variation in soybean varieties: II. Silver ion chromatographic analysis. Lipids 12:1037-1042.

Fatemi, S. H. and E. G. Hammond. 1980. Analysis of oleate, linoleate, and linolenate hydroperoxidic in oxidized ester mixtures. Lipids 15:379-385.

Fioriti, J. A., M. J. Kanuk and R. J. Sims. 1974. Chemical and organoleptic properties of oxidized fats. J. Am. Oil Chem. Soc. 51:219-223.

Frankel, E. N. 1980. Soybean oil flavor stability. p. 229-244. In D. R. Erickson, E. H. Pryde, O. L. Brekke, T. L. Mounts and R. A. Falk (eds.) Handbook of soy oil processing and utilization. American Soybean Association, St. Louis, MO.

Frankel, E. N. 1982. Volatile lipid oxidation products. Prog. Lipid Res. 22:1-33.

Frankel, E. N., W. E. Neff and E. Selke. 1983. Analysis of autoxidized fats by gas chromatography-mass spectrometry: VIII. Volatile thermal decomposition products by hydroperoxy cyclic peroxides. Lipids 18:353-357.

Gaddis, A. M., R. Ellis and G. T. Currie. 1959. Carbonyls in oxidizing fat. I. Separation of steam volatile monocarbonyls into classes. Food Res. 24:283-297.

Gardner, H. W. and E. Selke. 1984. Volatiles from thermal decomposition of isomeric methyl (12S, 13S)-(E)-12,13-epoxy-9-hydroperoxy-10-octadecenoates. Lipids 19:375-380.

Gensic, J. L., B. F. Szuhaj and J. F. Endres. 1984. Automated gas chromatographic system for volatile profile analysis of fats and oils. J. Am. Oil Chem. Soc. 61:1246-1249.

Gray, J. I. 1978. Measurement of lipid oxidation: A review. J. Am. Oil Chem. Soc. 55:539-546.

Halsbeck, F. and W. Grosch. 1983. Autoxidation of phenyl linoleate and phenyloleate: HPLC analysis of the major and minor monohydroperoxides as phenyl hydroxystearates. Lipids 18:706-713.

Hammond, E. G. and W. R. Fehr. 1975. Oil quality improvement in soybeans—*Glycine max* (L.) Merr. Fette Seifen Anstrichim. 77:97-101.

Hammond, E. G. and W. R. Fehr. 1983. Registration of A5 germplasm line of soybeans. Crop Sci. 23:192.

Hammond, E. G. and W. R. Fehr. 1984. Progress in breeding for low-linolenic acid soybean oil. (in press). In J. B. M. Rattray and C. Ratledge (eds.) Biotechnology for the oils and fats industry. Chemists Society, Champaign, IL.

Hammond, E. G., W. R. Fehr and H. E. Snyder. 1972. Improving soybean quality by plant breeding. J. Am. Oil Chem. Soc. 49:33-35.

Haverkamp-Bergeman, P. and K. deJong. 1959. Analytical applications of a Celite/dinitrophenylhydrazine column. Rec. trav. chim. Pays-Bas. 78:275-283.

Hildebrand, D. F. and T. Hymowitz. 1981. Two soybean genotypes lacking lipoxygenase-1. J. Am. Oil Chem. Soc. 58:583-586.

Hoffmann, G., J. B. A. Stronik, R. G. Polman and C. W. Van Ossten. 1973. The oxidative stability of some diacid triglycerides: Zesz. Probl. Postepow. Navk Roln. 146:93-98.

Horikx, M. M. 1964. Decomposition of methyl oleate hydroperoxide on the Celite/dinitrophenylhydrazine-hydrochloric acid column. J. Appl. Chem. 14:50-52.

Jackson, H. W. and D. J. Giacherio. 1977. Volatiles and oil quality. J. Am. Oil Chem. Soc. 54:458-460.

Keppler, J. G. 1977. Twenty-five years of flavor research in a food industry. J. Am. Oil Chem. Soc. 54:474-477.

Korycka-dahl, M. and T. Richardson. 1980. Initiation of oxidative changes in foods. J. Dairy Sci. 63:1181-1198.

Lau, F. Y. and E. G. Hammond. 1982. Effect of randomization on the oxidation of corn oil. J. Am. Oil Chem. Soc. 59:407-411.

List, G. R. and D. R. Erickson. 1980. Storage, handling and stabilization. p. 267-353. In D. R. Erickson, E. H. Pryde, O. L. Brekke, T. L. Mounts and R. A. Falb (eds.) Handbook of soy oil processing and utilization. American Soybean Association, St. Louis, MO.

Michalski, S. T. and E. G. Hammond. 1972. Use of labeled compounds to study the mechanism of flavor formation in oxidizing fats. J. Am. Oil Chem. Soc. 49:563-566.

Min, D. B. 1981. Correlation of sensory evaluation and instrumental gas chromatographic analysis of edible oils. J. Food Sci. 46:1455-1456.

Morita, M. 1981. Non-metal-requiring catalyst contained in recovered methyl linolenate from an autoxidizing sample. J. Agric. Food Chem. 45:2403-2407.

Mounts, T. L. and K. Warner. 1980. Evaluation of finished oil quality. p. 245-266. In D. R. Erickson, E. H. Pryde, O. L. Brekke, T. L. Mounts and R. A. Falb (eds.) Handbook of soy oil processing and utilization. American Soybean Association, St. Louis, MO.

Neff, W. E., E. N. Frankel and D. Weislender. 1982. Photosensitized oxidation of methyl linolenate. Secondary products. Lipids 17:780-790.

Neff, W. E., E. N. Frankel, E. Selke and D. Weislender. 1983. Photosensitized oxidation of methyl linolenate monohydroperoxides: hydroperoxy cyclic peroxides, dihydroperoxides, keto esters, and volatile thermal decomposition products. Lipids 18:868-876.

Pan, W. P. and E. G. Hammond. 1983. Stereospecific analysis of triglycerides of *Glycine max, Glycine soya, Avena sativa,* and *Avena sterilis* strains. Lipids 18:882-888.

Park, D. K., J. Terao and S. Matsushita. 1983. Influence of interesterification on the autoxidative stability of vegetable oil. Agric. Biol. Chem. 47:121-123.

Peers, K. W., D. T. Coxon and H. W.-S. Chan. 1984. Thermal decomposition of individual positional isomers of methyl linoleate hydroperoxides, hydroperoxy cyclic peroxide and dihydroperoxides. Lipids 19:307-313.

Pradel, G. and J. Adda. 1980. Peroxides as a source of error in the quantitative determination of monocarbonyls in cheese. J. Food Sci. 45:1058-1059.

Raghuveer, K. G. and D. C. Hammond. 1967. The influence of glyceride structure on autoxidation. J. Am. Oil Chem. Soc. 44:239-243.

Reindl, B. and S. Hans-Jurgen. 1982. Determination of volatile aldehydes in meat as 2,4-dinitrophenylhydrazones using reverse phase high performance liquid chromatography. J. Agric. Food Chem. 30:849-854.

Rice, R. D., L. S. Wei, M. P. Steinberg, and A. I. Nelson. 1981. Effect of enzyme inactivation on the extracted soybean meal and oil. J. Am. Oil Chem. Soc. 58:578-583.

Schieberle, P. and W. Grosch. 1981a. Decomposition of linoleic acid hydroperoxides II. Breakdown of methyl 13-hydroperoxy-cis-9-trans-11-octadecadienoate by radicals or copper II ions. Z. Lebensm. Unters. Forsch. 173:192-198.

Schieberle, P. and W. Grosch. 1981b. Detection of monohydroperoxides with unconjugated diene system as minor products of the autoxidation of methyl linoleate. Z. Lebensm. Unters. Forsch. 173:199-203.

Smouse, T. H. 1979. A review of soybean oil reversion flavor. J. Am. Oil Chem. Soc. 56:747A-750A.

Stone, R. and E. G. Hammond. 1983. An emulsion method for the sensory evaluation of edible oils. J. Am. Oil Chem. Soc. 60:1277-1281.

Swartz, D. P., H. S. Haller and M. Keeney. 1963. Direct quantitative isolation of monocarbonyl compounds from fats and oils. Anal. Chem. 35:2191-2194.

Waltking, A. E. 1982. Progress report of the AOCS flavor nomenclature and standards committee. J. Am. Oil Chem. Soc. 59:116A-120A.

Waltking, A. E. and H. Zmachinski. 1977. A quality control procedure for the gas liquid chromatographic evaluation of the flavor quality of vegetable oils. J. Am. Oil Chem. 54:454-457.

White, P. J. and E. G. Hammond. 1983. Quantification of carbonyl compounds in oxidized fats as trichlorophenylhydrazones. J. Am. Oil Chem. Soc. 60:1769-1773.

Wilcox, J. R., J. F. Cavins and N. C. Nielsen. 1984. Genetic alteration of soybean oil composition by a chemical mutagen. J. Am. Oil Chem. Soc. 61:97-100.

Williams, J. L. and T. H. Applewhite. 1977. Correlation of the flavor scores of vegetable oils with volatile profile data. J. Am. Oil Chem. Soc. 54:461-463.

Wong, W.-S. D. and E. G. Hammond. 1977. Analysis of oleate and linoleate hydroperoxides in oxidized ester mixtures. Lipids 12:475-479.

Yabuuchi, S., R. M. Lister, B. Axelrod, J. R. Wilcox and N. C. Nielsen. 1982. Enzyme-linked immunosorbent assay for the determination of lipoxygenase isoenzymes in soybean. Crop Sci. 22:333-337.

BREEDING AND GENETICS

Plenary Paper

BIOTECHNOLOGY AND ITS IMPACT ON FUTURE DEVELOPMENTS IN
SOYBEAN PRODUCTION AND USE

Robert M. Goodman and Jack Kiser

Modern biotechnology is based on scientific advances that make it possible to isolate and clone specific pieces of DNA containing genes, and to sequence the nucleotides in a DNA molecule (i.e., read the genetic code), so that the precise location and structure of genes can be studied at the molecular level. These advances were fueled by major improvements in technique, by the discovery and exploitation of specialized enzymes (restriction endonucleases, DNA polymerases, reverse transcriptases, and ligases) with which DNA is manipulated, and the exploitation of plasmids (originally identified as sex factors in bacteria) with which DNA from any source can be amplified (cloned). Related advances make it possible to transfer genes from organism to organism by means that by-pass the normal sexual processes governing intraspecific inheritance (National Research Council, 1984). The purpose of this contribution is to survey the ways in which modern biotechnology has already had, and may in the future have, an impact on our understanding of the soybean. We shall review the prerequisites for genetic engineering of a crop plant with some examples of recent accomplishments with other species, and summarize soybean breeding objectives that might be amenable in the future to genetic engineering. Finally, we will summarize the current state of the art as regards soybeans.

As with any new field, terminology tends to grow quickly. The careful definition of the terms lags behind. New terms are invented and used in different ways by different people, or old terms accumulate new meanings, again often not very precisely. Here, then, are some definitions of terms as we shall use them.

Biotechnology: The application of recombinant DNA, cell and tissue culture, classical plant breeding and other methods used to develop new and improved plants and plant products. Because they are useful in crop production, genetically-engineered plant disease diagnostic tools and agricultural microbes can also be considered products of plant biotechnology.

Genetic Engineering: Genetic manipulations that use recombinant DNA methods (gene-splicing) to change the genetic make-up of an organism.

Gene Expression: Control of when the newly introduced genes are turned on and off during the life cycle of the recipient plant, and the levels at which the genes are working when turned on.

Gene Isolation: The identification of specific genes (each encoding a specific protein) responsible for a given genetic trait, and the isolation of those genes, usually *via* a bacterial plasmid or phage vector cloning system.

Gene Transfer (Transformation): The coding regions of isolated genes, along with appropriate regulatory signals, are moved into a recipient organism (in this case a crop plant) on a vector in such a way that the genes are integrated into the chromosome and expressed.

Plasmid: A relatively small, autonomous circular piece of DNA, not part of the chromosome, most common in bacteria. Their convenient size and ability to be moved from organism to organism make plasmids very useful in cloning genes and as vectors for inserting genes into cells.

Regeneration: The creation of a complete plant from a single plant cell or small piece of plant tissue. A regenerated plant thus is a clone of another plant. Regeneration can be a complex process involving multiple hormone, nutrient and growing condition treatments.

Vectors: Pieces of DNA that can move or be moved from one organism to another and survive in both. Genes can be spliced into a vector to be carried into another organism. Certain plasmids and viruses are useful vectors. Alternatively, vector systems can be based on other means of moving genes, such as microinjection, liposomes, or direct uptake of DNA from the medium.

BIOTECHNOLOGY AND SOYBEANS

Recent Applications of Biotechnology Methods to Soybean Research

Modern biology, including recombinant DNA and other related methods, has begun to have a major impact on our understanding of soybeans. This section is intended to present three brief examples.

The Stress Response. When organisms are placed in stressful situations, the machinery of gene expression and protein synthesis responds rapidly and often dramatically. Overall protein synthesis is reduced, and new proteins, not present in the absence of the stress, are produced. One of the best characterized plant species in this regard is the soybean, due to the work of Joe Key and his colleagues over the past several years. Briefly, soybean seedlings, or excised hypocotyls (cv. Wayne), when transferred from 28 C to 40 C produce a new set of proteins, called heat-shock or hs proteins, and cease to produce proteins that are normally produced in unstressed tissue (Key et al., 1981). These proteins are of low molecular weight (15-18,000 kD) and present a much more complex picture than that found in the well-studied heat-shock response of *Drosophila* (Schoffl and Key, 1983).

From the start, the study of this response, including its kinetics, its complexity, and its possible role in conferring thermal tolerance in soybeans (Lin et al., 1984) relied heavily upon molecular biology methods. Several genes encoding heat-shock proteins in soybeans have now been cloned and their structures are being studied. It is known that heat-shock genes occur in multigene families and that the genes of one class of such proteins may be clustered on the chromosome (Schoffl and Key, 1983).

Gene Structure. Soybean genes encoding glycinin, β-conglycinin, and lectin (the seed storage proteins), leghemaglobin, RuBP carboxylase, and actin have been cloned and at least partly characterized (Kitamura et al., 1984; Goldberg, 1983). The Le1 gene for seed lectin in soybean (Vodkin et al., 1983) encodes a mature protein of 253 amino acids, plus a 32-amino acid signal sequence at the N-terminus; it is one of two known genes for lectin in the soybean genome (the other, Le2, apparently is not expressed in seeds). The mature protein is processed in-vivo after translation from the messenger RNA to remove the signal sequence. The signal sequence is presumed to be involved in the transport and/or compartmentalization of the mature protein into protein bodies in developing seeds. The Le1 gene contains no introns (i.e., the mRNA is an unprocessed exact copy of the DNA in the coding region of the gene). Thus, it shares several properties with other more abundant seed proteins (no or few introns, signal sequences), but differs in being encoded by a single gene rather than by a multigene family (Goldberg, 1983).

Several soybean and *Glycine soja* genotypes phenotypically lack lectin (Le⁻). Such plants contain the coding region for Le1 but it is interrupted (in the Le⁻ genotype Sooty) approximately in the middle of the gene by a 3.4 kilobase sequence that has structural features of a transposable element (Vodkin et al., 1983). Thus, the absence of lectin in Le⁻ plants is due not to the absence of the gene, but to its failure to function because of a large insertion in the coding sequence.

Nitrogen Fixation and the Rhizobium Symbiosis. Because molecular biology and recombinant DNA technology originated in the study of microorganisms and viruses, it is perhaps not surprising that the most rapid progress to date in agricultural biotechnology has been made in manipulation of plant-associated microorganisms. Of the 24 million hectares of soybeans grown annually in the U.S., some 12 million are inoculated with the nitrogen-fixing symbiont, *Rhizobium japonicum*. Strain selection and improvement has been an agricultural research goal for many years. Significant progress, due to biotechnology, may result from processes now under active investigation.

Strains presently in use often lack significant traits. For example, of the nine dominant subgroups in U.S. soils, three-fourths are missing the hydrogen uptake (Hup) trait; thus they waste energy that is evolved in the form of hydrogen gas. A 5 to 10% gain in nitrogen efficiency by strains carrying the hydrogen uptake genes has been calculated. The annual soybean crop value is ten billion dollars, so these genes could mean a one-half to one billion dollar increase in crop value, if improvement in the energetics of nitrogen fixation were to result in a corresponding percentage increase in yield.

Two strategies for incorporating Hup genes into competitive wild strains are being investigated. The first entails the use of recombinant DNA technology to transfer "homologous" genes from Hup⁺ to Hup⁻ *R. japonicum* strains (Cantrell et al., 1983). The second strategy is to bring in foreign Hup genes and their promoters from other bacterial species (Timblin and Kahn, 1984).

Another possibility for improvement is to make *R. japonicum* strains that are more tolerant of osmotic stress (Strom et al., 1983). Such strains may aid in improving the stress tolerance of soybean plants in the field, as nitrogen assimilation is intimately related to the plant's ability to respond favorably to osmotic stress.

Osmotic stress protection genes are completely missing in *Rhizobium*. Such genes confer tolerance to salinity and drought in free-living organisms. Genes found in *Salmonella typhimurium* that confer tolerance have been transferred to *Klebsiella pneumoniae* and function in this background (Strom et al., 1983). The contribution of these genes, when in *Rhizobium*, to plant osmotic stress tolerance in the symbiote is unknown, but will soon be testable, thanks to the contribution of modern biotechnology.

Finally, the nature of the symbiotic interaction itself, including recognition and the infection process, and the control of genes that are induced as a consequence of symbiosis are under active investigation in several laboratories.

Soybean Improvement Goals Amenable to Genetic Engineering

Among the major breeding objectives in the soybean that are amenable to biotechnological solutions are seed protein and oil quality, plant stress tolerance, pest and disease resistance, and herbicide tolerance. The potential for biotechnology in meeting these objectives is great. We will briefly examine each objective, pointing out the factor needing improvement, the desired result, and the possible biotechnological strategems for solution.

THE STORAGE PROTEINS

Seventy percent of the storage proteins in soybean seeds are accounted for by two components of the globulin fraction, glycinin and β-conglycinin (Kitamura et al., 1984). Glycinin is made up of two subunits and β-conglycinin of three. The quality deficiency of these proteins is due to very low levels of the sulfur containing amino acids, methionine and cysteine (Larkins, 1983). The subunits of these proteins are coded for by multigene families (Goldberg, 1983). Some of the genes code for subunits with better levels of sulfur containing amino acids than others (Larkins, 1983). Goldberg (1983) has studied gene expression in these proteins and has found that regulation occurs primarily at the transcriptional level. Thus, it may be possible to alter the balance of subunits made by various genes in favor of those with higher levels of sulfur containing amino acids, by small changes in the regulatory region of the DNA. However, further elucidation of the regulatory mechanism is needed before this can be realized. Another suggested method for increasing the quality of these proteins is to alter structurally the subunits (Larkins, 1983).

Another approach to changing seed protein quality is to increase the levels of a protease inhibitor found in seed. The Bowman-Birk inhibitor, and its related family of inhibitors accounts for a small percentage of the seed protein (6% in the cultivar Tracy), but contains 20% cysteine. These inhibitors make up over half of the cysteine found in soybean seed (Larkins, 1983). So, small changes in the inhibitor content of seed could result in large changes in the cysteine content. The most desirable use of this strategy would couple increases in the amount of the inhibitor polypeptides with genetic changes that inactivate the polypeptide as a protease inhibitor.

Questions remain as to the effects of these changes on the plant. Changes in the globulin storage protein could affect the processing or packaging of the proteins as well as seedling nutrition. The function of the protease inhibitors in seeds and seedling development is largely unknown; increases in concentration of these compounds may have a deleterious effect (Larkins, 1983). On the other hand, Kitamura et al. (1984) describe two soybean genotypes that lack certain of the glycinin and β-conglycinin subunits. This tends to indicate that changes in the balance of the subunits may not have deleterious effects.

Although there are uncertainties in the engineering of seed proteins, a large amount of information has been gathered through the use of molecular biology, and the prospects for genetic manipulation of the seed proteins is encouraging.

Fatty Acids. Soybean oil is one of the major edible oils (Burton and Brim, 1983). The fatty acid contents of soybean oil from twenty lines tested by Hawkins et al. (1983) were as follows: 10.2 to 11.9% palmitic, 2.9 to 4.2% stearic, 18.6 to 27.9% oleic, 50.0 to 58.6% linoleic and 6.8 to 9.3% linolenic. Linolenic acid can be oxidized to form chemical intermediates that cause poor flavor (Burton and Brim, 1983). For commercial uses, the linolenic acid in soybean oil is lowered by hydrogenation (Burton and Brim, 1983; Cramer and Beversdorf, 1984). This process is becoming increasingly expensive (Burton and Brim, 1983); it also lowers the levels of linoleic acid and causes formation of positional and geometrical isomers of linoleic and oleic acids (Cramer and Beversdorf, 1984). Thus, many breeders are attempting to lower the linolenic content by conventional breeding techniques (Burton and Brim, 1983; Hawkins et al., 1983; Cramer and Beversdorf, 1984). Less than 2% linolenic acid is the target to alleviate the flavor problem (Brossman and Wilcox, 1984). Brossman and Wilcox (1984) found a mutant line with 3.5% linolenic acid and state that the lowest percentage reported from selection experiments was 4.2%.

The pathway of saturated fatty acid biosynthesis in plants is relatively well understood. Less well-characterized is the pathway for desaturation of fatty acids. The biosynthesis of oleic acid (18:1) from stearic acid is clear. The biosynthesis of linoleic acid (18:2) from oleic acid and of linolenic acid (18:3) from linoleic acid has not been clearly defined (Stumpf, 1980). Elucidation of these processes, or at least the ability to screen for linoleoyl desaturase, is needed before cloning or molecular biological techniques can be applied to this problem. Once these reactions are defined, the gene and promoter for this enzyme could be used to reduce the activity of the gene by new techniques being developed for turning off genes (Izant and Weintraub, 1984).

There is considerable interest in changing the overall fatty acid complement of various crops to produce valuable nonedible oils. Soybean is a good target crop for some of these manipulations, as it is widely grown, the facilitiates are in place for handling the crop and for crushing, and successful development of new uses for soybean oil would improve the profitability of the crop. Some of the fatty acids of interest are lauric, palmitic and ricinoleic acid; also of interest are the long chain waxy esters as in jojoba. Ricinoleic acid and the long chain fatty acids are both biosynthetic products from oleic acid (Stumpf and Shimakata, 1983) and would require the transformation of soybean by foreign genes.

Environmental Stress. Drought and salt tolerances are important objectives in soybean breeding. The study of osmoregulation has provided some insight into the mechanisms of these tolerances. Osmoprotective compounds have been postulated for plants. Various organic compounds are known to accumulate in plants following osmotic stress. The direct link of these compounds with stress tolerance in plants has not been made, but the circumstantial evidence is attractive (Strom et al., 1983). Boyer and Meyer (1980) studied osmoregulation in soybean seedlings and found that internal osmotic adjustment amounting to a 70% increase in solute content occurred when seedlings were placed under water stress. The compounds responsible for the increase were mainly fructose, sucrose and free amino acids. Singh and Gupta (1983) report differences among soybean cultivars in proline accumulation and relative water content when put under the same stress. Glycine betaine and proline, found in many plants, will protect bacteria against osmotic stress when added to their growth medium. Genes encoding production of proline compounds in bacteria have been cloned. When introduced into non-tolerant *Escherichia coli* strains, the genes conferred tolerance to these strains. The use of these genes in engineering bacteria is being pursued, as in the *R. japonicum* example given previously (Strom et al., 1983). There is potential for genetic engineering to play a role in the widening of tolerance to osmotic stresses in plants, but more information is needed on the mechanisms of tolerance in plants.

Pests and Diseases. The major pest in soybeans is the soybean cyst nematode, *Heterodera glycines* Ich. It occurs in most of the major soybean growing areas of the world and poses a continuing threat to soybean production in the United States (Sinclair, 1982). Luedders (1983) discusses the genetics of the soybean-nematode relationship and the conflicting reports regarding physiologic races. The major fungal pathogen of soybeans is *Phytophthora megasperma* pv. *glycinea* Kaun and Erwin, which causes a stem and root rot. It can cause severe damage under certain conditions with losses in yield of up to 50% (Sinclair, 1982). Twenty-two races have been reported (Laviolette and Athow, 1983). There are several other pests and diseases of somewhat less significance.

The use of genetic engineering to help solve disease and pest problems is probably going to take longer to realize than some of the other objectives discussed here. Genetic engineering relies on an effective screening method of gene products in order to isolate and clone a particular gene. Although there are many genes for resistance to various diseases, the mechanism or product of these genes that controls resistance is unknown. This makes the task of gene isolation, out of the one hundred thousand or more genes present in a plant, difficult. An alternative approach to cloning known resistance genes into the host is to clone genes encoding a pesticidal compound, such as the toxin to phytophagous lepidopterans found in *Bacillus thurengiensis* (Comai and Stalker, 1984).

Herbicide Compatibility. Increased tolerance of soybeans to several herbicides would be of interest. The herbicide market for the soybean crop amounts to approximately $1 billion, but the major herbicides are all for preemergence use. Trifluralin is effective on both broadleaf and grass weeds, alachlor is effective on grasses, and metribuzin is effective on broadleaves. There is a trend toward greater use of postemergence herbicides. Bentazon and acifluorfen are broadleaf weed herbicides, and glyphosate is both a grass and broadleaf herbicide. Glyphosate is presently applied by a wick method; for example, to kill volunteer corn and several of the difficult broadleaf weeds.

The major herbicides, for which there would be interest in increased tolerance, are glyphosate, metribuzin and atrazine. Tolerance to glyphosate by soybeans would allow its use as a general postemergence herbicide for control of grasses and eliminate the need for wick application (John Callahan, Calgene, pers. comm.). Tolerance to metribuzin by soybean plants at present varies by cultivar and environment (Andersen, 1975). Increased tolerance would allow the use of metribuzin as a postemergence herbicide at higher rates for improved weed control. Atrazine is an herbicide used in corn (*Zea mays* L.) and a large number of other crops for the control of broadleaf weeds. Soybeans are relatively intolerant to atrazine. Atrazine has a slow decomposition rate in soils, and thus, use of atrazine in corn-soybean rotations can present problems (Andersen, 1975). Atrazine tolerance would allow for easier crop rotations and the use of a good broadleaf herbicide in soybean crops.

Genetic engineering for herbicide tolerance will probably be one of the earliest areas in which to see results. Because the mechanism of action of herbicides is generally well described, and bacterial or other microbial screening can often be used, isolation of genes for tolerance is straightforward. When the mode of action of the herbicide is as an inhibitor of enzyme function, the DNA encoding an altered enzyme not inhibited by the herbicide can be cloned and used to transform the plant. Alternatively, a plant could be transformed by a gene encoding an enzyme whose activity results in degradation or conjugation of the herbicide, thereby inactivating it before the herbicide can damage the plant (Comai and Stalker, 1984).

A gene has been cloned from a mutant *Salmonella typhimurium* that provides glyphosate tolerance in bacteria (Comai et al., 1983). Work is now underway to engineer crop plants, including the soybean, for glyphosate tolerance by transfer of this bacterial gene.

The target site for atrazine is a chloroplast protein involved in photosynthetic electron transport. An altered target approach to atrazine resistance thus would require engineering of the chloroplast. A detoxification mechanism for atrazine in corn has been fully described, and is an attractive alternative approach for making other species more atrazine tolerant. In tomato, metribuzin tolerance has been shown to be governed by a single dominant gene, and the biochemistry has been suggested. These examples show the great potential for biotechnological solutions to improving herbicide compatibility of elite soybean cultivars (Comai and Stalker, 1984).

GENETIC ENGINEERING OF SOYBEANS: A PROSPECTIVE VIEW

We have briefly summarized the technology and surveyed some aspects of soybean biology that have been advanced with the use of modern biotechnology methods. We now turn our attention to genetic engineering of soybeans. By this, we mean the insertion and expression of useful foreign genes in the soybean genome to derive useful elite cultivars with new traits.

An Overview of Plant Genetic Engineering

The ability to manipulate DNA at the molecular level enables agricultural biologists to make genotypic combinations not previously possible, as well as making combinations possible by conventional methods in much more rapid and precise ways. There are, however, major technological requirements that will need to be met before these tools can be used outside the laboratory. These prerequisites are discussed here in the order in which they are used; gene isolation, vectors, gene transfer, and gene expression.

Gene Isolation. Relatively few important genes from plants have been cloned and sequenced. In part this is because our knowledge of biochemical pathways in plants is limited (i.e., few important gene products have been isolated and purified to homogeneity for study). There is a major need for increased understanding of the genetic basis of important plant traits. This knowledge will only come from a concerted effort, by plant geneticists, cytogeneticists, biochemists and developmental biologists, to study the germplasm of major crop species and their relatives to define agriculturally important traits in both genetic and biochemical terms. In the several cases where plant and bacterial metabolic pathways are similar, and where mutants are available or can be made in the bacterial system, genes from bacterial sources may well find use in genetic engineering of plants. Fatty acid synthesis, aromatic amino acid synthesis, and carbon fixation are examples currently under investigation in a number of laboratories.

Vectors. Two naturally occurring systems in plants involve insertion of DNA sequences into chromosomes. The megaplasmids Ti and Ri, transmitted in nature by the soil bacteria *Agrobacterium tumefaciens* and *A. rhizogenes,* respectively, contain a small region of DNA (the T-DNA) which is transferred by an as yet unknown mechanism to the chromosome of the host plant. In nature, the results are the diseases crown gall (Ti) and hairy root (Ri). Once it was understood that insertion of DNA containing functional genes that were expressed in the plant was the basis of these diseases, it was fairly straightforward to adapt these systems to the first-generation plant genetic engineering experiments. Much more sophisticated use of vectors, based on the ability of T-DNA to insert into chromosomes, will be possible once the molecular mechanism of the transfer is understood.

The other class of sequences with the ability to insert DNA into plant chromosomes are the transposable elements. These elements, which may carry functional genes of their own also (e.g., possibly encoding an enzyme for their own transfer), are receiving increasing attention because of their potential for use as vectors for plants.

Viruses are very important as vectors for genetic engineering of bacteria and animals, but plant viruses have so far found limited use in plant genetic engineering. On the basis of current knowledge, it is unlikely that a plant virus vector system will be developed that results in integration of foreign genes into plant chromosomes. Development of viruses as cloning vectors for use in plants might, however, lead to an efficient system for producing large amounts of a particular gene product in an inexpensive alternative (farmers' fields) to cell cultures in fermenters.

Several other vector approaches are under investigation. Chief among these are microinjection and direct DNA uptake. Microinjection as a means of introducing DNA into the nucleus has been very successful in animal embryos. Fine glass needles can be inserted into plant cells and even into the nucleus and a few picoliters of fluid containing recombinant DNA injected. Such cells can then be cultured. To date, no confirmed transformation of a plant species by this approach has been reported, but such results are expected soon. Microinjection technology will also be very important for transfer of chromosomes in advanced cytogenetic manipulations and possibly also for gene transfer into organelles.

Direct DNA transfer refers to methods by which DNA is presented to cells under conditions in which the DNA is taken up and some portion of it integrated by unknown mechanisms into the chromosome. Such methods work in bacteria and animals. Similar approaches have so far proven less successful in plants, but that situation may be changing. It has long been known that plant viral RNAs and DNAs can be taken up in a biologically active form and at a low efficiency the same has been shown for Ti DNA. These latter methods are attractive because they may be of very general usefulness (e.g., across many different species) and may not require the use of DNA from a pathogenic agent.

Gene Transfer. Gene transfer requires, in addition to a vector, the ability to manipulate cells of the species of interest so as to introduce DNA in a way that does not kill the cells, and to obtain from that cell a viable, functioning plant that has not been altered in undesirable ways. Plant organ and tissue culture is a well established technology whose origins go back to the early part of the 20th century. Progress in manipulating cultures of major food crops (the cereals and legumes) has been much slower, potato (*Solanum tuberosum* L.), tomato (*Lycopersicon esculentum* Mill.) and some vegetable species excepted. Whether from somatic or gametophytic cells, in the major crop species the ability to willfully, reproducibly, and frequently regenerate whole plants from individual cells or even cell aggregates in culture continues to be a limitation. Because of this difficulty, attention is now being given to the possible transfer of genes to pollen, ovules, or recently fertilized embryos. This approach would have the added advantage of avoiding an in-vitro culture step altogether, and thus would avoid the often undesirable manifestation of what has come to be called somaclonal variability.

Gene Expression. Though limited progress has been made in comparative studies of gene structure that have yielded insights into the factors governing expression of plant genes, it is obvious that there is much more that we don't know about this important topic. Newly available methods that make efficient gene transfer experiments possible will quicken dramatically the pace of research in this area. Such experiments will enable scientists to dissect the DNA sequences flanking the coding regions of genes, to make very specific changes, even at the single-nucleotide level, and then to study the consequences of such changes by measuring expression of the gene when it is placed back into the chromosome.

Cytogenetic Manipulation. For many years, cytogeneticists have been able in certain species (notably wheat [*Triticum aestivum* L.]) to obtain novel gene combinations by crosses with different species of the same or a near-relative genus. Though a time-consuming process to develop new plant cultivars, much of importance has been learned using these methods.

The development of microinjection, together with refinement of the methods for isolating chromosomes, development of fluorescence-activated sorting technology, and the construction of artificial chromosomes, bring hope that complex genetic traits of agricultural significance will one day be transferred and expressed to yield novel genotypes. As experimental tools, these methods will lead to advances in our understanding of coordinated gene regulation. As practical tools, they will lead to more rapid product development and make possible genetic engineering of plants for complex quantitative traits, such as yield, horizontal disease resistance and production of important secondary products, such as flavors, fragrances and pharmaceuticals.

Soybean Genetic Engineering

Vectors. Though reported not to be crown gall susceptible (DeCleene and DeLey, 1976; Matthysse and Gurlitz, 1982) soybean plants are susceptible to transformation by T-DNA of both the nopaline and octopine strains of *A. tumefaciens* (Pedersen et al., 1983; Hood et al., 1984); crown gall symptoms, stable transformation and T-DNA gene expression were observed (Pedersen et al., 1983). In as yet unpublished work by our colleagues at Calgene (D. Facciotti, C. Shewmaker, A. Valentine and J. O'Neal, pers. commun.), this important observation has been confirmed and extended. Thus, the use of recombinant Ti plasmids as one approach to genetic engineering of soybean plants seems to be possible.

Other vector possibilities include microinjection, direct DNA uptake, and viruses. Protoplasts of soybean cells in culture, including those transformed by T-DNA, have been prepared (Xu et al., 1982; Gamborg et al., 1983b). Thus, application of microinjection or direct DNA transfer to somatic cells in culture is possible, though untested. Among the viruses, several geminiviruses (Goodman, 1981) infect soybean plants. Development of geminivirus-based vectors for plant genetic engineering, however, is several years off.

Gene Transfer and Expression. Though T-DNA genes were expressed in soybean crown galls in the experiments by Pedersen et al. (1983), there are no published reports of the transfer and expression of a recombinant foreign gene in soybeans. Facciotti, Shewmaker and colleagues (pers. commun.) at Calgene have recently succeeded in this important accomplishment. Their experiment made use of the bacterial gene that inactivates the antibiotic kanamycin, to which soybean cells in culture are very sensitive. The recombinant DNA molecule inserted into the soybean genome contained a promoter obtained from a light-regulated nuclear gene in soybeans fused to the bacterial coding region for aminoglycoside phosphotransferase. Transformed soybean cells in culture showed a higher level of resistance to kanamycin when grown in the light than when grown in the dark. Thus, the results indicated that the promoter and the gene both were working as expected.

Regeneration of Plants from Cells in Culture. Exciting as the availability of soybean transformation technology is, it is of little use until methods are devised to regenerate whole, fertile, plants from soybean cells in culture (or until the transformation systems are amenable, as they are not now, to transformation of gametes). Two reports published in late 1983 (Christianson et al., 1983; Gamborg et al., 1983a), and unpublished work from at least one industrial laboratory, suggest that progress is being made towards the goal of regenerating soybean plants from somatic cells in culture. Obviously, the next step is to combine the transformation technology with the ability to regenerate plants.

CONCLUSIONS

While advances in plant biotechnology have outpaced the application of these new methods to the soybean, the necessary tools for progress with soybeans are falling into place. By the next World Soybean Research Conference, there will undoubtedly be examples of genetically engineered soybeans to talk about. We will also know a lot more about soybeans at the molecular genetic level. These are indeed exciting times in agricultural biology. We all can look forward to important changes as our technical capabilities expand. One hopes that, as a result, farming will become a more profitable and successful enterprise, and that real progress will be made possible in reducing the chronic and acute shortages of food that continue to plague humankind.

NOTES

Robert M. Goodman, Vice President, Research and Development, and Jack Kiser, Research Associate (Plant Breeding), Calgene Inc., 1920 Fifth St., Davis, CA 95616.

REFERENCES

Andersen, R. N. 1975. Differential soybean variety tolerance to herbicides. Sci. J. Series, Minn. Agr. Exp. Stn. No. 9098.

Boyer, J. S. and R. F. Meyer. 1980. Osmoregulation in plants during drought. p. 199-202. In D. W. Rains, R. C. Valentine and A. Hollaender, (eds.) Genetic engineering of osmoregulation. Impact on plant productivity for food, chemicals, and energy. Basic Life Sciences 14. Plenum Press, New York.

Brossman, G. D. and J. R. Wilcox. 1984. Induction of genetic variation for oil properties and agronomic characteristics of soybean. Crop Sci. 24:783-787.

Burton, J. W. and C. A. Brim. 1983. Recurrent selection in soybeans. IV. Selection for increased oleic acid percentage in seed oil. Crop Sci. 23:744-747.

Cantrell, M. A., R. A. Hangland and H. J. Evans. 1983. Construction of a *Rhizobium japonicum* gene bank and use in isolation of a hydrogen uptake gene. Cell 80:181-185.

Christianson, M. L., D. A. Warnick and P. S. Carlson. 1983. A morphogenetically competent soybean suspension culture. Science 222:632-634.

Comai, L., L. C. Sen and D. M. Stalker. 1983. An altered aroA gene product confers resistance to the herbicide glyphosate. Science 221:370-371.

Comai, L. and D. M. Stalker. 1984. Impact of genetic engineering on crop protection. Crop Protection (in press).

Cramer, M. M. and W. D. Beversdorf. 1984. Effect of genotype environment interactions on selection for low linolenic acid soybeans. Crop Sci. 24:327-330.

DeCleene, M. and J. DeLey. 1976. The host range of crown gall. Bot. Review 42:389-466.

Gamborg, O. L., B. P. Davis and R. W. Stahlhut. 1983a. Somatic embryogenes in cell cultures of Glycine species. Plant Cell Rep. 2:209-212.

Gamborg, O. L., B. P. Davis and R. W. Stahlhut. 1983b. Cell division and differentiation in protoplasts from cell cultures of Glycine species and leaf tissue of soybean. Plant Cell Rep. 2:213-215.

Goldberg, R. B. 1983. Organization and expression of soybean seed protein genes. p. 137-150. In L. D. Owens, (ed.) Genetic Engineering: Application to Agriculture. Beltsville Symposium 7. Rowman and Allanheld, Totowa.

Goodman, R. M. 1981. Geminiviruses. J. Gen. Virol. 54:9-21.

Hawkins, S. E., W. R. Fehr and E. G. Hammond. 1983. Resource allocation in breeding for fatty acid composition of soybean oil. Crop Sci. 23:900-904.

Hood, E. E., G. Jen, L. Kayes, J. Kramer, R. T. Fraley and M. D. Chilton. 1984. Restriction endonuclease map of pTl Bo542, a potential Ti plasmic vector for genetic engineering of plants. Bio/Technology 6:702-709.

Izant, J. G. and H. Weintraub. 1984. Inhibition of thymidine kinase gene expression by anti-sense RNA: A molecular approach to genetic analysis. Cell 36:1007-1015.

Key, J. L., C. Y. Lin and Y. M. Chen. 1981. Heat shock proteins of higher plants. Proc. Nat. Acad. Sci. 78:3526-3530.

Kitamura, K., C. S. Davies and N. C. Nielsen. 1984. Inheritance of alleles for Cgy_1 and Gy_4 storage protein genes in soybean. Theor. Appl. Genet. 68:253-257.

Larkins, B. A. 1983. Genetic engineering of seed storage proteins. p. 93. *In* T. Kosuge, C. P. Meredith and A. Hollaender (eds.) Genetic Engineering of Plants. An Agricultural Perspective. Basic Life Sciences 26. Plenum Press, New York.

Laviolette, F. A. and K. L. Athow. 1983. Two new physiologic races of *Phytophthora megasperma* f.sp. *glycinea*. Plant Dis. 67:497-498.

Lin, C. Y., J. K. Roberts and J. L. Key. 1984. Acquisition of thermotolerance in soybean seedlings. Plant Physiol. 74:152-160.

Luedders, V. D. 1983. Genetics of the cyst-nematode soybean symbiosis. Phytopathology 73:944-948.

Matthysse, A. G. and R. H. G. Gurlitz. 1982. Plant cell range for attachment of *Agrobacterium tumefaciens* to tissue culture cells. Physiol. Plant Pathol. 21:381-387.

National Research Council. 1984. Genetic Engineering of Plants: Agricultural Research Opportunities and Policy Concerns. National Academy Press, Washington.

Pedersen, H. C., J. Christiansen and R. Wyndalle. 1983. Induction and *in vitro* culture of soybean crown gall tumors. Plant Cell Rep. 2:201-204.

Schoffl, R. and J. L. Key. 1983. Identification of a multigene family for small heat shock proteins in soybean and physical characterization of one individual gene coding region. Plant Mol. Biol. 2: 269-278.

Sinclair, J. B. 1982. Copendium of Soybean Diseases. The American Phytopathological Society, St. Paul, Minn.

Singh, B. B. and D. P. Gupta. 1983. Proline accumulation and relative water content in soya bean (*Glycine max*) varieties under water stress. Ann. Bot. 52:109-110.

Strom, A. R., D. LeRudulier, J. W. Jakowec, R. C. Bunnell and R. C. Valentine. 1983. Osmoregulatory (Osm) genes and osmo-protective compounds, p. 39-60. *In* T. Kosuge, C. P. Meredith, and A. Hollaender (eds.) Genetic Engineering of Plants. An Agricultural Perspective. Basic Life Sciences 26. Plenum Press, New York.

Stumpf, P. K. 1980. Biosynthesis of saturated and unsaturated fatty acids. p. 177-204. *In* P. K. Stumpf and E. E. Conn (eds.) The Biochemistry of Plants. A Comprehensive Treatise. Vol. 4 Lipids: Structure and Function. Academic Press, New York.

Stumpf, P. K. and T. Shimakata. 1983. Molecular structures and functions of plant fatty acid synthetase enzymes. p. 1-15. *In* W. W. Thomson, J. B. Mudd and M. Gibbs (eds.) Proceedings of the Sixth Annual Symposium in Botany. American Society of Plant Physiologists, Rockville, MD.

Timblin, C. R. and M. L. Kah. 1984. Lactose inhibits the growth of *Rhizobium meliloti* cells that contain an actively expressed *Escherichia coli* lactose operon. J. of Bacteriol. 158:1204-1207.

Vodkin, L. O., P. R. Rhodes and R. B. Goldberg. 1983. A lectin gene insertion has the structural features of a transposable element. Cell 34:1023-1031.

Xu, Z. H., M. R. Davey, and E. C. Cocking. 1982. Callus formation from root protoplasts of *Glycine max* (soybean). Plant Sci. Lett. 24:111-115.

Molecular Biology

USE OF TISSUE CULTURE AS A TOOL IN SOYBEAN IMPROVEMENT

E. J. Roth, K. G. Lark and R. G. Palmer

The increasing use of tissue culture and molecular biology in plant science has emphasized the expectation that these techniques may be used to formulate new approaches to agricultural problems and to develop new germplasm. More often than not, the emphasis is on the production of new plants, through alteration or selection of special attributes at the cellular level and subsequent regeneration of intact plants from these new cell types. However, there are many other uses of tissue culture, particularly as an adjunct to experiments with whole plants, in which it often provides a rational explanation for phenomena observed in the growing plant. For soybean, methodology of plant regeneration is still in its infancy, and therefore, the benefits to be derived from cell-culture research are not always evident. In this paper, we shall first review the "state of the art" and then discuss some specific experiments using somatic genetic techniques. We shall then indicate how these experiments have led to a reevaluation of results obtained from sexual crosses and the constraints that may be imposed on a plant breeding system as a result of evolutionary changes that accompany obligatory inbreeding.

TECHNIQUE

Preparation of Tissue Cultures

Callus or suspension-cell cultures have been prepared from a number of cultivars and species without any difficulty (Beversdorf and Bingham, 1977; Phillips and Collins, 1981; Gamborg et al., 1983). Cultures can be prepared from almost any part of the growing plant. Heterotrophic cultures usually are prepared by using sucrose or somtimes glucose as a carbon-energy source (Gamborg et al., 1968). Other carbon sources can be substituted, depending on the cultivar. Recently, autotrophic cultures have been prepared (Horn et al., 1983). A condition for their growth in cell culture seems to be an increased partial pressure of CO_2. Frequently, cultures prepared from different cultivars will display different nutritional requirements for optimal growth, usually requiring different concentrations of metals or auxin and, sometimes, a different temperature for optimal growth (Table 1).

Protoplast Formation and Cell Fusion

In tissue culture, somatic hybrids usually are prepared by fusion of protoplasts and regeneration of the cell wall. Soybean was used extensively in some of the first cell-fusion studies (Kao and Michayluk, 1974; Constabel et al., 1976). However, all these involved interspecific hybrids, usually of soybean with tobacco (*Nicotiana tabacum* L.), barley (*Hordeum vulgare* L.), corn (*Zea mays* L.), or pea (*Pisum sativum* L.). More extensive studies have shown that soybean can be protoplasted easily (Gamborg et al., 1974), but that these protoplasts are difficult to cultivate and do not regenerate cell walls readily. Unlike *Datura*, *Petunia*, or tobacco, fusions between soybean lines are not yet possible on a routine basis. Most likely, this reflects the low viability of soybean protoplasts (1-10%), which could reduce the probability of successful fusion to as much as 1/10,000. It may well be that the early success with interspecific fusions reflects the ability of other cells, as in tobacco, to regenerate cell walls. Until recently, genetically altered cell lines, such as nutritional auxotrophs, with good selective markers were not available. These are necessary to allow rapid and easy selection of somatic hybrids. As these have become available (to be discussed later), we can expect to see more progress in the improvement of cell-fusion techniques for soybean.

Genetic Transformation and Regeneration of Plants from Cultured Cells

Transformation of tobacco, *Petunia*, and *Datura*, by using the Ti plasmid of *Agrobacterium tumefasciens*, has made it possible to consider introducing a number of new genes into these plants. As yet, this system cannot be used with soybean. At least two factors restrict its use. First, there is the difficulty in producing tumors on soybean, although there is at least one report that this can be done (Pedersen et al., 1983). Second is the inability to get large numbers of viable protoplasts into which plasmic DNA can be introduced. Finally, even if transformation of individual cells could be accomplished, a reproducible system for regenerating plants from individual cells is needed. As yet, no such system has been published. Some progress has been made, however, in regenerating plants from tissue cultures prepared from embryos (Christianson et al., 1983) or from tissues grown in a type of organ culture (Kartha et al., 1981). Thus, genetic transformation of soybean must await at least two technical advances: regeneration of plants from protoplasts and/or the development of a suitable vector system if the Ti plasmic system cannot be adapted to soybean.

Production of Haploid or Partial-Haploid Cell Lines

Haploid cell lines have been produced by preparing cell lines from haploid plants (Weber and Lark, 1980; Vasil, 1980). More recently, we have developed a method of producing partial haploid cells by treating suspension cultures with the herbicide isopropyl-N-(3-chlorophenyl) carbamate (CIPC) (Roth et al., 1982). These cell lines are of use in both genetic mapping and production of recessive mutants, as well as in understanding the interactions between genes sharing a common genetic background. Because such lines can be produced rapidly and in large numbers, it is possible to use them for preparing a genetic map. This can be done as follows: Two parents are combined in a sexual cross to produce a heterozygote. Cell lines are prepared from the parental plants and the heterozygote. The DNA (or protein) from these three lines is extracted and examined carefully for the presence of any of an abundance of molecular markers. For DNA, these would be represented by restriction polymorphisms; for proteins, polymorphisms can be detected by isozyme staining and monoclonal antibody reactions. Monoclonal antibody reactions would be the easiest, because they can be used by any scientist or technician as easily as blood-typing tests. Markers are found that have a different polymorphism in each parent. Of course, both polymorphic forms will be present in the heterozygote. When chromosomes are lost from the heterozygote, polymorphisms present on that chromosome will disappear. Once such a map has been made, it can be used by breeders or field geneticists to examine segregating crosses. An advantage of tissue culture is the speed and the ability to control segregation distortion due to selective pressures (see later).

GENETIC VARIATION

Two sources of genetic variants are available: mutants produced by mutagenizing cells growing in culture and cells derived from different germplasm. Cell-culture techniques have been developed for the isolation of mutant or variant clones derived from single cells (Weber and Lark, 1979) and for the storage of these cell lines in liquid nitrogen (Weber et al., 1983). A detailed study of the parameters of mutagenesis has led to a routine procedure for the isolation of mutants (Weber and Lark, 1980; Zhou et al., 1981; Roth and Lark, 1982). Figure 1 summarizes this procedure.

In general, mutations that are isolated nonselectively by screening mutagenized cells seem to be stable (Zhou et al., 1981; Roth and Lark, 1982). However, mutants that were selected for resistance to a variety of selective agents have proved unstable, possibly due to reversible changes in the genome, such as gene rearrangements or gene amplification. Recently, we have had success in enriching particular classes of mutations by selections that kill wild type cells without injuring the mutants. These classes of mutations include nucleoside auxotrophs (Roth et al., 1982) as well as amino acid auxotrophs. For most of these studies, it has been essential to have haploid or partial haploid cell lines in order for recessive phenotypes to be expressed.

Because soybean is an obligatory inbreeding plant, variation in germplasm segregates rapidly. Cell lines isolated from widely separated plant introductions display different phenotypes for a variety of physiological markers as well as isozymes. The physiological markers include growth in culture at different temperatures, at different auxin concentrations, and on different nitrogen sources (Roth and Lark, 1984). The variants isolated in this way, both induced mutations and those found in different plant introductions, are useful for two types of study: the analysis of biochemical and physiological traits and the study of gene-product interactions that can have an effect on breeding strategies.

GENETIC BACKGROUND EFFECTS

Gene Expression

Possible effects of genetic background on growth physiology, which could result in segregation distortion, have been encountered (Roth et al., 1982; Roth and Lark, 1984). 'Minsoy' (PI 27890) is a Maturity Group 0 plant introduced into the United States from France in 1910. 'Noir' 1 (PI 290136), also of Maturity Group 0, was introduced into the U.S. from Hungary in 1963. Minsoy and Noir 1 are standard lines for $fr_1 fr_1$ and $fr_2 fr_2$ (nonfluorescent roots), respectively (Delannay and Palmer, 1982). Tissue-culture cell lines were prepared from these, and from their F_1 hybrid. The hybrid cell line was treated with CIPC to prepare partial-haploid cell lines that had lost one or more chromosomes. Growth of the parent, hybrid, and partial-haploid cell lines was then measured under different sets of controlled conditions. In addition, several isozyme polymorphisms were examined. Results demonstrated that the different partial haploids had lost from 7 to 18 chromosomes and that chromosome loss was occurring randomly. Isozyme data indicated (a) that partial haploids had phenotypes (isozyme patterns) that would be expected if markers behaved independently in the

276

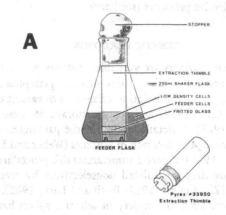

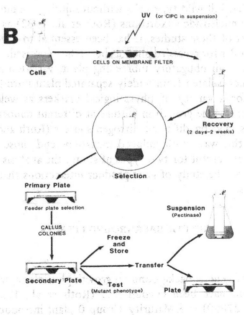

Figure 1. A schematic representation of the steps involved in mutant or partial-haploid production using soybean cell suspension cultures. (A) System for maintaining low-density cell suspensions in culture-sharing medium of high-density cultures. The low-density culture is maintained inside the glass extraction thimble with the high-density cells outside. When selections are operative, the cells on the outside, which may be dying, can be replaced with fresh, healthy, cells every two days. Medium from the outside diffuses in and feeds the cells on the inside. Toxic products from cells that die inside the thimble diffuse out and are removed every one or two days when the cells are changed. This system resembles the feeder plating system for obtaining callus clones from single cells (Weber and Lark, 1980). (B) Use of the system shown in A in conjunction with feeder plates to produce or select mutant and partial-haploid cell lines.

heterozygote and if chromosome loss had revealed the expected pattern for the parental genes that remained, and also (b) that other, new, isozyme patterns were created, suggesting that enzyme mobilities were altered by different patterns of protein modification dependent on the genetic backgrounds of the different partial haploids.

Physiological studies of the partial haploids revealed two aspects of chromosome loss: (a) parental phenotypes and (b) new, usually more restrictive, phenotypes (i.e., a quantitative change). Table 1 summarizes data of three growth phenotypes from this study (for the actual growth curves, see Roth and Lark, 1984).

It can be seen that partial haploid 8 is much more severely affected by 22 C than Noir 1, and that 2 is much more severely affected by growth in allantoin than either of the parent cell lines. Partial haploid 12 requires much more auxin than any other line (including Minsoy), and 5 is much more sensitive to high auxin concentrations than either parent. Partial haploid 8 seems to be exceptionally resistant to high auxin concentrations.

These results led us to hypothesize that the phenotypes of these cells depended on the interactions of several genes, and that the phenotype that characterized one parent was derived from a different combination of genes (or gene alleles) than the ones involved in the phenotype of the other parent. As a result, removal of chromosomes from one parent could not be compensated for completely by the remaining homologous chromosomes derived from the other parent. The basis for this supposition was that the genetic background of a particular soybean cultivar evolved independently from other cultivars and, as such, will have developed a number of genetic idiosyncrasies that have "fitted" it to the selective conditions under which it has been growing.

A prediction based on this hypothesis was that a healthy sexual hybrid between widely divergent cultivars might not always give rise to all possible genetic combinations in subsequent generations. This idea was tested by examining the progeny of

Table 1. Growth of parental and partial-haploid cell lines under different conditions.

Cell lines:[a]	M	Fl	Nr	2	5	8	12
Growth condition[b]							
22 C	G[c]	G	15	G	G	3	12
Allantoin	50	G	50	5[d]	G	G	G
Auxin:[e]							
0.5 mg/L	10	G	G	G	G	G	2
1 mg/L	10	G	G	G	G	G	10
2 mg/L	G	G	G	G	G	G	10
12 mg/L	G	G	G	G	1	G	G
24 mg/L	2	G	2	G	1	G	G
48 mg/L	1	1	1	1	1	300	2

[a]M = Minsoy; Nr = Noir 1; Fl = hybrid; 2, 5, 8, 12 are partial haploids derived from Fl.

[b]22 C—Cells were transferred from 33 C to 22 C, and the increase in cell mass was measured. Allantoin—Cells were transferred to medium containing allantoin as the sole source of nitrogen, after which packed cell volume (PCV) was measured.

[c]G indicates that cell lines lived and continued to grow exponentially. Numbers indicate that cell lines died after an increase in PCV equivalent to the initial cell mass multiplied by the number shown.

[d]This value corresponds to the increase in PCV in the absence of any added extracellular nitrogen.

[e]Cells were transferred into medium containing auxin concentration shown, after which PCV was measured.

sexual hybrids between Minsoy and Noir 1.

Selection Against Certain Gene Combinations

We have observed aberrant segregation of the Fr_2 allele among F_2 plants of the cross Minsoy x Noir 1, and in the reciprocal cross. Roots from F_2 seedlings derived from these crosses were examined under ultraviolet (UV) light. All seedlings with non-fluorescent roots were transplanted to the field. The identity of these plants was maintained, and they were used as male parents in cross-pollinations both to Minsoy and to Noir 1. The resulting hybrid seed were germinated, and their roots examined with UV light and classified as fluorescent or nonfluorescent.

F_2 plants of the original cross of Minsoy x Noir 1, and the reciprocal cross, segregated the expected 9 fluorescent: 7 nonfluorescent roots. Genotypically, the non-fluorescent class was composed of:

$$2 \; Fr_1 \; fr_1 \; fr_2 \; fr_2$$
$$1 \; Fr_1 \; Fr_1 \; fr_2 \; fr_2$$
$$2 \; fr_1 \; fr_1 \; Fr_2 \; fr_2$$
$$1 \; fr_1 \; fr_1 \; Fr_2 \; Fr_2$$
$$1 \; fr_1 \; fr_1 \; fr_2 \; fr_2$$

In our cross-pollinations of the nonfluorescent phenotype to Minsoy, we expected to identify three genotypic classes among the F_2 plants:

2   segregating 1 fluorescent: 1 nonfluorescent ($Fr_1 \, fr_1$)
1   all fluorescent ($Fr_1 \, Fr_1$)
4   all nonfluorescent ($fr_1 \, fr_1$)

Similarly, in our cross-pollinations of the nonfluorescent phenotype to Noir 1, we expected the 2:1:4 ratio. Table 2 gives the expected and observed ratios for fluorescent and nonfluorescent roots. It can be seen that, at the Fr_2 locus, segregation was nonrandom in the Noir 1 cross (Table 2). There were no differences between reciprocal crosses.

We then germinated additional seed of the cross, F_2 nonfluorescent phenotype x Noir 1. In addition to fluorescence and nonfluorescence of the roots, two additional genes were classified. Minsoy has $Pb \, Pb$ (sharp pubescence tip) and is $W_1 \, W_1$ (purple flower); Noir 1 has $pb \, pb$ (blunt pubescence tip) and is $w_1 \, w_1$ (white flower). Sharp pubescence tip (Pb --) and purple flower (W_1 --) nonfluorescent F_2 plants were used in crosses to Noir 1. The data on segregation of pubescence tip and flower color involved only nine plants and were consistent with random segregation of Pb and W_1. If segregation were as distorted as at the Fr_2 locus, we might have expected to see an effect,

Table 2. Backcross data with Minsoy and Noir 1 as female parents.[a]

Female parent		Phenotypic classes[b]			X^2	F
		1F:1NF	All F	All NF		
Minsoy	Expected	2	1	4		
	Observed	7	2	5	3.38	>0.10
Noir 1	Expected	2	1	4		
	Observed	0	1	10	5.56	>0.05

[a]Male parents were nonfluorescent F_2 plants of Minsoy x Noir 1, and Noir 1 x Minsoy. Data were combined.

[b]F = fluorescent roots; NF = nonfluorescent roots.

but lesser degrees of segregation distortion would not have been observed.

In other data, significant segregation distortion has been observed, albeit of a less pronounced degree. Larger samples were studied in the analysis of sexual crosses between *G. soja* and *G. max* (Gai et al., 1982) in which the distribution of a qualitative character, tawny (*T*) and gray (*t*) pubescence color, was examined. In one such experiment a ratio of tawny: segregating: grey of 90:60:90 was expected among F_3 plants, but the observed ratio of 90:87:58 showed a significant deviation from expected (see F_3 plants of PI 424001 x Amsoy 71, in Table 5 of Gai et al., 1982). In another series of experiments involving the same parents and the same alleles (Carpenter, 1984) segregation distortion again was observed that became less pronounced when a higher percentage of the Amsoy 71 parental genome was introduced through successive back-crossing. When these two sets of experiments are taken together, they indicate that segregation distortion is the result of unequal mixing of genetic background rather than the presence or absence of genes from a particular parent.

DISCUSSION

We conclude that certain gene combinations, not predicted from the phenotypes of the parental cell lines or the heterozygote from which the partial haploid was derived, can exert a detrimental effect on plant growth. This leads us to hypothesize that certain gene combinations could lead to the loss of progeny plants and, hence, give rise to distorted segregation ratios. Such an effect is of great importance to soybean breeding. To better understand this loss of certain progeny or restricted recombination, it will be necessary to undertake a detailed genetic analysis of wide crosses.

Tissue culture, coupled with modern molecular techniques, provides the methodology to obtain such an analysis quickly. It would now be possible to obtain a detailed genetic map of soybean within less than two years if sufficient resources were committed to the effort. Because of the natural isolation of germplasm that results from inbreeding, a multitude of molecular markers are available that could be characterized readily by using tissue-culture lines prepared from different plant introductions. Because of the short generation times of cultured cells, it should be possible to shorten the time required for such an analysis and then to confirm the results by a rapid survey of F_2 plants obtained by self-pollination of the F_1 hybrids from the sexual crosses.

NOTES

E. J. Roth, Graduate Student, K. G. Lark, Professor, Department of Biology, University of Utah, Salt Lake City, Utah 84112. R. G. Palmer, Research Geneticist, ARS USDA, Departments of Agronomy and Genetics, Iowa State University, Ames, Iowa 50011.

Joint contribution: Agricultural Research Service, U.S. Department of Agriculture, and Journal Paper No. J.11559 of the Iowa Agric. and Home Econ. Exp. Stn., Ames, Iowa 50011; Project 2471.

Mention of a trademark or proprietary product by the USDA, Iowa State University, or University of Utah does not imply its approval to the exclusion of other products that also may be suitable.

Research of EJR and KGL, described in this paper, was supported by grants from the NIEHS, NIES No. 01498 and the NSF, No. PCM-8010771.

REFERENCES

Beversdorf, W. D. and E. T. Bingham. 1977. Degrees of differentiation obtained in tissue cultures of *Glycine* species. Crop Sci. 17:307-311.

Carpenter, J. A. 1984. Segregation for agronomic traits in populations with varying percentages of *Glycine soja* germplasm. M.S. Thesis, Iowa State University Library, Ames, Iowa.

Christianson, M. L., D. A. Warnick and P. S. Carlson. 1983. A morphogenetically competent soybean suspension culture. Science 222:632-634.

Constabel, F., G. Weber, J. W. Kirkpatrick and K. Pahl. 1976. Cell division of intergeneric protoplast fusion products. Z. Pflanzenphysiol. 79:1-7.

Delannay, X and R. G. Palmer. 1982. Four genes controlling root fluorescence in soybean. Crop Sci. 222:278-281.

Gai, J., W. R. Fehr and R. G. Palmer. 1982. Genetic performance of some agronomic characters in four generations of a backcrossing program involving *Glycine max* and *Glycine soja*. Acta Genet Sin. 9:44-56.

Gamborg, O. L., F. Constabel, L. Fowke, K. N. Kao, K. Ohyama, K. Kartha and L. Pelcher. 1974. Protoplast and cell culture methods in somatic hybridization in higher plants. Can. J. Genet. Cytol. 16:737-750.

Gamborg, O. L., B. P. Davis and R. W. Stahlhut. 1983. Somatic embryogenesis in cell cultures of *Glycine* species. Plant Cell Rep. 2:209-212.

Gamborg, O. L., R. A. Miller and K. Ojima. 1968. Nutrient requirements of suspension cultures of soybean root cells. Exp. Cell Res. 50:151-158.

Horn, M. E., J. H. Sherrard and J. M. Widholm. 1983. Photoautotrophic growth of soybean cells in suspension culture. I. Establishment of photoautotrophic cultures. Plant Physiol. 72:426-429.

Kao, K. N. and M. R. Michayluk. 1974. A method for high-frequency intergeneric fusion of plant protoplasts. Planta (Berl.) 115:355-367.

Kartha, K. K., K. Pahl, N. G. Leung and L. A. Mroginski. 1981. Plant regeneration from meristems of grain legumes: soybean, cowpea, peanut, chick pea, and bean. Can. J. Bot. 59:1671-1679.

Pedersen, H. C., J. Christiansen and R. Wyndaele. 1983. Induction and in vitro culture of soybean crown gall tumors. Plant Cell Rep. 2:201-204.

Phillips, G. C. and G. B. Collins. 1981. Induction and development of somatic embryos from cell suspension cultures of soybean. Plant Cell Tissue Organ Cult. 1:123-126.

Roth, E. J. and K. G. Lark. 1982. Isolation of an auxotrophic cell line of soybean (*Glycine max*) which requires asparagine or glutamine for growth. Plant Cell Rep. 1:157-160.

Roth, E. J. and K. G. Lark. 1984. Isopropyl-N-(3-chlorophenyl) carbamate (CIPC) induced chromosome loss in soybean: A new tool for plant somatic cell genetics. Theor. Appl. Genet. (in press).

Roth, E. J., G. Weber and K. G. Lark. 1982. Use of isopropyl-N-(3-chlorophenyl) carbamate (CIPC) to produce partial haploid cells from suspension cultures of soybean (*Glycine max*). Plant Cell Rep. 1:205-208.

Vasil, I. K. 1980. Androgenetic haploids. p. 195-223. *In* G. H. Bourne and J. F. Danielli (eds.) International review of cytology, supplement 11A. Academic Press, New York.

Weber, G. and K. G. Lark. 1979. An efficient plating system for rapid isolation of mutants from plant cell suspensions. Theor. Appl. Genet. 55:81-86.

Weber, G. and K. G. Lark. 1980. Quantitative measurement of the ability of different mutagens to induce an inherited change in phenotype to allow maltose utilization in suspension cultures of the soybean *Glycine max* (L.) Merr. Genetics. 96:213-222.

Weber, G., E. J. Roth and H. G. Schweiger. 1983. Storage of cell suspensions and protoplasts of *Glycine max* (L.) Merr. by freezing. Z. Pflanzenphysiol. 109:29-39.

Zhou, J. P., E. J. Roth, W. Terzaghi and K. G. Lark. 1981. Isolation of sodium dependent variants from haploid soybean cell culture. Plant Cell Rep. 1:48-51.

MOLECULAR FEATURES OF STORAGE PROTEINS IN SOYBEANS

N. C. Nielsen

The principal storage proteins in soybeans have been given the trivial names glycinin and β-conglycinin. The properties of these proteins have been studied in a large number of laboratories and the work was recently reviewed (Nielsen, 1985; Beachy and Fraley, 1985). Together they account for about 70% of the total seed protein, and they both are located in specialized cell compartments called protein bodies. Since the proteins account for such a large proportion of total seed protein, they have a major influence on seed nutritional quality and the functionality of seed products. Opportunities undoubtedly exist that will permit genetic manipulation of these properties, but efforts in this area will be hampered because of a lack of information about the structure of the proteins, the genes which encode them, and the mechanisms that are operative during expression of these genes.

CHEMICAL COMPOSITION

My laboratory has concentrated on glycinin because it is generally present in a higher concentration and is a richer source of sulfur amino acids than β-conglycinin. Glycinin is devoid of sugar and does not undergo the ionic strength association/dissociation phenomena characteristic of β-conglycinin. When purified from dilute extracts, it is a 12.2 S hexamer of about 360,000 daltons. Each of the six subunits consist of two polypeptide components, one with an acidic and the other with a basic isoelectric point. The acidic and basic polypeptides are linked by a single disulfide bond, which is located at an homologous position in each subunit (Staswick et al., 1984a). The basic polypeptide components are all about 20,000 daltons, while the acidic polypeptide components are more variable in apparent molecular weight and range between 38,000 and 48,000 daltons.

The initial objective of our studies was to purify each major glycinin polypeptide and establish the structural relationships between them. The early studies (Moreira et al., 1979) revealed that the acidic and basic polypeptides each belonged to small families of proteins with homologous NH_2-terminal sequences (Table 1). Subsequent studies (Moreira et al., 1981; Staswick and Nielsen, 1983) showed that the high degree of homology extended into the interior portions of the polypeptides and permitted identification of six major acidic (A_{1a}, A_{1b}, A_2, A_3, A_4, A_5) and five major basic (B_{1a}, B_{1b}, B_2, B_3, B_4) polypeptides. It was concluded that the proteins were synthesized from a homologous gene family that has evolved from a common ancestral gene.

Differences in the first 10 to 20 residues of the NH_2-terminal sequence permitted unambiguous identification of each acidic and basic polypeptide component (Table 1).

Table 1. Comparison of the NH_2-terminal sequences of the acidic and basic polypeptides in glycinin subunits from breeding line CX635-1-1-1.

Group	Subunit	Mol. wt.[a]	No. met.[b]	Acidic	Basic
I	$A_{1a}B_2$	58,000	5-6	FSSREQPQQNECQIQKLNALKPD...GIDETICTMRLRQNIGQTSSPDIY	
I	$A_{1b}B_{1b}$	58,000	5-6	FSFREQPQQNLUQIQKLNALKPD...GIDETICTMRLRHNIGQTSSPDIY	
I	A_2B_{1a}	58,000	7-8	LREQAQQNECQIQKLNALKPD...GIDETICTMRLCRHNIGQTSSPDIF	
II	A_3B_4	62,000	3	ITSSKF NECQLNNLNALQPD...GVEENICTMKLHENIARPSWARFY	
II	$A_5A_4B_3$	69,000	3	ISSSKL NECQLNNLNALEPD...GVEENICTLKLHENIARPSWARFY	

[a]Apparent molecular weight estimated by SDS-electrophoresis. Molecular mass for A_2B_{1a} determined from primary structure is $31,600 \pm 100$ (A_2) + $19,900 \pm 100$ (B_{1a}) = $51,500 \pm 200$ (Staswick et al., 1984a).

[b]Variability in methionine content reflects heterogeneity among subunits (Staswick et al., 1984b).

We, therefore, separated the A-SS-B subunits and determined their identity by sequence analysis (Staswick et al., 1981). The principal conclusion drawn from this study was that, with the exception of A_4, each basic polypeptide was non-randomly linked to a specific acidic one (Table 1). It was subsequently shown that A_4 originated from the NH_2-terminal region of subunit $A_5A_4B_3$ (Staswick et al., 1984b; Kitamura et al., 1984; Scallon and Nielsen, unpublished). Evidently, a proteolytic cleavage occurs between A_4 and A_5, and this causes the subunit to dissociate into free A_4 and A_5-SS-B_3 upon exposure of the protein to denaturing conditions. The polypeptide assignments for subunits are based on our studies with breeding line CX635-1-1-1, and sequence heterogeneity will undoubtedly be observed among soybean cultivars.

Comparison of the physical properties of the glycinin subunits in CX635-1-1-1 prompted us to separate them into two distinct groups (Table 1). Subunits in Group-1 have more uniform molecular weights and contain more methionine than members of Group-2. This feature could be significant in efforts to manipulate seed nutritional quality, since legumes contain suboptimal levels of sulfur amino acids. The extent of sequence homology between subunits provides the second criterion which distinguishes the two subunit groups. NH_2-terminal and internal sequence homology among members of the same group exceeds 90%, whereas that between members of different groups is only 50 to 60% (Table 1 and Nielsen, unpublished).

Analytical isoelectric focusing showed that each glycinin polypeptide component exhibited charge microheterogenity (Moreira et al., 1981). To determine the basis for this heterogenity, the complete amino acid sequence of the A_2B_{1a} subunit was determined. Structural heterogeneity was observed at a number of locations in both A_2 (Figure 1) and B_{1a} (Figure 2). The fact that the heterogenity is not confined to the ends of the components, and that amino acids other than those which can undergo deamidation are involved, implies that several different coding sequences contribute to the A_2B_{1a} subunits. I consider it likely that the sequence heterogenity reflects a larger number of genes than has been detected by DNA solution hybridization (Fischer and Goldberg, 1982; Goldberg et al., 1981), although the possible involvement of residual genetic heterogeneity among the population of inbreds from which the seed polypeptides were purified cannot be excluded.

A₂ POLYPEPTIDE COMPONENT

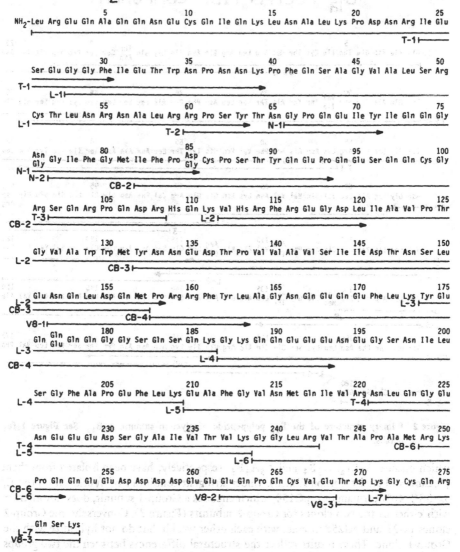

Figure 1. Primary structure of the A₂ polypeptides from glycinin subunit A_2B_{1a}. Standard three letter abbreviations are used to denote amino acids. Horizontal lines indicate regions sequenced for each peptide. The beginning and end of peptides are denoted with vertical bars. Arrowheads indicate the peptide continues, but was not sequenced beyond that point. Peptides were generated by cleavage with CNBr (e.g., CB-1, CB-2, etc.), NH₂OH (e.g., N-1, N-2, etc.), trypsin on citraconylated peptides (e.g., T-1, T-2, etc.), *Staphylococcus aureus* V-P protease (e.g., V8-1, V8-2, etc.) and endoproteinase Lys-C (e.g., L-1, L-2, etc.). Data from Staswick et al. (1984b).

GENE STRUCTURE AND EXPRESSION

Messenger RNA's isolated from midmaturation stage seeds have been used to construct cDNA libraries. Three glycinin clones, denoted pG154, pG23 and pG258,

B_{1a} POLYPEPTIDE COMPONENT

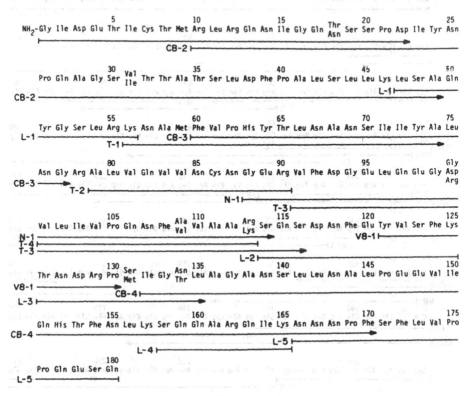

Figure 2. Primary structure of the B_{1a} polypeptide of glycinin subunit A_2B_{1a}. See Figure 1 for details.

which encode A_2B_{1a}, A_3B_4 and $A_5A_4B_3$, respectively, have been isolated from them (Scallon and Nielsen, unpublished). Under stringent conditions of hybridization (64 C, 5x SSC, 40% formamide), pG154, which encodes a Group-1 subunit, does not interact with either of the two clones for Group-2 subunits (Figure 3). Conversely, the Group-2 clones pG23 and pG258 interact with each other weakly but do not hybridize with the Group-1 clone. These results reflect the structural differences between the two groups of glycinin subunits, and provide a means to distinguish among certain glycinin coding sequences.

The three cDNA clones have been used as probes to identify genomic DNA fragments that encode glycinin subunits. In southern blots, the Group-1 probe, pG154, hybridizes to three Eco RI restriction fragments at moderate stringency (Table 2). The three fragments are about 5.3, 4.1 and 3.2 kb in size and correspond to G1, G2 and G3 described by Fisher and Goldberg (1982). They are derived from the 3'-end of genes that I refer to as glycinin-1, glycinin-2 and glycinin-3, respectively. The two Group-2 probes, pG23 and pG258, do not hybridize with G1, G2 or G3 at moderate or high stringency, but instead hybridize to 13 and 9 kb Eco RI restriction fragments. We (Scallon and Nielsen, unpublished) have since demonstrated that an intact $A_5A_4B_3$ gene (e.g., glycinin-4) is in the 13 kb fragment, and that the A_3B_4 gene

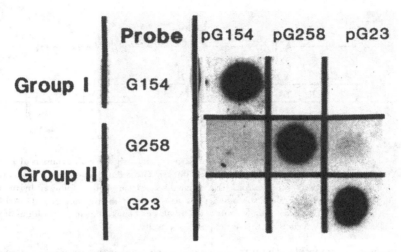

Figure 3. Cross hybridization among glycinin probes. About 5 ng of DNA was bound to nitrocellulose. It was denatured in 0.1 N NaOH at 50 C for 5 min and equilibrated in 10x SSC. Southern hybridization was at 40% formamide, 5x SSC at 64 C for 16 h and the blot was washed at 64 C in 0.2x SSC and 0.1% SDS.

(e.g., glycinin-5) is in the 9 kb fragment. Genomic reconstructions using the three probes suggests that there is about one copy of each glycinin gene per haploid genome (Fischer and Goldberg, 1982; Scallon and Nielsen, unpublished).

Genomic clones that encode the major glycinin subunits have been isolated from libraries of soybean DNA. The coding regions for glycinin-1, glycinin-2 and glycinin-3 have been sequenced in their entirety, whereas only the essential features of the sequences for the group-2 genes have been determined. The studies show that all of glycinin genes share common structural features (Figure 4). Each gene encodes a precursor with a short signal peptide, followed by the acidic component, a linker region and then the basic component. The coding sequence is interrupted three times by introns, twice in the region that encodes the acidic component and once in the region

Table 2. Comparison of glycinin genes in cultivar Dare.

Subunit group	Subunit ident.[a]	Eco RI fragment size (kb)[b]	Hybridization			Glycinin gene
			pG154	pG23	pG258	
I	$A_{1a}B_{1b}$	5.3	+	-	-	Gy_1
I	A_2B_{1a}	4.1	+	-	-	Gy_2
I	$A_{1b}B_2$	3.2	+	-	-	Gy_3
II	$A_5A_4B_3$	13.0	-	±	+	Gy_4
II	A_3B_4	9.0	-	+	±	Gy_5

[a]NH_2-terminal sequences of A_{1a}, A_{1b}, B_{1b}, B_2 are similar but not identical with those of CX635-1-1-1.

[b]Probes include about the 3'-terminal half of coding sequence for A_2B_{1a} (pG154), A_3B_4 (pG23) and $A_5A_4B_3$ (pG258). In the case of the Group-1 genes, other Eco RI fragments, which originate from the 5'-end of the gene, may hybridize to probes that contain additional coding sequences.

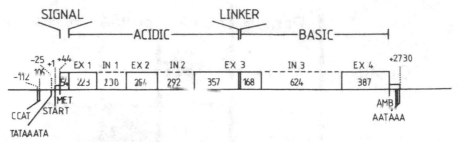

Figure 4. Schematic representation of the glycinin-2 gene. Transcriptional unit consists of 44 bp and
77 bp untranslated 5' and 3' regions (low boxes). The coding sequence (tall boxes) consists
of 4 exons (EX) and 3 introns (IN). It encodes a signal peptide, followed by the acidic,
linker and basic regions. The numbers of base pairs in each region are indicated by the
numbers in the boxes. Numbers preceded by plus or minus indicate the positions of potential regulatory regions relative to the putative cap site.

for the basic polypeptide. The introns occur in each gene at the same relative position,
but their length is variable. Not unexpectedly, the intron-exon junctions correspond
with the consensus GT/AG splicing rule for the 5' and 3' borders of other enkaryotic
genes (Breathnach and Chambon, 1978). When comparing corresponding introns
among genes, sequence homology is found around the intron borders, but successively
less homology occurs as the central parts of the introns are approached (Nielsen et al.,
unpublished).

The gene product encoded by the glycinin-2 gene corresponds to subunit A_2B_{1a}
(Figure 4). Only five differences exist between the amino acid sequence determined by
Edman degradation (Staswick et al., 1984b) and that predicted from the nucleotide
sequence of the gene (Marco et al., 1984; Nielsen et al., unpublished). The differences
observed are not unexpected considering the degree of sequence heterogenity observed
in the A_2B_{1a} subunit purified from seeds (Staswick et al., 1984b). S_1-nuclease mapping with glycinin mRNA was used to define the transcriptional unit. Transcription
begins about 44 bp upstream from the methionine initiation codon. A consensus
TATAAATA promotor site is located at -25 bp upstream from this putative "cap"
and a pair of CCAT sequences are located slightly further upstream (Nielsen et al.,
unpublished). Three potential polyadenylation sites are found in the DNA 3' from the
coding region. Comparison of this region with that for the trailer sequence of clone
pG154 revealed that the polyadenylated tail began 10-12 bp after the third consensus
sequence (Marco et al., 1984). Multiple polyadenylation signal sequences are features
of many eukaryotic genes and are used with different efficiencies (Rosenfeld et al.,
1983; Henikoff et al., 1983).

Each of the genes for the Group-1 subunits encodes a short signal peptide (Table
2). As anticipated from an analysis of glycinin precursors made in vitro (Tumer et al.,
1981, 1982), a high degree of homology among them is evident. The signal peptide of
the Gy_4 precursor of Group-II is longer than those of the Group-I precursors. In each
case, cleavage of the signal peptide occurs after a PheAla sequence and prior to the
hydrophobic NH_2-terminal amino acid of the mature subunits. As with secreted precursors of proteins in animal cells (Harwood, 1980), the plant signal peptides have a
central hydrophobic region followed by a more open region where cleavage occurs.
In analogy with animal systems, cleavage takes place co-translationally as the precursors are synthesized vectorally across rough endoplasmic reticulum.

Sequences in the precursors that correspond to the acidic and basic polypeptide components are covalently joined by peptide linkers of variable size (Tumer et al., 1981). The primary structure in this region has been deduced for the five major types of glycinin subunits (Table 3). In each case the tripeptide ArgAsn↑Gly is conserved at the junction between the linker and the basic polypeptide component. Since all basic polypeptides have NH_2-terminal glycine, cleavage occurs just before this residue. A proteolytic enzyme with specificity for Asn-Gly bands is to our knowledge unknown, although an enzyme with similar specificity would be required to cleave viral polyprotein precursors described by Kitamura et al. (1981). This post-translational cleavage probably occurs after the precursor enters the protein body (Chrispeels, 1983).

The site of cleavage between the acidic component and the linker region is not well conserved among subunits (Table 3). In the case of A_2B_{1a}, comparison of the sequence deduced from the nucleotide sequence with that determined from Edman degradation of the protein indicates that cleavage takes place between a lysine and an arginine residue. Paired basic amino acids such as these have been implicated as recognition signals for enzymes that modify prohormones (Docherty et al., 1982), but their significance in legume storage protein precursors remains to be evaluated. In this regard, some but not all of the glycinin precursors have paired basic residues that potentially could be involved in removal of the linker region.

MODIFICATION OF FUNCTIONAL AND NUTRITIONAL QUALITY

The glycinin complex consists of six nonidentical subunits, but as described earlier, composition of the subunits varies. For example, there is a three- to four-fold difference among them in methionine content (Table 1). This raises the possibility that seed quality can be altered by eliminating subunits with less desirable properties and replacing them with ones that have more suitable features. In this regard, mutants due to recessive alleles that result in the elimination of specific glycinin or β-conglycinin subunits have been identified (Kitamura et al., 1984). The mutants exhibit two important characteristics. First, the genetic elimination of the subunits has no apparent effect on physical properties of glycinin. Thus, the glycinin and β-conglycinin complexes will tolerate at least some degree of compositional polymorphism. Second, significant differences in percentage protein are not observed among seeds of genotypes that bear the null-alleles compared with those from normal genotypes (Kitamura et al., 1984). Since these mutants appear to alter the expression of individual subunit genes rather than the whole complex, they are less likely to have detrimental pleotropic effects than most of the regulatory mutations in cereals that produce high lysine phenotypes. The soybean mutants may, therefore, prove useful in altering the nutritional properties of legume seed proteins and the functionality of food products made from them.

It has been suggested that emerging biotechnology will permit modification of agronomically important traits. This approach might be applied to seed protein genes, although success will depend on identification of sites where changes can be made without adverse effects on subunit structure and function. With this in mind, computer-assisted alignment and secondary structure predictions have been made based on the amino acid sequences for a number of glycinin and β-conglycinin polypeptides (Argos and Nielsen, unpublished). The predictions reveal an interesting similarity between glycinin and β-conglycinin (Figure 5). Three similar domains are predicted for each of them. The NH_2-terminal amino acids from both kinds of subunits comprise a domain predominated by helices and turn structure. This region is followed by a slightly larger one with mixed β-stand, helix, and turns. It is noteworthy that areas of helix,

Table 3. Comparison of the amino acid sequences for 5 major glycinin subunits around the cleavage sites for the signal and linker peptides.

Gene	Signal Peptide	↑	Acidic Polypeptide	↑	Linker Region	↑	Basic Polypeptide[a]
Gy₁	MAKLVFSLCFLLFSQCCFA	↑	FSSREQ		GKDKHCQRPRGSQSK SRRN	↑	GID ....
Gy₂	MAKLVLSLCFLLFSGC.FA	↑	LREQAQ		PQCVETDKGCQRQSK↑RSRN	↑	GID ....
Gy₃	MAKLVLSLCFLLFSGCCFA	↑	FSFREQ		EECPDCDEKDKHCQS QSRN	↑	GID ....
Gy₄	MGKPFTLSLSSLCLLLLSACFA	↑	ISSSSK		QPRRPRQEEPRERGC ETRN	↑	GVE ....
Gy₅			....		SRPEQQEPRGRGC QTRN	↑	GVE ....

[a]Underlined amino acids in the signal region are hydrophobic. For Gy₂ the position of cleavage between the acidic polypeptide and linker region is known from the comparison of the primary structure of the mature protein (Staswick et al., 1984b) with that predicted for the precursor from its DNA sequence (Nielsen et al., unpublished). Paired basic residues in other sequences could act as recognition sites for processing. The NH_2-terminal region of the Gy₅ subunit precursor is unknown.

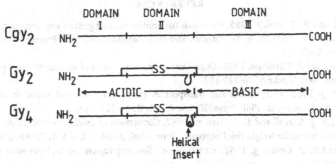

Figure 5. Schematic representation of conglycinin (Cgy) Group-1 (Gy₂) and Group-2 (Gy₄) structural alignments. The proteins are divided into three domains. Domain I contains mainly
predicted α-helical, turn structure, domain-II mixed α-helical, turn and β-sheet conformation, and domain-III predominantly β-sheet and turn structures. The glycinin subunits are
distinguished from the β-conglycinin by an insert of variable size between the second and
the third domains, which is predicted to assume mainly an α-helical conformation. The
position of the disulfide bridges (-SS-) that link the acidic and basic polypeptide components of glycinin are indicated.

turn and β-strand within the central domains in both kinds of subunits can be aligned,
whereas corresponding areas of structure cannot be aligned in the NH_2-terminal domains. The basic polypeptide component of glycinin and the 250 COOH-terminal
amino acids of β-conglycinin subunits appear equivalent and exhibit a high degree of
predicted β-strand and turn conformations. Thus, the COOH-terminal region is hydrophobic and probably is buried within the molecule. The highest degree of amino acid
sequence homology found between glycinin and β-conglycinin subunits is in this domain, which suggests it plays an important role in maintenance of subunit structure.
Modifications here would probably have an adverse effect on subunit structure and
function.

The most striking difference between the glycinin and β-conglycinin subunits
occurs in the middle of the glycinin molecules adjacent to the position of the linker in
the precursor. Large regions with predicted helical plus turn structures are located
there which are absent β-conglycinin subunits (Figure 5). This region has a higher concentration of aspartate, glutamate and other charged amino acids than other parts of
the molecule. It is this domain that also accounts for the size difference between
Group-1 and Group-2 glycinin subunits. The Group-2 subunits contain several 20-
amino-acid repeats in the central part of this insertion that are not found in the Group-
1 subunits. Because of its predicted high helical content, it is likely that at least part of
this domain is at the surface of the molecule. More importantly, the natural variation
found in this region indicates that it can accommodate changes in primary structure
without detrimental effects on other areas in the molecule that are critical for association of subunits into glycinin complexes. This region, therefore, appears to be a logical
target for efforts to "engineer" legume seed proteins.

NOTES

N. C. Nielsen, USDA/ARS Purdue University, West Lafayette, IN, USA.

The data summarized are the result of work by M. A. Moreira, P. E. Staswick, J. S. Mediros, N. E.
Tumer, Y. Marco, B. Scallon, C. Davies, L. Floener, T. J. Cho, C. Dickenson, K. Kitamura and T. Vu
while they were either graduate students or post-doctoral associates in the author's laboratory. Without their skillful efforts, the work would not have been possible.

REFERENCES

Beachy, R. N. and R. T. Fraley. 1985. Potentials for applications of genetic engineering technology to soybeans. p. (in press). *In* A. M. Altschul and H. L. Wilcke (eds.) New Protein Foods. Academic Press, New York.

Breathnach, R. and P. Chambon. 1981. Organization and expression of eukaroytic split genes for proteins. Ann. Rev. Biochem. 50:349-383.

Chrispeels, M. J. 1983. Biosynthesis and transport of storage proteins and lectins in cotyledons of developing legume seeds. Phil. Trans. R. Soc. Lond. B304-309-319.

Docherty, K., R. J. Carroll and D. E. Steiner. 1982. Conversion of insulin to proinsulin: involvement of a 31,500 molecular weight thiol protease. Proc. Natl. Acad. Sci. U.S.A. 79:4613-4617.

Fischer, R. and R. B. Goldberg. 1982. Structure and flanking regions of soybean seed protein. Cell 29:651-660.

Goldberg, R. B., G. Hoscheck, G. S. Ditta and R. W. Breidenbach. 1981. Developmental regulation of cloned super abundant embryo mRNAs in soybean. Developmental Biol. 83:218-231.

Harwood, R. 1980. Protein transfer across membranes: The role of signal sequences and signal peptidase activity. p. 3-52. *In* The Enzymology of Post-translational Modification of Proteins. Acad. Press, New York.

Henikoff, S., J. S. Sloan and J. D. Kelly. 1983. A drosophila metabolic gene transcript is alternatively processed. Cell 34:405-414.

Kitamura, K., C. S. Davies and N. C. Nielsen. 1984. Inheritance of null-alleles for the α'-subunit of β-conglycinin and the $A_5A_4B_3$ subunit of glycinin in soybean. Theor. Appl. Genet. 68:253-257.

Kitamura, N., B. Semler, P. Rothberg, G. Larsen, C. Adler, A. Dorner, E. Emini, R. Hanecak, J. Lee, S. van der Werf, C. Anderson and E. Wimmer. 1981. Primary structure, gene organization and polypeptide expression of poliovirus RNA. Nature 291:547-549.

Marco, Y. A., V. H. Thanh, N. E. Tumer, B. J. Scallon and N. C. Nielsen. 1984. Cloning and structural analysis of DNA encoding an A_2B_{1a} subunit of glycinin. J. Biol. Chem. (In press).

Moreira, M. A., M. A. Hermodson, B. A. Larkins and N. C. Nielsen. 1979. Purification and partial characterization of the acidic and basic subunits of glycinin. J. Biol. Chem. 254:9921-9926.

Moreira, M. A., M. A. Hermodson, B. A. Larkins and N. C. Nielsen. 1981. Comparison of the primary structure of the acidic polypeptides of glycinin. Arch. Biochem. Biophys. 210:633-642.

Nielsen, N. C. 1985. The structure of soy proteins. p. (in press). *In* A. M. Altschul and H. L. Wilcke (eds.) New Protein Foods. Vol. 5. Academic Press, New York.

Rosenfeld, M. G., J. J. Mermod, S. A. Amara, L. W. Swanson, P. E. Sawchenko, J. River, W. W. Vale and R. M. Evans. 1983. Production of a novel neuropeptide encoded by the calcitonin gene via tissue specific RNA splicing. Nature 304:129-135.

Staswick, P. S., M. A. Hermodson and N. C. Nielsen. 1981. Purification and characterization of the complexes of acidic and basic subunits of glycinin. J. Biol. Chem. 256:8752-8755.

Staswick, P. S. and N. C. Nielsen. 1983. Identification of a soybean cultivar lacking several glycinin subunits. Arch. Biochem. Biophys. 223:1-8.

Staswick, P. S., M. A. Hermodson and N. C. Nielsen. 1984a. Identification of the cystines linking the acidic and basic components of glycinin subunits. J. Biol. Chem. (in press).

Staswick, P. S., M. A. Hermodson and N. C. Nielsen. 1984b. The complete amino acid sequence of the A_2B_{1a} subunit of glycinin. J. Biol. Chem. (in press).

Tumer, N. E., V. H. Thanh and N. C. Nielsen. 1981. Purification and characterization of mRNA from soybean seeds: identification of glycinin and β-conglycinin precursors. J. Biol. Chem. 256:8756-8760.

Tumer, N. E., J. D. Richter and N. C. Nielsen. 1982. Structural characterization of the glycinin precursors. J. Biol. Chem. 257:4016-4018.

THE ORGANIZATION OF GENES INVOLVED IN SYMBIOTIC NITROGEN FIXATION ON INDIGENOUS PLASMIDS OF *RHIZOBIUM JAPONICUM*

Alan G. Atherly, R. K. Prakash, Robert V. Masterson,
Nancy B. DuTeau, and Kim S. Engwall

Rhizobium japonicum is a member of the family Rhizobiaceae that forms a symbiotic relationship with soybean plants and fixes atmospheric nitrogen in root nodules. *R. japonicum* is of particular interest, not only because of its symbiosis with the important agronomic crop, soybean, but also because of the recent discovery of fast-growing strains of *R. japonicum* that appear to be unrelated to traditionally used U.S. strains (Keyser et al., 1982; this volume). The ability of both slow- and fast-growing *R. japonicum* strains to nodulate and fix nitrogen on the same cultivar is unusual but not unique. Slow- and fast-growing *R. species* have been identified which nodulate *Lotus pedunculatus*, and also from the cowpea group of rhizobia which nodulate a wide variety of plants (Brockwell, 1980). The physiological differences between slow- and fast-growing *R. japonicum* strains are significant (Keyser et al., 1982; Yelton et al., 1983; Sadowsky et al., 1983) and reflect a divergent genetic background. It has, therefore, been of interest to investigate the genetic content of *R. japonicum* fast- and slow-growing strains in terms of sequence organization and conservation.

LARGE PLASMIDS IN *R. JAPONICUM*

A common feature of both slow- and fast-growing *R. japonicum*, as well as other rhizobia, is the presence of large plasmids (between 90 and 750 megadalton [Mdal]) (Casse et al., 1979; Gross et al., 1979; Prakash et al., 1981; Russell and Atherly, 1981). The slow-growing *R. japonicum* strains examined in this study usually contained one large plasmid ranging between 118 and 196 Mdal, whereas the fast-growing strains contained one to four plasmids ranging from 54 to 500 Mdal. Therefore, large plasmids are a consistent feature in both slow- and fast-growing *R. japonicum* strains, with the exception of strain USDA 110, a slow-growing strain of great interest. In addition, almost all fast-growing rhizobia examined have been found to possess very large plasmids. These plasmids have been referred to as megaplasmids. *Nif* and *Nod* genes have been found on megaplasmids in some strains (Denarie et al., 1981). However, megaplasmids have never been positively identified in slow-growing *R. japonicum* strains. Because of the location of nodulation and nitrogen fixation genes on large plasmids in rhizobia, it is of interest to determine the genetic content of plasmids in these strains to further our understanding of the *R. japonicum*-soybean symbiosis.

Table 1 contains the molecular weights of large plasmids separated from slow- and fast-growing *R. japonicum* strains, as well as the presence of *Nod* or *Nif* gene clusters

Table 1. Properties of large plasmids from *R. japonicum*.

Strain of origin	Approximate mol. wt.	Hybridization to:	
		nif probe	*nod* probe
	M daltons		
Slow-growing type			
61A76	178 ± 4	-	
3I1b31	160 + 11	-	
3I1b71a	164 ± 2	-	
3I1b74	195 ± 6	-	
3I1b94	58 ± 9	-	
	118 ± 6	-	
3I1b143	159 ± 6	-	
3I1b110	none		
Fast-growing type			
USDA191	30	-	-
	69	-	-
	195	+	+
	>500	-	-
USDA193	186	+	+
	>500	-	-
USDA194	76	-	-
	145	-	-
	>500	-	-
USDA201	117	-	-
	192	+	+
	>500	-	-
USDA205	57	-	-
	112	+	-
	192	-	+
	>500	-	-
USDA206	50	-	-
	60	-	-
	197	+	+
	>500	-	-

on the plasmid. Molecular weights were estimated by agarose gel electrophoresis, by using the procedure of Eckhardt (1978) with internal molecular weight standards. The presence of *Nif* or *Nod* gene sequences was determined by hybridization of blotted DNA to cloned fragments of *Nif* DNA (pRmR2; Ruvkun and Ausubel, 1981) or *Nod* DNA (pRmSL26, Long et al., 1982). Table 1 shows that, in every case examined, no slow-growing *R. japonicum* strain has *Nif* genes on the indigenous plasmid. In contrast, in every fast-growing *R. japonicum* strain, except USDA 194, *Nod* and *Nif* were both present on the same plasmid. USDA strain 194 seemingly has *nod* and *nif* on the chromosome.

NIF SEQUENCE CONSERVATION IN *R. JAPONICUM*

Slow-growing Strains

To obtain an estimate of the sequence conservation of *nif* genes in slow-growing strains, we hybridized pRmR2 DNA (containing *nif* DH genes from *R. meliloti*) to a

Southern blot of EcoR1 digested total DNA from nine different slow-growing strains. Six of the nine strains examined have identical hybridization patterns with relatively strong hybridization to a 11.0 Kb fragment. A 4.2 Kb fragment also hybridized, as did two relatively weak signals of 3.3 and 5.4 Kb (Figure 1, lanes 2-7).

The plasmidless strain 3I1b110 had a different hybridization pattern with the *nif* probe in which two bands of 2.8 and 6.1 Kb hybridized strongly and a 2.0 Kb band hybridized relatively weakly (Figure 1, lane 1). The two other strains examined, AA102 and 3I1b123, both contain plasmids. The hybridization pattern of the *nif* genes in each of these strains is distinctly different. Strain AA102 (Figure 1, lane 8) has two large bands of 12.0 and 13.7 Kb that hybridize and strain 3I1b123 (results not shown) contains a 2.8-Kb fragment, common to the same sized fragment in 3I1b110, and two other bands of 5.5 and 8.4 Kb. Therefore, a great amount of conservation of *nif* gene sequences is found in the nine strains examined.

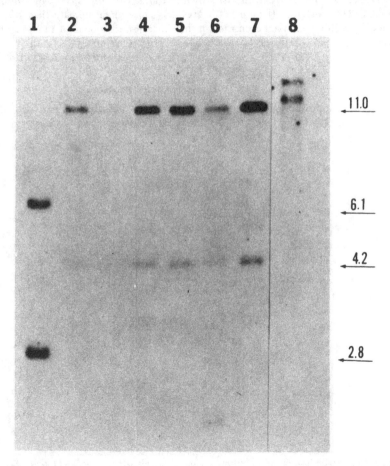

Figure 1. Hybridization of ^{32}P-labeled pRmR2 (containing *nif* DH from *R. meliloti*) to EcoRI-digested total DNA from slow-growing strains. Lane 1, 3I1b110; lane 2, 3I1b94; lane 3, 3I1b74; lane 4, 3I1b143; lane 5, 3I1b71a; lane 6, 61A76; lane 7, 3I1b31; lane 8, AA102. Molecular weights are in Kbp.

Fast-growing Strains

We have previously determined the location of structural *nif* genes to be on large plasmids in the fast-growing strains, except for strain USDA 194 (Table 1) (Masterson et al., 1982). The 76, 145 and 500 Mdal plasmids in strain USDA 194 did not contain structural *nif* genes. To estimate the sequence conservation of *nif* DNA in fast-growing *R. japonicum* strains, the Southern blot of EcoRI digested plasmid DNA of each strain was hybridized with ^{32}P-labeled pRmR2 DNA. The results are presented in Figure 2A. Every strain with structural *nif* genes on a large plasmid contained the same EcoRI size fragments, 4.2 and 4.9 Kb, which hybridized with the labeled *nif* probe. As expected, no hybridization was observed with the plasmid, pRja61A76, from the slow-growing strain (lane 1 of Figure 2A) or with the plasmids isolated from fast-growing strain USDA 194 (lane 2 of Figure 2A). Furthermore, when the ^{32}P-labeled *nif* DNA was hybridized to total DNA from fast-growing strains, the same sized bands hybridized to strain USDA 194 DNA (lane 6 of Figure 2B). These data indicate that *nif* structural genes are highly conserved even though their location may be on a 112-Mdal plasmid in strain USDA205, or plasmids larger than 190 Mdal in strains USDA191, USDA193, USDA201, and USDA206, as well as on the chromosome in strain USDA194.

NOD SEQUENCE CONSERVATION

Using costramid clone pRmSL26 containing a nodulation gene from *R. meliloti* (Long et al., 1982; the kind gift of F. Ausubel, Harvard U., and S. Long) allowed us to

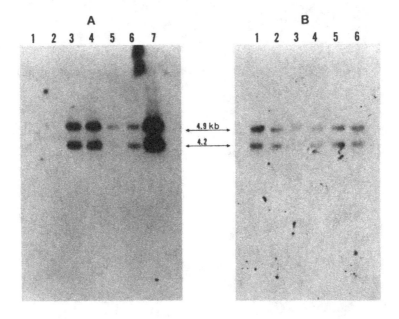

Figure 2. A. Hybridization of ^{32}P-pRmR2 (containing *nif* DH from *R. meliloti*) to EcoRI-digested plasmid DNA from fast-growing strains. Lane 1, pRja61A76; lane 2, pRjaUSDA194a,b; lane 3, pRjaUSDA201a,b; lane 4, pRjaUSDA205a,b,c; lane 5, pRjaUSDA206a,b,c; lane 6, pRjaUSDA191a,b; lane 7, pRjaUSDA193. B. ^{32}P-pRmR2 hybridized to EcoRI-digested total DNA from fast-growing strains. Molecular weights shown are the sizes of Kb of the EcoRI fragments that hybridize with the *nif* probe.

further isolate a 3.5 Kb fragment known to carry essential *nod* genes (S. Long, Stanford U., pers. commun.). When this *nod* fragment was hybridized to EcoR1-digested plasmid DNA from fast-growing strains, the result was that both common and unique bands of the large plasmids hybridized (Figure 3). Lane 1 of Figure 3 contains an EcoR1 digest of total DNA from strain USDA 194. Although *nif* genes are conserved in this strain, similarly to those of other fast-growing *R. japonicum* strains examined, the nodulation genes seem to have three bands in common (11.2 Kb, 5.3 Kb, and 2.8 Kb) with EcoR1 fragments of plasmids from strains USDA 201, USDA 191, and USDA 193 (Figure 3, lanes 2, 5, and 6). However, a smaller band (4.8 Kb) is common with bands of plasmic DNA isolated from USDA 206 and total DNA isolated from USDA 191 (Figure 3, lanes 1 and 4). This hybridization assay indicates that some diversity exists in the EcoR1 restriction enzyme sites surrounding the nodulation genes in fast-growing *R. japonicum*.

In contrast, the same 3.5-Kb fragment hybridized very weakly to total DNA from slow-growing strains and is seen only after long exposures with intensifying screens

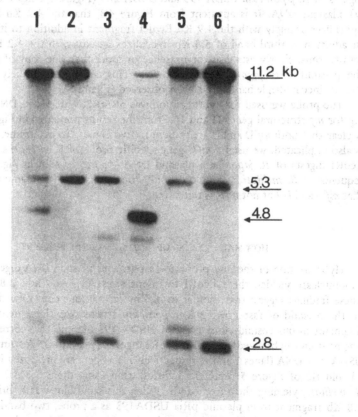

Figure 3. Hybridization of ^{32}P-labeled *nod* genes to EcoRI-digested plasmid DNA from fast-growing strains. A 3.5 Kb EcoRI-BamHI fragment, isolated from a clone known to carry *nod* genes (described in the text) was hybridized to a Southern blot containing: lane 1, total DNA of PRC194; lane 2, pRjaUSDA201a,b; lane 3, pRjaUSDA205a,b,c; lane 4, pRjaUSDA206a,b,c; lane 5, pRjaUSDA191a,b; lane 6, pRjaUSDA193.

(results not shown) and may or may not be related sequences. These results suggest that significantly different DNA sequences are involved in the symbiotic event of nodulation by slow-growing R. japonicum.

HOW MANY COPIES OF NIF GENES ARE PRESENT?

Hybridization of the nif probe (pRmR2) to fast-growing DNA digested with EcoRI endonuclease yielded two EcoRI fragment sizes, 4.2 and 4.9 Kb (Figure 2). These data suggest that two nif gene clusters may be present on the plasmid of each fast-growing R. japonicum strain. Additional evidence leading to the same conclusion comes from the finding that a cosmid clone bank, constructed from the plasmid of USDA 193, showed that the 4.2 and 4.9 Kb EcoRI fragments homologous to nif genes of R. meliloti were located on separate clones. To confirm this speculation, we isolated the 4.2 Kb EcoRI fragment that is homologous to nif genes of R. meliloti from the clone bank, labeled the DNA with ^{32}PdCTP and hybridized to blots containing EcoRI digests of R. japonicum USDA 193 and USDA 205 (Figure 4, lanes 1 and 2, respectively) plasmid DNA. It is apparent from Figure 4B, that the 4.2 Kb EcoRI fragment hybridizes strongly with the 4.9 Kb EcoRI fragment in addition to its own sequence. In addition, a third band of 5.8 Kb hybridizes weakly with the 4.2 Kb probe. These results, thus, clearly establish the presence of more than one copy of nif sequences on the plasmid of R. japonicum fast growers. This conclusion is likely common to all strains since multiple bands are always observed (Figure 2).

The probe we used for hybridization was pRmR2, which has a DNA fragment coding for nif structural genes D and H. From the results presented in Figure 4A and B, it is clear that both nif D and H are present in two copies. To determine, likewise, if nifK is also duplicated, we used a nifK gene-specific probe of R. meliloti and hybridized to EcoRI digests of R. japonicum plasmid DNA. As can be seen in Figure 4C, the nifK sequence of R. meliloti is located on only the 4.2 Kb EcoRI fragment, which implies that nifD and H but not nifK is reiterated.

HOW MANY COPIES OF NOD GENES ARE PRESENT?

Hybridization of the nod probe to fast-growing plasmid DNA digested with EcoRI endonuclease yielded three EcoRI fragment sizes; 11.3, 5.3 and 2.8 Kb (Figure 3). These findings suggest that, similar to nif, more than one copy of nod may be present on the plasmid of fast-growing R. japonicum strains. We, thus, identified the EcoRI fragments in our cosmid clone bank of USDA 193 plasmid that corresponded to each fragment and hybridized each to a EcoRI digest of plasmid DNA from USDA 193 and USDA 205 DNA (lanes 1 and 2, respectively). The data are presented in Figure 5. Lane B1 and B2 of Figure 5 represent hybridization using the 3.5 Mdal nod probe from R. meliloti; yielding the three bands. When this same blot was hybridized, using the 2.8 Kb fragment from plasmid pRja USDA193 as a probe, two bands are seen (lanes D1 and D2). These findings suggest at least two copies of the nod sequences are present. When the 11.3 Kb fragment is used as a probe (lanes C1 and C2) it only hybridizes to itself. However, after longer exposure (2 days), weak signals were observed with two additional EcoRI fragments of sizes 2.5 Kb and 3 Kb (data not shown). These results show the presence of more than one copy of the nod region on the plasmid of fast-growing R. japonicum strains.

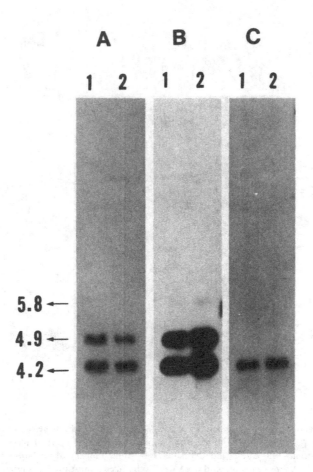

Figure 4. Hybridization of heterologous probe of *nif* genes with the plasmid DNA from fast-growing *R. japonicum* strains USDA 193 (Lane 1) and USDA 205 (Lane 2). Approximately 1 μg of plasmid DNA from *R. japonicum* USDA 193 and 205 was digested with restriction enzyme *EcoRI* and separated on 0.7 agarose gels. The restriction pattern of the plasmid DNA is shown in Figure 5A. DNA was transferred to nitrocellulose by the method of Southern and annealed with 1-2 x 10^6 c.p.m. of ^{32}P-labeled nick-translated probe. A. Autoradiogram of *EcoRI* digestion pattern of *R. japonicum* plasmid DNA obtained after hybridization against pRmR2, which contains *nif* structural genes D and H of *R. meliloti*. B. Hybridization pattern of plasmid DNA, when 4.2 Kb *EcoRI* fragment of *R. japonicum* 193 plasmid DNA was used as probe. C. Autoradiogram of *R. japonicum* plasmid DNA after hybridization with *nif* K probe of *R. meliloti*. The 1.7 Kb *EcoRI* fragment of plasmid pRmR8L2, which contains the *nif* structural genes K, was isolated from agarose gel.

TRANSFER OF SYMBIOTIC PLASMID TO OTHER BACTERIA STRAINS

For plasmid transfer or elimination it was first necessary to obtain an easily identifiable genetic marker on the plasmid. This was accomplished by insertion of Tn5 into the *nif-nod* plasmid of strains USDA 191 and 193, by using a "suicide" plasmid (Simon et al., 1983). Both strain 191::Tn5 and 193::Tn5 were exposed to high temperatures

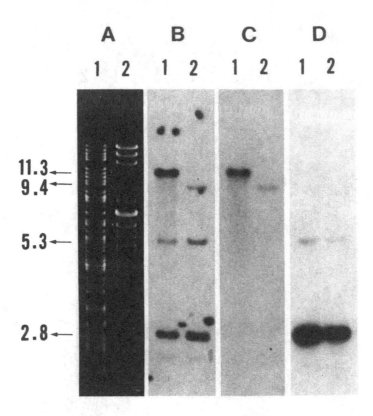

Figure 5. Hybridization of DNA fragment of *R. meliloti* nod genes with the plasmid DNA from fast-growing *R. japonicum*. A. Restirction endonuclease *EcoRI* digest of plasmid DNA from USDA strains 193 (lane 1) and 205 (lane 2). B. Autoradiogram of the blots with the EcoRI digests of *R. japonicum* plasmid DNA were hybridized against the 3.5 kb-sized fragment of *R. meliloti* containing the *nod* genes. The 3.5 kb size *nod* probe of *R. meliloti* was prepared by subcloning and *EcoRI-BamHI* fragment from plasmic pRmS126 into pBR322 cut with *EcoRI* and *BamHI*. C and D. Autoradiogram of *R. japonicum* plasmid DNA obtained after hybridization with 11.3 kb- and 2.8 kb-sized *EcoRI* fragments of plasmid DNA from *R. japonicum* USDA 193.

(37 C) in the presence of ethidium bromide (Zurkoski and Lorkiewicz, 1979) to eliminate the plasmid containing the Tn5-Kan gene. Following selection of plasmidless isolates, the plasmidless USDA 193 strain was tested on soybean plants for ability to form nodules. No nodules were ever observed; however, control plants inoculated with the strains 191::Tn5 and 193::Tn5 formed nodules. These data strongly suggest that the plasmid carrying the *nod* and *nif* genes (Table 1) are required for nodulation and have, thus, been designated *Sym* (for symbiotic) plasmids.

None of the plasmids in strain USDA 191 or USDA 193 are self transmissable. However, plasmid transfer is possible in the presence of the helper plasmid pRL180 (Hooykaas et al., 1982). Using pRL180 as a helper, we have successfully transferred the Sym plasmid of USDA 193 into *E. coli* strain HB101, *R. japonicum* USDA 3I1b110, *A. tumefaciens* strains A136 and LBA288, *R. leguminosarium* and *R. meliloti* where it is maintained.

However, pSym193 is unstable in all other fast-growing *Rhizobium* strains tested. On the other hand, pSym191 is unstable in all *Rhizobium* fast-growing strains, even *R. leguminosarium* and *R. meliloti*. Even though pSym193 is stable in *R. leguminosarium* and *R. meliloti*, its presence does not extend the host range to include soybean. In contrast, the presence of either pSym191 or pSym 193 in *A. tumefaciens* allows nodule formation on soybean plants, but not nitrogen fixation. Thus, *nod* genes are expressed in *A. tumefaciens* but not *nif* genes, perhaps because additional important genes are absent or promoters are not recognized.

CONCLUSION

It is interesting that *nif* genes appear mostly on the plasmids of fast-growing *R. japonicum* strains and, very likely, only on the chromosome or megaplasmids of slow-growing strains. Ruvkun and Ausubel (1980) speculate that plasmid-borne, nitrogen-fixation-related traits may be the result of recent radiation, perhaps by conjugative plasmids to other bacterial species. The acquisition of nitrogen-fixation genes would allow an organism new niches, including symbiotic interaction. Thus, it is possible that the fast-growing *R. japonicum* strains isolated from the People's Republic of China are relatively recent additions to the family of nitrogen-fixing organisms. The isolation and cloning of the plasmids in this study are in progress, and the characterization of *nif* and other genes present should produce valuable information about these symbionts.

NOTES

Alan G. Atherly, R. K. Prakash, Robert V. Masterson, Nancy B. DuTeau, and Kim S. Engwall, Department of Genetics, Iowa State University, Ames, IA 50011.

This investigation was supported by Iowa Agric. and Home Econ. Exp. Stn. Project No. 2622, grant 59-2191-0-1-494-0 from the U.S. Department of Agriculture, and funds from Land O'Lakes Corp.

REFERENCES

Brockwell, J. 1980. Experiments with crop and pasture legumes, principles and methods. p. 417-488. *In* F. J. Bergersen (ed.) Methods for Evaluating Nitrogen Fixation. John Wiley Press, N.Y.

Casse, F., C. Boucher, J. S. Julliot, M. Michel and J. Denarie. 1979. Isolation and characterization of large plasmids in *Rhizobium meliloti* using gel electrophoresis. J. Gen. Microbiol. 113:229-242.

Denarie, J., P. Boistard, F. Casse-Delbart, A. G. Atherly, J. O. Berry and P. Russell. 1981. Indigenous plasmids in *Rhizobium*. p. 227-234. *In* K. L. Giles and A. G. Atherly (ed.) Biology of the Rhizobiaceae. Academic Press Inc., New York.

Eckhardt, T. 1978. A rapid method for the identification of plasmid deoxyribonucleic acid in Bacteria. Plasmid. 1:584-588.

Gross, D. C., A. K. Vidaver and R. V. Klucas. 1979. Plasmids, biological properties and efficacy of nitrogen fixation in *Rhizobium japonicum* strains indigenous to alkaline soils. J. Gen. Microbiol. 114:257-266.

Hooykaas, P. J. J., H. den Dulk-Ras and R. A. Schilperoot. 1982. Method for the transfer of large cryptic, non-self-transmissible plasmids: *Ex planta* transfer of the virulence plasmid of *Agrobacterium* rhizogenes. Plasmid. 8:94-96.

Keyser, H. H., B. B. Bohlool and T. S. Hu. 1982. Fast-growing *Rhizobia* isolated from root nodules of soybeans. Science 215:1631-1632.

Long, S. L., W. J. Buikema and F. M. Ausubel. 1982. Cloning of *Rhizobium meliloti* nodulation genes by direct complementation of *nod* mutants. Nature. 298:485-488.

Masterson, R. V., P. R. Russell and A. G. Atherly. 1982. Nitrogen fixation (nif) genes and large plasmids of *Rhizobium japonicum*. J. Bacteriol. 152:928-931.

Prakash, R. K., R. A. Schilperoort and M. P. Nuti. 1981. Large plasmids of fast-growing *Rhizobia*: homology studies and location of structural nitrogen fixation (*nif*) genes. J. Bacteriol. 145: 1129-1136.

Sadowsky, M. J. and B. B. Bohlood. 1983. Possible involvement of a megaplasmid in nodulation of soybeans by fast-growing rhizobia from China. Appl. Environ. Microbiol. 46:906-911.

Russell, P. F. and A. G. Atherly 1981, Plasmids in *Rhizobium japonicum*. p. 256-262. In J. H. D. Stevens and K. Clarke (ed.) Proceedings of 8th North American Rhizobium Conference. University of Manitoba Press, Winnipeg.

Ruvkun, G. B. and F. M. Ausubel. 1980. Interspecies homology of nitrogenase genes. Proc. Nat. Acad. Sci. U.S.A. 77:191-195.

Ruvkun, G. B. and F. M. Ausubel. 1981. A general method for site-directed mutagenesis in prokaryotes. Nature 289:85-88.

Simon, R., U. Priefer and A. Puhler. 1983. A broad host range mobilization system for in vivo genetic engineering: transposon mutagenesis in gram negative bacteria. Biotechnology 1:784-791.

Yelton, M. M., S. S. Yang, S. A. Edie and S. T. Lim. 1983. Characterization of an effective salt-tolerant, fast-growing strain of *Rhizobium japonicum*. J. Gen. Microbiol. 129:1537-1547.

Zurkowski, W. and Z. Lorkiewicz. 1979. Plasmid mediated control of nodulation in *R. trifolii*. Arch. Microbiol. 123:195-201.

Germplasm and Breeding Methodology

STRATEGIES FOR CULTIVAR DEVELOPMENT IN THE TROPICS

Romeu Afonso de Souza Kiihl, Leones Alves de Almeida and
Amélio Dall'Agnol

Although soybean is cultivated today in many parts of the world, the most important areas are located at latitudes greater than 23°. Brazil is an exception because it has around 2 million hectares of soybean below 23°S, with an average yield close to two tons per hectare. In Brazil, there is a clear tendency for the soybean to move to areas closer to the equator.

There is no doubt that there will be a tremendous increase in world demand for soybeans. Tropical regions will certainly be responsible for meeting that demand, and we hope that people from those areas will be the first to benefit from that development. There is an expectation that soybeans will become the primary protein source for many people in the tropics. In several tropical areas where new soybean cultivars are planted, the crop has looked as good as in temperate regions. The establishment of the soybean as a major crop in many tropical areas will be dependent on the development of adapted cultivars. In the development of soybean cultivars for the tropics, several aspects should be considered: yield, flowering, height, maturity, disease resistance, seed quality, insect resistance, nematode resistance, tolerance to soil acidity and in some cases, promiscuous nodulation.

IMPORTANT TRAITS FOR TROPICAL CULTIVARS

Any group working on the development of soybean cultivars for the tropics should consider a minimum number of characteristics that would permit the crop to be successful in the area considered. For Africa, high seed quality and promiscuous nodulation is a necessity. In Southeast Asia resistance to soybean rust, caused by *Phakopsora pachyrizi*, is one of the most serious diseases receiving great emphasis (Shanmugasundaram et al., 1977). In Brazil, because of the possibility of off-season seed production, a good inoculant industry, and in the beginning, the absence of rust, work was concentrated mainly on the development of high yielding cultivars with determinate growth and adquate height for mechanical harvest. This strategy proved successful, and this year in a location at 12°S there is an area of 11,000 ha for off-season seed production. Table 1 shows some characteristics of the four soybean cultivars released in Brazil for areas at less than 15°.

In temperate regions, the soybean crop is normally planted 20 to 50 days before the longest day. Therefore, seed development takes place when days are shortening (Hinson, 1974). In tropical regions, date of planting usually will be determined by rainfall distribution. In some cases, off-season seed production will be desirable (Brazil),

Table 1. Time[a] to flowering and maturity, and height of four soybean cultivars at Imperatriz, MA, and their yields at Imperatriz and Brejo, MA.

Cultivar	Time to Flowering	Time to Maturity	Height at maturity	Seed yield Imperatriz	Seed yield Brejo
	d	d	cm	— kg/ha —	
BR-10 (Teresina)	52	120	109	3369	3766
Timbira	42	107	88	3037	3112
BR-11 (Carajás)	52	120	102	2531	3021
Tropical	44	107	94	2675	2857

[a]Days from emergence (21 Dec.).

or two crops can be grown the same year (Colombia). In Southeast Asia the crop may have to fit a certain period of the year, between two other crops (rice [*Oryza sativa* L.] mainly). It becomes clear, therefore, that the control of flowering, and as a consequence height, plays a major factor in developing soybean cultivars for the tropics, and that control of flowering is aimed toward the development of types less sensitive to variations in date of planting and movement across latitudes. The idea is to have late flowering (a long juvenile period and a long inductive phase), coupled with a critical maximum daylength longer than the longest day for the latitude considered. We will concentrate our discussion on the importance of controlling flowering in order to develop soybean cultivars adapted to the tropics, and also we will comment on how early selection can be done.

SELECTION CRITERIA

The soybean is a short-day plant (Garner and Allard, 1920). Besides daylength, temperature also influences time to flowering, and often there is a significant interaction between these two factors (Summerfield and Wien, 1980). Considerable effort has been devoted to the identification of day-neutral, or photoperiod insensitive, cultivars (Criswell and Hume, 1972; Polson, 1972; Nissly et al., 1981; Shanmugasundaram, 1981), but these genotypes appear to be too early to be used in the tropics.

Another approach to developing soybean cultivars for the tropics is to look for types with late flowering under short-day conditions. Some genotypes with this characteristic were identified (Miyasaka et al., 1970; Hartwig and Kiihl, 1979). This character is controlled by recessive genes (Kiihl, 1976; Hartwig and Kiihl, 1979). Two other sources for late flowering under short-day conditions have been identified: IAC 73-2736 and Paranagoiana. They are natural mutants from Hardee and Paraná, and were selected in 1972 and 1977, respectively. The mutants and their original parents differ in one pair of genes, with lateness being recessive (Gilioli et al., 1984). The simple genetic control, and the fact that lateness is recessive, permit most of the screening work to be done outside the anticipated area of production. More extensive testing is made only with the types that make good growth. In Londrina (23°S), where the National Soybean Research Center is located, early planting (September 20 to October 10) is adequate for screening for flowering, and consequently for height. As can be seen in Table 2, it is easy to differentiate between types. Tropical, a cultivar adapted for production at latitudes lower than 15°, makes good growth if sown in September and October, while UFV-1 will have adequate height only if sown in November. For

Table 2. Time to maturity and height of two soybean cultivars (UFV-1 and Tropical) grown at Londrina 23°S on nine planting dates in 1981-82 (Garcia, pers. comm.).

Cultivar		Emergence date								
		20 Sept.	5 Oct.	20 Oct.	5 Nov.	20 Nov.	5 Dec.	20 Dec.	5 Jan.	5 Feb.
JFV-1	maturity	204	187	174	169	158	142	131	113	99
	height, cm	38	45	54	68	72	79	75	52	32
Tropical	maturity	201	187	177	170	156	145	131	128	125
	height, cm	93	103	110	112	110	101	101	99	87

most planting dates, they mature at the same time.

In Table 3, we compare the mutant, Paranagoiana, and the original cultivar, Paraná. Interestingly, the mutation that permitted late flowering, thus resulting in a taller plant, also affected the fruiting period. Paranagoiana was released as a new cultivar and is adapted for production in a wide latitude belt, and permits planting in areas around 23° during September and October.

In a study with five late flowering genotypes (Paranagoiana, BR 80-7553, Tropical, Santa Maria and PI 159925), Kiihl (unpublished) verified that their late flowering was a combination of a longer than usual juvenile period and a somewhat longer period from the beginning of induction to opening of the flower, the variation in juvenile period being the greatest among the genotypes.

Table 3. Comparisons between Paraná and Paranagoiana sown at Londrina, 23°S, on seven planting dates in 1981-82 (Garcia, pers. comm.).

	Emergence date						
	20 Sept.	5 Oct.	20 Oct.	5 Nov.	20 Nov.	5 Dec.	20 Dec.
Number of days from emergence to flowering							
Paraná	39	36	40	40	38	39	35
Paranagoiana	68	67	69	64	61	62	56
Number of days from flowering to maturity							
Paraná	71	78	70	69	68	60	57
Paranagoiana	95	92	84	84	76	69	62
Number of days from emergence to maturity							
Paraná	110	114	110	109	106	99	92
Paranagoiana	163	159	153	148	137	131	118
Height of plants (cm) at maturity							
Paraná	31	27	57	64	51	55	46
Paranagoiana	85	90	109	106	115	90	85
Yield (kg/ha)							
Paraná	2036	2467	2489	3222	2586	2102	2058
Paranagoiana	3340	3114	3566	3513	3118	2646	2560

NOTES

Romeu Afonso de Souza Kiihl, Leones Alves de Almeida and Amélio Dall'Agnol, EMBRAPA/CNPS, C.P. 1061, 86100 Londrina, PR, BRAZIL.

REFERENCES

Criswell, J. C. and D. J. Hume. 1972. Variation in sensitivity to photoperiod among early maturing soybean strains. Crop Sci. 12:657-60.

Garner, W. W. and H. A. Allard. 1920. Effect of relative length of day and night and other factors of environment on growth and reproduction in plants. J. Agric. Res. (Belts.) 18:553-606.

Gilioli, J. L., T. Sediyama and N. S. Fonseca, Jr. 1984. Herança do número de dias para floração em quatro mutantes naturais em soja (Glycine max (L.) Merrill) estudada sob condições de dias curtos. III Seminário Nacional de Pesquisa de Soja. (in press).

Hartwig, E. E. and R. A. S. Kiihl. 1979. Identification and utilization of a delayed flowering character in soybean for short-day conditions. Field Crop Res. 2:145-151.

Hinson, K. 1974. Tropical production of soybeans. p. 38-54. In Proc. workshop on soybeans for tropical and sub-tropical conditions. INTSOY series No. 2. INTSOY, Univ. of Ill., Urbana, Illinois.

Kiihl, R. A. S. 1976. Inheritance of two characteristics in soybeans (Glycine max (L.) Merrill): I. Resistance to soybean mosaic virus, II. Late flowering under short-day conditions. Ph.D. Dissertation. Mississippi State Univ., Mississippi State, Miss.

Miyasaka, S., G. Guimarães, R. A. S. Kiihl, L. A. C. Lovadini and J. D. Dematte. 1970. Variedades de soja indiferentes ao fotoperiodismo e tolerantes a baixas temperaturas. Bragantia 29:169-174.

Nissly, L. R., R. L. Bernard and C. N. Hittle. 1981. Variations in photoperiod sensitivity for time of flowering and maturity among soybean strains of maturity Group III. Crop Sci. 21:833-836.

Polson, D. E. 1972. Day neutrality in soybeans. Crop Sci. 12:773-76.

Shanmugasundaram, S. 1981. Varietal differences and genetic behavior for the photoperiodic responses in soybeans. Bull. Inst. Trop. Agric. Kyushu Univ. (Japan) 4:1-61.

Shanmugasundaram, S., S. C. S. Tsuo and T. S. Toung. 1977. Bull. Inst. Trop. Agric. Kyushu Univ. (Japan) 2:25-40.

Summerfield, R. J. and H. G. Wien. 1980. Effects of photoperiod and air temperature on growth and yield of economic legumes. p. 17-36. In R. J. Summerfield and A. H. Bunting (eds.) Advances in legume science. Royal Botanic Gardens, Kew.

STRATEGIES FOR CULTIVAR DEVELOPMENT IN TEMPERATE CLIMATES

John Schillinger

Improvement strategies for soybeans must be geared to meeting current and future production demands. As cost of production continues to rise, the need for greater productivity per unit area becomes more acute. Thus, breeding strategies to enhance soybean yields are paramount.

Many non-beneficial biological systems interact with the soybean plant and often cause yield losses. These destructive biological agents are often dynamic in nature and readily give rise to new biotypes. Soybean breeding must involve the incorporation of defensive mechanisms to counter the destructive nature of these pests.

Soybean growers are adopting new management practices to increase yields. Narrow row-spacings, irrigation and no-tillage culture are becoming more common. Breeding strategies must include the incorporation of improved standability and rapid emergence into cultivars designed for high management conditions.

I will briefly deal with individual strategies to improve yield and pest resistance and to identify high management ideotypes. I will then suggest combined strategies that should provide for the development of improved soybean cultivars for the future.

OFFENSIVE STRATEGIES: YIELD IMPROVEMENT

Identifying new genetic recombinants that promote higher soybean seed yields is the primary goal of most soybean breeders. The strategies employed to obtain higher yields has changed in the last ten years and will continue to change as modern technologies enhance breeder efficiency.

Breeding Populations

Breeding populations derived from elite crosses continue to be the prime source of new genetic recombinants for improved yield. With intensified interest in soybean breeding in both private and public sectors in the last 20 years, there are now available a large number of elite cultivars to use as parents. The basic strategy followed by Asgrow is to make numerous crosses each year, as indicated in Table 1. The frequency of elite crosses is generally 75% each year. Even with this high frequency of elite breeding populations, less than 1% of the total populations synthesized between 1973 and 1978 produced an improved cultivar.

Combining good cultivars from diverse regions; for example, northern and southern U.S., has proven to be productive. Cooper (1981), Thorne (J. C. Thorne, pers. commun.) and Asgrow have released several distinguished cultivars from such crosses (Table 2). Cooper indicated that new genes were introduced from the southern cultivar,

306

Table 1. Relationships of number of breeding populations initiated to those yielding improved soybean cultivars in the Asgrow breeding program, 1973-1978.

Year	Populations initiated	Populations producing cultivars	%
1973	203	2	1.0
1974	291	2	0.7
1975	672	3	0.4
1976	711	6	0.8
1977	600	6	1.0
1978	734	10	1.4
Total	3211	29	0.9

Table 2. Improved cultivars derived from diverse crosses.

Cultivars released	Pedigree		Breeder
Elf, Gnome, Pixie, Sprite	Williams	X Ransom	Cooper (1981)
A2858	Beeson	X Mack	Asgrow[a]
A3127, A3659, A3860, A4268	Williams	X Essex	Asgrow
A4997	Harcor	X Essex	Asgrow
B242	Hampton 266	X Wells	Thorne[b]
S3031	B216	X Davis	Thorne
S4240	Williams	X Essex	Thorne

[a] Asgrow Seed Co., Ames, IA.
[b] Dr. John Thorne, Northrup-King Seed Co., Washington, IA.

Ransom, into the northern gene pool producing higher yielding, short statured cultivars adapted to northern U.S. from the cross of Williams X Ransom. Soybean cultivars and breeding lines vary in their ability to contribute to yields of their progenies. I have found that certain lines tend to be good "general combiners" and contribute to higher yielding progenies. For example, 16% of the lines entered in the 1983 Preliminary Yield Tests at Asgrow (Ames) had A3127 as a parent. After selecting for yield, 42% of the remaining lines had A3127 as a parent.

Exotic germplasm is a possible source of new yield genes. However, the process of identifying and transferring these genes to highly adapted genotypes will require long term studies conducted by individuals with patience and skill in detecting the effects of these new genes.

Early Generation Testing

Boerma and Cooper (1975a,b) described an early-generation testing procedure for identifying high yielding F_2-derived determinate plants in the F_3 generation. Elf, Sprite, Gnome and Pixie are cultivars developed by Cooper using this procedure.

Eby of Midwest Oilseeds, Inc. (pers. commun.) has also successfully used a version of early generation testing in his program. His procedure emphasizes high selection pressure for yield in F_3-derived F_4 and F_5 lines. Asgrow breeders are also actively utilizing selection for yield in F_3-derived F_4 lines. They are combining yield selection with simultaneous selection for lodging and disease resistance. Selection pressure for yield in the Midwest Oilseeds and Asgrow programs is approximately 5%. Table 3 contains an outline of these early generation testing systems.

Preliminary comparisons of methods of selection for yield within 100, F_4 lines from an elite X elite population are given in Table 4. Early-generation testing for yield and good agronomic type is obviously valuable when the number of lines advanced is considered. The correlation between yield in early-generation testing and the preliminary yield test was r = .56 in 1983 for the 16 populations studied. The correlation between a visual rating in F_4-progeny rows and preliminary yield test results was only r = .11 in 1983.

Further selection for yield and other traits among F_6 sublines from the highest yielding F_3-derived lines is a beneficial addition to early-generation testing. This is particularly true for populations derived from diverse parents.

DEFENSIVE STRATEGIES—PEST RESISTANCE

Any gain in yield potential through soybean breeding must be accompanied by the incorporation of mechanisms that protect the yield against losses from major disease

Table 3.   Two commercial breeding systems utilizing early-generation testing procedures.

	Midwest Oilseeds Inc.	Asgrow Seed Co.
Generation tested:	F_3-derived F_4 lines	F_3-derived F_4 lines
Plot type:	76 by 102 cm Hills (bordered) (10 seed/hill)	2 rows, 1.5 m long (35 seed/row)
Reps/location:	1	1
Locations:	3	2
Combine yield recorded by:	Volume	Weight
Selection intensity:	5% selected	8% selected
Selection criteria:	Yield, maturity	Yield, maturity, lodging
Next cycle of selection:	2-row plots, 2 reps. at three locations	4-row plots, 2 reps at three locations

Table 4.   Effectiveness of early-generation yield testing and visual selection among F_4 lines in elite X elite breeding populations of soybeans.

| | Yield of selected lines | | | |
Selection type:	1982 EGT[a]	1983 Prelim[b]	Gain over random	Lines advanced
	t/ha	t/ha	%	
Random	4.55	2.51	--	1
Visual	4.66	2.73	9	0
Yield	4.94	2.84	13	0
Visual & Yield	4.85	2.86	14	4

[a]EGT = Plot size: 2 rows, 1.5 M long, 1 replication at 2 locations.
[b]Prelim = Plot size: 2 rows, 4 M long, 2 replications at 2 locations.

pathogens, nematodes, insect pests, and soil factors. Fortunately, genetic variability for either resistance or tolerance has been found for many of the major biological yield deterrents of soybeans in temperate regions.

Diseases

Pathogens that have the capacity to reduce soybean productivity in temperate climates of North America are:

Phytophthora Root Rot
(U.S. and Canada)

Phytophthora megasperma var.
glycinea (Kuan & Erwin)

Brown Stem Rot
(U.S. - North)

Phialophora gregata
(Allington & Chambert)

Stem Canker
(U.S. - South)

Diaporthe phaseolorum SACC var.
caulivora (Athow & Caldwell)

A large yield testing program that samples many diverse environments will likely provide exposure to one or more of these destructive pathogens during the years of evaluation before release. However, an efficient strategy is to expose early-generation populations to artificial or natural disease epiphytotics for elimination of susceptible genotypes. In Asgrow's program, we use field disease nurseries that are planted to continuous soybeans, as well as laboratory screening to determine race or isolate specific resistance and tolerance. We screen annually over 40,000 lines or individual plants for disease resistance.

Nematodes

Soybeans are attacked by two major nematodes: a) soybean cyst nematode, *Heterodera glycines* (Ichinohe), of which at least five races are known, and b) root knot nematode, *Meloidogyne incognita* (Kofoid & White) and *M. arenaria* (Neal) Chitwood.

The soybean cyst nematode continues to spread across North America and is a major threat to soybeans. Resistance to races 1, 3 and 4 of cyst nematode has been incorporated into highly productive cultivars. Screening of F_2, F_3 and F_4 populations in the greenhouse can be done with infested soil, allowing approximately 30 days for the female cyst to develop. By utilizing two resistant parents and screening large numbers of F_2 and F_3 plants from resistant x susceptible crosses, a sufficient population of resistant plants should be available for yield selection. The necessity for very large populations (3,000+) is dictated by the multi-genetic nature (five to six genes are estimated) for resistance to cyst nematode in soybeans.

Insects

Breeding for insect resistance in soybeans is difficult. Progress toward improved cultivars is slow. Strategies for improving cultivars for insect resistance must include rearing of insect pests and screening large populations of plants under controlled conditions. Handling large numbers of plants at advanced growth stages for screening is a major obstacle to rapid gain in breeding for insect resistance.

Soil Factors

The development of improved cultivars of soybeans with tolerance to iron deficiency chlorosis has been reported by Fehr (1982). Methods for improving soybean cultivars in regions where iron deficiency chlorosis is a problem include utilizing parents with good tolernce, screening F_4 and F_5 lines in replicated hill plots under high alkali field conditions, and simultaneously testing these same lines in the standard yield program. Fehr (1982) emphasizes the need for backcrossing to increase the level of tolerance to iron chlorosis in soybean plants.

MEETING THE NEEDS OF MODERN SOYBEAN PRODUCTION PRACTICES

The increased use of new management practices for soybeans has stressed the need for an improved soybean ideotype. Practices such as narrower row-spacings, irrigation, reduced-tillage and increased use of postemergence herbicides has demanded extra breeding and selection for lodging resistance, rapid seedling emergence, and increased seedling vigor. Improvement in lodging resistance has occurred by selecting an indeterminate growth habit with shorter internodes, or by selecting for a semi-dwarf (dt_1, dt_1) growth habit in the northern U.S. Selection for seedling emergence can be accomplished by using the hypocotyl elongation procedure of Burris and Fehr (1971) in growth chambers and through field observations.

FUTURE STRATEGIES WITH BIOTECHNOLOGY

The role of the new biotechnologies in future breeding strategies is still unproven, but nevertheless promising. The major advantage may be to select for specific traits more efficiently through the use of cellular and seedling screens for disease and chemical resistance or electrophoretic analysis of protein bands associated with resistance. The identification, isolation and transfer of simply inherited genes that confer pest and chemical resistance have been discussed (National Research Council, 1984). These procedures, once perfected, offer great opportunities for improving breeding efficiency.

Use of pollen irradiation to transfer segments of DNA to a donor parent has been accomplished by Pandey (1978) in other crop species. If perfected, this procedure would greatly simplify or replace the backcross method of breeding.

COMBINED STRATEGIES FOR CULTIVAR IMPROVEMENT

The following trends will develop in soybean cultivar development programs:
1. Population Synthesis
    Greater use of parents from diverse origins will be possible with the advent of new cultivars from expanded breeding programs in the United States, Canada, Brazil, China, Argentina and Australia. Greater use and introgression of exotic germplasm will be possible through advances in genetic studies and biotechnology procedures.
2. Breeding Methodology
    Traditional methods of soybean breeding such as the pedigree and backcross methods will become less popular. Breeders will utilize more early-generation selection for:
    a. resistance to major diseases and nematodes
    b. yield
    Biotechnological procedures will enhance:
    a. screening of large numbers of plants or cells
    b. the incorporation of single genes in adapted germplasm in a reduced time
    The above procedures plus the use of an integrated winter nursery program will reduce the time from crossing to the finished cultivar to six years or less.

3. Yield Testing

Improvements in efficiency of research equipment and computerization will reduce labor requirements, and permit a great expansion in yield testing. It should be possible for breeders to plant and harvest many more plots than at present with the same resources and to provide the initial data summaries more quickly than occurs at present.

In summary, the future strategies for soybean cultivar development will encompass many new ideas and technologies. They will result in shorter response time to environmental stresses and greater improvements in cultivar performance.

NOTES

John Schillinger, Research Manager, Asgrow Seed Co., Ames, IA 50010.

REFERENCES

Boerma, H. R. and R. L. Cooper. 1975a. Comparison of three selection procedures for yield in soybeans. Crop Sci. 15:225-229.

Boerma, H. R. and R. L. Cooper. 1975b. Effectiveness of early-generation yield selection of heterogeneous lines in soybeans. Crop Sci. 15:313-315.

Burris, J. S. and W. R. Fehr. 1971. Methods for evaluation of soybean hypocotyl length. Crop Sci. 11:116-117.

Cooper, R. L. 1981. Development of short-statured soybean cultivars. Crop Sci. 21:127-131.

Fehr, W. R. 1982. Control of iron-deficiency chlorosis in soybeans by plant breeding. J. Plt. Nutr. 5:611-621.

National Research Council. 1984. Genetic engineering of plants: agricultural research opportunities and policy concerns. National Academy Press, Washington, D.C.

Pandey, K. K. 1978. Gametic gene transfer in *Nicotiana* by means of irradiated pollen. Genetica 49: 53-69.

THE APPLICATION OF QUANTITATIVE GENETICS THEORY TO PLANT BREEDING PROBLEMS

S. K. St. Martin

As in many other technological fields, the theory of genetics as applied to plant breeding developed concurrently with practical applications. In fact, many practices, such as the pedigree method, were firmly established before the first attempts to apply genetic principles to quantitative inheritance.

It seems proper that theory should be motivated by practical questions. Quantitative genetics theory should serve to increase the efficiency of plant breeding programs by increasing the genetic advance per dollar spent. This paper is intended to suggest issues for future quantitative genetics research that are based on practical questions that arise in breeding programs. "Quantitative genetics" is used in a broad sense to include statistical tools that are not derived from Mendelian principles.

GENE ACTION

The partition of genetic variation (σ_G^2) in a Hardy-Weinberg population is an outgrowth of the work of Fisher (1918):

$$\sigma_G^2 = \sigma_A^2 + \sigma_D^2 + (\sigma_{AA}^2 + \sigma_{AD}^2 + \sigma_{DD}^2 + \sigma_{AAA}^2 + ...)$$

where σ_A^2 = additive variance, σ_D^2 = dominance variance and the terms in parentheses represent interactions involving two or more loci (epistasis). The principal assumptions here are random mating and absence of linkage or maternal effects.

Most relevant to the development of pure lines is the partition of the genetic variance among lines derived by inbreeding without selection from a Hardy-Weinberg population,

$$\sigma_{G(inbred\ lines)}^2 = (1 + F)\sigma_A^2 + (1/4)(1 - F^2)\sigma_D^2 + (1 + F)^2\sigma_{AA}^2 + ...$$

where F is the inbreeding coefficient of the single plant parent (F = 0 for S_1 lines, etc.). The term involving σ_D^2 is correct only if there are two equally frequent alleles per segregating locus; otherwise the variance among lines cannot be expressed in terms of the variance components of the source population. The relative contribution of epistasis to σ_G^2 increases with inbreeding.

In particular, for pure lines F = 1 and therefore,

$$\sigma_{G(pure\ lines)}^2 = 2\sigma_A^2 + 4\sigma_{AA}^2 + ...$$

Thus, only the additive component and epistatic components involving additive effects contribute to variation among pure lines.

Studies of gene action in quantitative traits of the soybean have been limited in number and somewhat contradictory in result. The existence of heterosis suggests that dominance variance is present for seed yield. Epistatic effects probably exist, but are of less importance than the additive component. It remains an open question whether epistasis is of practical significance in selection among soybean lines.

PREDICTING GENETIC GAIN

When Selections Are Intercrossed

Prediction of genetic advance associated with a breeding method is one of the important goals of quantitative genetics. The need for expensive long-term selection experiments would be reduced if a reliable estimate of progress could be obtained by theoretical methods. Unfortunately, predicted gains and realized gains often disagree.

Galton (1889) observed that individuals at the extremes of a quantitative distribution transmit only a portion of their phenotype to the next generation. In Galton's terminology, mesurements on the progeny of individuals selected for high (or low) scores display *regression* toward the grand mean of the population. Plant and animal breeders are vividly aware of this phenomenon, and predict gain using regression equations.

The expected gain (ΔG) per year when individuals or families are tested and the selections are intercrossed is given by:

$$\Delta G = i(K\sigma_A^2 + C)/y\sigma_P \qquad (1)$$

where i = the standardized selection differential, K and C are constants, y = the number of years required to complete a cycle of selection and recombination, and σ_P = phenotypic standard deviation.

In (1), K depends on the type of family under evaluation (for example, K = 1/4 for half-sib families and K = 1 for S_1 lines). The constant C equals 0 if the individuals or families are not inbred or if the dominance variance is 0. If inbred lines are tested, then C is a function of the degree of dominance and allele frequencies, and may be positive or negative.

It is significant that the phenotypic standard deviation (σ_P) occurs in the denominator of (1) and includes nonadditive genotypic variance as well as the genotype x environment interaction. Replication over locations and years decreases σ_P.

Expression (1) estimates the difference in performance between successive cycles of Hardy-Weinberg populations [cycle (n)S_0 - cycle (n-1)S_0]. More pertinent to breeders who intend to use improved populations as source populations for line development is the difference between the means of pure lines derived from successive cycles [cycle (n)S_∞ - cycle (n-1)S_∞]. The formula for estimated gain in this case is similar to (1) except that the constant C is generally greater in absolute value. The complication posed by C in these formulas can be overcome by using estimates of variance components derived from inbred populations rather than from a Hardy-Weinberg source population.

When Selections are Reproduced Exactly

When selected families are reproduced exactly (e.g., by self-fertilization if they are pure lines) rather than intermated, then the predicted genetic gain per generation of selection includes in the numerator the total genotypic variation:

$$\Delta G = i \, \sigma_G^2 / \sigma_{P'} \tag{2}$$

where σ_G^2 = the genotypic component of variance among families and $\sigma_{P'}$ = the phenotypic component of variance among families.

If families are pure lines then the additive x additive epistatic component of variance is exploitable through selection and selfing, although it is not transmitted through crossing.

Many assumptions underlie (1) and (2) and therefore many factors affect the reliability of prediction. Perhaps the most important factors leading to poor prediction are as follows:

(a) Predicted gains are expected values. Gain achieved in any individual selection study may deviate considerably from the expected value, especially when the number of individuals selected is small.

(b) Standard errors of variance components are large. Thus, unless a large number of genotypes, replications, and environments is used to obtain estimates of σ_A^2, σ_G^2 or σ_P^2, the expected gain is likely to be imprecise. It is not unusual for the standard error of ΔG to be 30 to 50% as large as the estimate itself.

(c) The magnitude of the genotype x environment interaction is often underestimated. Failure to partition Genotype x Environment into its components Genotype x Year, Genotype x Location, and Genotype x Year x Location, such as when an experiment is conducted in a single year or when an unbalanced set of location-year combinations is used, can lead to biased estimates of σ_G^2 (Comstock and Moll, 1963).

(d) Non-normality in the distributions of phenotypic or genotypic value is a source of error in prediction. Major gene effects could cause deviations from normality.

(e) Linkage and nonadditive gene action are difficult to measure and could reduce the reliability of prediction formulas.

(f) The linear model used in statistical analysis; i.e., Phenotype = Genotype + Environment + Genotype x Environment + Error, may not be adequate. Some environments are more suited to the expression of genetic variation than others. In particular, there is likely to be a correlation between the genotypic variance in an environment and the environmental mean.

In summary, predicted gain formulas are generally imprecise and sometimes biased. Inaccuracy and bias are probably of less concern when comparing breeding methods using relative predicted gains than when absolute predictions are required. It seems unlikely that quantitative genetic theory alone will be able to surmount the obstacles posed in (a) through (d). Improved models might alleviate (e) and (f), but the difficulty in obtaining suitable parameter estimates remains.

Plant breeders achieve genetic gain by selection in genetically variable populations. Therefore, it is appropriate to discuss applications of genetic theory to problems related to genetic variation and selection. The common theme of these applications is optimizing the allocation of resources to hybridization and selection.

HYBRIDIZATION

The major source of genetic variation for quantitative traits is hybridization of selected parents and the resulting Mendelian segregation. Biparental crosses are probably the most frequently used for this purpose in soybean breeding. The opportunities for genetic recombination in the progeny of a biparental cross, however, are limited. Very few lines from a biparental cross can be expected to obtain more than 70% of their genetic material from one parent (St. Martin, 1982). Bailey and Comstock (1976) discussed at length the limitations of a two-line gene pool.

To broaden the range of genetic recombinations, a breeder can use more extensive intermating. Of course, this is a feature of formal recurrent selection, but the use of intermating need not be confined to that setting. Table 1 shows a scheme for using genetic male sterility to achieve intermating with very little effort. The scheme differs from formal recurrent selection in permitting introgression of new material (presumably superior lines selected from one's own and other breeding programs) in each cycle. This procedure is not suitable for controlled selection experiments, since the intermated lines do not derive from the population itself. It is similar to recurrent selection, however, in producing successive cycles of intermated populations with continual improvement in allele frequencies. It is, perhaps, more appropriately labelled "recurrent introgression" than recurrent selection. It is similar in concept to Jensen's (1970) diallel selective mating system.

The limitations associated with biparental crosses might be overcome in part by making many such crosses. The optimum allocation of resources for preliminary testing, as to numbers of crosses and of lines per cross, is an important question. Weber's (1979) argument that one line per cross is optimal was based on a simple model in which crosses fell into two classes: those that produced no superior genotypes and those in which the frequency of superior genotypes was some fixed value, $p > 0$. A continuous distribution of p over crosses would be more realistic, but assumptions concerning the nature of the distribution would be needed for a mathematical solution.

SELECTION

Use of Cross Information in Line Selection

Animal breeders, faced with the inability to replicate animal performance trials, have long used pedigree information to increase the effectiveness of selection. Plant breeders could make use of the same type of data. Specifically, the effectiveness of selection in the early stages of testing could be enhanced by using information on the cross mean performance to supplement line means. Cross means could be used in at least three ways:

Table 1. A scheme for recurrent introgression.

Season	Activity
1	Interplant composite of selected lines with a male sterile maintainer line in isolation. Harvest hybrid seed (genotype *Msms*) produced on male sterile (*msms*) plants.
2	Produce self-fertilized seed on *Msms* plants.
3	Repeat Season 1 activity using seed from Season 2 as male-sterile maintainer along with a new composite of selected lines as a pollen source.
	Also continue self-fertilizing seed from Season 2 for use in pure line development.

(a) By testing crosses as bulks or maturity group bulks and developing lines only from superior crosses (Harlan et al., 1940).

(b) By adopting an early-generation testing procedure like that described by Boerma and Cooper (1975) and Cooper (1982), where unreplicated F_2-derived line entries are used as "replicates" in estimating cross means.

(c) By use of cross information in the first test of a set of unselected lines derived by single-seed descent. Selection of lines for further testing could be based on an index $I = P_L + bP_C$ where P_L = line mean, P_C = cross mean, and b is a weighting factor chosen to optimize expected genetic gain. This method could be used in conjunction with little additional effort, except that the practice of discarding lines based on visual evaluation prior to harvest would conflict with the goal of obtaining an unbiased estimate of P_C. Weber (1982) developed a solution for weighting means of F_j-derived families for selection within crosses.

An element of selection among crosses is especially useful when additive x additive epistatic effects are large, causing significant "specific combining ability" effects that make prediction of cross means difficult. Research is needed to determine whether the value of cross mean data warrants the effort to obtain it by one of these methods. The use of cross means in selection is also an additional complication in the problem, mentioned earlier, of optimizing the number of lines per cross.

Allocation of Resources in Yield Testing

Yield testing is the most expensive aspect of most plant breeding programs and, therefore, the area in which the greatest potential for increased efficiency might be expected. Yield testing leading to cultivar release is a multi-stage process in which a decision is required at each stage as to whether to discard or retain each line. The later stages involve fewer lines but more extensive testing (i.e., more replications and locations) of each line. Allocating a large percentage of effort to the early stages increases the number of superior lines present, but the resulting high selection intensity decreases the probability of retaining a given superior line through subsequent stages. Thus, an optimum allocation is sought.

Theoretical research in this area is difficult because of the large number of variables (e.g., number of stages, selection intensity at each stage, genotypic and environmental variance components). Finney (1966) examined multiple stage selection at length and recommended that, in general, no more than three or four stages should be used. The greatest progress was made when the selection intensity and the proportion of resources employed were kept approximately constant from stage to stage. Finney did not, however, consider the existence of a genotype x stage (year) interaction.

St. Martin and Nagaraja (unpublished) used variance component estimates from the extensive yield evaluation of random soybean lines by Garland and Fehr (1981) to simulate selection with different allocations of resources (plots) to three stages (Table 2). Near-optimum gains were made over a broad range of allocation schemes, but it appeared that at least one-half of the available plots should be allocated to the first year's test. Discarding 80 to 90% of the initial group of lines, based on yield data from a single plot, was necessary for maximum progress. A similar selection intensity was optimal for the second stage.

Selection Based on Probabilities

Cochran (1951) suggested practicing selection to maximize the probability of obtaining a line superior to the best check cultivar. In this context it would be useful to

Table 2. Relative genetic gain from different allocations of a total of 2500 plots to three stages of testing.[a]

		30	35	40	45	50	55	60	65	70	75	80	85
	65	90											
	60	91	88										
%	55	88	94	88									
plots	50	88	97	90	90								
allo-	45	92	91	90	93	84							
cated	40	92	91	89	93	90							
to	35	85	88	89	95	97	92	92					
year	30	90	93	97	95	95	96	93	94				
2	25	88	90	90	86	90	95	97	97	97			
	20	90	90	90	91	93	91	90	95	92	100		
	15				93	87	96	94	94	94	95		
	10								89	88	95	95	93

% plots allocated to year 1

[a]Year (stage) 1: one location, one replication; year (stage) 2: two locations, two replications; year (stage) 3: four locations, three replications. Variance components from Garland and Fehr (1981).

estimate for each line in the testing program the probability, given its testing history and performance, that its genotypic value exceeded that of the check. This probability would be a function of the observed phenotypic difference between the line and the check, the number of replications and environments in the testing history, the magnitude of variance components associated with experimental error and genotype x environment interaction, and perhaps the total number of lines tested. The probability statement could be used to select lines for further testing or for the crossing block. The principal benefit of this approach is that it would permit comparison of lines at different stages in the program. For example, given a line that outyields the check by 10% averaged over two environments and another outyielding the check by 5% over four environments, which is more likely to prove successful in more extensive training?

CONCLUSION

Plant breeding programs have grown enormously in number, scope and size in recent years. The total expenditure of resources in plant improvement efforts is impressive. Quantitative and statistical genetics should contribute to the efficient deployment of these resources for maximum benefit.

NOTES

S. K. St. Martin, Assistant Professor, Department of Agronomy, Ohio State University, Columbus, Ohio.

REFERENCES

Bailey, T. B. and R. E. Comstock. 1976. Linkage and the synthesis of better genotypes in self-fertilizing species. Crop Sci. 16:363-370.

Boerma, H. R. and R. L. Cooper. 1975. Effectiveness of early-generation yield selection of heterogeneous lines in soybeans. Crop Sci. 15:313-315.

Cochran, W. G. 1951. Improvement by means of selection. p. 449-470. *In* Proc. Second Berkeley Symp. Math. Stat. and Prob., University of California, Berkeley.

Comstock, R. E. and R. H. Moll. 1963. Genotype-environment interactions. *In* Statistical Genetics and Plant Breeding, NAS-NRC Publ. 982. National Academy of Science, Washington, D.C.

Cooper, R. L. 1982. Early generation testing as an alternative to the single seed descent breeding method in soybean. Agron. Abstr., p. 62.

Finney, D. J. 1966. An experimental study of certain screening processes. J. Roy. Stat. Soc. Ser. B 28:88-109.

Fisher, R. A. 1918. The correlation between relatives on the supposition of Mendelian inheritance. Trans. Roy. Soc., Edinburgh 52:399-433.

Galton, F. 1889. Natural inheritance. Macmillan, London.

Garland, M. L. and W. R. Fehr. 1981. Selection for agronomic characters in hill and row plots of soybeans. Crop Sci. 21:591-595.

Harlan, H. V., M. L. Martini and Harland Stevens. 1940. A study of methods in barley breeding. USDA Tech. Bull. 720.

Jensen, N. F. 1970. A diallel selective mating system for cereal breeding. Crop Sci. 10:629-635.

St. Martin, S. K. 1982. Effective population size for the soybean improvement program in maturity groups 00 to IV. Crop Sci. 22:151-152.

Weber, W. E. 1979. Number and size of cross progenies from a constant total number of plants manageable in a breeding program. Euphytica 28:453-456.

Weber, W. E. 1982. Selection in segregating generations of autogamous species. I. Selection response for combined selection. Euphytica 31:493-502.

EQUIPMENT AND TECHNIQUES FOR FIELD RESEARCH IN SOYBEANS

John C. Thorne

Major advancements in mechanization of field research in the last two decades have allowed agronomists to improve dramatically their management of time and resources. In 1964, Mr.Egil Oyjord and other interested persons formed a group called, "The International Association on Mechanization of Field Experiments" (IAMFE), to provide a means of communication among researchers involved in mechanization. At that time, most soybean plots were planted with single-row push planters, end-trimmed with hand sickles, cut for harvest with sickles or weed mowers, and threshed with stationary threshers. A breeding program with 5000 yield-test plots was considered large.

During the 1960s people began putting cones on tractor-driven planters and discovered that it was not only easier, but faster and more labor efficient, to ride than to walk. Further refinements, such as seed dividers, electric trip mechanisms, check-wires, seed monitors, and automatic seed-feeding systems have led to even greater accuracy and efficiency.

Combine harvesting of small soybean plots began in the early 1970s. Some researchers modified existing small commercial combines, others began with European plot combines, such as the Hege 125 or Wintersteiger Seedmaster, or the American Chain SP50 (Table 1), which had been designed primarily for use in small grains. A few designed and built their own combines. Some of the problems encountered with the early combines involved lack of capacity, speed, and durability. Seed delivery and collection systems were often cumbersome or unreliable. Excessive field loss and seed damage, on one hand, or incomplete clean out between plots, on the other, were often encountered. (Clark and Fehr, 1976).

In recent years, equipment manufacturers have been building machines with bigger engines, heavier shafts, heavy-duty bearings, and other improvements, which have overcome many of the problems relating to capacity. Clark and Fehr (1976) reported a system for reducing field loss which, with modifications, has been incorporated into several plot combines. Thus, most of the major shortcomings of plot combines have been overcome, and most researchers are now comfortable with mechanized harvest.

If the 1960s was the decade for mechanizing planting, and the 1970s for harvest, probably the 1980s will be the decade for advances in computer handling of data. As graduate students in the mid-1960s, my colleagues and I dreamed of the day in the distant future when a computer on the plot combine would print the analysis of variance as the last plot in the trial was harvested. Since we were using mechanical calculators for most routine data analyses at the time, the idea seemed far-fetched. Today, it is not far from reality.

Table 1. Manufactuers of equipment mentioned in the text.[a]

Item	Manufacturer
Hege 125 Combine	Hans-Ulrich Hege, Saatzuchmaschinen, 7112 Waldenburg-Hohebuch, West Germany
Wintersteiger Seedmaster Combine	Walter and Wintersteiger K G, A-4910 Reid, Austria
K.E.M. (Chain) LD50 Combine	K.E.M. Corp., Box 471, Haven, KS 67534
Tye drill units	Tye Co., Box 218, Lockney, TX 79241
John Deere Flexi-Planter	John Deere Co., 1400 3rd Ave., Moline, IL 61265
K.E.M. spinning divider	K.E.M. Corp., Box 471, Haven, KS 67534
Zero-Max reduction gearbox	Zero-Max Co., 2845 Harriot Ave., South Minneapolis, MN 55408
L.G. 2007 planter	K.E.M. Corp., Box 471, Haven, KS 67534
Almaco planters and drills	Almaco, Box 296, Nevada, IA 50201
L.G. 2005 planter	K.E.M. Corp., Box 471, Haven, KS 67534
Almaco check head system	Almaco, Box 296, Nevada, IA 50201
Almaco CA500 automatic trip	Almaco, Box 296, Nevada, IA 50201
Almaco Hilldrop units	Almaco, Box 296, Nevada, IA 50201
Almaco Seed Tray Indexer	Almaco, Box 296, Nevada, IA 50201
John Deere Row Crop Header	John Deere Co., 1400 3rd Ave., Moline, IL 61265
Almaco Row Crop Header	Almaco, Box 296, Nevada, IA 50201

[a]The use of trade names in this publication does not imply endorsement of the products named or criticism of similar products not mentioned.

PLOT PLANTING

Almost all research programs that require large-scale planting of plots have developed a mechanized planting system. Sixteen articles in the 1980 Proceedings of Fifth International Conference on Mechanization of Field Equipment (Dijkstra and van Santen, eds., 1980) describe planters, drills, or planting systems, indicating world wide interest in this area.

Several recent trends in soybean production have resulted in the need for further sophistication in plot planters. The first is the increasing use of narrow rows (15 to 50 cm). Oplinger et al. (1983) developed a plot planting system using interchangeable tool bars, front-mounted on a small, rear-engine tractor. One tool bar is equipped with nine Tye drill double disc openers that plant 18-cm rows. The other tool bar, equipped with John Deere Flexi-Planter units, is used for planting 38- to 95-cm rows. A K.E.M. spinning divider with interchangeable heads allows for planting two to nine rows from one packet (Table 1). A Zero-Max, variable-reduction gearbox in the cone drive provides nearly infinite flexibility in row length. The planter is small enough to be easily transported for planting off-station plots.

At Northrup King, we were also concerned with planting various row spacings. Assistant soybean breeder, Leonard Goeke, working with K.E.M. Corporation, designed and built a front-mounted, rear-engine planter which will plant four, single-row plots spaced 76 cm apart, two-row plots, spaced up to 102 cm apart, with 61 cm between rows within the plot, three-row plots, spaced 38 cm apart, with borders spaced 76 cm from the plot, or five-row plots, spaced 19 cm apart, with borders spaced 76 cm from the plot (Table 1). The hydraulic system was plumbed so that runners can be raised and lowered independently. The K.E.M. spinning dividers and the Zero-Max gearbox were used for seed separation and row length flexibility as previously described.

A second trend in soybean production that has necessitated changes in planters is the interest in reduced-tillage or no-till planting systems. A part of this interest has come from the increasing hectarage of double-crop soybeans, much of which is planted no-till in narrow rows following wheat. Almaco offers ripple or fluted coulters, down pressure springs, and cast-iron closing wheels as options to equip plot planters and drills for no-till planting (Table 1).

The agronomic research group at Northrup King identified a need for a planter that could be used for both conventional and no-till planting, and which could plant either wide or narrow rows. Leonard Goeke and K.E.M. Corporation designed and built a pull-type 2- to 5-row planter to meet these requirements (Table 1). They mounted five, Tye, double-disc planter-units on a toolbar (Figure 1). Two cones, mounted above two spinning dividers, allow for planting one, two, three, or five rows from one or two seed packets. No-till equipment includes 2.5-cm, fluted coulters mounted in front of each disc opener, with downpressure from 300 kg springs on each unit (Figure 1). Up to 500 kg of additional weight may be added to the planter to hold it down in untilled soil. Weights and coulters are removable for conventional planting.

In the past, measuring and cross-marking aisles between plots has been a time-consuming operation that had to be completed before planting. Clark et al. (1978) suggested the use of a check-wire to signal the position of aisles so that prior marking would not be necessary. Almaco offers a similar system, which automatically trips seed cups, as optional equipment on its plot planters (Table 1). The check-wire system works effectively, but requires cables that are bulky for transport and handling. Further, failure to set the wire the same each time will result in crooked aisles.

Several researchers have experimented with ground-driven devices for tripping cones automatically based on distance traveled by a measuring wheel. We tried to use one with measurement based on revolution of the drive wheel of our self-propelled planter, but found that reliability was insufficient. Almaco recently developed a system that relies on an independent measuring-wheel tracking directly behind the gauge-wheel of the planter (Table 1). A chain drive from the wheel to a rotary encoder translates each 0.5 cm distance into a light pulse. A light sensor, which counts each light pulse, can be preset to trip seed cups for any desired row length from 0.5 to 27 m.

END-TRIMMING

End-trimming is one of the most labor intensive, tedious, and time-consuming operations in conducting soybean yield trials. Wilcox (1970) showed that trimming plots earlier than growth stage R5 would inflate yield estimates unless plots are corrected for end-plant effect. Genotypes varied for the amount of inflation, but these differences were associated with differences in maturity time. With this study and considerable personal bias as background, researchers have proceeded in three general ways with regard to end-trimming:

1. End-trim at least 0.5 m from each end of the plot at or near maturity to eliminate all end-plant effect.
2. Trim plots to harvest length shortly after emergence and either ignore end-plant bias or correct for it. Genotype x end-plant interaction can be reduced by restricting the range in time of maturity within a trial and/or by regression of yield on maturity date.
3. End-trim sometime between growth stages R1 and R6, when disposal of

Figure 1. Views of LG2005 planter showing placement of planter units and coulters (top) and down-pressure springs and additional weights (bottom).

discarded plant material is not difficult. Some compromise with end-plant bias is inherent with this approach.

Several techniques for reducing the labor required for mid- to late-season end-trimming have been developed. Kuell Hinson (pers. commun.) uses a small tractor equipped with a tool-bar and two disc-markers to mark harvest row length. Shortly after emergence, as the tractor is driven through the center of each aisle, the discs cut a 10-cm gap about 50 cm from the end of each row. Plants outside of the mark are removed shortly before harvest. The assumptions that row length is uniform in each range and that the tractor operator will drive straight are inherent in this system. J. H. Williams (pers. commun.) marks the ends of the plots by driving the tractor with marker across the center of the plot. With this system, uniform plot length is assured, since the plot length will be the length of the toolbar. Some damage occurs to the plants that are driven over by the tractor wheels, however. Freed et al. (1978) achieved the objective of uniformity by using two rotary mowers connected at a fixed distance equal to the plot length. Both ends of the plot are cut simultaneously. Charles Laible (pers. commun.) sprays end-plants with glyphosate at flowering to kill them. He applies metolachlor to the aisles at the same time for increased grass control. This system reduces, but does not eliminate, end-plant effect.

At NK, we trim plots to harvest length before flowering with a cultivator and simultaneously apply herbicide to the aisles. We use lattice designs, with blocks assigned to ranges, to adjust for minor variations in row length, if they exist. We assign entries to trials within as narrow a range of maturity time as possible, and use a regression of yield on maturity to reduce interactions based on maturity time.

HILL-PLOTS

The traditional plot used for yield evaluation consists of one or more rows from 3 to 10 cm in length. However, in recent years many soybean researchers have adopted the use of hill-plots (or rows less than 1 m) for yield trials. Breeders have found that using hill-plots for preliminary screening allows for testing large numbers of lines with minimal seed and land requirements. Similarly, hill-plots are useful for evaluation of herbicides, growth regulators, and other chemicals, and for disease and insect screening.

Garland and Fehr (1981) compared hill- and row-plots for evaluating unselected lines for yield and other agronomic characteristics. They also compared a randomized design for the hill-plots with a nonrandom arrangement whereby entries were planted in the same order in each replication. The nonrandom design was evaluated because its use would facilitate visual elimination of lines before harvest. Phenotypic correlations of performance in random hills, nonrandom hills, and rows were very high for all characters. Heritability estimates were similar for random hills and rows. Garland and Fehr concluded that "hill-plots were as effective as rows for phenotypic selection of agronomically desirable genotypes".

Many researchers have avoided using hill-plots because of the difficulty in planting and harvesting them. However, equipment now exists for mechanizing both planting and harvest. Almaco offers a hill-drop unit, consisting of a small cone for even distribution of a few seeds (Table 1). This unit can be used with their automatic-trip device and a seed-tray indexer for virtually completely automated planting. Carlson et al. (1975) described modifications of a Hege 125B combine that facilitates combine harvest of hill-plots. Thus, handling hill-plots in a program can be as convenient as handling row-plots.

PLOT HARVEST

At the 1976 World Soybean Conference, Quick (1976) listed several commercially available plot combines. Most of the combines available now are modifications of those listed or new models from the same manufacturers. Many of the changes in combines involve strengthening them for greater capacity, durability, and power. Although these changes represent major improvements, I will deal with modifications that represent new ideas or systems.

The NK soybean research group began combine harvest with the K.E.M. SP50 (Table 1). Goeke (1980) described several modifications of this machine, which have improved its usefulness for our program. The original machine had a forced-air seed delivery system, which often plugged when the seed moisture was above optimum, or damaged seed when moisture was too low. The air system was replaced with an elevator that carries seed to a double-sacker on top of the machine. To prevent seed mixtures resulting from incomplete cleanout of the elevator, a clam-type trap door was installed at the elevator boot (Figure 2). This trap door opens when the lever controlling the sacker is moved from one position to the other between plots.

We also believed that operators would be more diligent in checking and adjusting cylinder speed and concave clearance if they could do so easily. Therefore, a magnetic tachometer with a gauge on the control panel was installed, allowing the operator to monitor engine, cylinder, or fan speed by moving a switch. Also, a graduated scale for monitoring concave clearance was installed. Concave clearance can be adjusted hydraulically while the machine is in operation, if necessary. Cylinder speed can also be changed instantly by changing the adjustment of two variable-speed pulleys with a hydraulically-controlled rocker arm (Figure 3).

Finally, the combine is equipped with a load-cell scale and a moisture tester. Harvested seed can be weighed and discarded into a bulk bin or caught in a bag, weighed, tagged, and dropped into a bin, or holding sack if the seed is to be saved (Figure 3).

Several researchers have found that grain heads often do not operate properly in lodged soybeans. Several years ago, John Deere introduced a row crop header designed for harvesting soybeans with a commercial combine (Table 1). Lindahl (pers. commun.) adopted one of these for plot work. Almaco now offers a row-crop header as an option for their plot combines (Table 1).

DATA COLLECTION

While improvements in plot planters and combines may still be forthcoming, probably the major advancements in mechanization in these areas have now been made. However, advancements in the area of computer technology will have a major impact on data collection and handling systems in the next few years.

Konzak et al. (1983) described a system for combine harvest and data collection by one person. Plots are harvested into bags and weighed using an electronic balance system. The weight for each plot is transmitted to a portable, electronic data-terminal. Prior to harvest, plot numbers are down-loaded from a computer in which the field layout has been stored. As each plot is harvested, the plot number is displayed on the data terminal and checked for conformity with the harvest tag before the weight is recorded. When harvest is complete, the data are transferred to a computer for analysis. Additional data can be added by use of a bar-code system. Bar-codes associated with plot numbers are read by a wand, and additional data are entered manually.

Some soybean researchers prefer to dry harvested samples before weighing. Mason and Fehr (pers. commun.) devised a system, using a bar-code, to facilitate weighing samples in the lab. Bar-codes are established for each 12-digit plot number in the breeding program. Bagged plots are weighed on an electronic balance while the bar-code

Figure 2. Seed delivery elevator boot showing clam-type, trap-door in closed, operating position (top) and open, clean-out position (bottom).

is read with a wand. A printer prints the plot number and weight to provide a hard-copy backup for the system. At the same time, the data are automatically recorded in a portable data recorder. On completion of weighing, the data are down-loaded to the main-frame computer for analysis.

Figure 3. Plot combine showing position of rack for holding bags for use when plots are saved (top) and variable speed pulleys and rocker arm for cylinder speed control (bottom).

A CHALLENGE

At World Soybean Rsearch Conference I, Quick (1976) issued a challenge to manufacturers of plot equipment to provide more reliable plot combines for soybean research. Today, I believe the challenge is to the researcher to find the best uses of the equipment available, much of which was only imagined a few, short years ago.

NOTES

John C. Thorne, Soybean Research Department, Northrup King Co., P.O. Box 949, Washington, IA 52353-0949.

REFERENCES

Carlson, B. E., T. J. Bandstra, W. R. Fehr and R. C. Clark. 1975. Modification of a self-propelled plot combine for harvesting hill plots of soybeans. Crop Sci. 15:869-870.

Clark, R. C. and W. R. Fehr. 1976. Seed cleaning system for plot combines. Crop Sci. 16:880-881.

Clark, R. C., W. R. Fehr and J. C. Freed. 1978. Check-wire system for plot planters. Agron. J. 70: 357-359.

Dijkstra, J. and A. van Santen, eds. 1980. Proceedings of Fifth International Conference on Mechanization of Field Experiments.

Freed, J. C., J. B. Bahrenfus, T. J. Bandstra, W. R. Fehr and R. C. Clark. 1978. Mechanical system for end-trimming soybean plots. Crop Sci. 18:351-352.

Garland, M. L. and W. R. Fehr. 1981. Selection for agronomic characters in hill and row plots of soybeans. Crop Sci. 21:591-595.

Goeke, L. L. 1980. L.G. 3001 soybean plot harvester. p. 162-166. *In* J. Dijkstra and A. van Santen, eds. Proceedings of Fifth International Conference on Mechanization of Field Experiments.

Konzak, C. F., M. A. Davis and M. R. Wilson. 1983. Combine harvest and data acquisition by one person. Crop Sci. 23:1205-1208.

Oplinger, E. S., J. P. Wright and A. Klassy. 1983. A plot planting system utilizing a rear-engine tractor. Agron. J. 75:848-850.

Quick, Graeme R. 1976. Plot research equipment for soybeans. p. 321-341. *In* Lowell D. Hill (ed.) World Soybean Research. Interstate Printers and Publishers, Danville, Illinois.

Wilcox, J. R. 1970. Response of soybeans to end-trimming at various growth stages. Crop Sci. 10: 555-557.

OFF-SEASON NURSERIES ENHANCE SOYBEAN BREEDING AND GENETICS PROGRAMS

Silvia Rodriguez de Cianzio

Breeding and genetics programs use off-season nurseries for hybridization, generation advance, and seed increase during the winter when the crop cannot be planted in its area of adaptation. An effective off-season nursery reduces the number of years from the time selected parents are crossed to form a population until a superior segregate from the population is released as a cultivar. The efficiency of these programs may be increased further if selection for important characters can be practiced during the winter (Fehr, 1976).

The objective of this paper is to discuss the role of an off-season nursery in a breeding and genetics program. The nursery environment in Puerto Rico will be described as an example. Research conducted at the Iowa State University Soybean Breeding Nursery at the Isabela Substation of the University of Puerto Rico and at Iowa State University, Ames, will be the primary source of information.

Off-season nurseries play an important role in breeding and genetics programs. Our results indicate that an efficient off-season nursery can be used for selection of both quantitative and qualitative traits as well as hybridization and generation advance. The availability of multiple seasons during the year increases genetic gain per year. Management practices may be developed that will improve the handling of experimental material.

THE NURSERY ENVIRONMENT

Off-season nurseries may be established in the tropics or in a temperate location. A survey of public and private soybean breeders that I conducted in 1984 has shown that 80% of them are using off-season nurseries at some stage of their programs. Nurseries are located in eight countries of the northern hemisphere and in three of the southern hemisphere (Table 1). In the northern hemisphere, most of the countries are in the tropical region. In the southern hemisphere, most of the nurseries are in the temperate zone. Environmental conditions in temperate locations may be similar to those of the area where the genotypes are grown commercially and thus may present an advantage in terms of seed production.

Breeding material for U.S. breeders is grown at a nursery during the winter and early spring. The number of generations obtained during the off-season will depend on the latitude of the nursery location and on the maturity group of the experimental material. Breeding methodology and available resources will also determine the number of generations grown.

Table 1. Nursery location, by country and latitude, of public and private soybean breeding programs and approximate generation length.

| Nursery location | | Generation length[a] | |
Country	Latitude	Unlighted	Lighted
	deg-min	— days after planting —	
	Northern hemisphere		
Colombia	3-32	90-110	
Belize	16	100	130-150
Puerto Rico	18	75-110	100-140
Jamaica	18	75-100	
China	18	90-120	
Mexico	21	160	
U.S.A., Hawaii	21 to 22	90-100	120-150
Taiwan	23	100	
U.S.A., Florida	25	90-130	100-140
	Southern hemisphere		
Argentina	24	110	
Chile	33-34	120-135	
New Zealand	35 to 40	150	

[a]Approximate range in days, not taking into consideration maturity groups.

Fifty-two percent of soybean breeders that I surveyed indicated that they grow one generation per year at their off-season nursery locations and 48% obtain two generations per year. Breeders who located their nurseries in the temperate zone obtain one generation per year in the off-season, with an average generation length of approximately 130 days.

Soybeans are sensitive to photoperiod, and this sensitivity determines to a considerable degree their time of maturity (Garner and Allard, 1923; Parker and Borthwick, 1951). Short daylength in the tropics favors early flowering and seed maturation of cultivars adapted to temperate climates. In Puerto Rico, Maturity Groups 000 to IV flower in approximately 27 days and mature in an average of 82 days (Table 2). Two generations may be grown from November to May under these conditions. Seed production under natural daylength will be less than that obtained in temperate locations. The late-maturity groups adapted to lower latitudes generally will require longer periods to flower and mature in tropical regions. Seed numbers will be greater than for earlier maturity groups, but only one generation may be possible during a six-month period.

Hybridization of cultivars unadapted to tropical conditions may be impossible without the use of artificial lighting to extend daylength. Under natural daylength, flowers are few and not suitable for hybridization, because self-pollination occurs before they are large enough to manipulate (Fehr, 1980). To increase flower number and obtain flowers suitable for hybridization, breeders use artificial lights to extend natural daylength (Brim, 1974; Parker and Borthwick, 1951). At the ISU Soybean Breeding Nursery in Puerto Rico, lighting consists of 240-volt, 1,500-watt quartz-iodide bulbs installed on a pole approximately 7 m high. This provides sufficient light, for crossing purposes, in an area 18 by 18 m. In the lighted area, plants are exposed to continuous light for a period of 2 weeks after emergence, followed by 14.5-h days until crossing is completed. After crossing, plants are exposed to natural daylength for the

Table 2. Days from emergence to full bloom (R2) and maturity (R8) of cultivars of different maturity groups planted in Puerto Rico.

Maturity Group	Environment	Days from emergence to	
		R2	R8
000	Unlighted (UL)	25	80
00	UL	28	80
0	UL	27	80
I	UL	27	85
	Lighted (L)	35	94
II	UL	27	85
	L	35	100
III	UL	29	83
	L	36	101
IV	UL	29	83
V	UL	34	83
VI	UL	31	86
VII	UL	30	86
VIII	UL	31	86
IX	UL	41	102

remainder of the growing season. Generation length in the lighted area is longer, and more days are necessary to reach full bloom (R2) (Table 2). Two generations are still possible in 6 months, one under lights and the second under natural daylength, or vice versa. We have observed that the percentage of successful pollinations is greater in Puerto Rico than in Iowa (Walker et al., 1979).

In my survey, 66% of the breeders indicated that they use artificial lights in their off-season nurseries, and 42% of them plant under lights to facilitate hybridization. There were 58% of the breeders who used plantings under lights to obtain greater seed production.

USES OF THE NURSERY

Selection

The feasibility of using an off-season nursery to select for quantitative traits depends in part on the relative importance of genotype x environment interactions for the traits under consideration. To assess the practical significance of the interaction, the relative performance among lines at the off-season nursery and at the adapted location should be compared.

Evidence is available that indicates that the nursery can be used as a selection environment for chemical composition of the seed and seed size. We evaluated protein, oil, protein + oil, and fatty acid composition of the seed oil (Cianzio et al., 1985; Hawkins et al., 1983). Both studies were conducted during a period of at least 2 years in Iowa and Puerto Rico. In Puerto Rico, unlighted and lighted conditions were used. Similar experimental procedures were used in both studies. Variance components were estimated, phenotypic and rank correlations were calculated between the two locations, and the ranking of lines was compared for the adapted and unadapted conditions.

In both studies, genotype x environment interactions were found; however, the genetic components were larger than the interactions. Phenotypic and rank correlations for each trait between Iowa and Puerto Rico were positive, highly significant, and similar in value (Table 3). Ranking of lines differed as much from one another in Iowa environments as they did in Puerto Rico environments (Table 4). Similarity in chemical performance of lines in Iowa and Puerto Rico allowed identification of the same entries in each study. In the protein study, a selection intensity of 50% on an entry-mean basis in individual environments would include the top lines for protein, oil, and protein + oil percentage. In the fatty acid study, lines with the highest and lowest percentage of each fatty acid were similar in Iowa and Puerto Rico.

The studies on chemical composition of the seed indicated that selection among soybean lines adapted to Iowa is effective in Puerto Rico. To select for seed composition in tropical sites, plantings under either natural daylength or with daylength artificially extended could be used.

Selection for pod width is an efficient method of indirect selection for seed size (Bravo et al., 1980). Pod width and seed size (g/seed) of nine cultivars from Maturity Groups I, II, and III were evaluated in Iowa and Puerto Rico. Variance components were obtained for both traits, and ranking of cultivars was compared for the two locations. The rank of cultivars for seed weight fluctuated among environments more than for pod width (Table 5). Ranking of the nine genotypes for pod width was similar across environments. These results indicated that a reliable estimate of pod width can be obtained over a broad range of environments. Pod width can be measured in an adapted or unadapted environment when the green pods have reached full size. A pod has reached full size when the developing seed is at least half as wide as the pod (Cianzio et al., 1982).

Use of an off-season nursery to conduct selection for qualitative traits is practical because the environment generally does not influence the expression of a character. A nuclear-cytoplasmic interaction of a conditional lethal between y20-k2 and cyt-Y2 has been characterized for soybeans by using the Iowa and Puerto Rican environments (Palmer and Cianzio, unpublished).

Hybridization and Generation Advance

An off-season nursery fulfills an unavoidable requirement in any breeding program for hybridization and generation advance. Experimental material is exchanged between

Table 3. Phenotypic and rank correlations on an entry mean basis for eight chemical traits between Iowa and Puerto Rico.

| | Correlation coefficients | |
	Phenotypic	Rank
Protein[a]	0.89**	0.90**
Oil[a]	0.79**	0.78**
Protein + Oil[a]	0.77**	0.79**
Fatty Acids[b]		
Palmitic	0.89**	0.74**
Stearic	0.86**	0.82**
Oleic	0.74**	0.74**
Linoleic	0.71**	0.63**
Linolenic	0.84**	0.82**

[a]From Cianzio et al. (1985).
[b]From Hawkins et al. (1983).

Table 4. Ranking of top five lines for protein, oil, and protein + oil percentage on the basis of the standard, compared with rankings of individual environments in Iowa and in Puerto Rico, and ranking of the two highest and the two lowest lines for five fatty acids in Iowa and in Puerto Rico.

Rank of standard[b]		Rank in Puerto Rico Environments, Year 1[a]					
	Iowa	Nov. UL	Nov. L	Feb. UL	May L	May UL	Aug. UL
Protein							
1	2	2	1	1	1	1	2
2	1	1	17	3	2	2	1
3	4	3	2	4	5	5	5
4	5	9	6	6	3	3	3
5	8	10	3	2	6	4	8
Oil							
1	8	3	4	1	2	3	1
2	11	2	10	5	4	2	8
3	1	6	1	4	1	1	4
4	2	5	9	3	6	5	14
5	14	11	13	10	10	7	3
Protein + Oil							
1	1	2	1	2	2	1	2
2	2	1	14	3	1	2	1
3	7	11	2	1	10	3	6
4	5	4	3	4	4	4	4
5	10	3	16	13	5	5	3

Rank of Iowa	Rank in Puerto Rico				
	Palmitic	Stearic	Oleic	Linoleic	Linolenic
Fatty acid[c]					
1	1	1	3	1	1
2	4	2	4	2	6
19	18	20	14	19	20
20	20	18	18	20	18

[a]From Cianzio et al. (1985). Nov. UL = November unlighted; Nov. L = November lighted; Feb. UL = February unlighted; May L = May lighted; May UL = May unlighted; Aug. UL = August unlighted.

[b]Standard for year 1 (1979 to 1980 plantings) was the mean of all plantings in Iowa and Puerto Rico in 1980 to 1981.

[c]From Hawkins et al. (1983).

the nursery location and its area of adaptation. This type of arrangement requires careful coordination for continuous cycles of planting and harvest. Timing of planting and harvest is one of the most critical factors.

A practical solution to facilitate the exchange of experimental material may be the reduction in duration of a generation in the nursery. Two possibilities are the harvest of immature viable seed (Fehr, 1976) or the use of chemical compounds to hasten maturity. One of the most widely used breeding methods to advance generations in

Table 5. Ranking of nine cultivars for seed weight and pod width evaluated in Iowa and Puerto Rico environments (Bravo et al., 1980).

Character and genotype	Environments			
	Iowa[a]		Puerto Rico[b]	
	Agron. Res.	Burkey Farm	Oct. UL	Jan. L
Seed weight				
Prize	3	2	1	1
Disoy	1	1	2	2
Vinton	2	4	3	4
A74-201026	5	3	4	5
Beeson	6	6	6	6
A74-201020	4	5	5	3
A74-104030	7	7	7	7
A72-512	8	8	8	8
A73-19084	9	9	9	9
Pod width				
Prize	1	1	1	1
Disoy	2	2	2	2
Vinton	3	3	3	3
A74-201026	4	4	4	4
Beeson	5	5	5	5
A74-201020	5	6	6	6
A74-104030	7	6	7	7
A72-512	8	8	8	8
A73-19084	9	9	9	9

[a]Agron. Res. = Agronomy Research Center, Ames.
[b]Oct. UL = October unlighted; Jan. L = January lighted.

the nursery is single-seed descent (Goulden, 1941; Brim, 1966). Harvest of immature viable seeds would be particularly suitable for use with the single-seed descent method.

Research conducted at the Puerto Rico nursery has shown that it is possible to harvest immature viable seeds under tropical conditions (Ortiz et al., 1983). In this study, there were plantings in November and February under natural conditions, and in November under artificial lights. Pods were pulled from plants at weekly intervals from 24 to 59 days after flowering and were dried under ambient conditions. The percentage field-emergence of seeds harvested as early as 31 days after flowering was similar to that of seeds harvested at maturity. By harvesting immature seeds, there was a saving in time of approximately 2 weeks compared to harvesting at maturity (R8) (Table 6). To incorporate this harvest procedure efficiently into a breeding program, however, visual indicators are needed to identify pods 31 days old. Research is now in progress to develop such indicators.

The use of chemical treatments may be practical when seeds are harvested in bulk. Twenty-two percent of the breeders who answered my survey indicated that they apply defoliants in their off-season nurseries to hasten seed maturation. The chemical is applied at stages R6 or R7 (Fehr and Caviness, 1977). The estimated time saved in a generation ranges from 7 days to 2 weeks.

Plants have a shorter flowering period in the lighted area of the nursery than in their area of adaptation. A plant generally will have flowers suitable for crossing for only 6 to 7 days. This may reduce the number of hybridizations obtained on a plant.

Table 6. Average percentage field emergence of two cultivars whose seed was harvested from 24 to 59 days after flowering in three environments. (From Ortiz et al., 1983).

| | November | | February | |
	Unlighted	Lighted	Unlighted	Combined
Days after flowering		– % –		
24	32+[a]	87+	21+	47+
31	86@	90+	48@†	75@
38	91@	93+	52†	79@
45	93@	96+	46@	78@
52	90@	93+	38@	74@
59 (mature)	89@	92+	36@	72@

[a]Means within columns followed by different symbols are significantly different at the 0.05 probability level according to Duncan's Multiple Range Test.

Sequential plantings to extend the flowering period should be sown 7 to 8 days apart. The nursery provides the right set of conditions to obtain crosses among cultivars of widely different maturity groups. Different planting dates of the parents are required to match the flowering among these genotypes. The first planting of cultivars of the later maturity group can be approximately 2 weeks before the first date of the early cultivar, and weekly plantings of both parents follow for approximately 3 to 4 weeks.

CONCLUSIONS

Off-season nurseries play an important role in breeding and genetics programs for hybridization and generation advance. Results indicate that breeding for quantitative traits and basic genetic studies on qualitative characters may be conducted at the nursery location. The availability of multiple seasons each year can reduce the length of time required to conduct a breeding program for quantitative traits and increase genetic gain per year. The availability of year-round plantings to conduct hybridizations, advance generations, and progeny tests makes the nursery a convenient research tool for qualitative traits. Space generally is not a limiting factor in an off-season nursery, and irrigation and daylength may be controlled. Appropriate handling of the material can improve efficiency of the operation. Harvest of immature viable seed is possible to shorten the time of generation length. Increased number of hybridizations may be obtained by sequential plantings of parental material. New possibilities will arise as more knowledge is acquired of the environment and the performance of the material in those conditions. Other traits will have to be evaluated, improved testing techniques under laboratory and field conditions developed, and different management procedures will have to be tested.

NOTES

Silvia Rodriguez de Cianzio, Department of Agronomy, Iowa State University, Ames, IA 50011 and Department of Agronomy, University of Puerto Rico, Mayaguez, PR 00708.

Journal Paper No. J-11650 of the Iowa Agricultural and Home Economics Experiment Station, Ames, Iowa. Project No. 2475.

REFERENCES

Bravo, J. A., W. R. Fehr, and S. R. de Cianzio. 1980. Use of pod width for indirect selection of seed weight in soybeans. Crop Sci. 20:507-510.

Brim, C. A. 1966. A modified pedigree method of selection in soybeans. Crop Sci. 6:220.

Brim, C. A. 1974. Off-season nurseries. p. 65-69. *In* Proc. Workshop on Soybeans for Tropical and Subtropical Conditions. Univ. of Puerto Rico—INTSOY, Mayaguez.

Cianzio, S. R. de, S. J. Frank, and W. R. Fehr. 1982. Seed width to pod width ratio for identification of green soybean pods that have attained maximum length and width. Crop Sci. 22:463-466.

Cianzio, S. R. de, J. F. Cavins, and W. R. Fehr. 1985. Genotype x environment interactions for protein and oil content of soybean genotypes in temperate and tropical locations. Crop Sci. (in press).

Fehr, W. R. 1976. Description and evaluation of possible new breeding methods for soybeans. p. 268-275. *In* L. D. Hill (ed.) World Soybean Research. Interstate Printers and Publishers, Inc., Danville, Ill.

Fehr, W. R. and C. E. Caviness. 1977. Stages of soybean development. Iowa Agric. Exp. Stn., Iowa Coop. Ext. Serv. Spec. Rep. 80.

Fehr, W. R. 1980. Soybean. p. 589-599. *In* W. R Fehr and H. H. Hadley (eds.) Hybridization of crop plants. Am. Soc. of Agron., Madison, Wis.

Garner, W. W., and H. A. Allard. 1923. Further studies in photoperiodism, the response of the plant to relative length of day and night. J. Agric. Res. 23:871-920.

Goulden, C. H. 1941. Problems in plant selection. p. 132-133. *In* Proc. 7th International Genetical Congress, Edinburgh.

Hawkins, S. E., W. R. Fehr, E. G. Hammond, and S. R. de Cianzio. 1983. Use of tropical environments for selection of oil composition among genotypes adapted to temperate climates. Crop Sci. 23:897-899.

Ortiz, C., S. R. de Cianzio, P. Hepperly, and W. R. Fehr. 1983. Premature harvest of soybeans for rapid generation advance. Agron. Abstr. 1983:74.

Parker, M. W. and H. A. Borthwick. 1951. Photoperiodic response of soybean varieties. Soybean Dig. 11:26-30.

Walker, A. K., S. R. de Cianzio, J. A. Bravo, and W. R. Fehr. 1979. Comparison of emasculation and nonemasculation for hybridization of soybeans. Crop Sci. 19:285-286.

SOYBEAN CYTOGENETICS

Reid G. Palmer

Cytogenetics is the scientific study of the genetics of an organism, with particular reference to the chromosomes. This discussion will be limited to chromosome aberrations (interchanges and inversions), aneuploids, haploids, and polyploids. A recent review of male-sterile mutants in soybean is available (Graybosch and Palmer, 1984). A chapter on qualitative genetics and cytogenetics of the soybean is forthcoming in the revised soybean monograph to be published by the American Society of Agronomy.

The soybean, *Glycine max* (L.) Merr., may be a disomic polyploid. The chromosome number of the cultivated soybean is $2x = 2n = 40$; 20 pairs of chromosomes are observed during meiosis. Mitotic chromosomes of root-tip cells show two satellite chromosomes, with very few morphological differences among the remaining chromosomes. Establishment of a karyotype will be difficult, but a quantitative method has been applied to soybean chromosomes (Ahmad et al., 1983).

ANEUPLOIDS

Sources

Khush (1973) cites many references indicating that trisomics are common among progeny of triploid plants. Porter and Weiss (1948) and Sadanaga and Grindeland (1981) failed to produce triploid ($3x = 3n = 60$ chromosomes) soybean plants from crosses of diploids with tetraploids, or from the reciprocal cross. Triploids occur among the progeny of male-sterile (ms_1 ms_1) plants (Kenworthy et al., 1973; Beversdorf and Bingham, 1977; Chen et al., 1984). Buss and Autio (1980) observed one triploid among progeny of ms_1 ms_1 plants and one triploid among progeny of ms_2 ms_2 plants. Chen and Palmer (unpublished) indicate that progeny of triploids from ms_1 ms_1 plants have chromosome numbers distributed around 60. Aneuploids near 40 chromosomes were present among the progeny of these 60-chromosome plants and are a source of trisomics. Sadanaga and Grindeland (1979) obtained aneuploids among progeny of irradiated soybeans. However, these are not efficient ways to obtain trisomics in soybean.

Haploids generated via the ms_1 mutant were cross-pollinated by using 2n plants as the male parent. Aneuploids occurred among the F_1 plants (Sorrells and Bingham, 1979). Because of the low frequency of haploids and the difficulty in obtaining hybrid seed, this is not an efficient system for generating aneuploids.

Sadanaga and Grindeland (1981) and Graybosch (1984) obtained a low frequency of aneuploids among progeny of ms_2 ms_2 plants. This is somewhat surprising, because ms_2 ms_2 plants are highly female fertile and can be used successfully in cross-pollinations.

Synaptic mutants (e.g., st_2 st_2, st_3 st_3, st_4 st_4, and st_5 st_5) are a source of aneuploids (Palmer, 1974; Palmer and Heer, 1976; Palmer and Kaul, 1983). These genotypes are highly male sterile and female sterile, but an occasional seed is produced. A general scheme for obtaining aneuploids in soybean is shown in Table 1. The frequency of fertile aneuploids can be increased by growing the st st plants among a population of St St plants.

Identification

The three primary trisomics that have been characterized morphologically are similar to their respective diploid sibs (Palmer, 1976). Two more trisomics are being studied (Gwyn, 1984). Currently, we cross-pollinate our unknown trisomic plants with our known trisomics and analyze root-tip chromosome number of the F_1 plants. The 42-chromosome F_1 plants are examined meiotically to determine if they are double trisomic (i.e., two different extra chromosomes) or tetrasomic (i.e., two identical extra chromosomes). Gwyn (1984) has proposed that primary trisomics can be identified

Table 1. A general scheme for obtaining aneuploids in soybean.

Possible chromosome number of male gametes and female gametes from the st st genotype

			— Chromosome number —		
			Gametes (♀)		Gametes (♂)
st st	→	a)	20		Same as for female
		b)	)20		
		c)	(20		
		d)	40		

Self-pollination of a heterozygote for a male-sterile, female-sterile trait

St st	→	1)	St St	- Fertile
		2)	St st	- Fertile
		1)	st st	- Male sterile and female sterile

Self-pollination of the st st genotype and possible chromosome number of the progeny

	Chromosome number	Genotype
a)	40	st st
b)	)40	st st
c)	(40	st st^a
d)	80	restitution nucleus

Cross-pollination of the st st female parent with an St St male parent and possible chromosome number of the progeny

	Chromosome number	Genotype
a)	40	St st
b)	)40	St st
c)	(40	St st^b
d)	60	St st st^b

[a]Deficiency aneuploids never found.
[b]Deficiency aneuploids or 60-chromosome plants never found.

one from the other on the basis of the morphology of the 42-chromosome plants (tetrasomics), obtained by selfing the parent 41-chromosome plants. This method should have application for future studies with soybean trisomics.

Utilization

Aneuploids may be used to locate genes and linkage groups to specific chromosomes and to study the phenotypic and biochemical effects of individual chromosomes (Khush, 1973). A total of 29 mutants, representing 9 of the 13 linkage groups in soybean, has been crossed to trisomics A, B, and C (Palmer, 1984). The only example in soybean of trisomic inheritance is that of a chimera (gene symbol and genetic type collection number not yet assigned) that has been located to trisomic A (Newhouse et al., 1983).

Deficiency aneuploids; e.g., 39-chromosome plants, have been observed (Sadanaga and Palmer, unpublished). They have been identified among progeny of irradiated plants and plants trisomic for the ms_1 chromosomes. Progeny of 39-chromosome plants, either from self-pollination or cross-pollination, never gave deficiency aneuploids. If soybean is truly a disomic polyploid, deficiency aneuploids should be produced in greater frequencies (Khush, 1973). Clearly, future work with trisomics (addition and deletion) will involve their utilization in genetic studies. Generation of addition-substitution lines also should be feasible, if soybeans are a disomic polyploid.

POLYPLOIDS

Sources

Polyploids can be produced by the application of colchicine to young growing points (Sadanaga and Grindeland, 1981; Cheng and Hadley, 1983). The ms_1 ms_1 system can be used for producing haploids and polyploids (Kenworthy et al., 1973; Beversdorf and Bingham, 1977; Chen et al., 1984).

Utilization

Polyploids probably do not have any importance for direct soybean improvement because of generally lower fertility and seed yield. Polyploids of G. *max* may have application in crosses with the perennial soybeans, many of which have $4x = 4n = 80$ chromosomes. A further application might include using 80-chromosome soybean with the ms_1 allele in crosses with the perennial soybeans (Table 2). The ms_1 mutant may be utilized in crosses with the perennial soybeans to scale down the ploidy level in interspecific crosses. Alternatively, genotypes that produce 2n gametes might have application in hybridization programs between G. *max* and the perennial soybeans.

Genome segregation, chromosome segregation, and recombination in the hybrids between annual and perennial soybeans are only recently being investigated. Many obstacles need to be overcome before this germplasm can be used in soybean improvement programs. This area of research is deserving of more attention.

HAPLOIDS

Sources

Haploids, 20-chromosome plants, occur at a low frequency among progeny of ms_1 ms_1 plants. They may occur as monoembryonic seedlings or as members of polyembryonic seedlings (Kenworthy et al., 1973; Beversdorf and Bingham, 1977; Chen et al.,

Table 2. Possible use of 80-chromosome soybean plants with the ms_1 allele in crosses with perennial soybeans to scale down the ploidy level.

Female parent	Male parent
80 chromosomes	4x = 4n = 80 chromosomes
Ms_1 Ms_1 Ms_1 ms_1	(perennial *Glycine* species)
or Ms_1 Ms_1 ms_1 ms_1	
or Ms_1 ms_1 ms_1 ms_1	
or ms_1 ms_1 ms_1 ms_1	
F_1 Ms_1 Ms_1 -- --	
F_1 ms_1 ms_1 ms_1 ms_1	

These plants might be capable of producing 40-chromosome embryos via the mechanism(s) that give 20-chromosome embryos from 40-chromosome ms_1 ms_1 plants.

1984). Ahmad et al. (1975) found one haploid as a member of a twin from a non-ms_1 source. Anther culture of soybean has received some attention (Schaeffer, 1975), but no method has been reported for obtaining a haploid plant that proceeds to maturity.

Utilization

The low frequency and, perhaps more importantly, the presence of the ms_1 allele in soybean haploids is discouraging to plant breeders. One possible androgenic haploid has been produced via the ms_1 system. This haploid had the Ms_1 allele and the diploid produced was fully fertile and did not segregate the ms_1 allele (Palmer, unpublished).

INVERSIONS

Sources

Chromosome inversions have been reported in F_1 hybrids between *Glycine max* and land races of *G. max* (Ahmad et al., 1977; Delannay et al., 1982) and between *G. max* and *G. soja* (Delannay et al., 1982). These occur at a low frequency in *G. max* (Table 3), but at a much higher frequency among *G. soja* from Japan and Korea (Table 4). We have backcrossed the inversion from *G. soja* (PI 407179) to *G. max* and have identified the homozygous inversion. This genetic stock will represent our standard inversion line.

Table 3. Fertility of F_1 hybrid plants from crosses involving *Glycine max*.

	Maturity Groups 000-IV	Maturity Groups V-X
F_1 fertile	478	148
F_1 10-30% sterile	4[a]	3[c]
F_1 50% sterile	3[b]	3[d]

[a]Two lines from China, one from Japan, one from U.S.S.R.
[b]Two lines from China, one from France.
[c]Three lines from India.
[d]Two lines from Japan, one from India.

Table 4. Fertility of F_1 hybrid plants from crosses between *Glycine max* and *G. soja*.

Country of origin	Fertile	Number of accessions for each level of sterility in F_1			Total
		10-15%	20-40%	50%	
USSR	4	0	0	22	26
China	3	0	0	16	19
Korea	42	7	9	1	59
Japan	32	1	3	1	37
Unknown	0	0	0	1	1
					142

Utilization

An inversion in soybean first needs to be identified as pericentric or paracentric by cytological studies. Unknown homozygous inversions can be crossed to the standard line; then pollen fertility analyses and meiotic studies will be necessary to determine if the inersions are identical or not. Once the genetic map of soybean is more complete, inversions can aid in gene order placement. Inversions will be important in future cytogenetic research in soybean.

INTERCHANGES

Sources

Six standard interchanges have been identified (Sadanaga and Newhouse, 1982) and are described in Table 5. Interchanges also have been identified among accessions of *G. max* (Table 3) and of *G. soja* (Table 4). Table 6 lists 10 chromosome interchanges that are being investigated. These include all the interchanges reported by Delannay et al. (1982) from *G. max* and three from *G. soja* (one each from Korea and Japan, and one of unknown origin), plus several recently identified interchanges.

Identification

Sadanaga and Newhouse (1982), using the six standard interchanges, concluded that:

    a. one common chromosome is involved in an interchange in Clark T/T, KS 172-11-3, and KS 175-7-3.

Table 5. Origin of cytologically described interchange lines in soybean.

Interchange line	Origin
Clark T/T	A near-isogenic Clark line with the interchange from PI 101404B incorporated. PI 101404B is an introduction from northeastern China.
L75-0283-4	A spontaneous interchange in an F_4 progeny row of a Beeson x Amsoy 71 cross.
PI 189866	*Glycine gracilis* introduction from France.
KS172-31-2	Irradiated population of Hodgson. Selected by K. Sadanaga.
KS172-11-3	Irradiated population of Hodgson. Selected by K. Sadanaga.
KS175-7-3	Irradiated population of Steele. Selected by K. Sadanaga.

Table 6. Origin of suspected interchange lines in soybean.

Interchange line	Origin
PI 212239	*G. soja*, origin not known.
PI 274453	*G. max*, Okinawa, Japan.
PI 274454	*G. max*, Okinawa, Japan.
PI 323551	*G. max*, India.
PI 339735A	*G. soja*, Yee Cock Ri, Pyung Chang Myun, Pyung Chang Kun, Korea.
PI 407020	*G. soja*, Akita-ken, Japan.
PI 468917	*G. soja*, Shenyang Agricultural College, China.
PI 468919	Semi-wild line, Shenyang Agricultural College, China.
PI 468923	*G. max*, Beijing, China.
Progeny from MT-5	Progeny of a 42-chromosome plant that was the offspring of an st_5 st_5 plant (Palmer and Kaul, 1983).

  b. one common chromosome is involved in an interchange in L75-0283-4 and KS 171-31-2.
  c. one common chromosome (satellite chromosome) is involved in an interchange in KS 172-11-3 and KS 175-7-3.

Palmer and Newhouse (1984) have studied 14 *Glycine soja* accessions from the People's Republic of China (PRC) and 19 *G. soja* accessions from the USSR. Crosses were made between these accessions and cultivars of *G. max*, which had noninterchange chromosomes. Plants were classified for pollen fertility by using a solution of I_2KI. Fertile pollen grains were plump and stained red-brown; aborted pollen grains were shrunken, collapsed, unstained, or only very lightly stained. In a plant with interchange chromosomes, approximately half of the pollen is aborted; this condition is termed semisterility.

All 14 accessions from the PRC and all 19 accessions from the USSR were identical in chromosome structure to our Clark T/T standard interchange line. In addition, Palmer and Newhouse (1984) made eight different intercrosses between homozygous interchange accessions from these two countries. All F_1 hybrids were fertile. This indicates that all the accessions from the PRC and the USSR that have been examined, and have homozygous interchange chromosomes, have identical chromosome structure.

The basis for the difference in frequency of homozygous interchanges in *G. soja* accessions from the PRC and the USSR versus Korea and Japan is not known. These differences might suggest that the *G. soja* populations had different progenitors.

Utilization

Chromosome interchanges are useful in chromosome mapping studies. The Clark T/T interchange shows linkage with mutants of Linkage Group 6 and with Linkage Group 8 (Figure 1). Sadanaga and Grindeland (1984), using interchange KS 172-11-3, reported a recombination frequency of 1.9 ± 0.8% between w_1 and the interchange breakpoint. Furthermore, the w_1 locus was located on the satellite chromosome, and the gene order is w_1 *wm*, breakpoint, and ms_1 (Sadanaga, 1983).

Sacks and Sadanaga (1984), also using interchange KS 172-11-3, determined that the recombination frequency between the ms_1 and w_1 loci and the breakpoint were 18.4 ± 3.3% and 2.0 ± 1.2%, respectively. The linkage estimates between w_1 and the breakpoint were similar in the two experiments.

With six genetically and cytologically identified interchanges in soybean, progress

343

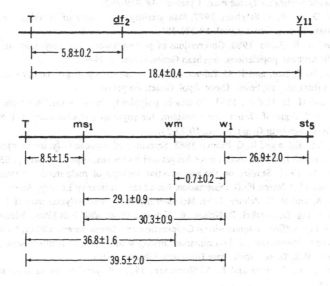

Figure 1. Linkage Group 6 (above) and 8 (below). T = interchange breakpoint; $\underline{df}_2$ = Lincoln dwarf; $\underline{y}_{11}$ = lethal yellow; ms_1 = male sterile, female fertile; wm = magenta flower; w_1 = white flower; and st_5 = synaptic mutant.

has been made in determining their relationship one with another. Chromosome mapping studies using interchanges will continue to have added importance in soybean research.

Linkage Groups

Twenty linkage groups are expected in soybean, but only 13 have been identified. Mutants and their linkage group with linkage estimates are given in the chapter on qualitative genetics and cytogenetics in the forthcoming revised soybean monograph. A compilation of locus-to-locus linkage data is available (Yee and Palmer, 1984). As discussed in the sections on Aneuploids and Interchanges, these chromosome types are very important in chromosome mapping studies. In the future, as additional aberrations are identified, they will provide the foundation for further linkage studies.

NOTES

Reed G. Palmer, USDA-ARS, Departments of Agronomy and Genetics, Iowa State University, Ames, IA 50011.

Joint contribution: Agricultural Research Service, U.S. Department of Agriculture, and Journal Paper No. 11606 of the Iowa Agriculture and Home Economics Experiment Station, Ames, IA, U.S.A. 50011. Project 2471.

REFERENCES

Ahmad, Q. N., E. J. Britten and D. E. Byth. 1975. A colchicine induced diploid from a haploid soybean twin. J. Hered. 66:327-330.
Ahmad, Q. N., E. J. Britten and D. E. Byth. 1977. Inversion bridges and meitoic behavior in species hybrids of soybeans. J. Hered. 68:360-364.

Ahmad, Q. N., E. J. Britten and D. E. Byth. 1983. A quantitative method of karyotypic analysis applied to the soybean, *Glycine max*. Cytologia 48:879-892.

Beversdorf, W. D. and E. T. Bingham. 1977. Male sterility as a source of haploids and polyploids of *Glycine max*. Can. J. Genet. Cytol. 19:283-287.

Buss, G. R. and W. R. Autio. 1980. Observations of polyembryony and polyploidy in ms_1 and ms_2 male-sterile soybean populations. Soybean Genet. Newsl. 7:94-97.

Chen, L. F. O., H. E. Heer and R. G. Palmer. 1984. The frequency of polyembryonic seedlings and polyploids from ms_1 soybean. Theor. Appl. Genet. (in press).

Cheng, S. H. and H. H. Hadley. 1983. Studies in polyploidy in soybeans: A simple and effective colchicine technique of chromosome doubling for soybean (*Glycine max* [L.] Merr.) and its wild relatives. Soybean Genet. Newsl. 10:23-24.

Delannay, X., T. C. Kilen and R. G. Palmer. 1982. Screening of soybean (*Glycine max*) accessions and *G. soja* accessions for chromosome interchanges and inversions. Agron. Abstr., p. 63.

Graybosch, R. A. 1984. Studies on the reproductive biology of male-sterile mutants of soybean (*Glycine max* [L.] Merr). Ph.D. Dissertation. Iowa State University Library, Ames.

Graybosch, R. A. and R. G. Palmer. 1984. Male sterility in soybean, *Glycine max* [L.] Merr. p. 232-253. *In* S. Wong, D. Boethel, R. Nelson, W. Nelson and W. Wolf (eds.) Proc. Second U.S./China Soybean Symp. Office of International Cooperation and Development, USDA, Washington, D.C.

Gwyn, J. J. 1984. Morphological discrimination among some aneuploids in soybean (*Glycine max* [L.] Merr.). M.S. Thesis. Iowa State University Library, Ames.

Kenworthy, W. J., C. A. Brim and E. A. Wernsman. 1973. Polyembryony in soybeans. Crop Sci. 13:637-639.

Khush, G. S. 1973. Cytogenetics of aneuploids. Academic Press, New York.

Newhouse, K. E., L. Hawkins and R. G. Palmer. 1983. Trisomic inheritance of a chimera in soybean. Soybean Genet. Newsl. 10:44-49.

Palmer, R. G. 1974. Aneuploids in the soybean, *Glycine max*. Can. J. Genet. Cytol. 16:441-447.

Palmer, R. G. 1976. Chromosome transmission and morphology of three primary trisomics in soybean (*Glycine max*). Can. J. Genet. Cytol. 18:131-140.

Palmer, R. G. 1984. Summary of trisomic linkage data in soybean. Soybean Genet. Newsl. 11:127-131.

Palmer, R. G. and H. E. Heer. 1976. Aneuploids from a desynaptic mutant in soybeans (*Glycine max* [L.] Merr.). Cytologia 41:417-427.

Palmer, R. G. and M. L. H. Kaul. 1983. Genetics, cytology, and linkage studies of a desynaptic soybean mutant. J. Hered. 74:260-264.

Palmer, R. G. and K. E. Newhouse. 1984. Evaluation of *Glycine soja* from The People's Republic of China and the USSR. Soybean Genet. Newsl. 11:105-111.

Porter, K. B. and W. G. Weiss. 1948. The effect of polyploidy on soybeans. J. Am. Soc. Agron. 40:710-724.

Sacks, J. M. and K. Sadanaga. 1984. Linkage between the male sterility gene (ms_1) and a translocation breakpoint in soybean, *Glycine max*. Can. J. Genet. Cytol. 26:401-404.

Sadanaga, K. 1983. Locating *wm* on linkage group 8. Soybean Genet. Newsl. 10:39-41.

Sadanaga, K. and R. Grindeland. 1979. Aneuploids and chromosome aberrations from irradiated soybeans. Soybean Genet. Newsl. 6:43-45.

Sadanaga, K. and R. Grindeland. 1981. Natural cross-pollination in diploid and autotetraploid soybeans. Crop Sci. 21:503-506.

Sadanaga, K. and R. L. Grindeland. 1984. Locating the w_1 locus on the satellite chromosome in soybean. Crop Sci. 24:147-151.

Sadanaga, K. and K. Newhouse. 1982. Identifying translocations in soybeans. Soybean Genet. Newsl. 9:129-130.

Schaeffer, G. W. 1975. Tissue and anther culture of soybeans. (Abstr.) Plant Physiol. Suppl. 56:37.

Sorrells, M. E. and E. T. Bingham. 1979. Reproductive behavior of soybean haploids carrying the ms_1 allele. Can. J. Genet. Cytol. 21:449-455.

Yee, C. C. and R. G. Palmer. 1984. Summary of locus-to-locus linkage data in soybeans. Soybean Genet. Newsl. 11:115-126.

COLLECTION AND UTILIZATION OF WILD PERENNIAL *GLYCINE*

A. H. D. Brown, J. E. Grant, J. J. Burdon, J. P. Grace and R. Pullen

Economically, soybean is one of the world's most important crops. In production of edible dry matter, it ranks sixth after wheat (*Triticum aestivum* L.), maize (*Zea mays* L.), rice (*Oryza sativa* L.), potato (*Solanum tuberosum* L.) and barley (*Hordeum vulgare* L.); and only maize and wheat surpass soybean in production of plant protein (Harlan and Starks, 1980). For a crop of such world significance, the future breeding of soybean rests on a relatively narrow base of genetic resources. The major collections of soybean germplasm, held by the U.S.D.A., amount to about 10,000 entries. This number contrasts with that of wheat for which the estimated total number world-wide exceeds 250,000. The International Board for Plant Genetic Resources (IBPGR) has recently coordinated efforts to augment the world soybean collection. However, much of the unsampled diversity occurs in relatively inaccessible regions of Asia, and may not be collected before extinction.

In such a situation, the wild evolutionary relatives of soybean are of great interest as potential genetic resources. Yet, in contrast to the world's major crops, soybean is relatively neglected. Only recently have significant steps been taken to collect, study and exploit the genetic resources of the more distant wild relatives of soybean (Hymowitz and Newell, 1977, 1980; Marshall and Broué, 1981). Here we recount recent, exciting developments—namely, the building of representative collections, the crossing of these species with soybean, and the evaluation of them as sources of disease resistance—that offer a major expansion of soybean germplasm.

THE WILD PERENNIAL *GLYCINE* SPECIES

The soybean genus, *Glycine*, consists of two subgenera, *Glycine* and *Soja* (see Hymowitz and Newell, 1981, for an historical account of *Glycine* taxonomy). The subgenus *Soja* includes soybean (*Glycine max* (L.) Merr), along with its annual wild progenitor (*G. soja* Sieb. & Zucc.). The other subgenus, *Glycine*, contains nine described perennial species, and at least three awaiting formal description. The nine named species are listed in Table 1, together with their somatic chromosome numbers.

Information on the geographic distribution of these species comes primarily from herbarium specimens. The major component of both interspecific and intraspecific diversity occurs in Australia. All seven diploid species are essentially restricted to Australia. All cytological races of the two complex species *G. tomentella* and *G. tabacina* also occur in Australia. Diploid and tetraploid *G. tomentella* (2n=40,80) occur in Papua New Guinea, and tetraploid strains of both species are found in South Pacific Islands, Philippines, Taiwan and Southern China.

345

Table 1.   Species of the subgenus *Glycine*, somatic chromosome numbers, the number of accessions collected in Queensland in 1983, and in the total CSIRO active collection.

Species	Chromosome Number	IBPGR Qld. trip 1983	CSIRO Collection
1. *G. clandestina* Wendl.	40	1	135
2. *G. canescens* F. J. Herm.	40	-	71
3. *G. latrobeana* (Meissn.) Benth.	40	-	12
4. *G. tomentella* Hayata	38,40,78,80	79	230
5. *G. falcata* Benth.	40	2	10
6. *G. latifolia* (Benth.) Newell / Hymowitz	40	-	27
7. *G. tabacina* (Labill.) Benth.	40	2	119
	80	2	143
8. *G. argyrea* Tind.	40[a]	-	6
9. *G. cyrtoloba* Tind.	40[a]	11	28
G. spp.	-	7	14

[a]Grant, unpublished data.

Within Australia, the subgenus is widespread, and various species occur in a diverse array of habitats. The ecological amplitude embraces most of the continent's climatic zones, from the monsoonal tropical north (*G. tomentella*), the central semi-arid expanses (*G. canescens*), the periodically flooded, heavy basaltic plains (*G. falcata*), the subalpine woodlands to altitudes of 1450 m (*G. clandestina*), to the cooler southerly latitudes (42°S) of Tasmania (*G. latrobeana*).

Typically, the perennial *Glycine* species are minor, inconspicuous components of semi-open *Eucalyptus* woodland. Often in such sites, more than one species or form can co-occur. They also can occur in more shaded, tall woodland, and in exposed, open grassland sites, such as stabilized coastal sand dunes or adjacent to roads, ditches, airstrips and recently disturbed sites.

As a group, the perennials display a wide variety of growth habit and morphological features (Hymowitz and Newell, 1980). Important life history characteristics which are common in the group are the deep, thick tap root, and the production of both chasmogamous and cleistogamous flowers. Variable life history traits include the growth habit, which may be twining or scrambling (in some forms with adventitious roots at the nodes), the timing and duration of flowering, and the remarkable setting of pods underground (*G. falcata*).

Thus, in broad outline the perennial *Glycine* exhibits a striking diversity of genotypes and ecological adaptation. Yet, until recently, they have been little studied, mainly because ample and representative collections did not exist. Over the last five years, increased efforts to acquire samples from several collectors, and in particular, joint U.S.A.-Australia collecting trips have significantly remedied this situation, improving the number and coverage of samples (Figure 1). The most recent steps have been collecting trips to remote North Queensland (104 accessions, Table 1) and the Kimberley district of Western Australia, both funded by IBPGR. As well as amplifying the range of variability within established species, these collecting missions continue to unearth new and rare taxa. This indicates that there is still much to be discovered about the taxonomic and genetic diversity within the genus. As of June 1984, the species composition of the active collection held by CSIRO, Division of Plant Industry is as listed in Table 1.

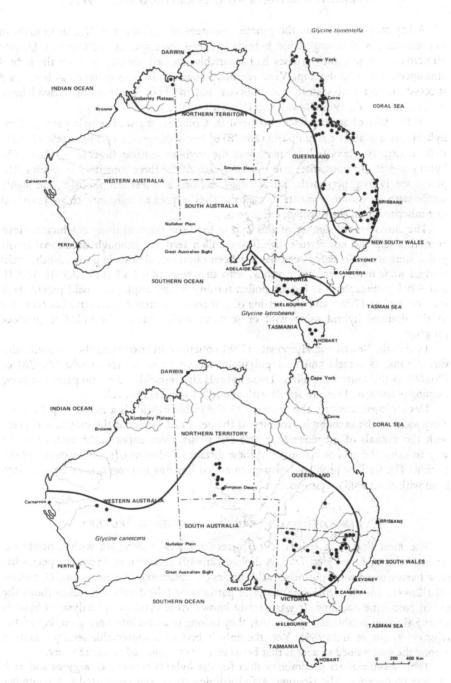

Figure 1. Origin of seed samples of *G. tomentella* and *G. latrobeana* (upper map), and *G. canescens* (lower map), compared to their reported Australian distribution. The limit of the distribution, based on herbarium specimens, is shown as a thick line.

HYBRIDIZATION BETWEEN SOYBEAN AND PERENNIAL *GLYCINE*

A key step to exploiting the genetic resources of the subgenus *Glycine* in soybean improvement is obtaining viable hybrids between these species and soybean. Despite numerous attempts, such crosses had invariably aborted, usually because the hybrid endosperm failed to develop. Very recently, however, *in vitro* culture has been used successfully to rescue such hybrid embryos, and to obtain viable plants in two laboratories (Broué et al., 1982; Newell and Hymowitz, 1982).

In the first of such crosses produced in Canberra, the wild female parent of the hybrid was a synthetic amphiploid (2n=78) of two wild species, *G. tomentella* (2n=38) x *G. canescens* (2n=40). The male was the soybean cultivar Lincoln or Hark. The hybrid had 59 chromosomes; one set from each of the three contributing species. The plants are twining perennials, intermediate in floral and leaf morphology, and highly sterile with only four bivalents at meiosis. They expressed resistance to soybean leaf rust inherited from the *G. canescens* parent.

The chromosome number of this hybrid has been doubled using colchicine. However, this step did not restore significant pollen fertility, although occasional fertile grains were seen. No selfed seed has yet been obtained. Meiosis in pollen mother cells showed some multivalent associations, with an average of 0.1 VI, 2 IV, 0.9 III, 50.6 II, and 4.6 I at metaphase 1. Indeed pollen fertility in the amphiploid wild parent itself was incomplete (70%), and the setting of self seed was erratic. Extensive backcrossing of the doubled hybrid using wild or soybean pollen has so far failed to produce progeny.

In Illinois, Newell and Hymowitz (1982) obtained hybrids using the soybean cultivar, Altona, as female and two polyploid accessions of *G. tomentella* (2n=78) or (2n=80) as the source of pollen. These hybrids also resembled the wild parent in being twining perennials. They are sterile with about 7.3 bivalents per cell.

Despite their sterility, the recovery of the viable hybrids is a major step forward. They validate the taxonomic grouping of the perennial species of the subgenus *Glycine* with the annuals of subgenus *Soja* into one genus. More importantly, they open the way to using the genetic resources of these distant relatives of *Glycine* through hybridization. The fertility problems have given added impetus to programs of crossing soybean with other wild accessions.

A NEW POTENTIALLY "BRIDGING" SPECIES: *G. ARGYREA*

The most promising results of more recent hybridizations are with a newly described species, *G. argyrea* Tind. A diploid (2n=40) accession of this new species has now been crossed with four other wild species (*G. canescens*, *G. clandestina*, *G. tomentella* 2n=40, and *G. latrobeana*) to yield partially or fully fertile hybrids without the aid of colchicine doubling. It was already known from cytological analysis of hybrids among the four established species that they belong to an evolutionary grouping within *Glycine* (Grant et al., 1984). Yet, the only hybrid with appreciable seed production among the four named species is that between *G. canescens* and *G. clandestina*.

Table 2 summarizes preliminary data for the hybrids between *G. argyrea* and each of the other species. The Genome Affinity Index (GAI) was computed as the number of bivalents divided by the basic chromosome number (x) for *Glycine*. This was taken to be x=10, assuming that soybean is a diploidized ancient tetraploid. The four values in Table 2 are close to the maximum value of 2.0 for regular meiosis. Comparable estimates of GAI for crosses among the four species had an average of 1.50 and a range

Table 2. Preliminary cytological and genetic data for interspecific hybrids between *G. argyrea*[a] and four diploid *Glycine* species.

Pollen Source:	*G. cland.*[b]	*G. can.*[c]	*G. tom.*[d]	*G. latro.*[e]
F_1 meiosis, GAI	1.99	2.00	1.89	1.93
Pollen fertility, %	77	84	54	51
F_2 isozyme segregation: Frequency of 'cooloola' allele in F_2 array				
Enp[g]	0.51	0.45	0.53	0.38
Ndh	0.70**[h]	0.47	0.81**	0.66
Gpi 2	0.75**	0.58	0.83**	0.45
Mdh 1	0.49	0.48	0.56	-
Pgm	0.48	0.57	0.50	0.50

[a]*G. argyrea* (2n=40) 1420, Cooloola National Park, Qld.

[b]*G. cland.* = *G. clandestina* (2n=40) 1145, Tingha, N.S.W.

[c]*G. can.* = *G. canescens* (2n=40) 1232, Cobham Lake, N.S.W.

[d]*G. tom.* = *G. tomentella* (2n=40) 1300, Mt. Garnet, Qld.

[e]*G. latro.* = *G. latrobeana* (2n=40) 1385, Inverleigh, Vic.

[f]GAI = Genome Affinity Index $\cong$ Number of bivalents $\div$ 10.

[g]Isozyme loci: Enp = Endopeptidase; Ndh = NADH dehydrogenase; Gpi = Glucosephosphate isomerase; Mdh = Malate dehydrogenase; Pgm = Phosphoglucomutase.

[h]Asterisks denote significant departure from 50% allele frequency.

of 1.03 to 2.0 (Grant et al., 1984). Pollen fertility in the *G. argyrea* hybrids is also relatively high.

The segregation of isozyme differences was studied in the F_2 progeny of these crosses, and Table 2 reports some examples. For three loci, the F_2 segregation did not differ significantly from the 1:2:1 expectations. However, for two loci (Ndh, Gpi 2) marked deviation in genotypic frequencies was found. In these cases, the allele originating from *G. argyrea* was over-represented in the progeny array. Distortions in F_2 segregation are a frequent feature of interspecies hybrids. The relationship between this phenomenon and hybrid fertility is under study.

Crosses between *G. max* and *G. argyrea* or its hybrid derivatives are now being attempted using in vitro embryo culture. Success was first obtained with the hybrid *G. argyrea* x *G. canescens*, when pollinated with *G. max* cv. Improved Pelican, and the new soybean hybrid plant is now established in the greenhouse. Like the earlier hybrids, it is a vigorous twining perennial, and sterile, with an average of 2.6 bivalents and 34.8 univalents at metaphase 1 of pollen mother cell meiosis. This recent hybrid appears to have an added advantage in responding to techniques (first worked out for *G. canescens*; Grant, 1984) that regenerate several plantlets from the proliferation of multiple shoots. Thus, several plantlets from the one original hybrid embryo have been recovered in this cross. This is an important development because of the added insurance it provides against loss during the critical step of transferring plants cultured in vitro to the greenhouse. Further, it is possible that induced variation in chromosomal constitution might arise during tissue culture, and this might promote variation in fertility of the hybrids or their chromosome-doubled derivatives.

WILD CONTRIBUTIONS TO SOYBEAN IMPROVEMENT

Several specific attributes present among the perennial *Glycines*, and of potential use in soybean improvement can be listed (Marshall and Broué, 1981). These include tolerance of adverse environments (frost, drought, salinity), pest and disease resistance, improved seed protein and oil quality, daylength insensitivity and reduced floral shedding. Our work has concentrated on resistance to soybean leaf rust, which is caused by the pathogen *Phakopsora pachyrhizi*.

This pathogen is widely distributed throughout the tropics and sub-tropics of the world and, in much of Asia, is one of the major constraints in soybean production. In that region, epidemics of the disease frequently result in yield losses of 20 to 30% (Bromfield, 1976). As a consequence, resistance is an important breeding goal, but is one that is severely hampered by the scarcity of known sources of resistance in either *G. max* or *G. soja* germplasm. In Australia, the pathogen has a wide distribution along the east coast and is found on a range of native species. Sporadic outbreaks on cultivated crops of soybean occur in favorable seasons and have resulted in serious and occasionally complete yield losses.

Fortunately, surveys of perennial *Glycine* accessions for resistance to soybean leaf rust have shown them to be a rich source of resistance genes (Burdon and Marshall, 1981a,b). Both qualitative and quantitative resistance occurs in the six species tested. Inheritance studies are in progress, but in at least two lines of *G. canescens* and one of *G. clandestina* the hypersensitive resistance segregates as a single dominant gene.

More recently, the variation in response, of a large number of accessions from four common species, to infection by eight Australian isolates of the pathogen has allowed the recognition of six virulence races (Burdon and Speer, 1984). In turn, this has led to the definition of a number of sets of differential hosts, or lines of *Glycine* that both discriminate among the races and maximize the chance of identifying further races. Such differential sets may be constructed either from a combination of all four species (*G. canescens*, *G. clandestina*, *G. tabacina* and *G. tomentella*) or from each of the species separately. If it is assumed that the reaction types observed in the four *Glycine* species are the product of a simple gene-for-gene interaction, then a minimum of 5, 3, 4 and 4 resistance genes must be present in these accessions of the four species, respectively. An example of one of these differential sets, that for *G. canescens* is shown in Table 3.

This development represents a significant new contribution from wild perennial germplasm. A set of differential lines was, until now, entirely lacking for this pathogen because of the very low levels of resistance found within *G. max*. The newly designated lines will allow a study of the genetic basis of resistance in the host species, the range of virulence found in populations of the pathogen, and more generally, features of the coevolutionary interaction of host resistance and pathogen virulence in natural

Table 3. *Glycine canescens* differentials for soybean leaf rust (after Burdon and Speer, 1984).

Glycine accession number and geographic origin		Soybean Leaf Rust Races					
		R1	R2	R3	R4	R5	R6
1232.1	Cobham Lake, N.S.W.	S[a]	S	S	S	S	S
1123	Young, N.S.W.	R	S	R	S	S	S
1112	Candoblin, N.S.W.	R	R	I	I	S	S
1113	Condoblin, N.S.W.	R	R	S	S	R	S
1302	Nymagee, N.S.W.	R	R	S	R	S	S
1340	Murray Downs, N.T.	R	R	R	R	R	S
1232.2	Cobham Lake, N.S.W.	R	R	R	R	R	R

[a]The reactions are classified as S = susceptible, I = intermediate, R = resistant.

communities. Such uses of the wild germplasm can proceed independently, and in addition to the direct use of the resistance genes in soybean breeding.

FUTURE PROSPECTS

Until recently the perennial *Glycine* species were the neglected poor cousins of soybean. The last five years have seen three major developments which have dramatically altered their significance. These developments are the assembly of a large and diverse collection, the achievement of hybrids between the perennials and soybean, and the uncovering of diverse sources of resistance to soybean leaf rust. It seems justified to expect that a new chapter in soybean breeding is about to open.

The importance of the germplasm of wild relatives in improving autogamous crops with a narrow genetic base is well illustrated by the case of tomato. In a recent review, Rick (1982) listed a truly impressive range of characters introgressed or available for introgression from wild and exotic taxa, after a history of hybridizations "dating back as far as five decades". Yet "great opportunities exist for further improvement of the cultivated tomato". The case of tomato may indeed be an archetype for that of soybean. Although in soybean hybrid fertility is presently a problem, the techniques of genetic engineering may offer another way of extracting desirable genes. Yet, whether transfer is to be accomplished by conventional or molecular means, thorough systematic, genetic and physiological research into the variation within the wild species will be needed as a basis for their exploitation in soybean improvement.

NOTES

A. H. D. Brown, J. E. Grant, J. J. Burdon, J. P. Grace and R. Pullen, CSIRO, Division of Plant Industry, Canberra A.C.T. 2601, Australia.

The work was supported in part by grants from the Rural Credits Development Fund of Australia, the Australian Oilseeds Research Committee and the I.B.P.G.R.

REFERENCES

Bromfield, K. R. 1976. World soybean rust situation. p. 491-500. *In* L. D. Hill (ed.) World soybean research, proceedings of the world soybean research conference. Interstate Printers, Danville, Ill.

Broué, P., J. Douglass, J. P. Grace and D. R. Marshall. 1982. Interspecific hybridisation of soybeans and perennial *Glycine* species indigenous to Australia via embryo culture. Euphytica 31:715-724.

Burdon, J. J. and D. R. Marshall. 1981a. Evaluation of Australian native species of *Glycine* for resistance to soybean rust. Plant Dis. 65:44-45.

Burdon, J. J. and D. R. Marshall. 1981b. Inter- and intra-specific diversity in the disease-response of *Glycine* species to the leaf-rust fungus *Phakopsora pachyrhizi*. J. Ecol. 69:381-390.

Burdon, J. J. and S. S. Speer. 1984. A set of differential *Glycine* hosts for the identification of races of *Phakopsora pachyrhizi* Syd. Euphytica (in press).

Grant, J. E. 1984. Plant regeneration from cotyledonary tissue of *Glycine canescens*—a perennial wild relative of soybean. Plant, Cell, Tissue and Organ Culture (in press).

Grant, J. E., J. P. Grace, A. H. D. Brown and E. Putievsky. 1984. Interspecific hybridization of *Glycine* Willd. subgenus *Glycine* (Leguminosae). Aust. J. Bot. (in press).

Harlan, J. R. and K. J. Starks. 1980. Germplasm resources and needs. p. 254-273. *In* F. G. Maxwell and P. R. Jennings (eds.) Breeding plants resistant to insects. John Wiley, New York.

Hymowitz, T. and C. A. Newell. 1977. Current thoughts on origins, present status and future of soybeans. p. 197-209. *In* D. S. Seigler (ed.) Crop resources. Academic Press, New York.

Hymowitz, T. and C. A. Newell. 1980. Taxonomy, speciation, domestication, dissemination, germplasm resources and variation in the genus *Glycine*. p. 251-264. *In* R. J. Summerfield and A. H. Bunting (eds.). Advances in Legume Science. Royal Botanical Gardens, Kew.

Hymowitz, T. and C. A. Newell. 1981. Taxonomy of the genus *Glycine*, domestication and uses of soybeans. Econ. Bot. 35:272-288.

Marshall, D. R. and P. Broué. 1981. The wild relatives of crop plants indigenous to Australia and their use in plant breeding. J. Aust. Inst. Agric. Sci. 47:149-154.

Newell, C. A. and T. Hymowitz. 1982. Successful wide hybridization between soybean and a wild perennial relative, *G. tomentella* Hayata. Crop Sci. 22:1062-1065.

Rick, C. M. 1982. The potential of exotic germplasm for tomato improvement. p. 1-28. *In* I. K. Vasil, W. R. Scowcroft and K. J. Frey (eds.) Plant improvement and somatic cell genetics. Academic Press, New York.

UTILIZING THE GENETIC DIVERSITY OF ANNUAL SOYBEAN SPECIES

Junyi Gai

BREEDING PROGRESS AND ANNUAL GERMPLASM

Gai (1983) reviewed soybean breeding progress in the U.S. After the first cycle of breeding effort in the northern U.S., cultivars released in the 1940s and 1950s had 26% greater yield compared to selections from within plant introductions made in the 1920s. The second cycle gave another 16% increase over the first cycle. Also, there was 20% genetic progress in yield from the 1940s to the 1970s in the southern U.S. Simultaneously, many economically important traits also were improved significantly. Similar results were obtained in China, Japan, Brazil, and other countries. Actually, the source of all breeding progress is annual species. No perennial has been used in a breeding program yet.

The taxonomy of soybean species is not fully agreed upon. Most soybean scientists accept two species in the subgenus *Soja*; i.e., *Glycine max* (L.) Merr. and *G. soja* Sieb. & Zucc. (Hadley and Hymowitz, 1973), while Russian scientists tended to treat the two species as one with five subspecies (Korsakov, 1979). The other two subgenera are basically perennials. However, annual soybean germplasn will continue to be the most important genetic resource for further breeding progress in raising yield level and stability, and in improving human nutrition.

There is ample evidence for the significance of annual soybean germplasm. In the U.S., where no land cultivars were available, the first success in utilizing germplasm was the introductions from Northeast China, and the second success was in screening for specific genes or genetic materials, especially genes lending resistance, and tolerance to diseases and stresses. Among 55 important ancestors of U.S. soybean cultivars, according to Delannay et al. (1983), 27 were from China, 7 from Japan, 8 from Korea, and 13 were of unknown origin. For the northern cultivars released after 1971, seven introductions made a cumulative genetic contribution of 0.721. For the southern cultivars released after 1971, one of these plus six different introductions, made a cumulative genetic contribution of 0.820.

In China, land cultivars were preferred at first, since each region had numerous native ecotypes with good adaptation to the local environments. Therefore, the ancestors of cultivars were quite different from region to region, and somewhat localized in comparison with the situation in the U.S. G. Zhang (1983) listed the cultivars released in Heilongjiang from the 1950s to the 1980s. They were divided into five groups according to major parentage: Mancangjin (including Dongnong 4 and Kejiao 56-4258, two derivations from Mancangjin), Jingshanpu, Zihua 4, Yuanbaojin, and Fenfdihuang. All five parents were native to the Northern Spring Region. It is supposed that more and more exotic germplasm will add to the parentage since breeding for

354

resistance and tolerance to diseases, insects and stresses have been emphasized recently.

One way to expand genetic diversity is to utilize the wild species *G. soja*. Resistance and tolerance to pests and stresses, and high protein might be the main reasons for using *G. soja*. Gai et al. (1982) indicated two or three backcross generations were enough to avoid the wild characteristics, including vining, non-abscission of leaves, shattering, colored seed coat, and lodging. However, studies are needed to determine the recovery of the yield potential of *G. max*.

GENETIC DIVERSITY OF AGRONOMIC TRAITS

Although a center of diversity is not the same as a center of origin, and crops did not necessarily originate in centers (Harlan, 1971), there is no doubt that centers of diversity do exist for most crop plants, and that China is both a center of diversity and a center of origin for soybeans. The fact that, in the U.S. soybean germplasm collection, the number of entries from eastern Asia dominates over the others is good evidence.

Hawkes (1981) has indicated that, by the 1950s, concern developed that the natural reservoirs of germplasm resources were rapidly being destroyed, or genetic erosion was taking place. Much effort has been made to maintain and collect soybean germplasm since 1950, especially since southern corn leaf blight spread across the U.S. in 1970 raising concern about narrowness of the genetic base of crops. Thus, the diversity of soybean germplasm was strongly emphasized and studied.

The early emphasis relating to genetic diversity was on maturity and a series of morphological and ecological characters. In the U.S., soybeans are classified into 12 maturity groups, from 00, 0, I, through X. Earlier and later ones could be added to that system when found. In eastern Asia, China and Japan, various cropping systems have been used, and a number of corresponding land cultivars have been selected as a consequence. Japanese scientists classified these cultivars into eight groups (Ia, Ib, IIa, IIb, IIIb, IIIc, IVc, and Vc) according to days to flowering and days from flowering to maturity. In China, environments vary greatly due to a wide range of latitutde, altitude and cropping systems, including spring, summer, fall, and winter planting. Two classification systems, similar to the American and the Japanese, have been proposed, but there are still some problems to be discussed and studied.

Table 1 shows the phenotypic frequencies of some agronomic and morphological traits in the Chinese native germplasm collection. In the whole country, summer- and spring-planting types, determinate and indeterminate stem termination, erect and semi-erect growth types, yellow seed coat, yellow cotyledon, and broad leaf shape were dominant over their counterparts, respectively. In the Northern Spring Region, 100% were of the spring-planting type, and 94.8% and 70.5% were of the summer-planting type in the Northern Summer and Southern Multiple regions, respectively, while there appeared 5.6% of the fall-planting type in the latter region. In the Northern Spring region, indeterminates were in majority, while in the other two regions, especially the Southern Multiple region, determinates were in majority. Except for yellow seed coat, black was in more abundance than other colors in the Northern Spring and Northern Summer regions, while green was more frequent than the others in the Southern Multiple region. The two northern regions tended to have more white flower and more gray pubescent strains, while more purple flower and brown pubescent strains were in the Southern Multiple region.

There has also been emphasis on genetic diversity for resistance to pests. In the U.S., soybean breeders have realized that requirement since the 1950s. Two nationwide

Table 1. The phenotypic frequencies (%) of agronomic and morphological characters of soybean land cultivars in China. (Summarized from historical records.)

		Northern Spring[a]	Northern Summer[a]	Southern Multiple[a]	Total
% of Strains		33.5	37.4	29.1	100.0
Planting	SP	100.0	5.2	23.9	42.4
Type[b]	SU	0	94.8	70.5	56.0
	FA	0	0	5.6	1.6
Stem	IN	71.8	32.5	13.2	40.0
Termination[c]	SD	5.3	15.3	8.3	9.9
	DE	22.9	52.2	78.5	50.1
Growth	ER	60.6	41.6	67.8	56.1
Type[d]	SE	27.8	39.5	22.5	31.0
	SP	3.5	10.8	2.7	5.0
	PR	8.1	8.1	7.0	7.9
Seed Coat	YW	61.9	62.2	59.4	61.3
Color[e]	GN	12.0	10.7	24.7	15.2
	BW	5.6	5.9	7.2	6.2
	BK	17.6	18.5	6.7	14.7
	BI	2.9	2.7	2.0	2.6
Cotyledon	YW	95.1	97.1	98.6	97.0
Color[f]	GN	4.9	2.9	1.4	3.0
Flower	PU	47.9	49.0	67.6	54.1
Color[g]	WH	52.1	51.0	32.4	45.9
Pubescence	GY	61.1	68.2	38.6	56.9
Color[h]	TW	38.9	31.8	61.4	43.1
Leaf Shape[i]	BD	93.9	96.1	95.6	95.4
	NW	6.1	3.9	4.4	4.6

[a]Planting region/season.
[b]Planting type: SP—Spring, SU—Summer, FA—Fall.
[c]Stem termination: IN—Indeterminate, SD—Semi-determinate, DE—Determinate.
[d]Growth type: ER—Erect, SE—Semi-erect, SP—Semi-prostrate, PR—Prostrate.
[e]Seed coat color: YW—Yellow, GN—Green, BW—Brown, BK—Black, BI—Bicolor.
[f]Cotyledon color: YW—Yellow, GN—Green.
[g]Flower color: PU—Purple, WH—White.
[h]Pubescence color: GY—Gray, TW—Tawny.
[i]Leaf shape: BD—Broad, NW—Narrow.

diseases, phytophthora root rot (*Phytophthora megasperma* Drechs. var. *sojae* A. A. Hildeb.) and cyst nematode (*Heterodera glycines*, Ichinohe), were controlled by transferring resistance genes to current cultivars. The sources of resistance genes were found in germplasm collections; viz., A.K., Arksoy, Blackhawk, CNS, Illini, Monroe and Mukden, which have a dominant gene for resistance to phytophthora root rot, and Peking, PI 90.763 and PI 88.788, which have recessive resistance genes to cyst nematode.

In China, Kwanggyo and Xicaohuang were found to be resistant to all six strains of soybean mosaic virus (SMV) from Jiangsu and Heilongjiang Province (Pu et al., 1983). Of 809 strains, Nongkang SCN781, Nongkang SCN782, Nongkang SCN791, and Nongkang SCN792 were found to have resistance to cyst nematode in Heilongjiang, but there was no specific resistance to races of the pathogen (Wu et al., 1983). Of 934 strains, 113 were resistant to frogeye leaf spot (*Cercospora sojina* Hana) in Heilongjiang. A single dominant gene for resistance was confirmed. Screening for resistance to soybean pod borer, *Leguminivora glycinivorella* (Mats), was conducted in Jilin. The original source of resistance was Tiejiasilihuang from which a number of commercial cultivars, such as Jilin 3 and Jilin 16, were released. Other sources were also reported. The mechanism of resistance was considered to be a comprehensive result of non-preference and antibiosis. The resistance was proved to be quantitatively inherited (Guo and Feng, 1983). Sources of resistance to the agromyzid beanfly (*Melanagromyza sojae*) were screened in Jiangsu. Non-preference and antibiosis are also significant factors in resistance to it (Xia et al., 1983).

In Japan, screening for sources and breeding for cultivars resistant to soybean mosaic virus, soybean stunt virus (SSV), cyst nematode, downy mildew (*Pernospora manshurica* [Naum.] Syd.), and purple seed stain (*Cercospera kikuchii* Mat. & Tomoy) were conducted. Nemashirazu and Harosoy were reported to be resistant to SMV-A, -B and SSV-A, -B, and SMV-A, -C, -D and SSV-A, -B, -C, respectively. Selections from their hybrids were resistant to all the strains of SMV and SSV in Tohoku (Nagasawa and Iisuka, 1977). Geden Shirazu and Nangun Takedate were the main sources in Japanese commercial cultivars tolerant to the cyst nematode. Of 923 strains, 105 were resistant to downy mildew, indicating a broad genetic base (Murakami et al., 1981). The sources reported to be resistant and tolerant to diseases, insects, and stresses were summarized by Gai (1983).

Land cultivars can be divided into groups according to performance of their major ecological traits relative to local, natural and agricultural conditions. Ma and Pei (cited by Fehr and Gai, 1980) reported on the genetic diversity of a population of land cultivars in the lower Yangtze and Hwai valleys of China. The Xu-Hwai group, whose environment was characterized by a short growing season and low rainfall, had cultivars with fewer days to flowering and maturity, lower seed weight, and a lower proportion of determinate types than cultivars of other groups. Seed of this group was used for commercial purposes; therefore, yellow seed coat was preferred. In the Jiang-Hwai and Jiangnan groups, the longer growing season and greater rainfall characteristic of their environment were associated with more determinate cultivars possessing more days to flowering and maturity, and heavier seed weight than in the Xu-Hwai group. Colored seed coats were prevalent in the Jiang-Hwai group because the soybeans often were consumed directly by the local people.

Genetic diversity for yield and yield-related traits mostly concerns breeders. Ma and Gai (1979) started a program to evaluate the genetic diversity for yield and other agronomic, quantitative characters of the land cultivar population in the lower Yangtze and Hwai valleys. The results showed appreciable variability for yield-related traits. The genotypic coefficient of variation, and 1% and 0.1% expected genetic advance for yield, were 14.5%, 35.2% and 44.4%, respectively. The estimates of genotypic multiple coefficient of variation for yield component characters, morpho-quantitative characters, developmental-period characters, and yield, plus the above three groups of characters, were 9.0, 17.2, 1.6, and 6.7%, respectively.

STRATEGY AND METHODS OF UTILIZING THE DIVERSITY

The strategy and methods for utilizing genetic resources depend on the breeding objectives and the base materials. There can be two kinds of germplasm relative to each specific station, local (adapted) and exotic. In some breeding programs, crosses of adapted x adapted, or good x good often are preferred to ensure that the selections will be adapted. Another kind of program tends to add some gene(s) or characteristic(s) onto an adapted strain, which may be advantageous in broadening genetic diversity of the selections. Therefore, it may become more important to breeders than the adapted X program.

However, no matter which kind of breeding program is preferred, the basic step is to screen for parents with good yield potential. Cooperative regional evaluations of a portion of the U.S. collection have been initiated. Nelson (pers. comm.) at Illinois yield-tested about 1700 new introductions for two years. Cregan and Kenworthy (pers. comm.) at Maryland yield-tested more than 1000 introductions, from which 15 of group II, 32 of group III, and 54 of group IV were selected. Most of those selections were from Northeast China, Japan, and Korea. They further evaluated the combining ability of 12 PIs from visual selection and 12 PIs from yield tests by using four cultivars as testers. Experience indicated that good cultivars often came from a few crosses, which suggested the importance of evaluating combining ability, especially specific combining ability.

Backcrossing has been very successful in transferring some major gene(s) to a commercial cultivar. When a polygenic trait is to be transferred, it is usually not easy to get both the trait and yield recovered in the two respective parents in a single backcrossing program. In this situation, a double backcrossing program may be helpful. The hybrids are backcrossed separately to the respective parents. In other words, both parents are used as donors and two separate backcrossing programs are started. Selections from the high-yield donor side should aim toward improving yield while keeping a high level of the trait, and vice versa for the transferred-trait donor side. Finally, crossing between the superior lines from both sides is made to combine both yield and the trait on the basis of inferior genes having been discarded. This procedure is essentially 'convergent improvement' proposed for improving inbred lines. Of course, if the trait is not genetically complicated, a normal backcrossing program with an appropriate sample size and yield-testing capacity might satisfy the breeders' purposes.

There is an increasing tendency in utilizing germplasm to transfer more than two traits from one parent to one adapted cultivar. A strategy of stepwise accumulation of favorable traits is recommended. Breeders can have several programs with individual goals rather than one program for all goals. In each program, there is a major goal, such as resistance to cyst nematode, or early maturity, in addition to yield. Yield is always included in a program. During the same period, different goals can be achieved in separate programs with yield as a common goal. Then the next step is to combine each of the two superior genotypes from separate programs. The same procedure is used for further steps to make more advanced combinations.

To explain the advantages, suppose there is a good yield cultivar to be improved in four traits, each controlled by a single dominant gene from a separate parent. In the first step, four backcrossing programs can be initiated to transfer the individual genes to the high-yield cultivar. If five backcrosses are needed for each program, 63/64 of the yield genes will come from the recurrent parent, and 1/4 of the genotypes in BC_5F_2 from the dominant BC_5F_1 will be homozygous resistant. That means 63/64 x 1/4 = 63/256 will be of the desired genotype at the BC_5F_2 generation. In other words, BC_5F_3 lines with YYAA, YYBB, YYCC, and YYDD genotypes, where YY represents yield genes (might be a large number of genes), AA, BB, CC, and DD, the single dominant genes, can be obtained from individual programs. In the next step, crosses such as YYAA x YYBB, are made to get YYAABB. The proportion of lines of that genotype in the F_3 generation will be 1/16. The same is used for the other crosses to get YYAACC, YYAADD. Next, YYAABBCC and YYAABBDD can be obtained at the same rate by the same procedure, and finally YYAABBCCDD.

Obviously, in this way the intermediate products, such as YYAA, YYAABB, and YYAABBCC, can be used commercially at each step. Meanwhile, the desirable genotype can be easily detected, since only the gene to be transferred is segregating, and the yield genes are basically homozygous.

In contrast, if several goals are considered in a single program rather than in step-wise improvement; for example, crossing two parents with all the required traits complementary to each other, the probability of obtaining desirable genotypes is very low, and they may be easily lost due to genetic drift during the segregating generations. Suppose for single dominant genes and five yield genes are involved, only $(1,4)^4 \times (16/1024) = 1/16384$ will be AABBCCDD in the F_2 with at least four homozygous yield genes. The probability of finding at least one desirable genotype will be only $1 - (1-(1/16384)^{2000}) = 0.115$ if 2000 F_2 plants are grown. This explains why one usually fails to get the desirable type when an introduction is directly used in a two-way cross.

When breeding aims for high-yielding ability, three kinds of philosophy have been considered: (1) removing factors restricting yield, such as disease, insect and stress problems, (2) developing an ideotype with some revolutionary traits, like dwarfism in cereal crops, to reveal yield potential, (3) accumulating high-yield genes. Different genetic mechanisms underlie each. Backcrossing and stepwise accumulation can be used for the first two approaches. Recurrent selection fits the third one.

In the soybean, to composite genes from a great diversity of parents into one population, three or four intermatings are needed. The first intermating may be a diallel cross, complete or partial, followed by a chain cross and a plant-to-plant cross; or two-way followed by four-way and multiple-way crosses can also be used (Fehr and Ortiz, 1975). This is tedious work in soybeans because it must be done by artificial hybridization. Genetic male sterility has been used to simplify intermating. Four nuclear male sterile genes ms_1, ms_2, ms_3, and ms_4, have been reported in the U.S. In addition, the partial male sterile gene msp was studied. Two germplasm populations using ms_1 maintainer N69-2774 also are available. Recurrent selection methodology needs to be studied. It includes types of families tested, number of years of testing before recombination, testing procedures, and number of generations between cycles in addition to the intermating procedure.

The strategy and methods of utilizing the diversity discussed here are rather traditional. Today, the progress in biotechnology and genetic engineering looks promising; however, it will not be considered in this paper.

FURTHER EXPLORING THE DIVERSITY

Breeding objectives or target traits will vary from environment to environment, and new ones will be raised for breeders as the soybean production and processing industry develops. That means there will be an increasing requirement to further explore soybean germplasm.

One approach is to explore the diversity in the present collections in the world. A cooperative program among countries and regions is required for the evaluation of soybean germplasm so that target genes can be recognized and used in breeding programs. A trait found to be of agronomic importance would be cataloged and evaluated. Consequently, such a cooperative program is possibly a long-term work.

A second approach is to make further explorations to enlarge the germplasm collection or add new genes to the world genebank. Every soybean producing area has old germplasm and/or newly developed germplasm to be collected. Among them eastern Asia seems very important. From that standpoint, soybean germplasm in China, Japan, Korea, and Southeast Asia has been collected several times. A program to further collect soybean germplasm in China is underway.

In addition to these two approaches, a third one would be to create new germplasm through mutation and recombination. It should be emphasized that important variation so obtained should be treated as a germplasm resource.

Soybean germplasm can be considered in two categories. The original one is basically natural, and the secondary one is somewhat artificial, developed through mutation and hybridization. The origin and original site should be kept on record for both kinds of germplasm, especially the former because of the high relationship of traits with original environment. Most of the original, and some of the secondary, are a mixture of genotypes, or a population, while the others are a pure line.

The reason for collecting artificial germplasm is that the mutant genes, recombined chromosomes, modified genomes, and mixtures of modified genotypes are genetically revised, just as the original germplasm was revised during history, and the recombinations might have new importance in breeding due to gene interaction.

The methods of conservation of population germplasm and pure-line germplasm should be different. For a population, a certain sample size of seeds should be maintained such that genetic diversity can be preserved. Hawkes (1981) suggested 2500-5000 seeds for one population. In addition, to cope with genetic drift, a certain sample size of plants is needed in reproduction. It might be a good idea to maintain a bulk sample and some major pure lines for population germplasm. Of course, for pure-line germplasm only a small sample is needed.

There is ample room open for scientists to work and study soybean germplasm. Moreover, there is an urgent need to avoid extinction of land cultivars and genetic drift of rare genes.

NOTES

Junyi Gai, Soybean Research Laboratory, Nanjing Agricultural College, Nanjing, People's Republic of China.

The author is grateful to Dr. R. H. Ma, Director, SRL, NAC, for his helpful comments on the manuscript.

This research was supported by IBPGR through Cooperative Agreement PR 3/11 IBPGR Soybean.

REFERENCES

Delannay, X., D. M. Rodgers and R. G. Palmer. 1983. Relative genetic contributions among ancestral lines to North American soybean cultivars. Crop Sci. 23:944-949.

Fehr, W. R. and Junyi Gai. 1980. Germplasm exchange and cooperative research with the People's Republic of China. p. 24-38. *In* Proc. 10th soybean seed res. conf. American Seed Trade Association, Washington, D.C.

Fehr, W. R. and L. B. Ortiz. 1975. Recurrent selection for yield in soybeans. J. Agr. Univ. P. R. 59: 222-232.

Gai Junyi, W. R. Fehr and R. G. Palmer. 1982. Genetic performance of some agronomic characters in four generations of a backcrossing program involving *Glycine max* and *Glycine soja*. Acta Genetica Sinica (China) 9:44-56.

Gai Junyi. 1983. The progress of soybean improvement and its potential direction in the United States. Soybean Sci. (China) 2:225-231, 327-341, 3:70-80.

Guo Shougui and Zhen Feng. 1983. Studies on resistance of soybean varieties to soybean pod. p. 345-347. *In* S. Wong et al. (eds.) Proc. Second US-China Soybean Symposium.

Hadley, H. H. and T. Hymowitz. 1973. Speciation and cytogenetics. p. 97-116. *In* B. E. Caldwell (ed.) Soybeans: Improvement, Production, and Uses. Amer. Soc. Agron., Madison, Wisconsin.

Hawkes, J. G. 1981. Germplasm collection, preservation, and use. p. 57-83. *In* K. J. Frey (ed.) Plant Breeding II. Iowa State Univ. Press, Ames, Iowa.

Harlan, J. R. 1971. Agricultural origins: centers and noncenters. Science 174:468-474.

Korsakov, N. I. 1979. Mobilization, conservation and utilization of soybean germplasm in the USSR. p. 241-247. *In* F. T. Corbin (ed.) World soybean research conference II: proceedings. Westview Press, Boulder, Colorado.

Ma, R. H. and Junyi Gai. 1979. Preliminary study on the local soybean varieties in the lower Yangtze and Hwai valleys II. Acta Genetica Sinica (China) 6:331-338.

Murakami, S. et al. 1981. Disease ratings of soybean varieties for downy mildew. Tohoku Natl. Agric. Exp. Stn. (Japan) Misc. Publ. 2:41-84.

Nagasawa, T. and N. Iizuka. 1977. Breeding of soybeans for resistance to virus diseases. Bull. Tohoku. Natl. Agric. Exp. Stn. (Japan) 55:75-80.

Pu Zhuqin, Cao Qi and Xi Baobi. 1983. Resistance of soybean varieties to different strains of the soybean mosaic virus. p. 389-390. *In* S. Wong et al. (eds.) Proc. second US-China soybean symposium. Dupont Far East Inc., Monsanto Far East Ltd., Pecteri Chemicals Inc., Pioneer Hi-Bred International, Inc., and Potash and Phosphate Institute in cooperation with OICD/USDA.

Wu Heli, Liu Hanqi, Shang Shaogang, Yao Zhenchun, Li Xiulon, and Mao Dougqing. 1983. Studies on the screening of resistant sources of soybean cyst nematode. p. 381-383. *In* S. Wong et al. (eds.) *ibid.*

Xia Jikang et al. 1983. Determination of the production loss caused by the agromyzid fly, *Melanagromyza sojae* (Zehntner), and proposal of criterion for evaluating the resistance of soybean cultivars to the agromyzid fly. p. 360-371. *In* S. Wong et al. (eds.) *ibid.*

Zhang Guodong. 1983. Analysis of the pedigree of soybean cultivars in Heilongjiang Province. Soybean Sci. (China) 2:184-193.

Quality and Utilization

BREEDING SOYBEANS FOR IMPROVED PROTEIN QUANTITY AND QUALITY

J. W. Burton

The development of near-infrared reflectance instruments has made it possible to easily measure protein, oil, and moisture in soybeans. These developments have increased the number of genotypes that can be screened in a season, and thus, made larger breeding efforts possible. These techniques have also made it possible for markets to monitor the constituents of the soybean crop. In the future, this could mean that soybeans will be priced on the basis of chemistry as well as seed grade. In that event, it should be to a producer's advantage to grow either high protein or high oil cultivars, depending upon the relative prices of meal and oil.

In addition to emphasis on increasing seed protein percentage, research efforts have also been directed toward the improvement of protein quality. Soybean protein ranks high compared to other vegetable protein sources, but it is still of lower quality than animal proteins. Sulfur containing cystine and methionine are the most nutritionally limiting amino acids in soybeans. Of these, methionine limits protein quality most. Methionine level in soybean protein of currently grown cultivars (1.0 to 1.3%) is about 43% of that in the FAO (Food and Agricultural Organization) standard, hen's egg (Wolf and Cowan, 1975). Thus, soybean protein needs to have double its current methionine content to provide the equivalent of egg protein.

The purpose of this paper is to review the plant breeding research that has been conducted to increase protein quantity and improve its quality in soybeans.

SELECTION FOR INCREASED PERCENTAGE PROTEIN

Cultivars currently being grown in the U.S. have approximately 40.5% seed protein and 21.0% seed oil on a dry weight basis (Hartwig, 1979). However, there is a wide range (approx. 15%) in percentage protein among lines of the U.S. soybean germplasm collection. About 10% have a protein percentage higher than 44.5% (Hartwig, 1979). The wild species, *G. soja*, with a protein content that is typically greater than 45%, also has been used as a source of high protein germplasm (Erickson et al., 1981). Thus, genetic resources are available for use in increasing protein percentage.

Percentage protein is a highly heritable trait. In six separate studies involving 13 populations of random lines from 2- and 3-way crosses, broad sense heritability estimates for percentage protein ranged from 0.39 to 0.92 (Table 1). The mean heritability for populations from crosses between adapted lines with average protein percentages (≤42%) was 0.71. Mean heritability of percentage protein was 0.82 for populations from crosses between adapted lines where one or more of the parents had above average protein percentages (>42%). There were also 10 populations which had exotic

Table 1. Summary of variance component heritability estimates for seed protein percentage and the response to selection for increased protein percentage from six separate studies.

Investigators	Populations		Number of test sites	Herita-bility	Unselected popula-tion mean	Pro-portion selected	Selection response[b]
	Type	Derivation[a]					
					%		%
Johnson et al. (1955a)	F_3 lines in the F_4 and F_5	1. AA X AA	4	0.39	42.2	.05	1.5
		2. AA X AA	2	0.83	43.7	.05	4.3
Thorne and Fehr (1970)	F_6 lines	3. AA X UH	1	0.91[c]	43.9[c]	.20	3.8[e]
	F_5 lines	4. (AA X UH) X AM	1	0.91[c]	44.3[c]		
Shannon et al. (1972)	F_4 lines in the F_5	5. AM X AM	2	0.88	44.5	.10	4.0
		6. AA X AM	2	0.89[d]	42.0[d]	.10	3.6
		7. AA X AA	2	0.92	41.1	.10	3.9
Shorter et al. (1976)	F_3 lines in the F_4	8. AA X UA	2	0.60	41.2	.10	2.2
		9. AM X UA	2	0.58	42.8	.10	3.2
		10. AA X AM	2	0.54	41.5	.10	1.5
Erickson et al. (1981)	F_3 lines	11. Composite of 4 G. max X G. soja crosses	2	0.78	45.3	.10	5.1
Openshaw and Hadley (1984)	F_3 lines	12. UH X AH	1	0.90	47.8	.10	3.4
		13. AM X AM	1	0.75	45.6	.10	1.9

[a] AH = adapted, high protein line (>46%); AM = adapted, moderately high protein line (42-46%); AA = adapted, average protein line (≤42%); UH, UM, UA = as above except that U = unadapted.
[b] Observed or predicted selection response as a percentage of the unselected population mean.
[c] Average for six populations.
[d] Average for four populations.
[e] Average for 12 populations.

germplasm (*Glycine max* or *G. soja* plant introductions) in their derivation. The average heritability for percentage protein in these populations was 0.83.

Depending on methods and the genetic population used, expected or observed progress from selection for increased percentage protein ranged from 1.5% to 5.1% of the population mean (Table 1). These were increases in the population mean of selected lines in the following generation, with no recombination among lines. Using recurrent selection for increased protein among S_1 progenies of a population derived from intermating 12 high protein and 12 high yielding lines, Miller and Fehr (1979) achieved a 3.4% increase over the base population mean in one cycle of selection. This was an actual change from 43.1% to 44.6%. With six cycles of recurrent selection in two populations (designated I and II), Brim and Burton (1979) averaged 0.7% and 1.3% increases per cycle over the initial population means. In Population I, which originated from a cross between two adapted lines with differing oil and protein percentages, this represented a change from 46.3% to 48.4%. In Population II, which was more genetically diverse (9 high protein PIs crossed to D49-2491, a sister line of the cv. Lee), the mean changed from 42.8% to 46.1%. Realized heritabilities in the two populations were 0.29 and 0.34, respectively.

A major difficulty in selection for increased protein in most soybean breeding populations has been the existence of negative genetic correlations between percentage protein and the two other economically important traits, yield and percentage oil. Thus, selection for increased protein usually results in decreases in percentage oil (Figure 1), and often in decreased yield. Percentage protein and percentage oil were found to be negatively correlated in 12 soybean populations investigated in five separate studies (Table 2). Most of these correlations had absolute values greater than 0.50. However, weaker relationships between the two traits do occur. In one population, derived from the cross Woodworth X PI 159,764, the genetic correlation between protein and oil was only -0.15 (Simpson and Wilcox, 1983). Negative correlations between percentage protein and yield, though frequent, were usually not great, only two having absolute values greater than 0.50. One population, derived from the cross, Woodworth X PI 159,764, had a positive genetic correlation of 0.54 between protein and yield (Simpson and Wilcox, 1983).

The correlations between percent protein and other agronomic traits have been less consistent in direction and magnitude than those previously discussed. Simpson and Wilcox (1983) found positive phenotypic correlations between maturity and percentage protein, and Miller and Fehr (1979) found that days to maturity increased with selection for increased protein percentage. Other studies have found negative correlations between maturity and percentage protein, ranging from -0.45 to +0.27 (Johnson et al., 1955b; Shorter et al., 1976). Similarly, weak correlations that are positive in some studies and negative in others have been found with other traits such as seed size, plant height, lodging and time to flowering.

Other investigations have sought relationships between percentage seed protein and vegetative traits. Schapaugh and Wilcox (1980) found no differences in dry matter accumulation between high and average protein lines. In a growth analysis of selected cycles from the Populations I and II previously described, Carter et al. (1982) found that selection for protein had indirectly increased total vegetative N in Population I and indirectly decreased total vegetative N in Population II. The changes in vegetative N were manifested primarily in leaves. Both populations showed increases in

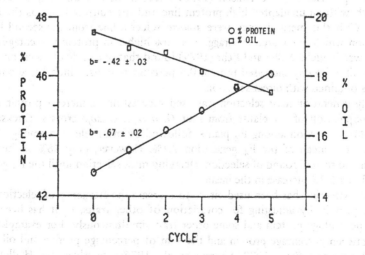

Figure 1. The response of Population II to selection for increased percentage seed protein and the accompanying changes in percentage seed oil (Brim and Burton, 1979).

Table 2. Genotypic correlations between seed protein and seed oil percentages and between protein percentage and yield.

Population	Johnson et al. (1955b) 1[a]	2[a]	Thorne & Fehr (1970) 3[a]	4[a]	Shorter et al. (1976) 9[a]	10[a]	Simpson & Wilcox (1983) 14[b]	15[b]	16[b]	17[b]	Openshaw & Hadley (1984) 12[a]	13[a]
% oil genotypic	-0.48	-0.70	-	-	-0.96	-0.35	-0.15	-0.96	-0.72	-0.36	-0.83	-0.68
% oil phenotypic	-0.48	-0.69	-0.66*	-0.58*	-0.79*	-0.24	-	-	-	-	-0.80*	-0.65*
yield genotypic	-0.64	-0.12	-	-	-	-	+0.54	-0.74	-0.40	-0.20	-	-
yield phenotypic	-0.33	-0.08	-0.21	-0.27	-	-	-	-	-	-	-	-

*Significant at $\alpha \leqslant .05$

[a]These populations correspond to those in Table 1.

[b]Random F_5 lines from crosses between unadapted high protein lines and adapted average protein lines (UH X AA).

percentage petiole N throughout the pod-filling period. These indirect responses to selection for protein were also of interest because yield did not change in Population I, whereas Population II showed a correlated decline in yield.

When considering the problems of genetically increasing the quantity of protein produced by a soybean crop, there must be a recognition of both the producer's desire for high yield and the soybean processor's desire for high protein percentage and acceptable oil levels. Thus, breeding methods have been varied depending upon the breeding goals. As already noted, recurrent selection has been used successfully to increase protein; however, in each case percentage oil declined significantly (Brim and Burton, 1979; Miller and Fehr, 1979). This negative relationship between protein and oil has led some investigators to attempt to increase protein indirectly by selection for low oil. This has some economic advantages in that percentage oil can be measured rapidly and nondestructively in soybean seeds by nuclear magnetic resonance (NMR) spectroscopy. Hartwig and Hinson (1972) selected for low oil using a backcrossing approach with an unadapted high protein line and the cultivar Bragg as the recurrent parent. With this method, they were able to select a line from the second backcross generation which, compared to Bragg, was 10% higher in protein percentage and only 3.5% lower in yield. Miller and Fehr (1979) found that one cycle of recurrent selection for low oil indirectly increased protein 0.8 percentage units, which was about half the progress obtained with direct selection.

Early generation mass selection was used successfully to increase protein in a composite population of F_2 plants from four G. max X G. soja crosses (Erickson et al., 1981). Mass selection among F_2 plants, followed by mass selection among F_3 plants, increased the mean of the F_4 generation 7.1%. However, only 28% of the gain was made in the second round of selection. Delaying mass selection until the F_3 generation resulted in a 7.7% increase in the mean.

Index selection has been used either to increase the efficiency of selection for percentage protein by adjusting for covariation of other traits, or it has been used to change percentage protein and some other trait simultaneously. For example, correlations between percentage protein and the sum of percentage protein and oil are positive (Thorne and Fehr, 1970; Shorter et al., 1976; Openshaw and Hadley, 1983). Thorne and Fehr (1970) found that selection for an increased sum of oil and protein increased protein by 3.4% while decreasing oil by only 0.5%. Heritability estimates for the sum of the two traits (protein + oil) were 0.55, 0.78, and 0.67, determined for populations of F_3 lines (Shorter et al., 1976; Openshaw and Hadley, 1983). Openshaw and Hadley (1983) predicted the gain in protein resulting from selection for oil + protein to be 1.4% of the midparent. The observed gain was only 0.3% of the midparent; however, percentage oil also increased slightly. Thus, it appeared that slow progress could be made in both traits by selecting for increased protein + oil. 'Desired gains' indexes were also used to increase percentage protein while changing percentage oil or holding it constant. These indexes were reasonably effective in controlling changes in both traits simultaneously. Brim et al. (1959) used selection indexes in which oil yield and protein yield were given relative economic weights of 1:1, 1:0.6, and 1:0.2. Results showed that index selection was superior to single trait selection for changing both traits simultaneously, but the difficulty of determining realistic economic weights was noted.

Increased protein yield can also be accomplished by selection for increased yield, provided percentage protein does not decline significantly. In this respect, recurrent restricted index selection could be used to hold protein constant while increasing yield. It might be possible to select for protein yield directly; however, with this approach there is the risk that percentage protein would decline. Other approaches include independent culling and tandem selection. A kind of tandem selection was used with some success by Sebern and Lambert (1984) to identify high, intermediate and low percentage protein groups in the F_2 or F_3 generations, followed by selection for yield within the groups in later generations.

INCREASING METHIONINE CONTENT OF SOYBEAN PROTEIN

Breeding efforts to increase percentage methionine in soybean protein have been limited, compared with those to increase protein quantity. This has been due primarily to the difficult and time-consuming nature of methionine assays. Thus, some of the research on this problem has been directed toward easier and more rapid methods for methionine assay, or toward finding a correlated trait that is easier to measure. Breeding research has followed traditional patterns; i.e., assessment of available genetic variation, estimation of heritability and identification of correlated traits.

Surveys of germplasm resources show that there is variability for methionine content. Examination of the published methionine levels in the soybean germplasm collection, Group V through Group X, reveals a range of levels from 1.1% to 1.6% of total protein (Hartwig and Edwards, 1975). Kaizuma and Fukui (1975) found a range of methionine content from 0.64% to 0.96% among a group of 54 Japanese cultivars. The range for cystine, the other major sulfur-containing amino acid, was from 0.59% to 1.21%. Their investigation of other *Glycine* species, *G. soja*, *G. gracilis*, *G. tomentella* and *G. clandestina*, showed no evidence that seed proteins of these species were higher in methionine content than *G. max*. Taira et al. (1976) screened 1110 *G. max* lines from 15 regions worldwide. In a single environment test, these ranged from 0.78% to 1.34% methionine in protein. The overall mean was 1.03%. Cystine ranged from 0.85% to 2.36%. Using a set of 34 cultivars grown in two years as the reference population, Kaizuma et al. (1974) estimated the heritabilities for methionine, cystine and total sulfur-containing amino acids to be 0.55, 0.67, and 0.67, respectively.

Correlations between percentage protein and methionine content of the protein generally have been nonsignificant, both in *G. soja* (Kaizuma and Fukui, 1975) and in *G. max* (Kaizuma et al., 1974). Burton et al. (1982) showed that increasing the percentage protein through five cycles of recurrent selection in two populations did not change concentrations of methionine in the protein. On the other hand, Taira et al. (1976) found a negative correlation (-0.21) between percentage protein and methionine (as a percentage of the protein) among 1110 *G. max* lines from various geographical regions. The correlation between cystine and percentage protein was -0.38.

Significant, positive correlations have been found between methionine and cystine and between methionine and total sulfur-containing amino acids (Kaizuma et al., 1974; Kaizuma and Fukui, 1975; Taira et al., 1976). This has led some to suggest that sulfur content or the nitrogen to sulfur ratio (N/S) could be used as a reasonable selection criterion for increasing the methionine content of seed protein. In a set of 13 *G. max* and *G. soja* lines representing a wide range of N/S ratios, Radford et al. (1977) found significant positive correlations (0.83, 0.83, 0.85) between percentage sulfur and concentrations of methionine, cystine, and methionine + cystine, respectively. The correlations between the three traits and N/S ratio were negative and significant (-0.79, -0.74, -0.79). Maranville et al. (1984) have shown that energy-dispersive, x-ray fluorescence spectrometry provides a rapid method of sulfur determination on a dry weight basis for soybean, corn (*Zea mays* L.), wheat (*Triticum aestivum* L.), and sorghum (*Sorghum bicolor* [L.] Moench.). In soybeans, they found that the correlation between sulfur values and methionine was 0.80. Similar correlations were obtained with cystine. This method may prove to be the economical means of screening genotypes for methionine content that plant breeders have been needing.

In mature soybean seeds, most of the protein is in storage globulins of which there are two main fractions, an 11S and a 7S, based on sedimentation. Because most of the sulfur containing acids have been shown to be present in the 11S fraction (Krober and Cartter, 1966), it has been proposed that decreasing the 7S/11S ratio would be another way to increase methionine content of protein. Variation in this ratio has been shown among cultivars. Taira and Taira (1972) found a range of 0.71 to 1.41 in 7S/11S ratio among 30 cultivars averaged over three environments.

CONCLUSION

Research findings suggest that both protein quantity and quality can be increased in soybeans without sacrificing agronomic performance. However, economic incentives such as constituent pricing, are probably needed before high protein cultivars will become widely available for production. If producers were paid a premium for producing high protein or high methionine soybeans, the demand for cultivars with those characteristics would increase and breeding efforts would surely increase to meet the demand. High protein lines (≥45%) are currently available and could be released as cultivars, but in most cases their yield would not equal the highest yielding cultivars now in production. Development of high yielding, high protein cultivars will require a long-term breeding effort. Research toward this goal should continue, even though, currently, there are no economic incentives for producing high protein percentage or quality soybeans.

NOTES

J. W. Burton, Research Agronomist, USDA-ARS and Associate Professor of Crop Science, N. C. State University, Raleigh, N.C. 27695-7631.

REFERENCES

Brim, C. A. and J. W. Burton. 1979. Recurrent selection in soybeans. II. Selection for increased percent protein in seeds. Crop Sci. 19:494-498.

Brim, C. A., H. W. Johnson and C. C. Cockerham. 1959. Multiple selection criteria in soybeans. Agron. J. 51:42-46.

Burton, J. W., A. E. Purcell and W. M. Walter, Jr. 1982. Methionine concentration in soybean protein from populations selected for increased percent protein. Crop Sci. 22:430-432.

Carter, T. E. Jr., J. W. Burton and C. A. Brim. 1982. Recurrent selection for percent protein in soybean seed—indirect effects on plant N accumulation and distribution. Crop Sci. 22:513-519.

Erickson, L. R., H. D. Voldeng and W. D. Beversdorf. 1981. Early generation selection for protein in Glycine max x G. soja crosses. Can. J. Plant Sci. 61:901-908.

Hartwig, E. E. 1979. Breeding productive soybeans with a higher percentage of protein. p. 59-66. In Seed protein improvement in cereals and grain legumes Vol. 2. International Atomic Energy Agency, Vienna.

Hartwig, E. E. and C. J. Edwards. 1975. Evaluation of soybean germplasm maturity group V to X. Germplasm report. USDA-ARS Soybean Production Research, Stoneville, Miss.

Hartwig, E. E. and K. Hinson. 1972. Association between chemical composition of seed and seed yield of soybeans. Crop Sci. 12:829-830.

Johnson, H. W., H. F. Robinson and R. E. Comstock. 1955a. Estimates of genetic and environmental variability in soybeans. Agron. J. 47:314-318.

Johnson, H. W., H. F. Robinson and R. E. Comstock. 1955b. Genotypic and phenotypic correlations in soybeans and their implications in selection. Agron. J. 47:477-483.

Kaizuma, N. and J. Fukui. 1975. Breeding for chemical quality of soybean in Japan. Trop. Agric. Res. 6:55-68.

Kaizuma, N., H. Taira, H. Taira and J. Fukui. 1974. On the varietal differences and heritabilities for seed protein percentage and sulfur-containing amino acid, contents in cultivated soybeans, Glycine max Merrill. Japan J. Breed. 24:81-87.

Krober, O. A. and J. L. Cartter. 1966. Relation of methionine content to protein levels in soybeans. Cereal Chem. 43:320-325.

Maranville, J. W., P. J. Mattern and R. B. Clark. 1984. Estimation of sulfur in grain by x-ray fluorescence spectrometry and its relation to sulfur and amino acids of field crops. Crop Sci. 24:303-305.

Miller, J. E. and W. R. Fehr. 1979. Direct and indirect selection for protein in soybeans. Crop Sci. 19:101-106.

Openshaw, S. J. and H. H. Hadley. 1984. Selection indexes to modify protein concentration of soybean seeds. Crop Sci. 24:1-4.

Radford, R. L. Jr., C. Chavengsaksongkram and T. Hymowitz. 1977. Utilization of nitrogen to sulfur ratio for evaluating sulfur-containing amino acid concentrations in seed of Glycine max and G. soja. Crop Sci. 17:273-277.

Schapaugh, W. T. Jr. and J. R. Wilcox. 1980. The accumulation and redistribution of dry matter and nitrogen in normal and high protein soybean genotypes. Agron. Abstr. p. 92.

Sebern, N. A. and J. W. Lambert. 1984. Effect of stratification for percent protein in two soybean populations. Crop Sci. 24:225-228.

Shannon, J. G., J. R. Wilcox and A. H. Probst. 1972. Estimated gains from selection for protein and yield in the F_4 generation of six soybean populations. Crop Sci. 12:824-826.

Shorter, R., D. E. Byth and V. E. Mungomery. 1976. Estimates of selection parameters associated with protein and oil content of soybean seeds (Glycine max [L.] Merr.). Aust. J. Agric. Res. 28:211-222.

Simpson Jr., A. M. and J. R. Wilcox. 1983. Genetic and phenotypic association of agronomic characteristics in four high protein soybean populations. Crop Sci. 23:1077-1081.

Taira, H. and H. Taira. 1972. Influence of location on the chemical composition of soybean seeds. III. Protein component content by disc electrophoresis. Proc. Crop Sci. Soc. Japan 41:235-243.

Taira, H., H. Taira, N. Kaizuma and J. Fukui. 1976. Varietal differences of seed weight, protein and sulfur-containing amino acid content of soybean seed. Proc. Crop Sci. Soc. Japan 45:381-393.

Thorne, J. C. and W. R. Fehr. 1970. Incorporation of high-protein, exotic germplasm into soybean populations by 2- and 3-way crosses. Crop Sci. 10:652-655.

Wolf, W. J. and J. C. Cowan. 1975. Soybeans as a food source. CRC Press, Inc., Cleveland, Ohio.

ANTI-NUTRITIONAL FACTORS IN SOYBEANS: GENETICS AND BREEDING

Theodore Hymowitz

Raw, mature soybean seeds contain anti-nutritional factors, such as protease inhibitors, lectins, goitrogens, cyanogens, anti-vitamin factors, phytic acid, saponins, and estrogens. These anti-nutritional factors must be destroyed, or their activity markedly reduced, before the two main products of the soybean, oil and the protein-rich meal, can be utilized effectively in foods or feeds. Obviously, soybean products would be less expensive if the cost of processing were reduced by removal of anti-nutritional factors via traditional plant breeding methods. Breeding for the removal of anti-nutritional factors in the soybean is an extremely complex problem. To be successful it must be accomplished without reducing yields, without lowering oil and protein content in seed, or increasing the plant's susceptibility to pests or pathogens. Nevertheless, investigations on the elimination of certain anti-nutritional factors in soybean seed have been successful and the results of those investigations are summarized in the following sections.

PROTEASE INHIBITORS

Almost 70 years ago, Osborne and Mendel (1917) reported that unheated soybean meal is inferior in nutritional quality to properly heated soybean meal. The trypsin inhibitor proteins in raw, mature soybean seed have been proposed to be one of the major factors responsible for the poor nutritional value of the unheated meal (Borchers et al., 1948). Physiologically, the ingestion of unheated soybean meal by monogastric animals causes pancreatic hypertrophy (Bray, 1964; Chernick et al., 1948) in addition to growth inhibition. These effects may be due to upset in the balance of methionine and cystine in the pancreas (Booth et al., 1960). Trypsin inhibitors make up ca. 6% of the total protein in soybeans.

Two classes of trypsin inhibitors exist in soybean seeds. Much of the soybean trypsin inhibitory (SBTI) activity is due to SBTI-A_2 (Rackis et al., 1962), which was purified by Kunitz (1945) and is generally called the "Kunitz soybean trypsin inhibitor." The complete amino acid sequence of the Kunitz inhibitor was reported by Koide and Ikenaka (1973). It has a relative molecular mass (Mr) of 21,384, consisting of 181 amino acid residues with two disulfide bonds. The active center of the Kunitz inhibitor is the Arg63-Ile64 bond, and it has specificity toward trypsin (Wolf, 1977).

The proteins in the second class are generally called the "Bowman-Birk type proteinase inhibitors." Actually, in soybean seed there may be as many as five different Bowman-Birk type inhibitors. The complete amino acid sequence of the Bowman-Birk inhibitor itself was reported by Odani and Ikenaka (1972). It has a Mr of 7,861, consisting of 71 amino acid residues with seven disulfide bonds. The protein has the unique property of inhibiting trypsin and chymotrypsin at independent binding sites. The trypsin inhibiting site is the Lys16-Ser17 bond and the chymotrypsin inhibiting site is the Leu44-Ser45 bond (Wolf, 1977).

Kunitz Trypsin Inhibitor

Seed from the U.S. Department of Agriculture soybean germplasm collection (2944 accessions) was screened by polyacrylamide gel electrophoresis for the presence or absence of electrophoretic forms of SBTI-A_2. Four electrophoretic forms of SBTI-A_2 were discovered. Three of the forms, designated Ti^a, Ti^b, and Ti^c are electrophoretically distinguishable from one another by their different R_f values of 0.79, 0.75, and 0.83, respectively (R_f is the mobility relative to a bromophenol-blue dye front in a 10% polyacrylamide-gel, anodic system, in which a Tris-glycine buffer, pH 8.3, is used). The three forms are controlled by a co-dominant multiple allelic system at a single locus. The fourth form of SBTI-A_2 that is found in soybean germplasm does not exhibit a protein band in the gels. The lack of a protein band is inherited as a recessive allele, and has been designated ti. In the homozygous recessive state ($ti\ ti$) no SBTI-A_2 band is produced in the seed. Soybeans without the Kunitz Trypsin inhibitor have about 50% less trypsin inhibitor activity per gram of protein than soybeans containing the Kunitz trypsin inhibitor.

Two accessions from Korea, PI 157440 and PI 196168, carry the ti allele. PI 157440 was introduced in 1947 under the name 'Kin du'. Possibly Kin du is the same as 'Kum du', which in English means golden bean. PI 196168 was introduced into the U.S. in 1951 under the name "Baik tae" (Hymowitz and Kaizuma, 1981; Orf and Hymowitz, 1979).

Near isolines of cv. Williams (BC5F4), carrying the ti allele from PI 157440, were yield tested in 1982 and 1983 at Urbana (Table 1). Preliminary data suggest that substitution of the Ti^a allele with the ti allele does not markedly affect yield. In addition, there were no obvious differences in other traits, such as germination, flowering time, maturity, plant height, and lodging. Analysis of the oil and protein content in seed from the 1982 trial also suggests no marked differences in these components of seed in the near-isolines of Williams (Bernard and Hymowitz, unpub.). Chick feeding trials

Table 1. Yield, protein and oil content of Williams (Ti^a) and near-isolines containing the ti allele for the Kunitz trypsin inhibitor, Urbana, 1982 and 1983 (Bernard and Hymowtiz, unpub.).

Strain	Yield		1982	
	1982	1983	Protein	Oil
	kg/ha		%	
Williams	3076	3188	42.5	17.9
L81-4583	3115	3320	42.5	17.9
L81-4584A	3122	2858	42.9	17.6
L81-4584B	3135	3010	42.9	17.8
L81-4585	3142	2831	42.4	17.8
L81-4590	3161	3260	42.7	18.0
L81-4593	3214	3340	42.4	18.4
L81-4594	3260	3128	42.9	17.8

comparing the near-isolates of Williams with Williams currently are being conducted at the University of Illinois.

Bowman-Birk Inhibitors

Investigators have separated five Bowman-Birk type inhibitors from soybean cultivars either by anion or cation exchange chromatography (Hwang et al., 1977; Ikenaka and Odani, 1978) Stahlhut and Hymowitz (1983) screened, electrophoretically, the U.S. Department of Agriculture cultivar and genetic type collections for Bowman-Birk inhibitor variants. They found accessions that seemed to be missing their Fraction III or V electrophoretic bands. Thus far, no genetic studies on the Bowman-Birk type inhibitors have been reported in the literature.

LECTIN

Among the components of soybean seed are a group of glycoproteins that cause the agglutination of certain red blood cells. These glycoproteins are called lectins or phytohaemagglutinins. At least four different lectin forms have been reported in soybean seed. Defatted soybean meal contains about 3% lectin (Liener and Rose, 1953). The major lectin present in the seed of most soybeans has a Mr of 120,000, and is composed of four subunits (Catsimpoolas and Meyer, 1969; Lotan et al., 1974). The lectin has specificity for N-acetyl-D-galactosamine and, to a lesser extent, for D-galactose (Lis et al., 1970).

Seed from the U.S. Department of Agriculture soybean germplasm collection (2784 accessions) was screened by polyacrylamide gel electrophoresis for the presence or absence of electrophoretic forms of soybean seed lectin (SBL); two were discovered. One form designated *Le* has an SBL band at R_f 0.48 (R_f is the mobility relative to the lysozyme protein band in a 10% polyacrylamide-gel cathodic system, in which a glycine-citric acid buffer, pH 3.7, is used). The second form of SBL found in the soybean germplasm does not exhibit a protein band in the gels. Orf et al. (1978) demonstrated that the presence of SBL is controlled by a simple dominant gene designated *Le*; the homozygous recessive allele *le le* results in the lack of SBL. Recently Goldberg et al. (1983) identified a 3.4 kb segment of DNA that is inserted into the *le* allele. This insertion blocks transcription and accounts for the absence of SBL mRNA.

Since 1953, lectin was believed to contribute to the poor nutritive properties of raw soybean meal. However, after conducting an experiment in which rats were fed soybean meal with and without lectin, Turner and Liener (1975) concluded that lectin contributed little to the deleterious effect of unheated soybean meal.

GOITROGENS

Raw mature soybeans contain a factor that causes goiter in animals and humans. This was first reported by McCarrison (1933), who found enlarged thyroids in rats fed soybeans. Wilgus and Gassner (1941) reported that raw soybean meal in the diet of poultry enlarged the thyroids and reduced the hatchability and fertility of the eggs. The effect of the goitrogenic factor in soybeans can be counteracted by the addition of iodide in soybean-based feeds and foods. Thus far, the goitrogenic factor has not been precisely identified in soybeans, and thus, no chemical screening studies have been conducted to determine if accessions are available without the goitrogenic factor. See Van Etten (1969) for a thorough review of goitrogen literature.

CYANOGENS

Soybeans, like many other legumes, contain glucosides from which hydrocyanic acid may be released by hydrolysis. Honig et al. (1983) report that they found 0.26 μg of hydrocyanic acid released per g of whole soybean meal. Hence the soybean cyanide levels do not appear to be of nutritional significance.

ANTI-VITAMIN FACTORS

Soybeans contain several anti-vitamin factors such as A, B_{12}, and D_3. These anti-vitamin factors have not been fully characterized and their mode of action is poorly understood. Thus, no screening studies have been conducted to determine if these factors can be eliminated from soybean cultivars (Liener, 1980; Rackis, 1972).

PHYTIC ACID

Phytic acid, *myo*-inositol 1,2,3,4,5,6-hexakisphosphate, accounts for about 70% of the total phosphorus in raw, mature soybean seed. It binds to cations of nutritional importance; e.g., calcium and zinc, thereby reducing their bioavailability. Raboy (1984) found that, in 38 soybean strains, phytic acid content ranged from 14 to 23 mg per g dry weight of seed. In 20 strains of *Glycine soja*, he found that phytic acid content ranged from 19 to 28 mg per g dry weight of seed. Breeding for reduced phytic acid content in seed will be extremely difficult, because it is intimately associated with phosphorus, which is an essential element for growth, and secondly, because phosphorus content in seed is greatly influenced by environmental factors affecting phosphorus nutrition.

SAPONINS

Saponins are glucosides characterized by their foaming in aqueous solutions and hemolyzing red blood cells. Sumiki (1929) first demonstrated the presence of saponins in soybeans. Saponin in soybeans has been separated into five fractions that differ in their aglycone moiety (sapogenin), commonly named soyasapogenol A, B, C, D, and E, and six carbohydrate moieties containing glucuronic acid, glucose, galactose, rhamnose, xylose, or arabinose (Birk et al., 1963; Gestetner et al., 1966a; Wolf and Thomas, 1971). Soybeans contain about 0.5% saponins (Gestetner et al., 1966b). Apparently, soybean saponins are harmless when ingested by chicks, rats, and mice at concentrations three times that normal in a soybean meal supplemented diet (Birk, 1969).

ESTROGENS

An estrogen is a substance capable of stimulating the growth of female reproductive organs and the development of female secondary characteristics in mammals. The isoflavones, genistein, and daidzein have been isolated from soybeans and shown to possess weak estrogenic activity (Carter et al., 1955; Nilson, 1962). Since the genistein and daidzein content of defatted soybean meal has been estimated to be about 0.10%, it is doubtful that there is sufficient to elicit a physiological response in animals or humans.

Recently, coumestrol was detected in soybeans by HPLC (Lookhart et al., 1978). Coumestrol possesses strong estrogenic activity. In seed of three soybean cultivars

(Amsoy 71, Clark 63, and Columbus) Lookhart et al. (1979) found that the concentration of coumestrol ranged from 0.09 to 0.02 μg per g dry weight. Thus, it may be possible to screen the soybean germplasm collection for low coumestrol containing accessions, if indeed, the coumestrol content in soybeans is demonstrated to be of physiological significance.

NOTES

Theodore Hymowitz, Department of Agronomy, University of Illinois, Urbana, IL 61801.

REFERENCES

Birk, Y. 1969. Saponins. p. 169-210. *In* I. E. Liener (ed.) Toxic Constituents of Plant Foodstuffs. Academic Press, New York.

Birk, Y., A. Bondi, B. Gestetner and I. Ishaaya. 1963. A thermostable hemolytic factor in soybeans. Nature 197:1089-1090.

Booth, A. N., D. J. Robbins. W. E. Ribelin and F. De Eds. 1960. Effect of raw soybean meal and amino acids on pancreatic hypertrophy in rats. Proc. Soc. Exp. Biol. Med. 104:681-683.

Borchers, R., C. W. Anderson, F. E. Mussehl and A. Mochl. 1948. Trypsin inhibitors. VIII. Growth inhibiting properties of a soybean trypsin inhibitor. Arch. Biochem. 19:317-322.

Bray, D. J. 1964. Pancreatic hypertrophy in layering pullets induced by unheated soybean meal. Poultry Sci. 43:382-384.

Carter, M. W., G. Matrone and W. G. Smart Jr. 1955. Effect of genistin on reproduction of the mouse. J. Nutr. 82:507-511.

Catsimpoolas, N. and E. W. Meyer. 1969. Isolation of soybean hemagglutinin and demonstration of multiple forms by isoelectric focusing. Arch. Biochem. Biophys. 132:279-285.

Chernick, S. S., S. Lepkovsky and I. L. Chaikoff. 1948. A dietary factor regulating the enzyme content of the pancreas. Changes induced in size and proteolytic activity of the chick pancreas by the ingestion of raw soybean meal. Am. J. Physiol. 155:33-41.

Gestetner, B., Y. Birk and A. Bondi. 1966a. Soya bean saponins. VI. Composition of carbohydrate and aglycone moieties of soya bean saponin extract and of its fractions. Phytochemistry 5: 779-802.

Gestetner, B., Y. Birk, A. Bondi and Y. Tencer. 1966b. Soya bean saponins. VII. A method for the determination of sapogenin and saponin contents in soya beans. Phytochemistry 5:803-806.

Goldberg, R. B., G. Hoschek and L. O. Vodkin. 1983. An insertion sequence blocks the expression of a soybean lectin gene. Cell 33:465-475.

Honig, D. H., M. E. Hockridge, R. M. Gould and J. J. Rackis. 1983. Determination of cyanide in soybeans and soybean products. J. Agric. Food Chem. 31:272-275.

Hwang, D. L.-R., K.-T. D. Lin, W.-K. Yang and D. E. Foard. 1977. Purification, partial characterization, and immunological relationships of multiple low molecular weight protease inhibitors of soybean. Biochim. Biophys. Acta 495:369-382.

Hymowitz, T. and N. Kaizuma. 1981. Soybean seed protein electrophoresis profiles from 15 Asian countries or regions: Hypotheses on paths of dissemination of soybeans from China. Econ. Bot. 35:10-23.

Ikenaka, T. and S. Odani. 1978. Structure function relationships of soybean double-headed proteinase inhibitors. p. 206-216. *In* S. Magnusson, M. Ottesen, B. Foltmann, K. Dano and H. Neurath (eds.) Regulatory proteolytic enzymes and their inhibitors. Pergamon Press, New York.

Koide, T. and T. Ikenaka. 1973. Studies on soybean trypsin inhibitors: 3. Amino-acid sequence of the carboxyl-terminal region and the complete amino-acid sequence of a soybean trypsin inhibitor (Kunitz). Eur. J. Biochem. 32:417-431.

Kunitz, M. 1945. Crystallization of a trypsin inhibitor from soybeans. Science 101:668-669.

Liener, I. E. 1980. Miscellaneous toxic factors. p. 429-467. *In* I. E. Liener (ed.) Toxic Constituents of Plant Food Stuffs. Academic Press, New York.

Liener, I. E. and J. E. Rose. 1953. Soyin, a toxic protein from the soybean. III. Immunochemical properties. Proc. Soc. Exp. Biol. Med. 83:539-544.

Lis, H., B. Sela, L. Sachs and N. Sharon. 1970. Specific inhibition by N-acetyl-D-galactosamine of the interaction between soybean agglutinin and animal cell surfaces. Biochim. Biophys. Acta 211: 582-585.

Lookhart, G. L., B. L. Jones, and K. F. Finney. 1978. Determination of coumestrol in soybeans by high performance liquid and thin-layer chromatography. Cereal Chem. 55:967-972.

Lookhart, G. L., P. L. Finney, and K. F. Finney. 1979. Note on coumestrol in soybeans and fractions at various germination times. Cereal Chem. 56:495-496.

Lotan, R., H. W. Siegelman, H. Lis and N. Sharon. 1974. Subunit structure of soybean agglutinin. J. Biol. Chem. 249:1219-1224.

McCarrison, R. 1933. The goitrogenic action of soybean and ground nut. Indian J. Med. Res. 21: 179-181.

Nilson, R. 1962. Estrogenic activity of some isoflavone derivatives. Acta Physiol. Scand. 56:230-236.

Odani, S. and T. Ikenaka. 1972. Studies on soybean trypsin inhibitors. IV. Complete amino acid sequence and anti-proteinase sites of Bowman-Birk soybean proteinase inhibitor. J. Biochem. 71:839-848.

Orf, J. H. and T. Hymowitz. 1979. Genetics of the Kunitz trypsin inhibitor: An antinutritional factor in soybeans. J. Am. Oil Chem. Soc. 56:722-726.

Orf, J. H., T. Hymowitz, S. P. Pull and S. C. Pueppke. 1978. Inheritance of a soybean seed lectin. Crop Sci. 18:899-900.

Osborne, T. B. and L. B. Mendel. 1917. The use of soybean as feed. J. Biol. Chem. 32:369-377.

Raboy, V. 1984. Phytic acid in developing and mature seed of soybean [*Glycine max* (L.) Merr.] and *Glycine soja* Sieb + Zucc. Ph.D. Dissertation, University of Illinois Library, Urbana.

Rackis, J. J. 1972. Biologically active components. p. 158-202. *In* A. K. Smith and S. J. Circle (eds.) Soybeans: Chemistry and Technology. Volume 1. Proteins. The AVI Publ. Co., Westport, Connecticut.

Rackis, J. J., H. A. Sasame, R. K. Mann, R. L. Anderson and A. K. Smith. 1962. Soybean trypsin inhibitors: isolation, purification and physical properties. Arch. Biochem. 98:471-478.

Stahlhut, R. W. and T. Hymowitz. 1983. Variation in the low molecular weight proteinase inhibitors of soybeans. Crop Sci. 23:766-769.

Sumiki, Y. 1929. Studies on the saponin of soybean. Bull. Agric. Chem. Soc. Japan 5:27-32.

Turner, R. H. and I. E. Liener. 1975. The effect of selective removal of hemagglutinins on the nutritive value of soybeans. J. Agric. Food Chem. 23:484-487.

Van Etten, C. H. 1969. Goitrogens. p. 103-142. *In* I. E. Liener (ed.) Toxic Constituents of Plant Foodstuffs. Academic Press, New York.

Wilgus, H. S., Jr. and F. X. Gassner. 1941. Effect of soybean oil meal on avian reproduction. Proc. Soc. Exp. Biol. Med. 46:290-293.

Wolf, W. J. 1977. Physical and chemical properties of soybean proteins. J. Am. Oil Chem. Soc. 54:112A-117A.

Wolf, W. J. and B. W. Thomas. 1971. Ion-exchange chromatography of soybean saponins. J. Chromatogr. 56:281-293.

BREEDING SOYBEANS FOR SPECIAL USES

O. G. Carter

Soybeans are a major source of protein in the human diet in Southeast Asia, but are not popular in many other parts of the world, in spite of a great need in the human diet for the nutrients they contain. A number of important quality characteristics contribute to this unpopularity; the main ones are flavor, cookability, flatus factors and anti-nutritional factors. Many of these problems can be overcome by processing, but this is costly and some countries have health regulations which prohibit hydrogenation of soybean oil. Complex processing facilities are frequently not available in developing countries.

Plant breeders have tended to emphasize yield, agronomic characters and factors such as pest and disease resistance. Quality aspects, in relation to human food use, have been largely neglected. Even protein and oil receive little emphasis when farmers are paid only on the amount of soybeans produced, with no premium for high protein or high oil content.

Food texture and flavor are of great importance in developing countries. New, high yielding wheat (*Triticum aestivum* L.) and rice (*Oryza sativa* L.) cultivars frequently are not grown for home use because of flavor problems, but are grown only when they are to be sold outside the village.

This paper will review recent studies on the use and improvement of soybean cultivars for manufacture of human food, and make brief reference to use of soybean oil as a liquid fuel for diesel engines.

OIL QUALITY

The major limiting factor in soybean oil quality for both human food and as a diesel fuel is the level of linolenic acid. Germplasm surveys of the soybean and related species have generally shown the absence of genotypes with less than the 2% linolenic acid needed to overcome oxidized flavor problems in soybean oil (Hymowitz et al., 1972; Hammond and Fehr, 1983). There is one report of a 'Line 64' with 0.0% linolenic acid (Tripathi et al., 1975).

It has been suggested that linolenic acid is an essential component of membranes in chloroplasts and that very low levels of linolenic acid are not biologically feasible in soybean seed (Fehr et al., 1971). Recent Australian studies with linseed (flax, *Linum usitatissimum* L.), using the chemical mutagen ethyl methanesulfonate (EMS), have reduced linolenic acid from 45% to 0.5% (Begg, pers. comm.). Linseed has extensive chloroplasts present in the developing seed. The chemical mutagen, EMS, has also been used to produce low linolenic acid lines of soybean (Hammond and Fehr, 1983; Wilcox et al., 1984).

Very low levels of linolenic acid, or its complete elimination, has great significance for soybean use as a human food in developing countries, where complex and expensive food processing technology to eliminate flavor problems is not available. A reduction in linolenic acid percentage also produces an oil quality better suited for use as a liquid fuel in diesel engines (Begg, pers. comm.).

There appears good reason to believe that linolenic acid in soybeans can be reduced below the critical 2% level, and possibly even eliminated, although development of high yielding commercial cultivars with this changed fatty acid composition is likely to take some years.

Many of the flavor problems with soybeans are considered to be due to the presence of linolenic acid. Oxidized flavors are produced by oxidization of linolenic acid by lipoxygenase. This enzyme is inactivated by heat and this problem can be partly, at least, overcome by heating, which is also necessary to destroy trypsin inhibitors during soybean processing. Methods have been described that allow food products to be developed from soybeans high in linolenic acid and lipoxygenase (Nelson et al., 1978), but removal of these factors from soybeans could greatly increase their flexibility as a human food and reduce processing costs.

PROTEIN

The amino acid composition in soy protein is well suited to both human and animal requirements, with some shortage of the sulfur containing amino acids cystine and methionine. Genetic variability has been identified and synthetic methionine can be added, but in most cases high methionine levels in cereals eaten in conjunction with soy protein overcome this problem, but genotypes with increased methionine levels would be helpful.

Food processors have developed a wide range of protein products that can be used directly or blended with other food products (Campbell, 1980). A major factor in processing is the removal of flatulence causing oligosaccharides. The two major food products, other than protein isolates, are tofu and soymilk, and a number of studies have been made to examine the suitability of soybean cultivars for manufacture of these products.

Tofu

This product is a curd resulting from precipitation of heated soymilk. Calcium sulphate or calcium chloride or magnesium chloride are used to bring about precipitation. Over one billion people in China and Southeast Asia are dependent on tofu as a major source of food protein (Watanabe, 1978). Differences in quality of tofu produced from different soybean cultivars has been reported, and the major difference is associated with texture of the curd (Skurray et al., 1980). Curd firmness is strongly influenced by concentration of calcium sulphate and the ratio of 11S to 7S protein in the soybean cultivar used (Skurray et al., 1980; Saio et al., 1969). In Japan, tofu manufacture is on a small scale, and a uniform soybean seed supply is important if flavor and texture are to be maintained. Adjusting calcium sulphate levels for different cultivars could be used under large scale manufacturing conditions. The U.S. cultivar, Beeson, has been recognized by Japanese importers as being particularly well suited to tofu manufacture. Conditions under which the soybean crop is grown, harvested and stored before use are also likely to be important. A detailed study of a range of temperature and humidity conditions in storage has shown that a combination of high temperature and relative humidity, 35C and 80%, causes rapid deterioration in soybean seed quality when used for manufacture of tofu or fermented soybean food products (Saio et al., 1980). This, and other studies by Mwandemele (1982), clearly show the importance of crop growing, harvest and storage conditions in influencing quality of soybeans for food manufacture, and they emphasize the need for great care in standardizing growing and storage conditions before evaluating soybean seed quality.

Breeding cultivars resistant to weather damage before harvest may need to be given higher priority where climatic conditions make this a problem in countries such as Indonesia. Storage conditions in tropical countries are likely to be very important, in relation to food use of soybeans, as they are for maintaining viability of sowing seed. Cultivars with yellow hilum are preferred for tofu manufacture.

Soymilk

Soymilk is becoming an increasingly important form of utilizing soybeans for human food use. Most commercial cultivars appear well suited to soymilk manufacture, but a yellow hilum is preferred. Favorable conditions during harvest and storage, to prevent seed deterioration, are essential (Saio et al., 1980). Heating to prevent development of off flavors due to oxidation of linolenic acid by lipoxygenase, is also an important part of soymilk manufacture (Hanene, 1980). Soymilk manufacture from ten soybean lines, with a range of chemical compositions, showed there was little correlation between seed protein content and protein in the soymilk, but there was a close relationship between oil content in soymilk and oil content in the seed (Mwandemele, 1982). These lines were derived from a cross between U.S. cultivars Big Jule and Bethel.

OLIGOSACCHARIDES

Legume seeds, including soybeans, cause discomfort following eating due to flatulence. The oligosaccharides raffinose and stachyose are not digested due to the absence of α-galactosidase in the digestive tract (Rackis et al., 1970; Murphy, 1972). When these sugars reach the lower ileum and ascending colon they are acted upon by anaerobic microflora with the release of carbon dioxide and hydrogen, causing flatulence (Calloway, 1972). Processing can be used to overcome the problem (Rackis, 1981), or there is a possibility of elimination by genetic selection (Murphy, 1972). A wide range of oligosaccharide levels in soybeans has been reported by Hymowitz and Collins (1974).

Studies on lines derived from three soybean crosses indicated a wide range of stachyose and raffinose levels (Mwandemele, 1982). Flatulence studies involving human beings are difficult and expensive, but providing oligosaccharide levels are a true indicator of flatulence, then chemical methods are readily available for genotype evaluation. Considerable care is needed, both in relation to storage conditions before analysis and during analysis, as enzyme systems which break down oligosaccharides are present in soybean seeds and can cause rapid changes in sugar composition (Mwandemele, 1982). It may be possible to use these same enzymes to eliminate oligosaccharides during processing.

There would appear to be good prospects for overcoming flatulence problems associated with use of soybeans as a human food, both through plant breeding and processing technology. Oligosaccharides remain in solution and are largely eliminated during tofu manufacture. They are also reduced markedly during manufacture of fermented soybean foods such as tempeh and miso.

COOKABILITY

Cooking time to achieve an acceptable level of softness is an important factor limiting use of soybeans as a human food. Cooking times of up to 3 h have been required to produce an acceptable food product from dry soybeans (Mueller et al., 1974).

A number of mechanical devices have been developed to obtain precise measurement of tenderness (Mitchell et al., 1961; Spata et al., 1973; Mwandemele, 1982). Cooking times can be modified by soaking and by use of 0.5% $NaHCO_3$ during boiling (Spata et al., 1974; Nelson et al., 1978). Hardness is believed to be caused by calcium pectate in the cell wall, and cooking softens the wall, with sodium replacing calcium in the cell wall, accelerating this process.

A study of a large number of recombinant lines of soybean indicated that there was considerable variation in cookability. Some lines combined rapid cooking with high protein content and satisfactory yield (Mwandemele, 1982). Cookability was also shown to be influenced by the environmental conditions under which the crop was grown.

Improved cookability would increase the attractiveness of soybeans as a human food and reduce energy costs, both in the food processing factory and in the home. There appear to be good prospects for breeding soybean lines with rapid cookability, high yield and other desirable characteristics as a human food (Chesterman, 1981).

ANTI-NUTRITIONAL FACTORS

Trypsin inhibitors have received greatest attention, but a number of other factors, including hemagglutinins, goitrogens and anti-vitamins have been identified and all are heat labile (Liener, 1980). Breeding cultivars without lipoxygenase appears a possibility (Hildebrand and Hymowitz, 1982; Kitamura et al., 1983). The use of *Glycine soja* Sieb. and Zucc. would appear to be helpful in reducing lipoxygenase activity in soybeans (Ahmad, 1980).

Heat stable anti-nutritional factors include the flatulence causing oligosaccharides discussed earlier, estrogens, saponins and lysinoalanine. Flatulence factors appear to be the only heat stable factors of any consequence, unless a diet solely consisting of soybeans is contemplated.

CONCLUSIONS

There are clearly some problems with using soybeans as a food for human consumption, with flavor associated with oxidation of linolenic acid by lypoxygenase being the most important. Cookability, flatulence and anti-nutritional factors such as trypsin inhibitor are also important. There seems to be good reason to believe that plant breeders will be able to overcome many of these problems, and food processors can overcome others. The conditions under which the crop is grown, harvested and stored before use are also very important for food quality.

This review has highlighted the need for a team approach involving plant breeders, agronomists, food technologists, home economists and sociologists who understand the religious and traditional customs associated with food use in the home. There are many lessons to be learned from the mistakes made in breeding new rice and wheat cultivars, where quality for food use was neglected in favor of yield.

With a team approach, many of the problems identified in this review can be solved and the soybean can make a much greater contribution to the nutritional needs of the world population. This will be achieved only by a team of specialists working together over a prolonged period.

NOTES

O. G. Carter, Assistant Principal, Hawkesbury Agricultural College, Richmond, New South Wales, 2753, Australia.

REFERENCES

Ahmad, Q. N. 1980. Soybean cytogenetics and isoenzymes. Ph.D. Dissertation. University of Queensland Library, Brisbane, Australia.

Calloway, D. H. 1972. Gas-forming property of food legumes. p. 263-270. In M. Milner (ed.) Nutritional improvement of food legumes by breeding. John Wiley and Sons, New York.

Campbell, M. F. 1980. Soy protein product characteristics. p. 713-719. In F. T. Corbin (ed.) World soybean res. conf. II: Proc. Granada.

Chesterman, C. 1981. The evaluation of soybean varieties for processing suitability for human foods. Graduate Diploma in Food Science Thesis. Hawkesbury Agricultural College, Richmond, Australia.

Fehr, W. R., J. G. Thorne, and E. G. Hammond. 1971. Relationship of fatty acid formation and chlorophyll content in soybean seed. Crop Sci. 11:211-213.

Hammond, E. G. and W. R. Fehr. 1983. Registration of A5 germplasm line of soybean. (Reg. No. GP44). Crop Sci. 23:192.

Hanene, E. M. 1980. Process and product development of soymilk from soy flour. Graduate Diploma in Food Science Thesis. Hawkesbury Agricultural College, Richmond, Australia.

Hildebrand, D. F. and T. Hymowitz. 1982. Inheritance of lipoxygenase-1 activity in soybean seeds. Crop Sci. 22:851-853.

Hymowitz, T., and F. I. Collins. 1974. Variability of sugar content in seed of G. max (L. Merrill) and G. soja (Sieb. and Zucc.). Agron. J. 66:239-240.

Hymowitz, T., R. G. Palmer, and H. H. Hadley. 1972. Seed weight, protein, oil and fatty acid relationships within the genus Glycine. Trop. Agric. 49:245-250.

Kitamura, K., C. S. Davies, N. Kaizuma and N. C. Nielsen. 1983. Genetic analysis of a null-allele for lipoxygenase-3 in soybean seeds. Crop Sci. 23:924-927.

Liener, I. E. 1980. Anti-nutritional factors as determinants of soybean quality. p. 703-712. In F. T. Corbin (ed.) World soybean res. conf. II: Proc. Granada.

Mitchell, R. S., D. J. Casimir and L. J. Lynch. 1971. The maturometer—instrumental test and redesign. Food Tech. 15:415-418.

Mueller, D. C., B. P. Klein and F. O. Van Dwyne. 1974. Cooking with soybeans. Coop. Ext. Serv. Circ. No. 1092, College of Agriculture, University of Illinois.

Murphy, E. L. 1972. The possible elimination of legume flatulence by genetic selection. p. 273-276. In M. Milner (ed.) Nutritional improvement of food legumes by breeding. John Wiley and Sons, New York.

Mwandemele, O. D. 1982. Genetic variation in quality characteristics of soybeans (Glycine max (L.) Merrill) that influence human acceptability. Ph.D. Dissertation, University of Sydney Library, Sydney, Australia.

Nelson, A. I., M. P. Steinberg and L. S. Wee. 1978. Whole soybean foods for home and village use. INTSOY Series NO. 14. International Agricultural Publications, University of Illinois, Urbana.

Rackis, J. L. 1981. Flatulence caused by soya and its control through processing. J. Amer. Oil Chem. Soc. 58:503-509.

Rackis, J. J., D. H. Honig, D. J. Sessa and F. R. Steggerda. 1970. Flavour and flatulence factors in soybean protein products. J. Agric. Food Chem. 18:977-982.

Saio, K., M. Kamiya and T. Watanabe. 1969. Food processing characteristics of soybean 11S and 7S proteins. Part I: Effect of different protein components among soybean varieties on formation of tofu gel. Agric. Biol. Chem. (Tokyo) 33:1301-1308.

Saio, K., I. Nikkuni, Y. Ando, M. Otsuru, Y. Terauchi and M. Kito. 1980. Soybean quality changes during model storage studies. Cereal Chem. 57:77-82.

Skurray, G., J. Cunich and O. Carter. 1980. The effect of different varieties of soybean and calcium ion concentration on the quality of tofu. Food Chem. 6:89-95.

Spata, J. A., A. I. Nelson and S. Singh. 1974. Developing a soybean dal for India and other countries. World Crops 26:82-84.

Spata, J. A., M. P. Steinberg and L. S. Wei. 1973. A simple shear press for measuring tenderness of whole soybeans. J. Food Sci. 38:722-723.

Watanabe, T. 1978. Traditional non-fermented soybean foods in Japan. p. 35. *In* Proc. Int. soya protein food conf. American Soybean Association, St. Louis.

Wilcox, J. R., J. F. Cavins and N. C. Nielsen. 1984. Genetic alteration of soybean oil composition by a chemical mutagen. J. Amer. Oil Chem. Soc. 61:97-100.

Tripathi, R. D., G. P. Srivostava and M. C. Misra. 1975. Note on the quality constituents of soybean (*Glycine max* (L.) Merrill) varieties. Indian J. of Agric. Res. 9:220-222.

BREEDING SOYBEANS FOR IMPROVED OIL QUANTITY AND QUALITY

J. R. Wilcox

The soybean is the most important oilseed crop in the world, accounting for 50% of all oilseeds produced and 30% of the total supply of all vegetable oils (USDA, 1982). It is, perhaps, paradoxical that virtually no improvement has been made in either oil quantity or quality of released cultivars of this important oilseed crop during the past 50 years. There have been some successes in genetic improvement of oil quantity and quality, but these successes have not been incorporated into improved cultivars.

OIL QUANTITY

Commonly grown soybean cultivars contain about 20% oil in the seed, and, with few exceptions, this value has remained relatively constant for the past 50 years. This consistency of oil content can be attributed to two factors. First, soybean breeders have put much greater emphasis on increasing seed yield, and on breeding for pest resistance, than on altering chemical composition of the seed. Soybean growers are paid for the quantity of seed, not the quantity of oil produced, so breeders have put their greatest emphasis on breeding for characteristics that contribute to increased yield, or that prevent yield losses. Second, and of less importance, the oil content of commonly grown cultivars is near the maximum oil content of germplasm accessions that were available to plant breeders at the time these cultivars were developed. The oil content of U.S. germplasm accessions acquired before 1970 ranged from a low of 13.2% in PI 82278 to a high of 23.5% in PI 79885 (U.S. Regional Soybean Laboratory, 1966; 1969). Since no accessions were available with markedly higher oil contents than those of released cultivars, there was no opportunity to incorporate exceptionally high oil content.

Two studies have demonstrated that oil content of soybean seeds is controlled by the maternal parent, rather than by the embryo. Brim et al. (1968) evaluated oil content of seeds from reciprocal crosses between low (14.5 to 14.6%) and typical (19.0 to 20.6%) oil lines of soybeans (Table 1). The oil content of F_1 seeds was not significantly different from that of selfed seeds produced on the maternal parent. Singh and Hadley (1968) compared the oil content of F_1 seeds from reciprocal crosses between low (13.0 to 15.8%) and typical (21.0 to 21.9%) oil content parents with that of selfed seeds on the parent plants. Oil contents of the F_1 seeds were much closer to those of selfed seeds from the maternal parents than of selfed seeds from the paternal parents. The mean effect of the male parent was to lower the oil content 1.1% when high oil parents were females and to raise oil content 1.3% when low oil parents were

Table 1. Oil content of selected parents and crosses (Brim et al., 1968).

Parent or cross	Seeds or plants	Oil content
	no.	%
PI 215,811	20	14.6 s[a]
PI 215,811 X PI 92,567	9	15.2 s
PI 92,567 X PI 215,811	25	18.9 t
PI 92,567	18	19.8 t
PI 201,421	25	20.6 s
PI 201,421 X PI 212,605	53	19.4 s
PI 212,605 X PI 201,421	11	14.5 t
PI 212,605	17	14.5 t

[a]Means within groups followed by the same letter do not differ at the 0.01 probability level.

females. Variability in oil content among seeds produced on F_1 plants was greater than the variability among seeds produced on the parent plants. The authors attributed the increased variability among F_2 seeds to interplant variability, since F_2 seeds were composited from several F_1 plants, rather than to differences among genotypes of F_2 embryos. These two studies indicate that selection for oil content based on individual F_2 seed analysis would be ineffective in soybeans.

Studies have shown that oil content in soybeans is controlled primarily by additive genetic effects (Brim and Cockerham, 1961; Hanson et al., 1967; Hanson and Weber, 1961; 1962). There is some evidence for epistatic components of variability for percentage oil (Hanson and Weber, 1961; 1962).

Oil content has a moderately high heritability, from 0.51 to 0.84 (Hanson and Weber, 1962; Johnson et al., 1955a; Kwon and Torrie, 1964; Shorter et al., 1976; Weber and Moorthy, 1952). Heritability for oil content is generally higher than for seed yield, but lower than for plant developmental characteristics, such as time of flowering or maturity.

Johnson et al.(1955b) reported that flowering time, fruiting period, and seed weight were all positively correlated with oil content, but that there was a strong negative correlation between oil and protein content of the seed. The use of a selection index composed of oil percentage plus flowering time, fruiting period, seed weight, and protein percentage, all genetically correlated with oil percentage, appeared to be no better than selecting for oil content alone.

Irradiation has been shown to be an effective way of increasing the genetic variability for oil content. Williams and Hanway (1961) irradiated seeds of cvs. Adams and Hawkeye with X-rays and thermal neutrons. Irradiation resulted in increased variability compared to the control population, and heritability of oil content in the irradiated populations (65 to 78%) was comparable to estimates from segregating populations derived from hybridization. However, in the irradiated populations the selection advantage was greater for low than for high oil content.

Progress has been made in breeding for high oil content. Burton and Brim (1981) evaluated three cycles of recurrent selection for oil content in a population segregating for male sterility (Table 2). The germplasm source was a set of 10 selected lines derived from crosses among six cultivars adapted to the southern U.S. Each of the ten lines was crossed to male-sterile plants (ms_1 ms_1) and seeds from five F_1 plants from each

Table 2. Progress in recurrent selection for oil content in soybeans (Burton and Brim, 1981).

Cycle	Oil	Yield	Total oil	Protein
	%	kg/ha	kg/ha	%
C_0	18.8	2,362	446	42.0
C_1 W	19.1	2,398	461	41.8
C_2 W	19.4	2,262	441	41.1
C_3 W	19.9	2,306	460	40.5
L.S.D.$_{.05}$	0.3	NS	NS	0.5

cross were composited to initiate the study.

Male-sterile plants in the recombination block were pollinated by insects, and those with hybrid seed were harvested. Seeds from each sterile plant were evaluated for oil content and the highest 10% were selected. Half-sib progeny of these selected plants were grown to maturity in the greenhouse, and the fertile individual with the highest oil percentage within each half-sib family was selected. Seeds from selected plants were composited to initiate the next cycle. Progenies from each cycle were evaluated in replicated tests at three locations. Data on the half-sib progenies, or the within-family selected populations, showed that oil percentage increased linearly at an average rate of 0.35 ± 0.3% per cycle. There was no significant change in seed yield or total oil produced per hectare with three cycles of recurrent selection. There was a significant decrease in percentage protein, but differences in total protein produced per hectare were not significant. The significant increase in percentage oil demonstrated that the oil concentration of a soybean population could be improved rapidly by recurrent selection.

Two soybean cultivars have been released that have exceptionally high oil content. Amsoy, a Maturity Group II cultivar and Ransom, a Group VII cultivar, both average 22 to 23% oil (Weber, 1966; Brim and Elledge, 1973). These two cultivars were readily accepted by growers and were extensively grown in their areas of adaptation. However, their popularity and persistence were not due to their high oil content, but to their yield advantage over other available cultivars at the time of their release.

Studies demonstrate that oil content in soybeans is controlled primarily by additive genetic effects, is more highly heritable than many other quantitative traits, and is responsive to selection. The lack of progress in developing soybean cultivars with high oil content appears to be due to a lack of priority rather than to a lack of opportunity. At present there is no economic incentive for developing high oil soybean cultivars.

OIL QUALITY

Soybean oil is the major vegetable oil produced and consumed in the U.S., with more than 90% of the oil going into edible products. The oil contains five fatty acids: palmitic, stearic, oleic, linoleic, and linolenic, typically occurring in the percentages shown in Table 3. The high linolenic acid content of the oil, 7 to 9%, has been associated with objectionable flavors and poor stability (Dutton et al., 1951). Crude soybean oil is refined, bleached, hydrogenated, and deodorized to produce a bland-tasting, light-colored oil acceptable to consumers. However, during storage the "grassy" or "beany" flavor of the oil may return, which has been referred to as flavor reversion

Table 3. Fatty acid components of soybean oil.

Common name	Systematic name	Structural formula	Amount in soy-bean oil
			%
Palmitic (16:0)	n-Hexadecanoic	$CH_3(CH_2)_{14}COOH$	11
Stearic (18:0)	n-Octadecanoic	$CH_3(CH_2)_{16}COOH$	3
Oleic (18:1)	cis 9-Octadecanoic	$CH_3(CH_2)_7CH = CH(CH_2)_7COOH$	22
Linoleic (18:2)	cis 9,12-Octadecadienoic	$CH_3(CH_2)_4CH = CHCH_2CH =$ $CH(CH_2)_7COOH$	56
Linolenic (18:3)	cis 9,12,15-Octadecatrienoic	$CH_3CH_2CH = CHCH_2CH =$ $CHCH_2CH = CH(CH_2)_7COOH$	8

(Smouse, 1979). Several theories have been proposed for the cause of flavor reversion and the undesirable taste of soybean oil. The high linolenic acid content of the oil, which can undergo oxidation, has been considered a major cause of the problem.

Breeding for improved oil quality has been concentrated on lowering the linolenic acid content of the oil. A soybean breeder's primary source of genetic variability for low linolenic acid would be the germplasm collections. The U.S. soybean collection contains strains with a minimum linolenic acid concentration of about 4.2% (Kleiman and Cavins, 1982). This is considerably higher than the 2% level considered a maximum for good oil flavor and stability.

Breeding programs have been initiated, using low linolenic acid germplasm accessions, to develop soybean strains with inherently low levels of this fatty acid. White et al. (1961) intercrossed low linolenic acid strains from the U.S. germplasm collection, and from progenies of these crosses identified an F_2 plant with 3.35% linolenic acid. However, the research was terminated because in succeeding generations the low linolenic acid content of the line was not maintained, and environmental effects strongly influenced the linolenic acid content of the line.

Hammond and Fehr (1975) attempted to decrease linolenic acid by intercrossing soybean accessions with the lowest linolenic acid content line available and selecting for transgressive segregates with still lower contents of this fatty acid. Using this procedure they were able to identify strains with 1 to 1.5% less linolenic acid than was present in the best parental strains.

Wilson et al. (1981) were successful in altering fatty acid composition of soybeans by selecting for high [18:1/(18:2 + 18:3)] ratios (R_1). Since the polyunsaturated fatty acids, 18:2 and 18:3, are believed to result from sequential desaturation of 18:1, selection for high R_1 should result in changes in the content of all three fatty acids. Comparisons of inbred selections from each of four cycles of recurrent selection demonstrated that R_1 values progressively increased from 0.4 in the initial population to 0.65 after four cycles of selection. An experimental line, N78-2245, developed during this recurrent selection study, contained 51% oleic acid (18:1), 31.9% linoleic acid (18:2), and 4.2% linolenic acid (18:3), in comparison with the representative cultivar, Dare, which contained 21.4% 18:1, 55.6% 18:2, and 8.9% 18:3.

Another potential source of variability for low linolenic acid is from species within the genus *Glycine* that are related to the cultivated soybean. However, accessions of related *Glycine* species that have been evaluated for oil composition all have higher linolenic acid contents than the cultivated soybean (Hymowitz et al., 1970).

A third opportunity available to the plant breeder is to create variability for fatty acid content using mutagenic agents. Hammond and Fehr (1975) used both X-rays and chemical mutagens in attempts to induce mutations that alter fatty acid content of soybean oil. Using these techniques, they have identified an M_4 plant selection, A5, from seeds treated with ethyl methanesulfonate (EMS) that averaged 2.9 to 4.3% linolenic acid, depending upon the environment where the line was grown (Hammond and Fehr, 1983a). A second mutant line, A6, was identified as an M_2 plant from treatment of seeds with sodium azide (Hammond and Fehr, 1983b). This line contained 28.1% stearic acid, compared with 4.4% in the parent line, and 19.8% oleic acid, compared with 42.8% in the parent line. Concentration of the other fatty acids, palmitic, linoleic, and linolenic, was unchanged in A6.

In the research by Hammond and Fehr, all selection efforts were put into producing lines with altered fatty acid composition; agronomic characteristics and disease resistance were ignored. This means that the altered fatty acid characteristics would have to be incorporated into lines or cultivars with good agronomic characteristics if they are to be competitive with existing cultivars.

Wilcox et al. (1984) attempted to alter the fatty acid composition of a high yielding, agronomically acceptable cultivar using EMS. Seeds of cv. Century were treated with EMS, and seeds from M_2 plants evaluated for fatty acid composition by gas chromatography. Treatment with EMS significantly increased the variability for each of the fatty acids. There was about a two-fold increase in variability of stearic and linolenic acid and a three- to four-fold increase in the variability of palmitic, oleic, and linolenic acid in the oil from seed of M_2 plants compared with the controls. One line was identified with 3.4% linolenic acid, and this low level of 18:3 has remained stable in successive generations grown in both greenhouse and field environments.

Reciprocal crosses were made between C1640, the low linolenic acid line, and Century. Fatty acid composition of the oil from reciprocally crossed F_1 seeds was compared with that of selfed seed produced on the same node of the maternal parents. The linolenic acid content of F_1 seeds was intermediate to that of selfed seeds from the two parents, and the values for the reciprocal crosses were essentially the same. These data demonstrate that for this mutant line, linolenic acid content is determined by the embryo rather than by the maternal parent. This is in contrast to previous reports that fatty acid composition of soybean oil is controlled by the maternal parent rather than by the embryo (Brim et al., 1968; Martin et al., 1983). Embryonic control of fatty acid composition would enable the breeder to evaluate the fatty acids in fragments of seeds produced on F_1 plants, then select and grow only those F_2 seeds with the desired composition.

These studies demonstrate that fatty acid composition of soybean oil is amenable to genetic manipulation. The wide changes in fatty acid composition that have resulted in just a few years research suggest that, with continued efforts, soybean cultivars could be developed with a range of fatty acid composition adapted to specific edible or industrial oil uses.

An alternative approach to improving oil quality by altering fatty acid composition is the genetic elimination of seed components that contribute to the instability of the oil (Kitamura et al., 1983). This approach would be feasible if specific components could be identified, and then, eliminated without adversely affecting seed germination or plant growth. Three lipoxygenase isoenzymes, lipoxygenase-1 (L-1), lipoxygenase-2 (L-2), and lipoxygenase-3 (L-3) are found in high concentrations in soybean seeds (Axelrod et al., 1981), and probably contribute to the instability of polyunsaturated fatty acids. Strains of soybeans have been identified that lack each of the lipoxygenase isoenzymes (Hildebrand and Hymowitz, 1981; Kitamura et al., 1983). Genetic analyses have demonstrated that the presence of L-1 and L-3 is controlled by single dominant genes, Lx_1 and Lx_3, respectively, that are inherited independently (Hildebrand and Hymowitz, 1982; Kitamura et al., 1983). The absence of both L-1 and L-3 has no apparent deleterious effects on the soybean plant. Incorporating null alleles for the lipoxygenase isoenzymes into soybean cultivars with either normal or low levels of linolenic acid is another potential method of improving soybean oil flavor and stability through breeding.

NOTES

J. R. Wilcox, Research Geneticist, USDA-ARS, and Professor of Agronomy, Purdue University, West Lafayette, IN.

REFERENCES

Axelrod, B., T. M. Cheesebourgh and C. Laakso. 1981. Lipoxygenase from soybeans. Methods Enzymol. 71:441-451.

Brim, C. A. and C. C. Cockerham. 1961. Inheritance of quantitative characters in soybeans. Crop Sci. 1:187-190.

Brim, C. A. and C. Elledge. 1973. Registration of Ransom Soybeans (Reg. No. 95). Crop Sci. 13:130.

Brim, C. A., W. M. Schutz and F. I. Collins. 1968. Maternal effect on fatty acid composition and oil content of soybeans, *Glycine max* (L.) Merrill. Crop Sci. 8:517-518.

Burton, J. W. and C. A. Brim. 1981. Recurrent selection in soybeans III. Selection for increased percent oil in seeds. Crop Sci. 21:31-34.

Dutton, H. C., C. R. Lancaster, C. D. Evans and J. C. Cowan. 1951. The flavor problem of soybean oil. VIII Linolenic acid. J. Am. Oil Chem. Soc. 28:115-118.

Hammond, E. G. and W. R. Fehr. 1975. Oil quality improvement in soybeans—*Glycine max* (L.) Merr. Fette Seifen Anstrichm 77:97-101.

Hammond, E. G. and W. R. Fehr. 1983a. Registration of A5 germplasm line of soybean, (Reg. No. GP44). Crop Sci. 23:192.

Hammond, E. G. and W. R. Fehr. 1983b. Registration of A6 germplasm line of soybean, (Reg. No. GP45). Crop Sci. 23:192-193.

Hanson, W. D., A. H. Probst and B. E. Caldwell. 1967. Evaluation of a population of soybean genotypes with implications for improving self-pollinated crops. Crop Sci. 7:99-103.

Hanson, W. D. and C. R. Weber. 1961. Resolution of genetic variability in self-pollinated species with an application to the soybean. Genetics 46:1425-1434.

Hanson, W. D. and C. R. Weber. 1962. Analysis of genetic variability from generations of plant-progeny lines in soybeans. Crop Sci. 1:63-67.

Hildebrand, D. F. and T. Hymowitz. 1981. Two soybean genotypes lacking lipoxygenase-1. J. Am. Oil Chem. Soc. 58:583-586.

Hildebrand, D. F. and T. Hymowitz. 1982. Inheritance of lipoxygenase-1 activity in soybean seeds. Crop Sci. 22:851-853.

Hymowitz, T., F. I. Collins, V. E. Sedgwick and R. W. Clark. 1970. The oil and fatty acid content of some accessions of *Glycine wiighti*. Trop. Agric. 47:265-269.

Johnson, H. W., H. F. Robinson and R. E. Comstock. 1955a. Estimates of genetic and environmental variability in soybean. Agron. J. 47:314-318.

Johnson, H. W., H. F. Robinson and R. E. Comstock. 1955b. Genotypic and phenotypic correlations in soybeans and their implications in selection. Agron. J. 47:477-483.

Kitamura, K., C. S. Davies, N. Kaizuma and N. C. Nielsen. 1983. Genetic analysis of null-allele for lipoxygenase-3 in soybean seeds. Crop Sci. 23:924-927.

Kleiman, R. and J. F. Cavins. 1982. Soybean germplasm evaluation-search for low linolenic lines. J. Am. Oil Chem. Soc. 59:305A.

Kwon, S. H. and J. H. Torrie. 1964. Heritability of and interrelationships among traits of two soybean populations. Crop Sci. 4:196-198.

Martin, B. A., B. F. Carver, J. W. Burton and R. F. Wilson. 1983. Inheritance of fatty acid composition in soybean seed oil. Soybean Genet. Newsletter 10:89-92.

Shorter, R., D. E. Byth and V. E. Mungomery. 1976. Estimates of selection parameters associated with protein and oil content of soybean seeds Glycine max (L.) Merr. Aust. J. Agric. Res. 28: 211-222.

Singh, D. B. and H. H. Hadley. 1968. Maternal control of soil synthesis in soybeans, Glycine max (L.) Merr. Crop Sci. 8:622-625.

Smouse, T. H. 1979. A review of soybean oil reversion flavor. J. Am. Oil Chem. Soc. 56:747A-751A.

United States Department of Agriculture. 1982. Agricultural Statistics. U.S. Govt. Printing Office, Washington, D.C.

United States Regional Soybean Laboratory. 1966. Evaluation of maturity groups I and II of the U.S.D.A. soybean collection. USDA, RSLM 230.

United States Regional Soybean Laboratory. 1969. Evaluation of maturity Groups III and IV of the U.S.D.A. soybean collection. USDA, RSLM 238.

Weber, C. R. 1966. Registration of Amsoy soybeans, (Reg. No. 57). Crop Sci. 6:611-612.

Weber, C.R. and B. R. Moorthy. 1952. Heritable and nonheritable relationships and variability of oil content and agronomic characteristics in the F_2 generation of soybean crosses. Agron. J. 44: 202-209.

White, H. B. Jr., F. W. Quackenbush and A. H. Probst. 1961. Occurrence and inheritance of linolenic and linoleic acid in soybean seeds. J. Am. Oil Chem. Soc. 38:113-117.

Wilcox, J. R., J. F. Cavins and N. C. Nielsen. 1984. Genetic alteration of soybean oil composition by a chemical mutagen. J. Am. Oil Chem. Soc. 61:97-100.

Williams, J. H. and D. G. Hanway. 1961. Genetic variation in oil and protein content of soybeans induced by seed irradiation. Crop Sci. 1:34-36.

Wilson, R. F., J. W. Burton, and C. A. Brim. 1981. Progress in the selection for altered fatty acid composition in soybeans. Crop Sci. 21:788-791.

BREEDING FOR RESISTANCE TO ROOT-KNOT NEMATODES

Kuell Hinson

CAUSAL ORGANISMS AND DISTRIBUTION

Four species of nematodes, *Meloidogyne incognita* (Kofoid and White) Chitwood; *M. arenaria* (Neal) Chitwood; *M. javanica* (Treub) Chitwood; and *M. hapla* Chitwood, cause root-knot of soybean. *M. hapla* thrives best in temperate regions centering near 40° latitude. The other three species are found in many other parts of the world between about 35°N and 35°S latitude. *M. javanica* thrives best in tropical regions. Four races of *M. incognita*, two of *M. arenaria*, and one each of *M. javanica* and *M. hapla* are recognized by Taylor and Sasser (1978).

Caviness and Riggs (1976) emphasized potential problems on soybean from variants within species of root-knot nematodes, similar to those found for races of soybean cyst nematode (*Heterodera glycines*, Ichinohe). Few serious problems have been recognized within the USA to this time. Boquet et al. (1975) found that five *M. incognita* populations indigenous to Louisiana produced a highly significant cultivar x population interaction when tested on 18 diverse soybean cultivars. One of the populations was the "Wartelle race", which is noted for its atypical effect on soybean plants. Some authorities consider it a subspecies. Differences in the response of soybean genotypes to two field populations of *M. arenaria* have been detected in South Carolina (H. L. Musen, pers. commun.). Other intraspecific variants of root-knot nematodes on soybean within the USA apparently are of minor importance. Data from Hussey and Boerma (1981) appear to describe the more general situation. They tested *M. incognita* collections from Georgia, South Carolina, and Florida on six soybean genotypes, noted apparent differences in aggressiveness, but obtained no significant soybean genotype x collection interaction for several traits studied.

A major shift in species recently occurred on many farms in the Gulf Coast production area of southeastern USA. To avoid damage from newly identified infestations of soybean cyst nematode Race 3, susceptible soybean cv. Bragg was replaced by resistant cv. Centennial. Within three years, damage caused by *M. arenaria* on many farms changed from unrecognized to severe. This illustrates the extent to which shifts in nematode populations can take place, but no such shifts have been reported for variants within root-knot nematode species.

Race differences from one continent to another may be more important. However, Brazilian soybean breeders have used U.S. genotypes resistant to *M. incognita* as sources of resistance there. Further, some genotypes classed as resistant to *M. javanica* in the U.S. have been classified similarly in Brazil.

GENETICS OF RESISTANCE

After more than 30 years of breeding for resistance to *M. incognita*, accurate information on the inheritance of resistance is lacking. This has not prevented the development of cultivars with satisfactory resistance to three root-knot nematode species, but better information on inheritance will result in improved breeding efficiency.

Resistance to *M. incognita* has received the most attention. In crosses between resistant and susceptible parents, resistance levels approximately equal to those of resistant parents are recovered in moderate size populations. This indicates that few major genes control resistance. The absence of discrete classes suggests that parents also differ in other genes having smaller effects. Individual plant scores for root galling have not been precise enough to distinguish small genetic differences. Data from replicated tests conducted over many years in Florida indicate that homozygous, distantly related cultivars form a continuum with respect to resistance to *M. incognita*. Few genes could still account for nearly all variation. Susceptible genotypes are sometimes found in progenies from crosses between two genotypes resistant to *M. incognita*. More than one major mechanism for resistance could be present. Combining mechanisms may result in still higher resistance levels.

These observations suggest that different parental combinations of resistant x susceptible genotypes can produce different frequencies of resistant and susceptible progenies. Boquet et al. (1976) arrived at the guarded conclusion that resistance to the Wartelle race of *M. incognita* was governed by one major gene and a second gene with a smaller effect. No other published information on the inheritance of resistance was found. After more than 30 years of breeding for resistance to *M. incognita*, E. E. Hartwig (pers. commun.) considers that two or three major genes usually must be assumed to account for the variation between parents resistant and susceptible to *M. incognita*. Less information is available on the inheritance of resistance to *M. arenaria*, *M. javanica*, and *M. hapla*. Whether one or more genes for resistance contribute protection against more than one species is not known.

MECHANISMS FOR RESISTANCE

In summarizing influences of resistance across crop species, Taylor and Sasser (1978) stated that larvae which invade roots of resistant plants may: 1) develop to maturity as females, but produce defective or no eggs, 2) develop to maturity as males, 3) have development arrested before completing the second, third or fourth molt, 4) be killed by an immune reaction, or 5) leave the root, and possibly enter another root. None of these reactions is entirely positive or negative. Individual nematodes may reproduce at low levels. Partial development may leave visible galls. The immune reaction may leave distorted or broken tissue in the root.

Immunity to root-knot nematodes is rare or absent in soybean. Resistance appears to depend on factors that reduce, rather than prevent, invasion and damage. Hague (1980) suggested breeding cultivars with less branched root systems, to reduce the number of infection sites, provided this could be done without adverse effects on nodulation or increased sensitivity to water stress.

deGuiran and Villemin (1980) reported that the arrest of egg development, considered diapause, occurred in all strains of the four *Meloidogyne* spp. important on soybean, when tested on tomato (species not identified). More limited tests on soybean produced similar results. Mean percentage diapause within egg masses depended on the host plant and the strain of the parasite.

Lim and Castillo (1978) found that *M. incognita* completed its life cycle in 43 to 47 days on susceptible cv. Clark 63, but the life cycle had not been completed after 53 days on resistant L-114. T. D. Wyllie (pers. commun.) observed similar relationships between length of life cycle and resistance about 20 years ago. Although Lim and Castillo found that nematode reproduction was four times greater on susceptible Clark 63, other evidence indicates that females that reach maturity on resistant plants generally are more fecund because of reduced competition. A longer nematode life cycle, and possibly lower reproduction, could result in lower mid-season inoculum densities in field plantings. Unanswered questions are: How is this mechanism inherited? Is the contribution to resistance major or minor?

Kaplan et al. (1980) reported that invasion of Centennial soybean roots by *M. incognita* larva caused increased localized production of the phytoalexin, glyceollin, within 48 h. Glyceollin greatly reduces the mobility of *M. incognita* larvae in the root and in-vitro. Invasion of Centennial roots by *M. javanica* larvae did not cause increased glyceollin. Centennial is resistant to *M. incognita* and susceptible to *M. javanica*. Pickett 71 is susceptible to both nematode species and did not produce glyceollin in response to invasion by either species.

It is likely that other important mechanisms have escaped detection. The importance or inheritance of those listed is not known; therefore, their relationships to the generally accepted hypothesis for inheritance of resistance to *M. incognita* (few genes with large effects, plus other genes with smaller effects) is unknown. The concept of identifying individual mechanisms, determining their importance and their inheritance, and assembling mechanisms in advantageous combinations seems important enough to pursue.

Even though major resistance mechanisms probably remain undetected, the presence of different potential mechanisms suggests that resistance may be achieved in different ways and that optimum combinations of mechanisms may further increase resistance levels. Also, the presence of different resistance mechanisms provides a rationale for unexpected results, such as the presence of susceptible progeny when both parents are resistant.

BREEDING TECHNIQUES

Sources of Resistance

Root-knot resistant cultivars must possess many other important characteristics; therefore the choice of parents and breeding procedures become important considerations. Most rapid progress is achieved when nematode-resistant parents have many other desirable characteristics. For long-term breeding work, however, several unrelated or distantly related sources of resistance should be used to accumulate factors for resistance and to meet unforeseen problems by means of genetic diversity. When resistant parents are deficient in many desirable characteristics, several cycles of crossing and selection usually are required to produce an acceptable cultivar.

Much of the primary germplasm for resistance to *M. incognita* in cultivars adapted to southern U.S. is considered to come from cvs. Palmetto and Laredo. Other parents with intermediate levels of resistance also occur in pedigrees. Palmetto was introduced from Nanking, China in 1927, and Laredo was introduced from Yangdingkiwan, China in 1914. Palmetto has better agronomic qualities; therefore was used more extensively. Diverse sources of resistance were important in the development of agronomically desirable genotypes resistant to the Wartelle race of *M. incognita*. The major gene for resistance evidently came from Laredo; the minor gene from cv. Hill (Boquet et al., 1976). Hill is not related to either Palmetto or Laredo.

Although moderately susceptible to Wartelle race, Bragg has demonstrated a high level of resistance to other populations of *M. incognita* in the U.S. and in other countries. It apparently obtained its resistance from Palmetto, through cv. Jackson (Figure 1A). Resistance has not diminished after nearly 20 years of extensive production on *M. incognita*-infested soils in southern U.S.

Cobb is also thought to derive resistance to *M. incognita* from Palmetto (Figure 1B). I obtained progeny susceptible to *M. incognita* from the cross Bragg x Cobb. E. E. Hartwig (pers. commun.) has obtained similar results from other crosses of resistant x resistant. Two different combinations of genes evidently produce adequate levels of resistance. Whether two mechanisms are involved is not known. Mode of gene action or exact source of all contributing genes is not known. Cultivars several breeding cycles removed from primary sources of resistance may have obtained contributing genes from parents not noted for high levels of resistance.

Forrest (Figure 1C) is among the U.S. cultivars most resistant to both *M. arenaria* and *M. incognita*. It is a progenitor of Kirby (Figure 1D), also resistant to both species. My data strongly suggest that Kirby is more resistant than Forrest to *M. arenaria*. The increased resistance, if present, is from Centennial or Cobb, both highly susceptible, or D68-216, an untested line closely related to Forrest.

Although Kirby equals or exceeds other U.S. cultivars in resistance to *M. arenaria*, it is better classed as moderately resistant. Still higher levels of resistance were indicated in my 1983 data. If these results are confirmed, contributing genes are from less resistant genotypes. Other U.S. cultivars with *M. arenaria* resistance nearly equal to that of Kirby include Braxton, Govan, and Gordon.

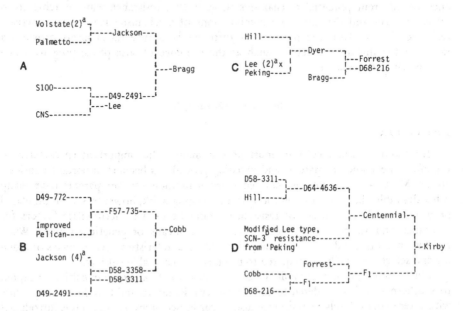

Figure 1. Pedigree of cultivars resistant to one or more species of *Meloidogyne*.
[a](2) and (4) indicate one and three backcrosses, respectively, to Volstate, Jackson, or Lee.
[b]SCN-3 is soybean cyst nematode, Race 3.

Kirby has demonstrated more resistance to *M. javanica* than any other cultivar tested in Florida. Some genes in Kirby may contribute resistance to more than one root-knot species. Other sources of resistance to *M. javanica* probably are available from Brazil, and possibly from other countries. Many U.S. cultivars in Maturity Groups V through VIII are resistant to *M. incognita*.

These examples identify the primary sources of resistance used successfully in the U.S. They do not identify adequately secondary sources of resistance, but rather suggest that some genes for resistance may have come from unexpected sources. Breeding procedures employed can be inferred from pedigrees. The modified backcross method was used in the development of Kirby. In modified backcrosses a recurrent phenotype is substituted for the recurrent genotype. In this example, the recurrent phenotype (resistance to both *M. incognita* and soybean cyst nematode Race 3) was present in both Forrest and Centennial. Soybean breeders in Brazil have used similar procedures and much of the same basic germplasm for resistance. Other breeding procedures; e.g., recurrent selection, can be used effectively also. For most rapid progress in new programs, parents with many desirable traits having confirmed resistance to the organism(s) in the area are recommended. For the long term, diverse sources of resistance are recommended.

Confirmed resistance to the organism(s) in the area is very important. In Florida, much published information on resistance of cultivars to specific root-knot nematodes is unreliable. This could be caused by variation within nematode species among test sites, or it could be associated with variation in inoculum densities within or among test sites. Soybean genotypes rated resistant at low inoculum densities, when compared to a more susceptible check, may be rated susceptible when inoculum densities are higher. Inoculum densities in our field plots produce large galls on susceptible genotypes. Soybean genotypes rated resistant to a particular nematode species under these conditions is seldom classed differently in other tests.

Evaluation Techniques

Methods used frequently for rating the reaction of soybean genotypes to root-knot nematodes are counts or indices of eggs, egg masses, or larvae; gall indices; plant vigor; and yield tests in infested fields. Indices usually discriminate five or six phenotypic classes, with zero or one representing the most resistant class. Combinations of criteria are often used. Some criteria are better suited to glasshouse work; others to field plots. Glasshouse-grown plants usually are examined 40 to 60 days from seeding. Field plot data, other than yield, usually are best when plants are approaching maturity. In all cases, however, data should be collected when maximum differences are evident. With very high inoculum density, delayed data collection often causes resistant or moderately resistant genotypes to appear susceptible. All properly used procedures produce acceptable results. The best procedure for a particular researcher may depend more on available facilities than on reliability of the method.

Hussey and Boerma (1981) outlined refined procedures for glasshouse evaluations. Their procedures and facilities are now being used for inheritance studies. Extensions of their techniques may result in the identification of other resistance mechanisms. They listed three advantages over field screening: continuous screening throughout the year, the specific organism need not be indigenous to the area, and uniform inoculum density. Refined glasshouse procedures provide for sterilized soil, surface sterilized eggs for inoculum, and easy standardization of inoculum. Their glasshouse results correlated well with gall scores of the same genotypes in west Florida field plots.

Field plots require less elaborate facilities, but establishment and maintenance of near uniform inoculum density is sometimes difficult. In Florida, I maintain separate nurseries for each of three species and usually plant one-half of each nursery to a susceptible cultivar in alternate years. Confining the test area to the smallest practical space results in more uniform inoculum densities. About 10 years ago, I reduced plot size from single rows 3 m long to single hills of six to eight plants spaced 45 cm apart within rows. Distance between rows remained 0.9 m. Adjusting scores of test lines to scores of a susceptible check, planted every third hill, increased discrimination under some conditions (Dall'Agnol et al., 1978). To be practical, however, precision must be increased by 50% to compensate for increased plot numbers. The method was practical only when islands of low and high inoculum density, determined from check plot scores, occurred within the test area. Plants of a morphologically distinct, highly susceptible genotype within plots can indicate relative inoculum density. Cultivar Pine Dell Perfection has served well in *M. incognita* nurseries.

Plant vigor scores in field plots can be used effectively to supplement data on gall scores or as a rapid screening technique. Vigor scores on nearly mature plants are more reliable than earlier scores (Williams et al., 1973). Late-season gall scores probably would favor identification of genotypes causing the nematode to require a longer life cycle.

Yield tests in uniformly infested fields have the advantage of integrating all factors associated with yield loss. When combined with gall scores, yield data can indicate tolerant genotypes. Williams et al. (1973) noted that two lines with intermediate gall scores produced higher seed yields than lines with lower gall scores. They stated that greenhouse screening would not have indicated this degree of tolerance.

Kinloch et al. (1985) found good agreement between gall scores and seed yield in *M. incognita* infested soil. Correlation coefficients were -0.82**, -0.82**, and -0.86** in three consecutive years. Their indices differ slightly from many others frequently used, and apparently describe effects on yield better (Table 1).

Similar data for *M. arenaria* and *M. javanica* are being tabulated (R. A. Kinloch and C. K. Hiebsch, pers. commun.). Kirby has yielded more than any other cultivar in nurseries infested with *M. arenaria* and *M. javanica*, and has equalled the yield of several other resistant cultivars in nurseries infested with *M. incognita*. Its yield is similar to yields of the better susceptible cultivars in fields that have no nematode problems.

These yield data illustrate the amount of yield loss that can be expected from growing susceptible cultivars in heavily infested soil, demonstrate that development of cultivars resistant to multiple species is practical, and suggest that programs to

Table 1. Gall indices and average effects on seed yield over three years (Kinloch et al., 1985).

Index	Root surface galled	Gall scores vs. seed yield	
	%	score	kg/ha
0	0		
0.2	0-5		
1	6-25	0-0.9	2483
2	26-50	1-1.9	2192
3	51-75	2-2.9	1226
4	75-100	3-4.0	539

breed for resistance should be started at the first indication of a potential root-knot problem.

NOTES

Kuell Hinson, Research Agronomist (Soybeans), U.S. Department of Agriculture, ARS, Gainesville, Florida 32611.

REFERENCES

Boquet, D. J., C. Williams and W. Birchfield. 1975. Resistance to five Louisiana populations of root-knot nematode. Plant Dis. Rep. 59:197-200.

Boquet, D., C. Williams and W. Birchfield. 1976. Inheritance of resistance to the Wartelle race of root-knot nematode in soybeans. Crop Sci. 16:783-785.

Caviness, C. E. and R. D. Riggs. 1976. Breeding for nematode resistance. p. 594-601. *In* L. D. Hill (ed.) World Soybean Research. The Interstate Printers and Publishers Inc., Danville, Illinois.

Dall'Agnol, A., K. Hinson and J. T. Johnson. 1978. Selecting soybeans for resistance to *Meloidogyne javanica.* Soil and Crop Sci. Soc. of Fla. Proc. 37:101-104.

deGuiran, G. and M. A. Villemin. 1980. Specificity of the embryonic egg diapause in *Meloidogyne* (Nematoda). Revue de Nematologie. 3:115-121.

Hague, N. G. M. 1980. Nematodes of legume crops. p. 199-205. *In* R. J. Summerfield and A. H. Bunting (eds.) Advances in Legume Science. Royal Botanic Gardens, Kew.

Hussey, R. S. and H. R. Boerma. 1981. A greenhouse screening procedure for root-knot nematode resistance in soybeans. Crop Sci. 21:794-796.

Kaplan, D. T., N. T. Keen and I. J. Thomason. 1980. Association of glyceollin with the incompatible response of soybean roots to *Meloidogyne incognita.* Physiol. Plant Pathol. 16:309-318.

Kinloch, R. A., C. K. Hiebsch and H. A. Peacock. 1985. Comparative root-knot galling and yield responses of soybean cultivars to *Meloidogyne incognita.* Plant Dis. 69:(In press).

Lim, B. K. and M. B. Castillo. 1978. Interactions of *Meloidogyne incognita* and *Rotylenchulus reniformis* with selected soybean varieties. Philippine J. Biol. 7:165-176.

Taylor, A. L. and J. N. Sasser. 1978. Biology, identification and control of root-knot nematodes (*Meloidogyne* species). North Carolina University Graphics, Raleigh, N.C.

Williams, C., W. Birchfield and E. E. Hartwig. 1973. Resistance in soybeans to a new race of root-knot nematode. Crop Sci. 13:299-301.

BREEDING PRODUCTIVE SOYBEANS WITH RESISTANCE
TO THE SOYBEAN CYST NEMATODE

E. E. Hartwig

SOURCES AND GENETICS OF RESISTANCE

A search for resistance to the soybean cyst nematode (SCN), caused by *Heterodera glycines* Ichinohe, in soybean was initiated soon after the nematode was identified in the U.S. A field screening of over 4,000 soybean germplasm lines in an SCN-infested field in North Carolina in 1957 revealed apparent resistance within about 0.3% of the lines. These lines were retested in the field in North Carolina during the same year, and in the greenhouse at Jackson, Tennessee. All lines identified as resistant were black-seeded and originated from northeast China. The cultivar Peking, introduced in 1906, was selected for use as a parent in a breeding program begun at Stoneville, Mississippi in 1957.

Caldwell et al. (1960) reported resistance to the soybean cyst nematode to be recessive and controlled by three major genes. From later studies, Matson and Williams (1965) concluded that resistance was determined by three recessive genes plus one dominant gene. The dominant gene was closely linked with the gene giving solid seed coat color. The difference between a ratio of 63:1 for three genes and a ratio of 253:3 for one dominant and three recessive genes is small and difficult to measure. Two of the three resistant plants would be expected to be heterozygous for reaction to SCN. This, along with linkage with a seed coat color gene, delayed recovery of yellow-seeded, homozygous-resistant lines. Once yellow-seeded, SCN-resistant lines were isolated, breeding programs could proceed with greater efficiency.

Foundation seed of cv. Pickett developed by Brim and Ross (1966) was distributed for increase in North Carolina, Tennessee, Missouri, and Arkansas in 1966. A back-crossing program with cv. Lee as the recurrent parent was used in the development of Pickett. A winter nursery in Puerto Rico, utilizing artificial light to delay flowering, was used to gain a year in making foundation seed available. Pickett averaged approximately 10% less in seed yield than Lee when grown in the absence of cyst nematodes, but was effective in increasing seed yields where SCN was present. Also, Pickett was more susceptible to phytophthora rot, caused by *Phytophthora megasperma* Drechs f. sp. *glycinea* Kuan and Erwin, than was Lee. Although phytophthora rot was more likely to cause problems in clay soils, where SCN was less likely to be a problem, both clay and sandy loam soils are present in some fields. Consequently, both problems could occur. This emphasized the importance of giving consideration to multiple pest problems in breeding programs. Pickett is similar to Lee in resistance to several of the foliar diseases.

The cultivar Dyer, developed by Hartwig and Epps (1968), was released in 1967. It matured nearly three weeks earlier than Pickett. Dyer has the same early generation parentage as Pickett, but Hill was the parent for the final crosses. Dyer was lower yielding in the absence of SCN, as was Pickett. These early releases demonstrated that with SCN resistant cultivars, soybean plants could be grown satisfactorily on SCN infested soils. As seed of the resistant cultivars became available for more widespread plantings, SCN injury was observed in localized areas within fields indicating variability with the SCN population. This led to the evaluation of the nematode populations and to the designation of four races by Golden et al. (1970). The original population found in North Carolina was designated Race 1, a population in southeastern Virginia was designated Race 2, the original population in western Tennessee, southeastern Missouri, and northeastern Arkansas was designated Race 3 and the secondary population in the SCN Race-3 area was designated Race 4. Peking was described as resistant to SCN Races 1 and 3.

Further screening of the soybean germplasm collection identified PI 88788 and PI 89772 as having high levels of resistance to SCN Race 4. Genetic studies conducted by Thomas et al. (1975), in which an F_2 population of Peking X PI 88788 was evaluated when infested with SCN Race 4, showed that Peking and PI 88788 differed by a single gene for reaction to SCN Race 4.

Second cycle cultivars have been developed which are resistant to SCN Race 3. These cultivars are highly productive in their area of adaptation and are resistant to several other major pests. Forrest, released in 1972 as SCN-Race-3 resistant (Hartwig and Epps, 1973), is also resistant to the foliar disease, bacterial pustule, caused by *Xanthomonas phaseoli* (E. F. Sm.) Dows. var. *sojensis* (Hedges) Starr & Burkh., two species of root-knot nematode, *Meloidogyne incognita* (Kofoid & White) Chitwood and *M. arenaria*, and the reniform nematode, *Rotylenchulus reniformis*. It has a twelve-year average seed yield of 3500 kg/ha on sandy loam soil at Stoneville, Miss. It is one of the highest yielding cultivars for this soil. Nematodes are not recognized as a problem in this soil.

Bedford, developed by Hartwig and Epps (1978), is an SCN-Race-3 and -4-resistant cultivar released in 1977. It has produced nearly as well as Forrest. Forrest, and a closely related line, were used as parents in the development of Bedford. PI 88788 was the source of SCN Race 4 resistance.

SCREENING FOR RESISTANCE

Techniques for screening are critical for conducting an efficient breeding program. The initial screening of germplasm to identify sources of resistance to SCN was made in a field in southeastern North Carolina that had a high population of cyst nematodes. Germplasm lines were planted in 3-m rows. A glabrous line, susceptible to the nematode, was grown 10 to 15 cm to the side of the germplasm lines. Since all germplasm lines were pubescent, they could readily be distinguished. Plants were dug approximately 30 days after emergence and roots inspected. If roots of a germplasm line were free of cysts, plants of the adjacent glabrous line were dug to verify the presence of nematodes. By this technique, researchers were confident that lines free of nematodes were truly resistant rather than escapes.

Greenhouse plantings have been used for most screenings of breeding material. Field soil from an infested field is used to fill individual pots. For preliminary screening of germplasm lines or breeding lines, 7 to 10 pots of each line are used with one plant per pot. Plants of a susceptible cultivar are inspected 26 to 32 days after planting. When these check plants show good development of white cysts, plants to be evaluated are removed from the pots and rated on a 0 to 4 basis. Plants which have 0, 1 to 5, 6 to 10, 11 to 20, and 21 to 40 cysts are scored 0, 1, 2, 3, or 4, respectively. Plants more heavily infested are frequently given a 4 plus. Within F_2 populations or breeding lines, plants having a score of 2 or less may be transplanted and grown to maturity. Progeny are evaluated in the field. With some SCN populations, where susceptible plants receive a score of 4 plus, plants with a score of 3 are saved for further evaluation of progeny. Progeny of F_2, F_3, or F_4 plants are evaluated in the greenhouse. Remnant seed is available for making field plantings of resistant or segregating lines.

Golden et al. (1970), in their discussion of SCN races, described resistance as plants or genotypes having no more than 10% of the number of cysts of the susceptible check. Cultivars, such as Forrest, which are resistant to SCN Race 3 have a near-immune reaction to this race. In a field comparison with Tracy (susceptible), Forrest had no cysts while a 250-cc sample of soil from Tracy plots averaged 410 cysts. Seed yield of Tracy was 61% of that for Forrest. However, in a study conducted by Epps et al. (1981) on SCN Race 4 infested soil, the three-year mean number of cysts per 250-cc of soil for Bedford and Tracy was 90 and 367, respectively. Bedford had a mean seed yield of 2960 kg/ha (33% greater than for Tracy). This difference was significant at the 1% level of confidence. Bedford gave no response to treatment by a nematicide while Tracy gave a statistically significant yield response from nematicide treatment. Although the cyst count for Bedford exceeds the 10% level described as necessary to qualify as resistant (Golden et al., 1970), it seems to provide an adequate level of protection. Similar responses have been observed in other fields. Cyst populations have remained at a very low level and seed productivity at a high level after eight years of continuous Bedford culture where initially there was a high population of SCN Race 4. In another three-year study conducted by Epps et al. (1981), cyst counts from Bedford and Forrest plots were 36 and 287, respectively, per 250-cc of soil. Although the cyst count for Bedford exceeded the 10% level of that for Forrest, seed yield of Bedford, non-treated, exceeded that for Forrest treated with each of five different nematicides.

From these studies, I have concluded that in breeding programs to develop productive germplasm resistant to SCN Race 4, I may use different criteria for selection after greenhouse screening, than when screening segregating populations for resistance to SCN Race 3. Field evaluation on SCN-infested soil is clearly needed to evaluate accurately breeding lines. I accept as resistant those lines that maintain cyst populations at low enough levels to avoid economic injury to the crop. This level will vary with different SCN populations.

Growing segregating populations in bulk on SCN Race 3 infested soil was an effective means of increasing the percentage of resistant plants in a population (Hartwig et al., 1982). An F_2 population of Forrest (resistant) X Tracy (susceptible) had only 62% as many cysts as Tracy per 250-cc of soil. Greenhouse screening of such an F_2 population would have shown that 98.8% of the plants were susceptible. It appears that heterozygous plants support fewer cysts to maturity than do homozygous, susceptible plants. The rate of shift to resistant plants in the population suggests that vigor and productivity of plants are related to the number of resistant genes carried by the plant. The difference in plant vigor did not appear to be adequate to permit visual selection.

BREEDING RESISTANT CULTIVARS

In the original screening of germplasm in a search for sources of resistance to SCN Race 4, Peking was classified as susceptible. Field observations indicate that Peking has a moderate level of resistance to SCN Race 4. A greenhouse study in which soybean strains were grown 60 days in pots having soil infested with SCN Race 4 verified the field observations. Pickett-71 and Forrest, resistant to SCN Race 3, produced as many cysts as did the susceptible Lee. Peking produced more cysts than PI 88788, but far less than Pickett-71 or Forrest.

Since Pickett-71 and Forrest derive their SCN Race 3 resistance from Peking, it is evident that Peking carries one or more genes influencing resistance to SCN that were not transferred in the original crosses. Preliminary studies suggest that Peking has an additional gene giving a moderate level of resistance to SCN Race 4. Later, an SCN isolate, similar to what was identified in Japan as Race 5, was found. Peking has a high level of resistance to SCN Race 5, whereas Forrest and Pickett-71 are susceptible.

A breeding program was initiated to transfer the additional gene from Peking to strains resistant to SCN Race 3, such as Forrest. The standard 30-day seedling screening program could not be used in screening F_2 plants for resistance to SCN Race 4, since Peking receives a susceptible rating. It was necessary to grow F_2 plants in infested soil long enough to permit two generations of this nematode to develop. Cyst counts between Peking and Forrest differed sufficiently to permit identifying Peking type plants within the F_2 population. After the isolation of the SCN Race 5 type, it was possible to screen 30-day seedlings and identify the Peking genotypes. Second backcross lines having the additional gene from Peking are now in the early stages of field evaluation.

A crossing program was also initiated to add the additional gene carried by Peking to SCN-Race-4 resistant strains, such as Bedford. SCN-Race-4-resistant lines were used as the recurrent parent, and progeny were screened for reaction to SCN Race 5. It was assumed that resistance to SCN Race 4 would be automatically retained. This did not prove to be the case. Nearly all lines resistant to Race 5 failed to have the Bedford level of resistance to SCN Race 4. The results could be expressed by multiple alleles at a single locus or linkage. We are inclined to believe that linkage is involved, since we have recovered a few lines which appear to be resistant to both Races 4 and 5. PI 88788 and PI 89772 were both classified as resistant to SCN Race 4 in the original screening. However, they differ in their reaction to SCN Race 5. PI 89772 is resistant and PI 88788 is susceptible.

PI 89772 was used as a parent with an SCN-Race-3-resistant line to develop more productive lines resistant to SCN Race 4. One of the lines selected as resistant to SCN Race 4 is J74-88. J74-88 has good resistance to SCN Race 5. This further indicates that linked genes control resistance to SCN Races 4 and 5.

D72-8927 was selected for resistance to SCN Race 2. PI 90763 was used as the source of resistance to SCN Race 2. The other parent was an SCN-Race-3-resistant line having Peking as the source of resistance. D72-8927 has a high level of resistance to Race 5 and a moderate level of resistance to Race 4. Approximately 25% of the plants in an F_2 population of Bedford X D72-8927 are resistant to Race 5. None has been homozygous resistant to SCN Race 4.

D79-5353 has D72-8927 as a parent, and appears to react similarly to D72-8927 when tested against either SCN Race 4 or 5. Progeny of 200 F_2 plants of Bedford X D79-5353 have been screened against the 5 races of SCN. All plants were resistant to SCN Race 3. Approximately 25% were resistant to either Race 1, Race 4, or Race 5, suggesting segregation at a single locus for reaction to each race. With independent distribution, three lines from a population of 200 would be expected to be resistant to the three races. Only five lines resistant to SCN Race 2 were found, suggesting that there must be recessive genes at three loci to give resistance. Combinations of resistance among F_3 lines are illustrated in Table 1. Homozygous resistant lines have been

398

Table 1. Reaction to soybean cyst nematode Races 1 to 5 of four selected F_3 lines from a population of Bedford X D79-5353.

Line No.	Race				
	1	2	3	4	5
	Classification[a]				
36	R	Seg	R	Seg	R
39	R	S	R	R	S
66	R	R	R	S	R
161	R	Seg	R	Seg	Seg

[a]R = resistant; Seg = segregating; S = susceptible.

isolated from lines that were originally classified as segregating. These lines have not been cross-checked to determine whether there are isolated lines having a high level of resistance to the five races. The results illustrate that one must work with large plant populations when attempting to develop soybean lines having a high level of resistance to the several races of SCN. Such a program becomes more practical after obtaining resistance to one or more races in adapted parents.

It appears that within the four soybean strains Peking, PI 88788, PI 89772, and PI 90763, genes for resistance to a wide range of SCN populations are present. Based upon the seedling test, some of these strains were classified as susceptible to individual SCN populations. However, when grown in the greenhouse for 60 days in soil infested with each of the five SCN races, each had developed a lower population of SCN than the susceptible cultivar Essex (Table 2). Each of these may have adequate field resistance to avoid or minimize economic injury. Bedford (Table 2) is susceptible to Race 5

Table 2. Number of cysts developing after 60 days on soybean strains grown in greenhouse pots in soil infested with five races of the soybean cyst nematode.

Soybean strain	Race				
	1	2	3	4	5
	— Number of cysts per 250-cc of soil —				
Essex	430	167	278	275	185
Peking	7	62	2	40	8
PI 88788	33	75	13	28	65
PI 89772	15	122	2	25	12
PI 90763	17	60	5	35	12
D72-8927[a]	13	130	7	68	17
J74-88[b]	33	42	7	48	15
D79-5353[c]	67	65	3	73	63
J81-116[d]	42	92	7	43	13
Forrest	93	205	3	272	98
Bedford	72	200	13	48	207

[a]Peking and PI 90763 are sources of resistance.
[b]Peking and PI 89772 are sources of resistance.
[c]Peking, PI 88788, and PI 90763 are sources of resistance.
[d]Peking, PI 88788, and PI 89772 are sources of resistance.

in greenhouse plantings. However, economic injury has not been recognized in fields where a population of this nematode is present. D72-8927, classified as susceptible to Race 4 in greenhouse tests, has been effective in reducing Race 4 to a low level in a five-year field study.

Breeding lines have been developed which have good multiple race resistance (Table 2). Lines such as J74-88 and J81-116 are superior in agronomic qualities to the original sources of resistance. Although inferior to Forrest in seed yield and seed holding, they can be used in field studies to evaluate SCN populations. They also provide superior lines that may be used as parents.

The soybean cyst nematode was first identified in the southern U.S. Consequently, initial emphasis was directed toward developing highly productive SCN-resistant culti-vars of Maturity Groups V and VI. Cultivars and improved germplasm having resistance to Race 3 from Peking are available in most Maturity Groups from I to VIII. Cultivars having resistance to Race 4 from PI 88788 are available for Maturity Groups III, V, and VI. Breeding techniques have included early-generation screening in the greenhouse with infested field soil, followed by evaluation in cyst infested fields.

NOTES

Edgar E. Hartwig, Research Agronomist, USDA, ARS, Soybean Production Research Unit, working in cooperation with the Delta Branch, Miss. Agric. and Forestry Expt. Stn., P.O. Box 196, Stoneville, MS 38776.

All field and greenhouse studies reported were conducted in cooperation with Lawrence Young, Research Plant Pathologist, USDA-ARS, who works in cooperation with the West Tennessee Experiment Station, Jackson, Tennessee.

REFERENCES

Brim, C. A. and J. P. Ross. 1966. Registration of Pickett soybeans. Crop Sci. 6:305.

Caldwell, B. E., C. A. Brim and J. P. Ross. 1960. Inheritance of resistance to the cyst nematode *Heterodera glycine*. Agron. J. 52:635-636.

Epps, J. M., L. D. Young and E. E. Hartwig. 1981. Evaluation of nematodes and resistant cultivars for control of soybean cyst nematode race 4. Plant Dis. 65:665-666.

Golden, A. M., J. M. Epps, R. D. Riggs, L. A. Duclos, J. A. Fox and R. L. Bernard. 1970. Terminology and identity of infraspecific forms of the soybean cyst nematode (*Heterodera glycine*). Plant Dis. Rep. 54:544-546.

Hartwig, E. E. and J. M. Epps. 1968. Dyer soybean. Crop Sci. 8:402.

Hartwig, E. E. and J. M. Epps. 1973. Registration of Forrest Soybeans. Crop Sci. 13:287.

Hartwig, E. E. and J. M. Epps. 1978. Registration of Bedford soybeans. Crop Sci. 18:915.

Hartwig, E. E., T. C. Kilen, L. D. Young and C. J. Edwards, Jr. 1982. Effect of natural selection in segregating soybean populations exposed to phytophthora rot or soybean cyst nematode. Crop Sci. 22:588-590.

Matson, A. L. and L. F. Williams. 1965. Evidence of four genes for resistance to the soybean cyst nematode. Crop Sci. 5:477.

Thomas, J. D., C. E. Caviness, R. D. Riggs and E. E. Hartwig. 1975. Inheritance of reaction to race 4 of the soybean cyst nematode. Crop Sci. 15:208-210.

RESISTANCE TO INSECT DEFOLIATORS

M. J. Sullivan

Major foliage feeders of soybean in the primary growing regions of North and South America include velvetbean caterpillar (*Anticarsia gemmatalis* Hubner) and soybean looper (*Pseudoplusia includens* Walker). Minor foliage feeders found in North America include Mexican bean beetle (*Epilachna varivestis* Mulsant), green cloverworm (*Platypena scabra* Fabricius), corn earworm (*Heliothis zea* Boddie) and cabbage looper (*Trichoplusia ni* Hubner). Others sometimes considered pests in both areas are various beetles including *Diabrotica* sp., *Cerotoma* sp., and *Colaspis* sp.

One strategy for controlling insect pests in soybean is the use of resistant cultivars; a desirable and ecologically sound control mechanism. At the first World Soybean Research Conference, two papers were presented involving plant resistance to soybean insect pests (Schillinger, 1976; Turnipseed and Sullivan, 1976). This paper will update research efforts conducted since 1976 concerning resistance to foliage feeders on soybean.

INITIAL RESISTANCE

The first report of resistance in soybean to insects was that of Hollowell and Johnson (1934), who indicated a correlation between pubescence type and potato leaf hopper injury. Coon (1946) reported differences in Japanese beetle (*Popellia japonica* Newman) attack on several soybean cultivars.

The 1960s provided the major breakthrough in insect resistance in soybean. Germplasm in world soybean collections from Maturity Groups VII and VIII were screened for Mexican bean beetle resistance (Van Duyn et al., 1971). Excellent resistance was found in three plant introductions, PI 171451; PI 227687; and PI 229358. In the early 1970s, Maturity Groups III-V were also screened for Mexican bean beetle resistance (Elden et al., 1974); they also listed three plant introductions with excellent resistance, PI 90481, PI 96089, and PI 157413.

Resistance sources discovered by Van Duyn were found to have resistance to other soybean insect pests: corn earworm, bean leaf beetle (*Cerotoma trifurcata* Forster) and striped blister beetle (*Epicouta vittata* Fabricius) (Clark et al., 1972). These sources of resistance have been utilized in breeding programs throughout the United States to develop advanced breeding lines with multiple insect resistance and good agronomic characteristics. Major efforts in developing insect resistant soybean cultivars have been concentrated in the United States with emphasis directed at foliage feeding insects.

400

VELVETBEAN CATERPILLAR

The most important defoliator in the western hemisphere is the velvetbean caterpillar (Turnipseed and Kogan, 1976). Advanced breeding lines, derived primarily from PI 229358, with resistance to Mexican bean beetle and other lepidopterous pests, have been screened in the U.S. and Brazil for velvetbean caterpillar resistance (Paschal and Minor, 1978; Turnipseed and Sullivan, 1976). Low levels of resistance were found in some lines. A germplasm line has been released, D75-10169, that has some degree of resistance to velvetbean caterpillar (Table 1).

The influence of resistant germplasm on the biology of velvetbean caterpillar also has been studied (Moscardi et al., 1981; Estes, 1983). Estes used advanced breeding lines, ED 73-371 and D75-10230, both derived from crosses of cv. Bragg X PI 229358. In single generation studies larvae exhibited greater mortality, longer developmental times, and reduced larval weights when compared to larvae reared on susceptible cultivars (Davis, Tracey, Bragg, and Braxton). D75-10230 had a greater effect than ED 73-371. Larvae reared through two consecutive generations on D75-10230 exhibited greater mortality, longer developmental time, and reduced fecundity. Larvae expressed less susceptibility to the effects of all genotypes in the second generation.

The three plant introductions initially identified for Mexican bean beetle resistance (PI 171451, PI 227687, PI 229358) are the best identified sources of resistance to velvetbean caterpillar. Germplasm Maturity Groups VII-IX have been screened for resistance to this pest in Florida. Eleven additional plant introductions have been found with resistance to velvetbean caterpillar; none are superior to the original three (Kuell Hinson, pers. commun.).

SOYBEAN LOOPER–CABBAGE LOOPER

After resistance was reported to Mexican bean beetle, PIs 171451, 227687, and 229358 were screened and found to have resistance to both soybean and cabbage

Table 1. Soybean germplasm releases with resistance to defoliators.

Germplasm release	Parentage	Insect resistance
L76-0038[a]	Williams X PI 171451	Mexican bean beetle
L76-0049[a]	Williams X PI 171451	Mexican bean beetle
L76-0132[a]	Beeson X PI 171451	Mexican bean beetle
L76-0272[a]	Williams X PI 229358	Mexican bean beetle
L76-0328[a]	Williams X PI 229358	Mexican bean beetle
N80-50232[b]	Forrest X [D68-216 X (Bragg X PI 229358)]	Corn earworm
		Mexican bean beetle
N80-53201[b]	Forrest X [D68-216 X (Bragg X PI 229358)]	Corn earworm
		Mexican bean beetle
N79-2282[b]	Forrest X [Govan X (Bragg X PI 229358)]	Corn earworm
		Mexican bean beetle
D75-10169[c]	Govan X (Bragg X PI 229358)	Soybean looper
		Velvetbean caterpillar
		Corn earworm
		Mexican bean beetle

[a]Joint release: Illinois, Maryland, and Purdue Agric. Exp. Stn.; and U.S.D.A., Sci. Educ. Admin., Agric. Res.

[b]Joint release: North Carolina Agric. Res. Serv. and U.S.D.A., Agric. Res. Serv.

[c]Joint release: U.S.D.A., Agric. Res. Serv., Mississippi Agric. and Forrest. Exp. Stn. and South Carolina Agric. Exp. Stn.

loopers (Kilen et al., 1977; Luedders and Dickerson, 1977). As with velvetbean caterpillar, advanced breeding lines have been developed with good levels of resistance to
soybean looper. The soybean looper is the more prevalent of the two species occurring
in soybean (Turnipseed and Kogan, 1976). For this reason, more emphasis has been
placed on developing resistance to this species.

Hatchett et al. (1979) evaluated five breeding lines derived from initial crosses of
Bragg X PI 229358 and determined that all the lines were as resistant as the parent,
PI 229358. In single-generation studies with two advanced breeding lines, Estes (1983)
found that mortality was significantly reduced in those larvae reared on resistant genotypes ED 73-371 and D75-10230. These lines were derived from the original cross of
PI 229358 X Bragg. Only larvae reared on D75-10230 weighed less and had longer
developmental times. Susceptible cultivars were Davis, Tracy, Bragg, and Braxton. In
two-generation studies, D75-10230 increased mortality, reduced larval weights, and
caused longer development times. In the second generation, only mortality and larval
weights were effected. ED 73-371 effected larval mortality in the first generation only.
Larvae reared on leaves of Davis, D75-10230, and ED 73-371 consumed less foliage
compared with those reared on the resistant genotype, and mortality was greater on
D75-10230 (Grant, pers. commun.). In a recent study, Yanes and Boethel (1983) re-
confirmed resistance in PI 227687 and found that both larval weights and leaf consumption were significantly reduced in loopers fed PI 227687 foliage compared with
Davis.

Breeding lines have been developed with good levels of resistance to soybean looper. One soybean germplasm line, D75-10169, has been released with soybean looper
resistance (Table 1).

At the present time, the Georgia program is screening 140 accessions from Maturity
Groups VII and VIII. These had previously shown levels of resistance in screening trials
conducted for velvetbean caterpillar in Florida. Soybean looper populations were not
sufficient for reliable resistance ratings in 1983; however, differential damage was seen
with velvetbean caterpillar (Todd, pers. commun.).

MEXICAN BEAN BEETLE

As previously mentioned, the best sources of resistance to this pest are PIs 171451,
227687, and 229358 from Maturity Groups VII and VIII (Van Duyn et al., 1971) and
PIs 90481, 96081, and 157413 from Maturity Groups III-V (Elden et al., 1974). These
sources have resistance to adults and larvae; both attack soybean foliage.

Smith and Brim (1979a) reported on resistance in backcross populations derived
from Bragg X PI 227687. Evaluations of progeny indicated 53% were similar in resistance as the resistant parent, PI 227687. Resistance to corn earworm revealed that only
7% had dual leaf feeding resistance. Leaf extracts of PI 229358 indicated high levels of
resistance when compared with Forrest (Smith et al., 1979).

After initial screening provided plant introductions with high levels of resistance,
several breeding programs proceeded to incorporate this resistance into acceptable
agronomic types. The results of these programs have provided several germplasm releases which have resistance to the Mexican bean beetle (Table 1).

CORN EARWORM–TOBACCO BUDWORM

Although the corn earworm poses a more serious threat as a pod feeder, it does
assume the role of a foliage feeder in many cases. Again, PIs 171451, 227687, and

229358 were tested and found also to be resistant to this pest (Clark et al., 1972; Hatchett et al., 1976). Leaf feeding resistance was also found for tobacco budworm, *Heliothis virescens* (Fabricius) (Hatchett et al., 1976).

Five breeding lines derived from crosses utilizing PI 229358 were evaluated for multiple insect resistance, including corn earworm and tobacco budworm. All were as resistant as PI 229358. Three were resistant to tobacco budworm. Progeny of three advanced backcross populations, with PI 277687 used as the resistance source, were tested for foliage feeding resistance (Smith and Brim, 1979b). Of progeny tested, 73% retained foliage resistance; the authors stated that evaluation of corn earworm leaf feeding resistance by visual estimate is a useful method of eliminating large amounts of susceptible germplasm. Further evidence of resistance in PIs 227687 and 229358 was demonstrated by grafting, using Davis as the susceptible cultivar (Lambert and Kilen, 1984). A resistance factor, which is translocated throughout the leaves, was indicated for 227657 and 229358.

As previously mentioned, scientists in Florida have screened Maturity Groups VII-IX for better and/or additional sources of resistance to velvetbean caterpillar. The Georgia program has 140 accessions from Florida to screen for soybean looper. At present, South Carolina scientists are screening Maturity Groups VII and VIII for additional and/or better sources of corn earworm resistance (Shipe, pers. commun.). This

Table 2. Current soybean insect resistance programs in the United States.

State	Insect	Source of resistance
Florida	Velvetbean caterpillar	PI 227687
		PI 229358
Georgia	Soybean looper	PI 171451
		PI 229358
Illinois	Mexican bean beetle	PI 171451
		PI 229358
Indiana	Mexican bean beetle	PI 171451
		PI 229358
Maryland[a]	Mexican bean beetle	PI 171451
		PI 229358
Maryland[b]	Corn earworm	PI 171451
		PI 229358
Mississippi	Soybean looper	PI 171451
	Corn earworm	PI 227687
	Tobacco budworm	PI 229358
	Velvetbean caterpillar	
	Beet armyworm	
North Carolina	Mexican bean beetle	PI 229358
	Corn earworm	
Ohio	Mexican bean beetle	PI 171451
		PI 229358
South Carolina	Corn earworm	PI 171451
		PI 227687
		PI 229358

[a]USDA and University of Maryland.
[b]University of Maryland, Eastern Shore.

will complement the efforts in Florida and Georgia. Germplasm releases with resistance to foliage feeding corn earworm are shown in Table 1.

PRESENT RESEARCH AND FUTURE PLANS

Various breeding programs have developed advanced lines with good resistance to major soybean defoliators. Although agronomic characteristics of these lines are generally good, all have fallen short in the most important facet, yield. Lines will not be released until they yield equal to or better than existing cultivars. Advanced lines that have yielded well in several different geographic locations are in yield trials at present. A commercial cultivar with resistance to foliage feeding insects may come from these.

Table 2 presents a list of public scientists involved in developing insect-resistant soybean cultivars. Additionally, several private companies are now including insect resistance as an objective in their breeding programs. The resistance sources used are PIs 171451, 227687, 229358, or advanced breeding lines developed in other programs that have these lines as the resistance source. Australia has begun a breeding program on the soybean (Rogers, pers. commun.). Sources of resistance include the three PIs previously mentioned.

The use of resistant cultivars offers an ideal insect control method that is compatible with other methods of insect control. Recent advances have been made in developing soybean genotypes that have multiple pest resistance; this includes insects, nematodes, and diseases. In the very near future there should occur cultivar releases with multiple pest resistance.

NOTES

M. J. Sullivan, Department of Entomology, Fisheries, and Wildlife, Clemson University, Edisto Experiment Station, Blackville, South Carolina.

REFERENCES

Clark, W. J., F. A. Harris, F. G. Maxwell and E. E. Hartwig. 1972. Resistance of certain soybean cultivars to bean leaf beetle, striped blister beetle, and bollworm. J. Econ. Entomol. 65:1669-1672.

Coon, B. F. 1946. Resistance of soybean varieties to Japanese beetle attack. J. Econ. Entomol. 39: 510-513.

Elden, T. C., J. A. Schillinger and A. L. Steinhauer. 1974. Field and laboratory selection for resistance in soybeans to the Mexican bean beetle. Environ. Entomol. 3:785-788.

Estes, A. L. 1983. Influence of susceptible and resistant soybean genotypes upon the biology of *Anticarsia gemmatalis* and *Pseudoplusia includens*. M.S. Thesis. Clemson University.

Hatchett, J. H., G. L. Beland and E. E. Hartwig. 1976. Leaf-feeding resistance to bollworm and tobacco budworm in three soybean plant introductions. Crop Sci. 16:277-280.

Hatchett, J. H., G. L. Beland and T. C. Kilen. 1979. Identification of multiple insect resistant soybean lines. Crop Sci. 19:557-559.

Hollowell, E. A. and H. W. Johnson. 1934. Correlation between rough-hairy pubescence in soybean and freedom from injury by *Emposca fabae*. Phytopathology 24:12.

Kilen, T. C., J. H. Hatchett and E. E. Hartwig. 1977. Evaluation of early generation soybeans for resistance to soybean looper. Crop Sci. 17:397-398.

Lambert, L. and T. C. Kilen. 1984. Insect resistance factor in PI's 229358 and 227687 demonstrated by grafting. Crop Sci. 24:163-165.

Luedders, V. D. and W. A. Dickerson. 1977. Resistance of selected soybean genotypes and segregating populations to cabbage looper feeding. Crop Sci. 17:395-396.

Moscardi, F., C. J. Barfield and G. E. Allen. 1981. Impact of soybean *Glycine-max* cultivar Bragg on phenology of velvetbean caterpillar (*Anticarsia gemmatalis*-Lepidoptera, Noctuidae) oviposition, egg hatch, and adult longevity. Can. Entomol. 113:113-120.

Paschal, E. H. and H. C. Minor. 1978. Velvetbean caterpillar resistance in soybean selections from crosses involving Mexican bean beetle resistant plants. Soybean Genet. Newsletter 5:24-27.

Shillinger, J. A. 1976. Host plant resistance to insects in soybeans. p. 579-583. *In* L. D. Hill (ed.) World Soybean Research. Interstate Printers and Publishers. Danville, Illinois.

Smith, C. M. and C. A. Brim. 1979a. Resistance to Mexican bean beetle and corn earworm in soybean genotypes derived from PI 227687. Crop Sci. 19:313-314.

Smith, C. M. and C. A. Brim. 1979b. Field and laboratory evaluations of soybean lines for resistance to corn earworm leaf feeding. J. Econ. Entomol. 72:78-80.

Smith, C. M., R. F. Wilson and C. A. Brim. 1979. Feeding behavior of Mexican bean beetle (*Epilachna varivestis*) on leaf extracts of resistant and susceptible soybean genotypes. J. Econ. Entomol. 72:374-377.

Turnipseed, S. G. and M. J. Sullivan. 1976. Plant resistance in soybean insect management. p. 549-560. *In* L. D. Hill (ed.) World Soybean Research Conference. Interstate Printers and Publishers, Danville, Illinois.

Turnipseed, S. G. and M. Kogan. 1976. Soybean entomology. Ann. Rev. Ent. 21:247-282.

Van Duyn, J. W., S. G. Turnipseed and J. D. Maxwell. 1971. Resistance in soybeans to the Mexican bean beetle: I. Sources of resistance. Crop Sci. 11:572-573.

Yanes, J. and D. J. Boethel. 1983. Effect of a resistant soybean genotype on the development of the soybean looper (Lepidoptera:Noctuidae) and an imported parasitoid, *Microplitis demolitor* Wilkinson (Hymenoptera:Braconidae). Environ. Entomol. 12:1270-1274.

Disease and Insect Resistance

BREEDING FOR MANAGEMENT OF SOYBEAN SEED DISEASES

E. A. Kueneman

Production and maintenance of good quality seed is very difficult in the warm, humid tropics due to rapid seed deterioration, and seed longevity becomes a critical factor in determining whether the crop can be successfully introduced. In this paper I put major emphasis on seed-borne diseases that have striking effects on seed longevity and seedling vigor, but I also discuss use of host-plant resistance for the control of several major seed-borne diseases that have pronounced effects on vegetative or reproductive growth. The symptomology and epidemiology of diseases being reviewed have been discussed adequately in the 'Compendium of Soybean Diseases' (Sinclair, 1982) and will not be repeated in detail.

The prospects are very good for controlling many seed diseases through genetic manipulation. However, I would like to emphasize that genetic improvement is a component of a production package. Both good crop management and proper seed handling also are essential to the production and maintenance of good quality seed.

POD AND STEM BLIGHT/SEED ROT DISEASE

Three related organisms are considered the disease agents of pod and stem blight: *Diaporthe phaseolorum* (Cke.&Ell.) var. *sojae* Wehm (*Dps*), *D. phaseolorum* var. *caulivora (Dpc)*, and the imperfect stage of *Phomopsis sojae* (Leh.). According to Schmitthenner and Kmetz (1980), *Dpc* is generally associated with canker-like symptoms while *Phomopsis* spp. and *Dps* are associated with pod and stem blight symptoms. All three organisms are highly virulent seed pathogens and frequently are associated with lower seed germination and poor stand establishment.

Sources of Resistance

According to Athow (1973) cultivars immune to pod and stem blight have not been identified, but cultivar differences in relative resistance occur: Harosoy 63 and Lindarin 63 are the least susceptible. Walters and Caviness (1973) reported that PI 82264 of Maturity Group IV was resistant to *Dps*. A number of other resistant sources were summarized by Tisselli et al. (1980). More recently, Higley and Tachibana (1982) screened 34 cultivars and breeding lines for resistance to stem canker disease. They reported that Midwest had the highest level of resistance; delayed disease development was observed on Peak and A79-331022.

There also appear to be cultivar differences for seed rot disease caused by these organisms. Pascal and Ellis (1978) found several germplasm accessions (PI 205912, PI 219653, PI 239235, PI 279088, and PI 341349) that had low incidence of *Phomopsis* spp. and good seed germinability following 4 weeks of delayed harvest. Miranda (1977), who studied three soybean lines, Mack, Dare, and D67-5677-13, found that infection by *Dps* during seed development had little effect on seed quality. However, the incidence of *Dps* after physiological maturity was related to seed germinability. Germination at 109 days after anthesis was 80, 62, and 32% for D67-5677-13, Dare and Mack, respectively, and the corresponding incidence of *Dps* infection was 50, 74, and 87%. Ndimande et al. (1981) reported that two germplasm lines, TGm 685 and Tgm 686, from Indonesia were resistant to field weathering, and the incidence of *Phomopsis* in seeds of these lines was less than that observed in lines susceptible to field weathering.

To the best of my knowledge, no research has been carried out to determine if the resistance identified for pod and stem rot disease or for stem canker provides resistance to seed rot disease or whether, by virtue of having lower inoculant load, cultivars resistant to pod and stem blight might escape seed rot diseases. There appear to be no reports of presence or absence of strains of these pathogens based on host specificity. Consequently, host plant resistance could likely provide stable disease control.

Screening Methods

Keeling (1982a) found that toothpick inoculation with *Dpc* could be used in screening for resistance to stem canker disease. The delayed harvest technique, which has been used (Pascal and Ellis, 1978; Miranda, 1979; Green and Pinnell, 1968) to identify lines with resistance to field weathering of seed, is not ideal when making comparisons among lines of different maturity. Dassou and Kueneman (1984) reported that screening precision was improved by removing pods at the yellow stage and storing them at 30 C and 95% RH for 7 days. This procedure is described elsewhere in this paper.

PURPLE SEED STAIN

Cercospora kikuchii (T. Matsu. & Tomoyasu) Gardner is generally considered the causal organism. However, other species of *Cercospora* have been shown to cause similar symptoms (Jones, 1959). The symptoms are conspicuous, irregular, purple blotches on seed which can occasionally cover the entire surface. The effects of the disease on seed germination are not clear because of an antagonistic interaction between *C. kikuchii* and *Diaporthe* spp. (Roy and Abney, 1976). *Diaporthe* causes major reductions in seed germination; inoculation of *C. kikuchii* lowers incidence of *Diaporthe*. Consequently, it is possible that when *Diaporthe* is prevalent, germination will be superior in seeds infected with *C. kikuchii* compared to uninfected seeds. Heavy infection of *C. kikuchii* can cause reduction in seed germination in the absence of *Diaporthe*, and weak seedlings are derived from seeds with 50% or more of the surface stained by *Cercospora* (Yeh and Sinclair, 1982).

Sources of Resistance

Results of early screening work are summarized by Tisselli et al. (1980) and will not be repeated here. Scientists at the Asian Vegetable Research and Development Center (AVRDC, 1979) screened 1200 accessions against purple seed stain and reported the following entries as resistant: Harosoy, Palmetto, Hidatsa, PI 248400, PI 181537, PI 200508, PI 200479, PI 238926, White Biloxi, Giant Sleeves, Kaoshiung No. 3, Lee, Sumbing, Aracadion, Pochal, No. 208, H15, L-206-4-M(2)-10-(M6), Fukuzu, Austin, Ross, Shin 2, Takiya (Waseshu), K0309, and Bikuni. Roy and Abney (1976) found PI 80837 to be resistant compared to Amsoy, Wayne, and Cutler.

Screening Methods

It may be difficult to compare cultivars that differ in maturity. Environmental conditions at flowering and early pod development can greatly influence the incidence of infection, and similarly, conditions during pod maturation can influence the expression of symptoms, which are manifested during the dry-down period (Roy and Abney, 1976). Large differences among cultivars for resistance can be identified successfully by weekly inoculations with C. kikuchii during flowering (Wilcox et al., 1975). To prepare inoculant Wilcox et al. cultured the organism on V-8 Juice-agar in petri dishes. When the fungus had overgrown the plates, the entire contents of 40 petri dishes were macerated in 2 L of water and strained through cheesecloth. This concentrate was diluted with 11 L of water before spraying plants until thoroughly wet in the late afternoon or early evening. Using this inoculation technique, Wilcox et al. studied the inheritance of resistance in a cross involving Amsoy, which is very susceptible, and a resistant germplasm accession, PI 80837. They assumed complete maternal plant influence, and their estimates of broadsense heritability were quite high: 0.91 in the F_2 and 0.51 in the F_3. The observed response to selection did not differ greatly from expected response, after adjustments were made to compensate for differences in overall incidence of disease (based on the susceptibility parent) that occurred in the different years. By use of these inoculation techniques, genetic differences among breeding lines can be accentuated, and progress in selection might be anticipated.

PHYTOPHTHORA ROT

Phytophthora rot, caused by *Phytophthora megasperma* (Drechs.) var. *sojae* A. A. Hildebrand, is primarily a seed and seedling disease associated with seed rot and pre- and post-emergence damping-off, all of which result in poor stand establishment. The disease is most frequently encountered on heavy textured soils in North America and Australia (Sinclair, 1982).

Sources of Resistance

Twenty physiological races (based on host-specificity) have been identified (Keeling, 1982c). At least five genes and four alleles control resistance. Because of the complexity of the host-parasite reactions, readers interested in specific cultivar reactions to different races should see reports by Tisselli et al. (1980) and Keeling (1982c).

Screening Methods

The two screening methods described by Keeling (1979) included a 'hypocotyl puncture' method and a 'hydroponic culture' method. In the former, a spear-shaped needle is dipped through a culture of fungus and then inserted through the hypocotyl of a 10-day-old seedling approximately 1 cm below the cotyledons. The inoculated seedlings are kept for 16 to 18 h in a moist chamber before being transferred to a greenhouse held at 22 to 24 C. Susceptible plants die in 4 to 5 days. In the hydroponic culture method, 5-day-old seedlings are transferred to holes punched in styrofoam sheets (25 cm thick). The sheets are floated on 25% Hoagland's nutrient solution. A fungus culture growing on a semisolid corn meal agar is added to the nutrient solution at the rate of 100 ml/10 L of nutrient solution. The plants are shaded for 3 to 5 days and scored as dead or alive after 7 days. The hydroponic culture method allows simultaneous screening of soybeans against several strains of the pathogen (Kilen and Keeling, 1977). More recently, Keeling (1982b) reported that distilled water can be used instead of nutrient solution, and infected seedlings (10 seedlings per 15 L of hydroponic solution) can be used instead of inoculum from cultured fungus.

Because of the great genetic variability in *Phytophthora*, breeding cultivars with stable resistance is difficult. The use of non-race-specific field tolerance, as reported for 'Forrest' (Hartwig et al., 1982) or 'Asgrow 2656' (Tooley and Grau, 1982), might be useful. Buzzel and Anderson (1982) suggested selection for tolerance based on low plant loss, followed by backcrossing to incorporate race-specific resistance. Tooley and Grau (1982) described a method for quantifying cultivar differences in non-race-specific field tolerance based on the calculation of a log LD_{50} value (the log of the spore concentration lethal to 50% of the plant population) after cotyledon inoculation. Their method may be very useful for selection of parental lines and for evaluating late-generation breeding lines.

ANTHRACNOSE

Anthracnose is caused by *Colletotrichum dematium* (Pers. ex Fr.) Grove var. *truncatum* (Schw.) Arx.; the perfect stage is *Glomerella glycines* (Hori) Lehman and Wolf. The symptoms include irregular shaped brown areas on stems, pods and pedicels. Pre-emergence and post-emergence damping-off may occur when infected seeds are planted (Sinclair, 1982).

Sources of Resistance

No major effort has been made to identify sources of resistance to Anthracnose in soybean. For a similar disease in beans (*Phaseolus vulgaris*) caused by C. *lindemuthianum*, host-plant resistance has been identified. The pathogen, however, possesses pathogenic variability and several races have been reported based on host-parasite interactions (Chaves, 1980); similar variability for Anthracnose might be expected in soybean.

Although not a confirmed source of resistance, Ndimande et al. (1981) observed that soybean germplasm accession, TGm 685, from Indonesia was highly resistant to field weathering, and that the incidence of Anthracnose in the seed was low compared to several other lines susceptible to field weathering. More basic research is required on cultivar reactions and on screening methods before an effective breeding project can be implemented.

BACTERIAL PUSTULE

Bacterial pustule disease caused by *Xanthomonas campestvis* pv. *phaseoli* (Smith) Dye is strongly seed borne and is commonly found when susceptible soybean cultivars are grown in a warm, humid environment. Localized leaf spot symptoms may coalesce and form brown areas. Under severe infection pressure, leaf defoliation may be observed on some cultivars. In Mississippi, yield losses of 8 to 11% were attributed to the disease (Hartwig and Johnson, 1953). Under conditions of moderate disease establishment, Weber et al. (1966) found that susceptible lines had 4.3% lower yields than resistant lines. In Indiana, no yield loss was associated with susceptibility (Laviolette et al., 1970). *Xanthomonas* infection can enhance infection of *Pseudomonas tabaci* (Chamberlain, 1956), so that in areas where wildfire disease is a problem yield losses may be substantial in cultivars susceptible to bacterial pustule.

Sources of Resistance

There are many sources of resistance (Sinclair, 1982), but Clemson Non-Shattering (CNS) is the source, direct or indirect, most commonly used in the southern USA. Hartwig and Lehman (1951) found that CNS and F.C. 31592 carried the same recessive gene for resistance (near immunity). There is no confirmed evidence for pathogenic variability resulting in the breakdown of disease resistance. Consequently, for now there is no compelling reason to use other sources of resistance. Hartwig and Lehman (1951) reported cultivar differences in degrees of susceptibility among susceptible lines; I have observed the same phenomenon at the International Institute of Tropical Agriculture (IITA). It could be argued that, to insure durable resistance, lines with field resistance should be developed followed by backcrossing to incorporate the single, recessive gene for total resistance.

Screening Methods

Jones and Hartwig (1959) reported a simple method for field inoculation of bacterial pustule. Leaves of infected plants were run through a food chopper and a food blender using tap water (10 leaflets/3.8 L). The suspension was allowed to stand for 2 h before filtering. The filtrate was sprayed using either low or high pressure. Infected leaves can be frozen and stored for use the following year. Preplanting spreader rows of a susceptible cultivar can reduce the incidence of escape, which can occur even when plants are inoculated (Kueneman and Allen, 1980). In crosses between resistant and susceptible parents, Hartwig postpones selection for resistance until the F_3 generation, at which time he can discern the homozygous susceptible families to be discarded. Resistant families (Homozygous) can be advanced, and resistant plants within segregating families can be identified (Hartwig, personal communication).

BACTERIAL BLIGHT

Bacterial blight, caused by *Pseudomonas syringae* pv. *glycinea* (Coerper) Young, Dye & Wilkie, is a very common disease of soybean plants, especially in cool, wet environments. Lesions on leaves are initially small, angular and water-soaked. The margin of the lesion is frequently yellow, giving a halo-like appearance. Lesions often tear away and leaves take on a ragged appearance. Seeds become infected and a slimy bacterial growth may develop (Sinclair, 1982).

Sources of Resistance

A number of races have been identified. Some cultivar-race reactions were summarized by Tisselli et al. (1980). The most thorough survey of pathogenicity of isolates (Cross et al., 1966) involved 104 isolates and 17 cultivars. Seven races were identified based on host specificity. Chippewa was resistant to six of the seven isolates and gave an intermediate reaction with the other race. It is surprising that no further assessment has been reported and that inheritance studies are lacking, with the exception of that reported by Mukhergee et al. (1966). They found that Norchef, Harsoy and PI 132207 carry a single dominant gene for resistance to Race 1. PI 189968 was resistant to both Race 1 and Race 2; resistance to Race 2 apparently involved more than one gene.

Screening Methods

The simple method developed by Jones and Hartwig (1959) and described under the section on bacterial pustule would likely be suitable for screening for resistance to bacterial blight. They noted that the method was useful for screening for *Pseudomonas tabaci* as well as for *Xanthamonas*. Mukhergee et al. (1966) used a similar technique and reported that low pressure spray was preferable to high (0.55 Pa; 80 psi) pressure. Spraying inoculum with a knapsack sprayer the evening after an overhead irrigation gave good results. If one desires to evaluate the reaction to a specific strain of the organism without confounding the effects of local strains, a 5 ml syringe with rubber tubing can be used. The end of the tube is held to the underside of the leaf and pressure is applied to gently infuse the suspension into the leaf mesophyll. Kennedy and Cross (1966) tested 16 methods of inoculation and reported that good results were obtained by use of an artist's paint sprayer and spraying the leaf until water congestion was evident in a distinct area approximately 5 mm in diameter. Best results are obtained on young, rapidly growing plants.

SOYBEAN MOSAIC

The disease, caused by soybean mosaic virus (SMV), is generally associated with leaf distortion (see Sinclair, 1982). However, SMV is seedborne, and infected seeds may fail to germinate or may produce diseased seedlings. SMV is spread in the field by aphids. Discoloration of seed, 'hilum bleeding', may occur when the virus infects cultivars with dark hila. Hartwig and Keeling (1982) inoculated a susceptible cultivar with SMV-1 and observed a yield loss of 12% averaged over 4 years.

Sources of Resistance

There are two approaches to breeding for control of SMV. One approach is to breed for resistance to the disease; the other is to breed for low levels of seed transmission.

Incorporation of genes for resistance to SMV is complicated by the existence of multiple strains. Seven strains have already been identified (Cho and Goodman, 1982), and it is probable that more will be found in the future. The variety Buffalo, selected in Zimbabwe, is resistant to six of the seven known strains. Suweon 86, Suweon 94, Suweon 95 and Suweon 106 of group III maturity are resistant to all seven strains; Suweon 97 of maturity group IV also was reported to be resistant (Cho and Goodman, 1982).

The genetics of resistance to all of the seven known strains has not been fully studied. Kiihl and Hartwig (1979) noted an allelomorphic series involving three alleles that controlled the disease reaction against two SMV strains. Resistance was dominant and the Rsv allele found in PI 96983 provided good protection against SMV-1 and SMV-1-B. Allele rsvt found in Ogden conditioned resistance to SMV-1, but allowed a necrotic reaction to develop when plants were inoculated with SMV-1-B. The allele, rsv, conditioned normal mosaic symptoms with both virus strains. Kwon and Oh (1980) found a recessive gene that conditioned resistance to a necrotic strain of SMV, and Lim (1982) reported a single dominant gene for Suweon 97 that conditioned resistance to strains G2 and G7. There is very little information concerning the deployment of genes for resistance to SMV and subsequent breakdown of resistance. One such case was reported (Cho et al., 1977) in Korea. Breeders making efforts to incorporate resistance to SMV should be mindful that there is considerable pathogenic variability within the SMV complex, and that resistance to SMV may not be stable across environments and over time.

Another approach to genetic control of SMV involves exploiting cultivar differences in the incidence of seed transmission. Goodman et al. (1979) screened 897 germplasm accessions for incidence of seed transmission using the 'Illinois severe' isolate of SMV (SMV-I1-S). Four tropically adapted accessions (PI 203406, PI 240664, PI 325779, and Arisoy) were classified as nontransmitting lines based on incidence of virus on 1000 seedlings taken from SMV infected plants. Fifteen temperate lines (FC 31678, PI 60279, PI 68680, PI 70019, PI 70036, PI 88303, PI 91115, PI 92684, PI 92718-2, PI 360835, Cloud, Manchu 2204, Merit, Mukden, and Virginia) also were classified as nontransmitting.

I believe that breeders wishing to exploit low seed transmission to control SMV should consider the following: (1) It is not yet known whether a cultivar with low incidence of seed transmission of a given strain will exhibit low seed transmission when infected with other strains. (2) There is no evidence that this approach will be more stable than the deployment of genes for resistance. (3) Selection will likely need to wait for late generations because a reasonable amount of seed will be needed to assess incidence of seed transmission. (4) It will be very difficult to employ resistance genes also if one is using the low seed transmission strategy. A line with low seed transmission could be used as a recurrent parent in a backcross program with a SMV resistant line as a donor parent. And, while it would be possible to have progenies with both mechanisms of disease control, without genetic markers it would be impossible to identify with certainty which resistant progenies also carried genes for low seed transmission. (5) In regions where SMV is a serious problem, cultivars with low seed transmission of SMV will likely require special promotional efforts for acceptance. Such lines are likely to be severely infected with SMV when sown in trials involving other susceptible lines. Consequently, it is my belief that unless we obtain additional evidence that deployment of resistance genes frequently leads to unstable disease control, and have clear evidence that low seed transmission provides stable control, it is preferable to use the strategy of breeding for resistance using SMV inoculum that is representative of the strains occurring frequently in the region. At IITA we currently use a local SMV strain that causes pronounced symptoms, but not necrosis, for screening in early generations. Use of such isolates minimizes the likelihood of escape. Selected progenies can then be tested across a wider range of local SMV isolates.

Screening Methods

In the method used by Cho and Goodman (1982), leaves showing symptoms were homogenized with 4 to 5 volumes of 0.01 M sodium phosphate (pH 7.0) in a Waring blender. The inoculum was filtered through a double layer of cheesecloth plus one layer of Miracloth. A small amount of Carborundum (600 mesh) was added and the inoculum was applied to primary leaves using a Type B Wren airbrush at 4.9 kg/cm^2 pressure supplied by an air compressor. They also inoculated by rubbing leaves with a cotton-tipped applicator that had been dipped into the inoculum.

OTHER PATHOGENS OF STORED SEED

Species of *Aspergillus, Penicillum, Macrophomina, Fusarium,* and *Rhizopus* are known to invade soybean seed in storage when seeds have moisture contents of 13% or greater. These fungal organisms and bacteria such as *Bacillus subtillis* can have very deleterious effects on seed longevity under warm, humid conditions.

Sources of resistance to these specific pathogens have not been clearly identified. York showed relative resistance to *Fusarium oxysporum* (Leath and Carrol, 1982). It is possible that lines showing superior levels of resistance to seed deterioration in storage may, in fact, carry some form of resistance to such organisms (Ndimande et al.,

1981; Dassou and Kueneman, 1984). Many lines with good seed longevity imbibe water slowly, suggesting that the integrity of the seed coat may play a role in maintenance of seed viability during storage. Seed coat pigments are frequently associated with better seed longevity. Whether the pigments *per se*, or other substances produced in the metabolic pathway of pigment synthesis, are bio-static is not known. Cultivar differences in seed longevity exist, but mechanisms of resistance to deterioration merit further study.

SCREENING FOR RESISTANCE TO SEED DETERIORATION

In 1978 we at IITA initiated a breeding program to develop cultivars with superior seed longevity. A first step was to identify parental lines with good seed longevity and then to develop screening methods that would give quick assessment of the storability of breeding lines, so that several generations could be advanced in a year.

Deterioration in Storage

In our first attempt we used the accelerated aging method developed by Byrd and Delouche (1971) in which seeds were stored for 72 h at 40 C and 99% RH. This test did not prove to be satisfactory. Fungal growth on the seed was prolific and there was extensive cross contamination of pathogens from one seed to the next (Wien and Kueneman, 1981). Lowering the relative humidity to 75% and extending the aging period to 6 weeks, however, resulted in a good correlation with ambient storage tests. This relationship was confirmed (Dassou and Kueneman, 1984) in other tests where the correlation of seedling emergence after modified-accelerated-aging with that after 8-months ambient storage was 0.82**. This method is now being used routinely to evaluate the potential seed longevity of advanced-generation (F_4 and greater) breeding lines.

Two other methods of screening were tested at IITA: hot water pre-germination stress and methanol stress (Kueneman, 1981). Both tests may be useful, but we have noted that a few genotypes resistant to deterioration under ambient storage are sensitive to these tests. There appears to be a strong, non-cytoplasmic maternal plant influence on the expression of seed longevity (Kueneman, 1983), and consequently, segregation is delayed for one generation. Early generation selection has not been very effective. The lines originally used as sources of good seed longevity (IITA accessions: TGm 737P, TGm 685, TGm 693) are low-yielding, lodge badly, are black-seeded, and are susceptible to bacterial pustule and pod shattering. Breeders interested in incorporating improved seed longevity into their breeding materials may wish to use breeding lines such as TGX 342-356D, TGX 536-01D, and TGX 709-01E as parental sources. These are recombinants with good seed longevity derived from crosses between the aforementioned parents and cultivars with superior agronomic characteristics.

Resistance to Field Weathering

The most common procedure for evaluating weathering resistance is to leave plants in the field beyond the normal harvest period. This 'delayed harvest' technique has several limitations: (1) Cultivars maturing at different times will be subject to different environmental weathering stress. (2) Since pods on a given plant do not all mature at the same time (the timing of pod maturation varies considerably with some cultivars), they will be subject to different periods of weathering. Dassou and Kueneman (1984) compared several screening methods. An 'incubator-weathering' method, where pods at physiological maturity were detached and kept in a plastic grid at 30 C and 90-95% RH for 10 days, provided better precision than the delayed harvest method. Furthermore, lines of different maturity can be compared without the confounding effects of fluctuating environment that occur in the field when using the delayed harvest procedure. Genotypes identified with the highest levels of resistance to field weathering were black-seeded (TGm 1171 [AVRDC 8457], TGm 106 [Lee A], TGm 46 [Fort Lamy], TGm 920 [Japan Black], and TGm 693 [Indo 153]). In general, lines with resistance to deterioration in the field were also resistant to deterioration during storage.

The heritability estimates for resistance to field weathering reported by Green and Pinnell (1968) were low, but these estimates were based on the delayed harvest technique, and the variation observed for parents was very large. Inheritance studies have recently been initiated at IITA using the incubator weathering technique. We expect that different heritability estimates will be observed depending on the mechanism of resistance. For example, if resistance is conditioned by hardseededness, a few major genes may condition resistance (Kilen and Hartwig, 1978). If disease resistance or membrane integrity are the mechanisms for resistance, then inheritance may be more complex.

RECOMMENDED FUTURE RESEARCH

Genetic variability in the pathogen for several seed-borne diseases; e.g., *Phytophthora*, bacterial blight, and soybean mosaic virus, has been observed. Comprehensive studies should be conducted to determine the gene action involved for race-specific reactions and for field resistance. Further studies on the mechanisms of resistance are required for all diseases discussed. For cases such as pod and stem blight, where a pathogen causes disease symptoms on vegetative tissues as well as on seed tissues, studies are required to determine the relationship between resistance for the vegetative disease and for the seed disease.

Large-seeded cultivars are nearly always prone to rapid seed deterioration whereas many, but not all, small-seeded lines are resistant. There is a need to clarify the reasons for poor longevity in large-seeded lines. Many lines with good seed longevity have colored seed, suggesting there may be a pigment-related factor that influences seed longevity. This, too, merits investigation.

NOTES

E. A. Kueneman, Soybean Breeder, Latin American Regional Program, International Institute of Tropical Agriculture, EMBRAPA/CNPAF, C.P. 179-74000, Goiania, Goias, BRAZIL.

REFERENCES

Athow, K. L. 1973. Fungal diseases. p. 459-489. *In* B. E. Caldwell (ed.) Soybeans: Improvement Production and Uses. Amer. Soc. Agron., Madison, Wisconsin.
Asian Vegetable Research and Development Center (AVRDC). 1979. Varietal development and germplasm utilization in soybean. AVRDC Tech. Bull. 13:78-102.
Buzzel, R. I. and T. R. Anderson. 1982. Plant loss response of soybean cultivars to *Phytophthora megasperma* f. sp. *glycinea* under field conditions. Plant Dis. Reptr. 66:1146-1148.
Byrd, H. W. and J. C. Delouche. 1971. Deterioration of soybean seed in storage. Proc. Assoc. Off. Seed Anal. 61:41-57.
Chamberlain, D. W. 1956. Methods of inoculation for wildfire of soybean and the effects of bacterial pustule on wildfire development. Phytopathology 46:96-98.
Chaves, G. 1980. Anthracnose. p. 39-54. *In* H. F. Schwartz and G. E. Galvez (eds.) Bean production problems. CIAT Series No. 09EB-1. CIAT, Cali, Colombia.

Cho, E. K., B. J. Chung and S. H. Lee. 1977. Studies on identification and classification of soybean virus diseases in Korea. II. Etiology of a necrotic disease of *Glycine max*. Plant Dis. Reptr. 61: 313-317.

Cho, E. K. and R. M. Goodman. 1982. Evaluation of resistance in soybeans to soybean mosaic virus strains. Crop Sci. 22:1133-1136.

Cross, J. E., B. W. Kennedy, J. W. Lambert and R. L. Cooper. 1966. Pathogenic races of bacterial blight pathogen of soybean *Pseudomonas glycinea*. Plant Dis. Reptr. 50:557-560.

Dassou, S. and E. A. Kueneman. 1984. Screening methodology for identification of soybean varieties resistant to field weathering of seed. Crop Sci. 24:774-779.

Goodman, R. M., G. R. Bowers Jr., and E. H. Paschal II. 1979. Identification of soybean germplasm lines and cultivars with low incidence of soybean mosaic virus transmission through seed. Crop Sci. 19:264-267.

Green, D. E. and E. L. Pinnell. 1968. Inheritance of seed quality. I: Heritability of laboratory germination and field emergence. Crop Sci. 8:5-11.

Hartwig, E. E. and S. G. Lehman. 1951. Inheritance of resistance to the bacterial pustule disease in soybeans. Agron. J. 43:226-229.

Hartwig, E. E. and H. W. Johnson. 1953. Effect of the bacterial pustule disease on yield and chemical composition of soybeans. Agron. J. 45:22-23.

Hartwig, E. E. and B. L. Keeling. 1982. Soybean mosaic virus investigations with susceptible and resistant soybeans. Crop Sci. 22:955-957.

Hartwig, E. E., R. C. Kilen, L. D. Young and C. R. Edwards Jr. 1982. Effects of natural selection in segregating soybean populations exposed to *Phytophthora* rot or soybean cyst nematodes. Crop Sci. 22:588-590.

Higley, P. M. and H. Tachibana. 1982. Resistance to stem canker of soybeans. Phytopathology 72: 1136.

Jones, J. P. and E. E. Hartwig. 1959. A simplified method for field inoculation of soybeans with bacteria. Plant Dis. Reptr. 43:946.

Jones, J. P. 1959. Purple stain of soybean seeds incited by several *Cercospora* species. Phytopathology 49:430-432.

Keeling, B. L. 1979. Research on phytophthora root and stem rot: Isolation, testing procedures, and seven new physiological races. p. 367-370. *In* F. T. Corbin (ed.) World soybean research conference. II. proceedings. Westview Press, Boulder, Colorado.

Keeling, B. L. 1982a. A seedling test for resistance to soybean stem canker by *Diaporthe phaseolorum* var. *caulivora*. Phytopathology 72:807-809.

Keeling, B. L. 1982b. Factors affecting the reaction of soybeans to *Phytophthora megasperma* var. *sojae* in hydroponic culture. Crop Sci. 22:325-327.

Keeling, B. L. 1982c. Four new physiologic races of *Phytophthora megasperma* f. sp. *glycinea*. Plant Dis. Reptr. 66:334-335.

Kennedy, B. W. and J. E. Cross. 1966. Inoculation procedures for comparing reaction of soybean to bacterial blight. Plant Dis. Reptr. 50:560-565.

Kiihl, R. A. S. and E. E. Hartwig. 1979. Inheritance of reaction to soybean mosaic virus in soybeans. Crop Sci. 19:372-375.

Kilen, T. C. and E. E. Hartwig. 1978. Inheritance of impermeable seed in soybeans. Field Crops Res. 1:65-70.

Kilen, T. C. and B. L. Keeling. 1977. Simultaneous screening soybeans against three races of *Phytophthora megasperma* var. *sojae*. Crop Sci. 17:185-186.

Kueneman, E. A. 1981. Genetic differences in soybean seed quality: Screening methods for cultivar improvement. p. 31-41. *In* Soybean seed quality and stand establishment. INTSOY Series No. 22. Univ. of Illinois, Urbana.

Kueneman, E. A. 1983. Genetic control of seed longevity in soybeans. Crop Sci. 23:5-8.

Kueneman, E. A. and D. J. Allen. 1980. A technique for screening populations of soybeans for resistance to bacterial pustule. Trop. Grain Legume Bull. (IITA) 19:34-37.

Kwon, S. H. and J. H. Oh. 1980. Resistance to necrotic strain of SMV in soybeans. Crop Sci. 20:403-404.

Laviolette, R. A., K. L. Ahtrow, A. H. Probst and J. R. Wilcox. 1970. Effect of bacterial pustule on yield of soybeans. Crop Sci. 10:150-151.

Leath, S. and R. B. Carrol. 1982. Screening for resistance to *Fusarium oxysporum* in soybean. Plant Dis. Reptr. 66:1140-1143.

Lim, S. M. 1982. A new source of resistance to soybean mosaic virus in a soybean line and its inheritance. Phytopathology 72:943.

Miranda, M. 1977. Influence of some seed-borne pathogens and field weathering on soybean (*Glycine max* (L.) Merrill) seed quality. M.S. Thesis. Mississippi State Univ. Library, Mississippi State, MS.

Mukhergee, D., J. W. Lambert, R. L. Cooper and B. W. Kennedy. 1966. Inheritance of resistance to bacterial blight (*Pseudomonas glycina* Coerper) in soybeans (*Glycine max* L.). Crop Sci. 6:324-326.

Ndimande, B. N., H. C. Wien and E. A. Kueneman. 1981. Soybean seed deterioration in the tropics. I. The role of physiological factors and fungal pathogens. Field Crops Research 4:113-121.

Pascal, E. H. and M. A. Ellis. 1978. Variation in seed quality characteristics of tropically grown soybeans. Crop Sci. 18:837-840.

Roy, K. W. and T. S. Abney. 1976. Purple seed stain of soybeans. Phytopathology 66:1045-1049.

Schmitthenner, A. F. and K. T. Kmetz. 1980. Role of *Phomopsis* sp. in the soybean seed rot problem. p. 355-366. *In* F. T. Corbin (ed.) World Soybean Research Conference II: Proceedings. Westview Press, Boulder, Colorado.

Sinclair, J. B. 1982. Compendium of soybean diseases. Amer. Phytopath. Soc., St. Paul, MN.

Tisselli, O., J. B. Sinclair and T. Hymowitz. 1980. Sources of resistance to selected fungal, bacterial, viral and nematode diseases of soybeans. INTSOY Series No. 18. Univ. of Illinois, Urbana.

Tooley, R. W. and C. R. Grau. 1982. Identification and quantitative characterization of rate-reducing resistance to *Phytophthora megasperma* f. sp. *glycinea* in soybean seedlings. Phytopathology 72:727-733.

Walters, H. J. and C. E. Caviness. 1973. Breeding for improved seed quality. Arkansas Farm Res. 23:5.

Weber, C. R., J. M. Dunleavy and W. R. Fehr. 1966. Effect of bacterial pustule on closely related soybean lines. Agron. J. 58:544-545.

Wilcox, J. R., F. A. Laviolette and R. J. Martin. 1975. Heritability of purple seed stain resistance in soybeans. Crop Sci. 15:525-526.

Wien, H. C. and E. A. Kueneman. 1981. Soybean seed deterioration in the tropics. II. Varietal differences and techniques for screening. Field Crops Res. 4:123-132.

Yeh, C. C. and J. B. Sinclair. 1982. Effects of *Cercospora kikuchii* on soybean seed germination and its interaction with *Phomopsis* sp. Phytopath. Zeit. 105:265-270.

BREEDING FOR THE MANAGEMENT OF SOYBEAN ROOT AND
STEM DISEASES

A. K. Walker

Breeding soybeans for resistance to root and stem diseases has served an important role in decreasing disease losses. This review presents the current account of breeding for the management of phytophthora rot, brown stem rot, stem canker, Fusarium blight and Sclerotinia stem rot.

PHYTOPHTHORA ROT

Phytophthora rot, caused by *Phytophthora megasperma* Drechs f. sp. *glycinea* Kuan and Erwin (Pmg), is a very widespread and destructive disease. Race-specific resistance conferred by single, major genes has been the most widely used method of control; however, Pmg is a variable pathogen and races are being found which will attack sources of resistance. Twenty-three physiologic races of Pmg have been reported since 1955. Races 21 to 23 were reported in 1983 (Laviolette and Athow, 1983; White et al., 1983). There is no doubt that new races will continue to be found.

Nine major dominant genes for resistance have been reported and were reviewed by Athow and Laviolette (1982). They are Rps_1, $Rps_1{}^b$, $Rsp_1{}^c$, $Rps_1{}^k$, Rps_2, Rps_3, Rps_4, Rps_5, and Rps_6. Resistance to most races is still possible with several single-gene and two-gene combinations. The combinations, $Rps_1{}^b$ Rps_4, $Rps_1{}^b$ Rps_6, $Rps_1{}^k$ Rps_4, $Rps_1{}^k$ Rps_6, provide resistance to races 1 to 22 (Laviolette and Athow, 1983). $Rps_1{}^b$ Rps_4 have been backcrossed into the cultivars Century and Pella (Athow, pers. commun.). In addition, several backcross cultivars have been developed with $Rps_1{}^c$ Rps_3. The cultivars, Keller and Miami, were developed by backcrossing Rps_3 into the $Rps_1{}^c$ cultivars, Beeson-80 and Wells-II, respectively. Keller and Miami are susceptible to Races 12, 19 and 22. Cultivar, Winchester, was developed by backcrossing $Rps_1{}^b$ Rps_3 into cv. Williams. Winchester is susceptible to Races 10, 12, 19 and 22. Cultivars, Tracy and Tracy-M, have $Rps_1{}^b$ Rps_3. A germplasm line, HW79149, released from the Ohio Agricultural Research and Development Center, has two resistance genes, $Rps_1{}^c$ and another unidentified gene, from PI 82263-2. Cultivars, Williams-82 and Century-84, which are susceptible to Races 12, 16, 19 and 20, were developed by backcrossing $Rps_1{}^k$ into Williams and Century.

Once race-specific resistance genes are incorporated into a cultivar, there exists a threat that new virulent races will increase in frequency because of selection pressure exerted by the resistant cultivars. Thus, additional sources of resistance must be identified. Moots et al. (1983) screened 85 cultivars from the germplasm collection to 14 races, and found 37 cultivars which were resistant to one or more races. Several cultivars, which were susceptible to Races 1 and 2, were resistant to other races. Plant introductions are being screened at agricultural experiment stations in Indiana, Minnesota, Mississippi, Ohio, and Ontario for new sources of resistance. Also, several inheritance studies are being conducted on plant introductions that have unidentified genes conveying multi-race resistance.

Several methods have been developed to screen populations for more than one gene. Lines with two genes for resistance can be selected with à minimum of effort, if the specific genes were in the parents and the proper races are used for inoculation (Wilcox, 1983). Hypocotyls of individual plants also can be inoculated wth several different races. Kilen and Keeling (1977) used the hydroponic inoculation technique to screen simultaneously soybean plants for reaction to several races. No antagonistic effects among races were found. Keeling (1982b) compared several variables of the hydroponic inoculation technique and reported the most rapid kill of the greatest number of plants was obtained using a 25% Hoagland's nutrient solution, shading the plants for three days after adding inoculum, and using infected fresh plants as inoculum. Currently, they are using zoospores as an inoculum source by adding 50 zoospores per plant to the hydroponic solution. This technique works very well for selecting Rps_2 genotypes, whereas the hypocotyl inoculation usually kills some Rps_2 plants. Also, herbicides that are highly soluble may be added to the solution to screen survivors of the Pmg inoculation for tolerance to the herbicides. For example, 125 $\mu g/g$ metribuzin or 60 $\mu g/g$ bentazon have been used for screening for tolerance to these herbicides.

Morrison and Thorne (1978) inoculated detached cotyledons to screen for reaction to two races. Cotyledons of resistant genotypes showed only a slight discoloration at the inoculation site, whereas cotyledons of susceptible genotypes became severely necrotic in 72 to 96 h.

Several researchers are injecting zoospores into the hypocotyls in place of mycelia (Moots et al., 1983; Schwenk et al., 1979). Zoospores of several races can be injected into the same plant. Zoospore concentration is controlled in order not to overload the plant. This technique also will detect intermediate responses (Moots et al., 1983).

The proportion of resistant plants in some heterogenous populations can be increased by growing the population on soils infested with Pmg. Buzzell and Haas (1972) determined the relative fitness of the $Rps\ Rps$ versus the $rps\ rps$ gene pairs by growing composites of lines differing in these alleles and in pubescence color. The average fitness of the rps allele, relative to the Rps allele, decreased from 0.784 to 0.372 during 3 years. Hartwig (1972) evaluated two populations in the F_6 generation after they were advanced four generations on infested soil. In the population from a cross of a highly susceptible by a resistant cultivar, natural selection was effective; however, natural selection was not effective in the population from the cross of a field-resistant by a resistant cultivar. Hartwig et al. (1982) planted an F_2 population of Forrest x Tracy and 10 backcross F_3 families of Forrest (2) x Tracy at a location where phytophthora rot caused injury to susceptible genotypes. The populations were advanced in bulk at the phytophthora location for four generations. Natural selection was not effective in increasing the frequency of resistant plants in those populations. This was attributed to the presence, in the segregating population, of field-resistant plants that competed effectively with plants having major genes for resistance.

With the increase in new races, many breeding programs have been redirected to identify soybean cultivars and lines with tolerance, which also is referred to as field-tolerance, field-resistance and rate-reducing resistance (Buzzell and Anderson, 1982;

Irwin and Langdon, 1982; Jimenez and Lockwood, 1980; Tooley and Grau, 1982). Tolerance is the relative ability of plants to survive root infection, either natural or artificial, without showing severe symptom development such as death, stunting or yield loss. Tolerance is race-nonspecific, and growing tolerant cultivars should not favor the build-up of one race relative to others.

Heritability estimates for tolerance to phytophthora rot range from 68 to 96% on an entry mean basis (Buzzell and Anderson, 1982; Walker and Schmitthenner, 1982). Walker and Schmitthenner (1982) showed that heritability estimates from a greenhouse study were not affected by major gene resistance; however, lines carrying resistance had a slightly higher mean tolerance rating than susceptible lines. This suggested that major gene resistance and tolerance were not completely independent. High heritability estimates indicate that progress should be expected from selection in early generations. Considerable genetic gain also can be made from selection. Walker and Schmitthenner (1982) released a tolerant population, PMGT(S1)C3, derived from recurrent selection using S_1-line evaluation.

Transgressive segregation for lower and higher tolerance levels than either parent has occurred, and continuous variation exists in populations. This evidence also suggests that tolerance is a quantitative trait.

There are two disadvantages of tolerance for phytophthora rot control. First, absolute control is not obtained. Under severe disease conditions, tolerant cultivars will suffer some yield loss. A second weakness of all known tolerant lines is their susceptibility to damping off disease (Schmitthenner and Walker, 1979). Stands can be severely reduced if conditions favorable for phytophthora rot occur before or during emergence.

Major gene resistance also has its disadvantages. First, with the build-up of new races, resistance with specific genes will not always hold. Secondly, the Vertifolia effect can occur. This has been observed with several released soybean cultivars. Buzzell and Anderson (1982) reported on the effect with the breeding line, 0X20-8, which was selected for Rps_1 resistance to Race 1, but in the presence of compatible races it is extremely susceptible. Walker and Schmitthenner (1982) reported an eightfold difference in yield between 0X20-8 and other more tolerant soybean lines. An integrated approach of selection for both tolerance and race-specific resistance should be used to avoid releasing new cultivars with race-specific resistance that are inherently highly susceptible to compatible races.

Another integrated approach is the planting of tolerant cultivars and protecting them from damping-off with *Phytophthora* fungicides. These fungicides will control Pmg during emergence and seedling growth until tolerance is established in the plant. Yield results have shown that this approach is equal to, or better than, multi-race resistance for control of phytophthora rot, without the selection pressure for development of new races (Schmitthenner and Walker, 1979).

If a field that is naturally infested with several races of Pmg is available, hundreds of lines can be visually screened in a small area using hill plots or short rows. A disadvantage of field evaluation is that tolerant and multi-race resistant genotypes cannot always be distinguished from each other. Multi-race genotypes can be screened for tolerance in the greenhouse by using a tester race for which the genotypes have no race-specific resistance. Several greenhouse techniques which differ in inoculum source and planting media are described by Irwin and Langdon (1982), Jimenez and Lockwood (1980), Tooley and Grau (1982), and Walker and Schmitthenner (1982).

BROWN STEM ROT

Brown stem rot, caused by *Phialophora gregata* (Allington and Chamberlain) W. Gams (Pg), was discovered on soybean plants in the United States in 1944. Brown stem rot occurs frequently in the North Central United States and Canada. Many years in which conditions are favorable for disease development, Pg can cause significant yield reductions.

Resistance for brown stem rot was found in PI 84946-2. This resistance has been transferred successfully into the cultivars, BSR201, BSR301, and BSR302, and the germplasms, A3 and A4. Resistance in PI 86150 is greater than resistance in PI 84946-2 (Tachibana and Card, 1972). They also reported resistance in PI 88820N, PI 90138, and PI 95769. The Japanese cultivar, Kosodefuri, is relatively resistant compared to other Japanese cultivars (Kobayashi et al., 1983). Currently, plant introductions are being screened for brown stem rot resistance by researchers in Illinois and Minnesota.

Differences in susceptibility also are observed. Cultivars Century and Weber are known to be more susceptible to brown stem rot than other currently grown cultivars (Nickell, Tachibana, pers. commun.).

BSR301 and BSR302 were developed by selecting single, F_4 plants from a field infested with Pg. The stem of each plant was split and each plant with little or no internal stem browning was saved. F_5 progeny from resistant F_4 plants were grown in rows, and the level of resistance was measured by stem splitting on 10 plants per row. The lines which showed the greatest resistance were grown in a yield test in hill plots the following year on infested soil. The highest yielding lines were advanced to yield tests in row plots (Fehr, pers. commun.).

The evaluation of resistance by stem splitting is a laborious procedure. In an attempt to reduce stem splitting, an alternate evaluation was conducted. Single, F_4 plants were evaluated in a field infested with Pg. The stem of each plant was split and each plant with little or no internal stem browning was saved. F_5 progeny from single, F_4 plants were grown directly in hill plots in infested soil, and it was assumed that resistant lines would have higher yields in hill plots than susceptible lines. BSR201 was developed using this procedure.

Ertl and Fehr (1983) compared 96, F_4-derived lines from a population formed by intermating high-yielding lines with brown-stem-rot-resistant lines. The lines were grown in hill plots in four Pg infested and four uninfested fields. Selection for yield in hill plots grown on Pg infested soil was not an effective method for identifying lines with brown stem rot resistance. The correlation between yield and percentage of stem browning was not significant (r=-0.19). There was a high correlation of yield of the 96 lines on infested and uninfested soil (r=0.80, p⟨0.01). They concluded that resistance to brown stem rot should be evaluated by growing lines on infested soil and splitting stems to determine the extent of infection.

Heritability of brown stem rot resistance on an entry-mean basis in three replications of hill plots at each of four locations was 0.82, but only 0.33 on a plot basis (Ertl and Fehr, 1983). Therefore, heritability on a single-plant basis is likely to be extremely low.

BSR201-, BSR301, and BSR302 are prescribed for fields where 75% of the plants had brown stem rot in any recent year (Tachibana, 1982). One recommendation is that, following a susceptible soybean crop, a farmer should plant one year of corn, followed by a resistant cultivar (BSR201, BSR301, or BSR302), followed by another year of corn, and then any non-resistant soybean cultivar.

Epstein et al. (1983) grew BSR301, and the susceptible cultivar, Oakland, continuously for five years on land that initially had 100% plant infection on a susceptible cultivar. Brown stem rot incidence and severity decreased with the growing of BSR301, but not with Oakland. This was shown by growing four susceptible cultivars (Cumberland, Oakland, Williams-79, and Williams-82) on the BSR301 and Oakland land. Average yields for the four susceptible cultivars was 10% greater on the BSR301 land than on land on which Oakland had been grown. The higher yields were attributed to reduction in brown stem rot.

Gray (1971) identified two types of pathogenic isolates, Type I, which caused typical leaf symptoms and defoliation in addition to vascular discoloration, and Type II, which caused only a less extreme vascular discoloration than Type I. He (Gray, 1975) ran experiments with extracts made from soybean stems infected with Type I and Type II isolates, in which detached leaves were placed. Leaves in the Type I extract wilted, whereas leaves in the Type II extract did not. Leaves from resistant PI 86150 did not wilt in either extract. A toxin in the Type I extract may be responsible for wilting of the soybean leaves.

A greenhouse method using visual ratings on 7-week-old plants has been developed by Sebastian et al. (1983). Roots of 14-day-old plants are dipped in inoculum derived from Type I isolate, then the plants are transplanted into pots and rated 5 weeks later for leaf and stem symptoms. Sebastian and Nickell (1983) screened 45 soybean genotypes for brown stem rot in an infested field and in the greenhouse. Yield in the field was highly correlated with visual leaf ratings in the greenhouse. Soybean genotypes can be screened sequentially in the greenhouse for resistance to both Pmg and Pg without an interaction (Sebastian et al., 1983).

STEM CANKER

Stem canker, caused by *Diaporthe phaseolorum* (Cke and Ell.) Sacc. var. *caulifora* Athow and Caldwell (Dpc), has been very destructive on certain cultivars. Prior to the late 1940s, stem canker was sporadic in occurrence and only rarely serious. Then, it became very serious, with severe yield losses occurring when the highly susceptible cultivars, Hawkeye and Blackhawk, were widely grown in the North Central United States and Canada. With the elimination of production of highly susceptible cultivars, it became unusual to find more than 1 to 5% infected plants (Wilcox, 1983). Resistance in most of the currently grown cultivars in the northern United States and in Canada may have been derived from the cultivar, Mandarin (Athow, 1973).

Stem canker was first observed in the southern United States in 1973. Since then the extent and severity of stem canker there has increased. It has been reported in Mississippi, Alabama, Tennessee, Arkansas, Louisiana, Georgia, South Carolina, and Florida. Increased incidence in the south can be attributed to widespread plantings (over four million hectares in 1983) of highly susceptible cultivars, such as Hutton, Wilstar 790, Bragg, Coker 237, Lee-74, Mack, Forrest, Bedford, and Nathan (Keeling, 1982a; Krausz and Fortnum, 1983; Weaver et al., 1983).

Weaver et al. (1983) compared 30 cultivars under natural field infestations over two years in Alabama. Seed yield was highly correlated with visual ratings. Tracy-M had the highest seed yield and the lowest visual rating of any cultivar. They concluded that the use of resistant soybean cultivars is effective in combating stem canker, and visual evaluations under field conditions is an effective method of selection.

Screening for stem canker in the field without artificial inoculation can be disappointing, because infection is dependent upon natural occurrence of the diseases.

Environmental stresses, particularly nutrient, herbicide, moisture stress, and other pest-induced stress, also affect disease severity.

Keeling (1982a) inoculated 10-day-old seedlings in the greenhouse by inserting toothpicks infested with Dpc into the stems, and compared the results with disease development on artificially inoculated field-grown plants and with field-grown plants subjected to natural disease development. All three methods showed good agreement on the eight genotypes tested. Thus, the seedling test can be used to evaluate accurately soybean lines for resistance.

Physiologic specialization has been found to occur in Dpc (Keeling, pers. commun.). At least six isolates with differing pathogenic capabilities have been tested. For example, cultivar Centennial is susceptible to some isolates of Dpc and resistant to others. Isolates from Iowa, Indiana, Ohio, and Mississippi have been found to differ in their reaction to a set of genotypes. Cultivar Kingwa has shown resistance to more isolates of Dpc than any other cultivar. Tracy appears to have two dominant genes controlling resistance (Keeling, pers. commun.). Tracy, Tracy-M, Harosoy, and Sanga may possess similar resistance genes, since they have similar reactions to the different isolates tested.

Breeders and pathologists in the south are screening cultivars and breeding lines so that no new cultivars with a high level of susceptibility will be released. Many breeders are cautious about developing large scale programs to breed for resistance to stem canker resistance because of the history of the disease in the northern United States. It is possible that, after the more susceptible cultivars are eliminated from production in the south, the disease will become less important.

FUSARIUM BLIGHT

Fusarium oxysporum Schlecht. causes a wilt and root rot disease called Fusarium blight. Severe losses from this disease have occurred in Delaware, Maryland, and Virginia in recent years (Leath and Carroll, 1982). The pathogen occurs in most soybean growing areas, and there is potential for widespread occurrence of the diesase. Fusarium blight appears to be worse when the early growing season is wet and cool, followed by periods of high temperatures and unusually dry weather later in the growing season.

Effectiveness in field screening depends upon both the presence of the pathogen and the proper environmental conditions. Differences in cultivar susceptibility have been found. Leath and Carroll (1982) grew 10 soybean cultivars in two fields where inoculum was applied. Control (uninoculated) plots were adjacent to inoculated plots. The cultivar, York, had the smallest percentage decrease in yield of any cultivar when inoculated plots were compared with uninoculated plots. Essex was very susceptible to Fusarium blight.

Leath and Carroll (1982) used a soil-drench inoculation method on plants grown in flats in the greenhouse. They found this method to be effective in differentiating reactions among cultivars. This method took 6 weeks to complete. They also developed a rapid test tube screening method that can determine differential reactions in 16 days. This method was effective in screening out very susceptible lines, but it was not as effective as field screening in identifying higher levels of resistance.

SCLEROTINIA STEM ROT

Sclerotinia stem rot, caused by *Sclerotinia sclerotiorum* (Lib.) de Bary has been considered a disease of minor importance on soybean. However, incidence is increasing

424

due to rotation of soybeans with susceptible crops (Cline and Jacobsen, 1983). The disease was first observed on soybean plants in the United States in 1924 and has since been reported in Brazil, Canada, Hungary, India, Nepal, South Africa, and other soybean-growing areas. The fungus causes the most severe damage in areas with a humid climate and abundant rainfall. Cool temperatures and moist conditions have been shown to favor Sclerotinia stem rot.

Soybean cultivars have been observed to differ in susceptibility to Sclerotinia stem rot, and cultivar selection could be useful in control of this disease. Grau and Radke (1984) compared six soybean cultivars under several cultural practices in naturally infested fields. Disease and severity, and consequently yield reductions, were greatest in narrow rows. Also, disease severity was greater under irrigated conditions. Yields were improved 10 to 22% by minimizing disease severity through reduced irrigation before beginning bloom. Hodgson and Corsoy had less disease than other cultivars.

Grau and Radke (1982) evaluated 23 soybean cultivars for their resistance to *Sclerotinia sclerotiorum* in naturally infested fields. Corsoy, Hodgson, and Hodgson-78 were less susceptible than other cultivars. Gnome and Weber were highly susceptible. Resistance or susceptibility was not related to plant architecture or to soybean maturity group.

Screening soybeans in naturally infested fields can be effective, but conditions for disease development do not occur every year. Therefore, the development of a reliable greenhouse technique would be useful. Cline and Jacobsen (1983) reported that the limited-term inoculation technique was effective for screening for resistance to Sclerotinia stem rot in the greenhouse. Limited-term inoculation was accomplished by attaching autoclaved celery pieces, colonized by the fungus, to the nodes of 4-week-old plants for 24 h. This method distinguished differences in disease susceptibility among the soybean cultivars evaluated. Corsoy, Williams and Union were much less susceptible to Sclerotinia stem rot than other cultivars tested. Elf was highly susceptible. Resistance to stem invasion was suggested to be related to factors associated with purple-flowered cultivars (Grau and Radke, 1982); however, Williams and Union are white-flowered and have moderate resistance.

NOTES

A. K. Walker, Research Scientist III, Asgrow Seed Co., Redwood Falls, MN 56283.

REFERENCES

Athow, K. L. 1973. Fungal diseases. p. 459-489. *In* B. E. Caldwell (ed.) Soybeans: Improvement, production, and uses. Am. Soc. Agron., Madison, WI.
Athow, K. L. and F. A. Laviolette. 1982. Rps6, a major gene for resistance to *Phytophthora megasperma* f. sp. *glycinea* in soybean. Phytopathology 72:1564-1567.
Buzzell, R. I. and J. H. Hass. 1972. Natural and mass selection estimates of relative fitness for the soybean rps gene. Crop Sci. 12:75-76.
Buzzell, R. I. and T. R. Anderson. 1982. Plant loss response of soybean cultivars to *Phytophthora megasperma* f. sp. *glycinea* under field conditions. Plant Dis. 66:1146-1148.
Cline, M. N. and B. J. Jacobsen. 1983. Methods for evaluating soybean cultivars for resistance to *Sclerotinia sclerotiorum*. Plant Dis. 67:784-786.
Epstein, A. H., J. D. Hatfield and H. Tachibana. 1983. Control of brown stem rot (caused by *Phialophora gregata*) with concomitant increase in yield by continuous cropping of resistant soybean. (Abstr.) Phytopathology 73:817.
Ertl, D. S. and W. R. Fehr. 1983. Simultaneous selection for yield and resistance to brown stem rot of soybean in hill plots. Crop Sci. 23:680-682.

Grau, C. R. and V. L. Radke. 1982. Resistance of soybean cultivars to *Sclerotina sclerotiorum*. Plant Dis. 66:506-508.

Grau, C. R. and V. L. Radke. 1984. Effects of cultivars and cultural practices on Sclerotinia stem rot of soybean. Plant Dis. 68:56-58.

Gray, L. E. 1971. Variation in pathogenicity of *Cephalosporium gregatum* isolates. Phytopathology. 61:1410-1411.

Gray, L. E. 1975. Evidence for toxin production by a strain of *Cephalosporium gregatum*. Phytopathology. 65:89-90.

Hartwig, E. E. 1972. Utilization of soybean germplasm strains in a soybean improvement program. Crop Sci. 12:856-859.

Hartwig, E. E., T. C. Kilen, L. D. Young and C. J. Edwards, Jr. 1982. Effects of natural selection in segregating soybean populations exposed to phytophthora rot or soybean cyst nematodes. Crop Sci. 22:588-590.

Irwin, J. A. G. and P. W. Langdon. 1982. A laboratory procedure for determining relative levels of field resistance in soybean to *Phytophthora megasperma* f. sp. *glycinea*. Aust. J. Agric. Res. 33:33-39.

Jimenez, B. and J. L. Lockwood. 1980. Laboratory method for assessing field tolerance of soybean seedlings to *Phytophthora megasperma* var. *sojae*. Plant Dis. 64:775-778.

Keeling, B. L. 1982a. A seedling test for resistance to soybean stem canker caused by *Diaporthe phaseolorum* var. *caulivora*. Phytopathology. 72:807-809.

Keeling, B. L. 1982b. Factors affecting the reaction to *Phytophthora megasperma* var. *sojae* in hydroponic culture. Crop Sci. 22:325-327.

Kilen, T. C. and B. L. Keeling. 1977. Simultaneous screening of soybeans against three races of *Phytophthora megasperma* var. *sojae*. Crop Sci. 17:185-186.

Kobayashi, K., N. Kondo, T. Ui, H. Tachibana and T. Aota. 1983. Difference in pathogenicity of *Phialophora gregata* isolates from adzuki bean in Japan and from soybean in the United States. Plant Dis. 67:387-388.

Krausz, J. P. and B. A. Fortnum. 1983. An epiphytotic of Diaporthe stem canker of soybean in South Carolina. Plant Dis. 67:1128-1129.

Laviolette, F. A. and K. L. Athow. 1983. Two new physiologic races of *Phytophthora megasperma* f. sp. *glycinea*. Plant Dis. 67:497-498.

Leath, S. and R. B. Carroll. 1982. Screening for resistance to *Fusarium oxysporum* in soybean. Plant Dis. 66:1140-1143.

Moots, C. K., C. D. Nickell, L. E. Gray and S. M. Lim. 1983. Reaction of soybean cultivars to 14 races of *Phytophthora megasperma* f. sp. *glycinea*. Plant Dis. 67:764-767.

Morrison, R. H. and J. C. Thorne. 1978. Inoculation of detached cotyledons for screening soybeans against two races of *Phytophthora megasperma* var. *sojae*. Crop Sci. 18:1089-1091.

Schmitthenner, A. F. and A. K. Walker. 1979. Tolerance versus resistance for control of phytophthora root rot of soybeans. p. 35-44. *In* H. D. Loden and D. Wilkenson (eds.) Proc. of the 9th Soybean Seed Res. Conf. Am. Seed Trade Assoc., Washington, D.C.

Schwenk, F. W., C. A. Ciaschini, C. D. Nickell and D. G. Trombold. 1979. Inoculation of soybean plants by injection with zoospores of *Phytophthora megasperma* var. *sojae*. Phytopathology. 69:1233-1234.

Sebastian, S. A. and C. D. Nickell. 1983. The relationship between yield and brown stem rot in soybeans. (Abstr.) Phytopathology. 73:795.

Sebastian, S. A., C. D. Nickell and L. E. Gray. 1983. Sequential screening of soybean plants for resistance to phytophthora rot and brown stem rot. Crop Sci. 23:1214-1215.

Tachibana, H. and L. C. Card. 1972. Brown stem rot resistance and its modification by soybean mosaic virus in soybeans. Phytopathology. 62:1314-1317.

Tachibana, H. 1982. Prescribed resistant cultivars for controlling brown stem rot of soybean and managing resistance genes. Plant Dis. 66:271-273.

Tooley, P. W. and C. R. Grau. 1982. Identification and quantitative characterization of rate-reducing resistance to *Phytophthora megasperma* f. sp. *glycinea* in soybean seedlings. Phytopathology. 72:727-733.

Walker, A. K. and A. F. Schmitthenner. 1982. Developing soybean varieties tolerant to phytophthora
   rot. p. 67-78. *In* H. D. Loden and D. Wilkenson (eds.) Proc. of the 12th Soybean Seed Res. Conf.
   Am. Seed Trade Assoc., Washington, D.C.

Weaver, D. B., P. A. Backman, B. H. Cooper and D. L. Thurlow. 1983. Susceptibility of soybean cultivars to field infestations of stem canker. Agron. Abstr. p. 84.

White, D. M., J. E. Partridge, and J. H. Williams. 1983. Races of *Phytophthora megasperma* f. sp.
   glycinea in eastern Nebraska. Plant Dis. 67:1281-1282.

Wilcox, J. R. 1983. Breeding soybeans resistant to diseases. p. 183-235. *In* J. Janick (ed.) Plant Breeding Rev. Vol. 1. AVI Publishing Co. Inc., Westport, CT.

BREEDING FOR RESISTANCE TO LEAF DISEASES

R. L. Bernard

In considering this topic, one must first decide just what constitutes a leaf disease. Many diseases affecting the root and stem may demonstrate their first obvious field symptoms in the leaves: wilting, yellowing, necrosis, etc. An example would be the often rather spectacular leaf symptoms (yellowing and necrosis) caused by brown stem rot (*Phialophora gregatum* [Allington and Chamberlain] W. Gams), which as its name implies, is primarily a stem disease. Likewise, many leaf diseases also affect pod tissue and consequently are important as seed diseases. Downy mildew, rust, anthracnose, and scab, which initially infect the leaf, are examples of leaf diseases which may have major effects on the seeds also. Virus diseases, too, often show their major symptoms in the tissue of the leaves. This presentation will cover non-viral diseases that are considered to affect leaves primarily.

Leaf diseases, because of their often very distinctive and prominent appearance in growing soybeans, have received considerable attention, perhaps more than their economic importance merits. Damage by root, stem, and seed diseases to yield and seed quality can be rather considerable, and yet, go unobserved. Since leaf diseases are so readily identifiable we, perhaps, know more about their frequency than other diseases.

Disease surveys and plots grown specifically to monitor the frequency of disease occurrence provide useful information on the prevalence of leaf diseases. A few are widespread and occur year after year in the major soybean growing areas of the United States, and these have received the most attention in breeding programs. In the U.S., bacterial blight, downy mildew, and brown spot are the most common leaf diseases and are observed in over half of the soybean fields in many areas in most years. They often can be found in virtually every field at some time during the growing season.

In the following sections major leaf diseases will be described from the breeding standpoint: cultivar differences, inheritance of resistance, economic loss, and how much progress breeding has made. Most of this is based on the situation in the U.S. Each of the major leaf diseases has a very different story, ranging from complete success to complete failure. Taken together they illustrate the various breeding methods that may be used in coping with diseases.

BACTERIAL PUSTULE

Bacterial pustule, caused by *Xanthomonas campestris* pv. *phaseoli*, illustrates the ideal breeding solution to a disease problem. Formerly widespread in the central and southern U.S., it is greatly reduced in occurrence now because of the use of resistant cultivars. It is still commonly observed whenever susceptible cultivars are grown. Its growth and spread are favored by warm, rainy weather, and therefore, it is more of a potential problem in the South. Varietal differences in response to bacterial pustule were studied as far back as the 1920s, but a high level of resistance found in the cultivar CNS was first reported in 1951 by Hartwig and Lehman and by Feaster, and was identified as being controlled by a single major recessive gene. This is a rather rare type of resistance, since it has not been found elsewhere in the U.S. germplasm collection, except in a few lines that are probably related to CNS. The gene from CNS exists in every major cultivar in the southern U.S. today (Maturity Groups V and later), beginning with the widely grown Lee cultivar released in 1954. By backcrossing and other breeding methods, it has also been transferred to many northern cultivars (Groups II, III, and IV), notably Clark-63, Wayne, Cumberland, and Williams, and this has greatly restricted frequency of disease occurrence in the southern Midwest.

Thus, we have a situation today where a once prevalent disease has, at present, been effectively controlled by a single resistance gene. More importantly, after 40 years of planting large commercial hectarages of resistant cultivars, no breakdown of resistance has been reported. How much is this worth to soybean production? Bacterial pustule is not a devastating disease, except perhaps on a few extremely susceptible cultivars. Studies using artificial inoculation, and other studies testing related lines with and without the resistance gene, have given yield loss estimates ranging from 0 to about 15%. A reasonable estimate would be that heavy infection in the South can cause 10% yield losses, whereas the lighter infections as occur farther north cause 0 to 5% losses. Considering the large area over which these rather small losses are prevented by the resistance gene, its value to the economy is quite great. This is an ideal plant breeding success story, but one which is not repeated with any of the other leaf diseases.

BACTERIAL BLIGHT

Bacterial blight, caused by *Pseudomonas syringae* pv. *glycinea*, in most years is the most prevalent leaf spot disease in the North Central Region. It is especially prevalent during the cooler, early part of the growing season, and is more common in the North. Its development is arrested by the onset of hot summer weather. Mid- and late-season plant growth, free of the disease symptoms, often overtops and obscures the damaged lower leaves. There are reports of a number of races of this pathogen (Cross et al., 1966; Thomas and Leary, 1980; Fett and Sequeira, 1981) based on differential responses of common U.S. cultivars, and a major gene for resistance to one of the races has been identified (Mukherjee et al., 1966). However, this information has not been the basis for much breeding effort. Rather, the approach has been to select for what might be described as horizontal resistance. Whereas a number of older U.S. cultivars and many strains in the germplasm collection are very highly susceptible to bacterial blight, with severe and long-lasting symptoms, U.S. commercial cultivars of today are only slightly to moderately susceptible and for the most part sustain little or no economic loss. Evidently, consciously or unconsciously, selection has proceeded away from highly susceptible types, and without selection for immunity or major-gene resistance the disease has been controlled. This moderate level of resistance (or low susceptibility) does not seem to be race-specific, and the hope is that this will continue to provide long-term, stable and adequate economic control of this very common disease.

OTHER BACTERIAL LEAF DISEASES

Several other lesser bacterial leaf spot diseases have been reported. Wildfire, caused by *Pseudomonas syringae* pv. *tabaci*, has caused serious damage in the past in the South, but its invasion of the soybean leaf is related to bacterial pustule infection and it has been controlled by the same gene that gives resistance to pustule. This gene is present in most cultivars grown where wildfire formerly occurred. Bacterial wilt and bacterial tan spot, caused by *Corynebacterium* spp. have been reported in Iowa in recent years and some cultivar differences occur, but these diseases have not received much attention from breeders as yet.

DOWNY MILDEW

Downy mildew, caused by *Peronospora manshurica* (Naoum.) Syd., is the most widespread fungus-caused disease in the U.S. In the Midwest, many years it may be observed in a majority of the soybean fields. It is transferred from place to place and crop to crop through spore encrustations on the seeds, which give rise to systemically infected plants, and the disease may then spread to the rest of the field through rain and wind-carried spores. Many races (over 30) have been reported for this fungus (Dunleavy, 1971, 1977b), and each commercial cultivar is resistant to some and susceptible to some. My unpublished results with yield-testing resistant isolines of susceptible cultivars, and the observations of others, suggest that downy mildew does not have much effect on yield as it normally occurs. However, in some fields, especially late-planted fields, very heavy infections have been observed, and this disease probably warrants continued attention from the breeder.

Based on Dunleavy's 1970 report of a few cultivars with resistance to all known races, we conducted a genetic study and found a major gene for this resistance in the vegetable cultivar, Kanrich (Bernard and Cremeens, 1972). Kanrich had appeared to be completely resistant to downy mildew in this country, although I had seen it heavily infected in an experimental plot in northeast China in 1974. The Group IV cultivar, Union, with the gene for resistance from Kanrich was released for the southern Midwest in 1977. At first no downy mildew was observed on Union, but as the commercial hectarage of this cultivar increased, not unexpectedly, in 1981 downy mildew was found on Union at several test locations in Illinois (Lim et al., 1984). It has appeared annually since then, although it often appears later or develops less intensively than on related susceptible cultivars. Other cultivars, which carry resistance to this new race, have already been identified, and so, the opportunity for further breeding exists. Three gene pairs controlling race-specific resistances were reported many years ago by Geeseman (1950), but not enough is yet known about the inheritance of specific race resistances. However, breeding for multigenic resistance would seem a likely option for the control of this pest.

The existence of a large number of races makes the use of this type of resistance of uncertain value, and a better approach may be available. Within the susceptible response there is a wide range in number and size of leaf spots that develop, and in my observations this is a rather consistent cultivar trait over a range of locations and years. Some cultivars never seem to be very heavily infected. Data from artificial inoculation (Dunleavy, 1970), however, indicate that these differences may also be race-specific. This needs to be further studied and its inheritance identified before it can be strongly recommended as a breeding procedure.

BROWN SPOT

A very different situation exists for brown spot, caused by *Septoria glycines* Hemmi, probably the second most commonly occurring leaf disease in the Midwest. The occurrence of this disease is often less noticeable because it concentrates its attack on the lower leaves of the plant, in contrast to downy mildew, bacterial blight, and pustule, all of which first infect and develop on the upper leaves of the plant canopy. Infection often occurs by splashing of spores that overwintered on plant debris from the previous soybean crop. Therefore, the disease has become more serious where soybeans are grown after soybeans in the same field, but it can also develop to a significant level in rotated fields. Data from inoculated plots demonstrated that significant yield losses occur in both the South (17% yield reduction, Young and Ross, 1979) and the North (12 to 34% reduction, Lim, 1980). Although some cultivar differences have been observed and PI 68708 has been described as showing some resistance (Dunleavy et al., 1960), no significant level of resistance has been found and little if any breeding work is underway. Considering the rather appreciable yield losses that seem to be occurring, the discovery of resistance to this disease should be a major breeding goal.

In my field observations, I have been impressed with the close relationship of plant lodging and intensity of brown spot development. Lodged plants or even the lodged part of plants are often much more heavily infected and defoliated than erect ones. It seems that the closer to the ground a leaf is the greater the infection and defoliation. Under these condtions breeding for lodging resistance will give a certain amount of protection from the disease.

FROGEYE LEAF SPOT

Frogeye leaf spot, caused by *Cercospora sojina* Hara, has occasionally severely attacked certain highly susceptible cultivars in both the North and the South, but is for the most part a rather rare disease in the U.S. today, presumably because commercial cultivars are relatively resistant. This has been accomplished evidently by simply avoiding highly susceptible cultivars like those on which frogeye epiphytotics have developed. Races of this disease organism have been reported by Athow et al. (1962) and Ross (1968), and many commonly grown cultivars were reported to be susceptible to one or more races. However, a rating of susceptibility to artificial inoculation does seem to mean that the cultivar will develop the disease in the field, and frogeye remains a relatively rare disease. Genetic studies have identified two major genes, occurring in commercial cultivars, which give resistance to two races of the frogeye pathogen (Athow and Probst, 1952; Probst et al., 1965).

TARGET SPOT

Target spot, caused by *Corynespora cassiicola* (Berk. and Curt.) Wei, has been shown to be capable of causing appreciable losses on susceptible cultivars in the South (Hartwig, 1959). It is a late-season disease, usually not beginning its development until August, but then advancing rapidly in warm, moist seasons and capable of causing near-complete defoliation. Fortunately, resistance is not rare and most cultivars now grown in the South are resistant. The breeder's objective should be to avoid releasing susceptible cultivars, which is not too difficult since resistance is in most of the adapted parent cultivars.

POWDERY MILDEW

Powdery mildew, caused by *Microsphaera diffusa*, occasionally develops in soybean fields and has been reported to cause appreciable yield loss (up to 26%) in Iowa. It is a common pest of greenhouse-grown soybean plants. A number of highly susceptible cultivars, Amsoy, Corsoy, Hark, and Wells, have become popular in the North and this has probably brought about the recent surge of interest in powdery mildew. There is a wide range of reaction expressed by commonly grown cultivars (Buzzell and Haas, 1975; Grau and Laurence, 1975; Dunleavy, 1977a) and some evidence of race differentiation (Grau and Laurence, 1975). Resistance is common among commercially grown cultivars, and a major dominant gene for resistance has been identified (Grau and Laurence, 1975; Buzzell and Haas, 1975; 1978). This is a disease that will bear watching, but for the present does not appear to be prevalent enough to merit much breeding attention.

RUST

Rust, caused by *Phakopsora pachyrhizi* Syd., is perhaps the most damaging leaf disease of soybeans. Under conditions favorable to the fungus, yield losses can range up to 50% (Bromfield, 1976; Yang, 1977) or more. It is most damaging in tropical and sub-tropical regions of eastern Asia. It is not known to occur in the U.S., but is widespread in South America and the Caribbean area. A few resistant cultivars have been identified as a result of germplasm screening efforts at AVRDC in Taiwan. The following strains from the USDA germplasm collection have shown some resistance to rust, based on observations in Taiwan and at the Plant Disease Research Laboratory in Frederick, Maryland: PI 200492, 224268, 227687, 230970, 230971, 339871, 459024, 459025. Some of these have been used in breeding programs, and cultivars with resistance from PI 200492 have been developed and released in Taiwan and the U.S. (Dowling, released in Texas in 1978). Because of the multiracial situation these resistance sources are inconsistent over location and time (Yeh, 1983), and no cultivar with complete resistance has been identified. Probably combining multigenic resistance and an ongoing search for new sources and new types of resistance should be breeding goals.

OTHER LEAF DISEASES

Important diseases such as anthracnose (*Colletotrichum dematium* var. *truncatum* and *Glomerella glycines*), scab (*Sphaceloma glycines*) (reported only in Japan), and Rhizoctonia aerial foliage and web blight (Sinclair, 1982) infect all above-ground plant parts and are not primarily leaf diseases, and so were not included in this report.

Alternaria and Phyllosticta leaf spots are not uncommon in U.S. soybean fields, but have not been observed at levels that would cause economic losses. In the tropics, a number of leaf diseases caused by various genera of fungi, *Choanephora, Dactuliophora, Drechslera, Leptosphaerulina, Pyrenochaeta*, and *Stemphylium* have been reported (Sinclair, 1982), but little is known about them. Where soybean production becomes more intensive it is to be expected that some of these may become recognized as production problems.

NOTES

R. L. Bernard, USDA/ARS and Department of Agronomy, University of Illinois, Urbana, Illinois.

REFERENCES

Athow, K. L. and A. H. Probst. 1952. The inheritance of resistance to frog-eye leaf spot of soybeans. Phytopathology. 42:660-662.

Athow, K. L., A. H. Probst, C. P. Kurtzman and F. A. Laviolette. 1962. A newly identified physiological race of *Cercospora sojina* on soybean. Phytopathology. 52:712-714.

Bernard, R. L., and C. R. Cremeens. 1972. A gene for general resistance to downy mildew of soybeans. J. Heredity. 62:359-362.

Bromfield, K. R. 1976. World soybean rust situation. p. 491-500. *In* L. D. Hill (ed.) World Soybean Research. Interstate Printers and Publishers, Danville, Illinois.

Buzzell, R. I. and J. H. Haas. 1975. Powdery mildew of soybeans. Soybean Genet. Newsletter 2:7-9.

Buzzell, R. I. and J. H. Haas. 1978. Inheritance of adult plant resistance to powdery mildew in soybeans. Can. J. Genet. Cytol. 20:151-153.

Cross, J. E., B. W. Kennedy, J. W. Lambert and R. L. Cooper. 1966. Pathogenic races of the bacterial blight pathogen of soybeans. Plant Dis. Rep. 50:557-560.

Dunleavy, J. M. 1970. Sources of immunity and susceptibility to downy mildew of soybeans. Crop Sci. 10:507-509.

Dunleavy, J. M. 1971. Races of *Peronospora manshurica* in the United States. Am. J. Bot. 58:209-211.

Dunleavy, J. M. 1977a. Comparison of the disease response of soybean cultivars to *Microspora diffusa* in the greenhouse and the field. Plant Dis. Rep. 61:32-34.

Dunleavy, J. M. 1977b. Nine new races of *Peronospora manshurica* found on soybeans in the Midwest. Plant Dis. Rep. 61:661-663.

Dunleavy, J. M., C. R. Weber and D. W. Chamberlain. 1960. A source of bacterial blight resistance for soybeans. Proc. Iowa Acad. Sci. 67:120-125.

Feaster, C. V. 1951. Bacterial pustule disease in soybeans; artificial inoculation, varietal resistance, and inheritance of resistance. Mo. Agric. Exp. Stn. Res. Bull. 487.

Fett, W. F. and L. Sequeira. 1981. Further characterization of the physiologic races of *Pseudomonas glycinea*. Can. J. Bot. 59:283-287.

Geeseman, G. E. 1950. Inheritance of resistance of soybeans to *Peronospora manshurica*. Agron. J. 12:608-613.

Grau, C. R. and J. A. Laurence. 1975. Observations on resistance to powdery mildew of soybeans. Plant Dis. Rep. 59:458-460.

Hartwig, E. E. 1959. Effect of target spot on yield of soybeans. Plant Dis. Rep. 43:504-505.

Hartwig, E. E. and S. G. Lehman. 1951. Inheritance of resistance to the bacterial pustule disease in soybeans. Agron. J. 43:226-229.

Lim, S. M. 1980. Brown spot severity and yield reduction in soybean. Phytopathology. 70:974-977.

Lim, S. M., R. L. Bernard, C. D. Nickell and L. E. Gray. 1984. New physiological race of *Peronospora manshurica* virulent to the gene *Rpm* in soybeans. Plant Dis. 68:71-72.

Mukherjee, D., J. W. Lambert, R. L. Cooper and B. W. Kennedy. 1966. Inheritance of resistance to bacterial blight in soybeans. Crop Sci. 6:324-326.

Probst, A. H., K. L. Athow and F. A. Laviolette. 1965. Inheritance of resistance to race 2 of *Cercospora sojina* in soybeans. Crop Sci. 5:332.

Ross, J. P. 1968. Additional physiological races of *Cercospora sojina* on soybeans in North Carolina. Phytopathology. 58:708-709.

Sinclair, J. B. (ed.). 1982. Compendium of soybean diseases (2nd ed.) Amer. Phytopathol. Soc., St. Paul.

Thomas, M. D. and J. V. Leary. 1980. A new race of *Pseudomonas glycinea*. Phytopathology. 70:310-312.

Yang, C. Y. 1977. Past and present studies on soybean rust incited by *Phakopsora pachyrhizi* SYD. Bull. Inst. Trop. Agric., Kyushu Univ. 2:78-94.

Yeh, C. C. 1983. Physiological races of *Phakopsora pachyrhizi* in Taiwan. J. Agric. Res. of China 32:69-74.

Young, L. D. and J. P. Ross. 1979. Brown spot development and yield response of soybean inoculated with *Septoria glycines* at various growth stages. Phytopathology. 69:8-11.

BREEDING FOR RESISTANCE TO VIRUSES IN SOYBEANS

G. R. Buss, C. W. Roane and S. A. Tolin

Over 50 different viruses have been reported to cause soybean diseases (Sinclair, 1982). Among the more commonly occurring and damaging viruses are soybean mosaic virus (SMV), tobacco ringspot virus (TRSV), bean pod mottle virus (BPMV), bean yellow mosaic virus (BYMV), peanut mottle virus (PMV), and cowpea chlorotic mottle virus (CCMV). Yield losses are determined by several factors, including the infecting virus, the soybean cultivar, the stage of growth at which plants are infected, and the proportion of plants that are infected. In addition to yield losses, quality of seed from infected plants is often reduced by mottling on the seed coat.

At this time, genetic resistance appears to be the most practical and effective means of controlling viruses of soybean, since virus infection is difficult to control by traditional disease control methods. No practical viricide treatments are presently available. Control of alternate hosts, such as weeds and other crops, and planting virus-free seed in the case of seed-borne viruses (SMV and TRSV) reduce the inoculum sources, but these methods often are not effective. Control of aphid (BYMV, PMV, SMV) or beetle (BPMV, CCMV) vectors is seldom practical. Aphids in particular are very transient and can quickly spread virus within and between soybean fields (Irwin and Schultz, 1981).

VIRUS IDENTIFICATION

If one is to conduct a successful breeding program for virus resistance, it is important to have positive identification of the virus that is causing losses. Collaboration with a plant virologist is very nearly essential in this step, and is very helpful in the whole breeding process.

Symptoms are unreliable for identifying viruses, since symptoms of many viruses are very similar in soybeans and will vary depending on the strain of the virus, the genotype of the host, and environmental conditions. In addition, a given plant can be infected by more than one virus.

Serological testing is the most reliable method for virus identification. The Ouchterlony, double-diffusion method (Tolin and Ford, 1983) and the enzyme-linked immunosorbent assay (ELISA) (Moore et al., 1982) are two that are commonly used. Reaction with antiserum, specific for a given virus, tests for the presence or absence of that virus and is very accurate if conducted properly. Serology is useful in distinguishing one virus from another, but will not differentiate strains of the same virus. In addition, these procedures require certain laboratory facilities that might not be generally available to plant breeders.

In situations where facilities are not available for serological testing, a virus can be identified by mechanical inoculation to differential host plants and observation of symptom development. Such biological tests are useful to verify serological tests. The reactions of selected differential host species can differentiate between viruses. For example, inoculation of Topcrop bean (*Phaseolus vulgaris* L.) results in large necrotic local lesions with PMV, small necrotic lesions with SMV, but systemic mosaic with BPMV or BYMV. Cowpea (*Vigna unguiculata* [L.] Walp.) seldom shows symptoms, if it is infected at all, with these four viruses, but shows a bright yellow mottle with CCMV and systemic necrosis with TRSV. *Chenopodium quinoa* Willd. is also a good differential host for these viruses. It is systemically infected by TRSV, but is immune to BPMV. CCMV causes tiny chlorotic lesions in 4 to 5 days. SMV and BYMV usually produce necrotic lesions in 8 to 10 days. The reaction of differential soybean culti-vars can be used to identify strains of a given virus and, in fact, is the only means presently available for identifying strains of some viruses such as SMV (Cho and Goodman, 1979).

GENETICS OF RESISTANCE

Resistance of soybean to a number of viruses has been reported. Tisselli et al. (1980) list a number of sources of resistance to SMV, CCMV, PMV and BYMV, but none for BPMV, TRSV or any others.

Resistance in soybean to viruses has been reported in most cases to be controlled by one or two host genes. The inheritance of resistance to SMV and PMV has been the most extensively studied; in all cases except one, resistance is conferred by a single dominant gene. Also, several strains of these viruses have been identified based on their reactions on soybean cultivars that have resistance to the common strain. It is clear from the SMV and PMV examples that multiple resistance genes are available and that they each offer specific combinations of resistance to the known strains that apparent-ly have different genes for pathogenicity. These two host-virus interactions will be con-sidered in detail below.

Soybean Mosaic Virus

Symbols have been assigned to three separate genes that independently confer resistance to SMV. Kiihl and Hartwig (1979) assigned the gene symbol *Rsv* to the dom-inant gene found in PI 96983. They assigned *rsv*$^+$ as the gene symbol for the gene in Ogden which is recessive to *Rsv* and allelic to it. Both alleles are dominant to the allele for susceptibility. Buzzell and Tu (1984) found that the dominant gene in Raiden was at a different locus and labelled it *Rsv*$_2$. A recessive gene for resistance was found in some experimental lines by Kwon and Oh (1980), but no gene symbols were assigned.

Ninety-eight isolates of SMV were classified by Cho and Goodman (1979, 1982) into seven strain classes based on disease reactions of a number of differential soybean cultivars (Table 1).

There is some association between the reported genes for resistance and these reac-tions. According to Cho and Goodman (1982), PI 96983 has the same reaction as Buffalo to the SMV strains. Since Buffalo is in a different reaction class from Ogden, there is good agreement with the conclusion of Kiihl and Hartwig (1979) that PI 96983 and Ogden have different alleles for resistance. Buzzell and Tu (1984) reported that Raiden is resistant to all seven strain groups. Thus, the three genes for which symbols have been assigned respond differently to the SMV strains. However, there are some inconsistencies in the reactions, which are difficult to explain. It is assumed that the gene for resistance in York is descended from Ogden via Hood, since Ogden is a parent

Table 1. Reaction of differential cultivars to SMV strains G1-G7 (Cho and Goodman, 1979; 1982).

Cultivar	Reaction to SMV Strain[a]						
	G1	G2	G3	G4	G5	G6	G7
Clark	M	M	M	M	M	M	M
Davis	-	-	-	N	M	M	M
Marshall	-	N	N	-	-	N	N
Ogden	-	-	N	-	-	-	N
Kwanggyo	-	-	-	-	N	N	N
Buffalo	-	-	-	-	-	-	N
Suweon 97	-	-	-	-	-	-	-

[a]- = symptomless; M = mosaic; N = necrosis.

of Hood and Hood is a parent of York. Hood is in the same reaction class as Ogden, but York is not (Cho and Goodman, 1982). The other parent of York is Dorman, which is susceptible to SMV. It would appear either that susceptible cultivars have some genes that modify the expression of the major genes for resistance, or that the resistance genes themselves can be modified as they are passed from one generation to the next. Cho and Goodman (1982) also cite examples of lines with a common resistant parent which have different reactions to the SMV strains. Obviously, resolution of this problem will require further research on soybean genetics.

An alternative to breeding for resistance is to breed for low incidence of seed transmission, since plants infected via seeds are considered to be the source of virus for the spread of SMV. Bowers and Goodman (1982) have identified 27 soybean lines that are infected by SMV, but appear to resist seed transmission. Until the inheritance of the trait is known, its value in a breeding program cannot be assessed.

Peanut Mottle Virus

Three genes for resistance to PMV have been studied. Boerma and Kuhn (1976) reported a single dominant gene in cultivars Dorman and DNS and assigned the gene symbol, *Rpv*. Later, Shipe et al. (1979b) found that a recessive gene in Peking was at a locus different from the dominant gene in Arksoy. It was assumed that Arksoy and Dorman have the same genes by descent so the Peking gene was labeled *rpv$_2$*. In a later study, Buss et al. (1985) showed that Arksoy and Dorman, as well as York and Shore, in fact, do have the same gene for resistance. They also found that the CNS gene is not at the *Rpv* locus, but deferred assigning a gene symbol until allelism tests with the *rpv$_2$* locus are completed. A number of other resistant genotypes have been identified, but have not been studied genetically (Shipe et al., 1979a; Boerma and Kuhn, 1976).

Five strains of PMV have been distinguished in soybean (Bays, 1983). Table 2 shows the reactions of the differential cultivars to those strains. The cultivar reactions to the strains are in agreement with the available genetic information. Arksoy, Dorman, York and Shore react similarly to the five strains and have the same gene for resistance, whereas Peking and CNS are resistant to all the strains and have resistance genes at other loci.

BREEDING FOR RESISTANCE

Breeding Objectives

The ultimate concern of the soybean breeder is to incorporate genetic resistance into high-yielding, adapted cultivars. Thus, the inheritance of resistance plays a major

Table 2. Reaction of some cultivars to PMV strains P1-P5 (Bays, 1983).

Cultivars	Reaction to PMV strain[a]				
	P1	P2	P3	P4	P5
Lee 68	S	S	M	S	M
Cumberland	S	M	M	M	-
Virginia	S	M	M	-	-
Arksoy, Dorman	-	-	-	-	M
Shore, York	-	-	-	-	M
Peking, CNS, Davis	-	-	-	-	-
Buffalo, Haberlandt	-	-	-	-	-
Kwanggyo	-	-	-	-	-

[a]- = symptomless; M = mild mottle; S = severe mottle and blistering.

role in how it is used in a breeding program. Resistance to viruses in soybeans is simply inherited, so it should be feasible to transfer the genes effectively by most methods already employed by the breeder, and perhaps by new genetic engineering or recombinant DNA methods if the genes can be isolated and cloned.

Since most reported virus resistance genes are dominant, the trait can be incorporated into adapted germplasm relatively quickly by backcrossing. This is particularly useful if resistance is found in unadapted germplasm. No genetic linkages between resistance genes and agronomically undesirable traits have been reported. However, the genes for SMV and PMV resistance in York are linked with 3.7 ± .8% recombination (Roane et al., 1983). This linkage can be used to transfer resistance to both viruses while using only one virus for inoculations. It would be necessary, ultimately, to verify the presence of both genes, but much effort could be saved by eliminating nearly half of the inoculations.

In situations in which resistance is available in adapted germplasm and virus resistance is just one of several objectives in the breeding programs, any breeding method, such as pedigree or single seed descent, is acceptable. The point in the process at which virus evaluation is conducted is largely determined by the priority of resistance in the overall objectives.

No sources of resistance to some viruses have been found in soybean. However, there are some partial solutions that can be used in a breeding program to lessen the impact of damage from those viruses. For example, with BPMV, Quiniones et al. (1971) reported yield losses of 10% from infection with BPMV and 18% from SMV, but infection by both viruses caused a 66% yield loss. While there is no known resistance to BPMV, the extreme losses from the synergistic effect can be avoided by cultivars having SMV resistance.

The mechanism by which different strains of a virus evolve is not understood, so it is difficult to devise a breeding strategy aimed at preventing development of more virulent strains. An interim strategy would seem to be one which avoids genetic uniformity, which may result in vulnerability to new strains or increased damage from these strains (Kwon and Oh, 1980). For example, the potential for loss is much greater for a host-virus combination that results in systemic necrosis rather than mosaic symptoms. Toward this end, more information is needed on the host-pathogen genotype interactions for the important soybean viruses. Research on the molecular basis of resistance of the host and pathogenicity of the virus might also be of value in this regard.

Inoculation Methods

In most situations, natural infections by virus are not of sufficiently high incidence for efficient screening of breeding materials. It is rare that a field will be uniformly infected with only a single virus or that the infection would occur within a short time. Thus, artificial inoculation is an important aspect of breeding for virus resistance.

Just as it is important to identify positively the virus with which one is working, it is also important to maintain pure cultures of the virus and to maintain that purity during production of inoculum. Since viruses can replicate only in a living host, inoculum must be increased on plants of a susceptible cultivar or a susceptible species. In order to maintain purity of the culture, care must be taken to use seeds that are not infected with any viruses. The best way to obtain clean seed is to grow a small increase block and several times through the season, remove any plants that show virus symptoms. Also, when the seeds are planted for inoculum increase, any seeds with seedcoat mottling should not be used, and any seedlings with virus symptoms should be removed before inoculation.

Seedlings can be inoculated any time after the unifoliolate leaves expand. Best results in the greenhouse are obtained following inoculation of rapidly growing plants before the first trifoliolate leaf unfolds.

Different methods of inoculation may be used, depending on the number of plants to be inoculated and the desired accuracy of the results. Inoculation by hand methods is the most effective and efficient for up to several hundred seedlings at a time. The procedure (Cho and Goodman, 1979; Roane et al., 1983) involves grinding about a gram of infected leaves with 10 mL of buffer solution in a mortar and pestle. The plants to be inoculated are dusted with a small amount of carborundum as an abrasive and a small amount of inoculum suspension is rubbed on a leaf of each plant with the pestle. The plants are rinsed with water and allowed to grow until symptoms appear, which is 7 to 10 days or more, depending on growing conditions. This method produces a high proportion of infected plants and is suggested for most greenhouse work and small field studies.

For inoculating large numbers of plants, as in field nurseries, mechanical devices are available to reduce the labor. When it is important that each plant is carefully inoculated, as in genetic studies, the air-brush technique is very useful (Shipe et al., 1979b; Kiihl and Hartwig, 1979; Roane et al., 1983). The preparation of inoculum is similar to the method used for smaller quantities except that grinding is done in a blender and carborundum is added directly to the suspension. A buffer containing additives to stabilize the virus for longer periods of time improves results (Tolin and Ford, 1983). The inoculum is applied to the under side of a leaf by a jet of compressed air from the air brush held 1 to 2 mm from a hand-held leaf.

For large nurseries, the pad inoculator is very efficient (Ross, 1978). The device consists of an open-ended box mounted on bicycle wheels. Foam rubber pads mounted on the inside of the box are kept moist with the inoculum and as the device is pushed along the soybean row, the inoculum is rubbed into the leaves. This method requires the plants to be somewhat larger than with the other methods. It is very useful in evaluating breeding lines for virus reaction and can be performed more rapidly and less laboriously than with the air brush.

Virus isolates can be maintained continuously by periodically transferring the virus to young seedlings. However, this is time-consuming and increases the risk of contamination by an alien virus or of genetic modifications of the virus through multiple passages through the host. The preferred method for long-term maintenance of inoculum is by desiccation of infected tissue over calcium chloride and storage at 4 C (Roane et al., 1983).

NOTES

G. R. Buss, Associate Professor, Department of Agronomy; C. W. Roane and S. A. Tolin, Professors, Department of Plant Pathology; Virginia Polytechnic Institute and State University, Blacksburg, VA 24061.

REFERENCES

Bays, D. C. 1983. Variability of the peanut mottle virus reaction in soybean (*Glycine max*). Ph.D. Diss. Virginia Polytechnic Institute and State University, Blacksburg.

Boerma, H. R. and C. W. Kuhn. 1976. Inheritance of resistance to peanut mottle virus in soybeans. Crop Sci. 16:533-534.

Bowers, G. R. and R. M. Goodman. 1982. New sources of resistance to seed transmission of soybean mosaic virus in soybeans. Crop Sci. 22:155-156.

Buss, G. R., C. W. Roane, S. A. Tolin and T. Vinardi. 1985. A second dominant gene for resistance to peanut mottle virus in soybeans. Crop Sci. (In press).

Buzzell, R. I. and J. C. Tu. 1984. Inheritance of soybean resistance to soybean mosaic virus. J. Hered. 75:82.

Cho, E. K. and R. M. Goodman. 1979. Strains of soybean mosaic virus: Classification based on virulence in resistant soybean cultivars. Phytopathology 69:467-470.

Cho, E. K. and R. M. Goodman. 1982. Evaluation of resistance in soybeans to soybean mosaic virus strains. Crop Sci. 22:1133-1136.

Irwin, M. E. and G. A. Schultz. 1981. Soybean mosaic virus. FAO Plant Protection Bull. 29:41-55.

Kiihl, R. A. S. and E. E. Hartwig. 1979. Inheritance of reaction to soybean mosaic virus in soybeans. Crop Sci. 19:372-375.

Kwon, S. H. and J. H. Oh. 1980. Resistance to a necrotic strain of soybean mosaic virus in soybean. Crop Sci. 20:403-404.

Moore, D. L., R. M. Lister, T. S. Abney and K. L. Athow. 1982. Evaluation of virus contents in soybean by enzyme-linked immunosorbent assay. Plant Dis. 66:790-793.

Quiniones, S. S., J. M. Dunleavy and J. W. Fisher. 1971. Performance of three soybean varieties inoculated with soybean mosaic virus and bean pod mottle virus. Crop Sci. 11:662-664.

Roane, C. W., S. A. Tolin and G. R. Buss. 1983. Inheritance of reaction to two viruses in the soybean cross 'York' x 'Lee 68'. J. Hered. 74:289-291.

Ross, J. P. 1978. A pad inoculator for infecting large numbers of plants with virus. Plant Dis. Rep. 62:122-125.

Shipe, E. R., G. R. Buss and C. W. Roane. 1979a. Resistance to peanut mottle virus (PMV) in soybean (*Glycine max*) plant introductions. Plant Dis. Rep. 63:757-760.

Shipe, E. R., G. R. Buss and S. A. Tolin. 1979b. A second gene for resistance to peanut mottle virus in soybeans. Crop Sci. 19:656-658.

Sinclair, J. B. 1982. Compendium of Soybean Diseases. Second Edition. American Phytopathological Society, St. Paul.

Tisselli, O., J. B. Sinclair and T. Hymowitz. 1980. Sources of resistance to selected fungal, bacterial, viral and nematode diseases of soybeans: INTSOY Series 18. INTSOY, University of Illinois, Urbana.

Tolin, S. A. and R. H. Ford. 1983. Purification and serology of peanut mottle virus. Phytopathology 73:899-903.

Physiological Traits

BREEDING SUPER-EARLY SOYBEAN CULTIVARS

Wang Jinling and Gao Fanglan

Development of extremely early soybean cultivars (Maturity Groups 00 and 000) has allowed the northern part of Heilongjiang province of China and Siberia, as well as northern U.S. and southern Canada, to become the newly developed soybean production regions. At the same time, several typical short-day habit cultivars, such as Tropical, Timbira and Jupiter, have permitted soybeans to march triumphantly into the tropical and subtropical regions.

In 1971 we made a cross between Logbeau from north Europe and Don Non 47-1D, a progeny line of a cross between very early cultivars, Kusun and Japanese Early Green. Several extremely early promising lines have been developed from the transgressive earlier segregates of this cross. Among the lines, Don Non 36 is 7 days earlier than Maple Presto (Maturity Group 000), and 11 days earlier than Portage (Maturity Group 00). It has 18-20 g per 100 seeds, light hilum color, and a possible yield of up to 1,623 kg/ha. This cultivar has been registered and released in the 50°N area of Heilongjiang province where the annual accumulated temperature above 10 C is only about 1,800 C, the frost-less season is 85 days, and formerly only spring wheat could be grown. This achievement prompted us to initiate a study about the potentiality of developing even earlier lines from crosses between parents of very early and extremely early cultivars of different geographical origin.

PARENTAL MATERIALS AND PROCEDURES

Ten extremely early to medium early soybean cultivars of different geographical origin (Table 1) were used as parents for 21 crosses. The crosses were made in 1979 at Harbin, and their F_1 (planted in 1980), F_2 (in 1981) and F_3 (in 1982) populations were also planted at Harbin. Some of the F_4 populations were planted at Swen We County (50°N) for breeding of super-early cultivars for this new soybean region.

During the course of the experiment, date of maturity was recorded every three days. Data were analyzed for the following items: mode of genetic segregation of maturity in the F_2 and F_3; amount of segregates earlier than their respective early parent through transgressive segregation; coefficient of variability (C.V.); broadsense heritability (h) and expected genetic advance (ΔG).

DISTRIBUTION OF MATURITY DATE IN THE HYBRID POPULATIONS

Genetic segregation in the F_1, F_2 and F_3 for 6 of the 21 crosses can be seen in Table 2. Maturity time for the F_1 plants was nearly equal to maturity time for the

Table 1. Soybean parental materials of different earliness and geographical origin.

Variety	Maturity date at Harbin	Maturity Group	Geographical origin
Don Non 4	Sept. 12	II	Harbin, Heilongjiang, PRC
Nanjing Green	Aug. 27	0	Nanjing, PRC
Bei Hu Do	Aug. 25	00	Bei An, Heilongjiang, PRC
Fiskeby	Aug. 21	000	Sweden
Early Heihe	Aug. 22	000	North Inner Mongolia, PRC
Toobin	Aug. 28	0	Soviet Union
Hime Kokane	Aug. 24	00	Japan
Sakamoto Wase	Aug. 23	00	Japan
Can. 840-7-3	Aug. 23	00	Canada
Can. 052-903	Aug. 23	00	Canada

respective parents. Segregation of maturity date in the F_2 is nearly normal in distribution, with transgressive segregation to both the early and late ends. Segregation in the F_3 is nearly the same as that of the F_2. The degree of transgressive segregation in maturity is different among crosses. It seems that super-early (earlier than Maturity Group 000) individuals can be discovered in the earlier transgressive segregates only when both or at least one of the parents of the cross is extremely early. Both parents of cross 79-56 belong to Maturity Group 000 and gave the largest number of super-early segregates in the F_3 (Table 3). Highly significant correlation coefficients (r) between parental difference in growth period and mean growth period of the hybrid population (F_1 = 0.60, F_2 = 0.67, F_3 = 0.75, n = 21) demonstrate that the greater the parental difference in growth period, the greater the tendency of the progeny to be late maturing.

For breeding of super-early cultivars, parents should not only both be very early, but also have similar maturity times and different geographical origins. Transgressive earliness in crosses 79-56, 79-60, 79-61, where Early Heihe was used as the female parent, is outstanding in segregating large numbers of early transgressive individuals, indicating that Early Heihe is a good germplasm source for earliness genes. On the other hand, transgressive segregation for late maturation in the F_2 and F_3 is prominent when Hime Kokane and Nanjing Green were crossed with other early parents. So, genes these two cultivars donated to the crosses are mostly for late maturity.

From Table 2 and Table 3, it can be seen that the mean growth period of every hybrid population remains almost the same in different generations, and parental means differed very little from the respective population means for the F_2 and F_3. All these denote that additive genetic variance is the major component of genotypic variance for maturity time (Croisiant and Torrie, 1971). Therefore, growth period of soybean parents influences greatly that of their offspring. The highly significant correlation coefficients between parental means and means of the F_1, F_2 and F_3 populations are 0.77, 0.87, and 0.90, respectively. Therefore, to obtain more super-early offspring, extremely early (Maturity Group 00 and 000) cultivars should be used as parents.

DIFFERENT RATE OF OCCURRENCE OF SUPER-EARLY SEGREGATES AMONG CROSSES

Wang and Zhu (1963) pointed out that transgressive inheritance of maturity time in soybean plants is a rather common phenomenon. The present study confirms this.

Table 2. Genetic segregation in maturity date for 6 of the 21 crosses, Harbin 1979-1982. (♀ and ♂ indicate data for the parents.)

Code	Cross	Gen-eration	Aug 7	Aug 10	Aug 13	Aug 16	Aug 19	Aug 22	Aug 25	Aug 28	Aug 31	Sep 3	Sep 6	Sep 9	Sep 12	Sep 15	Sep 18	Sep 21	Sep 24	Sep 27	Plants no.	X̄±s days	C.V. %
													— number of plants —										
79-56	Early Heihe x Fiskeby	F1				♀	♂7														7	88.0 0	
		F2		1	8	19	41	13	51	♀♂	9										142	85.4 4.5	5.38
		F3	2	14	33	129	♀114	♂79	31	2	1										405	83.3 3.8	4.51
79-60	Early Heihe x Can. 052-903	F1			♀	♂6															6	88.0 0	
		F2			2	4	10	13	♀36	♂25	1										91	89.0 4.8	5.39
		F3		2	16	25	105	90	♀♂88	11	1	2									350	83.1 3.8	4.57
79-76	Nanjing Green x Can. 840-7-3	F1			♂	17															17	88.0 0	
		F2					♀3	6	100	♂63	♀12										184	91.4 3.5	3.90
		F3				10	39	♂207	♀130	43	29	6									454	88.7 3.4	3.79
79-66	Hime Kokane x Can. 052-903	F1							♀♂7												7	91.0 0	
		F2						2	19	♀♂57	29	36	13	4	1						161	97.1 4.7	4.78
		F3				8	♂206	♀124	59	33	17	10	1								457	90.0 3.8	4.30
79-71	Bei Hu Do x Nanjing Green	F1										♂41									41	100.0 0	
		F2									♀55	14	♂20	140	23	19	26	38			335	104.7 6.3	6.10
		F3					8	♀88	♂84	26	71	39	40	62	46	9	26	2			510	97.3 8.4	8.66
79-42	Bei Hu Do x Don Non 4	F1													♂31						31	109.0 0	
		F2										26	11	4	♀5	9	17	15	♂14	1	211	116.3 4.7	4.00
		F3						18	42	♀38	35	83	133	77	♂52	38	18	5			539	101.3 6.7	6.63

Table 3. Early transgressive segregates for different crosses, Harbin 1979-1982.

Code	Cross	F₁ generation		F₂ generation					F₃ generation					
		Mean growth period of parents (days)	Mean growth period of F₁s (days)	Plants (no.)	Mean growth period of parents (days)	Mean growth period of F₂ population (days)	Transgressive segregates early[a] (%)	late[b] (%)	Plants (no.)	Mean growth period of parents (days)	Mean growth period of F₃ population (days)	Transgressive segregates early[a] (%)	late[b] (%)	Plants with growth period under 81 days (no.)
79-39	Early Heihe x Don Non 4	101.5	100.0	252	105.9	107.5	0	0	527	97.4	97.5	0	2.0	0
79-42	Bei Hu Do x Don Non 4	103.0	109.0	221	110.0	116.3	0	0	539	100.2	101.3	3.3	4.2	0
79-43	Fiskeby x Don Non 4	102.0	95.0	119	107.0	108.8	0	0	537	97.4	102.1	0	15.8	0
79-46	Toobin x Don Non 4	106.0	109.0	235	108.6	120.3	0	0.8	532	100.7	>105.5	13.5	35.7	0
79-48	Hime Kokane x Don Non 4	103.0	115.0	209	107.0	122.3	0	35.4	515	98.4	>106.9	0	38.4	0
79-50	Sakamoto Wase x Don Non 4	101.5	91.0	214	105.2	104.6	0	0	514	97.6	97.6	0	4.6	0
79-51	Can. 052-903 x Don Non 4	102.0	96.0	229	107.0	110.4	0	0	503	97.9	98.8	0	6.9	0
79-52	Can. 840-7-3 x Don Non 4	100.0	100.0	125	107.0	108.2	0	0	521	76.6	99.2	0	0	11
79-53	Nanjing Green x Don Non 4	107.5	105.0	245	111.5	110.7	4.0	0	510	99.7	99.8	12.6	1.9	0
79-56	Early Heihe x Fiskeby	86.5	88.0	142	91.2	85.4	57.7	0	409	85.4	83.3	43.9	8.3	178
79-58	Early Heihe x Hime Kokane	86.5	88.0	229	90.2	90.7	35.8	6.9	342	87.9	88.9	2.9	19.0	10
79-60	Early Heihe x Can. 052-903	86.5	88.0	91	94.9	89.0	17.5	0	350	87.0	83.1	45.1	0.8	158
79-61	Early Heihe x Can. 840-7-3	83.5	88.0	43	90.1	85.7	44.1	0	362	86.0	85.1	17.9	4.1	65
79-62	Early Heihe x Nanjing Green	91.0	100.0	95	93.4	102.2	0	32.6	499	88.9	92.5	0	32.4	0
79-71	Bei Hu Do x Nanjing Green	90.5	100.0	335	98.0	104.7	0	73.4	501	89.6	97.3	1.6	58.8	0
79-75	Nanjing Green x Sakamoto Wase	91.0	91.0	328	95.1	91.0	0	0.6	488	88.5	87.8	2.0	3.2	10
79-76	Nanjing Green x Can. 840-7-3	89.5	88.0	184	98.0	91.4	59.2	0	464	87.6	90.5	2.1	7.5	10
79-66	Hime Kokane x Can. 052-903	88.5	91.0	161	93.9	97.8	13.0	51.5	457	87.8	90.0	1.7	26.0	0
79-69	Hime Kokane x Fiskeby	86.5	94.0	73	92.1	90.2	0	2.7	369	87.4	83.6	1.6	6.5	6
79-70	Hime Kokane x Toobin	91.0	91.0	51	96.9	106.6	0	60.7	507	89.6	96.0	0.9	44.9	0
79-64	Can. 052-903 x Fiskeby	86.5	88.0	227	90.9	89.8	7.9	0.4	444	84.8	86.6	0	4.5	39

[a] Percentage of segregates earlier than earliest parent.
[b] Percentage of segregates later than latest parent.

In crosses 79-39 to 79-53 where Don Non 4 was a common parent, the large difference in maturity time between parents caused absence of transgressive segregates in the F_2, but in the F_3 segregates later than the late parent occurred in all crosses (Tables 2 and 3). For the purpose of obtaining segregates with growth periods less than 78 days, the parents should both be of Maturity Group 000, and 000 x 00 types would give the next best result. It is important to note that except for cross 79-62, the five crosses in which Early Heihe was the common parent had a large number of super-early segregates in the F_2 and F_3. Owing to the fact that the male parent, Nanjing Green, of cross 79-62 is from Maturity Group 0, the cross gave few super-early segregates through additive effects of genes for maturity. But, Nanjing Green combined well with Sakamoto Wase and Can. 052-903 to produce segregates maturing before Aug. 16. Cross 79-64, with parents of 00 and 000 maturity, gave a certain number of super-early segregates. This may be due to the difference in geographical origin of the parents. In short, earliness and difference in genotype of the combining parents are essential to obtain super-early offspring. In the present study, Early Heihe was an excellent germplasm source for earliness genes. Crosses in which Early Heihe was used as one of the parents gave not only more super-early segregates, but also more agronomically promising segregates. Early Heihe was a selection from Bei Hu Do, which was developed from Bei Liang 55 x Kusun, and the two parents of this cross are selections from local cultivars of north Heilongjiang province. It seems that Early Heihe is the Chinese North East type donor for earliness genes, while Fiskeby is the North Europe type donor. The combination of these two donors gave the best result in producing super-early and good agronomic performing offspring.

ESTIMATION OF GENETIC PARAMETERS

Broadsence heritability for growth period was quite high, and differences among crosses and generations are not obvious (Table 4). Difference in coefficient of variation (C.V.) among crosses is significant, and consequently, caused the expected genetic advance also to be different among crosses. In the nine crosses where Don Non 4 was the common parent, C.V. in the F_2 was comparatively small, and consequently expected genetic advances also were small in comparison with those of the other crosses. However, in the F_3 C.V. of the nine crosses increased, and consequently, the expected genetic advance in the nine F_3 populations also increased, while those of the crosses between extremely early parents remained the same. Therefore, when both or one of the parents is late in maturity, segregates earlier than the early parent can be obtained in the F_3 and later generations, though such segregates are scarce in the F_2. Segregation for maturity in cross 79-42 demonstrates this point.

Among the crosses between the extremely early parents, the C.V. values are considerably different, and the expected genetic advances then also are different among crosses. But, expected genetic advance in the F_2 and F_3 are quite similar. This means that segregation in crosses between the extremely early parents has already become fully expanded in the F_2. Thus, action of super-early segregates could be emphasized in this generation. The large expected genetic advance values of crosses 79-70, 79-71, and 79-62 (Table 4) are mainly due to the occurrence of a large number of late segregates. For the purpose of selecting super-early segregates from crosses between

Table 4. Genetic parameters for the 21 hybrid populations, Harbin 1979-1982.

Code	Crosses	F₂			F₃		
		C.V.	Herita-bility (h)	Expected genetic advance (ΔG)	C.V.	Herita-bility (h)	Expected genetic advance (ΔG)
		%		days	%		days
79-39	Early Hei x Don Non 4	6.8	0.96	14.7	8.2	0.98	16.1
79-42	Bei Hu Do x Don Non 4	4.0	0.93	8.9	6.6	0.96	13.2
79-43	Fiskeby x Don Non 4	6.4	0.96	13.9	8.7	0.98	17.9
79-46	Toobin x Don Non 4	3.7	0.92	8.4	9.5	0.99	20.5
79-48	Hime Kokane x Don Non 4	5.4	0.94	13.0	7.2	0.97	15.4
79-50	Sakamoto Wase x Don Non 4	6.8	0.97	14.3	8.2	0.98	16.2
79-51	Can. 052-903 x Don Non 4	5.2	0.94	11.2	8.5	0.99	17.1
79-52	Can. 840-7-3 x Don Non 4	6.1	0.95	12.9	8.6	0.97	16.6
79-53	Nanjing Green x Don Non 4	6.0	0.95	13.1	9.0	0.98	18.2
79-56	Early Heihe x Fiskeby	5.3	0.91	8.5	4.5	0.89	6.9
79-58	Early Heihe x Hime Kokane	7.4	0.96	13.3	4.1	0.88	6.7
79-60	Early Heihe x Can. 052-903	5.3	0.95	9.3	4.5	0.93	7.2
79-61	Early Heihe x Can. 840-7-3	6.3	0.95	10.3	3.6	0.80	5.1
79-62	Early Heihe x Nanjing Green	4.8	0.91	9.2	6.8	0.97	12.5
79-71	Bei Hu Do x Nanjing Green	6.1	0.96	12.4	8.6	0.97	16.8
79-75	Nanjing Green x Sakamoto Wase	3.8	0.88	6.1	2.9	0.81	4.3
79-76	Nanjing Green x Can. 840-7-3	3.9	0.90	6.5	3.7	0.89	6.1
79-66	Hime Kokane x Can. 052-903	4.7	0.90	8.6	4.2	0.87	6.8
79-69	Hime Kokane x Fiskeby	3.8	0.84	5.9	3.1	0.84	4.7
79-70	Hime Kokane x Toobin	8.0	0.97	17.0	8.4	0.98	16.1
79-64	Can. 052-903 x Fiskeby	3.1	0.96	12.4	3.2	0.79	4.6

extremely early parents, expected genetic advance is not a dependable parameter.

CONCLUSIONS

Breeding super-early soybean cultivars to push soybean production into high latitude areas of both hemispheres is exceedingly important for increasing world protein and oil supply for food and feed. Through the achievements already attained by soybean breeders from different countries, and the development of Don Non 36 in Heilongjiang province, PRC, we are convinced that there is still high potential lying ahead of us in this field. The key points of super-early soybean breeding are: primarily, collecting extremely early Maturity Group 000 to 00 parental materials with remote phylogenetic relationships from different geographical and ecological areas. In China, besides soybeans growing in northern Heilongjiang province and northern Inner Mongolia, early-sown spring soybeans in the middle part of China are also promising germplasm sources for this purpose. Soybeans from northern Europe, Siberia, Hokkaido and the 00 to 000 Maturity Group belt of the U.S. and Canada are also promising sources for genes of extreme earliness.

Taking as parents extremely early cultivars from places differing geographically and ecologically from each other would give a better chance of obtaining offspring earlier than their parents. With crosses of such combinations, selection for super-early segregates might properly begin in the F_2 generation. If one of the parents in the combination is more or less late in maturity, selection for super-early types should be postponed until the F_3.

NOTES

Wang Jinling and Gao Fanglan, North East Agricultural College, Harbin, Heilongjiang, Peoples Republic of China.

REFERENCES

Croisiant, G. L., and J. H. Torrie. 1971. Evidence of nonadditive effect and linkage in five hybrid population of soybeans. Crop Sci. 11:675-677.
Wang Jinling and Zhu Chichuang. 1963. A preliminary study of inheritance of growth period of soybeans. Acta Agronomica Sinica 2:333-336.

BREEDING FOR PHOTOSYNTHETIC CAPACITY

R. I. Buzzell and B. R. Buttery

Soybean breeding is concerned with increasing photosynthetic capacity as a possible means to an end: other things being equal, such as partitioning of photosynthate, adaptation, pest resistance, etc., increasing photosynthetic capacity will increase yield, and therefore is a worthwhile goal. Wells et al. (1982) found that yield differences among cultivars were partially accounted for by differences in photosynthetic capacity of the plant canopy during seed development.

CROP PHYSIOLOGY

Photosynthetic carbon assimilation is the basis of all plant growth. For crop production there are three main aspects to be considered: a) the specific leaf rate (net photosynthetic rate per unit leaf area), which depends on the biophysical and biochemical processes in the chloroplasts and on numerous internal and external control factors, such as temperature and CO_2 concentration; b) the quantity of leaves, and c) their duration. These components of photosynthesis form a hierarchical system from the biochemical to the canopy level, and obviously, involve a great many genes. Indeed, breeding for virtually any change in size, shape, rates of development, etc. is likely to have an effect on the photosynthetic capacity of the crop at some stage in the growing season.

Photosynthetic activity per unit of leaf area is affected by leaf mass, which is a function of leaf thickness and leaf density. Specific leaf mass (SLM = mass per unit area) seems to be more a function of leaf thickness than of leaf density (Dornhoff and Shibles, 1976; Lugg and Sinclair, 1980); other results suggest that differences in SLM are a consequence both of differences in leaf thickness and leaf density (Wiebold et al., 1981).

Leaf area is affected by leaf size, leaf number per plant, plant population, and the environment. Leaf area is not uniformly distributed at all levels of the canopy; in indeterminate cultivars the leaf area at the top of the canopy is relatively small. Leaf area index (LAI = leaf area per unit of land) varies with growth duration of the cultivar (Metz et al., 1984) and with environment; for example, it may be excessively high under irrigation (Paul et al., 1979).

The maximum photosynthetic rate varies with leaf position on the plant (Woodward, 1976). The early reproductive stage has a higher leaf rate than the vegetative stage (Larson et al., 1981; Gordon et al., 1982). Leaf rate has been found to increase at the beginning of seed-filling (Wittenbach et al., 1980; Boon-Long et al., 1983), but not in all cases (Lugg and Sinclair, 1980). Upper leaflets maintain a fast leaf rate for longer periods than do lower leaflets (Lugg and Sinclair, 1981; Secor et al., 1984). Leaves at upper nodes of indeterminate cultivars have greater SLM and leaf thickness (Lugg and Sinclair, 1980), and with different leaf positions, there is considerable variation in stomatal density (Lugg and Sinclair, 1979).

In the soybean, which is a C_3-type plant, photosynthesis (carboxylation) and photorespiration (oxygenation) are inter-related in that CO_2 and O_2 compete for ribulose 1,5-bisphosphate (RuBP) at the same or adjacent sites on RuBP carboxylase-oxygenase (RuBisCO) (Ogren, 1976). The conductance of CO_2, in its diffusion from the atmosphere to the site of carboxylation, affects leaf rate. Stomatal resistance is a function of the number and size of stomata and their degree of opening. Mesophyll resistance is a function of the size and arrangement of mesophyll cells and the dispersion of chloroplasts as carriers of RuBisCO (Buzzell and Buttery, 1984).

With leaf development, maximum leaf rate occurs at about the time that the maximum area of a leaflet is attained (Woodward, 1976; Lugg and Sinclair, 1981). Lincoln and Williams cultivars were similar in maximum leaf rate, but the rate of decline was faster in Lincoln than in Williams in the latter part of the reproductive stage (Boon-Long et al., 1983). Chlorophyll content declined concomitantly with photosynthesis during senescence (Wittenbach et al., 1980; Secor et al., 1983; 1984), but may not be the cause of declining photosynthesis (Mondal et al., 1978).

The above coverage has been brief; for further information on assimilation, partitioning, and respiration, the reader should refer to the review of carbon metabolism in soybeans by Shibles, Secor and Ford[1].

GENETIC VARIABILITY

Cultivars differ in the amount of leaf area per plant (Ecochard et al., 1979); leaf area per plant in the F_3 was significantly correlated with leaf area per row in the F_4 (Paul et al., 1979). In one of two crosses, late lines had significantly greater main-stem LAI than did early lines (Metz et al., 1984). There was significant variation for leaflet size and leaflet mass within maturity groupings.

Cultivars differ in leaf photosynthesis rate on an area basis (e.g. Buttery et al., 1981), but differences are not as great when expressed on a leaf mass basis (Ogren, 1976). There are cultivar differences in the persistence of high leaf rate (Sinclair, 1980).

Cultivar differences exist for SLM (Dornhoff and Shibles, 1976; Bhagsari et al., 1977; Buttery et al., 1981; Chang and Li, 1983). There are differences for chlorophyll content among cultivars and lines; i.e., excluding chlorophyll mutants (Juang et al., 1980; Buttery et al., 1981; Hesketh et al., 1981). Cultivars differ in % N in the leaves (Buttery et al., 1981; Kishitani and Tsunoda, 1982; Boon-Long et al., 1983) and in RuBisCO activity (Hesketh et al., 1981). Soybean leaves have higher carboxylase activity of RuBisCO than some C_3 species, but lower than others (Seemann et al., 1984). There are cultivar differences in photorespiration (Bhagsari et al., 1977) and in photo-respiratory metabolism (Latché et al., 1978) Cultivars differ for mesophyll resistance (Dornhoff and Shibles, 1976; Bhagsari et al., 1977) and for stomatal resistance (Dornhoff and Shibles, 1976). Cultivar differences exist for stomatal density (Ciha and Brun, 1975; Juang et al., 1980).

Flavonol glycoside genes may have an effect on photosynthetic capacity. Lower chlorophyll content, SLM, and leaf photosynthetic rate are associated with the combination of Fg_1 and Fg_3, especially in the presence of fg_2 and t (Buttery and Buzzell, 1976).

TRANSGRESSIVE SEGREGATION

Transgressive segregation has been observed for lower and higher canopy apparent photosynthesis (CAP) in each of two crosses (Harrison et al., 1981) and for greater SLM in one cross (Chang and Li, 1983). A comparison of cultivars/lines with their parents (Buttery et al., 1981) indicated that some were the result of transgressive segregation for higher leaf rate, SLM and chlorophyll; a few were the result of transgressive segregation for lower leaf rate, SLM and chlorophyll content.

HERITABILITY AND CORRELATION

Heritabilities for photosynthetic traits are low to moderate (Table 1). In some cases heritability for these traits may be greater than that for yield. Leaf photosynthesis rate is significantly correlated with a number of plant traits (Table 2). In terms of coefficient of determination, most of the values are low to moderate in magnitude. Genetic correlations have not been determined. Correlations within a given segregating population could differ from values that have been obtained.

SELECTION RESPONSE

Indirect

Comparison of cultivars/lines developed during 1940 to 1970 in the North Central Region of North America indicates that selection for yield and other agronomic characteristics resulted in the majority of the cultivars/lines having higher leaf photosynthesis rate, SLM and chlorophyll content than their parents (Buttery et al., 1981). Likewise Du et al. (1982) found the leaf rate of improved cultivars to be better than that of the best parents. Selection for yield also may have selected for cultivars with greater

Table 1. Heritabilities for traits contributing to photosynthetic capacity of the soybean.

Trait	Source	H%
Canopy apparent photosynthesis	1st cross[a]	41
	2nd cross[a]	65
Leaf photosynthesis rate	1st cross[b]	36
	2nd cross[b]	56
Specific leaf mass	Flowering, 1 cross[c]	18
	Pod-filling, 1 cross[c]	44
	1st cross[d]	7-70
	2nd cross[d]	15-69
Leaf area per plant	F₃ and F₄, 2 crosses[e]	52
Leaflet size	1st cross[d]	46-95
	2nd cross[d]	29-87
Leaflet mass	1st cross[d]	49-89
	2nd cross[d]	29-85

[a]Harrison et al. (1981).
[b]Wiebold et al. (1981).
[c]Chang and Li (1983).
[d]Metz et al. (1984).
[e]Ecochard et al. (1979).

Table 2. Phenotypic correlations of leaf photosynthesis rate with other soybean traits. (VS = vegetative stage; RS = reproductive stage.)

Trait	Source	r
Stomatal resistance	16 cv. VS upper; lower[a]	-0.67*; -0.32
	2 expts RS[b]	-0.65**
	29 cv. RS[c]	-0.30
Mesophyll resistance	16 cv. VS[a]	-0.93**
	2 expts RS[b]	-0.92**
Leaf area	29 cv.[c]	-0.66**
Specific leaf mass	29 cv. RS[c]	0.62**
	1st expt; 2nd expt[b]	0.71*; 0.01
	1st cross; 2nd cross[d]	0.46**; 0.65**
	48 cv. VS[e]	0.33*
	12 cv. VS; RS[e]	0.76**; 0.42
	1 cross RS[f]	0.19*
Leaf thickness	1st cross; 2nd cross[d]	0.51**; 0.29**
Leaf density	1st cross; 2nd cross[d]	0.12; 0.62**
Chlorophyll content	48 cv. VS[e]	0.67**
	12 cv. VS; RS[e]	0.81**; 0.55
	29 cv. RS[c]	0.59**
% Leaf N	1 cross, RS[f]	0.37**
	12 cv. VS; RS[e]	0.10; 0.75**
	32 lines RS[g]	0.55**
Leaf protein	29 cv. RS[c]	0.57**
	1 cross RS[f]	0.31**
RuBisCO	29 cv. RS[c]	0.79**
Photorespiration	2 expts[b]	0.68**
Yield	12 cv. VS; RS[e]	0.30; 0.78**
	32 lines RS[g]	0.41*
	9 rapid; 9 slow lines[h]	0.53; 0.18

[a] Kaplan and Koller (1977).
[b] Bhagsari et al. (1977).
[c] Hesketh et al. (1981).
[d] Wiebold et al. (1981).
[e] Buttery et al. (1981).
[f] Secor et al. (1982).
[g] Buttery (unpublished).
[h] Ford et al. (1983).

rooting density, which results in less midday water deficits and less reduction in photosynthesis (Boyer, 1983). Comparison of Williams and Essex with the older cultivars Lincoln and Dorman indicated that yield improvement was not the result of increased photosynthetic capacity for Williams and Essex (Gay et al., 1980).

Direct

Selection for either high or low leaf rate in the F_2 or F_3 was ineffective (Wiebold et al., 1981). In the same study, gain from hypothetical selection for SLM was about 5% in the F_2 and 5 to 11% in the F_3. Gain from hypothetical F_3 selection in leaf thickness was about 5%. There was very little gain for leaf density. Secor et al. (1982) concluded that selection for leaf rate is effective within a soybean population of homozygous lines of similar developmental stage. The selection of high and low CAP in two crosses was effective (Harrison et al., 1981). Expected genetic advance from selecting for leaf area per plant was 27.7% as an average over two crosses (Ecochard et al., 1979).

PRESENT STATUS AND FUTURE PROSPECTS

Effective divergent selection for high and low leaf rate did not significantly affect dry matter or bean yield (Ford et al., 1983). The effective selection of high and low CAP resulted in significant yield differences (Harrison et al., 1981; Boerma and Ashley, 1982). Selection for small leaflet size or mass was effective in increasing yield in the late-maturity groups of two crosses grown under narrow-row conditions (Metz et al., 1984).

Cultivar differences in a trait may not result in differences in other traits; compensation may occur. For example, barley (*Hordeum vulgare* L.) lines that differed in stomatal frequency compensated for these differences in the field by different degrees of stomatal opening, and did not differ in yield (Moss, 1980). There may be a negative correlation between traits; e.g. larger, thinner leaves tend to have lower leaf photosynthesis rates.

The literature indicates that considerable knowledge has been obtained on the physiology of the soybean plant. Current investigations will provide more information on the biological limitations of the soybean plant, but probably will not provide a "touchstone" for breeders. For example, even with fairly rapid measurement of photosynthetic leaf rate using a $^{14}CO_2$ technique, measuring a large number of lines is more difficult than testing them for yield. Furthermore, indirect selection for yield using physiological characters does not appear particularly promising—yield is the end-product integrating a number of factors which include photosynthetic capacity and harvest index. The application of modern technology makes the acquisition and handling of yield data from numerous environments increasingly more efficient and effective. Thus, it is encouraging to know that each and every soybean breeder can have an effect on photosynthetic capacity by selecting for yield. Being knowledgeable of C_3-type photosynthesis, the negative relationship between leaf area and leaf rate, the effect of LAI and harvest index, etc. will enhance the breeder's approach in developing improved cultivars suitable for specific environments and production systems.

Selection of mutant strains (using *Arabidopsis*) with lesions in enzymes of the photorespiratory pathway has been unsuccessful in increasing the efficiency of carboxylation (Ogren, 1983). For a review of the prospects of improving CO_2 fixation of crop plants through genetic engineering, see Buzzell and Buttery (1984).

NOTES

R. I. Buzzell and B. R. Buttery, Agriculture Canada, Research Station, Harrow, Ontario NOR 1GO Canada.

[1]See the American Society of Agronomy's forthcoming revision of their monograph on the soybean.

REFERENCES

Bhagsari, A. S., D. A. Ashley, R. H. Brown and H. R. Boerma. 1977. Leaf photosynthetic characteristics of determinate soybean cultivars. Crop Sci. 17:929-932.

Boerma, H. R. and D. A. Ashley. 1982. Canopy photosynthesis and its association with yield. p. 79-87. *In* Proc. 12th Soybean Seed Research Conference. American Seed Trade Assoc., Washington D.C.

Boon-Long, P., D. B. Egli and J. E. Leggett. 1983. Leaf N and photosynthesis during reproductive growth in soybeans. Crop Sci. 12:617-620.

Boyer, J. S. 1983. Environmental stress and crop yields. p. 3-7. *In* C. D. Raper and P. J. Kramer (eds.) Crop reactions to water and temperature stresses in humid, temperate climates. Westview Press, Boulder, Colorado.

Buttery, B. R. and R. I. Buzzell. 1976. Flavonol glycoside genes and photosynthesis in soybeans. Crop Sci. 16:547-550.

Buttery, B. R., R. I. Buzzell and W. I. Findlay. 1981. Relationships among photosynthetic rate, bean yield and other characters in field-grown cultivars of soybean. Can. J. Plant Sci. 61:191-198.

Buzzell, R. I. and B. R. Buttery. 1984. Breeding for improved CO_2 fixation. p. 87-112. *In* G. B. Collins and J. F. Petolino (eds.) Applications of genetic engineering to crop improvement. Martinus Nijhoff/Dr. W. Junk Publishers, The Hague.

Chang, Y. C. and J. Li. 1983. Inheritance and breeding for higher photosynthetic rate in soybeans. p. 157-160. *In* Soybean Research in China and the United States. Proc. of the 1st China/USA Soybean Symposium and Working Group Meeting. INTSOY Series No. 25. INTSOY, Univ. of Illinois, Urbana.

Ciha, A. J. and W. A. Brun. 1975. Stomatal size and frequency in soybeans. Crop Sci. 15:309-313.

Dornhoff, G. M. and R. Shibles. 1976. Leaf morphology and anatomy in relation to CO_2-exchange rate of soybean leaves. Crop Sci. 16:377-381.

Du Weiguang, Wang Yumin and Tan Kehui. 1982. Varietal differences in photosynthetic activity of soybean and its relation to yield. Acta Agron. Sinica 8:131-135.

Ecochard, R., A. Gallais, M. H. Paul and C. Planchon. 1979. Heritability and response to selection of physiological traits related to yield in soybean. Ann. Amélior. Plantes 29:493-514.

Ford, D. M., R. Shibles and D. E. Green. 1983. Growth and yield of soybean lines selected for divergent leaf photosynthetic ability. Crop Sci. 23:517-520.

Gay, S., D. B. Egli and D. A. Reicosky. 1980. Physiological aspects of yield improvement in soybeans. Agron. J. 72:387-391.

Gordon, A. J., J. D. Hesketh and D. B. Peters. 1982. Soybean leaf photosynthesis in relation to maturity classification and stage of growth. Photosynthesis Res. 3:81-93.

Harrison, S. A., H. R. Boerma and D. A. Ashley. 1981. Heritability of canopy-apparent photosynthesis and its relationship to seed yield in soybeans. Crop Sci. 21:222-226.

Hesketh, J. D., W. L. Ogren, M. E. Hageman and D. B. Peters. 1981. Correlations among leaf CO_2-exchange rates, areas and enzyme activities among soybean cultivars. Photosynthesis Res. 2:21-30.

Juang, Y. D., H. S. Chang and C. Y. Chen. 1980. Studies on photosynthesis, respiration and leaf anatomy of several soybean cultivars. J. Sci. Eng. 17:263-274.

Kaplan, S. L. and H. R. Koller. 1977. Leaf area and CO_2-exchange rate as determinants of the rate of vegetative growth in soybean plants. Crop Sci. 17:35-38.

Kishitani, S. and S. Tsunoda. 1982. Leaf thickness and response of leaf photosynthesis to water stress in soybean varieties. Euphytica 31:657-664.

Larson, E. M., J. D. Hesketh, J. T. Woolley and D. B. Peters. 1981. Seasonal variations in apparent photosynthesis among plant stands of different soybean cultivars. Photosynthesis Res. 2:3-20.

Latché, J. C., G. Viala, J. Calmés and G. Cavalié. 1978. Comparative study of photorespiratory metabolism in soybean (*Glycine max* L. Merr.) varieties. Ann. Amélior. Plantes 28:77-87.

Lugg, D. G. and T. R. Sinclair. 1979. Variation in stomatal density with leaf position in field-grown soybeans. Crop Sci. 19:407-409.

Lugg, D. G. and T. R. Sinclair. 1980. Seasonal changes in morphology and anatomy of field-grown soybean leaves. Crop Sci. 20:191-196.

Lugg, D. G. and T. R. Sinclair. 1981. Seasonal changes in photosynthesis of field-grown soybean leaflets. 1. Relation to leaflet dimensions. Photosynthetica 15:129-137.

Metz, G. L., D. E. Green and R. M. Shibles. 1984. Relationship between soybean yield in narrow rows and leaflet, canopy, and developmental characters. Crop Sci. 24:457-462.

Mondal, M. H., W. A. Brun and M. L. Brenner. 1978. Effects of sink removal on photosynthesis and senescence in leaves of soybean (*Glycine max* L.) plants. Plant Physiol. 61:394-397.

Moss, D. N. 1980. Role of physiology in soybean breeding. p. 225-232. *In* F. T. Corbin (ed.) World Soybean Research Conference II: Proceedings. Westview Press, Boulder, Colorado.

Ogren, W. L. 1976. Improving the photosynthetic efficiency of soybean. p. 253-261. *In* L. D. Hill (ed.) World Soybean Research. Interstate Printers & Publishers, Danville, Illinois.

Ogren, W. L. 1983. Soybean photosynthesis. p. 154-156. *In* Soybean Research in China and the United States. Proc. of the 1st China/USA Soybean Symposium and Working Group Meeting. INTSOY Series No. 25. INTSOY, University of Illinois, Urbana.

Paul, M. A., C. Planchon and R. Ecochard. 1979. The relationships between leaf area index, developmental phase durations and yield in soybean. Ann. Amélior. Plantes 29:479-492.

Secor, J., D. R. McCarty, R. Shibles and D. E. Green. 1982. Variability and selection for leaf photosynthesis in advanced generations of soybeans. Crop Sci. 22:255-259.

Secor, J., R. Shibles and C. R. Stewart. 1983. Metabolic changes in senescing soybean leaves of similar plant ontogeny. Crop Sci. 106-110.

Secor, J., R. Shibles and C. R. Stewart. 1984. A metabolic comparison between progressive and monocarpic senescence of soybean. Can. J. Bot. 62:806-811.

Seemann, J. R., M. R. Badger and J. A. Berry. 1984. Variations in the specific activity of ribulose-1,5-bisphosphate carboxylase between species utilizing differing photosynthetic pathways. Plant Physiol 74:791-794.

Sinclair, T. R. 1980. Leaf CER from post-flowering to senescence of field-grown soybean cultivars. Crop Sci. 20:196-200.

Wells, R., L. L. Schulze, D. A. Ashley, H. R. Boerma and R. H. Brown. 1982. Cultivar differences in canopy apparent photosynthesis and their relationship to seed yield in soybeans. Crop Sci. 22:886-890.

Wiebold, W. J., R. Shibles and D. E. Green. 1981. Selection for apparent photosynthesis and related leaf traits in early generations of soybeans. Crop Sci. 21:969-973.

Wittenbach, V. A., R. C. Ackerson, R. T. Giaquinta and R. R. Hebert. 1980. Changes in photosynthesis, ribulose bisphosphate carboxylase, proteolytic activity, and ultrastructure of soybean leaves during senescence. Crop Sci. 20:225-231.

Woodward, R. G. 1976. Photosynthesis and expansion of leaves of soybean grown in two environments. Photosynthetica 10:274-279.

BREEDING SOYBEANS TO PREVENT MINERAL DEFICIENCIES OR TOXICITIES

Rufus L. Chaney

PHILOSOPHY OF BREEDING PLANTS TO ADAPT TO SOIL PROBLEMS

It has been recognized for many years that plants differ remarkably in relative adaptation to limiting soil mineral problems. Research has shown that cultivars of soybean can differ in both ability to obtain nutrients from a soil and to utilize the absorbed nutrients to produce plant dry matter and yield (Brown and Jones, 1977). However, the mere existence of genetic diversity does not lead plant breeders to invest their efforts.

During the last decade it has become generally accepted that certain edaphic problems are best solved by breeding adapted cultivars (resistant, tolerant, efficient). If the soil fertility problem can be solved effectively, for the long term, by applying soil amendments, then the best solution is to "change the soil." If, on the other hand, the problem cannot be solved economically by changing the soil, the only hope for a solution is to "change the plant," or to grow other crops.

Public and private sector researchers play increasingly different roles in the breeding process. Public researchers establish the extent of a problem, and whether breeding improved cultivars is possible and needed. They publish ratings of available high-yielding cultivars to assist growers in their management decisions, and they identify germplasm with tolerance. Private companies then use the knowledge, methods, and germplasm to breed improved adapted cultivars, which are offered for sale to growers. Whether "changing the plant" is chosen over "changing the soil" depends on the progress of public and private research, and the competitive climate in seed purchases.

This review briefly summarizes soybean nutritional problems and important research progress. Several soybean nutrition problems have been identified as clearly requiring plant breeding to correct serious, extensive, yield loss—Fe-deficiency chlorosis on calcareous soils, and Al toxicity, which limits rooting into strongly acid subsoils. Great progress has been made in breeding soybean cultivars with improved Fe-chlorosis resistance, and research has progressed on Al tolerance. The Fe case is considered in greater detail to examine the many soil chemistry, agronomic, and plant breeding considerations which affect soybean management on Fe problem soils.

SOYBEAN NUTRITION PROBLEMS AND RESEARCH PROGRESS

Soybeans require P, K, S, Ca, Mg, Fe, Mn, B, Zn, Cu, Mo, Cl, Co, and Ni from soils for normal growth and reproduction. Growers normally apply P and K as fertilizers, and S and Cl accompany these. Liming materials supply Ca and Mg. Most soils have adequate levels of plant-available Fe, Mn, B, Zn, Cu, Co, and Ni for normal plant growth, even of modern high-yielding cultivars. However, deficiencies of microelements occur in some crops on local or regional problem soils where the soybean is often grown in rotation, and soybean plants suffer Mn, Fe, and possibly Mo and B on particular soils.

Nutrient Deficiencies of Soybean

Soybeans require little special attention with respect to micronutrient fertilizers—partly because they absorb soil nutrients effectively, partly because amounts needed are easily supplied (Cl, Co, Ni), and partly because elements which could be needed for soybeans are applied for other crops (Zn, Cu, B, Mo). Historically acidic, sandy soils of the Southeast have lower stores of adsorbed microelements than the near neutral loams of the midwest. However, all soils will require fertilization with some microelements over periods of centuries. Removal lowers availability, and nutrients must eventually be replaced. This information is a brief summary of research results. More complete information is available from plant nutrition textbooks (e.g., Menzel and Kirkby, 1982) and reviews by deMooy et al. (1973) and Anderson and Mortvedt (1982).

Soybeans suffer Fe-deficiency chlorosis on poorly-drained calcareous soils, and on over-irrigated calcareous soils. This problem was first identified in Iowa, Illinois and Minnesota, U.S.A. However, soybeans are now grown in other production areas with calcareous soils, and soybean Fe-chlorosis is a problem in many areas (Clark, 1982). Application of Fe fertilizers to the soil can correct the deficiency, but costs are prohibitive, and less expensive Fe fertilizers are unlikely to solve this need. Conceivably, genetic-engineered soil bacteria or mycorrhizae could produce natural Fe chelators and aid crop Fe uptake, but this is still a dream. Foliar sprays can correct Fe deficiency of soybean, but require careful timing, are required each year on only parts of a field, and are expensive (deMooy, 1972; Randall, 1976).

Cultivar differences were recognized early. The release of Wayne, Hark, Anoka, and other high-yielding, very chlorosis-susceptible, cultivars allowed wide expression of the chlorosis problem. The success of breeding programs to prevent yield loss due to Fe chlorosis is an important example, and is discussed more fully below.

The soybean is more sensitive to manganese deficiency than other crops grown in rotation. In the U.S., southeastern coastal plain soils and lake plain soils in Ohio, Michigan and Indiana cause Mn deficiency. The problem is more severe in sandier soils (lower initial Mn), and if soils are limed above pH 6. This often occurs when growers overlime surface soils to correct strongly acidic subsoils that contain toxic levels of exchangeable Al. A regional project studied methods to correct Mn deficiency on over-limed, light-textured soils (Anderson and Mortvedt, 1982). Application of 20 kg Mn/ha prevented deficiency. If Mn deficiency occurs, foliar Mn application can save that crop. Mn deficiency can be so severe as to cause zero yield; hence, liming and Mn must be managed wisely on these potential problem soils.

Nickel is the most recently identified essential nutrient for higher plants (Eskew et al., 1983), and is an exciting achievement of improved research technology. Nickel is essential for urease, and rhizobia. Phytotoxic levels of urea accumulate in leaves during Ni deficiency. It is extremely unlikely that soils are Ni deficient.

Nutrient Toxicities in Soybean

Although every element can be phytotoxic if present in large excess, only a few elements prove phytotoxic to soybean in the field. In some strongly acid (pH below

5.2) uncontaminated soils, natural Al or Mn can injure soybean. Although some cultivars are sensitive to P, P toxicity does not occur in the field (deMooy et al., 1973). Rather, P toxicity has confounded many nutrient solution studies on soybean N, Fe, salinity tolerance, etc. (Chaney, 1984; Ismande and Ralston, 1981; Grattan and Maas, 1984).

Chloride exclusion by soybeans was discovered during salinity research. Tolerant cultivars excluded Cl from shoots, and tolerance was dominant (Abel, 1969). Parker et al. (1983) found that poorly drained sandy soils could become Cl toxic because of Cl accumulation from KCl fertilizers. Injury was very severe during drought. Cultivar differences were pronounced, and sensitivity ratings are now provided to growers.

Even though surface soils are limed to reduce Al and Mn toxicity to plants and nodulation (Munns, 1980), roots cannot penetrate the acid subsoil unless they have adequate Al tolerance. Strongly acid subsoils limit soybean yields in southeastern U.S., Brazil, and elsewhere (Abruna, 1980). Armiger et al. (1968) found Chief to be relatively tolerant, and Perry, relatively sensitive cultivars. Screening for tolerance has been a problem that has slowed breeding efforts (Devine et al., 1979; Sartain and Kamprath, 1978; Sapra et al., 1982; Hanson and Kamprath, 1979). Some germplasm is more tolerant than Chief (Devine et al., 1979; Muzilli et al., 1978).

Several research efforts are needed before breeders will select for Al tolerance: 1) a better demonstration of the drought-by-Al tolerance interaction on yield in the field; 2) better screening methods that correlate with demonstrable field results; and 3) development of Al tolerance ratings for cultivars to better advise growers. Although soybean is appreciably Al tolerant, roots are limited to the plow layer if subsoils are strongly acid. During periods of drought, plant growth is impaired in proportion to Al-tolerance.

Cultivar differences have been shown for Mn and Zn toxicity (Anderson and Mortvedt, 1982; White et al., 1979). Toxicity from Mn occurs on soils so acid that yield is reduced due to ineffective nodulation more than because of Mn phytotoxicity (Munns, 1980). Liming corrects the problem.

BREEDING TO PREVENT Fe-CHLOROSIS OF SOYBEAN

Iron deficiency chlorosis of the soybean commonly occurs on calcareous soil areas in Iowa, Indiana and Illinois (Randall, 1976; Fehr and Trimble, 1982), and extends to other calcareous soil production areas. Often only the rim of depressional areas in the field cause chlorosis, although a much wider area is calcareous. Innskeep and Bloom (1984) found that soluble salts accumulate in these areas and alter soil chemistry. Although these salts are important in some fraction of the chlorotic fields, the fundamental cause of soybean chlorosis is bicarbonate in soil with low Fe availability (Chaney, 1984). Soil moisture, soil temperature, biodegradable organic matter, and soil compaction influence soybean Fe chlorosis by affecting bicarbonate and root development. Bicarbonate interferes with regulatory control of Fe-stress-response, the adaptive increase (up to 500-fold) in ability of roots to absorb and translocate Fe to shoots. Bicarbonate interacts strongly with the availability of soil Fe (Coulombe et al., 1984a), and causes chlorosis at concentrations found in problem soils.

Soil solution bicarbonate levels depend on CO_2 levels in the soil air. When soils are wet, gas exchange is reduced and CO_2 increases; CO_2 reacts with $CaCO_3$ to form $Ca(HCO_3)_2$. The poorly drained calcareous soils which cause soybean Fe chlorosis contain about 0.01 MPa water potential when chlorosis is most severe. Some locations dry more quickly and the crop recovers. Some locations have higher extractable Fe, and the crop recovers more easily as the soil dries. Other locations kill sensitive cultivars in most years.

Two principal problems delayed progress on understanding soybean chlorosis. First, bicarbonate chlorosis research in the 1950s was misinterpreted, and little further research was conducted until 1981 (Coulombe et al., 1984b). Second, researchers could not obtain chlorosis in the greenhouse by using soil from fields that caused severe chlorosis. The soil problem has now become better understood: 1) air-drying sharply increases available Fe, as does dry soil storage (Leggett and Argyle, 1983), 2) few researchers try to maintain water levels that occur in the field when chlorosis begins, and 3) a mulch is required on the soil surface to obtain conditions of gas exchange in the greenhouse similar to those in the field. When these factors are properly managed, soybean chlorosis severity in the greenhouse is equal to that in the field.

Plant breeding was considered soon after chlorosis was noted. Weiss (1943) studied very chlorosis susceptible and resistant lines and found a single gene difference. One of these lines (PI 54619, T203) has been used for many research studies and helped scientists recognize the importance of plant breeding to solve soil problems. However, no breeding effort occurred until severe chlorosis was found with Wayne, Hark, and Anoka, new high-yielding cultivars which rapidly displaced earlier more resistant cultivars. Breeders in Iowa, Illinois and Minnesota worked on this problem and achieved great progress.

One important achievement was the development of recommendations to farmers to grow chlorosis resistant cultivars on problem soils. Available cultivars were rated in problem fields, and results published in advisory bulletins (deMooy, 1972; Lambert et al., 1970). Rating and breeding remained difficult. Thus, of the 498 cultivars tested in the Iowa Soybean Yield Test in 1982 (Mason et al., 1982), no cultivars had a 1 rating (highly resistant); 3.6% had a 2 rating (moderately resistant); 26.1% had a 3 rating (moderately susceptible); 66.5% had a 4 rating (susceptible); and 3.8% had a 5 rating (highly susceptible), with an average 3.70 rating for all cultivars tested.

Fehr (1982, 1984) has reviewed soybean Fe research to establish the relationship between chlorosis rating of a cultivar and yield reduction on problem soils. Although full yield of susceptible cultivars could be recovered on problem soils by repeated spraying with Fe-chelates, yield loss was 20% for each unit in chlorosis scoring if not sprayed. Thus losses could be quite significant if farmers grew cultivars with 4 or 5 ratings (70.3% of available cultivars).

In Iowa, soybean Fe chlorosis is a common problem in the Clarion-Nicollet-Webster soil association. Based on soil survey reports, within this association in Iowa (2,775,000 ha), I have estimated that 733,000 ha are poorly drained, nearly level, calcareous soils (26% of area); within this area, 123,000 ha (17% of total calcareous) are Calciaquolls, soils with relatively lower Fe availability and more likely to cause chlorosis. Estimated annual losses depend on cultivar, soil type, rainfall, etc. Soil and crop specialists estimate that mapped Calciaquolls and other calcareous soils have approximately 70% and 20% chlorosis problem, respectively. About 40% of this land would be in soybeans each year. Using a chlorosis rating of 3.7 (average), one predicts 54% yield reduction. Presuming 3000 kg/ha grain in the absence of chlorosis, this predicted chlorosis loss would be 135,000 metric tons grain/yr, a substantial loss. The weakest link in this estimate is the portion of poorly drained calcareous soil on which chlorosis occurs. Research is badly needed to define soil properties in relation to chlorosis score of index cultivars to help growers in their selection of management practices.

Subsequent research has not fully supported the single-gene finding of Weiss (1943), since essentially continuous variation in chlorosis score is found rather than an "all or none" response. Cianzio and Fehr (1980; 1982) had to test F_3 lines as rows in the field, since individual plants could not be accurately scored. They found a strong influence of modifying genes. The recurrent selection program of Fehr (1982, 1984) allowed creation of a germplasm line (A7) with improved Fe-chlorosis resistance over available germplasm. Progress in recurrent selection supports the conclusion of additive gene effects, even though one primary gene confers susceptibility.

One conclusion of Cianzio and Fehr (1982) is that a backcrossing program can recover both high yield and good chlorosis resistance. By using a highly Fe-chlorosis-resistant line and a modern high-yielding but chlorosis susceptible line, one can obtain adequate chlorosis resistance and high yield. Progress in improving the chlorosis resistance of Fehr's breeding line has progressed until he can no longer obtain chlorosis in his breeding nurseries (Fehr, 1984). Other fields, perhaps with deliberate irrigation, and/or chopped straw or trifluralin herbicide amendments, may increase the severity of the soil problem and allow improved field screening. Similarly, carefully managed soil pots might allow scoring of individual seedlings and hasten progress in chlorosis resistance.

Another method is nutrient solution screening. Weiss (1943) used a sand culture system with controlled P and Fe to identify T203 types, highly susceptible. Brown and Jones (1976) introduced an iron limitation screen, which was further tested by Byron and Lambert (1983). This method does not cause chlorosis during the whole growth period, but only when Fe is depleted. When bicarbonate was shown to be the causal soil factor, Coulombe et al. (1984a; 1984b; 1985) developed an improved nutrient solution screening technique, which has shown good correlation with field ratings. Chaney and Coulombe (unpublished) have further improved this already useful technique by studying the role of each solution factor that affects chlorosis. They have improved the range of chlorosis severity score between A7 and Pride B216 from 2.4 units using the published method, to 4.5 units. Other factors remain to be tested. These two methods provide reproducible screening capability; scores on individual plants are reliable. These methods have been adopted by several private breeders because they provide needed screen results at low cost, compared to variable field screening (variable from point-to-point in the field, and year-to-year).

Other research may contribute to final solution of the Fe-chlorosis problem. Agronomic practices, such as use of cultivar mixtures, can help minimize economic loss (Trimble and Fehr, 1983). The new understanding about bicarbonate may lead to soil management practices that reduce CO_2 in the soil. In the future, selection for improved Fe uptake or bicarbonate resistance may be done by cell culture, if methods to regenerate plantlets are developed. Fortunately, great progress has already occurred, and growers have increasing numbers of chlorosis-resistant, high-yielding cultivars.

NOTES

Rufus L. Chaney, Research Agronomist, USDA-ARS, Biological Waste Management and Organic Resources Laboratory, Bldg. 008, BARC, Beltsville, MD 20705.

REFERENCES

Abel, G. H. 1969. Inheritance of the capacity for chloride inclusion and chloride exclusion by soybeans. Crop Sci. 9:697-698.

Abruna, F. 1980. Response of soybeans to liming on acid tropical soils. p. 35-46. *In* F. T. Corbin (ed.) World Soybean Research Conf. II. Proc., Westview Press, Boulder, Colorado.

Anderson, O. E. and J. J. Mortvedt (eds.) 1982. Soybeans: Diagnosis and correction of manganese and molybdenum problems. Southern Coop. Ser. Bull. 281. Univ. Georgia Agric. Exp. Stn., Athens.

Armiger, W. H., C. D. Foy, A. L. Fleming and B. E. Caldwell. 1968. Differential tolerance of soybean varieties to an acid soil high in exchangeable Al. Agron. J. 60:67-70.

Brown, J. C. and W. E. Jones. 1976. A technique to determine iron efficiency in plants. Soil Sci. Soc. Am. J. 40:398-405.

Brown, J. C. and W. E. Jones 1977 Fitting plants nutritionally to soils. I. Soybeans. Agron. J. 69: 399-404.

Byron, D. F. and J. W. Lambert. 1983. Screening soybeans for iron efficiency in the growth chamber. Crop Sci. 23:885-888.

Chaney, R. L. 1984. Diagnostic practices to identify iron deficiency in higher plants. J. Plant Nutr. 7:47-67.

Cianzio, S. R. and W. R. Fehr. 1980. Genetic control of iron deficiency chlorosis in soybeans. Iowa State J. Res. 54:367-375.

Cianzio, S. R. and W. R. Fehr. 1982. Variation in the inheritance of resistance to iron deficiency chlorosis in soybeans. Crop Sci. 22:433-434.

Clark, R. B. 1982. Iron deficiency in plants grown in the Great Plains of the U.S. J. Plant Nutr. 5: 251-268.

Coulombe, B. A., R. L. Chaney and W. J. Wiebold. 1984a. Use of bicarbonate in screening soybeans for resistance to iron chlorosis. J. Plant Nutr. 7:411-425.

Coulombe, B. A., R. L. Chaney and W. J. Wiebold. 1984b. Bicarbonate directly induces Fe-chlorosis in susceptible soybean cultivars. Soil Sci. Soc. Am. J. (In press).

Coulombe, B. A., R. L. Chaney, W. J. Kenworthy and W. J. Wiebold. 1985. Effectiveness of nutrient solution bicarbonate, macronutrients salts, and CO_2 in screening for resistance to Fe-chlorosis. Crop Sci. (In press).

deMooy, C. J. 1972. Iron deficiency chlorosis in soybeans—What can be done about it? Iowa State Univ. Coop. Ext. Serv. Pm-531.

deMooy, C. F., J. Pesek and E. Spaldon. 1973. Mineral nutrition. p. 267-352. *In* B. E. Caldwell (ed.) Soybeans: Improvement, production, and uses. Agronomy 16:267-352. Am. Soc. Agron., Madison, Wis.

Devine, T. E., C. D. Foy, D. L. Mason and A. L. Fleming. 1979. Aluminum tolerance in soybean germplasm. Soybean Genet. Newsletter 6:24-27.

Eskew, D. L., R. M. Welch and E. E. Cary. 1983. Nickel: An essential micronutrient for legumes and possibly all higher plants. Science 222:621-623.

Fehr, W. R. 1982. Control of iron-deficiency chlorosis in soybeans by plant breeding. J. Plant Nutr. 5:611-621.

Fehr, W. R. 1984. Current practices for correcting iron deficiency in plants with emphasis on genetics. J. Plant Nutr. 7:347-354.

Fehr, W. R. and M. W. Trimble. 1982. Minimizing soybean yield loss from iron deficiency chlorosis. Coop. Ext. Serv., Iowa State Univ. Pm-1059.

Grattan, S. R. and E. V. Maas. 1984. Interactive effects of salinity and substrate phosphate on soybean. Agron. J. 76:668-676.

Hanson, W. D. and E. J. Kamprath. 1979. Selection for aluminum tolerance in soybeans based on seedling root growth. Agron. J. 71:581-585.

Imsande, J. and E. J. Ralston. 1981. Hydroponic growth and the nondestructive assay for dinitrogen fixation. Plant Physiol. 68:1380-1384.

Innskeep, W. P. and P. R. Bloom. 1984. A comparative study of soil solution chemistry associated with chlorotic and nonchlorotic soybeans in western Minnesota. J. Plant Nutr. 7:513-531.

Lambert, J. W. 1970. Varietal trials of farm crops. Minn. Agr. Exp. Stn. Misc. Rept. 24.

Leggett, G. E. and D. P. Argyle. 1983. The DTPA-extractable iron, manganese, copper, and zinc from neutral and calcareous soils dried under different conditions. Soil Sci. Soc. Am. J. 47:518-522.

Mason, H. L., W. R. Fehr, J. B. Bahrenfus and B. K. Voss. 1982. 1982 Iowa soybean yield test report. Iowa State Univ., Coop. Ext. Serv. AG-18-2.

Menzel, K. and E. A. Kirkby. 1982. Principles of Plant Nutrition, 3rd Ed. Intnl. Potash Inst., Bern, Switzerland.

Munns, D. N. 1980. Mineral nutrition and nodulation. p. 47-56. *In* F. T. Corbin (ed.) World Soybean Research Conf. II. Proc. Westview Press, Boulder, Colorado.

Muzilli, O., D. Santos, J. P. Palhano, J. Manetti, A. L. Lantmann, A. Garcia and A. Cataneo. 1978. Soil-acidity tolerance in soybeans and wheat cultivars (in Portuguese). R. Bras. Ci. Solo 2:34-40.

Parker, M. B., G. J. Gascho and T. P. Gaines. 1983. Chloride toxicity of soybeans grown on Atlantic coast flatwoods soils. Agron. J. 75:439-443.

Randall, G. W. 1976. Correcting iron chlorosis in soybeans. Minnesota Agr. Ext. Serv. Soils Fact Sheet 27.

Sapra, V. T., T. Mebrahtv and L. M. Mugwira. 1982. Soybean germplasm and cultivar aluminum tolerance in nutrient solution and Bladen clay loam soil. Agron. J. 74:687-690.

Sartain, J. B. and E. J. Kamprath. 1978. Aluminum tolerance of soybean cultivars based on root elongation in solution culture compared with growth in acid soil. Agron. J. 70:17-20.

Trimble, M. W. and W. R. Fehr. 1983. Mixtures of soybean cultivars to minimize yield loss caused by iron-deficiency chlorosis. Crop Sci. 23:691-694.

Weiss, M. G. 1943. Inheritance and physiology of efficiency in iron utilization in soybeans. Genetics 28:253-268.

White, M. C., A. M. Decker and R. L. Chaney. 1979. Differential cultivar tolerance to phytotoxic levels of soil Zn. I. Range of cultivar response. Agron. J. 71:121-126.

SOYBEANS ADAPTED TO COOLER REGIONS

A. Soldati and E. R. Keller

CLIMATIC LIMITATIONS IN COOLER REGIONS

Soybeans are, for many reasons, a valuable crop for the cooler regions of middle and northern Europe as well as Canada (Beversdorf and Hume, pers. comm.; Soldati et al., 1983; Szyrmer and Szczepanska, 1982; Vidal and Arnoux, 1980). For years major efforts have been made in Europe and Canada to introduce the soybean into these areas. However, this has not yet proved to be an economically viable undertaking.

Daylength sensitivity was, for a long time, considered to be a limiting factor for soybean production in Europe, since delayed flowering often resulted in failure of the crop to reach maturity. Identification of daylength insensitive genotypes and development of such properties in Fiskeby-type material in Sweden (Holmberg, 1973), as well as within the breeding programs in Europe (Schuster and Böhm, 1981; Szyrmer and Szczepanska, 1982; Fossati and Uehlinger, pers. comm.) and Canada (Beversdorf and Hume, pers. comm.; Seitzer and Voldeng, 1978) have resulted in remarkable progress in eliminating the short day requirement for flowering of the soybean (Hadley et al., 1982; Summerfield and Roberts, 1983). There are numerous papers describing experiments in which photoperiod and temperature within a controlled environment were tested (reviewed by Hadley et al., 1982 and Shanmugasundaran et al., 1980). A limitation of such experiments may lie in the fact that other important factors such as the daily pattern of irradiation, temperature, air humidity, etc. are rarely, if ever, tested.

Research work conducted in cooler regions demonstrates that temperature is the main limiting factor. Cool temperatures in many areas of Europe and Canada continue to limit soybean yield and production. There are cultivars available that mature in cool areas, but poor vegetative and reproductive growth, and slow development, during cool periods result in reduced yields and yield instability (Table 1). As a result, soybean yields in cool areas are too low for successful competition of soybean with other crops that are presently better adapted to the prevailng climate. Other factors that may limit soybean yield include solar radiation (Gourdon and Planchon, 1982), water availability (Vidal and Arnoux, 1980), and plant nutrition, including nitrogen fixation. All of these factors contribute to plant stress which, during the critical flowering and pod filling periods, increases abortion of flowers and pods.

INFLUENCE OF CLIMATIC CONDITIONS ON GROWTH AND YIELD STRUCTURE

Since 1971 field experiments conducted in Switzerland, at locations having very different temperature conditions, revealed a strong influence of climatic conditions on yield components of soybean (Soldati, 1976). At locations with relatively cool

Table 1. Yield of different soybean cultivars in Zürich, Switzerland (Huber, personal communication.)

Cultivar	Grain yield							
	1976	1977	1978	1979	1980	1981	1982	1983
				− t/ha −				
Caloria (FRG)	2.22	1.39	0.88	2.17	1.49			
Gieso (FRG)	2.02	1.40	1.05		1.30	2.20	3.09	2.91
Altona (CAN)	1.50	1.54	0.85	1.83	0.67			
Evans (USA)		0.45	1.14	2.36	1.56	1.31	4.10	3.20
Maple Arrow (CAN)				2.19	1.83	2.13	2.63	2.76
Maple Presto (CAN)				2.44	1.28	1.29	2.44	2.26
ETH 187 (CH)					1.45	2.36	2.90	3.02
Heat units[a]	1217	1049	951	1124	1015	1121	1299	1299

[a]Sum of the daily temperature (May 1-Sept. 30; threshold 8 C).

temperatures, plants develop more nodes and branches due to the longer vegetative period. At locations with relatively warm temperatures, plants produce more pods, seeds per pod and greater yields (Table 2). In the cool area of eastern Switzerland, plants produce more leaves and flowers than in the warmer southern regions, but they abort a much higher proportion of their flowers. Reproductive development is strongly influenced by temperature. At the warmer locations plants develop flowers earlier, thus extending the period for pod development and grain filling. The specific response to temperature varies with genotype: Cultivars that mature later generally begin to flower later. Of the total heat units accumulated during the growing season, the portion available during reproductive stages (flowering to harvest) is much higher under the warmer than under the cooler growing conditions. This explains why soybeans at warmer locations produce higher yields.

The total biomass production (Table 3) shows that soybean plants are capable of

Table 2. Components of yield of four cultivars grown at two locations in Switzerland from 1971 to 1973 (selected data from Soldati, 1976).

Cultivar (maturity group)	Nodes per plant	Pods per node	Seeds per pod	Hundred seed weight	Grain yield per plant	Heat units[a]
	no	no	no	g	g	C
Zürich (cool climate)						
0-52-903 (000)	54.9	1.3	1.4	18.5	18.3	808
Gieso (00)	63.4	1.5	1.7	12.7	20.3	899
Clay (0)	58.6	1.1	1.7	15.1	17.2	889
Chippewa 64 (I)	60.5	1.0	1.6	10.2	14.8	910
Lugano (warm climate)						
0-52-903 (000)	51.3	1.8	1.6	20.9	31.7	1187
Gieso (00)	58.5	2.1	2.0	18.5	43.8	1320
Merit (0)	57.3	2.6	2.1	15.8	47.5	1379
Chippewa 64 (I)	54.6	2.2	2.4	15.4	41.8	1385

[a]Sum of the mean daily temperature (planting to maturity; threshold 10 C).

Table 3. Total dry matter production at stages R1 to R8[a] and grain yield of two soybean cultivars grown in Zürich.[b]

Year	Planting date	Total dry matter				Leaf Area Index (LAI) at			Grain yield[c]
		R1	R3	R5	R8	R1	R3	R5	R8
		— g/m^2 —							g/m^2
Fiskeby V (000)									
1980	13 May	160	405	520	360	2.1	3.5	3.7	225
	27 May	145	165	260	233	1.6	2.5	1.9	184
1981	9 May	160	430	880	647	2.3	5.1	7.3	304
	21 May	145	350	855	699	2.1	5.6	7.1	362
1982	28 April	155	335	565	.d	2.4	3.5	4.1	.d
1983	6 May	194	375	510	514	3.1	4.7	4.2	212
Gieso (00)									
1980	13 May	110	285	390	441	1.8	3.3	2.9	318
	27 May	105	205	350	263	1.3	2.3	2.6	198
1981	9 May	155	705	1320	921	2.5	9.0	11.7	366
	21 May	125	585	805	713	2.1	7.8	7.5	278
1982	28 April	140	320	495	.d	2.7	3.6	3.8	.d

[a]According to Fehr and Caviness, 1984.
[b]Density of 50 plants/m^2.
[c]Harvest by hand.
[d]Damaged by hail.

high production in cool regions. However, instability in partitioning biomass between vegetative and reproductive growth prevents the desired stabilization of yield. This partitioning is related both to interactions between temperature and photoperiod and to podding ability following cold stresses during early reproductive development.

Depending on the climatic conditions, the differences in the various parts of the plant are very large (Table 3). Under cool conditions, such as in 1981, the German cultivar Gieso reached an LAI of 11.7 at R5. However, we do not consider such a high LAI to be an advantage; harvestibility is not positively affected by this high value because of the extent of lodging and irregular maturity. Based on these results, we conclude that cultivars are needed that could be planted earlier and flower earlier. This would enable them to make better and longer use of the warm summer period, and consequently, would prolong the grain filling stage.

DIFFERENTIAL RESPONSES TO COOL TEMPERATURES

Numerous investigations have been conducted in the laboratory and in growth chambers with the aim of describing the cold tolerance of soybean plants. Warm temperatures are necessary for germination and emergence (Hopper et al., 1979; Seitzer and Voldeng, 1978; Spehar, 1977; Szyrmer and Szczepanska, 1982); cold tolerance is necessary during vegetative growth and during flowering. Schmid and Keller (1980) investigated cold tolerant and intolerant cultivars, in growth chambers and the glasshouse, and concluded that three factors are important in cold tolerance: (a) date of the cold stress during the vegetative and reproductive development; (b) duration of the cold stress; (c) temperature levels. A period of cool temperatures may stimulate some cultivars; e.g., the Hungarian lines I-1 and ISZ-7, so that they can better exploit their yield potential. Cold tolerance can improve yield in areas having unfavorable climatic conditions when genetically incorporated into high-yielding cultivars that come from areas having favorable conditions. Plant breeding for cold tolerance can only be efficient if simple and safe methods are employed, and when easily recognizable selection criteria can be identified. We have investigated the effects of cold stresses on shoots, roots and reproductive organs at distinct growth stages. Since plant organs differ in their temperature requirements for maximum growth, a cold stress affects the dry matter accumulation in leaves, stems and roots to varying degrees. A cool period of two weeks during emergence has a greater effect on dry matter accumulation in roots and stems than in leaves (Figure 1). Low temperatures cause an increase in the dry weight of stems, whereas a decrease in the dry weight of roots is found. The effects of a cold stress, thus, appear not only in the total dry matter accumulation, but also in the distribution pattern of the dry matter in the plant. The fact that the differences due to low temperatures have mainly quantitative and not qualitative expression makes selection of cold tolerant cultivars rather difficult.

Growth parameters during the seedling stage are rarely correlated with final yield. Thus, cold tolerance during the early stage of plant development is not necessarily indicative of cold tolerance during later stages of growth. The results show that Gieso, when exposed to a permanent cold stress, is less active photosynthetically than are I-1,

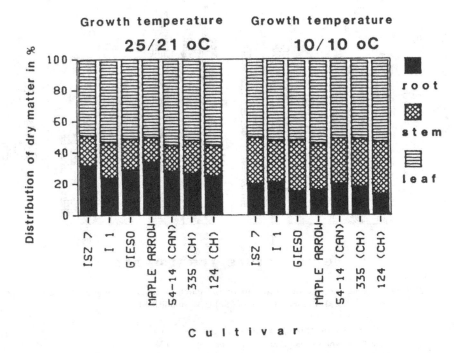

Figure 1. Dry matter content of roots, shoots and leaves, in percent of total dry matter, of 7 cultivars at two growth temperatures (harvested at growth stage V2) (Keller and Soldati, unpublished).

ISZ-7 and Maple Arrow. In contrast to other cultivars, Gieso does not show an increase in starch content of the leaves (Figure 2). We presume that the translocation rate of assimilates to the stem and roots is not higher, since the total dry matter accumulation of Gieso is relative low. The physiological activity of the different cultivars is certainly of great importance under cold stress conditions. For example, it would be impossible to use the starch content as an indirect measure of the photosynthetic activity or cold tolerance of the plant, because it varies greatly due to the source-sink capacities.

IMPORTANCE OF THE ROOT SYSTEM IN COLD TOLERANCE

Grafting techniques have been used to evaluate the importance of the root system for cold tolerance. The shoot of the line 54-14 (an early maturing line from the University of Guelph; Beversdorf, personal communication) was grafted onto rootstocks of I-1, ISZ-7 and Gieso. Grafting was done soon after germination, and control plants were self-graftings of 54-14. Apparently, plant height was not affected by the rootstock, since no differences were found between the various grafting combinations. Leaves and stems, which are directly affected by the cold stress, showed greater

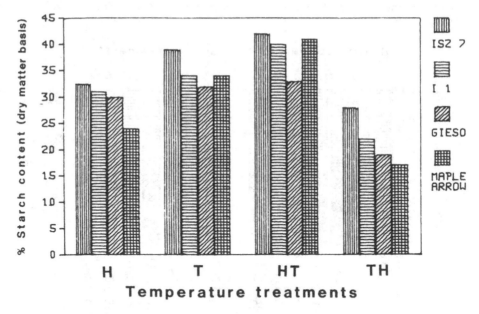

Figure 2.   Effect of four temperature treatments on the starch content of four cultivars (harvested at growth stage V 3.5) (Brenner, Keller and Soldati, unpublished).
Treatments:
H: 25/21 C (day/night temperature) from emergence to V 3.5
T: 12/8 C (day/night temperature) from emergence to V 3.5
HT: 25/21 C (day/night temperature) from emergence to V 2 and 12/8 C (day/night temperature) from V 2 to V 3.5
TH: 12/8 C (day/night temperature) from emergence to V 2 and 25/21 C (day/night temperature) from V 2 to V 3.5

damage than did the roots (Table 4). The rootstock had no direct effect on the reduced dry matter accumulation of the shoot. After termination of the cold stress, the stressed plants appeared to provide a better supply of assimilates to those plant parts that were most affected by the cold stress. This ability to compensate for cold stress is particularly evident in leaf area and leaf dry weight.

It seems that an exaggerated compensation occurred, which was influenced by the root system. The tested cultivars did not show a stable and temperature independent development of the reproductive organs. The cold stress led to an increased shedding of flowers and pods. At the same time, it delayed the formation of pods and grains considerably and affected yield and harvest index (Table 5). Grain yield was determined mainly by the shoot-line, 54-14, and was only slightly influenced by the rootstock. Temperature had an adverse effect on the relation between reproductive and vegetative growth, leading to a lower harvest index. The root system can be a limiting factor in the formation of yield. This depends not only on the size of the root system, but also on qualitative aspects such as the root's ability to take up water and nutrients.

REQUIREMENTS OF SOYBEANS ADAPTED TO COOLER REGIONS

The need to evaluate the natural environmental variability from historical weather data for each region of potential soybean production, and the response of genotypes to this variation, is the first step in defining requirements of adapted cultivars for cooler regions (Beversdorf et al., 1982). The Joint Field Trial, which is organized by the "Working Group for the Promotion of the Soybean Crop in Northern Europe and Canada", should provide information related to this matter. The main objective of the trial is the evaluation of growth and development of soybean plants under different climatic conditions (Soldati, 1983).

Until now it was not possible to identify selection criteria for cold tolerance. The

Table 4. Effect of cold stress on vegetative growth traits of grafted plants[a] (Keller and Soldati, unpublished). N: treatment without cold stress; S: treatment with cold stress (day temperature 12 C; night temperature 8 C); A: Greenhouse experiment; B: field experiment.

Trait		Immediately after cold stress		4 weeks after cold stress		At final harvest (R8)	
		N	S	N	S	N	S
Leaf		N	S	N	S		
area,	A	693	494 (146)[b]	1015	1262 (283)		
cm^2	B	2488	856 (114)	2438	2547 (275)		
Specific		N	S	N	S		
leaf weight	A	4.85	5.60 (.39)	6.24	7.44 (.69)		
mg/cm^2	B	4.15	5.16 (.33)	5.88	7.47 (.58)		
Root dry		N	S	N	S	N	S
weight,	A	1.67	1.38 (.31)	3.40	3.31 (.68)	2.80	3.19 (.73)
g	B	4.31	2.11 (.34)	6.90	6.04 (.99)	5.43	5.07 (.88)
Shoot:root		N	S	N	S	N	S
ratio	A	2.84	2.84 (.32)	4.92	4.57 (.47)	5.68	5.30 (.94)
	B	4.04	3.04 (.42)	6.59	4.86 (.57)	6.58	2.74 (.81)

[a]Mean of four cultivars.
[b]LSD 5%.

Table 5.   Effect of cold stress on grain yield per plant, harvest index and grain:root ratio[a] (Brenner, Keller and Soldati, unpublished). N: treatment without cold stress; S: Treatment with cold stress (day temperature 12 C; night temperature 8 C).

	Grain yield per plant		Harvest index		Grain:root ratio (dry weight)	
	− g −					
	N	S	N	S	N	S
Greenhouse	9.20	5.20 (1.70)[b]	0.48	0.38 (.12)	4.89	2.86 (.69)
Field	18.82	4.16 (2.33)	0.47	0.19 (.05)	3.58	0.77 (.47)

[a]Mean of four cultivars.
[b]LSD 5%.

utilization of greenhouses or growth chambers in a breeding program is too expensive. Results of many experiments show that the selection for cold tolerance must be based on both dry matter accumulation of the various plant parts and, increasingly, on physiological (qualitative) traits that reflect typical biochemical reactions of cold-stressed plants. In order to determine selection criteria for cold tolerance that are independent of cultivar, the overall carbohydrate economy of the plant must be studied. We believe that, in the future, we should look more closely at physiological reactions, such as photosynthetic activity and translocation of plants under cold stress. According to Brun and Setter (1980), soybean yield may be considered as the product of the rate of photosynthesis, integrated over time, and the partitioning of the resulting photosynthates between physiological and morphological yield components, often referred to assimilate sinks.

Cultivars for cooler regions should have an ideotype characterized as follows: (a) good germination and growth under cold temperature conditions; (b) tolerance of cold stress before and during the flowering period; i.e., no excessive flower and pod abortion; and (c) show high photosynthetic activity under moderate temperatures for maximum grain filling during a normal growing season. In searching for ideal cultivars, breeders should not neglect other important factors determining yield, such as a regular and sufficient water supply, improved inoculation methods and improved strains of *Rhizobium japonicum*, and optimum plant density in relation to growth habit.

CONCLUSIONS

By definition, a cold-tolerant ideotype should be able to overcome cool and inconsistent weather and respond with high and stable yields. Based on the results presented, we believe that there is considerable genetic variability for cold tolerance. We are aware, however, that cold tolerance may not be a simple trait. Many physiological and biochemical processes are undoubtedly influenced by temperature. In addition to cold tolerance, lines must be able to recover rapidly from cold periods in order to take advantage of warm periods during fluctuating temperature extremes typical of the continental climate.

NOTES

A. F. Soldati, Adjunct Research Associate; E. R. Keller, Professor of Agronomy; Swiss Federal Institute of Technology (ETH), Dept. of Crop Science, CH-8092 Zürich, Switzerland.

REFERENCES

Beversdorf, W. D., A. Soldati and E. R. Keller. 1982. Notes on the meeting "Research on soybeans for cooler regions of Europe". Soybean Genetics Newsletter 9:4-8.

Brun, W. A. and T. L. Setter. 1980. Effect of pod filling on leaf photosynthesis in soybean. *In* F. T. Corbin (ed.) World Soybean Research Conference II: Proceedings. Westview Press, Boulder, CO.

Fehr, W. R. and C. E. Caviness. 1977. Stages of soybean development. Iowa State Agric. Exp. Stn. Spec. Rep. 80.

Gourdon, F. and C. Planchon. 1982. Responses of photosynthesis to irradiance and temperature in soybean. Photosynthesis Res. 3:31-43.

Hadley, P., R. J. Summerfield and E. H. Roberts. 1982. Effects of temperature and photoperiod on reproductive development of selected grain legume crops. p. 19-41. *In* D. G. Jones and D. R. Davies (eds.) Temperate legumes: physiology, genetics and nodulation. Pitman Adv. Publishing Progr., London.

Holmberg, S. A. 1973. Soybean for cool temperate climates. Agric. Hort. Gen. 31:1-20.

Hopper, N. W., J. R. Overholt and J. R. Martin. 1979. Effect of cultivar, temperature and seed size on the germination and emergence of soyabeans. Ann. Bot. 44:301-308.

Schmid, J. and E. R. Keller. 1980. The behavior of three cold-tolerant and a standard soybean variety in relation to the level and the duration of a cold stress. Can. J. Plant Sci. 60:821-829.

Schuster, W. and J. Böhm. 1981. Experience in soyabean breeding in middle Europe. p. 158-171. *In* E. S. Bunting (ed.) Production and utilization of protein in oilseed crops. M. Nijoff Publishers, The Hague.

Shanmugasundaran, S., G. C. Kuo and A. Nalampang. 1980. Adaptation and utilization of soybeans in different environments and agricultural systems. p. 265-277. *In* R. J. Summerfield and A. H. Bunting (ed.) Advances in legume science. Royal Botanic Gardens, Kew.

Soldati, A. 1976. Abklärung von Komponenten des Ertragsaufbaues bei der Sojabohne unter verschiedenen klimatischen Bedingungen der Schweiz. Diss. ETH Nr. 5732, Zürich, Switzerland.

Soldati, A. 1983. Preliminary report of the joint field trial 1982. p. 86-99. *In* G. Robbelen (ed.) Proc. of the meeting of the section on oil and protein crops. EUCARPIA, Gottingen.

Soldati, A., E. R. Keller, H. Brenner and J. Schmid. 1983. Adaptation of soybeans to northern regions: a review of research work conducted at the Swiss Fed. Inst. of Technology (1971-1982). Eurosoya 1:27-34.

Spehar, C. R. 1977. Screening maturity groups 00 and 0 of the U.S. World soybean collection for germination at 10°C and field evaluation of selected lines. M.S. Thesis. University of Wisconsin Library, Madison.

Summerfield, R. J. and E. H. Roberts. 1983. The soyabean. Biologist 30:223-231.

Szyrmer, J. and K. Szczepanska. 1982. Screening of soybean genotypes for cold tolerance during germination. J. Plant Breeding 88:255-260.

Vidal, A. and M. Arnoux. 1980. Les problèmes posés par l'amélioration variétale du soja en France. Revue française des corps gras. 27:23-25.

BREEDING FOR DROUGHT AND HEAT RESISTANCE:
PREREQUISITES AND EXAMPLES

J. E. Specht and J. H. Williams

Much of the variability in annual soybean yields can be attributed to the annual fluctuation in seasonal precipitation and temperature (Runge and Odell, 1960; Thompson, 1970, 1975). There is, thus, a need for soybean cultivars with greater yield stability to seasonal weather variability, particularly drought and heat stress. However, relatively few breeders have as a breeding objective the direct selection of genotypes with greater drought and/or heat resistance per se. Many breeders would argue that selection for these traits occurs indirectly, in the sense that selection of genotypes is based on superior performance in a range of seasonal environments. Although these genotypes probably do possess an "adaptation" to the "mean environment" of the production region for which they are ultimately intended (Rosielle and Hamblin, 1981), it must be emphasized that any improvement in genetic stress resistance is coincidental rather than purposeful. Furthermore, severely stressful environments may not have occurred during the development of these genotypes, thereby providing minimal selection pressure for resistance to these stresses. Although breeding for adaptation per se will result in some enhancement in yield stability to weather variability, it is likely to be far short of what is truly needed for the infrequent severe stresses (e.g., 1983 in the American Midwest).

In this paper, we shall briefly discuss some prerequisites (and problems) inherent in developing strategies for the genetic improvement of drought and/or heat resistance in soybeans. This will be followed by some examples taken from our own research. Space does not permit a comprehensive treatment of the mechanisms of drought and heat resistance in plants; for this the reader is advised to consult some excellent recent publications (Christiansen and Lewis, 1982; Levitt, 1980; Mussell and Staples, 1979; Paleg and Aspinall, 1981; Specht and Williams, 1978; Turner and Kramer, 1980).

PREREQUISITES

The first and perhaps most important prerequisite is the need to characterize rigorously the environmental stresses for which greater resistance of a genetic nature is desired. This is best accomplished by a thorough review of annual weather records for the production region of interest to identify the kinds of drought and heat stress conditions that have historically had a serious impact on soybean yields. Key elements to consider in this regard are the stress intensity and duration, the rate of stress development, and the phenological timing of the stress, since soybean ontogenetic stages differ considerably in stress sensitivity. If the breeder has the ability to establish an "artificial stress nursery" (e.g., with rain shelters), then that nursery should mimic, as closely as possible, stress based on the climatological records. Even if the breeder cannot develop an artificial stress nursery, and has to rely on mechanistic rather than empirical screening techniques (discussed later), the characterization of the stress is no less important, because it provides a sound basis for choosing appropriate selection criteria and screening techniques.

The second prerequisite is the development of appropriate selection criteria, which can involve either "integrative" plant traits (i.e., seed yield, total dry matter, etc.) or "constitutive" plant traits (i.e., physiological, anatomical, or morphological traits). Breeders generally prefer integrative traits as selection criteria, since these traits are direct measures of performance. The constitutive traits, in contrast, are physiological and morphological features inherent in some stress-resistant ideotype; the latter formulated on the basis of theoretical considerations, speculative inferences, or ecological precedents. Such traits often have been proposed as selection criteria by physiologists, based on studies of phenotypic adaptations to stress (Turner and Kramer, 1980). The mechanisms by which plants can achieve some measure of stress resistance have been categorized into two general types: avoidance and tolerance (Levitt, 1980). Avoidance involves the ability of the plant to erect some sort of barrier, physiological or morphological, between it and the external stress condition, such that it can avoid the transduction of the external stress to its tissues. If the external stress is indeed transduced, either partially or totally, to the plant tissue, then tolerance involves the ability of the plant to maintain tissue integrity and function, perhaps via mechanisms that prevent or repair the injurious strain induced by the stress. This may allow the plant to continue adequate growth and development despite the stress.

The third prerequisite is the development of empirical or mechanistic screening techniques that can be used to detect the desirable stress-resistant genotypes in breeding populations. Ideally, the screening technique should be (1) reproducibly uniform when repeated during each selection cycle, (2) conveniently applicable to large breeding populations, (3) capable of detecting genotypic differences on an individual rather than line basis (for early-generation testing), and (4) precise enough to permit heritable variation to be expressed adequately and detected—i.e., have minimal environmental variation. Empirical techniques generally involve performance tests; i.e., the selection for integrative traits, wherein the breeder searches for genotypes exhibiting superior performance in field trials. The primary advantage of an empirical technique is that genotypic performance in yield trials ultimately provides the basis for a decision for cultivar release. The major disadvantage is that adequate control over a field testing environment is difficult to achieve, at least in a manner that would permit the selection pressure—i.e., stress—to be applied in a consistent and reproducible way. In contrast, mechanistic techniques involve selection for constitutive traits, physiological or morphological adaptations, that are purported to be fundamental components of a stress-resistant phenotype. If these traits are truly constitutive; i.e., do not require the presence of the stress for genetic expression, then the testing and selection environment need only be one that maximizes the heritable variation in these traits. This aspect is often a major advantage of mechanistic techniques, which can often be accomplished in the laboratory. The major disadvantage is the difficulty in experimentally verifying the thesis that mechanistic selection for a constitutive trait will indeed result in improved stress resistance. Genetic contrasts in constitutive traits are usually not available *a priori* to determine whether differences in the trait correlate with performance in controlled stress environments. Breeders are unlikely to allocate their limited resources to a mechanistic technique without such evidence.

The final prerequisite is the assemblage of appropriate germplasm upon which the screening technique can be applied. Genetic improvement in stress resistance cannot occur unless the germplasm contains significant heritable variation for the selected trait. Unlike pest resistance, for which qualitative variation, and often biological immunity, is often found, the traits associated with stress resistance are likely to be quantitatively inherited. Therefore, some sort of recurrent selection procedure would probably be effective, since the goal would be to improve the frequency of the favorable alleles at the many loci controlling the trait. However, the breeding method should not preclude a search for qualitative inheritance, because qualitative genes, if these exist, can be rapidly incorporated into commercial cultivars using backcrossing methods.

The four prerequisites described in the previous paragraphs are not easily satisfied relative to the development of effective breeding strategies for stress resistance. One way to illustrate this is to provide examples from our own research efforts involving breeding for improved drought and heat resistance in soybeans.

EXAMPLES

Thermostability

Several years ago, we became interested in a technique that a colleague had developed for evaluating heat tolerance in sorghum [*Sorghum bicolor* (L.) Moench.]. The technique involved a laboratory test using leaf disc samples collected from each genotype. It was based on the observation that when leaf tissue is exposed to an elevated temperature for a brief period of time, cellular membranes are 'injured' and become more permeable, such that electrolytes diffuse out of the leaf cells. If the heat-stressed tissue is subsequently bathed in deionized water for a specified period of time, the amount of electrolyte leakage can be quantitatively measured by electrical conductance of the bathing solution. The technique thus provides an indirect measure of membrane thermostability. We subsequently modified the technique and used it in soybeans (Martineau et al., 1979a, 1979b).

Consistent genotypic differences in cellular membrane thermostability were observed with this technique. These differences were reproducible at different sampling dates within the same season or across several seasons. For example, cultivars Corsoy, Williams and Harcor consistently ranked low in heat injury; i.e., they had high thermostability, whereas Bonus and Beeson ranked high. Encouraged by these data, we then decided to apply the technique to F_2-derived F_3 and F_4 lines obtained from crosses between genotypes ranked high and low in heat injury. The data were used to compute heritability estimates, phenotypic and genotypic correlations of heat injury with other traits, and estimates of predicted genetic gain. The heritability estimates obtained for heat injury were reasonably high and of the same order of magnitude as those for yield. Estimates of predicted genetic gain averaged 16 to 18%. Phenotypic and genotypic correlations of heat injury with seed yield were low and not statistically significant; however, these correlations were in the right direction; i.e., genotypes with greater membrane thermostability were generally higher yielding.

The membrane thermostability trait, thus, seemed to have heritable variation that could be exploited by selection. Yet, we decided not to continue this line of research for several reasons. First, the technique was time-consuming, laborious, and not conveniently applicable to large numbers of entries. Second, it could only be used to sample lines and not individuals, thereby restricting its use to advanced generations.

Third, the large experimental and analytical error associated with the technique necessitated considerable replication and sampling in order to reliably rank genotypes with some consistency relative to heat injury, thus restricting the number of entries that could be tested. Finally, and perhaps most importantly, we found it difficult to experimentally verify a relationship between this trait and field performance. These problems, and the uncertainty that selection for the trait would ultimately result in genotypes with a measurable and verifiable improvement in heat resistance, caused us to seek other lines of research.

Pubescence Density

Although the foregoing screening technique proved not to be a useful procedure, we still believed that there was merit in the mechanistic approach. However, we also recognized that it was important to first establish the value of a constitutive trait before embarking on a breeding program. Therefore, we examined the literature in order to choose a trait that was strongly implicated in drought and heat resistance. One such trait was plant pubescence. Ecological studies have demonstrated that a general increase in pubescence density occurs in closely related species as one traverses an environmental gradient of increasing aridity and/or thermal stress (Turner and Kramer, 1980). The adaptative value of increased plant pubescence in arid environments resides in the fact that greater pubescence density on the leaves increases leaf reflectivity. This in turn reduces the amount of radiant solar energy impacting upon the leaf, thereby lowering leaf temperatures and reducing transpiratory water loss. This ecological precedent led us to surmise that greater pubescence density on soybean leaves might lead to improvement in soybean drought and heat resistance.

Nearly all commercial soybean cultivars possess pubescence, but in some germplasm lines a gene exists that increases the number of trichomes per unit of epidermal surface by about four-fold (Bernard and Singh, 1969). Near-isogenic lines differing in this gene locus (Pd_1/pd_1) were available in the two older soybean cultivars Harosoy and Clark, and earlier research with these isolines had indicated that transpiration was lower with dense pubescence (Ghorashy et al., 1971). However, we felt more research was necessary to verify more critically the adaptive value of the increased pubescence in soybean response to drought and heat stress. With some grant funding and an interdisciplinary team of scientists, we conducted a comprehensive micrometeorological, physiological, and agronomic evaluation of these near-isogenic lines.

Space permits only a brief summarization of the results here. Baldocchi et al. (1983) observed significantly smaller fluxes of latent heat over the dense pubescence canopies, indicating that on a field scale, transpiration was reduced. The sensible heat flux was larger, however, indicating that denser leaf pubescence altered the partitioning of net radiation into latent and sensible heat. Net radiation penetrated deeper into the canopy of the dense pubescent isoline, due to reflective scattering of radiation intercepted by light-saturated sunlit leaves at the canopy surface to light-deficient shaded leaves in the lower canopy. Total canopy photosynthesis was thus improved, which coupled with the reduction in water vapor exchange, resulted in improved water-use efficiency. Clawson (1983) used classical growth analyses techniques and seasonal soil water data to compare the isolines. He observed that dense pubescence resulted in greater vegetative vigor as was evidenced by greater phytomass dry weight, leaf area index, leaflet number and dry weight, and plant height. Water-use efficiency per unit leaf area or ground area was greater for dense pubescence due to slightly less seasonal and daily evapotranspiration and greater phytomass production. Garay and Wilhelm (1983) observed that the dense pubescence isoline had a greater root density, foraged for water in deeper soil zones, and extracted more soil water during a drought than did the normal pubescence isoline. In his radiation balance studies, Nielsen (1983) observed that dense pubescence resulted in greater reflection of the incoming shortwave radiation of the canopy surface, including photosynthetically active radiation (PAR). However, no reduction in leaf photosynthesis was observed as a result of reduced PAR. Agronomic data collected in these and other studies (Hartung et al., 1980) indicated that dense pubescence also resulted in slightly higher yields, although not consistently so, because the extra vegetative vigor often increased lodging, which led to lower yields than the normal pubescence isoline.

The foregoing observations seemed to suggest that soybean cultivars with the gene for dense pubescence may be useful in high temperature, semi-arid production environments. We are now, in fact, crossing the lines containing this gene to new cultivars and experimental lines. Some caveats ought to be mentioned, however. First, there was an interaction of the dense pubescence gene with the genetic background into which it was introduced. The positive effects of dense pubescence were generally larger or evident only in the Harosoy, rather than the Clark genetic background, probably because Harosoy has a whitish-grey pubescence color (reddish-brown in Clark), which resulted in greater leaf reflectivity. Second, dew formation was greater on the dense pubescent leaves, which could conceivably offer an environment more conducive to greater disease incidence. Greater amounts of water from rainfall or sprinkler irrigation also tended to accumulate on the dense pubescence leaves, causing the plants to become heavy, and subsequently to lodge. Use of the dense pubescence gene may, thus, require greater emphasis on lodging resistance, or perhaps shorter plant stature. Finally, there is no certainty that increasing plant pubescence density of commercial cultivars will lead to improved drought and heat resistance, because in the final analysis, genotypes containing the dense pubescence gene will have to survive performance-based selection in stressful production environments.

Empirical Approaches

The final example from our research deals with an effort to use a line-source sprinkler system (Hanks et al., 1976) as an empirical technique for assessing genotypic performance across a range of soil moisture environments. The design of this system is such that it delivers a uniform amount of water to points along any given line parallel to the sprinkler pipeline. However, the delivered water amount also decreases in a uniformly linear fashion (in the absence of wind) at points along a line perpendicular to the pipeline, from a high water amount at the sprinkler line to a zero water amount at some point distant from the pipeline. A soil moisture gradient can be progressively developed during the course of the growing season by applying water amounts just sufficient to balance evapotranspiration in the plots located nearest the pipeline.

In 1983, we planted several cultivars in replicated plots perpendicularly arrayed to a line-source sprinkler system and evaluated the cultivar responses. Precipitation was extremely low during the 1983 growing season, with cultivar yields in the nonirrigated plots ranging from 0.8 to 1.6 Mg/ha, and those near the sprinkler line ranging from 3.0 to 4.1 Mg/ha. Seed yields averaged across the six plots were computed for each cultivar. In addition, a linear regression was performed for each cultivar, by regressing plot yield versus the amount of water applied (plus total precipitation). The regression coefficients were a measure of the linear change in seed yield per unit of water received. The R^2 values were 0.96 or better for each cultivar, indicating a good fit for the linear relationship between yield and the soil moisture gradient. A graph of the resulting data is presented in Figure 1, where the vertical axis is scaled in terms of

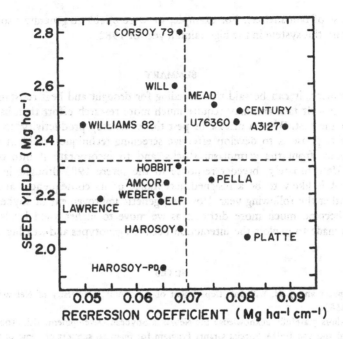

Figure 1. Mean seed yield (averaged over six plots) relative to the linear regression coefficient for seed yield versus the amount of water received per plot (irrigation plus precipitation). Irrigation amounts in the six plots (four replicates of each plot) ranged from zero to 36.7 cm, with precipitation of 24.9 cm from 1 June 1983 to 30 September 1983. Solid circles indicate cultivar coordinates and dashed lines represent averages of all cultivars.

cultivar mean yields (over the six plots) and the horizontal axis is scaled in terms of cultivar regression coefficients. Cultivars with a high mean yield and high yield stability (across the soil moisture gradient) should fall in the top part of the upper left quadrant of Figure 1. Whereas Williams-82 exhibited the best yield stability; i.e., lowest regression coefficient, Corsoy 79 gave the highest mean yield, and was, in fact, higher yielding than Williams-82 in all but the driest plots. Williams-82 and its previous near-isogenic releases were, and still are, popular cultivars grown widely throughout the north central USA, indirectly indicating a good measure of yield stability to a broad range of production environments. However, Corsoy-79 and its near-isogenic predecessor were popular with producers in a more restricted production region, indicating a narrower range of adaptation. Most of the other cultivars in Figure 1 fell along a line traversing the lower left and upper right quadrants, with recent releases, and a Nebraska experimental line, falling into the latter quadrant. A3127 had the largest regression coefficient, and was the highest yielding cultivar in the wettest plots. The steep response slope for Platte was due to its enhanced susceptibility to charcoal stem rot (*Macrophomina phaseolina* [Maub.] Ashby) in the drought-stressed plots. The lower yield and stability of Harosoy-Pd_1, relative to Harosoy, was due to a lodging differential that increased with greater amounts of applied water.

The line-source sprinkler technique, thus, seems to be useful for evaluating genotypic differences in response to an artificially created seasonal soil water gradient. However, drought stress can only develop in the outermost plots when the natural precipitation is below normal, unless these plots are protected from chance rainfall with the use of rain shelters. For example, we were unable to generate a soil moisture gradient with this system in the high rainfall year of 1982.

SUMMARY

In summary, it can be said that breeding for drought and heat resistance of soybean is a difficult task and may require much more research effort than has been expended in the past. When a disease or pest threatens crop productivity, the breeders' immediate response is to develop effective screening techniques, to identify and isolate resistance from the germplasm stocks, and to incorporate it into commercial cultivars. Unfortunately, breeder response to the severe 1983 drought in the north central USA is likely to be a resigned acceptance of its consequences and hope for better weather the following year. However, genetic improvement of soybean yield is likely to become much more difficult as we move to higher yield levels unless an attempt is made to exploit the interaction between genotypes and weather variability.

NOTES

J. E. Specht and J. H. Williams, Department of Agronomy, University of Nebraska, Lincoln, Nebraska 68583.

The authors gratefully acknowledge the Nebraska Soybean Development, Utilization, and Marketing Board and the USDA Special Grants Program for financial support of some of the research reported here.

REFERENCES

Bernard, R. L. and B. B. Singh. 1969. Inheritance of pubescence type in soybeans: Glabrous, curly, dense, sparse, and puberlent. Crop Sci. 9:192-197.

Baldocchi, D. D., S. B. Verma, N. J. Rosenberg, B. L. Blad, A. Garay and J. E. Specht. 1983. Leaf pubescence effects on the mass and energy exchange between soybean canopies and the atmosphere. Agron. J. 75:537-543.

Christiansen, M. N. and C. F. Lewis. 1982. Breeding plants for less favorable environments. John Wiley & Sons, Inc., New York.

Clawson, K. L. 1983. Physiological and agronomic responses of diverse pubescent soybean isolines to drought stress. Ph.D. Dissertation. Univ. of Nebraska-Lincoln Library.

Garay, A. F. and W. W. Wilhelm. 1983. Root system characteristics of two soybean isolines undergoing water stress conditions. Agron. J. 75:973-977.

Ghorashy, S. R., J. W. Pendleton, R. L. Bernard and M. E. Bauer. 1971. Effect of leaf pubescence on transpiration, photosynthetic rate, and seed yield of three near-isogenic lines of soybeans. Crop Sci. 11:424-427.

Hanks, R. J., J. Keller, V. P. Rasmussen and G. D. Wilson. 1976. Line source sprinkler for continuous variable irrigation-crop production studies. Soil Sci. Soc. Am. J. 40:426-429.

Hartung, R. C., J. E. Specht and J. H. Williams. 1980. Agronomic performance of selected soybean morphological variants in irrigation culture with two row spacings. Crop Sci. 20:604-609.

Levitt, J. 1980. Responses of plants to environmental stresses, (2nd Ed.). Acad. Press, New York.

Martineau, J. R., J. E. Specht, J. H. Williams and C. Y. Sullivan. 1979a. Temperature tolerance in soybeans. I. Evaluation of a technique for assessing cellular membrane thermostability. Crop Sci. 19:75-78.

Martineau, J. R., J. H. Williams and J. E. Specht. 1979b. Temperature tolerance in soybeans. II. Evaluation of segregating populations for membrane thermostability. Crop Sci. 19:79-81.

Mussell, H. and R. C. Staples. 1979. Stress physiology in crop plants. John Wiley & Sons, Inc., New York.

475

Nielsen, D. C. 1983. Influence of soybean pubescence type on radiation balance, photosynthesis and transpiration. Ph.D. Dissertation. Univ. of Nebraska-Lincoln Library.

Paleg, L. G. and D. Aspinall. 1981. The physiology and biochemistry of drought resistance in plants. Acad. Press, New York.

Rosielle, A. A. and J. Hamblin. 1981. Theoretical aspects of selection for yield in stress and nonstress environments. Crop Sci. 21:943-946.

Runge, E. C. A. and R. T. Odell. 1960. The relation between precipitation, temperature and the yield of soybeans on the Agronomy South Farm, Urbana, Illinois. Agron. J. 32:245-247.

Specht, J. E. and J. H. Williams. 1978. Testing soybeans for heat and drought tolerance. Proc. Eighth Soybean Seed Res. Conf. American Seed Trade Association, Washington, D.C.

Thompson, L. M. 1970. Weather and technology in the production of soybeans in the central United States. Agron. J. 62:232-236.

Thompson, L. M. 1975. Weather variability, climatic change, and grain production. Science 188:535-541.

Turner, N. C. and P. J. Kramer. 1980. Adaptation of plants to water and high temperature stress. John Wiley & Sons, Inc., New York.

NITRATE METABOLISM OF SOYBEAN—PHYSIOLOGY AND GENETICS

J. E. Harper, R. S. Nelson and L. Streit

The uptake and metabolism of nitrate has been extensively studied, particularly in cereal crops, as a possible key limitation to nitrogen input for protein production. Nitrate uptake and metabolism by plants is mediated by a unique complement of genetic information. This information prescribes limits within which biochemical and physiological reactions can operate. The expression of genetic potential is, however, partially regulated in many ways in plants. Much more is known about the environmental regulation of nitrate metabolism, by such factors as light, moisture, and temperature, etc., than about genetic limitations to nitrate metabolism.

VARIATION IN NITRATE METABOLISM

Although genetic variation in nitrate reductase (NR) activity has been found among soybean cultivars, no correlation is evident between NR activity and seed yield or seed protein content (Harper et al., 1972). This result is understandable in that symbiotic N_2 fixation by soybean likely supplements any lack of nitrogen input via nitrate metabolism. In addition, it appears that nitrate availability in the plant tissue is the primary limitation to nitrate metabolism, rather than enzyme level per se. Even those cultivars having low NR activity appear to have more than adequate NR enzyme to metabolize the available tissue nitrate; this is based on observations that all (over 200) soybean cultivars analyzed responded when additional nitrate was added to the *in vivo* NR assay (Harper, unpublished data). Furthermore, integration of *in vivo* NR activity in the presence of exogenous nitrate in the assay medium overestimated actual *in situ* reduced-N accumulation by about 50% (Harper, 1974). Therefore, efforts to select for enhanced nitrate reductase activity in soybean does not appear to be a viable research approach directed at increasing nitrogen input into seed protein.

Alternatively, selection for limited uptake and/or metabolism of nitrate, which may in turn allow enhanced symbiotic N_2 fixation, may provide a viable alternative in manipulating overall nitrogen metabolism of legumes. One approach has been to select mutants with altered NR activity. Although most NR mutants have been found in non-leguminous higher plant species (Oostindiër-Braaksma and Feenstra, 1973; Tokarev and Shumny, 1977; Warner et al., 1977; Müller and Grafe, 1978; King and Khanna, 1980; Strauss et al., 1981, Márton et al., 1982; Bright et al., 1983; Evola, 1983), NR mutants in pea (*Pisum sativum* L.) (Feenstra and Jacobsen, 1980; Warner et al., 1982) and soybean (Nelson et al., 1983) have also been identified. These legume mutants are useful in testing the hypothesis that symbiotic N_2 fixation in legumes is at least partially inhibited by preferential metabolism of NO_3^- into protein, thereby leading to a decrease in the number of carbon skeletons available as energy sources for nodule development and function. (This hypothesis will be referred to as the carbon deprivation hypothesis in a later section.)

ISOLATION AND SELECTION OF NR MUTANTS

In most studies, mutations have been induced either by chemical or irradiation treatment of seed. Commonly used mutagenic agents include ethylmethane sulfonate, N-nitro-N-methyl-guanosine, sodium azide, nitrosomethyl urea, gamma rays, X-rays, and unmoderated fission neutrons. Seed (M_1) which has been subjected to mutagenic treatment is usually advanced one generation and the resulting progeny (M_2) are then evaluated for mutants defective in nitrate metabolism.

One commonly used screen for NR-deficient mutants, which has been effectively used for *Arabidopsis* (Oostindiër-Braaksma and Feenstra, 1973) and soybean (Nelson et al., 1983), is to grow seedlings in the presence of chlorate. Chlorate can serve as an alternate substrate to nitrate for reduction by the NR enzyme. Reduction of chlorate to chlorite results in development of characteristic toxicity symptoms, generally attributed to the chlorite ion, which is highly reactive (Weaver, 1942; Aberg, 1947). Chlorate also competes with nitrate for uptake (Deane-Drummond and Glass, 1982; Doddema and Telkamp, 1979), and hence may also select mutants with altered ability for chlorate, and presumably nitrate, uptake. Uptake mutants have been isolated from *Arabidopsis* via chlorate screening (Oostindiër-Braaksma and Feenstra, 1973). A more direct screen for possible NR mutants is to analyze individual seedlings directly for *in vivo* NR activity. This approach has been successfully used to isolate NR-deficient mutants of barley (*Hordeum vulgare* L.) and pea (Warner et al., 1977; Warner et al., 1982). Although more time consuming than the chlorate screen, the *in vivo* NR assay has the advantage of being a more direct indicator of NR activity.

CHARACTERIZATION OF THE SOYBEAN NR MUTANT (nr_1)

Physiology

Nitrate reductase activity of most plant species is expressed only when nitrate is present in the growth media (Beevers and Hageman, 1983). Soybeans are an exception in that it has been shown that NR activity was present when no nitrogen source, other than N_2, was supplied (Harper, 1974). Likewise, Lahav et al. (1976) found that NR activity was present in soybean plants grown on urea in the absence of nitrate. This NR activity in soybean has been termed "constitutive" or "non-inducible". Nelson et al. (1983) and Ryan et al. (1983) presented the first evidence that two physiologically discrete isozymes of NR (nitrate inducible [iNR] and constitutive [cNR]) were present in soybean. This finding was based on selection of a soybean mutant (nr_1, formerly LNR-2) which lacked the constitutive form and retained the inducible form of NR.

Characterization of the nr_1 mutant line has provided insight into the nitrate metabolic capability of soybean. The constitutive NR activity is expressed in physiologically young tissue of the wild-type line. This includes cotyledons, unifoliolate leaves, and trifoliolate leaves (Nelson et al., 1983; Nelson and Harper, unpublished). Appearance of constitutive NR activity preceded the appearance of the inducible NR form (Figure 1). Constitutive NR activity peaked in the unifoliolate leaf at 8 to 9 days after planting at which time the leaf was essentially fully expanded. Constitutive NR activity then declined to essentially zero during the next 7 to 8 days. The inducible NR activity

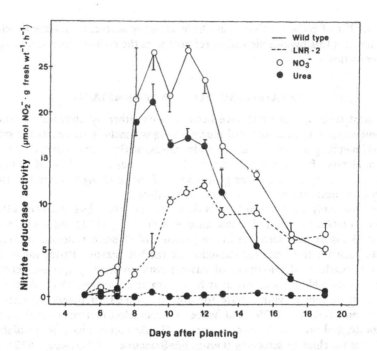

Figure 1. NR activity profiles for unifoliolate leaves from NO_3^-- or urea-grown wild-type and LNR-2 growth chamber plants. Data are means of three replicates ± SE, except for values from days 5 and 6 which were bulked samples. If not shown, the SE was within the data point for means from 7 to 19 DAP. Wild type is the cultivar Williams and LNR-2 is a low nitrate reductase mutant now designated nr_1 (Nelson et al., 1983).

peaked later (10 to 12 days after planting) and then gradually declined. However, inducible NR activity remained at approximately 70% of peak activity at the time constitutive NR had declined to zero. Each newly formed trifoliolate exhibited this profile, and therefore, only the uppermost leaf canopy had detectable constitutive NR activity at any time during the growing season (Nelson et al., 1983). Roots of both wild-type and nr_1 plants had no detectable NR activity when grown in the absence of nitrate; i.e., with urea nutrition. Considering the entire plant and the distribution of constitutive and inducible NR activities within the plant, it seems that the constitutive NR may account for approximately 12.5 to 20% of total NR activity.

The lack of constitutive NR activity (nr_1 plants) has not been shown to be detrimental to nitrogen metabolism. Nitrate uptake, accumulation, and incorporation into reduced-N compounds of nr_1 plants was not different from that of wild-type plants containing the constitutive NR activity (Ryan et al., 1983). Both plant types retained the substrate-inducible NR enzyme and it appears that this enzyme is sufficient to metabolize the incoming nitrate. In addition, nitrite reductase activity was normal in the mutant (Nelson et al., 1984b). The question remains as to whether the constitutive NR activity actually functions in the process of nitrate reduction *in situ*. It does seem that the constitutive NR activity is capable of reducing chlorate *in situ*. This is based on the observation that wild-type plants grown on urea (constitutive NR present) develop chlorate toxicity symptoms (Ryan et al., 1983). Since nitrate metabolism was unaffected in the nr_1 mutant, the carbon deprivation hypothesis would predict no effect on N_2 fixation in this plant. This was the result found for nr_1 (Ryan et al.,

1983). However, in a pea mutant (E_1) (Feenstra et al., 1982), which had lost considerably more of the measurable NR activity (80%), inhibition of N_2 fixation (acetylene reduction activity) by nitrate was decreased compared with the wild-type. Although more evaluation is necessary, these results are consistent with those that would be predicted by the carbon deprivation hypothesis.

Biochemistry

Most higher plants contain a single, nitrtate-inducible, NADH-preferred NR with a pH optimum of 7.5 (Beevers and Hageman, 1983). In addition all have been found to contain FAD, cytochrome b-557, and a molybdenum cofactor (MoCo) as prosthetic groups (Guerrero et al., 1981). Soybean is an exception to most higher plants in that it has a constitutive NR activity in addition to the inducible NR activity. The capability of obtaining inducible and constitutive NR activities without cross contamination, by growing nr_1 soybean plants on nitrate (inducible NR present) and wild-type plants on urea (constitutive NR present), has enabled biochemical characterization of these physiologically defined NR activities (Nelson et al., 1984b). The relationship between the constitutive and inducible NR activities, separated physiologically using the nr_1 mutant and wild-type plants (Nelson et al., 1983), and those separated and defined as NADPH- and NADH-NR activities on a biochemical basis (Jolly et al., 1976; Campbell, 1976), are shown in Table 1. The inducible NR from soybean is similar to other plant NRs in preferring NADH as its reductant source and in having a pH optimum of 7.5. The constitutive NR differs from the inducible NR in having greater capacity to use $FMNH_2$ and in having a pH optimum of 6.8. In addition, bicarbonate has been shown to be a competitive inhibitor of nitrate for constitutive NR but not for inducible NR (Nelson et al., 1984b), and cyanide and tungstate inhibit inducible NR

Table 1. Biochemical traits for physiologically and biochemically defined soybean NR isozymes.

	Physiologically defined[a]			Biochemically defined[b]		
				NADPH:	NADH:	
	cNR	iNR	Ref.[c]	NR	NR	Ref.[c]
Physical and kinetic traits						
pH optimum (NADH donor)	6.8	7.5	(1)	6.5	6.5	(4)
K_m (NO_3^-) NADH donor (mM)	0.16	0.14	(1)	5.8	0.11	(4)
Partial activities						
$FMNH_2$ or $FADH_2$ NR (as % of NADH NR)	52	25	(1)	211	79	(4)
Cyt. c reductase	+	+	(1)	+	+	(5)
Methyl viologen NR	+	+	(2)	+	+	(5)
Inhibitor	**% inhibition**			**% inhibition**		
Cyanide[d]	15	80	(3)	8	69	(4)

[a] cNR, constitutive nitrate reductase; iNR, inducible nitrate reductase.

[b] Based on primary electron donor specificity.

[c] Reference: (1) Nelson et al., 1984b; (2) Streit et al., 1984; (3) Nelson et al., 1984c; (4) Jolly et al., 1976; (5) Campbell, 1976.

[d] Cyanide concentration was 10 μM for cNR and iNR, and 100 μM for NADPH:NR and NADH: NR.

more strongly than constitutive NR (Harper and Nicholas, 1978; Aslam, 1982; Nelson et al., 1984c).

It is intriguing to propose a relationship between the constitutive and inducible NR activities, defined physiologically with the previously biochemically separated NADPH- and NADH-NR activities from soybean. Although some discrepancies exist when comparing the data (Table 1) (i.e., pH optima of NADH vs. inducible NR, $K_m(NO_3^-)$ of NADPH vs. constitutive NR when NADH is used as the electron donor), it appears to us that the constitutive NR is most like the NADPH-NR and inducible NR is most like the NADH-NR. The contention that constitutive NR and NADPH-NR activities are related has recently been confirmed by Western Blot techniques (Streit, Robin, Campbell, and Harper, unpublished). Duke and co-workers (Duke et al., 1982; Kakefuda et al., 1983) have also been active in characterizing NR isoforms from soybean cotyledons. Their findings, using a herbicide treatment to separate NR isoforms within the cotyledon, support our findings with soybean leaves regarding pH optima and Michaelis constants for nitrate for the two NR activities.

Since cyanide and tungstate both have been shown to affect the molybdenum co-factor (Solomonson, 1979; Notton and Hewitt, 1971), and results for soybean show differential sensitivity of the constitutive and inducible NR activities to cyanide and tungstate (Harper and Nicholas, 1978; Aslam, 1982; Kakefuda et al., 1983; Nelson et al., 1984c). it seems that constitutive NR may be altered at the MoCo site, compared with inducible NR. A MoCo is likely present in the wild type and the nr_1 mutant, based on existence of normal levels of xanthine dehydrogenase and inducible NR activity (Nelson et al., 1984b). [The existence of a MoCo common to the NR and xanthine dehydrogenase molybdoenzymes was suggested by Pateman et al. (1964), and reconstitution studies have shown that MoCo from various sources, except nitrogenase, are capable of supporting NR activity (see review by Kleinhofs et al., 1983).] The presence of a normal MoCo, however, indicates only that this cofactor is present if needed for the constitutive NR activity. Complementation experiments similar to those performed by Mendel (1983) with tobacco (*Nicotiana tabacum* L.) and fungal (*Neurospora crassa*) cell extracts, as well as electron paramagnetic resonance (EPR) studies, are in progress with the constitutive and inducible NRs and should determine whether the MoCo in constitutive NR is absent, altered, or incorporated differently due to an altered apoprotein.

Characteristics that are similar between the inducible and constitutive NR activities include 1) the presence of cytochrome c reductase activity, 2) susceptibility to pHMB inhibition, 3) thermal stability, and 4) K_m for NO_3^- (Nelson et al., 1984b and 1984c). The fact that the K_m for NO_3^- is similar for both activities (ca. 0.15 mM), and to that found for other plant species (Guerrero et al., 1984), supports the view that constitutive NR is active *in situ*. Definitive proof of this awaits selection of secondary mutants lacking inducible NR activity in which the metabolism of NO_3^- by constitutive NR can be evaluated.

Genetics

The constitutive NR activity is linked in some manner to the evolution of nitrogenous gases occurring during *in vivo* NR assays; evolution being absent in nr_1 and present in wild-type plants (Nelson et al., 1983). Whether gaseous nitrogen evolution occurs *in situ* remains to be established. It has been recently established that the gaseous nitrogen compound evolved is primarily acetaldehyde oxime (Mulvaney and Hageman, 1984). That this nitrogenous gas evolution is associated with constitutive NR activity is further supported by the lack of both gas evolution and constitutive NR activity from roots and cell suspension cultures of wild-type plants, where only inducible NR activity is prsent (Nelson et al., 1983; Nelson et al., 1984a).

Segregation of the F_2 of crosses between the wild-type and the nr_1 mutant indicated that absence of the constitutive NR activity and of nitrogenous gas evolution was controlled by a single recessive nuclear locus (Ryan et al., 1983). Joint inheritance of constitutive NR activity and nitrogenous gas evolution as a single dominant nuclear trait was confirmed in 16 F_3 lines of a cross between the wild type and the nr_1 line. Three lines were totally absent in both factors, four lines had both factors present in every plant, and the remaining nine lines were still segregating.

Although MoCo mutants and structural gene mutants of NR have been isolated and partially characterized in many plant species (Kleinhofs et al., 1983), no such mutants have been reported for soybean. The nr_1 mutant which has been isolated in our laboratory has not been fully characterized. However it is apparent that the gene coding for the apoprotein of the constitutive NR is either altered or not transcribed since cytochrome c reductase activity is decreased in the mutant (Nelson et al., 1984b). Determination of presence (in a nonfunctional state) or absence of the apoprotein will indicate whether the mutation is affecting the structural gene (apoprotein nonfunctional) or a regulatory gene (apoprotein not present). Although other interpretations of the results exist, these are the working parameters to be tested.

NOTES

J. E. Harper, USDA-Agriculture Research Service, and R. S. Nelson and L. Streit, Department of Agronomy, University of Illinois, Urbana.

REFERENCES

Aberg, B. 1947. On the mechanism of the toxic action of chlorate and some related substances upon young wheat plants. Kungl Lantbrukshögsk Ann. 15:37-107.

Aslam, M. 1982. Differential effect of tungstate on the development of endogenous and nitrate-induced nitrate reductase activities in soybean leaves. Plant Physiol. 70:35-38.

Beevers, L. and R. H. Hageman. 1983. Uptake and reduction of nitrate: Bacteria and higher plants. p. 351-375. In A. Läuchli and R. L. Bielski (ed.) Springer-Verlag, New York, NY.

Bright, S. W. J., P. B. Norbury, J. Franklin, D. W. Kirk and J. L. Wray. 1983. A conditional-lethan *cnx*-type nitrate reductase-deficient barley mutant. Mol. Gen. Genet. 189:240-244.

Campbell, W. H. 1976. Separation of soyean leaf nitrate reductases by affinity chromatography. Plant Sci. Lett. 7:239-247.

Deane-Drummond, C. E. and A. D. M. Glass. 1982. Nitrate reduction by roots of soybean (*Glycine max* L. Merr.) seedlings. Plant Physiol. 69:1298-1303.

Doddema, H. and G. P. Telkamp. 1979. Uptake of nitrate by mutants of *Arabidopsis thaliana*, disturbed in uptake or reduction of nitrate. II. Kinetics. Physiol. Plant. 45:332-338.

Duke, S. O., K. C. Vaughn and S. H. Duke. 1982. Effects of norflurazon (San 9789) on light increased extractable nitrate reductase activity in soybean *Glycine max* (L.) Merr. seedlings. Plant Cell Environ. 5:155-162.

Evola, S. V. 1983. Chlorate-resistant variants of *Nicotiana tabacum* L. I. Selection *in vitro* and phenotypic characterization of cell lines and regenerated plants. Mol. Gen. Genet. 189:447-454.

Feenstra, W. J. and E. Jacobsen. 1980. Isolation of a nitrate reductase deficient mutant of *Pisum sativum* by means of selection for chlorate resistance. Theor. Appl. Genet. 58:39-42.

Feenstra, W. J., E. Jacobsen, A. C. P. M. vanSwaay and A. J. C. deVisser. 1982. Effect of nitrate on acetylene reduction in a nitrate reductase deficient mutant of pea (*Pisum sativum* L.). Z. Pflanzenphysiol. Bd. 105:471-474.

Guerrero, M. C., J. M. Vega and M. Losada. 1981. The assimilatory nitrate-reducing system and its regulation. Ann. Rev. Plant Physiol. 32:169-204.

Harper, J. E. 1974. Soil and symbiotic nitrogen requirements for optimum soybean production. Crop Sci. 14:255-260.

Harper, J. E. and J. C. Nicholas. 1978. Nitrogen metabolism of soybeans. I. Effect of tungstate on nitrate utilization, nodulation, and growth. Plant Physiol. 62:662-664.

Harper, J. E., J. C. Nicholas and R. H. Hageman. 1972. Seasonal and canopy variation in nitrate reductase activity of soybean (Glycine max L. Merr.) varieties. Crop Sci. 12:382-386.

Jolly, S. D., W. Campbell and N. E. Tolbert. 1976. NADPH- and NADH-nitrate reductases from soybean leaves. Arch. Biochem. Biophys. 174:431-439.

Kakefuda, G., S. H. Duke and S. O. Duke. 1983. Differential light induction of nitrate reductases in greening and photobleached soybean seedlings. Plant Physiol. 73:56-60.

King, J. and V. Khanna. 1980. A nitrate reductase-less variant isolated from suspension cultures of Datura innoxia (Mill.). Plant Physiol. 66:632-636.

Kleinhofs, A., J. Taylor, T. M. Kuo, D. A. Somers and R. L. Warner. 1983. Nitrate reductase genes as selectable markers for plant cell transformation. p. 215-231. In P. F. Lurquin and A. Kleinhofs (eds.) Genetic engineering in eukaryotes. Plenum Press, New York.

Lahav, E., J. E. Harper and R. H. Hageman. 1976. Improved soybean growth in urea with pH buffered by a carboxy resin. Crop Sci. 16:325-328.

Márton, L., T. M. Dung, R. R. Mendel and R. Maliga. 1982. Nitrate reductase deficient cell lines from haploid protoplast cultures of Nicotiana plumbaginifolia. Molec. Gen. Genet. 182:301-304.

Mendel, R. R. 1983. Release of molybdenum co-factor from nitrate reductase and xanthine oxidase by heat treatment. Phytochemistry 22:817-819.

Müller, A. J. and R. Grafe. 1978. Isolation and characterization of cell lines of Nicotiana tabacum lacking nitrate reductase. Mol. Gen. Genet. 161:67-76.

Mulvaney, C. S. and R. H. Hageman. 1984. Acetaldehyde oxime, a product formed during in vivo nitrate reductase assay of soybean leaves. Plant Physiol. (In press).

Nelson, R. S., M. E. Horn, J. E. Harper and J. M. Widholm. 1984a. Nitrate reductase activity and nitrogenous gas evolution from heterotropic, photomixotrophic, and photoautotrophic soybean suspension cultures. Plant Sci. Lett. 34:145-152.

Nelson, R. S., S. A. Ryan and J. E. Harper. 1983. Soybean mutants lacking constitutive nitrate reductase activity. I. Selection and initial plant characterization. Plant Physiol. 72:503-509.

Nelson, R. S., L. Streit and J. E. Harper. 1984b. Biochemical characterization of nitrate and nitrite reduction in the wild-type and a nitrate reductase mutant of soybean. Physiol. Plant. (In press).

Nelson, R. S., L. Streit and J. E. Harper. 1984c. Inhibitor studies on crude and affinity purified constitutive and inducible nitrate reductases from soybean. Plant Physiol. Suppl. 74:53.

Notton, B. A. and E. J. Hewitt. 1971. The role of tungsten in the inhibition of nitrate reductase activity in spinach (Spinacea oleracea L.) leaves. Biochem. Biophys. Res. Comm. 44:702-710.

Oostindiër-Braaksma, F. J. and W. J. Feenstra. 1973. Isolation and characterization of chlorate-resistant mutants of Arabidopsis thaliana. Mutat. Res. 19:175-185.

Pateman, J. A., D. J. Cove, B. M. Rever and D. B. Roberts. 1964. A common co-factor for nitrate reductase and xanthine dehydrogenase which also regulates the synthesis of nitrate reductase. Nature 201:58-60.

Ryan, S. A., R. S. Nelson and J. E. Harper. 1983. Soybean mutants lacking constitutive nitrate reductase activity. II. Nitrogen assimilation, chlorate resistance, and inheritance. Plant Physiol. 72:510-514.

Solomonson, L. P. 1979. Structure of Chlorella reductase. p. 199-205. In E. J. Hewitt and C. V. Cutting (eds.) Nitrogen assimilation of plants. Academic Press, London.

Strauss, A., F. Bucher and P. J. King. 1981. Isolation of biochemical mutants using haploid mesophyll protoplasts of Hyoscyamus muticus. I. A NO_3 non-utilizing clone. Planta 153:75-80.

Streit, L., R. S. Nelson and J. E. Harper. 1984. Purification and properties of leaf nitrate reductase (NR) from nr_1-mutant and wild-type soybean plants. Plant Physiol. Suppl. 74:54.

Tokarev, B. I. and V. K. Shumnyi. 1977. Classification of barley mutants with lowered nitrate reductase activity after treatment of the grain with ethylmethane sulfonate. Sov. Genet. 13:1404-1408.

Warner, R. L., A. Kleinhofs and F. J. Muehlbauer. 1982. Characterization of nitrate reductase-deficient mutants in pea. Crop Sci. 22:389-393.

Warner, R. L., C. S. Lin and A. Kleinhofs. 1977. Nitrate reductase-deficient mutants in barley. Nature 269:406-407.

Weaver, R. J. 1942. Some responses of the bean plant to chlorate and perchlorate ions. Plant Physiol. 17:123-128.

HOST RANGE AND COMPATIBILITY OF SOYBEAN
WITH RHIZOBIAL MICROSYMBIONTS

T. E. Devine

EVOLUTION, MORPHOLOGY AND ANATOMY OF SYMBIOSIS

The nitrogen-fixing symbiosis seems to have evolved early in the phylogeny of the Leguminosae. Evidence for this rests upon the occurrence of this symbiosis in two major taxonomic subgroups of the Leguminosae, the Mimosoideae and Caesalpinioideae, both of which had developed by the end of the Cretaceous period (Polhill et al., 1981). The principles governing the coevolution of host and microsymbiont have yet to be discovered. In some cases the compatibility of host populations appears to be specific to a narrow range of rhizobial strains; in other cases the host appears to have a broad range of compatibility with a diverse array of rhizobial strains. Selection pressures in different environments, closed vs. clinal, or the tightness of reproductive barriers to introgression and gene flow, may all have played a role in the conservation of broad adaptability to rhizobial populations or the differentiation of specific affinity.

Soybean is in the sub-family Papilionoideae and forms desmodioid-type nodules similar to other genera in the tribe Phaseoleae, which is thought to be of tropical/subtropical origin. Desmodioid nodules are characterized as being round or oblate in shape and bearing obvious lenticels that facilitate gaseous exchange (Corby, 1981; Sprent, 1981). These nodules are determinate and have a continuous or closed vascular system. Sprent (1981) postulated that the combination of export of fixed nitrogen as ureides and determinate nodule morphology, characterized by a closed vascular system, are adaptations to the warm tropical environment. The export of fixed nitrogen as ureides is thought to be more economical in carbon requirement than export as amino acids. But, ureides are less soluble than amino acids. The association of warmer temperatures providing high solubility and high water flux through the nodule provided by a continuous vascular system are thought to compensate for the lower solubility. Efforts to adapt soybeans to cultivation in cool temperate condition may require modification of the nitrogen fixation system.

GENETIC CONTROL OF SYMBIOSIS

Rhizobial Genetics

Although caution should be observed in extrapolating from one species or genus to another, in view of the paucity of information on the genetics of the rhizobia that nodulate soybean, it is appropriate to consider other rhizobial species as potential model systems. The chromosomes of R. leguminosarum, R. phaseoli, and R. trifolii have very similar genetic maps, but differ in the extrachromosomal genetic material in their plasmids (Kondorosi and Johnston, 1981). Plasmid transfer experiments have established that host specificity for nodulation at the host species or genetic level among these three rhizobial taxa is controlled by genetic material in plasmids (Beringer et al., 1980). The chromosome map of R. meliloti is reported to differ sufficiently from the R. leguminosarum group to place it in a separate group. Sufficient information is not yet available to compare the chromosome homology and plasmid content of the fast- or slow-growing rhizobia that nodulate soybean with other rhizobia. However, Masterson et al. (1982) reported that slow-growing strains of R. japonicum do not have structural nif genes, homologous to nif D and nif H, on large plasmids, but that all the fast-growing strains that nodulate soybeans, except PRC 194, have structural nif genes on large plasmids. Kuykendall (1979) and Wells and Kuykendall (1983) described important steps in the development of a system of genetic analysis of R. japonicum: (a) introduction and genetic transfer of R factors to facilitate chromosomal genetic mapping, and (b) the isolation and characterization of amino acid auxotrophs to serve as genetic markers. R-factor-mediated chromosome transfer in R. japonicum has recently been accomplished (D. Kuykendall, pers. commun.).

Host-Plant Genes Controlling Rhizobial Interactions

In soybean, several host genes controlling nodulation response with individual rhizobial strains or groups of strains have been reported. The rj_1 allele, sometimes called the "nonnodulating gene", conditions a marked nodulation restriction with all rhizobial strains tested (Vest et al., 1973). In field tests, plants of the rj_1 rj_1 genotype yielded one nodule per 1,000 to 1,500 plants (Devine et al., 1979). In greenhouse studies, with sand or vermiculite as the growth medium, some rhizobial strains produced scattered nodulation on the root system of rj_1 rj_1 plants (Clark, 1957; Devine and Weber, 1977).

The dominant allele Rj_2, carried by the cultivar Hardee, conditions an ineffective nodulation response with rhizobial strains of the c1 and 122 serogroups (Vest et al., 1973). Plants with the Rj_2 Rj_2 genotype nodulate normally with most strains of R. japonicum, but form cortical proliferations or rudimentary nodules with strains of the c1 and 122 serogroups. The dominant allele Rj_3 was reported to produce an ineffective nodulation response with rhizobial strain USDA 33 (Vest et al., 1973). Cultivars, such as Hardee, carrying the Rj_3 allele produce cortical proliferation on the roots rather than nodules when inoculated with strain USDA 33, but nodulate normally with other strains of rhizobia. The dominant allele Rj_4, carried in the cultivars Hill and Dare, conditions an ineffective nodulation response with rhizobial strain USDA 61 (Vest et al., 1973). Plants carrying the Rj_4 gene form cortical proliferations and rudimentary nodules when inoculated with strain USDA 61, but nodulate normally with other strains of rhizobia, such as USDA 110. Strain USDA 61 is a strain with a strong propensity to induce rhizobitoxin-induced chlorosis.

Another mode of host-strain interaction is found in the combination of the cultivar Peking with rhizobial strain USDA 123 (Vest et al., 1973). In this combination, nodules of normal size and number are formed, but virtually no nitrogen is fixed. I have found that F_2 segregating populations from the cross of Peking X CNS, when inoculated with strain USDA 123, do not exhibit the bimodal distribution that is characteristic of segregation of a single genetic locus, suggesting that the genetic control of this reaction is multigenic. Peking has frequently been used as a source of nematode resistance in U.S. breeding programs. Since serogroup 123 is a major component of the rhizobial population in the northern part of the midwestern U.S., careful thought should be given to the possibility of inadvertent introgression of genes conditioning inefficient nitrogen fixation with these rhizobia when breeding for nematode resistance.

Peking also produces an interesting response with rhizobial strain USDA 110 (Caldwell and Vest, 1968). In a field test with soil containing a diversity of rhizobial strains, 0.3% of the nodules from Peking contained serogroup 110, although other cultivars had a range from 21.7% to 38.3% in the frequency of nodules containing serogroup 110. Yet, in a Leonard-jar test, in which Peking was inoculated by strain USDA 110 alone, nodulation was within the normal range with significant nitrogen fixation (Vest et al., 1973).

Keyser et al. (1982a) reported that several fast-growing rhizobial strains, obtained from east-central China, nodulated and fixed nitrogen with the G. max plant introduction Peking, but showed ineffective nodulation with several US-bred cultivars. Devine (1984) recently studied the cross of Peking with Kent, a cultivar displaying incompatible nodulation response with the fast-growing rhizobial strain USDA 205. The evidence indicated segregation at a major locus controlling nodulation response with strain USDA 205, with Peking carrying a recessive allele conditioning effective nodulation and Kent carrying a dominant allele conditioning ineffective nodulation (Devine, 1984).

A test of 285 plant introduction lines (PIs) from Asia established that ability to nodulate with strain USDA 205 was common (56%) in G. max plant introductions (Devine, 1985). Examples of ability to nodulate were found in all Maturity Groups, 00 through X, and in introductions from USSR, Korea, China, Japan, Burma, Thailand, Vietnam, Malaya, and Indonesia. The highest frequency (over 80%) of nodulating lines was found in PI lines from the southeast Asian nations of Thailand, Malaya, and Indonesia. Only 34% of the lines from northeast China nodulated with strain USDA 205. Many lines that were nodulated were also clearly nitrogen deficient, indicating that the total competence of the symbiosis (nodulation effectiveness and nitrogen fixation efficiency) is more complex genetically than a single gene controlling nodulation. Therefore, although it should be possible to transfer to US-bred cultivars the ability to nodulate with at least one fast-growing strain, the transfer of highly efficient fixation may be more difficult. However, some lines produced vigorous growth with dark green foliage with strain USDA 205, indicating an efficiency of fixation at least equal to that of Peking.

Significance of the *Rj* Genes

Of the five known genes governing nodulation response in soybeans (rj_1, Rj_2, Rj_3, and Rj_4, and the allele governing nodulation with the fast-growing strain USDA 205) in all cases except the rj_1 gene, the dominant allele conditions the ineffective nodulation response. The rj_1 gene was discovered in segregating lines of the Lincoln x Richland cross and would seem to be a mutation.

It is appropriate to ask why one encounters genes that cause the failure of a process considered beneficial to survival. The first explanation that comes to mind is that the alleles causing failure of nodulation are mutations—inborn metabolic errors. This explanation may fit the case of the rj_1 gene. However, since most mutations are recessive, the mutation explanation is not congruent with the dominant nature of the other four incompatible alleles. An alternative explanation is based on the history of soybean cultivation. The cultivated soybean and presumably its rhizobia, were introduced into the United States from Asia. It is reasonable to postulate that, within particular ecological areas in Asia, a natural selection pressure was exerted upon both the host and microsymbiont for mutual compatibility in symbiosis. Thus, mutually adapted and compatible ecotypes of the host and microsymbiont would evolve. When ecotypes of the host and microsymbiont that have not coevolved are artificially brought into association (either in laboratory experiments or in a field used for crop production), the symbiosis may be defective in either the effectiveness of nodulation, the efficiency of fixation, or both. The Rj_2, Rj_3, Rj_4 alleles and the gene controlling nodulation failure with the fast-growing strain USDA 205, may then be indicative of the association of ecotypes of the host that have not coevolved with ecotypes of the microsymbiont. For example, during soybean introduction into the United States, the reassortment of host and microsymbiont ecotypes in some instances may have resulted in the coupling of ecotypes *R. japonicum* from northern Asia with soybean host-plant germplasm from southern Asia.

These hypotheses, simple mutation vs. coevolution, were tested by Devine and Breithaupt (1981) by evaluating the nodulation response of a sample of Asiatic plant introductions with rhizobial cultures that elicit the ineffective nodulation response of the Rj_2 and Rj_4 phenotypes. If the Rj_2 and Rj_4 alleles were simply a mutation phenomenon, then the frequency of incompatible phenotypes should be low and their occurrence random with respect to geographic distribution. If the incompatible reactions result from coupling genotypes of the host and microsymbiont that have not coevolved, then the frequencies of incompatible phenotypes need not be low and should follow a geographic distribution pattern. A total of 847 PI lines were tested in growth trays (Devine and Reisinger, 1978) for the Rj_2 and Rj_4 phenotypes. The Rj_2 phenotype occurred with relatively low frequency (2% of the lines tested), but there appeared to be a clustering of the Rj_2 phenotype in the vicinity of the neighboring Chinese cities of Nanking and Hangchow. Thirty-one percent of the lines from this area had the Rj_2 phenotype. The Rj_4 phenotype occurred in 29.7% of the lines tested. The highest frequency of the Rj_4 phenotype occurred in lines from southeast Asia; i.e., from Burma, Malaya, Indonesia, Thailand, and Vietnam, and a lower frequency occurred in lines from northern Asia. Thus, there is a relationship between geographical sources of the PIs and frequencies of the Rj phenotypes.

The Rj genes produce phenotypes that are strikingly visible evidence of host-microsymbiont incompatibility. There are, however, more subtle, less easily distinguishable deficiencies in the efficiency of symbiosis following nodule formation. If the same principle of ecotypic coevolution that governs the effects of the Rj_2 and Rj_4 alleles on effectiveness of nodulation also governs the efficiency of nitrogen fixation, then coupling the proper ecotypes of the soybean host and rhizobial microsymbiont should result in improved efficiency of nitrogen fixation.

COWPEA RHIZOBIA AND SOYBEANS

Rhizobitoxin-Induced Chlorosis

Some strains of rhizobia that nodulate and fix nitrogen with soybean also produce rhizobitoxin in symbiosis (Johnson and Clark, 1958). Rhizobitoxin produces a foliar chlorosis. I interpret data from several reports as suggesting that the chlorosis-inducing strains represent a subpopulation of rhizobia not homologous with the cultivated soybeans grown in the USA, but with some other host, perhaps cowpea (*Vigna unguiculata* [L.] Walp.) (Devine et al., 1983). Rhizobitoxin-producing strains differ from other strains of *R. japonicum* in their nodulating ability. The rhizobitoxin-producing strains produce a noticeable degree of nodulation with the nodulation-restrictive soybean genotype rj_1 rj_1 in sand or vermiculite culture (Devine and Weber, 1977). These strains also induce nodule-like structures on peanut (*Arachis hypogaea* L.) (Devine et al., 1983).

I also interpret the reports of Hollis et al. (1981) and Agarwal and Keister (1983) as indicating that the rhizobitoxin-producing strains constitute a subpopulation. DNA/DNA hybridization analyses (Hollis et al., 1981) indicated that slow-growing rhizobial strains nodulating soybeans, and labelled as *Rhizobium japonicum*, could be separated into three DNA-homology groups. Two groups were classified as *R. japonicum* subspecies 1 and 2 and the third group was differentiated as *Rhizobium* sp. Although the rhizobitoxin-production characteristics are known for only some of the strains classified, none of the strains in subspecies 1 and 2 are known rhizobitoxin producing-strains, but four of the nine strains in the group classified as *Rhizobium* sp. are rhizobitoxin-producing strains. Agarwal and Keister (1983) studied 39 strains of rhizobia classified as *R. japonicum* and divided them into three groups, A, B, and C, on the basis of colony morphology, explanta acetylene reduction in a liquid medium, pellet characteristics and O_2 consumption under low O_2 tension. Six of the 17 strains classified in group A (positive explanta acetylene reduction) are known chlorosis-producing strains, whereas of the strains in groups B and C only one strain (USDA 117) in group C could be identified as a rhizobitoxin-producing strain. Even this strain is only known to produce rhizobitoxin in vitro, not in symbiosis.

Hydrogenase Activity

On the basis of chemical kinetics and energetics, it has been suggested that strains of rhizobia that have an uptake hydrogenase system (Hup$^+$) should be more efficient in N_2 fixation than strains lacking this system (Hup$^-$) (Evans, this volume). Lack of the uptake hydrogenase system, then, may be interpreted to be a defect in symbiosis. Some rhizobia classified as *R. japonicum* form an effective and efficient symbiosis and behave as Hup$^+$ strains with cowpea, but as Hup strains on soybean (Keyser et al., 1982b). Thus, strains thought to be lacking maximum efficiency with soybean were apparently not being evaluated with the proper host. The frequency of recovery of Hup$^-$ rhizobia is particularly high in samples from southeastern USA (Lim et al., 1981). This is also a traditional area of cowpea cultivation.

Thus, the evidence from the hydrogenase activity and rhizobitoxin production of rhizobial strains suggests to me that many of the rhizobial strains nodulating soybean in the southeastern U.S. are not truly homologous with soybeans, but more likely are homologous with cowpea or other legumes. These strains have an extended but limited symbiotic capability with soybean.

BREEDING FOR IMPROVED NITROGEN FIXATION

Rhizobial strains that produce increased soybean yields have been identified (Abel and Erdman, 1964; Caldwell and Vest, 1970) and efforts are underway to construct new highly-superior strains of *Rhizobium* (Williams and Phillips, 1983). Much of the world's soybean area contains an abundant heterogeneous and persistent population of rhizobia that range from poor to good in their N_2 fixation efficiency with current soybean cultivars. At present there is no economically feasible method of establishing introduced strains against the competition of indigenous strains (Ham, 1976; Vest et al., 1973). Therefore, two approaches to the problem of indigenous rhizobia may be taken: (a) breed the host plant for adaptation to the indigenous rhizobia, or (b) attempt to control the genotype of the rhizobia admitted to symbiosis.

Adapting the Plant to Fit the Microsymbiont

Selecting soybean germplasm for adaptation to indigenous rhizobia may have been the implicit practice of breeders whenever they selected for performance on low nitrogen soils, since nitrogen fixation would have been integrated with the other components of agronomic performance used as selection criteria. New techniques for measuring N_2 fixation directly may improve the efficiency of selection. A striking example of specific selection of plant germplasm for adaptation to indigenous rhizobia has been conducted by I.I.T.A. in Africa. Soybean is not an indigenous crop in Africa, and U.S.-bred soybean cultivars do not adequately nodulate and fix nitrogen with indigenous African "cowpea-type" rhizobia. The introduction of compatible rhizobia results in markedly improved fixation of the U.S.-bred lines in Africa. However, production and distribution of inoculum may be restricted in developing countries. Nangju (1980) reported that soybeans from southeast Asia nodulated successfully with indigenous African rhizobia and the resulting symbiosis was effective in N_2 fixation. U.S.-bred cultivars, however, have the advantage of higher grain yield potential and better resistance to lodging and pod shattering. Nangju suggested breeding to combine the desirable agronomic characteristics of U.S.-bred cultivars with the symbiotic potential for N_2 fixation with the indigenous rhizobia characteristic of the southeast Asian germplasm. Progress in developing such lines has been reported (Kueneman, This volume).

Controlling Rhizobial Specificity

Concepts. Devine and Weber (1977) proposed a genetic system for host specificity of the rhizobial strains admitted to symbiosis. The system requires (a) development of host cultivars that substantially exclude infection by the indigenous *Rhizobium* strains, (b) identification of development of *Rhizobium* strains having the genetic potential to infect these specific host cultivars, and (c) development of the technology to manipulate the genetic system of *Rhizobium* in order to couple this specific nodulating ability with high nitrogen fixation. Achievement of these three elements would provide farmers with a production package of cultivars that would not nodulate with indigenous strains and inoculum of several superior strains that would nodulate these cultivars.

The rj_1 Gene. By backcrossing the rj_1 gene, the nonnodulating gene, into our best agronomic lines of soybean, one can achieve the first requirement—exclusion of the indigenous *Rhizobium* strains from symbiosis. To achieve the second requirement, we at Beltsville are searching for genetic information in *Rhizobium* that will "break down the resistance" of the rj_1 genotype and permit nodulation. To obtain strains of *Rhizobium* with this ability, in one approach we plant a field containing a diverse population of rhizobia with the Clark-rj_1 soybean. After 6 to 10 weeks plants are dug and the roots examined for the presence of nodules. Pure cultures isolated from these nodules are tested for ability to nodulate Clark rj_1 plants in Leonard jars. Promising isolates are then tested for nodulation with Clark rj_1 plants in soil culture. In another approach, in cooperation with Dr. Dan Jeffers of the Ohio Agricultural Research and Development Center, we are using a field as a large petri dish in a manner analogous to the selection for bacterial mutants with resistance to antibiotics. First, corn (*Zea mays* L.) was grown in a field to lower the level of soil nitrogen. The following year, Clark rj_1 soybeans were planted. Because of the soil nitrogen deficiency, and the inability of the Clark rj_1 plants to nodulate, the plants displayed the pale green color typical of nitrogen deficiency. Each year, the field is replanted to Clark rj_1 soybeans. If rhizobial variants occur that nodulate rj_1 soybeans, then since nodulation favors the proliferation of nodulating rhizobia (Kuykendall et al., 1982), the rj_1-compatible rhizobia should increase in number at the site at which they occurred. The following season, plants in the vicinity should be nodulated, creating the appearance of dark green islands of plants in a field of pale green plants. Nodules from such sites should yield the desired rhizobia.

Can recombinant DNA technology be useful in solving the problem? If the rj_1rj_1 genotype lacks a biochemical entity necessary for nodulation that is present in the $Rj_1\ Rj_1$ genotype, would it be possible to identify and isolate the DNA of the Rj_1 allele by use of restriction enzymes, and then to transfer this allele to R. japonicum strains with superior nitrogen fixation? Would the plant gene be functional in the rhizobia? Would the genetically engineered rhizobia supply the entity lacking in the $rj_1\ rj_1$ plant and achieve nodulation with it?

Multigeneic Incompatibility Systems. A variation of the exclusion scheme utilizing incompatibility alleles other than rj_1 has been suggested (Devine and Breithaupt, 1980). The frequency of the Rj_2 and Rj_4 alleles reflects geographic distribution as discussed earlier. Then if, for example, the *Rhizobium* strains in a geographic area of the United States are derived from northern Japan or China, and these strains are incompatible with soybean lines from southeast Asia, it should be possible to avert nodulation with the indigenous U.S. strains by transferring the genes for this rhizobial-rejection response to adapted cultivars. In comparing this approach with the use of the rj_1 allele, it should be borne in mind that the single allele rj_1 may be readily transferred by backcrossing, whereas accumulation of several incompatible alleles, such as Rj_2, Rj_4, etc., is more difficult in a backcrossing program.

Efforts are also underway at the University of Minnesota to employ the system of host-plant exclusion of indigenous strains with selective receptivity to desirable strains (Kvien et al., 1981). Field screening of soybean plant introduction lines has identified lines that do not nodulate normally with the strains indigenous to the Minnesota test site. Some of these lines nodulate well with desirable *Rhizobium* strains, such as USDA 110. It will be important to determine the completeness and consistency of the exclusion of the indigenous strains in the soils of the upper Midwest. If the exclusion mechanism is controlled by a large complex of genes located at scattered loci in the genome, the task of transferring these genes from the agronomically unimproved plant introductions to agronomically acceptable cultivars will be difficult. Kvien also reported that poor nodulation with indigenous rhizobia did not assure high recovery of an introduced strain, nor did good nodulation with indigenous rhizobia preclude selection of introduced strains.

If a host plant can be found or developed that excludes nodulation by the slow-growing rhizobial strains found in the USA, but nodulates and fixes nitrogen very efficiently with the fast-growing rhizobial strains, a similar system could be constructed. If the host genotype has a strong preference for nodulation with the fast-growing rhizobia, this might be sufficient to achieve host specificity.

NOTES

T. E. Devine, ARS-U.S. Department of Agriculture, Nitrogen Fixation and Soybean Genetics Laboratory, Bldg. 011, H. H. 19, BARC-West, Beltsville, MD 20705.

REFERENCES

Abel, G. H., and L. W. Erdman. 1964. Response of Lee soybeans to different strains of *Rhizobium japonicum*. Agron. J. 56:423-424.

Agarwal, A. K. and D. L. Keister. 1983. Physiology of *ex planta* nitrogenase activity in *Rhizobium japonicum*. Appl. Environ. Microbiol. 45:1592-1601.

Beringer, J. E., N. J. Brewin and A. W. B. Johnston. 1980. The genetic analysis of *Rhizobium* in relation to symbiotic nitrogen fixation. Heredity 45:161-186.

Caldwell, B. E. and G. Vest. 1970. Effects of *Rhizobium japonicum* strains on soybean yields. Crop Sci. 10:19-21.

Caldwell, B. E. and B. Vest. 1968. Nodulation interactions between soybean genotypes and serogroups of *Rhizobium japonicum*. Crop Sci. 8:680-682.

Clark, F. E. 1957. Nodulation responses of two near-isogenic lines of the soybean. Can. J. Microbiol. 3:113-123.

Corby, H. D. L. 1981. The systematic value of leguminous root nodules. p. 657-669. *In* R. M. Polhill and P. H. Raven (eds.) Advances in Legume Systematics, Part 2. Royal Botanic Gardens, Kew, London, England.

Devine, T. E. 1984. Inheritance of soybean nodulation response with a fast-growing strain of *Rhizobium*. J. Hered. 75:359-361.

Devine, T. E. 1985. Nodulation of soybean (*Glycine max* L. Merr.) plant introduction lines with the fast-growing rhizobial strain USDA 205. Crop Sci. (in press, April issue).

Devine, T. E. and B. H. Breithaupt. 1980. Significance of incompatibility reactions of *Rhizobium japonicum* strains with soybean host genotypes. Crop Sci. 20:269-271.

Devine, T. E. and B. H. Breithaupt. 1981. Frequencies of nodulation response alleles, Rj_2 and Rj_4, in soybean plant introduction and breeding lines. USDA Tech. Bull., No. 1628.

Devine, T. E., L. D. Kuykendall and B. H. Breithaupt. 1979. Development of a genetic control system for establishing host cultivar symbiosis with agronomically preferred strains of *Rhizobium*. Agron. Abst. p. 60.

Devine, T. E., L. D. Kuykendall and B. H. Breithaupt. 1983. Nodule-like structures induced on peanut by chlorosis producing strains of *Rhizobium* classified as *R. japonicum*. Crop Sci. 23:394-397.

Devine, T. E. and W. W. Reisinger. 1978. A technique for evaluating nodulation response of soybean genotypes. Agron. J. 70:510-511.

Devine, T. E. and D. F. Weber. 1977. Genetic specificity of nodulation. Euphytica 26:527-535.

Ham, G. E. 1976. Competition among strains of rhizobia. p. 144-150. *In* L. D. Hill (ed.) World Soybean Research. Interstate Printers and Publishers Inc., Danville, Ill.

Hollis, A. B., W. E. Kloos and G. H. Elkan. 1981. DNA:DNA hybridization studies of *Rhizobium japonicum* and related *Rhizobiaceae*. J. Gen. Microbiol. 123:215-222.

Johnson, H. W. and F. E. Clark. 1958. Role of the root nodule in the bacterial-induced chlorosis of soybeans. Soil Sci. Soc. Amer. Proc. 22:527-528.

Keyser, H. H., B. B. Bohlool, T. S. Hu and D. F. Weber. 1982a. Fast-growing rhizobia isolated from root nodules of soybeans. Science 215:1631-1632.

Keyser, H. H., P. van Berkum and D. F. Weber. 1982b. A comparative study of the physiology of symbioses formed by *Rhizobium japonicum* with *Glycine max*, *Vigna unguiculata*, and *Macroptilium atropurpurum*. Plant Physiol. 70:1626-1630.

Kondorosi, A. and A. W. B. Johnston. 1981. Genetics of *Rhizobium*. *In* K. L. Giles and A. G. Atherly (eds.) Biology of the *Rhizobiaceae*. Academic Press, New York.

Kuykendall, L. D. 1979. Transfer of R factors to and between genetically marked sublines of *Rhizobium japonicum*. Appl. Environ. Microbiol. 37:862-866.

Kuykendall, L. D., T. E. Devine and P. B. Cregan. 1982. Positive role of nodulation on the establishment of *Rhizobium japonicum* in subsequent crops of soybeans. Curr. Microbiol. 7:79-81.

Kvien, C. S., G. E. Ham and J. W. Lambert. 1981. Recovery of introduced *Rhizobium japonicum* strains by soybean genotypes. Agron. J. 73:900-905.

Lim, S. T., S. L. Uratsu, D. F. Weber and H. H. Keyser. 1981. Hydrogen uptake (hydrogenase) activity of *Rhizobium japonicum* strains forming nodules in soybean production areas of the U.S.A. *In* J. M. Lyons, R. C. Valentine, D A. Phillips, D. W. Rains and R. C. Huffaker (eds.) Genetic engineering of symbiotic nitrogen fixation and conservation of fixed nitrogen. Plenum Press, New York.

Masterson, R. V., P. R. Russell and A. G. Atherly. 1982. Nitrogen fixation (*nif*) genes and large plasmids of *Rhizobium japoncium*. J. Bacteriol. 152:928-931.

Nangju, D. 1980. Soybean response to indigenous *Rhizobia* as influenced by cultivar origin. Agron. J. 72:403-406.

Polhill, R. M., P. H. Raven and C. H. Stirton. 1981. Evolution and systematics of the *Leguminosae*. p. 1-26. *In* R. M. Polhill and P. H. Raven (eds.) Advances in Legume Systematics, Part 1. Royal Botanic Gardens, Kew, London.

Sprent, J. I. 1981. Functional evolution in some papilionoid root nodules. p. 671-676. *In* R. M. Pohill and P. H. Raven (eds.) Advances in Legume Sytematics, Part 2. Royal Botanic Gardens, Kew, London, England.

Vest, Grant, D. F. Weber and C. Sloger. 1973. Nodulation and nitrogen fixation. p. 353 390. *In* B. E. Caldwell (ed.) Soybeans: Improvement, Production, and Uses. Am. Soc. of Agron., Madison, Wis.

Wells, S. E. and L. D. Kuykendall. 1983. Tryptophan auxotrophs of *Rhizobium japonicum*. J. Bacteriol. 156:1356-1358.

Williams, L. E. and D. A. Phillips. 1983. Increased soybean productivity with a *Rhizobium japonicum* mutant. Crop Sci. 23:246-250.

PATHOLOGY

PATHOLOGY

Diseases of Seeds

FACTORS AFFECTING THE SEVERITY OF PHOMOPSIS
SEED DECAY IN SOYBEANS

P. R. Thomison

Infection of soybean seed by the seed decay fungi belonging to the genera *Phomopsis* (Sacc.) Bubak and *Diaporthe* Nitschke is a major cause of poor seed quality in regions where the climate is warm and humid during and after crop maturity. Seed infection by these fungi may result in moldy seed at maturity or seed decay during germination. Seed exhibiting disease symptoms (discoloration, fissuring, superficial mycelial growth) are smaller in size and volume, lower in density, produce lower quality oil and flour, and have less durability than seed free of infection (Sinclair, 1982). Early detection of these seedborne pathogens is difficult because infections remain dormant and symptomless until plants begin to mature and senesce. Moreover, the absence of visual disease symptoms on pods and stems does not preclude the possibility of seed infection.

Researchers have implicated *Phomopsis* sp. (hereafter referred to as *Phomopsis*) as the pathogen primarily responsible for seed decay problems (Schmitthenner and Kmetz, 1980). *Phomopsis* has been found to account for a greater incidence of seed infection than *Diaporthe phaseolorum* (Cke. & Ell.) Sacc. var. *sojae* (Lehman) Wehm. and *D. phaseolorum* var. *caulivora* Athow & Caldwell.

ENVIRONMENTAL CONDITIONS

Climatic conditions (precipitation, humidity and temperature), during and after crop maturity, play a critical role in determining the severity of *Phomopsis* seed infection. However, until recently there was little known concerning the relative contribution of each of these weather related variables to overall disease development. Based on results of a three year survey, Shortt et al. (1981) concluded that rainfall rather than temperature was the dominant environmental factor influencing seed decay in Illinois. Spilker et al. (1981) reported that relative humidity was more important than temperature in determining *Phomopsis* infection during soybean maturation. Their findings, based on studies conducted in growth chambers, indicated that high temperatures (32 C day/24 C night), in combination with high humidity (90%), resulted in greater fungal growth (more moldy seed) and lower germination compared with seed from high humidity—low temperature (26 C day/16 C night) treatments. Balducchi and McGee (1984) showed that at high relative humidity, the rate of movement of *Phomopsis* from pod to seed increased with temperature.

In Kentucky, TeKrony et al. (1983) found that the incidence of seedborne *Phomopsis* was significantly correlated with air temperature and minimum relative humidity, but not with total precipitation or daily precipitation. A regression model,

495

employing environmental variables, developed to estimate *Phomopsis* damage indicated that precipitation and minimum relative humidity were more important than other environmental conditions in the development of *Phomopsis* infection.

The damage to germinating seed caused by *Phomopsis* may be modified by soil moisture conditions at planting. Gleason (1984) examined *Phomopsis* growth in the cotyledons of soybean seeds infected with comparable levels of *Phomopsis* and found that fungal colonization of cotyledons increased in dry soils (-0.1 nPa or less), but decreased in soil wetter than that. The greater fungal development in seeds planted in dry soil was associated with decreased seedling emergence and establishment.

The close association between environmental conditions and disease expression has been used to design disease forecast systems for use in scheduling foliar fungicide appli cations. In Illinois and Kentucky, checklists or 'point systems', based on weather variables, cropping history, cultivar selection and planting dates, have been developed to determine whether to apply foliar fungicides to soybeans grown for seed (Sinclair, 1982).

CROPPING PRACTICES AND RESIDUE MANAGEMENT

Seed infection by *Phomopsis* can be reduced and seed quality improved by practicing crop rotation. *Phomopsis* seed infection is lower following corn (*Zea mays* L.) than in continuously cropped soybeans. However, studies in Ohio (Jeffers and Schmitt-henner, 1981) found that burying crop residues, including soybean residue, by fall or spring plowing did not affect germination or disease incidence when compared with soybeans grown without tillage. In Iowa, *Phomopsis* was detected in seed harvested from soybeans grown in a field that had been previously cropped to corn continuously for 10 years (Garzonio and McGee, 1983). Infected seed are a minor source of inoculum where soybeans have been grown previously, and *Phomopsis* conidia blown in from adjacent fields or carried in by equipment represent the major inocula.

In certain situations, weeds such as velvet leaf (*Abutilon theophrasti* Medic.) may act as an alternate host for *Phomopsis*, and thereby serve as another source of disease inoculum in soybean fields (Hepperly et al., 1980).

CULTIVAR RESISTANCE AND SEED COAT IMPERMEABILITY

Presently, no satisfactory resistance to *Phomopsis* is available in widely used, adapted soybean cultivars, although varying degrees of resistance have been reported (Kueneman, 1982). The lower levels of diseased seed exhibited by soybean crops maturing later in the growing season have been misconstrued by some as resistance to *Phomopsis*. Differences in seed disease associated with varying maturity times are generally explained by environmental conditions during maturation, rather than by genotypic resistance (Schmitthenner and Kmetz, 1980). Some of the resistance to *Phomopsis* found in plant introductions may be related to hard, impermeable seed coats. Researchers have found that soybean lines that produce hard impermeable seed often have significantly less *Phomopsis* and *Diaporthe* infected seed and higher germination than lines with permeable seed coats (Miranda et al., 1980; Hill et al., 1984). The thicker seed coats of impermeable seed may limit fungal establishment and/or access to nutrients. However, it is questionable whether hard seedcoats can be used to control *Phomopsis*, since impermeable seed are frequently associated with uneven germination and poor stand establishment.

Hill and West (1982) detected naturally occurring pores that penetrated deeply into the palisade layer of the seed coat, providing passage into the hourglass layer. Fungal hyphae were observed to extend into these pores, suggesting that such pores could provide a means of fungal entry without the presence of visible seed coat defects. In light of this finding, it would be interesting to ascertain whether seed coat impermeability affects these pores and the fungal penetration associated with them.

FUNGICIDES AND THERMOTHERAPY TO REDUCE INFECTION

Fungicide foliar sprays and seed treatments can be used to reduce *Phomopsis* infection and increase germination of seed. Single or split applications of fungicide sprays at R3 and R5 have reduced, but not eliminated, seed infection. A single, high rate of benomyl at R6 (full seed) may also be effective in limiting *Phomopsis* damage and improving germination under certain conditions (Sinclair, 1982).

The efficacy of foliar fungicides in controlling *Phomopsis* may be determined, in large part, by the application method used. Jeffers et al. (1982b) evaluated the effectiveness of standard aircraft application and several techniques for ground application of benomyl. Aerial spraying of benomyl did not result in improvement of overall seed germination; whereas, application of benomyl with certain types of ground equipment, such as a mistblower or downdraft sprayer, was effective in both reducing fungal seed infection and improving germination.

Symptomless seed coat infections caused by *Phomopsis* may be reduced, without the use of fungicide, by aging seed or by thermotherapy. Zinnen and Sinclair (1982), using soybean oil as a medium for heating soybeans, demonstrated that thermotherapy of seed to control seedborne fungi was experimentally feasible. Treatments that reduced *Phomopsis* recovery from seed, while increasing germinable, pathogen-free seed, ranged from 5 min at 70 C to 10 s at 140 C.

INTERACTIONS WITH OTHER SEEDBORNE FUNGI OF SOYBEANS

Antagonistic interactions between *Phomopsis* and other seedborne fungi have received some attention as a possible mechanism of disease control. Roy and Abney (1977) reported that inoculation of soybean plants with *Cercospora kikuchii* (Mat. & Tomoy) Gardner reduced *Phomopsis* seed infection and improved seed germination. Studies have also indicated an inverse relationship of seed infection between *Alternaria* sp. and *Phomopsis*, suggesting that there may be an antagonism between these two organisms (Jeffers et al., 1982b; Thomison, 1983).

DESICCANT HERBICIDES

The use of desiccant herbicides, such as paraquat and glyphosate, have been proposed as a means of expediting maturation and accelerating dry-down of soybeans. However, spraying with these herbicides can reduce seed quality and actually increase *Phomopsis* infection under certain environmental conditions (Cerkauskas et al., 1982). The rapid death of plant tissues caused by paraquat and other desiccants provides more time for fungal colonization and sporulation.

This stimulatory effect of desiccant herbicides on *Phomopsis* development has been utilized as a method to detect *Phomopsis* in pods prior to seed infection. The amount of *Phomopsis* on pods is used to predict the severity of seed infection and the need for foliar fungicide control. In Iowa, McGee and Nyvall (1984) have developed a test in which pods are detached in the field, at R6, surface sterilized in sodium hypochlorite, and then treated with a desiccant herbicide to kill pod tissue rapidly and to enhance sporulation. Pods are examined after a 7-day incubation period at 95% relative humidity to determine infection, i.e. the presence of *Phomopsis* pycnidia. When more than 50% of the pods are infected, foliar fungicides are recommended.

PLANT MATURATION AND STRESS CONDITIONS

Little is known concerning the changes that occur in soybean plants which render them more susceptible to *Phomopsis* and *Diaporthe* infection at maturity. There has been speculation that various physiological and metabolic processes associated with plant senescence contribute to the transformation of the dormant *Phomopsis* infection into an active one.

Hepperly and Sinclair (1980a) inoculated detached soybean pods with *Phomopsis* mycelia at different reproductive stages and found a greater increase in seed infection in yellow soybean pods than in green pods. This increased seed infection was related to a breakdown of the mechanisms governing latency of infection. A lack of soluble substrates and the presence of fungistatic metabolites in immature pods were cited as possible factors causing *Phomopsis* latency.

Researchers have conjectured that, during the ripening of fruits, various compounds comprising the cell wall become more loosely bound and may only then be degraded by pectolytic enzymes of fungi (Verhoeff, 1974). An analogous situation may occur in soybean pods. The apparent failure of *Phomopsis* to degrade and develop in immature soybean tissue may be evidence that the fungus is limited in its invasion of immature pods by insufficient cell wall degrading enzyme activity.

Plant nutrition may also play a role in disease development. Changes in the nutrient content of soybean pods and stems during senescence coincide with the breaking of 'latent' *Phomopsis* infections and the subsequent appearance of pycnidia. Stress conditions such as low fertility and drought may increase plant susceptibility to pathogens by altering the nutrient composition of plant tissues.

Several studies have found significant relationships between potassium deficiency and increases in moldy seed caused by *Phomopsis* and *Diaporthe*. However, effects of K fertility on symptomless seed coat infections and seed germinability have usually not been considered. Jeffers et al. (1982a) reported that high levels of K (exceeding that necessary to maximize yields) generally reduced moldy levels. However, germination and yield responses to increased K fertilization were less consistent, and even where seed germination was improved by higher rates of K fertilization, the incidence of seed infection by *Phomopsis* and *Diaporthe* showed no response to K. Their results suggested that, whereas K fertilization might not influence *Phomopsis* and *Diaporthe* infection of seed directly, it might reduce the severity of infection, limiting fungal growth after infection has occurred.

A study was recently conducted in Ohio to determine whether differences in the nutrient composition of soybean pod tissue influence seed infection by *Phomopsis* (Thomison, 1983). Soybean plants with varying fruit loads were grown to generate different levels of carbohydrates and mineral nutrients in pods. Reducing pod numbers per node resulted in more severe *Phomopsis* seed infection, and a greater concentration of total non-structural carbohydrates (TNC), nitrogen and phosphorus in pod walls (Table 1). Results of moist chamber incubation tests on pods sampled from plants at different reproductive stages of growth indicated greater mycelial growth on pods from plants with reduced fruit loads. Similarly, in a study comparing *Phomopsis* seed infection under high and low moisture regimes, significantly greater *Phomopsis*

Table 1. Effect of reduced fruit load and pod position on *Phomopsis* seed infection and pod wall characteristics of Wells soybeans (Thomison, 1983).

Pod removal treatment	Pod position[a]	Moldy seed[b]	*Phomopsis*	Germ.	Pod wall[d]					
					TNC[c]	N	P	K	Ca	Mg
		– % –			– % of dry weight –					
1 pod/node	B	37	92	16	9.02	1.08	0.179	4.94	0.45	0.299
	T	4	25	82	13.24	1.63	0.163	2.66	1.08	0.543
No pods removed	B	4	57	66	6.38	0.76	0.059	5.04	0.47	0.350
	T	0	11	98	8.64	1.04	0.112	2.19	1.01	0.573
LSD (0.05)										
pod removal		3	11	11	0.16	0.21	0.020	NS	0.03	NS
position		2	6	4	0.11	0.07	0.008	0.20	0.08	0.027
pod remov. x pos.		4	12	8	0.29	0.13	0.014	0.35	NS	NS

[a] B, T—Bottom and top thirds of plants, respectively.
[b] Mean of six replications of 100 seeds each.
[c] Mean of six replications of 100 pod walls each.
[d] Nutrient levels of pod walls at R7 (physiological maturity).

infection and reduced germination of seed was associated with higher levels of TNC, N and P in pod walls of plants grown under high soil moisture conditions (Table 2). Although these results indicate that varying soil moisture and reduced fruit load did affect pod wall nutrient levels, which in turn, might have influenced *Phomopsis* development, other factors associated with these treatments also may have altered plant susceptibility to *Phomopsis*. Reducing pods per plant increased pod size, seed coat fissuring (hypodermal cracks), and delayed plant naturation. High soil moisture treatments resulted in greater lodging and canopy density, in addition to prolonging seed dry down and maturity.

Also, there may be a relationship between nutrient composition of pods and *Phomopsis* and *Diaporthe* infection of soybean seed in the upper and lower nodes of plants. Delaware researchers (Crittenden and Svec, 1974) observed that moldy seed

Table 2. Effect of soil moisture on *Phomopsis* seed infection and pod wall characteristics of Harosoy-63 soybeans (Thomison, 1983).

Soil moisture[a]	*Phomopsis*[b]	Germ.	Maturity[c] date	Pod wall[e]					
				TNC[d]	N	P	K	Ca	Mg
				– % of dry weight –					
Low (8%)	18	74	18 Sept	7.3	1.22	0.15	2.0	0.98	0.82
High (18%)	51	50	29 Sept	9.3	1.80	0.29	2.3	0.78	0.56
LSD (0.05)	6	4		1.1	0.09	0.04	NS	NS	0.18

[a] Average soil moisture content (% of dry weight) six weeks after water stress (plastic covers applied) at R1.
[b] Mean of six replications of 100 seeds each.
[c] Maturity = Date of harvest maturity.
[d] Mean of six replications of 100 pod walls each.
[e] Nutrient levels of pod walls prior to physiological maturity (R6.5).

levels were greater in pods from the upper nodes of plants and conjectured that seed infection was correlated with a decrease in K levels, which occurs in soybean pods late in the season. However, others have detected a greater incidence of *Phomopsis* in seed from the lower nodes of plants (Schmitthenner and Kmetz, 1980; Hepperly and Sinclair, 1980b; Jeffers et al., 1982a).

Recent studies indicate that infection of soybean seed within the plant is not related to K deficiency. Jeffers et al. (1982a) noted that , although K deficiency symptoms were exhibited primarily by upper pods and foliage, K fertilization had a more pronounced effect on reducing *Phomopsis* infection of seed in pods at the lower nodes than in those at upper nodes. Thomison (1983) found a lower incidence of *Phomopsis* seed infection and moldy seed in pods at the upper nodes of soybean plants to be associated with lower K and higher Ca levels in pod walls (Table 1). Conversely, seed infection was most severe at the lower nodes, where pod wall levels of K were greater and Ca lower. The inverse relationship between Ca concentration and *Phomopsis* infection warrants further attention, since calcium deficiencies play a significant role in a number of plant diseases.

Other factors such as time and duration of pod development may also contribute to differences in *Phomopsis* seed infection at varying pod positions. Kulik (1983) observed a reduction in sporulation by *Phomopsis* on soybean residue with the approach of autumn and the onset of cool weather. This decrease in spore production may limit the capacity of the fungus to infect pods and seeds formed at the upper nodes later in the growing season. Pods (and seed) from the lower halves of indeterminate soybeans begin development earlier and mature more slowly than pods set later at upper nodes. Consequently, pods at lower nodes are exposed to weathering conditions and pathogen attack over a longer duration than those at the top of the plant.

Virus diseases are another factor that may contribute to greater fungal seed decay. In Illinois, inoculation of 20 soybean cultivars with soybean mosaic virus (SMV) reduced germination and increased seed infection by *Phomopsis* (Hepperly et al., 1979). Ross (1977) found that *Phomopsis* was isolated with greater frequency from SMV susceptible lines than from SMV resistant lines. In Kentucky, Stuckey et al. (1982) reported that greater *Phomopsis* infection occurred when soybeans were inoculated with bean pod mottle virus (BPMV) than with SMV. Depending upon cultivar, the incidence of *Phomopsis* seed infection increased as much as five-fold, in plants inoculated with BPMV compared with uninoculated control plants. These virus inoculations did not appear to influence infection by *Cercospora kikuchii*, another seedborne fungus, which was also present.

How certain conditions such as virus infections and K deficiencies predispose soybeans to greater *Phomopsis* seed infection is not understood. There is evidence to suggest that delayed plant maturity may be a common underlying factor influencing *Phomopsis* seed infection. Virus infection and K deficiency often delay plant maturation. Reduced pod number and irrigation treatments, which prolong plant senescence and slow seed dry-down, also increase *Phomopsis* seed infections. Slower seed dry-down rates resulting from delayed maturity may provide *Phomopsis* infections in the pod wall more time to develop and colonize seed. Prolonged periods of high seed moisture are critical for fungal invasion of seed. Other factors associated with a slow rate of maturation, and therefore with greater levels of pod wall nutrients, may also stimulate greater seed infection in certain situations. Further work will be necessary to determine the actual contribution of these different factors to disease development.

NOTES

Peter R. Thomison, Department of Agronomy, University of Maryland, College Park, MD 20742.
Scientific Article No. A-3909. Contribution No. 6890 of the Maryland Agric. Exp. Stn., Dept. of
Agronomy, College Park, MD 20742.

REFERENCES

Balducchi, A. and D. C. McGee. 1984. Environmental factors affecting movement of a *Phomopsis*
spp. from soybean pods to seeds. Iowa Seed Sci. 6:3-4.

Cerkauskas, R. F., O. D. Dhingra, J. B. Sinclair and S. F. Foor. 1982. Effect of three desiccant herbicides on soybean seed quality. Weed Sci. 30:484-490.

Crittenden, H. W. and L. V. Svec. 1974. Effect of potassium on the incidence of *Diaporthe sojae* in
soybean. Agron. J. 66:696-697.

Garzonio, D. M. and D. C. McGee. 1983. Comparison of seed and crop residues as sources of inoculum for pod and stem blight of soybeans. Plant Dis. 67:1374-1376.

Gleason, M. 1984. Effects of soil moisture on seedborne *Phomopsis* sp. and soybean seedling establishment. *In* M. M. Kulik (ed.) Proceedings of the conference on the *Diaporthe/Phomopsis* disease
complex of soybean. USDA/ARS (In press).

Hepperly, P. R., B. L. Kirkpatrick and J. B. Sinclair. 1980. *Abutilon theophrasti*: Wild host for three
fungal parasites of soybean. Phytopathology 70:307-310.

Hepperly, P. R. and J. B. Sinclair. 1980a. Detached pods for studies of *Phomopsis sojae* pod and seed
colonization. J. Agric. Univ. Puerto Rico. 64:330-337.

Hepperly, P. R. and J. B. Sinclair. 1980b. Association of plant symptoms and pod position with
Phomopsis sojae seed infection and damage in soybean. Crop Sci. 20:379-381.

Hepperly, P. R., G. R. Bowers, Jr., J. B. Sinclair and R. M. Goodman. 1979. Predisposition to seed infection by *Phomopsis sojae* in soybean plants infected by soybean mosaic virus. Phytopathology.
69:846-848.

Hill, H. J., S. H. West and K. Hinson. 1984. *Phomopsis* infection of soybean seed differing in impermeability. *In* M. M. Kulik (ed.) Proceedings of the conference on the *Diaporthe/Phomopsis*
disease complex of soybean. USDA/ARS (In press).

Hill, H. J. and S. H. West. 1982. Fungal penetration of soybean seed through pores. Agron. J. 22:602-
605.

Jeffers, D. L. and A. F. Schmitthenner. 1981. Germination and disease in soybean seed as affected by
rotation, planting time, K fertilization and tillage. Agronomy Abstr. 73:119.

Jeffers, D. L., A. F. Schmitthenner and M. E. Kroetz. 1982a. Potassium fertilization effects on
Phomopsis seed infection, seed quality and yield of soybeans. Agron. J. 74:886-890.

Jeffers, D. L., A. F. Schmitthenner and D. L. Reichard. 1982b. Seed-borne fungi, quality, and yield
of soybeans treated with benomyl fungicide by various application methods. Agron. J. 74:589-
592.

Kueneman, E. A. 1982. Genetic differences in soybean seed quality: Screening methods for cultivar
improvements. p. 31-41. *In* J. B. Sinclair and J. A. Jackobs (eds.) Soybean seed quality and stand
establishment. INSOY, Series 22. Univ. of Illinois, Urbana.

Kulik, M. M. 1983. The current scenario of the pod and stem blight-stem canker seed decay complex
of soybean. Int. J. Trop. Plant Dis. 1:1-11.

McGee, D. C. and R. F. Nyvall. 1984. Pod test for foliar fungicides on soybeans. Iowa State Univ.
Ext. Serv. Pm 1136.

Miranda, F., C. H. Andrews and K. W. Roy. 1980. Seedcoat impermeability in soybean associated
with resistance to seedborne infections and weathering. Agronomy Abstr. 72:111.

Ross, J. P. 1977. Effect of aphid-transmitted soybean mosaic virus on yields of closely related resistant and susceptible soybean lines. Crop Sci. 17:869-872.

Roy, K. W. and T. S. Abney. 1977. Antagonism between *Cerospora kikuchii* and other seedborne
fungi of soybeans. Phytopathology 67:1062-1066.

Schmitthenner, A. F. and K. T. Kmetz. 1980. The role of *Phomopsis* sp. in the soybean seed rot problem. p. 355-370. *In* F. T. Corbin (ed.) World Soybean Research Conference II: Proceedings. Westview Press, Boulder, Colorado.

Shortt, B. J., A. P. Grybauskas, F. D. Tenne, and J. B. Sinclair. 1981. Epidemiology of *Phomopsis* seed decay of soybean in Illinois. Plant Dis. 65:62-64.

Sinclair, J. B. 1982. Compendium of soybean diseases. 2nd ed. Am. Phytopathol. Soc., St. Paul, MN.

Spilker, D. A., A. F. Schmitthenner and C. W. Ellett. 1981. Effects of humidity, fertility and cultivar on reduction of soybean seed quality by *Phomopsis* sp. Phytopathology 71:1027-1029.

Stuckey, R. E., S. A. Ghabrial and D. A. Reicosky. 1982. Increased incidence of *Phomopsis* sp. in seeds from soybeans infected with bean pod mottle virus. Plant Dis. 66:826-829.

TeKrony, D. M., D. B. Egli, R. E. Stuckey and J. Balles. 1983. Relationship between weather and soybean seed infection by *Phomopsis* sp. Phytopathology 73:914-918.

Thomison, P. R. 1983. *Phomopsis* seed infection and seed quality in soybeans as influenced by soil moisture, fruit load and nutrient accumulation. Ph.D. Dissertation. Ohio State Univ., Columbus.

Verhoeff, K. 1974. Latent infections by fungi. Ann. Rev. Phytopath. 12:99-110.

Zinnen, T. M. and J. B. Sinclair. 1982. Thermotherapy of soybean seeds to control seedborne fungi. Phytopathology 72:831-834.

PURPLE SEED STAIN AND CERCOSPORA LEAF BLIGHT

H. J. Walters

The fungus *Cercospori kikuchii* (T. Matsu & Tomoyasu) Gardner causes purple seed stain of soybeans and Cercospora leaf blight of the soybean plant. Most research pertaining to this fungus has focused on the purple staining of seed. The fungus was first reported in Korea in 1921 (Suzuki, 1921). Matsumoto and Tomoyasu (1925) indicated that the fungus attacked leaves, stems and pods to some extent, but primarily damaged seeds. Han (1959) reported wide distribution of the disease in Taiwan and described symptoms on leaves, stems, pods, and seed. Purple seed stain is now found world wide (Sinclair, 1982).

In the United States, purple seed stain was first observed in 1924 in Indiana (Gardner, 1926). In 1950, Lehman (1950) reported that seedlings were stunted or killed after emergence. Murakishi (1951) described symptoms on hypocotyls, stems, leaves, and petioles of greenhouse-grown soybean plants inoculated with mycelia of the fungus and on plants grown from infected seed. Cercospora leaf blight was reported in 1980 as a disease of economic importance occurring under field conditions in the United States (Walters, 1980).

The increased use of fungicide to control foliar diseases of soybeans has emphasized that Cercospora leaf blight is far more severe than previously recognized, and that it may be a much more important aspect of the damage caused by *C. kikuchii.*

SYMPTOMS

Purple Seed Stain

Seed discoloration occurs in irregular blotches, varying from light to dark purple, and ranging from tiny spots to the entire area of the seed coat. Seed discoloration is often accompanied by wide cracks in seed coats, usually transversely along the seed. Purple seed stain does not reduce yields, but may cause reduced stands (Sinclair, 1982).

Cercospora Leaf Blight

Cercospora leaf blight is an important economic disease of soybean plants in the southern United States. Estimated yield losses from the disease in 1978 were 173 kt in Arkansas and 3.56 Mt for the 15 southern states (Sturgeon, 1979).

Defoliation, beginning with the uppermost leaves and progressing down, has been observed on soybean plants for many years, and has often been mistaken for early maturity by growers. When efficient fungicides are used to control foliar pod and stem diseases, top defoliation usually does not occur.

Under field conditions, the first symptoms of the disease are observed during the late R5 (beginning seed) and R6 (full seed) soybean growth stages. Upper leaves exposed to the sun have a light purple, leathery appearance. Reddish purple, angular-to-irregular lesions later occur on both upper and lower leaf surfaces. Lesions vary from pinpoint spots to irregular patches up to 1 cm in diameter and may coalesce to form large necrotic areas. Veinal necrosis may also be observed. Numerous infections cause rapid chlorosis and necrosis of leaf tissue, resulting in defoliation starting with the young upper leaves. Green leaves often occur below the defoliated area. Lesions on petioles and stems are slightly sunken, reddish purple areas several millimeters long. Infection of petioles increases defoliation. The most obvious symptom is the premature blighting of the younger, upper leaves over large areas, even entire fields. On more susceptible cultivars round, reddish purple lesions, which later become purplish black, occur on pods.

CAUSAL ORGANISM

The fungus can be readily isolated from infected tissue of seed, stems, petioles and lamina by placing the infected tissue in a moist chamber for 48 to 72 h at 23 to 28 C. Abundant conidia are produced, which may be picked off with a drawn-out capillary tube with the aid of a dissecting microscope (Vathakos and Walters, 1979). Sporulation of *C. kikuchii* usually does not occur on potato dextrose agar (Murakishi, 1951; Vathakos and Walters, 1979). Abundant sporulation with all isolates tested occurred on dead soybean plant tissue agar (Vathakos and Walters, 1979) and on cleared V-8 juice agar (El-Gholl et al., 1982) with 12-h periods of alternating light and darkness.

EPIDEMIOLOGY

Cercospora kikuchii overseasons in infected leaf lamina, petioles, stems and seed of soybean plants. I have isolated the fungus from stems of the previous year's crop at weekly intervals from 1 June to mid-October in 1979 and from abscissed petioles as early as 4 weeks prior to blooming.

Infection depends on humidity and temperature and may occur at any time these factors are favorable. Laviolette and Athow (1972) showed that inoculations made during the flowering period resulted in almost the maximum amount of purple stained seed. Crawford (1979) showed that inoculations made by atomizing spores onto plants at R4 resulted in substantial purple seed stain, but did not do so at later growth stages. This agrees with the finding of Murakishi (1951) and Roy and Abney (1976) that *C. kikuchii* infects developing pods, and that wounding is not necessary for infection.

The effect of temperature and moisture on infection by *C. kikuchii* was evaluated with Forrest soybeans (Martin and Walters, 1982). Seedlings inoculated at the V1 growth stage were allowed a dew period of 24, 48 or 72 h at temperatures ranging from 16 to 36 C, and then placed in a growth chamber at 24 C for disease development. Although good infection occurred at 16 C, optimum infection occurred from 20 to 24 C, with less infection at 28 and 32 C, and none at 36 C. As the length of dew periods increased, infection increased at all temperatures. A cycle of 12 h of darkness followed by 12 h of light in the dew chamber resulted in significantly better infection compared with continuous darkness. The effect of alternate dew and dry periods on infection was determined by exposing inoculated plants to 1, 2 or 3 dew periods ranging from 0 to 24 h in 4-h increments, with one dew period every 24 h. Except for the

24-h dew period, plants were allowed to dry in a 24 C growth chamber for the remainder of the 24-h cycle. Following the specified dew periods, plants were returned to the growth chamber for 12 days to allow disease development. A dew period of at least 8 h was required to establish a significant amount of infection. The greatest amount of infection was obtained with the 24-h dew period. Disease ratings increased as exposure to a given dew period increased.

Conidia are borne by wind and splashing rains to other leaves and stems (Sinclair, 1982). My data, taken in Arkansas with Kramer-Collins Spore Samplers over a period of 3 years, show that conidia of *C. kikuchii* may be carried for relatively long distances by wind. The findings indicate that rotation with a non-host crop would lower the amount of early infection, but would not prevent infection during the reproductive stages of the soybean plant. Depending on the distance from other soybean fields, those rotated with rice or grain sorghum are usually as severely affected with Cercospora leaf blight as are non-rotated fields.

DISEASE RESISTANCE

Cultivars differ greatly in their susceptibility to purple seed stain and Cercospora leaf blight. Most appear to have some degree of resistance, and usually do not have a high percentage of purple-stained seed. Athow (1976) reported that PI 80837 possesses a high degree of resistance to seed stain. Roy and Abney (1976) obtained a low percentage of stained seed when plants of PI 80837 were inoculated at the R2, R3, R5

Table 1. Mean disease ratings of soybean cultivars inoculated with conidia of *Cercospora kikuchii* at the V1 growth stage.

| Cultivar | Disease ratings of cultivars tested by:[a] | |
	Vathakos and Walters (1979)	Crawford (1979)
Pickett 71	2.13	4.4
Bragg	2.64	4.4
Essex	–	4.3
Hood 75	4.94	4.3
Forrest	4.85	4.2
Bedford	–	4.0
Dare	4.61	3.7
Mack	3.73	3.0
Mitchell	–	2.7
Lee 74	2.43	2.3
Davis	2.77	2.2
Lancer	–	2.1
Centennial	–	2.1
Hill	2.10	2.1
Harsoy 63	–	1.9
Tracy	2.04	1.7
Pickett	4.57	–
Lee 68	2.36	–

[a]Disease ratings: 1 = No symptoms; 2 = Veinal purpling and a few leaf lesions; 3 = Moderate number of lesions with up to 25% of leaf area affected. Pod lesion may or may not be present; 4 = Severe infection with lesions covering up to 50% of pod; 5 = Greater than 50% of leaf area covered with lesions, large necrotic areas and/or leaf abscission. Pod lesions severe, covering over 50% of pod. Distortion and/or pod abscission.

and R8 growth stages.

Crawford (1979) showed that cultivars very susceptible to Cercospora leaf blight also developed pod lesions. These cultivars were most severely affected with purple seed stain. Seed stain was observed on Bragg, Forrest, Pickett-71, Essex, Bedford, Dare, Hood-75 and Mack when plants were inoculated at the R4 growth stage (Table 1). No purple stain was observed on seed of Mitchell, Lee-74, Davis, Lancer, Centennial, Hill, Harosoy-63 or Tracy.

Table 1 shows the relative differences in susceptibility of soybean cultivars to Cercospora leaf blight when inoculated with a spore suspension of *C. kikuchii* at the V1 growth stage in tests conducted by Vathakos and Walters (1979) and Crawford (1979). Disease ratings among cultivars by both investigators were similar except for Pickett-71. Crawford found no significant difference in disease ratings on cultivars inoculated in the V1 and R2, R3 or R4 growth stages. It is possible to predict the ratings of Cercospora leaf blight on adult plants with data from seedlings inoculated with *C. kikuchii.*

NOTES

H. J. Walters, Department of Plant Pathology, University of Arkansas, Fayetteville, Arkansas.

REFERENCES

Athow, K. L. 1976. Pathological factors affecting seed quality. p. 455-461. *In* L. D. Hill (ed.) World Soybean Research. The Interstate Printers and Publishers, Inc., Danville, Illinois.

Crawford, M. A. 1979. Comparative screening for determining resistance in soybean cultivars to the foliar blight disease caused by *Cercospora kikuchii* (Matsu. and Tom.) Chupp. MS Thesis, Dept. Plant Pathology, Univ of Arkansas, Fayetteville.

El-Gholl, N. E., S. A. Alfieri, Jr., W. H. Ridings and C. L. Schoulties. 1982. Growth and sporulation *in vitro* of *Cercospora apii, Cercospora arachidicola, Cercospora kikuchii* and other species of Cercospora. Can. J. Bot. 60:862-868.

Gardner, M. W. 1926. Indiana plant diseases. Proc. Indiana Acad. Sci. 35:237-257.

Han, Y. S. 1959. Studies on purple spot of soybean. J. Agric. and For. 8:1-32.

Laviolette, F. A. and K. L. Athow. 1972. *Cercospora kikuchii* of soybean as affected by stage of plant development. Phytopathology 62:771.

Lehman, S. G. 1950. Purple stain of soybean seeds. N.C. Agric. Exp. Stn. Bull. 369.

Martin, K. F. and H. J. Walters. 1982. Infection of soybean by *Cercospora kikuchii* as affected by dew temperature and duration of dew periods. Phytopathology 72:974.

Matsumoto, T. and R. Tomoyasu. 1925. Studies on purple speck of soybean seed. Am. Phytopathol. Soc. Jpn. 1:1-14.

Murakishi, H. H. 1951. Purple seed stain of soybean. Phytopathology 41:305-318.

Roy, K. W. and T. S. Abney. 1976. Purple seed stain of soybeans. Phytopathology 66:1045-1049.

Sinclair, J. B. 1982. Compendium of soybean diseases. 2nd ed., American Phytopathological Society, St. Paul.

Sturgeon, R. V., Jr. 1979. Southern United States disease losses. p. 14-17. *In* Proc. South. Soybean Dis. Workers, Hot Springs, Arkansas.

Suzuki, K. 1921. Studies on the cause of purple seed of soybeans. (In Japanese) Chosen Agric. Assoc. Rep. 16:24-28.

Vathakos, M. G. and H. J. Walters. 1979. Production of conidia of *Cercospora kikuchii* in culture. Phytopathology 69:832-833.

Walters, H. J. 1980. Soybean leaf blight caused by *Cercospora kikuchii.* Plant Dis. 64:961-962.

SOYBEAN MOSAIC VIRUS

Yves Maury

Soybean mosaic virus (SMV) is a potyvirus. It has a single structural protein sub-unit. Molecular weight estimates indicate a size of 28,300 daltons according to Hill and Benner (1980a) and of 32,150 according to Soong and Milbrath (1980). Infectious single stranded RNA, which is 5.3% of the virus particle, has an S value of 39.7 before and 25.4 after formaldehyde denaturation. Electrophoresis on polyacrylamide gels give a molecular weight value of 3.02×10^6 (Hill and Benner, 1980b). In vitro translation of SMV RNA suggested that protein processing might play a role in the expression of mature SMV proteins (Vance and Beachy, 1984).

DETECTION METHODS

A quantitative assay using *Phaseolus vulgaris* cv Top Crop was developed by Milbrath and Soong (1976). The sensitivity of this assay is about 100 ng/ml. This assay has been used frequently, but some SMV strains fail to induce necrotic lesions (Hunst and Tolin, 1982; Tsuchizaki et al., 1982). This assay was also used for verifying RNA infectivity (Hill and Benner, 1980b).

The minimum quantity of SMV detectable using purified SMV-specific IgG in ELISA is 2.5 ng/ml (Hill et al., 1981). Reaction of all representative isolates of the seven SMV strains G_1 to G_7 were positive in ELISA (Cho and Goodman, 1979). Bryant et al. (1983) developed a solid phase radioimmunoassay, SPRIA; values obtained were generally proportional to virus concentration with a detection limit of 25-50 ng/ml of purified SMV. This assay did not discriminate differences among isolates. Monoclonal antibodies, recently obtained (Hill et al., 1984) could help in studying serotypes.

VIRUS STRAINS AND SOYBEAN GENOTYPES

In search for maximum variability of SMV, Cho and Goodman (1979) used seeds of 98 accessions in the USDA soybean germplasm collection. All virus isolates were differentiated on the basis of the reactions of six SMV-resistant soybean cultivars. They could be grouped into seven strains (Table 1).

The main information drawn from this work was:
- the identification of new viral strains SMV G_3, -G_4, -G_5, -G_6, -G_7.
- the confirmation of previous observations by Han and Murayama (1970) and by Kühl and Hartwig (1979) and Hartwig (1979) that necrotic reactions occur in soybean

Table 1. Classification of soybean mosaic virus strains based on reaction of soybean cultivars (Cho, 1981).

Cultivars	Reactions of soybean cultivars to SMV strains						
	G_1	G_2	G_3	G_4[a]	G_5	G_6	G_7
Rampage	M[b]	M	M	M	M	M	M
Clark	M	M	M	M	M	M	M
Davis	-	-	-	N	M	M	M
York	-	-	-	N	M	M	M
Marshall	-	N	N	-	-	N	N
Ogden	-	-	N	-	-		N
Kwanggyo	-	-	-	-	N	N	N
Buffalo	-	-	-	-	-	-	N

[a]Results reported by Cho and Goodman (1979) were revised.

[b]Symbols for symptoms: - = symptomless; M = mosaic; N = necrosis.

lines that possess resistance to less virulent strains, after infection with virulent SMV strains.

- the absence of a cultivar resistant to all strains.

Cho and Goodman (1979) noted that there would be a risk of favoring an epidemic of SMV, similar to the outbreak observed in Korea (Cho et al., 1977), by widely growing soybean cultivars having an increased level of resistance to SMV.

After an evaluation of different soybean accessions reported by other workers to be SMV resistant, Cho and Goodman (1982) were successful in encountering, in five lines selected in Korea (Suweon -86, -94, -95, -97 and -106), a high degree of resistance to all seven SMV strains. Lim (1982), studying the inheritance of Suweon 97 resistance to G_2 and G_7, established that resistance to each was conditioned by a single dominant gene. In a collection of tropically adapted soybean lines, Bowers (1981) found a line, PI 424131, resistant to all seven strains. Resistance to G_1 and G_7 was shown to be conditioned by a dominant gene at the same locus as the allelomorphic series Rsv, rsvt, rsv previously shown by Kühl and Hartwig (1979) to be involved in the resistance of PI 96983 and Ogden to two strains, SMV_1 and SMV_{1-B}. In other lines Bowers found a dominant gene at a second locus, R_2, involved in resistance to strains G_5, G_6 and G_7.

Buzzell and Tu (1984) also were able to find, in lines derived from cv. Raiden, resistance to all strains. They showed that this resistance was controlled by a dominant gene different from Rsv; the name Rsv_2 was assigned to this gene. But Raiden was derived from Nemashirazu, which was described as susceptible to three out of five strains of SMV in Japan (Iizuka and Yunoki, 1983).

In conclusion several dominant resistance genes, and also a recessive one (Kwon and Oh, 1980), are known which, when combined, might now give a better chance of resistance to additional strains of SMV.

It would be very useful to know the mode of action of such resistance genes. In some Japanese resistant cultivars, Iizuka and Yunoki (1983) differentiated two types of resistance:

- resistance by hypersensitivity; seven cultivars, among them Nemashirazu, which are resistant to only some of the five Japanese strains, belong to this category.

- resistance by inhibition of the translocation of SMV from the inoculated leaves where it was shown to multiply; Peking, Cloud and Peking Krodaizu, resistant to all Japanese strains, behave in this way.

MAJOR FACTORS LINKED TO VIRUS SPREAD

SMV, having no known overwintering host, infected soybean seed is the most likely source of primary virus inoculum. Hill et al. (1980) verified this hypothesis by use of caged plots. As no aphid species are observed to colonize soybeans in Iowa migratory, winged aphids are probably most important in spread of SMV there. The distribution of SMV, when followed in an experimental field, suggested plant to plant spread from primary inoculum foci within the field.

Irwin and Goodman (1981) also monitored the movement of SMV from a point source into a surrounding, initially virus-free, field of Williams. Two plots, each 92 x 92 m were established within a large field. All plants in the center 0.5% of the area of the downwind plot were sap-inoculated with an isolate of SMV-G$_2$, which induces severe symptoms. The spread outward from the source was greater downwind than up-wind. Twenty-five percent of all spread occurred within 2 m of the source, and 95% of all infections were within 17 m of the source. Spread in the control plot was almost nil (two plants), whereas 209 plants in the inoculated plot were infected. No infection occurred more than 50 m from the source. This experiment again excluded any interference from sources outside the plot.

So, SMV transmission through soybean seeds plays, in this context, a pivotal role in the epidemiology of this virus. Two main factors are now shown to influence the incidence of seed transmission: the time of infection in relation to stages of growth and the plant genotype.

Seed Transmission

The effect of time of inoculation on incidence of seed transmission was followed in Williams (Bowers and Goodman, 1979). When the plant was inoculated before the onset of flowering, 18% transmission was recorded. This dropped to 11% for inoculation during R1 and then to 4% for later inoculations. Therefore, inoculation after flowering resulted in a reduction, but not elimination of seed transmission.

The effect of plant genotype was studied (Bowers and Goodman, 1979) using two soybean cultivars, Merit and Midwest, previously shown to exhibit low and high incidences, respectively, of SMV transmission through mature seed. Infectious virus was detected in the testas, cotyledons and axes dissected from immature seeds of both Merit and Midwest. The virus was present in higher proportion in testas than in embryos for both cultivars. As seeds matured and dried, the percentage of testas containing infectious virus decreased significantly; however viral antigen was detected by ELISA after infectivity was lost. In Merit, inactivation of SMV during seed maturation also occurred in embryos, confirming a previous report of such a phenomenon by Koshimizu and Iizuka (1963). So, SMV resembles the majority of seed transmitted viruses in that embryo infection is a prerequisite for seed transmission. However, the mechanism preventing seed transmission in certain genotypes is not caused by the resistance of the immature embryo to virus infection.

The possibility has been explored that improved cultivars with no seed transmission might be developed. Five tropical accessions and a high yielding tropical cultivar, Improved Pelican, were shown to have very low incidence (0.05 to 0.23%) of virus transmission through seed (Goodman and Oard, 1980). Twelve temperate lines also produced all healthy seedlings when 1000 seeds from SMV-infected plants were planted (Bowers and Goodman, 1982). After further work, with the goal of distinguishing zero transmission from a very low level of transmission, an evaluation using other isolates should be done. Induced variation in the frequency of seed transmission has been reported after testing different isolates (Takahashi et al., 1980); in particular an incidence of 14.8% was reported for Improved Pelican (Suteri, 1981).

As a short term solution to control the initial inoculum carried by seed, certification has been suggested (Bryant et al., 1982). A rapid test for the determination of level of SMV transmission in seed batches, useful also for experimental purposes, has been developed in several laboratories. Lima and Purcifull (1980) developed a simple hypocotyl disk immunodiffusion test that gave results similar to that obtained by a growth test. Bossennec and Maury (1978), Lister (1978) and Ma Defang et al. (1982) discussed the use of ELISA for detecting SMV in seeds and seed parts and also in germinated seeds. Their results were in good agreement with growth tests. This means that, in the cultivars examined, antigen was found only in embryos containing infectious virus. Lister also showed that the mean level of SMV in an infected seed is low, compared to the level of tobacco ring spot virus. Bryant et al. (1983) estimated, using SPRIA, the average antigen content in infected seeds of Williams to be 474 ng per seed.

Each of these techniques applied to individual embryos can be used to determine with accuracy high incidences of transmission. However, the lower the percentage of transmission in a seed lot, the higher the number of seeds that must be tested for an accurate result. Individual seeds can no longer be tested; the only economical way to test large numbers of seeds is to work with groups of seeds. Bryant et al. (1983) tried to establish, using SPRIA for detecting SMV in groups of 100 seeds, a relation between the quantitative result of this test and the percentage of seed transmission in the batch. A major difficulty in using this method arises from the considerable variation in virus content from one seed to another.

To determine the rate of transmission with more accuracy, the following methodology was proposed by Maury et al. (1983). Seeds of Altona, infected by SMV-1, were first analyzed after dissection into testas, cotyledons and axes and grinding at the same dilution (1/200 W/V). Figure 1 shows the diversity in virus titer from one embryo to another, and also in the axis and cotyledons of the same embryo. Curiously, for some embryos, SMV was found in the axis and not in the cotyledons, even at lower dilutions. Because the weight of cotyledons is about 30 times the weight of the axis, in some embryos virus content is very low. Knowing the O.D. (405 nm) of the less-infected embryo, 30-seeds was calculated to be, for this cultivar, the smallest sample compatible with detection of one infected embryo.

Testas contained a constant level of antigen. This antigen is likely to induce false positive reactions as the ratio of number of infected testas to number of infected embryos increases. The technological problem of working without the testas was solved in two ways (Maury et al., 1984). A procedure to prepare axes was developed by crushing dry seeds and sieving, and the test was shown to give good results using groups of 30 axes. But, the easiest way was to soak the seeds and to grind them in a Waring Blender. It was shown that, owing to their friability, embryos could be differentially ground, the antigen from the testas having no influence on the results.

Under these conditions, group testing using ELISA was found to be reproducible and to give accurate results. For example, a test with 1800 seeds, divided into 60 groups of 30 seeds each that would react positively in 6 of 60 reactions would mean that the rate of transmission of this seed lot is 0.4%, the 95% confidence interval being 0.2 to 0.8%. Such determinations have been verified in preliminary field experiments by following the spread of SMV using ELISA to test groups of leaves (Maury et al., unpublished).

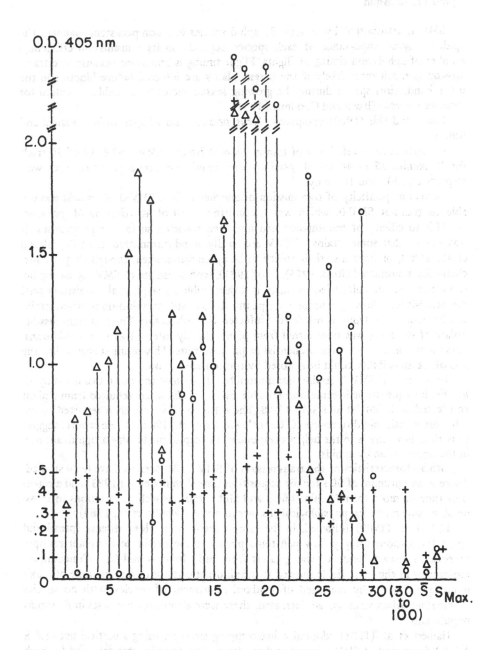

Figure 1. Distribution of SMV in 30 embryo infected seeds. On the right, results from embryos with a higher titer of SMV in cotyledons; on the left, results from embryos with a higher titer in axes (Maury et al., 1983).

△ = Axis. ○ = Cotyledons. + = Testas. $\overline{S}$ = Average values for healthy seeds. S_{Max} = Maximum values for healthy seeds. (30 to 100) = Average values for embryo healthy seeds collected from infected plants.

Aphid Transmission

SMV is transmitted by at least 31 aphid species in a non-persistent manner. The epidemiological importance of each species depends on its transmission efficiency, number of aphids and timing of flights. Flight timing is important because seed transmission is much more likely if the parent plants are infected before bloom. On the other hand, virus spread during the growing season increases available inoculum for secondary spread (Irwin and Goodman, 1981).

Lucas and Hill (1980) compared *Myzus persicae* and *Rhopalosiphum maidis* and found:

- a difference in efficiency of transmission of isolate, SMV 12 18, 14 of 40 trials for *R. maidis*, 35 of 40 for *M. persicae*. The acquisition access probe required was, respectively, 20 s and 10 s only.

- a vector specificity of transmission of another isolate, SMV-0; *R. maidis* was unable to transmit SMV-0, which was transmitted in 30 of 40 trials by *M. persicae*.

This specificity of transmission had not been reported so far for potyviruses. It was known that some strains of SMV are hardly aphid-transmitted at all (Takahashi et al., 1980), or have lost their ability to be aphid-transmitted, though they are mechanically transmitted (Ross, 1975). Cho (1981), with an isolate of SMV-G$_5$ having become non-transmissible, tried to make it transmissible by sequential acquisition test, the acquisition following a probe on a plant infected with an aphid-transmitted strain, and by acquisition from plants doubly infected by both strains. The non-transmissible isolate of strain G$_5$ was transmitted from plants doubly infected with an aphid-transmissible strain, but not from sequential acquisition tests. This result suggests that the loss of transmissibility could be restored by transencapsidation.

In the case of SMV-0, readily transmitted by *M. persicae* and not transmitted by *R. maidis*, infrequently it became transmissible by *R. maidis* when acquired from a plant coinfected with SMV-0 and SMV-12-18. These two strains were then reported to exhibit some antigenic differences (O'Connell-Ziegler et al., 1982). These reports suggest thus that, in addition to the helper component, the capsid might play a significant role in the transmission by aphids.

Other characteristics of the transmission of SMV by *M. persicae* have been studied. There is an optimum of 60 s for the acquisition time (Schultz et al., 1983) and a retention time of more than 75 min (Lucas and Hill, 1980), as well as variations in transmission associated with physiological changes in plants (Schultz et al., 1983).

Hill et al. (1980) followed, in the same experimental plots, disease spread and aphid fluctuations using yellow pan traps placed at the height of the plant canopy. Increase in disease incidence was greatest from early June to mid-July. Concomitant monitoring in the same fields showed that aphid populations during the time of maximum spread generally consisted of a mixed assortment of species, with no species dominating. When virus spread decreased, there were significant increases in *R. maidis* populations.

Halbert et al. (1981) adopted a live-trapping strategy, using a vertical net (1.6 x 4.6 m) downwind of SMV-infected soybean fields. This trapping was followed for each aphid by an assay for infectivity on a healthy soybean seedling. Trapping began about 2 weeks after inoculation and continued until the end of R7 several mornings per week. Their results show that *R. maidis*, due to its relative abundance, contributed a large percentage of the total transmission (26.4%) but did not transmit until the final week of July. The majority of early season transmissions were by *Aphis craccivora* and *Macrosiphum euphorbiae*. *A. craccivora* has been considered to be the most important

SMV vector in central Illinois because it flies in late spring and is a relatively efficient vector. *Aphis citricola* was also believed to play a significant role in SMV spread owing to its transmission efficiency in the laboratory and its flights in mid to late spring, but it could not be trapped and assayed.

NOTES

Yves Maury, I.N.R.A. Pathologie végétale, F7800 Versailles, France.

This report deals mainly with information published during the last five years. For more information the reader is invited to refer to a preceding synthesis "Ecology and Control of Soybean Mosaic Virus" by Irwin and Goodman (1981).

I thank Drs. Yong-Xuan Chen, E. K. Cho, S. A. Ghabrial, R. M. Goodman, J. H. Hill, N. Iizuka, M. E. Irwin, M. Iwaki, J. P. Ross, B. R. Singh and J. C. Tu for sending me recent information, and my colleagues J. M. Bossennec, G. Boudazin and E. Ron for reviewing my manuscript.

REFERENCES

Bossennec, J. M. and Y. Maury. 1978. Use of the ELISA technique for the detection of soybean mosaic virus in soybean seed. Ann. Phytopathol. 10:263-268.

Bowers, G. R. 1981. Inheritance of resistance to soybean mosaic virus in soybeans and studies on seed transmission. Diss. Abstr. Int. 41:3957 B.

Bowers, G. R. and R. M. Goodman. 1979. Soybean mosaic virus: infection of soybean seed parts and seed transmission. Phytopathology 69:569-572.

Bowers, G. R. and R. M. Goodman. 1982. New sources of resistance to seed transmission of soybean mosaic virus in soybeans. Crop Sci. 22:155-156.

Bryant, G. R., D. P. Durand and J. H. Hill. 1983. Development of a solid-phase radioimmunoassay for detection of soybean mosaic virus. Phytopathology 73:623-629.

Bryant, G. R., J. H. Hill, T. B. Bailey, H. Tachibana, D. P. Durand and H. I. Benner. 1982. Detection of soybean mosaic virus in seed by solid-phase radioimmunoassay. Plant Dis. 66:693-695.

Buzzel, R. I. and J. C. Tu. 1984. Inheritance of soybean resistance to soybean mosaic virus. J. Heredity 75:82.

Cho, E. K. 1981. Strains of soybean mosaic virus with emphasis on the mechanism of aphid transmisssibility. Korean J. Plant Prot. 20:177-180.

Cho, E. K., B. J. Chung and J. H. Lee. 1977. Studies on identification and classification of soybean virus diseases in Korea. II. Etiology of a necrotic disease of *Glycine max*. Plant Dis. Rep. 61:313-317.

Cho, E. K. and R. M. Goodman. 1979. Strains of soybean mosaic virus: classification based on virulence in resistance soybean cultivars. Phytopathology 67:467-470.

Cho, E. K. and R. M. Goodman. 1982. Evaluation of resistance in soybeans to soybean mosaic virus strains. Crop Sci. 22:1133-1136.

Goodman, R. M. and J. H. Oard. 1980. Seed transmission and yield losses in tropical soybeans infected by soybean mosaic virus. Plant Dis. 64:913-914.

Halbert, S. E., M. E. Irwin and R. M. Goodman. 1981. A late aphid (Homoptera : Aphididae) species and their relative importance as field vectors of soybean mosaic virus. Ann. Appl. Biol. 97:1-9.

Han, Y. H. and D. Murayama. 1970. Studies on soybean mosaic virus strains by differential hosts. J. Fac. Agric. Hokkaido Univ. 56:303-310.

Hill, E. K., J. H. Hill and D. P. Durand. 1984. Production of monoclonal antibodies to viruses in the potyvirus group: use in radioimmunoassay. J. Gen. Virol. 65 (in press).

Hill, J. H. and H. I. Benner. 1980a. Properties of soybean mosaic virus and its isolated protein. Phytopath. Z. 97:272-281.

Hill, J. H. and H. I. Benner. 1980b. Properties of soybean mosaic virus ribonucleic acid. Phytopathology 70:236-239.

Hill, J. H., G. R. Bryant and D. P. Durand. 1981. Detection of plant virus by using purified IgG in ELISA. J. Virol. Meth. 3:27-35.

Hill, J. H., B. S. Lucas, H. I. Benner, H. Tachibana, R. B. Hammond and L. P. Pedigo. 1980. Factors associated with the epidemiology of soybean mosaic virus in Iowa. Phytopathology 70:536-540.

Hunst, P. L. and S. A. Tolin. 1982. Isolation and comparison of two strains of soybean mosaic virus. Phytopathology 72:710-713.

Iizuka, N. and T. Yunoki. 1983. Evaluation of resistant soybean varieties to soybean mosaic and stunt viruses by sap or graft inoculation. Ann. Rept. Plant Prot. North Japan 34:115-117.

Irwin, M. E. and R. M. Goodman. 1981. Ecology and control of soybean mosaic virus, p. 182-215. In K. Maramorosch and K. F. Harris (eds.) Plant diseases and vectors. Ecology and Epidemiology. Acad. Press, N.Y.

Kühl, R. A. S. and E. E. Hartwig. 1979. Inheritance of reaction of soybean mosaic virus in soybeans. Crop Sci. 19:372-375.

Koshimizu, Y. and N. Iizuka. 1963. Studies on soybean virus diseases in Japan. Bull. Tohoku Natl Agric. Exp. Stn. 27:1-103.

Kwon, S. H. and J. H. Oh. 1980. Resistance to necrotic strain of SMV in soybeans. Crop Sci. 20: 403-404.

Lim, S. M. 1982. A new source of resistance to soybean mosaic virus in a soybean line and its inheritance (Abstract). Phytopathology 72:943.

Lima, J. A. A. and D. E. Purcifull. 1980. Immunochemical and microscopical techniques for detecting Blackeye Cowpea mosaic and soybean mosaic viruses in hypocotyls of germinated seeds. Phytopathology 70:142-147.

Lister, R. M. 1978. Application of the enzyme linked immunosorbent assay for detecting viruses in soybean seed and plants. Phytopathology 68:1393-1400.

Lucas, B. S. and J. H. Hill. 1980. Characteristics of the transmission of three soybean mosaic virus isolates by Myzus persicae and Rhopalosiphum maidis. Phytopathol. Z. 99:47-53.

Ma Defang, Xu Shaohua, Zhang Zuofang, Hu Weizhen, Zhang Chengliang, Song Shumin and Zhao Ruisbeng. 1982. Detection of soybean mosaic virus with enzyme linked immunosorbent assay (ELISA). Acta Phytopathologica Sinica 12:47-52.

Maury, Y., J. M. Bossennec, G. Boudazin and C. Duby. 1983. The potential of ELISA in testing soybean seed for soybean mosaic virus. Seed Science and Technology (in press).

Maury, Y., J. M. Bossennec, G. Boudazin and C. Duby. 1984. Group testing using ELISA: determination in soybean seed lots of the level of soybean mosaic virus transmission (submitted to Agronomie).

Milbrath, G. M. and M. M. Soong. 1976. A local lesion assay for soybean mosaic virus using Phaseolus vulgaris L. cv. "Top Crop". Phytopathol. Z. 87:255-259.

O'Connell-Ziegler, P., J. H. Hill and H. I. Benner. 1982. Characterization of two isolates of soybean mosaic virus and their transmission by aphids. Phytopathology 72:138.

Ross, J. P. 1975. A newly recognized strain of soybean mosaic virus. Plant. Dis. Reptr. 59:806-808.

Ross, J. P. 1983. Effect of soybean mosaic on component yields from blends of mosaic resistant and susceptible soybeans. Crop Sci. 23:343-346.

Schultz, G. A., M. E. Irwin and R. M. Goodman. 1983. Factors affecting aphid acquisition and transmission of soybean mosaic virus. Ann. Appl. Biol. 103:87-96.

Soong, M. M. and G. M. Milbrath. 1980. Purification, partial characterisation and serological comparison of soybean mosaic virus and its coat protein. Phytopathology 70:388-391.

Suteri, B. D. 1981. Effect of soybean mosaic virus on seed germination and its transmission through seeds. Indian Phytopath. 34:370-371.

Takahashi, K., T. Tanaka, W. Iida and Y. Tsuda. 1980. Studies on virus diseases and causal viruses of soybean in Japan. Bull. Tohoku Nath. Agric. Exp. Stn. 62:1-130.

Tsuchizaki, T., P. Thongmeearkom and M. Iwaki. 1982. Soybean mosaic virus isolated from soybeans in Thailand. JARQ 15:279-285.

Vance, V. B. and R. N. Beachy. 1984. Translation of soybean mosaic virus RNA in vitro. Evidence of protein processing. Virology 132:271-281.

SOYBEAN BUD BLIGHT: SEED TRANSMISSION OF THE CAUSAL VIRUS

R. I. Hamilton

Bud blight of soybean, as a symptom, can be induced by at least four viruses. Tobacco ringspot virus (TRSV) is the most common incitant, followed by tomato ringspot virus (TomRSV), tobacco streak virus (TSV) and the necrotic strain of soybean mosaic virus. All of these viruses, except TomRSV, have been isolated from natural field infections; TomRSV induced symptoms in experimental inoculations but it has not been associated with natural infections. A recent report indicates that a mycoplasma-like organism also induces some of the symptoms heretofore attributed to the TRSV-induced disease (Derrick and Newsom, 1984).

The disease, first reported in Indiana, USA in 1941 by Samson who established TRSV as the causal virus, has been reported since in Canada, in the eastern region of the USSR and in Himachel Pradesh, India. There may be doubt about the identification of the virus in Himachel Pradesh (Gupta, 1976). Bud blight has recently been observed in Turkey (J. B. Sinclair, pers. comm.), Sri Lanka (P. Shivanathan, pers. comm.), Egypt (E. K. Allam, pers. comm.) and in Karnataka, India (Muniyappa, pers. comm.). In North America, the disease was common in some soybean-producing areas of the U.S.; e.g., Iowa, Illinois, and Indiana and in Ontario, Canada (Hildebrand and Koch, 1947), between 1945 and 1960 with sporadic occurrences in the 1970s, but its incidence has decreased since. On the other hand, the disease is now appearing in countries where soybean has recently been introduced, probably via infected seed. The similarity in symptoms induced by TRSV and TSV (Sinclair, 1982) and the isolation of TSV, but not TSRV, from soybean with bud blight symptoms (Fagbenle and Ford, 1967) coupled with the seed-transmissibility of TSV in soybean (Ghanekar and Schwenk, 1974), indicates the necessity of identifying the causal virus by definitive methods (Hamilton et al., 1981) whenever bud blight is observed.

The epidemiology of the disease is not well understood. Transmission of the virus by the nematode, *Xiphinema americanum*, has been demonstrated, but its role in the epidemiology of bud blight is probably negligible because virus transmitted to the roots rarely infects the leaf and floral buds (Bergeson et al., 1964; Halk and McGuire, 1973). Increasing evidence suggests that an aerial vector is involved in introducing the virus to soybean plantings from reservoir hosts (Bergeson et al., 1964; Crittenden et al., 1966; Hill et al., 1973; Messieha, 1969); thrips appear to be a likely candidate (Irwin and Yeargan, 1980). Transmission by aerial vectors points to the importance of the wide range of reservoir hosts that are susceptible to TRSV (Stace-Smith, 1970), and to the distinct possibility that seed transmission of the virus plays a pivotal role in maintaining the virus in these hosts (Murant and Lister, 1967; Tuite, 1960).

The major causal virus, TRSV, is a member of the nepovirus group, which is composed of nematode-transmitted, isometric viruses containing a single-stranded RNA genome. The genome is bi-partite (Harrison and Murant, 1977); i.e. RNA-1 and RNA-2 (mol. wt. 2.4 and 1.4 x 10^6, respectively) are required for infectivity and expression of the genome. The bi-partite nature of the genome allows for genetic interaction (pseudo-recombination) (Harrison et al., 1972) with other isolates of the virus in mixed infection of a common host, thus providing an opportunity for evolution of new strains.

SEED TRANSMISSION OF PLANT VIRUSES

As previously mentioned, seed transmission of TRSV in reservoir hosts may play a key role in the epidemiology of bud blight. The basic aspects of seed transmission of plant viruses in general, and of TRSV in soybean in particular, will now be described.

Basic Aspects

Seed transmission occurs in many virus-host combinations. Not all viruses, however, are seed transmitted; about 100 of the more than 400 viruses described are transmitted in this way. The fact that seedlings originating from seeds produced by infected plants are not invariably infected clearly indicates that seed transmission results from a complex interaction between virus and host. The basic interaction involves infection of sporogenous tissues in the ovule and anther, which leads to the development of infected megaspore and microspore mother cells and the consequent development of an infected embryo. Continued development of the embryo through the stages of seed maturation and germination usually, but not always, leads to the transmission of virus to the seedling. It is possible that some viruses that are not seed-transmitted are able to infect sporogenous tissues but the infected cells are killed, thus effectively preventing the development of functional, infected gametes.

Development of functional infected gametes and consequent embryo infection, by definition, occurs before flowering is complete. The few reports (Eslick and Afanasiev, 1955; Crowley, 1959) suggesting invasion of embryos by viruses have not been confirmed, and such cases are likely explained by development of infected gametes in late flowers. Non-embryonic transmission does occur with some viruses; e.g., tomato mosaic virus as a consequence of infection or contamination of the seed coat. The recovery of southern bean mosaic virus from bean embryos in plants inoculated when pods were at a late developmental stage (Gay, 1973) may have been due to contamination with virus originating in the seed coat (McDonald and Hamilton, 1972).

Studies with seed-transmitted viruses indicate that there is considerable variability in seed-transmissibility that can be attributed to the host species, cultivar and virus strain. For example, the non-seed-transmitted strain of barley stripe mosaic virus does not spread systemically to the meristems in barley (*Hordeum vulgare* L.) as rapidly as the seed-transmitted type strain. No virions of the former were detected in developing pollen (Carroll and Mayhew, 1976a) or in eggs (Carroll and Mayhew, 1976b), although those of the seed-transmitted strain were readily detected in gametes. Studies with pseudorecombinants of two nepoviruses, raspberry ringspot virus (RRV) and tomato blackring (TBRV) (Hanada and Harrison, 1977), indicate that seed transmission is associated with RNA-1 and that speed of systemic infection in the experimental host, *Chenopodium quinoa*, is governed by RNA-1 (RRV) or by the interaction of RNA-1 and -2 (TBRV), suggesting that rate of systemic spread is associated with the genomic segment that governs seed transmission. More recently, in vitro translation of the genomic RNAs of two strains of tobacco mosaic virus that differed in their rates of cell-to-cell spread, indicates that the difference is associated with changes in a specific viral-coded protein (Leonard and Zaitlin, 1982). A similar protein has been detected in virus-infected plants (Dorokhov et al., 1983), and has been implicated in the transport of the infective principle in plants (Taliansky et al., 1982). Moreover, proteins of approximately the same molecular weight have been detected in the translation products of other viral RNAs, suggesting that a viral-coded "transport protein" is a general feature of virus infection. It is conceivable that such a protein could potentiate systemic spread, leading to development of infected gametes, and that a defective protein may inhibit such spread. Mobile cytoplasmic vesicles containing ds-RNA have been proposed as a means of facilitating systemic spread of pea enation mosaic virus (deZoeten and Gaard, 1983), and perhaps the "transport protein" is involved in some aspect of the stabilization of the RNA in the vesicles or in their construction.

Seed Transmission of TRSV in Soybean

TRSV is efficiently transmitted by soybean seed; rates of transmission are often in excess of 90%, especially in seed from plants infected at an early age or in seeds from plants arising from infected seeds (Athow and Bancroft, 1959; Crowley, 1959; Owusu et al., 1968). Infective virus has been recovered from embryos and seed coats of immature seeds (Yang and Hamilton, 1974) and from embryos, but not from seedcoats, of mature seeds (Athow and Bancroft, 1959). Infected seed retained high germinability and transmission to seedlings (70 to 90%) over a 5-year storage period (Laviollette and Athow, 1971).

Detailed studies on the effect of TRSV infection on gametogenesis in soybean (Yang and Hamilton, 1974) showed that pollen development was severely reduced and that seed transmission was essentially dependent upon infection of the megagametophyte; i.e., the egg. A significant reduction in pollen production was observed in flowers of infected plants. Partial to complete failure of pollen production was not uncommon in TRSV-infected flowers. Many anthers were necrotic and dehiscence of the abnormal anthers usually did not occur. The anther filaments in many infected flowers were shortened in comparison to those in virus-free flowers, thus rendering pollen deposition on the stigma virtually impossible.

The average number of pollen grains produced in each infected flower was less than 1000, whereas 4000-5000 were produced by the majority of virus-free flowers. Germination rates were about 45% and 75% for pollen from TRSV-infected and virus-free plants, respectively. Of considerable significance was the severe reduction of germ tube length in pollen from infected plants ($x = 252 \mu m$) compared to an average of 425 μm for germ tubes of pollen produced by virus-free flowers (Figure 1a,b). Very few pollen grains from infected plants were able to produce germ tubes long enough to fertilize even the nearest egg, which was about 1700 μm from the stigma. In controlled crosses, only one seed was produced when pollen from infected plants was used to pollinate virus-free flowers compared to 80 seeds which were produced in the reciprocal cross. Moreover, the majority of the pods from self-pollinated flowers on infected plants were single seeded, whereas the majority of pods on virus-free plants contained 2 or 3 seeds, demonstrating that pollen from virus-infected flowers rarely fertilized the distal eggs in the ovary.

Electron microscopy of pollen and ovules revealed TRSV in the inner layer (the intine) of the pollen wall (Figure 1c,d) as well as in the generative cell of the pollen and in the walls and cells of the embryo sac (Figure 2a,b), integuments and nucellus.

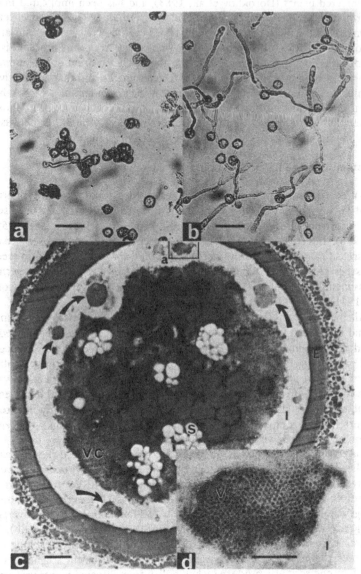

Figure 1. Effect of TRSV infection of Harosoy soybean on germination and ultrastructure of pollen. From Yang and Hamilton (1974). Reproduced with permission from Academic Press, Inc. 1a. Pollen from TRSV-infected soybean after germination for 18 h in 30% sucrose containing 120 ppm boric acid. Bar = 25 μm. 1b. Pollen from virus-free soybean after germination under same conditions as for Figure 1a. Bar = 25 μm. 1c. Thin section of pollen grain from TRSV-infected soybean showing virus aggregates arising from cytoplasmic protrusions in the intine (I) of the pollen wall. VC = vegetative cell; E = exine; and S = starch grain. Bar = 1 μm. 1d. Higher magnification of virus aggregate ("a" in Figure 1c) showing virions (V) in the intine. No such particles were seen in pollen from virus-free soybean. Bar = 250 μm.

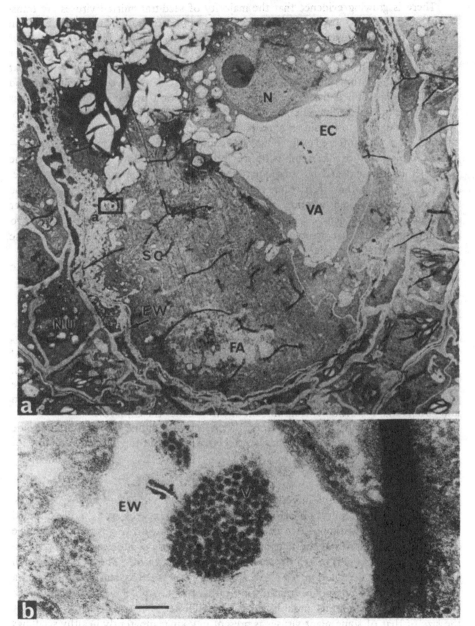

Figure 2. Thin section of a megagametophyte in TRSV-infected Harosoy soybean. 2a. The egg cell (EC) with nucleus (N) and vacuole (VA) situated next to a synergid cell (SC) and a filiform apparatus (FA). The embryo sac wall (EW) separates the megagametophyte from the nucellus (NU). A virus aggregate (arrow) is present in the cytoplasmic protrusion of the embryo sac wall. Bar = 4 μm. 2b. Higher magnification of virus aggregate ("a") in 2a, showing virions (V) in the embryo sac wall. Bar = 100 nm.

RE-EXAMINATION OF THE PROCESS OF SEED TRANSMISSION

There is growing evidence that the majority of seed-transmitted viruses are transmitted via infected embryos. This evidence has accumulated in the last two decades on the basis of experiments that depended upon cross pollination, infectivity assays of embryo extracts, and detection of viral antigen or virions in embryos by serology or electron microscopy. A major conclusion drawn from such experiments has been that the presence of virions in embryonic tissue is associated with seed transmission, and from this conclusion it has been inferred that the virion per se is the agent that is transmitted to the seedling; i.e., the virion is the agent responsible for seed transmission. However, it is difficult to accept the notion that the virion, which is an end-product of a virus-infected cell, should be required to undergo an infective process—i.e., rearrangement or disassembly of the protein coat, followed by release of the genomic nucleic acid and its subsequent translation into gene products that are required to support the infective process—as the requisite for seed transmission. The question can be asked whether the continued production of virions throughout gametogensis, embryogensis, seed maturation and germination is necessary in order to accumulate virions that are required to initiate transmission to the seedling. Is the virion the infective principle?

Recent advances in the molecular biology of viruses suggest other bases for the transmission of plant viruses via infected embryos.

Incorporation of Viral RNA in the Plant Genome

In this process single-stranded viral RNA would be used as a template for the in vivo transcription of a complimentary DNA, which in turn is converted to a double-stranded DNA and integrated into the host genome. The integrated viral genome would be among the complement of host genes that are transmitted during fertilization. Conversion of the integrated viral genome to its RNA form might then initiate the typical virus infection with its attendant production of progeny RNA, coat protein and virions and, incidentally, transmission of the virus via the embryo.

This process has been reported to be the basis for germ-line transmission of some sexually transmissible viruses of animals; i.e., retroviruses, but the phenomenon has not been observed with plant viruses.

Viral Replicative RNA

Most, if not all, embryo-infecting plant viruses contain a single-stranded RNA (ss-RNA) genome. The replication of ss-RNA plant viruses proceeds via a double-stranded RNA (ds-RNA) consisting of two complimentary strands with opposite polarity. The "negative" strand serves as the template for the enzymatic production of progeny ss-RNA ("positive" strand), much of which accumulates in virions. The ds-RNA is much more resistant to degradation by plant nucleases than ss-RNA, and it would seem a logical candidate to potentiate transmission of a virus through the seed. Recent work (Wakarchuk and Hamilton, unpublished) has established that two ds-RNA species of the size of that of some plant viruses is present in young, apparently healthy seedlings of bean (*Phaseoulus vulgaris* L., cv. Black Turtle Soup). The origin and function of these ds-RNAs is not known, but their presence in seedlings suggests that viral ds-RNAs may be transmitted via seed.

Protected Viral ss-RNA

If the mobile form of the virus during systemic infection of the plant is RNA rather than the virion, the "transport protein" may, in some way, protect the viral RNA from degradation by plant nucleases. The protected RNA may be able to withstand the vicissitudes of seed maturation and germination with consequent potentiation of viral replication and transmissions of the virus to the seedling.

CONCLUDING REMARKS

Recent developments in the molecular biology of viruses presage unusual opportunities for research in the molecular biology of viruses in seeds. When one considers the environment of the seed during the passage of viral genetic information from one host generation to the next, a central question is the nature of the transmissible phase. Large seeded legumes; i.e., bean and soybean, and the use of bi-partite and tri-partite seed-transmissible viruses and virus mutants provide useful research objects for research on the molecular events during seed transmission. Important topics include the role of viral genes in the events leading to infection of sporogenous cells, their survival and subsequent development of infected mother cells, and the nature of the transmissible phase of the virus during its passage in the seed. Tobacco ringspot virus, by virtue of its high transmissibility in soybean seed, would appear to be a particularly good model for such studies.

NOTES

R. I. Hamilton, Agriculture Canada Research Station, 6660 N.W. Marine Drive, Vancouver, Canada V6T 1X2.

REFERENCES

Athow, K. L. and J. B. Bancroft. 1959. Development and transmission of tobacco ringspot virus in soybean. Phytopathology 49:697-701.

Bergeson, G. B., K. L. Athow, F. A. Laviolette, and Sister M. Thomasine. 1964. Transmission, movement, and vector relationships of tobacco ringspot virus in soybean. Phytopathology 54:723-728.

Carroll, T. W. and D. E. Mayhew. 1976a. Anther and pollen infection in relation to the pollen and seed transmissibility of two strains of barley stripe mosaic virus in barley. Can. J. Bot. 54:1604-1621.

Carroll, T. W. and D. E. Mayhew. 1976b. Occurrence of virions in developing ovules and embryo sacs of barley in relation to the seed transmissibility of barley stripe mosaic virus. Can. J. Bot. 54:2497-2512.

Crittenden, H. W., K. M. Hastings, and D. M. Moore. 1966. Soybean losses caused by tobacco ringspot virus. Plant Dis. Rep. 50:910-913.

Crowley, N. C. 1959. Studies of the time of embryo infection by seed transmitted viruses. Virology 8:115-123.

Derrick, K. S. and L. D. Newsom. 1984. Occurrence of a leaf-hopper-transmitted disease of soybeans in Louisiana. Plant Dis. 68:343-344.

deZoeten, G. A. and G. Gaard. 1983. Mechanisms underlying systemic invasion of pea plants by pea enation mosaic virus. Intervirology 19:85-94.

Dorokhov, Y. L., N. M. Alexandrova, N. A. Miroschnichenko and J. G. Atabekov. 1983. Isolation and analysis of virus-specific ribonucleoprotein of tobacco mosaic virus-infected tobacco. Virology 127:237-252.

Eslick, R. F. and M. M. Afanasiev. 1955. Influence of time of infection with barley stripe mosaic virus on symptoms, plant yield and seed infection of barley. Plant Dis. Rep. 39:722-724.

Fagbenle, H. H. and R. E. Ford. 1970. Tobacco streak virus isolated from soybeans, *Glycine max*. Phytopathology 60:814-820.

Gay, J. D. 1973. Effect of plant variety and infection age on the presence of southern bean mosaic virus in floral parts and unripe seeds of *Vigna sinensis*. Plant Dis. Rep. 57:13-15.

Ghanekar, A. M. and F. W. Schwenk. 1974. Seed transmission and distribution of tobacco streak virus in six cultivars of soybeans. Phytopathology 64:112-114.

Gupta, V. K. 1976. Bud blight disease on soybean. Indian Phytopathol. 29:186-187.

Halk, E. L. and J. M. McGuire. 1973. Translocation of tobacco ringspot virus in soybean. Phytopathology 63:1291-1300.

Hamilton, R. I., J. R. Edwardson, R. I. B. Francki, H. T. Hsu, R. Hull, R. Koenig and R. G. Milne. 1981. Guidelines for the identification and characterization of plant viruses. J. Gen. Virol. 54: 223-241.

Hanada, K. and B. D. Harrison. 1977. Effects of virus genotype and temperature on seed transmission on nepoviruses. Ann. Appl. Biol. 85:79-92.

Harrison, B. D. and A. F. Murant. 1977. Nepovirus group. Descriptions of Plant Viruses No. 185. Commonw. Mycol. Inst. and Assoc. Appl. Biologists, Ferry Lane, Kew, U.K.

Harrison, B. D., A. F. Murant and M. A. Mayo. 1972. Evidence for two functional RNA species in raspberry ringspot virus. J. Gen. Virol. 16:339-348.

Hildebrand, A. A. and L. W. Koch. 1947. Observations on bud blight of soybeans in Ontario. Sci. Agric. 27:314-321.

Hill, J. H., A. H. Epstein, M. R. McLaughlin, and R. F. Nyvall. 1973. Aerial detection of tobacco ringspot virus-infected soybean plants. Plant Dis. Rep. 57:471-472.

Irwin, M. E. and K. V. Yeargan. 1980. Sampling phytophagous thrips on soybean. p. 283-304. *In* M. Kogan and D. C. Herzog (eds.) Sampling methods in soybean entomology. Springer-Verlag, New York.

Leonard, D. A. and M. Zaitlin. 1982. A temperature-sensitive strain tobacco mosaic virus defective in cell-to-cell movement generates an altered viral-coded protein. Virology 117:416-424.

Laviolette, F. A. and K. L. Athow. 1971. Longevity of tobacco ringspot virus in soybean seed. Phytopathology 61:755.

McDonald, J. G. and R. I. Hamilton. 1972. Distribution of southern bean mosaic virus in seed of *Phaseolus vulgaris*. Phytopathology 62:387-389.

Messieha, M. 1969. Transmission of tobacco ringspot virus by thrips. Phytopathology 59:943-945.

Murant, A. F. and R. M. Lister. 1967. Seed transmission in the ecology of nematode-borne viruses. Ann. Appl. Biol. 59:63-76.

Owusu, G. K., N. C. Crowley, and R. I. B. Francki. 1968. Studies on the seed-transmission of tobacco ringspot virus. Ann. Appl. Biol. 61:195-202.

Sinclair, J. B. 1982. Compendium of soybean diseases. American Phytopathological Society and University of Illinois, Urbana, IL.

Stace-Smith, R. 1970. Tobacco ringspot virus. Descriptions of Plant Viruses No. 17. Commonw. Mycol. Inst. and Assoc. Appl. Biologists. Ferry Lane, Kew, U.K.

Taliansky, M. E., S. I. Malyshenko, E. S. Pshennikova, and J. G. Atabekov. 1982. Plant virus specific transport function. II. A factor controlling host range. Virology 122:327-331.

Tuite, J. 1960. The natural occurrence of tobacco ringspot virus. Phytopathology 50:296-298.

Yang, A. F. and R. I. Hamilton. 1974. The mechanism of seed transmission of tobacco ringspot virus in soybean. Virology 62:26-37.

Nematology

CHEMICAL CONTROL OF NEMATODES ATTACKING SOYBEANS

M. E. Zirakparvar

Generally, nematodes are aquatic animals that may be found in almost any moist habitat. They can parasitize animals or plants, or they can live as saprobes in fresh water, in salt water and in the soil. These soil-borne species inhabit the water phase, moving in a film of moisture in the soil.

The amount of damage caused by plant-parasitic nematodes is related to many variables, including the nematode species, the size of the nematode population, the susceptibility of the host plant, and various environmental factors, such as temperature, duration of the growing season, availability of water and nutrients to the plant, and the presence of other organisms contributing to the total damage inflicted upon the crop. When the relationship between nematodes and a single plant is examined, the effect on the latter can range from death in some instances to absence of detectable damage in others.

About 50 species of plant parasitic nematodes have been reported to attack the soybean, the most important being *Meloidogyne* spp. and *Heterodera glycines*. Overall, these nematodes are estimated to cause a 10% yield reduction annually (Welch et al., 1978; Caviness and Riggs, 1976). Therefore, chemical control can be feasible, given certain circumstances.

MODE OF ACTION OF NEMATODES

There are several commercially available nematicides recommended as effective control measures for nematodes attacking soybeans. Generally, there are two categories of nematicides: fumigants and non-fumigants. Fumigants are developed as a liquid and are usually injected beneath the soil surface. Once in the soil, the chemical volatilizes to a vapor that kills nematodes. Non-fumigant nematicides are water soluble and are distributed through the soil by water percolation, entering the nematode bodies through the cuticle and/or ingestion. Some non-fumigant nematicides are systemic. The majority of the systemic compounds are applied to the soil. After solubilizing in soil water, the chemical is taken up by the plant roots and distributed throughout the plant, thereby affecting nematodes and possibly other pests feeding on the roots and foliage. A few systemic nematicides are applied to the foliage, then translocated to roots where they kill nematodes that feed on the roots. In addition to killing the nematodes through ingestion, these materials may also enter nematode bodies through the cuticle in contact with soil and plant tissue.

The nematicides that have been most commonly evaluated for their efficacy on soybean nematodes by scientists during the last four years include: Ethylene

Dibromide, Dichloropropene, Dibromochloropropane and the related chlorinated hydrocarbons aldicarb, carbofuran, ethoprop, fensulfothion, oxamyl, fenamiphos, and several other experimental compounds. Nematode species on which these chemicals were evaluated include: *Meloidogyne arenaria, M. javanica, M. incognita, M. hapla, Heterodera glycines, Pratylenchus penetrans, P. brachyurus, P. scribneri, P. hexincisus, Hoplolaimus columbus, H. galeatus, Paratrichodorus christiei, Belonolaimus longicaudatus, Tylenchorynchus claytoni, T. clarus,* and *Helicotylenchus* spp.

PLACEMENT METHODS

Proper placement of nematicides in relation to the soybean plants and target ports plays an important role in their efficacy to nematodes and toxicity to the soybean plants (Minton et al., 1979; Elliott et al., 1983; Rhoades, 1980). Based on the chemical and physical properties of a given nematicide, time of application and incorporation also may be critical. Nematicides that are easily photodegradable and have a very low water solubility will need to be properly distributed and incorporated to achieve expected performance. However, nematicides that are not photodegradable and have a very high water solubility can easily be distributed in the soil through irrigation or rainfall water, forming a continuous zone of protection. Through proper manipulation of application methods, dosage levels for acceptable pest control may be reduced. For example, TEMIK® aldicarb pesticide is federally labeled for soybean nematode control at the rate of 3.36 kg. active ingredient per hectare, to be applied in a 15 to 20 cm band and incorporated at planting. However, several years of experience have shown that a much reduced rate, 1.12 kg. active ingredient per hectare, will give acceptable control when it is applied in-furrow in close vicinity to the planted seed.

NEMATICIDES AND RESISTANT CULTIVARS

Although *H. glycines* and *Meloidogyne* spp. are probably the most important nematode species attacking soybeans worldwide, causing considerable yield loss (Caviness and Riggs, 1976), nematode damage to soybeans is not limited only to these species. Other nematode species such as *Pratylenchus* spp., *Hoplolaimus* spp., and *Belonolaimus* spp. have been demonstrated to cause significant yield losses and are quite devastating under certain conditions (Rodriguez-Kabana and Thurlow, 1980; Kinloch, 1980).

Due to the importance of *H. glycines* and *Meloidogyne* spp. as potential yield reducing agents, emphasis has been placed on the development of cultivars resistant to *H. glycines* races 1, 3, and 4 and to certain *Meloidogyne* spp. Physiological races of *H. glycines* have been shown to hybridize, and could therefore, give rise to new races capable of attacking the resistant soybean cultivars (Riggs et al., 1977; Triantaphyllou, 1975; Zirakparvar and Norton, 1981). With selection pressure shifting the gene composition of a given race, and with the presence of other damaging parasitic nematodes, yield losses on some resistant cultivars have been encountered. The use of nematicides under such circumstances has been beneficial (Kinloch, 1979; Bergeson, 1982; 1983; Phipps and Elliott, 1983).

AGROCHEMICAL INTERACTIONS

Several farm chemicals are often applied to a field of soybeans in a growing season to provide adequate nutrients and to control various pests, such as nematodes, fungi, insects, mites and weeds. Any of these chemicals may or may not interact with one or more of the other chemicals used.

Interactions between several nematicides and herbicides used in soybean production have been observed and studied by various scientists in recent years. Some herbicides, when applied in conjunction with nematicides to soybean fields, may alter the efficacy of the nematicide and influence the nematode population dynamics (Schmitt and Corbin, 1981; Schmitt et al., 1983). These interactions can be in the form of enhanced nematode control and subsequent yield increase (Kraus et al., 1982), or it may appear as an unusually high nematode population, phytotoxicity, reduction in soybean plant stand and yield loss (Schmitt and Corbin, 1981; Schmitt et al., 1983; Bostian et al., 1984; Rhoades, 1980). Examples of such phenomena are the interaction of metribuzin with some of the carbamate nematicides resulting in enhanced soybean cyst nematode control (Kraus et al., 1982) and with some of the organophosphate nematicides resulting in soybean cyst nematode population resurgence, phytotoxicity to soybeans, and yield reduction (Bostian et al., 1984; Rhoades, 1980).

The spectrum of activity for any given chemical is usually not broad enough to protect a crop from all attacking pests. Therefore, usually a combination of pesticides, both foliar and soil applied, are used in a growing season. There has been very little investigation (Kinloch and Schenck, 1978) of the possibility of interactions between these pesticides in terms of their effect on target pests, host plants, non-target organisms and the environment. This area deserves future attention.

Fertilizers are probably the most commonly used agricultural chemicals. However, their effect on nematodes parasitizing soybeans or the possibility of any interaction between nematicides and fertilizers used for soybean production has received limited attention. Further research on this subject is needed.

CONCLUSION

Several nematicides have been registered for control of soybean nematodes by the United States Environmental Protection Agency and by similar agencies in other countries. The recent action by the EPA to suspend the use of DBCP and EDB on all crops reduces the number of effective and economical soil fumigants that can be used for the control of nematodes on soybeans and emphasizes the fact that more comprehensive research should be conducted on non-fumigant nematicides to fill the gap.

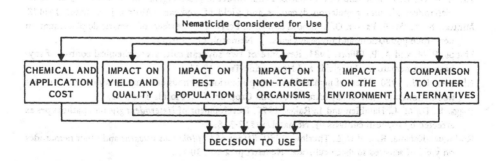

Figure 1. Parameters involved in selection of a nematicide for control of soybean nematodes.

Selection of a particular nematicide among all the available compounds should be based on a combination of factors, such as the expected cost/benefit ratio obtained by using the chemical, impact on target pests and non-target organisms, long and short term effects on the environment, and a cost/benefit analysis comparing all available alternative control measures (Figure 1). A unilateral decision without considering all the factors involved could lead to situations where not only the user will lose considerable benefit but the environment could suffer. The low cost of a nematicide is not always indicative of highest net profit. In contrast, the high price of an excellent nematicide may be offset by high yield return, resulting in highest net value to the user. Chemicals are proven to be very valuable, effective and sometimes the only tool available for battling nematodes, but their careless and indiscriminate use can lead to unpleasant results. Label directions and recommendations given by the manufacturer and local extension and university scientists for any given nematicide are the result of several years of scientific testing and should be followed carefully.

NOTES

M. E. Zirakparvar, Union Carbide Agricultural Products Company, Inc., Research Triangle Park, North Carolina 27709.

REFERENCES

Bergeson, G. B. 1982. Control of soybean cyst nematode with chemicals and a resistant variety. Fungicide and Nematicide Tests 37:198.

Bergeson, G. B. 1983. Evaluation of nematicides and resistant varieties for control of soybean cyst nematodes. Fungicide and Nematicide Tests 38:6-7.

Bostian, A. L., D. P. Schmitt and K. R. Barker. 1984. Early growth of soybean as altered by *Heterodera glycines*, phenamiphos and/or alachlor. J. Nematol. 16:41-47.

Caviness, C. E. and R. D. Riggs. 1976. Breeding for nematode resistance. p. 594-601. *In* L. D. Hill (ed.) World Soybean Research. Interstate Printers and Publishers, Danville, Illinois.

Elliott, A. P., P. M. Phipps, D. Babineau and J. Taylor. 1983. Effect of nematicides and application method on the soybean cyst nematode and soybean yield. Fungicide and Nematicide Tests 38:8.

Kinloch, R. A. 1979. Response of resistant soybean cultivar to fumigation at planting for control of soybean cyst and root-knot nematodes. Nematropica 9:27-32.

Kinloch, R. A. 1980. The control of nematodes injurious to soybeans. Nematropica 10:141-153.

Kinloch, R. A. and N. C. Schenck. 1978. Nematodes and fungi associated with soybeans growing under various pesticide regimes. Proc. Soil Crop Sci. Soc. Florida 37:224-227.

Kraus, R. G., G. R. Noel and D. I. Edwards. 1982. Effect of preemergence herbicides and aldicarb on *Heterodera glycines* population dynamics and yield of soybean. (Abstr.). J. Nematol. 14:452.

Minton, N. A., M. B. Parker, O. L. Brooks and C. E. Perry. 1979. Effects of nematicide placement on nematode populations and soybean yields. J. Nematol. 11:150-155.

Phipps, P. M. and A. P. Elliott. 1983. Response of two soybean cultivars to chemical control of soybean cyst nematode. Fungicide and Nematicide Tests 38:11.

Rhoades, H. L. 1980. Effect of nematicides and application methods on sting nematode control, root nodulations and yield of soybeans. Proc. Soil Crop Sci. Soc. Florida 39:90-92.

Riggs, R. D., M. L. Hamblen and L. Rakes. 1977. Development of *Heterodera glycines* pathotypes as affected by soybean cultivars. J. Nematol. 9:312-318.

Rodriguez-Kabana, R. and D. L. Thurlow. 1980. Effect of *Hoplolaimus galeatus* and other nematodes on yield of selected soybean cultivars. Nematropica 10:130-138.

Schmitt, D. P. and F. T. Corbin. 1981. Interaction of fensulfothion and phorate with preemergence herbicides on soybean parasitic nematodes. J. Nematol. 13:37-41.

Schmitt, D. P., F. T. Corbin and L. A. Nelson. 1983. Population dynamics of *Heterodera glycines* and soybean response in soil treated with selected nematicides and herbicides. J. Nematol. 15:432-437.

Triantaphyllou, A. C. 1975. Genetic structure of races of *Heterodera glycines* and inheritance of ability to reproduce on resistant soybeans. J. Nematol. 7:356-364.

Welch, E. R., R. D. May and R. D. Riggs. 1978. An economic analysis of controlling soybean cyst nematodes on soybeans. Cotton Gin and Oil Mill Press 79:(3) 12.

Zirakparvar, M. E. and D. C. Norton. 1981. Population characteristics of *Heterodera glycines* in Iowa. Plant Dis. 65:807-809.

STRATEGIES FOR RACE STABILIZATION IN SOYBEAN CYST NEMATODE

Robert D. Riggs

Soybean cyst nematode (*Heterodera glycines* Ichinohe) is a dynamic parasite that reproduces by amphimixis. Through its large number of genes for parasitic capability or through genetic recombination, or both, this nematode adapts to a wider host range than most species of cyst nematodes (Riggs and Hamblen, 1962; Riggs and Hamblen, 1966). Soybean cyst nematode (SCN) also has been able to adapt when previously resistant soybean cultivars were planted several consecutive years, and is eventually able to parasitize those cultivars.

Five races of SCN have been described (Golden et al., 1970; Inagaki, 1979) based on differential host range. However, other races have been observed, and if the number of differential hosts is expanded, the number of races is greatly increased. Miller (1971) demonstrated the variability of SCN when he selected seven isolates from one field that could be separated as races using soybean lines and cultivars. Slack et al. (1982) demonstrated the effects of planting resistant soybean cultivars several years in succession (Table 1). They found initially that the SCN juvenile population declined if it was high, or remained low if already low. However, after three years of planting a resistant cultivar, the juvenile population increased sharply. In plots where resistant cultivars were not grown, the population would not reproduce on the resistant cultivar.

When resistant cultivars became available, many growers planted them year after year. In some cases, after 4 to 5 years, damage to the resistant cultivar was observed. In other cases rice (*Oryza sativa* L.) was planted for a year, followed by two years of a resistant cultivar. In these cases damage to the resistant cultivar was observed after 6 years.

Caldwell et al. (1960) determined that three recessive genes controlled resistance (probably Race 1) in cv. Peking. A dominant gene was later found to be involved in resistance to what was probably Race 3 (Matson and Williams, 1965). Hartwig and Epps (1970) discovered another gene that conditioned resistance to SCN Race 4. Thomas (1974) and Thomas et al. (1975) studied the inheritance of resistance to all four races. Thomas proposed that 10 genes were involved with one race or another, but only part of the genes conditioned resistance against each race. These studies indicated that multiple genes are involved in resistance to SCN. If the gene-for-gene theory is applicable, there should be a like number of genes for parasitism in the nematode.

Triantaphyllou (1975) proposed that at least three genes for parasitism were present in SCN. Race 3 had none of the genes that would allow it to parasitize a cultivar or line with resistance. Race 1 had one gene that allowed it to parasitize PI 88788, Race 2 had two genes that allowed it to parasitize PI 88788 and Pickett, and Race 4

Table 1. Cropping sequence and *Heterodera glycines* juvenile counts (no/470 mL of soil) from rotation plots.

Year	Month	Plot 1		Plot 2	
1960	–	Cotton		Cotton	
1961	–	Soybean		Soybean	
1962	–	Cotton		Cotton	
1963	January	2		5	
		NC-55 Soybean		R58-82 Soybean	
	June	9		0	
1964		R58-82		Lee Soybean	
	June	4		2	
1965		A	B	A	B
		Lee[a]	R58-82[b]	Lee	R58-82
	June	162	42	825	490
1966		Lee	R58-82	Lee	R58-82
	June	960	47	320	11
1967		Lee	Pickett[b]	Lee	Pickett
	June	280	56	529	5
	September	570	120	414	20
1968		Lee	Pickett	Lee	Pickett
	June	1300	422	1160	400
1969	May	230	300	205	410

[a]Susceptible.
[b]Resistant.

had three genes that enabled it to parasitize Pickett, PI 88788 and PI 90763.

Limited studies by Price (1976) indicated that two loci controlled the parasitic capabilities of the four SCN races, with four alleles at one locus and two alleles at the other. Of course, both Triantaphyllou and Price recognized that they were working with a limited number of races, and that the number of races could be expanded greatly. If this occurred, the number of genes for parasitic capability would also increase.

Koliopanos (1976) crossed individuals from SCN populations (NC from North Carolina and VA from Virginia) that differed in ability to parasitize Pickett. The NC population did not reproduce, while the VA population did. The F_1 progeny from reciprocal crosses all parasitized Pickett better than the NC parent, but there was a difference depending on the female parent. Therefore, even though there was at least a gene difference between the two populations, there was also cytoplasmic inheritance that affected the parasitic capability.

The results of both Koliopanos (1976) and Price (1976) indicate that the ability to parasitize Pickett was dominant to the lack of such ability. However, when the SCN-susceptible cv. Lee was inoculated with a 1:1 mixture of eggs of Races 3 and 4, or 2 and 3, the results did not support dominance in the ability to parasitize Pickett (Price, 1976). After 1 month on Lee, only 7% of the progeny of the race 3 + 4 population were able to mature on Pickett; after 2 months, 2.1%; after 3 months, 1.4%; and after 4 months only 1.1%. When Races 2 and 3 were mixed, the subsequent maturation on Pickett was 22.2%, 10.5%, 9.7%, and 6.8% of the maturation of Lee after 1, 2, 3, and 4 months, respectively. When Races 3 and 4 were mixed 3:1 and 1:3, the results were similar (Riggs, unpublished). After 4 weeks, the female and cyst recovery from Pickett compared to Lee was 7.5 and 25.3% from the 1:3 and 3:1 mixtures, respectively, and after 8 weeks 4.0 and 33.9% from the 1:3 and 3:1 mixtures, respectively. A mixture of 1:9 Race 3 and Race 4 remained unchanged after 12 weeks on Lee.

Soybean cyst nematodes have shown a remarkable capacity for adjusting in the presence of resistant germplasm (Riggs et al., 1977; McCann et al., 1980; Young, 1982). This appears to be the result of mixtures of genotypes within SCN populations. Only in Florida has there been widespread use of SCN resistant cultivars with no occurrence of SCN races that will attack these cultivars (Kinlock, pers. commun.). However, soil from a Florida field where a Pickett soybean had been grown for two or more years was transferred to a greenhouse in Arkansas. The population was increased on Lee and then used to inoculate Pickett. In one generation on Pickett a few progeny were produced, and the next generation reproduced profusely. This indicates that conditions in Florida tend to be detrimental to the build-up of, at least, Race 4.

Through selection pressure provided by continuous reinoculation with those progeny that were produced on resistant cultivars, populations were selected from Race 3, which reproduced readily on Pickett, Peking, and PI 88788 and from Races 2 and 4 on Peking, PI 88788 and PI 90763 (Riggs et al., 1977). Subsequently, field populations were differentially selected by growing SCN-resistant soybean cultivars or lines in soil to which the nematodes had been added (McCann et al., 1980; Young, 1982). McCann et al. (1980) separated the field populations into three classes, as follows: Class 1, reproducing on cv. Cloud, Pickett-71, and PI 87631-1, PI 88788, PI 90763-R (a selection from PI 90763 by Miller in Virginia based on resistance to SCN) and PI 290332; Class 2, reproducing on Pickett-71 and PI 89772; and Class 3, reproducing only on Pickett. Young (1982) started with field populations of SCN Race 4, and after 11 selection generations high populations were present on cv. Bedford, and PI 88788, PI 89772, and PI 90763. The populations from Bedford and PI 88788 did not reproduce well on Peking, PI 89772, and PI 90763. On the other hand, populations which built up on Peking, PI 89772, or PI 90763 reproduced well on the other two, but did not reproduce well on Bedford or PI 88788. Luedders and Dropkin (1983) carried the selection process one step further. A population that had been selected on cv. Cloud, but would not reproduce well on PI 89772, and one that reproduced well on PI 89772, but did not reproduce well on Cloud, were re-selected on the resistant host. The population on Cloud was quickly re-selected on PI 89772 and vice versa, indicating that the original selection did not fix the parasitic capability of the population. However, growing a Race 4 population on Lee for several generations does not seem to alter its ability to parasitize Pickett (Riggs, unpublished; Miller, pers. commun.).

Rotation is suggested to manipulate the races of SCN. This includes the use of non-host crops and cultivars with resistance, but also involves the use of cultivars that are susceptible to all races of SCN. Slack et al. (1982) demonstrated that the absence of a host for 2 years reduces the SCN population to a level that will allow a susceptible cultivar to be grown the third year with little or no injury. Chinese scientists have reached the same conclusion (pers. commun.). The Chinese have not had resistant cultivars; therefore, they have established a practice of planting soybeans only one year of three, and they have little, if any, damage where this practice is followed. The recommended rotation includes a year of non-host, a year of a resistant cultivar, and a year of a susceptible cultivar. The year in a susceptible cultivar is to remove the selection pressure, which should allow the most competitive type to build up. Greenhouse tests have indicated that Race 3 is more competitive than either Race 2 or 4 (Price et al., 1976). In addition, the availability of two distinct sets of genes for resistance (McCann et al., 1980) allows for the rotation of cultivars with the different sets of genes to prevent the selection of races that attach these resistant cultivars.

NOTES

Robert D. Riggs, Department of Plant Pathology, University of Arkansas, Fayetteville, Arkansas 72701.

REFERENCES

Caldwell, B. E., C. A. Brim, and J. P. Ross. 1960. Inheritance of resistance of soybeans to the cyst nematode *Heterodera glycines*. Agron. J. 52:635-638.

Golden, A. M., J. M. Epps, R. D. Riggs, L. A. Duclos, J. A. Fox and R. L. Bernard. 1970. Terminology and identity of infraspecific forms of the soybean-cyst nematode (*Heterodera glycines*). Plant Dis. Rep. 54:544-546.

Hartwig, E. E. and J. M. Epps. 1970. An additional gene for resistance to the soybean-cyst nematode, *Heterodera glycines*. (Abstr.) Phytopathology 60:584.

Inagaki, H. 1979. Race status of five Japanese populations of *Heterodera glycines*. Jap. J. Nematol. 9:1-4.

Koliopanos, C. N. 1976. Hybridization between two populations of *Heterodera glycines*, Ichinohe, 1952. Ann. Inst. Phytopath. Benaki N.S. 11:168-175.

Luedders, V. D. and V. H. Dropkin. 1983. Effect of secondary selection on cyst nematode reproduction on soybeans. Crop Sci. 23:263-264.

Matson, A. L. and L. F. Williams. 1965. Evidence of a fourth gene for resistance to the soybean-cyst nematode. Crop Sci. 5:477.

McCann, J., V. H. Dropkin and V. D. Luedeers. 1980. The reproduction of differentially-selected populations of *Heterodera glycines* on different r-lines of soybean (*Glycine max*). (Abstr.) J. Nematol. 12:230-231.

Miller, L. I. 1971. Physiologic variation within the Virginia-2 population of *Heterodera glycines*. (Abstr.) J. Nematol. 3:318.

Price, M. 1976. Variability of *Heterodera glycines* races and genetics of parasitic capabilities on soybeans. Ph.D. Dissertation, University of Arkansas, Fayetteville.

Price, M., R. D. Riggs and C. E. Caviness. 1976. Races of soybean-cyst nematodes compete. Ark. Farm Res. 25(2):16.

Riggs, R. D. and M. L. Hamblen. 1962. Soybean cyst nematode host studies in the family Leguminosae. Ark. Agric. Exp. Stn. Rep. Ser. 110:1-18.

Riggs, R. D. and M. L. Hamblen. 1966. Further studies on the host range of the soybean-cyst nematode. Ark. Agric. Exp. Stn. Bull. 718:1-19.

Riggs, R. D., M. L. Hamblen and L. Rakes. 1977. Development of *Heterodera glycines* pathotypes as affected by soybean cultivars. J. Nematol. 9:312-318.

Slack, D. A., R. D. Riggs, and M. L. Hamblen. 1982. Nematode control in soybeans: Rotation and population dynamics of soybean cyst and other nematodes. Ark. Agric. Exp. Stn. Rep. Ser. 263:1-36.

Thomas, J. D. 1974. Genetics of resistance to races of the soybean-cyst nematode. M.S. Thesis, University of Arkansas, Fayetteville.

Thomas, J. D., C. E. Caviness, R. D. Riggs and E. E. Hartwig. 1975. Inheritance of reaction to race 4 of soybean-cyst nematode. Crop Sci. 15:208-210.

Triantaphyllou, A. C. 1975. Genetic structure of races of *Heterodera glycines* and inheritance of ability to reproduce on resistant soybeans. J. Nematol. 7:356-364.

Young, L. D. 1982. Reproduction of differentially-selected soybean cyst nematode populations on soybeans. Crop Sci. 22:385.

CONCEPT OF RACE IN SOYBEAN CYST NEMATODE

V. H. Dropkin

The soybean cyst nematode, *Heterodera glycines* Ichinohe, was identified as the cause of a disease of soybeans in Japan early in this century, and was named in 1952 (Ichinohe, 1952). The genus *Heterodera* contains many species of plant parasites that damage numerous plants. These nematodes are all characterized by striking sexual dimorphism. Males are typical elongate worms, but females develop into globular adults containing up to several hundred eggs. Mature eggs contain elongate infective larvae, about 0.4 mm in length, which emerge and invade growing roots. The larvae take up a position in the cortex or vascular cylinder, where the plant responds to their presence by developing a set of fused, enlarged cells upon which the nematode feeds. During development, the nematode remains in one position and becomes an elongate, motile male, or a swollen sessile female. These females are less than 1 mm in length, barely visible to the unaided eye on the surface of the root.

The soybean cyst nematode (SCN) completes its life cycle in just under a month in warm soils. As the female matures, its color changes from white to cream and then to brown when it dies. The term *cyst* refers to the thick-walled body of the female with its content of embryonated eggs. Unhatched larvae remain in a viable state for long periods enclosed in eggs within the cyst. They survive cold, drought, and other environmental hazards, for years in many cases. Some species of *Heterodera* have a mechanism for increasing the probability of successful infection of suitable hosts. Larvae remain unhatched until a signal diffuses from growing roots and stimulates the infective larvae to emerge.

SCN was first reported in the U.S. from a field of diseased soybeans in 1954 in North Carolina (Winstead et al., 1955). It is now known from 22 states along the Mississippi Valley, from Minnesota to Louisiana, as well as from the southeastern U.S. Losses attributable to this parasite are not uniform—they range from total crop failures to slight depression of yield. The severity of damage varies with soil type, weather, cultivar, and the numbers of nematodes present in the soil. These nematodes disturb root functions, including uptake and transport of water and minerals. Affected plants may show poor root development, depression of nodulation, inhibited top growth, yellowing of leaves, and reduced pod and seed production. These are symptoms of general root damage and are not diagnostic. Confirmation of cyst nematode involvement requires microscopic determination of the presence of the organism in significant quantity.

Search for genetic resistance began soon after discovery of SCN in North Carolina. Approximately 2800 selections and cultivars of soybeans were planted in an infested field. Among these, a few plant introductions appeared to show less damage than the rest and to have fewer adult females (Ross and Brim, 1957). The cultivar Peking was chosen for crosses with existing susceptible cultivars because it showed a high degree of resistance, together with favorable agronomic characteristics (Hartwig, 1981). Cultivars incorporating resistance from Peking were planted in infested soils. Soon these cultivars proved to be susceptible in some places and to retain resistance in others. Genetic variability of the pathogen was a likely contributor. This proved to be the case, and in 1970 a system of race classification was adopted (Golden et al., 1970).

The term "race", applied to SCN, refers to a population with characteristic abilities to reproduce on certain soybeans and not on others. In Europe, the term "pathotype" is used to refer to nematode populations that can be differentiated by a set of differential hosts. The word pathotype emphasizes the reproductive abilities of members of a population, but makes no reference to any other phenotypes. Zoologists use the term race to denote a subspecific population with a distinctive array of phenotypes and with a restricted flow of genes between it and other races of the species. The isolating mechanism may be geographic, partial sterility, behavioral, or other. In SCN, we make no reference to characteristics other than ability to reproduce in certain hosts, and no reference to gene flow between populations. We should, therefore, consider the adoption of the European usage of pathotype. The term "race" will be used, in this paper, as equivalent to "pathotype".

Differences in morphology and in host range of subspecific populations of phytonematodes are well documented. *Belonolaimus longicaudatus* is an ectoparasite distributed widely in Atlantic coastal states. Morphological differences measured by correlation of two characters were found in populations from various geographical areas (Rau and Fassuliotis, 1970). Another study on the ectoparasitic *Trichodorus* showed that variations exist in this species as well, related in part to geographical origin (Bird and Mai, 1967). These authors found considerable morphological variability among the offspring of a single female, suggesting heterozygosity, probably at more than one locus. Morphological variations were also noted in SCN, but they did not correlate with races determined by reproduction on differential hosts (Golden and Epps, 1965). Subspecific variation in host range has been known for many years in *Ditylenchus dipsaci*, an important pest of many crops in which approximately 20 races are recognized and designated by the name of the host on which each variant reproduces best (Hesling, 1966).

RACE CLASSIFICATION

Our concept of races in *Heterodera* sp. derives from the ability of the nematodes to reproduce in various hosts. On this basis, a number of races are recognized in the potato cyst, in the oat cyst, and in SCN. Potato cyst nematodes, formerly designated as one species, are now recognized as two. *Globodera rostochiensis* is separated into four races in Europe and *G. pallida* is separated into three. A test of Andean populations of *G. rostochiensis* revealed the same four races as those in Europe. But populations of *G. pallida* from South America were separated into six races (Canto Saenz and de Scurrah, 1977). Moreover, disc electrophoresis showed differences between a European and a Peruvian population, although both were classed as belonging to the same race according to the conventional host-range test (Franco, 1979). In this case, electrophoresis simply shows that two populations with the same reproductive performance on a restricted set of soybeans may have another set of phenotypes (proteins) that differ. Still other phenotypes, such as temperature optima or ability to survive drought, may differ among populations of SCN. The precise classification of subspecific populations, based on measurement of a given set of phenotypes, may be unique and unrelated to any other classification resulting from the use of other phenotypes.

The cereal cyst nematode, *H. avenae*, is separated into five races in Europe and one each in Australia and India. This species is of particular interest because there is evidence for differences in the ecology of races. This suggests that races may have evolved in response to selection for ecological variables. A French investigator found four races among 24 populations from France, Belgium, and Switzerland by measuring their reproduction on a limited set of differential oats (*Avena sativa* L.), barley (*Hordeum vulgare* L.), wheat (*Triticum* spp.), and rye (*Secale cerealae* L.). A second test on about 30 cultivars confirmed this classification (Rivaol, 1977). In a subsequent publication, he compared hatching and larval invasion of plants under varying soil temperatures in two of the four races, Fr 1 from southern France and Fr 4 from the north of France. The southern race hatched at 5 C and was inhibited by temperatures that rose to 10 C or higher, whereas the northern race hatched when soil warmed up to 10 C or higher. These differences were maintained when the populations were tested at an intermediate location, as well as in their native regions (Rivoal, 1978).

My last example to show that races occur in phytonematodes, both in sessile endoparasites and migratory ectoparasites, is the situation in *Meloidogyne*, the root-knot nematodes. Populations from many parts of the world have been assembled at Raleigh, NC. In 154 populations of *M. incognita*, 100 were classed by a differential host test as Race 1, 33 as Race 2, 13 as Race 3, and 8 as Race 4. Other species also contained races (Sasser and Taylor, 1978).

The existence of races of SCN in the U.S. became obvious soon after the original find (Brim and Ross, 1966; Epps and Duclos, 1970; Ross, 1962; Smart, 1964). As early as 1957, two Japanese sources of resistance were found to be susceptible to the North Carolina population. From the start, then, there was evidence of subspecific differences in the nematodes from separate geographic regions (Ross and Brim, 1957). To establish a classification, reproduction of a population on four soybean lines was compared with that on Lee-68 (Golden et al., 1970). Hosts were scored as susceptible when the number of cysts equalled 10% or more of the number on Lee-68 and resistant when cyst production fell below 10%. On this basis, the populations fell into four classes, called races. This terminology and system of classification is still used. It is an unwarranted simplification to assume that the four races delineated by five cultivars and PIs represent the total pathotype variability.

If the set of differentials is expanded, more differences among poulations should be demonstrated. The thorough study of Riggs and colleagues illustrates the great variability in reproductive ability present in SCN (Riggs et al., 1981). These workers assembled 33 populations from states in the U.S. with known infestations of SCN and five from Japan. Thirteen soybean kinds and five other host species (two *Lespedeza* spp., yellow sweet clover ([*Melilotus officinalis*] and *Cleome* sp.) constituted the differential host series. Reproduction on a host was scored in relation to reproduction on Lee, taken as the universal host. Separation of the populations into four races occurred in certain combinations of the test hosts, but not in others. The original set of differentials used by Golden et al. (1970) was less successful in separating races than use of more extensive and different sets of hosts. Golden's set of differentials separated six groups among the 38 populations. A population originally classed as Race 1 no longer fell into that category, nor did any other population of this series. Analysis of the results of the tests on 12 soybeans, rated according to one scheme of comparison with Lee, yielded 25 physiological groups; according to another scheme, 36 groups were distinguishable. This work establishes the great variability in reproductive capabilities of populations within the species *H. glycines*. Another investigator demonstrated a three-fold difference in several isolates in reproduction on Lee. He also found differences in reproduction on various hosts among populations from different locations, as well as in populations developed from single cysts collected in the same field (Miller, 1971).

Species of plant parasitic nematodes, like other species, are probably heterallelic at all loci and strongly so at many. This implies that gene frequencies will change in a population subjected to selection. There is good evidence that this is so in SCN. If the species carries multiple alleles for ability to reproduce on certain hosts at some of these loci and the nematodes are heterozygous at one or more loci, then selection should be effective. We may also expect to find adaptation to specific ecological situations in *H. glycines*. Evidence for response to selection is clear; that for ecotypes is almost lacking, because no investigations on this point have been reported. Race 1, according to the original classification, was reported from North Carolina; Race 2 from Virginia; Race 3 from the Mississippi Valley; and Race 4 from Arkansas, Missouri, and Tennessee. One bit of suggestive evidence that races might have differentiated on nonsoybean hosts is the finding of a field in Missouri in which lespedeza had grown for 25 years, followed by three years of susceptible soybeans. This field contained SCN classed as Race 4 (Epps and Duclos, 1970).

The effect of selection is dramatically shown by continuous cultivation on a resistant host. A population tested by inoculation of 500 larvae produced three cysts on Peking soybean and 108 cysts on Lee. After selection on Peking for seven consecutive generations, it then produced 103 cysts on Lee and 76 on Peking. Selection of another population on PI 88788 for six generations resulted in an almost steady increase of cyst production on the selecting host and no loss of reproduction on Lee (Triantaphyllou, 1975). Field experience confirms these results. In five years of continuous cultivation of Bedford, the ratio of cyst production on that cultivar to that on Essex, a susceptible host, rose from 0.10 to 0.94 (S. Anand, pers. commun.). However, Young (1983) found an interaction between selection and environmental factors. In fields cropped continuously to resistant Bedford, the rate of change of the ratio of SCN reproduction on Bedford to that on susceptible Essex varied from field to field and from year to year.

The zoological concept of race implies a restricted flow of genes between subspecific populations necessary to maintain the distinctions between races. Limited data of one study indicate that this may be so in SCN. Crosses within and between races were made by isolating lateral roots into vials, into each of which a single larva was introduced. After 15 days, 40 males of the appropriate type were added. Eggs per female averaged 133 in intraracial F_1s and 44 in interracial crosses. Although no field data are available, it appears that gene flow between races would be restricted (Price et al., 1978).

In the total interaction between SCN and soybeans, genetic controls may act at many points. Attraction to roots is the initial step. This aspect of behavior operates in many plant-nematode associations. Nematodes have complex sensory structures by which they respond to a variety of stimuli. Differences in invasion of roots occur in some associations of plant parasitic nematodes and plants, but they have not been documented in SCN-soybean interactions. Once at the plant surface, SCN larvae penetrate into the cortex and begin feeding. The action of nematode saliva and of feeding result in the formation of a set of enlarged, fused cells upon which the nematode feeds. Genetic incompatibility between host and parasite may disturb this process at any point from the initial penetration by larvae to the final stage of egg production at the conclusion of the nematode's life cycle. We (Acedo et al., 1984) have observed early blocks to the association manifested by a rapid necrotic response to invasion in some plants resistant to certain nematode populations and later blocks in other combinations. Sex ratios appear to be under genetic control. In some combinations, selection acts against males but in others there is no selection against males (J. Halbrendt, unpublished). The association of nematode-plant-nodulating bacteria is also under genetic control. Experiments in North Carolina showed that Race 1 inhibited nodulation and nitrogen fixation capacity in Lee soybean roots more than Races 2 and 4 (Lehman et al., 1971).

POSSIBLE ORIGINS OF RACES OF SCN

In any species distributed over a large area, we expect different gene frequencies to develop by mutation and selection in separate parts of the range. If these result in restrictions of gene flow, then further differences can accumulate. It is, therefore, not surprising to find that Japanese populations include Race 5, in addition to Races 1 and 3 known in the U.S. (Inagaki, 1979). Race 5 was not included in the original classification (Golden et al., 1970).

Clarke (1976) has proposed that hosts and parasites evolve together with the result that both develop polymorphisms of enzymes and other proteins. Parasites adapted to their hosts act as selective forces on the hosts leading to changes of frequency of alleles conferring resistance. The hosts, in turn, select parasites capable of reproducing on the altered hosts. Parasites, again, exert an altered selection leading to host resistance. Polymorphism in many loci may contribute to the wide host range of SCN.

The rapid spread of SCN throughout the soybean-producing areas of the U.S., together with the example of the field that contained a population of Race 4 nematodes after 25 years of lespedeza and three of a susceptible soybean, suggest that this parasite existed in the U.S. before soybean cultivation. Further, the demonstration that races differ in their reproduction on various weed hosts reinforces the suggestion that the various populations evolved on weeds (Miller and Duke, 1967). Price (1980) noted that the isolation of parasites in their hosts probably affects their evolution. In contrast to species of organisms with continuous distributions, such as herbivores, parasites tend to have patchy distributions. Parasites are selected to develop long-term survival mechanisms, in addition to adaptations for life on particular hosts. If *H. glycines* evolved in various legumes and some nonlegumes, then the species existed as partly isolated small demes, each adapted to its own host. Opportunities for exchange of genes must have been infrequent. The advent of commercial soybeans resulted in massive increases of nematode populations within which there was large-scale pooling of genes from the array of small populations, each surviving on its own weed host. Consequently, the amount of variability available to the parasites is large. The use of resistant cultivars is bringing shifting selection pressures on populations of SCN. As breeders continue to introduce new sources of resistance from the world collection of soybeans, selection pressures will increase the frequencies of those genes that permit SCN to adapt to each new resistant cultivar.

I shall conclude the paper with a brief review of a cooperative project between V. D. Luedders and me at UMC which seeks to develop understanding of the interorganismal genetics of SCN-soybean associations. We set out to produce populations of SCN with less variability than that present in the field in order to provide tools for the analysis of the genetics of resistance.

Our first step was to subject a field population of *H. glycines* to selection by growing it on six known sources of resistance. Cysts (a few at first) were harvested from each selecting host and inoculated back to that host. After two years of continuous selection, we had six populations originating from a small area of an infested field. Each population reproduced well on its selecting host (PI 209332, PI 87631-1, PI 88788, PI 89772, PI 90763-R, and Cloud). An additional selected population obtained from A. C. Triantaphyllou reproduced on Pickett-71. All possible combinations of populations x selecting hosts and cv. Williams were inoculated at a uniform level to obtain estimates of reproduction. Three different populations were identified. All populations reproduced as well on Pickett-71 as they did on Williams. The population selected on Pickett-71 reproduced poorly on five soybean lines but well on Pickett-71 and Williams. The population selected on PI 89772 reproduced poorly on five soybean lines but well on PI 89772, Pickett-71, and Williams. The remaining populations reproduced poorly on PI 89772, well on Pickett-71 and Williams, and moderately on the rest of the test plants.

These experiments confirm the efficacy of selection to change average reproduction of a population on "resistant" hosts. They also indicate that our field population had considerable variability (McCann et al., 1982). Similar results of selection have been reported by others (Anand and Brar, 1983; Young, 1982).

We then subjected two selected populations to secondary selection to determine whether the selected populations could be altered. The population selected on PI 89772 reproduced poorly on Cloud, and that selected on Cloud reproduced poorly on PI 89772. Reciprocal selections were made for four cycles. As a result of the secondary selection, each line lost its ability to reproduce on its original host and gained ability to reproduce on the second selecting host (Luedders and Dropkin, 1983). This established that the populations were not homozygous.

To increase the purity of populations, we produced a set of selected inbred populations. Cysts were selected from PI 89772 and PI 209332 plants grown in the gene pool. These were added singly to 100 seedlings of each kind. At every subsequent generation for nine generations, single cysts were transferred to seedlings to continue the selected inbred lines. We anticipated that many lines would be lost as recessive lethal genes reached homozygosity; this did not occur. Probably the breeding structure of SCN, resulting from their restricted movement in soil, continuously selects against recessive lethals in the field. After the ninth transfer, some populations were increased on the selecting hosts and tested on a set of 11 different soybean genotypes. As expected, we found differences between the lines selected on each host. Inoculations with nematodes selected on PI 209332 resulted in three groups of hosts: Williams, Essex, and Corsoy had the highest numbers of cysts per plant; PI 209332, Lines 1, 2, and 4 (obtained from Japan) and Ilsoy were intermediate but not significantly different from the highest group; and PI 89772, Pickett-71, and Peking were lowest. A different grouping was apparent in the inoculations with populations selected on PI 89772: Williams, Essex, Corsoy, Pickett-71, Lines 1, 2, and 4 all fell into the group with the highest numbers of cysts per plant; PI 89772 and Peking were intermediate; and Ilsoy and PI 209332 were lowest.

A few individual inbred populations were distinctive. Population 2 (selected on PI 209332) reproduced well on Japanese Line 1 but poorly on Lines 2 and 4, whereas other inbred populations reproduced equally well on these three soybean lines. The most striking differential soybean was Ilsoy. Three inbred populations selected on PI 209332 averaged 145 cysts on Ilsoy, while the remaining populations averaged 20. Three inbred populations selected on PI 89772 averaged 92 cysts on Ilsoy and the remaining populations averaged 34. Three populations selected on PI 209332 reproduced well on Pickett-71, but the others did not. We are currently moving toward controlled matings, rapid bioassays for reproduction, and methods of simplifying the handling of test plants.

SUMMARY AND CONCLUSIONS

Populations of SCN now reaching levels damaging to yields of soybean plants appear to have adequate variability to respond to changes in selection pressures resulting from the effects of breeders. Reproduction of this nematode in soybean plants is the end product of a multistep chain of events subject to regulation of processes in both partners. Genetic blocks to the completion of this association may occur at many points in the sequence.

The concept of race applied to nematodes was useful at the outset because it provided a framework within which to handle the observed differences among populations. It was also dangerous because it implied that each race had no or few components of the other races. I believe that this nematode species exists on soybean plants as a set of populations, each with its own distinctive distribution of gene frequencies and that these frequencies change in response to many selective pressures, including those imposed by soil, moisture, agronomic practices, climatic changes, and host cultivars. I also believe that genetic variability of the nematodes is continually maintained by mutation and immigration, as well as by Mendelian recombination.

How should we approach genetic control of SCN? We must address several problems.

1. Can we continue to rely on field populations of nematodes to screen the products of plant breeding? Gene pools and selected nematode populations assembled from different areas offer better screens.

2. Can we improve methods of handling plants and nematodes? Automated systems of producing synchronous infections are within reach. We have found that clipped seedlings with timed infections can be grown in hydroponics and scored for nematode reproduction 14 days after inoculation. This approach also lends itself to automation.

3. Should we rely on nematode reproduction as the sole phenotype to be measured? Other phenotypes may also be useful, such as the tolerance of soybean plants to SCN or the effect of SCN on nodulation. Yield reduction must result from the interaction of many organisms, including nematodes. We must take account of these interactions.

4. Can we develop methods of following the microevolution of populations of nematodes in the field? In addition to bioassays of individual cysts, we might explore the techniques of molecular biology to characterize nematode populations. Serology also offers promise.

5. It is time, I believe, to devote considerable effort to a broad-scale investigation of the association between soybeans and SCN. Both genetics and other aspects of the biology of the association merit attention. Until now, we have focussed on immediate practical solutions to field problems with great success. But we have operated in a vacuum of knowledge about a highly complex system in which soybeans, nematodes, fungi, bacteria, arthropods, viruses—all interact. We have neglected the whole subject of plant physiology in relation to nematodes. I believe that the shortest path to effective symbiosis between man and cyst nematodes is to understand the system in order to find those places that

are amenable to our management. I use the term symbiosis advisedly, because we will continue to live with these nematodes as long as we grow soybeans.

NOTES

V. H. Dropkin, Professor and Chairman, Department of Plant Pathology, University of Missouri-Columbia.

Experiment Station Journal Series: 9609.

Partial support from the American Soybean Association, Missouri Soybean Merchandising Council, and USDA Grants 901-15-36 and SEA 83 CRSR 22128 is gratefully acknowledged.

REFERENCES

Acedo, J. R., V. H. Dropkin and V. D. Luedders. 1984. Nematode population attrition and histopathology of *Heterodera glycines*-soybean associations. J. Nematol. 16 (in press).

Anand, S. and G. S. Brar. 1983. Response of soybean lines to differentially selected cultures of soybean cyst nematode *Heterodera glycines* Ichinohe. J. Nematol. 15:120-123.

Bird, G. W. and W. F. Mai. 1967. Morphometric and allometric variations of *Trichodorus christiei*. Nematologica 13:617-632.

Brim, C. A. and J. Ross. 1966. Relative resistance of Pickett soybeans to various strains of *Heterodera glycines*. Phytopathology 56:451-454.

Canto Saenz, M. and M. M. de Scurrah. 1977. Races of the potato cyst nematode in the Andean region and a new system of classification. Nematologica 23:340-349.

Clarke, B. 1976. The ecological genetics of host-parasite relationships. p. 87-103. *In* Symp. No. 14 British Society of Parasitology. Genetic Aspects of Host-Parasite Relationships.

Epps, J. M. and L. A. Duclos. 1970. Races of the soybean cyst nematode in Missouri and Tennessee. Plant Dis. Rep. 54:319-320.

Franco, J. 1979. Disc electrophoresis of female proteins of British and Peruvian potato cyst nematode populations, *Globodera* spp. Nematologica 25:32-35.

Golden, A. M. and J. M. Epps. 1965. Morphological variations in the soybean cyst nematode. (Abstr.) Nematologica 11:38.

Golden, A. M., J. M. Epps, R. D. Riggs, L. A. Duclos, J. A. Fox and R. L. Bernard. 1970. Terminology and identity of infraspecific forms of the soybean cyst nematode (*Heterodera glycines*). Plant Dis. Rep. 54:544-546.

Hartwig, E. E. 1981. Breeding productive soybean cultivars resistant to the soybean cyst nematode for the southern United States. Plant Dis. 65:303-307.

Hesling, J. J. 1966. Biological races of stem eelworm. p. 132-141. *In* Rep. Glasshouse Crops Res. Inst. Hurley, Berks., U.K.

Ichinohe, M. 1952. On the soybean nematode *Heterodera glycines* n. sp. from Japan. Mag. Appl. Zool. 17:1-4.

Inagaki, H. 1979. Race status of five Japanese populations of *Heterodera glycines*. Jap. J. Nematol. 9:1-4.

Lehman, P. S., D. Huisingh and K. R. Barker. 1971. The influence of races of *Heterodera glycines* on nodulation and nitrogen-fixing capacity of soybean. Phytopathology 61:1239-1244.

Luedders, V. D. and V. H. Dropkin. 1983. Effect of secondary selection on cyst nematode reproduction on soybeans. Crop Sci. 23:263-264.

McCann, J., V. D. Luedders and V. H. Dropkin. 1982. Selection and reproduction of soybean cyst nematodes on resistant soybeans. Crop Sci. 22:78-80.

Miller, L. I. 1971. Physiologic variation within the Virginia-2 population of *Heterodera glycines*. (Abstr.) J. Nematol. 3:318.

Miller, L. I. and P. L. Duke. 1967. Development of eleven isolates of the soybean cyst nematode on four species of the Scrophulariaceae. (Abstr.) Va. J. Sci. 18(4):143.

Price, M., C. E. Caviness and R. D. Riggs. 1978. Hybridization of races of *Heterodera glycines*. J. Nematol. 10:114-118.

Price, P. W. 1980. Evolutionary Biology of Parasites. Princeton Univ. Press XI, Princeton, NJ.

Rau, G. J. and G. Fassuliotis. 1970. Equal frequency tolerance ellipses for population studies of *Belonolaimus longicaudatus*. J. Nematol. 2:84-92.

Riggs, R. D., M. L. Hamblen and L. Rakes. 1981. Infra-species variation in reactions to hosts in *Heterodera glycines* populations. J. Nematol. 13:171-179.

Rivaol, R. 1977. Identification des races biologiques du nématode à kystes des céréales *Heterodera avenae* Woll., en France. Ann. Zool. Ecol. Anim. 9:261-272.

Rivoal, R. 1978. Biologie d'*Heterodera avenae* Wollenweber en France. I. Différences dans les cycles d'éclosion et de développement des deux races Fr1 et Fr4. Rev. Nematol. 1:171-179.

Ross, J. P. 1962. Physiological strains of *Heterodera glycines*. Plant Dis. Rep. 46:766-769.

Ross, J. P. and C. A. Brim. 1957. Resistance of soybeans to the soybean cyst nematode as determined by a double-row method. Plant Dis. Rep. 41:923-924.

Sasser, J. N. and A. L. Taylor. 1978. Biology, identification, and control of root-knot nematodes (*Meloidogyne* species). Dept. of Plant Pathology, North Carolina State Univ., Raleigh.

Smart, G. C. Jr. 1964. Physiological strains and one additional host of the soybean cyst nematode, *Heterodera glycines*. Plant Dis. Rep. 48:542-543.

Triantaphyllou, A. C. 1975. Genetic structure of races of *Heterodera glycines* and inheritance of ability to reproduce on resistant soybeans. J. Nematol. 7:356-364.

Winstead, N. N., C. B. Skotland and J. N. Sasser. 1955. Soybean cyst nematode in North Carolina. Plant Dis. Rep. 39:9-11.

Young, L. D. 1982. Reproduction of differentially selected soybean cyst nematode populations on soybean. Crop Sci. 22:385-388.

Young, L. D. 1983. Effects of continuous culture of resistant soybean cultivars on soybean cyst nematode reproduction. Plant Dis. 68:237-239.

PLANT-PARASITIC NEMATODES ASSOCIATED WITH SOYBEANS

D. P. Schmitt

Over 100 species of plant-parasitic nematodes are associated with soybean plants. Most of these are efficient parasites and cause little or no damage to this crop. However, a few are highly virulent and cause considerable damage. The soybean cyst nematode, *Heterodera glycines* Ichinohe, poses the greatest threat of any species to the crop. (This nematode will be discussed in two other papers in these Proceedings: "Concept of Race in Soybean Cyst Nematode" by V. H. Dropkin, and "Strategies for Race Stabilization in Soybean Cyst Nematode" by R. D. Riggs.) Root-knot nematodes (*Meloidogyne* spp.) are important on a worldwide basis. Other nematodes of economic importance, but generally more regional, include the reniform (*Rotylenchulus reniformis* Linford & Oliveira), sting (*Belonolaimus longicaudatus* Rau), lesion (*Pratylenchus* spp.), and lance (*Hoplolaimus columbus* Sher) nematodes. Many other nematodes are associated with soybean plants, but their relationships to soybean growth and yield need to be characterized.

Six species of root-knot nematode have been reported in soybean. *Meloidogyne incognita* (Kofoid & White) Chitwood is the most widespread, although *M. hapla* Chitwood, *M. javanica* (Treub) Chitwood, and *M. arenaria* (Neal) are also common. Two other species, *M. inornata*, and *M. bauruensis*, are known to occur in Brazil (Lordello, 1956).

Soybean is only a fair host of *Meloidogyne* spp.; however, *M. incognita*, *M. arenaria*, and *M. javanica* can cause severe damage. *Meloidogyne hapla* has a more subtle effect on plant growth but can suppress yield.

The characteristic symptom induced by this group of nematodes is the root gall. Plants infected with *M. incognita*, *M. arenaria*, or *M. javanica* produce relatively large galls, the size depending upon the amount of infection. Galls induced by *M. hapla*, however, are much smaller and may be overlooked. Roots proliferate at the infection site, may be stunted and have a "stubby root" appearance.

The relationship between numbers of *M. incognita* at planting and crop yield is negative in Florida (Figure 1) (Kinloch, 1982). At the northern limit of this nematode's distribution, severe damage occurs primarily on highly sensitive cultivars, whereas yields of moderately resistant to resistant cultivars are enhanced (Barker, 1982). In Florida, on the other hand, yields of resistant cultivars are still suppressed compared to yields of the same cultivar without nematodes (Kinloch, 1974).

Meloidogyne hapla may cause slight to moderate yield losses of soybean. Evidence for pathogenicity is based largely on nematicide trials. Greatest control was achieved with soil fumigants, which increased yields 2.1- to 2.5-fold (Table 1). Fenamiphos also gave a significant increase (1.6X) in yield.

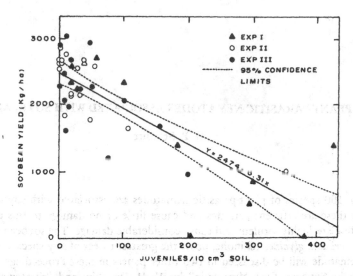

Figure 1. Relationship between soybean yield and numbers of southern root-knot nematode (*Meloidogyne incognita*) juveniles (from Kinloch, 1982).

Table 1. Effect of *Meloidogyne hapla* on soybean yield relative to control achieved by nematicides, Northhampton Co., NC, 1980.

Treatment	Nematodes at 66 days[a] per 500 cm³ soil	Yield[a]
	no.	kg/ha
1,3-D	13s	1920s
EDB	55s	1640st
Fenamiphos 15G	838st	1280t
Aldicarb 15G	1305st	850u
Turbofos 15G	1423s	580v
Control	1200st	780u

[a]Means followed by the same letter are not significantly different according to the Waller-Duncan K-ratio t-test.

Two other common root-knot species, *M. arenaria* (Rodriguez-Kabana and Williams, 1981) and *M. javanica* (Kinloch and Hinson, 1974) are more virulent than *M. incognita* or *M. hapla*. Losses due to *M. arenaria*, based on initial population density may be predicted by linear regression models (Rodriguez-Kabana and Williams, 1981). The slope of the regression line is greater for a susceptible cultivar (Ransom) than for Bragg, a moderately resistant cultivar. Relationships between initial numbers of *M. arenaria* and seed yield need to be established.

The reniform nematode (*Rotylenchus reniformis* Linford and Oliveira) is found on soybean plants, primarily in certain southeastern states in the USA. The juveniles do not feed, but the fourth stage juvenile penetrates the root with about one-third of the body to become the immature infective female (Rebois et al., 1970). As the female feeds, the posterior portion of her body swells and ruptures the epidermal tissue of the root. The cortical cells become distorted and devoid of cytoplasm (Rebois et al.,

1975). A syncytium is formed and the root system becomes stunted and chlorotic. Yield losses of dry seed have been as much as 33% (Rebois, 1971).

A few species of the lance nematode (*Hoplolaimus* spp.) are associated with soybean plants. The most virulent lance nematode is *Hoplolaimus columbus*, which feeds largely as an endoparasite, but also feeds as an ectoparasite (Lewis et al., 1976). It penetrates the root and migrates through the cortical cells, which rupture and form abnormally enlarged nuclei. The damaged cortical tissue sloughs off and does not produce secondary roots. Damage to the root results in an unthrifty, stunted plant which may also be chlorotic. Soil fumigation can effect a yield increase in some soybean lines (Table 2) (Nyczepir and Lewis, 1979).

Belonolaimus longicaudatus Rau is the sting nematode most economically significant to soybean. This species, however, may be a complex composed of two or more species (R. T. Robbins, pers. commun.). It has a very wide host range in important field crops and weeds (Robbins and Barker, 1973; Holdeman and Graham, 1953). It normally occurs in very sandy soils with a sand content between 84 and 94% (Miller, 1972; Robbins and Barker, 1973). Root damage by *B. longicaudatus* has been characterized on bean, *Phaseolus vulgaris* L. (Standifer, 1959). Feeding occurs just posterior to the apex of the root or at the base of lateral roots. A yellow area develops at the feeding site, root tips become swollen and roots may curve in the form of a "J". Lesions may extend from the cortex into the stele and may form a cavity encircling the injured cells. As a result, roots are stunted and fewer in number. Plants parasitized by this nematode are stunted, chlorotic, and rarely yield very well. Even a few individuals of this highly virulent pest can cause almost a complete loss of a soybean crop (Table 3). Excellent control of the sting nematode must be achieved in order to justify growing soybeans in infested fields.

Several species of lesion nematodes (*Pratylenchus* spp.) cause varying degrees of damage to soybean plants. These include *Pratylenchus penetrans* (Cobb) Filipjev & Schuurmans-Stekhoven, *P. coffeae* (Zimmerman) Filipjev & Schuurmans-Stekhoven,

Table 2.   Yield of 12 soybean genotypes in the field at Blackville, South Carolina, as affected by preplant soil fumigation with DBCP (from Nyczepir and Lewis, 1979).

Genotype	Yield		Mean difference[a]
	Fumigated	Unfumigated	
	g/3 m of row		g
Pickett-71	833	437	433s
Hardee	1053	752	145t
ED-371	763	607	193tu
Bragg	797	551	161tu
Coker 4504	977	898	106uv
W-4	898	829	71uvw
Forrest	508	551	16vwx
D71-9257	617	641	7vwx
Dyer	163	112	-18vwx
PI 90763	56	121	-43wx
PI 88788	75	201	-58wx
PI 89772	108	102	-94x

[a]The mean difference for yield was adjusted for plant number. Numbers followed by the same letter are not significantly different (P<0.05) by Duncan's Multiple-Range Test.

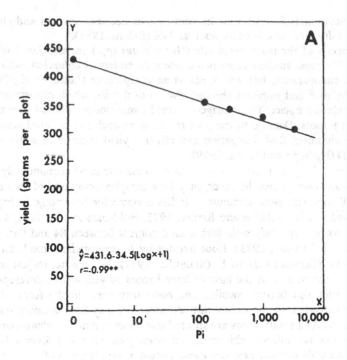

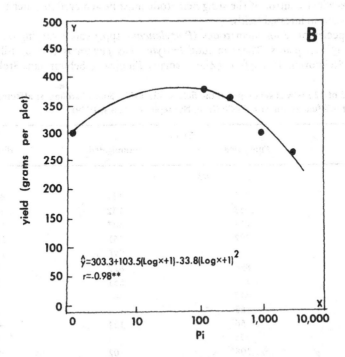

Figure 2. Regression of initial population densities (P$_i$) of *Pratylenchus brachyurus* and yield of Forrest soybean. A. Norfolk sandy loam. B. Appling sandy clay loam. Y = predicted yield in g/plot, X = initial numbers of nematodes/500 cm^3 soil (from Schmitt and Barker, 1981).

Table 3.   Soybean yields and numbers of the sting nematode (*Belonolaimus longicaudatus*)/500 cm³ soil at 29 days after planting as affected by nematicide treatments, Hertford Co., NC, 1977.

Treatment	Nematodes[a]	Yield[a]
	no.	kg/ha
Aldicarb 15G	3s	1100u
Carbofuran 10G	10s	650t
Fenamiphos 15G	8s	1330u
Control	21s	180s

[a]Means followed by the same letter are not significantly different.

P. hexincisus Taylor & Jenkins, *P. neglectus* (Rensch) Filipjev & Schuurmans-Stekhoven, *P. crenatus* Loof, *P. alleni* Ferris, *P. scribneri* Steiner, *P. agilis* Thorne & Malek, *P. zeae* Graham, and *P. brachyurus* (Godfrey) Filipjev & Schuurmans-Stekhoven (Schmitt and Noel, 1984). *P. scribneri* is probably the most widely distributed species in the United States, and is associated with soybeans in sub-temperate and temperate areas. Lesion nematodes are largely endoparasitic and they may complete their entire life cycle in root tissues. The nematodes either migrate into the roots or hatch within the root tissue and begin to feed. Lesions are typical of the feeding activity of *Pratylenchus* spp. (Acosta and Malek, 1981); root growth can be severely suppressed (Ferris and Bernard, 1962; Zirakparvar, 1982). Plants can be stunted and chlorotic as a result of the nematodes feeding activity.

The relationship of lesion nematode numbers to crop loss may be linear or curvilinear depending on soil type and cultivar (Figure 2). The yield loss of Forrest soybeans was negatively and linearly related to population densities of *P. brachyurus* in sandy soils, but was curvilinear in finer textured soils (Schmitt and Barker, 1981). Damping-off or air pollutant injury may be enhanced by some lesion nematodes (Lindsey and Cairns, 1971; Weber et al., 1979).

There are many other species of nematodes associated with soybean, but they do not have sufficient virulence to cause economic damage (Schmitt and Noel, 1984). These include such nematodes as *Tylenchorhynchus*, *Helicotylenchus*, *Scutellonema*, *Rotylenchus*, *Paratrichodorus*, *Criconemella*, and *Xiphinema*. *Xiphinema americanum*, however, may be a possible exception. This nematode caused significant loss in several cultivars in a sandy soil in North Carolina (Schmitt, unpublished).

Nematode control decisions will influence the profitability of producing soybeans. The numbers and species of nematodes should be determined in order to choose the proper control tactics. Nematode control can be achieved with one or more of the following tactics: nematicides, resistant cultivars, and cultural practices (primarily crop rotation) (Schmitt and Noel, 1984). Nematicides should be used in situations where they will provide the greatest profit and not simply for risk aversion. For maximum productivity, these tactics should be integrated to give long term control and yield good economic returns.

NOTES

D. P. Schmitt, Department of Plant Pathology, North Carolina State University, Raleigh, NC 27607.

REFERENCES

Acosta, N. and R. B. Malek. 1981. Symptomatology and histopathology of soybean roots infected by *Praytlenchus scribneri* and *P. alleni*. J. Nematol. 13:6-12.

Barker, K. R. 1982. Influence of soil moisture, cultivar, and population density of *Meloidogyne incognita* on soybean yield in microplots. (Abstr.) J. Nematol. 14:429.

Ferris, V. R. and R. L. Bernard. 1962. Injury to soybeans caused by *Pratylenchus alleni*. Plant Dis. Rep. 46:181-184.

Holdeman, Q. L. and T. W. Graham. 1953. The effect of different plant species on the population trends of the sting nematode. Plant Dis. Rep. 37:497-500.

Kinloch, R. A. 1974. Nematode and crop response to short-term rotations of corn and soybean. Proc. Soil and Crop Sci. Soc. Florida 33:86-88.

Kinloch, R. A. 1982. The relationship between soil population of *Meloidogyne incognita* and yield reduction of soybean in the coastal plain of Florida. J. Nematol. 14:162-167.

Kinloch, R. A. and K. Hinson. 1974. Comparative resistance of soybeans to *Meloidogyne javanica*. Nematropica 4:17-18.

Lewis, S. A., F. H. Smith and W. M. Powell. 1976. Host-parasite relationships of *Hoplolaimus columbus* on cotton and soybean. J. Nematol. 8:141-145.

Lindsey, D. W. and E. J. Cairns. 1971. Pathogenicity of the lesion nematode, *Pratylenchus brachyurus*, on six soybean cultivars. J. Nematol. 3:220-226.

Lordello, L. G. E. 1956. Nematoides que parasitam a soja na regaio de Bauru. Bragantia 15(6):55-64.

Miller, L. I. 1972. The influence of soil texture on the survival of *Belonolaimus longicaudatus*. Phytopathology 62:670-671.

Nyczepir, A. P. and S. A. Lewis. 1979. Relative tolerance of selected soybean cultivars to *Hoplolaimus columbus* and possible effects of temperature. J. Nematol. 11:27-31.

Rebois, R. V. 1971. The effect of *Rotylenchulus reniformis* inoculum levels on yield, nitrogen, potassium, phosphorus and amino acids of seeds of resistant and susceptible soybean (*Glycine max*). (Abstr.) J. Nematol. 3:326-327.

Rebois, R. V., J. M. Epps and E. E. Hartwig. 1970. Correlation of resistance in soybeans to *Heterodera glycines* and *Rotylenchulus reniformis*. Phytopathology 60:695-700.

Rebois, R. V., P. A. Madden and B. J. Eldridge. 1975. Some ultrastructural changes induced in resistant and susceptible soybean roots following infection by *Rotylenchulus reniformis*. J. Nematol. 7:122-130.

Robbins, R. T. and K. R. Barker. 1973. Comparisons of host range and reproduction among populations of *Belonolaimus longicaudatus* from North Carolina and Georgia. Plant Dis. Rep. 57:750-754.

Rodriguez-Kabana, R. and S. C. Williams. 1981. Assessment of soybean yields caused by *Meloidogyne arenaria*. Nematropica 11:105-113.

Schmitt, D. P. and K. R. Barker. 1981. Damage and reproductive potentials of *Pratylenchus brachyurus* and *P. penetrans* on soybean. J. Nematol. 13:327-332.

Schmitt, D. P. and G. R. Noel. 1984. Nematode parasites of soybeans. (In press) In W. R. Nickle (ed.) Plant and Insect Parasitic Nematodes, Academic Press, NY.

Standifer, M. S. 1959. The pathologic histology of bean roots injured by sting nematodes. Plant Dis. Rep. 43:983-986.

Weber, D. E., R. A. Reinert and K. R. Barker. 1979. Ozone and sulfur dioxide effects on reproduction and host-parasite relationships of selected plant-parasitic nematodes. Phytopathology 69:624-628.

Zirakparvar, M. E. 1982. Susceptibility of soybean cultivars and lines to *Pratylenchus hexincisus*. J. Nematol. 14:217-220.

Fungus Diseases

SOYBEAN ANTHRACNOSE

P. R. Hepperly

Hemmi (1920) and Lehman and Wolf (1926) were the first researchers to describe soybean anthracnose in Asia and the New World. The causal fungus was named *Colletotrichum glycines*, which later was found synonymous to *C. dematium*. Lehman and Wolf reported a sexual stage, but the relationship between asexual and sexual states was never proven. There has never been a confirmation of a sexual state for *C. dematium*, the major cause of soybean anthracnose.

Veinal leaf necrosis in soybeans can be caused by *C. dematium* or *Glomerella cingulata*. Unlike *C. dematium*, *G. cingulata* has never been found seriously damaging to soybean stems and seeds. Both *G. cingulata* and *C. dematium* infect a wide range of legumes and nonlegumes. One hundred synonyms are known for *C. dematium* and 600 for *G. cingulata* (Von Arx, 1970).

Athow (1973) and Tiffany (1951) considered soybean anthracnose a minor disease. Outside the temperate soybean zone, anthracnose causes important soybean losses (Backman et al., 1979; Hepperly et al., 1983; Sinclair, 1982). Soybean production has increased under tropical and sub-tropical climates. Prolonged periods of high rainfall, humidity, and temperature favor severe anthracnose. Calcium deficiency, which is common in leached, acid soils of the humid tropics and sub-tropics, predisposes soybeans to severe anthracnose (Muchovej et al., 1980). Besides reducing soybean yield (Backman et al., 1982; Roy and Miller, 1982; Whitney, 1982), *C. dematium* reduces soybean stand (Ellis et al., 1979; Hepperly et al., 1983; Nicholson and Sinclair, 1973). Inadequate soybean stand is a major constraint to soybean production in the tropics.

INFECTION

C. dematium depends on a film of water to penetrate soybeans. With free water, conidia of *C. dematium* germinate within 4 to 16 h at 20 to 30 C. Mycelial growth is sustained when relative humidities exceed 83%. Under desert and near desert conditions, anthracnose is restricted, based on its dependence on free water and high humidity (Neergaard, 1977).

The apical cell of the extended germ tube becomes swollen, thickened and darkened. This cell or appressorium possesses a sticky sheath that adheres to the plant cuticle and generates forces to rupture the plant surface. A thin hypha or infection thread breaches the cuticle and grows along the epidermal cells. In latent infections, the fungus becomes dormant after breaching the cuticle. In early seedlings and senescing plants, growth is uninterrupted and internal tissues are invaded. Intercellular portions of the thin walled cortical parenchyma are most easily attacked. Collapse of cortical tissues leads to characteristic lesions of anthracnose.

CULTURE

C. dematium, unlike some dry-spored conidial fungi, is not easily isolated in pure culture by transferring conidia. The slimy conidial matrix harbors sizable populations of bacteria. To reduce bacterial contaminants, the fungus should be isolated from newly formed local lesions that are not already overrun by secondary invaders or from latent infections. Best results are obtained by thoroughly rinsing plant material for 0.5 h in running water to remove soil, soaking materials in 0.5% NaOCl or CaOCl$_2$ for 2 to 4 min, and rinsing tissues in sterile, distilled water prior to plating. In isolations, emphasis should be placed on the margins of newly forming lesions. Acid potato dextrose agar (APDA) is a good general media for isolations. The fungus grows well on most common laboratory media, including PDA, V-8 agar, malt agar, and oatmeal agar. Growth and sporulation can often be enhanced on oatmeal and oatmeal sucrose agars. *C. dematium* grows optimally at from 26 to 32 C. The optimum for *G. cingulata* may be slightly lower. Conidia production can be stimulated in anthracnose fungi by growing them on sterile plant parts, such as soybean stems, petioles, or seeds. Large quantities of inoculum for field tests are conveniently produced on sterilized sorghum seeds. Fungus colonized sorghum seeds are easily distributed in test plot, and the fungus within the seeds can more easily survive harsh conditions.

IDENTIFICATION

Both *G. cingulata* and *C. dematium* have *Colletotrichum* asexual stages. *Colletotrichum* is characterized by sterile spines, setae, which are found interspersed in a compact palisade of conidiophores. The size and development of acervuli and setae of the soybean anthracnose fungi differ and can be used in their identification (see Neergaard, 1977). *C. dematium* has falcate or sickle shaped conidia which are readily differentiated from the cylindrical shape of the asexual stage of *G. cingulata*. Based on their asexual stages, both fungi belong to the Melanconiales. *G. cingulata* commonly produces perithecia and is classified under the Polystigmataceae of the Pyrenometeous Ascomycetes. No sexual stages have been documented for *C. dematium*.

A high degree of intraspecific variation is found in both *C. dematium* and *G. cingulata*. Rodriguez and Sinclair (1982) characterized four strains of *C. dematium* from soybeans in Illinois. Characteristics of *G. cingulata* are unstable in normal laboratory culture, and sectoring is extreme. In *G. cingulata* variations in colony growth, matrix color, and acervuli morphology are often striking. Pathogenicity of isolates of these fungi also varies. Srinivasan (1952) found that isolates of *C. dematium* from chili peppers (*Capsicum annum* L.) were nonvirulent on the same plant but highly virulent on *Sesbania*. Furthermore, *Sesbania* isolates were avirulent on the same plant but highly virulent on chili peppers.

DISTRIBUTION

Soybean anthracnose fungi are widely distributed and have been reported from most areas where soybeans are cultivated. The prevalence and severity of these fungi, however, appears to be greatly increased in humid tropical and sub-tropical environments. In Malaysia, Nik (1980) found *C. dematium* more common than other seed-borne fungi that are commonly reported from soybean. In India, Nicholson and Sinclair (1973) found up to 20% anthracnose stain on seed maturing in the early monsoon season. Hepperly et al. (1983) found up to 20% anthracnose seed stain in soybeans maturing under peak rains in May, but virtually none in a second rainy season peak in October. In Alabama, anthracnose disease ratings have been correlated with the amount of rainfall during soybean reproductive stages and with yield losses (Backman et al., 1979; 1982). Athow (1973) from Indiana and Schmitthenner and Kmetz (1980) from Ohio have found little economic development of anthracnose based on their experience in Midwestern soybeans. Hepperly et al. (Table 1) found that, in 25 soybean lines from Maturity Groups II, III, and IV, *C. dematium* seed infection ranged from 0 to 39% under heavy rainfall conditions, which favored anthracnose in central Illinois in 1977. In rainfed cropping areas, the anthracnose fungi have a wide distribution. This widespread distribution is probably related to their success as latent parasites and saprophytes of diverse plants. Shifts toward hotter and more humid world climates, unconscious incorporation of susceptible germplasm in soybean breeding programs, and adaptations of the anthracnose fungi themselves to soybeans or soybean fungicides may greatly increase the impact of soybean anthracnose in future crops.

DISEASE CYCLE AND SYMPTOMS

Soybean plants may be infected at any stage, but symptoms are generally found only at early seed and seedling stages and during soybean reproduction and senescence.

Table 1. Seed germination and infection percentages in 24 soybean lines under humid field conditions in central Illinois in 1977 (Hepperly et al., 1979).

Soybean line	Germination	Colletotrichum	Phomopsis
		– % –	
Williams	95.0	0.0	11.3
Midwest	88.0	15.0	9.3
Merit	52.3	6.8	60.8
AKHarrow	47.5	0.5	81.8
Cloud	61.8	14.8	31.0
Funman	61.8	0.0	54.0
Granger	72.8	0.0	46.3
Magna	41.5	1.3	64.8
Seneca	71.3	1.0	40.3
Manchu 2204	87.5	1.3	32.8
Mansoy	98.0	0.0	12.0
Mukden	69.5	0.0	61.3
FC 04.007A	70.3	1.0	61.3
PI 54.859	53.3	8.3	25.0
PI 60.279	90.0	15.3	17.3
PI 86.146	88.5	0.0	31.8
PI 360.835	98.5	0.0	1.0
PI 181.549	95.5	0.0	5.0
PI 70.019	81.0	5.3	37.3
Virginia	92.3	0.3	13.0
Hawkeye	96.0	0.0	23.0
PI 91.115	31.0	4.5	79.0
FC 31.678	12.8	39.0	60.5
PI 68.680	39.8	11.8	56.8
Range	12.8-98.5	0.0-39.0	1.0-81.8

Primary inoculum may come from infected soybean debris, alternate host debris, or from infected seed. Schneider et al. (1974) and Rodriguez and Sinclair (1978) have shown that seed infections are concentrated in the inner endodermal and mesodermal layers of the soybean seed coat before seed germination. During seed germination, *C. dematium* penetrates the cotyledons and often spreads to epicotyl and hypocotyl tissues. Cotyledon cankers are typically reddish-brown, sunken lesions concentrated on the outer surface of the seed leaves. Rot of the epicotyl and hypocotyl is more likely to cause pre- and post-emergence damping-off of soybean seedlings. Soybean seedlings with cotyledon cankers which do not spread to critical seedling parts may survive and after 3 to 4 weeks symptoms usually disappear. These plants, however, often lag in their development compared to plants which show no seedling disease, and their production is poor. Symptoms of anthracnose are usually absent in the vegetative stages, although the fungus can be isolated from surface disinfected plant tissues if the test plants are exposed to splashing rains in the field. A secondary cycle of inoculum production starts as lower leaves die from old age or shading. *C. dematium* acervuli are readily visible and abundant on the petioles of dead and dying lower leaves. Symptoms of anthracnose on leaves are varied and may be due to attack of the leaves, or indirectly result from stem damage. Necrotic local lesions are centralized on the underside of the leaflet veins. These types of veinal lesions are found for both *C. dematium* and *G. cingulata*. In response to severe aerial attacks of soybean anthracnose, leaves may cup downward and fall prematurely. Local lesions on stems, petioles, and pods are nearly rectangular in shape with a dark brown color. Stem tip die-back is associated with premature death of infected plants. In early stages of pod development, infection results in early pod abortion. Later infections cause local lesions on the outer surface of the pod and can lead to seed infections. Soybean seed are generally infected by *C. dematium* at and after physiological maturity. Infected seeds show a diffuse brown stain easily distinguished from the distinct hilum streaking sometimes found in seeds from soybean mosaic virus infected plants.

ASSAY

Direct assays of latent populations of pathogenic fungi may be useful in forecasting disease outbreaks. Cerdauskas et al. (1983) used desiccant-type herbicides to stimulate fruiting of *C. dematium* and *Phomopsis-Diaporthe* on soybean stems. Soybean disease forecasting has mostly been based on rainfall (Backman et al., 1979; 1982). By accounting for initial differences in inoculum, soybean anthracnose forecasting may be improved and unnecessary applications of fungicides may be reduced.

Assessment of soybean stem, pod, and leaf diseases have been standardized by using a scale based on the percentage of surface area with symptoms and/or fruiting of soybean pathogens (Walla, 1979). The scale varies from 0 to 9 with 0 = to no apparent disease and 9 = 90% disease coverage. Roy and Miller (1982), Stuckey and McCarter-Zorner (1982) and Whitney (1982) have shown the efficacy of this scale for evaluating fungicide sprays for control of soybean anthracnose.

Seed assays were used by Nicholson and Sinclair (1973) and Hepperly et al. (1983) to determine effects of seasonal tropical environments on soybean anthracnose in India and Puerto Rico. Hepperly et al. (1983) found that cellulose pad assays were more sensitive than agar plate assays in determining seed populations of *C. dematium*. Use of agar plate assays favored detection of *Phomopsis-Diaporthe* at the expense of *C. dematium*. Selective assays of soybean anthracnose will help improve studies of soybean disease epidemiology, control, and pathogen biology.

WEEDS

Dhingra and Silva (1978) first reported the increase of soybean anthracnose in weedy soybean fields. In Puerto Rico I have made a series of investigations to determine 1) whether weeds stimulate seedborne anthracnose and 2) how seasonal and cultural changes affect the crop-weed-disease interaction. Soybean anthracnose is not increased by early season weed competition. Nevertheless, late season and full-season weed competition greatly increases soybean anthracnose under humid conditions. Row spacing significantly affected the crop-weed-disease interaction with greater weeds being found at 60 cm and greater anthracnose found at 30 and 60 cm compared to 45 cm. Under dry season cultivation losses from weeds were much reduced and stimulation of disease was not noted (Tables 2 and 3).

Besides changing the physiological state of the soybean and influencing the microclimate for the pathogen, weeds can also serve as alternate hosts (Hepperly et al., 1980). Interactions of anthracnose with weeds deserves further study.

CONTROL

Anthracnose control can be directed at the host plant, the pathogen, or the environment. Only when several controls are exercised at the same time can we have much assurance of effectiveness of control and its stability over time. Most disease controls for soybean anthracnose are aimed at direct reduction of the pathogen population. These include: use of clean seed, treatment of seed with fungicides, crop rotation, residue destruction and the application of foliar fungicides. Use of weed control, fertilization, and selection of resistant cultivars of soybean are disease controls mostly focused

Table 2. Effects of weed competition and row spacing on the incidence of *Colletotrichum dematium* seed infection in Williams soybeans grown under humid season conditions in Puerto Rico in 1979.

Weed competition level[a]	*Colletotrichum dematium* Seed infection
	%
Full season	15.5
5 weeks early	7.6
4 weeks early	6.3
3 weeks early	7.1
2 weeks early	9.3
LSD 0.05[c]	6.7
Row spacing[b] cm	
30	10.4
45	4.1
60	13.0
LSD 0.05[c]	5.3

[a]Means based on 9 replicates of 50 seeds each assayed on potato dextrose agar (PDA).
[b]Means based on 15 replicates of 50 seeds each assayed on PDA.
[c]There were no significant interactions between Weed Competition Levels and Row Spacings. Weeds were increased at 60 cm row spacings. No statistical differences in any parameters were found in the same experiment repeated in the dry season.

Table 3. Influence of precipitation level and weed presence on the development of anthracnose in Cobb soybean grown under dry to wet and wet to dry growing seasons in Puerto Rico in 1980.

Growing season[a]	Rainfall during podding	Seedborne C. dematium	Visible seed damage	Germination PDA[b]	Germination Cellulose pads[c]
	cm		— % —		
March to July					
Dry to Wet	43.4	21.1	31.2	22.1	29.8
September to					
December (Wet to Dry)	19.3	1.1	9.7	45.3	88.5

	Seed presence	Days after inoculation[d] 27	34	41	48
		— % —			
Dry to Wet	+	13.2	17.5	47.7	60.2
	-	14.7	23.6	52.1	59.2
Wet to Dry[e]	+	28.5	34.3	38.0	48.2
	-	4.1	6.2	9.5	12.4

[a]Means based on 24 seedlots of 100 seeds each.
[b]Seeds were surface treated with 0.5% NaOCl and plated on sterile PDA at 27 C for 7 days.
[c]Seeds were plotted on Kimpac cellulose pads at 27 C for 7 days.
[d]Means are based on 4 replicates of 60 randomly selected plants following inoculation with C. dematium at V_5.
[e]In the wet to dry season stem anthracnose disease assessments were significantly different between with weeds and without weeds.

on the host itself. Finally, environment can be modified by planning and implementing shifts in planting dates. This can be particularly important in tropical areas characterized by distinctive dry and wet seasons. Planting in the dry season for maturation under the rainy season should be avoided.

FUNGICIDES

In southern U.S. fungicidal control of soybean anthracnose has resulted in increases in yield and seed quality (Backman et al., 1979; Roy and Buchring, 1981; Roy and Miller, 1982; Stuckey and McCarter-Zorner, 1982; Whitney, 1981; 1982). C. dematium isolates are more tolerant to benomyl than many other soybean seed pathogens (Rodriguez and Sinclair, 1982). Two to three µg/g of benzimidazole fungicides reduced C. dematium growth by half (Bong et al., 1983). Phomopsis-Diaporthe and Cercospora spp. are equally inhibited by benzimidazoles at concentrations less than 1 µg/g. Benlate significantly reduced Phomopsis-Diaporthe, but not C. dematium, in seeds of Tracy and Centennial soybeans (Roy and Buchring, 1981). In some instances, organotins appear to give greater control of C. dematium than benomyl. In Dare soybeans in Texas, Whitney (1981) found 15% stem and pod anthracnose with an organotin (Duter) and 25% with benomyl (Benlate). In 1982, he found 15% pod and stem anthracnose with the organotin, 35% for chlorothalonil and 68% in nontreated plots. For pod and stem blight and for *Cercospora* diseases on soybeans, benomyl has appeared to be superior to organotins. Soybean funigicides increasingly should be integrated with other control methods and with changes in cropping practices.

RESISTANCE

The relative susceptibilities of soybean cultivars and plant introduction to soybean anthracnose is not known. In Illinois, Hepperly et al. (1979) assayed 25 lines, finding seed anthracnose ranging from 0 to 39% depending on the line tested (Table 1). The cultivar Midwest appeared resistant to *Phomopsis*, but susceptible to anthracnose. PI 360835 appeared resistant to both *Phomopsis-Diaporthe* and *C. dematium* seed molds. Breeding programs may unconsciously use lines that are extremely susceptible to soybean anthracnose. This occurs because the disease is sporadic, and when it occurs it often goes undiagnosed. Buffalo, which has resistance to soybean mosaic, was extremely susceptible to anthracnose in Puerto Rico (Glen Bowers, pers. commun.). In breeding for virus resistance, anthracnose susceptibility was unconsciously increased in the Buffalo progeny.

Bong et al. (1983) and Backman et al. (1982) noted a range of susceptibilities and losses with cultivars tested in Malaysia and Alabama, respectively. In future work in soybean breeding programs, controlled inoculations that produce consistent epidemics should be emphasized, and should be helpful in identifying both sources of resistance and resistant progeny.

PROGNOSIS

Soybean anthracnose is a disease of increasing importance, partly because of increased cropping in humid subtropical and tropical zones. Even so, in temperate soybean areas anthracnose can be expected to increase if rotations continue to shorten. Stresses, such as weed competition during pod filling and calcium deficiency, predispose soybeans to severe anthracnose. Foliar fungicides can increase soybean yield and quality where diseases are limiting. For this reason, the role of foliar fungicides will likely increase in the near future, especially for production of high quality seeds under hot and humid environments. Increased monitoring and prediction of soybean anthracnose will be important in preventing unnecessary use of fungicides in soybeans. To accomplish this, refined detection techniques and their evaluation are needed. Long term control of soybean anthracnose will depend on increasing the resistance of commercial cultivars. We, therefore, must identify sources of disease resistance and determine their inheritance and stability. Finally, we must continue basic studies on the biology of the causal fungi, if we expect to control them in the long run.

NOTES

Paul R. Hepperly, Associate Professor of Plant Pathology, Department of Crop Protection, College of Agricultural Sciences, University of Puerto Rico, Mayaguez, Puerto Rico.

REFERENCES

Athow, K. L. 1973. Fungal Diseases. p. 459-489. *In* B. E. Caldwell (ed.) Soybeans: Improvement, Production and Uses, Agronomy 16. Am. Soc. Agron., Madison.

Backman, P. A., R. Rodríguez-Kabana, J. M. Hammond and D. L. Thurlow. 1979. Cultivar, environment, and fungicide effects on foliar disease losses in soybeans. Phytopathology 60:562-564.

Backman, P. A., J. C. Williams and M. A. Crawford. 1982. Yield losses in soybeans from anthracnose caused by *Colletotrichum truncatum*. Plant Dis. 66:1032-1034.

Bony, C. F. J., W. Z. Nik and T. K. Lim. 1983. Studies of *Colletotrichum dematium* f. sp. truncatum on soybean. Pertanika 6:28-33.

Cerdauskas, R. F., O. D. Dhingra and J. B. Sinclair. 1983. Effect of three desiccant-type herbicides on fruiting structures of *Colletotrichum truncatum* and *Phomopsis* spp. on soybean stems. Plant Dis. 67:620-622.

Dhingra, O. D. and J. F. da Silva. 1978. Effect of weed control on internally seedborne fungi in soybean seeds. Plant Dis Rep. 62:513-516.

Ellis, M. A., P. E. Powell and J. B. Sinclair. 1979. Internally seedborne fungi of soybean in Puerto Rico and their effect on seed germination and field emergence. Trop. Agric. 56:171-174.

Hemmi, T. 1920. Beitrage zur kenntnis der morphologie und physiologie der Japonischen Gloesporien. J. Coll. Agric. Hokkaido Imperial Univ. 9:1-159.

Hepperly, P. R., G. R. Bowers Jr., J. B. Sinclair and R. M. Goodman. 1979. Predisposition to seed infection by *Phomopsis sojae* in soybean plants infected by soybean mosaic virus. Phytopathology 60:846-848.

Hepperly, P. R., B. L. Kirkpatrick and J. B. Sinclair. 1980. *Abutilon theophrasti*: wild host for three fungal parasites of soybean. Phytopathology 70:307-310.

Hepperly, P. R., J. S. Mignucci, J. B. Sinclair and J. B. Mendoza. 1983. Soybean anthracnose and its assay in Puerto Rico. Seed Sci. Technol. 11:371-380.

Lehman, S. G. and F. A. Wolf. 1926. Soybean anthracnose. J. Agric. Res. 33:381-390.

Muchovej, J. J., R. M. C Muchovej, O. D. Dhingra and L. A. Maffia. 1980. Suppression of anthracnose of soybean by calcium. Plant Dis. 64:1088-1089.

Neergaard, Paul. 1977. Seed Pathology. MacMillan Press, London.

Nicholson, J. F. and J. B. Sinclair. 1973. Effect of planting date, storage conditions, and seedborne fungi on soybean seed quality. Plant Dis. Rep. 57:770-774.

Nik, W. Z. 1980. Seedborne fungi of soybean (*Glycine max* (L.) Merr.) and their control. Pertanika 3:125-132.

Rodríguez-Marcano, A. and J. B. Sinclair. 1978. Fruiting structures of *Colletotrichum dematium* and *Phomopsis sojae* formed in soybean seeds. Plant Dis. Rep. 62:873-876.

Rodríguez-Marcano, A. and J. B. Sinclair. 1982. Variation among isolates of *Colletotrichum dematium* var. truncata from soybeans and three *Colletotrichum* spp. to benomyl. J. Agric. Univ. Puerto Rico 66:35-43.

Roy, K. W. and N. W. Buchring. 1981. Effect of benlate on disease of soybeans grown in potassium deficient and fertile soils, 1979. Fung. Nemat. Test 36:99.

Roy, K. W. and W. A. Miller. 1982. Effects of fungicides on diseases and yield of soybean, 1980. Fung. Nemat. Test 37:106.

Schmitthenner, A. F. and K. T. Kmetz. 1980. Role of *Phomopsis* sp. in the soybean seed rot problem. p. 355-370. In F. T. Corbin (ed.) World Soybean Research Conference II: Proceedings. Westview Press, Boulder.

Schneider, R. W., O. D. Dhingra, J. F. Nicholson and J. B. Sinclair. 1974. *Colletotrichum truncatum* borne within the seed coat of soybean. Phytopathology 64:154-155.

Sinclair, J. B. 1982. Soybean Disease Compendium. Am. Phytopathol. Soc., St. Paul.

Srinivasan, K. V. 1952. Seedling blight of *Sesbania grandiflora*. Curr. Sci. 21:318.

Stuckey, R. E. and N. J. McCarter-Zorner. 1982. Effect of foliar fungicides on disease control and yield of Williams soybeans at Princeton, Kentucky, 1981. Fung. Nemat. Test 37:107.

Tiffany, L. H. 1951. Delayed sporulation of *Colletotrichum* on soybean. Phytopathology 41:975-985.

Von Arx, J. A. 1970. A Revision of the Fungi Classified as Gloeosporium. Lehre: Verlag J. Cramer, Berlin.

Walla, W. J. 1979. Soybean Disease Atlas. Southern Soybean Workers, Texas A&M University, College Station.

Whitney, N. G. 1981. Chemical control of soybean diseases, 1980. Fung. Nemat. Tests 36:102.

Whitney, N. G. 1982. Evaluation of fungicides for soybean disease control, 1981. Fung. Nemat. Test 37:107-108.

EPIDEMIOLOGY OF SOYBEAN DOWNY MILDEW

S. M. Lim

The epidemiology of plant disease deals with dynamic processes of host-pathogen interactions, which determine the prevalence and severity of the disease. Epidemic processes for most plant foliar diseases follow a series of steps: arrival of pathogens on plant surfaces, initial infection, incubation period, latent period, sporulation, dissemination of secondary inoculum, and infectious period. These complex biological processes are influenced by the environment. Man also often interferes with these processes by altering the host and pathogen populations and enviornment. Slowing or halting any of the epidemic processes can delay the development of the epidemic, so that serious disease loss does not occur.

Downy mildew of soybean, caused by *Peronospora manshurica* (Naoum.) Syd. ex Gaum, has been found in many soybean growing areas throughout the world (Dunleavy, 1981). Little is known about the epidemiology of downy mildew in relation to the occurrence of epidemics and the effects on seed quality and yield. Estimating cumulative effects of this disease on soybean yield during the entire growing season is difficult because downy mildew occurs sporadically in the field. Experiments that estimate yield losses caused by this disease have not been reported in the literature. Lack of information on downy mildew epidemiology probably exists also because *P. manshurica* is an obligate pathogen that cannot complete its life cycle apart from soybean plants.

PATHOGEN

In 1972, Miura first described the causal organism of soybean downy mildew as *Peronospora trifoliorum* de Bary var. *manshurica* (Naoum.). The occurrence of soybean downy mildew in the United States was first reported by Haskell and Wood (1923). Lehman and Wolf (1924) proposed that the causal organism was *P. sojae* Lehm. and Wolf. Gaumann (1923) placed *P. trifoliorum* var. *manshurica* into the rank of species and it became *P. manshurica* Syd. ex Gaum. In 1926, Wolf and Lehman verified that *P. sojae* and *P. manshurica* were the same causal organism of soybean downy mildew (Syn. *P. sojae* Lehman and Wolf).

The fungus overwinters either as oospores in soybean plant residue or encrusted on seeds (Johnson and Lefebvre, 1942). Oospores are globose with the exterior cell walls ranging from smooth to reticulate, and they are light brown. The size of oospores ranges from 20 to 36 μm in diameter (Lehman and Wolf, 1924). Reticulations, which are normally difficult to observe, can be easily observed by staining the sporangium with cotton blue (Dunleavy and Snyder, 1962). Encrusted seeds are a primary source of inoculum as the seed encrustation is composed of masses of oospores. Systemically infected seedlings result from encrusted seeds (Jones and Torrie, 1946; Koch and Hildebrand, 1946). Infection of seedling hypocotyls can also result from soil-borne oospores (Hildebrand and Koch, 1951). In either case, the fungus grows within the plant by establishing mycelium with haustoria. Sporangia are formed on sporangiophores on the underside of the leaf (Lehman and Wolf, 1924). Sporangia are ovoid to subglobose in shape, hyaline in color, and 20 x 24 μm in average size. Sporangia are the secondary inoculum which disseminate readily for local infections throughout the growing season.

The existence of physiological races in *P. manshurica* was first reported by Geeseman (1950a). He identified Races 1, 2, and 3 based on reactions of three differential soybean cultivars. Lehman (1953b; 1958) described Races 3A, 4, 5, 5A, and 6, and proposed a set of 10 differential cultivars for identifying these races. Grabe and Dunleavy (1959) described Races 7 and 8 using a set of 14 differential cultivars. Dunleavy (1971; 1977) made extensive surveys in the United States and identified Races 9 through 32 on the basis of reactions of 11 differential cultivars. Recently, Lim et al. (1984) reported the occurrence of a new race of *P. manshurica* in the Illinois soybean disease-monitoring plots. The new race was designated as Race 33 and was virulent on the soybean cultivar Union, which carries the gene *Rpm* for resistance to all races previously described.

The occurrence of *P. manshurica* over a wide geographic range throughout the world, and the diversity of the races in the United States, suggests that other physiological races may exist that have gone undetected and may appear in the future. Little is known, however, about the genetic potential of this fungus to produce races or the ability of different races to survive and increase in nature. Additionally, man is relatively unable to control downy mildew by controlling *P. manshurica* variation. Currently, the 33 described races of *P. manshurica* are not readily obtainable. Since *P. manshurica* is an obligate pathogen and preserving cultures can only be accomplished by inoculating susceptible soybean plants every 8 to 10 days, maintenance of *P. manshurica* races is a tedious task.

DISEASE REACTION

Reaction of soybeans to *P. manshurica* is classified by the phenotypic expression of soybean plants. Phenotypic expression is influenced by the interactions of soybean and pathogen genotypes and by the environment that may or may not be favorable for downy mildew development. Plant age, stage of development of the infected leaf (i.e., primary or trifoliolate), and the inoculum concentration significantly influence lesion development. Thus, a wide range of reactions can occur on a single soybean genotype (Wyllie and Williams, 1965). Soybean reaction to *P. manshurica* is both qualitative and quantitative. Qualitative reactions are characterized by symptom development: no symptom versus symptom or lesion types (i.e., small flecks versus chlorotic lesions). Quantitative reactions are characterized by lesion size, lesion number, or percentage of leaf area infected. In most rating systems, both infection type and lesion size are included.

Geeseman (1950a) used the severity rating system ranging from 0 to 4 to classify different types of soybean reactions to races of *P. manshurica*: 0 = no symptoms, 1 = 1-25%, 2 = 26-50%, 3 = 51-75%, and 4 = 76-100% of leaf area infected. Lehman (1958) used infection types to describe reactions to races: D = chlorotic lesions without necrotic center (1-2 mm in diam), F = chlorotic lesions without reddish or brownish flecks (2-4 mm in diam), H = chlorotic lesions, some having necrotic center (1-3 mm in diam), and N = yellowish lesions, some having small reddish or brownish flecks (2-4 mm in diam). Grabe and Dunleavy (1959) used a severity rating scale in which both infection type and lesion size were considered: 1 = immune, 2 = small flecks (0.5 mm in diam), 3 = small discrete chlorotic areas, irregular in shape (2 mm in diam), 4 = chlorotic areas (4 mm in diam), sometimes merging, often delimited by veins, presenting an angular appearance, and 5 = large confluent chlorotic area uniformly covering much of the leaf. Dunleavy (1971; 1977) used this rating system extensively in differentiating many races. He considered that ratings 1 and 2 were resistant and 3 through 5 were susceptible reactions. Bernard and Cremeens (1971) classified segregating progenies of soybean crosses into three discrete categories of resistant (R), intermediate (I), and susceptible (S) in an inheritance study of resistance to *P. manshurica*: R = no symptom, I = small chlorotic lesions (2 mm in diam), and S = large chlorotic area. Lim (1978) used a modified Horsfall and Barratt disease rating scale to quantify downy mildew severity in studying downy mildew gradients in a soybean field.

Use of reliable methods to classify soybean reactions to *P. manshurica* are important for identifying sources of resistance, differentiating physiological races and quantifying disease severity. In particular, a reliable method in quantifying downy mildew severity is needed to determine its effects on seed quality or yield.

SOURCES OF RESISTANCE

Identification of diverse sources and types of resistance to a particular disease is fundamental to a breeding program for disease resistance. The availability in soybean of a broad spectrum of resistance genes to *P. manshurica* provides alternatives to soybean breeders in developing resistant cultivars. Various sources of resistance to *P. manshurica* in soybeans have been identified based on immune reactions or small flecks on infected leaves. Dunleavy (1970) reported that three soybean cultivars, Mendota, Kanrich, and Kanro, were resistant to 14 races studied. Dunleavy and Hartwig (1970) also reported three additional sources which were resistant to 9 races from soybean plant introductions, PI 171443 and PI 201422, and from the cultivar Pine Dell Perfection. Prior to the occurrence of Race 33, these six sources were resistant to all known races and were immune to natural infection in Illinois fields. Cultivars Kanrich and Pine Dell Perfection were susceptible when tested against Race 33 (Lim et al., 1984). The other four soybeans have not been tested against Race 33. Lim et al. (1984) reported Fayette, Tracy, and PI 88788 were resistant to Races 2 and 33.

INHERITANCE OF RESISTANCE

A breeding program for disease resistance can be most effective if information on the inheritance of resistance is known. Inheritance can be monogenic or polygenic. Also, various gene actions can be estimated through genetic studies, which would provide valuable information in selecting a proper breeding method to develop effective resistant cultivars.

Resistance to *P. manshurica* in soybeans is mostly monogenic. Information on the polygenic resistance is not found in the literature. Although various sources of resistance to *P. manshurica* in soybeans are known, the nature of the inheritance of resistance has not been dealt with thoroughly. Bernard and Cremeens (1971) reported that resistance in the cultivars Kanrich and Pine Dell Perfection was controlled by a single gene, *Rpm*. F_3 lines from three backcross populations, involving Kanrich as the source of downy mildew resistance, segregated for monogenic control, with resistant, segregating, and susceptible progeny occurring in a 1:2:1 ratio. Resistant progeny showed no symptoms, whereas susceptible progeny developed large chlorotic areas. Segregating progeny fell into three discrete classes. Frequencies of the three classes were consistent with a 1 resistant : 2 intermediate (heterozygotes) : 1 susceptible ratio. The gene *Rpm* was shown to be different from the three genes that conferred differential reactions to Races 1, 2, and 3 of the fungus (Geeseman, 1950b), since soybean cultivars carrying these genes were highly susceptible to the races used in identifying the gene *Rpm*. Several studies indicate that other genes conferring resistance to *P. manshurica* races are yet to be identified (Dunleavy, 1970; Dunleavy and Hartwig, 1970; Lim et al., 1984).

DOWNY MILDEW EPIDEMICS

Downy mildew of soybean develops through systemic and local infections (Hildebrand and Koch, 1951). Systemic infection often occurs on soybean seedlings grown from oospore-encrusted seeds. Systemically infected seedlings are usually stunted. Leaves are small and tend to be curled downward at the edges. Light-green mottled areas appear on the infected leaves (Hildebrand and Koch, 1951; Athow, 1973). Sporangia are produced 1 to 2 weeks after emergence, depending upon temperature and moisture (Dunleavy, 1981). Sporulation is favored by high humidity and temperatures of 20 to 24 C (McKenzie and Wyllie, 1971). Lehman (1953a) reported that systemic infection of seedlings was affected by temperature and the rapidity of germination of oospore-encrusted seeds. At soil temperatures of 13 C, 40% of the seedlings grown from encrusted seeds were systemically infected, but no infection occurred above 18 C. Dunleavy and Snyder (1962) observed up to 40% systemic infection among volunteer soybean plants in fields that had plants infected with *P. manshurica* in the previous year.

Sporangia serve as secondary inoculum for local infections and are thought to be the chief sources of infection in the epidemic development of downy mildew. Sporangia are spread by wind and are produced repeatedly throughout the growing season through the polycyclic process of infections on younger leaves of the upper canopy of soybean plants. Rapidity of completing an infection cycle (i.e., the period from germination of a sporangium to production of a sporulating colony) is greatly influenced by environment and plant age (Hildebrand and Wolf, 1951). Sporulation occurs 5 to 6 days after infection at 23 C when other conditions such as moisture and leaf age are favorable (Dunleavy, 1981). Sporulation usually occurs between 10 and 25 C. Sporulation does not occur above 30 C or below 10 C (McKenzie and Wyllie, 1971). Under favorable conditions, an infection cycle can be completed in 8 to 10 days (Dunleavy, 1981). Several studies indicate that the susceptibility of soybean tissue to infection by *P. manshurica* differs with age (Hildebrand and Koch, 1951; Wyllie and Williams, 1965). The size of lesions decreases but the number of lesions tends to increase on older leaves. McKenzie and Wyllie (1971) observed that lesions were fewer in number but larger in size on young than on old plants, and abundant sporulation occurred on large lesions of young leaves. Pod infection may occur and not be evident externally. The interior of pods and seed coats are often covered with whitish masses of mycelium and oospores (Sinclair and Shurtleff, 1975).

Analysis of downy mildew epidemic development is scarce in the literature. Downy mildew develops through polycyclic infections within one season. Cumulative measurements of disease severity can be made. With such data, it is possible to plot the progress of downy mildew and calculate appropriate infection rates, which provide useful parameters for comparing epidemics. Measurements of cumulative disease severity are also very useful in estimating yield losses due to downy mildew.

Lim (1978) studied the disease severity gradient of downy mildew from an initial infection focus in a field plot of a susceptible soybean cultivar, Williams. The slopes of disease gradients from the infection focus were flattened at 58 days after inoculation of the focus. At this time, downy mildew severity at all distances from the infection focus (3.0, 6.1, 12.2, 18.3, and 27.3 m) ranged from 20% to 25%. Comparison of disease progress curves at each distance showed that initial severity of downy mildew was lower and downy mildew increase was more rapid at the areas more distant from the infection focus. Sporangia were collected at least once a week during the period of downy mildew ratings (June 29 to September 8). More sporangia were collected during periods of no precipitation. Frequent rainfalls, low wind speeds (0-8 km/h), and aging of soybean leaves resulted in trapping fewer spores despite favorable temperature and relative humidity. Heavy rains can cause "rain scrubbing" or wash out spores suspended in the air. On rainy days, more sporangia may have been carried from infected upper leaves to lower leaves (vertical spread within plants) rather than between plants (horizontal spread among plants).

While downy mildew can spread very rapidly and occurs frequently, it is usually not extremely severe. Downy mildew severity on susceptible cultivars in the soybean disease-monitoring plots in Illinois did not exceed 20% in any year since 1977. Most frequently, severity in the monitoring plots was about 10%. Downy mildew severity is usually higher in fields where soybeans are planted after wheat for double cropping. Hildebrand and Koch (1951) concluded that downy mildew did not cause yield reduction and is not a significant limiting factor in production of soybeans in Ontario, Canada. Bernard and Cremeens (1971) also indicated that, under natural infection in Illinois, there was no difference in yield between soybean lines that were closely related but differed for resistance and susceptibility to *P. manshurica*.

Attempts were made to estimate yield losses caused by downy mildew in field plots that were inoculated during the growing seasons for 1978-1980. Yields from inoculated plots of the susceptible cultivar Williams and a closely related resistant line L75-6141, carrying the *Rpm* gene, were compared to yields from uninoculated plots. Downy mildew did not develop on the resistant line L75-6141 in either inoculated or uninoculated plots, whereas downy mildew did develop on the susceptible cultivar Williams in both inoculated and uninoculated plots. The highest mean severity of downy mildew on Williams plants in inoculated plots was 22% in 1979. Mean severity on Williams plants in uninoculated plots was 9%. There were no significant differences in yields between inoculated and uninoculated plots of Williams in any of the 3 years of the study.

Although downy mildew has not been found to reduce soybean yield significantly, the need for more information about the occurrence of epidemics and their effect on yield is evident. The prevalence of this disease over a wide geographic range and the diversity of the races indicate its potential significance in soybean production.

NOTES

S. M. Lim, Agricultural Research Service, U.S. Department of Agriculture, and Department of Plant Pathology, University of Illinois, Urbana, Illinois.

REFERENCES

Athow, K. L. 1973. Downy mildew. p. 459-489. *In* B. E. Caldwell (ed.) Soybeans: improvement, production, and uses, Agronomy 16. Am. Soc. Agron., Madison, Wisconsin.

Bernard, R. L. and C. R. Cremeens. 1971. A gene for general resistance to downy mildew of soybeans. J. Hered. 62:359-362.

Dunleavy, J. M. 1970. Sources of immunity and susceptibility to downy mildew of soybeans. Crop Sci. 10:507-509.

Dunleavy, J. M. 1971. Races of *Peronospora manshurica* in the United States. Am. J. Bot. 58:209-211.

Dunleavy, J. M. 1977. Nine new races of *Peronospora manshurica* found on soybeans in the Midwest. Plant Dis. Rep. 61:661-663.

Dunleavy, J. M. 1981. Downy mildew. p. 515-528. *In* D. M. Spencer (ed.) The downy mildews. Academic Press, New York.

Dunleavy, J. M. and E. E. Hartwig. 1970. Sources of immunity from and resistance to nine races of the soybean downy mildew fungus. Plant Dis. Rep. 54:901-902.

Dunleavy, J. M. and G. Snyder. 1962. Inheritance of germination of oospores of *Peronospora manshurica*. Proc. Iowa Acad. Sci. 69:118-121.

Gaumann, E. 1923. Beitrage zu einer monographie der gattung *Peronospora* Corda. Beitr. Kryptogamenflora Schweiz., Bd. 5, Heft 4.

Geeseman, G. E. 1950a. Physiologic races of *Peronospora manshurica* on soybeans. Agron. J. 42:257-258.

Geeseman, G. E. 1950b. Inheritance of resistance of soybeans to *Peronospora manshurica*. Agron. J. 42:608-613.

Grabe, D. F. and J. M. Dunleavy. 1959. Physiologic specialization in *Peronospora manshurica*. Phytopathology 49:791-793.

Haskell, R. J. and J. I. Wood. 1923. Diseases of cereal and forage crops in the United States in 1922. Plant Dis. Rep. Suppl. 27:164-265.

Hildebrand, A. A. and L. W. Koch. 1951. A study of systemic infection by downy mildew of soybean with special reference to symptomatology, economic significance, and control. Sci. Agric. 31:505-518.

Jones, F. R. and J. H. Torrie. 1946. Systemic infection of downy mildew in soybean and alfalfa. Phytopathology 36:1057-1059.

Johnson, H. W. and C. L. Lefebvre. 1942. Downy mildew on soybean seeds. Plant Dis. Rep. 26:49-50.

Koch, L. W. and A. A. Hildebrand. 1946. Soybean diseases in southwestern Ontario in 1946. Ann. Rep. Can. Plant Dis. Survey 26:27-28.

Lehman, S. G. 1953a. Systemic infection of soybean by *Peronospora manshurica* as affected by temperature. J. Elisha Mitchell Sci. Soc. 69:83.

Lehman, S. G. 1953b. Race 4 of the soybean downy mildew fungus. Phytopathology 48:83-86.

Lehman, S. G. 1958. Physiologic races of the downy mildew fungus on soybeans in North Carolina. Phytopathology 49:791-793.

Lehman, S. G. and F. A. Wolf. 1924. A new downy mildew on soybeans. J. Elisha Mitchell Sci. Soc. 39:164-169.

Lim, S. M. 1978. Disease severity gradient of soybean downy mildew from a small focus on infection. Phytopathology 68:1774-1778.

Lim, S. M., R. L. Bernard, C. D. Nickell and L. E. Gray. 1984. New physiological race of *Peronospora manshurica* virulent to the gene *Rpm* in soybeans. Plant Dis. 68:71-72.

McKenzie, J. R. and T. D. Wyllie. 1971. The effect of temperature and lesion size on the sporulation of *Peronospora manshurica*. Phytopath. Z. 71:321-326.

Miura, M. 1922. Diseases of important economic plants in Manchuria. (English Abstr.). Jap. J. Bot. 1:9.

Sinclair, J. B. and M. C. Shurtleff. 1975. Compendium of soybean diseases. Am. Phytopath. Soc., St. Paul, Minnesota.

Wolf, F. A. and S. G. Lehman. 1926. Diseases of soybeans which occur both in North Carolina and the Orient. J. Agric. Res. 33:391-396.

Wyllie, T. D. and L. F. Williams. 1965. The effects of temperature and leaf age on the development of lesions caused by *Peronospora mansburica* on soybeans. Phytopathology 55:166-170.

SOYBEAN RUST

A. T. Tschanz and S. Shanmugasundaram

Soybean rust caused by *Phakopsora pachyrhizi* H. Syd. & P. Syd. is one of the major disease problems limiting soybean yield in the tropics and subtropics of Asia. The presence of *P. pachyrhizi* has been reported in both the eastern and western hemispheres, including Africa (Figure 1). However, the disease has caused significant economic losses only in Taiwan, Thailand, Indonesia, Philippines and Vietnam, and in parts of China, Japan, Australia, Korea, and India (Bromfield and Yang, 1976; Sinclair, 1982). But, rust should be regarded as a threat to most soybean producing regions. A number of reviews discuss in detail its distribution, host range and occurrence (Bromfield, 1980; Kochman, 1977; Yang, 1977; Sinclair, 1982; Tschanz and Wang, 1980),

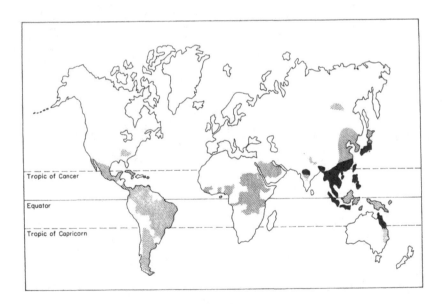

Figure 1. Areas in which soybean rust has been reported (dark shaded areas represent countries where soybean rust causes considerable economic loss).

as well as its epidemiology (Casey, 1979; Bromfield, 1980; Tschanz et al., 1984) and threat potential (Casey, 1979; Bromfield, 1980).

DISEASE SYMPTOMS

The disease begins as a small, water-soaked lesion on the abaxial surface of the leaves. Lesions gradually increase in size, and turn from gray to tan or reddish-brown. Young lesions may be easily mistaken for bacterial pustule (*Xanthomonas campestris* pv. *phaseoli* [Smith] Dye) until the uredia develop and discharge spores. Lesions eventually turn reddish-brown or dark brown, and assume a polygonal shape restricted by leaf veins. Although larger lesions have been observed, they usually attain a maximum size of less than 2 mm ^{2}. At 20 C chlorotic or tan-colored flecks will appear on infected plants about 5 days after inoculation. Uredia are usually differentiated in about 7 to 9 days and on a susceptible plant, spores are liberated by 9 to 10 days after inoculation (Bromfield, 1978).

Depending upon the soybean genotype and rust race, lesions develop into one of three different infection types RB, TAN, or O at 14 days after inoculation. The RB infection type is characterized by reddish-brown colored lesions, usually with 0-2 uredia per lesion. The TAN infection type is characterized by a tan-colored lesion usually with 2-5 uredia per lesion. The O infection type is characterized by the lack of macroscopically visible symptoms (Bromfield et al., 1980). Both RB and TAN type lesions may occur on a leaf of the same plant (Shanmugasundaram et al., 1980; Tschanz et al., 1980). The number of uredia per lesion may vary, but they are usually most abundant on the abaxial leaf surface. They are small and spherical with a central pore through which uredospores are exuded in columns or clumps. Although all above ground plant parts can be infected, symptoms are most abundant and characteristic on the leaves.

EPIDEMIOLOGY

The most intense rust epidemics have been observed in areas where the mean daily temperature is moderate, usually less than 28 C, and with precipitation and/or long periods of leaf wetness occurring frequently throughout the soybean growing season (Melching et al., 1979; Tschanz and Wang, 1980; Yeh et al., 1982; Tschanz et al., 1984). Rust development is inhibited by dry conditions and mean daily temperatures greater than 30 C and/or less than 15 C. Temperatures above 27 C for extended periods retard rust development even with adequate free moisture on leaf surface (Casey, 1979).

Field studies by Casey (1979) demonstrated that the development of a severe rust epidemic requires about 10 h/d of leaf wetness and mean daily temperatures of 18 to 26 C. In another field study, mean night temperatures consistently below 14 C prevented or greatly inhibited rust development, whereas mean night temperatures above 25 C had little effect on rust development when they occurred in conjunction with frequent, long leaf-wetness periods (Tschanz et al., 1984).

Free moisture on the plant surface is a prerequisite for uredospore germination and infection. Uredospore germination has been observed from 8 to 32 C (Kitani and Inoue, 1960), but the optimum has been variously reported as 19 to 23 C (Bromfield, 1980), 12 to 21 C (Casey, 1979) and 21 to 27 C (Kitani and Inoue, 1960). Although infection can occur with a 16 h dew period when temperatures are between 10 C and 26 C, temperatures above 28 C during the dew period appear to inhibit or prevent infection (Melching et al., 1979; Marchetti et al., 1976). At optimum temperatures, between 20 C and 25 C, infection of a susceptible host can occur with only 6 h of leaf surface wetness. Within this temperature range, maximal infection occurs within 10 to 12 h of leaf wetness. Increased periods of leaf wetness are necessary for infection when temperatures fall outside the optimum temperature range (Marchetti et al., 1976).

The ability of *P. pachyrhizi* to cause an epidemic in soybean is dependent on a number of factors. Two of these are temperature and leaf wetness duration, which together determine the suitability of infection periods. Although the frequency and duration of infection periods appear to be useful parameters in predicting soybean rust development, not enough data are yet available to accurately predict the effects of leaf wetness duration and temperature on the infection efficiency of soybean rust. Another factor affecting epidemic development, at least when soybeans are grown during the dry season, is the timing of the first rain and the amount of rainfall. Even when soybeans are grown with mist-type overhead or furrow irrigation, rainfall initiates the development of a rust epidemic. During low rainfall seasons the irrigation method can affect both when an epidemic begins and how rapidly it develops (Tschanz et al., 1984).

Recent studies have shown that rate of rust development is closely associated with development and maturation of the soybean plant. Rust develops later on plants grown under extended photoperiod, even though rust severities were similar in both natural and extended photoperiods at the beginning of the epidemic. Delay in rust development, therefore, results in a reduction in the rate of rust development. Therefore, the effect of soybean development and maturation on rust development has to be accounted for in epidemiological and host resistance studies (Tschanz and Tsai, 1982; Tschanz et al., 1984).

P. pachyrhizi telia and teliospore formation is induced by low temperatures and occurs when night temperatures are between 5 and 15 C and day temperatures are between 20 and 25 C (Bromfield, 1980; Kitani and Inoue, 1960; Hsu and Wu, 1968). Yeh et al. (1981) also induced formation of telia and teliospores on a number of other leguminous hosts. So far teliospore germination has not been observed, and its role in the epidemiology of the disease is unknown.

On the bases of epidemiological and climatological studies, it has been speculated that soybean rust is a potential threat in the U.S.A. (Casey, 1979; Bromfield, 1980). If viable uredospore inoculum were introduced early in the soybean growing season, it is likely that soybean rust could become established in some southern states. Host plants, temperature, humidity, and dew occurrence and duration would favor rust development in North Carolina, South Carolina, Georgia, Alabama, Mississippi, Tennessee, Louisiana, Florida, East Texas, and Kentucky (Casey, 1979; Bromfield, 1980). *P. pachyrhizi* has a wide host range including *Sesbaniz exaltata* and *Pueraria thunbergiana*, common weeds in the southern U.S.A.

HOST PLANT RESISTANCE AND PATHOGEN SPECIALIZATION

Many cultivars have been reported to have resistance to soybean rust (Tisselli et al., 1980). However, research at AVRDC indicates that most of the cultivars previously identified as resistant are susceptible in Taiwan. Many soybean cultivars reported as resistant are susceptible in Taiwan. Many soybean cultivars reported as resistant are suspected of carrying genes for specific resistance.

Considerable progress has been made on the inheritance of specific resistance. Singh and Thapliyal (1977) first suggested that their data from UPSS 3 x Clark 63 fitted a monogenic ratio with resistance being dominant, but that data from other crosses were inconsistent. McLean and Byth (1980) showed that the dominant resistant gene present in PI 200492, Tainung No. 3 and Tainung No. 4 was the same. However, since PI 200492 and Tainung No. 3 were susceptible to a new rust race but Tainung No. 4 was resistant, they suggested there was an additional gene or genes for resistance in Tainung 4. The resistance gene in PI 200492 was designated Rpp 1 (McLean and Byth, 1980). Another single gene for resistance was identified in PI 230970 and PI 230971 by Bromfield and Hartwig (1980).

Additional studies by Hartwig and Bromfield (1983) identified a dominant gene for resistance in PI 462312 (Ankur) and elucidated the interrelationship between the specific resistance genes in cultivars Ankur, PI 200492 and PI 230970. Each cultivar carried a different single dominant gene for resistance and these genes were located at different loci. The designations of the resistance genes in the three cultivars are Rpp1, Rpp2, Rpp3 in PI 200492, PI 230970 and PI 462312, respectively (Hartwig and Bromfield, 1983). Additional studies are underway to characterize and determine the interrelationship of other genes for specific resistance. Genes for resistance to rust are also suspected among the wild perennial *Glycine* species from Australia. Accessions in *G. canescens*, *G. clandestina*, *G. tabacina* and *G. tomentella* were ranked as highly resistant and are considered as potential sources of resistance genes for transfer to soybean once interspecific crosses are possible (Burdon and Marshall, 1981).

Symptoms on differential soybean cultivars and other hosts suggest the presence of a number of pathogenic races of *P. pachyrhizi*. Using four differentials (Wayne, PI 200492, Ankur and PI 230970), Bromfield et al. (1980) were able to identify four races among their worldwide collection of rust isolates. Yeh (1984) has identified three races in Taiwan using five differential cultivars (Ankur, Taita Kaohsiung No. 5, Tainung No. 4, PI 200492 and PI 230971), while a recent study at AVRDC has identified nine races of rust by utilizing 11 differential cultivars. A standardized set of soybean differentials would greatly facilitate the identification of rust races. We suggest that the following set be used initially: PIs 239871A and 239871B (*G. soja*), 230970, 230971, 200492, 459024, 459025, Taita Kaohsiung No. 5, Tainung No. 4, Wayne and Ankur (PI 462312).

All the differentials previously listed are known or suspected to have at least one gene for specific resistance. However, in the AVRDC study most races were compatible (TAN infection type) with 8 to 10 of the differentials (Srisuk and Tschanz, personal communication). In addition, two of Bromfield's (1981) races were compatible with three or more of their four differentials. Apparently, complex races of *P. pachyrhizi* are often compatible with a wide range of known or suspected specific resistance genes in different soybean cultivars. The presence of these complex races will limit the effectiveness of specific resistance, and there is, therefore, a need to identify and characterize other forms of resistance or tolerance which can minimize yield loss due to rust.

The presence in soybean of rate-reducing resistance to soybean rust was first identified and later confirmed at AVRDC (Tschanz et al., 1980; Tschanz and Wang, 1980; Tschanz et al., 1984). Rate-reducing resistance slows the rate of disease development and is usually effective against most races of the pathogen. A major factor influencing the identification and improvement of rate-reducing resistance is the influence of soybean development and maturation on rust development. However, a new evaluation methodology has been developed which partially corrects for the effect of variations in soybean development and maturity (Tschanz and Tsai, 1982; Tschanz et al., 1984).

Tolerance to soybean rust, as defined by the relative yielding ability of a soybean cultivar grown under a severe rust epidemic, has also been identified (Tschanz et al., 1984). Because of the difficulties associated with the identification and quantification of rate-reducing resistance, and because of the ineffectiveness of race specific resistance, tolerance is presently being utilized in AVRDC's soybean improvement program.

Levels of rust tolerance vary considerably among soybean cultivars with a range of 25 to 91% yield loss in one experiment (Tschanz et al., 1984). When stressed by a severe rust epidemic, newly developed rust tolerant lines at AVRDC yield 50 to 80% greater than the best commercial cultivar. Selecting for tolerance to rust appears to be more effective in reducing yield loss than selecting for resistance. To determine the tolerance level of soybean genotypes, all yield trials at AVRDC are conducted using a split-plot design, with main plots protected and unprotected by fungicides.

Further research is necessary before the range, complexity, distribution and genetics of either specific resistance genes or pathogen races can be fully understood. Research is also needed to determine the contribution of host plant resistance to the reduction in yield loss observed in tolerant cultivars and to elucidate the inheritance of both tolerance and resistance. Further characterization of rate-reducing resistance is necessary. The effect of environment and soybean development on rust development, resistance and tolerance also requires further study.

INTERNATIONAL COOPERATION

In 1976 and 1977 AVRDC, in cooperation with INTSOY and USDA, held three different conferences and workshops to discuss the importance of soybean rust. In order to exchange information on research results, techniques and approaches, an International Working Group on Soybean Rust (IWGSR) was organized in 1976 and has been publishing an annual newsletter since 1977. Through the proposed Asian Soybean Improvement Network, a series of soybean rust research activities will be initiated to consolidate the resources and information available in various countries. Such intensified research activities will, we hope, help to minimize the damage caused by soybean rust, both in countries where it is already a major problem and in countries where it is a potential threat.

NOTES

A. T. Tschanz, Associate Plant Pathologist and Head of Pathology; S. Shanmugasundaram, Plant Breeder and Legume Program Leader, the Asian Vegetable Research and Development Center (AVRDC); P.O. Box 42, Shanhua, Tainan 741, Taiwan, Republic of China.

REFERENCES

Bromfield, K. R. 1978. Review of research on soybean rust. p. 16-23. In N. G. Vakili (ed.) Proc. Workshop on soybean rust in the Western Hemisphere. ARS, USDA.

Bromfield, K. R. 1980. Soybean rust: some considerations relevant to threat analysis. Protection Ecol. 2:251-257.

Bromfield, K. R. 1981. Differential reaction of some soybean accessions to Phakopsora pachyrhizi. Soybean Rust Newsletter 4:2.

Bromfield, K. R., and E. E. Hartwig. 1980. REsistance to soybean rust and mode of inheritance. Crop Sci. 20:254-255.

Bromfield, K. R., J. S. Melching, and C. H. Kingsolver. 1980. Virulence and aggressiveness of Phakopsora pachyrhizi isolates causing soybean rust. Phytopathology 70:17-21.

Bromfield, K. R., and C. Y. Yang. 1976. Soybean rust: Summary of available knowledge. p. 161-164. In R. M. Goodman (ed.) Expanding the use of soybeans. INTSOY Series No. 10. University of Illinois, Urbana-Champaign.

Burdon, J. J., and D. R. Marshall. 1981. Evaluation of Australian native species of *Glycine* for resistance to soybean rust. Plant Dis. 65:44-45.

Casey, P. S. 1979. The epidemiology of soybean rust, *Phakopsora pachyrhizi* Syd. Ph.D. Thesis. University of Sydney, Australia.

Hartwig, E. E., and K. R. Bromfield. 1983. Relationships among three genes conferring specific resistance to rust in soybean. Crop Sci. 23:237-239.

Hsu, C. M., and L. C. Wu. 1968. Study on soybean rust. Sci. Agr. (China) 16:186-188 (in Chinese).

Kitani, K., and Y. Inoue. 1960. Studies on soybean rust and its control measures. I: Studies on the soybean rust. Shiloku Agric. Expt. Sta. (Japan) Bull. 5:319-342 (in Japanese).

Kochman, J. K. 1977. Soybean rust in Australia. p. 44-48. *In* R. E. Ford and J. B. Sinclair (eds.) Rust of soybean: The problem and research needs. INTSOY Series No. 12. University of Illinois, Urbana-Champaign.

Marchetti, M. A., J. S. Melching, and K. R. Bromfield. 1976. The effect of temperature and dew period on germination and infection by uredospores of *Phakopsora pachyrhizi*, the cause of soybean rust. Phytopathology 66:461-463.

McLean, R. J., and D. E. Byth. 1980. Inheritance of resistance to rust (*Phakopsora pachyrhizi*) in soybeans. Aust. J. Agric. Res. 31:951-956.

Melching, J. S., K. R. Bromfield and C. H. Kingsolver. 1979. Infection, colonization and uredospore production on Wayne soybean by four cultures of *Phakopsora pachyrhizi*, the cause of soybean rust. Phytopathology 69:1262-1265.

Shanmugasundaram, S., A. T. Tschanz, and K. R. Bromfield. 1980. RB and TAN infection types on G 8586 and G 8587 soybeans at AVRDC. Soybean Rust Newsletter 3:29.

Sinclair, J. B. 1982. Compendium of soybean diseases. The American Phytopathological Society, St. Paul, Minnesota.

Singh, B. B., and P. N. Thapliyal. 1977. Breeding for soybean rust in India. p. 62-65. *In* R. E. Ford and J. B. Sinclair (eds.) Rust of Soybean: The Problem and Research Needs. INTSOY Series No. 12. University of Illinois, Urbana-Champaign.

Tisselli, O., J. B. Sinclair and T. Hymowitz. 1980. Sources of resistance to selected fungal, bacterial, viral and nematode diseases of soybean. INTSOY Series No. 18, University of Illinois, Urbana-Champaign.

Tschanz, A. T. and B. Y. Tsai. 1982. Effect of maturity on soybean rust development. Soybean Rust Newsletter 5:38-41.

Tschanz, A. T. and T. C. Wang. 1980. Soybean rust development and apparent infection rates at five locations in Taiwan. Protection Ecology 2:247-250.

Tschanz, A. T., T. C. Wang, and L. F. Hu. 1980. Epidemic development of soybean rust and partial characterization of resistance to soybean rust. Soybean Rust Newsletter 3:35-39.

Tschanz, A. T., T. C. Wang, and B. Y. Tsai. 1984. Recent advances in soybean rust research at AVRDC. *In* S. Shanmugasundaram and E. W. Sulzberger (eds.) Soybeans in Tropical and Subtropical Cropping Systems. Asian Vegetable Research and Development Center, Shanhua, Taiwan, R.O.C. (in press).

Yang, C. Y. 1977. Soybean rust in Eastern Hemisphere. p. 22-33. *In* R. E. Ford and J. B. Sinclair (eds.) Rust of Soybean: The Problem and Research Needs. INTSOY Series No. 12. University of Illinois, Urbana-Champaign.

Yeh, C. C. 1984. Differential reaction of *Phakopsora pachyrhizi* on soybean in Taiwan. *In* S. Shanmugasundaram and E. W. Sulzberger (eds.) Soybeans in Tropical and Subtropical Cropping Systems. Asian Vegetable Research and Development Center, Shanhua, Taiwan, R.O.C. (in press).

Yeh, C. C., J. B. Sinclair, and A. T. Tschanz. 1982. *Phakopsora pachyrhizi*: Uredial development, uredospore production and factors affecting teliospore formation on soybeans. Aust. J. Agric. Res. 33:25-31.

Yeh, C. C., A. T. Tschanz, and J. B. Sinclair. 1981. Induced teliospore formation by *Phakopsora pachyrhizi* on soybeans and other hosts. Phytopathology 71:1111-1112.

POWDERY MILDEW, A SPORADIC BUT DAMAGING DISEASE OF SOYBEAN

Craig R. Grau

Powdery mildew of soybean was first reported in Germany in 1921 (Wahl, 1921) and since has been reported in Brazil, the People's Republic of China, India, Puerto Rico, South Africa and North America (Sinclair, 1982). Epiphytotics of powdery mildew occurred in many regions of North America in the 1970s. Many have speculated about what brought on the epiphytotics in the 1970s and why the incidence and severity of the disease has dropped in recent years. Dunleavy (1976) speculates that the shift from cultivars that eventually were proven to be resistant to the susceptible cultivar Harosoy-63, and other cultivars with Harosoy-63 parentage, was a primary factor in the sudden appearance of powdery mildew. However, these cultivars and their derivatives continue to be planted in the North Central states, and yet, a major epiphytotic has not occurred in recent years.

This disease has demonstrated the potential to cause significant yield loss and should be considered as a major disease of soybean. Although research on powdery mildew has increased, still there are many gaps in information about this disease. Epidemiology of powdery mildew is an aspect of this disease that is poorly understood. More research is needed to increase our understanding of this potentially important and interesting disease of soybean.

Although symptoms; i.e., changes in the host, are caused by the powdery mildew fungus, the disease is best recognized by the signs of the causal agent. White mycelium and conidia develop on leaves, but also are reported to occur on cotyledons, stems and pods (Dunleavy et al., 1966). Infection of stems and pods are not always reported and may be governed by cultivar, biotype of the pathogen and/or environmental factors. Chlorotic spots, veinal necrosis and reddening of leaf tissues are common symptoms associated with a specific form of resistance expressed by the host to the pathogen.

CAUSAL ORGANISM

Early reports on powdery mildew of soybean identified *Erysiphe polygoni* DC as the causal pathogen (Lehman, 1931; Wahl, 1921), but Lehman (1947) later reported that a species of *Microsphaera*, not *E. polygoni*, was the cause of powdery mildew. More recently, Roane and Roane (1976) reported finding cleistothecia of both *Microsphaera diffusa* Cke. and Pk. and *E. polygoni* on leaves of powdery mildew infected soybean plants. Most evidence supports *Microsphaera diffusa* as the cause of powdery mildew of soybean (Grau and Laurence, 1975; McLaughlin et al., 1977; Paxton and Rogers, 1974). Although it appears most workers are reporting *M. diffusa* to be the cause of powdery mildew of soybean, I consider this topic still open to scientific investigation. Inoculum has not been properly controlled in most experiments reported in the literature. Inoculum control is difficult for this air-borne pathogen, but such techniques are needed to resolve the issue of the cause of this disease. Also, it is vitally important that soybean researchers make an effort to correctly identify the powdery mildew fungus that they are using in their research. Although important for any research endeavor, the topic of correct identity of causal agents is especially critical for studies on host range and host resistance. Methods are available to purify and maintain isolates of *M. diffusa* (Mignucci, 1978), and offer possible control of inoculum that is needed for critical studies on host range, host resistance and physiological races.

It is not possible to identify *M. diffusa* by its conidiophores or conidia, for these structures are very similar to those of *E. polygoni*. *Microsphaera diffusa* produces conidia borne in chains on short, simple conidiophores. Conidia range in size from 17 to 21 μm in width by 27.7 to 54.1 μm in length, with a mean of 18 by 40 μm (Mignucci and Chamberlin, 1978).

Fully mature cleistothecia are needed to distinguish *M. diffusa* from *E. polygoni*. Cleistothecia of *M. diffusa* are initially white and progress to yellow, tan and finally black at maturity. Mature cleistothecia of *M. diffusa* are 18 to 125 μm in diameter and possess about 20 appendages that are 124 to 280 μm long and are dichotomously branched four to five times at their tips. Appendages are generally not dichotomously branched if less than 100 μm in length (McLaughlin et al., 1977; Paxton and Rogers, 1974). Cleistothecia contain several asci and each ascus contains up to six light yellow, ovoid ascospores measuring 9 x 18 μm (McLaughlin et al., 1977).

Although mature asci may be present, appendages may need more time to form their dichotomously branched tips. Cleistothecia of *M. diffusa* and *E. polygoni* are very similar until the dichotomous branching characteristic of *M. diffusa* occurs. Dichotomous branching at the tips of the appendages is a major taxonomic factor to distinguish *M. diffusa* from *E. polygoni*. Roane and Roane (1976) reported both pathogens to be simultaneously present on soybean leaves. They did not provide a scale for making size comparisons of the appendages presented in their figures. As stated earlier, McLaughlin et al. (1977) noted that appendages of *M. diffusa* needed to be 100 μm in length before they dichotomously branched and this was independent of ascus formation. I cannot overemphasize the importance of working with fully mature cleistothecia when distinguishing *M. diffusa* from *E. polygoni*.

DISEASE CYCLE AND EPIDEMIOLOGY

Disease Cycle

The disease cycle of powdery mildew of soybean has not been completely described. For example, the site and means by which *M. diffusa* survives in the absence of a host is not known. It can be assumed that *M. diffusa* survives as cleistothecia which liberate ascospores, the primary inoculum. The role or importance of other hosts, especially perennial hosts, has not been documented, but they could serve as an inoculum reservoir for soybeans. Conidia would likely be the primary inoculum in this latter case. Hosts of *M. diffusa* are listed in Table 1.

Once infection occurs, the asexual stage of the fungus becomes predominant. Conidia are abundantly produced. The fungus is capable of producing many cycles and enormous numbers of conidia per cycle on susceptible soybean cultivars. Under favorable environments, conidia may germinate within 3 h, after which appressoria are formed and an infection peg penetrates epidermal cells; after 8 h haustoria develop,

570

Table 1. Host range of *Microsphaera diffusa* (Johnson and Jones, 1961; Johnson et al., 1940; Luttrell and Samples, 1954; Mignucci and Chamberlain, 1978; Newell and Hymowitz, 1975; Thompson, 1951).

Host	Non-host
Arachis hypogaea	*Glycine canescens*
Cajanus cajan	*Lespedeza cuneata*
Cyamopsis tetragonaloba	*Lespedeza sericea*
Glycine clandestina	*Lotus uliginosus*
Glycine falcata	*Lotus pedunculatus*
Glycine max	*Lotus tenuis*
Glycine tabacina	*Melilotus alba*
Glycine tomentella	*Melilotus denata*
Lespedeza stipulacea	*Melilotus officinalis*
Lespedeza striata	*Psoralea glandulosa*
Lupinus albus	*Psoralea bituminosa*
Lupinus angustifolius	*Trifolium hybridum*
Lupinus hirustus	*Trifolium incarnatum*
Lupinus luteus	*Trifolium repenos*
Phaseolus aureus	
Phaseolus vulgaris	
Pisum sativum	
Psoralea physodes	
Psoralea tenax	
Psoralea tenuiflora	
Vigna unguiculata	

and a parasitic relationship is established by this obligate parasite. Mycelium develops on the leaf surface and mature conidia may be released by 144 h after initial deposition of inoculum on the leaf (Mignucci and Chamberlin, 1978). Thus, *M. diffusa* has the potential to be a very explosive pathogen and can uniformly infect soybeans on a regional basis.

Environmental Factors

The first signs of powdery mildew generally appear at or after flowering (growth stage R1). R1 generally occurs in early to late July in the North Central states, but this relationship may change with geographic location, time of planting and maturity group of the cultivar. Infection and mycelium development will occur from 18 to 30 C (Mignucci et al., 1977). However, mycelium develops more rapidly at 18 C compared with 24 or 30 C. Field observations also indicate that powdery mildew develops earlier and at a faster rate when air temperatures are below 30 C during the R1-R6 growth stages (Grau and Laurence, 1975; Leath and Carroll, 1982). No information was found on the effects of relative humidity, rainfall, solar radiation or other environmental factors.

Host

Soybean cultivars range from highly resistant to very susceptible. In the field, cultivars generally will appear resistant (little or no mycelium development) or susceptible. However, two forms of resistance are expressed when plants are inoculated and incubated in environmentally controlled facilities (Dunleavy, 1977; Grau and Laurence,

1975). Several Maturity Group 0 cultivars are reported to express a highly resistant reaction to *M. diffusa* (Dunleavy, 1977; Grau and Laurence, 1975). This highly resistant reaction produces no visible changes in the host, such as necrosis or chlorosis, and no signs of the pathogen. The nature of this reaction approaches immunity to *M. diffusa*, but has not been studied to determine its mode of inheritance. However, progeny of a Wilkin (highly resistant) x MN 63-194 (susceptible) cross expressed a reaction similar to Wilkin in the field. Thus, this trait appears to be heritable (Grau and Laurence, 1975). However, more detailed studies are needed to determine its actual mode of inheritance.

A second resistant reaction is characterized by degrees of mycelium development on the leaf surface that ranges from a trace to extensive coverage of the leaf surface. Although potentially extensive in terms of coverage, the density of the mycelium is commonly sparse when compared to a susceptible reaction. In addition, little or no sporulation occurs. This form of resistance is often characterized by mycelium development being confined to lower leaves. Cultivars expressing this second resistant reaction are often totally void of signs of powdery mildew when evaluated in the field. Mignucci and Lim (1980) described this reaction as adult plant resistance. This reaction is expressed in a quantitative manner in the greenhouse. However, this reaction is expressed very much like a qualitatively rather than a quantitatively inherited resistant reaction in the field (Arny et al., 1975; Demski and Phillips, 1974; Dunleavy, 1977; Grau and Laurence, 1975). This point of view is supported by the inheritance of this trait. Buzzell and Haas (1978) and Grau and Laurence (1975) found this resistant reaction to be inherited as a single, dominant gene in two cultivars, Blackhawk and Chippewa-64. A range of mycelium development is reported for cultivars with this resistant reaction (Dunleavy, 1977). Although limited mycelium development is important, a significant aspect of this form of resistance is the inhibition or suppression of sporulation. The lack of sporulation would be a great deterrent to natural epiphytotics. Cultivars express degrees of resistance under greenhouse conditions. However, such cultivars express high levels of resistance to *M. diffusa* in the field (Arny et al., 1975; Dunleavy, 1977; Grau and Laurence, 1975).

Most soybean cultivars express consistent reactions to *M. diffusa* (Arny et al., 1975; Demski and Phillips, 1974; Dunleavy, 1977; Grau and Laurence, 1975; Leath and Carroll, 1982). However, reactions of some cultivars vary within and between evaluations by specific researchers (Table 2). Such findings suggest the existence of physiologic races of *M. diffusa*. However, differences in reported reactions to *M. diffusa* may be due to other factors, such as interpretation of disease reaction and different causal fungi. Soybean cultivars may react differently to *E. polygoni* and *M. diffusa* assuming that *E. polygoni* may also cause powdery mildew of soybean as suggested by Roane and Roane (1976). The inheritance of resistance found in Chippewa-64 and Blackhawk has been documented and has remained effective. However, similar forms of resistance found in other cultivars have reacted differently to different populations of the pathogen (Arny et al., 1975; Dunleavy, 1977; Grau and Laurence, 1975). The highly resistant reaction to *M. diffusa* appears to be governed by different genes, of which not all are always effective (Table 2). Thus, physiologic races of *M. diffusa* appear to be associated with both forms of resistance. However, the presence of physiologic races of *M. diffusa* cannot be proven until experiments with more rigid control of inoculum are performed. However, evidence is available that supports the notion that physiologic races of *M. diffusa* are present in natural agro-ecosystems.

572

Table 2. Comparison of soybean cultivars and their reactions to powdery mildew (Arny et al., 1975; Dunleavy, 1977; Grau and Laurence, 1975).

Cultivar	Disease reactions Greenhouse[a]	Field[a]
Ada	HR	–
Anoka	HR-S	–
Altona	HR	R
Wilken	HR	HR
Clay	R-S	S
Steele	R	R-S
Chippewa-64	R	R
Wirth	R-S	R
Blackhawk	R	R
Dunn	R	R
Lincoln	R	R
Wayne	R	R
Rampage	R	R
Swift	R-S	R-S
Merit	R-S	R
Hodgson	R	S
Amsoy-71	S	S
Hark	S	S
Harosoy-63	S	S
Corsoy	S	S

[a]R = resistant, S = susceptible, HR = highly resistant.

IMPACT ON YIELD

Microsphaera diffusa invades epidermal cells and establishes a biotrophic relationship with the host by means of a haustorium. The presence of mycelium on the leaf surface of a susceptible cultivar is not accompanied by visible host symptoms until latter stages of the disease as the host approaches maturity. However, infection results in rates of photosynthesis and transpiration that are reduced to less than half those of non-infected plants (Mignucci and Boyer, 1979). Such an alteration of photosynthesis has the potential to reduce seed yield. Reduced photosynthesis would likely lead to earlier maturity, and both would contribute to a reduction of seed yield.

Several researchers have studied the impact of powdery mildew on seed yield. Phillips and Miller (1966; 1977) reported no yield loss in one study in Georgia, but a 35% yield loss caused by powdery mildew in a second study. Severity of powdery mildew was highly correalted (r = -.81) with yield. Both studies involved natural epidemics and determinant soybean cultivars.

Several attempts have been made to assess the impact of powdery mildew on indeterminate cultivars in the North Central states. Investigators have approached yield loss studies by comparing the yields of resistant and susceptible cultivars and by applying the fungicide benomyl to both groups of cultivars.

Mignucci and Lim (1977), using benomyl for control, reported an 11% yield reduction for Hark soybeans grown in an artificially induced epiphytotic, but no yield loss was measured in a natural epiphytotic. No yield increase was measured for a resistant cultivar treated with benomyl. Dunleavy (1978; 1980) approached the problem of measuring yield loss due to powdery mildew in a similar fashion, but he employed six cultivars, multiple locations, and studied the problem for four years. Also, he studied powdery mildew in natural epiphytotics that usually started in early to mid-July, and susceptible cultivars were severely diseased by the end of August. Yield loss was estimated by comparing the difference in yield between susceptible cultivars either treated or not treated with benomyl. Mean yield losses for three susceptible cultivars were 10%, 11%, 21% and 15% for 1975, 1976, 1977 and 1978, respectively. These data represent a 4-year mean of 14% and a range of 0 to 26% yield loss due to powdery mildew. The yield losses described above were associated with severe levels of powdery mildew at the end of the growing season, but final severity or rate of disease progress was not reported. In no case did benomyl treatments increase the yield of resistant cultivars.

CONTROL

Dunleavy (1980) reported that resistant cultivars produced yields that were 5 to 29% greater than susceptible cultivars over three years. Thus, resistant cultivars provide an effective means of control for powdery mildew. Circumstantial evidence suggests there are physiologic races of *M. diffusa* which would threaten the effectiveness of resistance. Resistance to *M. diffusa* needs to be better documented in terms of inheritance and its physiological basis in order to intelligently breed for resistance and deploy genes for resistance for control of powdery mildew.

Fungicides such as benomyl, thiophanate methyl, thiabendazole and chlorothalonil are registered for use on soybeans in the United States and other countries. Each of these fungicides are effective against *M. diffusa*. Dunleavy (1980) reported that susceptible cultivars treated with benomyl produced yields that were 11 to 21% greater than non-treated cultivars. Presented differently, benomyl treatments improved yields of susceptible cultivars 0.2 to 0.7 t/ha. However, these yield increases were the result of multiple applications of benomyl, which likely were not economically feasible. Thus, research is needed on how to use benomyl and other fungiicides economically for control of powdery mildew of soybean. Research is needed on epidemiology to better predict the need for control of powdery mildew and to determine the timing of fungicide application to enhance the economic benefits of their use.

Because so little is known about the disease cycle and its epidemiology, little can be done in terms of cultural practices to control this disease. Personal observation suggests that late-planted soybeans are more prone to develop severe infections.

It is well documented that powdery mildew is capable of reducing soybean yield. However, pathologists may be inclined to underestimate the importance of the disease because of its sporadic occurrence. Information on epidemiology and the genetics of resistance is needed to develop sound programs for control of powdery mildew.

NOTES

Craig R. Grau, Department of Plant Pathology, University of Wisconsin, Madison, Wisconsin.

REFERENCES

Arny, D. C., E. W. Hanson, G. L. Worf, E. S. Oplinger and W. H. Hughes. 1975. Powdery mildew on soybean in Wisconsin. Plant Dis. Rep. 59:288-290.
Buzzell, R. I. and J. H. Haas. 1978. Inheritance of adult plant resistance to powdery mildew in soybeans. Can. J. Genet. Cytol. 20:151-153.

574

Demski, J. W. and D. V. Phillips. 1974. Reactions of soybean cultivars to powdery mildew. Plant Dis. Rep. 58:723-726.

Dunleavy, J. M. 1976. A survey of powdery mildew of soybean in Central Iowa. Plant Dis. Rep. 60: 675-677.

Dunleavy, J. M. 1977. Comparison of the disease response of soybean cultivars to *Microsphaera diffusa* in the greenhouse and field. Plant Dis. Rep. 61:32-34.

Dunleavy, J. M. 1978. Soybean seed yield losses caused by powdery mildew. Crop Sci. 18:337-339.

Dunleavy, J. M. 1980. Yield losses in soybeans induced by powdery mildew. Plant Dis. Rep. 64:291-292.

Dunleavy, J. M., D. W. Chamberlin and J. P. Ross. 1966. Soybean Diseases. U.S. Dep. Agric. Handb. No. 302.

Grau, C. R. and J. A. Laurence. 1975. Observations on resistance and heritability of resistance to powdery mildew of soybean. Plant Dis. Rep. 59:458-460.

Johnson, H. W. and J. P. Jones. 1961. Other legumes prove susceptible to a powdery mildew of *Psoralea tenax*. Plant Dis. Rep. 45:542-543.

Johnson, H. W., C. L. Lefebvre and T. T. Ayers. 1940. Powdery mildew of lespedeza. Phytopathology 30:620-621.

Leath, S. and R. B. Carroll. 1982. Powdery mildew on soybean *Glycine max* in Delaware. Plant Dis. Rep. 66:70-71.

Lehman, S. G. 1931. Powdery mildew of soybean. J. Elisha Mitchell Sci. Soc. 46:190-195.

Lehman, S. G. 1947. Powdery mildew of soybean. (Abstrc.). Phytopathology 37:434.

Luttrell, E. S. and J. W. Samples. 1954. Mildew of lupines caused by *Microsphaera diffusa*. Plant Dis. Rep. 38:719-720.

McLaughlin, M. R., J. S. Mignucci and G. M. Milbrath. 1977. *Microsphaera diffusa* the perfect stage of the soybean powdery mildew pathogen. Phytopathology 67:726-729.

Mignucci, J. S. 1978. Development of soybean leaf cultures for maintenance and study of *Microsphaera diffusa*. Plant Dis. Rep. 62:271-273.

Mignucci, J. S. and J. S. Boyer. 1979. Inhibitions of phytosynthesis and transpiration in soybean infected with *Microsphaera diffusa*. Phytopathology 69:227-330.

Mignucci, J. S. and D. W. Chamberlain. 1978. Interactions of *Microsphaera diffusa* with soybeans and other legumes. Phytopathology 68:169-173.

Mignucci, J. S. and S. M. Lim. 1977. Effect of powdery mildew on soybean yield. Proc. Am. Phytopath. Soc. 4:148-149.

Mignucci, J. S. and S. M. Lim. 1980. Powdery mildew (*Microsphaera diffusa*) development on soybean with adult plant resistance. Phytopathology 70:919-921.

Mignucci, J. S., S. M. Lim and P. R. Hepperly. 1977. Effects of temperature on reactions of soybean seedlings to powdery mildew. Plant Dis. Rep. 61:122-124.

Newell, C. A. and T. Hymowitz. 1975. *Glycine canescens* a wild relative of the soybean. Crop Sci. 15:879-881.

Paxton, J. D. and D. P. Rogers. 1974. Powdery mildew of soybeans. Mycologia 66:894-896.

Phillips, D. V. and R. D. Miller. 1976. Control of powdery mildew of soybean in the field. Proc. Am. Phytopath. Soc. 3:277.

Phillips, D. V. and R. D. Miller. 1977. Soybean yield reductions caused by powdery mildew. Proc. Am. Phytopath. Soc. 4:95.

Roane, C. W. and M. K. Roane. 1976. Erysiphe and Microsphaera as dual causes of powdery mildew of soybeans. Plant Dis. Rep. 60:611-612.

Sinclair, J. B. 1982. Compendium of Soybean Diseases. 2nd Edition. American Phytopathological Society. St. Paul, Minnesota.

Thompson, G. E. 1951. *Microsphaera diffusa* on blue lupine. Plant Dis. Rep. 35:221.

Wahl, C. Von. 1921. Schadlinge an der Sojabohne. Z. Pflanzenkrankh 31:194-196.

PHYTOPHTHORA ROOT ROT OF SOYBEAN

Kirk L. Athow

Phytophthora root rot is one of the most destructive diseases of soybeans. The disease was first observed in northeastern Indiana and northwestern Ohio in the late 1940s, although the cause of the disease was not known at that time. The first published report of the disease was by Suhovecky and Schmitthenner (1955) in Ohio. Since then the disease has been reported in most of the soybean producing areas of the United States, and in Canada, Argentina, Australia, Hungary, Japan, and New Zealand. It has not been observed in Brazil, although some very susceptible cultivars have been grown there. The disease is more severe on heavy clay soils, which have small pore size, than on lighter soils. It is particularly prevalent on low, poorly drained areas, but it also appears on higher ground if the soil remains wet for several days. Soil compaction, which reduces porosity of the soil and aeration of the roots, increases incidences of the disease.

The disease became very destructive in the eary 1950s when the highly susceptible cultivars Harosoy, Hawkeye, and Lindarin occupied over 75% of the soybean acreage in the heaviest production areas of midwestern U.S.A. It was later learned that some of the older cultivars such as Illini and Mukden that were replaced by these susceptible cultivars had specific genes for resistance; and Dunfield and Lincoln were not highly susceptible although their yields were sometimes reduced 50% without death of plants. The resulting poor growth, yellowing, and reduction in yield often were attributed wholly to wet soil. Later, with resistant cultivars, it was possible to show that this was not entirely true, and that the condition was a disease whose development was favored by wet, poorly drained, heavy clay soils.

The disease may cause seed rot, pre- or post-emergence damping-off, or a more gradual killing or reduction of vigor of plants throughout the season. Young plants, however, are most susceptible and die most quickly. The disease is primarily a root rot, although stem rot (Kaufmann and Gerdemann, 1958) and infection of leaves and stems, when splashed by contaminated soil, has been reported.

SYMPTOMS

The first noticeable symptoms of the disease, except for a poor stand, often are yellowing and wilting of the leaves. The lateral roots are almost completely destroyed and the main root becomes dark brown. The dark discoloration generally extends up the stem several centimeters and occasionally to the fourth or fifth node. The main root and stem usually remain firm and there is no definite lesion other than the dark discoloration. Some cultivars are not readily killed, but show yellowing and poor vigor similar to that associated with wet soil and nitrogen deficiency. The condition is partly a result of nitrogen deficiency, because as the roots are destroyed there are fewer functional nodules. Yield reduction may approach 100% in some fields.

PATHOGEN

Suhovecky and Schmitthenner (1955) first reported the disease to be caused by a species of *Phytophthora*. Herr (1957) referred to the causal organism as *Phytophthora cactorum* (Leb. & Cohn) Schroet. Kaufmann and Gerdemann (1958) considered the organism distinct from previously described species of *Phytophthora* and suggested the name *P. sojae*. Hildebrand (1959) found that isolates of the soybean fungus from Illinois, North Carolina, and Ontario, Canada were morphologically similar to *P. megasperma*, but because of host specificity, suggested the trinomial *P. megasperma* Drechs. var. *sojae* A. A. Hildeb. Kuan and Erwin (1980) suggested the name *P. megasperma* Drechs. f. sp. *glycinea* Kuan & Erwin because host specificity was more correctly designated by formae speciales. They replaced *sojae* with *glycinea* to indicate the genus of the host.

The fungus belongs to a group of fungi known as the water molds (Phycomycetes means water loving). The optimum temperature for growth of the fungus is 25 to 28 C. It produces amphigynous and paragynous antheridia and thin-walled spherical or subspherical oogonia 28 to 59 μm in diameter. Oospores develop when an antheridium fertilizes an oogonium. The oospores have thick, smooth walls which protect them from desiccation and provides a means of survival during unfavorable soil conditions. After a dormant period, greatly influenced by temperature and moisture, the oospores germinate and produce a germ tube, which develops into either hyphae or a sporangium. The sporangia are vessels 42 to 65 by 32 to 53 μm in size and obpyriform in shape. A sporangium produces hundreds of minute zoospores which are released into the soil water that fills the pore spaces between the soil particles. The zoospores swim in the free water by means of two flagella—one anterior and one posterior.—They are attracted to the soybean roots where they attach themselves, encyst, germinate and infect the roots directly through the epidermis. The zoospores are considered the main source of root infection. Thus, if the pore spaces of the soil do not contain sufficient free water, the zoospores are unable to swim to and infect the soybean root. Without free soil water, infection can only occur if by chance a root grows in contact with the fungus; a rarity. After infection, fungal development in resistant cultivars is limited to the cortex and stele of lateral roots, whereas in susceptible cultivars ramification to tap root and hypocotyl occur (Beagle-Ristaino and Rissler, 1983). New oospores, sporangia, and zoospores are produced in diseased tissue.

PHYSIOLOGIC SPECIALIZATION

Physiologic specialization was reported by Morgan and Hartwig (1965) who isolated Race 2 from the soybean strain D60-9647. Race 2 was morphologically indistinguishable from Race 1, but the two races differed pathogenically. Race 1, isolated from D55-1492 and Lee, was nonpathogenic to D60-9647, Harrel, and Nansemond; whereas Race 2, isolated from D60-9647, was pathogenic to these soybean strains. Strains which were susceptible to Race 1 were also susceptible to Race 2.

Schmitthenner in 1972 isolated Race 3 in Ohio. The cultivars Arksoy, Higan, and Lee-68, which were resistant to Races 1 and 2, were also resistant to Race 3; whereas Harosoy-63, Amsoy-71, Beeson, Clark-63, and Mukden, which were resistant to Races

1 and 2, were susceptible to Race 3. D60-9647, which was resistant to Race 1 and susceptible to Race 2, was resistant to Race 3. Harosoy, Amsoy, Clark, Corsoy, and Cutler, which were susceptible to both Races 1 and 2, were susceptible to Race 3.

Race 4 was first identified in 1974 by Schwenk and Sim in Kansas. They were able to group host lines into four different reaction groups. The first group, which included Harosoy, Corsoy, Cutler, and Dare, was susceptible to Races 1, 2, 3, and 4. A group containing Amsoy-71, Beeson, Clark-63, and Cutler-71 was resistant to Races 1 and 2, and susceptible to Races 3 and 4. Arksoy and Mack were susceptible only to Race 4. D60-9647, Harrel, and Nansemond were susceptible only to Race 2. They did not report any lines resistant to all four races. However, Athow et al. (1974) found 95 lines, including Altona, resistant to Races 1, 2, 3, 4.

In 1976, Haas and Buzzell identified Races 5 and 6 in Ontario, Canada, and described the responses of Races 1 through 6 on an enlarged set of differential cultivars accepted by soybean researchers in 1975. These differentials consisted of Harosoy, Harosoy-63, Sanga, Mack, Altona, PI 103091, and PI 171442. Laviolette and Athow (1977) inoculated the above set of differentials with a group of isolates previously classified as Race 6 and identified new Races 7, 8, and 9 based on their reactions to the new differentials PI 103091 and PI 171442. In 1980, Keeling reported Races 10 through 16 from Mississippi, and in 1982 he reported Races 17 through 20 from the delta area of Mississippi and Arkansas. Laviolette and Athow (1983) reported races 21 and 22 from Indiana, and White et al. (1983) reported Race 23 from Nebraska. The reactions of the differential soybean cultivars to physiologic Races 1 through 23 are given in Table 1.

DISEASE CYCLE

Phytophthora root rot is a wet soil disease because the zoospores, the main source of root infection, move in soil water. Consequently, the disease is more prevalent in wet, poorly drained, fine textured clay soils. Soil compaction, which reduces soil porosity, also favors the dissemination of zoospores and the resulting disease. Reduced tillage systems, which tend to keep the soil cooler and wetter, increase disease incidence. D. H. Scott in Indiana (pers. commun.) observed 54% Phytophthora root rot in

Table 1.   Reaction of differential soybean cultivars to Physiologic races 1-23 of *Phytophthora megasperma* f. sp. *glycinea*.

Differential cultivar	Reaction[a] to physiologic race																						
	1	2	3	4	5	6	7	8	9	10	11	12	13	14	15	16	17	18	19	20	21	22	23
Harosoy	S	S	S	S	S	S	S	S	S	S	S	R	S	S	S	R	S	R	R	S	S	S	S
Harosoy-63	R	R	S	S	S	S	S	S	S	R	R	R	R	S[b]	R	R	R	R	R	S	S	S	S
Sanga	R	S	R	R	R	R	R	R	R	S	S	S	R	R	R	S	R	R	S	S	R	R	S
Mack	R	R	R	S	S	R	R	R	R	R	R	S	R	S	R	S	R	S	S	S	R	S	R
Altona	R	R	R	R	S	S	S	S	S	R	S	R	S	R	R	R	S	R	R	R	R	S	S
PI 171442	R	R	R	R	S	S	R	S	R	S	R	S	R	R	S	R	S	R	S	S	S	S	R
PI 103091	R	R	R	R	S	R	S	R	R	R	R	R	R	R	R	S	R	S	R	R	R	R	R

[a]Abbreviations: R = Resistant; S = Susceptible.

[b]Keeling (1982) reported Harosoy-63 resistant to Race 14. Moots et al. (1983) reported the majority of Harosoy-63 plants had killing lesions with Race 14. I have found Harosoy-63 to be uniformly susceptible to an isolate of Race 14 from Keeling.

replicated, no-till plots compared with 10% for chisel plowing and 1% for conventional, moldboard plowing. The use of the insecticide/nematocide carbofuran as a soil treatment increases the severity of the disease on soybean plants. Kittle and Gray (1979) reported that a soil temperature of 15 C, which is 10 to 15 C below the optimum for soybean root development, was most favorable for root infection. Although root infection may be favored by cooler soil temperatures, I have observed that disease development is much more rapid with an air temperature of 25 C or higher.

Primary inoculum comes from crop residue in the soil, where the causal fungus survives long periods without soybean roots, primarily as oospores. Oogonia and oospores are formed in infected root and stem tissues of susceptible, tolerant, and resistant cultivars. Many more oospores are formed in susceptible and tolerant cultivars than in resistant ones. It has not been determined how oospores germinate naturally in the soil. Sporangia form on infected plant debris and on the root surface of infected soybean seedlings. Zoospores are produced in abundance in the sporangia in flooded or waterlogged soils. Zoospores near soybean roots are attracted to the roots where they encyst and germinate. The germ tube penetrates directly through the epidermis and the resulting hyphae grow intercellularly in root tissues. In susceptible cultivars it ramifies the tap root and hypocotyl, but in resistant cultivars fungal development is limited to the cortex and stele of lateral roots. There is no evidence of cell necrosis in advance of the pathogen. Damage or death from the disease is probably the result of reduced water and nutrient uptake, caused by destruction of varying amounts of the root system.

CONTROL

Cultural

The trends in soybean culture, which emphasize larger equipment, earlier planting, reduced tillage, narrower rows, increased use of herbicides, and less rotation, all enhance the disease development. Improved soil tilth and drainage would do much to reduce phytophthora root rot. Overworking the soil, working the soil too wet, and repeated turning of heavy equipment in the same areas, which may cause soil compaction should be avoided. Sub-soiling or some form of deep tillage may help promote better drainage by breaking up a hard, plow-sole layer. Where practical, tile or surface drainage can greatly reduce the disease. It may be best not to plant soybeans in fields that are historically wet unless resistant cultivars are available.

Resistance

Resistant cultivars have provided the most effective control, even with the large number of physiologic races of the fungus. Thirteen or fourteen major, dominant genes, which condition resistance to various physiologic races of the fungus, have been identified. Bernard et al. (1975) reported Rps_1, which gave resistance to Races 1 and 2. Mueller et al. (1978) and Laviolette et al. (1979) reported $Rps_1{}^b$ from Sanga, PI 84637, or D60-9647; and $Rps_1{}^c$ from PI 54615, Arksoy, or Mack. Bernard and Cremeens (1981) reported the fourth allele at the Rps_1 locus, $Rps_1{}^k$, from the cultivar Kingwa. Rps_2 was reported from the cultivar CNS and in derived strains from root inoculation with Races 1 and 2 (Kilen et al., 1974). Rps_3 was reported in PI 171442 and PI 86972-1 (Mueller et al., 1978); Rps_4 was reported from PI 86050 (Athow et al., 1980); Rps_5 was reported by Buzzell and Anderson (1981) from Harosoy[6] x T240 (type collection strain); and Rps_6 was reported from the cultivar Altona (Athow and

Laviolette, 1982). Each of these genes gives resistance to several physiologic races Table 2) and is the basis for disease resistance. Recently, Layton, Ploper, Athow, and Laviolette (unpublished) have found evidence of four allelic genes at the Rps_3 locus, which tentatively have been designated $Rps_3{}^b$, $Rps_3{}^c$, $Rps_3{}^d$, and $Rps_3{}^e$. Also, a new gene, Rps_7, has been identified in PI 82312N (Athow and Laviolette, unpublished).

Rps_1 provided resistance to Races 1 and 2 in Clark-63, Harosoy-63, Hawkeye-63, Lindarin-63, Chippewa-64, Beeson, Calland, Protana, Amsoy-71, Cutler-71, Bonus, Harcor, Century, Oakland, Pella, Wells, Hardin, Evans, Hodgson-78, and others. The cultivars with Rps_1 are susceptible to Race 3 and many of the races found since 1972. The cultivars Vickery, Corsoy-79, Williams-79, Wells-II, Beeson-80, Vinton-80, Mack, Lee-78, Pickett-81, and Centennial have $Rps_1{}^c$ and are resistant to all races, except for 4 and 5, that have been found in the heavy production areas. Williams-82 has $Rps_1{}^k$ and is resistant to all races found in the Midwest. The new cultivars, Keller and Miami, have the two genes $Rps_1{}^cRps_3$, and Winchester has the two genes $Rps_1{}^bRps_3$. The genotype $Rps_1{}^cRps_3$ is susceptible only to Races 12, 19, 20, and 22, whereas the genotype $Rps_1{}^bRps_3$ is susceptible only to Races 10, 12, 19, and 20 (Table 2).

Although there are several 2- or 3-gene combinations that give resistance to all races of the fungus, there has been a reluctance by some to use specific gene resistance for fear that selection pressure would hasten the evolution of additional races. Several have advocated the use of so-called "tolerant" cultivars (Schmitthenner and Walker, 1979; Buzzell and Anderson, 1982; Walker and Schmitthenner, 1982). Schmitthenner and Walker (1979) defined tolerance as the ability of susceptible plants to survive infection without showing severe symptom development, such as death, stunting, or yield loss. In reality these "tolerant" plants actually have varying degrees of resistance to root infection, even though they are susceptible to pre- or post-emergence hypocotyl infection (susceptible to the seedling phase of the disease and inoculation into the hypocotyl). On the other hand, specific gene resistance gives resistance to both the seedling and root rot phases of the disease, as indicated by resistance to hypocotyl inoculation.

Even though the term "tolerance" has been badly used and much abused, resistance to root infection is a useful type of resistance. This resistance apparently is non-race specific, multigenic in nature, and could be expected to be more durable because

Table 2. Genes for resistance to specific races of *Phytophthora megasperma* f. sp. *glycinea* and their source.

Gene	Source	1	2	3	4	5	6	7	8	9	10	11	12	13	14	15	16	17	18	19	20	21	22	23
rps	Harosoy	S	S	S	S	S	S	S	S	S	S	S	R	S	S	S	R	S	R	R	S	S	S	S
Rps_1	Harosoy-63	R	R	S	S	S	S	S	S	S	R	R	R	R	S	R	R	R	R	R	S	S	S	S
$Rps_1{}^b$	Sanga	R	S	R	R	R	R	R	R	S	S	S	R	R	R	S	R	R	R	S	R	R	R	S
$Rps_1{}^c$	Mack	R	R	R	S	S	R	R	R	R	R	R	S	R	S	R	S	R	S	R	S	S	R	S
$Rps_1{}^k$	Kingwa	R	R	R	R	R	R	R	R	R	R	R	S	R	R	R	S	R	R	R	S	R	R	-
Rps_3	PI 86972-1	R	R	R	R	R	S	S	R	R	S	R	S	R	R	R	S	R	S	R	S	S	S	R
Rps_4	PI 86050	R	R	R	R	S	S	S	S	S	R	S	R	R	R	R	R	-	-	-	-	-	-	-
Rps_5	L62-904	R	R	R	R	R	S	S	R	R	S	R	S	R	R	R	S	R	-	-	-	-	-	-
Rps_6	Altona	R	R	R	R	S	S	S	S	S	R	S	R	S	R	R	R	S	R	R	R	R	S	S

[a]Abbreviations: R = Resistant; S = Susceptible.

of less selection pressure on individual genes. Walker and Schmitthenner (1982) reported that heritability was high and controlled by relatively few genes that were independent of the specific genes for resistance. Therefore, the two types of resistance together should complement each other. Selection for tolerance based on low plant loss, followed by backcrossing to include race-specific resistance, should provide effective long-term disease control.

Chemical

The systemic fungicide metalaxyl has activity against *Phytophthora*, *Pythium*, and downy mildews. The seed treatment formulation offers protection against the seedling phase of soybean phytophthora. It has been suggested (Schmitthenner, 1983) for use with those cultivars with a high degree of field tolerance but no seedling resistance. A granular formulation is intended as an in-furrow, band, or broadcast application. Varying degrees of effectiveness have been reported, but it may give season-long protection depending upon the rate used, the amount of rainfall, and the soybean cultivar. The cost-price relationship may be the deciding factor, particularly with broadcast application.

NOTES

K. L. Athow, Professor of Plant Pathology, Department of Botany and Plant Pathology, Purdue University, West Lafayette, IN 47907.

REFERENCES

Athow, K. L., F. A. Laviolette and T. S. Abney. 1974. Reaction of soybean germplasm strains to four physiologic races of *Phytophthora megasperma* var. *sojae*. Plant Dis. Rep. 58:789-792.

Athow, K. L. and F. A. Laviolette. 1982. Rps$_6$, a major gene for resistance to *Phytophthora megasperma* f. sp. *glycinea* in soybean. Phytopathology 72:1564-1567.

Athow, K. L., F. A. Laviolette, E. H. Mueller and J. R. Wilcox. 1980. A new major gene for resistance to *Phytophthora megasperma* var. *sojae* in soybeans. Phytopathology 70:977-980.

Beagle-Ristaino, J. E. and J. F. Rissler. 1983. Histopathology of susceptible and resistant soybean roots inoculated with zoospores of *Phytophthora megasperma* f. sp. *glycinea*. Phytopathology 73:590-595.

Bernard, R. L., P. E. Smith, M. J. Kaufmann and A. F. Schmitthenner. 1975. Inheritance of resistance to Phytophthora root and stem rot in the soybean. Agron. J. 49:391.

Bernard, R. L. and C. R. Cremeens. 1981. An allele at the rps$_1$ locus from the variety 'Kingwa'. Soybean Genet. Newslett. 8:40-42.

Buzzell, R. I., and T. R. Anderson. 1981. Another major gene for resistance to *Phytophthora megasperma* var. *sojae* in soybean. Soybean Genet. Newslett. 8:30-33.

Buzzell, R. I. and T. R. Anderson. 1982. Plant loss response of soybean cultivars to *Phytophthora megasperma* f. sp. *glycinea* under field conditions. Plant Dis. 66:1146-1148.

Haas, J. H. and R. I. Buzzell. 1976. New races 5 and 6 of *Phytophthora megasperma* var. *sojae* and differential reaction of soybean cultivars for race 1 to 6. Phytopathology 66:1361-1362.

Herr, L. J. 1957. Factors affecting a root rot of soybean incited by *Phytophthora cactorum*. Phytopathology 47:15-16.

Hildebrand, A. A. 1959. A root and stalk rot of soybean caused by *Phytophthora megasperma* Drechsler var. *sojae* var. nov. Can. J. Bot. 37:927-957.

Kaufmann, M. J. and J. W. Gerdemann. 1958. Root and stem rot of soybean caused by *Phytophthora sojae* N. sp. Phytopathology 48:201-208.

Keeling, B. L. 1980. Research on Phytophthora root and stem rot: Isolation, testing procedures, and seven new physiologic races. p. 367-370. *In* F. T. Corbin (ed.) Proc. World Soybean Res. Conf. II. Westview Press, Boulder, CO.

Keeling, B. L. 1982. Four new physiologic races of *Phytophthora megasperma* f. sp. *glycinea*. Plant Dis. 66:334-335.

Kilen, T. C., E. E. Hartwig and B. L. Keeling. 1974. Inheritance of a second major gene for resistance to Phytophthora rot in soybean. Crop Sci. 14:260-262.

Kittle, D. R. and L. E. Gray. 1979. The influence of soil temperature, moisture, porosity, and bulk density on the pathogenicity of *Phytophthora megasperma* var. *sojae*. Plant Dis. Rep. 63:231-234.

Kuan, Ta-Li and D. C. Erwin. 1980. Formae speciales differentiation of *Phytophthora megasperma* isolates from soybean and alfalfa. Phytopathology 70:333-338.

Laviolette, F. A. and K. L. Athow. 1977. Three new physiologic races of *Phytophthora megasperma* var. *sojae*. Phytopathology 67:267-268.

Laviolette, F. A. and K. L. Athow. 1983. Two new physiologic races of *Phytophthora megasperma* f. sp. *glycinea*. Plant Dis. 67:497-498.

Laviolette, F. A., K. L. Athow, E. H. Mueller and J. R. Wilcox. 1979. Inheritance of resistance in soybeans to physiologic races 5, 6, 7, 8, and 9 of *Phytophthora megasperma* var. *sojae*. Phytopathology 69:270-271.

Moots, C. K., C. D. Nickell, L. E. Gray and S. M. Lim. 1983. Reaction of soybean cultivars to 14 races of *Phytophthora megasperma* f. sp. *glycinea*. Plant Dis. 67:764-767.

Morgan, F. L. and E. E. Hartwig. 1965. Physiologic specialization in *Phytophthora megasperma* var. *sojae*. Phytopathology 55:1277-1279.

Mueller, E. H., K. L. Athow and F. A. Laviolette. 1978. Inheritance of resistance to four physiologic races of *Phytophthora megasperma* var. *sojae*. Phytopathology 68:1318-1322.

Schmitthenner, A. F. 1972. Evidence for a new race of *Phytophthora megasperma* var. *sojae* pathogenic to soybean. Plant Dis. Rep. 56:536-539.

Schmitthenner, A. F. 1983. Relative efficacy of metalaxyl seed and soil treatments for control of Phytophthora root rot of soybean. Phytopathology 73:801.

Schmitthenner, A. F. and A. K. Walker. 1979. Tolerance versus resistance for control of Phytophthora root rot of soybeans. p. 35-44. *In* Proc. 9th Soybean Seed Research Conf. Amer. Seed Trade Assoc., Washington, D.C.

Schwenk, F. W. and T. Sim. 1974. Race 4 of *Phytophthora megasperma* var. *sojae* from soybean proposed. Plant Dis. Rep. 58:352-354.

Suhovecky, A. J. and A. F. Schmitthenner. 1955. Soybean affected by early root rot. Ohio Farm Home Res. 40:85-86.

Walker, A. K. and A. F. Schmitthenner. 1982. Developing soybean varieties tolerant to Phytophthora rot. p. 67-78. *In* Proc. 12th Soybean Seed Research Conf. Amer. Seed Trade Assoc., Washington, D.C.

White, D. M., J. E. Partridge and J. H. Williams. 1983. Races of *Phytophthora megasperma* f. sp. *glycinea* on soybean in eastern Nebraska. Plant Dis. 67:1281-1284.

SCLEROTINIA STEM ROT OF SOYBEANS, ITS IMPORTANCE AND RESEARCH IN BRAZIL

José T. Yorinori and Martin Homechln

OCCURRENCE AND LOSSES

Sclerotinia stem rot (*S. sclerotiorum* [Lib.] de Bary) (= *Whetzelinia sclerotiorum* (Lib.) Korf & Dumont) is one of the most destructive soybean diseases in Brazil. It is found in the states of Minas Gerais, Paraná, Rio Grande do Sul and Santa Catarina. The disease is most severe in the states of Minas Gerais and Paraná at altitudes above 800 m where there is abundant rainfall during the crop season and temperatures are mostly in the range of 15 to 28 C (Figure 1), these conditions being optimum for apothecial production (Abawi and Grogan, 1975; Coley-Smith and Cooke, 1971). Losses are particularly severe where climate is best suited for production of high quality seed.

The first severe outbreak of the disease was recorded during the Paraná 1976-77 crop season. Assessment of losses on individual farms carried out in the 1980-81 crop season in three counties (Castro, Guarapuava and Palmeiras) in Paraná showed a yield reduction ranging from 71 to 92% in diseased fields (Homechin, 1981).

Assessment of yield losses carried out through the years often has been based on subjective observations. In 1983-84, an attempt was made to obtain more accurate data on yield losses caused by *S. sclerotiorum*. A method developed by Martinez et al. (1984) was used with slight modification. By determining the percentage of infected plants in a field, and the yield reduction of the infected plants, the percentage yield loss is calculated for a particular farm or region. By use of this method, seven farms surveyed at São Gotardo, MG, and three counties in Paraná had a disease incidence ranging from 8 to 40% with yield losses ranging from 7 to 22% (Table 1). Three of the seven farms surveyed did not have the yield loss determined because the plants were still in pod filling at the time of sampling. Although 1983-84 was considered unusually dry for most soybean production areas, incidence of *S. sclerotinia* was considered high.

Considering that the total soybean production area in Brazil is currently over 8 million ha and that about one-third is grown under conditions favorable for *S. sclerotinia* infection, an increasing incidence and loss can be expected in the future.

Observations made at São Gotardo, MG, have shown that, once *S. sclerotiorum* is introduced into a soybean field, it can quickly develop into epidemic proportions. Sclerotinia rot found at only trace levels in an area surveyed in 1981-82 (Yorinori, 1982) caused severe losses the following year (Shibuya, personal communication).

The increasing use of overhead irrigation in the flat lands of central and central-west Brazil may result in further outbreaks of Sclerotinia rot. The recent introduction of sunflower (*Helianthus annuus* L.) for a fall crop was responsible for the outbreak of

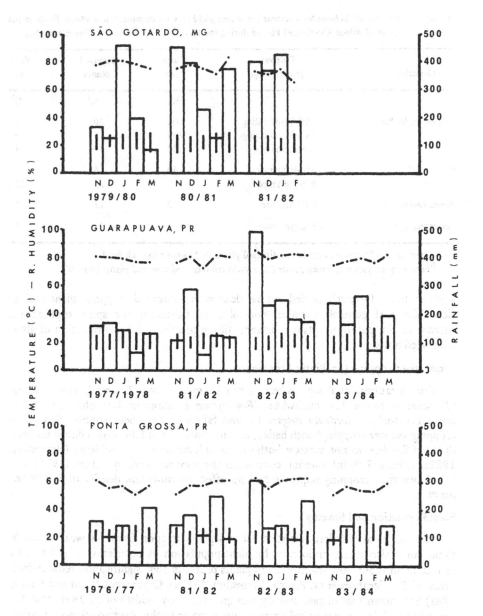

Figure 1. Minimum and maximum temperature (—), relative humidity (—·—·—) and rainfall (▢) during the soybean season in the counties with severe outbreaks of *Sclerotinia sclerotiorum*.

S. sclerotiorum on soybean, especially in western and southwestern Paraná, where the disease was not observed until a few years ago.

CONTROL OF *S. SCLEROTIORUM*

In the past 3 years, research work has been carried out at the National Soybean Research Center (CNPS) in an attempt to minimize the effect of *S. sclerotiorum* on

Table 1. Severity of *Scloerotinia sclerotiorum* and yield loss on commercial soybean fields in the states of Minas Gerais and Paraná during the 1983-84 crop season (Yorinori, unpub.)

Counties	Cultivar/ growth stage	Farm size	Diseased plants	Yield loss
		ha	%[a]	%[b]
Sao Gotardo, MG	IAC-8/pod filling	200	40	–
	IAC-8/pod filling	200	8	–
	Paraná/mature	50	20	9
Castro, PR	Pérola/mature	60	14	–
	Bragg/mature	2	25	22
Ponta Grossa, PR	FT-1/mature	30	12	7
Guarapuava, PR	Bragg/mature	4	25	7

[a]Average of 6-10 random counts of diseased plants in 2 m-rows in each field.
[b]Percent yield loss = diseased plants (%) x yield reduction of diseased plants (%)/100.

soybean crops. Research carried out has dealt with sequential cropping, plant population, biological control, chemical control (seed treatment and spray application), survival of the pathogen, plant resistance, tillage practices and identification of alternate weed hosts.

Sequential Cropping (or Winter Cropping)

Crop rotations involving soybean, lupin (*Lupinus luteus* L. and *L. albus*), wheat (*Triticum aestivum* L.), buckwheat (*Fagopyrum esculentum* Moench), oat (*Avena sativa* L.), barley (*Hordeum vulgare* L.) and fallowing have been studied. Sequential cropping (winter cropping) with barley, oat, buckwheat and fallowing reduced the incidence of *S. sclerotiorum*, whereas both species of lupin increased incidence (Homechin, 1982a, 1983a). This information contradicts the previous work by Adams (1975) who has shown that cropping sequence had no effect on inoculum density of *S. sclerotiorum*.

Plant Population and Spacing

The incidence of Sclerotinia stem rot is dependent upon favorable weather conditions, but is also greatly influenced by plant population. A population of 360,000/ha or less, in an 0.5-m row width with 20 plants or less/m row, significantly reduced incidence of *S. sclerotiorum* cv. Paraná (Homechin, 1983a). Other work (Grau and Radke, 1984) has shown that disease severity was greater at row widths of 25-28 cm than for row widths of 76 cm, but no information was given as to the plant population. Further research is needed to establish the most appropriate plant population and the best combination of row width and plant spacing for soybean cultivars of different maturity groups and plant architecture, and for soil fertility levels.

Biological Control

Preliminary studies on biological control of *S. sclerotiorum* have shown that a species of *Trichoderma* fungus applied in the soil or by seed treatment can help reduce incidence of Sclerotinia stem rot. An isolate of *Trichoderma* sp., obtained from colonized sclerotia of *S. sclerotiorum* and grown on autoclaved black oat (*Avena strigosa*

Schreb.), significantly reduced incidence of Sclerotinia stem rot (Homechin, 1983b). The antagonistic fungus was applied in four ways: (a) in furrow treatment by mixing 100 g of inoculum/m of row, (b) by spraying on the soil surface, (c) by spraying a spore suspension on the soil and/or on the plants at flowering, and (d) as a seed treatment. Disease incidence was reduced when the spore suspension was sprayed on the soil at planting, by spraying the plants at flowering, by a combination of both, and by seed treatment (Homechin, 1983b).

Chemical Control

Seed-borne *S. sclerotiorum* may be the main source of inoculum introduced into first-year soybean fields in the region of soybean expansion in the high plains of central Brazil, and from there to new areas of the northern regions. Currently, no restriction is imposed upon movement of soybean seed produced in fields where the pathogen is present. As an alternative measure for restricting the dissemination of seed-borne inoculum, fungicides have been tested for seed treatment. Fungicides and the dosage (active ingredient/kg of seed) which gave 100% control of *S. sclerotiorum* were: captan (1.5 g), benomyl (1 g), thiabendazol (0.2 g), and thiram (2.1 g) (Homechin and Henning, 1983).

In addition to seed treatment, fungicides also have been tested as a spray application to plants (Homechin, 1983a). Procimidone (1 and 0.75 kg a.i./ha), iprodione (1 and 0.2 kg a.i./ha) and vinclozolin (0.5 kg a.i./ha), applied at flowering and 15 days later, reduced the incidence of *S. sclerotiorum* infection. Further studies are needed to determine the economic feasibility of fungicide sprays. There also is need to adapt spray nozzles to allow proper coverage of stem and flower parts.

Survival of the Pathogen

The longevity of sclerotia of *S. sclerotiorum* is an important component to be considered in any soybean production system where the pathogen is already established. Gasparotto et al. (1982) studied the longevity of *S. sclerotiorum* in a field following a runner bean crop (*Phaseolus vulgaris* L.) that was severely damaged. Plots covered by *Brachiaria decumbens* stapf. and *Melinis minutiflora* Beauv. for 7 months showed great reduction in the number of sclerotia surviving at the depth of 0.5 cm. Nevertheless, it was not sufficient to eliminate Sclerotinia rot of lettuce (*Lactuca sativa* L.) grown subsequently on this area.

Work carried out in a soybean field severely damaged by Sclerotinia stem rot showed that, when sclerotia were buried at depths of from 5 to 50 cm, viability was unchanged after 18 months (Table 2). Sclerotia left on the soil surface were viable for 12 months, but were completely dead after 18 months. Four species of fungi were commonly associated with sclerotia at all burial depths: *Corynespora* sp., *Epicocum* sp., *Fusarium* sp., and *Trichoderma* sp. *Corynespora* sp. was the most frequent (Homechin, 1983a). During the period when the experiment was carried out, the area was covered by various weeds that grew naturally. The study is being continued.

Work by Cook et al. (1975) showed that nearly 75% of sclerotinia buried 5, 12.5 and 20 cm below the soil surface were recovered after 3 years. The majority were recovered from 5 to 10 cm. Survival was adversely affected by high soil moisture and temperature. According to these authors, the capacity of sclerotia to form secondary sclerotia and their longevity in soil insures that inoculum will be present during a 3-year crop rotation. Adams (1975) also found that sclerotia survived well over a 15-month period at 2.5, 15.0 and 30 cm depths, but poorly at 50 cm. Reduced survival of sclerotia in the field is thought to be due to the drying and remoistening of sclerotia

Table 2.   Survival of *Sclerotinia sclerotiorum* buried at 0 to 50 cm depths in a field (Homechin, 1983a).

Depth of burial	Viable sclerotia after		
	6 months	12 months	18 months
— cm —		— % —	
0	90[a]	80	0
5	97	98	94
10	94	96	95
15	93	95	96
20	76	97	87
25	77	98	98
30	54	97	94
35	59	95	93
40	59	95	99
45	49	99	98
50	48	95	98

[a]Average of 10 replicates of 100 sclerotia each.

near the soil surface (Adams, 1975). A very thorough review on survival and germination of sclerotia of various fungal species is presented by Coley-Smith and Cooke (1971).

Plant Resistance

Beginning with the 1980-81 crop season, over 800 soybean germplasm lines maintained by the National Soybean Research Center, Londrina, and 76 commercial cultivars were tested under field conditions (Homechin, 1982a, 1983a). Due to variation of climatic conditions from year to year, disease levels were very inconsistent, making selection very difficult.

Selections are being made based on rate of survival or percent of plants without disease symptoms. Accessions or cultivars with two-thirds or greater survival, or without infection in one meter of row, are being selected for further tests. None of the lines and cultivars has shown complete resistance, but many have consistently shown less disease. Among 76 commercial cultivars, 12 have been least affected: Bienville, Bossier, Coker 136, FT-2, Hampton, IAC-6, IAC-9, LC 72-749, Mineira, Numbaíra, UFV-4 and Viçoja (Homechin, 1982a, 1983a).

Previous work also has shown that some soybean cultivars are less susceptible to *S. sclerotiorum* infection: William, Union (Cline and Jacobsen, 1983); Corsoy (Cline and Jacobsen, 1983; Grau and Bissonette, 1974; Grau and Radke, 1984); Hodgson, Hodgson 78 (Grau et al., 1982); Ada, Clay, Beeson, Dunn, Norman and Portage (Grau and Bissonette, 1974). These may serve as possible sources of resistance in a breeding program. Studies by Grau et al. (1982) have indicated that resistance or susceptibility to *S. sclerotiorum* by soybean was not related to plant architecture or maturity group.

Tillage Practices

As shown by the studies on sclerotia survival (Table 2) (Homechin, 1983a), most sclerotia remain viable for 12 months when left on the soil surface. Thus, it would be expected that, with no-till, there would be a potential risk for a second crop of soybean established within 7-8 months after the previous crop. Studies on tillage practices, where no-till was compared with plowing (15-20 cm deep) + discing (twice) and deep plowing (40 cm deep) + discing (twice), showed that Sclerotinia stem rot under no-till was higher than on the other two treatments (Table 3).

Weeds as Alternate Hosts

Besides the sclerotia produced on infected soybean plants, several weed species are common hosts to *S. sclerotiorum* and may further increase sclerotia density in the field. In a survey made during a normal soybean season the following weed species were found infected by *S. sclerotiorum: Amaranthus* spp., *Bidens pilosa* L., *Brassica napus* L., *Borreria alata* DC., *Emilia sonchifolia* DC., *Euphorbia heterophylla* L., *Galinsoga parviflora* Cav., *Ipomoea* sp. and *Sida rhombifolia* L. (Homechin, 1982b).

No information was found on the effect of weeds on Sclerotinia stem rot incidence, but it would be expected that a soybean field infested with weeds would have higher disease incidence and a potentially greater inoculum density for the following crop. Thus, weed control should benefit in reducing Sclerotinia stem rot of soybean.

CONCLUSION

Early detection of the disease is especially important. This would allow for the adoption of cultural practices or crop rotations that could prevent the development of the disease to epidemic proportions. Further studies are needed on plant densities and spacings that would reduce disease incidence. The effects of rotation, sequential cropping, and tillage practices on sclerotia and disease incidence need further investigation in relation to precipitation and temperature. Use of overhead irrigation may predispose crops to greater disease severity. Irrigation management, especially during flowering, must be of primary concern where *S. sclerotiorum* is present.

Epidemiological studies are needed to better understand the mechanism and factors involved in the establishment and outbreak of the disease. Information is lacking on time of apothecial production during the crop season. If available, this information would allow for proper choice of sowing time or cultivar to prevent soybean flowering occurring at the time of apothecial production. By monitoring field weather data, especially temperature and relative humidity, a disease forecasting scheme could be developed. Information on the role of seed-borne inoculum would greatly contribute to the establishment of control strategies in areas where soybean is to be planted for

Table 3. Effect of tillage practices on the incidence of Sclerotinia stem rot of soybean (Homechin, 1982a, 1983a).

	1980-81	1981-82	1982-83
	—% Plants infected —		
No tillage	2.6[a]	2.1	12.9
Plowing (15-20 cm deep) + discing (2 x)	1.6	1.0	4.1b
Deep plowing (40 cm deep) + discing (2 x)	1.5	0.8	0.9b

[a]Average of four replicates of 5 x 10 m plots.

the first time. The selection and use of soybeans that are less affected by the disease could be an essential part in disease control strategy.

NOTES

José T. Yorinori and Martin Homechin, EMBRAPA/CNPS, C.P. 1061, 86 100 Londrina, PR, Brazil.

REFERENCES

Abawi, G. S. and R. G. Grogan. 1975. Source of primary inoculum and effects of temperature and moisture on infection of beans by *Whetzelinia sclerotiorum*. Phytopathology 65:300-309.

Adams, P. B. 1975. Factors affecting survival of *Sclerotinia sclerotiorum* in soil. Plant Dis. Reptr. 59:599-603.

Cline, M. N. and B. J. Jacobsen. 1983. Methods for evaluating soybean cultivars for resistance to *Sclerotinia sclerotiorum*. Plant Dis. 67:784-786.

Coley-Smith, J. R. and R. C. Cooke. 1971. Survival and germination of fungal sclerotia. Ann. Rev. Phytopathol. 9:65-92.

Cook, G. E., J. R. Steadman and M. G. Boosalis. 1975. Survival of *Whetzelinia sclerotiorum* and initial infection of dry edible beans in Western Nebraska. Phytopathology 65:250-255.

Gasparotto, L., G. M. Chaves, and A. R. Condé. 1982. Sobrevivência de *Sclerotinia sclerotiorum* em solos cultivados com gramíneas. Fitopatol. Bras. 7:223-232.

Grau, C. R. and H. L. Bissonette. 1974. Whetzelinia stem rot of soybean in Minnesota. Plant Dis. Reptr. 58:693-695.

Grau, C. R. and V. L. Radke. 1984. Effects of cultivars and cultural practices on Sclerotinia stem rot of soybean. Plant Dis. 68:56-58.

Grau, C. R., V. L. Radke and F. L. Gillespie. 1982. Resistance of soybean cultivars to *Sclerotinia sclerotiorum*. Plant Dis. 66:506-508.

Homechin, M. 1981. Avaliação de perdas de produção de soja devida a incidência da podridão branca da haste pelo fungo *Sclerotinia sclerotiorum*. p. 331-332. *In* Empresa Brasileira de Pesquisa Agropecuária. Resultados de pesquisa de soja 1980-81. Centro Nacional de Pesquisa de Soja, Londrina.

Homechin, M. 1982a. Epidemiologia e controle de *Sclerotinia sclerotiorum*. p. 181-200. *In* Empresa Brasileira de Pesquisa Agropecuária. Resultados de pesquisa de soja 1981-82. Centro Nacional de Pesquisa de Soja, Londrina.

Homechin, M. 1982b. Plantas daninhas hospedeiras de *Sclerotinia sclerotiorum* (Abst.). Fitopatol. Brasil. 7:472.

Homechin, M. 1983a. Epidemiologia e controle de *Sclerotinia sclerotiorum*. p. 164-177. *In* Empresa Brasileira de Pesquisa Agropecuária. Resultados de pesquisa de soja 1982-83. Centro Nacional de Pesquisa de Soja, Londrina.

Homechin, M. 1983b. Controle biológico de patógenos da soja. p. 184-185. *In* Empresa Brasileira de Pesquisa Agropecuária. Resultados de pesquisa de soja 1982-83. Centro Nacional de Pesquisa de Soja, Londrina.

Homechin, M. and A. A. Henning. 1983. Tratamento de sementes de soja com fungicidas para controle de *Sclerotinia sclerotiorum*. (Abst.). Fitopatol. Brasil 8:649.

Martinez, C. A., A. J. Ivancovich and G. Botta. 1984. Perdidas ocasionadas por *Sclerotinia sclerotiorum* (Lib.) de Bary en el partido de Pergamino (Buenos Aires, Argentina) durante três ciclos de cultivo de soja. *In* Seminário Nacional de Pesquisa de Soja, 3, Campinas, 1984. EMBRAPA-CNPS, Londrina. (In press).

Yorinori, J. T. 1982. Doenças da soja na região de São Gotardo, Minas Gerais. Congresso Brasileiro de Fitopatologia, 15 (Abst.). São Paulo, 1982.

ETIOLOGY, EPIDEMIOLOGY, AND CONTROL OF STEM CANKER

Paul A. Backman, David B. Weaver, and Gareth Morgan-Jones

Stem canker disease of soybean is one of a group of disorders caused by members of the *Diaporthe-Phomopsis* teleomorph/anamorph complex. The pathogen responsible for stem canker is the fungus *Diaporthe phaseolorum* (Cke. & Ell.) Sacc., var. *caulivora* Athow and Caldwell [Dpc] (Athow and Caldwell, 1954). It is related to the causal organism of pod and stem blight *D. phaseolorum* var. *sojae* (Lehman) Wehm. (Dps), and to a *Phomopsis* anamorph unconnected with an ascomycete state which also occurs on soybean. Both Dpc and Dps have *Phomopsis* anamorphs. Dps and the independent *Phomopsis* anamorph are particularly important in seed decay and poor seed quality (Schmitthenner and Kmetz, 1980). Although Dpc also reduces seed quality, it has become the subject of intensive research in the past five years, primarily because of its ability to kill the soybean plant well before harvest. Losses in the southeastern United States were estimated at 43 million dollars in 1983, with many legal actions initiated because of alleged spread by seed to noninfested fields.

HISTORY

The symptoms of stem canker were originally described by Crall (1950) in Iowa, when the causal fungus was identified as *D. phaseolorum* var. *batatatis* (Hunter & Field) Wehm. Later, Athow and Caldwell (1954) redescribed and named the organism *D. phaseolorum* var. *caulivora*. During the late 1940s and early 1950s the disease was prevalent in upper midwestern USA, but diminished in importance when the cultivars Hawkeye and Blackhawk were eliminated from production. Since then it has occurred sporadically, but with little impact on yield. However, that the pathogen is still endemic in the upper midwestern USA can be deduced from the frequent reports of Dpc occurrence when seeds are cultured (Kmetz et al., 1978).

During the past 10 years reports of significant damage to soybeans because of stem canker have been published. Earliest observations were made in Mississippi in 1973 (B. Moore, unpublished), followed by spread into Alabama in 1977, Tennessee in 1981, South Carolina and Georgia in 1982, and Florida, Louisiana, and Arkansas in 1983. In all states, some cultivars were almost destroyed (seed yields ‹100 kg/ha), while other cultivars were either unaffected or damaged at intermediate levels (Backman et al., 1981). Reports from Florida (Shokes and Sprenkle, unpublished) in 1983 found one county (Escambia) with 80% of the fields infected, while the counties on either side had less than 2%. These anomalies point to major differences in severity that are epidemiologically based and controlled by factors that are at best poorly identified. The recent report by Keeling (1984) of six races of Dpc indicates the potential for further adaptation and spread into previously un-affected areas.

MYCOLOGY

Within the *Diaporthe phaseolorum* (= *Sphaeria phaseolorum* Sacc.) species concept, four entities, which have been given varietal taxonomic rank, are recognized. These have, in large part, been based upon host parasite association. Morphologically, there are few, if any, differences among them. The lectotype variety, var. *phaseolorum* oc-curs on lima beans (*Phaseolus limensis* Macf.) while var. *batatatis* (= *Diaporthe batatatis* Hartner & Field) is the cause of dry rot of sweet potatoes (*Ipomoea batatas* [L.] Lam.). As stated above, vars. *caulivora* and *sojae* (= *Diaporthe sojae* Lehman) occur on soybean.

Wehmeyer (1933) recognized the similarity of the three teleomorphs—*D. phas-eolorum, D. batatatis* and *D. sojae*—when he transferred the latter two specific epithets to varietal rank within the former. He stated that all three were morphologically very similar and probably fully conspecific. Kulik (1984) concurred with this assessment and advocated the adoption of but one binomial for the three taxa. The valid name, based on nomenclatural date priority, is *D. phaseolorum. D. batatatis* and *D. sojae*, and the varietal entities, based on the recombination of these specific epithets, should therefore be treated as facultative synonyms of *D. phaseolorum. D. phaseolorum* is considered to be plurivorous in habit, but within it there are no stable morphological discontinuities sufficient to warrant separate, formal taxonomic status for biotypes occurring on different hosts and inducing differing disease syndromes.

Diaporthe phaseolorum var. *caulivora* was established by Athow and Caldwell (1954) and differentiated from var. *sojae* on the basis of the absence of an associated anamorph, occurrence of perithecia in caespitose clusters rather than singly, possession of shorter and more tapering perithecial beaks, and smaller asci and ascospores. Welch and Gilman (1948) had previously found isolates from soybean cankers (which, as mentioned before, was determined to be a strain of var. *batatatis*) to differ from var. *sojae* in pathogenicity and type of resultant disease induction, homothallism, caespi-tose perithecia, and absence of pycnidia. Although the validity of some of these dis-tinctions have been accepted, by and large, some question exists as to their taxonomic significance. Threinen et al. (1959) obtained radiation-induced mutants of both vars. *caulivora* and *sojae* that differed from parental types in morphology, pathogenicity, and genetic behavior. Whitehead (1966), who documented stem canker induction on soybean and birdsfoot trefoil (*Lotus corniculatus* L.) infected by var. *sojae*, reached a similar conclusion to Threinen et al. that the two varieties were insufficiently distinct to justify separate taxonomic rank. Schmitthenner and Kmetz (1980) added credence to this view when they reported that occasional isolates of var. *sojae* were able to pro-duce canker-like symptoms. Added to this, Hildebrand (1954) noted that occurrence of perithecia of var. *caulivora* in caespitose clusters was not always constant, and both he and Frosheiser (1957) found pycnidia in isolates from soybean stem canker.

There appear to be insufficient grounds for maintenance of var. *caulivora* as a separate taxonomic entity (Kulik, 1984), especially given that the other three varieties are considered superfluous and therefore redundant. The strains associated with stem canker, however, as Kmetz et al. (1978) have pointed out, can be consistently dis-tinguished from others of *D. phaseolorum*. To accommodate this fact, Kulik (1984) has advocated that soybean stem-canker-causing strains, at least provisionally, be re-ferred to as forma specialis *caulivora* to indicate association with a distinct disease condition. This recognizes that the distinction mainly involves physiological reaction, although this may be accompanied by slight morphological differences, especially in cultural characteristics *in vitro*.

RACES

Keeling (1984) has compared Dpc isolates from across the United States for their pathogenicity on various soybean cultivars. He suggests that there are six physiologic races that can be differentiated by the cultivars Kingwa, Tracy-M, Arksoy, Centennial, S-100, and J77-339. All northern isolates were classified as Races 4, 5, or 6, while southern isolates were Races 1, 2, or 3. Further, Keeling observed that northern cultivars were susceptible to southern isolates using the toothpick inoculation method. Morgan-Jones (1984) observed that there were cultural and/or some morphological differences, particularly ascospore shape between northern and southern races which was supported by Kulick's report (1984). Morgan-Jones and Backman (unpublished) indicated that observed differences might justify separation of northern and southern races into separate forma speciales of *D. phaseolorum*.

EPIDEMIOLOGY

Long distance movement of Dpc has been of particular concern since severe forms of the disease were found in the southeastern United States. There is ample evidence that the organism can be found on seed. However, typically seed from even the most severely affected southern fields will not exceed 5% detectable infestation (Crawford, 1984). This contrasts with studies of the northern stem canker organism which has been reported in seed at high frequencies, often ranging from 30 to 60% infestation (Kmetz et al., 1978). Observations made by numerous agronomists and plant pathologists working in the South often have revealed major differences in stem canker severity when comparing two seed lots of the same cultivar, planted side-by-side on the same day, and on land with no previous history of stem canker. This can only indicate that seed are involved in disease spread. The efficiency of spread by seed is a very controversial subject. There are many papers that indicate all members of the *Diaporthe-Phomopsis* complex greatly reduce seed germination (Schmitthenner and Kmetz, 1980). Even with low percentages of infested seed, and their reduced germinability, severe cases of stem canker have developed with seed as the only apparent inoculum source. This is supported by observations indicating a clustering of diseased plants in these same fields, and may indicate possible secondary disease cycles originating from primary inoculum introduced on seed (Backman and Crawford, unpublished). Observations indicating that soybean seed from the same lot planted on noninfested land at several different locations develop levels of stem canker ranging from 0 to 100% diseased plants (R. Burdett, personal communication), support the hypothesis that the fungus borne on seed infects the plant through a secondary cycle, not by direct penetration from the cotyledon into the plant. However, no physical evidence of a secondary cycle originating from seed (i.e., pycnidia or perithecia formed on shed cotyledons) has been reported. Typically, neither pycnidia nor perithecia are observed on infected plants during the summer season, although Krausz and Fortnum (1983), and Hildebrand (1956) have observed perithecia during the growing season. Our observations indicate that perithecia typically develop in late winter (February-March) in the southeastern states, and that ascospore release begins in late April and continues into June. It is these spores that are responsible for primary infections. Since ascospores are exuded in a sticky matrix, dispersal is primarily achieved by splashing of raindrops and windborne rain. These periods of rain not only serve for dispersal, they also supply the necessary moisture for infection. The conditions necessary for infection have not been defined. Other spread, either on a field-to-field basis or even on a regional basis, points to movement on debris contained in field equipment.

Athow (1954) found that practically all natural infection occurs through the leaves, and that removal of the first six trifoliolate leaves prevented stem canker development. Spray trials have indicated that fungicide applications must be made between V_2 and V_6 to control stem canker adequately (Backman, 1984; A. Chambers, unpublished). These facts, and evidence that reduced disease is associated with delayed planting (Weaver, 1984) indicate that spore release and plant infection occur early in the crop season, and probably relate to reduced spore dispersal due to exhaustion of perithecia. These observations open the door for research on prediction of spore release for more accurate timing of spray applications.

The possibility of alternative hosts for Dpc has been examined by several workers. Roy and Miller (1983) found *Diaporthe phaseolorum* on cotton (*Gossypium hirsutum* L.) that produced stem cankers when inoculated in soybean. Hildebrand (1956) was unsuccessful in establishing infections in several weed species common to his research location in Canada. However, Roy (pers. commun.) isolated fungi tentatively identified as *D. phaseolorum* that produced cankers when inoculated to soybeans.

DISEASE EVALUATION AND LOSSES

Several systems for the evaluation of stem canker severity have been utilized by scientists working independently. The only system we have found that provides a strong correlation between disease severity and yield is a pretransformed arcsine scale developed by our group (Backman, 1984; Weaver et al., 1984). This rating scale provides a string linear relationship between disease and yield, and is easily taught to the inexperienced evaluator. Data developed in 1983 spray trials relating treatment means for disase to treatment means for yield produced a linear regression line with an r^2 of 0.94, indicating that almost all of the yield variation was accounted for by the rating scale.

CONTROL

An understanding of the biology of Dpc has led to the development of tactics and recommendations for controlling stem canker of soybean. These have encompassed a broad range of methods ranging from sanitation, seed treatment and certification, rotation, time of planting, cultivar selection, and post-emergence fungicides.

Sanitation and Clean Seed

The prevention of the movement of Dpc to noninfested fields was first directed at the mechanisms of long distance movement of the organism. Any farm equipment that moves from infested to noninfested fields should be cleaned of all plant debris and soil. This recommendation must be coupled with a knowledge of the disease condition of the field in the previous season, and an ability to recognize the symptoms on plants when they occur in a field at frequencies of 0.1% or less. The second method of movement is on seed. No seed should be planted in noninfested fields that were harvested from fields with Dpc infected soybeans (Gazaway and Henderson, 1983; Moore et al., 1983). Seed sources often are unknown because seed from many fields are pooled when they are placed in commercial seed channels. Seedsmen should examine seed fields before harvest to determine the stem canker status. Treatment of seed with fungicides can greatly reduce stem canker, but does not eliminate it (Crawford, 1984). Carboxin-thiram and carboxin-thiram-captan were found to be best for Dpc infested seed. It is our opinion that fungicide should be used as an insurance against major levels of stem canker entering a previously noninfested field; that it should be used only on seed thought to be noninfested; and that it should be used with the understanding that it will not totally eliminate Dpc if the organism is present.

Post-emergence Chemicals

Early efforts to control stem canker relied on applications made just before symptom development (Backman et al., 1981). Typically, these applications were made in the early reproductive stages and indicated benefits only when systemic fungicides were applied. Efficacy was often enhanced by the addition of adjuvants (i.e., oil:surfactant blends), though our results were erratic between fields and trials run in different years. In 1982, our research trials indicated that benomyl applications during the vegetative period were more beneficial than those made during reproduction. Further, Chambers (unpublished) found that nonsystemic (contact) fungicides were efficacious when applied in the early vegetative period. Data from our 1983 fungicide spray trials support our previous observations of spore release and infection during May and June. Evaluations of fungicides for control of stem canker on cultivars of various levels of susceptibility indicate that control cannot be achieved on highly susceptible cultivars. Cultivars of intermediate susceptibility respond to fungicide treatment with yield increases commensurate with disease control, while no benefits were seen when resistant cultivars were treated (Backman, 1984). These data indicate that stem canker can be managed on cultivars of moderate susceptibility through the use of foliar fungicides. Spray applications should be applied in the early vegetative growth stages while spores are actively being produced. Fungicidal sprays applied during the early vegetative period were very economical because they could be banded (15 to 20 cm wide) over the small plants. Refinement of the treatment time could be achieved if there were more detailed knowledge of infection parameters.

Cultivar Reaction

Differential reaction of soybean cultivars to stem canker infection was reported by several investigators in the United States and Canada in the 1950s. Weiss (1952) noted the extreme susceptibility of Hawkeye. Athow (1954) compared Hawkeye with Harosoy under conditions of natural infestation and reported 5% infected plants (3% mortality) with Harosoy and 49% infected plants (33% mortality) in Hawkeye. He suggested that the buildup of stem canker in Indiana was largely due to widespread sowing of Hawkeye. Hildebrand (1956) reported a slight differential reaction of Blackhawk and Lincoln to artificial inoculation, and noted the general field susceptibility of Blackhawk and Hawkeye and the general field resistance of Harman and Harosoy. However, these cultivars tended to respond similarly when artificially inoculated in the seedling stage. Johnson (1955) reported that an extensive search had failed to reveal any cultivars with high levels of resistance.

Concurrent with reports of outbreaks of stem canker in southeastern USA, differential cultivar reaction was also noted. Backman et al. (1981) reported a complete range of cultivar reaction, with no disease observed on the cultivar Tracy-M to severe infection on Hutton and RA-800. Intermediate reaction was observed on most cultivars. Field susceptibility of Hutton, Bragg, and Coker 237 was reported by Krausz and

Fortnum (1983). Weaver et al. (1984) conducted extensive field experiments evaluating cultivars for disease development and seed yield under field conditions and confirmed the disease resistance of Tracy-M, but found other cultivars such as Braxton, Ransom, Davis, Wright, and Deltapine 105 to be equal to, or higher yielding than Tracy-M even though they showed higher levels of stem canker. Keeling (1982) established resistance of Tracy-M by inoculating seedlings in the greenhouse with toothpicks containing *D. phaseolorum* var. *caulivora*. Keeling also inoculated CNS as a suspected source of resistance, and Peking as a suspected source of susceptibility, because these cultivars tended to appear in the pedigrees of resistant and susceptible types, respectively. While CNS proved to be as resistant as Tracy-M to greenhouse inoculation, Peking was rated as moderately resistant. Thus, the source of susceptibility remains in doubt. Inheritance of genes controlling resistance are still being studied. Based on results of these studies and observations by workers in many states, cultivars are generally divided into four major groups (Table 1) on the basis of stem canker resistance (Gazaway and Henderson, 1983; Moore et al., 1983).

Crop Rotation

Benefits of crop rotation for control of stem canker have not been demonstrated in southeastern USA, probably because of the relatively recent nature of the problem and the long term nature of rotation studies. Because the fungus overwinters on crop debris, it would be logical to assume that rotation with a nonhost crop would lead to decomposition of the debris and would have some value in controlling the disease. Preliminary results by D. L. Thurlow (pers. commun.) indicate the disease is less severe in a susceptible cultivar (Hutton) when planted following the more resistant Ransom, than in Hutton planted continuously. Evidence by Roy and Miller (1983) that stem canker inoculum can come from other host crops such as cotton may limit the value of rotation as a control measure. Currently, rotation to a nonhost crop, such as corn (*Zea mays* L.) or grain sorghum (*Sorghum bicolor* [L.] Moench) is recommended for at least two years following conditions of severe disease infestation (Gazaway and Henderson, 1983; Moore et al., 1983).

Table 1. Relative stem canker resistance of selected soybean cultivars.[a]

Resistant

Tracy-M (VI)[b]; Braxton (VII); Dowling (VIII)

Moderately Resistant

Bay, Terra Vig 505, Deltapine 105, Deltapine 345, Wilstar 550 (V);
Centennial, Davis, RA 680, Coker 156 (VI); Wright, Ransom, Coker 317,
GaSoy 17 (VII); Coker 368, Coker 488, Cobb (VIII)

Moderately Susceptible

Bedford, Forrest, Essex, A 5474 (V); Jeff, Lee 74, S69-96,
Deltapine 506 (VI); Gregg (VII); Foster, Kirby (VIII)

Susceptible

A 5939 (V); RA 604, Brysoy 9, Bradley (VI); Coker 237, Bragg,
McNair 770, Wilstar 790, RA 701 (VII); Coker 338, Hutton, RA 801 (VIII)

---

[a]From Weaver et al. (1984) and Moore et al. (1983).
[b]Numbers in parentheses refer to maturity groups of preceding cultivars.

Planting Date

Benefits of delayed planting, particularly for mid- to full-season, moderately susceptible and susceptible cultivars, have been demonstrated (Weaver, 1984). In one year (1982) delayed planting benefited even moderately resistant cultivars, but the following year dry weather limited yields of late-planted plots to the extent that only the most susceptible cultivars showed a yield increase. Disease levels in these experiments were less than 50% infected plants, even on susceptible cultivars. Additional research is needed under higher levels of disease pressure to assess adequately the effect of late planting. Delayed planting appears to affect yield loss to stem canker in two ways. Plants are able to avoid the initial release of inoculum from crop debris that generally occurs during May, so fewer plants are infected. In those that do become infected, the shorter vegetative and reproductive period induced by late planting allows less time for the disease to develop to the point that plant death will occur before pod fill.

Tillage

Because Dpc overwinters on crop debris, the effects of tillage will depend largely upon whether the previous year's crop was infected or inoculum is introduced from some other source. Tyler et al. (1983) reported significantly more stem canker in no-till plots compared to four other tillage regimes, speculating that the lower incidence of brown spot (*Septoria glycines* Hemmi) in no-till plots caused plants to retain their lower leaves, providing an entry site for stem canker. It was not known whether inoculum was present at planting in this experiment.

BREEDING FOR RESISTANCE

When stem canker became prevalent in the Midwest in the 1950s, severity declined with the elimination of susceptible cultivars, so that continued breeding efforts to control the disease were not necessary (Athow, 1973). Many cultivars listed in Table 1 as susceptible or moderately susceptible have been among the most popular cultivars in their respective maturity groups. What effect eliminating these cultivars from production will have on future incidence of stem canker is not known. However, the current seriousness of the problem has led many breeders in affected areas to begin screening for resistance (E. E. Hartwig, pers. commun.). With the adaptation of the toothpick inoculation technique by Keeling (1982), breeders can screen effectively large numbers of lines in the seedling stage against known isolates of the pathogen. A major problem facing breeders in the future is the apparent association between stem canker susceptibility and cyst nematode resistance. Of the cultivars listed in Table 1, only two (Centennial and Coker 368) in the resistant or moderately resistant categories also have resistance to both cyst and root-knot nematodes. This observation led Keeling (1982) to suspect Peking, a source of cyst nematode resistance, as the source of stem canker susceptibility. Though Peking proved to be moderately resistant to toothpick inoculation, its field resistance to southern USA isolates is difficult to determine because of its extreme early maturity in southern latitudes.

Thus, there are two major goals in breeding for stem canker resistance. First, to screen all breeding lines in the greenhouse or field to avoid the release of susceptible types. Second, to develop types that have combined resistance to stem canker and the major nematode species. This second objective may be accomplished by backcrossing stem canker resistance into lines with multiple nematode resistance. Backcrossing appears to be the method of choice because of the ease of screening seedlings in the greenhouse and observations that genes for resistance are qualitatively inherited (Keeling, pers. commun.). Backcrossing would also be the most rapid way of introducing resistance genes into a wide range of agronomically acceptable genotypes. Evidence that physiological races of the pathogen may exist (Keeling, 1984) would present complications for plant breeders and make the utilization of a single source of resistance, such as Tracy-M, inadvisable.

CONCLUSIONS

Soybean stem canker has emerged from an obscure disease to a disease of major economic concern in a period of 10 years. Data presently being developed indicate that the aggressive strains of *Diaporthe phaseolorum* var. *caulivora*, found in the southern United States, are different from their northern counterparts, but a definitive study on their taxonomic status has not been made. The severity of stem canker in southern U.S.A. has already forced farmers to abandon some cultivars and to alter several aspects of their cultural practices. Unfortunately, there are not enough well adapted cultivars to deal with the problem solely through this method. Such practices as early season fungicide applications may be necessary while a broader selection of stem canker-resistant cultivars are being developed. The long-term status of resistant cultivars is in jeopardy because of the known races that have already developed. Further, these southern races have not yet indicated a northern limit of adaptability, and could threaten the soybean production areas of the midwestern United States.

NOTES

Paul A. Backman and Gareth Morgan-Jones, Professors, Department of Botany, Plant Pathology, and Microbiology; David B. Weaver, Assistant Professor, Department of Agronomy, Auburn University, Auburn, Alabama 36849.

The authors acknowledge the assistance of Mark A. Crawford, Barbara Cosper, and Elisa Smith in the development of much of the research information. We also acknowledge the aid of the Alabama Soybean Producers who support our research through their check-off program.

REFERENCES

Athow, K. L. 1954. Some midwest diseases. Soybean Dig. 14(8):19.

Athow, K. L. 1973. Fungal diseases. p. 459-489. *In* B. E. Caldwell (ed.) Soybeans: Improvement, Production and Uses. American Society of Agronomy, Madison, Wis.

Athow, K. L. and R. M. Caldwell. 1954. A comparative study of Diaporthe stem canker and pod and stem blight of soybeans. Phytopathology 44:319-325.

Backman, P. A. 1984. Effects of timing and rates of application of fungicides for control of stem canker of soybeans. *In* M. M. Kulick (ed.) Proc. Conf. on *Diaporthe/Phomopsis*, Ft. Walton Beach, Fla. (In press).

Backman, P. A., M. A. Crawford, J. White, D. L. Thurlow and L.A. Smith. 1981. Soybean stem canker: A serious disease in Alabama. Ala. Agric. Exp. Stn., Auburn Univ. Highlights Agric. Res. 28(4):6.

Crall, J. M. 1950. Soybean diseases in Iowa in 1949. Plant Dis. Rep. 34:96-97.

Crawford, M. A. 1984. Seed treatments and tillage practices as they affect spread and control of stem canker. *In* M. M. Kulick (ed.) Proc. Conf. on *Diaporthe/Phomopsis*. Ft. Walton Beach, Fla. (In press).

Frosheiser, F. I. 1957. Studies on the etiology and epidemiology of *Diaporthe phaseolorum* var. *caulivora*, the cause of stem canker of soybeans. Phytopathology 47:87-94.

Gazaway, W. S. and J. B. Handerson. 1983. Stem canker disease in soybeans. Alabama Coop. Ext. Timely Information Sheet.

Hildebrand, A. A. 1954. Observations on the occurrence of the stem canker and pod and stem blight on mature stems of soybean. Plant Dis. Rep. 38:640-646.

Hildebrand, A. A. 1956. Observations on stem canker and pod and stem blight of soybeans in Ontario. Can. J. Bot. 34:577-599.

Johnson, H. W. 1955. Soybean production research. Soybean Dig. 15(11):80-82.

Keeling, B. L. 1982. A seedling test for resistance to soybean stem canker caused by *Diaporthe phaseolorum* var. *caulivora*. Phytopathology 72:807-809.

Keeling, B. L. 1984. Evidence for physiologic specialization in *Diaporthe phaseolorum* var. *caulivora*. J. Miss. Acad. Sci. Suppl. 29:5.

Kmetz, K., C. W. Ellett and A. F. Schmitthenner. 1978. Soybean seed decay: Prevalence of infection and symptom expression of *Phomopsis* sp., *Diaporthe phaseolorum* var. *sojae* and *D. phaseolorum* var. *caulivora*. Phytopathology 68:836-839.

Krausz, J. P. and B. A. Fortnum. 1983. An epiphytotic of Diaporthe stem canker of soybean in South Carolina. Plant Dis. 67:1128-1129.

Kulik, M. M. 1984. Symptomless infection, persistence, and production of pycnidia in host and non-host plants by *Phomopsis batatatae*, *Phomopsis phaseoli*, and *Phomopsis sojae*, and the taxonomic implications. Mycologia 76:274-291.

Moore, W. F., J. A. Fox and F. Killebrew. 1983. Stem canker of soybean. Mississippi Coop. Ext. Plant Dis. Dispatch M-120, Mississippi State, Miss.

Morgan-Jones, G. 1984. The *Diaporthe/Phomopsis* complex of soybeans: morphology and ecology. *In* M. M. Kulick (ed.) Proc Conf. on *Diaporthe/Phomopsis*. Ft. Walton Beach, Fla. (In press).

Roy, K. W. and W. A. Miller. 1983. Soybean stem canker incited by isolates of *Diaporthe* and *Phomopsis* spp. from cotton in Mississippi. Plant Dis. 67:135-137.

Schmitthenner, A. F. and K. T. Kmetz. 1980. Role of Phomopsis sp. in the soybean seed rot problem. p. 355-366. *In* F. T. Corbin (ed.) Proc. World Soybean Research Congress II. Westview Press, Boulder, Colorado.

Threinen, J. T., T. Kommedahl and R. J. King. 1959. Hybridization between radiation-induced mutants of two varieties of *Diaporthe phaseolorum*. Phytopathology 49:797-801.

Tyler, D. D., J. R. Overton and A. Y. Chambers. 1983. Tillage effects on soil properties, diseases, cyst nematodes, and soybean yields. J. Soil Water Conserv. 38:374-376.

Weaver, D. B. 1984. Cultivar and planting date effects on field infestations of stem canker. *In* M. M. Kulick (ed.) Proc. Conf. on *Diaporthe/Phomopsis*. Ft. Walton Beach, Fla. (In press).

Weaver, D. B., B. H. Cosper, P. A. Backman and M. A. Crawford. 1984. Cultivar resistance to field infestations of *Diaporthe phaseolorum* var. *caulivora*. Plant Dis. 68:(In press).

Wehmeyer, L. E. 1933. The genus Diaporthe Nitschke and its segregates. Univ. Michigan Press, Ann Arbor.

Weiss, M. G. 1952. Agronomic research goals in soybean. Soybean Dig. 12:57.

Welch, A. W. and J. C. Gilman. 1948. Hetero- and homothallic types of *Diaporthe* of soybeans. Phytopathology 38:628-637.

Whitehead, M. D. 1966. Stem canker and blight of birdsfoot-trefoil and soybean incited by *Diaporthe phaseolorum* var. *sojae*. Phytopathology 56:396-400.

BROWN STEM ROT OF SOYBEANS

L. E. Gray

Brown stem rot (BSR) of soybean caused by *Phialophora gregata* (Allington and Chamberlain) W. Gams is a vascular disease of soybean plants that was first reported in 1944 in Illinois. Since that time it has been reported in the soybean producing states in midwestern USA as well as numerous states in the southern USA. This disease reduces soybean yields under certain environmental conditions (Gray, 1972; Gray and Sinclair, 1973; Weber et al., 1966).

DEFINITION OF FUNGUS ISOLATES PATHOGENICITY

Allington and Chamberlain (1948) emphasized the severe leaf necrosis caused by *P. gregata* when they first reported brown stem rot in Illinois. In their early work, they reported plant kill and leaf necrosis of soybeans inoculated with certain isolates of the fungus. Gray (1971) reported two pathogenic types of *Phialophora* isolates that could be differentiated by production of leaf symptoms on inoculated soybeans. Type 1 isolates caused typical leaf necrosis as well as stem vascular browning. Type 2 isolates caused only stem vascular browning and no leaf symptoms. Most reports in the literature have emphasized stem vascular browning with little mention of associated leaf symptoms. Consequently, determination of which type of isolate was used in these experiments is difficult.

Apparently, Type 2 isolates have little effect on soybean growth or yield (Gray, 1971). When soybeans were inoculated in replicated field plots with both Types 1 and 2 isolates, both isolates caused stem vascular browning, but only the Type 1 isolates reduced soybean yield (Gray, 1971). Type 1 isolates of the fungus have been recovered from infected soybean plants from Illinois, Indiana, Wisconsin, Minnesota, and Iowa.

SPORE GERMINATION AND FUNGUS SPORE PRODUCTION

Germination of *P. gregata* spores is affected by temperature. Allington and Chamberlain (1948) reported that only 0 to 9% of conidia germinated at 30 C, and 88% at 21 to 25 C. Little subsequent work has been done on determining the influence of air temperature on spore germination and sporulation of the fungus. Recent work in my laboratory has dealt with the effect of temperature on the rate of spore germination and sporulation of Types 1 and 2 isolates of the fungus. Spores readily germinate from 19 to 30 C, although germination is slower at 30 C for all isolates tested to date. One very important finding is that the fungus sporulates at temperatures from 19 to 25 C, but does not produce any spores at temperatures greater than 25 C. Where spores were subjected to an alternating temperature cycle of 30 C and 35 C for 7 h, followed by 17 h at 23 C, the spores germinated but no conidia were produced.

COLONIZATION OF SOYBEAN LEAVES BY PHIALOPHORA

My greenhouse studies have demonstrated that the fungus readily colonizes leaves of inoculated soybean plants that are showing symptoms. Leaf colonization generally progresses up the plant from the lower nodes to the upper nodes. Air temperature was found to influence leaf colonization of inoculated plants. Soybeans were inoculated by injecting petioles of individual leaves with spores (10^6 spores/mL). Replicated plants were then placed at two temperatures: 27 C for 8 h followed by 17 h at 23 C, and 7 h at 33 C followed by 17 h at 23 C. Three weeks after inoculation, leaves were removed from plants held at both temperature regimes. Leaf discs from veinal tissue of the leaves were plated onto water agar. The fungus was recovered from leaves of inoculated plants held at the lower temperature, but not from leaves of inoculated plants held at the higher temperature.

In additional work, plants were incoulated by a root dip technique (Sebastian et al., 1983) and held for 2 weeks at the first temperature regime. Then some of the plants were transferred to the high temperature regime. After 3 weeks the plants were sampled for leaf colonization by removing leaves of the plant and expressing the sap from the leaves with a leaf pressure chamber. The sap was collected aseptically and plated onto water agar. Leaf discs were also removed from leaves and plated onto water agar. The fungus was readily recovered from leaves of inoculated plants held at the lower temperature, but not from plants held at the high temperature. Spores of the fungus were also recovered from the expressed sap of inoculated plants held at the low temperature, but not from plants held at the high temperature.

SOURCES OF RESISTANCE

Chamberlain and Bernard (1968) reported that PI 84946-2 showed a low level of plant infection. Since that time, PI 86150 has also been reported to be resistant, based on low rates of plant infection under field conditions. Recently, several lines have been released (BSR 201 and BSR 302) that show a yield advantage in fields where brown stem rot is a chronic problem (Tachibana et al., 1983a,b). All of these lines result from the resistant parent PI 84946-2. Recently, from my own greenhouse tests, PI 437823 has shown a good level of resistance to defoliating strains of the fungus. This soybean plant introduction is of Group I maturity. Crosses have already been made between this introduction and numerous adapted commercial cultivars and will be evaluated for resistance and yield potential over the next growing seasons.

SCREENING PROCEDURES FOR RESISTANCE

Two important concepts that may help clarify assessment of soybean brown stem rot are the terms: disease incidence and disease severity. Disease incidence is a measure of the number of plants in a sample infected with the fungus. The percentage of plants infected at maturity, an index which has been used by numerous investigators, is an example of disease incidence. Disease severity is a measure of the amount of plant tissue infected; for example, the number of stem nodes which are colonized or in which vascular browning occurs (Phillips, 1971).

The relationship of the amount of disease to plant growth or yield can be based on incidence or severity. Some researchers have used the number of stem nodes with stem vascular browning (Phillips, 1971; 1973) or the percentage of the total length of stem with vascular browning (Tachibana and Card, 1979) as a measure of disease severity. These assessments have been used for measuring yield or for comparing pathogenicity of isolates. In most cases there have been poor correlations or erratic results in attempts to relate these assessments of BSR severity to yield. Similarly, the yield advantages of resistant over susceptible cultivars have been inconsistent. Tachibana and Card (1979) used the percentage of the stem with vascular browning as an indicator of soybean brown stem rot severity for evaluating soybeans for resistance and yield of resistant soybeans. In subsequent work (Ertl and Fehr, 1983) failed to show any significant correlation between percentage of stem browning and yield

Recent work at Illinois (Sebastian, 1984) has revealed significant negative correlations between soybean yield and leaf symptom severity attributed to *Phialophora*. Soybean lines that showed severe leaf symptoms, such as Century and Cumberland, yielded about 16% less than lines that showed no leaf symptoms. Further, there was very good agreement between greenhouse screening tests and severity of leaf symptoms and yield of plants in the field. Plants that developed severe leaf symptoms upon inoculation in the greenhouse also showed leaf symptoms in the field and were generally lower yielding than plants that showed resistance in greenhouse inoculation tests.

TOXIN PRODUCTION BY PHIALOPHORA ISOLATES

Kobayashi and Ui (1980) reported that an Iowa isolate of *P. gregata* produced a toxin that could be readily isolated in culture. They named the material "gregatin" and showed that at least five related materials were produced in culture filtrates. Gregatin A inhibited the growth of any bacteria and fungi in bioassays (Kobayashi and Ui, 1977). Gregatin A also caused vascular browning and leaf necrosis of soybeans that were exposed to 50 μg/g of material.

I have found that pathogenic isolates (isolates that cause leaf necrosis of inoculated plants) produce material in culture that, upon partial purification, causes leaf necrosis of soybean plants; however, nonpathogenic isolates of the fungus do not produce the same material in culture. The production of material is temperature sensitive. Preliminary work indicates that the material is produced at culture temperatures from 19 to 25 C, but not at 30 C. A rapid bioassay procedure for screening Phialophora isolates for toxin production has been developed using the alga *Chlamydomonas rheinhardii* as a bioassay organism. Future work is needed to determine the role of these metabolites of Phialophora in disease development.

NOTES

L. E. Gray, Plant Pathologist, USDA, ARS and Department of Plant Pathology, University of Illinois, Urbana, Illinois 61801.

REFERENCES

Allington, W. B. and D. W. Chamberlain. 1948. Brown stem rot of soybean. Phytopathology 38: 793-802.

Chamberlain, D. W. and R. L. Bernard. 1968. Resistance to brown stem rot in soybeans. Crop Sci. 8:728-729.

Ertl, D. S. and W. R. Fehr. 1983. Simultaneous selection for yield and resistance to brown stem rot of soybean in hill plots. Crop Sci. 23:680-682.

Gray, L. E. 1971. Variation in pathogenicity of *Cephalosporium gregatum* isolates. Phytopathology 61:1410-1411.

Gray, L. E. 1972. Effect of *Cephalosporium gregatum* on soybean yield. Plant Dis. Rep. 56:580-581.

Gray, L. E. and J. B. Sinclair. 1973. The incidence, development, and yield effects of *Cephalosporium gregatum* on soybeans in Illinois. Plant Dis. Rep. 57:853-855.

Kobayashi, Kiroku and Tadai Ui. 1977. Wilt inducing compounds produced by *Cephalosporium gregatum*. Physiol. Plant Pathol. 11:55-60.

Kobayashi, Kiroku and Tadao Ui. 1980. Production of gregatins by an Iowa isolate of *Cephalosporium gregatum*, the cause of brown stem rot of soybeans in U.S.A. Phytopathol. Soc. Japan 46: 256-257.

Phillips, D. V. 1971. Influence of air temperature on brown stem rot of soybeans. Phytopathology 61:1205-1208.

Phillips, D. V. 1973. Variation in *Phialophora gregata*. Plant Dis. Rep. 57:1063-1065.

Sebastian, S. A. 1984. The advantages and inheritance of brown stem rot resistance in soybeans and the implications to breeding for resistance. Ph.D. Dissertation. University of Illinois Library, Urbana.

Sebastian, S. A., C. D. Nickell and L. E. Gray. 1983. Sequential screening of soybean plants for resistance to *Phytophthora* rot and brown stem rot. Crop Sci. 23:1214-1215.

Tachibana, H., J. B. Bahrenfus and W. R. Fehr. 1983a. Registration of BSR 201 soybean. Crop Sci. 23:186.

Tachibana, H., J. B. Bahrenfus and W. R. Fehr. 1983b. Registration of BSR 302 soybean. Crop Sci. 23:186.

Tachibana, H. and L. C. Card. 1979. Field evaluation of soybeans resistant to brown stem rot. Plant Dis. Rep. 63:1042-1045.

Weber, C. R., J. M. Dunleavy and W. R. Fehr. 1966. Influence of brown stem rot on agronomic performance of soybeans. Agronomy J. 58:519-520.

ENTOMOLOGY

Plenary Paper

INTEGRATED PRODUCTION MANAGEMENT IN SOYBEAN SYSTEMS: A HOLISTIC VIEWPOINT

F. L. Poston, S. M. Welch, J. W. Jones, and J. W. Mishoe

Management of agricultural systems on the farm has become increasingly difficult during this century. Low commodity prices and rising production costs are limiting farm profits. Minor errors in decision-making can completely eliminate those profits. Today's farmer is faced with many inputs to manage and too few tools to accomplish the task. This has not, however, always been the case.

HISTORY OF AGRICULTURAL MANAGEMENT

During the late nineteenth and early twentieth centuries, a time of great agricultural expansion in the United States, crop production and marketing were substantially different than today. A farmer on the Great Plains tilled his field with a plow and draft animals, usually planted carry-over seed from the previous year, weeded with a mechanical cultivator and by hand, and hand-harvested when his crop was mature. Soil fertility and moisture were managed by a fallow system and by spreading manure. Dynamic occurrences of pests, diseases, and drought were not managed but simply endured, as were unstable or temporarily nonexistent markets. Realizing the need for an abundant and stable food supply, the United States Congress established (1) the Land-grant College system to conduct research and train students in agriculture and (2) the Federal Extension Service to provide information to farmers outside the formal college classroom (Elder, 1962).

Research and extension specialists at the beginning of this period can best be described as generalists. They had mastered the bulk of available knowledge concerning the production and marketing of various crops. Consequently, they were capable of integrating this information into appropriate production/marketing packages and extending them to farmers (Figure 1). As agricultural stability increased, so did the demand for relevant information. Agricultural disciplines began to segregate, and the roles of research and extension specialists were more clearly defined and divided. The structure that led to the future information explosion in U.S. agriculture was set in place.

From 1920 to 1950 large agricultural businesses developed and became a third information supplier to the farmer. The primary thrust of agriculture in the 1950s was the maximization of yield. The use of agricultural chemicals for fertility and pest control increased dramatically. During the 1960s the biological consequences of these practices surfaced and great conflict developed between farmers, scientists, private industry, environmentalists, conservationists, and public regulatory agencies. It is from

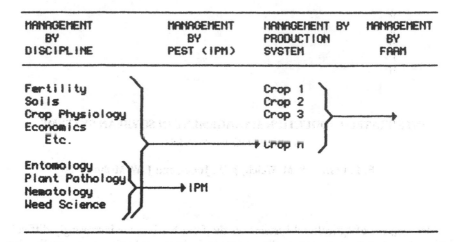

Figure 1. Evolution of the management concept in agriculture.

this caldron that integrated pest management (IPM) arose (Pedigo, 1985).

IPM represented an interesting development in agricultural management philosophy. It attempted to collate biological and limited economic information across well-defined disciplinary boundaries to form packages that could be implemented for the management of pests and diseases. The IPM concept was defined differently by various interest groups according to the impact that they desired to have on agriucltural production. Consequently, no clear group understanding of the concept emerged. Indeed, the diffusion of IPM to new groups is still occurring. For others, the concept has evolved to encompass some of the ideas expressed in this paper.

Other scientific disciplines have rich management histories worthy of discussion (fertility, plant breeding, water management, etc.). Each, undoubtedly, has contributed to our current and future management philosophy. All (including IPM), however, suffer from a critical flaw. Each management group attempts to optimize inputs and outputs for their part of the production or marketing process with little or no regard for the rest of the system. Likewise, recommendations that result from these management approaches often conflict. In short, these approaches are not integrated from the farmer's viewpoint. Consequently, the implementation of new production or marketing innovations in farming is slow. Farmers must, in essence, use empirical methods to determine the compatibility between an innovation in one area and the remainder of their production process.

As one examines current, discipline-oriented management, compared with well established, economically based, farm management programs, a clear gap emerges. Farm management, a multiple commodity management approach, often fails to consider the dynamic biological and physical relationships within a commodity in a manner allowing rapid response to changes in the crop environment. This of course is the strength of discipline management. Clearly, a commodity-based approach that incorporates the best of discipline management and interfaces with farm management programs is needed.

THE PRODUCTION SYSTEM CONCEPT

To develop a solution to the problems mentioned in the previous section, we begin by recognizing that the producer is the only person in agriculture today who attempts to integrate information to arrive at decisions. By using the farmer as a role model in program design, one can reduce fragmentation and improve technology transfer through enhanced compatibility between recommendations and current practices.

Clearly, if such compatibility is to exist, it must be based on a thorough knowledge of what current practices are. Most important is the decision-making process, which determines how, where, when, and why these practices are employed. The adoption of new technology is, after all, synonymous with the initiation of changes in decision-making. So important is this concept that we shall refer to the overall, geographically specific set of decisions, decision-making methods, and corresponding decision options (i.e., production and marketing practices) as the *production system* for that commodity.

An example will help to clarify the terminology. Soybean cultivar selection is an example of a particular decision a grower has to make. His decision-making method might be to consult yield trials and look for cultivars that perform well in his area. Alternatively, it might be simply to plant what he used last year, if it has proved adequate. For this decision, the individual soybean cultivars represent the decision options he has to choose between. To produce the crop, the farmer must make many decisions like this in some systematic fashion.

At an abstract level, at least, one can envision describing a production system simply by compiling a list of decisions, methods, and options. Without really discussing (for now) the practical details of how such a list might be compiled, consider what it might be used for. If the list included the best available decision-making methods and supporting educational materials, it could (with the appropriate packaging) be an important reference work for farmers and/or extension agents. Of course, all decisions would not be supported by equally complete research data, or be equally useful to growers. However, this variation would be more than offset by the level of integration achieved.

Also, such a list could be quite useful to administrators in establishing research needs and priorities. The list would have to be packaged differently and buttressed by data on (1) the efficacy of the decision methods, (2) the relative economic significance of various production operations, and (3) grower inputs. High priority needs could be identified by isolating economically significant decisions made by inadequate methods. On the other hand, an extension administrator could develop needs and priorities for improved technology transfer programs by comparing the best known management approaches (as listed) with actual current practices and grower evaluations. Finally, for professional staff, such a list would essentially be an agenda for interdisciplinary interaction because its basic elements are real world decisions, not narrow, disciplinary topics.

For these benefits to be realized, however, it must be practical to describe production systems. Superficially, the task is daunting. Complete lists of grower production decisions would (1) be very long, (2) incorporate many items relevant only to particular farms, (3) contain many inconsistencies because of geographic differences in production practices, and (4) be complicated by dependencies between decisions. The concept of the grower as a role model, however, results in tremendous simplification. First of all, there is a great deal of structure and organization to the decision-making process. At the crudest level, one can distinguish between within-season decisions (e.g., cultivar selection, planting date, etc.), between-season decisions (e.g., what crop to plant in which field), and long-term enhancement decisions (e.g., whether or not to terrace the field) (Stimac, pers. commun.).

Further structure emerges when considering within-season decisions. There is a regular pattern of production phases (i.e., clusters of decisions) which occur in a sequential order (Table 1). Each of these phases can be expanded into its major decisions and, if necessary, each of these can again be expanded as shown. By proceeding in this fashion from the general to the particular, a description of a given area's production system can be developed in essentially an outline format. By terminating the outline process at a level of resolution consonant with purposes, the description can be kept manageable; if further detail subsequently proves necessary, it can be added later.

In order to insure broad coverage, compilation of the production system description is best accomplished by an interdisciplinary team involving both research and extension. One or more workshops may be required to integrate the collective knowledge of the team. The result of this step will be a detailed description of each crop production decision made by producers. By way of example, Table 2 is a description for a millet (*Setaria italica* [L.] Beauv.) production system for a location in The Gambia. The table was produced as a part of an FAO funded study which used the production system approach to construct an IPM demonstration/research program in that country.

The specific information needed for each decision can be classified as follows: (1) the decision to be made plus the relevant decision variables (e.g., a pesticide choice might include pesticide type, application date, rate and method); (2) the person who makes the decision to implement a practice (not who may set overall policy); (3) the method or rules by which the decision-maker arrives at a choice; and (4) the options (including any problems or constraints) for each decision. In some cases (especially in lesser developed countries), physical circumstances or government policy

Table 1. Hypothetical list of production decisions for Central Iowa.

PREPLANT PHASE:
    Marketing
    Cultivar selection
    Tillage
    Planting
    Fertility
    Weed control
    Soil insect control
    Irrigation
VEGETATIVE PHASE:
    Painted lady control
    Replant
    Cultivation
    Irrigation
    Marketing
REPRODUCTIVE PHASE:
    Green cloverworm control
        Control economical?
        When?
        How?
    Fertility
    Irrigation
    Marketing
HARVEST PHASE:
    Harvest time
    Storage
    Marketing

Table 2. A summarized portion of the current millet production practices collated from three villages in The Gambia.

Decision	Decision maker	Decision method	Options
LAND PREPARATION:			
How?	Farmer	No choice	Weed, rake, heap and burn.
When?	Farmer	Before rain	No options.
PLANTING AND SEEDED PREPARATION:			
How?	Farmer	Available machinery	Carreau method, tracer method.
Spacing?	Farmer	Width of draft animals	80-100 cm between rows, 25-50 cm between hills.
When?	Farmer	Available machinery	Just after 1st rain, ca. June 15.
FERTILIZER APPLICATION:			
Application method?	Farmer	Availability of animals or chemicals	Manure or chemical fertilizer.
When?	Farmer	Application method	Manure—before planting; chemical—1-3 treatments.
Rate?	Farmer	Availability & soil test	Manure—tethering time; chemical—0-250 kg.
WEEDING AND THINNING:			
Weeding pattern?	Farmer	No choice	1-3 weedings starting at planting, within row and between row weeding patterns vary.
Thinning pattern?	Farmer	No choice	At 1st weeding, 2-14 days after emergence, 2-5 plants/hill.
SEED TO GERMINATION:			
Seed pest control?	Farmer	History of problem	Overplanting, deep planting, seed dressing, scaring.

may constrain certain decisions to a single option. At the other extreme, there may be several methods for making decisions, ranging from rules of thumb to some form of mathematical model. In either case the explicitly documented production system description can be used in an organized, improvement process.

Several projects, which use this method of organization on a number of commodities, are currently underway. In Kansas, a project funded by the W. K. Kellogg Foundation is developing a set of computerized decision aids for corn (*Zea mays* L.). This involves writing computer programs implementing the best available decision-making methods for each of a list of 27 important production decisions. In another effort the 12 North Central states are currently conducting a pest management research planning effort for confined cattle, corn, and potatoes (*Solanum tuberosum* L.) using the production system approach. This involves assembling descriptions of these production systems and then identifying and evaluating knowledge weaknesses. Another Kellogg funded project in Florida is assembling computerized databases of production-related educational materials. These materials cover the major management tasks for a number of crops. The first database to be completed was for soybean.

A second major problem in describing production decisions is that they are often interdependent. Because decisions occur in a sequential pattern, some early choices may influence the range of alternatives available later. For example, selection of a wide soybean row spacing (ca. 75 to 100 cm) increases the need for additional cultivation, because it decreases the inhibition of weed growth. Narrow rows inhibit weed growth and increase the difficulty of cultivation (Scott and Aldrich, 1970). Economics represents another form of dependence between decisions. Even if two decisions are not biologically dependent, they are related economically. For example, selection of a fertility program, superficially at least, is not biologically related to drying and storage decisions. Both areas are related, however, when the production budget for the crop is considered. Thus, economics is a tie binding all crop production decisions. It is, therefore, necessary to document dependencies between decisions. While such documentation can become complex, experience with the projects mentioned in the last paragraph indicates that it can be managed.

Another major problem in describing production systems stems from geographic variation. Not only do management options vary through space, but even the basic decisions can change (e.g., irrigation decisions do not need to be made in areas of dryland production). This variation may be continuous (e.g., along environmental gradients for region-wide monocultures) or discontinuous (e.g., for crops grown in geographically discrete patches). Clearly, some production features may be more spatially variable than others.

This variation can be accounted for by division of the crop range into distinct production areas based on site characteristics (soil, weather, economic, etc.). Production decisions can then be documented separately within each area. Production areas are mutually exclusive, but exhaustive geographic subregions are selected so that production practices are more similar within a given area than they are between areas. Obviously, this is a somewhat arbitrary definition. Production areas are units of convenience selected with an eye to ultimate purposes (crop management, research management, etc.). It may be appropriate beforehand to separate production systems by commodity use. For example, soybeans produced for oil may differ greatly from edible soybean production. The subdivision needs to be fine enough that rough uniformity of practices exist within an area but not so fine that the number of areas becomes unmanageable. In general, subdivision of a state-sized area into 5 to 10 pieces will be sufficient.

This subdivision may be accomplished in several ways. The first is by using a "Delphi approach" wherein a group of experts from various agricultural disciplines are asked individually to subdivide the region according to whatever subjective criteria they wish. After seeing each other's results the process may be repeatedly iterated until convergence of opinion is obtained. Another method is to use the mathematical techniques of cluster analysis to identify similar areas based on important production variables like soils, weather, pest factors, and market information. As an example, Kansas was separated into seven corn production areas using a cluster analysis technique. When given the same task, a committee using a Delphi approach yielded very nearly the same division.

RESEARCH AND IMPLEMENTATION PLANNING

When completed, a production system description establishes a basic framework for interdisciplinary cooperation and a direct link between research and implementation goals and actual production systems. We have synthesized a set of procedures (Figure 2) through experiences in applying the production system concept to soybean, corn, millet, and cotton production systems in Florida, Kansas, the Sahel in Africa (Gambia), and Sudan, respectively. While the entire procedure was not completed at

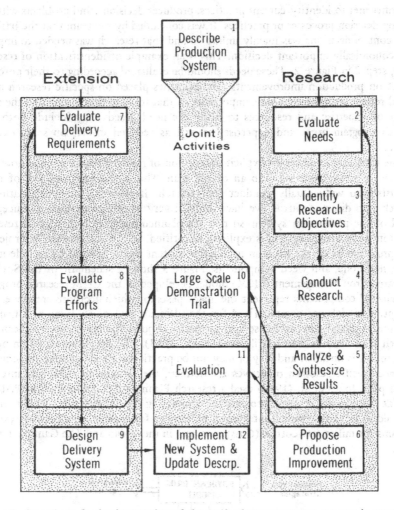

Figure 2. Procedures for implementation of the production management concept in research and extension programs. Steps are explained in the text.

any location, each site contributed its quota to our understanding of the complete process. At first glance, these procedures may seem obvious or even trivial. However, a discussion of each of the steps in Figure 2 will reveal its importance to the overall approach. The procedure can be divided into three parts: activities related to research, to extension, and those carried out jointly. None of the steps can be taken for granted in a successful integrated research and implementation program.

Research Activities

The first step in Figure 2 is essentially the development of the production system description as discussed in the preceding section. As an example, this procedure was used to identify four soybean production regions in Florida in the Consortium for Integrated Pest Management (CIPM) research program. Research and extension personnel met to identify current practices, producer decisions, and problems with the existing decision processes or practices. It was concluded by the team that the basis for weed control decisions was mainly unknown and that research was needed to improve this economically important decision. This is an example of identification of research needs, step 2 in Figure 2. These needs should be evaluated according to their potential impact on production improvements. The priorities placed on specific research needs should reflect both the economic importance of that topic to production and the availability of expertise and resources to fulfill the need. Needs may include both new research program areas and expertise, as well as reduced efforts on some existing programs.

The third step requires an explicit description of research objectives and how each research objective contributes to an overall plan. This is the responsibility of those scientists who will actually conduct the research. Because of the specialization of research into discipline areas, we have found it very helpful to develop a conceptual model of the production system, so that the planned integration of specific research results into the overall system is explicitly identified. In the soybean CIPM project, a conceptual model of the soybean production system was developed to include major insect, nematode, and weed components. Figure 3 shows a schematic of the Soybean Integrated Crop Management (SICM) model developed in the CIPM research program. The selected components represent the major pests for which decisions are made. This conceptual model was computerized for the ultimate goal of studying pest management practices and currently is serving as the focal point for integrating soybean pest research from various states (Wilkerson et al., 1983). The development of an operational model is optional and may or may not be practical for a given research program.

Once specific research objectives are defined, they should be organized into one overall plan. Jones et al. (1980) used a research Planning Activity Network (PAN) to organize the specific research projects for completing the SICM model. This tool was also used to identify research integration milestones for organizing suggested research programs for millet and cotton (*Gossypium* sp.) in the Sudan and The Gambia, respectively.

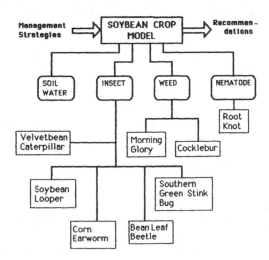

Figure 3. Schematic of the Soybean Integrated Crop Management model.

Step 4 in Figure 2 needs little explanation. Research may be conducted simultaneously by individuals and small teams. The analysis and synthesis of research (Step 5) is performed at two levels: independent experimental data analysis and integrated analysis of results by the interdisciplinary team. The integration may be highly empirical and subjective, again in a workshop format, or it may be through the use of production system models. Boggess et al. (1984) demonstrated the use of the Soybean Integrated Pest Management model for studying various pest management strategies for soybeans with three different insect pests. The interdisciplinary research team evaluates research results, empirically and/or using model simulations, and designs improved production decisions and practices (Step 6).

Until now we have discussed research activities that generate possible new management options and/or improved decision making methods for existing options. Before potential improvements can be made, however, they must be (1) evaluated under a range of actual production conditions and (2) integrated into ongoing education and support programs. These tasks fall under extension and joint extension/research activities as depicted in Figure 2.

Extension Activities

Educational and operational support needs for information delivery are generated in Step 7. In extension, methods of information delivery must be constantly evaluated for proficiency in supporting the current production system. Discrepancies between the best production system and current farmer practices should influence extension education programs. As an example, the Florida CIPM soybean workers identified a differential of one to two more sprays per season in non-scouted fields relative to scouted fields. This pointed out the need for additional educational and training programs.

While the identification of discrepancies is very useful, it is also desirable and possible to move beyond mere *post hoc* evaluations and operate in a timely anticipatory fashion. This can be done by evaluating all proposed changes to the production system in terms of their probable impact on methods of information delivery. Commonly, proposed changes will come from ongoing research programs, field demonstration trials (Step 10), agribusiness and/or grower innovations. Any or all of these may suggest needed changes in educational or support programs. In Step 8 these needs are evaluated. Priorities should reflect both the economic significance to production of the management methods being implemented as well as constraints on the resources (people, equipment, and funds) available to extension.

Step 9 involves two concurrent activities. The first is the design of information delivery methods associated with high-priority, new technologies that have been successful in large scale trials and are now ready for implementation. The design may incorporate the acquisition of data, timely delivery of that data in an appropriate format, and the educational support programs necessitated by the new technology. The second is the design of prototype delivery systems for use in large scale trials of the even newer methodologies that are currently being readied for large-scale testing. We are not suggesting that the format of information delivery systems should drive research innovation. It is important, however, to consider possible means of information delivery during the formulative stages of research.

Joint Research/Extension Activities

The large-scale demonstration (Step 10) and evaluation (Step 11) of a new production method has two purposes. The first is to provide feedback to research to delineate the limitations of the new method in production settings. In cases where unforeseen problems arise, enough data should be generated to indicate the nature of the difficulty and to suggest further lines of research where needed. The second objective is to provide information useful in determining large-scale implementation requirements (see Step 7 above). For example, the implementation of an insect pest model might well require scouting reports to establish initial conditions and daily weather data to calculate insect development rates. The trial might determine that a lapse in the availability of weather data is not a problem if it does not exceed a few days. Such a finding would have major significance for the design of the delivery system needed to support this model.

In Step 12, the new production system is implemented. This includes the management technology itself (as tested and evaluated in Steps 10 and 11) and the extension delivery system and training programs (as designed in Step 9). The production system descriptions are updated in this step so that ongoing planning and evaluation activities remain current. The iterative nature of this procedure prevents the stagnation of research and extension programs.

This set of procedures results in a continuum of possible improvements in decisions and practices, some of which can be tested and integrated into the production system over a short time period, and others over a much longer time. By analogy, this process can be compared to a pipeline in which some new technology is being implemented, some is being tested in demonstration plots, and some is in the early conceptual stages. The procedures developed in this paper constitute an orderly framework for interdisciplinary research and extension programs whose outputs can be readily integrated into existing production systems.

MAJOR DEFICIENCIES OF THIS APPROACH

A major deficiency of this approach is that it requires the coordination and cooperation of many people. In a sense, it is a structured exercise in group dynamics. Consequently, its potential for success is somewhat related to the desires and viewpoints of individuals within the group. Because of diversity of views, progress may be slowed or diverted by a number of different tactics or strategies (Keen, 1981). The strength of the approach, however, lies in the common delineation of the group task and the structured documentation of progress. An administrator responsible for the development of a research or implementation program can quickly look at an activity network, identify program delays, and correct them. This adds a new dimension to interdisciplinary research and implementation in agriculture. Using this approach, one may step beyond "promoting" interdisciplinary activities and actually "design" them.

Another weakness in this approach stems from its relative position within the management hierarchy. Just as IPM fails to integrate all features of crop production, so does this approach fail to capture all elements of farm management. Once the commodity level integration has been reached, however, the step to the farm level will be greatly facilitated. Much of the integration required at the farm level is economic in nature, which lends itself to quantitative approaches. Furthermore, many of the required decisions fall in the "between-season" or "long-term enhancement" categories. As a result, they affect production decisions in a relatively restricted fashion, compared with the multiplicity of interactions present in crop biology.

Undoubtedly, the production management concept developed in this paper will change as it is more broadly applied. The strength of the approach lies in its ability to organize and integrate a variety of information and to provide a common focal point for research and implementation activities. We believe that it closes a significant gap in existing management philosophy, and that it represents an important evolutionary step in agricultural operations.

NOTES

F. L. Poston and S. M. Welch, Department of Entomology, Kansas State University, Manhattan, KS.

J. Jones and J. Mishoe, Department of Agricultural Engineering, University of Florida, Gainesville, FL.

REFERENCES

Boggess, W. G., D. J. Cardelli and C. S. Barfield. 1984. A bioeconomic simulation approach to multi-species insect management. S. J. Agric. Econ. (in press).

Elder, C. R. 1962. People's colleges. p. 13-20. *In* A. Stefferud (ed.) The Yearbook of Agriculture, Washington, D.C.

Jones, J. W., J. W. Mishoe and J. S. Stimac. 1980. Systems analysis. *In* Research on pest management. ASAE Paper No. 80-1002. Am. Soc. Agric. Eng.

Keen, P. G. W. 1981. Information systems and organizational change. Comm. Assoc. Computing Mach. 24:24-33.

Pedigo, L. P. 1985. Integrated pest management. Yearbook of Science and Technology. McGraw Hill. (in press).

Scott, W. O. and S. R. Aldrich. 1970. Modern soybean production. S & A Publications, Champaign, Illinois.

Wilkerson, G. G., J. W. Mishoe, J. W. Jones, J. L. Stimac, D. P. Swaney and W. G. Boggess. 1983. SICM: Florida soybean integrated crop management model. Rept. AGE 83-1, Dept. Agric. Engineering, Univ. of Florida, Gainesville.

Plant Damage Syndromes, Yield Losses, and Economic Decision Indices

DAMAGE SIMULATIONS AS AN APPROACH TO UNDERSTANDING ECONOMIC LOSSES TO INSECTS

Gustave D. Thomas

Agronomists have used manual defoliation and depodding techniques to simulate hail damage to soybeans (Kalton et al., 1949; Camery and Weber, 1953; Weber, 1955; and McAlister and Krober, 1958). Entomologists have used hand defoliation and depodding to simulate insect damage: some have removed entire leaflets to accomplish various levels of defoliation (Begum and Eden, 1965; Turnipseed, 1972; Todd and Morgan, 1972; Thomas et al., 1974; 1976; 1978), whereas others have used cork borers or paper punches to remove portions of leaflets (Poston et al., 1976; Poston and Pedigo, 1976; Hammond and Pedigo, 1982).

In developing a management system for soybean insects, it is imperative to define how defoliation and pod loss affect soybean yield and quality and how this loss can be related to damaging populations of insect pests. My purpose is to examine how defoliation and depodding affect yield and further to relate defoliation to possible injury induced by caterpillar pests of soybean. I will discuss four different studies in this paper: 1) hand defoliation and depodding study, 2) sequential defoliation study, 3) hand defoliation versus insect defoliation study, and 4) defoliation study with the cabbage looper.

HAND DEFOLIATION AND DEPODDING STUDY

The major goal was to quantify the influence of leaf and pod loss on soybean yield at each of five stages of soybean growth. The experimental design was a split-split plot with five stages of soybean growth as whole plots, four levels of defoliation as sub-plots, and four levels of depodding as sub-subplots, with three replicates. Each sub-plot consisted of four, 4.9-m rows (of Clark-63), with a border row on each side that received the same level of defoliation as the subplot. The four levels of both defoliation and depodding were 0, 1/3, 2/3, and 3/3. Levels of defoliation were accomplished by removing 0, 1, 2, or 3 leaflets from each leaf. Levels of depodding were accomplished by counting the number of pods on each plant and removing the desired number evenly along the stem of the plant. The five stages of soybean growth were (Fehr et al., 1971): R3, R4, R5, R6, and R7.

Defoliation at R3 only affected yields when two-thirds of the leaves were removed (Table 1). Depodding at R3 never significantly reduced yields. Therefore, pod loss at R3 is relatively unimportant, probably because soybeans have sufficient time to recoup these losses. At R4, as at R3, removal of one-third of the leaves did not affect yields. However, one-third depodding did reduce yields significantly. At R5, one-third

Table 1. Mean percentage reduction in yield (from controls[a]) resulting from defoliation of Clark-63 soybean plants.

Growth stages	Levels of depodding	Levels of defoliation[b]			
		0	1/3	2/3	3/3
R3	0	0.0	5.0	15.1	59.7
	1/3	-2.1	1.8	9.7	55.8
	2/3	6.7	-3.7	16.8	52.7
	3/3	3.8	1.6	19.6	50.6
R4	0	0.0	7.4	25.4	83.5
	1/3	11.4	7.4	29.7	85.4
	2/3	12.3	17.5	34.8	85.4
	3/3	41.3	40.8	50.3	84.1
R5	0	0.0	13.8	24.8	74.8
	1/3	18.7	24.4	31.4	80.6
	2/3	34.5	38.6	46.5	83.5
	3/3	71.2	78.7	80.0	94.3
R6	0	0.0	22.8	21.2	52.0
	1/3	24.3	32.2	36.7	64.4
	2/3	45.8	51.6	57.6	77.8
	3/3	96.5	95.1	95.7	98.6
R7	0	0.0	6.3	1.4	9.6
	1/3	26.7	30.1	33.7	35.0
	2/3	63.3	64.9	66.1	63.5
	3/3	99.1	99.4	99.3	99.8

[a]Mean yields of the controls in kg/ha for the five whole plots were: R3, 2735; R4, 2963; R5, 2862; R6, 2903; and R7, 2903.

[b]Average of three replicates. LSD at 0.05 level of probability was 10.1.

defoliation or depodding reduced yield. One-third defoliation or depodding also reduced yield at R6. No level of defoliation at R7 caused significant reductions in yields. Leaf loss at this stage is unimportant, because the soybean plants have already matured. However, any pod loss at R7 will cause yield reductions.

Using the data from this experiment, regression equations were developed for percentage reduction in yield (Table 2). The equation with the best fit at R3 consists of

Table 2. Regression equations expressing percentage reduction in yield as functions of defoliation and depodding for each of five growth stages.

Stages of growth	Regression equations
R3	% reduction = 2.600708 - 0.390918 (% defoliation) + 0.009068 (% defoliation)^2
R4	% reduction = 2.033346 - 0.241863 (% defoliation) + 0.010688 (% defoliation)^2 + 0.003885 (% depodding)^2 - 0.003224 (% defoliation X % depodding)
R5	% reduction = 1.248382 + 0.007185 (% defoliation)^2 + 0.296637 (% depodding) + 0.004460 (% depodding)^2 - 0.004647 (% defoliation X % depodding)
R6	% reduction = 2.710017 + 0.237854 (% defoliation) + 0.002624 (% defoliation)^2 + 0.402883 (% depodding) + 0.005152 (% depodding)^2 - 0.004184 (% defoliation X % depodding)
R7	% reduction = -0.096083 + 0.083938 (% defoliation) + 0.819564 (% depodding) + 0.001775 (% depodding)^2 - 0.000835 (% defoliation X % depodding)

percentage defoliation only. However, at the other four growth stages, the equations with the best fit contain percentage defoliation, depodding, and defoliation X depodding. Defoliation was the most important factor at R3. The coefficient of determination (r^2) for the model with depodding only was 0.0 (Table 3). Although at R4, the equation with best fit contained both defoliation and pod removal, the model with defoliation only had a r^2 of 0.84. Defoliation and depodding were equally important at R5; r^2 for the model with defoliation only was 0.44, and that for depodding only was 0.43. At R6, pod removal was most important because the r^2 for the model with depodding only was 0.76. Depodding was the important factor at R7. The coefficient of determination for the model with defoliation only was 0.0.

How can these equations be translated into information a grower can use? By use of the regression equations, the average percentage defoliation (for fields yielding 1,680 to 2,688 kg/ha) that would induce a 33.6 kg/ha loss (the price of this amount of soybeans would about pay for the cost of the insecticide and application cost for insect control) was calculated for each of the five growth stages (Table 4). These damage

Table 3. Coefficients of determination (r^2) for each of the equations with best fit and those for equations for percent defoliation only and percent depodding only.

Stages of growth	Best fit	% defoliation only	% depodding only
R3	0.89	0.89	0.00
R4	0.96	0.84	0.10
R5	0.92	0.44	0.43
R6	0.92	0.13	0.76
R7	0.99	0.00	0.99

Table 4. Average number of larvae/m of row of four insect pests of soybean plants that would induce various levels of defoliation.

Plant stages	Defoliation[a] threshold	Average leaf area[b] Total	Damage	Pest species[c] T. ni 119.4[d]	P. includens 113.5	A. gemmatalis 85.2	S. exigua 52.3
	%	– cm^2 –		no.	no.	no.	no.
R3	40	1870	748	164.7	172.9	230.6	376.0
R4	19	1918	364	80.1	84.3	112.2	183.1
R5	6	1820	109	23.9	25.3	33.5	54.8
R6	6	1561	93	20.7	21.7	28.9	37.1
R7	20	871	174	38.4	40.3	53.8	87.6

[a]Estimated % defoliation that would provide a 33.6 kg/ha loss.

[b]Avg. leaf area per plant at density of 26 plants/m of row, damage computed from % threshold values.

[c]The common and full scientific names of the 4 species are: the cabbage looper, *Trichoplusia ni* (Hübner); the soybean looper, *Pseudoplusia includens* (Walker); the velvetbean caterpillar, *Anticarsia gemmatalis* (Hübner); and the beet armyworm, *Spodoptera exigua* (Hübner).

[d]Avg. leaf area consumed (cm^2) by each species in completing larval development (Boldt et al., 1975).

thresholds range from a high of 40% at R3, to a low of 6% at R5 and R6. Based on information about soybean leaf area at each of the growth stages, the amount of leaf loss/m of row that would produce the threshold level of damage was calculated. Also, based on laboratory studies (Boldt et al., 1975) that estimated the amount of soybean leaf feeding by caterpillars, preliminary economic injury levels for several insect pests were calculated (Table 4). It should be remembered that these levels are estimates.

SEQUENTIAL DEFOLIATION STUDY

Since the defoliation thresholds were calculated from values derived from the regression equations (Table 2), it was deemed necessary to field-test the derived threshold values. Furthermore, since caterpillars feed throughout several soybean growth stages, it was considered essential to measure defoliations over the entire five growth stages (i.e., R3 - 40% + R4 - 19% + R5 - 6% + R6 - 6% + R7 - 20%). Thus, the field design encompassed two objectives: to test the threshold values for a single defoliation and to determine the effects of sequential defoliations at threshold levels on yield and seed quality of soybeans.

The experimental field design was a randomized complete block at each of five stages of soybean growth with three replicates. Table 5 presents levels of defoliation at

Table 5.   Effect of sequential defoliations at threshold levels on yield of Clark-63.

Treatment no. and levels of defoliation	Yield[a]
	kg/ha
Initiated at growth stage R3	
1. Control	2589 r
2. R3-40%	2521 r
3. R3-40%+R4-19%	2447 rs
4. R3-40%+R4-19%+R5-6%	2400 rs
5. R3-40%+R4-19%+R5-6%+R6-6%	2252 st
6. R3-40%+R4-19%+R5-6%+R6-6%+R7-20%	2125 t
Initiated at growth stage R4	
1. Control	2474 r
2. R4-19%	2535 r
3. R4-19%+R5-6%	2636 r
4. R4-19%+R5-6%+R6-6%	2427 r
5. R4-19%+R5-6%+R6-6%+R7-20%	2326 r
Initiated at growth stage R5	
1. Control	2414 r
2. R5-6%	2454 r
3. R5-6%+R6-6%	2266 r
4. R5-6%+R6-6%+R7-20%	2333 r
Initiated at growth stage R6	
1. Control	2494 r
2. R6-6%	2528 r
3. R6-6%+R7-20%	2642 r
Initiated at growth stage R7	
1. Control	2320 r
2. R7-20%	2192 r

[a]Average of 3 replicates. Means within each growth stage followed by the same letter do not differ significantly at the 0.03 level of probability by Duncan's multiple range test.

each of the five growth stages. For example, treatment 6 of growth stage R3 shows levels of defoliation for each growth stage and the sequential defoliation procedure. Thus, treatment 6 included an initial 40% defoliation at R3, plus an additional 19% defoliation at R4, plus an additional 6% defoliation at R5, plus an additional defoliation of 6% at R6, plus an additional 20% defoliation at R7. Defoliation was accomplished by counting the number of leaflets per plant and then removing the desired percentage evenly along the plant. Numbers to be removed were rounded to the nearest 1/3 of a leaflet. Each treatment row was 4.9-m long, with a border row of the same length on each side that received the same level of defoliation as the treatment row.

The single defoliations at threshold levels did not have a significant effect on yield (Table 5). The only significant effect of sequential defoliations at threshold levels on yield occurred when defoliation was initiated at R3. Of these, only the two heaviest defoliations (R3 - 40% + R4 - 19% + R5 - 6% + R6 - 6% and R3 - 40% + R4 - 19% + R5 - 6% + R6 - 6% + R7 - 20%) caused significant yield reductions (13 and 17%, respectively). It would require heavy populations of caterpillars to cause these levels of sequential defoliations. The numbers necessary would vary from year-to-year because leaf surface area at any particular growth stage varies from year-to-year. For example, in the 7 years that records were kept, leaf surface area at R3 has varied from a high of 49089 to a low of 15798 cm^2/m row. Therefore, the number of cabbage looper larvae per m row necessary to cause the initial 40% defoliation at R3 would have varied from a high of 165 to a low of 53.

HAND-DEFOLIATION VERSUS INSECT DEFOLIATION STUDY

Of prime importance to the utilization of thresholds established by using manual defoliation methods is whether hand-defoliation actually simulates insect defoliation. Therefore, I compared the effects of hand and insect defoliation on yield and seed quality of soybeans.

The experimental design was a randomized complete block with three replicates. There were four treatments: 20.5 cabbage loopers/m of row (number calculated to cause 6% defoliation), 6% hand-defoliation, plants handled only, and control. Each plot consisted of two 2.44 m rows of caged Clark-63 soybeans. At R5 the desired numbers of early, 3rd-instar larvae (reared in the laboratory on artificial diet) were placed on the caged plants. Also, the desired numbers of leaves were removed in the hand-defoliation treatments. In the plants-handled treatments, the plants were handled in the same manner as the defoliated plants but no leaves were removed. After the cabbage looper larvae pupated, the amount of defoliation was determined.

Cabbage loopers caused 4.2% defoliation, which was close to the projected 6% defoliation level (Table 6). There were no significant differences in yield resulting from any of the four treatments. Therefore, hand-defoliation (at least at threshold levels) is an accurate estimate of the effect of caterpillar defoliation on soybean yield.

DEFOLIATION STUDY WITH THE CABBAGE LOOPER

The hand-defoliation and depodding study demonstrated which soybean growth stages and which levels of defoliation and depodding are the most important. The purpose of that study was to establish broad parameters. Hand removal of leaves and pods was used to simulate caterpillar damage. Based on the hand-defoliation and depodding study, field experiments were set up using caterpillars. By knowing the

Table 6.   Comparison of hand and insect defoliation at threshold levels at R5 on yield of Clark-63.[a]

Treatment	Defoliation	Yield
	%	kg/ha
Control	0	2009
Plants handled	0	2009
Hand defoliation	6	2143
20.5 T. ni/m of row	4.2	1982

[a]Average of three replicates. LSD at 0.05 level of probability for yield was 349.4.

Table 7.   Influence of six levels of *T. ni* on percentage defoliation and yield of Clark-63 soybeans treated at R5.[a]

T. ni/m of row	Projected defoliation	Actual defoliation	Yield
no.	%	%	kg/ha
0	0	0	3044
16.7	6	5.3	3064
31.8	12	9.8	2963
44.9	17	18.1	2573
70.5	25	21.6	2593
88.6	33	24.8	2540

[a]Average of three replicates. LSD at 0.05 level of probability for yield was 404.6.

amount of leaves an individual cabbage looper consumed in its development (Boldt et al., 1975) and the leaf surface area/plant in the field, it was possible to calculate how many loopers it would take to cause various levels of defoliation. Therefore, this study was conducted to determine the influence of leaf feeding by the cabbage looper on soybean yield.

The experimental design was a randomized complete block with six levels of cabbage looper infestation (0, 16.7, 31.8, 44.9, 70.5, and 88.6/m of row) and three replicates. Each experimental plot consisted of two, 2.44-m rows of caged Clark-63 soybean plants. The desired numbers of early, 3rd-instar larvae were placed on the caged plants at R5 and allowed to feed. After pupation, the amount of defoliation was determined.

Before the experiment, it was predicted that each level of the cabbage looper would cause a specific level of defoliation (Table 7). The actual and projected levels of defoliation were quite close for the four lowest levels of infestation, but differed significantly at the two highest levels of infestation. There was probably less feeding at the two highest levels because of larval competition. Yields between 0 and 31.8 loopers/m of row or 44.9 and 88.6/m of row were not different. Significant reductions in yield occurred, however, between 31.8 and 44.9 loopers/m of row.

NOTES

Gustave D. Thomas, Agricultural Research Service/USDA, Department of Entomology, University of Nebraska, Lincoln, NE 68583-0816.

REFERENCES

Begum, A. and W. G. Eden. 1965. Influence of defoliation on yield and quality of soybeans. J. Econ. Entomol. 58:591-592.

Boldt, P. E., K. D. Biever and C. M. Ignoffo. 1975. Lepidopteran pests of soybeans: consumption of soybean foliage and pods and development time. J. Econ. Entomol. 68:480-482.

Camery, M. P. and C. R. Weber. 1953. Effects of certain components of simulated hail injury on soybeans and corn. Iowa Agric. Exp. Stn. Res. Bull. 400:465-504.

Fehr, W. R., C. E. Caviness, D. T. Burmood and J. S. Pennington. 1971. Stage of development descriptions for soybeans, *Glycine max* (L.) Merrill. Crop Sci. 11:929-930.

Hammond, R. B. and L. P. Pedigo. 1982. Determination of yield-loss relationships for two soybean defoliators by using simulated insect-defoliation techniques. J. Econ. Entomol. 75:102-107.

Kalton, R. R., C. R. Weber and J. C. Eldredge. 1949. The effect of injury simulating hail damage to soybeans. Iowa Agric. Exp. Stn. Res. Bull. 359:736-796.

McAlister, D. F. and O. A. Krober. 1958. Response of soybeans to leaf and pod removal. Agron. J. 50:674-677.

Poston, F. L. and L. P. Pedigo. 1976. Simulation of painted lady and green cloverworm damage to soybeans. J. Econ. Entomol. 69:423-426.

Poston, F. L., L. P. Pedigo, R. B. Pearce and R. B. Hammond. 1976. Effects of artificial and insect defoliation on soybean net photosynthesis. J. Econ. Entomol. 69:109-112.

Thomas, G. D., C. M. Ignoffo, K. D. Biever and D. B. Smith. 1974. Influence of defoliation and depodding on yield of soybeans. J. Econ. Entomol. 67:683-685.

Thomas, G. D., C. M. Ignoffo and D. B. Smith. 1976. Influence of defoliation and depodding on quality of soybeans. J. Econ. Entomol. 69:737-740.

Thomas, G. D., C. M. Ignoffo, D. B. Smith and C. E. Morgan. 1978. Effects of single and sequential defoliations on yield and quality of soybeans. J. Econ. Entomol. 71:871-874.

Todd, J. W. and L. W. Morgan. 1972. Effects of hand defoliation on yield and seed weight of soybeans. J. Econ. Entomol. 65:567-570.

Turnipseed, S. G. 1972. Response of soybeans to foliage losses in South Carolina. J. Econ. Entomol. 65:224-229.

Weber, C. R. 1955. Effects of defoliation and topping simulating hail injury to soybeans. Agron. J. 47:262-266.

EFFECTS OF INSECT-PEST COMPLEXES ON SOYBEAN

James W. Todd and B. G. Mullinix

Pest management concepts are founded on principles of crop and pest ecology. The interrelationships among the animal, plant, and physical components of the soybean agroecosystem are just beginning to receive serious research attention. Although soybean researchers have generated considerable data on economic injury levels (EIL) for several pests, plant yield responses to damage by single arthropod pest species have not been firmly established. Poston et al. (1983) point out that the EIL concept may require some revision, reinterpretation or expansion as it is applied to pest complexes or to more complex pest control situations. Much of the early threshold work failed to produce definitive results; but rather raised more questions than it answered. The lack of success was due in part to questionable methodology and simulation techniques, and improper or inadequate timing and duration of treatments.

It has also been pointed out (Hinson et al., 1978) that vague or imprecise identification of plant growth stage at the time treatments were established could easily account for some of the early inconsistent results obtained in these studies. Other impediments to progress on EIL have arisen from difficulties in making the proper economic analysis of the cost/benefit factors of pest control and commodity prices. In any event, single species thresholds are of questionable value, since pests rarely occur in isolated pure populations exclusive of other pest species. When the added complications arising from coexistence of insect-weed, insect-nematode, insect-disease problems, and all possible three- or four-way combinations of these yield-limiting factors, along with a multitude of environmental factors such as of moisture stress and soil and nutritional problems are considered, it is easy to be overwhelmed by the magnitude of the problem. The obvious solution lies in a concentrated, multidisciplinary research effort where two or more key pest classes are evaluated simultaneously (McSorley and Waddill, 1982). Therefore, a logical approach is to conduct stepwise evaluations of the multiple pest damage-plant yield interactions to build a base from which further research on the multiplicity of yield-limiting factors can be performed.

ASSESSMENT OF THE DAMAGE-YIELD RELATIONSHIP

Very good reviews and discussions of the early research and development of EIL are presented by Kogan (1976), Kogan and Turnipseed (1980) and Poston et al. (1983). The amount of insect feeding damage sustained by crop plants is often positively correlated with pest population size (Kogan, 1976). Subsequent loss in crop yield is related to the amount of injury exceeding the tolerance level of the plant (Tammes, 1961; Davidson and Norgaard, 1973; and Bardner and Fletcher, 1974).

Southwood and Norton (1973) described procedures for economic analysis of the population-damage-yield relationships, which were used by Kogan (1976) to describe individual EIL for each of two groups of soybean pests; i.e., foliage feeders and pod feeders. Although these authors stressed monospecific thresholds, they all alluded to the extreme dearth of research data on multiple pest species that occur simultaneously or in sequence. Further development and utilization of pest management principles will be impeded until this research need is met.

In response to this need, a series of experiments was initiated in 1974 at the Georgia Coastal Plain Experiment Station to study the pest injury-plant yield relationship in relation to rainfall.

SOYBEAN YIELDS IN RELATION TO VARIOUS POPULATION DENSITIES OF VELVETBEAN CATERPILLAR AND SOUTHERN GREEN STINK BUG AT DIFFERENT REPRODUCTIVE GROWTH STAGES AND RAINFALL LEVELS

Infestations of the velvetbean caterpillar *Anticarsia gemmatalis* Hubner and the southern green stink bug, *Nezara viridula* (L.) were established in field cages on Coker 136 or Bragg soybean each year from 1974-1978 at Tifton, Georgia. Fourth and fifth stage larvae of *A. gemmatalis* and fifth stage nymphs of *N. viridula* were placed onto caged plants and allowed to remain for 14 to 21 days to represent a partial generation of each species initiated at each of the following soybean growth stages: R2 (full bloom), R3 (beginning pod), R4 (full pod), R5 (beginning seed), R5.5 (mid pod-fill) and R6 (full seed) as described by Fehr and Caviness (1977). Infestations were maintained at levels of 0.0, 29.25, 58.50 or 87.75 *A. gemmatalis* larvae and 0.0, 3.25, 6.50 or 9.75 *N. viridula* per meter of row by replacing missing or dead individuals on alternate days. Percentage defoliation resulting from *A. gemmatalis* feeding was rated visually and seed damage by *N. viridula* was rated and classified into damage groups as described by Todd (1976). Each year, rainfall was recorded daily and related to plant growth stage according to the R-stage designations referred to previously. Data were analyzed by the multiple regression (least squares), General Linear Models procedure (SAS, 1982). Variables included the relationship of soybean yield to the following: population densities and resultant damage by *A. gemmatalis* and *N. viridula*, amount of rainfall occurring through the end of R2, the amount of rainfall occurring during the stage in which the treatments (insect populations resulting in defoliation and pod-damage) were imposed, and all interactions.

Data collected during the four year periods were analyzed by use of forward-selection, stepwise regression to determine which of the collected predictor variables were most strongly related to yield. As a result, these related variables served as a basis for the model to analyze the data, whereby predicted damage-yield-rainfall relationships were derived. Three-dimensional plots (SAS, 1981) depict the best, reduced model, where we selected amounts of pretreatment rainfall (planting through R2) and rainfall during each of four soybean growth stages (R3, R4, R5, R6) to see how yield changed under those conditions and as a result of the damage inflicted by various population densities of *A. gemmatalis* and *N. viridula* present during a given stage.

An increase in rainfall from 0.6 cm to 3.2 cm at R3 (Figure 1) and from 0.0 cm to 6.4 cm at R4 (Figure 2) produced increased yields only at the higher combined population densities of *A. gemmatalis* and *N. viridula*. It is possible that, even at these early reproductive stages, interference among these insect species results in slower feeding (damage), enabling the plants to produce higher yields.

Increasing rainfall from 5 cm to 15 cm during R5 (Figure 3) and from 0.0 cm to

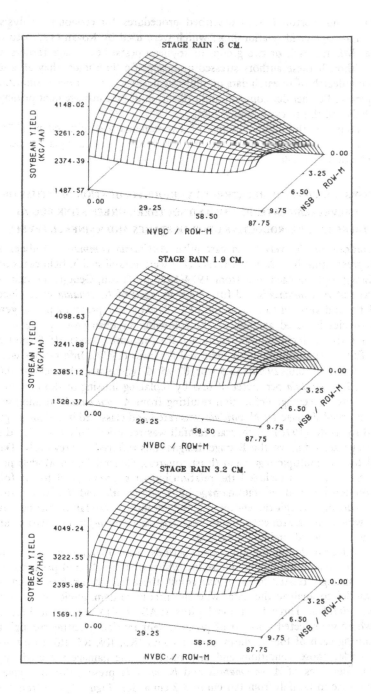

Figure 1. Response surface depicting the relationships between population densities of *A. gemmatalis* (NVBC) and *N. viridula* (NSB) during R3 with soybean yield at various rainfall levels and 53 cm of pre-treatment rainfall at Tifton, Georgia.

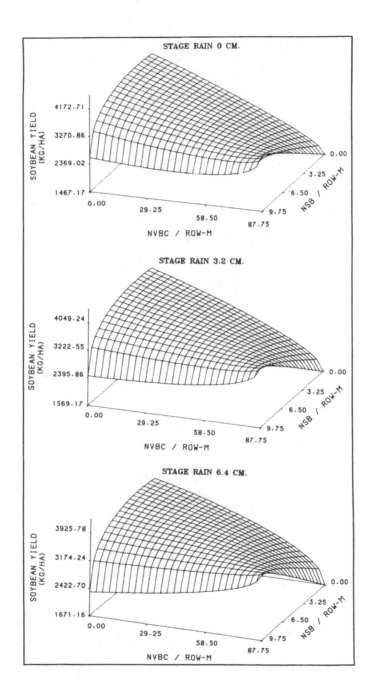

Figure 2. Response surface depicting the relationships between population densities of *A. gemmatalis* (NVBC) and *N. viridula* (NSB) during R4 with soybean yield at various rainfall levels and 53 cm of pre-treatment rainfall at Tifton, Georgia.

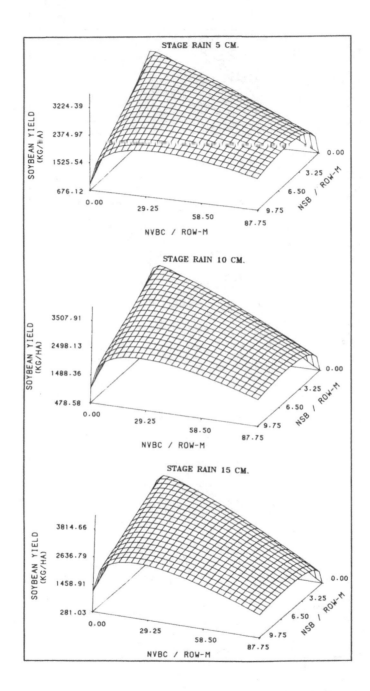

Figure 3. Response surface depicting the relationships between population densities of *A. gemmatalis* (NVBC) and *N. viridula* (NSB) during R5 with soybean yield at various rainfall levels and 53 cm of pre-treatment rainfall at Tifton, Georgia.

3.8 cm at R6 (Figure 4) tended to increase yield when *A. gemmatalis* and *N. viridula* populations were light, but did not offset yield losses from heavier damage resulting from denser populations. Plant damage resulting from higher populations of *A. gemmatalis* and *N. viridula* during R5 and R6 tended to produce larger yield differences compared to nondamaged plants as rainfall increased. Overall, there was a dramatic linear relationship between increasing *A. gemmatalis* populations (and thus, defoliation) and declining yields irrespective of rainfall level. The curvilinear effect was much less important, but still significant at the 1% level. The relative F-value for linear was of a second magnitude greater than for curvilinear.

Additionally, when considered across all stages and rainfall levels, defoliation tended to decrease yield sharply up to about the 53% level. Then, the rate of yield loss decreased with each additional increment in defoliation.

The relationship of increasing stink bug populations to yield was linear and highly significant. The curvilinear relationship was significant at the 5% level. As stink bug populations increased, the yield tended to decrease at lower rates. In the presence of a given population of *A. gemmatalis*, an increase in the *N. viridula* population tended to result in no further yield loss. At certain population densities of *A. gemmatalis* and *N. viridula* during R5 and R6, competition or interference between these two species tended to result in less yield reduction than when one or the other was present alone. Rainfall had less influence on this relationship than the plant growth stage at which the damage occurred (Figures 3 and 4).

SOYBEAN YIELD IN RELATION TO SIMULTANEOUS DAMAGE INFLICTED BY *A. GEMMATALIS* AND *N. VIRIDULA* BEGINNING AT R5.5 WITH VARIOUS RAINFALL LEVELS AT R3 THROUGH R6

It is important to remember that, although populations of the two insect species are plotted against soybean yields, the resultant damages from these populations are the primary yield-limiting factors. To more clearly illustrate the damage-yield relationships of *N. viridula* and *A. gemmatalis* on soybean during the pod-filling period, results of treatments imposed at R5.5 under varying rainfall levels during R3 through R6 are plotted in Figures 5, 6 and 7.

The dominant effect is the consistent drop in yield with increasing defoliation. The curvilinear effect, though significant, explained less yield loss. The consistent drop in yield due to defoliation became less pronounced as seed damaged by stink bug increased. This pattern appears consistently at all pretreatment and stage-rain levels. Rainfall did not change the destruction by the two pests studied. The relationship of plant damage inflicted by *A. gemmatalis* and *N. viridula* to yield accounted for 70% of the variation among the plots.

The interaction of damage by *A. gemmatalis* and *N. viridula* can be best seen by observing that, at given high defoliation levels ()50%), there was a general trend toward an increase in yield with increasing seed damage. Conversely, for defoliation levels up to ca. 50%, yield decreased as stink bug damage increased (Figures 5, 6 and 7). This clearly illustrates interference between the two species at denser populations. Also, it appears that *N. viridula* interferes more with *A. gemmatalis* than the reverse. Although defoliation rate (per day) was not a variable in this study, apparently it took *A. gemmatalis* longer to achieve a given level of defoliation when confined with the denser populations of *N. viridula*. It follows, then, that the plants could yield more under this lower defoliation rate.

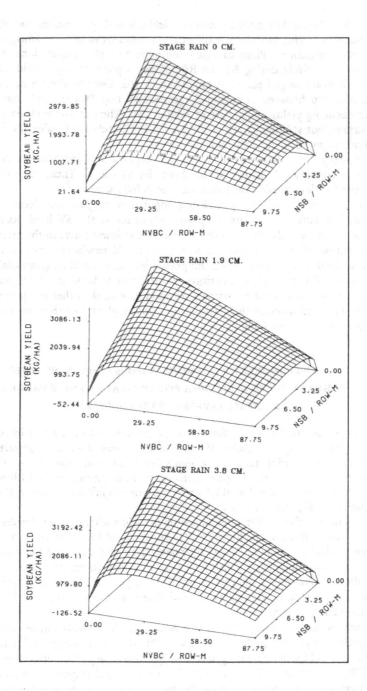

Figure 4. Response surface depicting the relationships between population densities of *A. gemmatalis* (NVBC) and *N. viridula* (NSB) during R6 with soybean yield at various rainfall levels and 53 cm of pre-treatment rainfall at Tifton, Georgia.

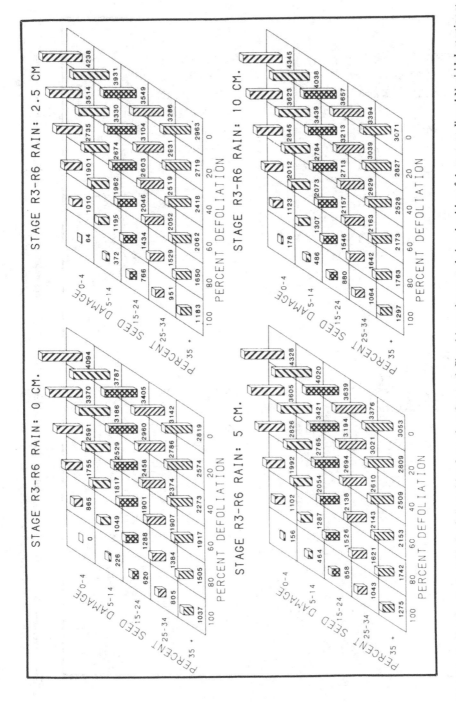

Figure 5. Predicted soybean yields (kg/ha) resulting from plant damage inflicted by various population densities of *A. gemmatalis* and *N. viridula* starting at R 5.5, at various rainfall levels from R3 through R6, and with 36 cm of pretreatment rainfall at Tifton, Georgia.

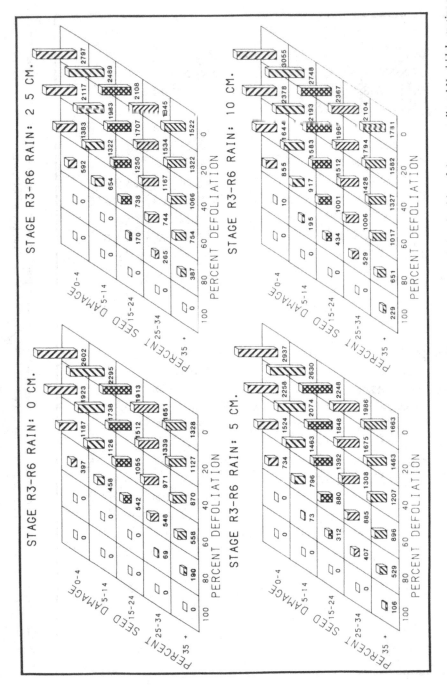

Figure 6. Predicted soybean yields (kg/ha) resulting from plant damage inflicted by various population densities of *A. gemmatalis* and *N. viridula* starting at R 5.5, at various rainfall levels from R3 through R6, and with 44.5 cm of pretreatment rainfall at Tifton, Georgia.

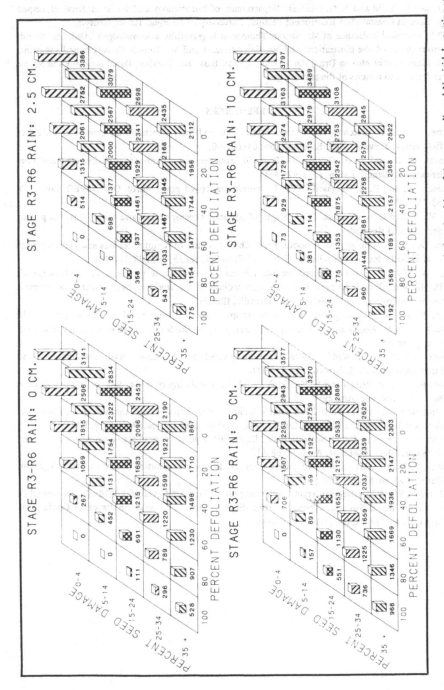

Figure 7. Predicted soybean yields (kg/ha) resulting from plant damage inflicted by various population densities of *A. gemmatalis* and *N. viridula* starting at R 5.5, at various rainfall levels from R3 through R6, and with 53 cm of pretreatment rainfall at Tifton, Georgia.

NOTES

James W. Todd and B. G. Mullinix, Departments of Entomology and Statistical Services, respectively, Georgia Coastal Plain Experiment Station, University of Georgia, Tifton, Georgia.

The technical assistance of Ms. Sheran Thompson is gratefully acknowledged. Also, Ms. Brenda Carpenter prepared the illustrations and Ms. Carol Ireland and Ms. Robbie Canady typed the manuscript. Many thanks also to Drs. John Woodruff, Max Bass, Jim Dutcher, David Isenhour and Myron Parker for critical reviews of the manuscript.

REFERENCES

Bardner, R. and K. Fletcher. 1974. Insect infestations and their effects on the growth and yield of field crops: a review. Bull. Entomol. Res. 64:141-60.

Caviness, C. D. and J. D. Thomas. 1980. Yield reduction from defoliation of irrigated and non-irrigated soybeans. Agron. J. 72:977-980.

Davidson, A. and R. B. Norgaard. 1973. Economic aspects of pest control. OEPP/EPPO Bull. 3(3): 63-75.

Fehr, W. R. and C. E. Caviness. 1977. Stages of soybean development. Iowa State Univ. Coop. Ext. Serv. Special Rep. 80.

Hinson, K., R. H. Nino and K. J. Boote. 1978. Characteristics of removed leaflets and yield response of artificially defoliated soybeans. Proc. Soil Crop Sci. Soc. Florida 37:104-109.

Kogan, M. 1976. Evaluation of economic injury levels for soybean insect pests. p. 515-533. *In* Lowell D. Hill (ed.) World Soybean Research: Proceedings of the world soybean research conference. Interstate Printers and Publishers, Inc., Danville, Illinois.

Kogan, M. and S. G. Turnipseed. 1980. Soybean growth and assessment of damage by arthropods. p. 3-29. *In* M. Kogan and D. Herzog (eds.) Sampling methods in soybean entomology. Springer-Verlag, New York.

McSorley, R. and V. H. Waddill. 1982. Partitioning yield loss on yellow squash into nematode and insect components. J. Nematol. 14:110-118.

Poston, F. L., L. P. Pedigo and S. M. Welch. 1983. Economic injury levels: Reality and practicality. Bull. Entomol. Soc. Am. 29:49-53.

SAS Institute Inc. 1981. SAS/Graph User's Guide: 1981 Edition. SAS Institute Inc., Cary, NC.

SAS Institute Inc. 1982. SAS User's Guide: 1982 Edition. SAS Institute Inc., Cary, NC.

Southwood, T. E. and G. A. Norton. 1973. Economic aspects of pest management strategies and decisions. Mem. Ecol. Soc. Aust. 1:168-84.

Tammes, P. M. L. 1961. Studies of yield losses. II. Injury as a limiting factor of yield. Tijdschr. Plantenziekten. 67:257-263.

Todd, James W. 1976. Effects of stink bug feeding on soybean seed quality. p. 611-618. *In* Lowell D. Hill (ed.) World Soybean Research Proceedings of the World Soybean Research Conference. Interstate Publishers, Inc., Danville, Illinois.

IMPACT AND ECONOMICS OF THREECORNERED ALFALFA
HOPPER FEEDING ON SOYBEAN

A. J. Mueller

The threecornered alfalfa hopper, *Spissistilus festinus* (Say), is a small triangular shaped insect in the family Membracidae of the Homoptera and, as the name implies, it is a pest of alfalfa (*Medicago sativa* L.). However, the first report of crop damage was on tomato (*Lycopersicon esculentum* Mill.) (Oemler, 1888). Favored host plants are members of the Leguminosae, but the threecornered alfalfa hopper has been reported to feed or reproduce on a variety of other plants, including Bermuda grass (*Cynodon dactylon* [L.] Pers.); Johnson grass (*Sorghum halepense* [L.] Pers.); wheat (*Triticum aestivum* L.); barley (*Hordeum vulgare* L.); oats (*Avena sativa* L.); sunflower (*Helianthus annuus* L.); cocklebur (*Xanthium* spp. L.); and cotton (*Gossypium hirsutum* L.).

DISTRIBUTION AND LIFE HISTORY

Caldwell (1949) restricted the distributional range of the threecornered alfalfa hopper to the southern United States and as far south as Costa Rica. Although this species may occur in northern USA, state records have not been reported in soybean north of Missouri (Mueller, 1980). The northern distributional range may fluctuate from year-to-year depending upon weather variations.

Developmental stages are egg, usually five nymphal stages and adult. The white, oblong egg is ca. 1 mm long and 0.35 mm in diameter and is oviposited beneath the plant epidermis. Mean incubation and nymphal developmental periods are reported to be 16.5 and 21.9 days, respectively (Meisch and Randolph, 1965). Females have a premating period of 8 to 10 days, but initiate oviposition immediately after mating (Kopp and Yonke, 1973). Mitchell and Newsom (1984a) reported that field collected females produced an average of 223.3 (±101.9) first instars during a mean reproductive life span of 38.6 (±25.1) days.

The threecornered alfalfa hopper overwinters as both adults and eggs (Wildermuth, 1915) and percentage survival is related to severity of the winter. Newsom et al. (1983) recently reported that in Louisiana the threecornered alfalfa hopper overwinters as an adult in a state of reproductive diapause and that the primary maintenance host is pine (*Pinus* spp.).

A four-year survey in Arkansas soybean fields indicates three overlapping generations per year (Mueller, 1980). Generation peaks occurred early to mid-July, mid-August, and September with each succeeding generation peak being of greater magnitude than the one before it. Four generations per year were reported on alfalfa in southern Arizona (Wildermuth, 1915) whereas only one generation per year was reported in the northern states (Kopp and Yonke, 1973).

NATURE OF DAMAGE

The threecornered alfalfa hopper is rather unique in its feeding behavior and no other Membracid feeds exactly like it. The most obvious feeding injury is a girdle at the base of the soybean plant. The insect girdles the stem by repeatedly inserting its piercing-sucking mouthparts, circumscribing the plant stem. Girdles low on the main stem are associated with lodging, breakage, and plant mortality affecting single plant yield (Mueller and Dumas, 1975). This type of feeding usually will occur before plants reach a height of 20 to 25 cm (Bailey et al., 1970; Davis and Laster, 1968). If soybeans are planted late, such as may occur in double-cropping, girdling incidence of young plants may be greater due to the increased numbers of threecornered alfalfa hopper present. This basal mainstem girdling usually is caused by immigrating adults shortly after plant emergence and/or the third-fifth stage nymphs of the first generation (Moore and Mueller, 1976). These girdles disrupt the vascular system of the plant, and frequently a swelling of the stem appears above the girdle. According to Mitchell and Newsom (1984b), the threecornered alfalfa hopper will preferentially feed at this swelling above the girdle, presumably exploiting the nutrient sink created by the phloem disruption and consequent blockage of translocate flow.

The first and second instars are not reported to girdle, but instead feed by single random punctures (Moore and Mueller, 1976). All others stages also may feed at times by random probing, and do not necessarily always cause girdles.

As the soybean plant increases in size, feeding sites are moved upward on the main stem, lateral branches, or onto the leaf petioles. The cause for the change in feeding sites is unknown, but possible explanations are that the physical hardening of the lower stem makes stylet penetration difficult and the increased stem thickness makes it difficult for insects to grasp the stem securely enough for the necessary leverage for stylet penetration. The leaf petioles are frequently girdled and occasionally the upper main stem and lateral branches are also girdled. During plant reproductive stages, the threecornered alfalfa hoppers may move to the racemes and also feed on the pedicels and peduncles; however, girdling of racemes is extremely rare. Mitchell and Newsom (1984b) suggest that since the developing pod is a nutrient sink, girdling of peduncles and pedicels is unnecessary.

PLANT RESPONSE AND AFFECT ON YIELDS

Plants girdled near the base respond in five major ways (Mueller and Jones, 1983). (1) Death, usually within two to three weeks after girdling. Occasionally the plant will survive as long as six weeks. Symptoms of this response are loss of vigor and growth, wilting from the top downward and reduction in size, and finally plant mortality. (2) Plants break at the girdle site and death occurs before harvest, or plants will not be harvestable by machine because they lie on the ground. In this category, breakage of the main stem may occur shortly after girdling up until maturity when plants have a full complement of pods. These plants do not contribute to yield. (3) The plant breaks at the girdle site, but remains alive and can be harvested. Many plants in this category are broken early, but the breaks are not complete. The lower portion of the plants lie flat on the ground, but as the plant continues to grow, new growth is vertical. (4) Plants remain upright (do not break), continue to grow in height, but become spindly and produce few, if any, pods. If pods are present, the number and size of the soybeans per pod are reduced. Since these plants compete with healthy, ungirdled plants for space and nutrients, their role is similar to a weed's. (5) The girdled area becomes "scabbed over", the plant remains upright, appears to recover, and produce a full or nearly full complement of pods. In a three-year field study in Arkansas (1978-1980), 34.4, 6.5, 2.5, 25.2, and 31.4% of the girdled plants were in categories 1 through 5, respectively (Table 1).

Caviness and Miner (1962) in a three-year field study simulated threecornered alfalfa hopper injury by reducing plant stands by 15, 30, and 45% at bloom and 2 weeks pre- and post-bloom. Significant yield reduction occurred only at 30% stand reduction 2 weeks post-bloom and 45% stand reduction at bloom and 2 weeks post-bloom.

Tugwell et al. (1972) regulated naturally occurring threecornered alfalfa hopper field populations by insecticide applications, which resulted in mean main stem girdling levels of 17, 24, 34, and 42%. Although there were no significant yield differences among girdling levels, the authors believed that differences might have occurred if girdled plants had lodged too late in the season for adjacent plants to compensate or had been unharvestable.

Mueller and Jones (1983), in a three-year small-plot field study, placed varying densities of laboratory reared third to fifth stage nymphs on susceptible stage plants (V3 to 6). In two of the three years, yield differences occurred when main stem girdling exceeded 5% (Table 2). Maximum girdling in the third year was only 50% and no differences were detected. Due to compensation by the ungirdled plants, plot yields were not reduced until ca. 65 to 70% girdling. Although this level may appear high, Wiggins (1939) and Probst (1945) demonstrated that only 9 to 12 healthy plants per m of row are required for maximum yields. Recommended seedling rates are ca. 39 seeds per m of row, and many farmers plant more. Since germination usually is above 80%, 9 to 12 plants per m of row still remain after a 65 to 70% girdling level occurs. Additionally, many of the girdled plants (ca. 30%) recover and have a full complement of pods. If our seeding rates were lower, a means suggested by some to conserve on planting costs, the threecornered alfalfa hopper probably would be a more serious pest.

Before 1981, the recommended control measure by most states that had a recommendation, was to treat when 10 to 15% of the plants were girdled and populations were still present. Since then, most recommendations have been changed to treat when fewer than 12 to 18 ungirdled plants per row m remain, plants are less than 25 cm tall, and threecornered alfalfa hoppers are present. This recommendation refers only to conventional row spacing (60 to 110 cm). No thresholds are presently recommended

Table 1. Plants in each girdled plant response category, 1978 through 1980 (Mueller and Jones, 1983).

Plant response	1978	1979	1980	Mean
	%	%	%	%
(1) Death within 6 wk	35.1	36.5	31.6	34.4
(2) Broken and unharvestable or death	3.4	6.0	10.1	6.5
(3) Broken, harvestable	1.8	1.9	3.8	2.5
(4) Upright, weak and spindly	26.8	26.5	22.2	25.2
(5) Plant recovered	32.9	29.1	32.3	31.4

Table 2. Mean yields from threecornered alfalfa hopper tests, 1978 through 1980 (Mueller and Jones, 1983).

Year	Girdling level[a]	Mean yields/plot[b]		Total
		Ungirdled plants	Girdled plants	
	%	g	g	g
1978	0	321 r	–	321 r
	22(19-23)	311 r	62 u	373 r
	31	313 r	36 v	379 r
	39(38-42)	274 rs	100 t	374 r
	52(50-54)	250 s	112 st	362 r
	65	203 s	132 s	355 r
	85	32 t	233 r	265 s
1979	0	130 r	–	130 r
	23	110 rs	7 s	117 r
	49(46-50)	83 st	34 rs	118 r
	69(65-77)	66 tu	41 r	107 r
	89(81-96)	22 u	38 rs	60 s
1980	0	167 r	–	167 r
	10(8-12)	186 r	10 s	197 r
	18(15-20)	183 r	8 s	191 r
	26(23-27)	144 r	13 s	158 r
	36(32-38)	136 r	22 s	159 r
	47(44-50)	145 r	36 r	181 r

[a]Values in parentheses indicate range in percentage girdling.

[b]Column means for each year followed by the same letter are not significantly different at P = 0.05, by Duncan's Multiple Range Test.

for narrow-row or solid-planted soybeans.

Recent studies in Louisiana (Layton, 1983) and Alabama (T. P. Mack, pers. commun.) indicate that threecornered alfalfa hopper feeding during plant reproductive stages reduces yields and may be more harmful to plants than early season main stem girdling. Layton (1983) used one to three week interval application of insecticides to manipulate populations of threecornered alfalfa hoppers, and she was able to maintain significant differences in populations using this technique. In one of her two field tests, she got a significant yield difference between the check and two insecticide regimes. Mean yields from the chlorpyrifos-treated plots were significantly lower and yields from the fenvalerate-treated plots were significantly higher. The chlorpyrifos plots had significantly higher seasonal numbers of threecornered alfalfa hoppers, while the fenvalerate plots had significantly lower seasonal numbers of hoppers. From these data, she suggests that late season threecornered alfalfa hopper feeding may be more harmful to plants and consequently to yields than early season main stem girdling injury.

In a two-year cage study in Alabama (T. P. Mack, unpublished) where 52 threecornered alfalfa hopper adults per meter were placed on caged plants at R2 and left undisturbed until harvest, yield losses averaged 46.5%.

Other sources of possible damage or reduced yields may be due to increased susceptibility to fungal infection or reduced nitrogen fixation. Herzog et al. (1975) studied the relationship of threecornered alfalfa hopper feeding damage and the fungus *Sclerotium folfsii* Saccardo. They concluded that disease transmission by this insect was of little importance, but that feeding sites served as a point of entry by the fungus. Hutchinson (1979) reported that stem girdling by the threecornered alfalfa hopper reduced nodulation and nitrogen fixation.

SUMMARY

Before 1981, action thresholds for threecornered alfalfa hopper main stem girdling damage ranged from no threshold to treatment when 10 to 15% of the plants were girdled with damaging populations still present. Since 1981, most states with three-cornered alfalfa hopper problems have revised their recommendations to treat when fewer than 12 to 18 ungirdled plants per row m remain, plants are less than 25 cm tall, and threecornered alfalfa hoppers are present. In most cases, early season populations will not be high enough to cause this level of main stem girdling. Therefore, when recommendations are followed, insecticide treatments are greatly reduced. Arkansas was the first to use this new recommendation, and extension personnel, consultants, and farmers have readily accepted it. One precaution is necessary. Although girdling injury appears to be random, under high population pressures a number of consecutive plants may be girdled. Therefore, damage assessment should be made by subsamples no greater than 0.5 m of row, because adjacent plants may not be able to compensate fully by filling in a spacing greater than 0.5 m. This recommendation is useful for conventional row spacing only. At present, no thresholds have been developed for narrow-row or solid-planted soybeans. In all probability, a greater incidence of mainstem girdling could be tolerated under these cultural regimes, because the hazard of lodging would be reduced.

Recently, several studies have suggested that late season feeding on the leaf petioles, racemes, pedicels, and peduncles may affect soybean yields. To date, no action thresholds have been developed and more research is needed to fully understand the impact this type of feeding has on yields.

NOTES

A. J. Mueller, Department of Entomology, University of Arkansas, Fayetteville.

REFERENCES

Bailey, J. C., L. B. Davis and M. L. Laster. 1970. Stem girdling by the threecornered alfalfa hopper and height of soybean plants. J. Econ. Entomol. 63:647-648.

Caldwell, J. S. 1949. A generic revision of the treehoppers of the tribe Ceresini in America north of Mexico, based on a study of the male genitalia. Proc. US Nat. Mus. 98(3234):491-521.

Caviness, C. E. and F. D. Miner. 1962. Effects of stand reductions in soybean simulating threecornered alfalfa hopper injury. Agron. J. 54:300-302.

Davis, L. B. and M. L. Laster. 1968. Stage of growth when soybeans are damaged by the threecornered alfalfa hopper. Miss. Agric. Exp. Stn. Information Sheet 1016. Mississippi State, Miss.

Herzog, D. C., J. W. Thomas, R. L. Jensen and L. D. Newsom. 1975. Association sclerotial blight with *Spissistilus festinus* girdling injury on soybean. Environ. Entomol. 4:986-988.

Hutchinson, R. L. 1979. The effects of insect damage on nitrogen fixation of soybeans. Ph.D. Dissertation, Louisiana State Univ. Library, Baton Rouge.

Kopp, D. D. and T. R. Yonke. 1973. The treehoppers of Missouri: Part 2. Subfamily Similinae, Tribes Acutalini, Ceresini, and Polyglyptini. J. Kans. Entomol. Soc. 46:233-242.

Layton, M. A. B. 1983. Effects of the threecornered alfalfa hopper, *Spissistilus festinus* (Say), on yields of soybean as determined by manipulation of populations by the use of insecticides. M.S. Thesis, Louisisana State Univ. Library, Baton Rouge.

Meisch, M. V. and N. M. Randolph. 1965. Life history studies and rearing techniques for the three-cornered alfalfa hopper. J. Econ. Entomol. 58:1057-1059.

Mitchell, P. L. and L. D. Newsom. 1984a. Seasonal history of the threecornered alfalfa hopper (Homoptera:Membracidae) in Louisiana. J. Econ. Entomol. 77 (In press).

Mitchell, P. L. and L. D. Newsom. 1984b. Histological and behavioral studies of threecornered alfalfa hopper (Homoptera:Membracidae) feeding on soybean. Ann. Entomol. Soc. Am. 77 (In press).

Moore, G. C. and A. J. Mueller. 1976. Biological observations of the threecornered alfalfa hopper on soybean and three weed species. J. Econ. Entomol. 69:14-16.

Mueller, A. J. 1980. Sampling threecornered alfalfa hopper on soybean. p. 382-393. *In* M. Kogan and D. C. Herzog (eds.) Sampling Methods in Soybean Entomology. Springer-Verlag, N.Y.

Mueller, A. J. and B. A. Dumas. 1975. Effects of stem girdling by the threecornered alfalfa hopper on soybean yields. J. Econ. Entomol. 68:511-512.

Mueller, A. J. and J. W. Jones. 1983. Effects of main stem girdling of early vegetative stages of soybean plants by threecornered alfalfa hopper (Homoptera: Membracidae). J. Econ. Entomol. 76:920-922.

Newsom, L. D., P. L. Mitchell and N. N. Troxclair, Jr. 1983. Overwintering of the threecornered alfalfa hopper in Louisiana. J. Econ. Entomol. 76:1298-1302.

Oemler, A. 1888. Extracts from correspondence regarding a new tomato enemy in Georgia, between A. Oemler and the Division of Entomology. USDA Div. Entomol. Insect Life 1:50.

Probst, A. H. 1945. Influence of spacing on yield and other characters in soybeans. J. Am. Soc. Agron. 37:549-554.

Tugwell, P., F. D. Miner and D. E. Davis. 1972. Threecornered alfalfa hopper infestations and soybean yield. J. Econ. Entomol. 65:1731-1733.

Wiggins, R. G. 1939. The influence of space and arrangements on the production of soybean plants. J. Am. Soc. Agron. 31:311-321.

Wildermuth, V. L. 1915. Threecornered alfalfa hopper. J. Agric. Res. 3:343-364.

APPROACHES TO STUDYING INTERACTIVE STRESSES CAUSED BY INSECTS AND WEEDS

Randall A. Higgins

Documenting whether combinations of stressors are synergistic, antagonistic, or additive in effect should aid efforts to develop cost-effective management strategies for pest complexes. Unfortunately, investigations combining weed and insect stress have been hindered by the unfamiliarity of weed scientists and entomologists with methods of establishing treatments in a manner acceptable to the other discipline. Review articles and policy statements (Allen and Bath, 1980; Norris, 1982; Pedigo et al., 1981) recognize the need for interdisciplinary investigations, but offer little advice on designing and implementing appropriate research programs. The primary objectives of this manuscript are to encourage quantitative studies of crop resilience to combinations of weeds and insects by reviewing the major approaches used to create crop stress and proposing standardized guidelines for data acquisition.

COMPONENTS OF THE EXPERIMENT

Proper selection of the component treatments is a prerequisite to successful interdisciplinary investigations. Although not all-inclusive, the following points should be considered in preliminary studies: (1) examine weeds and insects with relatively simple life cycles (discrete generations, etc.), (2) choose pests with substantial data bases (probably economically important pests), (3) select pests for which local rearing and establishment procedures have been confirmed, (4) combine species that can be maintained in spite of pest control measures required to minimize treatment confounding from other native weeds and insects, and (5) consider pest combinations for which there is a strong possibility that interdisciplinary management guidelines may be implementable. The interdisciplinary cooperator should be selected with equal care.

Published discipline-oriented studies have shown repeatedly that careful attention to detail must occur when any method of establishing biotic stress is employed. Whether studied individually or collectively, the importance of insects and weeds to the production system often depends on pest density and size, phenology of occurrence, influence of planting date, row widths, planting density, crop vigor, cultivar, and prevailing weather conditions. These influences must be appreciated and documented if effects on plant growth and yield will be accurately related to specific pest densities or combinations of density, dry matter, and related variables.

ESTABLISHING THE WEED COMPETITION STRESS COMPONENT
UNDER FIELD CONDITIONS

Recent studies of interference or competition by weeds in row crops have employed selective herbicides (e.g., safe-rate tables; Buchanan, 1977), manual weed removal, and artificially established pest populations to modify species composition, density, and spatial arrangement. This simplification of the natural situation (from species complexes occurring in irregular aggregations) may be even more crucial to obtain repeatable baseline data in interdisciplinary efforts. Also, plot size will probably be smaller than has been standard for competition studies, simply because techniques required to establish discrete insect populations over a large area are not generally available.

Weed scientists often locate competition trials where substantial natural weed populations have been confirmed. Mapping and augmenting known infested areas with propagules of local origin may be necessary to ensure that the heaviest infestation level required during the study will be attained.

Manual or mechanical planting of weed seed in continuous rows or in spaced hills on one or both sides of the crop row is also commonly used (Buchanan, 1977). Because the range of ideal planting depths is sometimes very narrow, several seeds per desired weed location often are sown to ensure that a few germinate. Dormancy-breaking treatments may be necessary, advisable, or deleterious depending on the species and current physiological state. Brief immersion in boiling water, acid baths, cold treatments, drying, or scarification substantially improves germination for some species.

Seeds may be planted in flats or peat pots and germinated in greenhouses, but the plants should be transferred to field plots soon after the seedlings emerge, preferably within a week, to minimize acclimation effects. Ideally, the weed and crop should be structurally and physiologically indistinguishable from plants developing in a typical crop production field.

For some tests, delayed planting may be desirable. However, if objectives include assessing competition duration in "equal start" situations, seeding or transfer of weeds must be concluded as crop planting or emergence is completed. In general, weed treatments should be thinned to desired densities soon after emergence and initial establishment have occurred. Selective removal of newly emerging weeds should be continued until the developing crop canopy eliminates further weed establishment.

A major problem resulting from the use of any pesticide to reduce a population is that a significant proportion of the survivors may be structurally or physiologically damaged. Thus, pesticides which may alter the experimental pest should not be used unless sublethal effects can be quantified or competition from the susceptible species is to be terminated. Inclusion and treatment of additional weed-free control plots to evaluate whether treatment-terminating herbicides cause significant crop injury may be advisable. Likewise, knowing the field herbicide history before planting and plot establishment should reduce unanticipated problems caused by carryover of persistent herbicides.

ESTABLISHING THE INSECT STRESS COMPONENT UNDER FIELD CONDITIONS

Approaches frequently used to manipulate existing insect populations and (or) to create insect damage include: (1) the use of unaltered natural infestations, (2) the modification of natural infestations with resistant isolines or selective insecticides,

643

(3) the use of caged or uncaged artificial infestations, and (4) the use of surrogate damage or injury simulation (Kogan, 1976; Poston et al., 1983). Methodology for documenting the effects of insect damage is less standardized than that for weed competition under field conditions.

Pest mobility and sampling difficulties often have hampered efforts to assess insect-induced stress quantitatively. Natural insect populations often are unpredictable, making duplication in time and space difficult or impossible to achieve, especially when the research area must be designated well in advance (as in weed-competition studies). Although true replication and randomization rarely occur, variation in pest population density across the infested area may develop with enough divergence to permit a limited assessment of several damage levels. Insect density gradients created by using genetic isolines of crops differing in factors that give the plant degrees of pest resistance may have promise in this regard. But the results may have broader practical application if equality in damage potential per insect among the isolines is maintained (e.g., only numbers and not feeding behavior have been altered).

Frequently, the stress history of a crop cannot be documented when late-season pest problems that developed fortuitously are studied. Unfortunately, several studies have shown that it may be very risky to extrapolate yield-loss findings to other situations without knowledge of the conditions under which the original response occurred. For instance, Higgins et al. (1984a) compared their data with those of Hammond and Pedigo (1982) and noted more than a twofold difference in loss per insect equivalent. Documented differences in moisture availability during reproductive development were cited as the most likely cause for the discrepancy, because the same techniques, cultivars, and geographic areas were employed. Higgins et al. (1983, 1984c) also reported that knowledge of leaf area remaining on stressed plants was much more informative than traditional measurements of insect damage, which frequently only assess tissue damage indirectly (e.g., percent defoliation).

Serial dilutions of broad-spectrum insecticides have been used on high-density insect populations to create a range of infestations under field conditions. Unfortunately, moribund individuals may substantially alter the relationship between pest density and damage-inflicting potential. Fewer problems may be encountered if selective insecticides rather than broad-spectrum products are used to eliminate unwanted species.

Artificial establishment of known insect densities has merit, especially if rearing procedures are well-established and not expensive. Synchronizing insect and host plant development plus achieving establishment may be difficult. In many instances, the more unrestricted the release environment, the more authentic the results. Some materials used to restrain insects in cage experiments can alter the microenvironment sufficiently to cause unnatural plant developmental responses. However, cage effects can be minimized by curtailing placement time. Non-screened barriers, isolation plots, and genetic traits restricting pest mobility have been proposed and used. Preventing native insects of the same species from altering the release densities may be difficult.

Ascertaining the actual number of insect equivalents sustained in any field study wherein real insects are employed over the course of a generation can be very difficult. Additional plot space should be allocated for each treatment combination to allow for frequent sampling capable of providing near-absolute population estimates. Otherwise, the yield loss realized might be attributed to a larger pest population (on an effective insect-equivalent basis) than actually existed for the duration of damage. If unaccounted for, this error could result in the recommended tolerance of denser infestations than are actually justified. Seemingly, this oversight would be especially serious where lepidopterous defoliators are released at known densities of young larvae and

644

left undisturbed until a yield analysis is conducted. Mortality caused by predators, parasites, pathogens, or weather after the insects have been released, but before the damage potential is realized, would artificially raise the control threshold established in the baseline study because 85% or more of the defoliation typically occurs in the final two larval stages. The deviation would be most serious if mortality similar to that overlooked during threshold development does not occur wherever these guidelines are applied.

Various types of surrogate injury or damage simulation have been used to obtain discrete levels of insect-like stress (e.g., Coggin and Dively, 1980; Hammond and Pedigo, 1982; Thomas, this volume). Initial attempts to quantify the relationship between foliage loss and insect numbers used a variety of techniques, such as leaf excision and leaf shredding with wire flails, to create damage that was assumed similar to injury caused by defoliating insects. However, comparative studies of simulated and actual insect feeding have shown that if the goal is to mimic a given insect damage pattern, appropriate methods must be selected (Poston et al., 1976; Hammond and Pedigo, 1981; Ostlie, 1984). Technique development and reasonable "validation" obviously require substantial supporting data on the damage syndrome of the species being simulated. Absolute validation that the simulation causes injury equal in effect to the real insect is probably not possible because the experiments are designed to test the null hypothesis that damage patterns are equal. Furthermore, it is common for simulation techniques to be very labor intensive (Higgins et al., 1983). Simulations may need to duplicate the phenological development of insect damage, and injury should be inflicted on the same plant stratum typically attacked by the insect. These latter precautions may be necessary because plants respond dynamically to many progressive stresses (i.e., they are "plastic"), and because insect damage is distributed over time rather than occurring on a single date. Another disadvantage is that for many types of insect damage no practical simulation method has been developed and tested. Elimination of the natural population is usually desirable to prevent undocumented damage.

Criticisms can be presented for all methods of producing insect damage if direct quantification of pest densities is desired. Nevertheless, most entomologists investigating crop loss readily admit that studies using these approaches have done much to advance practical insect management.

CONSIDERATIONS IN EXPERIMENTAL DESIGN AND ASSESSMENT
OF STRESS INTERACTION

Documentation of crop yield produced by gradients of stress has been a cornerstone in the development of practical pest management guidelines (Poston et al., 1983). Defining specific relationships between the stressor and the crop requires much more effort on the part of the investigator than simply creating weed, insect, or 'insect-like' stress successfully. A primary objective of damage-loss investigations must be quantitative interpretation of the relationship between degree of stress and altered growth, development, and yield of the crop regardless of the method chosen for imposing treatments. In contrast to insect damage potential, which often is assessed in terms of straightforward density or density and pest size or previous damage relationships (economic injury levels and economic thresholds corrected for crop stage), determining the importance of stress in situations where two plant species interact is usually more complex. Complications develop because competition is dynamic, with the stressors sometimes responding differently and dramatically to environmental influences. Therefore, records adequately describing the microclimate and edaphic environment should be maintained (Doraiswamy, 1982).

Confirmation that interaction exists between treatments is seldom possible unless each component reaches statistical significance independently. Simultaneously, the interdisciplinary experiment should examine treatment combinations representing actual field situations, so that management implications may be assessed (Higgins et al., 1984a; Schreiber, this volume; Wax and Stoller, this volume). Advanced studies of interactive stress should not simply combine full-season competition from weed densities capable of forming a continuous crop-shielding canopy along with severe insect damage restricted to the most yield-loss sensitive plant growth stage. Such an experiment would be useful for documenting extreme responses, but if interaction between components is substantial no information on the functional form of the response under more practical treatment combinations would be obtained.

Experimental designs employing factorial treatment arrangements have several advantages for assessing the extent of interaction between treatments. Unfortunately, a small increase in the number of treatments or the levels per treatment can create an unwieldy or unworkable experiment. The intentional confounding or elimination of some treatment combinations has been suggested as one approach to clarify interaction between treatments in complex intercropping experiments (Mead and Stern, 1980). Poston et al. (1983) also recognized the logistical problem created by traditional factorial treatment combinations and suggested an approach that makes greater use of existing single factor studies. They proposed the establishment of stressor equivalences based on similarity in physiological processes disrupted by the organisms stressing the crop. Substantial knowledge of plant processes is required during the experimental design phase because arriving at treatments involves assessing the role each stressor has in altering physiological aspects of crop health (Boote, 1981). This step is followed by experiments to determine the resilience response form exhibited by the crop to selected levels of stressors, and subsequently yield is established as a function of the physiological process being affected.

Clarifying the relationship between stressors and facilitating systems modeling require that simple studies of treatment effects based solely on economic yield be supplanted by response assessments documenting both the amount and rate of change over critical intervals. The primary objective then becomes accurately defining underlying relationships and not simply confirmation that insect A and weed B ultimately reduce profit. Season-long analyses of growth and development in interdisciplinary studies should focus on the crop, weed, and insect components simultaneously. Subsequently, statistical comparisons of the magnitude and form of the response curves established for single and combination treatments should prove very useful in assessing the significance of stress interaction during the growing season.

Plants from selected treatments periodically are sacrificed, partitioned into structural components, and then measured in studies of agronomic growth analysis (Radford, 1967; Patterson, 1982). Data on leaf area indices, biomass duration, morphological changes, and phenological development (Boote, 1981) are required in addition to the standard evaluation of dry weight. Enough additional plots or intraplot areas are required to facilitate the destructive analyses without eliminating the treated area. For soybean studies, samples should be obtained from plots of weed-free soybeans, soybean-free weeds, insect-free soybeans, and as many combinations of mixed stresses as labor and space permit while maintaining a statistically valid experimental design (Higgins et al., 1983; 1984b,c). Sampling frequency should be adequate to define the resilience response by documenting plant growth and development through each diverse developmental stage encompassed by the active stress period. Additional samples are required immediately before each stress treatment is imposed (for example, defoliation begins), immediately after the stress terminates (larvae cease feeding), and during the recovery period (e.g., two or more weeks after pupation). The latter samples will help document whether compensatory regrowth is occurring.

The temptation to exclusively use the 'regression approach' to growth analysis must be suppressed in some studies involving insects. Caution may be necessary because the 'smoothing' effect may mask significant, but transient changes caused by time-concentrated stresses. Because the 'traditional approach' (calculating rates based on two adjacent sampling dates) is less likely to mask these effects, but sometimes exhibits unacceptably high variability, a combination of methods may be needed. Degree of concern will probably depend on the species involved, type of damage inflicted, and the frequency of the sampling interval. Insects typically remove biomass that was manufactured at a cost to the plant rather than simply altering current production. Therefore, the term 'apparent' should probably precede all calculated growth rates unless the actual plant production within insect-damaged plots is somehow quantified. Recent applications also suggest that the expression of time should be changed from days to growing degree days (Russelle et al., 1984).

Partitioning of dry matter among structural and reproductive fractions can be studied through actual or apparent harvest indices. Path coefficient analyses have proven valuable for partitioning the relative influence of stress among each yield component so that concealed interactions can be assessed (Campbell et al., 1980).

Substitutive designs, including replacement series and mechanical diallel arrangements (Elmore et al., 1983), should be useful in some interdisciplinary stress evaluations. Assessing relative yield totals may be most appropriate when insect injury causes barren crop plants. Thus, without changing the absolute plant population, severely insect-injured crop plants may be considered 'functional weeds' with regard to their continuing use of resources and lack of yield contribution.

Variations of the 'critical duration of competition' and the 'critical weed-free requirement' formats used to study crop and weed competition are anticipated to dominate quantitative studies of weed and insect stress interaction. Most row crops are normally planted in a seedbed relatively free of competing vegetation. For a variable period of time, little interaction occurs between species, even at very high densities. Eventually, demands exceed the available supply of resources and plants exhibit altered growth and development. The 'critical period of competition' is reached when a crop can no longer tolerate a continued use of resources by weeds that emerged with the crop (Buchanan, 1977). Effective control measures should be applied before this threshold is exceeded because continued interference results in significantly decreased grain yields. The relatively scattered weeds that typically survive reasonably effective preemergence control efforts on well-managed farms cause greatest yield reductions when competition persists into the crop reproductive stages. Obviously, the weed treatments need to be established near the time of crop seeding in this type of experiment. In the other approach, seeds are planted or young seedlings transplanted at intervals after the crop was planted or has emerged. The 'critical weed-free requirement' is the interval after which future competitive losses to newly germinating weeds will be inconsequential, so that weed control efforts can be discontinued. In general, vigorously growing soybeans will provide adequate interspecific competition 4 to 6 weeks following emergence, if weed establishment was prevented before this interval. Crop canopy development gradually decreases the amount and quality of light available at the soil surface, thereby lessening the likelihood of additional weed establishment. Greater practical usage of the critical-duration-of-competition concept may result if anticipated yield loss is expressed as a function of density and duration (Barrentine and Oliver, 1977). The failure of these guidelines to account for harvest interference losses and an appreciation that control delay also may limit the suppression realized should be well understood by the user.

Factors known to influence the 'critical periods of competition' include weed species, density, duration, phenology, spacial arrangement, and many abiotic parameters. Insect-stressed crops may have an equally significant, albeit largely unappreciated, influence on the application of these management guidelines. Severe, early-season defoliation or root pruning of soybeans may substantially lengthen the weed-free requirement, in part, by extending the time required for canopy closure. Should weed competition and insect-induced stress interact, adequate documentation of the difference in time of crop and weed emergence may be important. Higgins et al. (1983; 1984a,b,c) noted only subtle evidence of interactions at the plot yield level in an interdisciplinary study combining realistic levels of simulated green cloverworm [*Plathypena scabra* (F)] defoliation with velvetleaf [*Abutilon theophrasti* (Medic)] competition. Identifiable treatment interactions detected during the growing season largely were transitory. Consistent interactive effects on yield at physiological maturity were restricted to strata within weed-proximate soybeans, where the effects of the sole treatments overlapped. The authors concluded that proximity-based analyses should become standard components of weed competition and insect-damage interaction studies where the weed density is sparse enough that it does not form a continuous overstory above the crop canopy. Infestations not exhibiting substantial interaction between adjacent weeds may be the most appropriate situation for yield-loss assessments based solely on insect and weed densities. Denser weed populations suffer mutual interference and often require biomass to be used as a covariate before reasonable predictions of yield loss are possible. Simple 'gain thresholds' (kg/ha of crop yield required to offset management costs, Higgins et al., 1984a) should still have value in hindsight determinations of whether or not pest-removal efforts were actually justified, however.

Lack of substantial interaction on yield does not mean that weed management is necessarily independent of insect management. Higgins (1982) determined that the minimal height differential required between crop and weed for proper clearance of rope wick applicators was exceeded earlier if severe crop defoliation occurred. Regardless of the economic advisability, this finding indicates that altering insect management may sometimes expand weed management opportunities. The converse has been reported where predators and parasites of pest insects use residual weed populations as alternative food sources.

Weed and insect interaction studies will contribute knowledge necessary to improving our understanding of pest complexes. Because no single treatment establishment method, experimental design, or evaluation protocol will apply to all pest combinations of interest, close cooperation between agronomists, entomologists, and weed scientists must be maintained. Statisticians probably will play an ever-increasing role if the experiments are to remain practical as treatment combinations become more elaborate and traditional analytical approaches prove inadequate. Team efforts are particularly important if the concepts of economic thresholds, economic injury levels, critical durations of competition, and critical weed-free requirements are to be expanded from limited discipline-specific tools to comprehensive pest management guidelines.

NOTES

Randall A. Higgins, Assistant Professor, Department of Entomology, 239 West Waters Hall, Kansas State University, Manhattan, KS 66506.

Contribution number 85-79-A of the Kansas Agricultural Experiment Station.

REFERENCES

Allen, G. E. and J. E. Bath. 1980. The conceptual and institutional aspects of integrated pest management. BioScience 30:658-664.

Barrentine, W. L. and L. R. Oliver. 1977. Cocklebur competition economics and control in soybeans. Miss. Agric. For. Exp. Stn. Tech. Bull. 83:1-27.

Boote, K. J. 1981. Concepts for modeling crop response to pest damage. ASAE Paper 81-4007, American Society of Agricultural Engineering, St. Joseph, MI.

Buchanan, G. A. 1977. Weed biology and competition. p. 25-41. *In* B. Trueblood (ed.) Research methods in weed science. S. Weed Sci. Soc.-Auburn Printing. Auburn,Ala.

Campbell, W. F., R. J. Wagenet, A. M. Bematraf and D. L. Turner. 1980. Path coefficient analysis of correlation between stress and barley yield components. Agron. J. 72:1012-1016.

Coggin, D. L. and G. P. Dively. 1980. Effects of depodding and defoliation on the yield and quality of lima beans. J. Econ. Entomol. 73:609-614.

Doraiswamy, P. C. 1982. Instrumentation and techniques for microclimate measurements. p. 43-63. *In* J. L. Hatfield and I. J. Thomason (eds.) Biometeorology in integrated pest management. Academic Press, New York.

Elmore, C. D., M. A. Brown and E. P. Flint. 1983. Early interference between cotton (*Gossypium hirsutum*) and four weed species. Weed Sci. 31:200-207.

Hammond, R. B. and L. P. Pedigo. 1981. Effects of artificial and insect defoliation on water loss from excised soybean leaves. J. Kans. Entomol. Soc. 54:331-336.

Hammond, R. B. and L. P. Pedigo. 1982. Determination of yield-loss relationships for two soybean defoliators by using simulated-insect defoliation techniques. J. Econ. Entomol. 75:102-107.

Higgins, R. A. 1982. Effects of simulated insect defoliation and annual weed competition on soybean and velvetleaf development. Ph.D. Dissertation. Iowa State Univ. Library, Ames, Iowa.

Higgins, R. A., L. P. Pedigo and D. W. Staniforth. 1983. Selected pre-harvest morphological characteristics of soybeans stressed by simulated green cloverworm defoliation and velvetleaf competition. J. Econ. Entomol. 76:484-491.

Higgins, R. A., L. P. Pedigo and D. W. Staniforth. 1984a. Effects of velvetleaf competition and defoliation simulating a green cloverworm (Lepidoptera:Noctuidae) outbreak in Iowa on indeterminate soybean yield, yield components, and economic decision levels. Forum: Environ. Entomol. 13:917-925.

Higgins, R. A., L. P. Pedigo, D. W. Staniforth and I. C. Anderson. 1984b. Partial growth analysis of soybeans stressed by simulated green cloverworm defoliation and velvetleaf competition. Crop Sci. 24:289-293.

Higgins, R. A., D. W. Staniforth and L. P. Pedigo. 1984c. Effects of weed density and defoliated or undefoliated soybeans (*Glycine max*) on velvetleaf (*Abutilon theophrasti*) development. Weed Sci. 32:511-519.

Kogan, M. 1976. Evaluation of economic injury levels for soybean insect pests. p. 515-533. *In* L. D. Hill (ed.) World soybean research. Interstate Printers and Publishers, Inc., Danville, Ill.

Mead, R. and R. D. Stern. 1980. Designing experiments for intercropping research. Exper. Agri. 16:329-342.

Norris, R. F. 1982. Interactions between weeds and other pests in the agro-ecosystem. p. 343-406. *In* J. Hatfield and I. Thomason (ed.) Biometeorology in integrated pest management. Academic Press, New York.

Ostlie, K. R. 1984. Soybean transpiration, vegetative morphology, and yield components following simulated and actual insect defoliation. Ph.D. Dissertation. Iowa State Univ. Library, Ames, Iowa.

Patterson, D. T. 1982. Effects of light and temperature on weed/crop growth and competition. p. 407-420. *In* J. Hatfield and I. Thomason (ed.) Biometeorology in integrated pest management. Academic Press, New York.

Pedigo, L. P., R. A. Higgins, R. B. Hammond and E. J. Bechinski. 1981. Soybean pest management. p. 417-537. *In* D. Pimentel (ed.) Handbook of pest management in agriculture. Vol. 3. CRC Press, Boca Raton, Fla.

Poston, F. L., L. P. Pedigo, R. B. Pearce and R. B. Hammond. 1976. Effects of artificial and insect defoliation on soyean net photosynthesis. J. Econ. Entomol. 69:109-112.

Poston, F. L., L. P. Pedigo and S. M. Welch. 1983. Economic-injury levels: Reality and practicality. Bull. Entom. Soc. Am. 20:529-553.

Radford, P. J. 1967. Growth analysis formulae—Their use and abuse. Crop Sci. 7:171-175.

Russelle, M. P., W. W. Wilhelm, R. A. Olson and J. F. Power. 1984. Growth analysis based on degree days. Crop Sci. 24:28-32.

Sampling, Ecology, and
Insect Population Dynamics

ADVANCES IN SAMPLING INSECTS IN SOYBEANS

D. C. Herzog

Before the appearance of *Sampling Methods in Soybean Entomology* (Kogan and Herzog, 1980), information on sampling soybean arthropods was largely fragmentary and scattered throughout the entomological literature. The publication of this sampling compendium brought together for the first time under one cover the great majority of information available on sampling soybean insect pests and their natural enemies. Yet, sampling of soybean arthropods remains a high research priority, as indicated by the 1984 report of the Soybean Research Advisory Institute to the U.S. Senate Committee on Agriculture, Nutrition and Forestry and the U.S. House of Representatives Committee on Agriculture (Quinn, 1984), which stated among its recommendations: "Develop more reliable sampling and assay techniques for pests, as a basis for determining their economic thresholds." There has been renewed interest in and considerable attention to sampling methodologies in the last several years. This has resulted in research that has filled certain technological and/or methodological voids, and allowed scientists rapidly and precisely to quantify or classify arthropod populations on soybean. The purpose and scope of this review is to serve as a brief supplement to Kogan and Herzog (1980) by briefly summarizing the sampling information that has become available since 1979.

HISTORICAL PERSPECTIVE

Sampling has always been a controversial issue. From the very first time that soybean entomologists met to compare notes, sampling has been a major issue. Probably the most provocative has been the choice of method for sampling of foliar arthropods: the shake/beat cloth or the sweep net. There has been no resolution of this issue. Both sampling methods, I am sure, will continue to be used, and staunchly supported by proponents.

A second controversial subject has been the need for quantification of absolute population densities. On the one hand, there are those who maintain that relative sampling methods provide all of the information necessary. On the other hand, there are those who feel very strongly that unless we know what proportion of the population is captured by a particular sampling method, no useful information can be acquired by sampling. It is my opinion that the truth lies somewhere between the two extremes. Certainly many, perhaps most, management-oriented sampling objectives can be satisfied through the use of relative sampling methods. However, as the questions to be answered become more complex, the more precise and accurate must be the information supplied to answer those questions. Whereas it would be beneficial to be able to measure absolute densities of target species in routine field sampling, for most situations this is impractical.

The derivation of absolute population densities for any species is extremely time-consuming and labor-intensive. Therefore, a realistic compromise is the development of calibration equations for conversion of captures by relative methods to absolute populations. This, of course, implies a standardization of relative sampling methodologies—another controversial subject which I will not discuss in detail.

NEW SAMPLING TECHNIQUES AND SPECIAL CASE PROBLEMS

Research and extension entomologists are constantly searching for new and improved sampling methodologies that will render their sampling tasks easier and more accurate or precise. The traditional shake cloth sampling method of Boyer and Dumas (1963) has been modified by Herbert and Harper (1983), a further modification of the method described in Shepard and Carner (1976). In this method, samples are placed in containers provided with a killing agent while in the field, and transported to the laboratory for storage, sorting and counting. This method thereby reduces error due to escape of highly mobile individuals or the more minute arthropods. While more accurate, this method is designed for research and would be impractical for the pest management practitioner.

A truly innovative modification of a relative sampling device has been developed by Drees and Rice (pers. commun.). The "vertical beat sheet," a modification of the shake cloth is designed such that plants are shaken against a vertically held galvanized metal sheet. Insects dislodged are collected in a tray at the bottom of the metal sheet, which is positioned at the base of the plants in a row. Insects may be counted either directly at eye level in the field or easily placed in a container for transport to the laboratory for storage and/or processing.

Throughout many of the chapters in Kogan and Herzog (1980), reference was made to difficulties encountered in sampling in soybean plantings made in other than the traditional wide-row configuration. Innovative methodologies are currently under development that will improve our capabilities for sampling insects in narrow row and broadcast soybean cropping systems. Attempts to overcome the difficulty encountered in sampling drilled (18 cm spacing) plantings were made by Buschman et al. (1981), who either created small clearings within the larger planting and beat the foliage over a shake cloth with a metal bar or shook foliage into a sweep net. Other modifications of sampling methods for broadcast and/or narrow row plantings are under study in North Carolina, Mississippi, and Louisiana (M. W. Braxton, H. N. Pitre and D. J. Boethel, pers. commun.).

Alternative methods of sampling green cloverworm eggs and pupae have been developed and evaluated by Buntin and Pedigo (1981) and Pedigo et al. (1982), respectively. A flushing technique for deriving absolute population estimates of green cloverworm, *Platthypena scabra* (F.), adults has been developed and evaluated by Pedigo et al. (1982). The flushing method of sampling adults has subsequently been adapted for use with the lesser cornstalk borer, *Elasmopalpus lignosellus* (Zeller) (J. E. Funderburk and D. C. Herzog, unpublished). Sampling of eggs of the velvetbean caterpillar, *Anticarsia gemmatalis* Hubner, has received extensive study (B. M. Gregory and C. S. Barfield, pers. commun.).

ABSOLUTE/RELATIVE SAMPLING METHOD CALIBRATION

Perhaps most notable among contributions to recent sampling research has been the development of precise, though time- and labor-intensive, methods of quantification of absolute populations of soybean arthropods, and the subsequent calibration of relative sampling methods to absolute population estimates. The effects of larval size, plant phenology and individual sampler variation on calibration ratios for shake cloth/ absolute were documented by Marston et al. (1979) for the velvetbean caterpillar, the soybean looper, *Pseudoplusia includens* (Walker), and the corn earworm, *Heliothis zea* (Boddie). They (Marston et al., 1982) subsequently calculated ratios for calibrating sweep net captures of a large number of species and groups to absolute population densities.

The most complete calibration among absolute and relative sampling methods was developed by Luna et al. (1982). They calibrated clam-trap (modified from Leigh et al., 1970) and sheet-arena (adapted from Marston et al., 1979) methods of absolute larval density measurement to known numbers of larvae released into a soybean canopy. One fact is certain: absolute sampling methods are not literally absolute. They further calibrated ground cloth and sweep net to absolute densities as estimated by the sheet arena sampler. High levels of correlation were obtained among all absolute and relative sampling methods studied. It is particularly encouraging that the calibration equations for sweep net to ground cloth derived by Luna et al. (1982) corresponded almost identically with those obtained by Rudd and Jensen (1977). Not so encouraging is the fact that similar efforts with the southern green stink bug, *Nezara viridula* (L.), by Menezes (1981) failed to corroborate those of Rudd and Jensen (1977). Perhaps the more highly contagious nature of phytophagous stink bug populations influenced the results of these two studies.

SEQUENTIAL SAMPLING AND PARAMETER ESTIMATION

Mathematical treatment of patterns of insect dispersion facilitate the determination of minimum and absolute numbers of samples necessary to achieve a prescribed level of precision or confidence in the sample estimate of some population parameter. The recognition of the mathematical distribution(s) represented by sample data of arthropod species under study is prerequisite to the development of sequential sampling or sequential count plans. Broad mathematical treatments of spatial or dispersion statistics are given in books by Poole (1974), Pielou (1977), Southwood (1980), Ripley (1981) and a recent review article by Taylor (1984).

Considerable disagreement apparently remains concerning the applicability of k (the index of dispersion) of the negative binomial distribution. Some contend that its lack of independence from the mean limits its usefulness. However, in his recent review, Taylor (1984) considers the k of the negative binomial distribution a valid population parameter. However, Taylor contends that Taylor's Power Law is universally applicable, while the parameter k of the negative binomial distribution is much more limited in its applicability. As more research is completed, it appears that very few insect populations conform to the Poisson distribution, and that most, if not all, display a clumped or aggregated pattern of dispersion. For example, Strayer et al. (1977) and Herzog and Todd (1980) based the development of sequential sampling plans for the velvetbean caterpillar on the premise that larvae display a random pattern of dispersion. However, recent research has revealed that, at least under north Florida field conditions, larvae of this species are somewhat aggregated, conforming to the negative binomial distribution (G. L. Cave and D. C. Herzog, unpublished). Situations such as this place the research entomologist in somewhat of a quandry. To analyze statistically experimental data, he must assume a normally distributed variance or choose from among several available transformations. It would be of tremendous benefit to entomology—to the whole of biological science—if applied statisticians would begin to address in earnest the question of non-normality of variances in "the real world."

Sequential sampling and sequential count plans have been developed for certain additional species of soybean insect pests, additional developmental stages of certain pests, and certain of their natural enemies that significantly reduce the time and sampling effort necessary to achieve sampling objectives. Sequential plans for either/or both parameter estimation and management decision-making for green cloverworm adults were developed by Pedigo et al. (1982), for eggs by Buntin and Pedigo (1981), for larvae by Bechinski et al. (1983) and for pupae in two additional papers by Bechinski and Pedigo (1983a,b).

A unique sequential sampling plan was developed by Bellinger et al. (1981) who based their decision plans on percent defoliation by the Mexican bean beetle, *Epilachna varivestis* Mulsant, on the binomial distribution. Because of the aggregated nature of stink bug populations, primarily the southern green stink bug, the possibility of developing sequential sampling plans based on the binomial distribution (presence/absence) in the sense of Wilson et al. (1983) and Wilson and Room (1983) is under study in Florida (D. C. Herzog, unpublished).

Another new, unique, and innovative use of sequential sampling recently has been developed by Pedigo and van Schaik (1984). This method applies the sequential probability ratio test through chronological time rather than through space, and decisions relative to the status of green cloverworm population trajectories are based on adult flushing samples made on a weekly basis, rather than, as normal, from sample to sample. In the case of the green cloverworm, the decision reached is whether the population in question is in either an endemic or an outbreak configuration.

QUANTIFICATION OF NATURAL ENEMY POPULATIONS AND PEST MORTALITY

Bechinski and Pedigo (1982) compared sweep-net, ground-cloth and vacuum-net sampling methods with absolute population measurement for several species and categories of natural enemies, and determined that plant shaking generally produced the most precise and cost-efficient estimates, although the sweep net proved superior for some groups. Bechinski and Pedigo (1981) derived population dispersion statistics and developed sequential estimation plans for *Orius insidiosus* (Say) and *Nabis* spp.

Sampling, in addition to quantification of natural enemy populations, often provides an aid in the estimation of mortality due to natural enemies. Samples of natural enemies collected in the field can be identified as having fed on a particular pest species through utilization of autoradiography (McCarty et al., 1980) or the enzyme linked immunosorbent assay technique (ELISA) (Ragsdale et al., 1981) developed by Ragsdale (1980). More precise estimates of predator-induced mortality were derived by Elvin (1983) through observation of introduced larval cohorts for visual confirmation of predation, reduction in number among the larval cohorts released, and intensive sampling of predatory species in surrounding soybean to determine overall mortality and proportions of that mortality attributable to each of the several abundant predatory species represented in the field.

Results of simulation model studies by Van Driesche (1983) supported cautionary statements made by Marston (1980) concerning the interpretation of percentage parasitism in insect samples. To alleviate problems associated with estimation of mortality, Van Driesche recommended precise definition of the susceptible stage, description of parasitoid and susceptible host phenologies, behavioral studies on parasitized hosts, delay of sampling until attack has been completed, and avoidance of summation of samples taken through space or time because of the inequity of percentage parasitism estimates among samples taken through space or time. Sweep-sampling of green cloverworm, soybean looper, and velvetbean caterpillar revealed that larvae infected with *Nomuraea rileyi* were more highly aggregated than were uninfected larvae (Fuxa, 1984). Thorpe (1984) placed tobacco budworm, *Heliothis virescens* (F.), egg cloths in the field to investigate *Trichogramma* spp. activity in soybean.

SPECIAL USES OF SAMPLING INFORMATION

Advances in the development of methods for quantifying lepidopteran adult abundance will provide means useful in the definition of patterns of short-range dispersal and long-range migration. An example is adult emergence traps, originally developed by Funderburk et al. (1983) for use with the seedcorn maggot, *Hylemya platura* (Meigen), under study for use for the lesser cornstalk borer. The age structure of samples of adults (Funderburk et al., 1984) captured in emergence traps and pheromone traps is being used to document population phenology and overwintering (J. E. Funderburk and D. C. Herzog, unpublished). An additional example is the work of Gregory and Barfield (B. M. Gregory and C. S. Barfield, pers. commun.), who are combining intensive sampling of adult velvetbean caterpillar populations with intensive egg sampling in order to relate adult numbers to egg populations produced in the field.

CONTINUING SAMPLING PROBLEMS

Narrow-row soybeans are the rule, rather than the exception, in Brazil, and perhaps other parts of the world. When asked how we in the U.S. sample in narrow rows, my answer had to be: "With great difficulty!" The question of sampling in narrow rows and broadcast plantings is being addressed in several states in this country and in Brazil. The vertical beat sheet of Drees and Rice (pers. commun.) is a big step in the right direction. But, we are a long way from having all the answers. Activity in this important area needs to be encouraged.

The soybean sampling area that has suffered most from lack of interest and activity is that of soil arthropods. Some excellent research has been conducted on bean leaf beetle larvae in Illinois and North Carolina, on the seedcorn maggot in Iowa, and is underway on the lesser cornstalk borer in several southeastern states. But, much remains to be done.

There are still those who enter the field to start sampling without having given thought to design of the sampling program. But, I'm not convinced that is worse than spending all time and effort in developing sampling programs with no thought given to acceptability to or implementation of the sampling programs by extension personnel, private consultants or producers. It is the responsibility of research to develop new information and methodologies. But, unless there is planning for and a mechanism of implementation, that information may be of use to no one other than those who developed it. How many sequential sampling plans for management have been developed? How many are actually implemented in the field?

A final problem that I would like to identify is the interaction between sampling method and spatial patterns of the species targeted by sampling. Population dispersion patterns are dynamic species characteristics. Those patterns observed in sampling result from a distortion of the population dispersion pattern by the sampling method. If, as

Taylor (1984) suggests, all species are more or less aggregated, the magnitude or extent of that aggregation can tremendously influence the number of samples that must be taken to satisfy a sampling objective. In the case of sampling for making management decisions, a sampling method should probably be chosen that minimizes the clumping observed. The best example that I know is the sweep net for sampling stink bugs. On the other hand, for intensive field studies of species biology, life history, mortality processes, etc., perhaps a method should be chosen that maximizes the aggregation or clumping observed in sampling. Because degree of aggregation results from interactions among individuals of both the same and different species, this aggregation has real biological meaning and should not be summarily dismissed. Perhaps, the time has arrived to reexamine choice of sampling method in light of sampling objective and insect spatial distributions or dispersion, and their biological meaning.

NOTES

D. C. Herzog, Department of Entomology and Nematology, University of Florida, North Florida Research and Education Center, Quincy, FL 32351.

REFERENCES

Bechinski, E. J., G. D. Buntin, L. P. Pedigo and H. G. Thorvilson. 1983. Sequential count and decision plans for sampling green cloverworm (Lepidoptera: Noctuidae) larvae in soybean. J. Econ. Entomol. 76:806-812.

Bechinski, E. J. and L. P. Pedigo. 1981. Population dispersion and development of sampling plans for *Orius insidiosus* and *Nabis* spp. in soybeans. Environ. Entomol. 10:956-959.

Bechinski, E. J. and L. P. Pedigo. 1982. Evaluation of methods for sampling predatory arthropods in soybeans. Environ. Entomol. 11:756-761.

Bechinski, E. J. and L. P. Pedigo. 1983a. Development of a sampling program for estimation of pupal densities of green cloverworm (Lepidoptera: Noctuidae) in soybeans and evaluation of alternative sampling procedures. Environ. Entomol. 12:96-100.

Bechinski, E. J. and L. P. Pedigo. 1983b. Microspatial distribution of pupal green cloverworms, *Plathypena scabra* (F.) (Lepidoptera: Noctuidae), in Iowa soybean fields. Environ. Entomol. 12:273-276.

Bellinger, R. G., G. P. Dively and L. W. Douglas. 1981. Spatial distribution and sequential sampling of Mexican bean beetle defoliation on soybeans. Environ. Entomol. 10:835-841.

Boyer, W. P. and B. A. Dumas. 1963. Soybean insect survey as used in Arkansas. USDA Coop. Econ. Insect Rep. 13:91-92.

Buntin, G. D. and L. P. Pedigo. 1981. Dispersion and sequential sampling of green cloverworm eggs in soybeans. Environ. Entomol. 10:980-985.

Buschman, L. L., H. N. Pitre and H. F. Hodges. 1981. Soybean cultural practices: Effects on populations of green cloverworm, velvetbean caterpillar, loopers and *Heliothis* complex. Environ. Entomol. 10:631-641.

Elvin, M. K. 1983. Quantitative estimation of rates of arthropod predation on velvetbean caterpillar (*Anticarsia gemmatalis* Hubner) eggs and larvae in soybeans. Ph.D. Dissertation. University of Florida Library, Gainesville.

Funderburk, J. E., D. C. Herzog and R. E. Lynch. 1984. Method for determining the age structure of adult populations of the lesser cornstalk borer (Lepidoptera: Pyralidae). J. Econ. Entomol. 77:541-544.

Funderburk, J. E., L. P. Pedigo and E. C. Berry. 1983. Seedcorn maggot (Diptera: Anthomyiidae) emergence in conventional and reduced-tillage soybean systems in Iowa. J. Econ. Entomol. 76:131-134.

Fuxa, J. R. 1984. Dispersion and spread of the entomopathogenic fungus *Nomuraea rileyi* (Moniliales: Moniliaceae) in a soybean field. Environ. Entomol. 13:252-258.

Herbert, D. A. and J. D. Harper. 1983. Modification of the shake cloth sampling technique for soybean insect research. J. Econ. Entomol. 76:667-670.

Herzog, D. C. and J. W. Todd. 1980. Sampling velvetbean caterpillar in soybean. p. 107-140. *In* M. Kogan and D. C. Herzog (eds.) Sampling methods in soybean entomology. Springer-Verlag, New York.

Kogan, M. and D. C. Herzog. 1980. Sampling methods in soybean entomology. Springer-Verlag, New York.

Leigh, T. F., D. Gonzalez and R. van den Bosch. 1970. A sampling device for estimating absolute insect populations on cotton. J. Econ. Entomol. 63:1704-1706.

Luna, J. M., H. M. Linker, J. L. Stimac and S. L. Rutherford. 1982. Estimation of absolute larval densities and calibration of relative sampling methods for velvetbean caterpillar, *Anticarsia gemmatalis* (Hubner) in soybean. Environ. Entomol. 11:497-502.

Marston, N. L. 1980. Sampling parasitoids of soybean insect pests. p. 481-504. *In* M. Kogan and D. C. Herzog (eds.) Sampling methods in soybean entomology. Springer-Verlag, New York.

Marston, N., D. G. Davis and M. Gebhardt. 1982. Ratios for predicting field populations of soybean insects and spiders from sweep-net samples. J. Econ. Entomol. 75:976-981.

Marston, N. L., W. A. Dickerson, W. W. Ponder and G. D. Booth. 1979. Calibration ratios for sampling soybean Lepidoptera: Effect of larval species, larval size, plant growth stage, and individual sampler. J. Econ. Entomol. 72:110-114.

McCarty, M. T., M. Shepard and S. G. Turnipseed. 1980. Identification of predaceous arthropods in soybeans using autoradiography. Environ. Entomol. 9:199-203.

Menezes, E. B. 1981. Population dynamics of the stink bug (Hemiptera: Pentatomidae) complex on soybean and comparison of two relative methods of sampling. Ph.D. Dissertation. University of Florida, Gainesville.

Pedigo, L. P., G. D. Buntin and E. J. Bechinski. 1982. Flushing technique and sequential-count plan for green cloverworm (Lepidoptera: Noctuidae) moths in soybean. Environ. Entomol. 11:1223-1228.

Pedigo, L. P. and J. W. van Schaik. 1984. Time-sequential sampling: A new use of the sequential probability ratio test for pest management decisions. Bull. Entomol. Soc. Am. 30:32-88.

Pielou, E. C. 1977. Mathematical ecology. Wiley Interscience, New York.

Poole, R. W. 1974. An introduction to quantitative ecology. McGraw-Hill, New York.

Quinn, P. J. 1984. U.S. Soybean production and utilization research: A report to the Senate Committee on Agriculture, Nutrition and Forestry and House Committee on Agriculture. Soybean Research Advisory Inst.

Ragsdale, D. W. 1980. The development and use of an enzyme linked immunosorbent assay (ELISA) for the quantitative assessment of predation of *Nezara viridula* (L.) within a soybean agroecosystem. Ph.D. Dissertation. Louisiana State University, Baton Rouge.

Ragsdale, D. W., A. D. Larson and L. D. Newsom. 1981. Quantitative assessment of the predators of *Nezara viridula* eggs and nymphs within a soybean agroecosystem using an ELISA. Environ. Entomol. 10:402-405.

Ripley, B. D. 1981. Spatial statistics. John Wiley & Sons, New York.

Rudd, W. G. and R. L. Jensen. 1977. Sweep net and ground cloth sampling for insects in soybeans. J. Econ. Entomol. 70:301-304.

Shepard, M. and G. R. Carner. 1976. Distribution of insects in soybean fields. Can. Entomol. 108:301-304.

Southwood, T. R. E. 1980. Ecological methods. Halsted Press, New York.

Strayer, J., M. Shepard and S. G. Turnipseed. 1977. Sequential sampling for management decisions on the velvetbean caterpillar. J. Georgia Entomol. Soc. 12:220-227.

Taylor, L. R. 1984. Assessing and interpreting the spatial distributions of insect populations. Ann. Rev. Entomol. 29:321-357.

Thorpe, K. W. 1984. Seasonal distribution of *Trichogramma* (Hymenoptera: Trichogrammatidae) species associated with a Maryland soybean field. Environ. Entomol. 12:127-132.

Van Diresche, R. G. 1983. Meaning of "percent parasitism" in studies of insect parasitism. Environ. Entomol. 12:1611-1622.

Wilson, L. T., C. Pickel, R. C. Mount and F. G. Zalom. 1983. Presence-absence sequential sampling
    for cabbage aphid and green peach aphid (Homoptera: Aphididae) on brussels sprouts. J. Econ.
    Entomol. 76:476-479.
Wilson, L. T. and P. M. Room. 1983. Clumping patterns of fruit and arthropods in cotton, with impli-
    cations for binomial sampling. Environ. Entomol. 12:50-54.

INFLUENCE OF TILLAGE PRACTICES ON SOIL-INSECT POPULATION DYNAMICS IN SOYBEAN

Ronald B. Hammond and Joseph E. Funderburk

A rich insect fauna inhabits the soil in soybean agroecosystems. Many are pests or beneficial organisms that are economically important to soybean production. Historically, soil insects have received less attention than above-ground insects from agricultural scientists and producers, for two reasons. First, soil insects and below-ground injury by soil pests are easily overlooked. Second, the sampling of soil insects can be laborious and costly, thereby inhibiting identification and quantification of soil insect populations in most research and scouting programs. As a result, this type of damage was often considered inconsequential, or not considered at all.

Injury to below-ground plant structures can greatly affect soybean growth and yield. However, these relationships have not been adequately determined or addressed for most soil pests. Reliable, practical control strategies that are compatible with pest management philosophy have not been developed and implemented for soybean soil pests. Economic injury levels that are based on actual soil-pest numbers are virtually useless in scouting programs, because sampling these pests is not feasible, and because the soil environment prevents reliable, much less, rapid suppression of a pest population. Injury from soil pests frequently occurs before rescue treatment can be achieved. Currently recommended control measures usually involve a prophylactic insecticidal application, which is either economically or ecologically unjustified in many situations.

The need for better information and methods has recently been recognized, partly because of concern about the effects of tillage on soil insect populations. Soybean growers are rapidly adopting reduced-tillage and no-tillage production systems; even though many are apprehensive about the potential for increased soil-pest problems. These concerns are probably legitimate, because tillage affects soil parameters which influence the behavior, development, and survival of soil insects.

TILLAGE EFFECTS ON SOIL PARAMETERS

Changes in soil parameters that influence insect biology and life history in the soil environment directly or indirectly result from the extent and type of tillage. We are concerned with tillage operations prior to planting, or the lack of them. These practices range from conventional tillage, either a fall or spring plowing followed by further disking or leveling, to what is collectively called conservation tillage. Simply stated, conservation tillage reduces the amount of soil disturbance by foregoing the use of the plow.

Varying the degree and/or the type of tillage influences the amount of plant residue that remains on the soil surface. Residue cover can range from none when the soil is plowed, to 100% coverage in no-tillage systems, with all levels in between depending upon the tillage method used. Residue cover, in turn, elicits modifications in both the abiotic and biotic parameters of the soil environment, providing for favorable or perhaps unsuitable habitats for insect fauna. Two abiotic soil parameters are greatly affected by the presence of a cover. First, the residue provides for greater moisture conservation through (1) better retention of water during rainfall, (2) less evapotranspiration, and (3) greater water holding capacity of the soil because of increased organic matter content. Secondly, a residue cover serves to lower soil temperature. While slowing arthropod development, this also slows the development of germinating soybean seeds, thereby increasing the time that they are subject to infestations by certain soil pests. In some situations, the lower temperatures might increase insect survivorship when the higher temperatures associated with no residue cover would have caused insect mortality.

Tillage operations also determine the placement of organic material. This organic matter, while decaying, may serve as an attraction and food source for various soil insect pests of soybean. From an integrated pest management viewpoint, the effects of tillage, manifested through changes in residue cover and abiotic soil parameters, may play an important role in the management of pest species through their influence on other biotic components in the agroecosystem. These parameters, which have an effect on the soil pest complex, can also influence beneficial organisms, viz. predaceous and parasitic arthropods, decomposers, and insect pathogens. Tillage operations can also modify the weed population occurring in fields, both in quantity and species. The presence of weeds can serve to modify the arthropod's habitat, serving as available oviposition and feeding sites. The following are examples where tillage influences on soybean pest and beneficial arthropods have been investigated.

SEEDCORN MAGGOT

Soil-inhabiting larvae of the seedcorn maggot, *Delia platura*, feed on cotyledons and plumules of germinating soybean seedlings, resulting in poor plant stands and damaged plants. The insect develops rapidly in cool temperatures and aestivates in warm temperatures (Funderburk et al., 1984). In the U.S.A., seedcorn maggot injury is restricted to northern latitudes where soybean often is planted in cool, wet soils.

Some forms of partially-buried crop residues have been observed to enhance seedcorn maggot oviposition. Soybean injury from this pest therefore was expected to increase as reduced-tillage systems were adopted. Recently, the influence of tillage practices on seedcorn maggot populations was evaluated in several studies involving soybean.

Funderburk et al. (1983) sampled seedcorn maggot populations in between- and within-row areas of four soybean tillage systems, including fall moldboard plow, fall chisel plow, till-plant, and no-till (Table 1). The cropping system was a corn (*Zea mays* L.)-soybean rotation, and soybean was planted each May in soil containing substantial amounts of corn residue. The amount of surface corn residue differed in each tillage system. Further, surface residue was equally distributed throughout between- and within-row areas of fall moldboard-plow and fall chisel-plow systems, but was greater in between- than in within-row areas of no-till and till-plant systems.

Tillage significantly affected seedcorn maggot numbers (Table 1). During soybean germination, the pest's numbers were greatest in the fall chisel-plow system, followed by the till-plant system. Seedcorn maggot numbers were lowest, and were statistically

Table 1. Estimates of seedcorn maggot populations during soybean germination in several cropping systems in the northern USA.

Treatment		Seedcorn maggots m/row[a]	
		Year 1	Year 2
		no	no
Soybean Tillage Systems, Iowa, Funderburk et al. (1983)[b]			
Till-plant	over rows	5.8 (5.6 s)[c]	1.3 (1.5 s)
	between rows	5.3	1.7
No-tillage	over rows	3.0 (2.9 t)	0.8 (0.8 s)
	between rows	2.7	0.8
Fall moldboard plow	over rows	4.2 (3.3 t)	1.0 (0.8 s)
	between rows	2.5	0.6
Fall chisel plow	over rows	9.2 (8.3 r)	5.5 (4.8 r)
	between rows	7.3	4.1
Soybean Cropping Systems, Ohio, Hammond and Jeffers (1983)[d]			
Intercropped in wheat		2.4	3.0
Following wheat, residue disked		11.8	34.0
No wheat		—	4.0
		FLSD = 5.0	FLSD = 9.8
Soybean Cropping Systems, Ohio, Hammond (unpublished data)[e]			
No rye:plant		1.3 u	2.2 u
No rye:disk-plant		1.2 u	16.9 t
Rye:disk-plant		46.9 s	75.6 r
Rye:paraquat-plant		2.1 u	11.6 tu
Rye:paraquat-disk-plant		20.0 t	50.9 s
Rye:plant-paraquat		3.2 u	10.3 tu
Rye:disk-plant-paraquat		47.4 s	62.3 s
Rye:plow-disk-plant		82.8 r	53.9 s

[a]Each study conducted between 1980-1983.
[b]Mean separations by orthogonal treatment comparisons.
[c]Numbers in parentheses are pooled means of over- and between-row areas.
[d]Mean separations by Fisher's least significant difference (FLSD).
[e]Mean separations by Duncan's multiple range test.

similar, in the no-till and fall moldboard-plow systems. Interestingly, populations were similar in between- and within-row areas in each system, thereby demonstrating that the species survives well as a saprophyte and that germinating soybean was not attractive for oviposition under field conditions. In this study, there was no relationship between seedcorn maggot numbers and the quantity of surface corn residue. Oviposition was not substantially enhanced by the corn residue, and seedcorn maggot was not an economic problem in any tillage system.

In another study, seedcorn maggot populations were sampled by Hammond and Jeffers (1983) in (1) soybean that was planted in disked soil with fresh wheat (*Triticum aestivum* L.) residue, (2) soybean that was intercropped with wheat, and (3) soybean that was planted in soil with no wheat residue (Table 1). Populations were small in the intercropped soybean and in the no-wheat residue soybean, but were 8.5 times greater when soybean was planted in disked soil with fresh wheat residue. The pest fed readily on the soybean in the disked-wheat system, greatly reducing plant density and subsequent yields.

Hammond (unpublished data) conducted another, more comprehensive investigation of the effects of tillage and fresh crop residues on seedcorn maggot populations. Wheat in this study was replaced by rye (*Secale cereale* L.), and treatments are shown in Table 1. Buried, fresh rye residues greatly enhanced oviposition in conventional tillage treatment, but not with no-till. Paraquat applications did not greatly alter the attractiveness of the partially-buried rye residues. As in the previous study, maggots fed readily on germinating soybean with yield reductions directly related to larval numbers.

Consequently, seedcorn maggot biology was affected by soil tillage. Populations were greatly enhanced in some reduced-tillage systems, but not in others. Cropping systems with a high risk for problems are those in which soybean is planted into soil containing freshly-decomposing, partially-buried crop residues.

LESSER CORNSTALK BORER

Larvae of the lesser cornstalk borer, *Elasmopalpus lignosellus*, are subterranean and inhabit the upper soil to a depth of 3 to 4 cm. The insect is polyphagous and damages numerous crops throughout temperate and tropical regions of the Western Hemisphere. In the USA, the lesser cornstalk borer is an occasional pest of soybean in southern latitudes, where outbreaks typically occur during periods of hot, dry weather (Tippins, 1982). Injury to soybean is primarily restricted to the period from emergence through early vegetative crop-growth stages. The larvae tunnel into the pith of the seedling, thereby damaging the vascular tissue and killing the plant. One larva usually destroys numerous seedlings.

Although soybean is a suitable larval host, it is much less preferred for oviposition under field conditions than other crops (R. E. Lynch and J. E. Funderburk, unpublished data). Severe stand reductions in soybean probably result from larval populations present before planting. These preplant populations apparently result in some cropping systems where oviposition is enhanced by the presence of host weeds (Tippins, 1982) and/or burned crop residues (Viana, 1981). Established populations then survive preplant tillage practices and feed on partially-buried plant residues (All, 1980). However, recent research has revealed that preplant tillage practices influence the feeding behavior of larval populations, thereby affecting the amount of crop injury by the pest.

Cheshire and All (1979), conducting greenhouse and laboratory studies, investigated the feeding behavior of lesser cornstalk borer larvae on crops planted in plowed soil and on crops planted no-till. The larvae fed on plant residues in the no-till situation, and fed directly on crops when the plant residues were buried by tillage. They hypothesized that this altered feeding behavior explained why injury under field conditions was greater to crops planted in plowed soils than to crops planted to no-till (All and Gallaher, 1977; All et al., 1979). In other field studies, plowing followed by a 3 to 4 week fallow period greatly reduced crop injury when preplant larval populations were present on weeds (All, 1980) or burned/unburned crop residues (B. Rogers, J. N. All, and J. M. Cheshire, pers. commun.).

Consequently, larval populations are expected to be greatly enhanced in the double-cropped soybean systems in the southern USA. In such systems, soybean usually is planted immediately following tillage operations in which burned or unburned small-grain residues are partially buried. Soybean damage in such cropping systems is being investigated by T. D. Reed and J. W. Todd (pers. commun.). Although preliminary, their results indicate, as expected, that injury was greatest to soybean plants seeded in soil containing burned weed and crop residues.

CARABID AND SPIDER PREDATORS

In soybean agroecosystems, arthropod predators play an important role in the population dynamics of soybean pests. Major components of the predator complex are the carabid beetles and predaceous spiders that inhabit the soil surface. House and All (1981) studied carabid beetles in conventional and conservation tillage fields where a winter wheat crop preceded soybean. In general, they found that carabid populations were several times greater in the conservation tillage system than under conventional tillage. Carabid fauna of conservation tillage systems was greater in both abundance and diversity. They hypothesized that this was a result of the increased number of available niches, owing in part to the mulch litter layer present in conservation tillage plots. They listed eight common and abundant carabid species that were several times greater in conservation-tillage fields, compared with only two species that were more numerous in conventional tillage (viz., *Apristus subsulcatus* and *Megacephala carolina*). The predominant species in the conservation system (*Harpalus pennsylvanicus*) was about six times more abundant than in the other system. This carabid, common to most soybean growing regions of the USA, has facultative phytophagus feeding habits. It has been found to be more abundant in soybean habitats with grasses and mixed broadleaf and grass weeds (Shelton and Edwards, 1983), suggesting the possibility of an indirect influence on this species from the lack of tillage if weed control is not adequate.

House and Stinner (1983) provided further evidence, using quadrat sampling, that carabid numbers and species diversity were significantly greater in no-tillage than in conventional tillage soybean seedings (Table 2). Pitfall-trap data indicated that two major soil macroarthropod guilds, spiders and predatory beetles (i.e., carabids and staphylinids), had greater densities and diversity in no-tillage. They reasoned that this was due to the greater structural diversity provided by the presence of weeds and litter in no-tillage systems.

Table 2. Estimates of predatory arthropod populations from several soybean growing regions.

Treatment	Predators/sample	
	Carabids	
	Individuals	Species
	no	no
Quadrat Samples from Georgia, House and Stinner (1983)		
Conventional tillage	0.38	0.44
No-tillage	17.56	5.63
	Carabids	Spiders
	no	no
Pitfall Samples from Ohio, Hammond and Stinner (unpublished data)[a]		
Conventional tillage	6.44 r	3.34 r
Reduced tillage	6.63 rs	3.84 s
No-tillage	7.28 s	4.55 t

[a]Mean separation by Duncan's multiple range test; represents combined totals from soybean and corn.

An ongoing study in Ohio is comparing soybean and corn soil arthropods in different tillage-management systems, including conventional (spring plow-disk), reduced (disking once) and no-tillage, where no winter cover crop is used. Also, included in the study as a factor is the influence of the previous crop. As expected, indications from 1982 data (Hammond and Stinner, unpublished data) are that the numbers of carabid beetles and spiders are significantly greater in no-tillage systems, followed by reduced tillage, and then conventional-tillage plots (Table 2). Percentage residue cover at planting time was ca. 70% in no-tillage, ca. 30% in reduced-tillage, and ca. 0% in conventional. This cover consisted of both the previous crop residue and weed biomass. The percentage cover was highest where soybean was the previous crop and no-tillage planting methods were used. Subsequently, more carabid beetles and spiders were collected from those plots as compared to when the previous crop was corn.

Whether this increase in number and diversity of predators can exert more control on pest species in soybean is still unclear and needs further study. However, House and Stinner (1983) showed that less leaf area was removed in the no-tillage systems than in the conventional tillage, suggesting a possible indirect influence of predators in controlling defoliators (e.g., bean leaf beetles, *Cerotoma trifurcata*).

OTHER ARTHROPOD FAUNA

Bean leaf beetles, whose females oviposit in the upper 3 to 4 cm of the soil, might be affected by tillage. More eggs are laid in moist and wet soil (Marrone and Stinner, 1983) and soils high in organic matter (Deitz et al., 1976). Both of these soil parameters are influenced by tillage practices. Troxclair and Boethel noted (Boethel, 1984) that beetles achieved denser populations in conventional-tillage soybean than in no-tillage soybean in Louisiana.

The larvae of the soybean nodule fly, *Rivellia quadrifasicita*, feed on soybean nodules. Because of its soil habitat during the larval stages, tillage might also affect its population dynamics. However, Koethe (1982) found overwintered adult fly emergence to be the same in no-tillage and conventionally planted corn plots that followed the previous year's soybean crop. It was proposed that plowing would have no overall effect because larvae would be below tillage level at the normal time of plowing in March or April. If tillage was late, however, it might have more of an effect because larvae pupate closer to the surface (<10.2 cm), and would be more vulnerable (Koethe, 1982).

There are many polyphagous insects, which are not considered primarily soybean pests, e.g., cutworms, wireworms, grasshoppers, and grubs, that will damage the crop under conditions suitable for the insects. Much of the knowledge and concern of possible problems in reduced or no-tillage soybean comes from information available from other crops grown in rotation with soybean. Black cutworm, *Agrotis ipsilon*, damage to soybean in Illinois, for example, may be related to elimination of fall plowing, allowing build-up of certain weeds attractive to ovipositing moths (Kogan, 1981). Slugs, the most serious non-insect pest in reduced and no-tillage corn, damaged no-tillage soybean in Ohio in the spring of 1983. In the past, these no-tilled fields would have been planted to corn. Because no-tillage soybean acreage is now expanding, damage from such pests might also increase. This demands further attention and monitoring.

Other fauna associated with the soybean agroecosystem also are affected by tillage. Insect detritivores were found to follow the same trends as carabids and spiders with greater numbers and diversity in no-tillage systems (House and Stinner, 1983; Hammond and Stinner, unpublished data). Aggregations of various microarthropods, including collembola, symphyla, diplura and acarina, found in tilled systems were smaller and less variable in area compared to no-tillage (Farrer and Crossley, 1983). They also found higher populations of these microarthropods in the no-tillage systems.

FUTURE NEEDS

The soil environment in the soybean agroecosystem acts as a dynamic and ever changing arthropod habitat. A basic understanding of the effects of tillage on the interactions of soil parameters and how they influence the insect's life history is needed if we are to solve or prevent pest problems.

Studying interactions in many different systems will be necessary. Soil environmental parameters (e.g., moisture, pH, soil type, temperature) can vary greatly from one geographical location to another, as do crop rotational-schemes, specific tillage practices and cover crops. Added to that are the differences in pest and beneficial insect fauna and weed populations that are unique to each area. Thus, to gain a full understanding, research will be needed on the effects of tillage on soil insects from numerous geographical locations.

An interdisciplinary approach to the research is a necessity. Agronomists will need to provide expertise in the areas of soil science, water hydrology, and crop management. For integrated pest management considerations, weed scientists, nematologists, and plant pathologists should be included to gain an understanding of the interactions between the biotic components.

There are many opportunities available to obtain further and better knowledge of the underlying principles involved in the interactions in the soil environment brought about by tillage and the subsequent physical changes in the system. Only when these principles are more fully understood can reliable and practical control strategies for insects be implemented.

NOTES

Ronald B. Hammond, Assistant Professor, Department of Entomology, Ohio Agricultural Research and Development Center, The Ohio State University, Wooster, OH; Joseph E. Funderburk, Research Assistant, North Florida Agricultural Research and Education Center, University of Florida, Quincy, FL.

The authors wish to thank the following persons who reviewed and made comments on the manuscript: Ben Stinner (Ohio), John Van Duyn (North Carolina), Tim Reed (Alabama), Joe Cheshire (Georgia), John All (Georgia) and Robert Treece (Ohio).

REFERENCES

All, J. N. 1980. Consistency of lesser cornstalk borer control with Lorsban in various corn cropping systems. Down to Earth 36(2):33-36.

All, J. N. and R. N. Gallaher. 1977. Detrimental impact of no-tillage corn cropping systems involving insecticides, hybrids, and irrigation on lesser cornstalk borer infestations. J. Econ. Entomol. 70:361-365.

All, J. N., R. N. Gallaher and M. D. Jellum. 1979. Influence of planting date, preplanting weed control, irrigation, and conservation tillage practices on efficacy of planting time insecticide applications for control of lesser cornstalk borer in field corn. J. Econ. Entomol. 72:265-268.

Boethel, D. J. 1984. Evolving crop protection systems for soybean: Impact on pest and beneficial arthropod species. In Proc. China/USA Soybean Science Symposium. Changchun, People's Republic of China. July 28-Aug. 2, 1983. (In press).

Cheshire, J. M., Jr. and J. N. All. 1979. Feeding behavior of lesser cornstalk borer larvae in simulations of no-tillage, mulched conventional tillage, and conventional tillage corn cropping systems. Environ. Entomol. 8:261-264.

Deitz, L. L., J. W. Van Duyn, J. R. Bradley, Jr., R. L. Rabb, W. M. Brooks and R. E. Stinner. 1976. A guide to the identification and biology of soybean arthropods in North Carolina. N. C. Agric. Exp. Stn. Tech. Bull. No. 238.

Farrar, Jr., F. P. and D. A. Crossley, Jr. 1983. Detection of soil microarthropods in soybean fields, using a modified tullgren extractor. Environ. Entomol. 12:1303-1309.

Funderburk, J. E., L. P. Pedigo and E. C. Berry. 1983. Seedcorn maggot (Diptera:Anthomyiidae) emergence in conventional and reduced-tillage soybean systems in Iowa. J. Econ. Entomol. 76: 131-134.

Funderburk, J. E., L. G. Higley and L. P. Pedigo. 1984. Seedcorn maggot (Diptera:Anthomyiidae) phenology in central Iowa and examination of a thermal-unit system to predict development under field conditions. Environ. Entomol. 13:105-109.

Hammond, R. B. and D. L. Jeffers. 1983. Adult seedcorn maggots in soybeans relay intercropped into winter wheat. Environ. Entomol. 12:1487-1489.

House, G. J. and J. N. All. 1981. Carabid beetles in soybean agroecosystems. Environ. Entomol. 10:194-196.

House, G. J. and B. R. Stinner. 1983. Arthropods in no-tillage soybean agroecosystems: Community composition and ecosystem interactions. Environ. Manage. 7:23-28.

Koethe, R. W. 1982. Descriptive studies of the biology and ecology of *Rivellia quadrifasciata* (Macquart) in eastern North Carolina. Ph.D. Dissertation. North Carolina State University, Raleigh.

Kogan, M. 1981. Dynamics of insect adaptations to soybeans: impact of integrated pest management. Environ. Entomol. 10:363-371.

Marrone, P. G. and R. E. Stinner. 1983. Effects of soil moisture and texture on oviposition preference of the bean leaf beetle, *Cerotoma trifurcata* (Forster) (Coleoptera:Coccinellidae). Environ. Entomol. 12:426-428.

Shelton, M. D. and C. R. Edwards. 1983. Effects of weeds on the diversity and abundance of insects in soybeans. Environ. Entomol. 12:296-298.

Tippins, H. H. 1982. A review of information on the lesser cornstalk borer *Elasmopalpus lignosellus* (Zeller). Georgia Agric. Exp. Stn. Spec. Pub. 17.

Viana, P. A. 1981. Effect of soil moisture, substrate color and smoke on the population dynamics and behavior of the lesser cornstalk borer, *Elasmopalpus lignosellus* Zeller 1848 (Lepidoptera: Pyralidae). M.S. Thesis. Purdue Univ., West Lafayette, Indiana.

ECOLOGICAL EFFECTS OF DOUBLE-CROPPING ON SOYBEAN INSECT POPULATIONS

Henry N. Pitre

Double-cropping is an agricultural system where two crops are produced on the same land during a 12-month period. Most double-cropped soybean plantings are made immediately after spring harvest of small grains, particularly wheat (*Triticum aestivum* L.) and barley (*Hordeum vulgare* L.), and in some areas after temporary winter pastures. However, soybean has been successfully double-cropped with corn (*Zea mays* L.), grain sorghum (*Sorghum bicolor* [L.] Moench), and sunflower (*Helianthus annus* L.) (Woodruff et al., 1981). Management of many pest species that attack soybean may be intensified by the cultural practices associated with effective double crop production.

Planting soybean no-till immediately after grain harvest, and as soon as the grain straw is burned, may be the most successful method of planting double-cropped soybean (Sanford, 1982). Burning practices contribute to mortality of specific stages of certain pest species, but careful consideration must be given to mortality of natural enemies of these pests.

The impact on pest and beneficial arthropod species of some nonconventional cultural practices in soybean crop production systems was recently reviewed by Boethel (1984). These practices may differ significantly from the normally recommended production practices for any one area, and must be examined carefully for use in integrated pest management schemes. This paper will consider the influence of double-cropping on insect populations in soybean, with particular emphasis given to the ecological effects of specific cultural practices associated with double-cropping systems.

TILLAGE

Double-cropping usually is associated with no-till or minimum tillage practices that favor conservation of soil moisture and development of weeds. Many pest and beneficial insect populations are influenced by this non-crop vegetation. Several insect groups; i.e., cutworms and armyworms, deposit their eggs on grasses in minimum tilled fields and the larvae cause damage to the cultivated crop.

In studies in North Carolina (Landis, 1983) the crop residue in no-till fields was associated with large predator populations; however, predation on corn earworm, *Heliothis zea* (Boddie) prepupae was greater in conventionally tilled fields. Landis observed that the microenvironment in no-till soil increased overwintering survival of corn earworm pupae. However, in some areas no-till soybean planted into grain stubble does not appear to be adversely affected by insects any more than does conventionally tilled soybean (Van Duyn, 1981). Corn earworms may seriously infest no-till soybean in North Carolina, but their occurrence is generally associated with late planting, not the no-till culture.

Soil insect pests; i.e., wireworms, seedcorn maggots, bill bugs, white grubs and lesser cornstalk borer (*Elasmopalpus lignosellus* [Zeller]) are frequently encountered in larger numbers in fields where conservation tillage is employed compared to conventionally tilled fields. The crop residue in conservation tillage fields serves as food and suitable habitat for these pests. This residue modifies the abiotic and biotic parameters that impact on insect populations (Musick, 1970; Hammond and Funderburk, this volume). Hammond and Funderburk include seedcorn maggot, lesser cornstalk borer, bean leaf beetle (*Cerotoma trifurcata* [Forster]), soybean nodule fly (*Rivellia quadrifasciata* [Macquart]), cutworms, and carabid and spider predators in their review.

Reduced tillage results in large amounts of decaying vegetative organic matter from crop residues, which provides ideal sites for development by insect pests like the seedcorn maggot (Miller and McClanahan, 1960). Although crop residues on the soil surface and soil moisture vary considerably among tillage systems, these variables were not considered to be important in influencing seedcorn maggot oviposition and development (Funderburk et al., 1983).

Funderburk and Herzog (pers. commun.) used pheromone traps to examine crop and non-crop habitats for lesser cornstalk borer adult activity in Florida. Adult activity was high in soybean during all vegetative growth stages in full-season and double-cropped plantings. The association of lesser cornstalk borers with grassy vegetation in soybean and other crops suggests that double-cropped soybean systems with limited tillage are potentially vulnerable to attack by this pest.

Soybeans under conservation tillage in Georgia supported larger and more diverse arthropod communities than conventionally tilled soybean (House and All, 1981; House and Stinner, 1983). Ground beetle populations were larger in the no-till fields and were associated with a degree of control of leaf-feeding species; i.e., bean leaf beetle. Denser threecornered alfalfa hopper (*Spissistilus festinus* [Say]) populations were reported in no-till soybean, whereas the bean leaf beetle and the banded cucumber beetle (*Diabrotica balteata* LeConte) increased to greater numbers in conventionally tilled soybean in Louisiana studies (Troxclair, 1984), which supports the idea proposed by House and Stinner (1983) that bean leaf beetle populations are influenced by ground beetle populations.

Green cloverworm (*Plathypena scabra* [F.]) larvae were less abundant in double-cropped soybean in Kentucky than in conventionally tilled soybean planted at the same time (Sloderbeck and Yeargan, 1983). The Mexican bean beetle (*Epilachna varivestis* Mulsant) was more abundant in tilled than no-tilled soybean, and red legged grasshopper (*Melanopus femurrubrum* [DeGeer]) populations were larger in late-planted, no-till soybean than in early conventionally planted soybean in Indiana (Sloderbeck and Edwards, 1979). The Mexican bean beetle developed two generations on early-planted soybean, but only one generation on double-cropped soybean.

Mechanical and chemical weed control procedures are used in conventionally tilled soybean fields to eliminate or reduce weed infestations that serve as reservoir hosts for some pest species, but in no-till double-cropped soybean there is adequate vegetation for establishment of early-season pest infestations. Todd et al. (pers. commun.) reported higher populations of predators, but lower populations of herbivores in plots with weeds than in weed-free plots. The populations of threecornered alfalfa hoppers and corn earworm larvae increased on soybean in conventional early plantings in fields kept weed free with toxaphene, compared with fields with some weeds; e.g., sicklepod (T. Mack, pers. commun.). This increase in abundance was attributed to reductions in populations of natural enemies in the toxaphene treated areas. Thus, the interactions of weed, pest and beneficial arthropod populations in production fields must be elucidated in order to understand the effects of cultural practices on insect populations on soybean in different ecosystems.

PLANTING DATE

In many years, soybean may be planted later than recommended due to factors delaying normal planting. The late-planted crop is frequently exposed to heavy infestations of arthropod pests and may be damaged more than early plantings. Velvetbean caterpillar (*Anticarsia gemmatalis* Hubner) and green cloverworm larvae are reported by Buschman et al. (1981) to develop large populations in late-planted, latematuring soybean in Mississippi. Early planting may be recommended in some areas to escape late-season populations of some lepidopterous pests, whereas early plantings may be damaged by Mexican bean beetle and bean leaf beetle in Illinois (Kogan and Kuhlman, 1982) and in Virginia (Deighan, 1983). Infestations may develop more quickly in early-planted fields than late-planted fields due to the large overwintering populations that emerge in the spring to attack early-planted soybean (M. Kogan, pers. commun.). The earliest-planted soybean fields in a production area serve as a trap crop by attracting the overwintering adults (Newsom and Herzog, 1977). Sloderbeck and Edwards (1979) reported that second generation Mexican bean beetle damage was greater on late-planted or double-cropped soybean because the late-planted fields are vulnerable to defoliation damage during pod developmental stages when the beetle population peaks, whereas the early plantings were generally physiologically mature at the time.

Late-planted soybeans double-cropped after wheat are particularly vulnerable to damage in late season by velvetbean caterpillar and stinkbugs, especially the southern green stink bug (*Nezara viridula* [L.]) in southeastern USA. Here, soybean plants are generally exposed to more than one generation of these pests each year (Boethel, 1984). In Virginia, soybean following wheat in a delayed-planting system contained high numbers of corn earworms (Deighan, 1983). Soybeans following wheat were the last fields to flower; thus when the corn earworm moved from corn the soybean fields in bloom were attractive to the moths for oviposition (Johnson et al., 1975; Hillhouse and Pitre, 1976). The early-planted soybean fields were past the peak of attractiveness to the moths. Populations of green cloverworms, cucumber beetle (*Diabrotica undecimpunctata howardi* Barber) adults, and spiders were highest in late-planted soybean fields, but numbers of green stinkbugs (*Acrosternum hilare* [Say]), and the beneficial nabids and geocorids were highest in early-planted soybean. In Louisiana, early plantings escaped economic thresholds of velvetbean caterpillar (Boethel, 1984). Buschman et al. (1981) reported denser populations of defoliating caterpillars in late-planted, late-maturing and narrow-row soybean fields, and early planting dates were associated with highest populations of whitefringed beetles, *Graphognathus* spp., soybean stem borers, *Dectes texanus texanus* LeConte, and early generation bean leaf beetles.

Natural enemies of insect pests of soybean have been observed to be influenced by cultural practices, including planting dates. Predators, primarily *Geocoris punctipes* (Say), were abundant in early-planted soybean fields but were not encountered in large numbers in double-cropped soybean fields (McPherson et al., 1982). Sprenkle et al. (1979) found that planting date had more effect than seeding rate or row spacing on insect populations on soybean in North Carolina. They reported high parasitism by the entomopathogen, *Nomuraea riley* (Farlow) Sampson, and predation on key pests on soybean planted late in narrow rows at a high seeding rate following small grains.

ROW SPACING

Increased interest in double-cropping, reduced tillage and erosion control combined with increased fuel costs have created a trend toward more narrow-row production of soybean. Understanding existing interactions between plant density and row spacing is essential for effective insect pest management.

Early-planted soybean is typically planted to standard row widths, whereas late-planted soybean may be planted no-till, narrow-row or drill. There are crop-pest management advantages to planting soybeans in narrow rows other than the agronomic benefits which include weed control and more effective use of soil moisture. For example, corn earworm moths have been reported to prefer open canopy, full bloom soybean plants for oviposition (Johnson et al., 1975; Hillhouse and Pitre, 1976), therefore soybean is planted with a grain drill in rows 15 to 30 cm to ensure early canopy closure. Bradley and VanDuyn (1980) associated lower populations of corn earworm larvae on soybean with the closed canopy system being less attractive for oviposition, less acceptable for larval development, and the larger natural enemy populations in the closed-canopy soybean plantings. Narrow-row planting can accomplish this effect when planting determinate soybeans late, or in a double-crop system, in southern USA (Boethel, 1984).

Populations of other herbivores and natural enemies of insect pests on soybean have been reported to respond to soybean row spacing. The potato leafhopper, *Empoasca fabae* (Harris), was found in greater numbers on soybean in wide rows (96 cm) than in narrow spacings in Illinois (Mayse, 1978). In southern USA populations of the green cloverworm, velvetbean caterpillar and loopers, *P. includens* and *Trichoplusia ni* (Hubner), were somewhat higher in drill (18 cm) than in wide-row (76 to 102 cm) soybean plantings (Bushman et al., 1981). Similar results for green cloverworm in narrow- and wide-row soybean were recorded in Tennessee (G. Lentz, pers. commun.). Mayse (1978) in Illinois and Sprenkel et al. (1979) in North Carolina also observed higher populations of defoliating caterpillars in narrow- than in wide-rows. Narrow-row spacing was associated with denser bean leaf beetle populations, but populations of other beetles and stink bugs did not appear to respond to row spacing in studies in Mississippi (Bushman et al., 1984).

Price (1976) indicates that soybean fields are colonized by predators only after plant canopy closure. Studies by Mayse (1978), Ferguson (1983), and Buschman et al. (1984) show that arthropod predators responded to row spacing, and that highest populations developed in narrow rows. Mayse (1978) attributes this observation to a more favorable microclimate in the closed canopy soybean, which occurred early, and thus there was ample time for predators to increase to high levels, as proposed by Price (1976). Ferguson (1983) lists factors contributing to enhancement of predators in double-cropped and drill-planted fields as: (1) increased plants per unit area in narrow-row spacings, thus, more habitats for predators and prey, (2) the increased levels of moisture, due to the dense canopy, attracting certain predators, (3) weedy fields providing more niches and food supply, (4) grain stubble providing refuges for predators, and (5) early predator colonizers building up to levels that are effective in keeping the pest from attaining economic status.

PESTICIDE INTERACTIONS

The complex of beneficial organisms in soybean may in many instances hold pest populations on the crop in check (Pitre et al., 1983), but if economic infestations develop, then proper insect and crop management tactics must be employed. Chemical insecticides that are highly effective in controlling pest outbreaks in conventional wide-row soybean may not be as effective when applied to the more dense soybean canopy of narrow-row plantings.

Hutchins and Pitre (1984) studied the influence of row spacing on spray penetration and efficacy of insecticides applied by aerial and ground equipment. Wide-row soybean systems allowed for superior control of the defoliating pest complex, including velvetbean caterpillar, green cloverworm, and loopers. Apparently, the dense foliage canopy inherent to narrow-row spacings, and thus to double cropping systems, limited droplet deposition into the canopy and consequently reduced larval mortality.

Since beneficial arthropods are influenced by double cropping practices, their population response to chemical pesticides is important in planning pest management systems. Herbicide-use programs may present certain problems to integrated pest management, foremost of which is the destruction of biological control agents in the agroecosystem (Newsom, 1979).

Since weeds can be the principal hindrance to good soybean yields in no-till soybean, timely use of pre-emergence and post-emergence herbicides is encouraged. Post-directed herbicides may be needed if earlier weed control measures were not completely successful. Studies by Farlow and Pitre (1983) on the effects of selected pesticide application routines, including herbicides, on pest and beneficial arthropods on soybean in Mississippi revealed that certain soil insecticides, herbicides, and fungicides may be used alone, or in routines as recommended in integrated pest management programs, without perturbation of the agroecosystem early in the season, as some have suggested.

Evidence that changes in cultural practices may increase insecticide usage on soybean has been presented (Boethel, 1984), and as stated, "it is imperative that research be initiated to determine whether the evolution in cultural practices currently underway is leading to increased dependence on pesticides and whether the current pest management technology will still be applicable to prevent this movement".

NOTES

Henry N. Pitre, Department of Entomology, Mississippi State University, Mississippi State, Mississippi 39762.

Mississippi Agricultural and Forestry Experiment Station Publication Number 5842.

The author expresses thanks to the following who provided published and/or unpublished information that was useful in the preparation of this review article: Drs. D. J. Boethel, J. R. Bradley, Jr., C. R. Edwards, N. C. Edwards, J. E. Funderburk, R. B. Hammond, M. Kogan, G. L. Lentz, T. Mack, R. M. McPherson, J. O. Sanford, J. Todd, and K. V. Yeargan.

REFERENCES

Boethel, D. J. 1984. Evolving crop production systems for soybean: Impact on pest and beneficial arthropod species. p. 348-359. *In* S. Wong, D. J. Boethel, R. Nelson, W. Nelson, W. Wolf (eds.) Proceedings of the Second U.S.-China Soybean Symposium. USAID-OICD, Washington, D.C.

Bradley, J. R. and J. W. VanDuyn. 1980. Insect pest management in North Carolina soybeans. p. 343-354. *In* F. T. Corbin (ed.) World Soybean Research Conference II: Proceedings. Westview Press, Boulder, Colorado.

Buschman, L. L., H. N. Pitre and H. F. Hodges. 1981. Soybean cultural practices: effects on populations of green cloverworm, velvetbean caterpillar, loopers, and *Heliothis* complex. Environ. Entomol. 10:631-641.

Buschman, L. L., H. N. Pitre and H. F. Hodges. 1984. Soybean cultural practices: effects on populations of *Geocoris*, nabids, and other soybean arthropods. Environ. Entomol. 13:305-317.

Deighan, John. 1983. The effect of cropping systems on the pest complex in Virginia soybeans and calibration of the sweep net and ground cloth sampling methods for use in these cropping systems. M.S. Thesis, Virginia Polytechnic Institute and State University, Blacksburg.

Farlow, R. A. and H. N. Pitre. 1983. Effects of selected pesticide application routines on pest and beneficial arthropods on soybean in Mississippi. Environ. Entomol. 12:552-556.

Ferguson, H. J. 1983. Effects of four cropping systems on arthropod predators of soybean insect pests in Virginia. M.S. Thesis, Virginia Polytechnic Institute and State University, Blacksburg.

Funderburk, J. E., L. P. Pedigo and E. C. Berry. 1983. Seedcorn maggot (Diptera: Anthomyiidae) emergence in conventional and reduced-tillage soybean systems in Iowa. J. Econ. Entomol. 76: 131-134.

Hillhouse, Thomas L. and Henry N. Pitre. 1976. Oviposition by *Heliothis* on soybeans and cotton. J. Econ. Entomol. 69:144-146.

House, G. J. and J. N. All. 1981. Carabid beetles in soybean agroecosystems. Environ. Entomol. 10: 194-196.

House, G. J. and B. R. Stinner. 1983. Arthropods in no-tillage soybean agroecosystems: community composition and ecosystem interactions. Environ. Manage. 7:23-28.

Hutchins, S. H. and H. N. Pitre. 1984. Effects of soybean row spacing on spray penetration and efficacy of insecticides applied with aerial and ground equipment. Environ. Entomol. (in press).

Johnson, M. W., R. E. Stinner and R. L. Rabb. 1975. Ovipositional response of *Heliothis zea* (Boddie) to its major hosts in North Carolina. Environ. Entomol. 4:291-297.

Kogan, Marcos and D. E. Kuhlman. 1982. Soybean insects: Identification and management in Illinois. Ill. Agric. Exp. Stn. Bull. 773.

Landis, D. A. 1983. Effects of no-tillage corn and soybean production on the behavior, development and survival of *Heliothis zea* (Boddie) prepupae and pupae. M.S. Thesis, North Carolina State Univ., Raleigh.

Mayse, M. A. 1978. Effects of spacing between rows on soybean arthropod populations. J. Appl. Ecol. 15:439-450.

McPherson, R. M., J. C. Smith and W. A. Allen. 1982. Incidence of arthropod predators in different soybean cropping systems. Environ. Entomol. 11:685-689.

Miller, L. A. and R. J. McClanahan. 1960. Life-history of the seedcorn maggot, *Hylemya cilicrura* (Rond.) and of *H. liturata* (Meig.) (Diptera: Anthomyiidae) in southwestern Ontario. Can. Entomol. 92:210-221.

Musick, G. J. 1970. Insect problems with no-till crops. Crops and Soil Magazine 23:18-19.

Newsom, L. D. 1979. Role of pesticides in pest management systems. *In* T. J. Sheets and D. Pimentel (eds.) Pesticides: Their Contemporary Roles in Agriculture, Health and the Environment. The Human Press, Clifton, N.J.

Newsom, L. D. and D. C. Herzog. 1977. Trap crops for control of soybean pests. Louisiana Agric. 20:14-15.

Pitre, H. N., C. R. Edwards, J. D. Harper and L. W. Bledsoe. 1983. Role of natural enemies in integrated pest management. p. 88-90. *In* H. N. Pitre (ed.) Natural Enemies of Arthropod Pests in Soybean. So. Coop. Ser. Bull. 285.

Price, P. W. 1976. Colonization of crops by arthropods: non-equilibrium communities in soybean fields. Environ. Entomol. 5:605-611.

Sanford, J. O. 1982. Straw and tillage management practices in soybean-wheat double-cropping. Agron. J. 74:1032-1035.

Sloderbeck, P. E. and C. R. Edwards. 1979. Effects of soybean cropping practices on Mexican bean beetle and redlegged grasshopper populations. J. Econ. Entomol. 72:850-853.

Sloderbeck, P. E. and K. V. Yeargan. 1983. Green cloverworm (Lepidoptera: Noctuidae) populations in conventional and double-crop, no-till soybeans. J. Econ. Entomol. 76:785-791.

Sprenkel, R. K., W. M. Brooks, J. W. VanDuyn and L. L. Dietz. 1979. The effects of three cultural variables on the incidence of *Nomuraea rileyi*, phytophagous lepidoptera, and their predators on soybeans. Environ. Entomol. 8:334-339.

Troxclair, N. N., Jr. 1984. The influence of tillage practices and row spacings on arthropod populations in soybean, *Glycine max* (L.) Merr. M.S. Thesis. Louisiana State Univ., Baton Rouge.

VanDuyn, J. W. 1981. Insect problems in no-till soybeans and corn. p. 12-13. *In* No-till Crop Production Systems in North Carolina—Corn, Soybeans, Sorghum, and Forages. North Carolina Agric. Ext. Serv. AG-273.

Woodruff, J. M., J. E. Elsner and W. D. Givan. 1981. Double-cropping soybeans in Georgia. The Univ. of Georgia Coop. Exp. Ser. Bull. 819.

DYNAMICS OF PHYTOPHAGOUS PENTATOMIDS ASSOCIATED WITH SOYBEAN IN BRAZIL

Antônio R. Panizzi

Soybean production in Brazil has increased dramatically since 1970. As cultivation of the crop spread, problems with insect pests have increased. Among the several insect pests, phytophagous pentatomids (stink bugs) are of special importance. These feed mainly on fruiting plants, causing severe reductions in seed yield and quality. Little information is available on factors influencing population dynamics of these insects. In this paper, I will discuss their dynamics in relation to the soybean crop and other host plants, and how control techniques influence their abundance. Opportunities for future research involving this group of insects and the soybean crop also will be discussed.

Species composition of phytophagous pentatomids associated with the soybean crop in Brazil has increased throughout the years. Silva et al. (1968) first reported four species as occurring in different states. Bertels and Ferreira (1973) listed eight species for Rio Grande do Sul and later Lopes et al. (1974) referred to 13 species for the same state. Presently, more than 25 species have been collected on soybean, mainly in Paraná (A. R. Panizzi and B. S. C. Ferreira, unpublished). Only three species are considered major pests, however. These are *Nezara viridula* (L.), *Piezodorus guildinii* (Westwood), and *Euschistus heros* (F.).

POPULATION TRENDS OF MAJOR PENTATOMID SPECIES

For characterization of a pest population it is important to know the synchronization of host plant and insect phenologies. Not many studies are available relating host plants and population trends of stink bugs on soybean. In the U.S., Todd and Herzog (1980) presented a generalized scheme for a typical seasonal host sequence of *N. viridula* in the southeastern coastal plain. They mentioned that five generations may occur per year and that adults move to soybean in late July and August. Soybean provides the only major source of food in late summer and early fall. Jones and Sullivan (1982), in South Carolina, found that the sequence black cherry (*Prunus serotina* Ehrhart) followed by elderberry (*Sambucus canadensis* L.) allowed large populations of *Acrosternum hilare* (Say) to develop before infesting soybean.

In the north of Paraná, Brazil, Ferreira and Panizzi (1982) carried out a survey on the incidence of eggs, nymphs and adults of *N. viridula*, *P. guildinii*, and *E. heros* on different host plants, including soybean (Figure 1). They found adults of the three species throughout the year, either on soybean or other host plants. In this region, *N. viridula* shows partial hibernation and some adults, as well as late instar nymphs, can

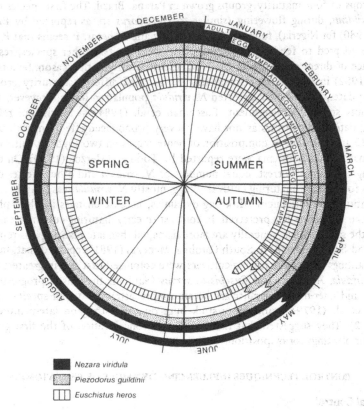

Figure 1. Annual occurrence of major soybean phytophagous pentatomids at different stages of development in the north of Paraná, Brazil (from Ferreira and Panizzi, 1982).

be found feeding on succulent plants throughout the year. Eggs were observed on soybean plants from mid-December to mid-May. Egg masses of *P. guildinii* and *E. heros* were observed earlier (late November), and nymphs from early December to late May (Figure 1). Reproduction for all three species occurs from late spring to mid-autumn, and it peaks during the summer when fruiting plants are most abundant.

North or south of this study area (23°S) a different situation is observed. In the northern area the pentatomids are reproductively active for a greater part of the year compared to the southern area. However, critical studies are needed to fully characterize the phenology of the pentatomid-pest population associated with the soybean crop in these areas.

Population Dynamics with Relation to Maturity Groups and Time of Planting

In general, pentatomids colonize soybean during its reproductive period, and their populations peak during the filling of pods. Early-maturing cultivars, in general, escape severe damage. However, stink bug populations do increase on them, and these then colonize successive later-maturing cultivars. As early-maturing cultivars senesce and become unattractive for feeding and reproduction, adults move to more acceptable hosts

(Todd and Herzog, 1980). This was confirmed by Ferreira and Panizzi (1982) for soybean crops of four maturity groups grown in Paraná, Brazil. The first species to appear is *P. guildinii*, during flowering (similar to *Piezodorus* sp. as reported by Ezueh and Dina (1980) for Nigeria), followed by *N. viridula* and *E. heros*. It seems that *P. guildinii* is more adapted to feed on flowering plants than the two other species, resulting in avoidance of direct competition for food resources early in the season. Schumann and Todd (1982) in Georgia mentioned that cultivars of different maturity groups, and planting dates, significantly affected *N. viridula* population levels; however, the main factor was stage of development. Buschman et al. (1984) in Mississippi referred to planting dates and cultivars as not having a very pronounced effect on stink bug abundance. Changes in species composition of pentatomids on two soybean cultivars of differing maturity in Louisiana were reported by Todd and Herzog (1980). On the early-maturing cultivar of Forrest, equal numbers of *N. viridula* and other species occurred, whereas for the later-maturing cultivar, Bragg, mostly *N. viridula* was found.

Planting time influences stink bug dynamics, because changes in plant phenology may make the plants less preferred. In contrast to early-maturing cultivars, early plantings of the same cultivar generally are more damaged than late plantings, as reported by Jones and Sullivan (1978) in South Carolina. Menezes (1981) reported that, in Florida, early plantings of early maturing-cultivars were colonized first and in greatest numbers by *N. viridula*, *A. hilare*, and *Euschistus servus* (Say). As the season progressed, later-planted and later-maturing cultivars became more attractive to all species. In Brazil, Panizzi et al. (1979) found more stink bugs on early- than on late-planted 'Bragg' (Figure 2). They suggested that this was due to concentration of the first generation on earlier plantings for oviposition.

CONTROL TECHNIQUES INFLUENCING DYNAMICS OF PENTATOMIDS

Chemical Control

Insecticides are the most effective control measure available to suppress stink bug populations on soybean crops on an emergency basis. However, because soybean is very tolerant of insect damage, relatively low amounts of insecticide are required to maintain these pests below the economic threshold levels.

Appropriate timing of applications, as well as use of selective insecticides at minimum effective rates, have been developed as a result of a large-scale insecticide screening program initiated in 1975 by the National Soybean Research Center of EMBRAPA (Gazzoni, 1980). Use of wide-spectrum insecticides (a mixture of methyl parathion and methomyl) early in the season was shown to cause an abnormal increase in *P. guildinii*, as compared to untreated areas (Figure 3). This was attributed to the destruction of natural enemies, which under normal conditions tend to maintain stink bug populations at lower levels.

Among the different insecticides, methyl parathion is most commonly reported as being effective against phytophagous pentatomids. However, different species may show differences in susceptibility to this insecticide (McPherson et al., 1979). In Brazil, *P. guildinii* is known to be more susceptible to carbaryl and less susceptible to methyl parathion, whereas with *N. viridula* the reverse was found. Thus, species composition will determine which material to use (Gazzoni, pers. comm.).

Biological Control

Several insects are known to attack phytophagous pentatomids. Parasites are probably the most important group of natural enemies and include both egg and adult

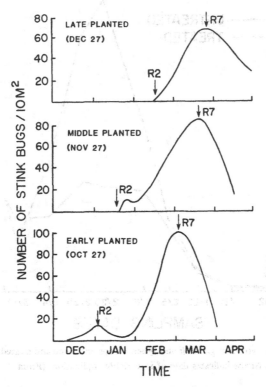

Figure 2. Population trend of *N. viridula*, *P. guildinii* and *E. heros* on soybean crops planted on three dates, Paraná, Brazil. R2 = full bloom; R7 = beginning maturity. (Modified from Panizzi et al., 1979).

parasites. Among the egg parasites, the most studied one is the scelionid *Trissolcus basalis* (Wollaston). This species occurs worldwide, and was first recorded in Brazil in 1979 from *N. viridula* eggs in soybean fields (Ferreira, 1980). Of 76 egg masses, 23% were found parasitized. Decrease in use of insecticides in the late 1970s, as well as the expanding soybean crop, are probably the main reasons for the presence of this parasite. Substantial parasitism has also been reported for *P. guildinii* eggs by *Telenomus mormideae* Lima in Paraná (Panizzi and Smith, 1976) and in Rio Grande do Sul (Link and Concatto, 1979). For the other major pentatomid pest, *E. heros*, as much as 40% of the eggs were parasitized by *T. mormideae* in field cages in Paraná (Villas Bôas and Panizzi, 1980).

Tachinid flies are the most common adult parasites of phytophagous pentatomids found in soybean agroecosystems. In North America, *Trichopoda pennipes* (F.) is the most common species. However, as many as nine other species of tachinids parasitize different species of phytophagous pentatomids that may occur on soybean plants in North America (Eger and Ables, 1981). In South America, the most common tachinid fly in soybean fields is *Eutrichopodopsis nitens* Blanchard. It is the principal parasite of *N. viridula*, and in one study 50.5% of the 2663 *Thyanta perditor* (F.) adults collected in the field were found to be parasitized by this fly in Paraná (Panizzi and Herzog, 1984).

Few references are found to predators and pathogens of phytophagous pentatomids in soybean agroecosystems. Probably due to the polyphagous habits of

678

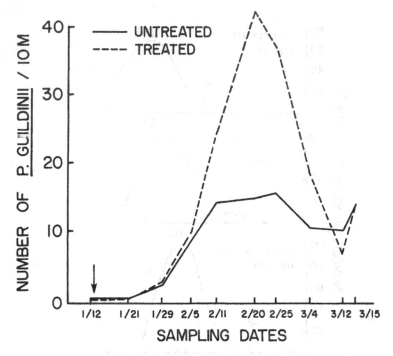

Figure 3. Population growth of *P. guildinii* on soybean crops untreated and treated with insecticides, Goiás, Brazil. Arrow indicates date of insecticide application. (From Panizzi et al., 1977).

predators, it is difficult to associate predators with their diets. Also, very little research has been conducted on diseases of stink bugs. The fungi *Beauveria bassiana* (Bals.) and *Metarhizium anisopliae* (Metsch.) attack *N. viridula* and *P. guildinii* in Brazil (F. Moscardi, pers. comm.). Further studies are needed to evaluate the impact of predators and pathogens on the stink bug complex.

Cultural Control

Methods of stink bug control used on soybean other than insecticides and biocontrol include the use of cultural practices, such as early-maturing cultivars, manipulation of planting date, and early-planted or/and early-maturing trap crops.

Early-maturing cultivars will escape severe damage done by stink bugs. The practice of concentrating on cultivars that mature early and within a narrow range to avoid stink bug damage is widespread in Brazil. This cultural practice has saved growers many insecticide applications. Early-maturing cultivars are also preferred because there is less chance of yield loss due to their shorter period of time in the field. However, this technique may be a risk to total production because, in years with low water availability, the whole crop will be susceptible to drought.

Planting date can also be used to change plant phenology, such that insect damage is reduced (see previous section). However, this practice is limited, because each cultivar has its optimum time for planting, and changes in planting time may result in a yield loss independent of stink bug damage. Manipulation of planting time, along with the use of early-maturing cultivars as trap crops, has been tried in Louisiana to control *N. viridula* (Newsom and Herzog, 1977). Early-maturing/early-planted soybean is highly attractive to *N. viridula* (Ragsdale et al., 1981).

In Brazil, effective control of *N. viridula, P. guildinii,* and *E. heros* using early-maturing cultivars as trap crops has been reported (Panizzi, 1980). This practice, however, has not been fully exploited. For instance, use of other host plants, perhaps more attractive than soybean, to concentrate stink bug populations in a particular area should be tried.

OPPORTUNITIES FOR FUTURE RESEARCH

Opportunities for future research include both applied and basic work. Of particular need are studies on the nutritional ecology of stink bugs. Information on the impact food may have on their biology, including the causes and timing of movement of stink bugs from native host plants to soybean, and how this switch in food source affects stink bug performance, should be generated. Identification of preferred host plants with potential use as trap plants is desired. Of special importance is the development of cultivars that may have one or more characteristics of resistance. The impact of resistant genotypes on stink bug biology should be assessed.

Cultural control involving time of planting, row spacing, crop rotation, and the effect of reduced tillage practices should be investigated. The trap crop technique is yet to be fully exploited as a means of manipulating stink bug abundance.

Perhaps the most important study area today is the biocontrol of stink bugs in soybean fields. In Brazil, the use of egg parasites has been greatly exploited since the discovery of *T. basalis* attacking *N. viridula* eggs. Introduction of new biocontrol agents should be promoted. For example, the introduction of *T. pennipes* to Brazil from North America, and the introduction of *E. nitens* to North America from Brazil should be tried to control *N. viridula*. If *P. guildinii* becomes an important pest species in southern USA, the egg-parasite *T. mormideae* is a potential candidate for introduction.

Finally, in view of the importance of stink bugs as a major pest component of soybean, pest management programs for these species should be constantly evaluated and updated based on the results of current research.

NOTES

Antônio R. Panizzi, entomologist, Centro Nacional de Pesquisa de Soja, EMBRAPA, Caixa Postal 1061, 86.100 - Londrina, Paraná, Brasil.

I want to express my gratitude to Dr. Frank Slansky, Jr., Dept. of Entomology and Nematology, University of Florida, for critically reviewing this manuscript.

REFERENCES

Bertels, A. and E. Ferreira. 1973. Levantamento atualizado dos insetos que vivem nas culturas de campo no Rio Grande do Sul. Univ. Católica Pelotas, Ser. Publ. Cient. n. 1.

Buschman, L. L., H. N. Pitre and H. F. Hodges. 1984. Soybean cultural practices: effects on populations of geocorids, nabids, and other soybean arthropods. Environ. Entomol. 13:305-317.

Eger, J. E., Jr., and J. R. Ables. 1981. Parasitism of Pentatomidae by Tachinidae in South Carolina and Texas. Southwest. Entomol. 6:28-33.

Ezueh, M. I. and S. O. Dina. 1980. Pest problems of soybean and control in Nigeria. p. 275-283. *In* F. T. Corbin (ed.) World soybean research conference II: proceedings. Westview Press, Colorado.

Ferreira, B. S. C. 1980. Ocorrência, no Brasil, de *Trissolcus basalis*, parasita de ovos de *Nezara viridula*. Pesq. Agropec. Bras. 15:127-128.

Ferreira, B. S. C. and A. R. Panizzi. 1982. Percevejos-pragas da soja no norte do Paraná: abundância em relação a fenologia da planta e hospedeiros intermediários. Ann. II Semin. Nac. Pesq. Soja v. II, p. 140-151.

Gazzoni, D. L. 1980. Selação de inseticidas para uso no programa de manejo de pragas da soja. An. VI Congr. Bras. Ent. p. 265-275.

Jones, W. A. and M. J. Sullivan. 1978. Susceptibility of certain soybean cultivars to damage by stink bugs. J. Econ. Entomol. 71:534-536.

Jones, W. A. and M. J. Sullivan. 1982. Role of host plants in population dynamics of stink bug pests of soybean in South Carolina. Environ. Entomol. 11:867-875.

Link, D. and L. U. Conuu(о) 1979. Hábitos de postura de *Piezodorus guildinii* em soja. Rev. Cent. Cien. Rur. 9:61-72.

Lopes, O. J., D. Link and I. V. Basso. 1974. Pentatomídeos de Santa Maria—lista preliminar de plantas hospedeiras. Rev. Cent. Cien. Rur. 4:317-322.

McPherson, R. M., J. B. Graves and T. A. Allain. 1979. Dosage-mortality responses and field control of seven pentatomids, associated with soybean, exposed to methyl parathion. Environ. Entomol. 8:1041-1043.

Menezes, E. B. 1981. Population dynamics of the stinkbug (Hemiptera: Pentatomidae) complex on soybean and comparison of two relative methods of sampling. Ph.D. Dissertation, University of Florida Library, Gainesville.

Newsom, L. D. and D. C. Herzog. 1977. Trap crops for control of soybean pests. Louisiana Agric. 20:14-15.

Panizzi, A. R. 1980. Uso de cultivar armadilha no controle de percevejos em soja. Trigo e Soja 47: 11-14.

Panizzi, A. R. and D. C. Herzog. 1984. Biology of *Thyanta perditor* (Hemiptera: Pentatomidae). Ann. Entomol. Soc. Amer. 77:646-650.

Panizzi, A. R., B. S. Corrêa, G. G. Newman and S. G. Turnipseed. 1977. Efeito de inseticidas na população das principais pragas da soja. An. Soc. Entomol. Bras. 6:264-275.

Panizzi, A. R., B. S. C. Ferreira, N. Neumaier and E. F. Queiróz. 1979. Efeitos da época de semeadura e do espaçamento entre fileiras na população de artrópodos associados a soja. Ann. I Semin. Nac. Pesq. Soja v. II: 113-125.

Panizzi, A. R. and J. G. Smith. 1976. Observações sobre inimigos naturais de *Piezodorus guildinii* (Westwood, 1837) (Hemiptera, Pentatomidae) em soja. Ann. Soc. Entomol. Brasil 5:11-17.

Ragsdale, D. W., A. D. Larson and L. D. Newsom. 1981. Quantitative assessment of the predators of *Nezara viridula* eggs and nymphs within a soybean agroecosystem using an ELISA. Environ. Entomol. 10:402-405.

Schumann, F. W. and J. W. Todd. 1982. Population dynamics of the southern green stink bug (Heteroptera: Pentatomidae) in relation to soybean phenology. J. Econ. Entomol. 75:748-753.

Silva, A. G. d'A., C. R. Conçalves, D. M. Galvão, A. J. L. Conçalves, J. Gomes, M. N. Silva and L. Simoni. 1968. Quarto catálogo dos insetos que vivem nas plantas do Brasil—seus parasitas e predadores. Min. Agric., Rio de Janeiro, Parte II, v. I: 1-622.

Todd, J. W. and D. C. Herzog. 1980. Sampling phytophagous Pentatomidae on soybean. p. 438-478. *In* M. Kogan and D. C. Herzog (eds.). Sampling methods in soybean entomology. Springer-Verlag, Berlin.

Villas Bôas, G. L. and A. R. Panizzi. 1980. Biologia de *Euschistus heros* (Fabricius, 1798) em soja (*Glycine max* (L.) Merrill). An. Soc. Entomol. Brasil 9:105-113.

DEPOSITION AND EFFICACY CHARACTERISTICS OF OIL
APPLIED PYRETHROIDS

S. G. McDaniel and L. D. Hatfield

The ability of pyrethroid insecticides to deliver a lethal dose at low rates allowed the expanded use of these problems in reduced volume applications (2.3 L/ha) of non-volatile, vegetable oils. This expanded use occurred in many agricultural crops, and is a result of increased efficiency due to characteristics involved in droplet size, canopy penetration, deposition uniformity and delivery volume (McDaniel, 1980; McDaniel et al., 1983). This application technique does require, however, that an aerial applicator either have two separate hydraulic delivery systems for oil and water sprays or purchase a more expensive rotary atomization delivery system which can accommodate both sprays. The use of this application technique has been moderated somewhat by this factor and, more recently, by increased expense of vegetable oil and more stringent regulatory policies.

Within the last two years, a potentially desirable intermediate application technique has arisen with the increasing availability of emulsified vegetable oils. This technique offers the opportunity to apply mixtures of water, emulsified oil and insecticide in a manner such that the advantages of nonvolatile, vegetable applications can be incorporated into lower volume water applications (9.2 L/ha). We investigated this application technique for cotton insect control and found that spray droplet volume median diameter, VMD, was significantly increased over that of water sprays when using a 1:3 ratio of vegetable oil to water in 9.2 L/ha applications of Ammo 2.5EC (cypermethrin). Equivalent efficacy of *Heliothis zea* Bodie and *Heliothis virescens* Fabricius was obtained with one less application of oil-water-cypermethrin treatments than with water-cypermethrin treatments at identical insecticide rates. We have also demonstrated that control of second generation southwestern corn borer, *Diatraea grandiosella* Dyar, with Ammo + emulsified oil + water at 9.2 L/ha was equivalent to that with Ammo + water at 27.6 L/ha.

Although use of vegetable oil-water mixtures in application of pyrethroids at 9.2 L/ha provides desirable efficacy, the 1:3 ratio of oil to water at this application volume is not an economic advantage over that of vegetable oil applications alone. This is because the same amount of vegetable oil (2.3 L/ha) is used in both techniques. Therefore, to obtain an economic advantage, vegetable oil quantity must be reduced in mixture applications. Although we briefly examined changes in oil-water ratio on spray droplet VMD and deposition, little or no information exists on the optimization of this ratio to provide adequate spray characteristics that would allow insect efficacy at a 9.2 L/ha application volume. Therefore, additional studies were conducted to examine the affect of various oil-water ratios on the spray characteristics of aerially applied oil-water-cypermethrin sprays.

APPLICATION AND DROPLET CHARACTERIZATION

In 1984, deposition characteristics were examined for various aerially applied treatments of water and oil-water mixtures. All treatments contained Ammo 2.5EC (cypemethrin) insecticde, manufactured by FMC Corporation, diluted to a rate of 55.4 g ai/ha, The aircraft used was a Thrush Commander fitted with eight mini-Micronair units calibrated to deliver the spray volume under normal application procedures. A Sorenson agricultural water soluble dye was included with the spray mixtures to enhance droplet imagery.

Treatment mixtures were applied at a total spray volume of 9.2 L/ha and consisted of modifications in the ratio of oil to water. The oil adjuvants used were refined soybean oil containing no emulsifiers and Bettakill, an emulsified vegetable oil product manufactured by Valley Chemical Company and containing 98% soybean oil + 2% emulsifier. All treatments were applied during a 2 h period, with constant meteorological conditions at 29.5 C, 60% relative humidity, and a 3.8-4.8 m/s wind speed.

Deposition characteristics were determined by placing 0.025 mm thick plastic coated kromecote cards at 0.92 m intervals over a total of 10 m. Each 77.2 mm X 127 mm card was positioned 43.2 cm above the soil surface and oriented slightly into the wind to ensure perpendicular impact of the spray droplets. Three replicate card lines were placed at 22.9 m intervals perpendicular to the wind direction and flight of the aircraft. This procedure ensured that a portion of each swath would be captured from three areas within the field. A total of five cards were chosen from each replicate following a single flight of the aircraft over the cards. Each card was evaluated for VMD of the resulting spray using the D-max technique (Maksymiuk, 1978). The number of drops per square centimeter was determined also.

VMD AND DROPLET DEPOSITION OF OIL/WATER SPRAYS

The results of aerial deposition data between Ammo + water alone and Ammo + water + emulsified oil at various ratios can be seen in Table 1. The Ammo + water treatment applied at 9.2 L/ha resulted in a VMD of 123 μm and the deposition of 55 drops/cm^2. When emulsified oil was added to the Ammo + water mixture at a one to three ratio, and the spray volume was kept constant, the VMD increased significantly to 155 μm. Thus, the addition of a non-volatile adjuvant increased the

Table 1. Deposition characteristics for aerially applied mixtures of oil and water using Ammo 2.5EC (cypermethrin) (55.4 g ai/ha) at a total spray volume of 9.2 L/ha.

Treatment[a]	Oil/water ratio	VMD[b]	Drops
		μm	no/cm^2
Ammo + Water	0 to 1	123 r	55 r
Ammo + EC Oil + Water	1 to 3	151 s	50 r
Ammo + EC Oil + Water	1 to 6	141 s	81 s
Ammo + EC Oil + Water	1 to 15	142 s	31 t

[a]EC Oil = Bettakill (98% vegetable oil, 2% emulsifier).

[b]Means in a column followed by the same letter are not significantly different (P=.95), as determined by Duncan's New Multiple Range Test.

overall droplet size of the aerial spray compared with water alone. However, the addition of emulsified oil at the one to three ratio did not change the number of drops impacting on the target.

When emulsified oil was added to the spray solution at a one to six oil-water ratio, the VMD of the spray was 141 μm. This ratio significantly increased the VMD over that of the Ammo + water treatment, and was statistically equivalent to the VMD produced by the higher emulsified oil dilution. The one to six ratio also significantly increased the drops/cm^2 to 81. This is well above both the Ammo + water and the one to three ratio of Ammo + emulsified oil + water. Thus, it appears a one to six oil-water ratio treatment would optimize the life expectancy of the spray by increasing droplet size and allowing for increased atomization. This procedure would also allow for lower application costs than the one to three oil-water mixture. These factors are highly important in the pursuit of maximum *Heliothis* spp. control in cropping situations using Ammo 2.5EC insecticide.

The treatment of Ammo + emulsified oil + water at the 1:15 ratio continued to provide a significantly higher VMD over that of the Ammo + water treatment, but a significantly lower number of drops/cm^2 was collected (31 drops/cm^2; Table 1). It is apparent there was sufficient oil to maintain the larger drops, but this mixture may have been deficient in the ability to maintain surface tension of the spray compared with higher oil ratios. Regardless, the 1:15 ratio of emulsified oil-water resulted in significantly fewer drops deposited when compared with higher oil-water mixtures or water alone. This could result in less coverage of the plant foliage and risk of less than desirable efficacy.

Based on the information in Table 1, the questions arise whether the emulsifiers in the treatments affected the deposition characteristics, and would a non-emulsified oil provide equivalent results without creating mixing problems? Table 2 shows the results of spray characteristics obtained from water sprays of Ammo containing emulsified oil or non-emulsified, refined soybean oil adjuvants at a one to three oil-water ratio. All treatments went into solution without difficulty. This indicates sufficient emulsifiers are contained in the Ammo 2.5EC formulation to prevent invert emulsions of these spray solutions. The deposition data obtained from these sprays shows that Ammo + non-emulsified oil + water has a significantly higher VMD (170 μm) than that of the Ammo + emulsified oil + water mixture (151 μm; Table 2). However, a difference of 19 μm in this VMD range would likely not affect field performance of the chemical.

The Ammo + non-emulsified oil + water treatment also had significantly more

Table 2. Deposition characteristics for aerially applied mixtures of oil and water using Ammo 2.5EC (cypermethrin) (55.4 g ai/ha) at a total spray volume of 9.2 L/ha.

Treatment[a]	Oil/water ratio	VMD[b]	Drops
		μm	no/cm^2
Ammo + Water	0 to 1	123 r	55 r
Ammo + EC Oil + Water	1 to 3	151 s	50 r
Ammo + Soy Oil + Water	1 to 3	170 t	72 s

[a]EC Oil = Bettakill (98% vegetable oil, 2% emulsifier); Soy Oil (non-emulsified refined soybean oil).

[b]Means in a column followed by the same letter are not significantly different (P=.95), as determined by Duncan's New Multiple Range Test.

drops/cm^2 than the Ammo + emulsified oil + water treatment (Table 2). Thus, it seems that the addition of emulsifiers to the soybean oil can impact the spray characteristics for oil-water dilutions.

NOTES

S. G. McDaniel, Southern Regional Manager and L. D. Hatfield, Research Biologist, FMC Corporation, Market Development, 2964 Terry Road (A-4), Jackson, MS 39212 USA.

REFERENCES

Maksymiuk, B. 1978. Field assessment methods—determining volume median diameter. USDA Tech. Bull. 1596:41-47.

McDaniel, S. G. 1980. Aerial application: Effects of formulation, volume and delivery on cotton insect control. p. 76-77. In Proc. Beltwide Cotton Conf.

McDaniel, S. G., B. M. McKay, R. K. Houston and L. D. Hatfield. 1983. Aerial drift profile of oil and water sprays. Ag. Aviation 10:25-29.

BIOLOGICAL CONTROL OF THE MEXICAN BEAN BEETLE: POTENTIALS FOR AND PROBLEMS OF INOCULATIVE RELEASES OF *PEDIOBIUS FOVEOLATUS*

Robert V. Flanders

Biological control strategies are divided into importation, augmentation, and conservation (DeBach, 1964). Historically, natural enemy importations have received the greatest emphasis, and have consequently resulted in the greatest number of biological control successes. However, augmentation and conservation have equally great potentials for the suppression of pest densities. The differential attempt and success rates between importation and the other strategies, especially augmentation, appear to be due to the differences in depths and breadths of biological data required to implement them. These differences are illustrated by biological control attempts against the Mexican bean beetle (MBB), *Epilachna varivestis* Mulsant (Coleoptera: Coccinellidae). The purpose of this paper is to review these attempts in regard to recent efforts to develop and implement inoculative releases of *Pediobius foveolatus* (Crawford) (Hymenoptera: Eulophidae) in soybean fields.

MEXICAN BEAN BEETLE BIOLOGY, DISTRIBUTION, AND PEST STATUS

The Mexican bean beetle is native to southern and central Mexico where it feeds on various species of legumes, especially *Phaseolus* spp. (Gordon, 1975). Its geographic distribution expanded northward into the western United States apparently with the cultivation of *Phaseolus* beans by native Indians. It was first observed as a pest of snap and lima beans in New Mexico in 1850 (Chittenden and Marsh, 1920). Before 1918, the MBB was restricted to the western United States, but in that year it was accidentally introduced into the east, where it spread rapidly. It presently occurs in an area extending from Florida to Maine, and as far west as Missouri. Eastern and western distributions are separated by the central plains, where climatic conditions apparently do not allow establishment (Gordon, 1975).

Since its introduction into the eastern United States, the MBB has been a perennial pest of cultivated snap beans, lima beans, and cowpeas, *Vigna unguiculata* (L.) Walp. When first introduced, it only was occasionally observed feeding on soybean plants (Davis, 1925). High population densities did not develop in soybean fields until the 1960s, coinciding with expanding soybean cultivation. Severe MBB infestations presently are limited to certain states on the Atlantic coast, especially Delaware, Maryland, and Virginia, and the states of Indiana, Kentucky, and Ohio in the Midwest. This geographic limitation is in contrast to frequently high MBB densities on cultivated *Phaseolus* spp. throughout the east and midwest.

The biology of the MBB has been reported by several authors, including Chittenden and Marsh (1920), Douglass (1933), and Graf (1925). Egg masses of 20 to 80 eggs are deposited on the undersides of leaves. There are four larval instars and pupation occurs on the plant. In those areas where the MBB is a pest of soybean plants, it exhibits two generations per year. The first generation, that arises from overwintered adults, usually exhibits low densities, but second generation densities typically are higher.

Larvae and adults of the MBB feed by scraping softer tissues primarily from the undersides of leaves, leaving tougher vascular tissues intact (Howard, 1941). Defoliated leaves characteristically appear net-like. Such defoliation reduces the photosynthetic areas of the plants, which ultimately affects yield. The defoliation potential of an MBB population depends on its density and age-class composition (i.e., small larvae eat less than large larvae and adults). The effect of defoliation on plant yield depends on the growth stage of the plant, and is most critical during reproductive growth (i.e., flowering, pod set, and pod fill). The co-occurrence of reproductive soybean growth stages and high second generation densities of MBB may reduce soybean yield by 40 to 50% in untreated fields. Economic thresholds for determining the need for control strategies are based on field estimates of percent defoliation at specific soybean growth stages.

PARASITE IMPORTATIONS

Early studies on natural control of MBB populations indicated that weather conditions, especially drought, were of primary importance in affecting MBB population dynamics (Graf, 1925; Howard, 1931). The recognized absence of effective MBB natural enemies in the eastern United States (Howard and Landis, 1936) prompted attempts at classical biological control. In the 1920s, a larval endoparasite, *Aploymyiopsis epilachnae* (Aldrich) (Diptera: Tachinidae), was imported from central Mexico, but failed to establish in the United States (Smyth, 1923; Landis and Howard, 1940). In the 1940s, another tachinid endoparasite of MBB larvae, *Lydinolydella metallica* Townsend, was imported from Argentina and Brazil, but also failed to establish (Berry and Parker, 1949). Establishment failures apparently were due to inabilities to overwinter in the United States. In 1966, the gregarious larval endoparasite *P. foveolatus* was imported from India and released in several eastern states beginning in 1967 (Angalet et al., 1968). It caused high rates of MBB parasitism when released in infested soybean fields, but also failed to overwinter (Stevens et al., 1975a).

Inabilities to establish the introduced parasites, despite their apparently high parasitism potentials during the growing seasons, were attributed to either lack of climatic tolerances or the absence of susceptible MBB stages during the winter months. To test the former hypothesis, a more cold-tolerant strain of *P. foveolatus* was imported from Japan in 1978, and released from Mississippi to Wisconsin during 1980 (Schaefer et al., 1983). Surveys during 1981 failed to detect the parasite in any release area, supporting the hypothesis that the absence of susceptible MBB larvae during the winter months probably was the primary obstacle hindering parasite establishment in the United States. To utilize these parasites, they consequently must be cultured with MBB in the laboratory during the winter and released in the field during the spring and summer.

Parasite releases are divided into two types: inoculation and inundation (DeBach and Hagen, 1964). Inundation involves the release of a large number of laboratory reared parasites into a field to suppress the density of a pest immediately. The progeny of the released parasites usually are not relied upon for continued control of the pest.

In inoculation, relatively small numbers of parasites are released, whose progeny then increase in the field to suppress the density of the pest eventually. Inoculative parasite releases against MBB in soybean fields is the most appropriate release strategy because of the high costs of rearing parasites in the laboratory and the seasonal changes in density of MBB populations (i.e., low density first generation, high density second generation). However, numerous biological, logistical, and economic considerations are involved in the development and implementation of such a release program.

DEVELOPMENT OF *P. FOVEOLATUS* INOCULATIVE RELEASE STRATEGIES

Because *P. foveolatus* exhibited potential for high rates of MBB parasitism following field releases, studies were initiated in Maryland in 1972 to develop and evaluate an inoculative release program (Stevens et al., 1975a,b). Basic biology studies on *P. foveolatus* and its relationship with MBB were conducted to facilitate mass rearing operations and field evaluations (Bledsoe et al., 1983; Flanders, 1984; Stevens et al., 1975b). Females prefer to oviposit in third and fourth instar MBB, but may also parasitize second instars and prepupae (Angalet et al., 1968). Following parasite oviposition, MBB larvae continue to feed for approximately six days at 25 C. Parasitized larvae then attach themselves to plant surfaces as if to pupate, but succumb to the internal feeding of the immature parasites. The attached larvae become abnormally swollen and their integuments turn a characteristic brown (mummy stage). When development is complete, parasite adults emerge from mummies by chewing small exit holes through their hosts' dried integuments.

Stevens et al. (1975a) reported results of field trials conducted on the DEL-MARVA (Delaware, Maryland, and Virginia) Peninsula from 1972 to 1974. Late spring and early summer releases of *P. foveolatus* caused from 90 to 100% parasitism in release and surrounding soybean fields by the end of the growing season (September). To foster the establishment and increase of parasites following release, early plantings of small plots of snap beans adjacent to soybean fields were recommended. MBB populations developed more rapidly in the snap bean plots than in soybean fields, allowing earlier releases and an apparently extra generation of the parasite. High numbers of parasites then dispersed from plots to surrounding soybean fields. Based on high parasitism rates in soybean fields, and observations that insecticide usage in the release area during the studies had decreased compared to previous years, Stevens et al. concluded that annual inoculative releases of *P. foveolatus* were capable of suppressing Mexican bean beetle in soybean fields. Subsequently, Reichelderfer (1979) analyzed the economic feasibility of inoculative releases in soybean fields. Assuming that such releases were as effective as insecticide applications, Reichelderfer concluded that adoption of the release program would beneficially reduce insecticide use on soybeans.

However, several questions and problems arose from the initial studies conducted by Stevens et al. (1975a) when further studies were attempted. First, the reported parasitism estimates were derived from timed searches in individual fields where only parasitized and nonparasitized MBB larvae were collected. No data were presented on age class distributions of MBB populations, on defoliation levels in the sampled fields, or on the growth stages of the soybean plants when the samples were taken. Since MBB generations are relatively discrete, and parasitized larvae and mummies remain visible longer than do nonparasitized larvae, the high rates of parasitism that were reported may have been due to mummies accumulating in fields while nonparasitized MBB larvae continued to develop to nonsusceptible and nonsampled stages (i.e., pupae and adults). In addition, parasitism was highest during September when the soybean plants probably were near or beyond the stage where defoliation would have significantly affected yield. Since no soybean defoliation data were reported, relationships between parasitism rates, MBB densities, and defoliation levels could not be evaluated. Additionally, the costs of planting and maintaining nurse-plots were not evaluated relative to increased probabilities of parasite establishment and subsequent MBB suppression, or to any other economic or logistic considerations. Other unanswered questions included parasite release rates, prediction of the need for MBB suppression in specific years, areas, or fields to justify releases when host densities are noneconomic, and the number of release sites or nurse plots required per area of soybean plants.

INOCULATIVE RELEASE EVALUATIONS IN INDIANA

Based on questions that arose from previous investigations, studies were conducted from 1980 to 1982 in southern Indiana to evaluate and develop a *P. foveolatus* release program in midwestern soybean fields. However, the objectives were not completely satisfied because of adverse weather conditions that arose during the first two years of the study. Despite these problems, considerable data were obtained on the effectiveness of parasite releases in suppressing MBB densities and defoliations, and on the functioning and use of nurse-plots.

1980 and 1981 Studies

In 1980, studies were formulated to evaluate parasite releases in fields with nurse-plots planted 2 to 4 weeks before soybean planting, with nurse-plots planted at the time of soybean planting, and without nurse-plots. The early plots were planted by broadcasting snap bean seed in a 23 by 23 m area of the field and tilling them into the soil. The plots planted at the time of soybean planting had snap bean seed interplanted in the soybean rows in the same sized area as the early plots. Assuming that parasite dispersal is primarily by wind and since the dominant wind direction is from the southwest, all plots were planted in the southwest corners of soybean fields. Thirty-two commercial soybean fields were studied and each nurse-plot type was replicated eight times. Sampling began when the first trifoliolate leaves appeared on plants (vegetative stage, V1; Fehr et al., 1971) in a field or nurse-plot, and continued weekly until soybean pods were filled (reproductive stage, R6). The sampling unit was a single soybean plant. Sample numbers varied from 30 to 150 per field depending on soybean growth stage, MBB density, and field size (10 to 50 ha). The number of MBB in each life stage, number of parasitized larvae (mummies), soybean growth stage, and percentage defoliation (visual estimate) were recorded for each sample. When third and fourth instar (large) larvae were found in a field or nurse-plot, 1000 to 5000 parasites were released. Releases were made in only half the fields of each plot type; the remaining served as controls.

Parasite establishment was confirmed in each release field. However, drought conditions during mid-July and August reduced MBB densities in all fields, and the impacts of the parasite releases could not be ascertained. Mexican bean beetle populations did not reach high enough densities to cause economically significant defoliations in any of the fields.

In 1981, 24 fields were studied. Investigations were to be the same as in 1980, except that all fields were to possess nurse plots planted either at or before soybean planting. However, frequent rains during May disrupted these plans. Soybean and nurse-plot plantings were delayed by 2 to 6 weeks. This delay significantly affected

MBB densities in soybean fields and snap beans. Six early nurse-plots were planted during the first week of May, before the wet weather began. In the absence of surrounding soybean plants, these plots were heavily invaded by MBB adults emerging from overwintering sites. Subsequent feeding damage destroyed the plots before parasites could be released and before the adjacent soybean fields were planted. The remaining nurse-plots, and soybean fields that were planted after the first of June, had very few MBB invade them. Parasite releases either could not be made or when made did not result in establishment because of low MBB densities. MBB densities in August and September were the lowest observed in southern Indiana for several years.

Although the impact of *P. foveolatus* releases on MBB population densities in soybean fields could not be evaluated during 1980 and 1981, several conclusions about the functioning and use of snap bean nurse-plots were possible. Since snap bean plants are preferred to soybean plants by ovipositing MBB adults, higher MBB densities developed in nurse-plots than in surrounding soybean fields. Parasite establishment and increase were aided by high MBB densities on snap bean plants, apparently because susceptible hosts were more easily located by ovipositing females than if releases were made directly into sparsely infested soybean fields. Additionally, the earlier a nurse-plot was planted, the higher the density of MBB that developed in it. The emergence of overwintering MBB adults is stimulated by climatic conditions, and does not necessarily correspond with the availability of host plants (Douglass, 1933). If adults did not locate host plants within a certain time after emergence, they died without ovipositing. In southern Indiana, MBB emergence began about 1 May and continued until about 1 June. Therefore, beans planted after 1 June were invaded by fewer MBB than those planted during the first of May. Consequently, nurse-plots planted before soybean fields exibited much higher MBB densities. Although nurse-plots planted at the same time as soybeans had lower MBB densities than earlier planted plots, the densities in the plots corresponded more closely to those in the adjacent soybean fields. Consequently, densities in nurse-plots planted at the time of soybean planting may indicate the subsequent damage potential of MBB in adjacent soybean fields, and these densities might be used to predict the need for parasite releases. Additionally, nurse-plots planted at the same time as soybeans reduced various logistic, weed control, and equipment problems that plagued earlier plantings.

1982 Studies

In 1982, 24 soybean fields in southern Indiana were used to evaluate parasite releases. Eighteen fields were planted before June 1, and the remaining were double-cropped (planted during late June following harvest of winter wheat, *Triticum aestivum* L.). Nurse-plots were planted by broadcasting snap bean seed in the southwest corners of all fields immediately before the planting of soybeans. The sampling design was similar to previous years, except that the sampling unit consisted of 0.306 m of soybean row. Twelve of the fields received parasite releases. In early-planted release fields, 5,000 to 7,000 parasites were released per nurse-plot when MBB populations primarily consisted of second and third instar larvae (late June). Because insufficient numbers of MBB invaded nurse-plots of the double-cropped fields during the first MBB generation, parasites were released during the second generation (late July and early August). Weather conditions during the spring and summer of 1982 were more "normal" than experienced during the previous two years, and MBB densities in fields and nurse-plots increased accordingly.

Parasite releases in the double-cropped soybean fields were ineffective, and all of these fields received insecticide treatments by late August. The failure of the parasite to suppress MBB densities economically was a result of the way these fields were invaded by the pest relative to the attributes of the parasite. There were very low MBB densities in these fields during June and July, due to the relationship between planting date and emergence of overwintering MBB adults. Invasion of these fields from late July to early August were by first generation adults that developed in surrounding soybean fields. Since the parasite only attacks large MBB larvae, it had insufficient time to increase in the nurse-plots and disperse across the fields to suppress MBB densities adequately. Observations appeared to indicate that such high invading densities may be suppressed by distributing large numbers of parasites uniformly across fields (i.e., no nurse-plots), but the costs for such high parasite numbers greatly exceeds those of applying equivalently effective insecticides.

In the early-planted fields, the parasites appeared to suppress MBB densities, compared with the control fields. Approximately half of the control fields, but none of the release fields, required insecticide treatments in late August. In release fields, the suppression potential of the parasite appeared to depend on the density of the first MBB generation. When first generation densities were high in release fields, parasitism rates during the second MBB generation were higher than in fields with lower first generation densities. This relationship is illustrated in Figures 1 to 3 for three representative fields. In all of these fields, small MBB larvae increased from the first to the second generation (Figure 1). However, densities of large larvae decreased in Field 23, increased moderately in Field 14, and greatly increased in the control field (Figure 2). Between release fields (14 and 23), density differences in large larvae between generations appeared to depend on differences in rates of parasitism (Figure 3). It appeared that the higher the MBB density, the greater the density increase and impact of the parasite.

Parasite releases had less of an impact on MBB defoliations than on densities (Figure 4). Soybean defoliation trends corresponded with occurrences of large MBB larvae and adults. Since parasitized MBB larvae continued to feed, parasitism rates and defoliation levels were poorly related. However, some effects of the parasite on MBB defoliation were discernible. Defoliation in Field 23 was higher throughout most of the season than in the other two fields, despite the higher parasitism rates. However, percentage defoliation in the control field greatly increased at the end of the season, and eventually exceeded those of the release fields. High defoliation of the control field on the last sampling date dictated the need for an insecticide application, but neither of the release fields required such a treatment. Without the parasite releases, the release fields probably would have been sufficiently defoliated to require insecticide applications. Therefore, parasite releases appeared to suppress MBB defoliations, but this was not as clear as density suppressions.

CONCLUSIONS

Inoculative releases of *P. foveolatus* may be effective in suppressing MBB densities and defoliations in soybean fields, but considerably more research is required to refine such releases for commercial use. A major problem is the inability to predict second generation MBB densities and defoliation potentials during the first generation. To justify parasite releases during the first MBB generation, such predictive capabilities must be developed.

The need for nurse-plots also must be studied further, as results in 1982 indicated that MBB densities in soybean fields that appear sufficient to justify parasite releases also may be sufficient to release and establish parasites directly into soybean fields.

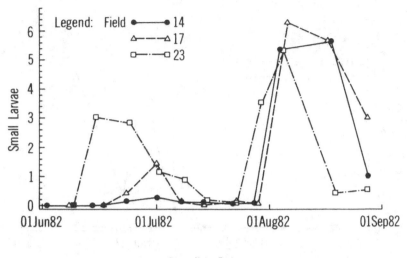

Figure 1. Density of small Mexican bean beetle larvae (first and second instars) per 0.306 m of soybean row in two *Pediobius foveolatus* release fields (14 and 23) and a nonrelease control field (17). Owen County, Indiana, 1982.

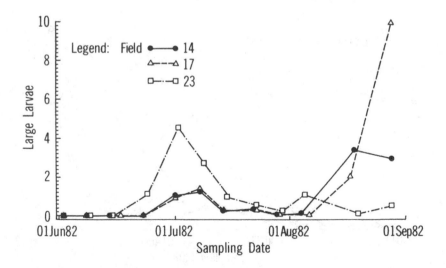

Figure 2. Density of large Mexican bean beetle larvae (third and fourth instars) per 0.306 m of soybean row in two *Pediobius foveolatus* release fields (14 and 23) and a nonrelease control field (17). Owen County, Indiana, 1982.

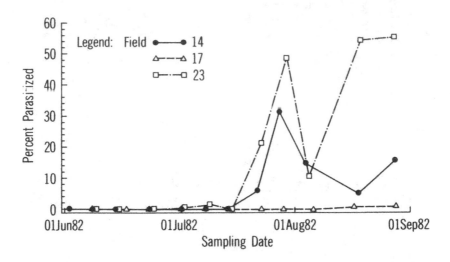

Figure 3. Percentage of large Mexican bean beetle larvae parasitized (mummified) by *Pediobius foveolatus* in two parasite release fields (14 and 23) and a nonrelease control field (17). Owen County, Indiana, 1982.

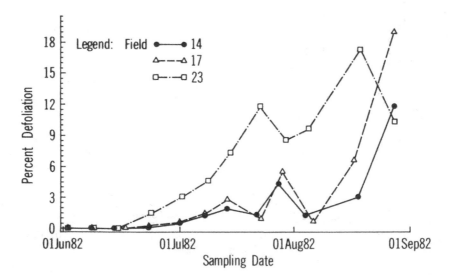

Figure 4. Percentage soybean defoliation by Mexican bean beetle in two *Pediobius foveolatus* release fields (14 and 23) and a nonrelease control field (17). Owen County, Indiana, 1982.

Soybean field releases, without nurse-plots, would nullify temporal lags that occur when parasites are initially established in nurse-plots and then must subsequently disperse through the field. If nurse-plots were eliminated, the economic costs and other problems associated with *P. foveolatus* releases would be significantly reduced, making the program more economically competitive with insecticide applications.

NOTES

Robert V. Flanders, Department of Entomology, Purdue University, W. Lafayette, IN 47907. This research was partially supported by USDA-APHIS(PPQ) Cooperative Agreement No. 12-16-5-2246. Authorized for publication as Journal Paper No. 10,212 by the Purdue Univ. Agric. Exp. Stn. Critical reviews of this manuscript by M. C. Wilson and R. J. O'Neal were appreciated.

REFERENCES

Angalet, L. M., L. W. Coles and J. A. Stewart. 1968. Two potential parasites of the Mexican bean beetle from India. J. Econ. Entomol. 61:1073-1075.

Berry, P. A. and H. L. Parker. 1949. Investigations on a South American *Epilachna* sp. and the importation of its parasite *Lydinolydella metallica* TNS. into the United States. Washington Entomol. Soc. Proc. 51:93-103.

Bledsoe, L. W., R. V. Flanders and C. R. Edwards. 1983. Morphology and development of the immature stages of *Pediobius foveolatus* (Hymenoptera: Eulophidae). Ann. Entomol. Soc. Am. 76:953-957.

Chittenden, F. H. and H. O. Marsh. 1920. The bean ladybird. USDA Agric. Bull. 843.

Davis, J. J. 1925. The Mexican bean beetle in Indiana. Ind. Agric. Exp. Stn. Circ. 126.

DeBach, P. (ed.). 1964. Biological Control of Insect Pests and Weeds. Chapman and Hall, London.

DeBach, P. and K. S. Hagen. 1964. Manipulation of entomophagous species. p. 429-458. *In* P. DeBach (ed.) Biological control of insect pests and weeds. Chapman and Hall, London.

Douglass, J. R. 1933. Habits, life history, and control of the Mexican bean beetle in New Mexico. USDA Tech. Bull. 376.

Fehr, W. R., C. E. Caviness, D. T. Burmood and J. Pennington. 1971. Stage of development descriptions for *Glycine max* (L.) Merrill. Crop Sci. 11:929-931.

Flanders, R. V. 1984. Comparisons of bean varieties currently being used to culture Mexican bean beetle (Coleoptera: Coccinellidae). Environ. Entomol. 13:995-999.

Gordon, R. D. 1975. A revision of the Epilachninae of the western hemisphere (Coleoptera: Coccinellidae). USDA Tech. Bull. 1493.

Graf, J. E. 1925. Climate in relation to Mexican bean beetle distribution. J. Econ. Entomol. 18:116-121.

Howard, N. F. 1931. Mexican bean beetles spread checked in 1930 by drought and heat. p. 375-376. *In* USDA yearbook of agriculture. U.S. Government Printing Office, Washington, D.C.

Howard, N. F. 1941. Feeding of the Mexican bean beetle larva. Ann. Entomol. Soc. Am. 34:766-769.

Howard, N. F. and B. J. Landis. 1936. Parasites and predators of the Mexican bean beetle in the United States. USDA Agric. Circ. 418.

Landis, B. J. and N. F. Howard. 1940. *Paradexodes epilachnae*, a tachinid parasite of the Mexican bean beetle. USDA Tech. Bull. 721.

Reichelderfer, K. H. 1979. Economic feasibility of a biological control technology: using a parasitic wasp, *Pediobius foveolatus*, to manage Mexican bean beetle on soybeans. USDA Agric. Econ. Report 430.

Smyth, E. G. 1923. A trip to Mexico for parasites of the Mexican bean beetle. Wash. Acad. Sci. J. 13:259-260.

Schaefer, P. W., R. J. Dysart, R. V. Flanders, T. L. Burger and K. Ikebe. 1983. Mexican bean beetle (Coleoptera: Coccinellidae) larval parasite *Pediobius foveolatus* (Hymenoptera: Eulophidae) from Japan: Field release in the United States. Environ. Entomol. 12:852-854.

694

Stevens, L. M., A. L. Steinhauer and J. R. Coulson. 1975a. Suppression of Mexican bean beetle on
soybeans with annual inoculative releases of *Pediobius foveolatus*. Environ. Entomol. 4:947-952.
Stevens, L. M., A. L. Steinhauer and T. C. Elden. 1975b. Laboratory rearing of the Mexican bean
beetle and the parasite, *Pediobius foveolatus*, with emphasis on parasite longevity and host-
parasite ratios. Environ. Entomol. 4:953-957.

PESTICIDE COMPATIBILITY IN SOYBEAN PEST MANAGEMENT

Kenneth V. Yeargan

Herbicides represent, by far, the most frequently used class of pesticides on soybean. During the past decade, the proportion of soybean plantings treated with herbicides increased from about 70% to more than 90% (USDA, 1983). During that same period, the proportion of soybean hectares treated with insecticides increased slightly, from about 8% to 12%. Use of fungicides and nematicides on soybean remains relatively low.

If one conservatively estimates an annual soybean production of 25 million hectares in the United States, then about 2.7 million ha are likely to receive applications of at least two classes of pesticides; namely herbicides and insecticides (25 million hectares X 0.9 X 0.12). Furthermore, herbicide applications frequently consist of two or more different compounds applied in combination or succession. For example, the Kentucky Cooperative Extension Service's 1984 recommendations for weed control in soybean lists 27 different options for preplant incorporated or preemergence herbicide use; 17 of those options consist of combinations of two or more different herbicides. Weed scientists have long been concerned with the study of the relative efficacy and safety to the crop plant of such herbicide combinations. More recently, weed scientists, entomologists, and others have begun to address questions concerning possible interactions between different classes of pesticides, such as herbicides, insecticides, and nematicides. Most of this work to date has concentrated on possible adverse effects to the crop plant from use of combinations of pesticides which were known to cause little adverse effect when used alone or in certain combinations with other products within their pesticide class.

Another area of concern regarding pesticide use on soybean is that of possible effects on nontarget organisms. Since the earliest use of modern pesticides, it has been obvious that an acceptable pesticide must control the target organism(s) and must not harm the crop being protected from pests. Likewise, dangerous levels of pesticide residue on the harvested product were unacceptable. Even some of the more subtle effects of usage of certain pesticides, such as pest resurgence or secondary outbreaks due to disruption of natural biological control, were recognized long ago by the more perceptive scientists, and are widely appreciated today. Recently, crop protection scientists have begun to investigate the repercussions of pesticide applications beyond traditional boundaries implied by disciplinary units, such as entomology or plant pathology. For example, while it is clear that broad spectrum insecticides are likely to kill beneficial insects along with pests, scientists have only recently addressed questions, such as the possible effects of postemergence herbicides on beneficial insects or the potential for fungicides directed at plant pathogens to adversely affect beneficial

696

entomopathogens that occur in soybean fields.

In a review of pesticide interactions in soybean, which covered information published through 1978, Newsom (1979) concluded that "the known and possible interactions within and between complexes of pests and tactics used for their control overwhelm the imagination" and that interdisciplinary research efforts will be required to resolve many of these questions. It is the purpose of this article to review progress made in this area since 1978, with primary emphasis on effects of interactions between classes of pesticides on the soybean plant and effects of pesticides on nontarget organisms.

RESPONSE OF SOYBEAN PLANTS TO COMBINATIONS OF PESTICIDES

As noted earlier, weed scientists, entomologists, and others have long been aware of potential crop phytotoxicity from combinations of pesticides. Much of the research in this area, however, has dealt with combinations of pesticides within a particular class, particularly herbicides. More recently, it has been shown that combinations of pesticides from different classes (e.g., herbicides with insecticides) may cause phytotoxicity to soybean, even when each pesticide is used at rates that cause no apparent problems when they are used separately. Those studies have concentrated on potential synergistic interactions between selected herbicides and pesticides which have insecticidal, nematicidal, or combined insecticidal-nematicidal activity. Selection of those classes of pesticides for study probably reflected their greater potential for combined use as compared with other types of pesticides (e.g., fungicides). Also, earlier studies on other crops had demonstrated phytotoxic, synergistic interactions between certain herbicides and organophosphate insecticides (Putnam and Penner, 1974).

Herbicide application is most likely to coincide with insecticide or nematicide application prior to the emergence of the soybean plant, because the majority of herbicides used in soybean production are applied as preplant incorporated or preemergence treatments. Insecticides or nematicides that may be applied at that time are primarily granular formulations of certain systemic carbamate or organophosphate compounds. Another possible, but less frequent, scenario is the concurrent application of postemergence herbicides (e.g., bentazon) and foliar insecticides. The effects of both types of combinations have been studied in recent years.

During 1975 and 1976, investigators in Kentucky (Hayes et al., 1979) found that combinations of metribuzin herbicide with either phorate or disulfoton interacted synergistically to significantly reduce both soybean plant population and yield in field plots. This occurred even at recommended rates of phorate and metribuzin. They also found that similar combinations of those insecticides with another widely used herbicide, linuron, did not adversely affect the crop when used at recommended rates. That study demonstrated the risks for soybean injury and yield reduction if soybean growers unwittingly followed weed and insect control recommendations which had been developed independently by scientists in different disciplines. It also signaled the lack of sufficient information on possible pesticide interactions and the need for an interdisciplinary approach to the problem.

In 1976, an interdisciplinary team of researchers at North Carolina State University began to study pesticide interactions in soybean (Bradley et al., 1979), and similar investigations have been reported recently by scientists in several other states. A common thread in most of those reports has been the occurrence of synergistic interactions of metribuzin with various organophosphate insecticides or nematicides, coupled with the lack of such interactions between metribuzin and carbamate insecticides, or between most other herbicides studied and organophosphate pesticides. Before considering specific results from those studies, it is worthwhile to summarize briefly the status of metribuzin use in soybean plantings.

Metribuzin is an effective herbicide for many weeds, and it is widely used; in 1982 metribuzin was second only to trifluralin in terms of the number of hectares of soybean treated (USDA, 1983). Nonetheless, soybean tolerance to metribuzin can be marginal under certain conditions. For example, coarse textured soils having low organic matter content, rainfall immediately following herbicide application, cultivar susceptibility, and planting depth have been implicated as possibly affecting soybean injury from metribuzin (Coble and Schrader, 1973). These factors should be kept in mind when evaluating interactions of metribuzin with other pesticides.

Specific organophosphate insecticides or nematicides which have been shown to produce adverse synergistic interactions in soybean are listed in Table 1, along with specific pesticides which have generally failed to produce such interactions. These results represent studies in five states (Georgia, Kentucky, North Carolina, Ohio, and Tennessee) involving several different soybean cultivars and soil types. With the exception of phenamiphos, all of the organophosphate insecticides or nematicides interacted with metribuzin to produce adverse effects on soybean. Conversely, the carbamate pesticides, aldicarb and carbofuran, generally did not cause adverse plant responses when used with metribuzin. The only exception involved a carbofuran-metribuzin combination in Georgia. In that study, several other factors may have influenced the results, including the sandy loam soil, the soybean cultivar (Tracy, which is known to be sensitive to metribuzin), and the application of 1.5 cm of irrigation water within 12 h of herbicide application (Dowler and Minton, 1979). Those investigators found

Table 1. Insecticides, nematicides, or insecticide-nematicides whose potential interactions with metribuzin in soybean have been reported.

Pesticide studied	Adverse response [a]	Type of study [b]	Reference
DBCP	+ [c]	F and G	Dowler and Minton, 1979
disulfoton	+	F and G	Hayes et al., 1979
	+	F and G	Hammond, 1983
ethoprop	+	F	Hayes et al., 1980
fensulfothion	+	F	Bradley et al., 1979
phorate	+	F	Bradley et al., 1979
	+	F and G	Hayes et al., 1979
	+	G	Hammond, 1983
terbufos	+	F	Hayes et al., 1980
	+	G	Hammond, 1983
aldicarb	-	F	Bradley et al., 1979
	-	F and G	Hammond, 1983
carbofuran	+ [c]	F	Dowler and Minton, 1979
	-	F	Hayes et al., 1980
	-	G	Hammond, 1983
phenamiphos	-	F	Hayes et al., 1980
	-	F and G	Hammond, 1983

[a] Adverse response greater than that caused by individual pesticides; + = yes, - = no.
[b] F = field study; G = greenhouse study.
[c] Adverse response obtained only on one of two cultivars studied and only during one of the two years of the study.

that another cultivar (GaSoy-17) did not show adverse response to carbofuran-metribuzin combinations in either of the two years of their study, and that Tracy showed no adverse response to that pesticide combination when the soybeans germinated prior to application of irrigation water or occurrence of significant rainfall.

Bradley et al. (1979), Hayes et al. (1979), Hayes et al. (1980), and Hammond (1983) independently studied several pesticide combinations involving metribuzin, and their results were generally in agreement when similar combinations were involved. For example, metribuzin-aldicarb combinations did not cause apparent soybean phytotoxicity in field studies by Bradley et al. (1979) in North Carolina or by Hammond (1983) in Ohio.

At-planting applications of insecticides and nematicides are not widely used in soybean production in the USA at this time. Use of preventative applications of insecticides is seldom justified economically, given the sporadic nature of insect infestations on this crop and the availability of effective alternative insecticides that can be applied as foliar sprays when needed. The availability of alternative control measures for nematodes (e.g., resistant cultivars) and economic considerations have suppressed the use of nematicides on soybean crops. Changes in current insect or nematode pest status, however, could lead to increased frequency of at-planting insecticide or nematicide applications, and knowledge of potential interactions of these pesticides with herbicides could prevent unnecessary direct crop injury from pesticide interactions.

In addition to the previously mentioned herbicide applications before soybean seedling emergence, increased use of postemergence herbicide applications is anticipated as new products become available. One of the postemergence herbicides most commonly used in soybean is bentazon. In a recent study, Campbell and Penner (1982) found that postemergence applications of bentazon in combination with certain organophosphate insecticides (malathion, parathion, or diazinon) caused severe injury to soybean plants even when applications were separated by 48 h. Similar treatments involving bentazon and carbamate insecticides (carbaryl or carbofuran) did not cause injury. Although the mechanism of the bentazon-organophosphate insecticide interaction was not determined, those investigators suggested that the insecticide probably interfered with metabolism of bentazon by the soybean plant.

From these studies, it is clear that several insecticides and nematicides are compatible with certain herbicides, whereas other combinations can cause direct injury to soybean. Although many potential combinations of pesticides from different classes remain to be studied, currently available data provide a general pattern that should be useful in selecting compatible combinations. If either preemergence applications of metribuzin or postemergence applications of bentazon are used, one would be well advised to avoid use of organophosphate insecticides or nematicides. The organophosphate nematicide, phenamiphos, appears to be an exception. Given the current availability of alternative carbamate and synthetic pyrethroid insecticides, and carbamate nematicides, many options exist for the grower who chooses to use metribuzin or bentazon herbicides. It is important that soybean growers be made aware of these alternatives and the possible consequences of ignoring them.

EFFECTS OF PESTICIDE APPLICATIONS ON NONTARGET ORGANISMS

In addition to avoiding possible adverse effects of pesticide combinations on the soybean plant, compatibility of pesticides in soybean pest management also implies minimizing undesirable impact of pesticide applications on other control tactics, such as biological control. Whether applied singly or in combinations, pesticides may affect nontarget organisms. The potential effects of broad-spectrum insecticides on beneficial arthropods are well known, but only recently have investigators begun to explore possible effects of pesticides on organisms that are not closely related to the target pests (e.g., effects of herbicides on nematodes or insects).

Resurgence of insect pest populations has been documented in soybean crops following foliar application of certain insecticides. Shepard et al. (1977) reported increased populations of the green cloverworm, *Plathypena scabra* Fabricius, the corn earworm and its congeners, *Heliothis* spp., the velvetbean caterpillar, *Anticarsia gemmatalis* Hubner, the soybean looper, *Pseudoplusia includens* (Walker), and the Mexican bean beetle, *Epilachna varivestis* Mulsant, in plots treated with foliar applications of methomyl plus methyl parathion, which they compared with untreated plots, in South Carolina. A similar application of monocrotophos resulted in little or no resurgence of pest populations in that study. In North Carolina, soybean plots treated with a foliar application of methyl parathion had approximately seven times more corn earworms, *Heliothis zea* (Boddie), than did untreated plots 2 weeks after treatment (Morrison et al., 1979). Those investigators attributed this resurgence to the drastic reduction of hemipteran predator populations that occurred in the methyl parathion-treated plots. Farlow and Pitre (1983b) did not observe substantially greater populations of lepidopteran larvae in methyl parathion-treated soybean plots in Mississippi, but that result seemed to be due to low overall populations of lepidopteran larvae at the study site. Thus, the possibility of pest population resurgence following foliar insecticide application in soybean has been demonstrated, but its occurrence is dependent on several factors, including the specific insecticide(s) used and the magnitude of the pest population available to colonize the plots following treatment.

It has also been demonstrated that applications of systemic insecticides or nematicides at planting may result in increased soybean insect pest populations later in the season. Morrison et al. (1979) showed that when the insecticide-nematicide aldicarb was applied in-furrow at planting, corn earworm populations were as much as seven times greater than in untreated soybean plots or those treated with certain other soil-applied pesticides, such as phorate or carbofuran. They also found that aldicarb was the only soil-applied pesticide in the study that drastically reduced hemipteran predator populations, and that its effect was as great as that of foliar applications of methyl parathion. Furthermore, they found that the soybean yield in aldicarb-treated plots was significantly lower than in untreated plots. Lentz et al. (1983) failed to obtain similar results in Tennessee, but noted that in their study early season predator population reductions due to soil-applied pesticides occurred well before expected late season increases in lepidopteran pest populations. Thus, they cautioned that their results might not apply to late-planted soybean, because in those cases predator populations might not have time to recover and prevent outbreaks of late-season insect pests. Farlow and Pitre (1983b) also failed to observe substantial late-season insect pest resurgence in aldicarb-treated soybean plots in Mississippi, but lepidopteran pest populations were generally quite low at their study site. In a similar study in Kentucky I found that aldicarb application at planting led to increased green cloverworm populations later in the season in only 1 of 3 years in which this was studied; this also apparently was attributable to generally low populations of potential colonizing lepidopteran pests. As has been observed with foliar insecticide applications, it appears that several factors may affect the mid- to late-season response of pest populations to soil-applied insecticides or nematicides. These include the type of pesticide used, the size of the pest population available for colonizing the field, and the amount of time elapsing between pesticide application and pest colonization. One scenario in which this type of pest resurgence may be likely is that of aldicarb application to late-planted soybean fields which are subsequently colonized by substantial lepidopteran pest populations.

Scientists have recently investigated the possibility that herbicides, alone or in combination with other types of pesticides, might affect nematode or insect populations in soybean. In a field study, Schmitt and Corbin (1981) reported that the herbicide alachlor, alone and in combination with phorate or fensulfothion, caused increases in nematode populations. The cause is unknown, but those authors noted that the effect may be complex, possibly involving alteration of bacterial populations, stimulation of nematode egg hatch, alteration of plant metabolism, or other factors. Farlow and Pitre (1983a) found that the predaceous hemipteran, *Geocoris punctipes* (Say), oviposited significantly greater numbers of viable eggs when exposed to plants treated with the herbicides bentazon or acifluorfen. They suggested that this may be an example of hormoligosis, a phenomenon in which certain biological processes are stimulated by sublethal quantities of a stressing agent. In that same study, no increases in field populations of *G. punctipes* were noted following herbicide application, perhaps due to increased oviposition being offset by increased mortality.

The introduction and increased use of the fungicide benomyl for plant pathogen control in soybean during the past decade raised concern among entomologists that the fungicide might adversely affect naturally-occurring fungal entomopathogens, possibly leading to increased populations of certain insect pests. As Newsom (1979) pointed out, initial studies generally did not bear this out. In a more recent investigation, Livingston et al. (1981) found that an epizootic of *Entomophthora gammae* (Weiser) in a soybean looper population in Arkansas was not affected by two or three applications of benomyl, even though those treatments were timed to maximize the possible suppressive action of the fungicide on the entomopathogen. Therefore, there seems to be little basis for concern over this particular fungicide-entomopathogen interaction.

The use of the insecticide toxaphene as a herbicide for the control of sicklepod, *Cassia obtusifolia* L., also caused concern among entomologists for a number of reasons. Newsom (1979) indicated that this pesticide could adversely affect insect parasite and predator populations in soybean crops. He also noted that unacceptably high residues of this organochlorine compound in fish had led to the closing of fishing in certain areas of Louisiana. The recent cancellation by the U.S. Environmental Protection Agency of the registration of toxaphene for most uses has perhaps rendered moot some of the concerns over its use on soybean crops in this country. However, it is worth noting the results of a recent study on this subject. Huckaba et al. (1983) reported that two applications of toxaphene on soybean plants in North Carolina had only a "modest effect upon population levels of the arthropod complex sampled." They noted that predator numbers were similar in treated and untreated plots, and they attributed observed differences in corn earworm populations to other factors. If toxaphene is used on soybean crops outside the United States in the future, either as a herbicide or insecticide, its potential effects on beneficial arthropods deserve further study in those countries. Even if its negative impact on biological control agents proves to be minor, scientists should not overlook other possible adverse effects, such as those which led to the cancellation of most of its uses in the United States.

CONCLUSIONS

Considerable progress has been made recently in identifying potential adverse effects on the soybean plant of interactions among selected pesticides belonging to different classes, particularly interactions between herbicides and insecticides or nematicides. The effective and popular herbicide metribuzin has been most frequently implicated in interactions with soil-applied, organophosphate insecticides or nematicides. Additional studies of these and other potential pesticide combinations are warranted in order to identify potentially damaging combinations and the mechanisms underlying the observed interactions.

Repercussions of pesticide application beyond the soybean plant and target pests have been documented. In addition to population resurgence of insect pests resulting from foliar applications of particular insecticides, similar resurgence has been observed under certain conditions following soil applications of aldicarb at planting. Also, certain herbicides have been shown to influence insect reproduction and the size of nematode populations.

Thus, in recent years, several previously unsuspected effects and interactions of pesticides have been demonstrated, some of which are potentially detrimental to soybean production. Conversely, several hypothesized interactions and effects of pesticides on nontarget organisms have proven to be unfounded, of limited significance, or important only under specific circumstances. Pesticides will continue to play a vital role in soybean production in the United States and many other parts of the world. Scientists engaged in the development and implementation of integrated pest management programs must continue to cross disciplinary boundaries in their studies of the compatibility of all pest management tactics, including use of pesticides. Information that is already available should be widely disseminated to growers, so that they can avoid needless losses due to the inadvertant selection of incompatible pesticide combinations or the use of single pesticides which may, under certain circumstances, adversely affect their soybean production system.

NOTES

Kenneth V. Yeargan, Department of Entomology, University of Kentucky, Lexington, KY 40546-0091.

REFERENCES

Bradley, J. R., Jr., J. W. Van Duyn, F. T. Corbin and D. P. Schmitt. 1979. Pesticide interactions: an IPM dilemma. p. 56-64. *In* H. D. Loden and D. Wilkinson (ed.) Report of the Ninth Soybean Seed Research Conference. American Seed Trade Assoc., Washington, D.C.
Campbell, J. R. and D. Penner. 1982. Enhanced phytotoxicity of bentazon with organophosphate and carbamate insecticides. Weed Sci. 30:324-326.
Coble, H. D. and J. W. Schrader. 1973. Soybean tolerance to metribuzin. Weed Sci. 21:308-309.
Dowler, C. C. and N. A. Minton. 1979. Selected pesticide interactions on two soybean varieties. p. 332. *In* Proc. 32nd Ann. Meeting Southern Weed Sci. Soc.
Farlow, R. A. and H. N. Pitre. 1983a. Bioactivity of the postemergence herbicides acifluorfen and bentazon on *Geocoris punctipes* (Say). J. Econ. Entomol. 76:200-203.
Farlow, R. A. and H. N. Pitre. 1983b. Effects of selected pesticide application routines on pest and beneficial arthropods on soybean in Mississippi. Environ. Entomol. 12:552-557.
Hammond, R. B. 1983. Phytotoxicity of soybean caused by the interaction of insecticide-nematicides and metribuzin. J. Econ. Entomol. 76:17-19.
Hayes, R. M., K. V. Yeargan, W. W. Witt and H. G. Raney. 1979. Interactions of selected insecticide-herbicide combinations on soybeans. Weed Sci. 27:51-53.
Hayes, R. M., G. L. Lentz and A. Y. Chambers. 1980. Response of soybeans to combinations of metribuzin and selected nematicides. p. 52. *In* Proc. 33rd Ann. Meeting Southern Weed Sci. Soc.

Huckaba, R. M., J. R. Bradley and J. W. Van Duyn. 1983. Effects of herbicidal applications of toxaphene on the soybean thrips, certain predators, and corn earworm in soybeans. J. Georgia Entomol. Soc. 18:200-207.

Lentz, G. L., A. Y. Chambes and R. M. Hayes. 1983. Effects of systemic insecticide-nematicides on midseason pest and predator populations in soybean. J. Econ. Entomol. 76:836-840.

Livingston, J. M., W. C. Yearian, S. Y. Young and A. L. Stacey. 1981. Effect of benomyl on an *Entomophthora* epizootic in a *Pseudoplusia includens* population. J. Georgia Entomol. Soc. 16:511-514.

Morrison, D. E., J. R. Bradley, Jr. and J. W. Van Duyn. 1979. Populations of corn earworm and associated predators after applications of certain soil-applied pesticides to soybean. J. Econ. Entomol. 72:97-100.

Newsom, L. D. 1979. Interactions of control tactics in soybeans. p. 261-273. *In* F. T. Corbin (ed.) World Soybean Res. Conf. II: Proceedings. Westview Press, Boulder, Colorado.

Putnam, A. R. and D. Penner. 1974. Pesticide interactions in higher plants. Residue Rev. 50:73-110.

Schmitt, D. P. and F. T. Corbin. 1981. Interaction of fensulfothion and phorate with preemergence herbicides on soybean parasitic nematodes. J. Nematol. 13:37-41.

Shepard, M., G. R. Carner and S. G. Turnipseed. 1977. Colonization and resurgence of insect pests of soybean in response to insecticides and field isolation. Environ. Entomol. 6:501-506.

U.S. Department of Agriculture. 1983. Pesticides. p. 4-13. *In* Inputs, outlook, and situation. IOS-2, October, 1983, Econ. Res. Serv., Washington, D.C.

BIOLOGICAL CONTROL OF SOYBEAN CATERPILLARS

Flávio Moscardi and Beatriz S. Corrêa Ferreira

Major soybean insect pests in Brazil are represented by defoliating caterpillars, especially the velvetbean caterpillar (VBC) (*Anticarsia gemmatalis* Hbn.) and stink bugs, mainly *Nezara viridula* (L.), *Piezodorus guildinii* (Westwood) and *Euschistus heros* (F.). These insects are distributed throughout the country's soybean growing area and frequently reach damaging levels, demanding control measures. The great majority of insecticide usage on the crop is directed against VBC and the stink bug complex.

Different integrated pest management (IPM) strategies, such as cultural practices, host plant resistance, and biological control have been investigated in Brazil, with the objective of implementing a soybean IPM program. Among the IPM strategies, considerable research effort has been concentrated on the biological control of insects. Efforts in this area have been dedicated especially to work on a nuclear polyhedrosis virus of *A. gemmatalis* (AgNPV).

STUDIES WITH THE AgNPV

Specificity

Laboratory trials with AgNPV on lepidopterous species such as *Bombyx mori* (L.) (silkworm), *Chlosyne lacinia saundersii* (Doubleday & Hewitson) (sunflower caterpillar), *S. latifascia* (pod worm), and *Trichoplusia ni* (Hbn.) (cabbage worm) showed that these species were infected only at very high doses, while the natural host (*A. gemmatalis*) presented high susceptibility to the pathogen (120- to 250,000-fold differences compared to the other species). The silkworm was the least susceptible, with 2.0 to 3.7% mortality at 2.5 to 3.0×10^6 polyhedron inclusion bodies (PIB)/larva, indicating that AgNPV would not represent a risk for the silk industry, even if used extensively on soybean crops near silkworm rearing facilities (Moscardi and Corso, 1981). Other studies, conducted in the U.S., also indicated low susceptibility of non-host species, such as *P. includens*, *Heliothis* spp., *Spodoptera* spp., and the sugarcane borer, *Diatraea saccharalis* (F.) to AgNPV (Carner et al., 1979; Pavan and Boucias, 1981).

Field Dosages

Previous field trials conducted on natural VBC larval populations in Florida, indicated that all virus doses tested, varying from 75 to 287.5 larval equivalents (LE)/ha, were sufficient to suppress VBC populations below damaging levels, with only one application during the soybean season (Moscardi, 1977; Moscardi et al., 1981). Efficient control was also obtained in South Carolina at 49 LE/ha (Carner and Turnipseed, 1977).

Trials conducted in Brazil, during the 1979/80 soybean season, with virus dosages varying from 10 to 320 LE/ha (1 LE = ca. 1.3 to 1.5×10^9 PIB), showed that substantial mortality (ca. 70%) was obtained even at the lowest dose. At 40 LE/ha, mortality was over 80% and approached 100% at the highest doses, if applications were directed against small VBC larvae (<1.5 cm) (Moscardi, 1983).

Leaf Consumption by Diseased Larvae

Even though high VBC mortality may be obtained with AgNPV, average time to death varies from 6 to 8 days, resulting in some leaf consumption during this period. Laboratory experiments with third instar VBC larvae indicated that mean leaf area consumption by diseased larvae was much reduced (27 cm^2) compared to consumption by healthy larvae (108 cm^2) (Moscardi, 1983). This reduction is related to the fact that larvae practically cease feeding by the 4th day after infection (Moscardi, 1977; Moscardi, 1983).

Effect of Host Age and Density on Virus Efficacy

Laboratory trials have shown a marked reduction in VBC susceptibility to AgNPV as the insect progresses in larval development (Table 1). Susceptibility decreased by as much as 40- to 50-fold from the first to fifth instar, with a very sudden decrease occurring after the 4th instar, indicating that field application of AgNPV should be directed against small VBC larvae (<1.5 cm) (Moscardi, 1983).

Since diseased larvae continue feeding for some days after application, it is important to define the maximum VBC larval population for virus use, in order to avoid soybean defoliation above critical levels. Artificial infestation of field-caged soybean plants by different VBC population densities, followed by virus applications, resulted in no significant soybean leaf area reduction for VBC levels up to 30 larvae/m of row, whereas at 40 larvae/m of row defoliation was significant (Moscardi, unpublished). For natural field situations the maximum level of 20 small VBC larvae/m of row was considered safe for virus use at the farmer level.

AgNPV Persistence on Soybean Leaves

Experiments on persistence on leaves of three AgNPV preparations indicated that the half-life of activity of purified virus was less than four days, whereas that of a crude

Table 1. Median Lethal Concentration (LC_{50}) of *Anticarsia gemmatalis* nuclear polyhedrosis virus at different larval instars of the host (Moscardi, 1983).

Larval instar	Larval size	Median Lethal Concentration (LC_{50})	
		PIB/ larva	PIB/mm^2 diet surface
	cm		
1st	<0.5	-	<2.0
2nd	0.5	9.3	3.1
3rd	1.0	28.0	13.0
4th	1.5	70.0	18.0
5th	2.0	445.0	121.0

preparation and a purified preparation with a clay-adjuvant was approximately 6 and 7 days, respectively, for a virus dose of 50 LE/ha (Moscardi, 1983). Although the half-life of the crude virus preparation is only about 6 days on leaves, experiments conducted at the farm level indicated that one AgNPV application was sufficient to maintain VBC populations below damaging levels for the remainder of the season. Persistence seems due to an increase in virus load on the crop, resulted from high quantities of PIB liberated from dead larvae (Moscardi, 1983).

AgNPV and *Nomuraea rileyi* Interactions

Since the fungus *N. rileyi* has an important role in natural regulation of VBC larval populations on soybean, laboratory and field studies were carried out to determine the potential for AgNPV application to interfere with development of *N. rileyi* epizootics. Seven doses of AgNPV (0.1 to 80 PIB/mm^2 of leaf surface) and four doses of *N. rileyi* (2.0 to 80 conidia/mm^2 of leaf surface) were tested singly or in combinations, in the laboratory, against 1.5 cm-VBC larvae. In field plots, virus was applied at 1.0×10^{11} PIB/ha and *N. rileyi* at 5.0×10^{11} conidia/ha, singly or in combinations against natural VBC populations.

Figure 1 shows results obtained in the laboratory with one dose of *N. rileyi* and different AgNPV doses (similar tendencies were observed for other *N. rileyi* doses). An apparent antagonism between the pathogens was observed, mainly in the simultaneous applications at the three lowest doses of the virus, because total mortality was lower than expected from additive effects of the pathogens. Furthermore, the virus substantially predominated over the fungus in effectiveness with simultaneous applications, larval mortality caused by *N. rileyi* being drastically reduced with increase in virus dose (Figure 1A). However, when the virus was offered to larvae 24 h after inoculation with *N. rileyi*, the fungus tended to cause mortalities comparable to that caused when it alone was applied, resulting in a consequent reduction in virus contribution to total mortality (Figure 1B). Similar results were obtained with applications of the two pathogens in field plots (Figure 2), indicating that virus applications in a large area may lead to a reduction of *N. rileyi* inoculum in situations where the virus is applied before or at the beginning of a *N. rileyi* epizootic on VBC larval populations on soybean. However, this effect may be reduced when a fungus epizootic is underway, especially in seasons with favorable climatic conditions for *N. rileyi*. Due to the importance of *N. rileyi* in preventing VBC outbreaks on soybean, these AgNPV and fungus relationships deserve further study.

Role of Predators on Virus Dissemination

Many predators have been observed feeding on virus-diseased VBC larvae, which are easy prey due to their loss of mobility by the fourth day after infection. Furthermore, phytophagous insects, such as healthy VBC larvae, crysomelid beetles (*Diabrotica speciosa* (Germ.) and *Colaspis* sp.), and stink bugs (*N. viridula* and *P. guildinii*), among others, also feed on moribund VBC larvae (personal observations). These facts may help to explain the sudden increase in virus-killed VBC in untreated areas that are near virus-treated fields, as has been observed in many instances in Brazil.

With the objective of better understanding the development of AgNPV epizootics in soybean fields, a study on the potential for virus dissemination by predators is underway at CNPS/EMBRAPA. Initial experiments have shown that virus PIB passing through the digestive tract of *P. maculiventris*, *Callida* spp., *L. concinna*, and *C. granulatum* were active against VBC larvae. Experiments in which predators, that previously fed on diseased larvae or were sprayed with a virus suspension, were liberated in

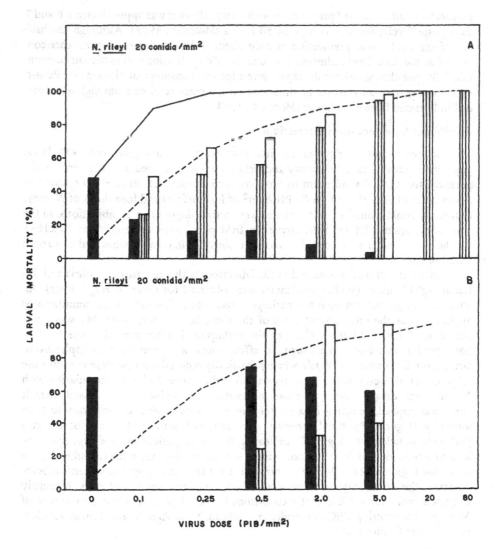

Figure 1. Mortality of third instar larvae of *Anticarsia gemmatalis* caused by one dose of *Nomuraea rileyi*, different doses of *A. gemmatalis*, nuclear polyhedrosis virus (AgNPV), and their combinations. A = simultaneous inoculation; B = AgNPV inoculated 24 h after *N. rileyi*; ■ = mortality by *N. rileyi*; ▥ = mortality by AgNPV; □ = total mortality; ---- = single effect of AgNPV; ——— = expected additive effect of the pathogens.

laboratory cages containing VBC larvae on potted soybean plants showed that *C. granulatum* and *L. concinna* induced ca. 75% and 45% virus mortality, respectively, on the insects. Virus mortality on VBC larvae was also induced when *L. concinna* ($10/m^2$) was liberated in 24 m^2-field cages. Studies under natural field conditions are to be started in the 1984-85 season.

Comparison of AgNPV Isolates

Initial experiments are being carried out at CNPS/EMBRAPA with five AgNPV isolates from distinct soybean-producing regions of Brazil, one isolate from Argentina

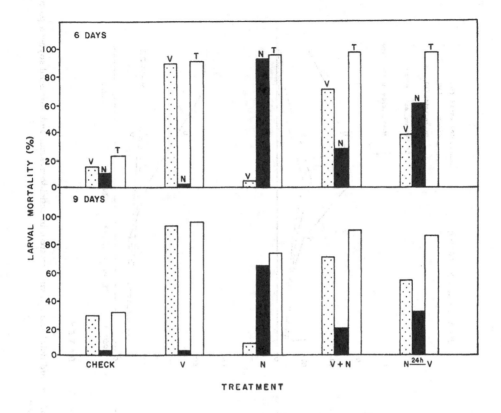

Figure 2. Mortality of *Anticarsia gemmatalis* larvae collected in soybean field plots after treatment with *Nomuraea rileyi* at 5.0×11^{11} conidia/ha, with *A. gemmatalis* NPV at 1.0×10^{11} PIB/ha, or with pathogen combinations. ▣ = mortality by AgNPV (V); ■ = mortality by *N. rileyi* (N); □ = total mortality (T).

and one from Uruguay. So far, these isolates have been tested at three doses against laboratory-reared VBC larvae, with no substantial differences in virulence being detected (Moscardi and Quintela, unpublished).

VIRUS UTILIZATION AT THE FARMER LEVEL

During the 1980-81 and 1981-82 soybean seasons, CNPS/EMBRAPA, in collaboration with the Extension Service (EMATER) and farmer cooperatives in the State of Paraná, carried out a pilot program with the objective of evaluating the feasibility of AgNPV use by farmers in lieu of chemical insecticides. The program consisted of using paired fields (virus treated, insecticide treated, and a check whenever possible) of at least a ha each on soybean farms in different regions of the state. Virus was applied at 50 LE/ha as a crude preparation when the majority of VBC larvae were still small (<1.5 cm), whereas insecticides were applied according to the IPM program. In the remainder of each soybean farm, the farmer himself would apply insecticide according to his own practice.

Results concerning VBC larval populations and defoliation occurring after application of the treatments are shown for two locations in Figure 3. In Ibiporá, the VBC

708

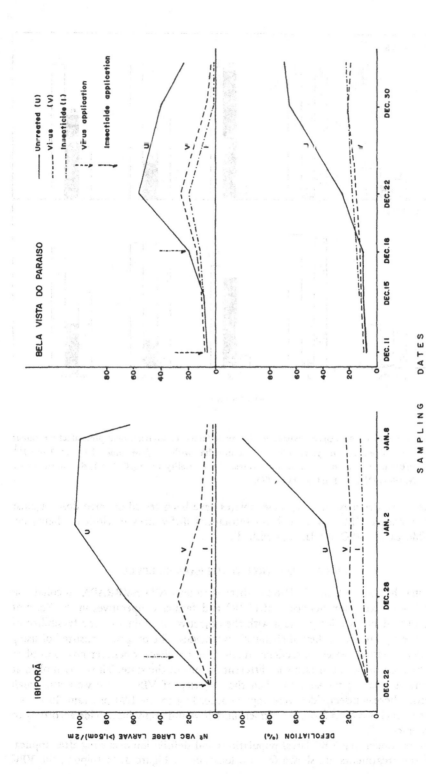

Figure 3. Number of velvetbean caterpillar (VBC) large larvae/2m of row and defoliation in virus-treated, insecticide-treated, and untreated soybean plots in Ibiporã and Bela Vista do Paraiso during the 1981-82 season.

population of large larvae ($>$1.5 cm) built up to high numbers, reaching over 100 specimens/2 m of row in the untreated area, whereas in virus- and insecticide-treated areas larval populations were much lower, remaining below the damage level (40/2 m of row). Furthermore, defoliation reached 100% in the untreated area compared to 10-20% in virus- and insecticide-treated areas. Similar results can be observed in Bela Vista do Paraiso.

Seed yield evaluation, as shown in Table 2 for some of the locations, reveals that there were not, in all cases, substantial differences between virus- and insecticide-treated areas. On the other hand, untreated areas in Ibiporã and Bela Vista do Paraiso showed significant yield reductions. Furthermore, in virus-treated areas the pathogen was applied only once, whereas one insecticide-treated area received two applications, and areas where the farmer applied insecticide according to his decision received two applications.

Results showed that the utilization of AgNPV at the farmer level is a viable alternative to chemical insecticides for controlling VBC larval populations on soybean in Brazil. The pilot program was conducted in other soybean-producing states with similar results.

During the pilot program in Paraná, demonstration areas distributed over the state were used by extension specialists for farmer education. The main objective was to show that the virus could be as effective as chemical insecticides in preventing yield

Table 2.   Number of applications and seed yield of soybean in virus-treated, insecticide-treated, and check areas for some locations in the State of Parana, Brazil.

Place	Treatment[a]	Number of applications	Seed yield
			kg/ha
Sertanópolis	V	1	2413
(1981)	I (IPM)	0	2560
	C	0	2535
	F	2	--
Primeiro de Maio	V	1	3120[b]
(1981)	I (IPM)	1	3120[b]
	F	2	--
Ibiporã	V	1	3913
(1982)	I (IPM)	1	3849
	C	0	949
	F	2	--
Bela Vista do Paraíso	V	1	3845
(1982)	I (IPM)	1	3840
	C	0	2848
	F	2	--
Rolândia	V	1	3250
(1982)	I (IPM)	2	3086
	F	2	-

[a]V = virus-treated area; I (IPM) = insecticide-treated area according to IPM recommendation; C = check area; F = insecticide-treated area according to farmer judgment.

[b]Farmer's estimate of yield.

loss by VBC, even though the pathogen would take longer to kill the host. Due to a favorable response by farmers, a simple and economical program for virus use was established in the 1981-82 season, in the state of Paraná.

The program consisted of expanding an existing VBC colony at CNPS/EMBRAPA for virus production, with the objective of providing extension service personnel and farmer cooperatives with an initial supply for further multiplication of the pathogen on naturally occurring VBC populations. Freshly dead larvae collected in virus-treated areas would be macerated, filtered and the crude suspension sprayed over much larger areas in the same season, or kept frozen for use the following season. Farmers, assisted by extension specialists, would learn the procedures so that virus use could progress rapidly in each region. Even though on a small scale, virus use in this way succeeded well in Paraná, and the program was adopted by other soybean-producing states, starting in Rio Grande do Sul, the largest producer, in 1982. A simple monitoring program was devised for testing the quality of the material being multiplied and collected under field conditions.

In the last season (1983-84) initial virus supplies were sent to different regions of all soybean producing states by CNPS/EMBRAPA. It is estimated that at least 11,000 ha of soybean plants were treated with the pathogen last season, with the majority of the area concentrated in Paraná and Rio Grande do Sul. Due to a high incidence of the host, large quantities of virus-killed larvae were collected under field conditions and stored. Virus sufficient for application to approximately 58,000 ha have been stocked in grower cooperatives or by farmers themselves. It is expected that the AgNPV-treated area in Brazil may reach 300,000 ha next season if VBC larval populations occur in sufficient numbers to allow virus multiplication at the farmer level.

STRATEGIES FOR INCREASING VIRUS USAGE

With the objective of increasing AgNPV usage on soybean in Brazil, regional virus-producing units are being developed in Paraná and other states. CNPS/EMBRAPA has provided training and technical assistance for institutions interested in continuous VBC mass rearing on an artificial diet for virus production. In Paraná, several grower cooperatives are already producing AgNPV, under laboratory conditions, for use next season. Similarly, in Rio Grande do Sul the State Extension Service, in conjunction with research institutions, is setting up facilities for production of 15,000 to 20,000 VBC-infected larvae a day.

Due to prospects for rapid expansion of virus use in Brazil, research efforts at CNPS/EMBRAPA are being directed towards developing or adapting virus formulation processes that would be simple and economical to adopt by the regional AgNPV producing units, in order to improve quality, storage and transport of the pathogen.

NOTES

Flávio Moscardi and Beatriz S. Corrêa Ferreira, EMBRAPA/CNPS, C.P. 1061, 86 100 Londrina, PR-Brazil.

REFERENCES

Carner, G. R., J. S. Hudson and O. W. Barnet. 1979. The infectivity of a nuclear polyhedrosis virus of the velvetbean caterpillar for eight Noctuidae hosts. J. Invertebr. Pathol. 33:211-216.

Carner, G. R. and S. G. Turnipseed. 1977. Potential of a nuclear polyhedrosis virus for control of the velvetbean caterpillar in soybean. J. Econ. Entomol. 70:608-610.

Moscardi, F. 1977. Control of *Anticarsia gemmatalis* Hübner on soybean with a Baculovirus and selected insecticides and their effect on natural epizootics of the entomogenous fungus *Nomuraea rileyi* (Farlow) Samson. MSc. Thesis. University of Florida, Gainesville.

Moscardi, F. 1983. Utilização de *Baculovirus anticarsia* para o controle da lagarta da soja, *Anticarsia gemmatalis*. EMBRAPA/CNPS, Com. Tec. 23. Londrina, Brasil.

Moscardi, F., G. E. Allen and G. L. Greene. 1981. Control of the velvetbean caterpillar by nuclear polyhedrosis virus and insecticides and impact of treatments on the natural incidence of the entomopathogenic fungus *Nomuraea rileyi*. J. Econ. Entomol. 74:480-485.

Moscardi, F. and I. C. Corso. 1981. Ação do *Baculovirus anticarsia* sobre a lagarta da soja (*Anticarsia gemmatalis*, Hübner, 1818) e outros lepidópteros. pp. 51-57. *In* EMBRAPA/CNPS. Anais do Seminário Nacional de Pesquisa de Soja, 2, vol. 2, Londrina, Brasil.

Pavan, O. H. and D. G. Boucias. 1981. Virus de poliedrose nuclear de *Anticarsia gemmatalis*: métodos de inoculação e especificidade. pp. 191-197. *In* EMBRAPA/CNPS. Anais do Seminário Nacional de Pesquisa de Soja, 2, vol. 2, Londrina, Brasil.

PHYSIOLOGY

PHYSIOLOGY

Plenary Paper

PARTITIONING OF CARBON AND NITROGEN IN N_2-FIXING GRAIN LEGUMES

J. S. Pate

Multicellular plants exhibit a complex division of labor among their tissues and organs, a feature patently essential to their growth and survival. Nowhere is this more evident than with respect to the plant's economy of carbon and nitrogen. This paper draws attention to the processes whereby these two elements are partitioned during growth, and discusses how the various transport phenomena involved impinge upon the growth and metabolism of the whole plant and its individual parts. The paper discusses studies conducted in the author's laboratory on the three annual grain legumes, cowpea (*Vigna unguiculata*, L. Walp.), pea (*Pisum sativum* L.) and white lupin (*Lupinus albus* L.), and will consider principally the situation in which these legumes are deriving their nitrogen entirely from symbiotic fixation in root nodules.

Differential partitioning of C and N is essential for three basic reasons. First, C and N are tkaen up from the environment at essentially opposite ends of the plant—C as CO_2 in photosynthesizing leaves, N below ground in roots or, in the case of legumes, in N_2-fixing nodules. The predominant flow of C is, therefore, inwards in phloem from leaves, that of N upwards in xylem from root or nodules. Without extensive subsequent mixing of these opposing streams, plant organs would obviously fail to be supplied adequately with one or other of the two elements. Second, relative requirements for C and N differ widely with age of plant organ, as shown for example in the legumes white lupin (Pate and Layzell, 1981) and cowpea (Pate et al., 1983; Peoples et al., 1983).

By using information on percentages of N and C in dry matter of a series of organs that act as significant sinks for C and N, and knowing the extents to which these organs are autotrophic for C and N, estimates can be made of each organ's utilization of carbon relative to nitrogen. These consumption ratios (C:N w/w) are then found to vary more than ten-fold between one class of organ and another; e.g., in lupin from as low as 12:1 for young leaves or shoot apices and fruits—organs all engaging in massive synthesis of protein, to 140:1 or more for old parts of stems currently involved in massive secondary thickening of their tissues with wall polymers rich in C relative to N. Third, the inputs of fixed C and N effected by annual grain legumes change continuously in both absolute and relative terms during plant growth, and often not in a manner reflecting the current demands of the plant for the two elements. For instance, much less N_2 is fixed relative to CO_2 in late than in early growth (Pate and Minchin, 1980), thus necessitating a much greater need for internal mobilization of reserves of N than of C during later growth, particularly when the developing seeds are making peak demands for N when laying down their reserves of protein. Indeed, N fixed before flowering is mobilized from vegetative organs to fruits with efficiencies approaching 70%, largely because the N is bound to an easily dismantled material, protein. By contrast, C fixed before flowering is available to seeds with only 2 to 5% efficiency, since the carbon in question is bound mostly to non-degradable cell wall materials of stem and root tissues (see Pate, 1983).

I now turn to consider a number of basic constraints which relate to the mechanisms whereby C and N are partitioned within the plant. In the first place, bulk long distance transfer of C and N between organs is, it is believed, confined almost entirely to the conducting elements of phloem and xylem. It, therefore, follows that if an organ, such as a young leaf, fruit or shoot apex, is importing consistently through both of these channels, the final levels of C and N in its dry matter will reflect accurately the ratio of supply in xylem and phloem, making allowances, of course, for any additions or losses of C by the organ through photosynthesis or respiration. Since the carbon:nitrogen weight ratio of the solutes of phloem streams are usually much higher (20 to 140:1) than those of xylem (1 to 3:1) (see Pate, 1980) it follows that even a slight increase or decrease in proportional intake through one or other channel will grossly affect the current C:N balance of an importing organ, with possible immediate repercussions on its growth and functioning. This principle is vividly illustrated, for example, in experiments in which lupin or cowpea seedlings have the air around their nodulated roots replaced by $Ar:O_2$ (80:20% v/v). Fixation ceases immediately, and within a few hours N levels in xylem of the $Ar:O_2$ treated plants have declined to less than 10% of those evident in control plants retained in air (Pate et al., 1984a). Within 24 h of the $Ar:O_2$ treatment the N contents of young leaves and the apical region of the shoot have fallen from 5 to 7% to less than 2% N in dry matter, and shoot growth accordingly declines abruptly. Then, if after several more days of $Ar:O_2$ treatment the seedlings are returned to air, fixation immediately resumes, N levels in dry matter of young organs are restored to their former levels, and growth of new shoot parts is quickly resumed.

The second class of constraint refers to specific architectural features of the transport system of an individual species. Certain legumes (e.g., pea and soybean) possess a horizontally-stratified pattern of photosynthate distribution in which upper leaves feed the apex, whereas middle and lower leaves feed the roots (see Pate and Farrington, 1981; Pate, 1983). When reproduction of such species commences, assimilate flow becomes even further restrained because fruits monopolize the photosynthate supply of their subtending nurse leaves. In other legumes (e.g., cowpea) (Pate et al., 1983), a less rigid partitioning system exists in which any sink has the capacity to acquire assimilates from virtually any source, with little evidence of preferential flow between sources and specific adjacent sinks. There are basic logistic problems in partitioning C and N within each of the above types of assimilate distribution pattern. As will be discussed later, their resolution within the plant requires the operation of a number of short-distance secondary exchange processes between and within xylem and phloem, each directed at ameliorating imbalances in C and N supply to specific regions of the plant.

Working in conjunction with the above constraints, one finds a series of important modifying influences whose apparent function is to redress any transient or long-term imbalances in the assimilatory inputs of C and N. Such influences become most obvious when plants are stressed, particularly if their photosynthetic input of C is changed relative to that of N_2 fixation. For example, in the $Ar:O_2$ experiment previously mentioned, prolonged inhibition of N_2 fixation will induce premature senescence of leaves, thus releasing leaf-bound N to the phloem stream and thereby reducing the elevated C:N ratios of phloem translocate associated with early stages of this treatment (Pate et al., 1984a). When applied to flowering or fruiting plants (cowpea and lupin), treatment with $Ar:O_2$ sets in train a fascinating complex of compensatory influences that work collectively in such a manner that seed yield is maximized within the context of a suddenly curtailed resource of N (M. B. Peoples, J. S. Pate and C. A. Atkins, unpublished data). The responses involved include a tailoring of seed number by abortion, reductions in size and percentage protein of surviving seeds, and earlier and more efficient mobilization of N from foliage to reproductive parts during fruiting. Conversely, when fruiting plants have their atmosphere enriched with CO_2, the extra photosynthate is, in part, supplied to nodules, thereby redressing the plant's elevated C:N balance by increasing rates of N_2 fixation. Plants then produce higher yields of larger seed of slightly lower than normal protein content. Finally, when photosynthetic inputs of C are artificially restricted during fruiting by placing plants in an atmosphere deficient in CO_2 or retaining them at low light, the reproductive effort of the plant is reduced by fruit and seed abortion and by the production of seed of high content of N relative to C.

THE ROOT AND ITS NUTRITIONAL RELATIONSHIPS WITH THE SHOOT

Whether nodulated or not, roots depend entirely on the shoot for carbon, except for small amounts they might acquire by dark CO_2 fixation or through symbiotic association with microorganisms. It is, therefore, possible to assess fairly accurately the requirements of a root for carbon simply by measuring its gain of the element in dry matter and its release of CO_2 to the rooting medium through respiration (Minchin and Pate, 1973; Pate, 1983). Measurements of root carbon balance become difficult, however, when roots are growing in soil, since it is extremely difficult to distinguish the true respiratory activities of the tissues of the root from that of soil microorganisms, and to identify what proportions of the total C budget of the rhizosphere are attributable to carbon lost as dead root tissue or root exudates.

To complicate matters further, soil-grown roots can vary greatly in their degree of autotrophy for N. Some legumes; e.g., pea (Wallace and Pate, 1965) and lupin (Atkins et al., 1979; Pate et al., 1979a), possess an active nitrate reductase system in their roots and therefore have the potential to supply reduced N directly to growing parts of the root close to where the nitrate has been absorbed and reduced. The roots of these species may still be only partly autotrophic for N, since they are likely to receive additional reduced N when importing carbohydrate from the shoot of the phloem. Other legume species, which possess only weak nitrate reductase activity in their roots; e.g., cowpea (Atkins et al., 1980a), will obviously be heavily dependent on shoot-derived N when growing on nitrate-N. And finally, there is the situation in the effectively nodulated legume growing in a rooting medium devoid of inorganic N. In this case, as shown by $^{15}N_2$ feeding studies (see Oghoghorie and Pate, 1972; Pate, 1973), fixed N is released from nodules directly to the shoot in the xylem, so that roots distal to nodules have access to this N only after it has cycled through the shoot and returned to the root with phloem translocate.

Aware of the above complications, most of our experimental studies on C and N exchange between root and shoot have dealt with nodulated plants grown in minus-nitrogen water or sand culture, thus avoiding problems due to extraneous inputs of C and N from soil-based media. The measured C and N consumption of the root then relates directly to its translocatory input, and provided that information is available on the C:N ratios of stem base phloem sap and root xylem sap, the relative amounts of C and N arriving in and cycling through the root can be computed (see Pate et al., 1979b). Data for lupin (Pate et al., 1979b; Pate et al., 1980) have shown that from mid-vegetative growth until flowering, the root receives from the shoot slightly more N in phloem translocate than it incorporates into dry matter. This extra N is assumed to return to the shoot via the same xylem stream that is exporting the newly-fixed N of the nodules. Comparisons of the level of N in root-bleeding xylem sap collected above and below the nodulated zone of the root (see Pate, 1980; Pate et al., 1984a) indicate that approximately 10% of the N leaving the root in xylem is recycled extra N provided from the shoot in the phloem, whereas the remaining 90% represents newly fixed N exported from nodules via the xylem.

Comparisons of the cost effectiveness of roots of different species can be made using data on carbon consumption of the nodulated root relative to its activity in fixing N. Our own studies in this connection suggest that there may be quite large differences between tap-rooted and slender-rooted legumes (Pate and Minchin, 1980; Pate (in press a)).

THE ROOT NODULE AND ITS EXCHANGES WITH THE PARENT PLANT

The developing nodules and roots of legume seedlings growing in a medium lacking in inorganic nitrogen share a common source of carbon and nitrogen, the N exclusively from cotyledons, the C initially from cotyledonary reserves and then increasingly from seedling photosynthesis. The cotyledonary N becomes progressively diluted within the enlarging body of the seedling during this pre-fixation stage, leading to noticeable N deficiency symptoms, such as reduced growth, yellowing of seedling leaves, and minimum percentage values for N in dry matter of seedling organs. In cowpea, for example, this 'nitrogen hunger' state is most pronounced over the period 12 to 16 days after germination, and continues for several days after the nodules have developed hemoglobin pigmentation and give positive assays for nitrogenase using the acetylene reduction technique (I. Matthews, C. A. Atkins and J. S. Pate, unpublished data).

Detailed balance sheets for N in cowpea seedlings during the transition period between cotyledon- and fixation-based nutrition indicate that the stem and first seedling (unifoliolate) leaves actually lose some of the N they gained initially from the cotyledons, and that this mobilized N is shared between root, nodules and shoot apex. During this starvation period, nodules commence to fix N, and over the subsequent two or three days more than double the concentration of N in their dry matter. Two days later xylem sap analyses first reveal the presence of ureides, indicating export of fixed N from nodules. Shortly after this, all seedling parts increase rapidly in N content as the newly-fixed N becomes generally available. Eventually, a week or so after the beginning of N_2 fixation, regreening and renewed expansion of the unifoliolate provides visual proof of the end of the seedling's N hunger state (I. Matthews, J. S. Pate, C. A. Atkins and P. Stanford, unpublished data).

A novel technique for studying how early-fixed N is partitioned in young legume seedlings involves the use of uniformly ^{15}N-enriched seed, obtained by raising a parent, non-nodulated generation of plants on nitrate of constant ^{15}N enrichment. Inoculated with Rhizobium and cultured on a mineral medium deficient in inorganic N, seedlings grown from these isotopically-enriched seeds at first develop organs of ^{15}N enrichment identical to that of their cotyledon N, but once nodules have started to fix N, the pools of ^{15}N in seedling parts become progressively diluted with atmospherically-derived ^{14}N. ^{15}N dilution data from such an experiment demonstrate conclusively that the first fixation products are preferentially sequestered by the insoluble (mainly bacteroid) fraction of the nodule, to the extent that by 5 or 6 days after fixation has commenced, over 75% of this fraction consists of newly-fixed N (Atkins et al., 1984a). Comparably high levels of dilution of cotyledonary ^{15}N are also evident at this time in the plumule and first trifoliolate leaf of the seedling, indicating that these actively expanding parts have prime access to any newly-fixed N that is released from the nodule. By contrast, the root, older parts of the stem and the unifoliolate leaves, show slower, less pronounced dilution profiles for the ^{15}N, partly because of the large size of their initial pools of cotyledonary N and partly due to their poor competing power for newly-fixed N (J. S. Pate, P. Stanford and C. A. Atkins, unpublished data).

Nodule growth is, thus, partly autocatalytic for N, enabling the newly-fixing nodule to establish immediately a much higher level of N in its dry matter (6 to 7% N) than in the dry matter of the roots (1.5 to 2% N) on which it is borne. Indeed, if nodulated seedlings are prevented from commencing N_2 fixation, by exposing their below-ground parts to $Ar:O_2$ (80:20 v/v) rather than to air, bacterial development is arrested, nodules show signs of senescence, and N in nodule dry matter fails to increase above the normal pre-fixation level of 2 to 3%. These deteriorative changes occur even if the $Ar:O_2$-treated seedlings have access to nitrate (Atkins et al., 1984b,c), suggesting that the ammonia released from bacterioids may have a specific regulatory role in the N metabolism of the nodule and in possibly protecting it against premature senescence (Atkins et al., 1984c).

Throughout their later growth and functioning, nodules remain continuously dependent for C on phloem translocate from the plant shoot, using imported carbohydrate partly as an energy source for nitrogenase functioning, partly to provide acceptor molecules for ammonia assimilation, and partly to synthesize dry matter during nodule growth. Budgets for mature nodules, based on their import, utilization and export of C and N, have been constructed for symbiotic associations of pea, cowpea and lupin (Pate et al., 1981; Pate (in press a)). Cowpea nodules, which synthesize N mainly as ureides, turn out to be more economical in usage of imported C in fixing N than in the case of the amide-producing species lupin and pea. Thus, fixation of 1 g N_2 requires import of 5.4 g C in cowpea nodules, compared with 6.5 g in pea nodules and 6.9 g in nodules of lupin. The better conversion efficiency of the cowpea nodule is, in part, due to a lower requirement for C in synthesizing and exporting ureides (C:N::1:1) as opposed to amides (C:N::2 to 2.5:1) in pea or lupin. Differences in efficiency between cultivars or between different *Rhizobium*-cultivar associations may also relate to whether or not H_2 is evolved by the nodule and the extent to which the nodule may engage in anapleurotic inputs of respired C. However, these differences tend to be relatively small proportions of the nodule's total budget for C (Pate et al., 1981; Rainbird et al., 1983).

There is good evidence that growing nodules continue to incorporate newly-fixed N directly into their bacterial tissues long after seedling stages have passed. For example, if $^{15}NO_3$ is provided to nodulated plants, ^{15}N enrichment of the insoluble N of nodules is much lower than that of other plant parts, indicating preferential incorporation of fixed N into nodule tissues (Oghoghorie and Pate, 1971). Estimates based on ^{15}N studies of this kind indicate that the nodule may provide itself with half or more of its N, a finding supporting the concept that insufficient N enters the nodule alongside sugar in the phloem. Evidence of this is also apparent in the empirical models of nodule C and N economy based on C:N ratios of phloem sap delivered to the nodule (Layzell et al., 1979).

Using experimentally derived consumption rates for C by nodules, one can assess the proportion of the plant's total net photosynthate that nodules monopolize during different periods of plant growth (see data for lupin, Pate and Herridge, 1978; Pate [in press b], and pea, [Minchin and Pate, 1973]). The amounts involved range from 5% to as much as 22% of the plant's current photosynthate (see Pate, in press c). In most cases allocation to nodules is greatest at or shortly after flowering, when nodules comprise a maximum proportion of plant fresh weight and are still highly active in fixation.

THE CARBON AND NITROGEN ECONOMY OF THE LEAF

Each leaf commences its life as an essentially heterotrophic organ, depending entirely on the plant for all its N, and in early stages of expansion, for all but a very small fraction of its requirements for C. After an initial phase of import through both phloem and xylem, the leaf starts to fix C photosynthetically faster than it currently requires C for growth. Export of C then commences. This graduation to self-dependency for C is usually reached when a leaf is one-third to one-half of its full size, after which photosynthesis and associated export of sugar rise rapidly to a maximum, usually as the leaf achieves full expansion and reaches peak content of N (see literature reviewed by Pate and Atkins, 1983; Pate, 1983). Near maximum rates of export of C may then continue for some time, but deteriorative changes eventually become evident. These are marked by losses of chlorophyll, total N and dry matter and concomitant losses in photosynthetic performance. By this time, of course, the leaf may occupy a low position in the plant canopy where its photosynthetic rate may be lowered as much by reduced light level as by deteriorative internal changes within its tissues (Pate and Farrington, 1981). With the approach of senescence, degeneration of physiological processes and associated losses of chlorophyll, N and dry matter accelerate, until the leaf finally enters negative C balance. Death follows quickly.

Applied to annual grain legumes, the above generalized picture of the functional life of a leaf has special significance in terms of N, since the bulk of the N fixed in nodules is processed by leaves before passing to the developing flowers and fruits (McNeil et al., 1979; Atkins et al., 1980b). Also, during later seed filling, a substantial part of the N provided to the grain is derived from breakdown of protein within the leaf canopy (Peoples et al., 1983). The timing of leaf senescence during fruiting is critical. On the one hand, prolonged maintenance of effective leaf surface allows net photosynthate to be produced as a general source for plant growth and seed filling, and particularly to implement fixation of further N in root nodules. On the other hand, induction of leaf senescence in early fruiting would provide immediately a bulk source of N for filling of seeds. The significance of the deployment of these events during reproduction will be elaborated upon later when the N nutrition of seeds is being considered.

To assess quantitatively the contribution of the leaf to the carbon economy of the whole plant, data are first required on the changes in total C that take place in the leaf's dry matter and the day-by-day exchanges of CO_2 that it makes with the atmosphere. The amount of C available for export in the phloem can then be determined. The picture relating to nitrogen is more complicated, since once the phase of import through xylem and phloem has passed, the leaf begins to release N to the parent plant as phloem translocate. Some of this represents N recently imported through the xylem; some represents mobilization of already existing pools of soluble and insoluble N from within the leaf (Atkins et al., 1983). To determine rates of cycling of N through the leaf, information is required on the ratios of C:N in both the entering xylem stream and the exiting phloem stream, necessitating use of a species whose phloem contents can be readily collected. White lupin, with its capacity to bleed spontaneously from the phloem, is an ideal grain legume in this regard, enabling a series of definitive studies on the functional economy of C and N in its leaves to be undertaken (Atkins et al., 1983; Pate and Atkins, 1983). The resulting models of leaf functioning (Pate and Atkins, 1983) depict the amounts of C and N exchanged through xylem and phloem with the plant during four distinct phases of development of the uppermost leaf on the main stem.

The first phase (1 to 11 days after leaf initiation) is characterized by net import through xylem and phloem. Rates of net photosynthesis are only slightly greater than losses in night respiration, and the leaf accordingly is dependent on the parent plant for the bulk (82%) of the C incorporated into its dry matter. Xylem furnishes more than four times as much N to the leaf as does phloem. Phase 2 (11 to 20 days) is the time of most rapid leaf growth and ends with the leaf being of maximum area, photosynthetic rate and contents of N and C. It is marked by massive import of N through xylem and export of C through phloem. Net photosynthesis supplies C at 20 times the rate of respiratory loss at night, and the equivalent of 77% of the leaf's net photosynthate is exported. By contrast, only 18% of the N imported by the leaf through the xylem is exported in the phloem, the C:N ratio of phloem sap being accordingly very high (100:1). That is, phloem is high in sucrose relative to amino compounds (Atkins et al., 1983). Phase 3 (20 to 38 days) represents the mid-age of the leaf when it is still highly active in photosynthesis, is almost maintaining its content of C and N in dry matter, and is particularly active in translocation of C and cycling of N. Average C:N ratio of the leaf's phloem translocate is 74:1 and, as in Phase 2, night respiration is equivalent to only 5% of the net daily gain of C from the atmosphere. Phase 4 (38 to 66 days) is the stage of leaf growth preceding senescence. The leaf is still green, though losing chlorophyll and N and C of dry matter. Photosynthetic activity is less, but N export is greater than in Phase 3, leading to a much lower C:N ratio in the outgoing phloem stream (35:1) than in Phase 3 (74:1).

By adding together the C and N fluxes of the above four phases of its development, the net total demands and contributions made during the life of the white lupin leaf may be determined (Pate and Atkins, 1983). An initial investment of 67 mg C and 7 mg N is made into structural dry matter, 51% of this C and all of the N derived from the parent plant. A total of 807 mg C is generated as net photosynthate, an amount some 15 times greater than the total C lost by the leaf in respiration at night, and representing a 12-fold return of photosynthate per unit of C initially invested in leaf dry matter. Viewed in terms of its balance of N, the leaf imports 18 mg N through xylem, and, during early growth, 0.4 mg N through phloem. Almost 80% of this imported N is returned to the plant as phloem translocate, most cycling immediately through the leaf after its receipt in the xylem, the remainder released later as the leaf approaches senescence.

THE C AND N RELATIONSHIPS OF DEVELOPING FRUIT AND SEEDS

The crop of fruits on a modern cultivar of grain legume contains upwards of 30% of the plant's final content of C and as much as 75 to 85% of its final N (see Pate and Minchin, 1980; Pate, in press a). Over 90% of the mature fruit's N is likely to be located in the reserve protein of seeds. To meet this especially large demand for N during seed filling a heavy premium must be placed on mechanisms mobilizing N quickly and effectively to the developing fruit. This applies particularly to rapidly maturing, determinate cultivars in which near-synchronous maturation of fruits will generate a climacteric in demand for N during the final weeks of plant growth.

Because fruits have limited access to pre-anthesis stores of carbon from vegetative organs, virtually all of the C bound into pod and seed must be derived from photosynthate generated while fruits are developing, and because of the relatively poor photosynthetic capacity of fruits, the bulk of their C supply has to be imported. Fruits, therefore, must compete directly with nodulated roots and growing vegetative parts of the shoot for photosynthate, and their success in doing this depends particularly on the intrinsic priorities that exist within the parent plant for partitioning of photosynthate. For instance, deep tap-rooted species, such as lupin, exhibit a continued high percentage allocation of net photosynthate to roots almost to the end of fruiting, enabling the plant to fix N at a high rate and pass this N directly to the developing fruits (Pate and Herridge, 1978; Pate, in press b). Other slender-rooted species, such as cowpea and pea, display the reverse of this trend, such that fixation ceases altogether or attenuates to insignificant levels by mid- or late-fruit filling. These grain legumes consequently rely very heavily on previously stored N when filling their fruits (Pate and Flinn, 1973; Peoples et al., 1983).

Bearing in mind the extent of variation between species indicated above, it comes as no surprise to find that the amount of pre-anthesis N mobilized to seeds of grain legumes may represent the equivalent of from less than one-tenth to almost two-thirds of the N requirements of the seeds (Peoples et al., 1983). This N may well have cycled through several age sets of leaves and shoot parts before reaching the seeds (Pate and Flinn, 1973; Pate and Minchin, 1980). The second source of nitrogen—that assimilated after flowering—also may be first incorporated into leaf protein before being finally released to seeds, but much of it may be transferred directly from nodule to seed, using pathways within the stem and leaf veins that effectively circumvent the mesophyll of the leaf of pea (Lewis and Pate, 1973), lupin (Atkins et al., 1975), and cowpea (Pate et al., 1984b).

As already pointed out, productivity of seed-bound N is highly dependent upon the maintenance during fruiting of effective nutritional relationships between leaf, fruit and nodule. However, the insoluble N of leaves, bound largely into chloroplastic proteins, represents by far the largest single reservoir of vegetative plant N, and if catabolized immediately, would provide a sizable fraction (e.g., 25 to 35% in cowpea; Peoples et al., 1983) of the fruit's total requirement for N. However, if the same pool of leaf N were reserved as actively photosynthesizing tissue until late in fruit filling, it would retain the capacity to synthesize sufficient photosynthate to fuel the fixation of at least an equivalent amount of nitrogen by the nodules. By augmenting the plant's total resource of N in this manner, the requirement for withdrawal of N from other vegetative structures would be correspondingly reduced, and the whole process of plant senescence delayed until much later in the growth cycle.

Two contrasting hypothetical strategies are envisaged for partitioning N optimally to fruits. The first would apply especially to grain legumes with high seed protein content (35 to 50% by weight), slow rates of fruit maturation and filling, and relatively small starting capitals of N accumulated in the plant before flowering. In this instance, the strategy for producing maximum seed yield would involve a retention of photosynthetic potential—and hence of nitrogen fixation—until as late as possible in the reproductive cycle, followed by mobilization of N from root, stem, and then finally, from leaves in the penultimate stages of fruit ripening. This may well approximate to the ancestral pattern of development in the progenitors of many grain legumes, and clearly resembles the strategy adopted by white lupin (see Pate and Herridge, 1978; Pate et al., in press b), and probably also by late maturing cultivars of other crop legumes, including soybean.

The second, alternative partitioning model is deemed generally appropriate for determinate, quick-maturing, large-fruited cultivars of grain legumes, such as mung bean (*Vigna radiata* L.), pea, cowpea and snap bean (*Phaseolus vulgaris* L.), all with seeds of relatively low (18 to 33%) protein content. The strategy in this case would be to establish a substantial reserve of plant N from symbiosis or from uptake of soil N before and shortly after flowering, and then to meet the sudden peak demand of the quickly maturing fruits by high proportional allocation of photosynthate to seeds and rapid and efficient mobilization of N from all vegetative parts of the parent plant.

The quantitative relationship between leaf N and seed-bound N should be noted in relation to the two partitioning strategies mentioned above. To fill a unit dry weight of seed, species with high seed protein level (e.g., lupin and cowpea, 35 to 50%) would have to senesce almost twice the amount of leaf chloroplastic N as would species such as cowpea and pea which have seed protein levels of only 18 to 25%. Duration of seed filling also would be critically important, since a species with high protein seeds, pursuing the strategy of fixing most of its N during fruiting, would have to maintain nodule activity for twice the length of time, or at twice the absolute rate, to effect the same yield of seed as would a similarly performing species with seeds of low protein content. Slow maturation of seeds (e.g., up to 12 weeks in lupin versus only 3 weeks in cowpea) would be a decided advantage in such circumstances.

The previous discussion leads one to consider how legume fruits effect such high inputs of N relative to C, and how even greater amounts of N relative to C are finally mobilized from pod reserves to seeds during late fruit ripening. Some of the processes involved can be demonstrated in lupin and cowpea, since in both species it is possible to sample phloem and xylem streams entering the fruit, and thus, assess how changing demands for C and N are met by import through vascular channels (Pate et al., 1977; Pate, 1984; Pate et al., 1984b). Both fruits exercise measurable economy of C by conserving some or all of their daytime respired CO_2 photosynthetically in green tissues of the pod wall (cowpea) and pod and seed (lupin) (Pate, 1984). The savings effected through such activity represent up to 15% of the total C requirement of the fruit, so are likely to have a considerable impact on the overall budget of the fruit. In the first half of fruit growth of the two species, when the C-rich, cellular fabric of the pod is the principal repository for dry matter, the fruit's import ratio of C:N through xylem plus phloem is relatively high, it being 15:1 for lupin (Pate et al., 1977); and 16:1 for cowpea (Peoples et al., in press). Phloem provides virtually all (97 to 98%) of the C (lupin and cowpea), and the bulk of N (83% in lupin, 59% in cowpea). Measurements of solute concentrations in phloem and xylem sap suggest that, at this stage, the mixture of xylem and phloem streams estimated to meet precisely the carbon and nitrogen requirements of the fruits also matches closely the fruit's current water usage in tissue growth and transpiration. In the lupin fruit, 66% of the water entering during the first half of fruit development comes in through phloem, in cowpea the corresponding figure is 76%. These estimates are based on the assumption that unidirectional mass flow operates in xylem and phloem, and that phloem is bringing in water and solutes in the amounts suggested from phloem sap analysis.

The second half of fruit growth shows very different patterns of behavior between the species. In lupin, the seed filling period (6 to 12 weeks after anthesis) records an intake of 9 C:1 N, a feature reflecting the high final content of protein in its seeds. Import of extra N is due mainly to a progressive lowering of the C:N ratio of phloem sap, and this is achieved partly by release of N during organ senescence, and partly by an increasing proportional transfer of recently fixed N directly from the xylem to the phloem stream serving the fruit. Indeed, direct intake through xylem at this stage provides only 7% of the fruit's net input of N. Mobilization of previously acquired N from the pod at this time provides the seeds with a significant proportion (up to 23%) of their N requirement (Pate, in press a). As in the first half of development, the phloem:xylem mixture calculated to bring C and N into the fruit in amounts meeting the consumption of these elements by the fruit, brings in an amount of water roughly balancing the transpiration and net changes in tissue water of the fruit (Pate et al., 1978).

The second half of development of the cowpea fruit shows extreme economy in water usage, the fruit recording a transpiration ratio of less than 7 ml water transpired/g dry matter accumulated. When the measured values for C and N intake through phloem and xylem are compared with estimates of the water requirements of the fruit, a gross anomaly becomes evident. Assuming that the phloem stream is operating as a mass flow system with a sugar concentration similar to that recorded for fruit phloem exudate, phloem import is shown to meet all of the fruit's requirements for carbon and water, but only 80% of its requirements for N (Peoples et al., in press). To gain entry of the extra N, a quite massive intake of xylem sap would be required, leaving the fruit with an estimated oversupply of 2 to 3 ml of water per day (Peoples et al., in press). This discrepancy has been resolved in a recent series of labeling studies (Pate et al., in press a) in which it is demonstrated that fruits are capable of engaging in backflow of water through xylem to the parent plant at times during the day when they are deriving a net surplus of water in the phloem. Conversely, at night and in certain circumstances during the early morning, when fruit transpiration is more than able to consume all phloem-delivered water, intake of water through xylem from the parent plant is in evidence. Presumably the 'extra' N required for fruit growth and other essential phloem immobile nutrients enters the fruit through xylem under these latter circumstances (Pate et al., in press a).

As well as possessing the aforementioned potential for increasing their uptake of N by reversible xylem exchanges of water with the plant, the fruits of certain cultivars of cowpea appear capable of directly increasing their supply of N relative to C by triggering senescence of adjacent foliage, particularly of their subtending nurse leaf. Such 'fruit induced senescence' is well authenticated for other legumes, including soybean (Nooden and Murray, 1982), and is indeed central to the 'self-destruction' hypothesis suggested by Sinclair and de Wit (1976) to apply generally to high protein seed crops. Hormonal influences have been implicated in mediating such senescence, although the compounds involved and their mode of transmission from fruit to leaf have not been ascertained.

In the large-fruited cowpea cultivar, Vita 3, the alleged 'senescence signal' from the first-formed fruit appears to be released early in its growth, and has the highly specific effect of inducing loss of protein and chlorophyll in the subtending nurse leaf. These events take place some 12 to 15 days after anthesis of the fruit; that is, a week or so before its maximum demand for C (Pate et al., 1983; Peoples et al., 1983). While such early senescence of a blossom leaf may release N to the subtended fruit, the overall loss of photosynthetic capacity at that node forces the subtended fruit to encroach upon the assimilate supplies from other foliar surfaces previously committed to roots, nodules, shoot apex and other younger fruits and inflorescences. Progressive starvation of these sub-dominant sinks ensues, as shown by greatly reduced rates of N fixation, abscission or retarded growth of young fruits, and cessation of shoot apical growth (Pate et al., 1983). Thus, the premature senescence of foliage at a first-formed reproductive node appears to be strongly counterproductive to yield.

THE STEM AND ITS ROLE IN PARTITIONING OF C AND N

The empirical models of C and N flow constructed for the white lupin (Pate et al., 1979b; Pate and Layzell, 1981) visualize a highly important role for the tissues of the stem in the diversion of xylem N from older parts of the shoot towards apical regions with high requirements for N. Two distinct processes are involved. The first, occurring in the lower part of the stem, consists of the abstraction of N from the xylem of departing leaf traces and the progressive transfer of this laterally withdrawn N to xylem streams passing farther up the shoot (Layzell et al., 1981). As a result, lower mature leaves receive less N than might be expected from their transpirational capacity to attract xylem fluids, while young upper leaves, tapping a xylem stream significantly enriched with N, benefit in terms of N intake to a greater extent than to be expected from their transpiration.

The second differential partitioning process resides in the topmost part of the stem, and consists of the transfer of N from the xylem to phloem streams passing from upper leaves to the shoot apex or developing inflorescences and fruits (Pate et al., 1979b). These upper leaves export a sugar-rich phloem stream (C:N::60 to 120:1), which were it to be the sole source of N, would grossly undernourish shoot apex and fruit parts with N. Xylem to phloem exchange in the upper part of the stem involves mainly the transfer to the phloem of the nitrogen-rich amides, glutamine and asparagine (Pate, 1980), thereby lowering the C:N ratio of the phloem stream to 15 to 20:1 (Pate et al., 1979b), a value close to that of the requirements of the recipient fruit (C:N ratio 12:1) or the shoot apex (13:1) (Pate and Layzell, 1981). These latter organs are thus able to gain much more N than they would ever acquire through their weak transpiration activity or their translocatory input from leaves.

FINAL COMMENTS

These essentially descriptive comments on the processes of C and N partitioning in grain legumes offer little indication of how supply and demand processes are modulated by and integrated within the whole plant as older organs age and senesce and new organs are added. Some indication of how nutrition of interacting organs might be regulated comes from those sections of the paper devoted to plant responses to artificially altered inputs of C and N. This is an obvious area where further studies should be directed, particularly if examined in sufficient detail to depict time courses of responses of donor and receptor organs to specific types of stress. By observing the extent and nature of the resilience of grain legume genotypes to such imbalances, information useful to crop physiologists and plant breeders may well emerge.

NOTES

J. S. Pate, Department of Botany, University of Western Australia, Nedlands WA 6009, Western Australia.

REFERENCES

Atkins, C. A., J. S. Pate, G. J. Griffiths and S. T. White. 1980a. Economy of carbon and nitrogen in nodulated and non-nodulated (NO_3-grown) cowpea (*Vigna unguiculata* (L.) Walp). Plant Physiol. 66:978-983.

Atkins, C. A., J. S. Pate and D. B. Layzell. 1979. Assimilation and transport of nitrogen in non-nodulated (NO_2-grown) *Lupinus albus* L. Plant Physiol. 64:1078-1082.

Atkins, C. A., J. S. Pate and D. L. McNeil. 1980b. Phloem loading and metabolism of xylem-borne amino compounds in fruiting shoots of a legume. J. Exp. Bot. 31:1509-1520.

Atkins, C. A. , J. S. Pate, M. B. Peoples and K. W. Joy. 1983. Amino acid transport and metabolism in relation to the nitrogen economy of a legume leaf. Plant Physiol. 71:841-848.

Atkins, C. A., J. S. Pate and P. J. Sharkey. 1975. Asparagine metabolism—key to the nitrogen nutrition of developing legume seeds. Plant Physiol. 56:807-812.

Atkins, C. A., J. S. Pate and B. J. Shelp. 1984a. Effects of short-term N_2 deficiency on N metabolism of legume nodules. Plant Physiol. 76:59-64.

Atkins, C. A., R. M. Rainbird and J. S. Pate. 1980c. Evidence for a purine pathway of ureide synthesis in N_2-fixing nodules of cowpea (Vigna unguiculata (L.) Walp). Z. Pflanzenphysiol. 97:249-260.

Atkins, C. A., B. J. Shelp, J. Kuo, M. B. Peoples and J. S. Pate. 1984b. Nitrogen nutrition and the development and senescence of nodules on cowpea seedlings. Planta 162:316-326.

Atkins, C. A., B. J. Shelp, P. J. Storer and J. S. Pate. 1984c. Nitrogen nutrition and the development of biochemical functions associated with nitrogen fixation and ammonia assimilation of nodules on cowpea seedlings. Planta 162:327-333.

Layzell, D. B., J. S. Pate, C. A. Atkins and D. T. Canvin. 1981. Partitioning of carbon and nitrogen and the nutrition of root and shoot apex in a nodulated legume. Plant Physiol. 67:30-36.

Layzell, D. B., R. M. Rainbird, C. A. Atkins and J. S. Pate. 1979. Economy of photosynthate use in nitrogen-fixing legume nodules: observations on two contrasting symbioses. Plant Physiol. 64:888-892.

Lewis, O. A. M. and J. S. Pate. 1973. The significance of transpirationally derived nitrogen in protein synthesis in fruiting plants of pea (Pisum sativum L.). J. Exp. Bot. 24:596-606.

McNeil, D. L., C. A. Atkins and J. S. Pate. 1979. Uptake and utilization of xylem-borne amino compounds by shoot organs of a legume. Plant Physiol. 63:1076-1081.

Minchin, F. R. and J. S. Pate. 1973. The carbon balance of a legume and the functional economy of its root nodules. J. Exp. Bot. 24:259-271.

Nooden, L. D. and B. J. Murray. 1982. Transmission of the monocarpic senescence signal via the xylem in soybean. Plant Physiol. 69:754-756.

Oghoghorie, C. G. O. and J. S. Pate. 1971. The nitrate stress syndrome of the nodulated field pea (Pisum arvense L.). Techniques for measurement and evaluation in physiological terms. Plant Soil, Special Volume:185-292.

Oghoghorie, C. G. O. and J. S. Pate. 1972. Exploration of the nitrogen transport system of a nodulated legume using ^{15}N. Planta 104:35-49.

Pate, J. S. 1973. Uptake, assimilation and transport of nitrogen compounds by plants. Soil Biol. Biochem. 5:109-119.

Pate, J. S. 1980. Transport and partitioning of nitrogenous solutes. Ann. Rev. Plant Physiol. 31:313-340.

Pate, J. S. 1983. Chapter 4: Distribution of metabolites. p. 335-401. In R. G. S. Bidwell and F. C. Steward (eds.) Plant Physiology. A Treatise, Volume 8—Nitrogen Metabolism. Academic Press, London.

Pate, J. S. 1984. The carbon and nitrogen nutrition of fruit and seed—case studies of selected grain legumes. p. 41-81. In D. R. Murray (ed.) Seed Physiology, Volume 1—Development. Academic Press, New York.

Pate, J. S. (in press a). Physiology of pea—a comparison with other legumes in terms of economy of carbon and nitrogen in plant functioning. (in press). In Proceedings of the 40th Easter School in Agricultural Science: The Pea Crop—A Basis for Improvement. University of Nottingham.

Pate, J. S. (in press b). Nitrogen metabolism of the white lupin (Lupinus albus L.). (in press). In Proc. 3rd International Lupin Congress.

Pate, J. S. (in press c). Economy of symbiotic nitrogen fixation. (in press). In T. J. Givnish (ed.) Proceedings of Cabot Symposium—Evolutionary Constraints on Primary Productivity: Adaptive Strategies of Energy Capture in Plants. Cambridge University Press.

Pate, J. S. and C. A. Atkins. 1983. Xylem and phloem transport and the functional economy of carbon and nitrogen of a legume leaf. Plant Physiol. 71:835-840.

727

Pate, J. S., C. A. Atkins, D. B. Layzell and B. J. Shelp. 1984a. Effects of N$_2$ deficiency on transport and partitioning of C and N in a nodulated legume. Plant Physiol. 76:59-64.

Pate, J. S., C. A. Atkins and R. M. Rainbird. 1981. Theoretical and experimental costing of nitrogen fixation and related processes in nodules of legumes. p. 105-116. In A. H. Gibson and W. E. Newton (eds.) Current perspectives in nitrogen fixation: Proc. 4th Int. symp. on nitrogen fixation. Griffin Press, South Australia.

Pate, J. S. and P. Farrington. 1981. Fruit set in Lupinus angustifolius cv. Unicrop. II. Assimilate flow during flowering and early fruiting. Aust. J. Plant Physiol. 8:307-318.

Pate, J. S. and A. M. Flinn. 1973. Carbon and nitrogen transfer from vegetative organs to ripening seeds of field pea (Pisum arvense L.). J. Exp. Bot. 24:1090-1099.

Pate, J. S. and D. F. Herridge. 1978. Partitioning and utilization of net photosynthate in a nodulated annual legume. J. Exp. Bot. 29:401-412.

Pate, J. S., J. Kuo and P. J. Hocking. 1978. Functioning of conducting elements of phloem and xylem in the stalk of the developing fruit of Lupinus albus L. Aust. J. Plant Physiol. 5:321-326.

Pate, J. S. and D. B. Layzell. 1981. Carbon and nitrogen partitioning in the whole plant—a thesis based on empirical modelling. p. 94-134. In J. D. Bewley (ed.) Nitrogen and carbon metabolism: Proc. of a symp. on the physiol. and biochem. of plant productivity. Martinus Nijhoff/Junk, The Hague.

Pate, J. S., D. B. Layzell and C. A. Atkins. 1979a. Economy of carbon and nitrogen in a nodulated and non-nodulated (NO$_3$-grown) legume. Plant Physiol. 64:1083-1088.

Pate, J. S., D. B. Layzell and C. A. Atkins. 1980. Transport exchange of carbon, nitrogen and water in the context of whole plant growth and functioning—case history of a nodulated annual legume. Ber. Deutsch. Bot. Ges. 93:243-255.

Pate, J. S., D. B. Layzell and D. L. McNeil. 1979b. Modelling the transport and utilization of carbon and nitrogen in a nodulated legume. Plant Physiol. 63:730-737.

Pate, J. S. and F. R. Minchin. 1980. Comparative studies of carbon and nitrogen nutrition of selected grain legumes. p. 105-114. In R. J. Summerfield and A. H. Bunting (eds.) Advances in legume science. Royal Botanic Gardens, Kew.

Pate, J. S., M. B. Peoples, A. J. E. Van Bel, J. Kuo and C. A. Atkins. (in press a). Diurnal water balance of the cowpea fruit. Plant Physiol. (in press).

Pate, J. S., P. J. Sharkey and C. A. Atkins. 1977. Nutrition of a developing legume fruit. Functional economy in terms of carbon, nitrogen, water. Plant Physiol. 59:506-510.

Pate, J. S., M. B. Peoples and C. A. Atkins. 1983. Post-anthesis economy of carbon in a cultivar of cowpea. J. Exp. Bot. 34:544-562.

Pate, J. S., M. B. Peoples and C. A. Atkins. 1984b. Spontaneous phloem bleeding from cryopunctured fruits of a ureide-producing legume. Plant Physiol. 74:499-505.

Pate, J. S., W. Williams and P. Farrington. (in press b). Chapter 10: Lupins. (in press). In R. J. Summerfield and E. H. Roberts (eds.) Grain legume crops. Granada Press, London.

Peoples, M. B., J. S. Pate and C. A. Atkins. 1983. Mobilization of nitrogen in fruiting plants of a cultivar of cowpea. J. Exp. Bot. 34:563-578.

Peoples, M. B., J. S. Pate, C. A. Atkins and D. R. Murray. (in press). Economy of water, carbon and nitrogen in the developing cowpea fruit. Plant Physiol. (in press).

Rainbird, R. M., C. A. Atkins, J. S. Pate and P. Sanford. 1983. Significance of hydrogen evolution in the carbon and nitrogen economy of nodulated cowpea. Plant Physiol. 71:122-127.

Sinclair, T. R. and C. T. de Wit. 1976. Analysis of the carbon and nitrogen limitations to soybean yield. Agron. J. 68:319-324.

Wallace, W. and J. S. Pate. 1965. Nitrate reductase in the field pea (Pisum arvense L.). Ann. Bot. 29: 655-671.

Partition and Transport of Assimilates

ASSIMILATE PARTITIONING IN SOYBEAN LEAVES DURING SEED FILLING

Robert T. Giaquinta, B. Quebedeaux, N. L. Sadler and V. R. Franceschi

The partitioning of assimilates from their sites of production in photosynthesizing source leaves to their sites of utilization in harvestable regions of crops is a major determinant of yield (Giaquinta, 1980; Gifford et al., 1984). Identifying and understanding the mechanisms operating within source and sink regions that control assimilate distribution are central to the eventual manipulation of partitioning by chemical or genetic means. The pathways and mechanisms of assimilate unloading in a variety of sinks have been reviewed recently (Thorne and Giaquinta, 1984), and assimilate unloading in developing soybean seeds is addressed elsewhere in this book (Thorne, this volume).

In source leaves, the subject of this study, several processes influence the availability of photosynthate for export. These include: photosynthesis rate; carbon partitioning between starch and sucrose; sucrose synthesis; intra- and inter-cellular compartmentation of assimilates; assimilate transfer to the phloem; and phloem loading (Geiger and Giaquinta, 1982; Giaquinta, 1983). This study characterizes several of these processes that operate in soybean leaves during seed filling.

PHOTOSYNTHESIS DURING SEED FILLING

The rate of seed growth in determinate Wye soybeans grown in a controlled environment reaches a maximum of 15 mg dry weight/seed·d at 25 days postanthesis (approximately one-half final seed size) (Figure 1). Source leaf photosynthesis

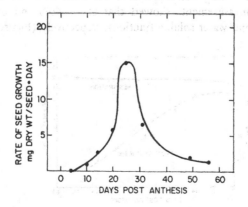

Figure 1. Soybean seed growth rate.

(measured by a 15 s $^{14}CO_2$ pulse) declines during seed filling with 50% inhibition occurring when the seeds are one-half final size (Figure 2). Leaf respiration, measured by $^{12}CO_2$ efflux, remains essentially unchanged during this period, although there appears to be a marginal increase during leaf senescence.

LEAF STARCH AND SUCROSE LEVELS

The decline in photosynthesis during seed filling was accompanied by an increase in leaf starch level. Starch, measured at 5 h into the light period, increased by three- to four-fold at the mid-seed filling stage compared with levels in fully expanded leaves at preanthesis (Figure 3). Starch decreased during the later stages of seed growth, a time when leaf photosynthesis rate also decreased, suggesting that starch mobilization contributes, in part, to the later stage of seed growth. Leaf sucrose levels were relatively constant during most of seed growth, but they did increase during early senescence. That the activity of leaf sucrose phosphate synthetase decreases in parallel with photosynthesis during seed-fill (Figure 4), suggests that the increase was due to a retention of sucrose within the source leaf rather than an increase in sucrose synthesis.

BASIS FOR INCREASED STARCH LEVELS

The increase in starch levels could result either from a preferential partitioning of photosynthetic carbon into starch versus water solubles/insolubles, or from a decrease in starch degradation leading to an accumulation. Figure 5 shows that there is no preferential partitioning of carbon, from either exogenous ^{14}C-glucose or from $^{14}CO_2$-derived assimilates, into starch during the first 30 days postanthesis, a time during which starch levels increase three-fold. Figure 5 (right) shows the ^{14}C-metabolite distribution following uptake of 2 mM glucose for 1 h into leaf discs obtained from soybeans at various stages of seed filling. The ^{14}C uptake rate and the percentage of label entering the water soluble and starch fractions were constant during 0 to 30 days postanthesis. The insoluble fraction contained 8 to 10% of the total ^{14}C accumulated at anthesis and decreased to about 2% by 30 days postanthesis. A similar trend (i.e., no increased partitioning of ^{14}C into starch) was noted for $^{14}CO_2$-derived material. For example, Figure 5 (right) shows the decline in leaf photosynthesis rate and increase in starch levels in soybean leaves during the first 30 days of seed growth. Experiments in which $^{14}CO_2$ was supplied to the same leaves used for the photosynthesis and starch measurements showed that about 40% and 60% of the label entered the insoluble and water soluble fractions, respectively (Figure 5, right). Thus,

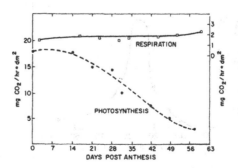

Figure 2. Soybean leaf photosynthesis and respiration rates during seed filling.

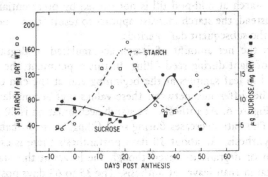

Figure 3. Soybean leaf starch and sucrose levels.

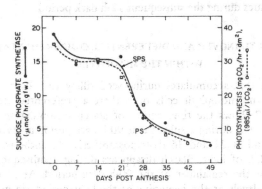

Figure 4. Sucrose phosphate synthetase activity and soybean leaf photosynthesis.

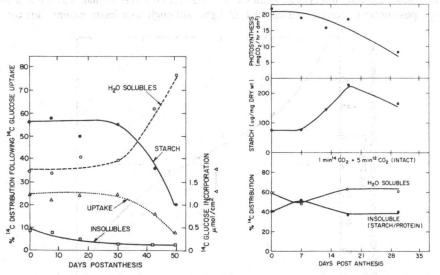

Figure 5. ^{14}C-metabolite distribution following ^{14}C-glucose accumulation for 1 h in soybean leaves (left); photosynthesis rate, starch levels, and distribution of ^{14}C-metabolites following $^{14}CO_2$ uptake (right).

732

the increase in leaf starch at mid-pod fill is not caused by preferential partitioning of carbon to starch. Instead, the starch increase appears to result from a decrease in starch degradation during the subsequent dark periods.

Figure 6 shows the net amount of starch accumulated or degraded in soybean leaves during 12 h of light during seed filling. In this experiment, the mass amount of starch was measured in leaf samples at the beginning and at the end of a 12-h photosynthetic period. The difference between these values (net accumulation or degradation) is plotted in Figure 6. At anthesis, net starch accumulation is 80 μg/mg dry wt·h in soybean leaves. This rate decreases during seed filling, most probably because of a decrease in photosynthesis. At about 30 days postanthesis there is essentially no net starch accumulation or degradation even though the leaves at this stage have two to three times more starch than leaves at anthesis. The 35 to 63 days postanthesis period is characterized by a net loss of starch from the leaves during the light period. This net loss of starch continues during the subsequent 12-h dark period.

STARCH TURNOVER AND DIFFERENTIAL COMPARTMENTATION
WITHIN THE MESOPHYLL

Does the starch that accumulates during seed filling turn over? Is starch evenly distributed among the mesophyll cells or is there a preferential storage within the mesophyll? Figure 7 shows the time course of starch accumulation and degradation during 12 h of light and during a subsequent 90 h dark period in leaves from plants at anthesis (0 day), and at 21 and 36 days postanthesis. At anthesis, starch increases threefold during 12 h of light. Starch disappears during a subsequent 12 h of dark and remains low for the remainder of the 90-h dark period. At 21 or 36 days postanthesis, the starch levels at the beginning of the light period are much greater than the levels observed in leaves at anthesis. The leaf starch levels in plants at 21 and 36 days postanthesis increase during 12 h of light, although to a lesser extent, but total

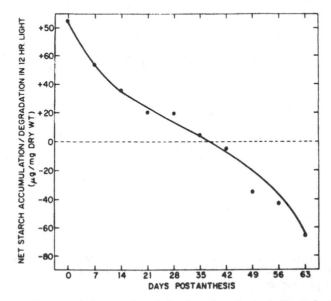

Figure 6. Net starch accumulation and degradation in soybean leaves during 12 h of light for days during seed filling.

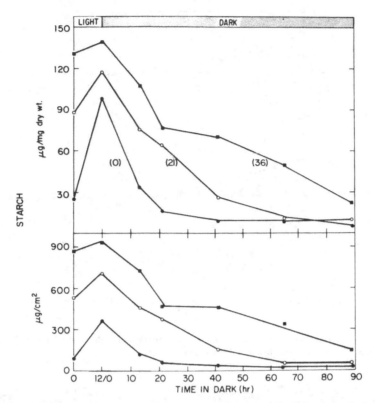

Figure 7. Time course of leaf starch accumulation and degradation at 3 times during seed filling. Upper panel is expressed as μg/mg dry wt; lower panel is expressed as μg/cm² leaf area.

starch remains high after 12 h of darkness. The starch in these plants can be completely degraded, but only after 65 to 90 h in darkness. Thus, it appears that the recently accumulated starch is mobilized, but a longer term starch pool also exists.

Starch in leaves of plants at 35 days post-anthesis is not degraded under the reduced sink demand caused by depodding, even after 140 h in darkness (Figure 8). The areas between the curves in Figure 7 represent the amount of leaf starch that is not turning over on a daily basis. Interestingly, this starch is compartmented preferentially within the mesophyll (Figure 9). Figure 9A shows a micrograph of a soybean leaf at the start of a light period soon after flowering. Most of the starch is in the second palisade parenchyma layer with very little in the upper layer of palisade parenchyma. Figure 9B is from the same plant, but at the end of a 12 h light period. Note the accumulation of starch in the upper palisade parenchyma. Thus, starch in the upper palisade layer is synthesized and degraded daily, whereas starch in the second palisade layer represents a longer term storage pool. The starch in the second palisade layer is available for mobilization during periods of extended dark (Figure 8) or during the later stages of seed filling, and during senescence (Figures 3, 6).

PHOSPHATE-INDUCED ALTERATIONS OF LEAF STARCH AND STARCH: SUCROSE RATIO

Studies on chloroplasts have shown that phosphate influences the distribution of fixed carbon between starch and sucrose. Under low phosphate concentrations, triose

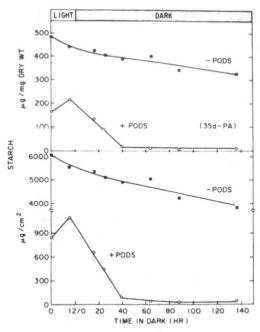

Figure 8. Time course of leaf starch accumulation and degradation in podded and depodded (flowers removed at anthesis) soybeans at 35 days post-anthesis.

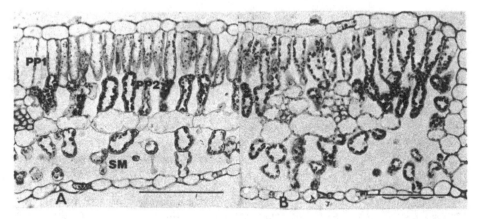

Figure 9. Light micrographs of soybean leaves at the start (A) and end (B) of a photoperiod. Note starch accumulation in PP2. Bar is 1 μm. Leaves were 3 μm thick, and were stained with basic fuchsin. PP1 is the upper palisade parenchyma; PP2 is the second palisade parenchyma; SM level is spongy mesophyll.

phosphate formed from photosynthesis remains within the chloroplast for starch synthesis. At high external phosphate concentrations, triose phosphate is exported across the chloroplast membrane to the cytoplasm for sucrose synthesis. The export of triose phosphate is coupled to the counter-exchange of phosphate via the "phosphate translocator" present in the chloroplast envelope (Heldt et al., 1977). We performed a series of experiments in an attempt to extend these findings to whole plants; that is, to soybeans during the seed filling stage. Our objective was to learn whether the starch and sucrose levels in soybean leaves could be altered by phosphate nutrition. Soybeans were grown in washed sand to the flowering stage using one-half strength of complete Hoagland's solution. At flowering, the sand was repeatedly washed with distilled water, and then different sets of soybeans were grown to maturity using one-half strength of Hoagland's solution in which the phosphate concentration was varied from phosphate deficiency (0.001 mM) to phosphate toxicity (100 mM). Leaf starch and sucrose levels were determined during seed filling. Figure 10 shows the leaf starch and sucrose levels, and the calculated leaf starch:sucrose ratio, from soybean plants that received 0.001, 0.1, 1, 100 mM phosphate. Both the amount of starch and the starch:sucrose ratio declined with increasing phosphate concentration. Since 0.001 and 100 mM phosphate were deficient and phytotoxic, respectively, we examined the effects of intermediate phosphate levels (0.5 and 5 mM) on these parameters (Figure 11). Phosphate concentrations of 0.5 and 5 mM had no appreciable affect on the decline of photosynthesis that occurred during seed filling, but 5 mM phosphate resulted in a lower starch level and lower starch:sucrose ratio. Figure 12 summarizes several experiments on the effects of phosphate concentrations on the leaf starch:sucrose ratio determined at mid-pod fill. It is evident that increasing phosphate concentrations supplied to the roots of sand-cultured plants markedly decrease the leaf starch:sucrose ratio.

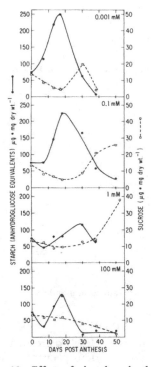
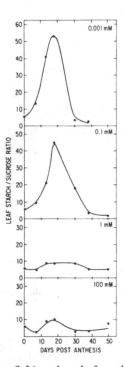

Figure 10. Effect of phosphate level on leaf starch and sucrose (left), and on leaf starch:sucrose ratio (right) during seed filling.

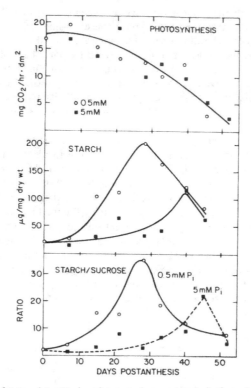

Figure 11. Effect of 0.5 and 5 mM phosphate on photosynthesis, leaf starch and starch:sucrose ratio.

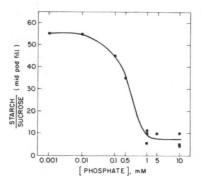

Figure 12. Effect of phosphate level on soybean leaf starch:sucrose ratio at mid-pod fill.

EFFECTS OF PHOSPHATE ON SEED GROWTH

Is the decrease in leaf starch induced by higher phosphate levels correlated with a change in seed growth? Figure 13 shows the dry weight of various plant parts during seed growth at 0.005, 0.05, 0.5 and 5 mM phosphate. Seed dry weight/plant was greater for the soybeans grown at 0.5 and 5 mM phosphate compared with the lower

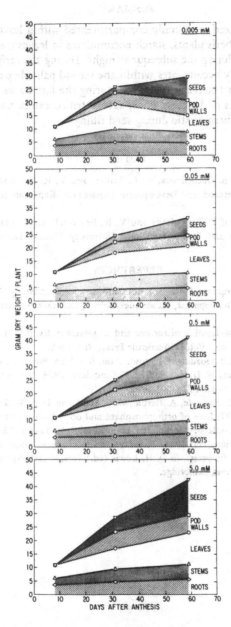

Figure 13. Effect of phosphate on dry matter accumulation in soybean plants.

phosphate concentrations. Since these were determinant plants that received a complete nutrient mix up to flowering, vegetative growth was not markedly affected by the phosphate regime. It is difficult to determine whether the increased seed growth at higher phosphate concentrations was due to more leaf carbohydrate availability for export to the seeds or from a phosphate enhancement of processes within the seeds.

SUMMARY

Starch is spatially and temporally compartmented within leaves during reproductive growth. In preanthesis plants, starch accumulates in leaves during the day and is completely mobilized during the subsequent night. During the early stages of seed filling, starch preferentially accumulates within the second palisade parenchyma layer for longer term storage. This starch is mobilized during the late stages of seed growth and during extended periods of darkness. Phosphate nutrition can be used to experimentally alter the leaf starch:sucrose ratio during seed filling.

NOTES

Robert T. Giaquinta, B. Quebedeaux, N. L. Sadler, and V. R. Franceschi, E. I. Du Pont de Nemours & Co., Central Research and Development Department, Experimental Station, Wilmington, Delaware 19898.

The present addresses of B. Quebedeaux and V. R. Franceschi are University of Maryland (Department of Horticulture) and Washington State University (Botany Department), respectively.

REFERENCES

Geiger, D. R. and R. T. Giaquinta. 1982. Translocation of photosynthate. p. 345-386. *In* Govindjee (ed.) Photosynthesis: Development, carbon metabolism, and plant productivity, Vol. 2. Academic Press, New York.

Giaquinta, R. T. 1980. Translocation of sucrose and oligosaccharides. p. 271-320. *In* J. Preiss (ed.) The biochemistry of plants. Vol. 3. Academic Press, New York.

Giaquinta, R. T. 1983. Phloem loading of sucrose. Ann. Rev. Plant Physiol. 34:347-387.

Gifford, R. M., J. M. Thorne, W. D. Hitz and R. T. Giaquinta. 1984. Crop productivity and photoassimilate partitioning. Science 225:801-808.

Heldt, H. W., C. J. Chon, D. Maronde, A. Herold, Z. S. Stankovic, D. A. Walker, A. Kraminer, M. R. Kirk and U. Heber. 1977. Role of orthophosphate and other factors in the regulation of starch formation in leaves and isolated chloroplasts. Plant Physiol. 59:1146-1155.

Thorne, J. H. and R. T. Giaquinta. 1984. Pathways and mechanisms associated with carbohydrate translocation in plants. (in press). *In* D. H. Lewis (ed.) Storage carbohydrates in vascular plants. Cambridge University Press, Cambridge.

ASSIMILATE TRANSPORT AND SOYBEAN SEED DEVELOPMENT

John H. Thorne

The first days of soybean seed development are marked by rapid cell division in the embryo. Vascularization also develops in the seed coat during this period (Carlson, 1973). Beyond about 20 days, seeds utilize external assimilates supplied via the pod from the vegetative canopy. The anatomical, physiological and biochemical aspects of the import process have been reviewed recently (Thorne, 1983). Briefly, these processes involve the transfer of assimilates between maternal (seed coat) and embryonic (cotyledon) tissues that are not cellularly linked. At least two of these steps involve carrier-mediated, energy-dependent processes that are sensitive to both environmental and chemical perturbations. The genetic and environmental controls of assimilate import and utilization by developing seeds appear to influence markedly soybean yield.

ROUTE OF ASSIMILATE IMPORT

Biochemical studies of oil and protein deposition during soybean seed growth indicate that large quantities of assimilates are transported to the embryonic cotyledons as they develop (Yazdi-Samadi et al., 1977). Anatomical and kinetic studies have revealed the following route of assimilate import in the developing pod: assimilates enter the pods via two major pod vascular bundles to which seeds are attached. Assimilates pass directly into the seed coat, essentially bypassing storage in the pod wall. A network of interconnected minor veins within the seed coat rapidly and evenly distributes the assimilates around the seed (Thorne, 1980; 1981). Assimilates exist (unload) from the phloem in the seed coat and subsequently emerge in the freespace separating the seed coat and the embryo (Figure 1). Assimilates emerging from the seed coat must diffuse across this freespace and into the cotyledons of the embryo; there are no cellular connections between the maternal seed coat and the embryo (Carlson, 1973; Thorne, 1981). Uptake by the cotyledons occurs from the cell walls following a lag created by the intercellular movement of assimilates from the phloem to the inner seed coat surface.

THE NUTRITIVE ROLE OF THE SEED COAT

Studies with intact pods have demonstrated the environmental sensitivity of the processes associated with the import of assimilates. Environmental regulation of import may occur at both the embryo level and at the phloem terminals within the seed coat (Thorne, 1982a). At higher ambient temperatures, the rate of oxygen diffusion

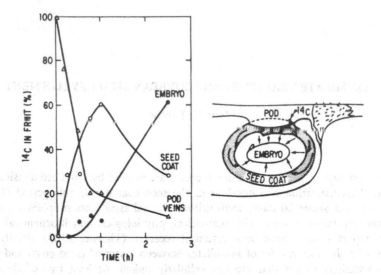

Figure 1. Kinetics of ^{14}C-photosynthate accumulation in field-grown soybean fruit following the application of a 60-s pulse of [14]CO$_2$ to the subtending leaf. Values represent time after arrival of the pulse in the fruit. The schematic illustrates the import and release of ^{14}C-photosynthate by the seed coat and the absorption by the cotyledons of the embryo. Adapted from Thorne (1980).

into the seeds may limit assimilate transport into seeds (Figure 2). Within the seed, unloading and efflux from the seed coat appear to be much more sensitive to oxygen availability than is subsequent import by the cotyledons; anoxia completely inhibits

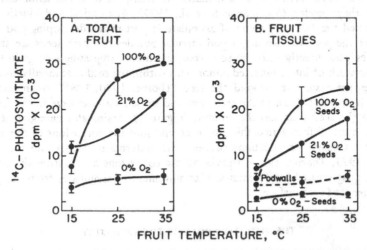

Figure 2. Accumulation of ^{14}C-photosynthate in field-grown soybean fruit following the labeling of the entire plant with [14]CO$_2$. Intact pods were contained in small plastic chambers maintained at temperatures of 15, 25, and 35 C and oxygen concentrations of 0, 21, and 100%. (Adapted from Thorne (1980).

unloading in the seed coat, but only reduces uptake by isolated cotyledons about 35% (Thorne, 1982a). Consistent with an inhibition of assimilate import and utilization, localized subambient oxygen treatments have been shown to prevent seed development (Quebedeaux and Hardy, 1975). Both seed coat and embryo processes are also temperature dependent, consistent with published reports of temperature influences on soybean yield components: harvest index, seed number, rate and duration of seed growth, final seed size, and nutritional quality (Egli et al., 1978, 1980; Hesketh et al., 1973; Howell and Cartter, 1953; Saito et al., 1970; Seddigh and Jolliff, 1984; Thorne, 1979).

The direct study of the nutritive role of the seed coat has begun recently with the development of an *in vivo* surgical technique in soybean (Thorne and Rainbird, 1983) and in *Vicia faba* (Wolswinkel and Ammerlaan, 1983). Access to the seed coat phloem for experimental manipulation of the unloading process can be gained by taking advantage of the lack of cellular connections between the maternal seed coat and the developing embryo. This technique involves making an incision in the pod wall, cutting the distal (unattached) halves of the seeds away, and gently removing the remaining embryo halves from the remainder of the attached seed coat. The exposed seed coat forms a cup (Figure 3) which maintains near-normal assimilate import, phloem unloading and efflux into a trap solution or agar matrix (agar allowed to harden in the cup).

The sites of phloem unloading, buried within the seed coat tissue, are accessible via the cell wall freespace and can be challenged with inhibitors, solutes, buffers, etc. to characterize the unloading process. PCMBS (p-chloromecuribenzoylsulfonate), a nonpenetrating sulfhydryl reagent known to bind reversibly to solute carriers of leaf phloem and other transporting cells, markedly inhibits assimilate release by the seed coat. A brief PCMBS pretreatment effectively inhibits phloem unloading and assimilate

Figure 3. Photograph of a surgically altered soybean pod and the attached seed coat cup in which assimilates may be trapped. The pod is attached to the plant, but held in an upright manner. Agar or buffer solution may be added to trap assimilates. From Thorne and Rainbird (1983).

release to the seed coat trap, consistent with the involvement in unloading of membrane-bound carriers. Dithiothreitol partially reverses the PCMBS inhibition of phloem unloading. Treatment with the membrane-penetrating metabolic inhibitors sodium fluoride, sodium arsenite or dinitrophenol completely prevents sucrose import (Thorne and Rainbird, 1983).

Unloading of assimilates into agar traps buffered with nonpenetrating zwitterionic buffers is maximal (Figure 4) under acid conditions (Thorne, 1983). The pH relationships of assimilate delivery are not resolved, however, for unloading into liquid buffer appears less sensitive to pH changes (Gifford and Thorne, unpublished). Unloading is stimulated 300 to 400% by micromolar concentrations of the fungal toxin, fusicoccin.

The presence of greater than 5 mM EGTA or EDTA in the seed coat trap profoundly enhances the release of assimilates, probably by chelating membrane-bound divalent cations from the seed coat phloem. Physiological unloading is apparently overwhelmed by massive phloem leakage or exudation in the presence of EGTA. EGTA provides a means for greatly stimulating the import and, thus 'sink strength', of an altered seed to determine the effect of a particular developing seed on unloading in adjacent seeds sharing vascular connections in the same pod. When 25 mM EGTA is present in the agar trap of the seed coat cup of one seed position, however, the 30-fold stimulation of import there has absolutely no effect on the unloading and release of assimilates in other seed positions (Figure 5), regardless of location in the pod (Thorne and Rainbird, 1983).

These data and others indicate that the delivery of assimilates to the embryo by the seed coat is a controlled process, and apparently is energy-dependent, carrier-mediated, and extremely sensitive to treatments that alter membrane integrity and permeability. The unloading mechanism is apparently a one-way process, since radiolabeled sucrose

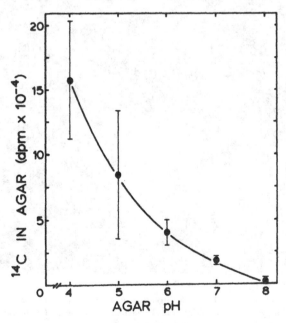

Figure 4. Effects of seed coat freespace pH on unloading of radiolabeled assimilates into the cup of surgically altered soybean seed coats. Zwitterionic buffers of an appropriate pKa were incorporated in the agar trap (4% agar containing 5 mM buffers).

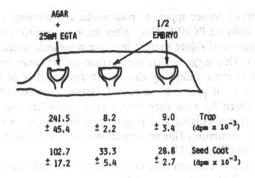

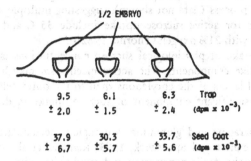

Figure 5. Autonomy of assimilate import among adjacent seeds in soybean pods. Three-seeded pods (n=5/treatment) were cut as described in the text. The embryo in one seed was replaced with agar containing 25 mM EGTA (pH 6). All other seeds were cut but with the half embryo left in place as a trap of ^{14}C-assimilates following labeling of the plant. Described in Thorne and Rainbird (1983).

added to the seed coat cup is not phloem transported to other parts of the plant (Bennett et al., 1984).

ASSIMILATE ACCUMULATION IN ISOLATED COTYLEDONS

Although it has been possible to evaluate the kinetics of assimilate accumulation and metabolism with embryos of intact seeds (Figures 1 and 2), mechanistic transport studies of specific solutes require the use of isolated cotyledons (or protoplasts isolated from the cotyledons). The lack of cellular connections between the maternal seed coat and the developing embryonic cotyledons facilitates the surgical removal of the embryos for in vitro culture.

Recent studies of the absorption of sucrose (the major assimilate transported to the embryos in vivo) by isolated soybean embryos show that much of the uptake of dilute exogenous sucrose results from a carrier-mediated cotransport of sucrose and protons (Lichtner and Spanswick, 1981a,b; Thorne, 1982b). Exogenous sucrose is accumulated both by saturable, carrier-mediated and by linear, diffusion-like transport mechanisms, resulting in a biphasic dependence on external sucrose concentration. PCMBS can be used to totally inhibit the active, carrier-mediated transport component and reveal the linear component. Metabolic poisons and proton ionophores are also effective inhibitors of active sucrose transport.

Very young embryos do not appear to possess the active transport mechanism and, as such, are not sensitive to PCMBS. Only after attaining approximately 60 mg fresh weight does the transport of dilute sucrose increase substantially and become sensitive to PCMBS (Figure 6). This suggests that the sucrose carriers may be inserted into the cotyledon cell plasmalemma following the switch from a diet of liquid endosperm (rich in hexoses) to one of primarily sucrose (Thorne, unpublished data). Although glucose is rapidly utilized by even very young cotyledons, its transport shows little evidence of being carrier mediated and is generally insensitive to free space pH and PCMBS (W. Lin, pers. commun.; Thorne, unpublished data). Hexoses are minor assimilate components in vivo (Thorne, 1980).

The active transport mechanism utilizes ATP to establish the necessary pH gradient across the cotyledon-cell plasmalemma to drive sucrose transport. Although anoxia and low temperatures reduce active transport somewhat (Figure 7), metabolic poisons severely inhibit this process (data not shown), suggesting multiple sources of energy. Optimum conditions for active sucrose uptake include 35 C and pH 6 incubation medium equilibrated with 21% oxygen (Thorne, 1982b).

Much of the uptake at physiological sucrose concentrations is due to the non-saturable, diffusion-like components. The active mechanism may be more important to the innermost cell layers of the cotyledons than to the outer cell layers. Nevertheless, it represents a significant component of sucrose uptake in developing soybean embryos.

The amides, asparagine and glutamine, are rapidly accumulated by developing soybean embryos (Bennett and Spanswick, 1983; Rainbird et al., 1984) in a manner similar to glucose uptake, facile penetration and rapid metabolism within the cells of the cotyledon. Sensitivity to metabolic poisons is due primarily to inhibition of this metabolism. Ureides and amino acids such as aspartate and glutamate are only slowly accumulated (Rainbird et al., 1984).

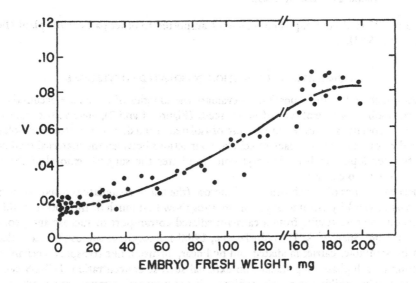

Figure 6. Rate of uptake of 1 mM sucrose by embryos of various size. V (velocity) = μmoles sucrose/h 0.1 g fresh weight. Uptake is in the presence of 5 mM MES, pH 6.

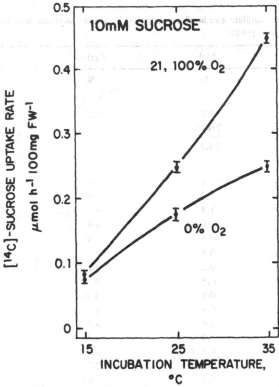

Figure 7. Rates of ^{14}C-sucrose uptake by excised immature soybean embryos from 10 mM solutions buffered to pH 6 with MES. Solutions were in equilibrium with 0, 21, or 100% oxygen. Data for 21 and 100% oxygen were indistinguishable. Metabolic poisons substantially reduce all values (data not shown). From Thorne (1982a).

SOURCE OF ASSIMILATES FOR STORAGE PRODUCT FORMATION

The developing soybean embryo is totally dependent on the surrounding maternal seed coat for all aspects of nutrition: assimilates, inorganic ions, water and oxygen. For the most part the nutritive role of the seed coat is one of delivery; however, recent studies indicate that the seed coat may regulate the synthesis or metabolism of specific phytohormones (M. Brenner, pers. commun.) and nitrogenous assimilates (Rainbird et al., 1984; in *Pisum sativum*; Murray, 1979).

Assimilates delivered by surgically accessed seed coats (Figure 3) are primarily sucrose and amino acids: 90% of the total carbon is sucrose and the carbon skeletons of the nitrogen solutes constitute the remaining 10% (Table 1). Glutamine is the principal nitrogenous solute released by the seed coat, comprising 52% of the nitrogen present. A further 19% of the nitrogen is asparagine. Methionine, a component of the 11S storage protein and regulator of the synthesis of the 7S storage protein (Holowach et al., 1984) is released only in trace amounts. Similarly, the otherwise ubiquitous ureides, allantoin and allantoic acid, are released by the seed coat in only trace amounts (Thorne and Rainbird, 1983; Rainbird et al., 1984). Other studies confirm that the ureides are extensively metabolized in the seed coat during import (Coker, pers. commun.). Ammonia, perhaps from ureide breakdown, represents nearly 7% of the nitrogen released by the seed coat (Rainbird et al., 1984).

Table 1. Components of assimilate exudate[a] from surgically altered soybean seed coats. Adapted from Rainbird et al. (1984).

Component	Mol	Carbon	Nitrogen
	%	%	%
Sugars:			
Sucrose	78.7	90.7	
Hexoses	Trace		
Total	70.7		
Amino Acids:			
Glutamine	10.4	5.0	52.6
Asparagine	3.7	1.4	18.8
Ammonia	1.3	0	6.9
Serine	0.8	0.4	2.2
Histidine	0.6	0.3	4.4
Proline	0.5	0.3	1.4
Threonine	0.5	0.2	1.3
Arginine	0.5	0.3	4.9
Valine	0.5	0.2	1.2
Alanine	0.4	0.1	1.1
Aspartate	0.4	0.2	1.1
Leucine	0.4	0.2	0.9
Glutamate	0.3	0.1	0.7
Isoleucine	0.3	0.1	0.7
Glycine	0.2	0.04	0.6
Phenylalanine	0.2	0.2	0.5
Tyrosine	0.2	0.2	0.5
Methionine	Trace		
Totals	21.2	9.24	100.0
Ureides:			
Allantoin	Trace		
Allantoic Acid	Trace		

[a]C:N ration = 30.8 mg C/mg N.

UNRESOLVED QUESTIONS

It seems apparent that photosynthate transport in developing soybean seeds is under two levels of control: active uptake of sucrose in cotyledon cells of the embryo, and controlled release by the seed coat phloem. Clearly, the latter provides an upper limit on the assimilate availability to, and growth of, the embryo. Yet, it is conceivable that the embryo could, in some manner, regulate the biochemistry and physiology of the seed coat, and in turn, alter the rate of assimilate release and even the chemical form of the incoming assimilates.

Other unresolved questions quickly come to mind: an understanding is needed of the determinants of physiological maturity and the cessation of assimilate import, if we hope to influence profitably the duration of seed growth. Further work is needed to understand environmental effects of embryo cell number and potential seed size and growth rate; and the roles of light and phytohormones in controlling soybean seed development surely deserve careful consideration. All may be important in searching for strategies for improving soybean yields.

747

NOTES

John H. Thorne, Plant Sciences Group Leader, Central Research and Development Dept., Experimental Station, E.I. DuPont de Nemours, Inc., Wilmington, Del. 19801.

REFERENCES

Bennett, A. B. and R. M. Spanswick. 1983. Derepression of amino acid-H^+ cotransport in developing soybean embryos. Plant Physiol. 72:781-786.

Bennett, A. B., B. L. Sweger and R. M. Spanswick. 1984. Sink to source translocation in soybean. Plant Physiol. 74:434-436.

Carlson, J. B. 1973. Morphology. p. 17-95. In B. E. Caldwell (ed.) Soybeans: Improvement, Production, and Uses. Am. Soc. Agron., Madison, Wis.

Egli, D. B., J. E. Leggett and J. M. Wood. 1978. Influence of soybean seed size and position on the rate and duration of filling. Agron. J. 70:127-130.

Egli, D. B. and I. F. Wardlaw. 1980. Temperature response of seed growth characteristics of soybean. Agron. J. 72:560-564.

Hesketh, J. D., D. L. Myhre and C. R. Willey. 1973. Temperature control of time intervals between vegetative and reproductive events in soybeans. Crop Sci. 13:250-254.

Holowach, L. P., J. F. Thompson and J. T. Madison. 1984. Effects of exogenous methionine on storage protein composition of soybean cotyledons cultured in vitro. Plant Physiol. 74:576-583.

Howell, R. W. and J. L. Cartter. 1953. Physiological factors affecting composition of soybeans. 1. Correlation of temperatures during certain portions of the pod filling stage with oil percentage in mature beans. Agron. J. 45:526-528.

Lichtner, F. T. and R. M. Spanswick. 1981a. Electrogenic sucrose transport in developing soybean cotyledons. Plant Physiol. 67:869-874.

Lichtner, F. T. and R. M. Spanswick. 1981b. Sucrose uptake by developing soybean cotyledons. Plant Physiol. 68:693-698.

Murray, D. R. 1979. Nutritive role of the seedcoats during embryo development in *Pisum sativum* L. Plant Physiol. 64:763-769.

Rainbird, R. M., J. H. Thorne and R. W. F. Hardy. 1984. Role of aides, amino acids, and ureides in the nutrition of developing soybean seeds. Plant Physiol. 74:329-334.

Saito, M., T. Yamamoto, K. Groto and K. Hashimoto. 1970. The influence of cool temperature before and after anthesis, on pod-setting and nutrients of soybean plants. Proc. Crop Sci. Soc. Japan 39:511-519.

Seddigh, M. and G. D. Jolliff. 1984. The effects of night temperature on dry matter partitioning and seed growth of indeterminate field-grown soybean. Crop Sci. (in press).

Thorne, J. H. 1979. Assimilate redistribution from soybean pod walls during seed development. Agron. J. 71:812-816.

Thorne, J. H. 1980. Kinetics of ^{14}C-photosynthate uptake by developing soybean fruit. Plant Physiol. 65:975-979.

Thorne, J. H. 1981. Morphology and ultrastructure of maternal seed tissues of soybean in relation to the import of photosynthate. Plant Physiol 67:1016-1025.

Thorne, J. H. 1982a. Temperature and oxygen effects of ^{14}C-photosynthate unloading and accumulation in developing soybean seeds. Plant Physiol. 69:48-53.

Thorne, J. H. 1982b. Characterization of the active sucrose transport system of immature soybean embryos. Plant Physiol. 70:953-958.

Thorne, J. H. 1983. Transport of photosynthate into developing soybean seeds. p. 43-53. In D. A. Randall (ed.) Current Topics in Plant Biochemistry and Physiology. Univ. Missouri, Columbia.

Thorne, J. H. and R. M. Rainbird. 1983. An *in vivo* technique for the study of phloem unloading in seed coats of developing soybean seeds. Plant Physiol. 72:268-271.

Quebedeaux, B. and R. F. W. Hardy. 1975. Reproductive growth and dry matter production of *Glycine max* (L.) Merr. in response to oxygen concentration. Plant Physiol. 55:102-107.

Wolswinkel, P. and A. Ammerlaan. 1983. Phloem unloading in developing seeds of *Vicia faba* L. The effects of several inhibitors on the release of sucrose and amino acids by the seed coat. Planta 158:205-215.

Yazdi-Samadi, B., R. W. Rinne and R. D. Seif. 1977. Components of developing soybean seeds: oil, protein, sugar, starch, organic acids, and amino acids. Agron. J. 69:481-486.

AMINO ACID EFFLUX FROM CELLS AND LEAF DISCS

J. Secor and L. E. Schrader

Whether, or to what extent, soybean yield is limited by carbon or by nitrogen is an issue as yet unresolved. Vegetative tissue, principally leaf lamina, contributes about half the carbon and nitrogen to mature seeds (Hume and Criswell, 1973). Translocated nutrients arise primarily from mesophyll cells. The identification of translocated nutrients, their translocation routes, and loading and unloading mechanisms are all areas of current, active research.

Sucrose has been established as the principal sugar translocated in soybean phloem sap; however, principal forms of translocated nitrogen in the phloem sap are less firmly established. Recent work has indicated that substantial amounts of amino acids are in the petiolar phloem sap, regardless of the form of nitrogen provided to roots (McNeil and LaRue, 1984; Layzell and LaRue, 1982). Even when xylem sap has high levels of ureides, petiolar phloem sap contains more amino acids than ureides. The relative consistency of phloem sap composition suggests that it is controlled by some regulatory process(es). Regulation of phloem sap composition is important because developing soybean embryos nutritionally depend upon amino acids, rather than ureides, for nitrogen (Rainbird et al., 1984; Hsu et al., 1984).

The location of selective regulatory mechanisms that control phloem sap composition is unknown. Compositional changes have been reported to occur in the phloem pathway. Layzell and LaRue (1982) showed that the major forms of nitrogen changed from 44% amino acids and 36% ureides in the petiole to 26% amino acids and 55% ureides in the fruit tip phloem sap. This suggests that metabolic conversions and/or selective withdrawal and/or deposition occurred. Ureides are now believed not to be taken up readily into the seeds (Rainbird et al., 1984); therefore, the changes in phloem sap composition reported by Layzell and LaRue could have been a consequence of selective withdrawal of amino acids by seeds. Because xylem-to-phloem transfers are quantitatively similar among treatments having different xylem sap compositions, and because phloem sap composition is similar among those treatments (McNeil and LaRue, 1984), it is likely that one point of control is where phloem loading occurs. That is, it is very likely that ureides are not readily loaded into the phloem. Thus, plants having a plethora of ureides in their xylem sap must somehow transfer the nitrogen from the ureides to carbon skeletons in order to synthesize the amino acids found in the phloem. The likely site of this action is the leaf lamina.

Within the leaf, there are several sites at which phloem sap composition can be regulated. First, loading of the phloem may involve specific carriers of different affinities. Second, the availability of compounds for loading may be important. Availability can be controlled by flux through the plasmalemma and/or organellar membranes (such as amino acid flux through the chloroplastic inner membrane).

Housley et al. (1977) established that amino acid transport occurs in the phloem. The uptake of amino acids into the phloem occurs from the apoplast and loading is dependent on metabolic energy, is distinct from sucrose, and is nonselective among several amino acids (Housley et al., 1979; Servaites et al., 1979; Schrader et al., 1980). Furthermore, the predominant, photosynthetically derived, radioactively labelled amino acids in petioles differ from the predominant, radioactively labelled ones in the leaf from which they had arisen (Schrader et al., 1980). Thus, because phloem loading is nonselective among amino acids, and because amino acid compositions of petiolar phloem and leaf lamina differ, it may be that the availability of amino acids in the apoplast dictates what is loaded into the phloem. The availability is apparently controlled by what is exported from mesophyll cells. Therefore, because experimental evidence points strongly to nonvascular tissue as being an important regulatory site, we investigated efflux of amino acids from mesophyll cells.

NET AMINO ACID EFFLUX FROM ISOLATED MESOPHYLL CELLS

Isolated cells were chosen as experimental material for two reasons: (1) they provide a convenient means to control experimental conditions, and (2) export from cells represents efflux strictly into the apoplast, whereas efflux from intact tissue occurs through both the apoplasm and symplasm. Thus, with isolated cells, there are no confounding effects of the two transport pathways.

Cells from reproducing plants were isolated using a mechanical-enzymatic technique, and subsequently were incubated in a buffered solution (Secor and Schrader, 1984). Efflux rates were determined by removing aliquots of solution at various times, then separating the cells from the bathing solution, and analyzing the cell-free fractions for total amino acid (ninhydrin-reactive compounds) content. An important consideration is that we measured net efflux, which is a consequence of simultaneous efflux and influx.

Net amino acid efflux does not seem to require concurrent photosynthetic energy. Net efflux was not significantly promoted when cells were aerated or when exposed to a photosynthetic photon flux density of 900 μmol quanta/s$\cdot$m (Table 1). Furthermore, DCMU did not enhance net amino acid efflux. These data suggest that diffusion, rather than a process directly dependent upon photosynthetic energy, is responsible for amino acid efflux. The effect of temperature supports the concept that net amino acid efflux may be a diffusional process. Q_{10} values ranged from 1.4 for the 10 to 20 C range to 2.7 for the 30 to 40 C range. However, the Q_{10} from 10 to 30 C, the physiological range for soybean plants, was 1.6. Because a Q_{10} of 2 is normally representative of biological activity, the Q_{10} of 1.6 for net efflux may represent a diffusional process that must overcome a barrier, which in this case is the plasma membrane.

No evidence existed that net amino acid efflux was mediated by an energy-requiring proton-ion pumping mechanism. Neither K^+ nor pH had consistently significant effects (Table 1). Thus, the experimental evidence suggests that net amino acid efflux is a diffusional process. The observations that an uncoupler of phosphorylation, FCCP, and an inactivator of proteins containing functional sulfhydryl groups, PCMBS, increased net amino acid efflux were of interest. It was unlikely that membrane permeability was increased because we have found that PCMBS reduces sucrose efflux. Further studies were undertaken to investigate the efflux of individual amino acids.

Table 1. Summary of experimental results characterizing net amino acid efflux from isolated cells (Secor and Schrader, 1984). Values are means ± SE (p = 0.68).

Experimental objective	Treatment	Efflux rate
		nmol/h·10^6 cells
Effect of aeration	aerated	8.4±0.7
	not aerated	7.2±0.5
Effect of light	light	16.6±3.9
	dark	15.8±5.1
Effect of temperature	10 C	9.1±1.7
	20 C	12.6±0.8
	30 C	24.1±5.2
	40 C	65.8±9.8
Effect of inhibitors	control	17.4±2.3
	50 μM FCCP	95.5±3.1
	1 mM PCMBS	38.6±2.2
	1 mM DCMU	18.7±0.9
Effect of K^+	0 mM	15.0±1.6
	1 mM	10.7±2.0
	10 mM	13.5±4.3
	100 mM	12.6±3.0
Effect of pH	pH 4	14.5±2.0
	pH 5	8.5±2.6
	pH 6	10.8±1.0
	pH 7	10.5±1.8
	pH 8	13.0±4.8

NET EFFLUX OF INDIVIDUAL AMINO ACIDS FROM
ISOLATED MESOPHYLL CELLS

Intra- and extracellular amino acid pools were analyzed at various intervals over a 210 min incubation period (Figure 1). The efflux pattern for each amino acid (except ornithine) was linear, but export rates varied greatly among amino acids. Additionally, some (e.g., aspartic acid, serine) had a lag period before export commenced. The amino acids that were exported fastest were alanine, lysine, leucine, and glycine. In fact, these four amino acids constituted about 50% of the external amino acid pool (Table 2). That these amino acids also constituted about 50% of the intracellular amino acid pool would suggest that efflux rate, or extracellular accumulation, is highly dependent upon a concentration gradient. This is not entirely true, because temporal changes of individual export rates showed no clear relationships with respective intracellular amino acid pools (Figure 1), or with lipid solubility (membrane permeability) according to the Merck Index. Internal amino acid pools may have changed because of both intracellular metabolic conversions and export during incubation periods. Perhaps, the strongest evidence for a mediated export is that the nonprotein amino acid gamma-aminobutyric acid (Gaba) constituted about 15% of the internal amino acid pool, while it had a very slow export rate. When the extra- and intracellular amino acid pool sizes are represented on a volume (concentration) rather than a cell number basis, it can be seen that the concentration gradient strongly favors efflux, yet the efflux rate of Gaba

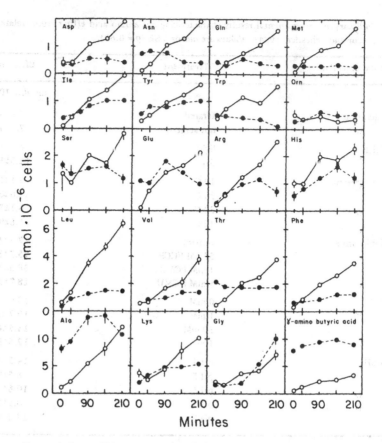

Figure 1. Changes in intracellular (●) and extracellular (○) amino acid content over a 210 min incubation period (Secor and Schrader, 1984).

Table 2. Predominant amino acids outside and inside cells after a 210 min incubation period (modified from Secor and Schrader, 1984).

Amino acid	Extracellular		Intracellular	
	nmol/10^6 cells	%	nmol/10^6 cells	%
Alanine	9.0±3.2	15	13.7±2.8	23
Lysine	8.1±2.0	14	5.0±0.3	9
Glycine	7.2±0.1	12	10.2±0.2	17
Leucine	4.8±1.7	8	2.0±0.5	3
Gaba	2.8±0.5	5	8.8±0.3	15
Others	26.8±6.6	46	19.1±5.7	33

is relatively slow (Table 3). Whether this indicates that Gaba is not available for export
(sequestered in an organelle or vacuole), or that it is selectively excluded from export,
or is rapidly taken up from the external solution is unknown.

Further evidence that selective export occurs is indicated by changes in the distribution of the internal and external pools of individual amino acids when cells are

Table 3. Concentration gradients of some amino acids outside and inside isolated cells after a 210 min incubation period.

Amino acid	Outside	Inside
	– μM –	
Alanine	9.7±3.4	195.8±40.0
Lysine	8.7±2.2	71.5± 4.3
Glycine	7.7±0.1	145.8± 2.9
Leucine	5.2±1.8	28.6± 7.1
Gaba	3.0±0.5	125.8± 4.3

subjected to various inhibitors (Secor and Schrader, 1984). For example, the partitioning of alanine and Gaba changes considerably when cells are incubated in 1 mM PCMBS (Figure 2). The addition of PCMBS, which inactivates proteins having functional sulfhydryl groups, increased the external partitioning of both alanine and Gaba, but perhaps via different mechanisms. On the one hand, there was little change in the total (intra- plus extracellular) amount of alanine, thus, alanine efflux occurred at the expense of the internal pool. On the other hand, unlike alanine the increase in export of Gaba was accompanied by an increase in its internal pool. Apparently, PCMBS increased the amount of Gaba accessible for export. In contrast, alanine export was independent of internal pools and its export may have depended upon other factors, such as a balance between simultaneous import and export. Further support that alanine export may be independent of its internal pool size is that 50 μM FCCP increases alanine export without affecting internal pool size (Secor and Schrader, 1984).

In summary, the data suggest that amino acids are exported from soybean cells via

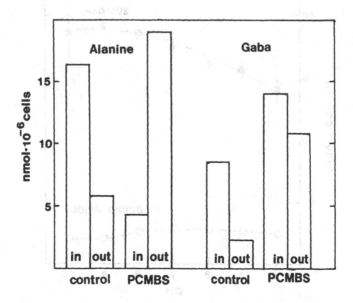

Figure 2. Effect of 1 mM PCMBS on the intra- and extracellular partitioning of alanine and Gaba after a 210 min incubation period (modified from Secor and Schrader, 1984).

a selective, diffusional process. Except for FCCP, treatments that reduced photo-synthesis or changed the proton-ion balance of cells did not significantly affect net amino acid efflux. However, intracellular amino acid concentration was not necessarily an indicator or a direct determinant of the rate of efflux. A good example is Gaba, which normally constituted a major portion of the total intracellular amino pool, yet normally had a slow net export rate. When compared with Gaba, alanine, which always had a fast net export rate and a large intracellular pool, seemed to have a different export control mechanism. The reasons for selective efflux are unclear; possibly, there are selective, nonenergetic export carriers present in the plasmalemma, or selective up-take carriers in the plasmalemma, or some amino acids are sequestered within cells and are not available for export.

SUCROSE AND AMINO ACID EFFLUX FROM LEAF DISCS

Recent reports (Franceschi and Giaquinta, 1983a; Franceschi et al., 1983) have indicated that paraveinal mesophyll cells may be important regulators of nutrient transport in leaf blades. Our investigations were undertaken to determine export characteristics of tissue containing intact cell layers. Net export of sucrose and total amino acids from leaf discs from leaves of reproducing plants was studied. Net export of total amino acids was about 50 to 90% slower than sucrose export and had physiological characteristics different from those of sucrose efflux. For example, net sucrose efflux increased to a plateau at about pH 8, whereas total net amino acid efflux hardly varied with pH, or perhaps, slightly declined above pH 8 (Figure 3). Also, we have demonstrated that PCMBS inhibits sucrose efflux while, in the same tissue, it promotes total

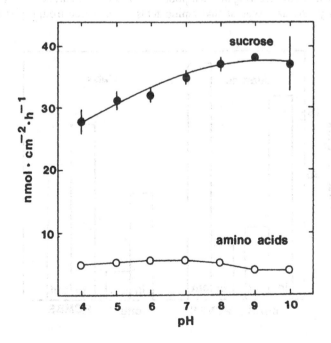

Figure 3. Effect of pH on net sucrose and net amino acid efflux from leaf discs after a 150 min incubation period.

amino acid efflux. This leads us to hypothesize that both the sucrose export carrier and the amino acid import carrier(s) are blocked simultaneously. Further studies are necessary to test this hypothesis.

NOTES

J. Secor, Dow Chemical USA, P.O. Box 9002, Walnut Creek, CA 94598-0902. L. E. Schrader, Department of Agronomy, University of Wisconsin—Madison, Madison, WI 53706.

Part of this research was supported by the College of Agricultural and Life Sciences, University of Wisconsin, Madison, by American Soybean Association Research Foundation Grant 80383, and U.S. Department of Agriculture, Competitive Research Grant No. 59-2551-0-1-445-0.

REFERENCES

Franceschi, V. R. and R. T. Giaquinta. 1983a. The paraveinal mesophyll of soybean leaves in relation to assimilate transfer and compartmentation. I. Ultrastructure and histochemistry during vegetative development. Planta 157:411-421.

Franceschi, V. R. and R. T. Giaquinta. 1983b. The paraveinal mesophyll of soybean leaves in relation to assimilate transfer and compartmentation. II. Structure, metabolic and compartmental changes during reproductive growth. Planta 157:422-431.

Franceschi, V. R., V. A. Wittenbach and R. T. Giaquinta. 1983. The paraveinal mesophyll of soybean leaves in relation to assimilate transfer and compartmentation III. Immunohistochemical localization of specific glycoproteins in the vacuole after depodding. Plant Physiol. 72:586-589.

Housley, T. L., D. M. Peterson and L. E. Schrader. 1977. Long distance translocation of sucrose, serine, leucine, lysine, and CO_2 assimilates. I. Soybean. Plant Physiol. 59:217-220.

Housley, T. L., L. E. Schrader, M. Miller and T. L. Setter. 1979. Partitioning of [^{14}C] photosynthate, and long distance translocation of amino acids in preflowering and flowering, nodulated and non-nodulated soybean. Plant Physiol. 64:94-98.

Hume, D. J. and J. G. Criswell. 1973. Distribution and utilization of ^{14}C-labelled assimilates in soybeans. Crop Sci. 13:519-524.

Hsu, F. C., A. B. Bennett and R. M. Spanswick. 1984. Concentrations of sucrose and nitrogenous compounds in the apoplast of developing soybean seed coats and embryos. Plant Physiol. 75:181-186.

Layzell, D. B. and T. A. LaRue. 1982. Modeling C and N transport to developing soybean fruits. Plant Physiol. 70:1290-1298.

McNeil, D. L. and T. A. LaRue. 1984. Effect of nitrogen source on ureides in soybean. Plant Physiol. 74:227-232.

Rainbird, R. M., J. H. Thorne and R. W. F. Hardy. 1984. Role of amides, amino acids, and ureides in the nutrition of developing soybean seeds. Plant Physiol. 74:329-334.

Schrader, L. E., T. L. Housley and J. C. Servaites. 1980. Amino acid loading and transport in phloem. p. 101-109. In F. T. Corbin (ed.) World soybean research conference II. Proceedings. Westview Press, Boulder, Colorado.

Secor, J. and L. E. Schrader. 1984. Characterization of amino acid efflux from isolated soybean cells. Plant Physiol. 74:26-31.

Servaites, J. C., L. E. Schrader and D. M. Jung. 1979. Energy-dependent loading of amino acids and sucrose into the phloem of soybean. Plant Physiol. 64:546-550.

EXPORT OF NITROGENOUS COMPOUNDS FROM SOYBEAN ROOTS

John G. Streeter

When cut off, the shoot of a soybean plant exudes liquid from the cut surface above the root; i.e., the root continues to develop a small positive pressure within the vascular system, technically within the xylem. This liquid is variously referred to as "bleeding sap", "xylem sap", "xylem exudate", or "stem exudate". The term "root exudate" is occasionally used, but this is incorrect; root exudate should be used to refer only to materials which are lost from roots to the rhizosphere. Since exudate may be collected from other organs (petioles, pods), and since the term "sap" is often used to refer to liquid obtained by crushing plant tissue, the term "stem exudate" may be least ambiguous and most descriptive and will be used in this review.

Soybean stem exudate can be collected in quantities large enough to permit analysis of individual carbon and nitrogen compounds. The changing spectrum of compounds can be very informative as to the response of the plant (and especially the root system) to various environmental influences, such as nitrogen nutrition.

Following the discovery of the ureides allantoin and allantoate in soybean stem exudate (Matsumoto et al., 1976), there was an explosion of interest in these nitrogen-rich compounds. Since the last review of this subject (Israel and McClure, 1980) several major papers, which enlarge the understanding of stem exudate composition, have been published. In addition, there has been significant recent progress on the biochemistry and localization of ureide synthesis in soybean nodules. Unfortunately, the space allotted for this review will not permit a discussion of biosynthetic pathways.

Many different lines of evidence have established that ureides in stem exudate originate in nodules and represent the major export products of N_2 fixation in soybean nodules (Fujihara and Yamaguchi, 1980; Streeter, 1979; Herridge, 1982a; Israel and McClure, 1980). This realization has spawned a great deal of interest in the possibility of using ureide analysis to estimate N input from N_2 fixation. In view of the practical importance of this possibility and the appearance of a number of recent papers on this subject, I have extended my review of root export to include this topic.

STEM EXUDATE COMPOSITION

It should first be acknowledged that stem exudate contains a wide variety of carbon and nitrogen compounds. The most complete inventory published thus far has been reported by Layzell and LaRue (1982), and a portion of their data is shown in Table 1. Sugars, cyclitols, and organic acids comprised a significant portion (15 to 26%) of total carbon. Even though ureides, which have a C/N ratio of 1.0, were the predominant compounds in exudate from plants grown without combined N, the

Table 1. Carbon and nitrogen compositions of soybean stem exudate (Layzell and LaRue, 1982).

Class of compound (major component)	-N plants[a]			+N plants[b]		
	Concentration	% of total C or N		Concentration	% of total C or N	
	μmol/mL	– % –		μmol/mL	– % –	
Sugars (sucrose)	0.64	8	-	0.53	5	-
Cyclitols (pinitol)	1.30	13	-	0.69	7	-
Organic acids (malonate)	1.82	5	-	0.58	3	-
Ureides (allantoate)	10.04	60	91	2.99	19	30
Amides (asparagine)	1.33	9	5	9.28	60	44
Amino acids (aspartate)	0.87	5	4	0.67	6	3
Nitrate	0.20	-	-	9.17	-	23
Totals	16.20	100	100	23.91	100	100
Total carbon (μg/mL)	848			752		
Total nitrogen (μg/mL)	620			568		
C/N ratio	1.37			1.32		

[a]Samples obtained 40 to 50 days after planting (DAP).

[b]Plants supplied with 15 mM NH_3NO_3 for 4 days (47-51 DAP) then 2.5 mM NH_4NO_3 for 3 days (51-54 DAP) and then sampled on day 54.

overall C/N ratio was 1.37. When plants were supplied with NH_4NO_3 for 7 days before sampling stem exudate, the shift from the predominance of allantoate to asparagine and nitrate signaled the depression of N_2 fixation in nodules (Table 1). Total nitrogen + carbon concentration in exudate was reduced about 10% by this treatment, but the C/N ratio was not significantly altered. Among the compounds shown in Table 1, the organic acids, allantoate, amides, amino acids, and nitrate all bear a negative charge at physiological pH values (5.5-8.5). Changes in the concentration of these compounds is, in part, governed by the need to balance cation charges in the xylem stream (Israel and Jackson, 1982).

Seasonal changes in nitrogen compounds in stem exudate collected from field-grown soybean plants are shown in Figure 1. In this experiment, nitrate supply could not be controlled, but it was relatively low since no fertilizer was applied. Note that nitrate was a major component of exudate until flowering, after which it declined to very low levels during pod growth. This occurred in spite of only small changes in NO_3^- concentration in the soil (Streeter, 1972). Amino acid concentration showed a similar pattern except that during seed growth (78-92 DAP), amino acid concentration increased relative to concentrations found during flowering and early pod growth. From the beginning of flowering to the last time when stem exudate could be obtained, ureides were the predominant nitrogenous compounds present.

The seasonal means from this experiment are shown in Table 2, along with some more recent results for greenhouse grown plants (see also Table 1). Several general conclusions can be drawn from the combined results: (1) Asparagine is the predominant amino acid and may represent a major portion of N in stem exudate from vegetative plants (and, perhaps, plants in the late seed-filling stage); (2) allantoate is the predominate compound on a seasonal mean basis; (3) allantoate comprises 65% to perhaps 95% of total ureide, depending on growth conditions.

As already mentioned, nitrogen nutrition of plants may significantly alter stem

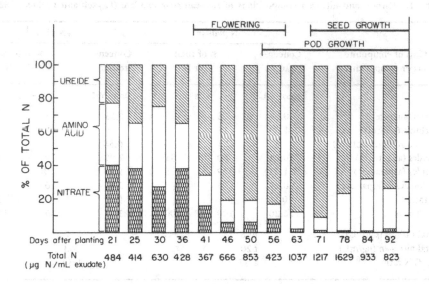

Figure 1. Proportional composition of stem exudate from field grown soybean plants. Amino acid N was calculated from the sum of individual compounds. Asparagine N comprised 61% of amino acid N on a seasonal mean basis. The soybean cultivar was Harosoy-63 and plants were grown in a soil which tested high in P and K and they received no fertilizer application during the year of the experiment. Nitrate N content of the soil was 10 to 13 μg N/g dry wt for the first 3 harvests and averaged about 5 μg N/g dry wt for the remainder of the season. Results shown are based on 4 replicates. Adapted from Streeter (1972, 1979).

Table 2. Relative concentration of individual ureides and amino acids in soybean stem exudate under various conditions.

| Growth conditions | Growth stage | % of total N in: | | | | | Allantoate as % of total ureide |
		Nitrate	Aspar-agine	Other amino acids	Allan-toate	Allan-toin	
	days after planting						
Harosoy-63 grown in field with no N application (Streeter, 1971, 1979)	seasonal mean (13 sample dates)	14	13	6	44	20	69
Davis grown in perlite in a greenhouse, no N supply (McClure et al., 1980)	50	0	9	10	64	17	75
Amsoy-71 grown in perlite in a greenhouse, no N supply (Schubert, 1981)	15	0	27	13	40	20	66
	45	0	1	1	94	4	95

exudate composition, and an exception to conclusion (2) must be emphasized. Namely, when plants are grown with combined nitrogen, the proportion of N in ureides decreases markedly, reflecting decreased N_2 fixation activity (McClure and Israel, 1979; McClure et al., 1980; Israel and Jackson, 1982; Layzell and LaRue, 1981;

Streeter, 1982). This effect has generally been demonstrated for nitrate, and some data for a wide range of nitrate concentrations are shown in Table 3. Note that, with increasing NO_3^- supply, the decrease in ureide-N concentration was not compensated by the increase in nitrate-N concentration.

INTERPRETATION OF STEM EXUDATE COMPOSITION

There are several precautions that are important in the acquisition and interpretation of data on stem exudate composition. Since problems of interpretation associated with nitrate-supplied plants are especially difficult, these will be discussed separately.

General Problems

1. Exudate composition may reflect composition of cells ruptured when severing the stem. This problem can be avoided by blotting or washing the surface of the cut stem prior to exudate collection.

2. Exudate flow rate will not be the same as in the intact plant. This is undoubtedly true at least for illuminated plants where transpiration will accelerate flow in the xylem. Thus, results should be interpreted only as the relative amounts of various compounds supplied to shoots and not as actual quantities supplied/unit time.

3. Exudate composition changes after severing the shoot. The exudate collection period should be kept as short as possible, preferably ⟨30 min, although changes may occur even within this short a period (McClure and Israel, 1979).

4. There is diurnal variation in exudate composition (McClure et al., 1980). Exudate should be collected at the same time each day.

Problems Associated with Nitrate-grown Plants

1. Exudate composition reflects recently absorbed nitrate, N (especially nitrate) from storage "pools" in the root, nitrate assimilated in roots, and possibly, N_2 fixation and nitrate assimilation in nodules. It is difficult to quantify the contribution of each of these potential sources. Nitrate assimilation in nodules is probably insignificant, but nitrate assimilation in roots may be substantial (Ohyama, 1983). Brandner and Harper (1982) have estimated that 30% of total plant nitrate reductase is in the roots.

Table 3. Effect of nitrate in nutrient solution supplied to soybean plants on nitrogen composition of stem exudate and N transport rate (Streeter, 1982)[a].

Nitrate in nutrient solution	Exudate composition			Exudate flow rate	Total N transport[b]
	Nitrate	Amino N	Ureide		
mg N/L	μg N/mL exudate			μL/plant·h	μg N/plant·h
0	0	62	572	11	6.9
15	16	52	406	26	12.2
30	155	43	204	67	26.8
60	237	55	82	106	39.8
100	247	55	42	76	26.0
150	249	49	34	98	32.7

[a]Beeson soybeans grown in sand culture in a greenhouse with a continuous supply of nitrate and sampled 41 days after planting. Data represent averages of two replicates, with 10 to 15 plants/rep.
[b]Sum of nitrate-N, amino-N, and ureide-N.

However, examination of ^{15}N in stem exudate after 1.5 to 3 h treatments with $^{15}NO_3^-$ indicated that only a small portion (perhaps 10%) of recently absorbed nitrate was reduced in the roots. Using a similar approach, Rufty et al. (1982) have estimated that 5 to 17% of recently absorbed nitrate was reduced in the root.

2. Exudate composition may also reflect "recycled" N; i.e., nitrate may be reduced in the shoot, transported as amino acids in the phloem to the root, and recycled in the xylem as reduced N. For plants grown with high nitrate (15 mM), it appears that a large proportion of reduced N in stem exudate is actually recycled N (Rufty et al., 1982; Brandner and Harper, 1982).

3. The amount of nitrate reduced in roots may vary with cultivar and with growth stage (Hunter et al., 1982).

4. Exudate flow rate increases with increasing nitrate supply (Table 3 and Israel and Jackson, 1982). This again emphasizes the importance of not interpreting results in terms of rate of supply of N to shoots.

Summary

There are numerous precautions which should be taken in gathering and interpreting data on stem exudate composition. Where nitrate supply is not high, these problems will generally not limit the utility of the results. Because of multiple sites for storage and assimilation of nitrate, interpretation of exudate composition of nitrate-grown plants is complex and should be undertaken with considerable caution.

USE OF UREIDE CONCENTRATION TO ESTIMATE N_2 FIXATION

Field evaluation of host and *Rhizobium* effects, and effects of various management practices, on N_2 fixation require an accurate and facile method for measuring N_2 fixation. Acetylene reduction assays were widely used in the 1970s. However, batch assays of dug roots suffer from serious errors due to loss of and damage to nodules. Systems for in situ acetylene reduction assays are complex and expensive. More importantly, there is growing evidence that the acetylene reduction assay seriously underestimates actual N_2 fixation (Patterson and LaRue, 1983a; Herridge, 1982b) and that it is subject to errors not previously recognized (Beringer, 1984).

Various ^{15}N methodologies may give good estimates when all the correct controls are included, but this approach is too expensive for a large number of comparisons, and the required analytical equipment is not available to many researchers. On the other hand, ureides are known to be synthesized in nodules and the analysis of ureides is easy. A method which is better adapted to large numbers of samples has recently been reported (Patterson et al., 1982), and the analysis of ureides can even be automated (VanBerkum and Sloger, 1983).

Because of seasonal changes in exudate composition (Figure 1), not all of which can be explained by changes in soil nitrate, it might be predicted that exudate [ureide] and N_2 fixation would not be highly correlated. This prediction would be supported by the problems previously mentioned for plants grown with nitrate. However, as outlined later, high correlations have been obtained, especially where N fractions in addition to ureides are accounted for.

McClure et al. (1980) have reported very high correlations between [ureide] in stem exudate and N_2 fixation activity of soybeans grown in a greenhouse (Figure 2). It should be emphasized that these high correlations were for growth stages preceding pod and seed growth. They also reported little influence of cultivar or *Rhizobium* strain on ureide concentration.

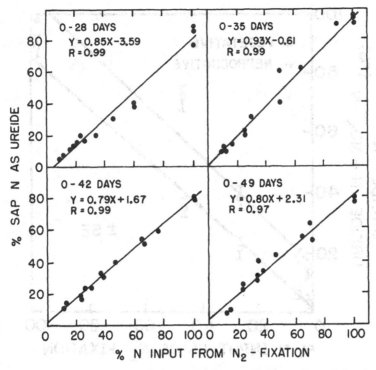

Figure 2. Relationship between the relative ureide content of stem exudate and the relative N input from N_2 fixation for greenhouse-grown Ransom at four sampling dates. Variation in % N from N_2 fixation was achieved by irrigating plants with nutrient solution containing 0, 2.5, 5, 7.5 or 10 mM KNO_3. Nitrate supplied was enriched with ^{15}N, and N input from N_2 fixation was estimated from ^{15}N enrichment of plants (McClure et al., 1980).

Herridge (1982a) related N_2 fixation to [ureide] of shoots or roots, expressing ureide-N as a % of ureide-N + nitrate-N ("relative [ureide]"). This is a reasonable approach because of the variable contributions of nitrate to total exudate N (Figure 1). In a greenhouse study, Herridge (1981a) found high correlations between relative [ureide] of shoot or roots and N_2 fixation, but because ureide content of plants in vegetative and reproductive stages was so different, close relationships were obtained only when data for the two growth stages were treated separately (Figure 3). Attempts to correlate N_2 fixation with [ureide] in leaflets was unsuccessful because of very large differences between growth stages in ureide content of leaf lamina. In field studies at two locations, measurement of ureides + nitrate in shoots and roots, and application of the "standard curves" developed in the greenhouse, did not accurately predict N input from N_2 fixation, which was estimated by two different methods (Herridge, 1982b).

Patterson and LaRue (1982b) also compared relative ureide content of field-grown plants with estimates of N_2 fixation based on acetylene reduction. In an attempt to establish a non-destructive assay, they focused most of their attention on young stem tissue, which permitted reproductive development to be completed on the remainder of the plant. They found that acetylene reduction activity was generally better correlated with relative [ureide] than with (simple) [ureide] in stem tissue. The relative [ureide] in young stem tissue at R3 showed the highest correlation with

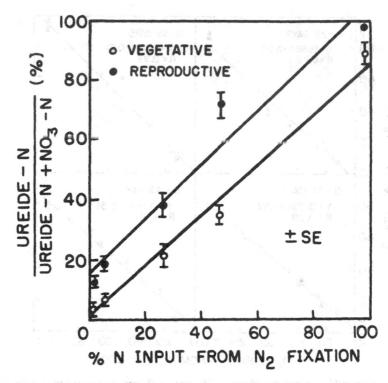

Figure 3. Relationship between N input from N_2 fixation and the relative abundance of ureides in the shoots of greenhouse-grown Bragg. Variation in % N from N_2 fixation was achieved by irrigating plants with solutions containing 0, 1.5, 3, 6, or 12 mM nitrate. N input from N_2 fixation was estimated from acetylene reduction assays of nodulated roots, assuming a ratio for acetylene reduced:N_2 fixed of 1.5:1 (Herridge, 1982a).

integrated seasonal acetylene reduction activity. However, the single ureide determination was not as closely correlated with the estimated seasonal fixation as a single acetylene reduction assay at the R2 or R3 stage.

Herridge (1984) has very recently reported an approach that appears to provide an accurate and reliable estimate of N_2 fixation for greenhouse-grown plants. He obtained stem exudate by the conventional method and compared it with exudate obtained by applying a vacuum to the lower end of the stem of the severed shoot. He also analyzed total amino N in exudate and expressed relative [ureide] as % of ureide-N + nitrate-N + amino-N. Using these innovations, R^2 values of 0.98 to 1.00 were obtained for 8 equations, each for a different growth stage, relating relative [ureide] and % of N input from N_2 fixation. Conventional exudate and "vacuum" exudate composition were equally accurate in their prediction of N_2 fixation. The inclusion of amino-N in the relative [ureide] calculation may explain the high predictive power of this variable but, if amino-N analysis is required, the approach is more difficult to exploit for large numbers of comparisons. The use of vacuum to obtain stem exudate may represent a technique that could be used to good advantage in obtaining samples more reliably from field-grown plants.

An ideal system for using ureide concentration to predict N_2 fixation should include rapid and reliable sampling and should require only a few simple chemical analyses. In addition, it would be desirable to have a non-destructive approach for some applications; e.g., the sampling of F_2 populations with a very limited number of plants. Research efforts to date have not established a system that has these qualities and predicts N input from N_2 fixation with high accuracy. However, good progress has been made toward an ideal system in just the past 2 or 3 years. The recent use of multiple regression (Herridge, 1984) seems like a logical direction for future work. Even if these expressions are empirical, they should be useful because of the increasing availability of computers for routine use.

NOTES

John G. Streeter, Department of Agronomy, Ohio State University/Ohio Agricultural Research and Development Center, Wooster, OH 44691.

REFERENCES

Beringer, J. E. 1984. Measuring nitrogen fixation. p. 31-44. *In* C. Veeger and W. E. Newton (eds.) Advances in nitrogen fixation research. Martinus Nijhoff, The Hague.

Brandner, S. J. C. and J. E. Harper. 1982. Nitrate reduction by roots of soybean (*Glycine max* (L.) Merr.) seedlings. Plant Physiol. 69:1298-1303.

Fujihara, S. and M. Yamaguchi. 1980. Nitrogen fixation and allantoin formation in soybean plants. Agric. Biol. Chem. 44:2569-2573.

Herridge, D. F. 1982a. Relative abundance of ureides and nitrate in plant tissues of soybean as a quantitative assay of nitrogen fixation. Plant Physiol. 70:1-6.

Herridge, D. F. 1982b. Use of the ureide technique to describe the nitrogen economy of field grown soybeans. Plant Physiol. 70:7-11.

Herridge, D. F. 1984. Effects of nitrate and plant development on the abundance of nitrogenous solutes in root-bleeding and vacuum-extracted exudates of soybean. Crop Sci. 25:173-179.

Hunter, W. J., C. J. Fahring, S. R. Olsen and L. K. Porter. 1982. Location of nitrate reduction in different soybean cultivars. Crop Sci. 22:944-948.

Israel, D. W. and W. A. Jackson. 1982. Ion balance, uptake, and transport processes in N_2-fixing and nitrate- and urea-dependent soybean plants. Plant Physiol. 69:171-178.

Israel, D. W. and P. R. McClure. 1980. Nitrogen translocation in the xylem of soybeans. p. 111-127. *In* F. T. Corbin (ed.) World soybean research conference II: proceedings. Westview Press, Boulder, Colorado.

Layzell, D. B. and T. A. LaRue. 1982. Modeling C and N transport to developing soybean fruits. Plant Physiol. 70:1290-1298.

Matsumoto, T., Y. Yamamoto and Y. Yatazawa. 1976. Role of root nodules in the nitrogen nutrition of soybeans. II. Fluctuation in allantoin concentration of the bleeding sap. J. Sci. Soil Manure 47:463-469.

McClure, P. R. and D. W. Israel. 1979. Transport of nitrogen in the xylem of soybean plants. Plant Physiol. 64:411-416.

McClure, P. R., D. W. Israel and R. J. Volk. 1980. Evaluation of the relative content of xylem sap as an indicator of N_2 fixation in soybeans. Greenhouse studies. Plant Physiol. 66:720-725.

Ohyama, T. 1983. Comparative studies on the distribution of nitrogen in soybean plants supplied with N_2 and NO_3 at the pod filling stage. Soil Sci. Plant Nutr. 29:133-145.

Patterson, T. G., R. Glenister and T. A. LaRue. 1982. Simple estimate of ureides in soybean tissue. Anal. Biochem. 119:90-95.

Patterson, T. G. and T. A. LaRue. 1983a. Nitrogen fixation by soybeans: Seasonal and cultivar effects, and comparison of estimates. Crop Sci. 23:488-492.

Patterson, T. G. and T. A. LaRue. 1983b. N_2 fixation (C_2H_2) and ureide content of soybeans: Ureides as an index of fixation. Crop Sci. 23:825-831.

Rufty, T. W., Jr., R. J. Volk, P. R. McClure, D. W. Israel and C. D. Raper, Jr. 1982. Relative content of NO_3^- and reduced N in xylem exudate as an indicator of root reduction of concurrently absorbed $^{15}NO_3^-$. Plant Physiol. 69:166-170.

Schubert, K. R. 1981. Enzymes of purine biosynthesis and catabolism in *Glycine max* I. Comparison of activities with N_2 fixation and composition of xylem exudate during nodule development. Plant Physiol. 68:1115-1122.

Streeter, J. G. 1972. Nitrogen nutrition of field-grown soybean plants: I. Seasonal variations in soil nitrogen and nitrogen composition of stem exudate. Agron. J. 64:311-314.

Streeter, J. G. 1979. Allantoin and allantoic acid in tissues and stem exudate from field-grown soybean plants. Plant Physiol. 63:478-480.

Streeter, J. G. 1982. Synthesis and accumulation of nitrite in soybean nodules supplied with nitrate. Plant Physiol. 69:1429-1434.

VanBerkum, P. and C. Sloger. 1983. Autoanalytical procedure for the determination of allantoin and allantoic acid in soybean tissue. Plant Physiol. 73:511-513.

SEASONAL AND DIURNAL CHANGES IN RIBULOSE BISPHOSPHATE CARBOXYLASE ACTIVITY IN FIELD-GROWN SOYBEANS

Dayle K. McDermitt and Carolyn A. Zeiher

It has been proposed that, under conditions of high light and moisture, carboxylation of ribulose-1,5-bisphosphate (RubP) is the rate limiting step in photosynthesis (Jensen and Bahr, 1977; Perchorowicz et al., 1981). The kinetic mechanism of ribulose-1,5-bisphosphate carboxylase/oxygenase (RubPC) has been extensively studied, and has been shown to occur in two stages (see Lorimer, 1981). The first stage involves the slow, reversible addition of carbon dioxide (C) to an activator binding site on inactive RubPC (E) forming a binary complex (EC), which in turn, is stabilized by rapid reversible binding of Mg ion (M) to form an active ternary complex (ECM). In the second stage, the ternary complex binds RubP and a second molecule of carbon dioxide at a separate site (Lorimer, 1979) and catalysis occurs. The binding of activator carbon dioxide and subsequent activation depends upon the concentration of carbon dioxide and magnesium ion, and upon pH, and probably involves a conformational change in the enzyme.

The slow reversion of EC back to E makes it possible to assay the proportion of RubPC in the active form (i.e., ECM + EC). Freshly isolated RubPC, or RubPC in intact chloroplasts (Jensen and Bahr, 1977) can be maintained in the active form in the absence of additional carbon dioxide or magnesium by incubation at 0 C. If plant leaves are chilled to 0 C, or frozen in liquid nitrogen and then extracted at 0 C, the initial activity which is expressed after extraction is stable for one to several hours in crude homogenates (McDermitt et al., 1983; Servaites and Torisky, 1984). In addition, it has been shown that the activity of RubPC expressed immediately after extraction is not determined by the concentrations of magnesium ion or carbon dioxide in the homogenate, but by the light the leaf received before extraction. This activity, which is measured prior to any in vitro activation, is termed "initial activity" (IA) and is presumed to reflect the physiological state of the enzyme before isolation (Perchorowicz et al., 1981; McDermitt et al., 1983; Servaites and Torisky, 1984). "Total activity" (TA) refers to the activity expressed when all RubPC is in the ECM form, and it is measured after an appropriate incubation period (6 minutes in the experiments reported here) at 25 C in assay buffer under optimal conditions for activation (McDermitt et al., 1983). After incubation, the reaction is initiated by addition of RubP.

Using a wheat (*Triticum aestivum* L.) seedling system, Perchorowicz et al. (1981) demonstrated a light-dependent increase in the fraction of RubPC activated (IA/TA), and showed that the increased RubPC activity correlated with the carbon dioxide exchange rate at high PPFD; however, maximal IA was only about 60% of TA at the highest PPFD (about 1500 $\mu mol/s \cdot m^2$). In field-grown soybeans, however, McDermitt et al. (1983) found that IA was about 95% of TA over much of the day when samples were taken diurnally late in the season (17 Sept.). Surprisingly, total activity was also low when samples were collected early in the morning. That work was first reported in 1982, and has since been confirmed by other workers (Vu et al., 1983; Servaites et al., 1984).

The variation in RubPC total activity has important implications, because it suggests there are physical or chemical regulatory mechanisms that act in addition to magnesium and carbon dioxide mediated activation. In view of the apparent complexity of RubPC regulation, it seemed desirable to measure activity of the enzyme over the season and within the soybean canopy under field conditions. To do that it was necessary to measure RubPC specific activity rather than activity per unit chlorophyll, to avoid the complication of changing RubPC/chlorophyll ratios.

METHODS

RubPC was purified by polyethylene glycol precipitation (Hall and Tolbert, 1978), followed by gel filtration (BioGel A5m) and ion exchange chromatography (DEAE-Sephacel). The purified protein gave a single band on 5% and 7% PAGE gels, and was used as antigen for the production of rabbit anti-RubPC. RubPC concentration in crude homogenates was measured by rocket immunoelectrophoresis, using purified soybean RubPC as standard. All activities are expressed in units of activity (1 unit = 1 mol/min·kg RubPC).

Williams-82 was planted 27 May 1983 and thinned by hand to 256,000 plants/ha on June 26. Pod development, R3, began 26 July and full seed (R6) was about 22 August. Maximum LAI was 5.3.

A record of node number was kept over the season, and a curve was fitted to the data so that the relative depth in the canopy of any node could be calculated for any sampling date. Samples for biochemical analysis were systematically removed over the season from nodes 7, 12, 14, and 16.

SEASONAL PATTERN OF RubPC ACTIVITY

Preliminary experiments, and results published by Vu et al. (1983) indicated that RubPC activity measured per unit chlorophyll was stable for long periods of time in liquid nitrogen or at -80 C. We expected our specific activity samples to be equally stable, but Figure 1 shows that this was not the case. Both initial (not shown) and total activity declined with time in storage at -80 C. That made it necessary to analyze the data by multiple regression techniques to correct for time in storage.

When the total activity data were corrected for storage time, no residual systematic variability remained. There was no evidence for variation with canopy position; however it is not possible to rule out crop or leaf age variability in total activity from those data because there was a highly significant correlation between crop age and storage time: later samples were stored longer. Consequently, correction for storage time also corrects for age. To address the question of crop age, an independent data set is required. We will return to this point later.

In contrast to total activity, initial activity showed a highly significant dependence both on canopy position and on leaf age. Figure 2 shows the dependence of initial activity (after correction for storage) on relative depth in the canopy for data collected over the entire season. Initial activity declines in an approximately log-linear manner from the top of the canopy, and approaches an asymptote at the bottom of the canopy.

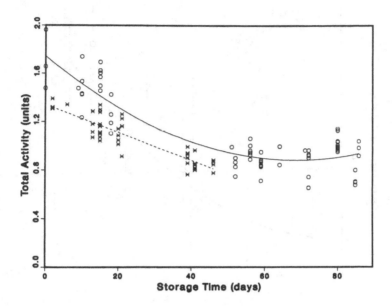

Figure 1. Loss of RubPC total activity with time in storage in either liquid nitrogen or at -80 C. Solid line: samples taken systematically over the season at nodes 7, 12, 14 and 16. Seasonal samples stored less than 20 days were all taken within 60 DAP. Broken line: samples taken from upper canopy, fully exposed leaves over the interval from 17 August to 2 September 1983 during diurnal and shading studies.

This behavior is expected because intial activity does not go to zero even at night, but varies with PPFD from about 0.2 units to, at most, about 1.4 units. If the initial activity data are corrected for both storage time and canopy position, no significant linear trend exists in the residuals over the season; however, a small but highly significant quadratic trend remains, such that the average level of activation is elevated both early and late in the season.

This point is illustrated in Figure 3. In the previous discussion, correction for time in storage was accomplished by regression methods. However, because both initial and total activities decline with time, and total activity does not depend upon position in the canopy, correction for time in storage can be accomplished by calculating the activation ratio (IA/TA) for each time point. Figure 3 shows the activation ratio for nodes 7, 12, 14, and 16 plotted against time over the season. (A similar plot is obtained if storage adjusted initial activities are used.) Both initial activity and the activation ratio increase for nodes 12 and 16 at the end of the season (Table 1). Full (100%) activation was observed only at node 16 at the end of the season. Those findings are supported by the diurnal data.

DIURNAL VARIATION IN RubPC ACTIVITY

Diurnal samples were taken twice during August and once in September. Sampled leaves were fully expanded, fully exposed upper canopy leaves that were relatively free

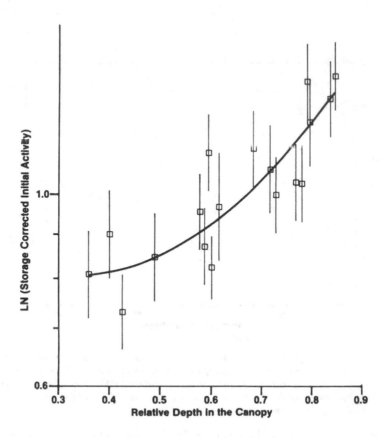

Figure 2. Decline in RubPC initial activity with relative depth in the canopy. Relative depth is defined as number of the sampled node divided by the total number of nodes present on that sampling date. Data are means of four replications with 95% confidence intervals.

Table 1. RuBPC initial and total specific activity from upper canopy leaves at two dates late in the growing season (22 Aug and 2 Sept).

Days after planting	Node[a]	Initial activity (IA)	Adjusted initial activity[b]	Total activity (TA)	Adjusted total activity[c]	IA/TA	Adjusted IA/ Adjusted TA
				— units —			
87	12	0.50	0.87	0.85	1.67	0.59	0.52
87	16	0.66	1.03	0.84	1.67	0.79	0.62
102	12	0.75	1.12	0.95	1.76	0.79	0.64
102	16	0.99	1.36	1.00	1.80	0.99	0.76
LSD (.05)		0.15	0.15	0.10	0.10	--	--

[a]Storage times. 87 DAP: 59 days. 102 DAP: 56 days (7 samples) and 51 days (1 sample).
[b]Adjusted IA = IA - 0.0112 STOR + 0.0000818 $STOR^2$.
[c]Adjusted TA = TA - 0.0293 STOR + 0.0001687 $STOR^2$.

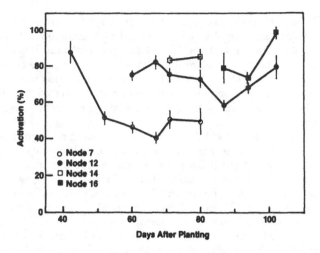

Figure 3. Variation in activation ratio for nodes 7, 12, 14 and 16 over the season. Activation ratio is defined as (initial activity)/(total activity), both uncorrected for storage time. Data are means of four replications plus or minus one SE.

from insect damage. Data taken 26 August are shown in Figure 4. Both initial and total activities increased sharply with light level until about 1000 h. After 1000 h, initial and

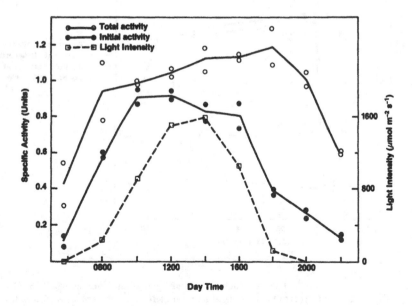

Figure 4. Variation in RubPC initial and total activities with time over the day (CDT). Data were taken 26 August 1983, and were not corrected for storage time.

total activities changed more slowly until about 1600 h when initial activity began to decline with light intensity. Total activity remained high until about 2000 h and then declined. It is evident that initial activity tracks light level more closely than total activity, and as observed in the seasonal samplings, initial activity was about 80% of total activity over most of the day. By contrast, apparent activation was virtually 100% over much of the day when samples were taken one week later on 2 September, but initial and total activities varied with light in a similar manner to those observed in the August samplings.

High activation was also observed in a second experiment performed on 2 September. In that experiment, shades were placed on two plots before dawn. Leaves were sampled at intervals until 1220 h when one of the shades was removed (Figure 5). Both initial and total activities increased sharply in response to light, with initial activity virtually identical to total activity. The shades were replaced after 75 min in the light, whereupon initial activity declined sharply but total activity remained high. Thus, in three separate lines of experiments (diurnal time course, shading study, seasonal sampling) late season apparent activation was nearly 100%, whereas at other times it appeared to be lower (80% to 85%). The period from 26 August to 2 September was very important in that senescence sharply increased during that interval.

A summary of initial activity response to light from all experiments in which it was measured is given in Figure 6. Data were corrected for storage time so that different sets could be compared. In all cases, light saturation was reached by 800

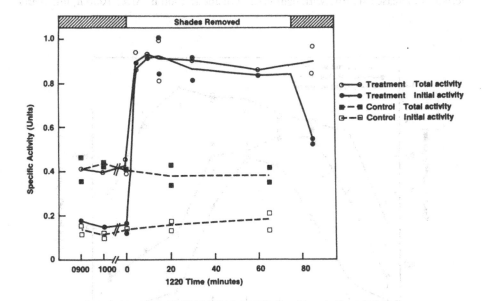

Figure 5. Effects of shades and shade removal on RubPC initial and total activities. Shades were applied at 0545 (CDT) before sunrise, and were left on until 1220 when shades from the treated plot (solid line) were removed. The control plot (broken line) was continuously shaded. Data were not corrected for storage time.

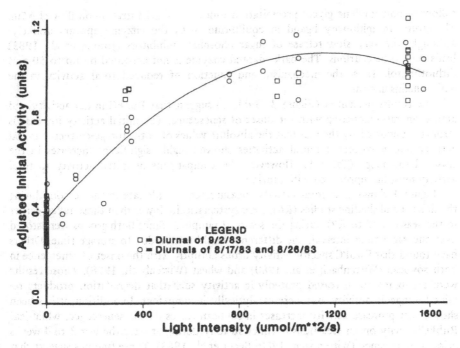

Figure 6. Variation in RubPC initial activity with light intensity combined over three different sampling dates, 17 and 26 August, 2 Sept. Data were corrected for storage time.

μmol/s·m^2, but the data from 2 September saturated at about one-half that level.

Total activity was found to increase sharply with light level, but it stayed high even after light declined. This has been observed in every diurnal time course we have performed, and it has been seen by others (Vu et al., 1983; Servaites et al., 1984). It is further demonstrated in the shading study of Figure 5. The seasonal data show a dependence of initial activity on canopy position, as expected from the light data of Figure 6, but no dependence of total activity on canopy position could be detected. On 17 August, initial and total activities were measured from lower canopy leaves before and after opening the canopy to expose the leaves to full sun. Initial activity sharply increased with kinetics similar to those in Figure 5, albeit to a level only about 75% of total activity, but there was no change in total activity. This means that either the PPFD in the lower canopy (about 40 μmol/s·m^2 at ground level in full shade) is saturating for total activity, or that total activity reaches maximal levels due to sun flecks (about 600 μmol/s·m^2 at ground level) and then stays high during periods of lower PPFD. We do not have sufficient data to construct a good light saturation curve for total activity, so the possibilities cannot yet be distinguished.

The physiological significance of the light dependence of total activity is not known. Variation in the carboxylation rate constant (initial activity) with light level makes sense in terms of maintaining a high RubP pool under conditions of varying light, but variation in the capacity for activation under optimal conditions of pH, carbon dioxide and magnesium ion concentration is more difficult to explain. However, it does not appear to be an artifact of the assay procedures. Apparent reduced total activity is not due to slow activation kinetics of dark-adapted RubPC. The dark-adapted enzyme is not activated by gel filtration through G-25, nor is it activated following polyethylene glycol precipitation followed by gel filtration on BioGel A5m. Therefore, an inhibitory ligand in equilibrium with the enzyme appears unlikely; although, the very slow release of sugar phosphate inhibitors (Jordan et al., 1983) leads one to be cautious. The dark-adapted enzyme is not activated by up to 50 mM dithiothreitol. Thus, the mechanism and function of reduced total activity in the dark remains unclear.

The results presented (Figure 3, Table 1) suggest that RubPC initial activity and activation ratio increased with the onset of senescence. The initial activity increase is strongly supported by the fact that the absolute values of both storage-corrected initial activity and uncorrected initial activities showed highly significant increases in the seasonal samplings (Table 1). However, the comparisons of initial activity to total activity must be approached with caution.

Figure 1 shows that total activity measurements made late in the season, during the diurnal and shading studies (R6), were systematically lower than those taken earlier in the season (V7 to R2) during the seasonal sampling. Since both groups were stored over the same time interval, the difference could not be due to storage time. Others have found that RubPC specific activity drops abruptly with the onset of senescence in both soybean (Wittenbach et al., 1980) and wheat (Wittenbach, 1978). Those results were attributed to increased proteolytic activity such that degradation products retained antigenic activity, but were catalytically incompetent. In addition, it has been shown that protease activity increases in soybean leaves during senescence, while leaf RubPC activity on an area basis declines steadily during at least the last 3 to 4 weeks prior to senescence (Wittenbach, 1980; Secor et al., 1983). Those findings suggest that the reduced apparent total activity seen in Figure 1 may be due to proteolytic reduction of catalytically active RubPC.

RubPC proteolysis occurs in vivo, but it may also occur during the assay activation period when crude homogenates from late season samples are measured. Preliminary studies showed that RubPC total activity was substantially stable in reaction mixtures at 25 C (total activity declined only 13% over a 40 minute incubation period); however, those studies were not performed using late season field samples. If it is true that proteolysis of RubPC accounts for reduced late season total activity, then it is not surprising to find that, while initial activity and total activity declined with storage in a similar manner, the rate of loss was greater for total activity than for initial activity. In that case the regression equation would not only correct for loss in total activity during storage, but also for loss due to proteolysis in older samples which happened also to have long storage times. Storage time over-correction due to a correlated but hidden variable may account for the observation that activation ratios calculated from storage adjusted variables followed the same pattern as corresponding ratios calculated from raw data, but were systematically lower for samples taken beyond R4 (e.g., Table 1). An artifact caused by in vitro proteolysis would not be expected to affect initial activity, because assays were always performed immediately after extraction and no 25 C incubation occurred. All operations prior to assay were performed at 0 C.

A question that remains is why apparent initial activity increases at the end of the season. That observation does not necessarily imply that the physiological activities of inidividual RubPC molecules increase, but may be explained equally well by an increase in the leaf average population specific activity. If the photosynthetic activity of relatively shaded spongy parenchyma cells in the lower portion of the leaf is less than that of fully exposed pallisade parenchyma cells in the upper portion, then physiological RubPC activity may also be reduced in spongy parenchyma cells.

Chlorophyll relative fluorescence yield from the upper surface is only about one half that from the lower surface, implying higher photosynthetic fluorescence quenching from upper surface cells (unpublished data). If RubPC remobilization is not uniform across the leaf but occurs first from lower leaf cells with reduced photosynthetic activity, then the leaf average population would be enriched in active RubPC from the upper surface pallisade parenchyma. Such a mechanism would lead to increased apparent initial activity.

NOTES

Dayle K. McDermitt and Carolyn A. Zeiher, Monsanto Agricultural Products Co., 800 N. Lindbergh Dr., St. Louis, Missouri 63167.

REFERENCES

Hall, N. P. and N. E. Tolbert. 1978. A rapid procedure for the isolation of ribulose bisphosphate carboxylase/oxygenase from spinach leaves. FEBS Lett. 96:167-169.

Jensen, R. G. and J. T. Bahr. 1977. Ribulose 1,5-bisphosphate carboxylase-oxygenase. Ann. Rev. Plant Physiol. 28:379-400.

Jordan, D. B., R. Chollet and W. L. Ogren. 1983. Binding of phosphorylated effectors by active and inactive forms of ribulose-1,5-bisphosphate carboxylase. Biochemistry 22:3410-3418.

Lorimer, G. H. 1979. Evidence for the existence of discrete activator and substrate sites for CO_2 on ribulose-1,5-bisphosphate carboxylase. J. Biol. Chem. 254:5599-5601.

Lorimer, G. H. 1981. The carboxylation and oxygenation of ribulose 1,5-bisphosphate: the primary events in photosynthesis and photorespiration. Ann. Rev. Plant Physiol. 32:349-383.

McDermitt, D. K., C. A. Zeiher and C. A. Porter. 1983. Physiological activity of RubP carboxylase in soybeans. p. 230. In D. D. Randall, D. G. Blevins and R. Larson (eds.) Current topics in plant biochemistry and physiology. Univ. of Missouri, Columbia.

Perchorowicz, J. T., D. A. Raynes and R. G. Jensen. 1981. Light limitation of photosynthesis and activation of ribulose bisphosphate carboxylase in wheat seedlings. Proc. Natl. Acad. Sci. USA 78:2985-2989.

Secor, J., R. Shibles and C. R. Stewart. 1983. Metabolic changes in senescing soybean leaves of similar ontogeny. Crop Sci. 23:106-110.

Servaites, J. C. and R. S. Torisky. 1984. Activation state of ribulose bisphosphate carboxylase in soybean leaves. Plant Physiol. 74:681-686.

Servaites, J. C., R. S. Torisky and S. F. Chao. 1984. Diurnal changes in ribulose 1,5-bisphosphate carboxylase activity and activation state in leaves of field-grown soybeans. Plant Sci. Lett. 35:115-121.

Vu, C. V., L. H. Allen, Jr. and G. Bowes. 1983. Effects of light and elevated atmospheric CO_2 on the ribulose bisphosphate carboxylase activity and ribulose bisphosphate level of soybean leaves. Plant Physiol. 73:729-734.

Wittenbach, V. A. 1979. Ribulose bisphosphate carboxylase and proteolytic activity in wheat leaves from anthesis through senescence. Plant Physiol. 64:884-887.

Wittenbach, V. A., R. C. Ackerson, R. T. Giaquinta and R. R. Hebert. 1980. Changes in photosynthesis, ribulose bisphosphate carboxylase, proteolytic activity, and ultrastructure of soybean leaves during senescence. Crop Sci. 20:225-231.

POTENTIAL FOR CONTROLLING PHOTORESPIRATION IN SOYBEANS

William L. Ogren

An analysis prepared for the First World Soybean Research Congress concluded that the genetic variability of soybean photosynthesis and photorespiration was small, and that no substantial improvement in photosynthetic efficiency could be expected from genetic manipulation of existing cultivars (Ogren, 1976). A corollary to this conclusion is that significant genetic variability in soybean photosynthetic efficiency will arise only if it is created. With current technology, there are two ways this objective might be pursued. The approach detailed in this paper describes attempts to create variability by induced mutagenesis. An alternative approach is to isolate genes encoding rate-limiting enzymes and then improving these enzymes by in vitro mutagenesis.

RUBISCO—THE TARGET ENZYME

Several studies have indicated that a pivotal enzyme in photosynthetic productivity in soybean and other C_3 species is ribulose 1,5-bisphosphate carboxylase/oxygenase (Rubisco). This enzyme catalyzes both the carboxylation reaction that initiates photosynthesis and an oxygenation reaction that yields glycolate-P, the first substrate of the photorespiratory pathway. Analysis of the properties of Rubisco, and of gas exchange by leaves of soybean in atmospheres of varying CO_2 and O_2 concentrations, showed that the regulation of leaf photosynthesis and photorespiration can be described by the kinetic properties of Rubisco (Laing et al., 1974; Farquhar et al., 1980). The ratio of these two processes is given in equation 1:

$$\text{Phosynthesis/photorespiration} = V_c K_o C / t V_o K_c O, \qquad (1)$$

where V_c and V_o are the maximum velocities of carboxylation and oxygenation, K_o and K_c are the Michaelis constants for O_2 and CO_2, C and O are the internal CO_2 and O_2 concentrations, and t is the ratio of oxygenase activity to photorespiratory CO_2 release. At normal atmospheric conditions ($320\ \mu L\ CO_2/L$ air, 21% O_2) at 25 C, the ratio of carboxylation to oxygenation is about 4:1. Thus, the objective of attempts to modify Rubisco with respect to reducing photorespiration is to increase this ratio. Complete elimination of oxygenase activity will increase photosynthetic productivity in air by about 45%. Because all Rubisco enzymes possess oxygenase activity, leading to the suggestion that it is an inevitable consequence of the carboxylation reaction mechanism (Andrews and Lorimer, 1978), experiments designed to improve the Rubisco enzyme are first being carried out on model organisms, which

774

have characteristics permitting rapid selection. If the procedures described here are successful, they can be suitably modified for use on soybean.

INDUCED MUTATION—THE APPROACH

Induced mutagenesis of specific traits requires an ability to examine large numbers of plants for the desired phenotype. One difficulty in screening for increased photosynthetic efficiency is that assays of photosynthesis are time-consuming, so it is not practicable to examine large plant populations. To circumvent this problem, two-stage selection schemes were devised that permit positive selection for mutant strains with a more agronomically efficient Rubisco. In one case, mutants of a C_3 plant were selected in which photorespiration was transformed from a merely wasteful process into a lethal one. These mutants permit the direct selection of second-site revertants that are defective in ribulose bisphosphate (RuBP) oxygenase activity and thus in the ability to initiate the photorespiratory process (Somerville and Ogren, 1982). The model C_3 plant *Arabidopsis thaliana* was used to demonstrate the feasibility of the selection procedure.

In the second case, mutations that rendered Rubisco inactive were selected from the green alga *Chlamydomonas reinhardii*. Then, revertants were sought that restored Rubisco activity. If the original mutation occurs at an appropriate site, it is possible that the reverted enzyme will contain an amino acid change that improves the kinetic characteristics of Rubisco (Spreitzer et al., 1982).

PROCEDURE WITH *ARABIDOPSIS THALIANA*

Arabidopsis thaliana is a C_3 plant, well-suited to genetic studies because it is small, is a profligate seed producer, and has a life cycle of about 6 weeks. Thus, large numbers of plants and several generations can be examined in a small space in a short time. In the first stage of the procedure, strains with defects in the photorespiratory pathway were sought by locating plants that did not grow in air but which grew normally in air supplemented with 1% CO_2. The rationale behind this selection was that in air, carbon would enter the photorespiratory pathway and accumulate to lethal concentrations as a substrate of the defective or missing enzyme. The 1% CO_2 in air atmosphere was used as the permissive condition because it was thought that the high CO_2 concentration would completely inhibit RuBP oxygenase activity, since CO_2 and O_2 are mutually competitive substrates on Rubisco (Laing et al., 1974). This phase of the procedure was successful, and mutants with defects in the following photorespiratory enzymes were recovered: 1) glycolate-P phosphatase, 2) glycine decarboxylase, 3) serine transhydroxymethylase, 4) serine-glyoxylate amino transferase, 5) glutamateoxyglutarate aminotransferase GOGAT), and 6) the chloroplast dicarboxylate transporter (Figure 1; Somerville and Ogren, 1982). In each case the mutant strains grew normally in 1% CO_2 but died when grown in air.

In the second stage of the procedure, seed of two of the mutants was again mutagenized, the F_2 seed was collected and planted in air. The rationale in this case was that if one of the genes encoding Rubisco subunits was altered so that there was little or no RuBP oxygenase activity, then the plants would grow in air because little or no carbon would enter the photorespiratory pathway. When this experiment was carried out, some plants were recovered that grew in air, but in all revertants the original lesion had been repaired so that the activity of the defective enzyme was restored (Somerville and Ogren, 1980). To reduce this background of true revertants a double mutant was

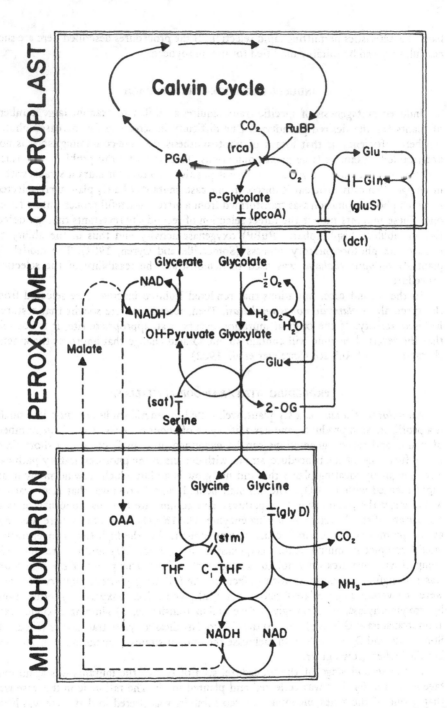

Figure 1. C$_3$ photosynthesis and the photorespiratory carbon and nitrogen cycles. *Arabidopsis thaliana* mutants with defects in the indicated enzymes have been isolated and characterized: Glycolate-P phosphatase (pcoA), glycine decarboxylase (glyD), serine transhydroxymethylase (stm), serine-glyoxylate aminotransferase (sat), glutamate synthase (gluS), and the chloroplast dicarboxylate transporter (dct). From Somerville and Ogren (1982).

constructed and mutagenized seeds screened for surviving plants. About five million double-mutant plants were screened, but none survived in air.

The native Rubisco enzyme consists of eight pairs of large (55,000 dalton) and small (15,000 dalton) subunits (Ellis, 1979). The larger subunit, containing the substrate binding sites, is encoded by the chloroplast genome. The smaller subunit, whose function is not yet known, is encoded in the nucleus. All of the photorespiration mutants isolated followed Mendelian inheritance patterns, and therefore, the lesions are probably in the nuclear DNA. Thus, it is possible that Rubisco enzymes with reduced oxygenase activity will require changes in the chloroplast DNA. Because higher plants contain several chloroplasts and each chloroplast contains several copies of the chloroplast genome, the recovery of specific chloroplast mutants will likely be a rare event. Thus, this procedure may not permit recovery of mutants with an altered large subunit of Rubisco. However, Spreitzer and Mets (1980) were able to recover a Rubisco large subunit mutant in the green alga *Chlamydomonas reinhardii*, so the quest for a Rubisco enzyme with altered carboxylase/oxygenase activity was pursued with this organism.

Procedure with *Chlamydomonas reinhardii*

Unlike higher plants, *Chlamydomonas reinhardii* has a single chloroplast. Although there are several copies of chloroplast DNA, the number of copies can be reduced by treating the alga with 5-fluorodeoxyuridine (Wurtz et al., 1979). Thus, it is possible to create and isolate desired mutations in the chloroplast DNA, if a suitable selection screen can be devised. The procedure followed in the *Chlamydomonas* study was to first create mutants in which Rubisco was rendered inactive. Such mutants would be incapable of photoautotrophic growth, but since *Chlamydomonas reinhardii* will grow on acetate as a carbon source, could be maintained on plates or in culture in the dark. The second stage of the procedure was to mutagenize the Rubisco mutant and search for revertants that had regained the capacity to grow on CO_2 in the light. Because Rubisco is essential for photoautotrophic growth, such mutants would also possess Rubisco with restored catalytic capacity. If an appropriate amino acid in the enzyme were changed, it might possess an altered ratio of carboxylase/oxygenase activity.

Several revertants of the *Chlamydomonas* Rubisco mutant were recovered, but in all cases the kinetic properties of the revertant enzymes were identical to wild type (Table 1). Thus, the procedure selected for restored Rubisco function, but the revertant

Table 1. Analysis of photoautotrophic revertants of a *Chlamydomonas* mutant with defective Rubisco (Spreitzer et al., 1982).

Strain	$Km(CO_2)$	Vmax	CO_2/O_2 specificity (V_cK_o/V_oK_c)
	$\mu M\ CO_2$	$\mu mol\ CO_2/mg\ prot \cdot h$	
Wild type	29	2.03	61
Mutant 10-6C	--	0	--
Revertants			
R4-7	28	2.01	61
R6-2	29	1.95	62
R6-3	28	1.99	61
R7-2	31	2.13	58
R7-2	28	2.07	60
R7-5	29	2.11	60
R7-6A	29	2.13	60
R7-6B	29	2.09	59

enzymes were not different from wild type. Subsequently additional Rubisco mutants were selected (Spreitzer and Ogren, 1983), but revertants of these mutants were also identical to wild type. Thus, although both the *Arabidopsis thaliana* and *Chlamydomonas reinhardii* procedures worked as designed, they did not yield an altered Rubisco enzyme.

RUBISCO–NATURAL VARIATION

It has been suggested that the oxygenase activity of Rubisco is an inherent characteristic of the carboxylation reaction mechanism (Andrews and Lorimer, 1978). Thus there is a question as to whether or not photorespiration can be reduced by preferentially altering the RuBP carboxylation/oxygenation ratio to favor carboxylation. However, a survey of diverse species in the plant kingdom demonstrated that substantial natural variation occurs (Table 2). The variation in CO_2/O_2 specificity, determined from the ratio of carboxylation to oxygenation at several CO_2 and O_2 concentrations, and expressed as the kinetic parameter V_cK_o/V_oK_c, was found to vary from 9 in Rubisco isolated from the photosynthetic bacterium *Rhodospirillum rubrum* to 80 in soybean and other C_3 plants. In this assay, higher values indicate a greater specificity for carboxylation relative to oxygenation. Differences in this specificity were found to occur in discrete steps, and except for the C_4 plants examined, specificity increased in the order of the presumed appearance of the various species during evolution. Photosynthetic bacteria were the first photosynthetic organisms, and arose when the atmosphere contained a high CO_2 concentration and a low O_2 concentration. The capability of the bacterial enzyme to catalyze high rates of RuBP oxygenation would be irrelevant in such an atmosphere because of the lack of substrate O_2. Photosynthetic bacteria are to this day found in anaerobic environments. Oxygen-producing photosynthesis evolved in cyanobacteria, initiating an irreversible change in the atmosphere from high CO_2 and low O_2 concentrations to low CO_2 and high O_2 concentrations. The increase in Rubisco specificity in cyanobacteria, green algae and higher plants was suggested to be driven by this change in atmospheric composition (Jordan and Ogren, 1981).

Also listed in Table 2 are the ratios of the rates of carboxylase to oxygenase activities that occur in the presence of air levels of CO_2 and O_2 (320 µL CO_2/L air, 21% O_2). This ratio ranges from 0.45:1 in *Rhodospirillum rubrum* to 4:1 in higher plants. Thus, if the *R. rubrum* enzyme were found in C_3 plants such as soybean, the rate of oxygenase activity would be more than twice the rate of carboxylase activity. From the stoichiometry of O_2 uptake to CO_2 evolved in photorespiration (Laing et al.,

Table 2. CO_2/O_2 specificity of Rubisco isolated from diverse species (Jordan and Ogren, 1981; 1983).

Species	CO_2/O_2 specificity (V_cK_o/V_oK_c)	Carboxylation:oxygenation in air (ratio)
Soybean, C_3 plants	80	4:1
Zea mays (C_4)	80	4:1
Sorghum bicolor (C_4)	70	3.5:1
Setaria italica (C_4)	58	2.9:1
Green algae	62	3.1:1
Euglena gracilis	54	2.7:1
Cyanobacteria	50	2.5:1
Rhodospirillum rubrum	9	0.45:1

1974), a ratio of 0.5 occurs at the CO_2 compensation concentration. The carboxylase/oxygenase ratio of 0.45:1 in *R. rubrum* Rubisco indicates that if this enzyme were present in soybean or any other C_3 plant, the rate of photorespiration in the present atmosphere would exceed the rate of photosynthesis, and C_3 plants could not exist. Thus, the observed changes are very significant in terms of crop productivity.

CONCLUSION

Although photorespiration has not yet been modified in soybean or other C_3 plants, procedures to select for reduced RuBP oxygenation, the first step in photorespiration, have been devised. Analyses of Rubisco enzymes from several diverse species demonstrate that natural selection has produced changes in Rubisco in the desired direction, providing optimism that application of the selection techniques described can lead to further improvements in the enzyme. Additional studies with the model organisms, *Arabidopsis* and *Chlamydomonas*, will demonstrate the feasibility of the induced mutation. If these studies prove to be successful, they can then be extended to soybean and other C_3 crop plants.

NOTES

William L. Ogren, U.S. Department of Agriculture, Agricultural Research Service, and Department of Agronomy, University of Illinois, Urbana, IL 61801.

REFERENCES

Andrews, T. J. and G. H. Lorimer. 1978. Photorespiration—still unavoidable? FEBS Lett. 90:1-9.

Ellis, R. J. 1979. The most abundant protein in the world. Trends Biochem. Sci. 4:241-244.

Farquhar, G. D., S. von Caemmerer and J. A. Berry. 1980. A biochemical model of photosynthetic CO_2 assimilation in leaves of C_3 species. Planta 149:78-90.

Jordan, D. B. and W. L. Ogren. 1981. Species variation in the specificity of ribulose bisphosphate carboxylase/oxygenase. Nature 291:513-515.

Jordan, D. B. and W. L. Ogren. 1983. Species variation in kinetic properties of RuBP carboxylase/oxygenase. Arch. Biochem. Biophys. 227:425-433.

Laing, W. A., W. L. Ogren and R. H. Hageman. 1974. Regulation of soybean net photosynthetic CO_2 fixation by the interaction of CO_2, O_2, and ribulose 1,5-diphosphate carboxylase. Plant Physiol. 54:678-685.

Ogren, W. L. 1976. Improving the photosynthetic efficiency of soybean. p. 253-261. *In* L. D. Hill (ed.) World Soybean Research. Interstate Printers & Publishers, Danville, Illinois.

Somerville, C. R. and W. L. Ogren. 1980. Photorespiration mutants of *Arabidopsis thaliana* deficient in serine-glyoxylate aminotransferase activity. Proc. Natl. Acad. Sci. USA 77:2684-2687.

Somerville, C. R. and W. L. Ogren. 1982. Genetic modification of photorespiration. Trends Biochem. Sci. 7:171-174.

Spreitzer, R. J., D. B. Jordan and W. L. Ogren. 1982. Biochemical and genetic analysis of an RuBP carboxylase/oxygenase deficient mutant and revertants of *Chlamydomonas reinhardii*. FEBS Lett. 148:117-122.

Spreitzer, R. J. and L. Mets. 1980. Non-mendelian mutation affecting ribulose 1,5-bisphosphate carboxylase structure and activity. Nature 285:114-115.

Spreitzer, R. J. and W. L. Ogren. 1983. Rapid recovery of chloroplast mutations affecting ribulose-bisphosphate carboxylase/oxygenase in *Chlamydomonas reinhardii*. Proc. Natl. Acad. Sci. USA 80:6293-6297.

Wurtz, E. A., B. B. Sears, D. K. Rabert, H. S. Shepherd, N. W. Gillham and J. E. Boynton. 1979. A specific increase in chloroplast gene mutations following growth of *Chlamydomonas* in 5-flurodeoxyuridine. Mol. Gen. Genet. 170:235-242.

FACTORS INFLUENCING CROP CANOPY CO_2 ASSIMILATION OF SOYBEAN

K. J. Boote, J. W. Jones, and J. M. Bennett

Recent studies have shown a good relationship of soybean seed yield with canopy CO_2 assimilation, especially if measured during seed filling. Wells et al. (1982) computed the integrated area under the apparent canopy photosynthesis (ACP) curve between R4 and R7 and found it to be positively correlated with seed yield (r = 0.59 and 0.66) in two different seasons. Harrison et al. (1981) reported that selecting for higher ACP in a breeding program resulted in increased seed yield. Larson et al. (1981) measured mid-day ACP for a range of cultivars and found the maximum relative difference to be 16%. Recently released, higher-yielding cultivars, such as Corsoy and Amsoy, had the highest rates, and older cultivars the lower rates.

We in Florida, and Christy and Porter (1982) in Missouri, have demonstrated that seed yield is closely related to seasonal canopy assimilation under a range of treatments (drought-defoliation and row spacing-shading treatments, respectively). Christy and Porter (1982) reported a very close relationship (r=0.86) of soybean seed yield with relative seasonal canopy photosynthesis under widely varying row spacing and shade treatments. For the Florida study (Bovi, 1983), daily integrated total canopy photosynthesis (TCP) data were put into a season-long soybean growth simulator (Wilkerson et al., 1983), which on a daily time basis, partitions assimilates among plant parts and assigns growth and maintenance respiration costs based on biochemical composition and theoretical considerations (Penning de Vries et al., 1974; Penning de Vries, 1975). Simulated dry matter accumulation was only 12% greater than field-measured dry matter accumulation. This illustrates, not only a close relationship of seasonal TCP to phytomass accumulation, but a direct coupling of carbon flow processes.

CO_2 ASSIMILATION: MEASUREMENT AND DEFINITIONS

Three methods have been used to measure carbon dioxide exchange rate (CER) of intact soybean canopies. The open system measures air flow rate and CO_2 concentration difference between incoming and outgoing air (Jeffers and Shibles, 1969; Kanemasu and Hiebsch, 1975). A second method used is a closed, controlled environment system that records the amount of CO_2 injection needed to maintain a constant concentration (Sakamoto and Shaw, 1967; Egli et al., 1970). A third approach is a transient method which measures the rate of CO_2 depletion in a closed system over a 1 to 2 min period (Boote and Bennett, 1982; Boote et al., 1983; Bovi, 1983; Christy and Porter, 1982; Harrison et al., 1981; Ingram et al., 1981; Larson et al., 1981; Peters et al., 1974; Wells et al., 1982). All three methods seem to give equivalent CER rates.

Most typically, soybean researchers have measured the apparent CER of an intact field canopy in a transparent chamber with the soil surface exposed in the chamber. Such a measurement of CER with the soil exposed will be defined here as apparent canopy photosynthetic (ACP) rate (see reports of Harrison et al., 1981; Jeffers and Shibles, 1969; Kanemasu and Hiebsch, 1975; Wells et al., 1982). ACP is the "apparent" result of three CO_2 flux components occurring concurrently: CO_2 uptake by photosynthesizing tissue, CO_2 efflux from above-ground tissue that is respiring or below the light compensation point, and thirdly, CO_2 efflux from the soil-root-microbial system.

One can eliminate the component of CO_2 efflux from the soil by covering the soil or by measuring it separately. The resultant CO_2 flux can be defined as net canopy photosynthesis (NCP) (see reports of Boote and Bennett, 1981; Egli et al., 1970; Ingram et al., 1981; Larson et al., 1981; Sakamoto and Shaw, 1967).

We propose that canopy assimilation reported as ACP or NCP fails to measure the total photosynthetic uptake, depending upon the degree of CO_2 efflux from the soil and from shoot tissues below light compensation. The CO_2 efflux from pods, seeds, stems, and from vegetative meristems can be considerable (Boote et al., 1983; Ingram et al., 1981; Mkandawire, 1984). Therefore, we suggest a third way to report canopy assimilation. Total canopy photosynthesis (TCP) rate will be defined here as total (not net) CO_2 uptake by photosynthesizing tissue. We have estimated TCP by measuring ACP, and adding back the absolute value of measured CO_2 efflux from plants and soil placed temporarily in darkness (DARKFLUX) (Boote et al., 1983). The dark measurement includes dark respiration by leaves as well as other tissues, and thus, corrects for the energy requirements of foliage which are normally supplied by ATP and reductant provided directly by the light reactions. This energy utilization is assumed equal to subsequent respiration of leaves in darkness. Because growth and maintenance respiration are dependent on temperature and current carbohydrate supply, TCP must be computed from ACP and dark respiration measured at nearly the same time of day. TCP is the desired measurement for soybean models and should also relate well to canopy assimilation simulated by light interception-photosynthesis models, such as that of Duncan et al. (1967).

PPFD RESPONSE OF ACP, NCP, AND TCP

Figures 1A and 1B illustrate the response of ACP, NCP, and TCP to photosynthetic photon flux density (PPFD) for soybean canopies at 21 and 58 days after planting (DAP) and having a leaf area index (LAI) of 0.27 and 5.08, respectively. The light response data were fit to the following asymptotic exponential models:

$$ACP = [PGMAX] [(1 - EXP_e(-EFF \cdot PPFD/PGMAX)) + DARKFLUX] \qquad [1]$$

$$NCP = [PGMAX] [(1 - EXP_e(-EFF \cdot PPFD/PGMAX)) + CROPRESP] \qquad [2]$$

$$TCP = [PGMAX] [1 - EXP_e(-EFF \cdot PPFD/PGMAX)] \qquad [3]$$

where PGMAX corresponds to the maximum rate of CO_2 (mg m^{-2}s^{-1}) assimilation in saturating light and EFF corresponds to the initial light-use efficiency (mg CO_2/μmol PPFD) for fixing CO_2. DARKFLUX is equal to SOILFLUX (CO_2 efflux from soil-root complex) plus CROPRESP (CO_2 efflux from shoot tissue). DARKFLUX is the y-intercept for ACP, whereas CROPRESP is the y-intercept for NCP. TCP has a zero intercept at zero PPFD. SOILFLUX corresponds to the vertical distance from the ACP line to the NCP line.

The asymptotic exponential model gives a slightly better fit than the rectangular

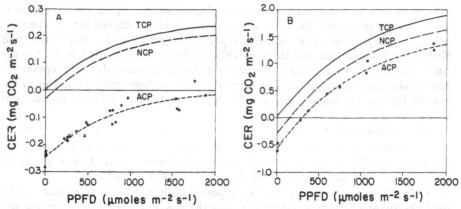

Figure 1. Response of TCP, NCP, and ACP to photosynthetic photon flux density (PPFD) for Bragg soybean canopies at mid-day on two dates in 1979: (A) at 21 days after planting, (B) at 58 days after planting. There were four replicate plots at 21 days and two at 58 days.

hyperbola model suggested by Goudriaan (1982). The photosynthetic capacity of the 21-day-old canopy (Figure 1A) was so small relative to SOILFLUX that ACP was negative even in full sunlight. The "apparent" light compensation point for ACP decreased as the canopy increased in LAI, because photosynthetic capacity became larger relative to CO_2 efflux from crop and soil. For the 58-day-old canopy, the EFF term was 0.00225, 0.00225 and 0.00216 mg CO_2/μmole PPFD for models 1, 2, and 3, respectively. This corresponds to an "apparent" quantum efficiency (QE) of incident PPFD of approximately 0.05 moles CO_2/mole PPFD. Assuming 90% absorption of PPFD by an LAI of 5, then QE = 0.055, which closely matches single leaf values of QE.

COMPARATIVE VALUES REPORTED FOR CANOPY ASSIMILATION AT MID-DAY

Values of ACP, NCP, and TCP for soybean canopies in mid-day sunlight are quite consistent across the 9 or 10 researcher-locations when compared for full canopies at early reproductive stages (Table 1). There seem to be greater differences due to the way of reporting assimilation (ACP, NCP, or TCP) than there are differences among locations, researchers, and cultivars. Mid-day CER rates reported as ACP, NCP, or TCP averaged 1.25, 1.57, and 2.02 mg CO_2/m^2s. Respiration from soil and plant components, respectively, account for the differences between the average values for ACP, NCP, and TCP. Wells et al. (1982) reported that genotypic differences were small during early reproductive stages (R4) and become important only during mid-to-late-seed development stages. Larson et al. (1981) reported up to a 16% difference in NCP between cultivars when measured during July and August in Illinois. The range among presently-grown, adapted cultivars was 8% or less, except for Wayne which had the lowest NCP but yielded well. Small cultivar differences in canopy assimilation were reported by Jeffers and Shibles (1969) and Egli et al. (1970).

SEASONAL PROFILE OF ACP, NCP, TCP, AND CO_2 EFFLUX

Figures 2A and 2B show the seasonal profile of ACP, NCP, TCP, and the CO_2 efflux components at mid-day for irrigated Bragg soybean grown at Gainesville in 1979. Although DARKFLUX, CROPRESP, and SOILFLUX are negative as used in equations

Table 1. Comparison of canopy assimilation rates measured at mid-day on full soybean canopies during early reproductive growth.

Researcher	Location	Cultivar	LAI	ACP	NCP	TCP
					$- mg\ CO_2/m^2s -$	
Sakamoto & Shaw, 1970	Ames, IA	Hawkeye	4-7		1.78	
Egli et al., 1970	Urbana, IL	Wayne			1.43	
Larson et al., 1981	Urbana, IL	Corsoy			1.32	
Christy & Porter, 1982	St. Louis, MO	Williams			1.31	
Boote & Bennett, 1982	Gainesville, FL	Williams	3.0		1.78	
Jeffers & Shibles, 1969	Ames, IA	Amsoy	8.0	1.39		
Kanemasu / Hiebsch, 1975	Manhattan, KS	Clark 63		1.00		
Wells et al., 1982	Athens, GA	MG 5-8		1.28		
Boote et al., 1983	Gainesville, FL	Bragg	4.7	1.31	1.65	2.11
Ingram et al., 1981	Quincy, FL	Bragg	5.5		1.72	2.06
Bovi, 1983	Gainesville, FL	Cobb				1.90
		Average		1.25	1.57	2.02

1, 2, and 3, they are hereafter presented as positive because most literature presents respiration as being positive. The ACP, NCP, and TCP increased as the canopy developed and achieved values as high as 1.24, 1.54, and 1.81 mg/m2s between 55 to 85 days of age. Thereafter, canopy assimilation declined as the canopy aged and as N was re-mobilized to seeds (Figure 2C). These results show no evidence that assimilation increased as the plants developed their pod load (sink). This is consistent with data of Christy and Porter (1982), Larson et al. (1981), and Wells et al. (1982). The CO_2 efflux from soil was initially fairly high, probably because a wheat cover crop was plowed under before planting on June 19 (Figure 2B). This resulted in negative mid-day ACP values at least up to 21 DAP when LAI was 0.27 (Figures 2A and 2C). ACP became negative again near the end of the season. NCP was close to TCP during early season, but diverged as the canopy developed and CROPRESP increased. During the first half of the season, CROPRESP increased with canopy assimilation rate and canopy size. During seed filling, CROPRESP remained constant or increased slightly, while assimilation declined. CROPRESP began to decline precipitiously after R7. CROPRESP averaged 0.295 mg CO_2/m^2s over 11 dates between R4 and R7 (Figure 2B). These values and trends are very similar to those of Ingram et al. (1981), but about 20% less than mid-day values measured by Mkandawire (1984) on Amsoy and Ford at R4.5 and R6.0 in Iowa in 1982. Larson et al. (1981) reported that average nighttime CROPRESP in Illinois was 0.306 mg/m2s.

Although early season SOILFLUX was confounded by decomposition of previous crop residue, later increases in SOILFLUX are consistent with expected root growth and nodule activity. The subsequent decline in SOILFLUX from about 0.30 to 0.07 mg/m2s appeared related to decreased root and nodule activity during seed fill as well as to slightly lower soil temperature. Ingram et al. (1981) reported a similar threefold decline in SOILFLUX, from 0.25 to 0.08 mg/m2s, during seed filling of Bragg. Mkandawire (1984) reported that respiration from soil and roots decreased from 0.26 to 0.13 mg/m2s from R4.5 to R6.0.

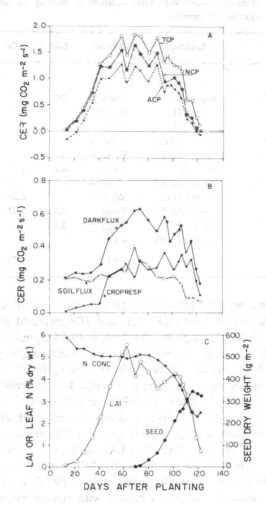

Figure 2. Seasonal trends in TCP, NCP, ACP and CO_2 efflux components, and canopy characteristics of Bragg soybean measured at mid-day at Gainesville in 1979: (A) TCP, NCP, and ACP, (B) DARKFLUX, SOILFLUX, and CROPRESP, (C) Leaf area index, leaf N concentration, and seed weight.

LAI AND DEFOLIATION

Most of the previously described seasonal trends in canopy CERs are related to development of LAI (see Figure 2C). If all data for TCP between 21 and 87 DAP are plotted versus their respective LAIs, the relationship in Figure 3A is obtained. The initial slope of midday TCP versus LAI is 0.874 mg CO_2/m^2s which approximates single leaf CER in high light. The response initially appears to be linear and becomes asymptotic above an LAI of 5. Larson et al. (1981) observed a similar response for NCP versus LAI. Jeffers and Shibles (1969) noted an interaction between LAI and response to PPFD. Their results showed that canopies having LAI below 4 showed

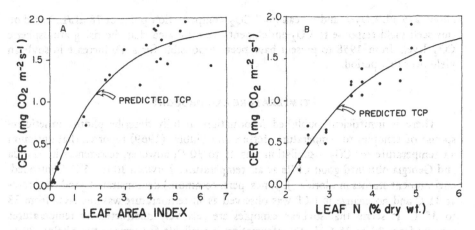

Figure 3. Mid-day TCP of Bragg soybean: (A) Response to LAI during 21 to 87 days after planting. TCP = [1.914] [1 - EXP$_e$(-0.4569·LAI)]. (B) Response of leaf N concentration between 80 to 119 days after planting. TCP = [1.996] [1 - EXP$_e$(-0.500(%N - 1.964))].

light-saturation at less than full sun.

Photosynthetic response to insect defoliation is proposed to act primarily via reductions in LAI. Ingram et al. (1981) reported response to natural insect defoliation over a 2-week period. A 49% reduction in LAI (5.5 to 2.8) reduced mid-day light interception by 17%, mid-day NCP by 15%, and seed yield by 16% compared to the yield of non-defoliated plots. In these and other unpublished data (Boote et al., 1979), the data points for NCP versus LAI for defoliated plots fall on the same curve as NCP versus LAI for non-defoliated plots.

SEASONAL DECLINE IN ASSIMILATION

Much of the decline in canopy assimilation during seed filling is associated with N remobilization from leaf tissue. Figure 3B shows mid-day TCP of Bragg soybean from 80 to 119 days plotted as a function of canopy leaf N concentration. Although the data were not covariance adjusted for concurrent decline in seasonal PPFD or LAI, the seasonal decline in TCP was much better associated with N concentration than with LAI. LAI on day 115 was still as high as 2.5, but TCP had decreased to one-third of the mid-season rates. The value of canopy leaf N concentration at which TCP became zero was 1.96%.

EFFECT OF CO$_2$-ENRICHMENT ON CANOPY ASSIMILATION

Atmospheric CO$_2$ has increased from about 315 μL/L in 1958 to ⟩340 μL/L at present (Keeling et al., 1982), and presently is increasing at about 1 μL/L a year. Thus, canopy assimilation response to CO$_2$ is of more than academic interest. Egli et al. (1970) reported that increasing CO$_2$ from 300 to 600 μL/L caused NCP, averaged over three cultivars, to increase 71% whereas transpiration decreased 18%. Results obtained from outdoor, enclosed, controlled-environment, plant growth chambers show that a doubling of CO$_2$ concentration (340 to 680 μL/L) can be expected to increase midday soybean NCP by about 60%, crop biomass by about 40%, and seed yield by about 30% (L. H. Allen, pers. comm.). Midday NCP response to CO$_2$ concentration had a K$_m$ of about 279 μL CO_2/L and a "canopy" CO_2 compensation point of 78 μL/L. Based on this seed yield response to CO_2-enrichment, we estimated that the rising atmospheric CO_2 levels from 1958 to present have been responsible for a 4% increase in soybean yield over that period.

TEMPERATURE AND DROUGHT

There is insufficient published information to fully describe photosynthetic response of canopies to temperature. Jeffers and Shibles (1969) reported that optimum air temperature for CO_2 assimilation was 25 to 30 C; however, researchers in Florida and Georgia obtained good CERs at air temperatures between 30 to 35 C. Controlled-environment studies in Florida (J. Jones, pers. commun.) demonstrated good NCP rates at 31 C and no change in NCP was observed as air temperature was increased from 28 to 35 C. It seems that soybean canopies are relatively unaffected by temperatures ranging from 20 to 35 C. Little information is available for temperatures below 20 or above 25 C.

The influence of soil water deficit on TCP, ACP, and CO_2 efflux components is illustrated in Figure 4. Wilting began on the 6th day after the last irrigation, and the diurnal measurements were made on the 7th day. Assimilation of irrigated plants responded predictably to the diurnal cycle in PPFD with no mid-day depression. By contrast, assimilation of water-stressed canopies began to diverge from irrigated values by 0900 h, reaching a maximum depression by mid-day with scarcely any recovery in late afternoon.

SOILFLUX was drastically reduced by soil water deficit, but showed only minor diurnal changes (Figure 4B). Total DARKFLUX (and by difference CROPRESP) showed a considerable diurnal trend, which appeared to be highly correlated with air temperature (Figures 4C and 4D). Most of the apparent drought reduction of DARK-FLUX was from reduced SOILFLUX. CROPRESP (DARKFLUX minus SOILFLUX) for stressed plants was maintained nearly equal until midday, by which time water stress had reduced both the carbon supply and the turgor needed for vegetative expansion.

A strong temperature effect on concurrent crop respiration may play a role in the diurnal hysteresis loop of ACP or NCP plotted versus PPFD (reported by Larson et al., 1981, and observed by us in these data). In our data, the canopy light compensation point and the dark respiration intercept for ACP and NCP continue to change throughout the day as temperature changes. When we plotted diurnal TCP versus PPFD, much of the hysteresis loop disappeared. We suspect a considerable proportion of the hysteresis loop of NCP versus PPFD reported by Larson et al. (1981) is associated with a diurnal temperature effect on crop respiration. Those authors also reported a linear effect of temperature on nighttime crop respiration.

CONCLUSIONS

There is a need to properly report canopy assimilation as ACP, NCP, and TCP, depending on whether the carbon exchange measurements include CO_2 efflux from soil-roots-nodules (SOILFLUX) and/or CO_2 efflux from non-photosynthesizing shoot tissue (CROPRESP).

Responses of ACP, NCP, TCP, and efflux components are influenced by light, temperature, CO_2, LAI, N-remobilization, root-and-nodule growth, canopy biomass, and soil water. Response of soybean TCP to PPFD is well described by an asymptotic

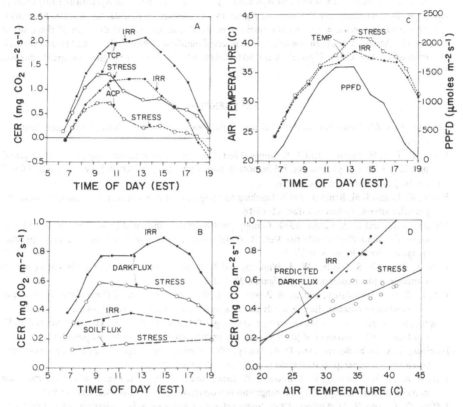

Figure 4. TCP, ACP, and CO_2 efflux components for irrigated (●) and water-stressed (○) Bragg soybean canopies on 5 August 1980, 53 DAP: (A) TCP and ACP, (B) DARKFLUX and SOIL-FLUX, (C) Chamber air temperature and PPFD throughout the day, (D) DARKFLUX versus measurement temperature. Equations of DARKFLUX versus temperature:
(1) Stressed: DARKFLUX = -0.213 + 0.01949 TEMP; R^2 = 0.73.
(2) Irrigated: DARKFLUX = -0.656 + 0.04027 TEMP; R^2 = 0.93.

exponential equation. Minimal effects of air temperature on TCP are observed in the range of 20-35 C, although temperature has a strong linear effect on CROPRESP. Water stress reduces assimilation and CO_2 efflux components, especially SOILFLUX.

Future atmospheric CO_2 increase will result in increases in assimilation, water use efficiency, and seed yield.

There are only minimal effects of cultivar and location on mid-day CERs of healthy, irrigated, full soybean canopies if CERs are compared within ACP, NCP, or TCP categories and are measured during early reproductive growth. Significant cultivar differences in assimilation during mid-to-late-seed growth are likely associated with a prolonged rate of N remobilization from vegetative tissue to seeds. Potential for genetic improvement of canopy assimilation appears only moderate and may be obtained by slower N mobilization, a longer seed filling period, and increased specific leaf weight.

NOTES

K. J. Boote, Associate Professor of Agronomy; J. W. Jones, Professor of Agricultural Engineering, and J. M. Bennett, Associate Professor of Agronomy, University of Florida, Gainesville, FL 32611.

This research was supported by the Institute of Food and Agricultural Sciences of the Univ. of Florida, the American Soybean Association Research Foundation (Project No. 78223), and the U.S. Department of Agriculture (Special Grant 801-15-58). Florida Agric. Exp. Stn. Journal Series No. 5715.

REFERENCES

Bovi, O. A. 1983. Estimating carbon balance of field-grown soybeans. Ph.D. Dissertation, Univ. of Florida.

Boote, K. J., J. M. Bennett, and J. W. Jones. 1983. Canopy photosynthesis: Apparent, net, or gross? p. 121-124. *In* C. Sybesma (ed.) Advances in photosynthesis research Vol. IV. Martinus Nijhoff/ Dr. W. Junk Publishers, The Hague.

Boote, K. J., and J. M. Bennett. 1982. Teaching techniques in crop water relations, gas exchange, and growth analysis. J. Agron. Educ. 11:13-16.

Christy, A. L., and C. A. Porter. 1982. Canopy photosynthesis and yield in soybean. p. 499-511. *In* Govindjee (ed.) Photosynthesis: Vol. II, Development, carbon metabolism, and plant productivity, Academic Press, New York.

Duncan, W. G., R. S. Loomis, W. A. Williams, and R. Hanau. 1967. A model for simulating photosynthesis in plant communities. Hilgardia 38:181-205.

Egli, D. B., J. W. Pendleton, and D. B. Peters. 1970. Photosynthetic rate of three soybean communities as related to carbon dioxide levels and solar radiation. Agron. J. 62:411-414.

Goudriaan, J. 1982. Potential production processes. p. 98-113. *In* F. W. T. Penning de Vries and H. H. van Laar (eds.) Simulation of plant growth and crop production, PUDOC, Wageningen.

Harrison, S. A., H. R. Boerma, and D. A. Ashley. 1981. Heritability of canopy-apparent photosynthesis and its relationship to seed yield in soybeans. Crop Sci. 21:222-226.

Ingram, K. T., D. C. Herzog, K. J. Boote, J. W. Jones, and C. S. Barfield. 1981. Effects of defoliating pests on soybean canopy CO_2 exchange and reproductive growth. Crop Sci. 21:961-968.

Jeffers, D. L., and R. M. Shibles. 1969. Some effects of leaf area, solar radiation, air temperature, and variety on net photosynthesis in field-grown soybeans. Crop Sci. 9:762-764.

Kanemasu, E. T., and C. K. Hiebsch. 1975. Net carbon dioxide exchange of wheat, sorghum, and soybean. Can. J. Bot. 53:382-389.

Keeling, C. D., R. B. Bacastow, and T. P. Whorf. 1982. Measurements of the concentration of carbon dioxide at Mauna Loa Observatory, Hawaii. p. 377-385. *In* W. C. Clark (ed.) Carbon dioxide review, Oxford Univ. Press, New York.

Larson, E. M., J. D. Hesketh, J. T. Woolley, and D. B. Peters. 1981. Seasonal variations in apparent photosynthesis among plant stands of different soybean cultivars. Photosyn. Res. 2:3-20.

Mkandawire, A. B. C. 1984. Respiration rates of soybean cultivars in the field. M.S. Thesis, Iowa State Univ. Library, Ames.

Penning de Vries, F. W. T. 1975. The cost of maintenance processes in plant cells. Ann. Bot. 39:77-92.

Penning de Vries, F. W. T., H. M. Brunsting, and H. H. van Laar. 1974. Products, requirments, and efficiency of biosynthesis: A quantitative approach. J. Theor. Biol. 45:339-377.

Peters, D. B., B. F. Clough, R. A. Garves, and G. R. Stahl. 1974. Measurement of dark respiration, evaporation, and photosynthesis in field plots. Agron. J. 66:460-462.

Sakamoto, C. M. and R. H. Shaw. 1967. Apparent photosynthesis in field soybean communities. Agron. J. 59:73-75.

Wells, R., L. L. Schulze, D. A. Ashley, H. R. Boerma, and R. H. Brown. 1982. Cultivar differences in canopy apparent photosynthesis and their relationship to seed yield in soybeans. Crop Sci. 22:886-890.

Wilkerson, G. G., J. W. Jones, K. J. Boote, K. T. Ingram, and J. W. Mishoe. 1983. Modeling soybean growth for crop management. Trans. Am. Soc. Agric. Eng. 26:63-73.

PHYSIOLOGY OF GENOTYPIC DIFFERENCES IN PHOTOSYNTHETIC RATE

John D. Hesketh, Joseph T. Woolley and Doyle B. Peters

Light manipulation studies with shades and reflectors over crops in the field (Schou et al., 1979; Kokubun and Watanabe, 1983) suggest that differences in canopy photosynthesis during pod set and seed growth should correlate well with differences in yield. Such a relationship would depend upon the percent of photosynthate partitioned into seed being constant. Obtaining an integrated, whole-plant value for the supply of photosynthate during reproductive growth is not a simple procedure and presents sampling difficulties. An alternative is to study leaf and organelle photosynthetic behavior within the crop to learn how canopies behave. The ultimate objective is to understand yield physiology and how various relevant plant physiological processes interact to control yield. Once the yield development process is understood, it should be easier to search and test for simple inexpensive assays for yield-correlated parameters.

After some 20 years of studying genotypic differences in leaf photosynthetic behavior, most crop physiologists are pessimistic about manipulating variations in leaf characteristics for yield improvement; in fact, this pessimism began to appear in the early 1970s (Elmore, 1980). New information about leaf behavior and its interactions with other plant processes is accumulating rapidly, necessitating frequent reanalyses of the problem. The soybean literature is extensive and is growing rapidly, justifying frequent literature reviews. We were limited in the number of references we should cite; therefore this will not be an historical presentation. R. Shibles, J. Secor and D. Ford kindly provided us with a well documented analysis of the history of research in this subject area; the reader is advised to consult their paper[1] and our literature citations for such detail.

C_4-C_3 PHENOMENON

Soybean leaves possess the C_3 photosynthetic carbon fixation pathway, and photosynthetic CO_2 exchange rates (CER) can be enhanced by some 40% by depleting O_2 from the air around a leaf. Vegetative dry matter production rates also are enhanced in plants exposed to O_2-depleted air, but pod set and bean growth are repressed by such conditions (Quebedeaux and Hardy, 1975). C_4 plants exhibit greater leaf CER values under intense light and 30 to 40 C, and their growth does not respond to O_2-depleted air.

Soybean geneticists should be aware that examples of both C_4 and C_3 behavior occur in some 18 genera. Naturally occurring or artificially produced plants possessing C_4-C_3 intermediate behavior occur in some five of these genera. All types of behavior have been reported in the same plant of *Mollugo ruicaulis*: C_3 behavior in leaves at the first to fourth mainstem positions, C_4-C_3 intermediate behavior in leaves at the 5th and 10th nodes and C_4 behavior in leaves above the 11th node. Study of leaves showing intermediate C_4-C_3 behavior has revealed much about the phenomenon; the details of such research have been reviewed elsewhere (Raghavendra, 1980).

GENOTYPIC DIFFERENCES IN CER IN SOYBEAN

The present status of information about the physiology of genotypic differences in photosynthesis among soybean strains reflects that for all species. Elmore (1980) listed many studies of genotypic variability in CER among crop cultivars, including 16 for soybean. He emphasized the frustrations encountered when attempting to correlate a simple photosynthetic assay with yield. There have been many reviews before (cf. Elmore, 1980) and since, reaching the same general conclusion.

Recent Research

We will review here research efforts in progress since 1980 of which we are aware. The group at Iowa State University, Ames (Secor et al., 1982; Ford et al., 1983) screened genotypes for photosynthetic characteristics and then studied in detail leaves from genotypes exhibiting the two extremes in CER values, searching for CER-correlated leaf parameters. They then made crosses between such genotypes and analyzed photosynthetic characteristics in successive generations. The wide range of CER values encountered were not correlated with Specific Leaf Mass (SLM, mg leaf dry weight per cm^2 leaf area), which they and others had found earlier when screening cultivars. Most recently (Ford et al., 1983), they calculated a mean photosynthetic rate for 30 days after the R5 stage and found that the range of 20 to 27 μ mol CO_2/m²·s CER values encountered in their breeding stock was not correlated with the range in crop yields measured.

The group at Agriculture Canada, Harrow, (cf. Buttery et al., 1981) first screened spaced plants of the same cultivars used in the early Ames studies for growth analysis attributes, as calculated from dry weights and leaf area measurements. Their whole-plant values correlated well with comparable Ames leaf values, both studies being carried out during vegetative growth and flowering. These results seemed encouraging at the time. They later found a good correlation between leaf photosynthetic rates and chlorophyll concentrations during vegetative growth, but this correlation did not hold up during reproductive growth in their most recent report (Buttery et al., 1981). However, leaf photosynthetic rates and percent nitrogen values during reproductive growth did correlate well with seed yields.

The group at Cornell University, Ithaca (Lugg and Sinclair, 1980; 1981a,b) studied many of the same cultivars, but also included strains selected for high and low SLM values. They followed SLM over the season and measured CER every 10 min, with 30 chambers in place at the same time, and reported frequency distributions of maximum daily CER values during August and September. Strains selected for large SLM values did not always exhibit rapid CER values.

A mainland Chinese group (Chang and Li, 1982) made crosses between strains exhibiting low and high CER values and measured SLM at flowering and pod fill through the F_4 generation. They were optimistic that lines exhibiting high SLM and photosynthetic values could be developed, but they acknowledged the many other factors that must be selected for when developing a new cultivar. These workers followed the dynamics of SLM and CER over the growing season.

The group at the University of Illinois, Urbana (Gordon et al., 1982; Boyer et al., 1980) screened some of the same cultivars but also included the maturity group and leaf shape and area isolines of Clark and Harosoy (Bernard and Weiss, 1973). Their correlations among leaf photosynthesis (CER), SLM and N were high, with all three parameters varying greatly within the backcross isolines, suggesting maturity group (stage of growth) and leaf dimension effects. All three parameters were inversely correlated with area per foliolate from which a leaflet was sampled. Supply and demand considerations for N and C during leaf expansion suggested an inverse relationship between N and C per unit area and expanded area.

In the reports previously discussed and in earlier reports from other groups, cultivars have tended to rank the same with respect to CER values and associated characteristics. We will discuss some general trends later.

Technical Problems

Before proceeding further, we need to discuss the many factors that can interfere with a study of photosynthesis, and would be correlated leaf characteristics. SLM at times may be associated with substantial starch levels in the leaf, which either have nothing to do with photosynthetic capacity of the leaf or may be inhibitory to photosynthetic processes. Mutants low in chlorophyll but with normal levels of P700 chlorophyll, an indicator of the number of photosynthetic units per unit leaf area, can exhibit very rapid photosynthetic CO_2 exchange rates in full sunlight. Under these conditions photosynthetic unit density is the factor controlling the supply of reductant.

The photosynthetic response to a parameter may be curvilinear. For example, Lugg and Sinclair (1981b) showed that leaf percentage N was saturating above levels that are common in leaves of field-grown plants (15 mg N/m^2).

A positive correlation between photosynthesis and some parameter assumes a constant stomatal conductance, or positive correlations between conductance and the attribute in question. Boyer et al. (1980) reported evidence for genotypic differences between old and new cultivars in afternoon leaf conductances that were associated with differences in leaf water potentials even for well watered plants. Bunce (1982) reported that increasing the leaf chamber humidity above ambient levels enhanced photosynthetic rates by a factor of 1.2. We found greater effects than Bunce reported for stressed leaves, but no effect for leaves on well watered plants. Boyer et al. (1980) reported 50% of the normal conductance at a leaf water potential of -1.4 MPa, which we have also measured under extremely successful conditions. Leaves on well watered plants easily approach the threshold water potential for stomatal closure by midday; stressed leaves maintain a slightly greater negative potential (-0.2 MPa) by closing stomata, in an apparent attempt to slow transpirational water loss. We have other evidence (Larson et al., 1981) that transpiration is not slowed at these water potentials and conductances because of increased leaf temperatures.

Effects of light during leaf expansion (Bunce, 1983), leaf senescence (Secor et al., 1983), starch buildup (Nafziger and Koller, 1976) and mineral nutrition (Chang and Li, 1982) have been described in detail. These factors must be controlled in a genetic study taking into account that genotypic differences may exist in how a leaf acclimates to such factors (see later).

State of the Art

We have attempted to summarize the status of research on genotypic differences in photosynthesis in an outline in Table 1.

Table 1. Summary of physiological factors controlling photosynthetic differences among genotypes.

A. Available genetic isolines
1. Chlorophyll-deficient mutants
2. Leaf shape and size: multiple and narrow leaflets, area per foliate
3. Flowering-maturity genes
4. Determinate, semideterminate and indeterminate genes

B. Characteristics of the leaf apparatus
1. Thickness
2. SLM
3. Protein N: 'insoluble' and 'soluble' N
4. Chlorophyll: Photosynthetic Unit Density
5. Stomatal Conductance: stomatal numbers and lengths

C. Characteristics of the canopy apparatus
1. Rate of leaf area expansion: area per foliolate, SLM, C or N requirements
2. Leaves: shape, angle, LAI light interception characteristics
3. Flowering: see A, 3 and 4 above, control of LAI, canopy structure
4. Pod fill: source-sink effects, N-redistribution

D. 1. Leaf and plant age or developmental stage
2. Light or photosynthate supply during leaf expansion
3. Water stress, soil and atmospheric
   a. Excess water
   b. Deficient water: stomatal closure, leaf flagging, restricted leaf expansion; root-shoot ratio, root and stem diseases, beneficial mycorrhiza
4. Cold nights: sky back-radiation
5. Global CO_2 increase
6. Fertility: beneficial mycorrhizal interactions
7. Measurement techniques

Genetic Isolines. Bernard and Weiss (1973) and Hartung et al. (1981) have described the available collection of backcross isolines. The flowering-maturity genes, including those controlling apical stem termination, are available in Clark and Harosoy backgrounds. Clark is representative of a group of 'large-leafed' genotypes which tend to have lower photosynthetic rates than Harosoy and associated 'small-leafed' genotypes (cf. Gordon et al., 1981); this trend exists throughout the relevant literature. Much of what might be concluded to be a physiological genotypic difference can be found within these isolines. Understanding such apparent genotypic differences within such a group of isolines is essential to isolating and studying real physiological genotypic differences unassociated with effects of maturity or stem termination genes. All this applies to LAI as well as CER behavior.

The Leaf Photosynthetic Apparatus. Leaf thickness, SLM, protein and chlorophyll (Table 1-B) are apt to be equally sensitive as indicators of the numbers of photosynthetic cells or chloroplasts in a leaf. The development of an enriched photosynthetic apparatus capable of a rapid CER value requires extra photosynthetic energy and nitrogen. It may be more efficient during seedling growth for the plant to develop less costly leaves, in terms of energy and N required, in which the photosynthetic apparatus is 'diluted' over a larger area. In such a case, a larger leaf area would result in lower leaf but higher canopy CER values early in the growing season. When comparing wild species related to many of the major crop plants, it appears that man has selected for larger leaves and lower photosynthetic rates (cf. Gordon et al., 1981).

However, as discussed below, one can also manipulate the canopy LAI behavior by controlling row width and reproductive behavior.

Generally speaking, there has been little progress in selecting for greater stomatal conductance by increasing stomatal numbers and lengths per unit leaf area. CER is greater for plants with stomata in both surfaces of the leaf.

The Canopy Photosynthetic Apparatus. The leaf apparatus, as discussed above, is an important part of the canopy apparatus. Leaf size, shape and angle affect the way the canopy intercepts light. As emphasized by Hartung et al. (1981), the genetic isolines discussed above (Table 1-A) can reveal how the maturity and stem termination genes control canopy architecture as well.

Three stages of growth are listed under Table 1-C: (1) early seedling leaf area expansion, (2) leaf area expansion and N-fixation during flowering and pod set, and (3) pod filling. Each stage has unique source:sink relationships. In the early phase the plant is developing a crop canopy for supplying photosynthate and a root system for supplying nitrogen. In the second phase the canopy is being completed, N is being stored in leaves for future redistribution, and the reproductive mechanism is being established. In the final phase the beans draw upon supplies and reserves of carbohydrate and nitrogen to the detriment of leaf activity. The soluble fraction of nitrogen associated with the dark reaction aspects of photosynthesis is available for redistribution to the seed; the insoluble fraction associated with the light reactions of photosynthesis is not redistributed. During the second phase, given the right supply of photosynthate and nitrogen and the right leaf expansion rate, leaves with an enriched photosynthetic apparatus at the top of the plant are possible (Lugg and Sinclair, 1980; Gordon et al., 1981). The isoline maturity and stem termination genes control the whole plant supply-demand ratio for carbohydrate and nitrogen, and thereby the existence of such enriched leaves.

Acclimation Processes. Table 1-D lists the various acclimation processes which may well vary with genotype. We have already discussed work of Boyer et al. (1980), which suggested that newer cultivars maintained less negative water potentials and greater leaf conductances than older cultivars, in the afternoon. The newer cultivars have greater resistance to root diseases, although Boyer et al. (1980) could not detect a disease effect in their studies. The biological control of root diseases, through the release of resistant cultivars, represents, of course, one of the major success stories of agricultural research. Because pest resistance is usually race-specific, and there usually are so many races of the pest in the natural system, the associated breeding effort in most crop species is unending and takes precedence over developing aggressive breeding programs for improving physiological traits controlling yield (cf. Chang and Li, 1982). Many physiologists fail to recognize this point. The issue for soybean could be readily cleared up by repeating the Boyer et al. (1980) study with the near isoline collections of Williams, Corsoy, Harosoy and Clark backgrounds carrying various combinations of genes associated with root resistance to a specific race of a disease.

The effect of carbohydrate supply during leaf expansion is too great to permit a meaningful study of genotypic differences in the leaf apparatus of plants grown in growth cabinets or greenhouses (Bunce, 1983). Leaves left in an enclosure for any length of time (days) will acclimate to the enclosure conditions; enclosure conditions may mask the cold night effect that occurs under natural conditions (Larson et al., 1981).

The acclimation issue amplifies the complexity of the system described in previous sections. An investigator must put considerable thought into research plans so as to control the factors we list, as well as others, such as temperature, before he can hope to study genotypic differences in leaf or canopy photosynthesis. We have presented some concepts as to how various factors might be expected to interact to control photosynthesis; it is best that investigators think and plan at the whole-plant level, and take advantage of the genetic isolines available.

THE OCCURRENCE OF EXCEPTIONALLY RAPID CER VALUES

Maximum reported photosynthetic CO_2 exchange rates vary from 22 to 44 μmol CO_2/m² · s at different stages of growth and among maturity group isolines in August. Rates for leaves smaller than 100 cm² per trifoliolate have approached 38 μmol/m³ · s. Such rates occur in leaves at nodes at the top of the mainstem of indeterminate plants when beans can be detected in pods at those nodes. Lugg and Sinclair (1980) reported 3 layers of palisade cells in such leaves. Such rates only occur for some 10 days and are very sensitive to stressful conditions induced by cold nights of 10 to 15 C, dry air associated with a cold front, or a drought. Such rates were detected in four out of five growing seasons; the July-August average maximum daily temperatures exceeded 32 C in the off season. These rates were induced by short days in early August. We have never measured such rates for Group III strains under natural daylengths at Urbana. These plants produced leaves of similar lengths and widths, but such leaves were never as thick or green as those in earlier cultivars. Such plants, of course, had a much larger leaf area per plant at this stage of growth over which to distribute their supply of C and N. We never measured such rates in determinate plants under natural photoperiods at Urbana.

At Urbana we place leaf chambers on leaves and obtain a CER value in about 10 min; such an approach gives an immediate feedback as to how to avoid excessive variability in CER or how to measure maximum CER values. At Ithaca, leaf chambers were left in place for some time and may have created an environment favorable for high CER values. By deduction, such high rates reflect what is possible under optimum conditions.

Counting trifoliolates 1 cm or longer down from the main shoot apex, the 6th or 7th trifoliolate was most likely to exhibit maximum photosynthetic rates. This occurred after maximum expansion, and was associated with an apparent thickening and greening process. Rapid rates were easily measured on most of the leaves on the plant on very humid days after a rain, as one might predict from Bunce (1982).

LAI INCREASE

Snyder and Bunce (1983) and Sinclair (1984) recently addressed the problem of leaf area expansion in a canopy. They looked at the leaf appearance and expansion rate components of the process, using the plastochron index concept. This appears to be a fruitful approach to the study of LAI development and may soon lend support to some of the concepts discussed above. Sinclair (1984) has observed genotypic differences in the minimum temperature at which leaves begin to appear.

CONCLUSIONS

The many factors affecting CER within a genotype present technical problems. Soybeans are grown over a wide range of latitudes; interactions between genotype, year, location can be considerable. Of the five years that CER was studied at Urbana, one season was excessively wet, one excessively hot, one excessively cold, and there were numerous droughts. We did irrigate, but atmospheric stresses can still be a problem. Added to these difficulties, CER was dynamic with leaf age and stage of growth, with low CER values occurring early in the season. However, we have some evidence that high CER values can be induced in seedlings by clear weather and cool nights.

Despite these technical difficulties, an incredible information base has been built that is fairly consistent among locations. Photosynthesis scientists should be impressed, not depressed by the progress reported in the past few years. Future studies at the canopy, leaf and organelle level, and interactions with other plant processes, should lead to an understanding of factors limiting yield and how these factors can be manipulated genetically and culturally.

NOTES

John D. Hesketh, Joseph T. Woolley and Doyle B. Peters, U.S. Department of Agriculture, Agricultural Research Service and Department of Agronomy, University of Illinois, Urbana, Illinois 61801.

[1]See the American Society of Agronomy's forthcoming revision of their monograph on the soybean.

REFERENCES

Bernard, R. L. and M. G. Weiss. 1973. Qualitative genetics. p. 117-154. In B. E. Caldwell (ed.) Soybeans: Improvement, production and uses. Am. Soc. Agron., Madison, WI.

Boyer, J. S., R. R. Johnson and S. G. Saupe. 1980. Afternoon water deficits and grain yields in old and new soybean cultivars. Agron. J. 72:981-986.

Bunce, J. A. 1982. Photosynthesis at ambient and elevated humidity over a growing season in soybean. Photosyn. Res. 3:307-311.

Bunce, J. A. 1983. Photosynthetc characteristics of leaves developed at different irradiances and temperatures: an extension of the hypothesis. Photosyn. Res. 4:87-97.

Buttery, B. R., R. I. Buzzell and W. I. Findlay. 1981. Relationships among photosynthetic rate, bean yield and other characters in field-grown cultivars of soybean. Can. J. Plant Sci. 61:191-198.

Chang, Y. C. and J. Li. 1982. Inheritance and breeding for higher photosynthesis rate in soybeans. p. 157-161. In B. J. Irwin, J. B. Sinclair and W. Jin-ling (eds.) Soybean Research in China and the United States. Proc. 1st China/USA Symposium and Working Group Meeting. INTSOY Series No. 25, Univ. Illinois, Urbana.

Elmore, C. D. 1980. The paradox of no correlation between leaf photosynthetic rates and crop yields. p. 155-167. In J. D. Hesketh and J. W. Jones (eds). Predicting photosynthesis for ecosystem models. Vol. II. CRC Press, Boca Raton.

Ford, D. M., R. Shibles and D. E. Green. 1983. Growth and yield of soybean lines selected for divergent leaf photosynthetic activity. Crop Sci. 23:517-520.

Gordon, A. J., J. D. Hesketh and D. B. Peters. 1982. Soybean leaf photosynthesis in relation to maturity classification and stage of growth. Photosyn. Res. 3:81-93.

Hartung, R. C., J. E. Specht and J. H. Williams. 1981. Modification of soybean plant architecture by genes for stem growth habit and maturity. Crop Sci. 21:51-56.

Kokubun, M. and K. Watanabe. 1983. Analysis of the yield-determining process in field-grown soybeans in relation to canopy structure. VII. Effects of source and sink manipulations during reproductive growth on yield and yield components. Japan J. Crop Sci. 52:215-219.

Larson, E. M., J. D. Hesketh, J. T. Woolley and D. B. Peters. 1981. Seasonal variations in apparent photosynthesis among plant stands of different soybean cultivars. Photosynthesis Res. 2:3-20.

Lugg, D. G. and T. R. Sinclair. 1980. Seasonal changes in morphology and anatomy of field-grown soybean leaves. Crop Sci. 20:191-196.

Lugg, D. G. and T. R. Sinclair. 1981a. Seasonal changes in photosynthesis of field-grown soybean leaflets. 1. Relation to leaflet dimensions. Photosynthetica 15:129-137.

Lugg, D. G. and T. R. Sinclair. 1981b. Seasonal changes in photosynthesis of field-grown soybean leaflets. 2. Relation to nitrogen content. Photosynthetica 15:138-144.

Nafziger, E. D. and H. R. Koller. 1976. Influence of leaf starch concentration on CO_2 assimilation in soybean. Plant Physiol. 57:560-563.

Quebedeaux, B. and R. W. F. Hardy. 1975. Reproductive growth and dry matter production of *Glycine max* (L.) Merr. in response to oxygen concentration. Plant Physiol. 55:102-107.

Raghavendra, A. S. 1980. Characteristics of plant species intermediate between C_3 and C_4 pathways of photosynthesis: Their focus on mechanism and evolution of C_4 syndrome. Photosynthetica 14:271-283.

Secor, J., D. R. McCarty, R. Shibles and D. E. Green. 1982. Variability and selection for leaf photosynthesis in advanced generations of soybeans. Crop Sci. 22:255-259.

Secor, J., R. Shibles and C. R. Stewart. 1983. Metabolic changes in senescing soybean leaves of similar plant ontogeny. Crop Sci. 23:106-110.

Schou, J. B., D. L. Jeffers and J. G. Streeter. 1979. Effects of reflectors, black boards, or shades applied at different stages of plant development on yield of soybeans. Crop Sci. 19:29-34.

Sinclair, T. R. 1984. Leaf area development in field-grown soybeans. Agron. J. 76:141-146.

Snyder, F. W. and J. A. Bunce. 1983. Use of the plastochron to evaluate effects of light, temperature and nitrogen on growth of soya bean (*Glycine max* L. Merr.). Ann. Bot. 52:895-903.

Nitrogen Metabolism

OXYGEN LIMITATION TO NITROGEN FIXATION IN SOYBEAN NODULES

Thomas R. Sinclair, P. Randall Weisz, and Robert F. Denison

Nitrogen fixation has a high energy requirement, with the theoretical biochemical requirement equal to 12 moles ATP for every mole of molecular nitrogen reduced. However, the data of Tjepkema and Winship (1980) suggested an ATP requirement in soybean at least double the theoretical value. Not surprisingly then, soybean bacteriodal suspensions have been found to have rates of acetylene reduction linearly related to the ATP/ADP ratio (Appleby et al., 1975). Such an intimate association between energy status in the nodule and nitrogen fixation rate has caused many to conclude that carbohydrate supply is the limiting factor in nitrogen fixation. However, it must be remembered that oxygen is also an essential "substrate" for respiration in tissues producing ATP. In this paper, we review the physiology of the soybean nodule with respect to the oxygen requirement, examine the experimental observations on the effects of oxygen on nitrogen fixation, and consider the potential agronomic significance of an oxygen limitation.

NODULE PHYSIOLOGY

An unusual feature of nitrogen fixation is that the key enzyme, nitrogenase, is inactivated by oxygen. Nitrogenase consists of two metalloproteins, both of which are irreversibly destroyed by oxygen (Robson and Postgate, 1980). Burris (1979) found from particulate preparations of *Azotobacter vinelandii* that oxygen was an uncompetitive inhibitor of nitrogen reduction with a K_i of 0.014 pO_2. The Fe-protein is particularly sensitive to oxygen and has been shown to be inactivated in only a few minutes of exposure to air.

The apparent paradox of a large oxygen requirement for respiration, combined with the potential inactivation of nitrogenase by oxygen, has been resolved by microorganisms in several ways (Robson and Postgate, 1980). It seems possible that the development of a nodule on soybean might be the solution adopted by *Rhizobium*. A key feature of the nodule must be that it offers an isolated atmosphere in which low oxygen partial pressures (pO_2) are maintained. Additional features that would enhance the ability of the nodular system to reduce nitrogen would be mechanisms to supply oxygen at the reducing site at high rates, and to regulate the oxygen supply rate with changing conditions. We examine each of these three criteria for a successful solution to the nitrogenase paradox by a develoing nodule system.

Low Oxygen Atmosphere

The anatomy of the soybean nodule is well suited for potentially providing a low oxygen environment. The bacteroids are contained in cells in the interior of the nodule

(Figure 1). The bacteriod-containing volume is surrounded by a cortex that is roughly 150 μm thick. The cortex can be further divided into an outer cortex, which contains many air spaces, and an inner cortex, which appears to be devoid of air spaces (Sprent, 1972). Between the two cortical regions is a single layer of scleroid cells.

Tjepkema and Yocum (1974) obtained direct evidence that the nodule interior was reduced in oxygen. They inserted oxygen microelectrodes into intact nodules, observing pO_2 with depth of insertion. Oxygen partial pressure in the outer cortex was near atmospheric but showed a drastic decline in the inner cortex and remained low throughout the bacteroidal volume. Tjepkema and Yocum proposed that the inner cortex acted as a barrier to oxygen diffusion greatly impeding the flow of oxygen into the bacteroidal-containing volume.

Sinclair and Goudriaan (1981) examined some of the mathematical considerations in regards to an oxygen-diffusion barrier in the nodule cortex. They calculated the permeability required of a diffusion barrier to cause an oxygen decrease of 0.2 pO_2. They concluded that a continuous shell of water around the bacteroid-containing volume with a thickness of about 45 μm would give the desired oxygen concentration decrease. Therefore, an inner cortex devoid of air spaces would offer an oxygen-diffusion barrier and provide nitrogenase with a low oxygen environment.

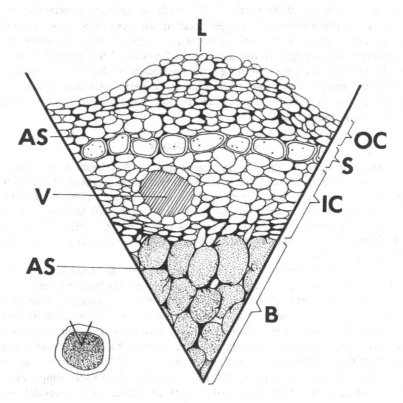

Figure 1. Schematic drawing of a cross-section of a soybean nodule. L = Lentical, OC = Outer Cortex, S = Sclerid cell layer, IC = Inner Cortex, B = Bacteroid containing volume, AS = Air Spaces, V = Vascular bundle.

Oxygen Transport

With low pO_2 inside the inner cortex, reasonably high nitrogen fixation rates can be obtained only if oxygen is readily transported to the bacteroids. Continued oxygen diffusion through liquid would greatly constrain fixation rates. Sprent (1972) and Bergersen and Goodchild (1973) observed that the bacteroid-containing volume of the nodule is permeated with continuous airspaces. Bergersen and Goodchild estimated these airspaces in soybean nodules occupied 2.5 to 5% of the volume. Since oxygen diffusivity in gas is more than four orders of magnitude greater than in liquid, Sinclair and Goudriaan (1981) showed that the observed airspaces would result in a fairly uniform oxygen distribution in the nodule interior.

Oxygen diffusion from the airspaces into the individual bacteroidal-containing cells also is potentially a large impediment to high nitrogen fixation rates. Large oxygen flux rates into these cells cannot be achieved from a combination of oxygen diffusion in liquids and a small oxygen gradient. An apparent adaptation to overcome this problem is the development of leghemoglobin. Leghemoglobin has a very high affinity for oxygen, with a half-saturated partial pressure of 5×10^{-6} MPa (Appleby et al., 1976). This high affinity for oxygen coupled with a fairly high leghemoglobin concentration, allows oxygenated leghemoglobin to transport oxygen across the liquid to the vicinity of an oxidase for respiration. Considerable evidence has been accumulated to demonstrate the supportive role of leghemoglobin in facilitating oxygen transport in the bacteroid-containing cells (e.g., Appleby et al., 1976).

Regulation of Oxygen Supply

If nodule respiration rate is limited by any factor other than pO_2, it is critical for the preservation of functioning nodules to restrict oxygen supply. Without such regulation pO_2 could build within the nodules and inactivate nitrogenase.

In a theoretical analysis of facilitated diffusion by oxygenated leghemoglobin, Stokes (1975) concluded that one important asset of the leghemoglobin system was the stabilization of the free oxygen concentration at the delivery site. This buffering capacity of leghemoglobin would be especially important to accommodate short-term fluctuation of pO_2 inside the nodule. Data presented by Appleby et al. (1976) showed that leghemoglobin in a bacteroid suspension not only increased the acetylene reduction rate at all oxygen concentrations, but also resulted in similar rates between 0.0026 and 0.0105 MPa pO_2.

However, leghemoglobin has a limited capacity to buffer against fluctuating oxygen partial pressure, so a mechanism to accommodate more long-term changes in nitrogen fixation ability is also required. Pankhurst and Sprent (1975) suggested from dehydration studies of nodules that it was possible for the magnitude of the oxygen-diffusion barrier in the cortex to change. They presented data that clearly showed reductions in the oxygen permeability of soybean nodules as they were dehydrated. Sheehy et al. (1983) reported that nodule permeability in white clover (*Trifolium repens* L.) responded to changes in pO_2 of the atmosphere outside the nodule. They found nodule permeability apparently decreased at higher pO_2 so, while some increase in acetylene reduction was observed, the increase relative to 0.02 MPa pO_2 was not large.

In our own studies of field-grown soybeans, we have attempted to make extensive measurements of nodule gas permeability and conductance (permeability x total nodule surface area). Using in situ, open-ended root chambers that allow continuous flow of various air-acetylene mixtures (Denison et al., 1983a), we have developed a data analysis technique that allows the conductance of the nodules in the chamber to be calculated (Denison et al., 1983b). Under a wide range of experimental conditions, we have consistently found that variations in the acetylene-reduction rate for saturating acetylene were very closely correlated with the conductance. Studies of variability in acetylene reduction rates among plants at Ithaca, NY and at Gainesville, FL both showed very high correlation coefficients with nodule gas conductance. Analysis of seasonal acetylene reduction rates also showed that the seasonal changes were closely correlated with nodule conductance. Finally, drought studies showed a very close correlation between nodule conductance and acetylene reduction rate at all stages of the drying cycle. An intriguing aspect of the drought study was that the flux density of acetylene for saturating acetylene was linearly related to nodule permeability (Figure 2), regardless of the severity of the stress or the magnitude of the nodule permeability. The results from this drought stress study suggested a very close relationship between the ability of oxygen to diffuse into the nodule and the nitrogen fixation potential. In fact, all our results for acetylene reduction rates at saturating acetylene can be explained by differences in conductance arising from variability in nodule gas permeability and/or surface area.

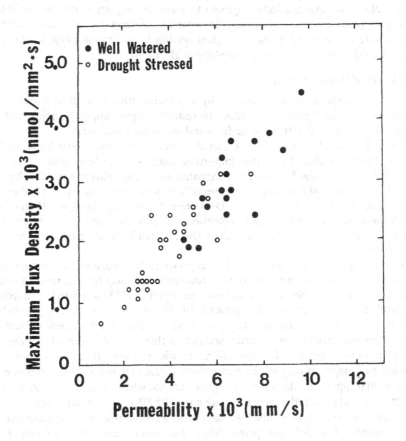

Figure 2. Maximum flux density of acetylene reduction versus nodule permeability for field-grown Biloxi plants subjected to both well-watered and drought-stressed conditions.

NODULE RESPONSE TO OXYGEN

In the previous section the hypothesis was presented that, due to the problem of nitrogenase inactivation by oxygen, the nodule functions to provide an oxygen-protected environment for the enzyme. Oxygen will remain at a low level, only if oxygen is the limiting factor in nodule respiration. If photosynthate was limiting, then the nodule must either have an energy-independent mechanism for consuming oxygen or suffer the fate of nitrogenase inactivation as pO_2 increases within the nodule.

In fact, considerable evidence has accumulated that shows that pO_2 in short-term experiments does limit nitrogen fixation rates. In 1940, Allison et al. concluded, from studies on the effects of increased oxygen partial pressures upon soybean nodule respiration, that oxygen levels were normally low and limiting within nodules. Bergersen (1962) showed that nitrogen fixation (^{15}N uptake) of intact nodules increased with increasing oxygen up to 0.03 to 0.05 MPa pO_2, above which fixation was inhibited. With the advent of the acetylene reduction assay, Bergersen (1970) and Mague and Burris (1972) found that rates increased substantially as oxygen was increased to 0.06 MPa pO_2 and remained high at higher oxygen levels.

Pankhurst and Sprent (1975) found that by raising the pO_2 around dehydrated nodules the acetylene reduction rate met or exceeded the rate of unstressed nodules at 0.02 MPa pO_2, depending on the degree of dehydration. Nodules that were severely stressed continued to have increasing acetylene reduction rates to 0.01 MPa pO_2, in contrast to the unstressed nodules, which had peak reduction rates at 0.04 MPa pO_2. The retarded acetylene reduction rates by the unstressed nodules above 0.04 MPa pO_2 could be a consequence of nitrogenase inactivation. Pankhurst and Sprent concluded from their results that an oxygen-diffusion barrier exists in soybean nodules, which can be compensated by increasing external pO_2. Their interpretation of the decreased permeability in response to nodule dehydration was a loss of turgor in the vacuolated cells cells of the cortex, and thus a reduction in the airspaces available for diffusion through the cortex.

Ralston and Imsande (1982) offered further support for the existence of a dynamic oxygen-diffusion barrier in soybean nodules. Attached nodules exposed to a range of oxygen partial pressures showed peak acetylene reduction rates at 0.03 MPa pO_2. Upon detachment the peak activity was shifted upwards to 0.05 MPa pO_2. They concluded that the water relations of the nodule was altered by detachment and, as a consequence, the permeability of the diffusion barrier was reduced.

We have confirmed the oxygen response in drought stress soybeans under field conditions. Using the continuous flow root chambers, pO_2 was increased and the short-term response in acetylene-reduction rate was measured. The plants studied were part of the drought response experiment referred to previously. In Figure 3 are plotted the mean acetylene reduction rates observed on 21 May 1982 for six plants that had been well watered, and on 9 June 1982 for four plants that had been rewatered for 2 days after 5 days of drought, and five plants that had been droughted for 7 days. While no observations were made at 0.6 pO_2 for the well-watered plants, all oxygen enrichment treatments resulted in a positive response in acetylene reduction rate. These results again demonstrated the existence of an oxygen-diffusion barrier that limits nitrogen fixation rate, and that the permeability decreases in response to dehydrating conditions.

AGRONOMIC IMPLICATIONS

Short-term exposures to atmospheres enriched in oxygen have clearly resulted in stimulations of acetylene-reduction rates, suggesting strongly that nitrogen fixation is

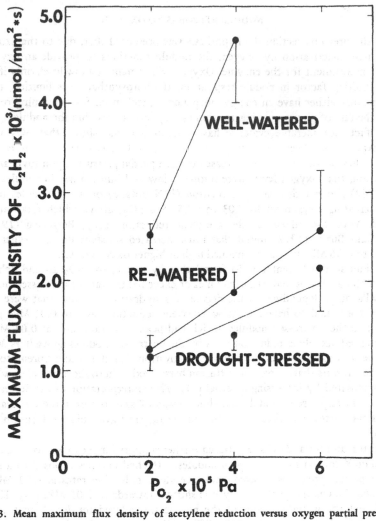

Figure 3. Mean maximum flux density of acetylene reduction versus oxygen partial pressure for field-grown Biloxi plants. Bars are standard error of the mean.

limited by oxygen. An intriguing question with important agronomic implications is: Will long-term oxygen enrichment of nodules result in greater seasonal nitrogen fixation and stimulated plant growth?

We attempted to answer this question by establishing long-term exposure treatments to altered oxygen concentrations through two cropping seasons. These experiments were performed in Gainesville, FL during the spring, 1983 and summer, 1983. The cultivar, Williams, was seeded on 25 March and Bragg was seeded on 29 June. The seeds were sown into the open-ended chambers used for acetylene reduction measurements (Denison et al., 1983a). Gas of various air-oxygen mixtures was continuously introduced at the bottom of the chambers (15 cm depth) at a rate of 1.7 mL/s. No lids were placed around the plant stems at the top of the chamber, as was done when measuring acetylene reduction.

During the spring experiment the gas mixtures were adjusted to 0.014, 0.021, 0.028 and 0.035 MPa pO_2 and flowed to ten chambers each. Half the chambers at each pO_2 were exposed continuously for 7 weeks from 7 April to 25 May, at which time they had developed to the beginning of podfill. The other five chambers at each pO_2 were exposed for shorter periods: 7 to 21 April, 21 April to 5 May, and 5 to 25 May. At the end of each exposure period the plants were harvested, nodules in the chambers counted and weighed, and plant dry weight determined.

For the summer experiment with Bragg only three pO_2 were tested: 0.012, 0.021, and 0.030. The two exposure periods studied were 18 July to 4 August and 4 to 24 August. During the second exposure period, a number of treatments were obtained by differing the pO_2 from those the plants had been subjected to during the first exposure period. Therefore, many plants harvested at the end of the second exposure period had been supplied with various gas mixtures for a total of 6 weeks. The plants at the end of this study were at beginning pod.

No significant difference in plant dry weight resulted from any of the treatments on Williams (Table 1). The only response was an increase in nodule fresh weight and number for the 7 to 21 April exposure to 0.014 MPa pO_2. No significant differences were found in the study with Bragg.

The failure to observe a response to elevated oxygen in long-term exposure suggests that the nodules could not sustain the elevated nitrogen fixation rates observed in short-term studies. Such an observation is consistent with the results of Criswell et al. (1976), whereby after a 4-h exposure to 0.006 MPa pO_2 the initially low rates of acetylene reduction returned to rates equivalent to plants kept at 0.021 MPa pO_2. The explanation of a long-term equality in nitrogen fixation rates subjected to differing pO_2, even though the instantaneous rates differ, may be associated with changing nodule permeability. As discussed earlier, Sheehy et al. (1983) have demonstrated a sensitivity in nodule permeability to pO_2. One explanation for the reduction in nodule permeability under elevated oxygen is that the catastrophic circumstance of exhaustion of carbohydrate reserves is anticipated, and the rate of oxygen entry to the interior of the nodule is limited further by a decrease in permeability. Certainly the mechanism regulating nodule permeability appears to be in intimate synchrony with the biochemical activity in the interior of the nodule.

If under well-watered conditions the permeability of the nodule is regulated to maintain an excess of carbohydrate reserves in the nodule, then the *de facto* limitation of nitrogen fixation rate would be carbohydrate supply. Consequently, conditions leading to enhanced carbohydrate supply would be expected to result in greater nodule activity. However, recent studies that examined the response of nodule acetylene reduction of plants whose shoots had been exposed to atmospheres enriched in carbon dioxide failed to show any increase in nodule specific activity (Finn and Brun, 1982; Williams et al., 1982).

The explanation for the lack of response to the carbon dioxide treatment may be that the anatomical characteristics of the nodule cortex may allow permeability only to increase to some maximum value. This may be a consequence of an anatomical configuration insuring that the structural integrity of the diffusion barrier is maintained. In our experiments under field conditions the greatest permeability we measured was 0.01 mm/s, which translates into a liquid layer approximately 20 μm thick. Since the cells of the inner cortex are roughly 10 μm wide, it may be that two layers of cells are joined together in an essentially permanent arrangement to form a minimum oxygen-diffusion barrier. In any event, the plant response to the carbon dioxide enrichment treatments was the development of more nodules, rather than a stimulation of nitrogen fixation by the nodules existing at the beginning of the treatments (Finn and Brun, 1982; Williams et al., 1982).

Table 1. Growth response of soybean plants grown with continuous exposure to various oxygen partial pressures around the nodules. Those data denoted with differing letters within a treatment date are significantly different ($\alpha=0.05$) by Duncan's multiple range test.

Treatment dates	pO_2	Plant mass	Nodule fresh weight	Nodules
	MPa	g	g	no
		Williams		
7-12 Apr	0.014	0.83 r	0.14 r	35 r
	0.021	0.82 r	0.05 r	9 s
	0.028	0.88 rs	0.08 r	18 s
	0.035	0.96 rs	0.06 r	8 s
	check	1.12 s	0.10 r	10 s
21 Apr-5 May	0.014	2.20 r	0.45 rs	38 r
	0.021	1.52 s	0.22 rs	28 r
	0.028	1.58 s	0.38 rs	53 r
	0.035	1.55 s	0.16 r	18 r
	check	1.76 rs	0.47 s	40 r
5-25 May	0.014	8.67 r	1.15 r	49 r
	0.021	11.38 r	1.73 r	99 s
	0.028	9.57 r	1.21 r	68 r
	0.035	11.60 r	1.49 r	67 r
	check	10.81 r	1.57 r	75 r
7 Apr-25 May	0.014	13.13 r	1.23 rs	74 r
	0.021	14.31 r	1.26 rs	74 r
	0.028	12.45 r	1.82 r	87 r
	0.035	11.61 r	0.87 s	57 r
	check	10.81 r	1.57 rs	75 r
		Bragg		
18 July-4 Aug	0.012	2.00 r	0.49 r	63 r
	0.021	2.51 r	0.56 r	56 r
	0.030	1.95 r	0.45 r	48 r
	check	1.76 r	0.35 r	40 r
(18 July-4 Aug)	0.012-0.021	12.28 r	2.45 r	137 r
to (4 Aug-24 Aug)	0.012-0.030	11.83 r	2.57 r	148 r
	0.021-0.021	10.29 r	2.18 r	130 r
	0.030-0.012	13.47 r	2.59 r	144 r
	check-0.012	9.39 r	2.01 r	105 r
	check-0.030	11.52 r	2.12 r	144 r
	check-check	11.06 r	2.52 r	150 r

The apparent great sensitivity to dehydration of nodules and the permeability of the diffusion barrier may have considerable impact on the agronomic performance of soybeans. Pankhurst and Sprent (1975) found in their nodule dehydration studies that any moisture loss relative to the most turgid nodules (estimated water potential of 0.1 MPa) resulted in substantial decreases in acetylene reduction. In our studies under field conditions, we have observed decreases in acetylene reduction as quickly as three to four days after irrigation. Such observations suggest that loss of turgor in nodules can readily reduce the nitrogen fixation potential. Since most nodules reside in the upper soil layers, which dry first, it seems likely that nodule dehydration may occur well before any drought effects influence carbohydrate supply. Therefore, under many field conditions reductions in nitrogen fixation potential due to nodule dehydration may be a common occurrence. The sensitivity of nitrogen fixation to soil drying, and its impact on crop productivity, appear to be very important areas for future investigation to fully understand the significance of the oxygen limitation in soybean nodules.

NOTES

Thomas R. Sinclair, P. Randall Weisz, and Robert F. Denison, Environmental Physiology Project, USDA-ARS, University of Florida, Gainesville, FL 32611.

REFERENCES

Allison, F. E., C. A. Ludwig, S. R. Hoover and F. W. Minor. 1940. Biochemical nitrogen fixation studies. I. Evidence for limited oxygen supply within the nodule. Bot. Gaz. 101:513-533.

Appleby, C. A., F. J. Bergersen, P. K. MacNicol, G. L. Turner, B. A. Wittenberg and J. B. Wittenberg. 1976. Role of leghemoglobin in symbiotic N_2 fixation. p. 274-292. *In* W. E. Newton and C. J. Nyman (eds.) Nitrogen fixation. Washington State University Press, Pullman.

Appleby, C. A., G. L. Turner and P. K. MacNicol. 1975. Involvement of oxyleghaemoglobin and cytochrome P-450 in an efficient oxidative phosphorylation pathway which supports nitrogen fixation in *Rhizobium*. Biochem. Biophys. Acta 387:461-474.

Bergersen, F. J. 1962. The effects of partial pressure of oxygen upon respiration of nitrogen fixation by soybean root nodules. J. Gen. Microbiol. 29:113-125.

Bergersen, F. J. 1970. The quantitative relationship between nitrogen fixation and the acetylene-reduction assay. Aust. J. Biol. Sci. 23:1015-1025.

Bergersen, F. J. and D. J. Goodchild. 1973. Aeration pathways in soybean root nodules. Aust. J. Biol. Sci. 26:729-740.

Burris, R. H. 1979. Inhibition. p. 569-604. *In* R. W. F. Hardy, F. Bottomley and R. C. Burns (eds.) A Treatise on dinitrogen fixation. John Wiley and Sons, New York.

Criswell, J. G., U. D. Havelka, B. Quebedeaux and R. W. F. Hardy. 1976. Adaptation of nitrogen fixation by intact soybean nodules to altered rhizosphere pO_2. Plant Physiol. 58:622-625.

Denison, R. F., T. R. Sinclair, R. W. Zobel, M. M. Johnson and G. M. Drake. 1983a. A non-destructive field assay for soybean nitrogen fixation by acetylene reduction. Plant and Soil 70:173-182.

Denison, R. F., P. R. Weisz and T. R. Sinclair. 1983b. Analysis of acetylene reduction rates of soybean nodules at low acetylene concentrations. Plant Physiol. 73:648-651.

Finn, G. A. and W. A. Brun. 1982. Effect of atmospheric CO_2 enrichment on growth, nonstructural carbohydrate content, and root nodule activity in soybean. Plant Physiol. 69:327-331.

Mague, T. H. and R. H. Burris. 1972. Reduction of acetylene and nitrogen by field-grown soybeans. New Phytol. 71:275-286.

Pankhurst, C. E. and J. I. Sprent. 1975. Effects of water stress on the respiratory and nitrogen-fixing activity of soybean root nodules. J. Exp. Bot. 26:287-304.

Ralston, E.J. and J. Imsande. 1982. Entry of oxygen and nitrogen into intact soybean nodules. J. Exp. Bot. 33:208-214.

Robson, R. L. and J. R. Postgate. 1980. Oxygen and hydrogen in biological nitrogen fixation. Ann. Rev. Microbiol. 34:183-207.

Sheehy, J. E., F. R. Minchin and J. F. Witty. 1983. Biological control of the resistance of oxygen flux in nodules. Ann. Bot. 52:565-572.

Sinclair, T. R. and J. Goudriaan. 1981. Physical and morphological constraints on transport in nodules. Plant Physiol. 67:143-145.

Sprent, J. I. 1972. The effects of water stress on nitrogen-fixing root nodules. II. Effects on the fine structure of detached soybean nodules. New Phytol. 71:443-450.

Stokes, A. N. 1975. Facilitated diffusion: The elasticity of oxygen supply. J. Theor. Biol. 52:285-297.

Tjepkema, J. D. and L. S. Winship. 1980. Energy requirements for nitrogen fixation in actinorhizzal and legume root nodules. Science 209:279-281.

Tjepkema, J. D. and C. S. Yocum. 1974. Measurement of oxygen partial pressure within soybean nodules by oxygen microelectrodes. Planta 119:351-360.

Williams, L. E., T. M. DeJong and D. A. Phillips. 1982. Effect of changes in shoot carbon-exchange rate on soybean root nodule activity. Plant Physiol. 69:432-436.

THE INTERACTION OF OXYGEN WITH NITROGEN FIXATION
IN SOYBEAN NODULES

Jay B. Peterson

The availability of usable forms of nitrogen is limiting to cereal-grain crop growth in field situations (Havelka and Hardy, 1976). The continuous removal of soil nitrogen as a result of agricultural practices has created an ever-increasing demand for regular input in the form of nitrogen fertilizer. This fertilizer is produced largely from atmospheric N_2 in the Haber-Bosch process, and requires large amounts of energy. The distribution and application of fertilizer require further expenditures of energy. Furthermore, the increasing world population will require substantial increases in food production.

In contrast to the industrial production of fertilizer, biological nitrogen fixation does not require the use of fossil fuel energy reserves. In field crops, the energy is provided by the plants through the production of photosynthate. For these reasons, biological nitrogen fixation has received a great deal of interest as an alternative technology.

The contribution of biological nitrogen fixation is, however, limited by a number of factors that are not clearly understood. These factors include the following: 1) nitrogen fixation occurs only within cells of some species of bacteria and algae (Stewart, 1982); 2) symbiotic nitrogen fixation in crop plants is, with the exception of associative symbioses, limited to the legumes (Stewart, 1982); 3) under normal field situations, soybeans fix only about 25% of the nitrogen they use (Havelka and Hardy, 1975); 4) the enzyme nitrogenase, which converts N_2 to the biologically useful NH_3, is rapidly and irreversibly inhibited by O_2 (Winter and Burris, 1976; Robson and Postgate, 1980). The O_2 sensitivity might be the most important factor limiting the range of organisms with N_2 fixing capability, the range of symbiotic relationships, and the range of environmental "niches" where a nonsymbiotic organism can fix nitrogen.

This discussion is centered on aspects of the last point (4). Aerobic diazotrophs depend on O_2 for the synthesis of ATP, which is a required substrate for nitrogenase. However, the inactivation of nitrogenase by O_2 is apparently prevented. The *Rhizobium* endosymbiont of soybean root nodules is one such aerobic diazotroph. Its metabolism and that of the surrounding plant cytoplasm are geared for the reduction and assimilation of N_2. The interaction of oxygen with these processes and how the interaction is controlled are the subjects of these discussions. Recent developments are emphasized.

THE NODULE

Root nodules form on soybeans as a result of invasion by the soil bacterium *Rhizobium japonicum*. In the mature soybean nodule, the central region of the nodule consists mostly of infected cells and is surrounded by the nodule cortex. The rhizobia, now called bacteroids, are observed as membrane-enclosed structures usually with several bacteroids per structure. This membrane is referred to as the peribacteroid membrane. The volume occupied by the bacteroids and peribacteroid membranes can be 80% of the volume of a nodule plant cell, and estimates of the numbers of bacteroids per plant cell range as high as 44,000 (Bergersen, 1982). As a result, the symbiosis is faced with the problem of maintaining sufficient O_2 and carbon flux to support the respiration of a large number of bacteroids, while simultaneously protecting nitrogenase in the bacteroids from exposure to O_2. There are a variety of mechanisms to accomplish these goals. Most of these mechanisms are not completely understood.

LEGHEMOGLOBIN AND DIFFUSION OF OXYGEN TO THE BACTEROIDS

The role of leghemoglobin is well documented. A brief discussion here is sufficient. Leghemoglobin is an oxygen-binding protein with properties similar to those of myoglobin. Leghemoglobin is located in the infected nodule plant cells where it binds oxygen firmly, maintaining a free O_2 concentration of about 10 nM outside the bacteroids (Bergersen, 1962). Researchers generally agree that leghemoglobin in vivo is located in the plant cell cytoplasm, but disagree as to its presence inside the peribacteroid membrane (Bergersen, 1982). Estimates of the concentration in the cytoplasm are about 3 mM (Bergersen, 1982). The leghemoglobin is normally about 20% oxygenated (Appleby, 1969), giving a bound O_2 concentration of about 600 μM. This is more than twice the concentration of O_2 in air-equilibrated water (28 C). Addition of oxygenated leghemoglobin to unstirred bacteroid suspensions allows maximal nitrogenase activity at a lower free O_2 concentration than without the protein (Bergersen and Turner, 1975). Therefore, the apparent functions of leghemoglobin are to keep the free O_2 concentration low and to promote the diffusion of O_2 to the bacteroids (Bergersen, 1982).

The nodule cortex provides a barrier that slows the diffusive entry of O_2 into the bacteroid-containing central tissue. The central tissue apparently has a uniform oxygen distribution, and this is thought to be due to the presence of intercellular spaces for rapid gas diffusion (Tjepkema and Yocum, 1974).

OXYGEN AND SOYBEAN NODULE N_2 FIXATION

There has been a good deal of experimentation on the effects of sub- and supraambient O_2 concentrations on nodule nitrogenase activity. The results must be interpreted with a good deal of caution. Nitrogenase itself is clearly O_2 labile, and the two components, nitrogenase reductase (Fe protein) and nitrogenase (MoFe protein), differ markedly in their sensitivity to O_2 (Robson and Postgate, 1980). Oxygen could, in addition, affect an array of other plant and bacterial reactions either directly or indirectly. These oxygen effects might also be influenced by the presence of leghemoglobin.

In two publications, Criswell et al. (1976, 1977) described the effects of continuous exposure to altered soil pO_2 on nitrogenase (C_2H_2) activity of intact soybeans. An increase to 0.032 MPa from 0.021 MPa gave a slight increase in nitrogenase activity. A very high O_2 concentration (0.089 MPa) inhibited the activity 98% within 1 h and 86% after 48 h. None of the activity at 24 h was recovered by returning to 0.021 MPa O_2. Alternatively, when intact root systems were continuously exposed to a subambient concentration of 0.006 MPa O_2, nitrogenase activity initially decreased, but recovered to control levels in about 24 h. This suggested that there were major structural and/or metabolic changes in the nodules. The effects of 0.02 MPa O_2 were "permanent"; activity was initially 27% of controls and increased to only 68% by 71 h. Short-term (30 min) incubations at near-ambient concentrations of 0.006 and 0.033 MPa O_2 had essentially no effect on nitrogenase activity when measured immediately after return to ambient O_2. In contrast, after 30 min exposure to 0.089 MPa O_2, only 56% of control activity was immediately recovered. Excised roots did not have the capacity for recovery. The general conclusions were: 1) at all O_2 concentrations tested, the nitrogenase activity showed a degree of adaptation with continuous exposure; 2) at the more extreme (high and low) pO_2 treatments, the ability to adapt declined; 3) once adapted, return to ambient pO_2 inhibited activity, 4) short-term (30 min) exposure to high (0.089 MPa) O_2 caused changes that were only partially reversed when the O_2 concentration was lowered; 5) since nodules would be exposed to 0.021 MPa O_2 or less under field conditions, N_2 fixation would be inhibited by short-term declines in O_2 concentration.

The effects of short-term exposure to a high O_2 concentration were recently studied in greater detail by Patterson et al. (1983). The investigators used a respirometer to monitor nitrogenase (C_2H_2) activity and respiration (CO_2 evolution) of the nodulated root system of intact soybeans. The root system was continuously flushed with air (with 5% C_2H_2) and this was interrupted with a 6 min flush with 100% O_2. There was a 90% reduction in nitrogenase activity when measured within 30 min after the O_2 treatment. The root system respiration decreased almost 40% as a result of this treatment. Two hours after O_2 treatment, nitrogenase had recovered to almost 50% of the pretreatment level. There was a slower increase thereafter that was continuous over the period measured (21 h). Respiration recovered to near-beginning values with a similar time course. The rapid recovery of significant nitrogenase activity within 2 h after treatment with 100% O_2 suggests that the initial loss with O_2 treatment was not due to permanent inactivation of the enzyme. That is, there appears to be a proportion of the enzyme that is "protected" from destruction by the O_2. The second and much slower phase of recovery might be more readily explained by protein synthesis.

SOYBEAN BACTEROIDS AND FREE-LIVING *RHIZOBIUM JAPONICUM*

Studies on nitrogenase activity with isolated soybean bacteroids have generated results that could help explain some of the phenomena observed with intact soybeans. Early studies (Bergersen and Turner, 1967) with $^{15}N_2$ as nitrogenase substrate showed that anaerobically isolated bacteroids could reduce nitrogen with energy supplied by an endogenous energy (carbon) source. Experiments were done in closed shaken vials with one (starting) level of O_2, which was about 5 to 6%. Additions of pyruvate, fumarate, or succinate enhanced nitorgenase activity. These compounds also enhanced O_2 uptake (as a measure of respiration) by bacteroids.

The development of the acetylene reduction technique as an assay for nitrogenase activity greatly accelerated the bacteroid studies. With this technique, a far greater number of experiments could be done in the same period of time. Researchers were able to develop a broader picture of the effects of oxygen on nitrogenase activity.

Wittenberg et al. (1974) and Emerich et al. (1979) established a method for bacteroid assays that is widely used. The bacteroids are isolated and washed anaerobically or in the presence of low O_2. Assays are performed in serum-stoppered vials that are rotated continuously to equilibrate the gas (usually 10% C_2H_2 in Ar, with O_2) and liquid (bacteroid suspension) phases. A range of initial O_2 concentrations is tested.

Nitrogenase activity increases from zero at zero O_2 to a peak value at some characteristic O_2 concentration. Above that concentration, the activity drops off rapidly to zero. The optimum O_2 concentration is dependent on the carbon source (or H_2 for bacteria with uptake hydrogenase). Isolated bacteroids apparently have an endogenous carbon source; substantial activities are observed without additions. With additions of an energy-yielding substrate, there is a marked upward shift in the O_2 optimum and sometimes a noticeable increase in the maximum nitrogenase activity (Emerich et al., 1979; Peterson and LaRue, 1981). As an example, addition of 10 mM sodium succinate shifted the observed O_2 optimum from approximately 0.0002 to about 0.0016 MPa (Emerich et al., 1979), and from about 0.0009 to about 0.0025 MPa (Peterson and LaRue, 1981). There were corresponding increases in nitrogenase activity of about 2.5- and 1.5-fold, respectively. The usable carbon sources or H_2 (Emerich et al., 1979) also stimulate O_2 uptake by the bacteroids.

The nitrogenase activity of isolated bacteroids can recover from inhibitory effects of O_2 over a short period of time. Recovery was first demonstrated with French bean bacteroids (Trinchant and Rigaud, 1979). Bacteroids were exposed to a supraoptimal (inhibitory) pO_2, and addition of succinate allowed recovery of activity. The same effect of delayed succinate addition was observed by Patterson et al. (1983) with soybean bacteroids. Patterson et al. extended the observations by using a method for quickly lowering the pO_2 in the assay vials. No carbon substrates were added in these experiments. Activity was recovered by lowering from inhibitory (0.003 MPa) to near-optimal (0.001 MPa) pO_2. With longer exposures to the inhibitory level of oxygen, less recovery was observed.

Recovery of bacteroid nitrogenase activity after inhibitory O_2 treatment apparently correlates with the rapid phase of recovery from supraoptimal O_2 treatment in intact plants (Patterson et al., 1983). The rapid recovery observed with bacteroids suggests again that protein synthesis is not responsible. A protecting protein similar to those found in *Azotobacter* spp. (Robson, 1979; Scherings et al., 1983) could possibly bind to the oxidized nitrogenase, protecting it from inactivation.

It is likely that respiration has a major role in protecting the nitrogenase. Soybean bacteroids have a series of high-affinity terminal oxidases (Bergersen and Turner, 1980) that vary in their affinity for O_2. The O_2 saturation for these oxidases is well above the free O_2 concentration in the nodule (Bergersen and Turner, 1980; Trinchant et al., 1981), and the oxygen concentrations supporting nitrogenase activity in isolated bacteroids (Trinchant et al., 1981). Since the nitrogenase activity of isolated bacteroids and intact nodules is not completely destroyed by supraoptimal O_2, the losses in activity are likely due to oxidative phosphorylation electron transport out-competing electron transport to nitrogenase for available reductant. Both electron transport processes may compete for the same "pool" of reductant, and the higher potential process (respiration) could easily be favored in such a competition. The increasing respiration would keep the internal pO_2 very low and protect nitrogenase. At substantially higher O_2 concentrations, the nitrogenase would be irreversibly inactivated. At first, this line of thought would not seem to explain the observation by Patterson et al. (1983), who found whole root nitrogenase activity *and respiration* to decrease after O_2 treatment. However, Patterson et al. (1983) used CO_2 evolution as an indicator of respiration. The CO_2 evolution likely reflects reductant generated for both oxidative phosphorylation and electron transport to N_2 and energy for production for nitrogen assimilatory reactions. If oxidative phosphorylation in bacteroids is not tightly coupled to nitrogenase activity the return to 0.021 from 0.1 MPa O_2 might result in the same level of O_2 uptake (and the same internal O_2 concentration) by the bacteroids as before the O_2 treatment. Therefore, nodule O_2 uptake may not have been inhibited as greatly as CO_2 evolution by the O_2 treatment. Similarly, bacteroid experiments should be performed measuring both CO_2 evolution and O_2 uptake as indicators of respiration during increases in O_2 concentration and after O_2 treatment.

PROTEIN SYNTHESIS AND OXYGEN

Low oxygen concentrations are required for expression of nitrogenase in free-living *Rhizobium japonicum* (Agarwal and Keister, 1983). Oxygen might, therefore, repress synthesis of the enzyme. A nif-specific repressor is known to mediate the O_2 repression of nitrogenase synthesis in *Klebsiella pneumoniae* (Hill et al., 1981; Merrick et al., 1982). The genetic control has not been worked out for rhizobia. Repression of nitrogenase synthesis by oxygen has, however, been demonstrated in nitrogenase-derepressed cultures of *R. japonicum* (Scott et al., 1979). An inhibition of total cell protein synthesis was also observed. Cyclic GMP repressed nitrogenase, hydrogenase, and nitrate reductase synthesis and this correlated with an observed rise in cyclic GMP levels in O_2-treated cultures (Lim et al., 1979). The treatment, however, consisted of increasing the O_2 concentration from 0.1% (gas phase above the bacterial suspension) to 20%. This may have little relevance to the physiology experiments of Patterson et al. (1983) where 3% was used to inhibit activity and 1% was near optimum in isolated bacteroids.

NODULE PLANT CELL METABOLISM

Nodule plant cells supply the bacteroids with carbon compounds to support nitrogen fixation. The assimilation of ammonia, which also requires carbon substrates, takes place outside the bacteroids in the plant cell cytoplasm (cytosol). Little is known, however, of how O_2 interacts with this metabolism. The idea was recently advanced (DeVries et al., 1980; Peterson and LeRue, 1981) that legume nodule plant cells metabolize a portion of this carbon through anaerobic pathways as a result of the low O_2 concentration. Key enzymes of such pathways have been found in soybean nodules. Phosphoenolpyruvate carboxylase (Peterson and Evans, 1979) for oxalo-acetate (and malate) production and pyruvate decarboxylase and alcohol dehydrogenase (Tajima and LaRue, 1982) for acetaldehyde and ethanol production were found in the soybean nodule cytosol. Carbon flow to these compounds is thought to be an alternative to further aerobic metabolism by the plant cells. It could also be an alternative to pyruvate production, since pyruvate does not support significant nitrogenase activity or O_2 uptake in isolated bacteroids (Ruiz-Argueso et al., 1979; Peterson and LaRue, 1981). Nitrogenase activity and O_2 uptake by isolated bacteroids are promoted by malate, ethanol, and acetaldehyde (Ruiz-Argueso et al., 1979; Peterson and LaRue, 1981). The ethanol-acetaldehyde pathway has been studied in somewhat more detail than the malate pathway. Ethanol and acetaldehyde were detected in the vapor phase surrounding detached nodules when incubated under anaerobic or aerobic conditions (VanStraten and Schmidt, 1974a,b). Much higher levels accumulated under anaerobic conditions. Similar experiments were performed (J. B. Peterson and T. A. LaRue, unpublished) with untreated nodules and nodules pretreated with 100% O_2. Nitrogenase activity was eliminated and aerobic CO_2 evolution decreased as a result of the oxygen pretreatment. Ethanol accumulated in the vapor phase around all anaerobically-incubated nodules but not aerobically-incubated nodules (20% O_2, 80% Ar), regardless of whether they were pretreated with oxygen. If acetaldehyde and/or ethanol from nodule cytosol metabolism supports nitrogenase activity under normal aerobic conditions, then they should accumulate after the O_2 pretreatment that inactivates nitrogenase. The lack of accumulation suggests that this pathway does not accommodate substantial carbon flux, at least in this soybean/*Rhizobium* combination (cv. Wilkin/strain USDA 138).

QUESTIONS AND DIRECTIONS

So what does all this mean to soybean nitrogen fixation? Since the soil pO_2 would never be higher than 21%, only the experiments with subambient pO_2 treatments would have meaning to current symbioses. Soybean nodules adjust to continuous subambient oxygen (down to about 0.006 MPa) by some unknown mechanism(s) and near-maximal rates are eventually obtained. Very low oxygen concentrations. 0.002 MPa O_2 or below (Criswell et al., 1976), would probably have a significant and lasting inhibitory effect on nitrogen fixation. Such conditions would probably correlate with the low nitrogenase activity in isolated bacteroids at suboptimal pO_2. Furthermore, low oxygen might increase the carbon flow through anaerobic pathways. Ethanol and acetaldehyde are volatile products of plant cell fermentative metabolism and might be lost. This loss would reduce the overall efficiency of carbon use by the nodule. Alternatively, these compounds might be toxic to the nodule (VanStraten and Schmidt, 1974a,b). No bacteroid nitrogenase activity is observed in the absence of O_2. Conditions of water logging in fields would, therefore, have a severe effect on soybean nitrogen fixation.

The effects of supra-atmospheric oxygen concentrations on nodule and bacteroid nitrogen fixation have implications for artificial transfer of symbiotic capabilities or transfer of nitrogenase genes to aerobic bacteria or to plants. Although progress has been made in understanding the effects of supraoptimal O_2 on nitrogen fixation, there is a need for a closer examination of the mechanism(s) controlling the interactions. Respiration can apparently provide some protection for nitrogenase. Oxygen at high levels can repress nitrogenase synthesis. But to what degree do each of these operate at the various O_2 concentrations? Are there other interactions, such as metabolic reactions or conformational or covalent modification, to protect nitrogenase? Is nitrogenase synthesis altered during the bacteroid or whole nodule O_2 treatments, and how are other genes affected as well? Is nitrogenase synthesis "directly" regulated by O_2, or is the synthesis a result of indirect effects through alterations of levels of certain cellular metabolites? These are the types of questions that will be answered with future research.

NOTES

Jay B. Peterson, Assistant Professor, Department of Botany, Iowa State University, Ames, Iowa 50011.

REFERENCES

Agarwal, A. K. and D. L. Keister. 1983. Physiology of *ex planta* nitrogenase activity in *Rhizobium japonicum*. Appl. Environ. Microbiol. 45:1592-1601.
Appleby, C. A. 1969. Properties of leghaemoglobin *in vivo* and its isolation as ferrous oxyleghaemoglobin. Biochim. Biophys. Acta. 188:222-229.
Bergersen, F. J. 1962. Oxygenation of leghemoglobin in soybean root nodules in relation to the external oxygen tension. Nature 194:1059-1061.
Bergersen, F. J. 1982. Root nodules of legumes: structure and functions. Research Studies Press, New York.

Bergersen, F. J. and G. L. Turner. 1967. Nitrogen fixation by the bacteroid fraction of soybean root nodules. Biochim. Biophys. Acta. 141:507-515.

Bergersen, F. J. and G. L. Turner. 1975. Leghemoglobin and the supply of O_2 to nitrogen-fixing root nodules: studies of an experimental system with no gas phase. J. Gen. Microbiol. 89:31-47.

Bergersen, F. J. and G. L. Turner. 1980. Properties of terminal oxidase systems of bacteroids from root nodules of soybean and cowpea and of N_2-fixing bacteria grown in continuous culture. J. Gen. Microbiol. 118:235-252.

Criswell, J. G., U. D. Havelka, B. Quebedeaux and R. W. F. Hardy. 1976. Adaptation of nitrogen fixation by intact soybean nodules to altered rhizosphere O_2. Plant Physiol. 58:622-625.

Criswell, J. G., J. D. Havelka, B. Quebedeaux and R. W. F. Hardy. 1977. Effect of rhizosphere pO_2 on nitrogen fixation by excised and intact nodulated soybean roots. Crop Sci. 17:39-44.

DeVries, G. E., P. InT'Veld and J. W. Kijne. 1980. Production of organic acids in Pisum sativum root nodules as a result of oxygen stress. Plant Sci. Lett. 20:115-123.

Emerich, D. W., T. Ruiz-Argueso, T. M. Ching and H. J. Evans. 1979. Hydrogen-dependent nitrogenase activity and ATP formation in Rhizobium japonicum bacteroids. J. Bacteriol. 137:153-160.

Havelka, U. D. and R. W. F. Hardy. 1975. Nitrogen fixation research: a key to world food? Science 188:633-643.

Havelka, U. D. and R. W. F. Hardy. 1976. Legume N_2 fixation as a problem in carbon nutrition. p. 456-475. In W. E. Newton and D. J. Nyman (eds.) Symposium on Nitrogen Fixation. Wash. State Univ. Press, Pullman.

Hill, S., C. Kennedy, E. Kavanagh, R. B. Goldberg and R. Hanau. 1981. Nitrogen fixation gene (nifL) involved in oxygen regulation of nitrogenase synthesis in K. pneumoniae. Nature 290:424-426.

Lim, S. T., H. Hennecke and D. B. Scott. 1979. Effect of guanosine 3',5'-monophosphate on nitrogen fixation in Rhizobium japonicum. J. Bacteriol. 139:256-263.

Merrick, M., S. Hill, H. Hennecke, M. Hahn, R. Dixon and C. Kennedy. 1982. Mol. Gen. Genet. 185:75-81.

Patterson, T. G., J. B. Peterson and T. A. LaRue. 1983. Effect of supra-ambient oxygen on nitrogenase activity (C_2H_2) and root respiration of soybeans and isolated soybean bacteroids. Plant Physiol. 70:695-700.

Peterson, J. B. and H. J. Evans. 1979. Phosphoenolpyruvate carboxylase from soybean nodule cytosol. Evidence for isoenzymes and kinetics of the most active component. Biochim. Biophys. Acta. 567:445-452.

Peterson, J. B. and T. A. LaRue. 1981. Utilization of aldehydes and alcohols by soybean bacteroids. Plant Physiol. 68:489-493.

Robson, R. L. 1979. Characterization of an oxygen-stable nitrogenase complex isolated from Azotobacter chroococcum. Biochem. J. 181:569-575.

Robson, R. L. and J. R. Postgate. 1980. Oxygen and hydrogen in biological nitrogen fixation. Ann. Rev. Microbiol. 34:183-207.

Ruiz-Argueso, T., D. W. Emerich and H. J. Evans. 1979. Characteristics of the H_2 oxidizing system in soybean nodule bacteroids. Arch. Microbiol. 121:199-206.

Scherings, G., H. Haaker, H. Wassink and C. Veeger. 1983. On the formation of an oxygen-tolerant three-component nitrogenase complex from Azotobacter vinelandii. European J. Biochem. 135:591-599.

Scott, D. B., H. Hennecke and S. T. Lim. 1979. The biosynthesis of nitrogenase MoFe protein polypeptides in free-living cultures of Rhizobium japonicum. Biochim. Biophys. Acta 565:365-378.

Stewart, W. D. P. 1982. Nitrogen fixation—its current relevance and future potential. Israel J. Bot. 31:5-44.

Tajima, S. and T. A. LaRue. 1982. Enzymes for acetaldehyde and ethanol formation in legume nodules. Plant Physiol. 70:388-392.

Tjepkema, J. D. and C. S. Yocum. 1974. Measurement of oxygen partial pressure within soybean nodules by oxygen electrodes. Planta 119:351-360.

Trinchant, J. C., A. M. Birot and J. Rigaud. 1981. Oxygen supply and energy-yielding substrates for nitrogen fixation (acetylene reduction) by bacteroid preparations. J. Gen. Microbiol. 125:159-165.

Trinchant, J. C. and J. Rigaud. 1979. Sur les substrats energetiques utilises, lors de la reduction de C_2H_2, par les bacteroids extraits des nodosites de *Phaseolus vulgaris* L. Physiol. Veg. 17:547-556.

VanStraten, J. and E. L. Schmidt. 1974a. Diminuation of acetylene reduction during assay of detached soybean nodules. Soil Biol. Biochem. 6:231-234.

VanStraten, J. and E. L. Schmidt. 1974b. Volatile compounds produced during acetylene reduction by detached soybean nodules. Soil Biol. Biochem. 6:347-351.

Winter, H. C. and R. H. Burris. 1976. Nitrogenase. Ann. Rev. Biochem. 45:409-426.

Wittenberg, J. B., F. J. Bergersen, C. A. Appleby and G. L. Turner. 1974. Facilitated oxygen diffusion. The role of leghemoglobin in nitrogen fixation by bacteroids isolated from soybean root nodules. J. Biol. Chem. 249:4057-4066.

CARBON METABOLISM IN SOYBEAN ROOTS AND NODULES: ROLE OF DARK CO_2 FIXATION

Karel R. Schubert and George T. Coker III

Many nonphotosynthetic plant tissues, including roots (Jackson and Coleman, 1959) and nodules of N_2-fixing leguminous and actinorhizal plants, possess an active system for assimilating CO_2 (Table 1). Lowe and Evans (1962) demonstrated that CO_2 was required for growth of *Rhizobium* and infection of the leguminous host. Likewise, high concentrations of CO_2 apparently stimulated nodule and root growth (Mulder and Van Veen, 1960). Davies (1979) has reviewed the role of dark CO_2 fixation; i.e., phosphoenolpyruvate (PEP) carboxylase activity, in plants. Recent studies have focused on establishing the significance of dark CO_2 fixation on the

Table 1. Dark CO_2 fixation in nodules of N_2-fixing plants.

Plant species	Major nitrogenous products	Enzymes involved	Reference
Alnus glutinosa (European black alder)	Citrulline	PEP Carboxylase Carbomyl phosphate synthetase	Gardner and Leaf (1960); McClure et al., (1983)
Glycine max (soybean)	Ureides (Asparagine, Glutamine)	PEP Carboxylase; phosphoribosylamino-imidazole carboxylase	Coker and Schubert (1981)
Lotus corniculatus (birdsfoot trefoil)	Asparagine	PEP Carboxylase	Vance et al. (1983)
Lupinus species (lupins)	Asparagine	PEP Carboxylase	Christeller et al. (1977); Layzell et al. (1979)
Medicago sativa (alfalfa)	Asparagine	PEP Carboxylase	Vance et al. (1983)
Pisum sativum (garden pea)	Asparagine	PEP Carboxylase	Minchin and Pate (1973)
Phaseolus vulgaris (dwarf French bean)	Ureides (Asparagine)	PEP Carboxylase	Cookson et al. (1980)
Vicia faba (broad bean)	Asparagine	PEP Carboxylase	Lawrie and Wheeler (1975)
Vigna angularis (Adzuki bean)	Ureides (Asparagine)[a]	PEP Carboxylase	Vance et al. (1983)
Vigna unguiculata (cowpea)	Ureides (Asparagine)[a]	PEP Carboxylase	Layzell et al. (1979)

[a]Ureide synthesis and export decreases and amide synthesis and export increases in the presence of nitrate or ammonium.

815

carbon economy and function of nodules and roots.

Nonphotosynthetic CO_2 fixation may play a key role in the synthesis of nitrogenous organic solutes transported from nodules and roots, respiratory substrates, and counterions. Layzell et al. (1979) suggested that the reassimilation of respired carbon in nodules of N_2-fixing plants may actually increase the efficiency of these symbionts. Label from $^{14}CO_2$ is incorporated into organic acids, amino acids, amides, and ureides. PEP carboxylase appears to be the primary enzyme involved in CO_2 fixation, based on the time-course of incorporation of label from $^{14}CO_2$ into malate. Several other carboxylating enzymes are involved in CO_2 fixation in ureide-exporting plants (McClure et al., 1983; Boland and Schubert, 1982a).

Christeller et al. (1977) have proposed that in lupin (*Lupinus* sp.) nodules dark CO_2 fixation functions primarily in the production of carbon skeletons for the synthesis of asparagine, the major product of ammonium assimilation in nodules. In lupins, estimated rates of CO_2 fixation are equivalent to those for N_2 fixation. Similar results have been obtained for other amide-exporters, including the perennial legume, alfalfa (*Medicago sativa* L.) (Vance et al., 1983).

Although soybean plants synthesize and transport the amides, asparagine and glutamine, when grown on nitrate or ammonium, the primary products of N_2 fixation exported from the nodule are the ureides, allantoin and allantoic acid. The synthesis of the latter does not require oxalacetate, the product of the PEP carboxylase reaction. Therefore, the importance and function of CO_2 fixation in soybean nodules growing totally on symbiotically fixed N_2 might differ from that for plants grown on combined nitrogen (nitrate or ammonium). The following is a summary of the results of studies on carboxylation reactions in soybean roots and nodules, including a comparison of the role of CO_2 fixation in soybeans grown symbiotically or on combined nitrogen.

CARBOXYLATION REACTIONS IN SOYBEAN PLANTS GROWN SYMBIOTICALLY

CO_2 Fixation During Nodule Development

Coker and Schubert (1981) monitored the rates of CO_2 fixation during early plant development (Figure 1). Prior to nodule development, secondary roots exhibited high rates of CO_2 fixation. As immature nodules developed, the rate of dark CO_2 fixation in roots declined about 12-fold. This precipitous decline corresponded to the onset of N_2 fixation, as measured by the C_2H_2 reaction assay, and the increased rate of $^{14}CO_2$ incorporation by nodules. During this period of time prior to the formation of functional nodules, the predominant nitrogenous compounds in the root xylem exudate were amino acids, especially asparagine (Coker and Schubert, 1981; Schubert, 1981). Shortly after N_2 fixation began, the relative content of the ureides in the sap increased and amino acids decreased. At that time the specific activity of CO_2 fixation in nodules declined in parallel with a decrease in the ratio of amino acids to ureides in the xylem exudate, while total CO_2 fixation per plant leveled off. In contrast, CO_2 fixation and in vitro PEP carboxylase activity paralleled C_2H_2 reduction activity throughout nodule development in lupins (Christeller et al., 1977). Thus, CO_2 fixation plays an important role in young roots and nodules. These tissues are dependent on amino acids for growth, and are involved in amino acid synthesis and transport. These results are consistent with the proposed role of CO_2 fixation in amino acid biosynthesis.

Initial Products of CO_2 Fixation

The initial products of $^{14}CO_2$ fixation were analyzed by high voltage electrophoresis or HPLC methods (Coker and Schubert, 1981; Schubert and Coker, 1981;

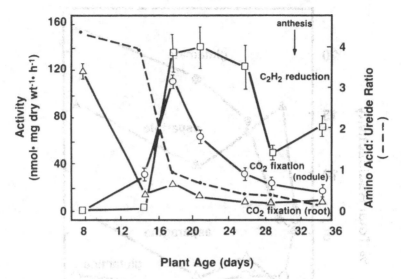

Figure 1. Rates of dark CO_2 fixation and C_2H_2 reduction during development of soybeans. CO_2 fixation of roots (△) and nodules (○), C_2H_2 reduction activity (□), and the ratio of the concentration of amino acids to ureides in the xylem exudate were measured during early nodule development. Figure modified from Coker and Schubert (1981).

Coker, 1982). The organic acid fraction accounted for two-thirds or more of the total acid stable radioactivity after exposure to labeled CO_2 (Coker and Schubert, 1981; Coker, 1982). The organic acid fraction accounted for a similar proportion of the total label incorporated in nodules of other N_2-fixing plants (Cookson et al., 1980; McClure et al., 1983; Vance et al., 1983; Christeller et al., 1977; Lawrie and Wheeler, 1975). The organic acid fraction was further separated either by chromatography on thin-layer plates, by gas chromatography or by ion exchange chromatography (Coker and Schubert, 1981; Coker, 1982). Label accumulated initially in malate with lesser amounts of ^{14}C in succinate, fumarate and citrate after longer incubations (Coker, 1982). Although attempts to trap labeled oxalacetate were unsuccessful (Coker, 1982), the labeling pattern and results of pulse-chase experiments support that PEP carboxylase is a major carboxylating enzyme in nodules and roots. The oxaloacetate formed is rapidly converted into malate through the action of malate dehydrogenase.

Role of CO_2 Fixation in Amino Acid Biosynthesis

Amino acids, primarily aspartate and glutamate, are labeled rapidly after exposure to $^{14}CO_2$ with lesser amounts of ^{14}C detected in alanine, asparagine, glutamine, serine, glycine, and other amino acids (Coker, 1982) (Figure 2). Very little radio-activity (less than 5%) accumulated in asparagine within nodules of plants grown symbiotically (Coker, 1982). The amount of ^{14}C in aspartate and glutamate increased after pulse-labeling, while the label decreased in the organic acid fraction (Coker and Schubert, 1981). These results substantiate the role of CO_2 fixation in amino acid biosynthesis in soybean roots and nodules.

In very young roots and nodules, a higher percentage of the ^{14}C incorporated is present as asparagine (Coker, 1982). Thus, CO_2 fixation may play a major role in amino acid synthesis during the very early stages of plant growth. As the plants develop and ureide synthesis, which does not require a supply of C_4 acids, increases,

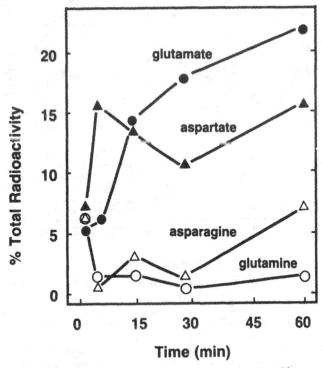

Figure 2. Distribution of label in amino acids of nodules exposed to $^{14}CO_2$. Values presented are for % total radioactivity incorporated into glutamate (●), aspartate (▲), asparagine (△), and glutamine (○) after exposure of nodulated root systems to $^{14}CO_2$. Figure modified from Coker (1982).

the role of CO_2 fixation in amino acid biosynthesis declines. This is consistent with the developmental pattern observed (Coker and Schubert, 1981).

Synthesis of Respiratory Substrates

Assuming that CO_2 fixation was necessary for asparagine biosynthesis, Christeller et al. (1977) suggested that the molar ratio of CO_2 fixed per N_2 fixed should be at least one. In fact, they estimated a value of about 2 for lupin nodules. Coker and Schubert (1981) estimated a value of 3.4 at the peak of CO_2 fixation in soybean nodules, even though the synthesis of asparagine only accounted for a small percentage of the N_2 fixed during this period. The higher than predicted rates of CO_2 fixation are consistent with alternative uses for the carbon skeletons produced via PEP carboxylase activity. Since the C_4 acids are very effective substrates for the respiratory activity of bacteroids and mitochondria, Coker and Schubert (1981) proposed that at least part of the organic acids synthesized in nodules and roots from recently fixed CO_2 were used to support the respiratory activities within the nodule. In fact, high levels of CO_2 stimulated C_2H_2 reduction activity and respiration in detached nodules (Coker and Schubert, 1981), consistent with the utilization of C_4 acids produced via CO_2 fixation as respiratory substrates. In pulse-chase experiments approximately 50% of the acid stable radioactivity was lost from detached nodulated roots during an 18-min chase (Coke and Schubert, 1981). This loss is the direct result of a decrease in radioactivity in the organic acid fraction. Lawrie and Wheeler (1975) reported a similar loss in total radioactivity in nodules of *Phaseolus vulgaris* L.

Synthesis of Carbon Skeletons to Maintain Ion Balance

High levels of malate are found in the xylem exudate of a number of legumes and other plants. The levels of malate in the xylem vary depending upon the source of nitrogen used for growth, with the highest levels of malate present in the sap of symbiotically-grown plants (Israel and Jackson, 1980). Dark CO_2 fixation may play a role in the synthesis of malate for export as a counterion for cation transport. Vance et al. (1983) reported that 60 to 70% of the ^{14}C in the root bleeding sap from alfalfa was in the form of organic acids. Indeed, labeled malate was detected in the xylem exudate and stems of nodulated soybean plants in which the root systems were exposed to $^{14}CO_2$ (Coker, 1982). With time, the label was metabolized in the leaves (unpublished results).

Carboxylation Reactions and Ureide Biosynthesis

Cookson et al. (1980) reported that ^{14}C from $^{14}CO_2$ was incorporated into allantoin and allantoic acid in dwarf French bean (*Phaseolus vulgaris* L.) nodules. They suggested that ^{14}C was first incorporated into C_4 acids, which were subsequently used via the glyoxylate cycle for the production of serine and glycine, essential precursors for ureide biogenesis. The amounts of ^{14}C incorporated into glycine and serine in dwarf French beans were higher than in soybean plants. In soybean nodules the ureides account for 2 to 3% of the total radioactivity incorporated from $^{14}CO_2$ (Schubert and Coker, 1981; Coker, 1982). Although ^{14}C was incorporated into serine and glycine in nodules, the levels were low (Figure 3). Soybean nodules apparently do not possess a

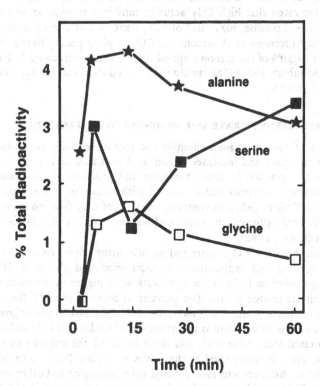

Figure 3. Distribution of label in alanine, serine, and glycine after incubation of nodules with $^{14}CO_2$. Figure modified from Coker (1982).

functional glyoxylate cycle (Johnson et al., 1966; Hanks, Schubert and Boland, unpublished results). Therefore, the most likely pathway for the incorporation of label from $^{14}CO_2$ into glycine and serine is via the reversal of glycolysis, with subsequent formation of labeled 3-phosphoglycerate. Boland and Schubert (1982a,b) have reported enzymological evidence for the synthesis of serine and glycine from 3-phosphoglycerate in soybean nodules. Ureide biogenesis in soybean nodules involves *de novo* purine biosynthesis (Schubert, 1981; Reynolds et al., 1982; Boland and Schubert, 1982a; Boland and Schubert, 1983). Boland and Schubert (1982a) demonstrated that purine biosynthesis in soybean nodules involves the incorporation of ^{14}C from $^{14}CO_2$ into C_6 of the purine ring in a reaction catalyzed by phosphoribosylaminoimidazole carboxylase. Thus, nodules contain a second very active carboxylating enzyme. The ^{14}C incorporated into the purine ring in this manner, however, is lost during the oxidation of the ring to form allantoin in the reaction catalyzed by uricase. Thus, ^{14}C incorporated via this enzyme is present only at low amounts in purine nucleotides, nucleosides, and free bases. Normally, the pools of these compounds are very low (Boland and Schubert, 1982a).

Thus, dark CO_2 fixation appears to play a key role in the synthesis of carbon skeletons for amino acid biosynthesis, respiratory substrates, and counterions, and in the *de novo* synthesis of purines for ureide biogenesis in symbiotically-grown soybeans. Layzell et al. (1979) suggested that dark CO_2 fixation also increases the efficiency of N_2 fixation. Dark CO_2 fixation, however, does not improve the efficiency of N_2 fixation *per se*. The high levels of CO_2 present within nodules and in the surrounding rhizosphere may decrease non-essential respiratory activity of roots and nodules. Mahon (1979) indicated that high CO_2 actually inhibited respiration of pea (*Pisum sativum* L.) nodules. Likewise, high rates of CO_2 fixation give rise to an apparent increase in efficiency (decrease in the amount of CO_2 respired per N_2 fixed). In the soybean, at least 10 to 30% of the carbon respired by nodules is refixed via PEP carboxylase (Coker and Schubert, 1981). This would result in a decrease in the apparent cost of N_2 fixation of up to 25%.

EFFECTS OF NITRATE AND AMMONIUM ON CO_2 FIXATION

The presence of nitrate or ammonium in the rooting medium decreases N_2 fixation and ureide transport and increases amino acid synthesis and transport. On this basis, combined nitrogen should affect the rates and products of CO_2 fixation. The rates of CO_2 fixation in soybean nodule treated with nitrate or ammonium (Table 2) decreased by 50% (Coker, 1982). In contrast, the rates of CO_2 fixation in roots (Table 2) was not significantly affected by nitrate and was increased by 50% in plants treated with ammonium (Coker, 1982).

After incubation with $^{14}CO_2$, roots and nodules from plants treated with nitrate or ammonium accumulated radioactivity in asparagine and glutamine (Figure 4). The rates of asparagine and glutamine synthesis were higher in ammonium-treated plants than in nitrate-treated plants. Ten percent or more of the ^{14}C fixed accumulated in asparagine and/or glutamine (Coker, 1982). In the case of plants grown in the presence of ammonium, ammonium is not transported (Polayes, 1983) and, therefore, must be incorporated into amino acids. On the other hand, the majority of the nitrate taken up by the plant is transported to the leaves as nitrate. Only 10 to 20% of the nitrate is reduced in the roots and incorporated into asparagine and other amino acids (Polayes, 1983). This would account for the apparent differences in the effects of nitrate and ammonium on the rate of dark CO_2 fixation and amide biosynthesis in

Table 2. Effects of nitrate and ammonium on the rate of dark CO_2 fixation[a] in soybean roots and nodules. Modified from Coker (1982).

Treatment	Roots		Nodules	
	7 days	14 days	7 days	14 days
	nmol/mg dry wt·h			
Control	16 ± 2 r	16 ± 3 r	51 ± 6 r	30 ± 2 r
KNO₃ (10 mM)	12 ± 1 r	14 ± 1 r	20 ± 3 t	10 ± 1 s
KNO₃ (3 mM)	15 ± 1 r	17 ± 2 r	41 ± 3 rs	12 ± 1 s
(NH₄)₂SO₄ (1.5 mM)	17 ± 2 r	24 ± 1 s	34 ± 3 st	14 ± 2 s

[a]Values followed by the same letter were not significantly different at the 5% level.

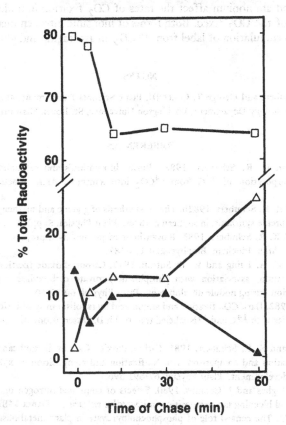

Figure 4. Distribution of label in roots of nitrate-treated soybean plants after pulse labeling with $^{14}CO_2$. Soybean roots were pulse labeled with $^{14}CO_2$ and the distribution of label in the organic acid fraction, asparagine (△), and aspartate (▲) were measured. Data taken from Coker (1982).

roots.

Nitrate and ammonium had very little effect on the distribution of label from $^{14}CO_2$ in the organic acid fraction of roots and nodules (Coker, 1982). Nitrate, however, did significantly alter the level of C_4 acids in roots. The levels of malate decreased approximately twofold in roots of plants grown in the presence of nitrate (Coker, 1982).

SUMMARY

An active system for incorporating CO_2 exists in soybean roots and nodules. This nonphotosynthetic CO_2 fixation is catalyzed by PEP carboxylase and phosphoribosylaminoimidazole carboxylase. The former is apparently important for the synthesis of carbon skeletons, which are used for the synthesis of amino acids, respiratory substrates, and counterions. The labeled products that accumulate in nodules and roots (organic acids, amino acids, and low levels of ureides) are labeled as a direct result of PEP carboxylase activity. Radioactivity incorporated through the action of phosphoribosylaminoimidazole carboxylase is lost in the synthesis of allantoin from uric acid. Nitrate and ammonium affect the rates of CO_2 fixation in nodules as well as the final products of the CO_2 fixed. Both forms of inorganic nitrogen increase the rates of synthesis and accumulation of label from $^{14}CO_2$ in asparagine and glutamine.

NOTES

Karel R. Schubert and George T. Coker III, Plant Sciences Department, Monsanto Agricultural Products Co., and Biology Department, Washington University, St. Louis, Missouri.

REFERENCES

Boland, M. J. and K. R. Schubert. 1982a. Purine biosynthesis and catabolism in soybean root nodules: Incorporation of ^{14}C from $^{14}CO_2$ into xanthine. Arch. Biochem. Biophys. 213: 486-491.

Boland, M. J. and K. R. Schubert. 1982b. The biosynthesis of glycine and methenyl tetrahydrofolate, precursors of ureide synthesis, in soybean nodules. Plant Physiol. Suppl. 69:618.

Boland, M. J. and K. R. Schubert. 1983. Biosynthesis of purines by a proplastic fraction from soybean nodules. Arch. Biochem. Biophys. 220:179-187.

Christeller, J. T., W. A. Laing and W. D. Sutton. 1977. Carbon dioxide fixation by lupin nodules. I. Characterization, association with phosphoenolpyruvate carboxylase and correlation with nitrogen fixation during nodule development. Plant Physiol. 60:47-50.

Coker III, G. T. 1982. Dark CO_2 fixation and amino acid metabolism in symbiotic N_2-fixing systems. Labeling studies with ^{14}C and ^{13}N-labeled tracers. Ph.D. Dissertation. Michigan State University, East Lansing.

Coker III, G. T. and K. R. Schubert. 1981. Carbon dioxide fixation in soybean roots and nodules. I. Characterization and comparison with N_2-fixation and composition of xylem exudate during early nodule development. Plant Physiol. 67:691-696.

Cookson, C., H. Hughes and J. Coombs. 1980. Effects of combined nitrogen on anaplerotic carbon assimilation and bleeding sap composition in Phaseolus vulgaris L. Planta 148:338-345.

Davies, D. D. 1979. The central role of phosphoenolpyruvate in plant metabolism. Ann. Rev. Plant Physiol. 30:131-158.

Gardner, I. C. and G. Leaf. 1960. Translocation of citrulline in Alnus glutinosa. Plant Physiol. 35: 948-950.

Israel, D. W. and W. A. Jackson. 1980. The influence of nitrogen source on ionic and carbon-nitrogen balance of soybean xylem sap. Plant Physiol. 65 Suppl.: 319.

Jackson, W. A. and N. T. Coleman. 1959. Fixation of carbon dioxide by plant roots through phosphoenolpyruvate carboxylase. Plant Soil. 11:1-16.

Johnson, G. V., H. J. Evans and T. M. Ching. 1966. Enzymes of the glyxylate cycle in Rhizobia and in nodules of legumes. Plant Physiol. 41:1330-1336.

Lawrie, A. C. and C. T. Wheeler. 1975. Nitrogen fixation in the root nodules of *Vicia faba* L. in relation to the assimilation of carbon. II. The dark fixation of carbon dioxide. New Phytol. 74: 437-445.

Layzell, D. B., R. M. Rainbird, C. A. Atkins and J. S. Pate. 1979. Economy of photosynthate: Use in nitrogen-fixing legume nodules. Plant Physiol. 64:888-891.

Lowe, R. H. and H. J. Evans. 1962. Carbon dioxide requirement for growth of legume nodule bacteria. Soil Sci. 94:351-356.

Mahon, J. D. 1979. Environmental and genotypic effects on the respiration associated with symbiotic nitrogen fixation in peas. Plant Physiol. 63:892-897.

McClure, P. R., G. T. Coker III and K. R. Schubert. 1983. Carbon dioxide fixation in roots and nodules of *Alnus glutinosa* I. Role of phosphoenolpyruvate carboxylase and carbamyl phosphate synthetase in dark CO_2 fixation, citrulline synthesis, and N_2 fixation. Plant Physiol. 71:652-657.

Minchin, F. R. and J. S. Pate. 1973. The carbon balance of a legume and the functional economy of its root nodules. J. Exp. Bot. 24:295-308.

Mulder, E. G. and W. L. Van Veen. 1960. The influence of carbon dioxide on symbiotic nitrogen fixation. Plant Soil 13:265-278.

Polayes, D. A. 1983. Nitrogen metabolism in soybeans. The biosynthesis of ureides in seedlings and the partitioning of N into vegetative and reproductive tissue. Ph. D. Dissertation. Michigan State Univ., East Lansing.

Reynolds, P. H. S., M. J. Boland, D. G. Blevins, K. R. Schubert and D. D. Randall. 1982. Enzymes of amide and ureide biogenesis in developing soybean nodules. Plant Physiol. 69:1334-1338.

Schubert, K. R. 1981. Enzymes of purine biosynthesis and catabolism in *Glycine max* I. Comparison of activities with N_2 fixation and composition of xylem exudate during nodule development. Plant Physiol. 68:1115-1122.

Schubert, K. R. and G. T. Coker III. 1981. Nitrogen and carbon assimilation in N_2-fixing plants. Short-term studies using ^{13}N and [11]C. Adv. Chem. Ser. 197:317-339.

Vance, C. P., S. Stade and C. A. Maxwell. 1983. Alfalfa root nodule carbon dioxide fixation I. Association with nitrogen fixation and incorporation into amino acids. Plant Physiol. 72:469-473.

INTERACTIONS BETWEEN CARBON AND NITROGEN DURING PODFILLING

D R Nelson, R J Bellville, C A Zampini and C A Maxwell

Two papers, consistent in their theme, were published independently in 1975. Each proposed a conceptual model to explain the role of carbon and nitrogen in reproductive development of soybean. The relationships defined in the model were attractive and consistent with the information at that time; as a result, they persisted.

Sinclair and de Wit (1975) stated simply that the nitrogen requirements of soybean were so great that the crop was forced to 'self-destruct', i.e. destroy its own productivity, in order to sustain seed growth. Even with redistribution during senescence, they state that yield is inhibited by inadequate nitrogen.

Thibodeau and Jaworski (1975) found that nitrate reductase (NR) activity in the leaves decreased to zero by early podfilling (Figure 1). These data and a reported lack

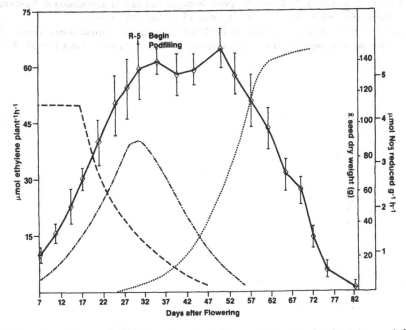

Figure 1. Patterns of nitrogen assimilation. Detached root acetylene reduction (AR) (–·–·–), leaf nitrate reductase activity (––––), and seed weight (······) profiles; components of a nitrogen assimilation/yield model presented by Thibodeau and Jaworski (1975). Solid line is seasonal nondestructive (AR) profile from Nelson et al. (1984).

of response of soybean to combined nitrogen late in the season (Lathwell and Evans, 1951), led to the conclusion that soybean is incapable of taking up or assimilating nitrate during podfilling. Their profile of acetylene reduction (AR) increased as NR declined, consistent with the recognized antagonism between the processes. AR then peaked before stage R5 (beginning seed growth; Fehr and Caviness, 1977), and had all but disappeared by mid-podfill, consistent with a report by Harper (1974). They concluded that dinitrogen fixation was inadequate to provide nitrogen during pod-filling. They further observed that AR declined as seed weight increased, and con-cluded that carbohydrate directed toward the roots is intercepted by the pods, ulti-mately starving the nodules. Finally, they stated that, as the supply of nitrogen from the nodules declines (NR having long since disappeared), the soybean must 'scavenge' nitrogen from leaf protein, which leads to senescence. The theme was remarkably similar to that of Sinclair and de Wit.

SEASONAL PATTERNS OF NITROGEN ASSIMILATION

These results were consistent with the available evidence. However, studies con-ducted since then support a different perception. The non-destructive field plot AR system of Nelson et al. (1984) demonstrated that nodule activity could increase until 30 days after R5 (Figure 2). Profiles from three seasons show the consistency of the data from year-to-year. Plants were grown in sand-nutrient culture with a complete nutrient solution. However, nitrogen was excluded from flowering until maturity. The yields of plots that depended on nitrogen fixation alone were comparable to those of soil grown companion plots. Integrating the area under the curve, 78% of seasonal

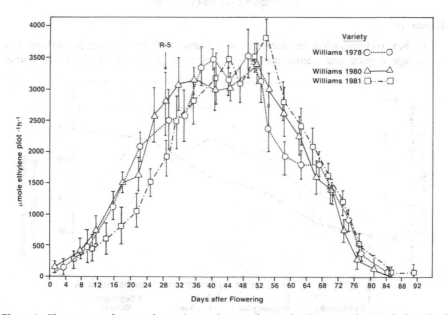

Figure 2. Three years of seasonal non-destructive acetylene reduction control data. Each point is the mean of three replications, 65 plants per replicate; profiles are plotted over one an-other using stage R2 as zero time. Bars represent the standard errors. From Nelson et al. (1984).

nitrogen is fixed after stage R5, demonstrating that dinitrogen fixation is capable of contributing to N needs during podfilling.

As further evidence of nitrogen assimilation during podfill, five studies reporting nitrogen analyses at both R5 and maturity demonstrated that 60% or more of seasonal nitrogen was accumulated after stage R5 (Nelson et al., 1984). Even more interesting, nonnodulating isolines accumulated 68% of their seasonal nitrogen after R5 (Pal and Saxena, 1975), after NR is absent from the leaves. Thibodeau and Jaworski disclaim NR activity in the roots and nodules; however, we conclude that sufficient activity exists in an organ other than the leaves. There is evidence that NR activity begins to increase in the roots at flowering (Hunter et al., 1982). It is peculiar that NR abandons a source of 'free' energy in the leaves to reside in another part of the plant, prior to the stage when carbon becomes limiting and nitrogen is in greatest demand.

In the context of the inverse relationships proposed by Thibodeau and Jaworski, it is also probable that AR is not increasing in response to declining NR, but due to the increased sink demand. Further, when the nondestructive AR profile is compared (Figure 1), nodule activity appears to be coincident with, rather than inversely related with seed growth. When AR begins to decline, seed growth does also. Evidence for prolonged seasonal activity can be reconciled with the AR profile of Thibodeau and Jaworski. In the nondestructive system, fixation of dinitrogen was the sole source of N after flowering; in the soil system, the contribution from NR likely resulted in the more rapid decline in nodule activity. The extent to which NR contributes will determine the stage at which AR declines.

COMPENSATION MECHANISMS

Transport

Carbon compensation mechanisms are orchestrated through photosynthesis rate, percentage export and redistribution. The plots in Figure 3 illustrate the carbon

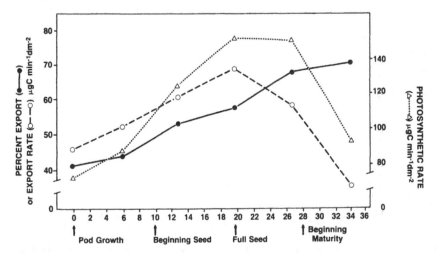

Figure 3. Carbon assimilation and transport during podfilling. Fiskeby was grown in the growth chamber in sand-nutrient culture at 1000 $\mu E/m^2 \cdot s$. Photosynthesis and percentage export (partitioning) from a source leaf were measured following pulse labeling with $^{14}CO_2$. Export rate is computed from the measured values. The three replications were repetitively labeled to develop the profiles.

strategy during podfilling. Note that during podfilling, percentage export is constantly increasing to maintain a high export rate (carbon flow) until maturity. The developing pods do not starve the roots, the leaves simply increase the supply of carbon to accommodate both seed growth and the enhanced nitrogen assimilation. An example of this is shown in Figure 4. When photosynthesis and nitrogen fixation were measured on the same plots, photosynthesis declined to nearly 50% of its peak rate before AR activity began to decline. This was initially hypothesized as evidence for a surplus of carbon during podfilling. However, it is likely that the decline in nitrogen fixation was coincident with the decline in export rate. Until then, no decline in carbon flow was perceived by the nodules.

Conceptually, one might expect an analogous mechanism in the roots, to insure an uninterrupted supply of nitrogen. When we compared the seasonal profiles of ureide content and AR, ureide was found to be relatively independent of fluctuations in AR (Figure 5). This suggests a regulated system to provide the shoot with a dependable source of nitrogen until the end of the season; a buffer against moderate environmental fluctuations.

H₂ Evolution

Another example of carbon economy after R5 is hydrogen evolution. It is recognized that nodules can evolve more than 40% of their energy as H_2, yet strains that eliminate this loss provide no yield advantage. If analogy with pea (*Pisum sativum* L.) is possible, this apparent inconsistency may result because, after R5, an inefficient strain becomes completely efficient (Bethlenfalvay et al., 1978). Analogous with this,

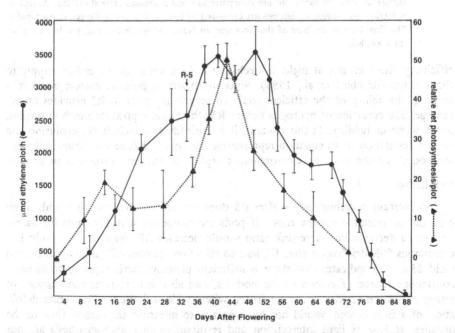

Figure 4. Seasonal canopy photosynthesis (dotted line) and acetylene reduction (solid line) profiles. Both measurements were made on the same plots; each point was the mean of three replicates, 65 plants per replicate. Canopy photosynthesis was measured in saturating light. Bars represent standard errors.

828

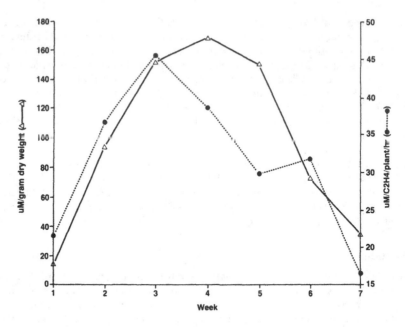

Figure 5. Seasonal acetylene reduction and ureide content. Fiskeby was grown in sand-nutrient culture at 1000 $\mu E/m^2 \cdot s$. Nitrate concentration was decreased after flowering. Acetylene reduction and ureide measurements consisted of four, one plant replicates at each point. The first 2 cm at the base of the stem was analyzed for ureide content. The last sampling stage was R6.

efficiency also increases at night, a mechanism which stabilizes the carbon supply to the nodules (Rainbird et al., 1983). Since both strains preserve carbon when it is limiting, the ability of the efficient strain to recycle H_2 prior to R5 provides no advantage. The evolution of hydrogen before R5, therefore, represents a carbon surplus, not a waste or liability. In this context, it is misleading to include H_2 evolution and nodule respiration in an equation representing the 'cost' of reducing various nitrogen sources, if, at the stage or environmental regime of interest, carbon is in surplus.

Defoliation

An increase in carbon supply after R5 does not address the issue that pods intercept carbon intended for the roots. If pods are competing with the roots for carbohydrate, a decrease in source:sink ratio should decrease AR. As shown in Table 1, a continuous 60% defoliation after R2 had no effect on seasonal AR and only decreased yield 23%. This indicates that there is sufficient photosynthetic capacity to insure a continuous source of carbon to the nodules, and also demonstrates the support of nitrogen fixation activity under conditions where yield is limited. A moderate defoliation of a field canopy would not be expected to interrupt the carbon flow to the nodules, as long as light interception and resultant canopy photosynthesis are not decreased. Declines in nodule activity from a defoliation have been reported, however. In one case, nodule activity of untreated plants declined before flowering (Lawn and Brun, 1974), and in the other, plants were grown at less than optimum light level (Bethlenfalvay et al., 1978). Overall, any environmental regime that shortens the seasonal profile of nitrogen fixation or decreases the carbon available to the nodules,

Table 1. Effect of defoliation after R5 on seasonal acetylene reduction and yield.

Treatment	Yield		Integrated acetylene reduction	
	% of control	SD Mean	% of control	SD Mean
Control	100	0.03	100	0.11
60% Defoliation[a] from R2 to R8	77	0.04	103	0.03
60% Defoliation from R5 to R8	64	0.10	91	0.18
40% Defoliation from R5 to R8	96	0.07	98	0.08

[a]Defoliation consisted of removing the lateral (60%) or terminal (40%) leaflets from each tri-foliolate twice weekly. Each treatment had three replications.

will influence the response of nodule activity to defoliation. Finally, if yields are un-affected when 40% of the leaves are irretrievably lost (Table 1), redistribution would not seem to be a primary resource. Warembourg et al. (1982) showed that 43% of leaf material was redistributed during rapid seed development. From the point of maximum dry weight, however, overall this amounted to less than 10% of the vege-tative biomass. The weight of seed produced after R5, however, was equal to the total vegetative weight accumulated in the first 70 days. This emphasizes the importance of current photosynthesis and nitrogen fixation. Environmental conditions can result in more or less redistribution. However, the key factor is the amount of nitrogen redistributed.

Shading and Elevated CO_2

Shading is another adaptation. Orcutt and Fred (1935) and Trang and Giddens (1980) demonstrated a stimulation of nodule activity with a moderate shade treat-ment. Each further showed a strong relationship between plant dry weight and total nodule activity. Trang and Giddens found that percentage N remained relatively constant with moderate shade. However, total non-structural carbohydrate declined and nodule specific activity increased. We conducted a study in 1983 that examined the interaction between shade and elevated CO_2 treatments (Table 2). Fiskeby, a determinate cultivar, was used. Unshaded plots accumulated more than 50% of their seasonal dry weight after R5, of which 70 to 80% was partitioned into the seeds. With 50% shade, partitioning increased to essentially 100%, and there was a pronounced effect of CO_2 concentration on redistribution. In both shaded and unshaded treat-ments, the increase in yield from elevated CO_2 was associated with an increase in partitioning to the seeds. This suggests that a decrease in vegetative:fruiting ratio may be a major component in the observed enhancements from elevated CO_2.

Elevation of CO_2 has been recognized for its stimulation of nodule activity for 50 years (Wilson et al., 1933); as much as a 500% enhancement has been reported (Hardy and Havelka, 1975). The study reported in Table 2 also examined the interaction between nitrogen source and elevated CO_2. An apparent difference in yield strategy between the nodulated plants and plants grown on 400 µg nitrate/mL was evident even before harvest. The nodulated plants progressed gradually from a pale green to a golden yellow. The nitrate-grown plants on elevated CO_2 began to senesce very rapidly and were mature 10 days before those with nodules. The high-nitrate, ambient-CO_2 treat-ment remained green throughout most of the experiment, and was the last to senesce. Despite the marked visible effects, there were only small differences in yield between the nodulated plants given elevated CO_2 and the nitrate-grown plants in either CO_2 regime (Table 2). Statistically, yield of both nitrogen treatments was enhanced by

Table 2. CO_2 effects during pod filling.

Treatments			Vegetative effects			Reproductive effects		
CO_2[a]	Nitrogen[b]	Shade	Total plant wt.	Δ Plant wt.	Wt. accum. during treatment	Seed yield per plant	100 seed wt.	Δ Harvest index seed wt/ Δ plant wt.
			g	g	%	g	g	
Low	Zero	–	21.0	9.4	42.9	8.67	12.94	0.928
High	Zero	–	29.2	12.5	51.7	10.57	13.96	0.851
Low	High	–	26.5	14.5	54.7	10.14	12.88	0.699
High	High	–	25.9	13.9	53.7	10.80	14.88	0.778
Low	High	50%	22.4	7.4	33.0	7.75	11.24	1.065
High	High	50%	21.8	5.4	24.8	8.58	13.20	1.337

[a]Low = ambient, high = 300 μL CO_2/L air.

[b]Zero = Complete nutrient without nitrogen after R5. High = Complete nutrient with 400 μg nitrogen/mL.

elevated CO_2. However, high nitrate appeared to attenuate the CO_2 response. The effect of elevated CO_2 on nodulated plants appears to be, in essence, satisfaction of a nitrogen stress by increasing the flow of carbon to the nodules.

Controls grown on zero nitrogen partitioned more of their weight into seeds after treatment than those growing on nitrate, suggesting a coupling between the nodules and seeds. In this context, Ishizuka (1983) found, using ^{15}N, that nitrate was processed primarily in mature organs (roots, stems, leaves), while nitrogen as amides or ureides was processed in developing organs (developing seeds after R5). Further, Bethlenfalvay and Phillips (1978) showed that vegetative weight was lower when plants depended on nodule activity than when grown on nitrate; however, photosynthesis per leaf area was higher. Results of Hardman and Brun (1971) showed that elevated CO_2 enhanced pod number of nodulated plants, but not 100-seed weight. Differences in utilization strategy between nitrogen sources requires further attention.

Grafting

Grafting of multiple roots on shoots, and vice versa, has been used as evidence that root or shoot activities can be pushed by enhanced source or sink activity. The most cited of these studies is that of Streeter (1974), who found that two shoots on the same root produced a transitory enhancement of nodule activity. In a lesser known article, Sanders and Brown (1976) grafted up to three shoots on a root and three roots on a shoot. A 3:1 root to shoot ratio enhanced nitrogen content 72% compared with 23% for a 1:3 ratio, shoot weight 56% compared with zero, and yield 32% compared with 15%. This again suggests that a carbon limitation can be demonstrated to the extent that it alleviates a nitrogen limitation.

SUMMARY (CONCEPTUAL MODEL)

Contribution of Assimilation

Under high-yielding, field conditions, nitrogen fixation and nitrate reduction both have major roles during podfilling. During this period, the nitrogen demand is such that nitrate is less inhibitory to nodule activity. However, nitrate still determines the contribution the nodules will make. Redistribution normally accounts for a minor portion of the seed nitrogen.

Resource Management

Carbon and nitrogen seem to be allocated as follows: Photorespiration exists in the leaves to prevent photooxidation of the machinery. Environmental influences on photosynthesis are buffered by changes in export rate. Carbon is transported to all the sinks; however, an excess of carbon goes to the roots. Prior to R5, energy beyond the needs of the root is released from the nodules as hydrogen, less at night than during the day. This insures an uninterrupted source of carbon to support nodule activity. Environmental influences on nitrogen fixation are buffered by a regulated flow of ureides to the shoot. Nitrate in the environment will inhibit nodule activity and decrease partitioning to the seed.

Conservative Nature

A moderate defoliation, shade or water stress treatment can be imposed without a detrimental effect on yield, because the soybean can respond with an increase in photosynthesis per unit area, partitioning, percentage export, or redistribution in order to maintain the carbon supply. In the absence of these environmental stimuli, this 'surplus' carbon is not produced. The absence of a carbon surplus after R5 is illustrated by the decline in H_2 evolution and near maximum percentage export. If it were possible to manipulate the compensatory mechanisms and fill the pods present at R5, yields would increase 40%. This demonstrates the importance of fully understanding the components of the environment and how they influence photosynthesis, transport and nitrogen fixation during podfilling.

C/N Limitations to Yield

The high correlations between nitrogen fixation or accumulation and yield (Lathwell and Evans, 1951; Nelson et al., 1984) strongly suggest that nitrogen is limiting. Processes that influence redistribution will also affect nitrogen use efficiency and thereby the correlation. The yield enhancement by elevated CO_2, which has been construed as evidence of a carbon limitation, may result from improved nitrogen status, resulting from an increase in transport of current photosynthate to the nodules. Referring to the hypothesis of Thibodeau and Jaworski, nodule activity is, in fact, adequate for higher yields. However, carbon transport to the nodules is inadequate. There are differences in yield strategy with regard to nitrogen source. Plants grown on zero nitrogen during podfilling partition a greater proportion of their weight to the seeds.

Overall, nitrogen and carbon are both limiting, in the sense that a stimulation of yield can occur through either system, though a stimulation by carbon appears to be coupled with alleviation of a nitrogen stress. Both carbon and nitrogen are in surplus, in the sense that excess capacity can be called upon to prevent a decline in yield in the presence of a moderate stress.

NOTES

D. R. Nelson, R. J. Bellville, C. A. Zampini and C. A. Maxwell, Monsanto, St. Louis, MO, USA.

REFERENCES

Bethlenfalvay, G. J., S. S. Abu-Shakra, K. Fishbeck and D. A. Phillips. 1978. The effect of source-sink manipulation on nitrogen fixation in peas. Plant Physiol. 43:31-34.

832

Fehr, W. R. and C. E. Caviness. 1977. Stages of soybean development. Special Report 80, Iowa Co-operative Extension Service. Iowa State University, Ames.

Hardman, L. L. and W. A. Brun. 1971. Effect of atmospheric carbon dioxide enrichment at different developmental stages on growth and yield components of soybean. Crop Sci. 11:886-888.

Hardy, R. W. F. and U. D. Havelka. 1975. Photosynthate as a major factor limiting N_2 fixation by field grown legumes with an emphasis on soybeans. p. 421-439. In R. S.Nutman (ed.) Symbiotic Nitrogen Fixation in Plants. Cambridge University Press, London.

Harper, J. E. 1974. Soil and symbiotic nitrogen requirements for optimum soybean production. Crop Sci. 14:255-259.

Hunter, W. J., C. J. Fahring, S. R. Olsen and L. K. Porter. 1982. Location of nitrate reduction in different soybean cultivars. Crop Sci. 22:944-948.

Ishizuka, J. 1983. ^{15}N study of the partitioning and metabolism of the nitrogen derived from atmospheric N_2 and combined N in soybeans. p. 552. In C. Veeger and W. E. Newton (eds.) Advances in Nitrogen Fixation Research. Nijhoff/Junk Publishers, The Hague.

Lathwell, D. J. and C. E. Evans. 1951. Nitrogen uptake from solution by soybeans at successive stages of growth. Agron. J. 43:264-270.

Lawn, R. J. and W. A. Brun. 1974. Symbiotic nitrogen fixation in soybeans: 1. Effect of photosynthetic source:sink manipulations. Crop Sci. 14:11-25.

Nelson, D. R., R. J. Bellville and C. A. Porter. 1984. Role of nitrogen assimilation in seed development of soybean. Plant Physiol. 74:128-133.

Orcutt, F. S. and E. B. Fred. 1935. Light intensity as an inhibitor in the fixation of atmospheric nitrogen by Manchu soybeans. J. Am. Soc. Agron. 27:550-558.

Pal, U. R. and M. C. Saxena. 1975. Contribution of symbiosis to the nitrogen needs of soybean. Acta Agron. Sci. Hungaricae 24:430-437.

Rainbird, R. M., C. A. Atkins and J. S. Pate. 1983. Diurnal variation in the functioning of cowpea nodules. Plant Physiol. 72:308-312.

Sanders, J. L. and D. A. Brown. 1976. Effect of variations in source:sink ratio upon chemical composition and growth of soybeans. Agron. J. 68:713-717.

Sinclair, T. R. and C. T. de Wit. 1975. Photosynthetic and nitrogen requirements for seed production by various crops. Science 189:565-567.

Streeter, J. G. 1974. Growth of two soybean shoots on a single root. J. Exp. Bot. 25:189-198.

Thibodeau, P. S. and E. G. Jaworski. 1975. Patterns of nitrogen utilization in the soybean. Planta (Berl) 127:133-147.

Trang, K. M. and J. Giddens. 1980. Shading and temperature as environmental factors affecting growth, nodulation and symbiotic nitrogen fixation by soybeans. Agron. J. 72:305-308.

Warembourg, F. R., D. Montange and R. Bardin. 1982. The simultaneous use of $^{14}CO_2$ and $^{15}N_2$ labelling techniques to study the carbon and nitrogen economy of legumes growing under natural conditions. Physiol. Plant 56:46-55.

Wilson, P. W., E. B. Fred and M. R. Salmon. 1933. Relation between carbon dioxide and elemental nitrogen assimilation in leguminous plants. Soil Sci. 35:145-165.

Developmental Aspects of
Growth and Productivity

SOYBEAN SEEDLING GROWTH AND VIGOR

J. S. Burris

Considerable research has dealt with the basic and applied aspects of soybean seedling growth and vigor. Symposia have been devoted to discussions of vigor, and handbooks have been published on methodology related to testing for seedling vigor (AOSA, 1983). A considerable volume of evidence deals with the basic physiological principles involved in the transition of the seed from the resting state. Germination encompasses the processes beginning with water uptake and culminating in the rupture of the seed coat, at which time seedling growth begins. This paper deals with both the germination and the growth phase of seedling development and stand establishment as influenced by seed vigor.

GERMINATION PHASE

The studies of Obendorf and Hobbs (1970) indicate significant reductions in seedling performance following rapid imbibition. They suggested that the imbibitional damage was due to the formation of stellar lesions associated with the physical stresses imposed by rapid water uptake. The soybean seed coat or testa is the primary regulator of water uptake. Disruptions of the testa, either because of mechanical damage or because of imperfect seed coat development, can result in substantial differences in performance. The imperfect seed coat trait (etching) has been described by several authors, and its genetics has been studied (Liu, 1949; Stewart and Wentz, 1930). Although these breaks in the seed coat constitute only a very small surface area, their occurrence allows the uptake of moisture at rates similar to those of the naked seed. Indalsingh (1981) compared the rate of water uptake of etched and nonetched seed with seed that had the testa removed. Similar rates of water uptake were exhibited by the naked seed and the etched seed. In a recent study of temperature effects on water uptake, Duke et al. (1983) reported substantial differences in the leakage of soluble contents from the seed in response to imbibition temperature. Although the effects of temperature on conductivity are quite modest, the differences in malate dehydrogenase (MDH) activity were quite large. Thus, the rapid leakage of cellular constituents because of the rapid water uptake may be associated with failure of enzyme systems resulting in impaired seedling development. Reduced enzymatic activity, coupled with increased microbial activity in response to the increased substrate leakage, may often result in reduced seedling performance or stand failure.

In a recent investigation of cold-tolerance, Mortensen (1984) compared the effects of oxygen concentrations and germination temperatures on soybean seedling development. Reduced germination, percentage dry-matter transfer, and reduced vigor index resulted from germination at reduced oxygen concentrations and/or reduced temperatures. Seeds grown at 0.2% oxygen did not germinate, although they remained completely viable and, when transferred to ambient oxygen conditions, germinated readily. She also reported greater oxygen requirements at warmer temperatures than at 10 C. The inhibition of growth at low O_2 was similar to a dormancy caused by an O_2 impermeable seed coat. Overcoming this dormancy by returning to ambient O_2 levels resulted in normal seedling development, indicating little if any damage due to the reduced O_2 level. The extension of this line of research may provide considerable insight into the biochemical events associated with the dormancy imposed by low temperatures and flooded soils.

GROWTH PHASE

Hypocotyl Development

The elongation of the soybean hypocotyl can be limiting to seedling growth, resulting in stand failure. The factors affecting hypocotyl elongation include soil or substrate temperature, substrate resistance, and atmospheric components. Work in my laboratory has dealt with a number of these factors and their physiological and agronomic implications (Burris and Knittle, 1975; Gilman et al., 1973; Saminy and LaMotte, 1976; Seyedin et al., 1982).

Gilman et al. (1973) described the effects of temperature on hypocotyl elongation, which depicted a classical bimodal response with maximum elongation occurring at 20 or 30 C and a minimum at 25 C. The soil environment can have a significant effect on hypocotyl development through both temperature and gaseous composition (Saminy and LaMotte, 1976). Knittle and Burris (1979) studied soybean hypocotyl development in the laboratory as effected by cultivar, seed size and downward force (Table 1). They reported that, as the force applied to the hypocotyl was increased from 0 to 100, the length of the hypocotyl decreased. However, when seed size was reduced in Corsoy, hypocotyl length decreased or remained constant as pressure was increased, whereas hypocotyls of Amsoy-71 seedlings increased in length as seed size was reduced and seedlings were exposed to increased pressure. The responses were less pronounced at 100 g pressure. Decreased length is generally associated with an increased hypocotyl diameter, as indicated by the increased swelling index. Changes in hypocotyl dimensions are required because the narrow column represented by the hypocotyl must thicken to provide a physical base to support penetration of a compacted or crusted soil. Failure to expand, as well as excessive expansion, can result in failure to emerge. Knittle et al. (1979) developed regression equations for the rate of the soybean hypocotyl elongation under field conditions. Hypocotyl elongation-rate regression equations were developed separately for the germination phase (Figure 1) and the elongation phase (Figure 2) and were substantially different. Soil temperature was shown to be a limiting variable during the initial stages of soybean germination and hypocotyl elongation. In addition, if soil moisture was included as a variable in the regression model it confirmed earlier models developed by Wanjura et al. (1973). The significance of soil resistance as a variable in hypocotyl growth has not been previously included in the models for soybean hypocotyl growth.

Although extremes in soil moisture and compaction were not included in this

Table 1. Effect of seed size and downward force on soybean hypocotyl growth at 25 C.[a]

Measurement	Cultivar	Force (g)	Seed size[b] Large	Small	Half[c]
Hypocotyl	Corsoy	0	17.5 r	19.3 s	18.5 rs
length (cm)		25	12.4 r	11.9 r	11.2 r
		100	6.1 r	6.0 r	5.5 r
	Amsoy 71	0	6.8 r	8.9 s	12.6 t
		25	5.4 r	7.9 s	9.0 s
		100	2.3 r	3.3 rs	4.1 s
Swelling	Corsoy	0	41.0 s	29.0 r	30.5 r
index (mg/cm)		25	61.7 s	44.1 r	47.7 r
		100	98.8 s	79.7 r	74.6 r
	Amsoy 71	0	54.4 s	40.3 r	32.1 r
		25	77.7 s	52.1 r	43.4 r
		100	110.2 s	86.5 r	80.0 r

[a]Seedlings were grown for 96 h under the given force.

[b]Values within cultivar and force followed by the same letters are not significantly different at P = 0.05 using Duncan's Multiple Range Test.

[c]50% retention of large seed after cotyledon excision as described by Burris and Knittle (1975).

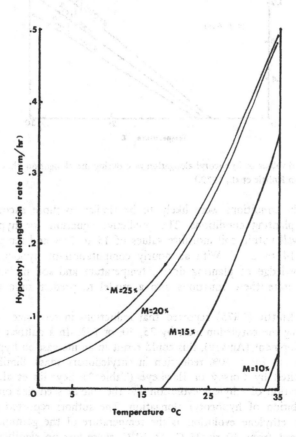

Figure 1. Predicted values of hypocotyl elongation rate during the germination phase using Equation 1. (From Knittle et al., 1979.)

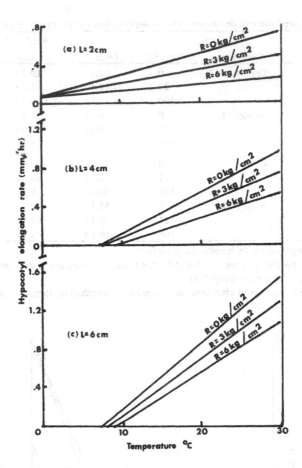

Figure 2. Predicted values of hypocotyl elongation rate during the elongation phase using Equation 2. (From Knittle et al., 1979.)

study, the study conditions were likely to be similar to those encountered under most soybean planting conditions. The predictive equation for hypocotyl elongation worked well within soil moisture values of 15 to 25% and temperatures from approximately 14 to 22 C. With an hourly computation of hypocotyl elongation rate and a knowledge of planting depth, temperature and soil moisture, it is possible to incorporate these equations into a model to predict time to 50% emergence.

Burris and Knittle (1975) reported that reductions in seed size could be mimicked by reducing the cotyledon mass by 25, 50, or 75%. In a cultivar with inhibited hypocotyl development (Amsoy), this could result in an increase in hypocotyl length from 13 to 19 cm for a 50% reduction in cotyledonary mass. Similar differences were not exhibited by Corsoy or Hawkeye (Table 2). Seyedin et al. (1982) dealt with the localization of ethylene evolution and the role of seed size on temperature-dependent inhibition of hypocotyl elongation. The authors reported a nearly two-fold increase in ethylene evolution as the temperature of the germination environment was reduced from 30 to 25 C. At 30 C, there was no significant difference in ethylene evolution among seedling portions. In contrast, those grown at 25 C

Table 2. Hypocotyl lengths of 7-day-old soybeans with excised cotyledons at 25 C.

Cultivar	Retention of whole seed[a], %			
	100	75	50	25
	– cm –			
Amsoy	13.4 r	16.0 s	19.0 t	15.3 s
Beeson	15.1 r	16.6 s	18.0 t	16.3 rs
Corsoy	21.3 s	21.8 s	21.7 s	18.5 r
Hawkeye	21.0 s	20.4 s	19.7 s	17.4 r

[a]Values within the same cultivar with different letters are significantly different at the 5% level according to Duncan's new multiple range test.

showed the hypocotyl section to be producing large amounts of ethylene during early stages of germination. Cultivar differences were quite dramatic, with the inhibited cultivars Amsoy-71 and Cutler-71 showing substantially greater ethylene evolution at 25 compared to 30 C. In contrast, Corsoy exhibited a nonsignificant, but slightly lower, ethylene evolution from seedlings grown at 25 C. The results of an experiment dealing with the differences in seed size and percentage cotyledon retention are given in Figure 3. The control seedlings grown at 25 C show a greater ethylene evolution than those at 30 C, while the removal of 50% of the cotyledon resulted in ethylene volumes similar to those generated by seedlings produced by small seeds. Their study also demonstrated that the highest rate of ethylene evolution occurs immediately below the hypocotyl arch, which may be the seedling portion that senses temperature and/or soil resistance. The ethylene involved in

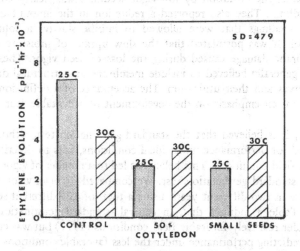

Figure 3. Effects of temperature, 50% cotyledon excision at planting time, and small seed size on rate of ethylene evolution during each 3-h interval in excised hypocotyl segments from Amsoy-71 seedlings. SD is standard deviation in picoliters per gram fresh weight per h.

hypocotyl growth is localized in the hypocotyl tissue and to some extent is controlled by unknown factors supplied by the cotyledons. The cotyledonary effect may be a response to substrate availability or to the reduction in a precursor of ethylene.

SEEDLING VIGOR

Seedling vigor is of considerable interest to the seed producer, marketer, and ultimately has a bearing on the performance of soybean seed planted by the farmer (Burris, 1976; Edje and Burris, 1970). The AOSA Vigor Handbook (AOSA, 1983) defines seed vigor as: "Seed vigor comprises those seed properties which determine the seed's potential for rapid uniform emergence and development of normal seedlings under a wide range of field conditions." The authors of this definition did not include plant performance after emergence. Reductions in seed vigor can be due to numerous environmental, cultural, and genetic causes. Few genetic effects, other than those previously described for the hypocotyl elongation, have been incorporated into current breeding programs. Reduced tillage and cold tolerance have received considerable study. A recent report by Unander et al. (1983) indicates that location of production may have a greater effect on cold tolerance than does genotype. This finding may seriously limit screening for this and other seed-quality traits until the location effect is further clarified.

Plant development after emergence is an integrated process and, therefore, more difficult to improve genetically. Numerous physiological processes have been associated with reductions in seedling growth rate and vigor, including: mitochondrial function, membrane integrity, reduced oxygen uptake, higher respiratory quotients, imbalances between the TCA and glycolytic pathways, and excessive levels of ethanol and acetaldehyde. Woodstock and Taylorson (1981), in a recent paper, indicated that the acetaldehyde evolution by low-vigor seedlings was nearly twice that of the high-vigor seedlings. They also reported a reduction in the amount of acetaldehyde produced by seedlings that were allowed to imbibe slowly in solutions of polyethylene glycol. It was postulated that the slow uptake of moisture allowed the repair of membrane damage caused during the loss of seed vigor. These deteriorative processes are generally believed to include membrane and structural damage, changes in stored reserves and their utilization. The acceptance of a definition for seed vigor was followed by an emphasis on the development of acceptable reproducible testing procedures.

In general, it is believed that the standard germination test, although valuable in describing seed lot performance under ideal conditions, fails to describe performance under adverse field conditions. This belief has led to a number of vigor tests as supplements to the standard germination test. We compared soybean vigor test results and field emergence in two different years, using a total of 109 different seedlots. Results presented in Table 3 indicate that, in general, standard germination predicts field emergence under relatively favorable field conditions (1977) but was considerably less accurate in predicting performance under the less favorable conditions encountered in 1978. The predictive value of the cold test was consistent across years, whereas the accelerated aging test was superior in predicting emergence under favorable conditions. The sand emergence test predicted field emergence relatively well in both years. Although the literature contains numerous correlative studies, much of the improved correlation depends upon the range in quality of the seedlots used and the range in field planting conditions. Wall et al. (1983) studied the emergence and yield of soybean

Table 3. Correlation coefficients between vigor tests and average field emergence.

Test	Average emergence	
	1977	1978
Acceleraged Aging 30 h	.828	.606
Accelerated Aging 72 h	.498	.584
Soil Cold Test	.771	.758
Vermiculite Cold Test	.787	.742
Standard Germination	.903	.581
Shoot Weight Mg/Seedling	.057[a]	-.167[a]
Root Weight Mg/Seedling	.531	-.318[a]
Seedling Vigor Classification	.893	.307
Hypochlorite Soak	-.262[a]	-.699
Emergence—Sand	.938	.810

[a]Not significant at 5% level of probability. All other values are statistically significant at the 5% level or better.

seed of differing quality over two years. The authors could not demonstrate yield differences due to quality except in seed with >50% *Phomopsis* infection. The lack of yield differences in soybean crops with differing stand levels may be explained by the results presented in Table 4. The study compares the agronomic development of high and low vigor seed in the field. As the season progresses, the initial differences in vigor (caused by aging) decrease and, as vegetative development approaches flowering, the differences are essentially nonexistent with no difference in yield. Although seedling vigor can be a valuable quality control parameter, which may be measured with some degree of accuracy, its expression in the field is difficult to predict.

The period of plant development from germination through to stand establishment has been a difficult growth period to predict and model. This would seem understandable on the basis of the large number of variables that affect the developing seedling. As our knowledge of seedling physiology improves, it is hoped that better models for soybean seedling development can be developed.

Table 4. Agronomic performance of high and low vigor Amsoy seed measured at several times after planting through to final yield.

Days from planting	Vigor level	Height dm	Leaves s/dm^2	Stem s/dm	Pods per plant	Seeds per pod	Yield
		cm	g/m^2	g/m^2	no	no	g/m^2
40	Hi	13.3	196	95			
	Lo	12.5	172	85			
60	Hi	51.8	715	716			
	Lo	46.5	613	584			
70	Hi	72.8	1001	1269			
	Lo	71.5	988	1271			
80	Hi	97.0	1590	2415			
	Lo	94.0	1558	2312			
Harvest	Hi				36.1	2.5	383
	Lo				39.7	2.5	381

NOTES

Professor J. S. Burris, Plant Pathology Seed and Weed Science, Iowa State University, Ames. Journal Paper No. J-11740 of the Iowa Agriculture and Home Economics Experiment Station, Ames, Iowa 50011. Project 2525.

REFERENCES

AOSA. 1983. Vigor Testing Handbook. Vol. 32 of Seed Testing Handbook. Association of Official Seed Analysts.

Burris, J. S. 1976. Seed/seedling vigor and field performance. J. Seed Technol. 1(2):58-74.

Burris, J. S. and K. H. Knittle. 1975. Partial reversal of temperature-dependent inhibition of soybean hypocotyl elongation of cotyledon excision. Crop Sci. 15:461-462.

Duke, S. H., G. Kukefuda and T. M. Harvey. 1983. Potential leakage of intracellular substances from imbibing soybean seeds. Plant Physiol. 72:919-924.

Edje, O. T. and J. S. Burris. 1970. Seedling vigor in soybeans. Proc. Assoc. Off. Seed Anal. 60:149-157.

Gilman, D. F., W. R. Fehr and J. S. Burris. 1973. Temperature effects on hypocotyl elongation of soybeans. Crop Sci. 13:246-249.

Indalsingh, J. T. 1981. Soybean seed quality as influenced in the physical integrity of the seed coat. M.S. Thesis, Iowa State Univ. Library, Ames, IA.

Knittle, K. H. and J. S. Burris. 1979. Effect of downward force on soybean hypocotyl growth. Crop Sci. 19:47-51.

Knittle, K. H., J. S. Burris and D. C. Erbach. 1979. Regression equations for rate of soybean hypocotyl elongation by using field data. Crop Sci. 19:41-46.

Liu, H. L. 1949. Inheritance of defective seed coat in soybeans. J. of Hered. 40:317-322.

Mortensen, J. J. 1984. Soybean seed performance under low temperature and low oxygen concentrations. M.S. Thesis, Iowa State Univ. Library, Ames, IA.

Obendorf, R. L. and P. R. Hobbs. 1970. Effect of seed moisture on temperature sensitivity during imbibition of soybean. Crop Sci. 10:563-566.

Saminy, C. and C. E. LaMotte. 1976. Anomalous temperature dependence of seedling development in some soybean cultures. Role of ethylene. Plant Physiol. 58:786-789.

Seyedin, N., J. S. Burris, C. E. LaMotte and I. C. Anderson. 1982. Temperature-dependent inhibition of hypocotyl elongation in some soybean cultures. I. Localization of ethylene evolution and role of cotyledons. Plant Cell Physiol. 23:427-431.

Stewart, R. T. and J. B. Wentz. 1930. A defective seedcoat character in soybeans. J. Am. Soc. Agron. 22:657-662.

Unander, D. W., J. W. Lambert and J. H. Orf. 1983. Effect of seed production environment on genetic differences in cold tolerance during germination. Soybean Genet. Newsl. 10:59-62.

Wall, M. T., D. C. McGee and J. S. Burris. 1983. Emergence and yield of fungicide treated soybean seed differing in quality. Agron. J. 75:969-973.

Wanjura, D. F., D. R. Buxton and H. N. Stapleton. 1973. A model for describing cotton growth during emergence. Trans. Am. Soc. Agric. Eng. 16:227-231.

Woodstock, L. W. and R. B. Taylorson. 1981. Soaking injury and its reversal with polyethylene glycol in relation to respiratory metabolism in high and low vigor soybean seeds. Physiol. Plant 53:263-268.

GROWTH AND DEVELOPMENT OF SOYBEAN ROOT SYSTEMS

T. C. Kaspar

The soybean usually is described as being weakly taprooted. Principal components of a soybean root system are a taproot, secondary roots, tertiary and other higher-order roots, adventitious roots originating from the hypocotyl, and root hairs. Soil physical conditions strongly influence growth and development of root systems, but plant genotype and the growth, development, and functioning of the shoot also are controlling factors. This paper presents a generalized description of soybean root systems and discusses briefly the influence of plant genotype and carbohydrate supply on root growth and development.

ROOT SYSTEM MORPHOLOGY

Development of a soybean root system begins when the radicle emerges from the seed. The taproot, which develops from the radicle, is strongly geotropic and usually grows downward at a rate of 25 to 50 mm/day (Mitchell and Russell, 1971; Taylor et al., 1978; Kaspar, 1982). Taproots can reach depths of more than 1.5 m in the field, but often their growth terminates much sooner because of unfavorable soil conditions or damage to the apex (Kaspar and Taylor, unpublished). Tangential division of cambial cells causes enlargement of the upper 10 to 15 cm of a taproot. As secondary growth proceeds, the epidermis, cortex, and endodermis on the enlarged upper portion often break apart and slough-off (Carlson, 1973). Even after secondary thickening, taproots usually are less than 7.5 mm in diameter at the hypocotyl-taproot junction, and decrease to approximately one-third of that 60 mm below the junction (Zobel, 1980).

Lateral or secondary roots first appear on taproots 3 to 7 days after germination (Mitchell and Russell, 1971). Lateral root primordia are initiated in the pericycle opposite the four protoxylem ridges, and usually do not occur within 30 mm of a taproot apex (Sun, 1955). Because soybean taproots have tetrarch xylem, lateral roots arising from them usually are found in whorls of four roots and are spaced at 90° intervals around the taproot circumference (Mitchell and Russell, 1971). As a result, laterals are oriented in four rows along the taproot axis, but this symmetry is sometimes difficult to observe in soil-grown plants because taproots often twist as they push through soil. Initiation of laterals proceeds acropetally at irregular intervals along the entire length of the taproot (Sun, 1955). Average diameter of secondary roots is 0.65 to 0.75 mm, which is about half of taproot diameter before secondary thickening takes place (Carlson, 1969).

A large portion of the root system consists of four to seven extensively branched, second-order roots originating from the basal portion of the taproot. These lateral roots, sometimes called major laterals or basal roots, are much longer and have larger diameters (2.0 to 3.0 mm) than lateral roots originating from lower portions of the taproot (Zobel, 1980). Generally, major laterals are plagiotropic; that is, they tend to grow out from the taproot at a characteristic angle (Raper and Barber, 1970; Kaspar et al., 1981). Additionally, Mitchell and Russell (1971) observed that, after major laterals had extended horizontally 20 to 36 cm, they suddenly turned downward and began rapid growth to depths of 1.80 m or more.

Secondary roots have three or four protoxylem ridges, but tertiary, quatenary, and other higher-order roots have either a triarch or diarch xylem (Carlson, 1973). Thus, higher-order roots usually are arranged in two, three, or four rows along the root axis, depending on the number of protoxylem ridges in their source root. Higher-order root primordia can be initiated whenever soil conditions are favorable, and unlike secondary roots, they are often found within 30 mm of a source-root apex (Carlson, 1973). Diameters of higher-order roots are less than half of second-order root diameters (Carlson, 1969). These small roots contribute greatly to the dynamic nature of a soybean root system. Depending on surrounding soil conditions, higher-order roots have an average functioning life span of 10 to 20 days (Huck and Davis, 1976). As soon as water or nutrients are no longer available in their immediate vicinity, these roots begin to die and decay even though their source root often remains intact. When dry soil that contains source roots is rewetted, or when a taproot or a major lateral root grows into favorable soil, new higher-order roots are initiated and grow into the surrounding soil (Huck and Davis, 1976; Sivakumar et al., 1977).

Root hairs first appear soon after the taproot begins to elongate. Each root epidermal cell is capable of producing one root hair; and so, root hairs may be found on all portions of the root system, except where the epidermis has sloughed-off or is very close to an apical meristem (Carlson, 1969). Size of root hairs varies, depending on size of the source root. Root hairs on tertiary or higher-order roots are the smallest, and those on major laterals or a taproot usually are the largest. Root hairs of most plant species usually die and decay on mature portions of root systems, but as long as the epidermis is intact soybean root hairs can persist until plant maturity (Carlson, 1969). Like small, higher-order roots, root hairs die when moisture or nutrients in surrounding soil is exhausted. Their primary function seems to be water and nutrient uptake, but exactly how much water passes through them is not known. They can, however, account for as much as 85% of a soybean root system's surface area (Carlson, 1969).

Very little is known about the adventitious roots, which originate from the underground portion of the soybean hypocotyl (Tanaka, 1977). In appearance and function, they seem to be similar to second-order laterals and may attain the same length and diameter as major laterals. Furthermore, like lateral roots, adventitious roots usually are initiated in the vicinity of differentiating vascular tissue. They normally appear during early vegetative growth, but they also can be initiated much later. Length of hypocotyl underground (Tanaka, 1977) and temperature (Stone and Taylor, 1983) partly determine the number of adventitious roots.

GROWTH AND DEVELOPMENT OF ROOT SYSTEM

Development of the root system in the field is strongly linked to shoot development. Mitchell and Russell (1971) described three general phases of soybean root development, which they related to shoot development: early vegetative growth, late vegetative growth to pod-set, and seed-set to maturity. Similarly, Sanders and Brown

(1979) related four phases of root growth to shoot development by adding a late-flowering to seed-set phase. In both studies, root systems were sampled on only four dates. Sivakumar et al. (1977) and Mason et al. (1980) sampled field-grown soybean root systems on five and seven dates, respectively. Additionally, Kaspar et al. (1978) observed soybean root growth in a rhizotron at three-day intervals throughout the growing season. These studies, except for that of Sanders and Brown (1979), were conducted in Iowa using indeterminate cultivars. Hence, the relationship between root growth and stage of development can be described in general terms for indeterminate cultivars grown in North Central USA.

Early Vegetative Growth

Soybeans planted in mid-May in Iowa normally reach growth stage V5 or V6 (Fehr and Caviness, 1977) in 40 to 45 days. By this time, taproots may extend to a depth of 0.8 to 1.0 m (Mason et al., 1980; Sivakumar et al., 1977) in a barrier-free soil. Major lateral roots have elongated 25 to 35 cm horizontally and, depending on soil temperature and moisture, may be within 3 cm of the soil surface (Mitchell and Russell, 1971). Second- and third-order roots have already begun to proliferate in the 0- to 15-cm soil layer, mostly under the row. Mason et al. (1980; Table 1) found 67% of root dry weight and 46% of root length in the upper 15 cm of soil at this time. On the other hand, Mitchell and Russell (1971) reported that 93% of soybean root dry weight could be found in the 0- to 15-cm soil layer 31 days after planting, and that more than 85% was present there throughout the growing season.

Preflowering

Before flowering begins (V6 to R1), taproot extension slows. Sivakumar et al. (1977) reported no increase in rooting depth between growth stages V6 and V10-R1. Mason et al. (1980), however, found that rooting depths increased by 30 to 40 cm during late vegetative growth. Typically, major lateral roots reach their maximum horizontal extension (30 to 50 cm) by this time (Mitchell and Russell, 1971; Bohm, 1977), and second- and third-order roots begin to grow throughout the inter-row soil volume. Total root length and root dry weight continue to increase (Mason et al., 1980; Table 2), and the 0- to 15-cm soil layer contains 68% of total root dry weight and 28% of root length by R1 (Table 1). Mason et al. (1980) found that percentage of total root

Table 1. Shoot-to-root dry weight ratio and percentage total root dry weight and root length per unit ground surface area in the 0- to 15-cm soil layer at various sampling times (Mason et al., 1980).

	Date (days after planting) and growth stage					
	June 21 (37) V3.4	June 27 (43) V5.5	July 12 (58) V9.1,R0.2	July 25 (71) V13.9,R2.3	Aug 7 (83) V18.3,R4.2	Aug 20 (97) V19.7,R5.1
Shoot to root dry weight ratio	5.5	5.1	6.8	7.8	8.7	9.9
Root dry weight in 0- to 15-cm layer, %	72	67	68	67	67	62
Root length in 0- to 15-cm layer, %	52	46	28	30	33	29

Table 2. Soybean root dry weight and root length per unit ground surface area for various depths and times, as measured with the soil monolith method (Mason et al., 1980).

Depth	Date (days after planting) and growth stage											
	June 21 (37) V3.4		June 27 (43) V5.5		July 12 (58) V9.1,R0.2		July 25 (71) V13.9,R2.3		Aug. 7 (83) V18.3,R4.2		Aug. 20 (97) V19.7,R5.1	
cm	g/m^2	m/m^2	g/m^2	m/m^2	g/m^2	m/m^2	g/m^2	m/m^2	g/m^2	m/m^2	g/m^2	m/m^2
0-15	4.05	1.56	8.22	5.16	20.12	2.28	36.39	6.56	44.79	8.38	47.60	7.97
15-30	0.89	0.71	2.63	2.57	3.93	2.58	4.04	2.89	4.37	2.42	7.53	3.68
30-60	0.15	0.25	0.90	0.74	2.45	1.67	4.76	3.85	4.65	3.31	5.62	3.67
60-90	–	–	0.30	0.25	1.96	0.95	4.56	3.84	4.36	3.70	5.33	3.42
90-120	–	–	–	–	0.47	0.32	2.95	2.69	5.02	4.24	5.32	3.95
120-150	–	–	–	–	0.02	0.02	0.61	0.81	2.40	2.35	3.02	2.73
150-180	–	–	–	–	–	–	0.07	0.06	0.29	0.35	1.08	0.61
Loose	0.53	0.31	0.17	0.14	0.49	0.33	0.74	0.93	0.79	0.86	1.31	1.44
Total	5.60	2.63	12.25	6.86	29.44	8.15	54.12	21.63	66.67	25.61	76.81	27.47

dry weight in the 0- to 15-cm layer remained fairly constant, but that percentage of total root length in this layer decreased drastically until flowering. Because of a prolonged drought, Sivakumar et al. (1977) observed that percentage of both total root dry weight and total root length in the 0- to 15-cm layer declined until the soil was rewetted. Shoot-to-root ratio increases to 6.8 (Table 1) by R1 and continues to increase throughout the growing season because of rapid reproductive growth.

Flowering

Between first flowering and early pod set (R1 to R3), several important events occur. First, shoots continue to grow at near maximum rates and approach maximum levels of light interception (Mason et al., 1980). Second, major lateral roots, which initially grew horizontally, turn downward and begin rapid vertical growth. Raper and Barber (1970) speculated that intraspecific competition or antagonism among soybean plants may cause major laterals to turn downward. This seems unlikely, however, because Bohm (1977) showed that soybean roots do penetrate soil that is occupied by roots of neighboring plants. Another possibility is that higher soil temperatures or lower soil moisture near the surface may cause major laterals to turn downward (Mitchell and Russell, 1971).

Third, there is a great increase in the root growth rate. After accounting for reductions in root length due to inadequate soil moisture, data of Sivakumar et al. (1977) showed that soybean roots elongated at their fastest rates during flowering. Similarly, Kaspar et al. (1978) observed that maximum rates of downward root growth of seven soybean cultivars occurred between R1 and R3. Lastly, Mason et al. (1980; Table 2) found that total root dry weight increased 84% and root length by 165% during flowering. Sixty-four percent of new root length was found below 30 cm, whereas 67% of the dry weight increase occurred above the 30-cm depth. Some of the dry-weight increase above 30 cm must have been the result of secondary thickening of taproots and major laterals.

Pod Set and Growth

Between R3 and R4, pod set and growth continues especially at the uppermost nodes, while seed growth begins at lower nodes. Vegetative shoot growth slows and most of the leaf area has been produced. Total root length and dry weight continue to increase (Table 2), but at a much reduced rate compared with the flowering period. Most root elongation takes place below 90 cm, but some new root growth does occur in the 0- to 15-cm layer if the soil is moist (Mason et al., 1980).

Seed Growth and Maturity

By R6 (mid-seed growth), leaf senescence has reduced leaf area and vegetative shoot dry weight. During early seed growth (R4 to R5), root length and dry weight continue to accumulate above 60 cm and below 1.35 m, but at a reduced rate compared to earlier periods (Table 2). In the 0- to 15-cm layer the increase in root dry weight is primarily due to secondary thickening of the taproot and large laterals. Mitchell and Russell (1971) reported that all eight of the cultivars they tested showed an increase in root dry weight during early seed growth. One of the cultivars they tested, however, accumulated dry matter faster during this period than it did earlier, two nonnodulating isolines maintained the same rate of growth, and the root systems of the other four grew at much reduced rates. Rooting depth also continues to increase during this period. Kaspar (1982) reported an average rooting depth of 1.72 m and a maximum depth of 2.30 m before R6 for eight cultivars.

Some root growth, although very little, continues until physiological maturity (Mitchell and Russell, 1971; Kaspar et al., 1978; Sanders and Brown, 1979). Total root dry weight and root length, however, usually decrease near maturity because of death and decomposition of old roots.

In summary, soybean root growth accelerates throughout the vegetative stage of development. Maximum rates of root growth coincide with attainment of near-maximum levels of light interception and vegetative shoot growth, which occur during flowering and early pod set. After seed growth begins, root growth rate usually declines and ceases altogether at physiological maturity. Extension of the root system is often limited by soil physical characteristics, but soybean roots have the potential to extend 50 cm or more horizontally and over 2.0 m vertically.

INFLUENCE OF CARBOHYDRATE SUPPLY ON ROOT GROWTH

Reduced root growth during seed development implies that carbohydrate supply might limit root growth during that stage. Evidently, developing seeds create a strong sink for carbohydrates, and their proximity to leaves reduces the relative amount of photosynthate partitioned to roots (Stephenson and Wilson, 1977). Alternatively, depodding often stimulates root growth (Brun, 1976). In general, any treatment or condition that decreases carbohydarte demand by the shoot, or increases photosynthate production, increases root growth relative to shoot growth. For example, longer daylength favors root growth, resulting in a lesser shoot-to-root dry weight ratio (Chatterton and Silvius, 1979). Similarly, increasing CO_2 concentration (Finn and Brun, 1982) or photosynthetic irradiance (Bunce, 1978) also increases root growth relative to shoot growth.

GENOTYPIC VARIATIONS

Growth and development of soybean root systems also are strongly influenced by genotype. For example, Zobel (1980) found that lines differed in taproot diameter, number of basal roots, and basal root diameter. Similarly, Taylor et al. (1978) and Kaspar (1982) showed that taproot elongation rates of cultivars varied by as much as

13 mm/day. As a result, root systems of soybean genotypes differ in size, composition, and development even when grown under similar environmental conditions. Mitchell and Russell (1971) reported that eight soybean cultivars differed in their relative rates of root dry matter accumulation during late stages of development. Similarly, Kaspar et al. (1978) observed different rates of downward growth among seven cultivars. Differential growth rates can result in differences in volume of soil penetrated by root systems, total surface area of the root system, and root density (Raper and Barber, 1970; Sanders and Brown, 1979). Furthermore, response of soybean genotypes to specific environmental conditions, such as soil temperature, also can vary, resulting in altered root growth and distribution (Stone and Taylor, 1983; Kaspar et al., 1981).

NOTES

T. C. Kaspar, Agricultural Research Service, USDA, and Dept. of Agronomy, Iowa State University, Ames, IA 50011.

Contribution from Agricultural Research Service, USDA, and Journal Paper J-11472 of the Iowa Agriuclture and Home Economics Experiment Station, Ames; Project 2659.

REFERENCES

Bohm, W. 1977. Development of soybean root systems as affected by plant spacing. Z. Acker Pflanzenbau. 144:103-112.

Brun, W. A. 1976. The relation of N$_2$ fixation to photosynthesis. p. 135-143. In L. D. Hill (ed.) World soybean research: Proc. of the World Soybean Res. Conf. Interstate Printers and Publishers, Danville, IL.

Bunce, J. A. 1978. Effects of shoot environment on apparent root resistance to water flow on whole soybean and cotton plants. J. Exp. Bot. 29:595-601.

Carlson, J. B. 1969. Estimating surface area of soybean root system. J. Minn. Acad. Sci. 36:16-19.

Carlson, J. B. 1973. Morphology. p. 17-96. In B. E. Caldwell, R. W. Howell, and H. W. Johnson (eds.) Soybeans: Improvement, production, and uses. American Society of Agronomy, Madison, Wis.

Chatterton, N. J. and J. E. Silvius. 1979. Photosynthate partitioning into starch in soybean leaves. I. Effects of photoperiod versus photosynthetic period duration. Plant Physiol. 64:749-753.

Fehr, W. R. and C. E. Caviness. 1977. Stages of soybean development. Iowa Agric. Home Econ. Exp. Stn. Spec. Rep. 80.

Finn, G. A. and W. A. Brun. 1982. Effect of atmospheric CO$_2$ enrichment on growth, nonstructural carbohydrate content, and root nodule activity in soybean. Plant Physiol. 69:327-331.

Huck, M. G. and J. M. Davis. 1976. Water requirements and root growth. p. 16-27. In L. D. Hill (ed.) World soybean research: Proc. of the World Soybean Res. Conf. Interstate Printers and Publishers, Danville, IL.

Kaspar, T. C. 1982. Evaluation of the taproot elongation rates of soybean cultivars. Ph.D. Dissertation. Iowa State University Library, Ames. University Microfilms (82-24332), Ann Arbor, MI.

Kaspar, T. C., C. D. Stanley and H. M. Taylor. 1978. Soybean root growth during the reproductive stages of development. Agron. J. 70:1105-1107.

Kaspar, T. C., D. G. Woolley and H. M. Taylor. 1981. Temperature effect on the inclination of lateral roots of soybeans. Agron. J. 73:383-385.

Mason, W. K., H. M. Taylor, A. T. P. Bennie, H. R. Rowse, D. C. Reicosky, Y. Jung, A. A. Righes, R. L. Yang, T. C. Kaspar and J. A. Stone. 1980. Soybean row spacing and soil water supply: Their effect on growth, development, water relations, and mineral uptake. Adv. Agric. Technol. AAT-NC-5. Agric. Res., North Central Region, SEA, USDA, Peoria, IL.

Mitchell, R. L. and W. J. Russell. 1971. Root development and rooting patterns of soybean [Glycine max (L.) Merr.] evaluated under field conditions. Agron. J. 63:313-316.

Raper, C. D. and S. A. Barber. 1970. Rooting systems of soybeans. I. Differences in root morphology among varieties. Agron. J. 62:581-584.

Sanders, J. L. and D. A. Brown. 1979. Measurement of rooting patterns for determinate and inde-terminate soybean genotypes with a fiber-optic scope. p. 369-379. *In* J. L. Harley and R. S. Russell (eds.) The soil-root interface. Academic Press, London.

Sivakumar, M. V. K., H. M. Taylor and R. H. Shaw. 1977. Top and root relations of field-grown soybeans. Agron. J. 69:470-473.

Stone, J. A. and H. M. Taylor. 1983. Temperature and the development of the taproot and lateral roots of four indeterminate soybean cultivars. Agron. J. 75:613-618.

Stephenson, R. A. and G. L. Wilson. 1977. Patterns of assimilate distribution in soybean at maturity. I. The influence of reproductive development stage and leaf position. Aust. J. Agric. Res. 28: 28:203-209.

Sun, C. N. 1955. Growth and development of primary tissues in aerated and non-aerated roots of soybean. Bull. Torr. Bot. Club 82:491-502.

Tanaka, N. 1977. Studies on the growth of root systems in leguminous crops. Agric. Bull. Saga Univ. (Japan) 43:1-82.

Taylor, H. M., E. Burnett and G. D. Booth. 1978. Taproot elongation rates of soybeans. Z. Acker Pflanzenbau. 146:33-39.

Zobel, R. W. 1980. Rhizogenetics of soybeans. p. 73-87. *In* F. T. Corbin (ed.) World soybean research conference II: Proceedings. Westview Press, Boulder, CO.

PHOTO-THERMAL REGULATION OF FLOWERING IN SOYBEAN[1]

R. J. Summerfield and E. H. Roberts

Soybean is cultivated in a remarkably diverse range of climates, and the timing of developmental events, particularly the initiation of floral primordia and their subsequent expansion into open flowers, differs markedly throughout this wide range of environments, depending on the genotype grown and the date on which it is planted. Flowers may appear as early as 25 days after sowing or not until after 50-60 days, or much later (Carlson, 1973). Time to flowering is an important determinant of productivity in this crop (Summerfield et al., 1984).

Rates of phenological development most often have been thought of as being *dictated* by relative responsiveness to prevailing photoperiod. Certainly photoperiod *is* a major factor influencing time to flowering of most genotypes (Summerfield and Roberts, 1984). However, *equally clear* is the fact that *temperature also* influences the rate of reproductive development, that interactions occur between photoperiodic and temperature effects, and that previous research has probably inclined us to a too dogmatic view of the need for a specific photoperiodic stimulus for flower evocation (Evans, 1969). Indeed, in *virtually every case* where studies have been sufficiently extensive, in domesticated or wild species, photoperiodic responses have been readily modified by temperature (Salisbury, 1981). Moreover, as long ago as 1936, Steinberg and Garner found that warmer temperatures hastened flowering of Biloxi, Peking and Mandarin cvs., at least up to an optimum of 28 C, above which flowering could be delayed depending on prevailing daylength.

An ability to explain, quantify and predict photo-thermal effects on flowering, and to screen large populations of diverse genotypes reliably in the field, might be expected to ensure a more rapid development of soybean genotypes well adapted to the regions and seasons for which they are intended.

PREDICTING PHOTO-THERMAL EFFECTS

Notwithstanding the apparent complexity of photoperiod x temperature x genotype interactions in the progress of soybean cultivars towards maturity, we now believe that flowering responses to wide ranges each of photoperiod and temperature are amenable to truly quantitative description by a number of simple general equations. These relations have been recently described in detail elsewhere (Hadley et al., 1984) and will be summarized only briefly here.

Factorial combinations of five photoperiods (8 h 20 min, 10 h, 11 h 40 min, 13 h 20 min and 15 h) and three night temperatures (14, 19 and 24 C) combined with a single day temperature (30 C) were imposed on nine diverse genotypes grown in pots in growth cabinets. The times to first appearance of open flowers were recorded. For a photoperiod-insensitive cultivar (Fiskeby V), and for the remaining eight photoperiod-sensitive genotypes in photoperiods shorter than their critical daylength, the rates of progress towards flowering (the reciprocals of the times taken to flower) were linear functions of mean diurnal temperature (Figure 1a). Thus, flowering responses in these genotype-environment combinations can be described by the following equation:

$$1/f = a + b\bar{t} \qquad (1)$$

where f denotes the number of days from sowing to the onset of flowering and a and b arc constants. As discussed elsewhere (e.g., Hadley et al., 1984), Equation (1) represents a thermal sum relation in that it infers that flowering will occur in any photo-thermal environment after a constant number of thermal units, T_f (in degree days above a base temperature of t_o), have accumulated after sowing. The value of the base temperature can be calculated from the constants of regression Equation (1) using the expression:

$$t_o = -a/b \qquad (2)$$

Furthermore, the thermal sum required for flowering is given by:

$$T_f = 1/b \qquad (3)$$

We have compared our relations for Fiskeby V with those calculated from field data in Australia (Lawn et al., 1984), and have also derived comparable relations from data published by Tanner and Hume (1976) from research in growth cabinets (Table 1). The remarkable similarity between field and controlled environment data, we suggest, is powerful evidence that Equations (1), (2) and (3) are validated.

For all photoperiod-sensitive genotypes, typified by TGm 39 in Figure 1b, a photoperiodic response became evident when it called for a time to flower longer than that otherwise determined by the basic temperature response. Thus, flowering was delayed in longer photoperiods, and the shortest photoperiod in which the delay becomes apparent may be defined as the critical photoperiod (P_c). However, even in photoperiods longer than P_c rates of progress towards flowering are still a linear function of mean temperature, but a different function from that which describes the basic temperature response in Equation (1). When photoperiods are longer than the critical value, and thus delay flowering, there is no interaction between photoperiod and temperature on the rate of progress towards flowering, and so the response in these photo-thermal regimes can be described by the following equation:

$$1/f = a' + b'p + c'\bar{t} \qquad (4)$$

where a', b' and c' are constants.

Finally, where our results are complete (i.e., flowering occurred in all photo-thermal regimes before the experiment was terminated) they suggest (e.g., Figure 1b) that for any genotype there is a minimum rate of progress towards flowering irrespective of photo-thermal environment; i.e., flowering cannot be delayed beyond a given maximum number of days. This maximum varies among genotypes, though it is conceivable that it might be infinity in very sensitive ones (and see Major, 1980).

The critical photoperiod, P_c, varies according to temperature. This is because its

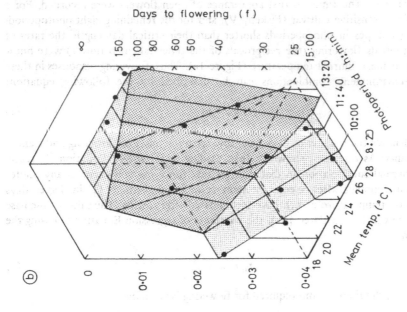

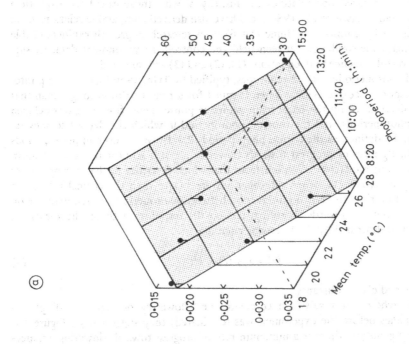

Figure 1. Effects of photoperiod (h) and mean diurnal temperture (t̄ C) on rates of progress towards flowering (1/f) in (a) the photoperiod-insensitive soybean, Fiskeby V and (b) photoperiod-sensitive TGm 39 (solid circles denote the experimental mean values in each of 10 (a) and 15 (b) environments, respectively).

Table 1. Estimates of the thermal responsiveness of flowering in the photoperiod-insensitive soybean, Fiskeby V.

Aspect[a]	Source of information		
	Hadley et al., 1984	Lawn et al., 1984	Tanner and Hume, 1976
$1/f = a + b\bar{t}$	$-0.0143 + 0.00181\,\bar{t}$	$-0.0127 + 0.00175\,\bar{t}$	$-0.0139 + 0.00178\,\bar{t}$
R^2	0.86	0.93	0.95
$t_o = -a/b$	7.9	7.3	7.8
$T_f = 1/b$	552	571	562

[a]Rate of progress towards flowering $(1/f)$ in relation to mean temperature $(\bar{t})$; coefficient of multiple determination (R^2) for $1/f$; base temperature (that value of t_o which, in theory, is sufficiently cool to arrest any progress towards flowering); and the thermal sum (T_f) required from sowing to flowering (Cd above a base temperature of t_o), respectively.

value is determined by the position of the intersection of the two response surfaces representing the basic temperature response (the only response of photoperiod-insensitive genotypes), which is described by Equation (1), and the photoperiod-temperature response described by Equation (4). The value of P_c for each genotype can be estimated for any photo-thermal regime by inserting the calculated values of the constants derived from Equations (1) and (4) into the following equation:

$$P_c = a - a' + \bar{t}(b - c')/b' \qquad (5)$$

For the eight photoperiod sensitive genotypes tested hitherto (Hadley et al., 1984), P_c increased from about 9 h 50 min at a mean temperature of 18 C to about 11 h 34 min at 27 C (i.e., by about 11 min/C).

Thus, for photoperiod-sensitive genotypes we conclude that, when the times taken to flower are transformed to rates of progress towards flowering, it is apparent that the entire response surface can be described by three planes (Figure 1b): One, in photoperiods shorter than the critical daylength, where rates of progress towards flowering are solely a positive linear function of mean temperature; another, in photoperiods longer than the critical value, where the rates are a positive linear function of temperature and a negative linear function of photoperiod with no interaction between them; and, finally, a third plane, defining the minimum rate for the genotype and which is affected by neither temperature nor photoperiod. Thus, although powerful statistical interactions between photoperiodic and temperature effects on rates of development appear if the treatments transgress any of these planes, within none of the three areas is there any interaction between the two factors.

We have re-examined other published data in light of these quantitative relations. Figure 2a shows our recalculations for: (A) 12 American genotypes adapted to latitudes north of about 37°N (MG 000 through III) grown in a constant photoperiod (12 h) with different diurnal thermal regimes and artificial lighting in growth cabinets (Tanner and Hume, 1976); (B) Wayne (MG III) grown in various diurnal thermal regimes in growth cabinets with artificial light and a photoperiod of 12 h (Tanner and Hume, 1976) or (C) 16 h in a controlled-temperature glasshouse with natural daylength extended with artificial illumination (Shibles et al., 1974); (D) 16 photoperiod-insensitive genotypes (75% from Europe) and (E) 7 photoperiod-sensitive ones (mostly from

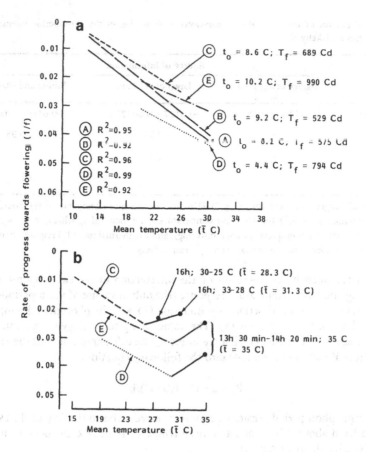

Figure 2. (a) Relations between rates of progress towards flowering (1/f) in diverse genotypes of soybean and mean diurnal temperature ($\bar{t}$ C) calculated from previously published data (A and B, Tanner and Hume, 1976; C, Shibles et al., 1974; D and E, Inouye et al., 1979), and (b) as in (a) over the widest range of $\bar{t}$ C investigated.

Asia) grown under natural daylengths (13 h 30 min to 14 h 20 min) with different constant temperatures in naturally-lit rooms (Inouye et al., 1979).

Several points emerge from these recalculations: First, that best-fit regression lines as described by Equation (1) explain at least 92% of the variations in rates of progress towards flowering in response to mean temperature; second, that differences between artificial and natural light, or between constant and changing photoperiods, do not detract from the precision of this relation; and, third, that base-temperatures and thermal sums for flowering calculated from Equations (2) and (3) seem entirely reasonable and plausible when related to the regions to which different genotypes are best adapted. For example, European genotypes and those adapted to the northern States in America or to southern Canada have cool average base temperatures (4.4 and 6.2 C, respectively) whereas the average for several genotypes from Asia (Taiwan, China and Japan) is much warmer (10.2 C). Then again, independent data for Wayne yield almost identical base teperatures (8.6 and 9.2 C) when Equation (2), derived from Equation (1), is used (and see Brown, 1960). Thus, although our own and other previous work on photothermal effects on flowering in soybean has emphasized differential effects of day and night temperatures (e.g., Huxley et al., 1976; Tanner and Hume, 1976; Shanmugas-undaram and Lee, 1980), it is now clear that genotypes are sensitive to mean diurnal temperature rather than to day or night temperature per se, *at least over wide ranges of each of these factors*. This qualification is necessary because, as Figure 2b shows, very hot days ()30 C) and/or very large diurnal variations in temperature (e.g., 33-19 C; 26.7-10 C) can delay flowering significantly compared with values predicted for respective mean temperatures from Equation (1) (and see Brown, 1960). We suggest that such extreme diurnal thermal regimes, which are relatively rare in the field, are likely to impose severe heat and/or water stress on the plants (as reflected by their very poor yields; Huxley et al., 1976). However, since these relations refer to 'times to flowering' rather than 'times to floral initiation' (see later), we do not yet know, for example, if these effects of hot (and, usually, dry) air reflect direct effects on the abortion/abscission of immature floral organs or indirect effects on attributes such as plant water status, or both. Notwithstanding, the relations described above seem to have wide applicability across both genotypes and photo-thermal regimes.

Other earlier publications have generated more extensive data than those considered hitherto in that the effects on flowering of factorial combinations of temperature and photoperiod have been studied (e.g., Steinberg and Garner, 1936; van Schaik and Probst, 1958; Lu and Yen, 1975). Our reanalyses of these data are shown in Figure 3. These response surfaces describe rates of progress towards flowering in different photo-thermal regimes for: U.S. cultivars of diverse adaptaticn (MG I, Mandarin; MG IV, Peking, Clark and Midwest; and MG VIII, Biloxi and PI 181698); for determinate and indeterminate forms; and for genotypes originating from or selected in Asia (Sangoku, Chung-Hsing 2 and Nung-Yuan 1), when all were grown in artificial light and in either constant or diurnally-changing thermal regimes. It is unfortunate that data for Clark and Midwest (van Schaik and Probst, 1958) relate to only single plants in each photo-thermal regime, and that the number of replicates for the Asian material is ambiguous (Lu and Yen, 1975). Nevertheless, it seems clear to us from Figures 3a-i that Equations (1), (2), (3), (4) and (5) offer alternative and more plausible explanations for these data than the often tortuously complex and incomplete interpretations originally proposed. Any apparent inconsistencies from our model in Figures 3e to 3i can be largely accounted for once it is recognized that these experiments included supra-optimal temperatures (31 to 32 C; see Figure 2b) and/or exceedingly short photoperiods (5 to 9 h) not experienced by soybean crops. In suggesting that the 'model' based on Equations (1)-(5) may have some unifying validity we recognize, of course, that the photoperiodic mechanism is not yet well understood (Vince-Prue, 1983), and that the sheer diversity in flowering behavior among annual seed plants, combined with seemingly complex interactions between internal and external 'cues', might be seen to mitigate strongly against a common basis of floral regulation (Marx, 1983). On the other hand, the 'model' seems remarkably appropriate, not only for soybean but also for other short- *and* long-day species of grain legumes (Hadley et al., 1983), over wide ranges each of photoperiod and mean temperature.

SCREENING GERMPLASM

An objective of many research programs, including our own[3], has been to develop techniques for screening large numbers of genotypes that may be more or less sensitive to photoperiod and temperature and which, therefore, are likely to be well adapted either to a relatively few or a wide range of environments, respectively. In order to develop appropriate techniques, it is important first to consider the range of natural

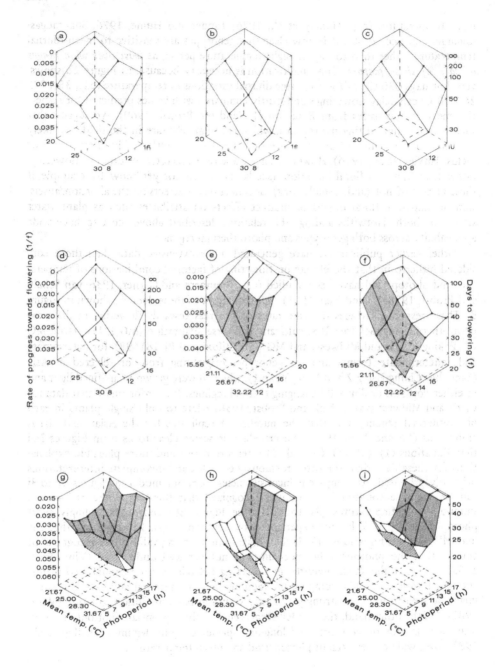

Figure 3. Relations between rates of progress towards flowering (1/f) in diverse genotypes of soybean and mean diurnal temperature (t̄ C) and photoperiod (h) calculated from previously-published data. Panels a-d, tropical and sub-tropical genotypes (Lu and Yen, 1975); e-f, Clark and Midwest (van Schaik and Probst, 1958); g-i, Mandarin, Peking and Biloxi (Steinberg and Garner, 1936).

climates in which the crop is typically cultivated. Tropical daylengths (including civil twilight) vary from about 11 to 14 h. In more temperate latitudes soybean crops grown during spring and summer months will most often experience daylengths longer than 14 h. Consequently, it seems likely that soybean crops will almost always experience photoperiods longer than the critical value (Hadley et al., 1984; and see Figures 1b and 3). It follows then that the environments chosen for screening should be located in a region defined by the photo-thermal response plane in photoperiods longer than the critical value. Since the position of this plane is defined by both temperature and photoperiod, then the minimum number of environments required to establish it is three (providing they include two photoperiods *and* two temperatures). Results obtained from these environments would establish the magnitude of both the photoperiodic and temperature responses of photoperiodically-sensitive genotypes and, in the case of photoperiod-insensitive ones, the results would determine the basic temperature response. In photoperiod-insensitive genotypes it would be possible to estimate the values of a and b in Equation (1) and, from these, could be calculated the base temperature (Equation 2) and thermal sum (Equation 3)—a concept that can be applied directly to the varying diurnal and seasonal thermal environments in the field.

What might be the most appropriate environments to choose in order to determine the photo-thermal response surfaces of photoperiod-sensitive genotypes? Clearly, it would be preferable if there were little probability of these environments falling on either of the other two planes (the basic temperature response and the maximum period to flower), which combine with the photo-thermal response to define the entire response surface (Figure 1b). Within these constraints the response would be shown most accurately if differences between the two temperatures and the two photoperiods were as large as possible. Our data hitherto (Hadley et al., 1984) suggest three environments (mean temperature-photoperiod combinations) that meet these criteria, viz. 21 C/12 h, 25 C/12 h and 25 C/14 h. It would also be advantageous to include a fourth treatment in order to define the maximum period to flowering. Conveniently, it seems that an appropriate combination for this purpose would be 21 C/14 h. Thus, a simple 2 x 2 factorial experiment, in the field or in glasshouses, should establish the principal features of the photo-thermal response and the maximum period to flower of photoperiod-sensitive genotypes, and the basic thermal response of photoperiod-insensitive ones.

Several authors have discussed the merits of using photoperiod-insensitive cultivars, particularly in areas of the tropics and sub-tropics where more than one soybean crop can be grown each year (e.g., Shanmugasundaram, 1981). Such genotypes are now available and could presumably be incorporated into cultivar development programs; their rates of phenological development can be reliably predicted by thermal integration above an appropriate base temperature. On the other hand, we cannot yet find evidence that truly photoperiod-insensitive soybean lines can also be of late maturity, and so photoperiod-sensitive cultivars may still be important where longer duration types are required to match more exactly the time available for growth and reproduction.

The prediction of phenological development of photoperiodically-sensitive types, based on climatic variables prevailing in the field, still presents difficulties, although some models based on field experiments have been attempted (e.g., Major et al., 1975a,b; Jones and Laing, 1978). Information in two areas is required before concepts such as those described here can be expected to predict with reliable precision the phenological behavior of field-grown soybean crops in non-inductive photoperiods: in particular, our model (which is accurate in constant environments) does not take into account changes in relative photoperiod sensitivity that might occur during development, and how the responses described here relate to seasonal changes in climate. Experiments in which reciprocal transfers of plants between long and short photoperiods and between warm and cool temperatures are made at regular intervals during the development of diverse genotypes may well provide this information. When these data are combined with current understanding, they may well lead to the development of fundamental models that will predict reliably the phenological behavior of germplasm in widely different climatic regions and growing seasons.

CONCLUDING REMARKS

We have consciously concentrated on the photothermal regulation of the discrete event of floral anthesis, taken to be the time of appearance of the first open flower (corolla color visible). We recognize, of course, that floral development includes the initiation of floral primordia and their subsequent period of expansion, culminating in anthesis. Initiation and anthesis are regulated independently by photoperiod, and floral initiation of axillary racemes is independent of that at stem apices (Thomas and Raper, 1983). Then again, photo-thermal effects on carbon metabolism and nitrogen nutrition post anthesis (Cure et al., 1982), as well as those on continued phenological progress towards maturity, senescence and death, are likely to be significant aspects of 'adaptability to environment and cropping season duration' in this species. Our ultimate aim is to quantify the relative significance of each of these diverse facets among soybean lines, and to incorporate the information into a reliably predictive scheme that can be exploited, with confidence, by plant breeders.

NOTES

R. J. Summerfield, Reader in Agriculture, University of Reading, Dept. of Agriculture and Horticulture, Plant Environment Laboratory, Shinfield Grange, Cutbush Lane, Shinfield, Reading, RG2 94D, Berkshire, England.

E. H. Roberts, Professor of Crop Production, University of Reading, Dept. of Agriculture and Horticulture, Earley Gate, Whiteknights Road, Reading, RG6 2AT, Berkshire, England.

[1]Prepared during the tenure of a research award from the UK Overseas Development Administration (ODA). The financial support of ODA and of the Organizing Committee for World Soybean Research Conference III, which combined to allow RJS to present this paper at Ames, is gratefully acknowledged.

[2]The engineering and technical assistance of Messrs. D. Dickinson, A. C. Richardson, M. Boxall, K. Chivers, M. Craig, S. Gill, C. Billing, Dr. S. Rawsthorne, Miss C. Chadwick and Mrs. B. Whitlock is gratefully acknowledged. We are also glad to acknowledge the scientific contribution of Dr. P. Hadley, Sub-Department of Horticulture, University of Reading for the formulation of Equations (1)-(5), as reflected in the original publication which first described these relations in detail (Hadley et al., 1984).

REFERENCES

Brown, D. M. 1960. Soybean Ecology. 1. Development-temperature relationships from controlled environment studies. Agron. J. 52:493-496.

Cure, J. D., R. P. Patterson, C. D. Raper and W. A. Jackson. 1982. Assimilate distribution in soybeans as affected by photoperiod during seed development. Crop Sci. 22:1245-1250.

Evans, L. T. 1969. The nature of flower induction. p. 457-480. In L. T. Evans (ed.) The induction of flowering. Macmillan, Melbourne.

Hadley, P., R. J. Summerfield and E. H. Roberts. 1983. Effects of temperature and photoperiod on reproductive development of selected grain legumes. p. 19-41. *In* D. R. Davies and D. Gareth-Jones (eds.) The physiology, genetics and nodulation of temperate legumes. Pitmans, London.

Hadley, P., E. H. Roberts, R. J. Summerfield and F. R. Minchin. 1984. Effects of temperature and photoperiod on flowering in soyabean (*Glycine max* (L.) Merrill): a quantitative model. Ann. Bot. 3:669-682.

Huxley, P. A., R. J. Summerfield and A. P. Hughes. 1976. Growth and development of soyabean cv. TK5 as affected by tropical daylengths, day/night temperatures and nitrogen nutrition. Ann. Appl. Biol. 82:117-133.

Inouye, J., S. Shanmugasundaram and T. Masuyama. 1979. Effects of temperature and daylength on the flowering of some photo-insensitive soybean varieties. Jap. J. Trop. Agric. 22:167-171.

Jones, P. G. and D. R. Laing. 1978. Simulation of the phenology of soybeans. Agric. Systems 3:295-311.

Lawn, R. J., J. D. Mayers, D. F. Beech, A. L. Garside and D. E. Byth. 1984. Adaptation of soybeans to subtropical and tropical environments in Australia. *In* S. Shanmugasundaram and E. Sulzberger (eds.) Proc. first international symposium on soybeans in tropical and sub-tropical farming systems. AVRDC, Taiwan (in press).

Lu, Y. C. and H. Yen. 1975. Photoperiod and temperature responses of soybean varieties observed in a phytotron. SABRAO J. 7:171-182.

Major, D. J. 1980. Photoperiod response characteristics controlling flowering of nine crop species. Can. J. Plant Sci. 60:777-784.

Major, D. J., D. R. Johnson and V. D. Luedders. 1975a. Evaluation of eleven thermal unit methods for predicting soybean development. Crop Sci. 15:172-174.

Major, D. J., D. R. Johnson, J. W. Tanner and I. C. Anderson. 1975b. Effects of daylength and temperature on soybean development. Crop Sci. 15:174-179.

Marx, G. A. 1983. Developmental mutants in some annual seed plants. Ann. Rev. Plant Physiol. 34:389-417.

Salisbury, F. B. 1981. Responses to photoperiod. p. 135-168. *In* O. L. Lange, P. S. Nobel, C. B. Osmond and H. Ziegler (eds.) Physiological plant ecology, Volume 1. Responses to the physical environment. Springer-Verlag, New York.

Shanmugasundaram, S. 1981. Varietal differences and genetic behaviour for the photoperiodic responses in soybeans. Bull. Inst. Trop. Agric. Kyushu Univ. (Jpn.) 4:1-61.

Shanmugasundaram, S. and M. S. Lee. 1980. Influence of night temperature on the flowering of the photoperiod sensitive and day-neutral soybeans. p. 53-66. *In* Legumes in the tropics. Universiti Pertanian Malaysia, Selangor, Malaysia.

Shibles, R., I. C. Anderson and A. H. Gibson. 1974. Soybean. p. 151-189. *In* L. T. Evans (ed.) Crop physiology: some case histories. Cambridge University Press, Cambridge.

Steinberg, R. A. and W. W. Garner. 1936. Response of certain plants to length of day and temperature under controlled conditions. J. Agric. Res. 52:943-960.

Summerfield, R. J. and E. H. Roberts. 1984. Soybean. *In* A. H. Halevy (ed.) A handbook of flowering. CRC Press, Boca Raton, Florida (in press).

Summerfield, R. J., S. Shanmugasundaram, E. H. Roberts and P. Hadley. 1984. Adaptation of soybeans to photo-thermal environments and implications for screening germplasm. *In* S. Shanmugasundaram and E. Sulzberger (eds.) Proc. first international symposium on soybeans in tropical and sub-tropical farming systems. AVRDC, Taiwan (in press).

Tanner, J. W. and D. J. Hume. 1976. The use of growth chambers in soybean research. p. 342-351. *In* L. D. Hill (ed.) World soybean research. Interstate Printers and Publishers, Danville, Illinois.

Thomas, J. F. and C. D. Raper. 1983. Photoperiod and temperature regulation of floral initiation and anthesis in soyabean. Ann. Bot. 51:481-489.

van Schaik, P. H. and A. H. Probst. 1958. Effects of some environmental factors on flower production and reproductive efficiency in soybeans. Agron. J. 50:192-197.

Vince-Prue, D. 1983. Photoperiodic control of plant reproduction. p. 73-97. *In* W. J. Meudt (ed.) Strategies of plant reproduction. Allanheld, Osmun and Co., Totowa, New Jersey.

CORRELATIVE EFFECTS OF FRUITS ON PLANT DEVELOPMENT

I. A. Tamas, P. J. Davies, B. K. Mazur and L. B. Campbell

Growing fruits regulate a number of developmental characteristics in plants, all of which affect plant productivity. Among these are the number of branches (and thus number of leaves) (Tamas et al., 1979a), the size and number of fruits and seeds (Tamas et al., 1979b), and plant senescence (Lindoo and Nooden, 1977; Wareing and Seth, 1967; Davies et al., 1977).

In field beans (*Phaseolus vulgaris* L.) the presence of developing fruits has a pronounced effect on several developmental phenomena. The fruits impose dormancy on axillary buds, which varies with cultivar (Tamas et al., 1979a). When fruits are removed axillary bud growth resumes, leading to a two-fold increase in leaf and shoot number. The ratio of reproductive to total biomass is affected by the extent to which the axillary buds are suppressed in the different cultivars. Developing fruits, therefore, have a direct effect on the partitioning of organic matter between reproductive and vegetative organs by altering their respective rates of development.

Fruits also affect the development of other fruits on the plant (Tamas et al., 1979b). When older fruits at the base of racemes were removed, the abortion rate of the younger ones was reduced. Competition, thus, exists among developing fruits.

Varying the fruit load on bean plants results in a dosage effect of fruits on photosynthesis (Tamas et al., 1981). When plants were fully defruited, the CO_2 exchange rate (CER) doubled in two weeks. A smaller increase in CER occurred after partial fruit removal. Leaf senescence and the concomitant loss of photosynthetic capacity was delayed progressively with increasing number of fruits removed. The removal of seeds from fruits had a dramatic effect on the chlorophyll content of leaves. Whereas the leaves of control plants maintained a steady chlorophyll concentration, those of the deseeded plants doubled their chlorophyll content within two weeks following seed removal.

In soybeans and the pea (*Pisum sativum* L.) the effect of fruits on senescence also has been shown to be mediated by the developing seeds rather than the carpels. The surgical removal of the developing seeds from the pod delayed leaf senescence in soybeans (Lindoo and Nooden, 1977) and apical senescence in peas (Gianfagna and Davies, 1981).

CAN COMPETITION FOR NUTRIENTS ACCOUNT FOR
CORRELATIVE FRUIT EFFECTS?

Removal of flowers or fruits will usually delay or prevent leaf or shoot senescence. The association of reproductive growth with senescence led Molisch (1928) to propose that the fruits induce senescence by mobilizing the nutrients required for growth, leaving the remaining parts of the plant, from which the nutrients have been withdrawn, in a state of deficiency. The finding in soybean plants that the senescence signal intensifies towards the time of fruit maturation and does not coincide with the time of greatest seed growth, indicates however, that nutritional exhaustion is unlikely to be the cause of leaf senescence (Lindoo and Nooden, 1977). In addition the effects of fruits in promoting leaf senescence in soybean appear to be quite localized (Lindoo and Nooden, 1977). If all the leaves near the fruit are removed, the leaves distal to the fruit appear to be unaffected even though they still supply nutrients to the fruit. In G2 peas Gianfagna and Davies (1981) have calculated from seed growth rates that nutrient drain on the plant imposed by the developing seed is approximately equal under long and short photoperiods, conditions under which the plants do and do not senesce, respectively. Nutrient exhaustion can, therefore, be ruled out as the cause of plant senescence. It is still possible, however, that competition for nutrients and hormones between developing sinks may result in the growth of one to the detriment of the other.

HORMONAL CONTROL

The photoperiodic control of senescence in G2 peas appears to be hormone mediated. In short days a non-senescing G2 plant can prevent senescence of a photoperiod insensitive scion grafted to it. In long days senescence of G2 plants can be prevented or delayed by applications of several gibberellins and the gibberellin content, notably GA_{19} and GA_{20}, is greater in short days than in long days (Davies et al., 1982).

Apical senescence occurs in peas only in the presence of fruit. It is the early stages of seed development that correlate with the development of senescence symptoms. The presence of a fruit signal above a certain threshold for a certain period of time is needed for the death of the apex (Proebsting et al., 1976; Davies et al., 1977; Gianfagna and Davies, 1981). This fruit signal could be the export of a senescence factor or a competitive effect of the fruit at the expense of the developing vegetative tissues. Fruits allowed to photosynthesize in $^{14}CO_2$ exported a small percentage of their labeled material to the shoot apex and to the seed of young pods. The compounds exported by the fruits differed from those exported to the apex by leaves (Gianfagna and Davies, 1983). One or more of the compounds could be a senescence factor. The transported compounds have not been identified.

Correlative effects of fruits may alter the level of several growth substances in the target organ. In senescing soybean leaves the cytokinin level was found to decrease and the abscisic acid (ABA) level increase before the onset of chlorophyll loss, suggesting that these hormonal changes contributed to leaf senescence (Lindoo and Nooden, 1978). However, foliar application of cytokinin was able to prevent fruit-induced leaf yellowing only if used in conjunction with a-naphthaleneacetic acid (NAA). When ABA was sprayed on leaves it accelerated senescence in podded plants, but it did not induce senescence in depodded plants (Lindoo and Nooden, 1978). These results suggest that correlative fruit effects involve complex hormonal interactions, and that changes in several growth factors may be required to alter the course of events in the target organ.

INDOLEACETIC ACID AS A POSSIBLE CORRELATIVE SIGNAL

Several lines of evidence suggest that indoleacetic acid (IAA) is involved in fruit-induced correlative phenomena. Wareing and Seth (1967) showed that the senescence of bean leaves was retarded upon excision of seeds, but was restored by the application of IAA to the deseeded pods. In field bean fruit-applied IAA caused an almost complete inhibition of axillary bud growth. Senescence of bud tissue also occurred. In contrast, axillary buds resumed growth when deseeded fruits were not treated with IAA. IAA treatment also caused a nearly 60% loss of leaves, and over 70% loss of chlorophyll from the remaining leaves, indicating extensive leaf senescence (Tamas et al., 1981). These results show that IAA can replace the correlative effect of seeds on axillary bud growth and leaf senescence. The same IAA treatments have been tried in deseeded fruits of G2 peas, but with no effect on senescence of the apical bud. Thus, differences in the correlative mechanism may exist between species, or factors such as uptake of the hormone may vary.

The seed replacement treatment described here has recently been used to test the role of auxins in fruit-to-fruit competition in beans. When seeds of dominant fruits were replaced with 0.1% NAA or IAA instead of plain lanolin, the number and size of younger fruits (those in existence on the day of treatment) were greatly reduced (Tamas, unpublished). Development of another category of young fruits, those on newly formed axillary shoots, also was inhibited by the same treatment. The older, dominant fruits of bean plants, thus, control the development of younger fruits—and thereby overall fruit load—by two distinct means. They cause abortion of already existing young fruits and prevent the formation of new fruiting branches by repressing axillary bud growth. The source of the inhibiting signal for both effects is the seed of dominant fruits, and auxins can replace both effects.

The foregoing results support the hypothesis that IAA is the correlative signal that originates from dominant fruits in field bean and controls development in buds, leaves and other fruits. Seeds of many plants are rich in IAA (Bandurski and Schulze, 1977) and could be the source of IAA transported to other organs. The inability of deseeded fruits to maintain dominance over other organs may be due to impaired IAA export. There is ample evidence indicating that IAA is involved in the dominance of the growing shoot apex over axillary buds (Tucker, 1978; White et al., 1975). That IAA itself is transported from dominant to target organs is implied by the results of Sorrells et al. (1978). These authors found that injection of the IAA-transport inhibitor N-1-naphthylphthalamic acid (NPA) into the maize (*Zea mays* L.) stem between two ears stimulates development of the lower ear. Thus, IAA transport seems to be required for upper ear dominance. We found a similar effect on bud growth in bean where NPA, applied as a 1% lanolin dispersion on the peduncles of intact fruits, relieved axillary bud growth inhibition (Figure 1). The inhibition of bud growth apparently involved IAA transport through fruit peduncles but not through petioles because treatment of petioles with NPA had no effect. The increase in bud growth, due to NPA treatment of peduncles, was nearly as great as that caused by deseeding of fruits. Bud growth stimulation by NPA appeared to decrease with age suggesting perhaps a greater flux of IAA in younger plants. Resumption of bud growth was not caused by decreased competition for nutrients between buds and fruits because on NPA treated plants the fruits did not show any decrease in development. In fact, on older NPA-treated plants the number and size of NPA treated fruits were greater than those of controls (Figure 2, D, E, F). On younger plants there was no significant difference among fruits of different treatments (Figure 2, A, B, C). NPA treatment of fruit peduncles also prevented the senescence of leaves. The foregoing indicates that inhibition of IAA transport out of dominant fruits can prevent the correlative effect of fruits on target organs. The results are, therefore, consistent with the hypothesis that IAA can serve as a correlative signal released by dominant fruits.

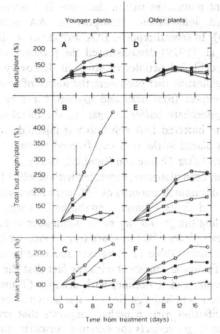

Figure 1. Effect of NPA treatment of peduncles or petioles on the growth of axillary buds in bean plants, cv. Redkloud (Tamas et al., 1981). NPA was 1% (w/w) in lanolin; 0.05 ml was applied as a ring on each peduncle or petiole. Plant age on day of treatment: A, B, C, 45 days; D, E, F, 53 days. ▲, intact control; ○, defruited plants; ■, NPA on peduncles; □, NPA on petioles. Vertical bar is minimum significant difference at α = 0.05.

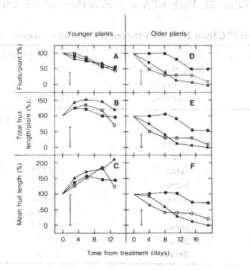

Figure 2. Effect of NPA on fruit development. See Figure 1 for details.

The distance between the dominant fruits and the affected organs is often considerable. Therefore, efficient movement within the plant is a necessary attribute of the correlative signal. Rapid, long-distance transport of IAA occurs in phloem tissue (Goldsmith et al., 1974). In several reports, IAA was shown to be transported in conjugated form. Hein et al. (1979) demonstrated the presence of conjugated IAA in phloem exudates from debladed petioles of soybean. Recovery was much less when plants were depodded, indicating that pods were the source of IAA.

We examined IAA export from bean and soybean fruits by using 2-^{14}C-IAA applied, either in citrate-phosphate buffer or agar, to the distal seed of a single donor fruit on each plant. The buffered IAA solution was injected into the seed. The agar preparation of IAA was placed in the seed cup formed by surgical removal of the embryo from the donor seed (after Thorne and Rainbird, 1983).

When donor fruits on bean plants were injected with labelled IAA and left attached to the plant for 1 or 2 days, small amounts of ^{14}C were recovered from nearby organs including leaves, buds and fruits (Tamas, unpublished). Large fruits exported relatively more ^{14}C paralleling the greater degree of dominance exhibited by these fruits. In both bean and soybean plants ^{14}C transport from donor fruits was predominantly to receivers at the same node (Table 1). Receivers above or below the donor showed much less activity. This finding is consistent with the fact that the degree of dominance is greatest within a raceme, and it declines with distance (Tamas, unpublished). The middle leaflet of receiver leaves had significantly less activity than the other two. The consistency of the distribution pattern perhaps suggests that transport of IAA from the donor to a particular target reflects the extent of vascular connection between the two organs.

We tested the effect of NPA and chlorflurenol (methyl-2-chloro-9-hydroxyfluorene-(9)-carboxylate on ^{14}C-IAA export by fruits. Both of these substances inhibited the movement of ^{14}C to seeds and pods of neighboring fruits when applied as a 1% lanolin dispersion to the peduncles of donors. The effect was greatest on transport to young fruits (Table 2). Both NPA and chlorflurenol are known to inhibit auxin

Table 1. Transport of ^{14}C from 2-^{14}C-IAA-treated fruits to other organs.

Receiver		^{14}C in Receiver Organ	
Organ	Node Position Relative to Donor Node	Field Bean[a]	Soybean[b]
		— Bq x 10/g dry wt ± SE —	
Seed	Above	-	12±4
Carpel	Above	-	8±4
Seed	Same	128±70	745±201
Carpel	Same	49±14	497±138
Left leaflet	Same	93±32	-
Middle leaflet	Same	31±8	-
Right leaflet	Same	121±73	-
Seed	Below	17±3	13±4
Carpel	Below	21±6	15±3

[a]Donors on bean plants were injected with 2-^{14}C-IAA in buffer, 1.3 x 10^4 Bq/fruit. Plants were 48 days old.

[b]Donors on soybean plants were treated with 2-^{14}C-IAA in agar, 1.85 x 10^4 Bq/seed cup. Plants were 103 days old.

Table 2. Effect of NPA and chlorflurenol on ^{14}C transport from 2-^{14}C-IAA-treated fruits to neighboring fruits in field beans.

Receiver Organ	Donor[a] Peduncle Treated with		
	Plain Lanolin	NPA; 1% in Lanolin	Chlorflurenol; 1% in Lanolin
	— Bq x 10/g dry wt ± SE —		
Seed	82±15	24±4	51±12
Carpel	97±14	38±14	57±15
Young Fruit	610±197	12±6	136±54

[a]Donors were treated with 2-^{14}C-IAA in agar, 1.85 x 10^4 Bq/seed cup. Plants were 61 days old.

movement in plants (Thomson and Leopold, 1974; Watkins and Cantliffe, 1980). Therefore, our results suggest that fruits of field bean plants export IAA to other organs and that the amount of ^{14}C recovered from receivers represents transported IAA.

The mechanism through which IAA, released by fruits, controls the development of other organs is not known. According to several studies, correlative effects of IAA, imposed from distant organs, resulted in growth inhibition and senescence of the target organ, whereas direct treatment with IAA or a rise in its endogenous level gave the opposite effect. The endogenous auxin level in the lateral buds of *Vicia faba* (Thimann and Skoog, 1934) was found to increase rather than decrease when the buds were released from the growth-suppressing effect of the apex by decapitation. The lateral buds of these plants, however, remained inhibited if the cut stump was treated with IAA. When IAA was applied directly to buds of pea plants, growth stimulation rather than inhibition was observed (Sachs and Thimann, 1967). Similarly, direct application of IAA to axillary buds of bean plants stimulated their growth, whereas fruit-applied IAA caused bud inhibition (Tamas et al., 1981). The way auxin affects leaf senescence seems to depend also on the site of auxin application. Leaves of bean plants were found to show increased senescence when fruits were treated with IAA (Wareing and Seth, 1967), whereas senescence of soybean leaves was retarded when the leaves were sprayed with NAA (Lindoo and Nooden, 1978). Furthermore, when deseeded older fruits of bean plants were treated with IAA, senescence in younger fruits was enhanced. However, senescence was delayed in the treated fruit itself (Tamas, unpublished).

The foregoing evidence suggests that IAA may be a link in mediating correlative control by fruits over other organs, but it does not directly inhibit development in the correlatively affected organ. IAA may accomplish this task indirectly through the agency of another growth regulating substance, such as ethylene. Ethylene is well known for its role in causing growth inhibition and senescence. Because IAA promotes ethylene synthesis, especially at high IAA concentrations (Kang et al., 1971), it is likely that ethylene plays a role in IAA-mediated correlative effects. A possible second factor is ABA. There is now some evidence that ABA is involved in correlative events along with IAA. In field bean plants we found that, shortly after fruit removal, axillary buds resumed growth and their ABA content declined substantially, but direct treatment of axillary buds with a solution of ABA inhibited their growth (Tamas et al., 1979a). The presence of older fruits also increased the rate of abortion and ABA content of younger fruits (Tamas et al., 1979b). The older fruits, therefore, seemed to have a direct controlling influence over the ABA content of younger fruits, suggesting, in turn, an active regulatory role for ABA in this correlative interaction. In the soybean, ABA concentration increased just prior to or during the senescence of pods (Quebedeaux et al., 1976). In leaves, the concentration of endogenous ABA increased during senescence, and applied ABA enhanced chlorophyll loss (Lindoo and Nooden, 1978), suggesting that ABA participated in the regulation of leaf senescence. ABA release from leaves appears to be under the control of fruits. Recent results (Hein et al., 1979; Setter et al., 1980) showed ABA accumulation in leaves of depodded soybean plants as a result of decreased export. Depodding also decreased the flux of IAA (transported in its conjugated form) from the plant into debladed petioles. Therefore, ABA export from leaves of adult plants may be stimulated by an IAA signal released by fruits.

The action of IAA and ABA in fruit-induced correlative phenomena seems to resemble that occurring in apical dominance. Tucker (1978) has shown that the auxin transport inhibitor, 2,3,5-triiodobenzoic acid, relieved the growth suppressing effect of the apical meristem over the lateral buds, but this suppression was restored by treatment of these buds with ABA. That IAA can stimulate ABA synthesis was demonstrated by Eliasson (1975) in pea stem sections incubated in IAA solution. More recently, Knox and Wareing (1984) showed that decapitation of bean plants decreased the ABA content of axillary buds. This change was prevented by the application of IAA to the decapitated stem. These results support the suggestion that the correlative inhibition of axillary buds in apical dominance results from the stimulation by auxin of inhibitor formation. A similar mechanism may be involved in the regulation of bud growth by fruits. Therefore, correlative control of axillary bud growth by the apical meristem, and also by fruits, may be viewed as two essentially similar phenomena mainly differing in the source organ of the correlative signal. In young plants the growing vegetative apex is the dominant structure, whereas in mature plants the fruits become the source of correlative influence. Control is eventually extended over the development of several organs including axillary buds, leaves, and other fruits. The extent to which these phenomena share a common hormonal mechanism requires clarification.

NOTES

Imre A. Tamas, Britta K. Mazur and Laura B. Campbell, Biology Department, Ithaca College, Ithaca, NY 14850. Peter J. Davies, Section of Plant Biology, Cornell University, Ithaca, NY 14853.

This work was supported in part by research grant PCM82-16159 from the National Science Foundation.

The authors thank Dr. Roger M. Spanswick for the use of the sample oxidizer.

REFERENCES

Bandurski, R. S. and A. Schulze. 1977. Concentration of indole-3-acetic acid and its derivatives in plants. Plant Physiol. 60:211-213.

Davies, P. J., E. Emshwiller, T. J. Gianfagna, W. M. Proebsting, M. Noma and R. P. Pharis. 1982. The endogenous gibberellins of vegetative and reproductive tissue of G2 peas. Planta 154:266-272.

Davies, P. J., W. M. Proebsting and T. J. Gianfagna. 1977. Hormonal relationships in whole plant senescence. p. 273-281. In P. E. Pilet (ed.) Plant growth regulation, Springer-Verlag, Berlin.

Eliasson, L. 1975. Effect of indole acetic acid on the abscisic acid level in stem tissue. Physiol. Plantarum. 34:117-120.

Gianfagna, T. J. and P. J. Davies. 1981. The relationship between fruit growth and apical senescence in the G2 line of peas. Planta 152:356-364.

Gianfagna, T. J. and P. J. Davies. 1983. The transport of substances out of developing fruits in relation to the induction of apical senescence in *Pisum sativum* line G2. Physiol. Plant. 59:676-684.

Goldsmith, M. H., D. A. Cataldo, J. Karn, T. Brenneman and P. Trip. 1974. The rapid nonpolar transport of auxin in the phloem of intact *Coleus* plants. Planta 116:301-317.

Hein, M. B., M. L. Brenner and W. A. Brun. 1979. Source/sink interactions in soybeans. II. A possible role of IAA. Plant Physiol. Suppl. 63:S43.

Kang, B. G., W. Newcomb and S. P. Burg. 1971. Mechanism of auxin-induced ethylene production. Plant Physiol. 47:504-509.

Knox, J. P. and P. F. Wareing. 1984. Apical dominance in *Phaseolus vulgaris* L.: The possible roles of abscisic and indole-3-acetic acid. J. Exp. Bot. 35:239-244.

Lindoo, S. J. and L. D. Nooden. 1977. Studies on the behavior of the senescence signal in Anoka soybeans. Plant Physiol. 59:1136-1140.

Lindoo, S. J. and L. D. Nooden. 1978. Correlation of cytokinins and abscisic acid with monocarpic senescence in soybeans. Plant Cell Physiol. 19:997-1006.

Molisch, H. 1928. Die Lebensdauer der Pflanze (The Longevity of Plants, transl. by E. H. Fulling, NY.) 266 pp. (1938).

Proebsting, W. M., P. J. Davies and G. A. Marx. 1976. Photoperiodic control of apical senescence in a genetic line of peas. Plant Physiol. 58:800-802.

Quebedeaux, B., P. B. Sweetser and J. C. Rowell. 1976. Abscisic acid levels in soybean reproductive structures during development. Plant Physiol. 58:363-366.

Sachs, T. and K. V. Thimann. 1967. The role of auxins and cytokinins in the release of buds from dominance. Am. J. Bot. 54:136-144.

Setter, T. L., W. A. Brun and M. L. Brenner. 1980. Effect of obstructed translocation on leaf abscisic acid, and associated stomatal closure and photosynthesis decline. Plant Physiol. 65:1111-1115.

Sorrells, M. E., R. E. Harris and J. H. Lonnquist. 1978. Response of prolific and nonprolific maize to growth-regulating chemicals. Crop Sci. 18:783-787.

Tamas, I. A., C. J. Engels, S. L. Kaplan, J. L. Ozbun and D. H. Wallace. 1981. Role of indoleacetic acid and abscisic acid in the correlative control by fruits of axillary bud development and leaf senescence. Plant Physiol. 68:476-481.

Tamas, I. A., J. L. Ozbun, D. H. Wallace, L. E. Powell, C. J. Engels. 1979a. Effect of fruits on dormancy and abscisic acid concentration in the axillary buds of *Phaseolus vulgaris* L. Plant Physiol. 64:615-619.

Tamas, I. A., D. H. Wallace, P. M. Ludford and J. L. Ozbun. 1979b. Effect of older fruits on abortion and abscisic acid concentration of younger fruits in *Phaseolus vulgaris* L. Plant Physiol. 64:620-622.

Thimann, K. V. and F. Skoog. 1934. On the inhibition of bud development and other functions of growth substances in *Vicia faba*. Proc. R. Soc. Lond. B. Biol. Sci. 114:317-339.

Thomson, K. S. and A. C. Leopold. 1974. *In-vitro* binding of morphactins and 1-N-naphthylphthalamic acid in corn coleoptiles and their effects on auxin transport. Planta 115:259-270.

Thorne, J. H. and R. M. Rainbird. 1983. An *in-vitro* technique for the study of phloem unloading in seed coats of developing soybean seeds. Plant Physiol. 72:268-271.

Tucker, D. J. 1978. Apical dominance in the tomato: The possible roles of auxin and abscisic acid. Plant Sci. Lett. 12:273-278.

Wareing, P. F. and A. K. Seth. 1967. Aging and senescence in the whole plant. Soc. Symp. Exp. Biol. 21:543-558.

Watkins, J. T. and D. J. Catliffe. 1980. Regulation of fruit set in *Cucumis sativus* by auxin and an auxin transport inhibitor. J. Am. Soc. Hort. Sci. 105:603-607.

White, J. C., G. C. Medlow, J. R. Hillman and M. B. Wilkins. 1975. Correlative inhibition of lateral bud growth in *Phaseolus vulgaris* L. Isolation of indoleacetic acid from the inhibitory region. J. Exp. Bot. 26:419-424.

THE PHYSIOLOGY OF REPRODUCTIVE ABSCISSION IN SOYBEANS

W. A. Brun, J. C. Heindl, and K J Betts

Soybean plants flower profusely, but set only a limited number of pods and seeds. Several authors (Van Shaik and Probst, 1958a; Hansen and Shibles, 1978; Wiebold et al., 1981) have documented flower and pod abscission varying from 32% to 83% in determinate and indeterminate cultivars.

The agronomic significance of this high incidence of pod and flower abscission is not fully understood. Some authors (Streeter and Jeffers, 1979) conclude, from analysis of total non-structural carbohydrate content, that soybean plants are capable of maintaining higher reproductive loads than those currently achieved.

The physiological mechanism of reproductive abscission is also poorly understood. It appears to be unrelated to pollen viability (Van Shaik and Probst, 1985b) or lack of fertlization (Abernethy et al., 1977). Several authors speculate that the incidence of reproductive abscission is related to the availability of photosynthetic assimilates to the young flowers and pods (Hardman and Brun, 1971; Schou et al., 1978; Streeter and Jeffers, 1979). However, since reproductive abscission occurs at a stage of plant development when photosynthetic rate is relatively high and reproductive growth rate is relatively low, it is difficult to envision that the abscission is caused by a limitation in the contemporary supply of photoassimilates. Others have inferred that reproductive abscission is at least in part hormonally mediated (Heindl and Brun, 1983; Huff and Dybbing, 1980), although the specific hormone system involved is unclear.

Recent work in our laboratory has attempted to elucidate the involvement of carbon assimilation and partitioning in reproductive abscission in soybean by examining the effects of light on the process, and by examining the sink relations of setting and abscising soybean flowers.

THE EFFECT OF LIGHT ON REPRODUCTIVE ABSCISSION

Field experiments were conducted in 1981 and 1982 to study the effects of light on flower and pod abscission of soybean plants of cultivar, Evans (Heindl and Brun, 1983). Cool-white and red fluorescent lights illuminated the lower part of the canopy during daylight hours for 3 weeks late in the flowering period. Naturally-lit canopies were the controls. Flowers and young pods on the main stems of half of the plants were shaded with aluminum foil (Table 1). Reproductive abscission was monitored by tagging each flower on its day of anthesis, and then, observing if and when that flower or resulting pod abscised.

Responses to red and white light were similar and light had no effect on the number of flowers produced (Table 2). In the natural, white, and red light treatments,

Table 1. Light and shade treatments used in the 1981 abscission study. Treatments were applied for 3 weeks during flowering and early pod growth.

Light regime	Shade treatment	Designation
Natural light	No shade	(NL)
	Shade nodes	(NLS)
Supplemental white light[a]	No shade	(WL)
	Shade nodes	(WLS)
Supplemental red light[b]	No shade	(RL)
	Shade nodes	(RLS)

[a] Westinghouse F96T12/Cool White lamps supplied 65 $\mu E/m^2$ s PPFD measured 30 cm from the source.

[b] General Electric F96T12/Red lamps supplied 20 $\mu E/m^2$ s PPFD measured 30 cm from the source. Maximum intensity at 660 nm.

Table 2. Effects of light and shade treatments on flowering and abscission (Heindl and Brun, 1983).

Treatment[a]	Flowers produced	Flower abscission	Pod abscission	Combined flower and pod abscission
		%	% remaining flowers	%
NL	4.0	35.4 s[b]	79.9 rs	90.5 r
NLS	3.8	48.8 r	85.6 r	92.2 r
WL	4.2	32.1 s	71.3 s	80.4 s
WLS	4.3	56.7 r	90.3 r	95.4 r
RL	4.6	29.8 s	72.3 s	82.6 s
RLS	4.1	49.2 r	90.9 r	93.1 r

[a] Treatments were applied to and data collected from trifoliolate nodes 1 to 5. See Table 1 for treatment designations.

[b] Values in the same column followed by the same letter are not significantly different according to Duncan's New Multiple Range Test ($\alpha=0.05$).

shading of the reproductive structures significantly increased both flower and pod abscission, except for pod abscission in natural light.

Seed yield per node was decreased in each light treatment by shading the nodes. This effect was attributable to pod number and not to weight per seed or number of seeds per pod (Table 3). These results suggested that the abscission responses to light were photomorphogenic, not photosynthetically, mediated. Red and white light were equally effective, and they were perceived in the reproductive structures, not in the leaves, as would have been expected had the response been photosynthetic.

To further elucidate the effects of light on the carbon nutrition of flowers and pods, a field experiment was performed in which a 30 min pulse of 1.11 MBq of $^{14}CO_2$ was administered to single leaves as described by Heindl and Brun (1983). After a 3 h chase, flowers and pods were collected separately from the pulsed node, the two nodes below, and the two nodes above the pulsed node. The plant material was dried at 60 C, weighed and analyzed for radioactivity. The samples were combusted

Table 3. Effects of light and shade treatments on pod and seed yield in 1981 (Heindl and Brun, 1983).

Treatment[a]	Mature pods per node	Seed yield per node	Weight per seed	Seeds per pod	Seed weight per pod
	no	g	mg	no	mg
NL	0.4 s[b]	0.11 s	95	2.6	248
NL3	0.3 s	0.06 s	93	2.3	221
WL	1.0 r	0.22 r	90	2.5	217
WLS	0.2 s	0.04 s	91	2.6	233
RL	1.0 r	0.22 r	89	2.6	227
RLS	0.4 s	0.10 s	114	2.2	260

[a,b]See Table 2.

on a sample oxidizer (Packard Model B-306) the exhaust from which was collected in a CO_2 absorber (Carbosorb) and subjected to scintillation spectroscopy (Beckman, Model LS 8000).

Flowers at the node of the pulsed leaf were shaded from light with aluminum foil. This light varied from midday, full sun to midday full cloud cover. The shades were installed on the pulsed node 0, 24, 72, or 120 h before labeling.

The sink strength (as defined by Warren-Wilson, 1972), expressed as the percentage distribution of ^{14}C in the reproductive structures at each of the three nodes, was calculated. Sink strengths were further partitioned into the components sink size (mg dry weight) and sink intensity (% dpm/mg dry weight). Sink intensity, thus, expressed the competitive ability of a given sink to accumulate current photoassimilate per unit mass.

In the control, the treated node contained 59.0% of the radioactivity and the two nodes below 40.2% (Table 4). The remaining 0.8% was in the two nodes above (data

Table 4. Distribution of exported ^{14}C-assimilates among reproductive structures. $^{14}CO_2$ pulse-chase administered to single leaf 1 week post anthesis. Values are means of 20 replicates. Al-foil shades covered flowers and pods at the pulsed node (Heindl and Brun, 1983).

Hours of shade before $^{14}CO_2$ pulse	Flowers and pods at pulsed node			Flowers and pods two nodes below pulsed node		
	Sink strength	Sink size	Sink intensity	Sink strength	Sink size	Sink intensity
	% dpm	mg dry wt	% dpm/mg dry wt	% dpm	mg dry wt	% dpm/mg dry wt
Nonshaded control	59.0 r[a]	29	2.2 r	40.2 s	71	0.6
0	40.4 s	31	1.3 s	59.0 r	97	0.7
24	47.3 rs	28	1.8 rs	52.2 rs	84	0.7
72	41.1 s	28	1.6 rs	57.9 r	85	0.8
120	50.4 rs	29	1.9 rs	49.1 rs	81	0.7
		NS			NS	NS

[a]Values in the same column followed by the same letter are not significantly different according to Duncan's New Multiple Range Test ($\alpha=0.05$).

not shown). Such a partitioning of photoassimilate is in agreement with that observed by Blomquist and Kust (1971).

When the node subtending the pulsed leaf was shaded, the sink strength of its flowers was clearly reduced, while that of the flowers at the two nodes below was increased (Table 4). These effects were independent of the length of time the shade had been in place when the pulse was administered, indicating that it was the light regime of the sink at the time of the partitioning that was important.

Sink size was not affected by the treatments (Table 4), so that the sink strength effect was clearly expressed by sink intensity. These results are consistent with those of Mor and Halevy (1980), who darkened young rose shoots and reported a resulting reduction in partitioning of recently fixed carbon to such shoots.

The mechanism by which light, perceived by the reproductive organs, regulates their capacity to accumulate photoassimilates is not known. Mor et al. (1980) postulated that the effect in roses may be on membrane transport, and Lichtner and Spanswick (1981) showed that darkness rapidly inhibited the transient membrane depolarization associated with carrier-mediated sucrose uptake in excised soybean cotyledons. Although regulation of membrane transport is an attractive mechanism, such carrier-mediated transport systems have not been demonstrated in soybean reproductive tissues at the very early stages of development when the shade effects here reported were most effective.

Regardless of the mechanism of the shade effect, the relationship between photoassimilate accumulation and reproductive abscission needs to be explored further. To do this, we undertook a series of experiments utilizing some soybean genetic material in which flower set or abscission could be reliably predicted.

SINK RELATIONS OF ABSCISING FLOWERS

The objectives of these experiments were to examine the partitioning of recently fixed carbon to normally setting and normally abscising flowers within a given soybean raceme. By examining this partitioning pattern as a function of time after anthesis of individual flowers, and in relation to the time of their abscission, we hoped to provide a better understanding of the mechanism of abscission.

Soybean plants of cv., Clark isoline E_1t, were grown as previously described (Brun and Betts, 1984) in a growth chamber from seed obtained from Dr. R. L. Bernard, USDA-ARS, Urbana, Illinois. This isoline of Clark produces 2 to 5 cm long axillary racemes, on which individual flowers develop sequentially from the base to the top. The first three or four flowers on the raceme normally set pods, while later flowers abscise soon after anthesis. If the first three or four flowers are removed, then the next three or four flowers will set pods (Huff and Dybing, 1980).

Individual flowers in racemes at mainstem nodes 7 to 13 were tagged at anthesis with color-coded embroidery thread. These flowers were then inspected daily until they abscised or until the raceme was harvested.

Sink activities of abscising and non-abscising flowers were compared in two experiments. In the first experiment, a comparison was made of normally setting flowers at raceme positions I-IV, with normally abscising flowers at raceme positions V-X. In the second experiment, half of the plants were treated by removing the flowers at raceme positions I-IV on the day of anthesis of the flower at position I. Comparisons were then made of setting and abscising flowers by comparing: a) raceme positions I-IV with positions V-VII of untreated plants; b) raceme positions V-VIII with positions IX-XVI of treated plants; and c) raceme positions V-VIII of treated plants with positions

V-VIII of untreated plants.

At different dates, ranging from 0 to 30 days after anthesis of the first flower at node 10, groups of three plants each were pulse-labeled with $^{14}CO_2$. The three plants were covered with a transparent plastic chamber within the growth chamber, and exposed to 1.48 MBq of $^{14}CO_2$ for 3.5 h. After a 0.5 h chase, the racemes were harvested and frozen in liquid N_2. The reproductive structures at various raceme positions were separated and dried for 2 days at 60 C. After dry weight determination, the samples were combusted (Packard Sample Oxidizer, model B-306) and the resulting CO_2 trapped in a CO_2 absorber (Carbosorb) and counted by liquid scintillation spectroscopy (Beckman LS 8000).

Individual sinks (flowers) within the raceme were characterized by expressing their radioactivity as a percentage of the total activity in the raceme (i.e., sink strength). The sink strength was further partitioned into its components, sink size (mg dry weight) and sink intensity (% dpm/mg dry weight).

The pattern of reproductive abscission at the various floral positions in control and treated racemes is shown in Table 5. At positions I-IV, only 17% of the flowers abscised, while 47% and 75% abscised at positions V-VIII and above, respectively. When flowers at positions I-IV were removed, abscission at positions V-VIII decreased from 47% to 21%. These observations confirm the earlier report by Huff and Dybing (1980), who used the same genetic material.

Comparing the normally setting positions I-IV with the normally abscising positions V-X in the first experiment, showed that the sink strengths (Figure 1A) of the two groups of flowers were not greatly different from shortly before until shortly after anthesis. Later, after set had occurred, the sink strength of the two groups of flowers diverged. Sink size of the setting flowers increased sharply after day 6, while that of the few remaining normally abscising flowers remained very low, and quite variable (Figure 1B). Pre-anthesis sink intensity was very high in both normally setting and normally abscising flowers (Figure 1C), indicating high rates of metabolic activity. Sink intensity then dropped sharply on or just before the day of anthesis in both groups of flowers. In normally setting flowers, sink intensity reached a minimum on the day after anthesis, but then recovered slightly on the next day and remained somewhat higher than in abscising flowers until day 6 when growth started and sink intensity dropped sharply.

In the second experiment, abscising and non-abscising flowers were compared in three ways. Comparing floral positions I-IV with floral positions V-XII gave very similar results (Figure 2) as in the first experiment, except that growth of the setting flowers was somewhat slower in the second experiment (i.e., sink size [Figure 2B] did not start to increase until day 18). There was a 3-4 day period just after anthesis when sink strength (Figure 2A) and sink intensity (Figure 2C) were similarly low in both

Table 5. Total reproductive abscission at various positions in 12 axillary racemes of Clark isoline $E_1 t$, with and without flowers at positions I-IV (basal) removed (Brun and Betts, 1984).

Floral position	% Abscission	
	Control	With flowers I-IV removed
I-IV	17	–
V-VIII	47	21
IX-Up	75	70

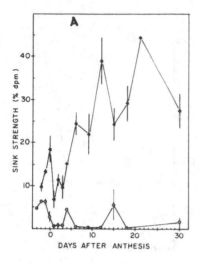

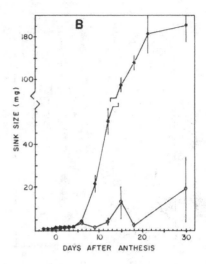

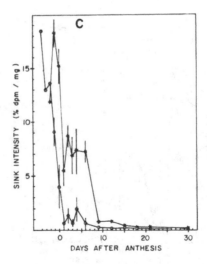

Figure 1. A: Sink strength (% dpm), B: Sink size (mg dry wt.), and C: Sink intensity (% dpm/mg dry wt.) of tissues at various floral positions within axillary racemes of Clark isoline E_1t as a function of time after anthesis of each flower. ✱—normally setting positions I-IV. ○—normally abscising positions V-X. Standard errors are shown where treatment means differed at the 5% level. Data from Experiment No. 2.

setting and abscising flowers.

Very much the same picture emerged when abscising and non-abscising flowers were compared within treated racemes (i.e., racemes with positions I-IV removed) (Figure 3A), and when the comparison was made between positions V-VIII of treated and non-treated racemes (Figure 3B). In each case, sink intensity dropped sharply

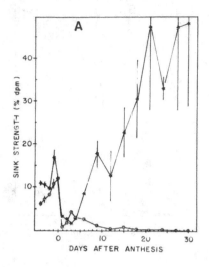

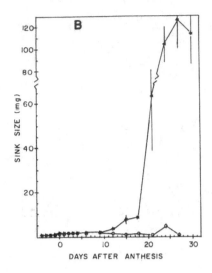

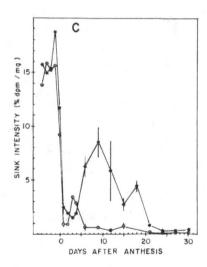

Figure 2. A: Sink strength (% dpm), B: Sink size (mg dry wt.), and C: Sink intensity (% dpm/mg dry wt.) of tissues at various floral positions in axillary racemes of Clark isoline $E_1 t$ as a function of time after anthesis of each flower. •—normally setting positions I-IV. ○—normally abscising positions V-XII. Standard errors are shown where treatment means differed at the 5% level. Data from Experiment No. 2.

on the day of anthesis, then remained low for 2 days, after which it partly recovered in the setting flowers, but not in the abscising flowers.

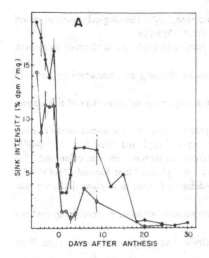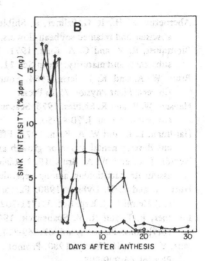

Figure 3. Sink intensity (% dpm/mg dry wt.) of tissues at various floral positions within axillary racemes of Clark isoline E_1t as a function of time after anthesis of each flower. •—normally setting flowers. o—normally abscising flowers. A: Comparison between positions V-VIII (I-IV removed) and positions IX-XVI (I-IV removed). B: Comparison between V-VIII (I-IV removed) and positions V-VIII (I-IV intact). Standard errors are shown where treatment means differed at the 5% level.

CONCLUSIONS

From these experiments, we concluded that soybean reproductive abscission is determined at or very near the date at which the individual flower reaches anthesis. The determination appears to be sensitive to light striking the flowers or the associated nodal tissues. This light response seems not to be a photosynthetic response, although it does seem to be associated with an increased ability of the reproductive tissue to accumulate recent photoassimilate from the leaves. More specifically, reproductive abscission appears to be associated with a failure of sink intensity to increase after its temporary decline just after anthesis. This lack of intensity indicates a reduction in the flowers' competitive ability to attract current photoassimilate from the leaves. Whether there is a cause and effect relationship between the lack of sink intensity and the abscission response remains to be determined, as does the question of possible hormonal mediation of the observed responses.

NOTES

W. A. Brun and K. J. Betts, Professor and Assistant Scientist, respectively; Department of Agronomy and Plant Genetics, University of Minnesota, St. Paul, MN 55108. J. C. Heindl, Monsanto Agricultural Products Co., 800 N. Lindbergh Blvd., St. Louis, MO 63167.

Supported in part by the U.S. Department of Agriculture under Grant 59-2271-0-2-020-0 from the Competitive Research Grants Office, and in part by a grant from the Minnesota Soybean Research and Promotion Council. Contribution from the Minnesota Agricultural Experiment Station, St. Paul, MN 55108. Paper No. 13,906, Scientific Journal Series.

REFERENCES

Abernethy, R. H., R. G. Palmer, R. Shibles and I. C. Anderson. 1977. Histological observations on abscising and retained soybean flowers. Can. J. Plant Sci. 57:713-716.

Blomquist, R. V. and C. A. Kust. 1971. Translocation pattern of soybeans as affected by growth substances and maturity. Crop Sci. 11:390-393.

Brun, W. A. and K. J. Betts. 1984. Source/sink relations of abscising and non-abscising soybean flowers. Plant Physiol. 75:In Press.

Hansen, W. R. and R. Shibles. 1978. Seasonal log of the flowering and podding activity of field-grown soybeans. Agron. J. 70:47-50

Hardman, L. L. and W. A. Brun. 1971. Effects of atmospheric carbon dioxide enrichment at different developmental stages on growth and yield components of soybeans. Crop Sci. 11:886-888.

Heindl, J. C. and W. A. Brun. 1983. Light and shade effects on reproductive abscission and ^{14}C-assimilate partitioning among reproductive structures in soybean. Plant Physiol. 73:434-439.

Huff, A. and C. D. Dybing. 1980. Factors affecting shedding of flowers in soybean [*Glycine max* (L.) Merrill.]. J. Exp. Bot. 31:751-762.

Lichtner, F. T. and R. M. Spanswick. 1981. Electrogenic sucrose transport in developing soybean cotyledons. Plant Physiol. 67:869-874.

Mor, Y. and A. H. Halevy. 1980. Promotion of sink activity in developing rose shoots by light. Plant Physiol. 66:990-995.

Mor, Y., A. H. Halevy and D. Porath. 1980. Characterization of the light reaction in promoting the mobilizing ability of rose shoot tips. Plant Physiol. 66:996-1000.

Schou, J. B., D. L. Jeffers and J. G. Streeter. 1978. Effects of reflectors, black boards, or shades applied at different developmental stages on growth and yield components of soybeans. Crop Sci. 18:29-34.

Van Shaik, P. H. and A. H. Probst. 1958a. The inheritance of inflorescence type, peduncle length, flowers per node and percent flower shedding in soybeans. Agron. J. 50:98-102.

Van Shaik, P. H. and A. H. Probst. 1958b. Effects of some environmental factors on flower production and reproductive efficiency in soybeans. Agron. J. 50:192-197.

Warren-Wilson, J. 1972. Control of Crop Processes. p. 7-30. *In* A. R. Rees, K. E. Cockshull, D. W. Hand and R. G. Hurd (eds.) Crop processes in controlled environments. Academic Press, London.

Wiebold, W. J., D. A. Ashley and H. R. Boerma. 1981. Reproductive abscission levels and patterns for eleven determinate soybean cultivars. Agron. J. 73:43-46.

Streeter, J. G. and D. L. Jeffers. 1979. Distribution of total nonstructural carbohydates in soybean plants having increased reproductive loads. Crop Sci. 19:729-734.

INFLUENCE OF DURATION AND RATE OF SEED FILL ON
SOYBEAN GROWTH AND DEVELOPMENT

Robert P. Patterson and C. David Raper, Jr.

Final dry matter accumulation in a soybean seed is the product of duration and rate of seed fill. Both of these components of seed growth are influenced by genotype (Egli et al., 1978; 1984) and environment (Cure et al., 1982; 1983; Egli and Leggett, 1973). While duration of seed fill, perhaps, has the greater correlation with differences in seed yield among cultivars (Egli, 1975; Egli and Leggett, 1973; Gbikpi et al., 1981; Hanway and Weber, 1971; Kaplan and Koller, 1974), the effects of environment on duration involve consideration of rate of seed fill. In this discussion we shall consider some possible relationships between duration and rate of seed fill with growth and development of the whole plant.

Duration and rate of seed fill are influenced by genotype (Egli et al., 1984), photoperiod and planting date (Boote, 1981; Cure et al., 1982; Jones and Laing, 1978; Raper and Thomas, 1978; Williams et al., 1979), temperature (Dunphy and Hanway, 1976; Egli and Leggett, 1973; Raper and Thomas, 1978), and water deficit (Cure et al., 1983; Meckel et al., 1984). These factors may have an effect directly on the developing seed. Additionally, they may indirectly alter seed growth by effects on assimilate availability and translocation from vegetative organs to the seed. In either event, it seems reasonable to document some of the observed effects of cultivar and environment on rate and duration of seed fill and then to explore the possible relationships brought about by changes in source strength of vegetative organs and sink demand of developing seeds.

EFFECTS OF GENOTYPE AND ENVIRONMENT

Genotype and Cultivar

Yield differences among cultivars are associated with differences in the length of the seed-filling period (Boote, 1981; Dunphy et al., 1979; Gay et al., 1980; Hanway and Weber, 1971; McBlain and Hume, 1980). A wide range in duration of seed fill exists within genotypes of the U.S. Soybean Germplasm Collection (Reicosky et al., 1982). Distinctions among genotypes were consistent across years, which supports the conclusion that duration of seed fill is under genetic control.

A number of methods are used to measure the length of seed fill. Some are based on growth characteristics of individual seed (Fraser et al., 1982; Gbikpi and Crookston, 1981), while others are determined on a whole-plant basis (Boote, 1981; Gay et al., 1980). Since the individual seed is the basic unit of yield, duration of filling of the individual seed provides the most quantitative description of yield. The effective filling period, which is calculated as the quotient of final seed weight and rate of accumulation of dry matter by seed during the linear phase of seed growth, was originally proposed for corn (Daynard and Kannenberg, 1976), but also is useful for soybeans. Use of the effective filling period avoids the problem of determining the beginning and end of seed growth. The linear filling phase of seed growth, which occurs between lag and plateau phases, is the period of greatest contribution to final seed yield.

For a number of soybean cultivars, the linear filling phase, or the effective filling period, is restricted to as little as 20 to 25 days (Hanway and Weber, 1971). Significant differences in effective filling period during each of three years was found (Egli et al., 1984) among the genotypes from the U.S. Soybean Germplasm Collection (Table 1). Genotypic differences remained apparent when the estimates of filling period were expressed in growing degree days; thus, the genotypic differences were not simply responsiveness to temperature. The correlation between final seed yield and effective filling period, however, was not very strong. Nonetheless, with more study of the relationship between rate and duration of seed growth, genotypes with long filling periods may prove to be useful in developing higher yielding cultivars.

Photoperiod and Planting Date

Each distinct reproductive event from floral induction to seed maturity is affected by the current photoperiod (Raper and Thomas, 1978; Thomas and Raper, 1983). Successively shorter days accelerate the rate of reproductive growth and shorten the duration of the effective filling period. Beginning with an optimum planting date a few weeks before the summer solstice, successive delays in planting result in reductions in days between planting and maturity. Part of this shortened growth period with late planting is caused by a reduction in the time to floral initiation. However, part also can result from the acceleration of rate of seed fill. The sensitivity to effects of planting date and photoperiod regime differ among cultivars. Late planting of early-maturing cultivars results in greater yield reductions than late planting of later maturing cultivars (Hartwig, 1954; Parker et al., 1981).

Some effects of photoperiod during seed development on rate and duration of seed fill and associated dry matter and nitrogen partitioning were investigated in a phytotron study (Cure et al., 1982). Significantly greater rates of accumulation of both dry matter and nitrogen occurred under short-day than long-day photoperiods (Figure 1). Plants under long-day photoperiods during seed development had a greater

Table 1. Characteristics of seed growth for selected genotypes compared with Essex. Adapted from Egli et al. (1984).

Genotype (Maturity group[a])	Effective filling period[b]	R5 to R7	Seed size	Seed yield
	days	days	mg/seed	g/m^2
Essex (V)	29	44	127	283
Columbus (IV, +2)	35	42	222	256
PI424.217a (+2)	47	48	241	195
PI398.401 (+2)	57	55	350	214

[a]Maturity groups compared with Essex, 1982.
[b]Individual seed measurements.

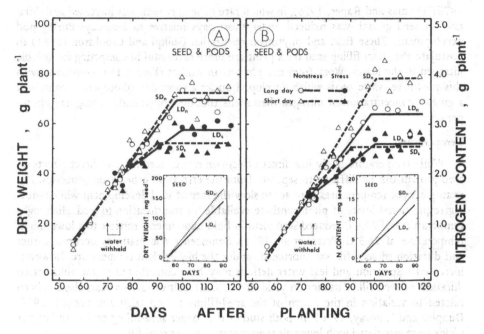

Figure 1. Effect of photoperiod on rate of accumulation of dry matter and nitrogen in seed and pods following a cycle of drying and recovery imposed at R5. Adapted from Cure et al. (1983).

concentration of nitrogen in vegetative parts, but a lower concentration in seed, than plants under continued short-day photoperiod. The long-day photoperiod during seed growth promoted greater leaf area and total vegetative growth than occurred under the short-day photoperiod.

Flowering and initiation of pods for soybean occurs over a relatively long time interval, while the span during which seeds on a plant reach physiological maturity is relatively more narrow. Thus, seed resulting from later-formed flowers would be expected to have a relatively shorter duration of seed fill than seed from earlier-formed flowers. Seeds from late-formed flowers normally experience shorter photoperiods than do seeds from early-formed flowers on the same plant. Under field culture, seeds from late-formed flowers had faster rates of dry matter and protein accumulation (Table 2) than seeds from early-formed flowers (Gbikpi and Crookston, 1981). These field observations are consistent with results of phytotron studies (Raper and Thomas,

Table 2. Effect of time of soybean pod set on seed characteristics for a Maturity Group I cultivar grown in Minnesota. Adapted from Gbikpi and Crookston (1981).

Pod position on stem	Rate of accumulation		Effective filling period	Seed size	Protein concentration in seed
	Dry matter	Pro- tein			
	— mg/seed·day —		days	mg	%
Early (lower)	4.3	1.7	45	181	36.4
Late (upper)	6.0	2.6	32	188	40.1

1978; Thomas and Raper, 1976), in which rate of seed growth was increased and duration of seed growth was reduced under short-days relative to long-days during pod development. These field and phytotron results led Gbikpi and Crookston (1981) to postulate that later-filling seed on a plant are more successful in competing for soluble nitrogen in the leaves. Work from our phytotron studies (Figure 1) is consistent with this postulate, since seeds which developed under a short-day photoperiod contained a greater concentration of nitrogen than seeds that developed under a long-day photoperiod.

Temperature

While temperature likely has direct effects on rate of seed fill, the direct effects on seed growth can be difficult to separate from the effects on other plant processes. For example, cool temperatures may act to slow the rate of seed development while reducing respirational losses of photosynthate available for translocation to seed. High night temperatures of 29.4 C reduced seed yield in field experiments relative to a lower night temperature of 18.3 C (Peters et al., 1971). Senescence and maturity occurred earlier and duration of seed fill was shortened, under the higher night temperature. However, increased respiration and leaf water deficits may have contributed to the abbreviated duration of seed fill. Variation in yield of the same cultivar across years also has been related to variation in the length of the seed-filling period (Egli and Leggett, 1973; Dunphy and Hanway, 1976). In both studies the longer seed-filling period and higher yields were associated with lower air temperatures during seed fill.

A number of phytotron studies suggest that seed growth is affected directly by temperature. Temperatures between 22 and 30 C have only slight effects on in vivo rate of seed growth, pod and seed weights at maturity, and days between anthesis and first brown pod (Hesketh et al., 1973; Thomas and Raper, 1976), although at the higher end of the range accelerated leaf senescence is associated with reduction in duration of seed fill by several days (Egli and Wardlaw, 1980). When day/night temperature during seed growth was lowered to 18/13 C, rate of seed growth was reduced for seed developing on the plant (Table 3). In vitro culture of cotyledons, however, indicates that over the range of 18 to 30 C rate of seed growth increases with temperature (Raper and Patterson, unpublished data; Egli and Wardlaw, 1980).

While it is important to understand the direct effects of temperature on seed growth, the interaction between direct and indirect effects in determining final seed yield, as is the case for duration of seed development and seed size (Hartwig, 1973), may be under partial genetic control. In field studies, slower rates of seed growth, a longer duration of seed growth, and higher seed yield for two cultivars occurred in a

Table 3.   Effect of day/night temperature during seed growth on characteristics of seed growth of Fiskeby V soybeans. Adapted from Egli and Wardlaw (1980).

Day/night temperature	Seed growth rate	Seed dry weight	Physiological maturity
C	mg/seed·day	mg/seed	
18/13	6.1	150	–
24/19	7.0	196	26 Apr
27/22	7.9	196	26 Apr
30/25	7.2	209	26 Apr
33/28	6.6	153	23 Apr

season with a relatively low temperature during seed fill (Egli et al., 1978). A third cultivar, however, had a higher yield in association with a shorter duration of seed fill during a season having a higher temperature during seed fill. In reality, temperature is not the single environmental factor that varies in field culture. The possible genetic differences in response of duration and rate of seed fill to temperature need to be investigated under controlled-environment conditions.

Water Deficit

Water stress during reproductive growth is a major factor in limiting yield of soybeans. The occurrence of a stress during seed filling causes greater reductions in yield than at other reproductive stages (Doss et al., 1974; Shaw and Laing, 1966; Sionit and Kramer, 1977). The reduction in yield caused by a water stress during seed fill is associated with a reduction in duration of seed fill and early senescence of leaves (Sionit and Kramer, 1977). In fact, the effects of water stress on yield appear to be related more to limited availability of photosynthate and nitrogen for translocation to seed (Boyer, 1976; Silvius et al., 1977) than to effects directly on growth rate of the individual seed. Unfortunately, there is little data available on effects of moisture stress on rate of growth for individual seeds.

Water stress reduced both vegetative growth and yield of soybeans in a field experiment (Meckel et al., 1984). Growth rate of individual seeds was not affected (Table 4), although duration of seed filling was reduced when a severe stress was imposed from planting to maturity. The reduction in duration of seed fill can be attributed to the effects of reduced photosynthetic activity of leaves (Figure 2) during the stress and failure of recovery of photosynthetic activity following rewatering (Cure et al., 1983). That rate of seed growth is not appreciably slowed by water stress may accentuate the reduction in duration. If rate of seed growth were reduced, reserves of photosynthate and nitrogen required to support both seed growth and continued photosynthetic and N_2-fixation activity would not be as rapidly depleted as occurs when rate of seed growth is not slowed.

SOURCE-SINK RELATIONSHIPS AFFECTING SEED GROWTH

The process of seed growth involves partitioning of assimilates from the vegetative source organs to the reproductive sink. Details of the translocation mechanism are not well understood (Wardlaw, 1974; Moorby, 1977); however, under a normal range of growth conditions translocation of assimilates between source and sink should be

Table 4. Effect of moisture stress on characteristics of seed growth when water stress was applied early (planting to R5), late (R5 to R7), or severe (planting to maturity) under field culture. Adapted from Meckel et al. (1984).

Irrigation treatment	Seed growth rate	Effective filling period	Seed size
	mg/seed·day	days	mg/seed
Well-watered	7.2	25	176
Early stress	6.4	26	160
Late stress	6.6	25	165
Severe stress	7.3	21	150

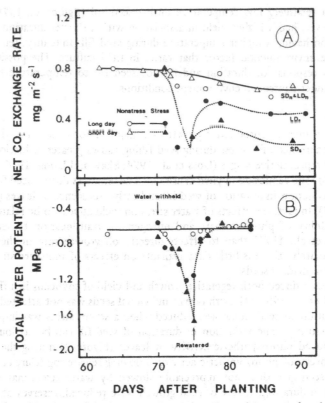

Figure 2. Effect of water deficit imposed at R5 on net CO_2 exchange rate for plants grown under long-day and short-day conditions during seed development. Adapted from Cure et al. (1983).

sufficient to sustain growth and functioning of the receiving organ, and when availability of assimilate in the source organ is decreased, translocation to the sink should be reduced accordingly. Based on this postulate (Wann and Raper, 1979; 1984), source strength can be represented by the concentration of assimilates in the exporting (vegetative) organ and sink demand can be represented by metabolic activity of the receiving (seed) organ. Translocation between source and sink is thus responsive to such factors as temperature, water deficit, etc., only indirectly as they affect assimilate concentration of the source or rates of growth and respiration of the sink (Christy and Swanson, 1976; Cook and Evans, 1976).

Based on these assumptions about translocation between source and sink, we conducted a study in the phytotron to examine the relationship between rate and duration of seed growth by using photoperiod, which was extended with incandescent lamps with little contribution to photosynthetically active radiation, to alter rate of seed growth (sink demand), and water stress to alter assimilate production and availability (source strength). Presumably, photoperiod directly affected only growth rate of the seed and not assimilate production, while the water stress directly affected only assimilate production and not seed growth. A change in photoperiod from short-day to long-day conditions during seed fill reduced the rate of dry matter and nitrogen accumulation in seed (Figure 1). Application of a water stress reduced photosynthetic activity of leaves (Figure 2) as well as N_2-fixation activity of roots (data not shown).

Under the nonstress conditions, the effect of the long-day photoperiod on reducing rate of seed growth had no effect on extending the duration of seed growth. However, photosynthetic and N_2-fixation rates under the nonstress conditions were sufficient to maintain a modest concentration and content of nitrogen in leaves throughout seed filling, and to actually increase the concentration and content of nonstructural carbohydrates (Figure 3). These plants did not undergo senescence at the end of seed fill but continued to accumulate dry matter and nitrogen in vegetative organs. Thus, we assume that, for the nonstressed plants, availability of neither photosynthate nor nitrogen was limiting in the source pool, and when source strength was not limiting, a reduction in sink demand (growth rate of seed) did not increase seed mass by prolonging duration of seed fill.

When photosynthetic activity was reduced by application of a water stress, the effect of the long-day photoperiod on slowing rate of seed growth did prolong duration of seed fill and resulted in increased seed mass. Apparently, the source-sink relationships for nitrogen were more important than for nonstructural carbohydrates. For the stressed plants, content of carbohydrates in leaves declined only slightly and concentration actually increased (Figure 3). Clearly, source strength for carbohydrates was not limiting for seed growth. The concentration of nitrogen in leaves declined rapidly after stress (Figure 3), but since the sink demand for nitrogen was less under the long-day than the short-day photoperiod (Figure 1), nitrogen reserve in leaves declined at a slower rate under the long-day photoperiod. The nitrogen concentration in leaves of stressed plants under long-day photoperiod remained above critical levels for continued photosynthetic activity (Boote et al., 1978) while, in response to the

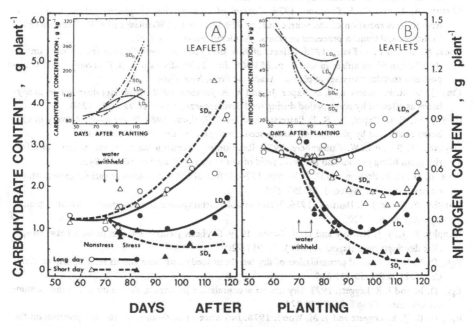

Figure 3. Effect of photoperiod during seed development on contents of nitrogen and soluble carbohydrates in soybean leaf tissue following a cycle of drying and recovery imposed at R5. Unpublished data from the experiments of Cure et al. (1983).

greater sink demand under the short-day photoperiod, nitrogen concentration for stressed plants dropped below the critical level.

The experimental relationship between rate and duration of seed fill for stressed plants concurs with the analysis by Sinclair and de Wit (1976) that a reduction in rate of seed growth reduces the rate of nitrogen depletion from vegetative reserves, delays senescence of leaves, and prolongs duration of seed filling. These results emphasize the potential of exploring genotypic sensitivity to photoperiodic regulation of rate of seed fill. Distinctions among cultivar and planting dates for rate of seed fill suggest that a range of genetic sensitivity exists. Certainly, a slower rate of seed fill enhances the capacity of soybeans to recover from stress-related interruptions in photosynthetic and N_2-fixation activity and helps maintain yield levels.

NOTES

Robert P. Patterson, Department of Crop Science, and C. David Raper, Jr., Department of Soil Science, North Carolina State University, Raleigh, North Carolina 27695-7620.

REFERENCES

Boote, K. J. 1981. Response of soybeans in different maturity groups to March plantings in Southern USA. Agron. J. 73:854-859.

Boote, K. J., R.N. Gallaher, W. K. Robertson, K. Hinson and L. C. Hammond. 1978. Effect of foliar fertilization on photosynthesis, leaf nutrition, and yield of soybeans. Agron. J. 70:787-791.

Boyer, J. S. 1976. Water deficits and photosynthesis. p. 151-190. *In* T. T. Kozlowski (ed.) Water deficits and plant growth. Vol. IV. Soil water measurement, plant responses, and breeding for drought resistance. Academic Press, New York.

Christy, A. L. and C. A. Swanson. 1976. Control of translocation by photosynthesis and carbo-hydrate concentration of the source leaf. p. 329-338. *In* I. F. Wardlaw and J. B. Passioura (eds.) Transport and transfer processes in plants. Academic Press, New York.

Cook, M. G. and L. T. Evans. 1976. Effects of sink size, geometry and distance from source on the distribution of assimilate in wheat. p. 393-400. *In* I. F. Wardlaw and J. B. Passioura (eds.) Transport and transfer processes in plants. Academic Press, New York.

Cure, J. D., R. P. Patterson, C. D. Raper, Jr. and W. A. Jackson. 1982. Assimilate distribution in soy-beans as affected by photoperiod during seed development. Crop Sci. 22:1245-1250.

Cure, J. D., C. D. Raper, Jr., R. P. Patterson and W. A. Jackson. 1983. Water stress recovery in soy-beans as affected by photoperiod during seed development. Crop Sci. 23:110-115.

Daynard, T. B. and L. W. Kannenberg. 1976. Relationships between length of the actual and effec-tive grain filling periods and the grain yield of corn. Can. J. Plant Sci. 56:237-242.

Doss, B. D., R. W. Pearson and H. T. Rogers. 1974. Effect of soil water stress at various growth stages on soybean yield. Agron. J. 66:297-299.

Dunphy, E. J. and J. J. Hanway. 1976. Water-soluble carbohydrate accumulation in soybean plants. Agron. J. 68:697-700.

Dunphy, E. J., J. J. Hanway and D. E. Green. 1979. Soybean yields in relation to days between spe-cific developmental stages. Agron. J. 71:917-920.

Egli, D. B. 1975. Rate of accumulation of dry weight in seed of soybean and its relationship to yield. Can. J. Plant Sci. 55:215-219.

Egli, D. B. and J. E. Leggett. 1973. Dry matter accumulation patterns in determinate and indetermin-ate soybeans. Crop Sci. 13:220-222.

Egli, D. B., J. E. Leggett and J. M. Wood. 1978. Influence of soybean seed size and position on the rate and duration of filling. Agron. J. 70:127-130.

Egli, D. B., J. H. Orf and T. W. Pfeiffer. 1984. Genotypic variability for duration of seedfill in soy-bean. Crop Sci. 24:587-592.

Egli, D. B. and I. F. Wardlaw. 1980. Temperature response of seed growth characteristics of soybeans. Agron. J. 72:560-564.

Fraser, J., D. B. Egli and J. E. Leggett. 1982. Pod and seed development in soybean cultivars with differences in seed size. Agron. J. 74:81-85.

Gay, S., D. B. Egli and D. A. Reicosky. 1980. Physiological aspects of yield improvement in soybeans. Agron. J. 72:387-391.

Gbikpi, P. J. and R. K. Crookston. 1981. Effect of flowering date on accumulation of dry matter and protein in soybean seeds. Crop Sci. 21:652-655.

Hanway, J. J. and C. R. Weber. 1971. Dry matter accumulation in eight soybean [*Glycine max* (L.) Merrill] varieties. Agron. J. 63:227-230.

Hartwig, E. E. 1954. Factors affecting time of planting soybeans in the Southern states. USDA Circular 943.

Hartwig, E. E. 1973. Varietal development. p. 187-210. *In* B. E. Caldwell (ed.) Soybeans: improvement, production, and uses. American Society of Agronomy, Madison, Wisconsin.

Hesketh, J. D., D. L. Myhre and C. R. Wiley. 1973. Temperature control of time intervals between vegetative and reproductive events in soybeans. Crop Sci. 13:250-254.

Jones, P. G. and D. R. Laing. 1978. The effects of phenological and meteorological factors on soybean yield. Agric. Meteorol. 19:485-495.

Kaplan, S. L. and H. R. Koller. 1974. Variation among soybean cultivars in seed growth rate during the linear phase of seed growth. Crop Sci. 14:613-614.

McBlain, B. A. and D. J. Hume. 1980. Physiological studies of higher yield in new, early-maturing soybean cultivars. Can. J. Plant Sci. 60:1315-1326.

Meckel, L., D. B. Egli, R. E. Phillips, D. Radcliffe and J. E. Leggett. 1984. Effect of moisture stress on seed growth in soybeans. Agron. J. 76:647-650.

Moorby, J. 1977. Integration and regulation of translocation within the whole plant. p. 425-454. *In* D. H. Jennings (ed.) Integration of activity in the higher plant. Cambridge University Press, London.

Parker, M. B., W. H. Marchant and B. J. Mullinix, Jr. 1981. Date of planting and row spacing effects on four soybean cultivars. Agron. J. 73:759-762.

Peters, D. B., J. W. Pendleton, R. H. Hageman and C. M. Brown. 1971. Effect of night air temperature on grain yield of corn, wheat, and soybeans. Agron. J. 63:809.

Raper, C. D., Jr. and J. F. Thomas. 1978. Photoperiodic alteration of dry matter partitioning and seed yield in soybeans. Crop Sci. 18:654-656.

Reicosky, D. A., J. H. Orf and C. Poneleit. 1982. Soybean germplasm evaluation for length of the seed filling period. Crop Sci. 22:319-322.

Shaw, R. H. and D. R. Laing. 1966. Moisture stress and plant response. p. 73-94. *In* W. H. Pierre, D. Kirkham, J. Pesek and R. Shaw (eds.) Plant environment and efficient water use. American Society of Agronomy, Madison, Wisconsin.

Silvius, J. E., R. R. Johnson and D. B. Peters. 1977. Effect of water stress on carbon assimilation and distribution in soybean plants at different stages of development. Crop Sci. 17:713-716.

Sinclair, T. R. and C. T. de Wit. 1976. Analysis of the carbon and nitrogen limitations to soybean yield. Agron. J. 68:319-324.

Sionit, N. and P. J. Kramer. 1977. Effect of water stress during different stages of growth of soybean. Agron. J. 69:274-278.

Thomas, J. F. and C. D. Raper, Jr. 1976. Photoperiodic control of seed filling for soybeans. Crop Sci. 16:667-672.

Thomas, J. F. and C. D. Raper, Jr. 1983. Photoperiod and temperature regulation of floral initiation and anthesis in soya bean. Ann. Bot. 51:481-489.

Wann, M. and C. D. Raper, Jr. 1979. A dynamic model for plant growth: Adaptation for vegetative growth of soybeans. Crop Sci. 19:461-467.

Wann, M. and C. D. Raper, Jr. 1984. A dynamic model for plant growth: Validation study under changing temperatures. Ann. Bot. 53:45-52.

Wardlaw, I. F. 1974. Phloem transport: Physical, chemical or impossible. Annu. Rev. Plant Physiol. 25:515-539.

Williams, W. A., C. O. Qualset and S. Gerg. 1979. Ridge regression for extracting soybean yield factors. Crop Sci. 19:869-873.

NITROGEN MOBILIZATION DURING SEEDFILL IN SOYBEANS

D. B. Egli and J. E. Leggett

The soybean seed contains approximately 40% protein; thus, it is not possible to understand the yield production process in soybean crops without considering N metabolism. Soybean plants can obtain N by absorption of soil solution NO_3^- and by the fixation of atmospheric N_2 by *Rhizobium japonicum* in the nodules. In a normal field environment NO_3^- is the primary source of N during the early stages of vegetative growth, whereas N_2 fixation becomes the predominant source of N in the later stages of vegetative growth and the early stages of reproductive growth (Thibodeau and Jaworski, 1975). However, a portion of the N in the seed at maturity comes from N that is mobilized or redistributed from nonseed plant parts and not directly from NO_3^- accumulated to N_2 fixed during seed filling. In this manuscript we want to consider the factors influencing N redistribution and its involvement in the production of yield.

N MOBILIZATION

The realization that plants redistribute mineral nutrients, including N, from vegetative plant parts to the developing seed is not new, and is not restricted to plants producing seed with a high N content (Molisch, 1928; Williams, 1955). During reproductive growth dry weight accumulates in the seed at a constant rate (Egli and Leggett, 1973). However, canopy photosynthesis declines during much of seedfill (Wells et al., 1982), which suggests that the redistribution of previously accumulated assimilates must occur to maintain the constant rate of seed growth. The accumulation of N by the seed also occurs at a relatively constant rate since the concentration of protein in the seed does not vary during seed development (Yazdi-Samadi et al., 1977). The rate of NO_3^- accumulation and N_2 fixation declines during seed filling (Thibodeau and Jaworski, 1975), suggesting that the redistribution of N is required to maintain the constant rate of N accumulation in the seed.

Early work with soybeans demonstrated clearly that the N concentration in vegetative plant parts and the pod walls declined during seed filling, which suggested that N was being redistributed from these plant parts to the seed (Borst and Thatcher, 1931; Hammond et al., 1951; Hanway and Weber, 1971a). Recent work has confirmed that the total N concentration in individual leaves starts declining early in seed filling, and this reduction continues at a relatively constant rate until the leaf abscises from the plant (Boote et al., 1978; Lugg and Sinclair, 1981; Boon-Long et al., 1983). Boon-Long et al. (1983) demonstrated that the decrease in leaf N started earlier in older leaves than it did in younger leaves, although the amount of N in the leaf when it abscised from the plant was similar. Similar patterns have been shown for leaf protein

(Boon-Long et al., 1983; Wittenbach et al., 1980; Thibodeau and Jaworski, 1975; Sesay and Shibles, 1980) and for ribulose,1-5,bisphosphate carboxylase RubPCase (Wittenbach et al., 1980). Decreasing concentrations of N have also been shown in the stems, petioles and pod walls during seed filling (Hanway and Weber, 1971a).

Leaf photosynthetic activity is closely related to changes in the N status when the leaf is losing N (Figure 1) (Boote et al., 1978; Sesay and Shibles, 1980; Wittenbach et al., 1980; Lugg and Sinclair, 1981; Boon-Long et al., 1983).

Thus, it is clear that N is lost from vegetative plant parts and the pod walls during seed filling. This loss of N from the leaves is associated with a decrease in photosynthetic activity of the leaves which continues until leaf abscission. It has been generally assumed that the N lost from the vegetative plant and pod walls is redistributed to the seed, and this assumption has been confirmed with the use of ^{15}N (Morris and Weaver, 1983). It has been reported that gaseous losses of N can occur from soybean leaves (Stutte et al., 1979), suggesting that all of the N lost from the vegetative plant parts and pod walls during seed filling may not be redistributed to the seed. However, only limited data are available on gaseous N losses and it is difficult to relate estimates of gaseous N losses to the quantity of N redistributed on a quantitative basis.

CONTRIBUTION OF REDISTRIBUTED N TO SEED N

Estimates of the quantity of N that is redistributed are obtained by measuring the change in N content of all plant tissues during seed filling, and this quantity is then compared with the amount of N in the seed at maturity. This technique requires the collection of all abscised leaves and petioles for analysis and usually neglects changes in the N status of the root system. The proportion of the N in the seed at maturity that came from redistribution has been reported to vary widely. Estimates obtained using field-grown soybeans in Iowa ranged from 50 to 64% (Hammond et al., 1951; Hanway and Weber, 1971b). Zeiher et al. (1982), using cultivars of varying maturity and growth habit, reported that the contribution of redistributed N to seed N at maturity ranged from a low of 33% to a maximum of essentially 100% (Table 1). A similar range was reported by Egli et al. (1983) for a single cultivar subjected to varying levels of moisture stress during vegetative growth, flowering and pod set, or during seed filling. Contributions ranging from 20 to 60% were reported for soybeans that

Table 1. Nitrogen redistribution of seven soybean cultivars in 1977 and 1978 (Zeiher et al., 1982).

Cultivar	Maturity Group	Leaf	Petiole	Stem	Pod wall	1977	1978	Mean
		\multicolumn{4}{Source of redistributed N[a]}	RN/SN[b]		
					– % –			
Corsoy	II	64	5	19	11	55	39	47
Williams	III	45	6	23	27	83	41	62
Lincoln	III	50	4	19	27	100	56	78
Elf	III	58	4	19	19	78	40	59
Kent	IV	62	3	19	15	61	33	47
Essex	V	46	5	27	21	82	62	72
York	V	46	2	22	29	100	111	105
L.S.D. (0.05)						29	34	

[a]Amount of redistributed N contributed by each plant part, as a percentage of the total amount redistributed, average of 1977 and 1978.
[b]Proportion of the total seed N (SN) that came from redistribution (RN).

were not nodulated and grown in nutrient culture with varying levels of NO_3^- in the medium (Egli et al., 1978). Leaves are the major source of redistributed N, followed by stems and pod walls, which are nearly equal, and the petioles contribute less than 5% of the total quantity of N redistributed (Table 1) (Zeiher et al., 1982). These data suggest that redistributed N can vary from being the primary source of N for the developing seed to the point where it represents only a relatively minor source of N. Obviously, the variation in the contribution of redistributed N to the seed requires an inverse variation in the supply of N to the seed from the uptake of NO_3^- or by N_2 fixation. These processes must remain at relatively high levels longer when redistributed N is only a small portion of the seed N at maturity; however, they must decline early in seed fill as the contribution of redistributed N approaches 100%. The timing of the decrease in N_2 fixation during seed filling should be highly variable across environments and cultivars and, in fact, considerable variation has been reported (Thibodeau and Jaworski, 1975; Nelson et al., 1984).

FACTORS INFLUENCING N REDISTRIBUTION

The variation in the contribution that redistributed N makes to seed N suggests that the process of N redistribution may be influenced by genetic and environmental factors. Hanway and Weber (1971a) measured plant N contents in field experiments that included varying levels of fertilizer (N, P, K) applications. The loss of N from vegetative plant parts was not prevented by high levels of N fertilizer, and they concluded that redistribution occurred even though nutrients were readily available in the soil solution. The application of foliar nutrient sprays containing N, P, K and S during seed filling had only limited effects on N concentrations in the leaves, and did not prevent the decline in leaf N during seed filling (Boote et al., 1973; Sesay and Shibles, 1980). Inoculation of soybean seed with rhizobial strains that varied in effectiveness had only a minor effect on redistribution of N from the leaves (Morris and Weaver, 1983).

Egli et al. (1978) investigated the effect of varying the level of NO_3^- in the nutrient medium on N redistribution of nonnodulated soybean plants grown in the greenhouse. Removing all of the NO_3^- from the medium during seed filling increased the rate of leaf senescence, and increased the contribution that redistributed N made to seed N to approximately 60%. However, N redistribution also occurred in plants receiving a normal or elevated level of NO_3^- during seed filling, although the quantity was reduced to approximately 20% of the final seed N. Moisture stress treatments applied during vegetative growth, flowering and pod set or during seed filling did not have a consistent effect on proportion of seed N coming from redistribution (Egli et al., 1983). Thus, although N redistribution was influenced by severe N stress during seed filling, it occurred regardless of the supply of N to the plant during seed filling. Zeiher et al. (1982) measured N redistribution on eight cultivars of varying maturity (Maturity Group II to V) and growth habit and reported cultivar differences in the proportion of seed N coming from redistribution (Table 1). There was a trend for late maturing cultivars to get more of their seed N from redistribution than early maturing cultivars did. Jeppson et al. (1978) evaluated the harvest N index of a number of soybean cultivars and concluded that there were cultivar differences in the efficiency with which N was mobilized to the developing seed.

RELATIONSHIP OF N REDISTRIBUTION TO YIELD

Although it is clear that a variable proportion of the N in soybean seed comes from N that is redistributed from above ground vegetative plant parts and pod walls during seed development, the importance of N redistribution in the yield production process is not clearly understood. The utilization of already reduced N as a source of N for the developing seed would seem to be an energy efficient process. The utilization of reduced forms of N redistributed from vegetative plant parts for seed growth would reduce the use of photosynthetically derived energy for N_2 fixation. However, Sinclair and de Wit (1975, 1976) hypothesized that the redistribution of N from the leaves was triggered by the inability of the plant to take up enough N to meet the requirements of the developing seed and caused a loss in the physiological activity of the leaf and ultimately caused leaf senescence. They characterized the soybean as being self-destructive and suggested that the self-destructive characteristic could limit yield by limiting the length of the seed filling period. The correlation between leaf N status and photosynthetic activity during seed filling (Figure 1) (Boote et al., 1978; Sesay and Shibles, 1980; Wittenbach et al., 1980; Lugg and Sinclair, 1981; Boon-Long et al., 1983) is consistent with the self-destruction hypothesis. Subjecting soybean plants to N stress during seed development has been shown to increase the rate of N loss from the leaves causing earlier leaf senescence and reductions in yield, which is also consistent with the self-destruction hypothesis (Egli et al., 1978).

Israel (1981) reported differences in the contribution of redistributed N to seed N between two cultivars that produced equal yields. Zeiher et al. (1982) measured the proportion of seed N that came from redistribution in eight cultivars in which the duration of seed fill varied from 27 to 46 days. Although the proportion of seed N coming from redistribution varied from 33 to essentially 100%, there was no relationship (r = 0.09) between the proportion of seed N coming from redistribution and the duration of seed fill. Similar results were obtained when a single cultivar was subjected to moisture stress during vegetative growth, flowering and pod set or during seed filling (Egli et al., 1983). The proportion of seed N that came from redistribution varied from 33 to 100% and the stress treatments created significant differences in yield; however, there was no relationship between the proportion of seed N coming from redistribution and yield (r = 0.07). In both of these experiments, a relatively constant proportion of the N in the vegetative plant and pod walls at beginning seed fill (R5) was redistributed during seed fill (Figure 2), and the amount of N in the plant at R5 was determined primarily by plant size (dry weight), which varied among cultivars and across moisture stress treatments. Vegetative plant size at beginning seed fill was not closely related to yield; however, the amount of N redistributed was associated with vegetative plant size, which accounted for the variation in the proportion of seed N coming from redistribution. We concluded that the contribution that redistributed N makes to the N in the seed at maturity is more closely related to the amount of N available for redistribution, than it is to the ability of the plants to obtain N from the soil or via N_2 fixation during seed filling (Zeiher et al., 1982; Egli et al., 1983).

Nooden et al. (1979) were able to prevent normal senescence of soybean leaves during seed filling by applying foliar sprays containing auxin and cytokinin. The concentration of N and starch increased in leaves on the treated plants, but there was no increase in seed yield. Examination of the duration of seed fill of 13 species of crop plants indicated a wide range in the length of the seed filling period; however, there was no relationship between the duration of seed fill and the seed N content (Egli, 1981). Schapaugh (1980), utilizing normal and high protein genotypes, could not demonstrate a causal relationship between N demand and premature leaf senescence.

Thus, it is not clear from the available data that the process of N redistribution

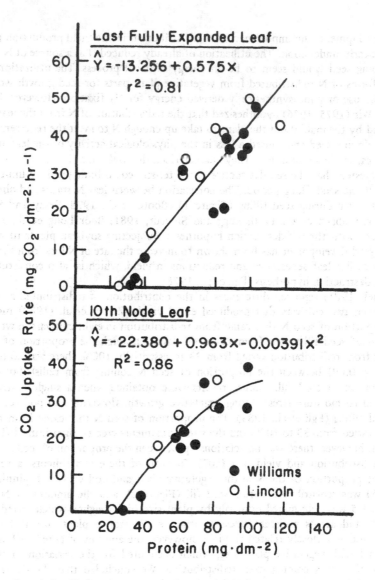

Figure 1. The relationship between CO_2 uptake rate and leaf protein during reproductive growth of two soybean cultivars. (From Boon-Long et al., 1983.)

has a negative or limiting effect on yield. It is clear that N redistribution is associated with leaf senescence and can be a significant source of N for the developing seed. If leaf senescence is initiated by the inability of the plant to meet the N requirement of the seed via NO_3^- uptake or N_2 fixation, as suggested by Sinclair and de Wit (1975, 1976), then N redistribution may be an important determinant of yield. If, however, leaf senescence is triggered by some other factor, such as a senescence signal originating in the seed as suggested by Lindoo and Noodén (1976), then it would seem that N redistribution would occur as a result of senescence and may not be important in determining final yield.

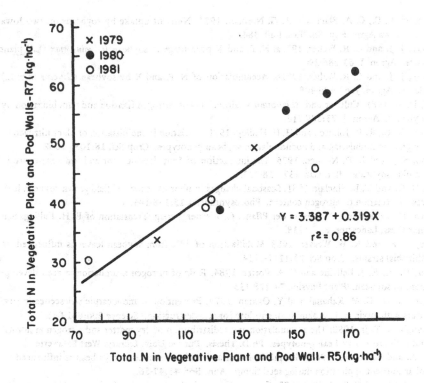

Figure 2. Relationship between total N in the plant at growth stages R5 and R7. Total N at R7 includes the N in abscised leaves and petioles. The cultivar Williams was subjected to moisture stress all season or either from planting to R5 or from R5 to maturity to create differences in vegetative plant size and yield. (From Egli et al., 1983.)

NOTES

D. B. Egli and J. E. Leggett, Department of Agronomy, USDA-ARS, University of Kentucky, Lexington, KY 40546-0091.

REFERENCES

Boon-Long, Preeda, D. B. Egli and J. E. Leggett. 1983. Leaf N and photosynthesis during reproductive growth in soybeans. Crop Sci. 23:617-620.

Boote, K. J., R. N. Gallaher, W. K. Robertson, K. Hinson and L. C. Hammond. 1978. Effect of foliar fertilization on photosynthesis, leaf nutrition, and yield of soybeans. Agron. J. 70:787-791.

Borst, H. L. and L. E. Thatcher. 1931. Life story and composition of the soybean plant. Ohio Agric. Exp. Stn. Bull. 494.

Egli, D. B. 1981. Species differences in seed growth characteristics. Field Crops Res. 4:1-12.

Egli, D. B. and J. E. Leggett. 1973. Dry matter accumulation patterns in determinate and indeterminate soybeans. Crop Sci. 13:220-222.

Egli, D. B., L. Meckel, R. E. Phillips, D. Radcliffe and J. E. Leggett. 1983. Moisture stress and N redistribution in soybean. Agron. J. 75:1027-1031.

Egli, D. B., J. E. Leggett and W. G. Duncan. 1978. Influence of N stress on leaf senescence and N redistribution in soybeans. Agron. J. 70:43-47.

Hammond, L. C., C. A. Black and A. G. Norman. 1951. Nutrient uptake by soybeans on two Iowa soils. Iowa Agric. Exp. Stn. Res. Bull. 384.

Hanway, J. J. and C. R. Weber. 1971a. N, P, and K percentage in soybean (*Glycine max* (L.)) plant parts. Agron. J. 63:286-290.

Hanway, J. J. and C. R. Weber. 1971b. Accumulation of N, P, and K by soybean (*Glycine max* L.) plants. Agron. J. 63:406-408.

Israel, D. W. 1981. Cultivar and *Rhizobium* strain effects on nitrogen fixation and remobilization by soybeans. Agron. J. 73:509-516.

Jeppson, R. C., R. R. Johnson and H. H. Hadley. 1978. Variation in mobilization of plant nitrogen to the grain in nodulating and nonnodulating soybean genotypes. Crop Sci. 18:1058-1062.

Lindoo, S. J. and L. D. Noodén. 1976. The interaction of fruit development and leaf senescence in 'Anoka' soybeans. Bot. Gaz. 137:218-223.

Lugg, D. G. and T. R. Sinclair. 1981. Seasonal changes in photosynthesis of field-grown soybean leaflets. 2. Relation to nitrogen content. Photosynthetica 15:138-144.

Molisch, H. 1928. Die Lebensdauer der Pflanz. G. Fischer Verlag. Translation of E. H. Fulling. Science Press, Lancaster, PA. 1938.

Morris, D. R. and R. W. Weaver. 1983. Mobilization of ^{15}N from soybean leaves as influenced by Rhizobial strains. Crop Sci. 23:1111-1114.

Nelson, D. R., R. J. Bellville and C. A. Porter. 1984. Role of nitrogen assimilation in seed development of soybean. Plant Physiol. 74:128-133.

Noodén, L. D., G. M. Kahanak and Y. Okatan. 1979. Prevention of monocarpic senescence in soybeans with auxin and cytokinin: an antidote for self-destruction. Science 206:841-843.

Schapaugh, W. T. Jr. 1980. The accumulation and redistribution of dry matter and nitrogen in normal and high protein soybean genotypes. Ph.D. Thesis. Purdue Univ. Library, West Lafayette.

Sesay, A. and R. Shibles. 1980. Mineral depletion and leaf senescence in soya bean as influenced by foliar nutrient application during seed filling. Ann. Bot. 45:47-56.

Sinclair, T. R. and C. T. de Wit. 1975. Comparative analysis of photosynthesis and nitrogen requirements in the production of seeds by various crops. Science 189:565-567.

Sinclair, T. R. and C. T. de Wit. 1976. Analysis of the carbon and nitrogen limitations to soybean yield. Agron. J. 68:319-324.

Stutte, C. A., R. T. Weiland and A. R. Blem. 1979. Gaseous nitrogen loss from soybean foliage. Agron. J. 71:95-97.

Thibodeau, P. S. and E. G. Jaworski. 1975. Patterns of nitrogen utilization in soybeans. Planta 127:133-147.

Wells, R., L. L. Schulze, D. A. Ashley, H. R. Boerma and R. H. Brown. 1982. Cultivar differences in canopy apparent photosynthesis and their relationship to seed yield in soybeans. Crop Sci. 22:886-890.

Williams, R. F. 1955. Redistribution of mineral elements during development. Ann. Rev. Plant Physiol. 6:25-42.

Wittenbach, V. R., R. C. Ackerson, R. T. Giaquinta and R. R. Herbet. 1980. Changes in photosynthesis, ribulose bisphosphate carboxylase, proteolytic activity and ultra-structure of soybean leaves during senescence. Crop Sci. 20:225-231.

Yazdi-Samadi, Bahman, R. W. Rinne and R. D. Seif. 1977. Components of developing soybean seeds: Oil, protein, sugars, starch, organic acids, and amino acids. Agron. J. 69:481-486.

Zeiher, Carolyn, D. B. Egli, J. E. Leggett and D. A. Reicosky. 1982. Cultivar differences in N redistribution in soybeans. Agron. J. 74:375-379.

REGULATION OF SOYBEAN SENESCENCE

L. D. Noodén

SENESCENCE: THE CONCEPT AND THE PROCESS

Senescence is an internally programmed degenerative process leading to death of an organism, organ or cell (Leopold, 1961; Noodén and Leopold, 1978). Many physiological and biochemical changes are known to precede death, but it is not clear which are central (primary) to the senescence process and which are peripheral, a result rather than cause of senescence (Noodén and Thompson, 1984). Even though the biochemistry of the senescence process is quite uncertain, senescence is a useful concept. Moreover, peripheral changes (e.g., decreased N_2 fixation) may, nonetheless, be very important theoretically and economically.

As a whole plant, such as a soybean, degenerates preceding its death, coordinated changes occur in all parts (Noodén, 1980b; 1984). Understanding the timing and correlative controls of this syndrome is prerequisite to unraveling the senescence process (Noodén, 1984). The physiological decline (monocarpic senescence) of the soybean plant during the reproductive phase is coupled with development of the reproductive structures (Leopold et al., 1959; Lindoo and Noodén, 1977; Noodén, 1980a,b). The developing seeds ultimately cause death of the plant, and this influence—the senescence signal—is exerted relatively late in podfill (Lindoo and Noodén, 1977; Noodén et al., 1978). The leaves are the main target of that influence: that is, the plant dies because the leaves are lost (Noodén, 1980a). The nature of this senescence signal is unknown; however, the simplest explanation that passes a variety of tests holds this signal to be a senescence-inducing hormone (Lindoo and Noodén, 1977; Noodén, 1980a,b). The younger pods also act directly or indirectly on the vegetative parts, including the roots, to produce changes that seem to be prerequisite to the later action of the senescence signal (Noodén, 1984); these might be termed early or preparatory changes. In addition to their effects on the vegetative parts, the developing reproductive structures influence each other. Together with all of the changes occurring in the various parts of the soybean plant, this adds up to a very complex series of changes in the maturing soybean plant (Noodén, 1984).

Inasmuch as the physiological decline occurs in all parts of the maturing soybean plant, pod development and monocarpic senescence must consist of parallel processes; these appear to be integrated but not rigidly coupled (Noodén, 1984). This parallelism no doubt accounts for some of the variability in the relative decline of different processes during the whole plant senescence—and probably also the cell senescence—process. Thus, not only can the relative timing of changes differ a bit with conditions and cultivar, but even the sequence of visible events may vary.

Under field conditions, stress is a fact of life and response to stress represents an important aspect of soybean physiology; however, studies aimed at understanding senescence mechanisms need to avoid the complications introduced by stress. This is a particularly difficult task under field conditions.

MEASURES OF SENESCENCE

A practical and immediate upshot of the complications and uncertainties described above is the problem of what parameter(s) to use as a measure of senescence. Ideally, that parameter would also be central to the senescence process. Chlorophyll loss is the most obvious manifestation of soybean leaf senescence and it reflects a loss of photosynthetic capacity. There are, however, genetic lines (mutants?) whose leaves do not turn yellow; green leaves are abscised (Bernard and Weiss, 1973; Kahanak et al., 1978). Loss of nitrogen from the leaves seems to be an important aspect of senescence (Noodén, 1980b), yet pod-induced senescence may occur without this loss if the phloem connection through the petiole is destroyed with heat (Noodén and Murray, 1982). Either these—chlorophyll or nitrogen loss from the leaves—are not essential components in the senescence syndrome or individual components can be uncoupled— perhaps because leaf senescence consists of loosely coupled, parallel processes—from the senescence process without preventing death (Noodén, 1984). Depending on the circumstances, any of several parameters such as chlorophyll loss, leaf abscission, foliar N loss, increased stomatal resistance, decreased photosynthesis, altered membrane properties, or any of the components of the senescence syndrome, may serve as measures of senescence, but some cross checks against second or additional criteria at key points seem warranted. Photosynthetic rate, however, is influenced by sink demand, which may be an important complicating factor in its use as a measure of leaf senescence (Noodén, 1980b). Nondestructive procedures; e.g., visual scoring (Lindoo and Noodén, 1976; Okatan et al., 1981), which allow the very same leaves or pods to be followed through development, offer very substantial advantages where they can be used.

In the particular case of soybean monocarpic senescence, studies on the effects of any treatment to alter senescence *should include some measure of treatment effects on pod development*, for alteration of pod development could secondarily affect leaf senescence. Simple quantitative measures such as the rate of seed dry weight accumulation are helpful though incomplete (Lindoo and Noodén, 1976; Noodén, 1984).

ROLE OF MINERAL REDISTRIBUTION AND A DECLINE IN ROOT FUNCTIONS

A prominent feature of monocarpic senescence in the soybean (Figure 1) and most, but not all, monocarpic species is the redistribution of mobile mineral nutrients, particularly N, from the vegetative structures to the reproductive structures (Noodén, 1980b). This correlation has given rise to the idea that soybean pods must draw their N from the leaves, the main depots of N; however, if monocarpic senescence is prevented through any of several means, genetic, surgical or hormone treatments, thereby maintaining the plant's assimilatory capacity, the pods are supplied by current assimilation rather than withdrawal from the leaves (Abu-Shakhra et al., 1978; Noodén et al., 1978; 1979; Noodén, 1980a,b). Thus, the pods can develop without necessarily drawing on the accumulated nutrients in the vegetative parts. Similarly, foliar senescence may occur in the absence of nutrient withdrawal from the leaves, for blocking of the exodus of N and other mineral nutrients from a leaf by phloem destruction does not prevent senescence of that leaf (Noodén and Murray, 1982).

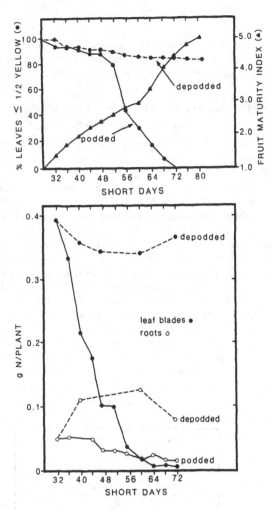

Figure 1. Foliar and root nitrogen in relation to leaf yellowing. Effect of depodding. Redrawn from Derman et al. (1978). Visual scoring of pod development and leaf yellowing as described by Lindoo and Noodén (1976). Pods were removed as they reached full extension (short day 32) which is before the start of pod fill.

During the reproductive phase, the roots of a soybean plant slow their growth and mineral assimilation (see Shibles et al., 1975; Hicks, 1978; Noodén, 1984). This is reflected in the mineral and N content of the xylem sap, which after some fluctuation generally declines during pod development (Streeter, 1972; 1979; McClure and Israel, 1979; Noodén, unpublished data). Net redistribution of the mobile minerals from the leaves to the seeds begins during pod extension (Derman et al., 1978). As shown in Figure 1, the decline in foliar N can be prevented by depodding; however, the N level in the roots (here nonnodulated) is increased by the depodding. This suggests that depodding, which inhibits loss of chlorophyll and associated photosynthetic capacity (Noodén, 1980b; 1984), may prevent, or at least retard, the decline of the root system. The recovery, or maintenance, of the root system of plants depodded at full pod extension is also indicated by an increase in mineral nutrients in the xylem sap (Noodén, unpublished data). Though the exact cause of the decline in the root system is not known (Noodén, 1984), this change seems distinct from the above-mentioned senescence signal, which occurs later, but the decline in root assimilate production could be a major factor in the preparatory phase previously discussed.

SUBSTITUTING DEFINED SOLUTIONS FOR THE ROOT SYSTEM:
THE USE OF EXPLANTS

The root system, especially the young root tips, is the major site of mineral assimilation and the major, if not sole, source of cytokinins for the leaves (Letham, 1978; van Staden and Davey, 1979). Use of explants which are excised stem sections with one node, its leaf and associated pods, offers an opportunity to study the role of a decreased mineral and cytokinin supply, and to even evaluate the contribution of individual minerals, from the roots to leaf senescence and pod development. Withdrawal of the mineral and cytokinin supply, that is culturing the explants on water alone, hastens both foliar senescence and pod development (Neumann et al., 1983). As reported earlier (Neumann et al., 1983) and shown in Figure 2, explants with pods at the same stage senesce at identical rates even though the leaves differ in age; however, an explant with a younger pod (very early podfill) is delayed. Thus, leaf senescence in a reproductive soybean is regulated primarily by the pods and is independent of age. In a whole plant, unlike explants, the pods at other nodes could induce senescence in a leaf at a node with less advanced pods. Although the progressive senescence of lower leaves, characteristic of vegetative soybeans, may halt or at least be retarded during the reproductive phase (Noodén, 1984), the lowest leaves may start monocarpic senescence earlier than the upper leaves (Lindoo and Noodén, 1977).

Mineral nutrients, roughly approximating xylem sap, supplied via the transpiration stream delay both pod development and leaf senescence, as measured by a variety of

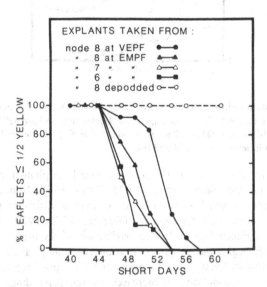

Figure 2. Leaf yellowing and pod development in explants from various nodes in the middle of the pod-bearing region of the soybean plant and depodded explants. Plantlets excised with a single, three-seeded pod at very early podfill (VEPF) or early-midpodfill (EMPF) on short-day 40 and cultured in water. Visual scoring as in Figure 1.

parameters (Mauk et al., 1983; Neumann et al., 1983 and Figure 3). While the minerals do delay pod development (4 days), this is not sufficient to account for their effect on most leaf senescence parameters. Of the individual mineral nutrients, N may have greatest effect (Neumann and Noodén, 1983; Mauk et al., 1982). Since depodded explants or leaf disks cultured on water do not show visible senescence (Neumann et al., 1983 and Figure 2), the pods, and not the decrease in mineral flux from the roots, cause senescence of the foliage. Nonetheless, a decline in the mineral supply through the xylem (see previous discussion) could be important in the preparatory phase discussed above.

Cytokinin (zeatin, zeatin riboside, etc.) delays both pod development and leaf senescence, especially abscission in the explants (Neumann et al., 1983; Garrison et al., 1984 and Figure 3). Perhaps most striking is the positive synergistic effect of the

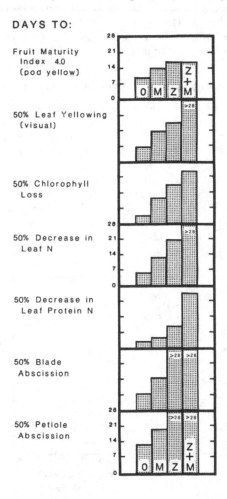

Figure 3. Effects of mineral nutrients and cytokinin on several senescence-related parameters in soybean explants. (From Mauk et al., 1983). Excised at early-mid podfill. Mineral nutrients modified from Neumann et al. (1983). Cytokinin as 4.6 μM zeatin.

896

minerals and cytokinin together (Neumann and Noodén, 1983; Mauk et al., 1983 and Figure 3). This synergism is also reflected in the dose-response curve for zeatin riboside with and without mineral nutrients, the threshold for an effect is about 10^{-9} M in the presence of minerals, but 3×10^{-7} M without the minerals (Noodén, unpublished data). In the intact soybean plant, the foliar cytokinin activity (Lindoo and Noodén, 1978) and cytokinin flux from the roots (Heindl et al., 1982) decline during early pod development, thereby suggesting that a decreased level of cytokinin plays a role in foliar senescence. As with the minerals discussed above, a decline in cytokinin flux seems to be necessary but not in itself sufficient to cause leaf yellowing, for depodding inhibits chlorophyll loss even if the explants are kept on water alone. Thus, the pods, but not a decrease in cytokinin or mineral flux, cause leaf yellowing. Though the lack of mineral and cytokinin flux into the leaves does not seem to be the primary cause of foliar senescence in the explants, cytokinin, at concentrations (10^{-5} M) well above the reported physiological level (Heindl et al., 1982) in combination with the minerals, can override whatever influence (senescence signal) the pod exerts, thereby keeping the foliage green (Noodén, unpublished data).

ON THE ROLE OF CYTOKININ: CONTROL OF MINERAL REDISTRIBUTION

The redistribution of N from the leaves to the seeds is an important aspect of soybean maturation and is influenced by the developing pods (Derman et al., 1978; Noodén, 1984). N redistribution occurs in pod-bearing explants even when the minerals are supplied through the xylem; however, cytokinin can block net redistribution (Figure 4). Bioassays of partially purified extracts indicate that the foliar cytokinin titer decreases at about the time that N is lost from the leaves (Oritani and Yoshida, 1973; Derman et al., 1978; Lindoo and Noodén, 1978). This suggests that a drop in foliar cytokinins may also control the release of N and other minerals in the leaves of intact plants, as in Figure 1, or conversely that an increase in cytokinin could inhibit N redistribution.

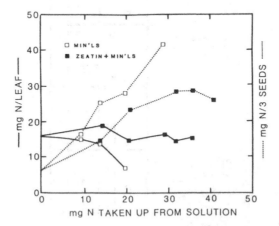

Figure 4. Influence of cytokinin (4.6 μM zeatin) on nitrogen levels in leaf lamina and seeds of explants supplied with mineral solution (4.2 mM NH_4NO_3). (From Mauk and Noodén, 1983.) Conditions similar to those in Figure 3.

TRANSPIRATION AS A FACTOR IN FOLIAR SENESCENCE AND
PARTITIONING OF ASSIMILATES

The xylem is the major, if not the sole, supplier of minerals and cytokinin to the leaves in soybean. Given the influence of these materials on foliar senescence and the role of transpiration in driving movement through the xylem, it seems that transpiration could play some role in controlling movement of minerals and cytokinin to the leaves in general, or even the partitioning between individual leaves. Insofar as stomatal resistance is a factor regulating transpiration and flux of xylem sap, alterations in stomatal resistance (r_s) offer possibilities for controlling the flux and partitioning of minerals and cytokinin, and thereby influencing foliar senescence. It is, however, well known that several treatments such as depodding and petiole cincturing, which increase r_s (Kriedemann et al., 1976; Koller and Thorne, 1978; Setter et al., 1980; Noodén, 1980b), do not cause foliar senescence and may in fact inhibit senescence (Leopold et al., 1959; Lindoo and Noodén, 1977; Noodén and Murray, 1982; Noodén, 1984). During the normal course of senescence in unstressed plants, a large, sustained rise in r_s occurs relatively late, about the time of visible yellowing (Neumann et al., 1983; Garrison et al., 1984). Thus, the large, sustained rise in r_s seems not to cause foliar senescence, at least not the earlier stages. Looking at the opposite side, since cytokinin or minerals can suppress r_s (Neumann et al., 1983; Garrison et al., 1984), a decreased flux of cytokinin or minerals into the leaves, as outlined above, could influence the rise in r_s; however, the cytokinin and minerals appear to decline long before r_s increases.

Even if r_s does not regulate whole-plant senescence, it could control partitioning of minerals and cytokinin by favoring the flux of xylem sap to leaves with low r_s and away from leaves with high r_s. Phloem cincturing of a single leaf on a depodded plant increases the r_s of that leaf, but does not cause it to senesce (Figure 5 and Noodén and

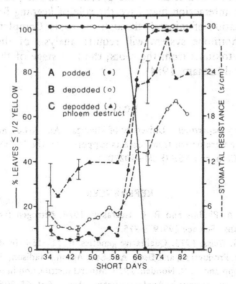

Figure 5. Leaf yellowing and stomatal resistance (r_s) for comparable, central leaves of plants treated as indicated at early-mid podfill. Procedures described by Neumann et al. (1983) and Noodén and Murray (1982).

Murray, 1982). A similar situation exists in "Y"-shaped plants (Lindoo and Noodén, 1977), which are depodded on one branch of the Y; the pod-bearing branch yellows normally, while the depodded branch stays green.

While the role of the stomata and r_s in long-term regulation of the distribution of xylem sap and also photosynthesis is unclear, if not doubtful at least in the context of soybean monocarpic senescence, perhaps they provide some rapid, short-term regulation of these processes in addition to their obvious prevention of excessive water loss. One wonders what regulates xylem flux and partitioning of xylem sap if not r_s. Probably, the answer lies mainly in the hydraulic architecture of the plant (Zimmermann, 1983), and one component which deserves special consideration is possible changes in the conductivity of the petiole, especially in the abscission zone. Despite the existence of a vascular framework, movement between sectors (orthostichies) can occur when necessary (Murray et al., 1982), and new connections may form as needed.

CONCLUSIONS AND OUTLOOK

It appears that the decline in leaf levels and root production of cytokinins may be an important factor in monocarpic senescence of soybean; however, this appears not to be a primary cause, rather an early or preparatory change. A similar statement can be made about the decline in mineral nutrient assimilation by the roots. The pods seem to exert some influence (senescence signal), which acts on leaves with a decreased level of cytokinin and mineral influx, but this requires further investigation. In particular, it seems important to determine whether or not depodding, which inhibits monocarpic senescence, also causes an increase in foliar cytokinin and root production of cytokinin. The striking synergism of cytokinin and minerals could be explained by decreased metabolism of the cytokinin in the presence of minerals. In the intact plant, this interaction may have the role of keeping foliar function in balance with the assimilation of minerals through the roots. Complete elucidation of this cytokinin-mediated hormone system will require analysis of the translocation and metabolism of the cytokinins from the roots; the first stage of that study is reported elsewhere (Noodén and Letham, 1984).

NOTES

L. D. Noodén, Botany Department, University of Michigan, Ann Arbor, MI 48109-1048.

The data shown here were drawn from projects supported by the USDA (CSRS grant No. 416-15-79), Allied Chemical Corp. and NSF (PCM-8302707).

REFERENCES

Abu-Shakhra, S. S., D. A. Phillips and R. C. Huffaker. 1978. Nitrogen fixation and delayed leaf senescence in soybeans. Science 199:973-975.

Bernard, R. L. and M. G. Weiss. 1973. Qualitative genetics. p. 117-154. *In* B. E. Caldwell (ed.) Soybeans: Improvement, Production, and Uses. Am. Soc. Agron., Madison, Wisconsin.

Derman, D. B., D. C. Rupp and L. D. Noodén. 1978. Mineral distribution in relation to fruit development and monocarpic senescence in Anoka soybeans. Am. J. Bot. 65:205-213.

Garrison, F. R., A. M. Brinker and L. D. Noodén. 1984. Relative activities xylem-supplied cytokinins in retarding leaf senescence and sustaining pod development. Plant Cell Physiol. 23:213-224.

Heindl, J. C., D. R. Carlson, W. A. Brun and M. A. Brenner. 1982. Ontogenetic variation of four cytokinins in soybean root pressure exudate. Plant Physiol. 70:1619-1625.

Hicks, D. R. 1978. Growth and development. p. 27-44. *In* A. G. Norman (ed.) Soybean Physiology Agronomy and Utilization. Academic Press, New York.

Kahanak, G. M., Y. Okatan, D. C. Rupp and L. D. Noodén. 1978. Hormonal and genetic alteration of monocarpic senescence in soybeans. Plant Physiol. Suppl. 61:26S.

Koller, H. R. and J. H. Thorne. 1978. Soybean pod removal alters leaf diffusion resistance and leaflet orientation. Crop Sci. 18:305-307.

Kriedemann, P. E., B. R. Loveys, J. V. Possingham and M. Satch. 1976. Sink effects on stomatal physiology and photosynthesis. p. 404-414. *In* I. F. Wardlaw and J. B. Passioura (eds.), Transport and Transfer Processes in Plants. Academic Press, New York.

Leopold, A. C. 1961. Senescence in plant development. Science 134:1727-1732.

Leopold, A. C., E. Niedergang-Kamien and J. Janick. 1959. Experimental modification of plant senescence. Plant Physiol. 34:570-573.

Letham, D. S. 1978. Cytokinins. p. 205-263. *In* D. S. Letham, P. B. Goodwin and T. J. V. Higgins (eds.) Phytohormones and Related Compounds—A Comprehensive Treatise Vol. I. Elsevier/North Holland Biomedical Press, Amsterdam.

Lindoo, S. J. and L. D. Noodén. 1976. The interrelation of fruit development and leaf senescence in 'Anoka' soybeans. Bot. Gaz. 137:218-223.

Lindoo, S. J. and L. D. Noodén. 1977. Studies on the behavior of the senescence signal in Anoka soybeans. Plant Physiol. 59:1136-1140.

Lindoo, S. J. and L. D. Noodén. 1978. Correlation of cytokinins and abscisic acid with monocarpic senescence in soybean. Plant Cell Physiol. 19:997-1006.

Mauk, C. S., A. Brinker and L. D. Noodén. 1983. Synergistic action of cytokinin and mineral nutrients on foliar senescence and pod development in soybean explants. Hort Sci. 18:597.

Mauk, C. S., P. M. Neumann and L. D. Noodén. 1982. Response of leaf senescence and seed yield in soybean explants to mineral nutrient combinations and single deletions. Plant Physiol. Suppl. 69:8.

Mauk, C. S. and L. D. Noodén. 1983. Cytokinin control of mineral nutrient distribution between the foliage and seeds in soybean explants. Plant Physiol. Suppl. 72:167.

McClure, P. R. and D. W. Israel. 1979. Transport of nitrogen in the xylem of soybean plants. Plant Physiol. 64:411-416.

Murray, B. J., C. Mauk and L. D. Noodén. 1982. Restricted vascular pipelines (and orthostichies) in plants. What's New in Plant Physiol. 13:33-36.

Neumann, P. M. and L. D. Noodén. 1983. Interaction of mineral and cytokinin supply in control of leaf senescence and seed growth in soybean explants. J. Plant Nut. 6:735-742.

Neumann, P. M., A. T. Tucker and L. D. Noodén. 1983. Characterization of leaf senescence and pod development in soybean explants. Plant Physiol. 72:182-185.

Noodén, L. D. 1980a. Regulation of senescence. p. 139-152. *In* F. T. Corbin (ed.) World Soybean Research Conference II: Proceedings. Westview Press, Boulder, CO.

Noodén, L. D. 1980b. Senescence in the whole plant. p. 219-258. *In* K. V. Thimann (ed.) Senescence in Plants. CRC Press, Boca Raton, FL.

Noodén, L. D. 1984. Integration of soybean pod development and monocarpic senescence. A minireview. Physiol. Plant. 62:273-284.

Noodén, L. D., G. M. Kahanak and Y. Okatan. 1979. Prevention of monocarpic senescence in soybeans with auxin and cytokinin: An antidote for self-destruction. Science 206:841-843.

Noodén, L. D. and A. C. Leopold. 1978. Phytohormones and the endogenous regulation of senescence and abscission. p. 329-369. *In* D. S. Letham, P. B. Goodwin and T. J. V. Higgins (eds.) Phytohormones and Related Compounds—A Comprehensive Treatise. Vol. II. Elsevier/North Holland Biomedical Press, Amsterdam.

Noodén, L. D. and D. S. Letham. 1984. Translocation of zeatin riboside and zeatin in soybean explants. J. Plant Growth Regul. 2:265-279.

Noodén, L. D. and B. J. Murray. 1982. Transmission of the monocarpic senescence signal via the xylem in soybean. Plant Physiol. 69:754-756.

Noodén, L. D., D. C. Rupp and B. D. Derman. 1978. Separation of seed development from monocarpic senescence in soybeans. Nature 271:354-357.

Noodén, L. D. and J. W. Thompson. 1984. Aging and senescence in plants. Chap. 5. *In* D. E. Finch and E. L. Schneider (eds.) Handbook of the biology of aging. 2nd Edition. Van Nostrand Reinhold Publishers, New York.

Okatan, Y., G. M. Kahanak and L. D. Noodén. 1981. Characterization and kinetics of soybean maturation and monocarpic senescence. Physiol. Plant. 52:330-338.

Oritani, T. and R. Yoshida. 1973. Studies on nitrogen metabolism in crop plants. XII. Cytokinins and abscisic acid-like substance levels in rice and soybean leaves during their growth and senescence in the crop plants. Crop Sci. Soc. Japan Proc. 42:280-287.

Setter, I. L., W. A. Brun and M. L. Brenner. 1980. Stomatal closure and photosynthetic inhibition in soybean leaves induced by petiole girdling and pod removal. Plant Physiol. 65:884-887.

Shibles, R., I. C. Anderson and A. H. Gibson. 1975. Soybean. p. 151-189. *In* L. T. Evans (ed.) Crop Physiology. Cambridge University Press, London.

van Staden, J. and J. E. Davey. 1979. The synthesis, transport and metabolism of endogenous cytokinins. Plant Cell Environ. 2:93-106.

Streeter, J. G. 1972. Nitrogen nutrition of field-grown soybean plants: I. Seasonal variations in soil nitrogen and nitrogen composition of stem exudate. Agron. J. 64:311-314.

Streeter, J. G. 1979. Allantoin and allantoic acid in tissues and stem exudate from field-grown soybean plants. Plant Physiol. 63:478-480.

Zimmerman, M. H. 1983. Xylem Structure and the Ascent of Sap. Springer-Verlag, Berlin.

RHIZOBIUM

EFFECT OF SOIL ENVIRONMENTAL FACTORS ON RHIZOBIA

R. J. Roughley

Rhizobia are a very minor component of the soil microflora, and while not restricted to rhizosphere soil, reach their maximum numbers in association with plant roots. They are stimulated by carbon compounds either leaked by or sloughed off plant roots and are controlled by microbial competition, antagonism, lysis and predation as well as by their physical and chemical environment. Their ability to infect legume roots and multiply within the resulting root nodule, protected from the soil environment, provides a special advantage over their competitors.

Knowledge of their behavior in soil in response to changing environmental conditions is surprisingly limited. This may be attributed to their general nondescript behavior, which makes their recognition and enumeration difficult. Attempts by Pattison and Skinner (1974) to develop a selective agar medium using six antibiotics illustrated difficulties inherent in this approach, because rhizobia were often inhibited at levels of the antibiotics necessary to suppress the soil microflora. Counting indigenous populations of rhizobia is still restricted to the MPN method based on the presence or absence of nodules on a trap host plant at particular dilutions of soil. This technique is limited and may not be appropriate for populations $<10^2$/g soil due to interference from the soil at low dilutions (Thompson and Vincent, 1967). More recently two methods have been developed to allow direct counting of particular strains when introduced into soil. The first utilizes fluorescent antibody to detect the distribution of a strain in the rhizosphere or to count rhizobia following extraction of cells from soil if populations are greater than 10^3 cells/g (Schmidt, 1974).

Numbers of introduced rhizobia may be counted, provided they carry an antibiotic marker, using agar containing antibiotics. Such a method was developed by Bushby (1981a) in which cells, naturally-resistant to streptomycin, were selected from the parent strain and counted on agar containing streptomycin, rifampacin, pimufucin and cyclohexamide. The method has been used successfully by Bushby (1981b) to describe changes in numbers of rhizobia during the growing season of *Vigna mungo* and by Brockwell et al. (1984) for soybeans.

Identification of particular strains is possible using one or a combination of serological techniques, phage typing, antibiotic marking and recently polyacrylamide gel electrophoresis (PAGE) (Noel and Brill, 1980). Using a combination of the counting and identification methods, several recent reports have followed population dynamics and the fate of introduced soybean rhizobia.

Mahler and Wollum II (1982) counted soybean rhizobia in soils of North Carolina and found they were fewest in early spring before soybeans were sown. Thereafter numbers increased and remained high until early winter. As expected, numbers were highest in soils where soybeans had been planted, reaching populations up to 10^5/g soil. It is important to bear in mind that this population consists of a wide diversity of strains providing a range of characteristics. Thus, the balance between them will change as conditions change in the soil. Noel and Brill (1980) identified 19 and 18 indigenous gel types of *R. japonicum* at two sites in Wisconsin using PAGE. This diversity of strains already occupying an ecological niche in the soil is part of the barrier confronting the establishment of inoculum strains.

The population reached and maintained by rhizobia in soil is a resultant of the physical, chemical and biological components of the environment. How these affect cowpea and soybean rhizobia is described below.

PHYSICAL COMPONENTS OF THE ENVIRONMENT

Soil Temperature

Survival of *Rhizobium* in soil is adversely affected by high soil temperatures. The effect is modified by the soil type; survival is better in heavy or organic soils than in light soils. Survival in light soils may be improved by amendments including montmorillonite and illite, which probably reduce the amount of water loss from the bacterial cell (Marshall, 1964). Rhizobia are generally less tolerant of high temperatures in moist than in dry soils, an exception was the improved survival of cowpea rhizobia at 35 C in moist compared with saturated or dry soil (Boonkerd and Weaver, 1982); however, the temperature stress was not severe and the result may express their response to moisture. As would be expected, there are differences between strains in tolerance to high temperatures even when selected from soils of the same region subjected to similar soil temperatures.

Although populations of rhizobia may be quite small in the surface layers of the soil, numbers may reach high levels down the profile depending on the ground cover and soil type. Day et al. (1978) counted the number of cowpea rhizobia in the profile of soils at Samaru, northern Nigeria, where bare soil surface temperatures can exceed 60 C. In the upper 5 cm there were 5 to 50/g soil, but they increased with depth reaching 18,000/g soil at 20 to 25 cm. There was considerable variation in the survival of six cowpea-type strains on germinating cowpea (*Vigna unguiculata* [L.] Walp.) seeds exposed to temperatures of up to 45 C, illustrating the possibility of selecting for improved tolerance.

The proportion of nodules formed by strains of *R. japonicum* from different serogroups was shown to be temperature dependent (Weber and Miller, 1972). Such differences were examined in broth culture using 42 strains of *R. japonicum* from different environments (Munévar and Wollum II, 1981). Strains differed in response to temperatures above 27.4 C; rises of only 1.5 C had noticeable effects. The response of strains to temperature in these studies was related to their symbiotic response.

There is little information on the effect of low root temperatures on rhizobia in soil. Strain differences between clover rhizobia have been clearly demonstrated in their ability to infect, nodulate and fix nitrogen. Ek-Jander and Fåhraeus (1971) found isolates of clover rhizobia were adapted to conditions in sub-arctic Scandinavia. Roughley et al. (1970) reported *R. trifolii* strain TA1 grew well in the rhizosphere of subterranean clover (*Trifolium subterraneum*) at 7 C yet failed to infect the root hairs at populations equivalent to those which could infect at 19 C, suggesting rhizobia may not be the limiting factor.

Soil Moisture

Soil water affects the number of rhizobia in soil, their distribution down the profile and the susceptibility of plant root hairs to infection. Although the importance of soil moisture on rhizobia has been long recognized, there are few reports describing their response. Fewer still have expressed this response in terms of water potential rather than percent moisture to allow more general application of the studies.

Mahler and Wollum II (1980) studied the population dynamics of 10 strains of *R. japonicum* in a loamy sand adjusted to water potentials between -1.5 and -0.01 MPa. Numbers of all strains were reduced in proportion to the moisture content. In general, isolates from serogroups 110, 122 and 138 survived better than other strains at -1.5 MPa. Those from serogroups 24, 31, 94 and 123 were particularly sensitive; their numbers declined more than 4 log numbers after 9 weeks. A potential of -0.5 MPa was the critical value; at more negative potentials, numbers of *R. japonicum* were reduced. They later investigated the effects of water potential and soil texture on the survival of *R. japonicum* and *R. leguminosarum* (Mahler and Wollum II, 1981). Numbers were always highest at 0.03 MPa and again fell proportionately to -1.5 MPa. They were also highest in the sandy loams, silt loams and sandy clay loam soils, and lowest in sands and clay loam soils.

The importance of the soil characteristics on the survival of rhizobia undergoing drying was confirmed by Wei-Liang and Alexander (1982). Survival of rhizobia was improved when they were established in soil before drying began. Mahler and Wollum II (1981) found that numbers in soil fell most rapidly in the first week after inoculation. The stress posed by low moisture could therefore discriminate between the naturalized population and newly introduced rhizobia. Strains of cowpea rhizobia forming most extra-cellular polysaccharide were most susceptible to desiccation (Osa-Afiana and Alexander, 1982).

Al-Rashidi et al. (1982) distinguished between the effect of matric and osmotic moisture potentials on four strains of *R. japonicum* in three soils. Two strains, CC709 and USDA 110, were tolerant of both matric and osmotic stress. The sensitive strains, CB1809 and 123, absorbed more water at all water activities, a factor which Bushby and Marshall (1977) found was related to susceptibility to desiccation.

Despite the adverse effect of dry soil on rhizobia, results of Worrall and Roughley (1976) indicated that infection of subterranean clover was restricted at soil water potentials that did not affect the number of *R. trifolii* in the rhizosphere. In this case the host's root hairs were distorted preventing root hair curling and infection.

The effect of waterlogging on rhizobia in soil has been neglected since Vandecaveye (1927) reported that two weeks flooding drastically reduced numbers of *R. leguminosarum*.

An understanding of the movement of microorganisms in soil is of particular significance to studies of *Rhizobium*, as their spatial distribution is a major factor determining the onset and pattern of nodulation on legume roots. Active movement may be neglected when water flow is considerable. Griffin and Quail (1968) suggest that, for active movement, bacteria need a continuous path of water-filled pores of sufficient size to allow flagellar movement. In experiments measuring the lateral movement of *R. trifolii*, Hamdi (1971) found that discontinuous pathways were more important in restricting lateral movement than was pore size. Lateral movement of *R. trifolii* ceased at -1.1 MPa in coarse sand, at -2.2 MPa in fine sand and at -0.04 MPa in a silt loam, although pore size was 60 nm, 11.5 nm and 6.8 nm, respectively.

R. trifolii when applied to seedling roots did not move through the profile of dry sandy soil unless flushed by water (Worrall and Roughley, 1976); a growing root had no effect. In saturated soil, or when unsaturated soil was watered and allowed to drain, rhizobia moved rapidly down the profile. Similar findings for *R. japonicum* were reported by Madsen and Alexander (1982). Therefore, when rhizobia are introduced into soil by inoculation they are likely to remain at the depth of seeding and be subject to desiccation and wide fluctuations in temperature, unless distributed by rain or irrigation.

CHEMICAL COMPONENTS OF THE ENVIRONMENT

Acid Soil Factors

Acid soils have long been implicated in nodulation problems, especially with temperate legumes. Slow-growing rhizobia from tropical legumes are usually more tolerant of acidity, but not invariably so (Bushby, 1982). Variation in tolerance exists within *Rhizobium* including *R. japonicum* (Ham et al., 1971), and many workers have now attempted to exploit this by screening strains for tolerance by use of laboratory media. Two problems immediately arise. Tolerance to soil acidity may require tolerance to hydrogen, aluminum and manganese ions in the presence of low concentrations of phosphorus and calcium. Testing for tolerance of the above factors requires the pH of laboratory media to be controlled. Many reports indicate these problems have either not been recognized or overcome, but attention was drawn to them by Keyser and Munns (1979a).

These authors, together with Cassman and Beck in a series of studies (Keyser and Munns, 1979a,b; Munns and Keyser, 1981; Cassman et al., 1981a,b) addressed both these problems using cowpea and soybean rhizobia. They used calcium, 50 μM to represent its low range in soil solution, manganese, 200 μM found to be toxic to soybean and cowpeas, aluminum, 50 μM a concentration likely to occur in acid soils, and 5 to 10 μM phosphorus. At these levels they considered calcium or manganese to impose little stress to either group of rhizobia grown in broth. None of the strains tolerant to 50 μM aluminum were sensitive to the high manganese or low calcium.

Tolerance of low pH (4 to 5) was not necessarily linked to tolerance of high concentrations of aluminum. Cowpea rhizobia tended to be more tolerant of aluminum than soybean rhizobia, but within these groups there was strain-to-strain variation in tolerance. However, even for tolerant strains, aluminum reduced growth rate in broth and proved a more severe stress than low phosphorus and low pH. Dividing cells were more sensitive to high aluminum. Tolerance was a stable character and was not enhanced by selecting for tolerant mutants or by prolonged culturing with aluminum stress.

R. japonicum, when grown on phosphate-rich (1 mM) laboratory medium, stored polyphosphate as granules. These would provide sufficient phosphorus for only 1 to 2 generations. Cells of *R. japonicum* also contained up to 1.5% phosphorus when grown in solutions containing 6 μM phosphorus, equivalent to that in a fertile soil. This would support up to 2.5 generations. They are, therefore, unlikely to be able to store sufficient phosphorus to colonize plant rhizospheres where levels of phosphorus may be 0.1 μM or lower. At this concentration some strains were stressed. Further strains of *R. japonicum* were screened at concentrations between 0.05 μM and 2 mM. Strain USDA 110 was unaffected within this range, whereas USDA 142 was particularly inefficient at taking up phosphorus and may, therefore, have difficulty colonizing soybean rhizospheres in low phosphorus soils.

Our knowledge of the mineral nutrition of rhizobia is very restricted, and is based almost entirely on studies on laboratory media or in broth. These findings need both expanding and testing in glasshouse and field experiments.

Combined Nitrogen

The effects of combined nitrogen on the legume symbiotic system has been described, almost exclusively, in relation to nodulation and nitrogen fixation. However, Nutman et al. (1978) showed that heavy dressings of nitrogenous fertilizer to pastures could reduce numbers of R. trifolii despite application of lime to counter excess acidity. Herridge et al. (1984) found that increasing the number of R. japonicum strain CB1809 added to high NO_3 (30 µg/g, top 30 cm) rhizobia-free soil increased the rhizosphere population in the 42 day period following inoculation and resulted in earlier nodulation. In the following year the subsequent established soil population colonized the new seedling rhizospheres better than freshly-introduced rhizobia of the same strain (Brockwell et al., 1984), suggesting the possible selection of adapted cells.

McNeil (1982) demonstrated variation in the ability of 16 strains of R. japonicum to nodulate soybeans under nitrate stress. He considered that random testing of strains was an inefficient selection method compared with applying a selection pressure to populations of R. japonicum.

Salinity

Sodium chloride adversely affects growth and survival of a number of Rhizobium species (Steinborn and Roughley, 1974). Eleven isolates of rhizobia all were adversely affected when the conductivity of the medium was raised from 1.2 to 6.7 mS/cm; 4 were very resistant (Singleton et al., 1982); R. japonicum USDA 110 was very sensitive (Singleton and Bohlool, 1984). Colonization of roots by strain 110 was not affected between 0.0 and 79.9 nM NaCl, but within this range nodulation was reduced. The progeny of the cells that nodulated soybeans at 79.9 nM NaCl were not better adapted to salt stress than the parent population. They attribute reduced nodulation to salt sensitivity of root infection sites.

THE MICROBIOLOGICAL COMPONENTS OF THE ENVIRONMENT

Microbial Interactions

There are numerous reports of microbial antagonism to rhizobia in pure culture; e.g., with R. japonicum (Pugashetti et al., 1982). The demonstration of similar effects in the field is more difficult. Chatel and Parker (1972) demonstrated that water extracts of soils in which clovers nodulated poorly was inhibitory to R. trifolii but not R. lupini when both were seeded on agar. The origin of the initial toxin remains unknown. Rhizobia persist better in sterile than in non-sterile soil, suggesting that biological agents are involved. Antagonism from fungi and actinomycetes, lysis by bacteriocins and bacteriophage, predation by protozoa and parasitism by Bdellovibrio have all been implicated. Much of the evidence is based on the responses of strains of rhizobia to soil isolates of fungi, actinomycetes, bacteria and protozoa in laboratory media. There is little evidence that any antagonism demonstrated will occur in soil where the toxic principle must be produced, must not be absorbed on the organic or colloidal fraction and must come in contact with the rhizobia. It is unlikely that numbers of rhizobia would be reduced to levels likely to cause nodulation failures.

Dobereiner et al. (1981) solved a nodulation problem in virgin Cerrado soils, Brazil by using *R. japonicum* resistant to 80-160 µg/ml streptomycin. These resistant strains were used after finding that of 218 isolates of *R. japonicum*, the progeny of 3-4 years inoculation with sensitive rhizobia, 86% were resistant to at least 80 µg/ml streptomycin. This suggests the introduced strains were exposed to the antibiotic in the rhizosphere where actinomycetes and the percentage of streptomycin resistant bacteria increased markedly after the application of lime.

The effect of protozoa on populations of rhizobia in soil has been described by Danso et al. (1975). They found that the population of rhizobia varied inversely with that of the protozoa. Rhizobia were never eliminated but were reduced in number until the energy used by the protozoa in hunting for survivors equalled that obtained from feeding. In their experiments numbers of rhizobia were not reduced below 10^6/g soil.

Rhizobiophages are common in the rhizosphere of legumes. Their practical significance has not been demonstrated beyond doubt in the field but Barnet (1980) has shown that when added to two strains differing in phase sensitivity, numbers of the sensitive strain in the rhizosphere declined. Perhaps, a more significant effect may result from genetic changes in the population resulting from phage attack.

Production of bacteriocins, substances produced by bacteria that are bactericidal against specific strains, is often under plasmid gene control. Schwinghamer et al. (1973) suggest they may be of ecological significance in regulating numbers of specific strains. This suggestion has been confirmed in experiments in non-sterile soil containing bacteriocin sensitive rhizobia. When an effective resistant strain was added together with a bacteriocin-producing strain, the proportion of nodules formed by the resistant strain was increased (Hodgson et al., 1984).

NOTES

R. J. Roughley, Horticultural Research Station Gosford, N.S.W. Department of Agriculture, P.O. Box 720, Gosford, N.S.W. 2250, Australia.

REFERENCES

Al-Rashidi, R. K., T. E. Loynachan and L. R. Frederick. 1982. Desiccation tolerance of four strains of *Rhizobium japonicum*. Soil Biol. Biochem. 14:489-493.

Barnet, Y. M. 1980. The effect of rhizobiophages on populations of *Rhizobium trifolii* in the root zone of clover plants. Can. J. Microbiol. 26:572-576.

Boonkerd, N. and R. W. Weaver. 1982. Survival of cowpea rhizobia in soil as affected by soil temperature and moisture. Appl. Environ. Microbiol. 43:585-589.

Brockwell, J., R. J. Roughley and D. F. Herridge.1984. Impact of rhizobia established in a high NO_3^- soil on a soybean inoculant of the same strain. p. 328. *In* C. Veeger and W. E. Newton (eds.) Advances in nitrogen fixation research. Martinus Nijhoff/Dr. W. Junk Publishers, The Hague.

Bushby, H. V. A. 1981a. Quantitative estimation of rhizobia in non-sterile soil using antibiotics and fungicides. Soil Biol. Biochem. 13:237-239.

Bushby, H. V. A. 1981b. Changes in the numbers of antibiotic-resistant rhizobia in the soil and rhizosphere of field grown *Vigna mungo* cv. Regur. Soil Biol. Biochem. 13:241-245.

Bushby, H. V. A. 1982. Ecology. p. 36-70. *In* W. J. Broughton (ed.) Nitrogen fixation. Vol. II. *Rhizobium*. Clarendon Press, Oxford.

Bushby, H. V. A. and K. C. Marshall. 1977. Water status of rhizobia in relation to their susceptibility to desiccation and to their protection by montmorillonite. J. Gen. Microbiol. 99:19-27.

Cassman, K. G., D. N. Munns and D. P. Beck. 1981a. Phosphorus nutrition of *Rhizobium japonicum*: Strain differences in phosphate storage and utilization. Soil Sci. Soc. Am. J. 45:517-520.

Cassman, K. G., D. N. Munns and D. P. Beck. 1981b. Growth of *Rhizobium* strains at low concentrations of phosphate. Soil Sci. Soc. Am. J. 45:520-523.

Chatel, D. L. and C. A. Parker. 1972. Inhibition of rhizobia by toxic soil-water extracts. Soil Biol. Biochem. 4:289-294.

Danzo, S. K. A., S. O. Keya and M. Alexander. 1975. Protozoa and the decline of *Rhizobium* populations added to soil. Can. J. Microbiol. 21:884-895.

Day, J. M., R. J. Roughley, A. R. J. Eaglesham, M. Dye and S. P. White. 1978. Effect of high soil temperatures on nodulation of cowpea, *Vigna unguiculata*. Ann. Appl. Biol. 88:476-481.

Dobereiner, J., M. Scotti, N. Sa and M. Vargas. 1981. Resistance to streptomycin of *Rhizobium* isolates from Cerrado and Amazon soils. p. 434. *In* A. H. Gibson and W. E. Newton (eds.) Current perspectives in nitrogen fixation. Aust. Acad. of Sci., Canberra.

Ek-Jander, J., and G. Fåhraeus. 1971. Adaption of *Rhizobium* to sub-arctic environment in Scandinavia. Plant Soil Special Volume: 129-137.

Griffen, D. M. and G. Quail. 1968. Movement of bacteria in moist particulate systems. Aust. J. Biol. Sci. 21:579-582.

Ham, G. E., L. R. Frederick and I. C. Anderson. 1971. Serogroups of *Rhizobium japonicum* in soybean nodules sampled in Iowa. Agron. J. 63:69-72.

Hamdi, Y. A. 1971. Soil-water tension and the movement of rhizobia. Soil Biol. Biochem. 3:121-126.

Herridge, D. F., R. J. Roughley and J. Brockwell. 1984. Effect of rhizobia and soil nitrate on the establishment and functioning of the soybean symbiosis in the field. Aust. J. Agric. Res. 35:149-161.

Hodgson, A., W. P. Roberts and J. S. Waid. 1984. Use of bacteriocin-production to increase the proportion of nodules formed by desired strains of *Rhizobium trifolii*. p. 43. *In* Proc. 7th Aust. Legume Nodulation Conf. Australian Institute of Agricultural Science, Sydney.

Keyser, H. H. and D. N. Munns. 1979a. Effects of calcium, manganese, and aluminum on growth of rhizobia in acid media. Soil Sci. Soc. Am. J. 43:500-503.

Keyser, H. H. and D. N. Munns. 1979b. Tolerance of rhizobia to acidity, aluminum, and phosphate. Soil Sci. Soc. Am. J. 43:519-523.

McNeil, D. L. 1982. Variations in ability of *Rhizobium japonicum* strains to nodulate soybeans and maintain fixation in the presence of nitrate. Appl. Environ. Microbiol. 44:647-652.

Madsen, E. L. and M. Alexander. 1982. Transport of *Rhizobium* and *Pseudomonas* through soil. Soil Sci. Soc. Am. J. 46:557-560.

Mahler, R. L. and A. G. Wollum II. 1980. Influence of water potential on the survival of rhizobia in a Goldsboro loamy sand. Soil Sci. Soc. Am. J. 44:988-992.

Mahler, R. L. and A. G. Wollum II. 1981. The influence of soil water potential and soil texture on the survival of *Rhizobium japonicum* and *Rhizobium leguminosarum* isolates in the soil. Soil Sci. Soc. Am. J. 45:761-766.

Mahler, R. L. and A. G. Wollum II. 1982. Seasonal fluctuations of *Rhizobium japonicum* under a variety of field conditions in North Carolina. Soil Sci. 134:317-324.

Marshall, K. C. 1964. Survival of root-nodule bacteria in dry soils exposed to high temperatures. Aust. J. Agric. Res. 15:273-281.

Munévar, F. and A. G. Wollum II. 1981. Growth of *Rhizobium japonicum* strains at temperatures above 27°C. Appl. Environ. Microbiol. 42:272-276.

Munns, D. N. and H. H. Keyser. 1981. Response of *Rhizobium* strains to acid and aluminum stress. Soil Biol. Biochem. 13:115-118.

Noel, K. D. and W. J. Brill. 1980. Diversity and dynamics of indigenous *Rhizobium japonicum* populations. Appl. Environ. Microbiol. 40:931-938.

Nutman, P. S., M. Dye and P. E. Davis. 1978. The ecology of *Rhizobium*. p. 404-410. *In* M. W. Louitt and J. A. R. Miles (eds.) Microbial ecology. Springer Verlag, Berlin.

Osa-Afiana, L. O. and M. Alexander. 1982. Differences among cowpea rhizobia in tolerance to high temperature and desiccation in soil. Appl. Environ. Microbiol. 43:435-439.

Pattison, A. C. and F. A. Skinner. 1974. The effects of antimicrobial substances on *Rhizobium* spp. and their use in selective media. J. Appl. Bacteriol. 37:239-250.

Pugashetti, B. K., J. S. Angle and G. H. Wagner. 1982. Soil microorganisms antagonistic towards *Rhizobium japonicum*. Soil Biol. Biochem. 14:45-49.

Roughley, R. J., P. J. Dart, P. S. Nutman and C. Roderiguez-Barrueco. 1970. The influence of root temperature on root-hair infection of *Trifolium subterraneum* L. by *Rhizobium trifolii* Dang. p. 451-454. *In* M. J. T. Norman (ed.) Proc. XI int. grassland cong. University of Queensland Press, Brisbane.

Schmidt, E. L. 1974. Quantitative autecological study of micro-organisms in soil by immunofluorescence. Soil Sci. 118:141-149.

Schwinghamer, E. A., C. E. Pankhurst and P. R. Whitfild. 1973. A phase-like bacteriocin of *Rhizobium trifolii*. Can. J. Microbiol. 19:359-368.

Singleton, P. W., S. A. El-Swaify and B. B. Bohlool. 1982. Effect of salinity on *Rhizobium* growth and survival. Appl. Environ. Microbiol. 44:884-890.

Singleton, P. W. and B. B. Bohlool. 1984. Effect of salinity on nodule formation by soybean. Plant Physiol. 74:72-76.

Steinborn, Julia and R. J. Roughley. 1974. Sodium chloride as a cause of low numbers of *Rhizobium* in legume inoculants. J. Appl. Bacteriol. 37:93-99.

Thompson, J. A. and J. M. Vincent. 1967. Methods of detection and estimation of rhizobia in soil. Plant Soil 26:72-84.

Vandecaveye, S. C. 1927. Effect of moisture, temperature and other climatic conditions on *R. leguminosarum* in the soil. Soil Sci. 23:355-362.

Weber, D. F. and V. L. Miller. 1972. Effect of soil temperature on *Rhizobium japonicum* serogroup distribution in soybean nodules. Agron. J. 64:796-798.

Wei-Liang, C. and M. Alexander. 1982. Influence of soil characteristics on the survival of *Rhizobium* in soils undergoing drying. Soil Sci. Soc. Am. J. 46:949-952.

Worrall, V. S. and R. J. Roughley. 1976. The effect of moisture stress on infection of *Trifolium subterraneum* L. by *Rhizobium trifolii* Dang. J. Exp. Bot. 27:1233-1241.

RHIZOBIUM STRAIN COMPETITION FOR HOST NODULATION

M. J. Trinick

For a highly effective *Rhizobium* strain to be a successful inoculation it must be able to compete successfully for host nodulation against a background of less effective resident soil rhizobia and other soil bacteria. The outcome of competition is the summation of the influences of many microbiological, plant and environmental factors. This paper reviews some factors that affect *Rhizobium* competition for nodulation with emphasis given to the soybean-*Rhizobium* symbiosis.

COMPETITION AND SOIL ESTABLISHMENT OF INOCULANT STRAINS

The success of a highly effective inoculum strain depends on its ability to become established in the host rhizosphere in competition with the indigenous microflora, to compete with resident soil rhizobia for nodule sites and to establish itself in the new soil environment as the major component of the rhizobial population.

Because the soybean is grown throughout the world its *Rhizobium* symbionts have become readily established as part of the soil microflora. The occurrence of soils with few or no resident strains of *R. japonicum* is diminishing. Significant responses to inoculation are obtained only when the soil population of rhizobia is very low. The numbers of rhizobia present in the soil, their effectiveness, and their often superior ability to compete with an inoculant strain, determine the success of the inoculum. Inoculated soybeans sown into soils containing appreciable levels of *R. japonicum* do not usually result in significant differences in seed yield. Generally, only about 5 to 10% of the nodules are formed by the inoculant applied to the seed (Kvien et al., 1981), even when the rhizobia are supplied at numbers manyfold higher than usual. Even when 50% of the nodules result from inoculation, an inoculum response may not occur. The rhizobia in these soils are usually effective and their soil numbers are frequently in excess of 10^4/g which minimizes the chances of detecting significant responses (Weaver et al., 1972).

Soybeans grown on soils containing 1000 or more rhizobia/g are not likely to be extensively nodulated by an inoculum applied at normal rates. Weaver and Frederick (1974) and Nelson et al. (1978) predicted that, for inoculum strains to form 50% or more of the nodules, an inoculum rate of at least 1,000 times the soil population must be used. However, any benefits would be modified by the 'competitive' nature of the inoculum strain.

Plants with nodules formed by effective and ineffective strains of rhizobia usually have lower yields than those nodulated only with an effective strain (Ireland and Vincent, 1968; Franco and Vincent, 1976). Singleton and Stockinger (1983) found that the relationship between percent effective nodules and shoot nitrogen in soybean plants was nonlinear, so that 95% of maximum nitrogen accumulation was obtained when only 75% of the nodules were effective. Near maximum yields have also been reported for subterranean clover (*Trifolium subterraneum* L.) when 25% of the nodules were ineffective (Robinson, 1969).

FACTORS AFFECTING COMPETITION

Influence of Host Genotype

The outcome of competition between strains of *Rhizobium* can be influenced by the legume species and even cultivar (Vincent and Waters, 1953; Diatloff and Brockwell, 1976; Materon and Vincent, 1980). Soybean cultivar differences are well established and common, and their influence on the outcome of competition between *Rhizobium* strains is receiving increased attention (Kuykendall and Weber, 1978; Materon and Vincent, 1980; Kosslak et al., 1983). Nigerian and Indonesian cultivars nodulate effectively with indigenous rhizobia in Nigerian soils (Pulver et al., 1982), but high producing U.S. cultivars respond to inoculation against the background of resident soil rhizobia, which are poor nodulators of the U.S. cultivars. Kvien et al. (1981), after testing 1600 lines, found two that recovered an introduced inoculum strain (USDA 110) at enhanced levels (55% of plant nodules) and responded with increased seed yield at both test locations.

Similar observations of genotypic influence have been made for *R. trifolii* nodulating white (*Trifolium repens* L.), red (*Trifolium pratense* L.), and subterranean clovers (Robinson, 1969; Masterson and Sherwood, 1974). Effective *Rhizobium* strains isolated from the non-legume *Parasponia* are ineffective (except for CP 283) on tropical legumes; only some strains isolated from tropical legumes, growing in the same soils, are effective on *Parasponia*. Competition between the effective isolates from the markedly different host genotypes always resulted in all nodules on *Parasponia* being formed by the *Parasponia* isolates (Trinick and Appleby, 1984). Little is known about the selection of specific strains of *Rhizobium* by most promiscuous tropical legumes, but legumes with degrees of specificity (e.g., *Centrosema*) are considered to favor, like *Parasponia*, certain members of the soil rhizobial population, if present.

Rhizobium Strain Representation in the Rhizosphere

Vincent and Waters (1953) found that different strains of the same *Rhizobium* species yielded unequal rhizosphere populations depending upon host genotype. The dominant strain varied depending upon the *Trifolium* species used, and the distribution of the strains in the nodules did not appear to be related to rhizosphere populations. Similarly with soybean, the rhizosphere colonization was not necessarily related to nodule representation (Materon and Vincent, 1980; Reyes and Schmidt, 1979, 1981; Moawad et al., 1984). Interestingly, the indigenous strain of *R. japonicum* common in the soils of North Central U.S., 123, and which generally nodulates soybeans in this region irrespective of inoculation practices, is only mildly stimulated in the rhizosphere (Ham et al., 1971b; Reyes and Schmidt, 1979, 1981). Paired effective, paired ineffective and paired effective-with-ineffective strains of *R. trifolii* and *R. meliloti* on their respective hosts were found by Marques-Pinto et al. (1974) to form a proportion of nodules that was related to their proportional representation on the root surface. They defined the term "competitive index" as the ratio of nodules originating from each strain under conditions of equal representation on the root. This index was found to be not always a true indicator of competitive success between strains of *R. trifolii*

(Labandera and Vincent, 1975; Franco and Vincent, 1976). Labandera and Vincent (1975) obtained evidence suggesting that *Rhizobium* strains can colonize roots independently. Trinick et al. (1983) found that the slow-growing strain, CB 756, responded to higher temperatures and produced larger rhizosphere populations on cowpeas (*Vigna unguiculata* [L.] Walp.), compared with the fast-growing cowpea *Rhizobium*, NGR 234, and this was reflected in greater nodule representation by CB 756. It would appear that the outcome of competition cannot be predicted with certainty from information on rhizosphere colonization and representation.

Delaying the inoculation of a second strain of *Rhizobium* greatly influences the outcome of competing rhizobia. When less competitive strains of *R. japonicum* were inoculated before the more competitive strain (USDA 110), their nodule occupancy was significantly increased (Kosslak et al., 1983). Thus, when the USDA 110 was delayed for 6, 48 and 148 h, the incidence of USDA 138 (primary inoculum) nodules increased from 6% (zero time) to 28, 70 and 82%, respectively. Thus, it is important that inoculum strains are presented in high numbers to the emerging root to assist in their competition with soil rhizobia.

Influence of Other Soil/Rhizosphere Microorganisms

Soil microorganisms, including other *Rhizobium* strains, can interfere with the competition between strains of *Rhizobium* for nodulation. This is probably achieved by altering the proportional numbers of the competing strains on the root surface. Microbial interactions in the soil/rhizosphere, and their influence upon the numbers of rhizobia, have been reviewed elsewhere (Trinick, 1982).

Inhibitory effects among *Rhizobium* strains on laboratory media have been shown (Trinick, 1982), but their ecological significance, particularly in competition for nodule sites, has received only scant attention. Schwinghamer and Belkengren (1968) isolated a polypeptide-type of antibiotic from an ineffective *R. trifolii* strain that strongly suppressed nodulation by effective sensitive strains. The ecological significance of antibiotic producing *Rhizobium* strains has not been fully investigated. Schwinghamer and Brockwell (1978) found that bacteriogenic and lysogenic strains of *R. trifolii*, grown with sensitive strains in broth and moist peat cultures, resulted in the sensitive strains being suppressed, and they speculated that bacteriocinogenic or lysogenic rhizobia might have a competitive advantage over sensitive rhizobia in the complex field environment. Rhizobiophages have received very little attention. Evans et al. (1979) and Barnet (1980) showed a significant decrease in number of susceptible *R. trifolii*, frequently accompanied by an increase in resistant types, and an increase in the proportion of ineffective variants but no effect on nodule numbers was noted in the presence of rhizobiophage. Recently, it has been shown that non-nodulating strains may block nodulation by effective strains (Broughton et al., 1982).

Temperature

Changing the temperature can change nodule occupancy markedly. Roughley et al. (1980) found strains of other *Rhizobium* species were poor competitors with *R. japonicum* on the promiscuously nodulating *G. max* cv. Malayan between 24 and 33 C, but at 36 C they formed more nodules (74 to 88%) than *R. japonicum*. In the natural soil environment, Caldwell and Weber (1970) found different serogroups were able to dominate nodule samples according to planting date, an effect later attributed to root temperature differences (Weber and Miller, 1972). Competition studies (Trinick et al., 1983) between fast- and slow-growing strains of *Rhizobium* on cowpea showed that the fast-grower was a superior competitor for nodule sites at 25/23 C (day/night) compared with three slow-growing strains, but at 30/26 C the slow-growing rhizobia dominated the nodules formed.

Soil Factors

Soil type influences competition between *Rhizobium* strains. Some *Rhizobium* strains have been shown to be well adapted to living as a soil saprophyte, and hence frequently are found in different soils; Ham et al. (1971a) and Damirgi et al. (1967) found the indigenous strain 123 to be widely distributed in soils of the north central U.S. Introduced rhizobia that form a high proportion of nodules on clovers and medics (*Medicago* sp.) in the first year frequently decline progressively in their soil population levels and nodule representation in the second and subsequent years (Chatel et al., 1968). Similar studies on persistence of introduced *R. japonicum* in different soils in various climatic zones have not been reported. However, poor nodule representation in the first year suggests that introduced strains do not compete and establish themselves in the presence of resident soil rhizobia. Soil characteristics that alter the population of one strain in relation to another have the potential to influence the outcome of competition.

The influence of soil pH on *Rhizobium* growth and survival is well documented, but its influence on *Rhizobium* competition has received little attention. Damirgi et al. (1967) found that *R. japonicum* serogroups 123 and 135 were related to soil pH; serogroup 123 dominated in nodules in soils of low pH (Clarion, pH 5.9), and 135 dominated in a Harps soil at pH 8.3. Russell and Jones (1975) found that neutral and alkaline conditions favored an ineffective strain. The effective strain was more competitive under acid conditions.

INHERENT NATURE OF THE SYMBIONT

Effectiveness and Infectivity with Host

Effectiveness and infectivity are both measures of host compatibility with *Rhizobium* strains, but neither attribute can be depended upon to predict the outcome of competition. Highly effective rhizobia vary in their ability to compete for host nodulation, and some effective strains are very poor competitors (Ireland and Vincent, 1968; Robinson, 1969; Labandera and Vincent, 1975; Franco and Vincent, 1976; Trinick and Appleby, 1984). Gibson (1968) found that a poorly nodulating strain of *R. trifolii* could lower the proportion of nodules formed by a second strain in competition with the native soil rhizobia. Diatloff and Brockwell (1976) reported that nodulation by an effective strain of *R. japonicum* on Hardee was suppressed to such an extent by an ineffective strain that plant growth was as poor as the uninoculated controls. In a soil containing 10^5/g of rhizobia ineffective on subterranean clover, Ireland and Vincent (1968) found that approximately 10^6 effective rhizobia per seed was required to obtain 90% effectively nodulated plants.

The time taken for a *Rhizobium* strain to infect its host and to form a visible nodule may be considered an aspect of host-*Rhizobium* compatibility. There does not seem to be a consistent relationship between this characteristic of *Rhizobium* strains and their competitive ability. Marques-Pinto et al. (1974) observed that competitiveness did not correlate with the relative speed with which strains produced nodules, whereas Franco and Vincent (1976) found with *Stylosanthes*, but not *Macroptilium* (Siratro), that the marked dominance of an ineffective over an effective competitor was correlated with the faster nodulating ability of the ineffective strain. When studying the competition between fast- and slow-growing rhizobia for nodulation of cowpeas, Trinick et al. (1983) found that time taken for the appearance of the first nodule was related to the success of competing strains. Effective strains of rhizobia isolated from legumes take much longer to nodulate the non-legume *Parasponia* than do isolates from *Parasponia*, and in competition studies between the two groups of rhizobia, the *Parasponia* isolates exclude other slow-growing rhizobia (Trinick and Appleby, 1984). Perhaps, the apparent discrepancy between reports can be accounted for if the time for the appearance of the nodule after root infection differs between strains of *Rhizobium*. Perhaps, the outcome of competing strains is more likely to depend on which strain is able to first infect the root, irrespective of the time required for the appearance of the nodule.

Growth Rate and Physiological Age

Franco and Vincent (1976) compared the competitive ability of a fast-growing *Leucaena* isolate (ineffective on *Macroptilium atropurpureum*) with an effective slow-grower for root surface colonization and nodulation representation. The fast growth rate of the *Leucaena* isolate did not ensure successful competition. The fast-growing Lotus *Rhizobium* strains were not found to be better competitors than slow-growers for nodules on *L. pedunculatus* (Pankhurst, 1981). Trinick et al. (1983) found that fast growth rate on laboratory media did not ensure a competitive advantage on cowpea.

The physiological age of an inoculum culture may greatly affect the outcome of competition between *Rhizobium* strains. Infection thread counts produced by *R. trifolii* were influenced by the physiological condition of the rhizobial cells, and there was an effect due to cell age and growth media (Napoli and Hubbell, 1976). Similar effects with other *Rhizobium* species and with strains of *R. japonicum* have been reported (Bhagwat and Thomas, 1982; Bhuvaneswari et al., 1980).

CONCLUSION

Development of inoculum cultures to benefit soybean plants and other legumes, grown in soils having a high resident population of less effective rhizobia, demands an understanding of the mechanisms that influence the outcome of its competition with native soil strains. The mechanisms determining the competitive ability of rhizobia are poorly understood and appear to vary greatly between different legume/cultivar-*Rhizobium* associations, as well as between various competing *Rhizobium* strains. No definite guidelines can be given to predict the outcome of competition, but a knowledge of the factors discussed should be considered when selecting new strains of *Rhizobium*. The *Rhizobium* strain selected for an inoculum should not only be highly effective over the range of conditions of the growing season, but more competitive than the indigenous population under the conditions (e.g., temperature, pH, moisture, etc.) prevailing at the time of inoculation and host seed germination.

NOTES

M. J. Trinick, Division of Plant Industry, CSIRO, G.P.O. Box 1600, Canberra, A.C.T. 2601, Australia.

REFERENCES

Barnet, Y.M. 1980. The effect of rhizobiophages on populations of *Rhizobium trifolii* in the root zone of clover plants. Can. J. Microbiol. 26:572-576.

Bhagwat, A. A. and J. Thomas. 1982. Legume *Rhizobium* interactions: Cowpea root exudate elicits faster nodulation response by *Rhizobium* species. Appl. Environ. Microbiol. 43:800-805.

Bhuvaneswari, T. V., B. G. Turgeon and W. D. Bauer. 1980. Early events in the infection of soybean (*Glycine max* L. Merr.) by *Rhizobium japonicum*. 1. Localization of infectible root cells. Plant Physiol. 66:1027-1131.

Broughton, W. J., U. Samrey and B. B. Bohlool. 1982. Competition for nodulation of *Pisum sativum* cv. Afghanistan requires live rhizobia and a plant component. Can. J. Microbiol. 28:162-168.

Caldwell, B. E. and D. F. Weber. 1970. Distribution of *Rhizobium japonicum* serogroups in soybean nodules as affected by planting dates. Agron. J 62:12-14.

Chatel, D. L., R. M. Greenwood and C. A. Parker. 1968. Saprophytic competence as an important character in the selection of *Rhizobium* for inoculation. p. 65-73. *In* Trans. 9th Int. Cong. Soil Sci., vol. 2. Angus and Robertson, Sydney.

Damirgi, S. M., L. R. Frederick and I. C. Anderson. 1967. Serogroups of *Rhizobium japonicum* in soybean nodules as affected by soil types. Agron. J. 59:10-12.

Diatloff, A. and J. Brockwell. 1976. Ecological studies of root-nodule bacteria introduced into field environments. 4. Symbiotic properties of *Rhizobium japonicum* and competitive success in nodulation of two *Glycine max* cultivars of effective and ineffective strains. Aust. J. Exp. Agric. Anim. Husb. 16:514-521.

Evans, J., Y. M. Barnet and J. M. Vincent. 1979. Effect of a bacteriophage on colonization and nodulation of clover roots by paired strains of *Rhizobium trifolii*. Can. J. Microbiol. 25:974-978.

Franco, A. A. and J. M. Vincent. 1976. Competition among rhizobial strains for the colonization and nodulation of two tropical legumes. Plant Soil 45:27-48.

Gibson, A. H. 1968. Nodulation failure in *Trifolium subterraneum* L. cv. Woogenellup (syn. Marrar). Aust. J. Aric. Res. 19:907-918.

Ham, G. E., V. B. Caldwell and H. W. Johnson. 1971a. Evaluation of *Rhizobium japonicum* inoculants in soils containing naturalized populations of rhizobia. Agron. J. 63:301-303.

Ham, G. E., L. R. Frederick and L. C. Anderson. 1971b. Serogroups of *Rhizobium japonicum* in soybean nodules sampled in Iowa. Agron. J. 63:69-72.

Ireland, J. A. and J. M. Vincent. 1968. A quantitative study of competition for nodule formation. p. 85-93. *In* Trans. 9th Int. Cong. Soil Sci., vol. 2. Angus and Robertson, Sydney.

Kosslak, R. M., B. B. Bohlool, S. Dowdle and M. J. Sadowsky. 1983. Competition of *Rhizobium japonicum* strains in early stages of soybean nodulation. Appl. Environ. Microbiol. 46:870-873.

Kuykendall, L. D. and D. F. Weber. 1978. Genetically marked *Rhizobium* identifiable as inoculum strain in nodules of soybean plants grown in fields populated with *Rhizobium japonicum*. Appl. Environ. Microbiol. 36:915-919.

Kvien, C. S., G. E. Ham and J. W. Lambert. 1981. Recovery of introduced *Rhizobium japonicum* strains by soybean genotypes. Agron. J. 73:900-905.

Labandera, C. A. and J. M. Vincent. 1975. Competition between an introduced strain and native Uruguayan strains of *Rhizobium trifolii*. Plant Soil 42:327-347.

Marques-Pinto, C., P. Y. Yao and J. M. Vincent. 1974. Nodulating competitiveness amongst strains of *Rhizobium meliloti* and *R. trifolii*. Aust. J. Agric. Res. 25:317-329.

Masterson, C. L. and M. T. Sherwood. 1974. Selection of *Rhizobium trifolii* strains by white and subterranean clovers. Ir. J. Agric. Res. 12:91-99.

Materon, L. A. and J. M. Vincent. 1980. Host specificity and interstrain competition with soybean rhizobia. Field Crops Res. 3:215-224.

Moawad, H. A., W. R. Ellis and E. L. Schmidt. 1984. Rhizosphere response as a factor in competition among the serogroups of indigenous *Rhizobium japonicum* for nodulation of soybeans. Appl. Environ. Microbiol. 47:613-615.

Napoli, C. A. and D. H. Hubbell. 1976. Growth characteristics of *Rhizobium trifolii* in yeast-extract mannitol broth, soil extract and root exudate. Abstr. Ann. Meet. Am. Soc. Microbiol. 76:170.

Nelson, D. W., M. L. Swearingin and L. S. Beckham. 1978. Response of soybeans to commercial soil-applied inoculants. Agron. J. 70:517-528.

Pankhurst, C. E. 1981. Effect of plant nutrient supply on nodule effectiveness and *Rhizobium* strain competition and nodulation of *Lotus pedunculatus*. Plant Soil 60:325-339.

Pulver, E. L., F. Brockman and H. C. Wien. 1982. Nodulation of soybean cultivars with *Rhizobium* spp. and their response to inoculation with *R. japonicum*. Crop Sci. 22:1065-1070.

Reyes, V. G. and E. L. Schmidt. 1979. Population densities of *Rhizobium japonicum* strain 123 estimated directly in soil and rhizospheres. Appl. Environ. Microbiol. 37:854-858.

Reyes, V. G. and E. L. Schmidt. 1981. Populations of *Rhizobium japonicum* associated with surfaces of soil-grown roots. Plant Soil 61:71-80.

Robinson, A. C. 1969. Competition between effective and ineffective strains of *Rhizobium trifolii* in the nodulation of *Trifolium subterraneum*. Aust. J. Agric. Res. 20:827-841.

Roughley, R. J., E. S. P. Bromfield, E. L. Pulver and J. M. Day. 1980. Competition between species of *Rhizobium* for nodulation of *Glycine max*. Soil Biol. Biochem. 12:467-470.

Russell, P. E. and D. G. Jones. 1975. Variation in the selection of *Rhizobium trifolii* by varieties of red and white clover. Soil Biol. Biochem. 7:15-18.

Schwinghamer, E. A. and R. P. Belkengren. 1968. Inhibition of rhizobia by a strain of *Rhizobium trifolii*: some properties of the antibiotic and of the strain. Arch. Mikrobiol. 64:130-145.

Schwinghamer, E. A. and J. Brockwell. 1978. Competitive advantage of bacteriocin- and phage-producing strains of *Rhizobium trifolii* in mixed culture. Soil Biol. Biochem. 10:383-387.

Singleton, P. W. and K. R. Stockinger. 1983. Compensation against ineffective nodulation in soybean. Crop Sci. 23:69-72.

Trinick, M. J. 1982. Biology. p. 76-146. *In* W. J. Broughton (ed.) Nitrogen fixation, vol. 2: *Rhizobium*. Oxford Univ. Press, Oxford.

Trinick, M. J. and C. A. Appleby. 1984. The rhizobia of *Parasponia*. *In* I. R. Kennedt and L. Copeland (ed.) The Seventh Australian Legume Nodulation Conference. Aust. Inst. Agr. Sci. Occasional Publication No. 12.

Trinick, M. J., R. L. Rhodes and J. H. Galbraith. 1983. Competition between fast- and slow-growing tropical legume rhizobia for nodulation of *Vigna unguiculata*. Plant Soil. 73:105-115.

Vincent, J. M. and L. M. Waters. 1953. The influence of the host on competition amongst clover root-nodule bacteria. J. Gen. Microbiol. 9:357-370.

Weaver, R. W. and L. R. Frederick. 1974. Effect of inoculum rate on competition on nodulation of *Glycine max* L. Merrill. II. Field studies. Agron. J. 66:233-236.

Weaver, R. W., L. R. Frederick and L. C. Dumenil. 1972. Effect of soybean cropping and soil properties on numbers of *Rhizobium japonicum* in Iowa soils. Soil Sci. 114:137-141.

Weber, D. F. and V. L. Miller. 1972. Effect of soil temperature on *Rhizobium japonicum* serogroups distribution in soybean nodules. Agron. J. 64:796-798.

BACTEROIDS IN THE SOYBEAN: *BRADYRHIZOBIUM JAPONICUM* SYMBIOSIS

J. C. Zhou, Y. T. Tchan and J. M. Vincent

The root-nodule bacteria of legumes occur in two forms: free-living and symbiotic. Establishment of a nitrogen-fixing symbiosis depends on a series of steps from initial root infection to the establishment of an active nitrogen-fixation system, variously affected by host and rhizobial genome, and the environment. It is generally accepted that bacteria in infection threads and, in some cases, in the intercellular spaces are similar to the free-living form. Bacteroids are found within host cells, surrounded by the peribacteroid membrane. Their form and number within the membrane varies from multiple, rod-like cells (as in soybean) to those that are single, markedly enlarged and distorted (as in clover).

The term bacteroid has been variously defined but for the present discussion it will be applied to all rhizobia found within the nodule cells of the host legume.

BIOCHEMICAL AND CYTOCHEMICAL FEATURES

Several attempts have been made to differentiate bacteroids and cultured rhizobia biochemically (Rantanen and Saubert [cited by Sutton, 1974]; Bergersen, 1958; Bisseling et al., 1977). Ching et al. (1977), using sucrose density gradient centrifugation, separated the soybean nodule homogenate into three fractions: the lightest contained mature bacteroids, the intermediate fraction composed of transforming cells, and the heaviest fraction consisted of apparently unchanged bacteria. Paau and Cowles (1978), using velocity sedimentation, found that alfalfa (*Medicago sativa* L.) bacteroids could be fractioned into four groups. The results obtained by using simultaneously ethidium bromide fluorescence and light scattering signals indicated that the slow fraction contained small rods which were similar in size and in nucleic acid content to cultured bacteria; the fast sedimenting fraction contained cells which were two to three times longer and had a nucleic acid content three to four times greater than cultured cells. They also reported positive correlation between bacteroid size and nucleic acid content per cell.

Sutton and Paterson (1983) hypothesized that infected cells of many legumes produce substances that inhibit the synthesis of one or more rhizobial cell wall components, such as peptidoglycan (Sutton et al., 1981), while permitting cell expansion, and DNA and protein synthesis. These authors saw in the correlation between detergent sensitivity and viability of bacteroids a relationship to the integrity of their cell wall. They went on to explain decreasing viability with nodule age (which characterized their lupin (*Lupinus* sp.) bacteroids) as due to progressive weakening of the cell wall to the point where the bacteroids had become highly sensitive to detergent and were fully non-viable. No such progression towards reduced viability was seen in our experience with soybean bacteroids; in fact, the reverse was clearly demonstrated (discussed later).

MORPHOLOGY

The morphology of bacteroids is largely plant-host dependent. Dart (1977) dramatically showed that the strain CB 2364 formed multiple rod-shaped bacteroids within the peribacteroid membrane when the host plant was *Vigna (phaseolus) angularis* Willd, Ohai & Ohashi. The same strain produced large spherical, singly enclosed bacteroids in *Arachis hypogaea* L. Sutton and Paterson (1983) indicated that bacteroids of *Phaseolus vulgaris* L. and *Vigna radiata* (L.) Wilczek (*P. aureus*) had slight or low sensitivity to detergent and high viability, whereas the bacteroid of *A. hypogaea* was highly sensitive to detergent and was low in viability. Sutton and Paterson (1980) demonstrated that, irrespective of strain of rhizobia (four strains for *Lotus pedunculatus* Cav. and three strains for *Medicago sativa* L.), the bacteroids produced in *L. pedunculatus* were always detergent resistant, whereas with *M. sativa* all bacteroids were detergent sensitive. It is not known whether bacteroids formed in soybean by the recently discovered fast-growing rhizobia (Keyser et al., 1984) belong to the detergent-resistant or -sensitive groups. Table 1 illustrates the relationship between detergent sensitivity, viability, morphology of bacteroids and host plant.

SEROLOGY

Bacteroids commonly share one or more surface antigens with the cultured bacteria, which permits strain recognition with agglutination by antisera developed against the latter form (Means et al., 1963), or by immunofluorescence (Trinick, 1969). Such reactivity does not prove antigen identity between the two forms. It seems significant, in the light of our experience with soybean bacteroids from different aged nodules (discussed later), that suspensions from very young nodules were negative to cultured cell antiserum, becoming positive as the nodule matured (Means et al., 1963). Another indication of differences between bacteroid and cultured cell is the observation that antisera prepared against a bacteroid suspension of soybean rhizobia agglutinated poorly, or not at all, the same strain of cultured cell (Skrdleta, cited by Bergersen, 1974). Recently we have produced the more specific monoclonal antibodies from mice and tested their reactivity by immunofluorescence (Table 2). It is clear that cultured

Table 1. Relationship between viability and detergent sensitivity, morphology and multiplicity of bacteroids (adapted from Sutton and Paterson, 1983, and Sutton et al., 1981).

Host plant	Detergent sensitivity	Viability	Size and shape	Bacteroids per envelope
Arachis hypogaea	high (H)	low (L)	branched (B), large round (Lr)	single (Si)
Medicago sativa	H	L	B, large pear shaped (Lp)	Si
Pisum sativum	H	L	B, Lp	Si
Trifolium repens	H	L	Lr, Lp	Si
Lupinus corniculatus	slight (S)	H	rod (R)	multiple (M)
Lupinus pedunculatus	S	H	R	M
Phaseolus vulgaris	S	H	R	M
Glycine max	L	H	R	M

Table 2. Immunofluorescence reaction of soybean bacteroids

	Immunofluorescence[a] with		
	Soybean bacteroids from different aged nodules		Cultured cells
	2 wk	8 wk	
Polyclonal serum[b]	I	+	+
Monoclonal serum 1[c]	+	-	+
Monoclonal serum 2[c]	-[d]	+	+

[a]All with *Bradyrhizobium japonicum* CB 1809.
[b]Prepared against cultured *B. japonicum* CB 1809.
[c]Prepared against bacteroids of *B. japonicum* CB 1809 from 2 week-old nodules.
[d]Clone culture subsequently became positive with 2 week-old bacteroids.

bacteria possessed surface antigens specific to young and older bacteroids, but the bacteroids of different age had specific antigens whose presence varied with age of nodule. Because immunofluorescence detects surface antigens, these results imply that the surface antigens of bacteroid underwent changes during the progress of nodule development. Such an approach could be very important as it offers means by which to investigate the transformation of bacteria to bacteroid and vice-versa. This changing nature of the bacteroid's surface antigen, as it develops into the older form, parallels morphological development (Zhou et al., 1984) as well as the transition from non-reacting to reactivity in agglutination already noted (Means et al., 1963).

VIABILITY OF BACTEROIDS

The capacity of bacteroids to reproduce on laboratory media has long been a controversial subject. Almon (1933), using direct observation on the morphologically differentiated bacteroids of pea (*Pisum sativum* L.), red clover (*Trifolium pratense* L.) and alfalfa (*Medicago sativa* L.), concluded that "Bacteroids, if they reproduce at all, do not do so with as great facility as the rod forms...". On the other hand, Means et al. (1963) found that viable cells from crushed soybean nodules amounted to 10-30% and Tsien et al. (1977) reported a high percentage recovery with bacteroids of the same host, as did Gresshoff and Rolfe (1978). Gresshoff et al. (1977) had earlier reported very high recovery from clover bacteroids. The last authors used a single protoplast isolation technique to exclude rhizobia from infection threads and intercellular spaces, and attributed their success with both hosts to the use of osmoprotected isolation, dilution and plating media. Tsien et al. (1977), on the other hand, were able to secure their high percentage recovery from soybean without this extra precaution. Their success, although irregular, was of a dimension that could not be attributed to intercellular rhizobia.

We have documented reproductive growth of the intracellular forms of clover and soybean by direct observation and video-recording of the contents of protoplasts that had been carefully freed of any external bacteria by micromanipulation (Zhou et al., 1984). A single protoplast was isolated and burst and the total and viable plate counts made. Direct observation of growth in a microchamber was carried out to determine the reproductive capacity of individually identified cells.

Table 3 illustrates the viability of clover and soybean intracellular forms. The

Table 3. Colony forming ability of the bacterial population per protoplast.

Age of nodule[a]	Clover			Soybean	
	Direct microscopy observation		Viable count.[b] Number of colonies	Growth: Direct microscopy[c] observation	Viable count.[b] Number of colonies
	Bacteroidal forms	Rod-like forms			
	Present Growth				
First appearance	- -	+	304 ± 123		
2 days	+ -	+	74 ± 27		
2 weeks	+ -	±[d]	0-5	-	⟨25
3 weeks	+ -	±	0-5	+	⟨25
4 weeks	+ -	±	0-5	+	30
6 weeks	+ -	±	0-5	+	1800
8 weeks	+ -	±	0-5	+	8000
11 weeks	+ -	±	0-5	+	22000

[a]For clover the age of nodule was denoted from the day of first appearance. For soybean (grown in sand culture), the age of the nodule was denoted from the date of inoculation. For soybean, the appearance of very young nodules was obscured from observation by the sand used in the Leonard jar. Therefore, information relating to the viability of rhizobia at the early stage of nodulation was unavailable.

[b]RMM was used for clover and BMM for soybean (Gresshoff et al., 1977; Gresshoff and Role, 1978).

[c]Bacteroids and rods are morphologically indistinguishable.

[d]±: In some cases a few small rods were found and showed positive growth in microchamber.

contrast between these two legumes regarding the reproductivity of their bacteroids was remarkable. Direct observation showed that morphologically differentiated clover bacteroids failed to show any reproductivity, and all the colonies produced on laboratory medium came from intracellular rods which decreased in number as the nodule aged. The relatively undifferentiated soybean bacteroids showed an opposite trend. The intracellular bacteria in young nodules were practically unable to reproduce in laboratory medium, whereas a large proportion of the bacteroids contained in old nodules were capable of *in vitro* multiplication.

FACTORS AFFECTING THE COLONY-FORMING CAPACITY

Age of Nodule

Table 3 also shows the proportionate colony-forming ability of soybean bacteroids from different aged nodules. The intracellular bacteria in very young nodules were practically unable to reproduce in laboratory medium, whereas a large proportion of these in older nodules were capable of *in vitro* multiplication. This corresponded with an observed change in bacteroid morphology from long thin rods to fatter rods, more like those of the cultured form.

Soybean Nodule Extract

Nodule extract added to the washing and suspending solutions, and thereby carried over to the plating medium, promoted the recovery of colonies from washed bacteroids at all stages of nodule development (Table 4 and Figure 1). This growth promoting

Table 4. Effect of washing and suspending solutions on percentage colony-forming ability[a] of bacteria from soybean nodule of different age.

| | Washing solution (bacterial suspension) | | | | | |
| Suspending solution | Distilled H$_2$O (DW) | | Protoplast dilution buffer, (PDB) (plain) | | Protoplast dilution buffer (PDB$^+$) (with mannitol) | |
	DW	DW+X[b]	PDB$^-$	PDB$^-$+X	PDB$^+$	PDB$^+$+X
Age of nodule (weeks)						
2-4	2.1	14.4	0.3	5.4	0.018	2.4
5-7	23.4	34.7	14.3	29.9	4.9	29.1
8-10						
8-13	34	54	37.9	53	18.6	50

[a]Plate count expressed as a percentage of total count.
[b]X = nodule extract.

effect was also demonstrated as a function of relative concentration of extract by adding extract in wells. In this case, relative to the most concentrated solution of extract used (0.02 g nodule/10 ml medium), half-strength produced approximately 70%, and quarter-strength approximately 50% of the colonies. On the same basis, only 10% of the colonies were produced when the extract was omitted. The relatively low percentage recovery reported by Means et al. (1963) could be attributed to the absence of nodule extract in their case. The nodule extract could not be replaced by a range of vitamins (thiamine, biotin, vitamin B$_{12}$, Ca pantothenate or riboflavin), nor by yeast

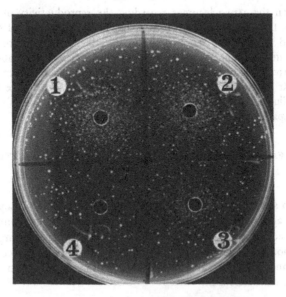

Figure 1. Influence of nodule extract on the colony forming ability of soybean bacteroids. 1. Full strength of nodule extract. 2. Half strength of nodule extract. 3. Quarter strength of nodule extract. 4. Control without nodule extract.

extract.

The growth requirement of the nodule extract by soybean bacteroids was only transitory. By use of the "replica-plating technique" it was possible to locate the nodule-extract requiring colonies. Such colonies, after a second transfer on nodule-extract medium, acquired enough biosynthetic ability to grow in its absence. It must be emphasized that without nodule extract up to 90% of cells could remain undetected during reisolation of soybean rhizobia from nodules less than 6 weeks old.

Osmolarity of Suspending and Plating Media

Unlike Gresshoff et al. (1977), we have not found that osmoprotected suspending and plating media secured the reproductivity growth of morphologically differentiated clover bacteroids. With soybean, colony-forming ability of bacteroids was never enhanced by the use of high osmolar solutions (Table 4). This result also differs from those of some other workers (Gresshoff and Rolfe, 1978), who reported that soybean bacteroids would not form colonies on plain BMM (see Table 3), but require the addition of 0.2 M mannitol, purportedly for osmoprotection. It agrees however with the earlier experience of Means et al. (1963) and Tsien et al. (1977), who were able to secure a moderate to large degree of colony formulation on unfortified media.

GENERAL CONSIDERATIONS

Without the inclusion of nodule extract in the plating medium the commonly used techniques for the isolation of rhizobia from young soybean nodules may only recover a small part of the total viable bacteroids (10% or less). Although such a potentially serious exclusion would be less when older nodules were involved, the possibility remains that, in its present form, such a technique may not be adequate to investigate the recovery of an inoculated soybean rhizobial strain in a field trial, especially in the presence of indigenous competing rhizobia, or for a survey of naturally occurring rhizobia. It will be even more deficient in the case of legumes whose bacteroids have an even lower colony-forming ability, clover being an extreme case.

The routine method of recovering rhizobia from nodules could also be a serious handicap for the rhizobial geneticist studying the interaction of mutants within the nodule, and in the detection of mutation that may occur inside the nodule. Discussion on methodology indicated here, showed the importance of the addition of nodule extract to dilution solutions or cultural media. On the other hand, the recommended osmoprotection was not only unnecessary but had a detrimental effect. The relationship between detergent sensitivity, bacterial morphology and viability and other biochemical information, indicates that rhizobia may have the surface modified during the process of transformation into bacteroids in the nodule cells. The preliminary monoclonal antibody study provided direct indication of such antigenic changes at the surface of the bacteria, and this technique opens up the possibility of a more penetrating serological analysis.

PERSPECTIVE

It was found that distilled water was the best washing and suspending solution for the recovery of soybean bacteroids. In fact, bacteroids of G. *max* washed with low osmolarity media respire at maximal rate when supplied with organic acids including malate and succinate (Stumpf and Burris, 1979). They also indicated that these two constitute the major organic acids in G. *max*. One should, therefore, consider the use of organic acids and other substances, including precursors, for cell-wall synthesis in future experimentation to ascertain their value as additive to media for recovery of bacteroids from soybean nodule suspension.

No detailed information on the transformation from rod to bacteroid in nodule cells is available. Many factors are known to be able to produce bacteroid-like cells *in vitro* cultures (Humphrey and Vincent, 1962; Vincent and Colburn, 1961; Skinner et al., 1977; Pankhurst and Craig, 1978). Monoclonal antibodies would be the tool for such investigations.

Now that different methods of separating bacteroids of different stages are available (Ching et al., 1977; Paau and Cowles, 1978), the highly specific monoclonal antibody could be used to identify the cytochemical changes of bacteroids. A comparative investigation of soybean and other rhizobia may help in the understanding of the biological differences between them.

Affinity chromatography with monoclonal antibody could also provide useful techniques not only in the isolation of differentially modified nodule bacteroids and simulated forms found in culture, it also could be used to probe the genetics of soybean rhizobia both in free-living and symbiotic forms. The powerful Fluorescence Activated Cell Sorter (FACS) would facilitate such research even further.

Finally, it cannot be stated too strongly that the behavior of bacteroids from hosts such as clover and soybean indicates that they are so different that results obtained with one cannot be transposed to a consideration of the other. It is apparent, however, that the host can exercise a major influence in where a particular symbiosis lies in relation to these extremes, and it will be of great interest to determine how the bacteroids formed by a fast-growing rhizobium in soybean behave.

As these improved techniques are now available, specialized soybean rhizobia research programs should be encouraged and could in the future improve the understanding of this symbiosis, so as to increase the nitrogen fixation efficiency of this system, to the benefit not only of fundamental research but agricultural production also.

NOTES

J. C. Zhou, Y. T. Tchan and J. M. Vincent, Department of Microbiology, Univeristy of Sydney, New South Wales 2006, Australia. J. C. Z., present address: Department of Soil and Agricultural Chemistry, Huachung Agricultural College, Wuhan, People's Republic of China.

We wish to acknowledge the assistance of Dr. B. R. Rolfe for the supply of cultures used in some experimentation reported here.

REFERENCES

Almon, L. 1933. Concerning the reproduction of bacteroids. Zentralbl. Bakteroil. II Abt. Bd. 87:289-297.

Bergersen, F. J. 1958. The bacterial component of soybean root-nodules. Changes in respiratory activity, cell dry weight and nucleic acid content with increasing nodule age. J. Gen. Microbiol. 19:312-323.

Bergersen, F. J. 1974. Formation and function of bacteroids. p. 473-498. *In* A. Quispel (ed.) The Biology of Nitrogen Fixation. North Holland Publishing Co., Amsterdam.

Bisseling, T., R. C. van den Bos and A. Van Kammen. 1977. Cytofluorometrical determination of the DNA contents of bacteroids and corresponding broth cultured *Rhizobium* bacteria. J. Gen. Microbiol. 101:79-84.

Ching, T. M., S. Hedtke and W. Newcomb. 1977. Isolation of bacteria, transforming bacteria and bacteroids from soybean nodules. Plant Physiol. 60:771-774.

Dart, P. J. 1977. p. 367-472. *In* W. S. Silver (ed.) Section 3: Biology, of R. W. F. Hardy (ed.) Treatise on Dinitrogen Fixation. John Wiley and Sons Inc., New York.

Gresshoff, P. M. and B. G. Rolfe. 1978. Viability of *Rhizobium* bacteroids isolated from soybean nodule protoplasts. Planta. 142:329-333.

Gresshoff, P. M., M. L. Skotnicki, J. F. Eadie and B. G. Rolfe. 1977. Viability of *Rhizobium trifolii* bacteroids from clover root nodules. Plant Sci. Lett. 10:299-304.

Humphrey, B. and J. M. Vincent. 1962. Calcium in cell walls of *Rhizobium trifolii*. J. Gen. Microbiol. 29:557-561.

Keyser, H. H., M. J. Sadowsky and B. B. Bohool. 1984. Fast-growing soybean rhizobia. p. 6. *In* R. Shibles (ed.) World soybean res. conf. III: Abstracts. WSRC-III. Ames, Iowa.

Means, U. M., H. W. Johnson and R. A. Date. 1963. Quick serological method of classifying strains of *Rhizobium japonicum* in nodules. J. Bacteriol. 87:547-553.

Paau, A. S. and J. R. Cowles. 1978. Studies on bacteroid size and nucleic acid content of alfalfa bacteroids fractionated by velocity sedimentation. Can. J. Microbiol. 24:1283-1287.

Pankhurst, C. E. and A. S. Craig. 1978. Effect of oxygen concentration, temperature and combined nitrogen on the morphology and nitrogenase activity of *Rhizobium* sp. strain 32H1 in agar culture. J. Gen. Microbiol. 106:207-219.

Skinner, F. A., R. J. Roughley and M. R. Chandler. 1977. Effect of yeast extract concentration on viability and cell distortion in *Rhizobium* spp. Appl. Bacteriol. 43:287-297.

Stumpf, D. K. and R. H. Burris. 1979. A micromethod for the purification and quantification of organic acids of the tricarboxylic acid cycle in plant tissues. Anal. Biochem. 95:311-315.

Sutton, W. D. 1974. Some features of the DNA of *Rhizobium* bacteroids and bacteria. Biochem. Biophys. Acta 36:1-10.

Sutton, W. D., C. E. Pankhurst and A. S. Craig. 1981. The *Rhizobium* bacteroid state. Int. Rev. Cytology Suppl. 13:149-177.

Sutton, W. D. and A. D. Paterson. 1980. Effect of the plant host on the detergent sensitivity and viability of *Rhizobium* bacteroids. Planta 148:287-292.

Sutton, W. D. and A. D. Paterson. 1983. Further evidence for a plant host effect on *Rhizobium* bacteroid viability. Plant Sci. Lett. 30:33-41.

Trinick, M. J. 1969. Identification of legume nodule bacteria by the fluorescent antibody reaction. J. Appl. Bacteriol. 32:181-186.

Tsien, H. C., P. S. Cain and E. L. Schmidt. 1977. Viability of *Rhizobium* bacteroids. Appl. and Envir. Microbiol. 34:854-856.

Vincent, J. M. and J. R. Colburn. 1961. Cytological abnormalities in *Rhizobium trifolii* due to a deficiency of calcium or magnesium. Aust. J. Sci. 23:269-270.

Zhou, J. C., Y. T. Tchan and J. M. Vincent. 1984. Reproductive capacity of bacteroids in nodules of *Trifolium repens* L. and *Glycine max* (L.) Merr. Planta (in press).

FAST-GROWING SOYBEAN RHIZOBIA

H. H. Keyser, M. J. Sadowsky, and B. B. Bohlool

Rhizobium are the bacteria capable of forming a nitrogen-fixing symbiosis with leguminous plants. Traditionally, their classification has been based on specific host infectivity with the recognition of two broad groups of fast- and slow-growing strains (Jordan and Allen, 1974). In addition to growth rate, the two groups are distinguished by several important differences, which has led to the proposed establishment of a separate genus for slow growers (*Bradyrhizobium* gen. nov.), while retaining fast growers in the genus *Rhizobium* (Jordan, 1982).

The soybean-nodulating bacteria, *R. japonicum*, have been classified as slow growers. Recently, however, Keyser et al. (1982) reported the existence of fast-growing isolates obtained from soil and nodules collected in China. Examination of the 11 fast-growing isolates for microbiological, genetic, and symbiotic properties has shown them to be distinct from the slow growers. They represent a unique and valuable addition of *Rhizobium* germplasm. In this article we present an overview of microbiological and symbiotic properties of the fast-growing soybean rhizobia. Results from genetic studies are presented elsewhere in this volume by Atherly et al. Several investigators are currently examining these fast growers for various attributes, and knowledge of them should increase rapidly.

ISOLATION

The 11 fast-growing strains were obtained from three separate collection expeditions in China (Table 1). Strain USDA 191 was isolated by J. C. Burton (Nitragin Co., Milwaukee) and the others by T. S. Hu and S. N. May in the USDA Nitrogen Fixation and Soybean Genetics Laboratory at Beltsville, Md. The isolates were authenticated by three cycles of plant passage through *G. max* cv. Peking and reisolation of single colonies. Careful examination showed no slow-growing colonies that could have been masked by the fast-growing type. Subsequent use of fluorescent antibodies confirmed their purity.

CULTURAL CHARACTERISTICS

The fast growers have generation (doubling) times of 2.9 to 4.8 h at 28 C in yeast extract-mannitol-salts (YEM) broth with a concomitant acidic reaction (Keyser et al., 1982). This is in contrast to slow-growing *R. japonicum*, which have generation times of 6 or more hours and a neutral to alkaline reaction in YEM media (Vincent, 1974). On YEM agar plates the fast growers gave 1 to 5 mm diameter colonies after 6 to 7

Table 1.   Strain history of fast-growing soybean rhizobia from China.

USDA accession no. and synonyms	Geographic origin of isolate	Source material from which isolation was made
191 (=440)	Shanghai, Jiangsu Province	Soil: isolated from nodule of trap host *G. max* cv. Jupiter[a]
192 (=0B2=PRC 192)	Tsinan, Shantung Province	Desiccated root system of *G. soja*: isolated from nodule of trap host *G. max* cv. Peking[b]
193 (=0B3=PRC 193)	Taiyuanchen, Shansi Province	Desiccated root system of *G. soja* PI 468397: isolated from nodule of trap host *G. max* cv. Peking[b]
194 (=PRC 194)	Chengchou, Honan Province	Desiccated nodule of *G. max*: direct isolation[c]
201 (= PRC 201)	Same as 194	Same as 194
205 (=PRC 205)	Same as 194	Same as 194
206 (=PRC 206)	Sinsiang, Honan Province	Same as 194
208 (=PRC 208)	Same as 206	Same as 194
214 (=PRC 214)	Same as 206	Same as 194
217 (=PRC 217)	Same as 206	Same as 194
257 (=PRC 257)	Wukung, Shensi Province	Same as 194

[a]Collected by J. Tanner, Univ. of Guelph, 1979.
[b]Collected by T. S. Hu, Chinese Academy of Agric. Sciences, Peking, 1980.
[c]Collected by W. Fehr, Iowa State Univ., and K. Hinson, Univ. of Florida, 1979.

days, whereas colonies of slow growers are only 0.5 to 1 mm. The fast growers produce typical circular, convex, white colonies on YEM agar, with little production of extracellular polysaccharide (Sadowsky et al., 1983). In liquid media, cells from late-log-phase to stationary-phase became enlarged and exhibited marked pleomorphism, while slow growers remained as typical rods. The growth rate difference between the two types of soybean rhizobia is maintained in sterile soil (Sadowsky, 1983). This information indicates that the cultural characteristics of the fast-growing soybean rhizobia are similar to other fast-growing *Rhizobium* species (e.g., *R. leguminosarum* and *R. trifolii*).

GROWTH RESPONSES

The fast-growing soybean rhizobia do not require vitamins (biotin, thiamine, pantothenate) for growth in a defined medium (Stowers and Eaglesham, 1984), a characteristic shared by slow-growing rhizobia. However, unlike slow growers they are tolerant of pH 9.5 and sensitive to pH 4.5 (Sadowsky et al., 1983). As with other fast growers, they are more sensitive to antibiotics than are the slow growers (Sadowsky, 1983; Stowers and Eaglesham, 1984). Also, in contrast to slow-growing *R. japonicum*, the fast growers are tolerant of high levels of NaCl (Sadowsky et al., 1983; Yelton et al., 1983; Stowers and Eaglesham, 1984). Likewise, such tolerance is also found among the fast-growing *R. meliloti* (Graham and Parker, 1964). Yelton et al. (1983) demonstrated that tolerance of 0.4 M NaCl is accompanied by increased levels of K$^+$ and glutamate in USDA 191. Stowers and Eaglesham (1984) found the fast growers, with one exception, to be tolerant of 0.17 M NaCl, but completely sensitive to 0.34 M NaCl. Sadowsky et al. (1983) found that all 11 of the fast growers made growth at 0.34 M NaCl, although some strains had growth limited to a few tolerant colonies after 14 days. None of the slow-growing *R. japonicum* tested by Sadowsky et al. (1983) produced any colonies at 0.34 M NaCl.

CARBOHYDRATE UTILIZATION

Fast-growing species of *Rhizobium* can use many sugars and organic acids for growth, whereas the slow growers are more specialized and prefer pentoses (Fred et al., 1932). Studies of the carbohydrate utilization by the fast-growing soybean rhizobia have shown they match the pattern of other fast-growing rhizobia (Sadowsky et al., 1983; Yelton et al., 1983; Stowers and Eaglesham, 1984). Unlike the slow-growing *R. japonicum* the fast growers utilized cellobiose, inositol, lactose, maltose, raffinose, glucitol, sucrose, and lactose. The ability to use lactose for growth requires sufficient activity of β-Galactosidase. Sadowsky et al. (1983) demonstrated large differences between the fast- and slow-growing soybean rhizobia in levels of β-Galactosidase activity (Table 2).

Consistent with the division of the rhizobia into fast- and slow-growing groups based on carbohydrate utilization is the division of these organisms based on the

Table 2. Enzyme activities of fast- and slow-growing soybean rhizobia (Sadowsky et al., 1983).

Strain	Enzyme activities	
	6-PGD[a]	β-Galactosidase[b]
Slow-growing rhizobia		
USDA 31	⟨0.5	7.1
USDA 110	⟨0.5	3.8
USDA 123	⟨0.5	3.0
USDA 136	⟨0.5	7.6
USDA 138	⟨0.5	3.4
PRC 005	⟨0.5	4.2
PRC 113-2	⟨0.5	3.3
PRC 121-6	⟨0.5	4.8
PRC B15	⟨0.5	4.8
Fast-growing rhizobia		
USDA 191	97	–
USDA 192	181	459
USDA 193	140	183
USDA 194	126	101
USDA 201	50	345
USDA 205	159	97
USDA 206	59	261
USDA 208	147	383
USDA 214	160	245

[a]6-phosphogluconate dehydrogenase, expressed as nanomoles of reduced NADP produced per min per mg of protein.

[b]Expressed as micromoles of o-nitrophenol produced per min per mg of protein.

presence and absence of enzymes of the pentose phosphate pathway. Only fast-growing rhizobia have NADP-linked 6-phosphogluconate dehydrogenase (6-PGD) activity, a key enzyme of that pathway (Keele et al., 1969; Martinez-de Drets and Arias, 1972). Table 2 shows the clear differences between the two types of soybean rhizobia in 6-PGD activity, again linking these fast growers to other fast-growing rhizobia. Studies by Yelton et al. (1983) and Stowers and Eaglesham (1984) also showed the presence of 6-PGD activity in these fast-growers.

Of the 30 strains examined from 6 species of *Rhizobium*, only the fast-growing soybean rhizobia were capable of substantial growth in ethanol (Sadowsky, 1983). While the fast growers grew to final cell densities of 10^8-10^9 cells/mL on 0.2% ethanol, the slow-growing *R. japonicum* never reached densities greater than 10^7 cells/mL on any concentration of ethanol. No activity of NAD- or NADP-linked alcohol dehydrogenase was found.

BIOCHEMICAL ATTRIBUTES

In comparing the 11 fast-growing soybean rhizobia with 7 isolates of slow-growing *R. japonicum*, Sadowsky et al. (1983) found both groups to be positive for catalase, oxidase, urease, penicillinase, and nitrate reductase. Neither group produced H_2S or 3-Ketolactose. Bernaerts and DeLey (1963) found that production of 3-Ketolactose from lactose is limited to species of *Agrobacterium*, a genus closely related to the fast-growing *Rhizobium*. Graham and Parker (1964) found that production of penicillinase was more common in slow-growing rhizobia, though some isolates of the fast-growing *R. leguminosarum* possessed this attribute.

Gelatinase activity clearly differentiated the two groups. The slow growers were negative while the fast growers and an isolate from *Leucaena* were positive (Sadowsky et al., 1983).

In a test of litmus milk reactions, 20 of 21 slow growing *R. japonicum* produced an alkaline pH change with no peptonization (Sadowsky et al., 1983). By comparison, 10 of 11 fast growers produced peptonization of the medium, though pH changes were variable. This response of the fast-growing soybean isolates is typical of other fast-growing *Rhizobium* species (Graham and Parker, 1964).

The fast growers apparently lack hydrogenase activity, as determined in culture and in bacteroid preparations. All seven of the fast growers tested for hydrogenase in free-living culture were negative (S. Uratsu, pers. commun.). Yelton et al. (1983) found USDA 191 to lack hydrogenase activity in culture as well as in symbiosis with *G. max* cv. Clark and cowpea (*Vigna unguiculata* [L.] Walp.) cv. Calif. Blackeye 5. We have found all 11 of the fast growers also to be negative in symbiosis with *G. max* cv. Peking and *G. soja* PI 468398.

SEROLOGICAL RELATIONSHIPS

The fast-growing soybean rhizobia are serologically related (Sadowsky, 1983). Tests with whole-cell antisera indicated that the fast growers shared at least one heat-labile antigen in common. Examination with the more specific somatic antigens showed at least three serological groups among the fast growers; USDA 192 (3 strains), USDA 194 (2 strains), and USDA 205 (6 strains).

In agglutination tests with somatic antisera from 14 non-cross-reacting strains of slow-growing *R. japonicum*, the only cross reaction we found with the fast growers occurred between the USDA 194 group and the USDA 122 group (representing the slow growers USDA 122, USDA 136, CB 1809, and others). Sadowsky (1983) found the same degree of difference between the fast and slow growers. He also found that while USDA 136 cross-reacted with USDA 194 in immunofluorescence reactions, this was not detected in immunodiffusion reactions using somatic antisera.

Immunodiffusion tests with species of *Rhizobium* other than *R. japonicum* with antisera from three of the fast growers showed some very interesting relationships. Fast-growing isolates of *R. meliloti*, an isolate from *Sesbania*, and an isolate from *Lablab purpureus* (L.) Sweet had whole cell and somatic antigens in common with the fast-growing soybean rhizobia (Sadowsky, 1983). Slow growers from *R. lupini* and the 'cowpea' groups did not cross react. The fast-growing soybean rhizobia appear to be fairly distinct among *Rhizobium* in their serological relatedness to both slow growers (*R. japonicum*) and fast growers (*R. meliloti* and 'cowpea' types), though Trinick (1980) reported the same for isolates of *Leucaena*.

SYMBIOSIS WITH LEGUMES

The fast-growing soybean rhizobia nodulate and fix N_2, to varying degrees of effectiveness, with *G. max* and *G. soja* genotypes and with some legumes of the cowpea cross inoculation group. Similar host range is found with the slow-growing *R. japonicum* (Leonard, 1923; Van Rensburg et al., 1976). The fast growers do not nodulate hosts of other fast-growing species (*R. meliloti; R. trifolii; R. phaseoli; R. leguminosarum*) (Keyser et al., 1982; unpublished data). Many adapted genotypes of North American soybeans form ineffective symbioses with the fast growers (Keyser et al., 1982). Table 3 shows the response of four cultivars. In most cases the fast-growers

Table 3. Top dry weight of soybean cultivars inoculated with fast- and slow-growing soybean rhizobia.[a]

Strain	Cultivars of *G. max*			
	Wilson-6	Bedford	Clark	Hardee
	$-$ g/jar[b] $-$			
Slow growers				
USDA 110		4.8 r		6.1 r
USDA 122	6.4 r		4.6 r	
Fast growers				
USDA 191		1.9 t		2.4 s
USDA 192	1.9 u		2.7 s	2.4 s
USDA 193	1.5 uv		2.2 s	
USDA 194	1.7 uv	3.0 s	2.9 s	
USDA 201			2.8 s	2.5 s
USDA 205	2.6 t	1.9 t	2.8 s	2.2 s
USDA 206		2.0 t		
USDA 208	3.3 s		3.1 s	2.4 s
USDA 217	1.8 uv		2.3 s	2.6 s
Uninoculated	1.3 v	1.8 t	2.6 s	1.9 s

[a]Tests conducted in vermiculite-Leonard jar assemblies, grown for 5-6 weeks in a growth chamber.

[b]Each value is a mean of 3 replicates, 3 plants per replicate. Values followed by the same letter within a column are not significantly different (P < .05) according to Duncan's New Multiple Range Test.

were not significantly better than an uninoculated control, as well as being significantly inferior to the reference slow grower. Nodule production on these cultivars ranges from normal (with cultivars Bedford and Hardee) to sparse and atypical (with cultivars Clark, Kent, and Lee). Similar ineffective responses by North American soybeans have been reported by Van Rensburg et al. (1983) and Stowers and Eaglesham (1984). This general ineffective response is not without exception as Yelton et al. (1983) reported USDA 191 to be effective with Clark, and Dowdle (pers. commun.) found three of the fast growers to be effective with other cultivars, based on acetylene-reduction.

Peking forms an effective symbiosis with the fast growers (Table 4). Peking is an unimproved selection of a line from China. Also, Stowers and Eaglesham (1984) found the fast growers to be very effective with three "Asian-type" soybeans. Van Rensburg et al. (1983) found an effective symbiosis between the fast growers and two South African soybeans (Table 5). Cultivar Geduld is a selection from a cross of Dixie and a line from China, while Usutu was selected from crosses in which Geduld was involved. Perhaps, the low frequence of effectiveness with the U.S. cultivars is atypical of soybean germplasm in general and simply results from the narrow germplasm base of U.S. cultivars, which may be built upon accessions with a very low frequency of effectiveness with the fast growers.

Cowpea forms an effective symbiosis with the fast growers (Table 4). This has been confirmed by Stowers and Eaglesham (1984) and Yelton et al. (1983). While they are

Table 4. Shoot nitrogen content in three legumes inoculated with fast- and slow-growing soybean rhizobia.[a]

	Host legume		
Strain	G. max cv. Peking	G. soja PI 468398	Vigna ungumiculata cv. Calif. Blackeye-5
		– mg N/jar[b] –	
Fast growers			
USDA 191	65.3 r	6.9 rs	97.6 rstu
USDA 192	37.7 u	8.5 r	88.8 rstuv
USDA 193		6.8 rs	103.0 rst
USDA 194	46.3 stu	7.1 rs	63.5 vw
USDA 201	63.8 r	7.0 rs	78.3 tuv
USDA 205	50.4 s	6.0 rs	93.6 rstu
USDA 206	38.7 tu	5.8 s	49.5 w
USDA 208	45.9 stu	6.9 rs	78.8 tuv
USDA 214	45.3 stu	7.6 rs	73.2 uvw
USDA 217	49.1 st	7.0 rs	82.9 stuv
USDA 257		7.0 rs	104.9 rs
Slow growers			
USDA 122	52.6 s	6.2 rs	
176A22[c]			116.1 r
Uninoculated	6.9 v	1.1 t	8.7 x

[a]See footnote a, Table 3.

[b]See footnote b, Table 3.

[c]A slow-growing reference strain of R. sp. 'cowpea miscellany'.

Table 5. Dry mass and nodulation of cultivars of G. max inoculated with slow- and fast-growing strains of R. japonicum (from Van Rensburg et al., 1983).

	Cultivar of American origin	Cultivars of South African origin		
Strain	Forrest	SSS3	Geduld	Usutu
		— g/plant[a] —		
Slow grower				
WB1	2.72	2.66	2.11	1.87
Fast growers				
USDA 193	0.56[b]	0.58	2.14	1.46
USDA 194	0.64[b]	0.52	1.59	1.33
USDA 201	0.67[b]	0.49	2.07	1.79
USDA 205	0.67[b]	0.80	2.29	1.38
USDA 206	0.71[b]	0.52	1.57	1.47
Uninoculated	0.82[b]	0.68[b]	0.56[b]	0.54[b]

[a]Average of 3 replicates LSD (P ⟨ 0.01) = 0.52; c.v. = 28.4%.
[b]No nodules formed.

also effective with pigeon pea (*Cajanus cajan* [L.] Huth) the fast growers are ineffective with mung bean (*Vigna radiata* [L.]) and species of *Sesbania* and *Macroptilium* (Keyser et al., 1982; Stowers and Eaglesham, 1984).

The most interesting comparison of host preference between the fast- and slow-growing soybean rhizobia is found in their reaction with different genotypes of wild soybean (*G. soja*). Keyser et al. (1982) reported the fast growers to be effective at N_2-fixation with a line from China (Table 4). Subsequent tests with seven other *G. soja* genotypes demonstrated significant effectiveness interactions with both rhizobia groups (Keyser and Cregan, 1984a). The response of two genotypes with opposite group preferences is shown in Figure 1. PI 468397 exhibits a general specificity for effectiveness with the fast growers as opposed to slow growers. This very unusual reaction has not yet been found in *G. max*, and indicates there is a large range of symbiotic adaptations in *G. soja* germplasm.

CONCLUSIONS

The fast-growing soybean rhizobia from China are distinct from the slow-growing *R. japonicum*. Several distinguishing bacteriological properties used to separate the slow-growing *Bradyrhizobium* from the fast-growing *Rhizobium* place these new isolates in the *Rhizobium* genus. However, the similar host range with the slow growers points out the difficulty of including both the bacteriological and the host related properties in the classification system of root nodule bacteria.

China is considered to be the gene center of the *G. max* and *G. soja*, and apparently it is also the gene center of two types of soybean rhizobia. The occurrence of the fast growers in China is not rare, as they were found in five provinces. In the collection made by Fehr and Hinson, 8 of 69 nodules contained fast-growing isolates. In contrast, the occurrence of the fast growers in U.S. soils is apparently very low at best, though most workers would probably discard such isolates as suspected contaminants.

Genetic studies of the fast growers have also shown them to be distinct from the

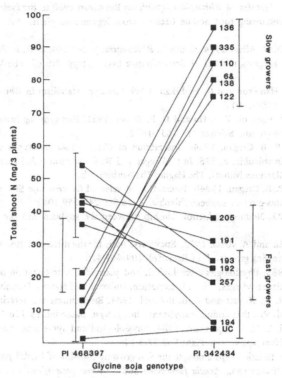

Figure 1. Total shoot N in two *G. soja* genotypes in symbiosis with fast- and slow-growing strains of rhizobia. Strains are listed by USDA number, UC = uninoculated control. Values within a *G. soja* genotype encompassed by the vertical bar are not significantly different (P ≤ 0.05) as tested by Duncan's New Multiple Range Test from Keyser and Cregan, 1984b).

slow growers (see the chapter by Atherly et al., this volume). Their potential for genetic manipulation, combined with their differential symbiotic reactions with *G. max* and *G. soja* genotypes, provides researchers with valuable organisms for use in increasing our understanding of the soybean-rhizobia symbiosis.

NOTES

H. H. Keyser, USDA, ARS, Nitrogen Fixation and Soybean Genetics Laboratory, Beltsville, Maryland 20705; M. J. Sadowsky, Department of Biology, Genetic Manipulation Research Group, McGill University, Montreal, Canada H3A 1B1; B. B. Bohlool, Department of Microbiology, University of Hawaii, Honolulu 96822.

REFERENCES

Bernaerts, M. J. and J. DeLey. 1963. A biochemical test for crown gall bacteria. Nature (London) 197:406-407.

Fred, E. B., I. L. Baldwin and E. McCoy. 1932. Root nodule bacteria and leguminous plants. University of Wisconsin Studies in Science. No. 5, University of Wisconsin, Madison.

Graham, P. H. and C. A. Parker. 1964. Diagnostic features in the characterization of the root-nodule bacteria of legumes. Plant Soil 20:383-396.

Jordan, C. D. 1982. Transfer of *Rhizobium japonicum* Buchanan 1980 to *Bradyrhizobium* gen. nov., a genus of slow-growing root nodule bacteria from leguminous plants. Int. J. Syst. Bacteriol. 32:136-139.

Jordan, D. C. and O. N. Allen. 1974. Genus II. *Rhizobium*, p. 262-264. *In* R. E. Buchanan and N. E. Gibbons (eds.) Bergey's manual of determinative bacteriology. 8th ed. The Williams & Wilkins Co., Baltimore.

Keele, B. B., P. B. Hamilton and G. H. Elkan. 1969. Glucose catabolism in *Rhizobium japonicum*. J. Bacteriol. 97:1184-1191.

Keyser, H. H., B. B. Bohlool, T. S. Hu and D. F. Weber. 1982. Fast-growing rhizobia isolated from root nodules of soybean. Science 215:1631 1632.

Keyser, H. H. and P. B. Cregan. 1984a. Interaction of *Glycine soja* genotypes with fast and slow growing soybean rhizobia. p. 598. *In* C. Veeger and W. E. Newton (eds.) Adv. in nitrogen fixation research. Martinus Nijhoff, The Hague, The Netherlands.

Keyser, H. H. and P. B. Cregan. 1984b. Interaction of selected *Glycine soja* Sieb & Zucc. genotypes with fast- and slow-growing soybean rhizobia. Crop Sci. 24:1059-1062.

Leonard, L. T. 1923. Nodule-production kinship between the soybean and the cowpea. Soil Sci. 15:277-283.

Martinez-de Drets, G. and A. Arias. 1972. Enzymatic basis for the differentiation of *Rhizobium* into fast- and slow-growing groups. J. of Bacteriol. 109:467-470.

Sadowsky, M. J. 1983. Physiological, serological, and plasmid characterization of fast-growing rhizobia that nodulate soybeans. Ph.D. Dissertation, University of Hawaii, Honolulu.

Sadowsky, M. J., H. H. Keyser and B. B. Bohlool. 1983. Biochemical characterization of fast- and slow-growing rhizobia that nodulate soybeans. Int. J. Syst. Bacteriol. 33:716-722.

Stowers, M. D. and A. R. J. Eaglesham. 1984. Physiological and symbiotic characteristics of fast-growing *Rhizobium japonicum*. Plant Soil 77:3-14.

Trinick, M. J. 1980. Relationships amongst the fast-growing rhizobia of *Lablab purpureus, Leucaena leucocephala, Mimosa* spp., *Acacia farnesiana* and *Sesbania grandiflora* and their affinity with other rhizobial groups. J. Appl. Bacteriol. 49:39-53.

Van Rensburg, H. J., B. W. Strijdom and M. M. Kriel. 1976. Necessity for seed inoculation of soybeans in South Africa. Phytophylactica 8:91-95.

Van Rensburg, H. J., B. W. Strijdom and C. J. Otto. 1983. Effective nodulation of soybeans by fast-growing strains of *Rhizobium japonicum*. South African J. Sci. 79:251-252.

Vincent, J. M. 1974. Root-nodule symbiosis with *Rhizobium*. p. 265-347. *In* A. Quispel (ed.) Biology of nitrogen fixation. North-Holland Publishing Co., Amsterdam.

Yelton, M. M., S. S. Yang, S. A. Edie and S. T. Lim. 1983. Characterization of an effective salt-tolerant, fast-growing strain of *Rhizobium japonicum*. J. Gen. Microbiol. 129:1537-1547.

HYDROGEN RECYCLING IN NODULES AFFECTS NITROGEN FIXATION AND GROWTH OF SOYBEANS

Harold J. Evans, F. Joe Hanus, Richard A. Haugland, Michael A. Cantrell, Liang-Shu Xu, Sterling A. Russell, Grant R. Lambert and Alan R. Harker

Each nitrogenase that has been examined from different sources is composed of a molybdenum-iron protein and an iron protein. Together these two components catalyze the N_2 fixation reaction, which may be represented as follows:

$$N_2 + 8e + 8H^+ + 16\,ATP \rightarrow 2\,NH_3 + 16\,ADP + 16\,Pi + H_2$$

Hydrogen appears to be an indispensable product of the reaction and under optimum conditions H_2 evolution utilizes about 25% of the electron flux through the nitrogenase reaction (Evans et al., 1982; Robson and Postgate, 1980). When the amount of reductant is limiting in reactions or the proportion of iron protein to molybdenum iron protein is less than optimum an increase in the allocation of electrons to protons and a corresponding decrease in allocation of electrons to N_2 is observed. Under such circumstances, the rate of H_2 evolution may be considerably greater than 25% of the rate of electron flow in the nitrogenase reaction (Eisbrenner and Evans, 1983).

The efficiency of energy utilization in the nitrogenase reaction has some very practical implications that are relevant to N_2 fixation and growth of the soybean and other legumes. The necessary energy to provide the reductant and ATP for N_2 fixation must be derived from the oxidation of products of photosynthesis supplied to the nodule bacteroids by the legume. On the basis of electrons transferred, the reduction of protons to H_2 requires approximately the same amount of ATP as the reduction of N_2 to NH_3. The evolution of H_2 during N_2 fixation, therefore, consumes photosynthates and may be viewed as a side reaction in which energy from the plant is wasted.

Not all legumes lose H_2 during N_2 fixation. About 22% of the strains of *R. japonicum* possess a capacity to synthesize the H_2 recycling system, and as a consequence nodules formed by such strains internally recycle the H_2 produced by the bacteroid nitrogenase reaction (Evans et al., 1982). So far, no effective H_2 recycling system has been identified in nodules formed by *Rhizobium meliloti* or *Rhizobium trifolii* (Eisbrenner and Evans, 1983). The relationship of nitrogenase catalyzed H_2 evolution and H_2 oxidation via a membrane bound hydrogenase coupled to the respiratory electron transport chain of rhizobial cells is shown in Figure 1. This scheme indicates that H_2 oxidation may lead to O_2 consumption and ATP generation both of which may be beneficial to the function of the nitrogenase system (Emerich et al., 1979). In addition, Dixon et al. (1981) have proposed that the removal of H_2 within nodules by a

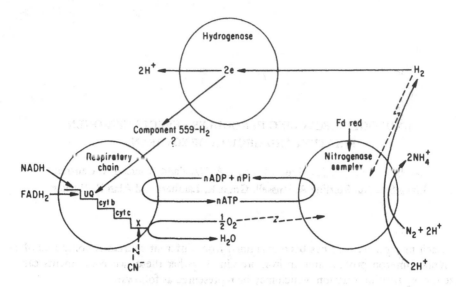

Figure 1. A scheme showing possible relationships of nitrogenase, hydrogenase and respiratory electron transport in *Rhizobium japonicum*. (From Eisbrenner and Evans, 1983).

recycling mechanism may benefit the N_2 fixation process by preventing H_2 inhibition of nitrogenase. Also, substantial evidence has accumulated showing that nodules formed by strains of *Rhizobium* that utilize H_2 in the generation of ATP respire CO_2 at decreased rates, indicating they have a lower rate of consumption of respiratory substrates than nodules formed by Hup⁻ strains (Drevon et al., 1982; Rainbird et al., 1983).

Of probable practical significance is the recent work of Klucas et al. (1983), who demonstrated that nickel is an essential element for the expression of hydrogenase activity in *R. japonicum* and for growth of this bacterium in a medium in which H_2 was the sole energy source. More recently, Harker in our laboratory has purified hydrogenase from *R. japonicum* cultured chemolithotrophically in a medium containing ^{63}Ni and found that the isotope was bound to the purified enzyme. It is now apparent that the synthesis of a functional hydrogenase by *R. japonicum* requires a supply of available nickel.

Eisbrenner and Evans (1983) summarized results of 13 different experiments in which several researchers have compared Hup⁺ and Hup⁻ *Rhizobium* strains as inoculants for legumes. Among these trials, eight showed statistically significant increases (10 to 49%) in dry weight and/or total N content of plants from use of Hup⁺ inocula. In five experiments no significant effect of the inoculum could be measured. Most of these experiments are open to the criticism that plants were grown in pot cultures for periods of only 28 to 45 days and Hup⁺ and Hup⁻ field isolates or poorly defined Hup⁺ parent strains and Hup⁻ mutants were used as inoculants. Undoubtedly, such strains differ in many genetic characteristics other than their capacity to recycle H_2. In a more recent evaluation of possible benefits from H_2 recycling, Nelson (1983) reported that pea (*Pisum sativum* L.) inoculated with groups of Hup⁺ and Hup⁻ *R. leguminosarum* strains and grown for 35 days in pot cultures at three irradiance levels in some cases, showed a negative effect of Hup on dry weights and N contents of plants. However, the rates of C_2H_2 reduction and H_2 loss from nodules formed by coupled Hup⁺ inocula revealed that 31% of the nitrogenase electron flow was consumed in H_2 evolution. This report and several others (Eisbrenner and Evans, 1983) have failed to demonstrate that the Hup$^+$ inocula utilized in experiments produced nodules that recycled a great majority of the H$_2$ that was evolved during N$_2$ fixation. Despite all of the imperfections of short-term experiments that compared poorly defined Hup$^+$ and Hup$^-$ strains, eight of the experiments have revealed significant increases in favor of the Hup$^+$ inocula (Eisbrenner and Evans, 1983; Nelson, 1983).

Assuming that H$_2$ recycling capability is a desirable trait, the techniques of molecular genetics may be useful in the construction of improved strains and the transfer of the determinants for Hup into other species that lack Hup activity. Cantrell et al. (1983) have constructed a gene bank from a Hup$^+$ strain of *R. japonicum* and selected a cosmid (pHU1) with 25 kb of insert DNA that complements three out of six Hup$^-$ *R. japonicum* mutants. Data obtained by transposon mutagenesis of pHU1 and transfer of the mutated genes into the genome of a Hup$^+$ strain of *R. japonicum* indicates that *hup*-specific sequences span about 15 kb in pHU1 (Haugland et al., 1984). Since pHU1 gives complementation of only three of six Hup$^-$ *R. japonicum* mutants and does not confer Hup activity upon Hup$^-$ *R. japonicum* wild-type strains, new cosmids have been isolated that contain part of pHU1 and an additional 20 kb of DNA adjacent to it. It is hoped that these new cosmids will include all the determinants for H$_2$ uptake expression.

For more comprehensive discussions of the relationship of H$_2$ loss during N$_2$ fixation to efficiency of energy utilization, see reviews by Eisbrenner and Evans (1983) and Robson and Postgate (1980).

EVALUATIONS OF HUP USING GENETICALLY DEFINED STRAINS

To make valid assessments of advantages from H$_2$ recycling capability, Hup$^+$ and Hup$^-$ strains of *Rhizobium* are needed that are identical with the exception of a point mutation in a Hup determinant. Hup$^-$ mutants and revertants that occur at frequencies that would be expected for point mutations should be appropriate for tests of the two inoculant types. Also, Hup$^-$ mutants that have been complemented by incorporation of Hup DNA by recombinant techniques should be useful. The experiments to be described have utilized strains of this type. Some of the experiments were conducted during relatively short growth periods whereas others were carried out in large cylinders where plants could be grown to maturity.

Materials and Methods

Inoculant Strains. The *R. japonicum* strains PJ17-1 and PJ18-1 are Hup$^+$ revertants of the Hup$^-$ mutants PJ17 and PJ18, respectively, which were selected by their ability to grow chemolithotrophically as described by Lepo et al. (1981). PJ17-19 and PJ17-1-20 are reisolates of PJ17 and PJ17-1, respectively. Strain PJ18nal (Cantrell et al., 1983) is a spontaneous nalidixic, acid-resistant derivative of strain PJ18.

The Hup$^+$ strain PJ18nalHR was obtained by transferring the *hup* cosmid pHU1 into PJ18nal by conjugation as described by Cantrell et al. (1983). PJ18nal (pHU1) transconjugants were identified by the tetracycline resistance marker carried on pHU1 and then were tested for Hup activity by an H$_2$-dependent methylene blue colony assay (Haugland et al., 1983). A resultant Hup$^+$, PJ18nal (pHU1) transconjugant was cultured in yeast extract mannitol broth in the absence of tetracycline to allow growth of segregants that no longer carried the cosmid. Cells were then plated on HUM plates containing nalidixic acid at 50 μg/mL. The resultant colonies were tested for Hup activity as described above and for tetracycline sensitivity by replica plating onto an H$_2$ uptake medium containing tetracycline at 100 μg/mL. Strain PJ18nalHR was one of the Hup$^+$, tetracycline-sensitive isolates identified by the procedure.

On the basis of genetic evidence, PJ18nalHR is a Hup$^+$ derivative of PJ18nal and arose as a result of marker exchange between the wild-type *hup* sequence on the cosmid pHU1 and the mutant *hup* sequence in the genome of PJ18nal, followed by segregational loss of pHU1. Strains PJ18nal and PJ18nalHR should, therefore, be identical with the single exception of the mutated *hup* gene in PJ18nal.

Leonard Jar Experiment (Table 1). Wilkin soybean seed were surface disinfected with 2% sodium hypochlorite, germinated at 28 C, then inoculated by immersing in a suspension containing about 7.0 x 10^8 cells per mL. Germinated seeds were transferred to sterilized Leonard jar assemblies (20-cm upper pot) containing sterile washed river sand (Vincent, 1970). Each assembly was provided with 1500 mL of a one-half strength nitrogen-free nutrient solution (Harper and Nicholas, 1976) supplemented with NiCl$_2$ and CoCl$_2$, each at 0.05 μM. The solution was replaced weekly. After 7 days, plants were thinned to three per pot. Eight replicate cultures of each treatment were arranged in randomized complete blocks in a glasshouse and grown during November and December, 1982. They were provided with day light and supplemental light (16 h per day) the total of which at mid-day was about 350 μE/m^2·h. Plants were harvested after 40 days.

Concrete Cylinder Experiment (Table 2). Surface-disinfected Wilkin seed were planted on 31 May 1983 in concrete cylinders (106 cm diameter by 48 cm height) each containing 522 kg of a mixture of 90% sand, 10% peat. The mixture received sufficient CaCO$_3$ to adjust the pH to 6.8 and was fertilized with the following (kg/ha): P as K$_2$HPO$_4$, 45; K as K$_2$HPO$_4$, 67: Mg as MgSO$_4$·7H$_2$O, 22; Na$_2$MoO$_4$·2H$_2$O, 1.2; CoCl$_2$·6H$_2$O, 0.28. Plants were watered twice weekly by a mist irrigation system and after 17 June each cylinder was supplied twice weekly with 1 L of an N-free nutrient solution (see Leonard jar experiment above). At planting, seeds in each two 1-m rows in each cylinder were inoculated with 60 mL of a broth culture of either PJ18nal or PJ18nalHR containing about 6.9 x 10^8 cells per mL. After one week, plants were thinned to 5-cm intervals and inoculated again as described above. The experiment was arranged in randomized complete blocks with 10 replicates of each treatment. Two additional cylinders were planted without inoculation. Plants were harvested after 141 days. Prior to harvest, senescent leaves were collected when they became severely chlorotic. At 90 days after planting, two plants outside each 1-meter row were harvested for nodule measurements.

Analyses. C$_2$H$_2$ reduction activity, H$_2$ evolution and H$_2$ uptake by bacteroids were measured as described or referred to by Hanus et al. (1981). All plant material from each replicate culture or plot was harvested and dried at 74 C. After grinding the entire sample, a sub-sample of 30 g was removed, ground to pass a 40-mesh screen, dried overnight at 74 C and a 200 mg sample weighed for micro-Kjeldahl analysis using the titration procedure described by Bergersen (1980). Nitrogen analyses presented are means of duplicate titrations on all of the replicate samples from each experiment. Analyses of variance were performed by standard procedures.

Results

Leonard Jar Experiment (Table 1). Several short term experiments have been conducted in pot cultures to compare mutants and revertants of *R. japonicum* as inoculants for soybeans. The results of the experiment presented in Table 1 are typical of those obtained. Both Hup$^+$ revertant strains (PJ17-1-20 and PJ18-1) produced nodules which recycled essentially all the H$_2$ from the nitrogenase reaction, whereas nodules formed by the Hup$^-$ mutants PJ17-19 and PJ18 lost H$_2$ at rapid rates. In the PJ17-19

Table 1. Comparison of Hup⁻ mutants of *Rhizobium japonicum* with their Hup⁺ revertants as inoculants for soybean plants grown in modified Leonard jars.[a]

Strain	Pheno- type	H_2 evolu- tion	Nodule wt	Shoot wt	Root wt	Total wt	N as % of total	Total N
		$\mu mol/h \cdot g^b$	g	g	g	g	%	g/culture
PJ17-19	Hup⁻	1.83	0.64	9.00	3.04	12.68	3.37	0.43
PJ17-1-20	Hup⁺	0.03	0.85	11.50	3.49	15.84	3.38	0.54
PJ18	Hup⁻	1.45	0.57	8.03	2.78	11.38	3.45	0.39
PJ18-1	Hup⁺	0.03	0.67	9.19	3.18	13.04	3.38	0.44
Strain effect significance (P≤):		0.001	0.01	0.005	0.10	0.001	NS	0.005
LSD ≤ 0.05		0.64	0.15	1.70	–	1.89	–	0.067
LSD ≤ 0.01		0.88	0.20	2.32	–	2.57	–	0.092

[a] All data are expressed (dry wt basis per culture unless otherwise indicated) as means of measurements on eight replicate cultures harvested 40 days after planting.

[b] Expressed on a fresh wt basis.

and PJ17-1-20 comparison, use of the Hup⁺ strain resulted in significant increases in weights of nodules, shoots, total plant material and in total N per culture. The Hup⁺ inoculant in the PJ18 and PJ18-1 comparison also produced plants with increased weights and N contents, but with the exception of nodule weights, other differences were not significant at $P \leq 0.05$. In this experiment variability among replicates was considerable as evidenced by coefficients of variation ranging from 15 to 21%. In other comparisons of Hup⁻ mutants and their Hup⁺ revertants as inoculants for soybean plants in pot cultures grown for 35 to 40 days, some have shown increases in growth and N contents from use of Hup⁺ strains, whereas no effects could be measured in others. However, Dr. Daniel Israel (pers. commun.) has observed statistically significant increases in total N fixed by use of Hup⁺ strains PJ17-1-20 and PJ18-1 when compared with Hup⁻ strains PJ17-20 and PJ18 as inoculants in experiments in which large pots containing soybean plants were grown to maturity.

Cylinder Experiment (Table 2). This experiment was designed to compare a Hup⁻ *R. japonicum* mutant with a Hup⁺ stable derivative obtained by recombinant techniques. Use of large cylinders made it feasible to allow the plants to reach maturity. As shown in Table 2, there was no significant difference from use of the two strains on numbers, weights, or C_2H_2 reduction rates of nodules. Rates of H_2 evolution and bacteroid H_2 uptake after 90 days of growth clearly demonstrated that nodules formed by the Hup⁺ PJ18nalHR recycled essentially all H_2 produced. These data, together with the fact that the weight of plants produced on the non-inoculated plots was 1.5% of that of inoculated plots, show that cross contamination was a negligible factor. At harvest, plants inoculated with the Hup⁺ PJ18nalHR, when compared with plants inoculated with the Hup⁻ PJ18nal, produced the following increases in weight: Seed, 5.2%; leaves, 14%, roots, 16%; and total plant material, 9.0%. Differences in leaf weight and total weight were significant at $P \leq 0.05$, whereas increases in weights of seed and roots were significant at $P \leq 0.07$ and 0.1, respectively. Furthermore, use of the Hup⁺ PJ18nalHR produced seed, leaves and total plant material with significantly greater percentage N and total N than plants inoculated with the Hup⁻ PJ18nal. Although the total weight of roots of plants nodulated with the Hup⁺ strain was 16%

Table 2. Comparison of a Hup⁻ mutant of *Rhizobium japonicum* with a Hup⁺ derivative of the mutant as inoculants for soybeans in a cylinder experiment.[a]

	Strain		Strain effect significance	Increase over PJ18nal
	PJ18nalHR Hup⁺	PJ18nal Hup⁻		
Nodule Function				
Nodulation				
no/plant	22.7	21.1	NS	
g/plant	0.56	0.48	NS	
C_2H_2 reduction, μmol/h·g[b]	9.4	7.4	NS	
H_2 evolution, μmol/h·g[b]	0.11	7.18	0.001	
Bacteroid H_2 uptake, nmol/ min·mg protein	115	0.1	0.001	
Yield				
Seed, g/plot	514.9	489.6	0.07	5.2
Leaves, g/plot	377	332	0.05	14
Stems and hulls, g/plot	420	389	NS	–
Roots, g/plot	100	86	0.1	16
Total biomass, g/plot	1412	1296	0.05	9.0
N content				
Seed N				
%	6.87	6.65	0.005	3.3
g/plot	35.4	32.6	0.02	8.6
Leaf N				
%	1.86	1.66	0.04	12
g/plot	7.01	5.51	0.03	27
Stem and hull N				
%	1.15	1.07	NS	–
g/plot	4.83	4.16	NS	–
Root N				
%	1.40	1.60	0.09	-13
g/plot	1.40	1.38	NS	–
Total plant N				
%	3.42	3.35	NS	–
g/plot	48.3	43.4	0.03	11

[a]All data are expressed on a dry weight basis (unless otherwise indicated) and are means of determinations on materials from each of 10 replicate plots.
[b]Expressed per g of fresh nodules.

greater than that of plants inoculated with the Hup⁻ mutant, the total N contents of roots from plants inoculated with the different strains were about the same. From the results of the cylinder experiment, it was concluded that the Hup⁺ PJ18nalHR was superior to the Hup⁻ mutant PJ18nal as an inoculant for soybeans.

Discussion

We have pointed out that strong evidence supports the view that H_2 recycling within nodules of legumes results in potential benefits to the legume-*Rhizobium* symbiosis. Whether the potential advantages from H_2 recycling that have been demonstrated in vitro actually function in vivo and increase growth and N_2 fixation of nodulated legumes has to be determined by growth trials. Unfortunately, the great majority of investigators have utilized poorly defined Hup$^+$ and Hup$^-$ wild-type or field isolates for their evaluations of in vivo benefits of H$_2$ recycling (Eisbrenner and Evans, 1983). Furthermore, these experiments have compared Hup$^-$ strains with Hup$^+$ strains possessing an inefficient H$_2$ recycling system. As a consequence, pea nodules formed by these Hup$^+$ strains have lost H$_2$ at rates that are as high as 31% of the nitrogenase electron flux (Nelson and Child, 1981; Nelson, 1983). In the experiments with pea and our own pot culture experiments with soybean (Table 1), plants have been harvested after relatively short periods; for example, 23 to 35 days for pea and 35 to 40 days for soybean. The time course of N$_2$ fixation by field grown soybean plants was examined by Hardy et al. (1971) who reported that N$_2$ fixation was initiated at a measurable rate near 25 days after planting. Total N fixed proceeded logarithmically and began to plateau near 125 days after planting. From the curve showing cumulative N fixed during growth (Hardy et al., 1971) it is apparent that the amount of N fixed by soybeans after 40 days was less than 10% of the total N$_2$ fixed during growth for 125 days. Assuming that the shape of the growth curve for pea plants during a growth period of 70 days is similar to that of soybeans, we have calculated that the amount of N fixed by pea plants in 35 days would be less than 15% of the amount of N fixed during a 70 day growth period. It is apparent that the evaluation of in vivo beneficial effects of H$_2$ recycling requires carefully selected strains, the Hup$^+$ ones of which should efficiently recycle H$_2$. Also, well replicated experiments need to be conducted that allow plants to reach maturity. Such conditions are essential for the compounding of the beneficial effect of H$_2$ recycling during the logarithmic phase of growth. Our experiment in large cylinders, and also the experiment by Rainbird et al. (1983), has indicated that use of Hup$^+$ inocula substantially increased weights of roots. In controlled experiments where adequate nutrients and water are available the advantages of a more extensive root system may be less than that expected under field conditions where water and nutrients may be limiting.

Three experiments have been conducted with Hup$^+$ and Hup$^-$ *Rhizobium japonicum* as inoculants for soybean plants in which the plants were allowed to grow to maturity. They include a comparison of groups of Hup$^+$ and Hup$^-$ field isolates (Hanus et al., 1981); the cylinder experiment with genetically defined strains described in this paper; and a large pot experiment comparing the Hup mutants and revertant strains listed in Table 1 (Dr. Daniel Israel, pers. commun.). All of these tests have shown statistically significant increases from use of the Hup$^+$ strains.

NOTES

Harold J. Evans, F. Joe Hanus, Richard A. Haugland, Michael A. Cantrell, Liang-Shu Xu, Sterling A. Russell, Grant R. Lambert and Alan R. Harker, Laboratory for Nitrogen Fixation Research, Oregon State University, Corvallis, OR 97331.

We express our appreciation to Mr. David Dalton for reading this paper and to Mrs. Sheri Woods-Haas for typing the manuscript. The work was supported by NSF Grant PCM 81-18148, U.S. Department of Agriculture CRGO Grant 82-CRCR-1-1073, and The Oregon Agricultural Experiment Station from which this is paper number 7186.

REFERENCES

Bergersen, F. J. 1980. Measurement of nitrogen-fixation by direct means. p. 65-111. *In* F. J. Bergersen (ed.) Methods for evaluating biological nitrogen fixation. John Wiley and Sons, New York.

Cantrell, M. A., R. A. Haugland and H. J. Evans. 1983. Construction of a *Rhizobium japonicum* gene bank and use in the isolation of a hydrogen uptake gene. Proc. Natl. Acad. Sci. 80:181-185.

Dixon, R. O. D., E. A. G. Blunden and J. W. Searl. 1981. Inter-cellular space and hydrogen diffusion in root nodules of pea (*Pisum sativum*) and lupine (*Lupinus alba*). Plant Sci. Lett. 23:109-116.

Drevon, J. J., L. Frazier, S. Russell and H. J. Evans. 1982. Respiratory and nitrogenase activities of soybean nodules formed by hydrogen uptake negative (Hup⁻) and revertant strains of *Rhizobium japonicum* characterized by protein patterns. Plant Physiol. 70:1341-1346.

Eisbrenner, G. and H. J. Evans. 1983. Aspects of hydrogen metabolism in nitrogen-fixing legumes and other plant-microbe associations. Ann. Rev Plant Physiol. 34:105-137.

Emerich, D. W., T. Ruiz-Argueso, T. M. Ching and H. J. Evans. 1979. Hydrogen dependent nitrogen ase activity and ATP formation in *R. japonicum* bacteroids. J. Bacteriol. 137:153-160.

Evans, H. J., G. Eisbrenner, M. A. Cantrell, S. A. Russell and F. J. Hanus. 1982. The present status of hydrogen recycling in legumes. Israel J. Bot. 31:72-88.

Hanus, F. J., S. L. Albrecht, R. M. Zablotowicz, D. W. Emerich, S. A. Russell and H. J. Evans. 1981. The effect of the hydrogenase system in *Rhizobium japonicum* inocula on the nitrogen content and yield of soybean seed in field experiments. Agron. J. 73:368-372.

Hardy, R. W. F., R. C. Burns, R. R. Hebert, R. D. Holsten and E. K. Jackson. 1971. Biological nitrogen fixation: a key to world protein. p. 561-590. *In* T. A. Lie and E. G. Mulder (eds.) Plant and Soil, Special Volume. Martinus Nijhoff, The Hague.

Harper, J. E. and J. C. Nichols. 1976. Control of nutrient pH with an ion exchange system: Effect on soybean nodulation. Plant Physiol. 38:24-28.

Haugland, R. A., M. A. Cantrell, J. S. Beaty, F. J. Hanus, S. A. Russell and H. J. Evans. 1984. Characterization of *R. japonicum* Hup gene cosmids. p. 684. *In* C. Veeger and W. E. Newton (eds.) Advances in nitrogen fixation research. Nijhoff/Junk, The Hague.

Haugland, R. A., F. J. Hanus, M. A. Cantrell and H. J. Evans. 1983. A rapid colony screening method for identifying hydrogenase activity in *Rhizobium japonicum*. Appl. Environ. Microbiol. 45:892-897.

Klucas, R. V., F. J. Hanus, S. A. Russell and H. J. Evans. 1983. Nickel: A micronutrient element for hydrogen-dependent growth of *Rhizobium japonicum* and for expression of urease activity in soybean leaves. Proc. Natl. Acad. Sci. 80:2253-2257.

Lepo, J. E., R. E. Hickok, M. A. Cantrell, S. A. Russell and H. J. Evans. 1981. Revertible hydrogen-uptake deficient mutants of *Rhizobium japonicum*. J. Bacteriol. 146:614-620.

Nelson, L. M. 1983. Hydrogen recycling by *Rhizobium leguminosarum* isolates and growth and nitrogen contents of pea plants. Appl. Environ. Microbiol. 45:856-861.

Nelson, L. M. and J. J. Child. 1981. Nitrogen fixation and hydrogen metabolism by *Rhizobium leguminosarum* isolates in pea (*Pisum sativum*) root nodules. Can. J. Microbiol. 27:1028-1034.

Rainbird, R. M., C. A. Atkins, J. S. Pate and P. Sanford. 1983. Significance of hydrogen evolution in the carbon and nitrogen economy of nodulated cowpea. Plant Physiol. 71:122-127.

Robson, R. L. and J. R. Postgate. 1980. Oxygen and hydrogen in biological nitrogen fixation. Ann. Rev. Microbiol. 34:183-207.

Vincent, J. M. 1970. A Manual for the Practical Study of Root-Nodule Bacteria. Blackwell Scientific Publications, Oxford.

ENVIRONMENTAL INTERACTIONS INFLUENCING INNOVATIVE
PRACTICES IN LEGUME INOCULATION

J. Brockwell

As an act of applied ecology, legume inoculation invariably involves interactions with biotic and abiotic features of the environment. The host plant is the major interactant, and the microflora, especially other rhizobia, have a significant role in determining the fate of the inoculant. The nature of carriers for inoculants and the physical characteristics of the soil into which they are introduced are important also. The progress made since World Soybean Research Conference II in further understanding some of these interactions is considered in this paper. Emphasis is placed on *Rhizobium japonicum* and its symbiosis with the soybean.

INOCULANT PREPARATION

While peat remains the favored base for legume inoculants, the search for alternative carrier materials continues. Substances recently promoted include polymers and vegetable oils. Dommergues et al. (1979) reported that an inoculant in which *R. japonicum* was entrapped in polyacrylamide gel survived and nodulated soybeans as well as a peat-based inoculant. The concept has been extended to the use of other polymers (Jung et al., 1982). Good survival of bacteria depends upon the maintenance of moisture in the gels. Kremer and Peterson (1982) prepared inoculants, by resuspending lyophilized cultures of *Rhizobium* in vegetable oil, that survived on seed as well as or better than peat cultures, and performed well in field trials (Kremer and Peterson, 1983). The daunting task of evaluating these novel inoculants in diverse environments with many species is before us.

The usual means of making peat inoculant uses broths containing large populations of rhizobia to inoculate the peat (e.g., Roughley and Vincent, 1967). Somasegaran and Halliday (1982) found that this practice may be unnecessary. Populations of rhizobia that developed in their peat cultures reached similarly high levels within a week of inoculation with broths of *Rhizobium* spp. varying between 10^6 and 10^8 bacteria/ml. This indicated that the peat itself might be the factor limiting populations of rhizobia in inoculants, not the concentration of bacteria in the broths used to make the inoculants.

Poor quality inoculants continue to find their way to the market. Evaluation of quality by enumeration of viable rhizobia is an accurate index of inoculating potential (Hiltbold et al., 1980). Vincent and Smith (1982) concluded that many retail samples of preinoculated seed carried less than optimum levels of viable inoculant, but that the quality of preinoculated seed could be improved by seed pelleting. Supplementation of inoculants with fungicide and/or Mo was associated with low numbers of viable rhizobia (Skipper et al., 1980). Gault and Brockwell (1980) observed that, of four Mo compounds, only sodium molybdate was harmful to R. meliloti and R. trifolii.

The peat itself may also determine inoculant quality. Increasing the gamma irradiation dosage for sterilizing peat resulted in larger populations of rhizobia in inoculants for soybean and alfalfa (Medicago sativa L.) (Strijdom and van Rensburg, 1981). In Australia, an unidentified inhibitor in a particular peat, which is specific against R. lupini inoculant strain WU425, is posing problems for manufacturers and the quality control authority (Roughley, pers. comm.). Schiffmann (1982) overcame a similar difficulty with slow-growing strains of rhizobia by culturing them in the peat together with unnamed, fast-growing, slime-producing bacteria which, he postulated, exerted a 'protective influence' on the slow-growers.

ECOLOGICAL INTERACTIONS

Sensitivity of soybean inoculants to contact and systemic fungicides applied to seed influences the location of root nodules (Chamber and Montes, 1982). On the other hand, Lennox and Alexander (1981) demonstrated that a thiram-resistant strain of R. phaseoli enhanced nodulation of beans (Phaseolus vulgaris L.) when the seed had been treated with thiram. This was attributed to thiram-induced suppression of protozoa, especially ciliates which are predatious on rhizobia, without any concomitant reduction in numbers of the inoculant.

Synergism between rhizobia and mycorrhiza is well known (e.g., Smith and Daft, 1977). More recently, enhancement of legume nodulation by mixed cultures of Rhizobium and Azotobacter has been described by Burns et al. (1981) and Maurya and Sanoria (1982). Burns et al. suggest that a non-excretable protein produced by A. vinelandii is involved in the improved nodulation. The beneficial effect of one strain of rhizobia upon another has been reported by Rolfe et al. (1980), who made the curious observation that a non-invasive mutant of R. trifolii, when mixed with an ineffective mutant, gave rise to effective nodules. In some way the presence of the invasive organism permitted entry into the plant of a non-invasive strain.

Interactions of Rhizobium spp. involving antagonism generated by other soil microorganisms, physical attributes of the environment, and factors influencing mobility in soil are dealt with by Roughley (1984) in a companion paper in these proceedings.

Recent evidence indicates that the soybean-R. japonicum association is more complex than previously believed. Bromfield and Roughley (1980) found that 'tropical' soybean cv. Malayan nodulated equally well with both R. japonicum and Rhizobium sp. isolates from cowpea (Vigna unguiculata [L.] Walp.), but that the Rhizobium sp. isolates failed to nodulate 'American' soybeans. Keyser et al. (1982) located fast-growing strains of rhizobia that were ineffective (or poorly effective) with 'American' soybean cultivars, but effective for black-seeded Peking. Keyser (pers. comm., 1983) also reported interactions in nitrogen fixation between the associations of different lines of Glycine soja Sieb. and Zucc. (same subgenus, Soja, as soybean) with fast- and slow-growing strains of R. japonicum. Other species of Glycine (subgenus Glycine) fix nitrogen with cowpea rhizobia, but not with slow-growing R. japonicum (Brockwell, unpublished). These findings represent a significant extension to the known genetic bases of R. japonicum (Keyser et al., 1982) and soybeans.

RATES OF INOCULATION

The act of legume inoculation consists of introducing a fragile organism in the vegetative state into an environment that is usually physically, chemically and biotically hostile. Even with strains of rhizobia that have been selected for tolerance of particular environmental hazards, it is well recognized that mortality levels of the inoculant are likely to be high. Accordingly, it has become normal practice to inoculate at high rates.

Rhizobia resident in soil pose a competitive threat to inocula. It is not generally appreciated to what extent inoculant rhizobia may be grossly outnumbered by apparently small populations of soil rhizobia (Brockwell, 1981). A population of R. japonicum CB1809 established in the soil for one year colonized the rhizospheres of soybeans better than an inoculant comprising a streptomycin-resistant mutant of CB1809, and the dominance increased with time (Brockwell et al., 1984). Compensation for an overall numerical disadvantage can be made by strategic placement of high rates of inoculant into the seedbed close to the point where root infection will occur.

Establishment of inocula in soil previously unoccupied by R. japonicum has received some recent attention. Smith et al. (1980), using increasing rates of frozen broth culture delivered into the seedbed, concluded that inoculation levels above log. 5.0 bacteria/cm of row were needed to establish effective nodulation in a R. japonicum-free soil. The data are consistent with a very high mortality rate of the rhizobia comprising the inoculum, but whether this was due to the form of the inoculum or to environmental factors is unclear. Similar work in Australia showed that, when sampled 2 h after inoculation, suspensions of peat inoculant sprayed into the seedbed were irregularly distributed through the soil profile (Herridge et al., 1984), and that survival of inocula was strongly influenced by weather conditions at sowing and immediately afterwards (Brockwell, unpublished data).

Although inoculation of a prior crop of a non-legume with R. japonicum can be used to establish rhizobia in the soil, providing inoculum for a subsequent crop of soybeans (Gaur et al., 1980), Kuykendall et al. (1982) have demonstrated that, when nodulating and non-nodulating isolines of soybeans and other legumes are planted in soil inoculated with R. japonicum, best establishment of the bacteria occurs via the nodulating line of soybean. This suggests that rhizobia multiply more prolifically within the nodule than elsewhere, and/or that 'nodule' rhizobia (recently released from nodules) are better equipped for survival than 'rhizosphere' or 'soil' rhizobia.

Once established in soil, R. japonicum appears to persist even in the absence of soybean crops (Crozat et al., 1982), although there may be seasonal fluctuations in the size of the populations (Mahler and Wollum II, 1982). However, appreciable growth of rhizobia in soil apparently occurs only in the presence of germinating seeds, growing roots and decomposing nodules (Pena-Cabriales and Alexander, 1983).

In certain field soils, mineralization of organic N and nitrification may provide levels of NO_3^- that satisfy the N requirements of young soybeans, but inhibit nodulation. When the NO_3^- is exhausted, the plants may enter a N-deficient phase, during which sufficient nodules are being formed to compensate for loss of soil N. If this period is prolonged, yield may be reduced. Herridge et al. (1984), working with one such soil, found that increasing the rates of inoculant applied into the seedbed led to more extensive distribution, numerically and in depth, of rhizobia through the soil. Also, there were greater numbers of rhizobia in the rhizosphere, improved nodulation and nitrogen fixation, and larger residual populations in the soil after harvest. They argued that concentrations of NO_3^- in soil water were not uniform, that those parts of the root system exposed to low concentrations of NO_3^- would nodulate first, and that these conditions would most likely be satisfied when large populations of rhizobia were extensively distributed through the soil by means of heavy rates of inoculation.

In another field experiment, the natural abundance of ^{15}N was measured in nodulated (various rates of inoculation) and non-nodulated soybeans at six intervals during growth (Turner et al., 1984). Good agreement was obtained between figures for nitrogen fixed, calculated from $\delta^{15}N$ values, and those calculated from differences in N yield between nodulated and non-nodulated treatments. Substantially more nitrogen was fixed where soil N had been depleted by a prior cereal crop. The proportion of plant N derived from atmospheric nitrogen increased with time as soil N was further depleted and reached >80% between 98 and 114 days with the highest rate of inoculation.

STRATEGIC INOCULATION AND PLANT RESPONSE

The placement of seed inoculant in the soil is usually relatively remote from the location of infection foci on the seedling roots, a circumstance likely to be exacerbated by inefficient natural transport of rhizobia. This is one of several reasons that justify the use of alternative means of inoculation (Brockwell et al., 1980). Granular inoculant applied into the seed row is a well-established and successful commercial practice (Muldoon et al., 1980). Liquid inoculants sprayed into the seedbed (Hely et al., 1980), less widely used, are equally successful in promoting nodulation (Hale, 1981), although some loss of viability may occur should inoculant circulate in the pumping equipment for >4 h (Gault et al., 1982).

Hydroseeding mixtures containing legume seed and rhizobia are a form of liquid inoculant. Satisfactory survival of rhizobia in those mixes that contain fertilizer depends on a pH value >6.0 (Brown et al., 1983). Survival of rhizobia in such slurries might be improved further by suspending the inoculant in vegetable oil (e.g., Kremer et al., 1982).

Post-emergence inoculation of uninoculated legumes was successful in inducing satisfactory nodulation on birdsfoot trefoil (*Lotus corniculatus* L.) and alfalfa (Rogers et al., 1982) and on soybeans (Boonkerd et al., 1984). Under different conditions, there was no benefit of post-emergence inoculation for non-nodulated soybeans in either glasshouse or field (Gault et al., 1984). It seems clear that, for post-emergence inoculation to be successful, soil moisture must be satisfactory. In addition the susceptibility of legume roots is a transient phenomenon (Bhuvaneswari et al., 1981), which suggests that infection foci on a rapidly growing root system might soon become too remote from the site of post-emergence inoculation for adequate infection and nodulation to occur.

Interesting work has been done on the means by which soybeans compensate for symbiotic deficiencies. Singleton and Stockinger (1983) inoculated soybeans with various mixtures of effective and ineffective strains of *R. japonicum*. As anticipated, they found that the proportion of effective nodules formed increased as the ratio of effective to ineffective bacteria became greater in the inoculant. However, the total volume of effective nodule tissue remained approximately constant throughout. This was regarded as a 'compensatory mechanism' for keeping the amount of effective nodule tissue constant even as the proportion of effective nodules declined.

In soybeans, the nitrogenous products of nitrogen fixation are mainly ureides (e.g., Herridge, 1982a), and the proportion of nitrogenous solutes as ureides can be used as an index of nitrogen fixation (e.g., Herridge, 1982b). This procedure was used to compare the N economies of nodulated and non-nodulated soybeans grown in a soil that initially contained high levels of NO_3^- throughout the profile but no *R. japonicum* (Herridge et al., 1984). The results indicated that non-nodulated plants compensated for symbiotic deficiencies by more efficient exploitation of soil N and by more efficient redistribution of vegetative N into grain N. Thus, the act of inoculation leading to effective nodulation was a means of conserving soil N. The physiology of these compensatory mechanisms is not understood.

CONCLUDING REMARKS

The development of suitable carriers for inoculants, particularly for those parts of the world where no deposits of peat occur naturally, remains a high priority. Much promise attaches to oil-based inoculants. The task of exhaustive testing across a range of environments is formidable indeed. A world-wide series of tests along the lines of the INLIT trials (Davis et al., 1984) suggests itself.

The recognition that the genetic bases for effective nodulation in *G. max* (and other *Glycine* spp.) and *R. japonicum* are more complex than previously believed has important implications for soybean improvement programs. In particular, it widens the options for soybean breeders concerned with the development of cultivars suited to the tropics. Meanwhile, the extent and nature of symbiotic diversity within the *G. max-R. japonicum* association remains to be properly defined.

The reputation often attributed to the soybean that it makes little contribution to total soil N may be undeserved. There are indications that, in some circumstances at least, soybean crops may add significantly to the total pool of soil N. Estimation of nitrogen fixation in the field is difficult and resolution is far from satisfactory (e.g., LaRue and Patterson, 1981; Bergersen and Turner, 1983). The N input due to soybean cropping warrants much more investigation.

The 1980s have heralded a surge of interest in the molecular biology of *Rhizobium*. On the practical front, this has been directed towards the construction of organisms with wider host range and greater N_2-fixing capacity. The question arises whether some of this emphasis has been misplaced. It can be argued that (i) populations of rhizobia occurring in nature are so diverse that there is ample scope for natural selection of highly effective strains; and (ii) expression of nitrogen fixation in the field is less constrained by level of effectiveness than by other limiting factors, such as temperature, soil moisture, nutrient balance, photosynthesis, management practices, etc., as well as the ability of inoculant to compete successfully with resident rhizobia. A wealth of literature on strain competition attests to the difficulties that have been experienced in introducing new strains. Understanding the basis of competition and exploiting this in strain construction is, therefore, of major importance (Roughley et al., 1984). There is an indication that the expression of competitiveness in rhizobia depends on a limited number of determinants (Brockwell et al., 1982). Identification of the steps preceding legume root infection by rhizobia is a matter of urgency. Some of these steps may be (i) initial attraction of rhizobia to legume root (chemotaxis), (ii) non-specific multiplication of rhizobia in the rhizosphere, (iii) mutual recognition of host and bacteria, (iv) specific bacterial multiplication in the rhizoplane adjacent to infection foci, (v) adhesion of the bacteria to the root hair surface; followed by further steps leading to nodulation and nitrogen fixation as defined by Vincent (1980). The utilization of this information for constructing strains that are strongly competitive would represent a major breakthrough towards more effective use of legume inoculants. Let us hope

for substantial progress in this area before World Soybean Research Conference
IV.

NOTES

John Brockwell, Division of Plant Industry, CSIRO, G.P.O. Box 1600, Canberra, A.C.T. 2601,
Australia.

REFERENCES

Bergersen, F. J. and G. L. Turner. 1983. An evaluation of ^{15}N methods for estimating nitrogen fixation in a subterranean clover-perennial ryegrass sward. Aust. J. Agric. Res. 34:391-401.

Bhuvaneswari, T. V., A. A. Bhagwat and W. D. Bauer. 1981. Transient susceptibility of root cells in four common legumes to nodulation by rhizobia. Plant Physiol. 68:1144-1149.

Boonkerd, N., C. Arunsri, W. Rungrattanakasin and Y. Vasuvat. 1984. Effects of postemergence inoculation on field grown soybeans. p. 327. In C. Veeger and W. E. Newton (eds.) Advances in nitrogen fixation research. Nijhoff/Junk, The Hague.

Brockwell, J. 1981. A strategy for legume nodulation research in developing regions of the Old World. Plant Soil 58:367-382.

Brockwell, J., R. R. Gault, D. L. Chase, F. W. Hely, M. Zorin and E. J. Corbin. 1980. An appraisal of practical alternatives to legume seed inoculation: field experiments on seed bed inoculation with solid and liquid inoculants. Aust. J. Agric. Res. 31:47-60.

Brockwell, J., R. R. Gault, M. Zorin and M. J. Roberts. 1982. Effects of environmental variables on the competition between inoculum strains and naturalized populations of Rhizobium trifolii for nodulation of Trifolium subterraneum L. and on rhizobia persistence in the soil. Aust. J. Agric. Res. 33:803-815.

Brockwell, J., R. J. Roughley and D. F. Herridge. 1984. Impact of rhizobia established in a high nitrate soil on a soybean inoculant of the same strain. p. 328. In C. Veeger and W. E. Newton (eds.) Advances in nitrogen fixation research. Nijhoff/Junk, The Hague.

Bromfield, E. S. P. and R. J. Roughley. 1980. Characterisation of rhizobia from nodules on locally-adapted Glycine max grown in Nigeria. Ann. Appl. Biol. 95:185-190.

Brown, M. R., D. D. Wolf, R. D. Morse and J. L. Neal. 1983. Viability of Rhizobium in fertilizer slurries used for hydroseeding. J. Environ. Qual. 12:388-390.

Burns Jr., T. A., P. E. Bishop and D. W. Israel. 1981. Enhanced nodulation of leguminous plant roots by mixed cultures of Azotobacter vinelandii and Rhizobium. Plant Soil 62:399-412.

Chamber, M. A. and F. J. Montes. 1982. Effects of some seeds disinfectants and methods of rhizobial inoculation on soybeans (Glycine max L. Merrill). Plant Soil 66:353-360.

Crozat, Y., J. C. Cleyet-Marel, J. J. Giraud and M. Obaton. 1982. Survival rates of Rhizobium japonicum populations introduced into different soils. Soil Biol. Biochem. 14:401-405.

Davis, R. J., J. Halliday and F. B. Cady. 1984. Preliminary data from legume inoculation trials. p. 333. In C. Veeger and W. E. Newton (eds.) Advances in nitrogen fixation research. Nijhoff/Junk, The Hauge.

Dommergues, Y. R., H. G. Diem and D. Divies. 1979. Polyacrylamide-entrapped Rhizobium as an inoculant for legumes. Appl. Environ. Microbiol. 37:779-781.

Gault, R. R., L. W. Banks, D. L. Chase and J. Brockwell. 1984. Salvage of soybean crops lacking nodulation. p. 105-106. In I. R. Kennedy and L. Copeland (eds.) AIAS Occ. Publ. No. 12. Aust. Inst. Agric. Sci., Sydney.

Gault, R. R. and J. Brockwell. 1980. Studies on seed pelleting as an aid to legume inoculation. 5. Effects of incorporating molybdenum in the seed pellet on survival, seedling nodulation and plant growth of lucerne and subterranean clover. Aust. J. Exp. Agric. Anim. Husb. 20:63-71.

Gault, R. R., D. L. Chase and J. Brockwell. 1982. Effects of spray inoculation equipment on the viability of Rhizobium spp. in liquid inoculants for legumes. Aust. J. Exp. Agric. Anim. Husb. 22:299-309.

Gaur, Y. D., A. N. Sen and N. S. Subba Rao. 1980. Improved legume-*Rhizobium* symbiosis by inoculating preceding cereal crop with *Rhizobium*. Plant Soil 54:313-316.

Hale, C. N. 1981. Methods of white clover inoculation—their effect on competition for nodule formation between naturalised and inoculated strains of *Rhizobium trifolii*. N. Z. J.Exp. Agric. 9:169-172.

Hely, F. W., R. J. Hutchings and M. Zorin. 1980. Methods of rhizobial inoculation and sowing techniques for *Trifolium subterraneum* L. establishment in a harsh winter environment. Aust. J. Agric. Res. 31:703-712.

Herridge, D. F. 1982a. Relative abundance of ureides and nitrate in plant tissues of soybean as a quantitative assay of nitrogen fixation. Plant Physiol. 70:1-6.

Herridge, D. F. 1982b. Use of the ureide technique to describe the nitrogen economy of field-grown soybeans. Plant Physiol. 70:7-11.

Herridge, D. F., R. J. Roughley and J. Brockwell. 1984. Effect of rhizobia and soil nitrate on the establishment and functioning of the soybean symbiosis in the field. Aust. J. Agric. Res. 35:149-161.

Hiltbold, A. E., D. L. Thurlow and H. D. Skipper. 1980. Evaluation of commercial soybean inoculants by various techniques. Agron. J. 72:675-681.

Jung, G., J. Mugnier, H. G. Diem and Y. R. Dommergues. 1982. Polymer-entrapped *Rhizobium* as an inoculant for legumes. Plant Soil 65:219-231.

Keyser, H. H., B. B. Bohlool, T. S. Hu and D. F. Weber. 1982. Fast-growing rhizobia isolated from root nodules of soybean. Science 215:1631-1632.

Kremer, R. J. and H. L. Peterson. 1982. Effect of inoculant carrier on survival of *Rhizobium* on inoculated seed. Soil Sci. 134:117-125.

Kremer, R. J. and H. L. Peterson. 1983. Field evaluation of selected *Rhizobium* in an improved legume inoculant. Agron. J. 75:139-143.

Kremer, R. J., J. Polo and H. L. Peterson. 1982. Effect of suspending agent and temperature on survival of *Rhizobium* in fertilizer. Soil Sci. Soc. Am. J. 46:539-542.

Kuykendall, L. D., T. E. Devine and D. B. Cregan. 1982. Positive role of nodulation on the establishment of *Rhizobium japonicum* in subsequent crops of soybean. Curr. Microbiol. 7:79-81.

LaRue, T. A. and T. G. Patterson. 1981. How much nitrogen do legumes fix? Adv. Agron. 34:15-38.

Lennox, L. B. and M. Alexander. 1981. Fungicide enhancement of nitrogen fixation and colonization of *Phaseolus vulgaris* by *Rhizobium phaseoli*. Appl. Environ. Microbiol. 41:404-411.

Mahler, R. L. and A. G. Wollum II. 1982. Seasonal fluctuations of *Rhizobium japonicum* under a variety of field conditions in North Carolina. Soil Sci. 134:317-324.

Maurya, B. R. and C. L. Sanoria. 1982. Effectiveness of rhizobial strains with and without co-inoculants and phosphate on Bengal gram (*Cicer arietinum*). J. Agric. Sci. (Camb.) 99:239-240.

Muldoon, J. F., D. J. Hume and W. D. Beversdorf. 1980. Effects of seed- and soil-applied *Rhizobium japonicum* inoculants on soybeans in Ontario. Can. J. Plant Sci. 60:399-409.

Pena-Cabriales, J. J. and M. Alexander. 1983. Growth of *Rhizobium* in unamended soil. Soil Sci. Soc. Am. J. 47:81-84.

Rogers, D. D., R. D. Warren Jr. and D. S. Chamblee. 1982. Remedial postemergence legume inoculation with *Rhizobium*. Agron. J. 74:613-619.

Rolfe, B. G., P. M. Gresshoff, J. Shine and J. M. Vincent. 1980. Interaction between a non-nodulating and an ineffective mutant of *Rhizobium trifolii* resulting in effective (nitrogen-fixing) nodulation. Appl. Environ. Microbiol. 39:449-452.

Roughley, R. J. 1984. Effect of soil environmental factors on rhizobia. Proc. World Soybean Research Conf. III (this volume).

Roughley, R. J., J. Brockwell and R. A. Date. 1984. Utilization of *Rhizobium* in agriculture. Aust. Microbiol. 5:150.

Roughley, R. J. and J. M. Vincent. 1967. Growth and survival of *Rhizobium* spp. in peat culture. J. Appl. Bact. 30:362-376.

Schiffmann, J. 1982. Biological and agronomic aspects of legume inoculation in Israel. Isr. J. Bot. 31:265-281.

Singleton, P. W. and K. R. Stockinger. 1983. Compensation against ineffective nodulation in soybean. Crop Sci. 23:69-72.

Skipper, H. D., J. H. Palmer, J. E. Giddens and J. M. Woodruff. 1980. Evaluation of commercial soybean inoculants from South Carolina and Georgia. Agron. J. 72:673-674.

Smith, R. S., M. A. Ellis and R. E. Smith. 1980. Effect of *Rhizobium japonicum* inoculant rates on soybean nodulation in a tropical soil. Agron. J. 72:505-508.

Smith, S. E. and M. J. Daft. 1977. Interactions between growth, phosphate content and N_2 fixation in mycorrhizal and non-mycorrhizal *Medicago sativa*. Aust. J. Plant Physiol. 4:403-413.

Somasegaran, P. and J. Halliday. 1982. Dilution of liquid *Rhizobium* cultures to increase production capacity of inoculant plants. Appl. Environ. Microbiol. 44:330-333.

Strijdom, B. W. and H. J. van Rensburg. 1981. Effect of steam sterilization and gamma irradiation of peat on quality of *Rhizobium* inoculants. Appl. Environ. Microbiol. 41:1344-1347.

Turner, G. L., R. R. Gault and F. J. Bergersen. 1984. Measurements of natural abundance of ^{15}N in a soybean crop. p. 7-8. *In* I. R. Kennedy and L. Copeland (eds.) AIAS Occ. Publ. No. 12. Aust. Inst. Agric. Sci., Sydney.

Vincent, J. E. and M. S. Smith. 1982. Evaluation of inoculant viability on commercially inoculated legume seed. Agron. J. 74:921-923.

Vincent, J. M. 1980. Factors controlling the legume-*Rhizobium* symbiosis. p. 103-129. *In* W. E. Newton and W. H. Orme-Johnson (eds.) Nitrogen fixation. II. Univ. Park Press, Baltimore.

PROBLEMS OF SOYBEAN INOCULATION IN THE TROPICS

Peter H. Graham

There are still many countries in the tropics and subtropics for which the soybean is a relatively new crop. Soils in these regions will usually contain few rhizobia infective or effective with introduced soybean cultivars (Zengbe, 1980; IITA, 1982), so inoculation is often needed for a reasonable crop yield. Positive response to inoculation is the norm (Table 1), but inoculation failures do occur (Chesney et al., 1973; Jiminez and Villalobos, 1980; Salema and Chowdhury, 1980; Awai, 1981). Moreover, soil, plant and environmental factors can limit the degree of response, with plant-growth and yield in inoculated treatments less than achieved by N-fertilized controls. Major limiting factors vary with region, but include moisture stress, soil acidity, high temperatures during inoculant transportation and in the soil, and mineral deficiencies and toxicities (Ayanaba, 1977; Dommergues et al., 1980; Freire, 1982). Cultivar specificity in response to inoculation is also a factor.

VARIETAL AND STRAIN SPECIFICITY IN INOCULATION RESPONSE

Modern soybean cultivars derive mostly from the USA, come from a very narrow germplasm base, and show only limited variation in ability to fix N_2 in symbiosis. Whereas some host x strain interaction has been reported previously, it is only with recent interest in Chinese and unimproved third-world landraces, that major differences between cultivars in inoculation response have become apparent. Thus, in studies at the International Institute of Tropical Agriculture (IITA) in Nigeria, uninoculated American cultivars, such as Bossier and Jupiter, remained essentially without nodules, while the agronomically inferior Asiatic cultivars, Malayan and Orba, showed excellent nodulation with native soil rhizobia, and benefited little from inoculation (IITA, 1978; 1979; Nangju, 1980; Pulver et al., 1982; Ranga Rao et al., 1982). When Bosser, the symbiotically promiscuous soybean line TGm 344 and cowpea (*Vigna unguiculata* [L.] Walp.) were used as hosts for MPN counts of rhizobia in three West African soils, Bossier recovered only 3-134 rhizobia/g, whereas counts with TGm 344 and cowpea were 208-1949 and 740-24162 cells/g, respectively. Because of concern over inoculant quality and availability in Africa, breeders at IITA and elsewhere have tried to incorporate symbiotic promiscuity into agronomically superior soybean lines. Unfortunately, several of the promiscuous lines have proved ineffective or only partially effective with a majority of the soil strains against which they were tested (IITA, 1978). As in the USA (Uratsu et al., 1982), a number of soil organisms can be expected to be Hup⁻, and inefficient in N_2 fixation.

As with cultivars, it is mainly US strains and inoculants that have been evaluated

951

Table 1. Response of soybean to inoculation in the tropics and subtropics.[a]

Country	Yield increase[b]	Reference
	%	
Australia	22-63	Bushby et al. (1983)
Bangladesh	38	Sobhan (1978)
Benin	126	Dumont (1981)
Brazil	61	Freire (1978)
	56	Vargas and Suhet (1980)
Cuba	67	Sistachs (1976)
India	14-235	Subba-Rao and Balasundaram (1971)
	446	Pal and Saxena (1975)
	84-92	Dube (1976)
	69	Kaul and Sekhon (1977)
	25-76	Singh and Tilak (1977)
Madagascar	335	Denaire (1968)
Mexico	78	Ramirez-Gama (1982)
Nigeria	88	Kang et al. (1975)
	20-50	IITA (1979)
	93-192	Rango Rao et al. (1982)
Philippines	130-160	Paterno et al. (1979)
	30	Halos et al. (1982)
Puerto Rico	30	Smith et al. (1981)
Senegal	26	Wey (1980)
	71	Wey and Saint Macary (1982)
Sierra Leone	279-360	Haque et al. (1980)
Tanzania	102	Chowdhury (1977)
Thailand	29-603	Boonchee and Schiller (1979)
USA	64	Scudder (1974)

[a]Results of field studies only included. Additional, earlier references are given by Ayanaba (1977).

[b]Response of best strain or treatment compared to the uninoculated control.

in the tropics, and these also are not of diverse origin (USDA, 1979). The recent isolation of fast-growing rhizobia from soybean nodules collected in China (Keyser et al., 1982) shows again the greatest variation in *Rhizobium* to be obtained at the center of origin of a species. These organisms have proved non-effective or ineffective with US soybean cultivars (Keyser et al., 1982; Eaglesham, pers. commun.), but effective symbioses have been reported between them and some tropical and PI lines (van Rensburg et al., 1983; Devine, pers. commun.; Eaglesham, pers. commun.). Further problems of specificity are to be expected as breeders expand the genetic base with which they work and as more national programs initiate breeding activities. Additional *Rhizobium* collection from China and from secondary gene pools in Asia is definitely necessary.

SOIL ACIDITY

Soil acidity is a constraint to agricultural productivity in from 35-70% of tropical soils, and is particularly important for soybean production in Brazil, where more than one million hectares of this crop are grown on initially acid "cerrado" soils. Poor plant growth on acid soils can result from pH effects per se, from Al or Mn excess, from P, Ca or Mo deficiency, or from a combination of these factors.

For the inoculated legume grown in acid soil, problems can include death or failure to multiply of the inoculant strain, poor root-hair or root development, nodulation failure or reduced N_2 fixation. For soybean, the critical pH below which nodulation does not occur has been reported as from pH 4.2-5.0, with nodule number and weight/plant near normal at pH 5.5 and above (Albrecht, 1933; Munns et al., 1977b; Mengel and Kamprath, 1978).

Whereas soil amendment, using lime to raise pH and reduce Al and Mn availability in the soil, is a traditional practice in the tropics and has been a major factor in the successful establishment of soybeans on "cerrado" soils, the current emphasis is more toward low-input technologies, including the selection of acid-tolerant cultivars able to grow at relatively low soil pH. Soybean lines tolerant of acid pH and aluminum (Sartain and Kamprath, 1978; Sapra et al., 1982) or manganese (Carter et al., 1975; Okhi et al., 1980) excess have been identified, though usually in studies which failed to consider the *Rhizobium* component. This is unfortunate because variation in strain sensitivity to soil acidity can complicate the interpretation of apparent host differences, and plants supplied fertilizer N are usually less sensitive to soil acidity than plants dependent on N_2 fixation (Munns et al., 1981). Döbereiner et al. (1965) did contrast nodule development and yield in nine cultivars grown on acid soil. Highest yield was obtained with the cultivar having greatest nodule mass, and nodule mass and yield were reasonably correlated under these conditions.

While the tendency is to group soybean rhizobia with strains from cowpea, as acid tolerant, many soybean strains do not grow well in low pH media. Ayanaba and Wong (1982) found 12 of 75 strains able to grow in pH 4.5 medium with 50 μM Al, while Keyser and Munns (1979) reported 23% of strains capable of growth under these conditions. In this study and in that reported by IITA (1980), soybean strains were clearly less tolerant of acid conditions than were strains from cowpea. Despite this, excellent responses to soybean inoculation have been reported in acid soils (Bromfield and Ayanaba, 1980), though somewhat higher than normal rates of inoculation have sometimes appeared desirable (Munns et al., 1981). Strain persistence during fallow has also appeared adequate (Rango Rao et al., 1982).

The pelleting of inoculated seed to protect rhizobia from acid soil conditions has not been used to any extent with soybean. This is, in part, because of the controversy as to whether lime or rock-phosphate should be used as the pelleting material with tropical legumes. Experimentally, Elkins et al. (1976) in Brazil found yields increased from 3278 kg/ha with inoculation to 4019 kg/ha when seeds were inoculated and lime pelleted. Similar results were obtained by Kang et al. (1977) in Nigeria.

TEMPERATURE

High temperatures contribute to the rapid death of rhizobia during transport and storage, and in the soil. Smith et al. (1980) found 36% of inoculants they shipped to the tropics and subtropics exposed to temperatures in transit greater than 38 C, while 8% were exposed to more than 49 C. Most probable number (MPN) counts in shipped peats dropped as low as 2 x 10^6 cells/g. High temperatures during transport and storage have been blamed for the poor performance of imported soybean inoculants in Ecuador (Graham and Hubbell, 1975), and have led Freire (1982) to stress the need for national inoculant capability.

The rapid death of rhizobia under hot conditions in the soil can necessitate higher than normal rates of inoculation for successful *Rhizobium* establishment. Thus, Smith and del Rio Escurra (1982) in Puerto Rico found greatest tap root and total nodulation when granular soil inoculants were applied at 10 times the normal rate, while Smith et al. (1981) found no nodules formed below $10^{3.59}$ cells applied/cm of row, with nodule number increasing from $10^{4.59}$ to $10^{9.59}$ cells applied/cm of row. Similarly, Wey and Saint Macary (1982) showed nodule number maximized when 10^{13} cells of USDA 138 were applied/ha, while Vargas and Suhet (1980) found yield increased 11% when inoculation rate was raised from 500 to 200 g/ha.

Strains show considerable variation in high temperature tolerance (Galletti et al., 1971; Munevar and Wollum, 1981a,b; 1982). Munevar and Wollum (1981a) compared the temperature tolerance of 42 strains of soybean rhizobia. Three strains showed an optimum temperature for growth of 35.2 C, but less than 10% of the strains survived exposure to 45.5 C for 96 h. Strains 587, Tal 102, 184, and 649, and NC 1005, 1010, 1030 and 1033 appeared most tolerant of temperature, while USDA 123 and 76 were relatively sensitive. Performance of strains in this trial was closely correlated with their ability to nodulate and fix N_2 at 33 C (Munevar and Wollum, 1981b). From these results, it is not surprising that high soil temperatures can also affect the proportion of nodules derived from particular inoculant strains (Weber and Miller, 1972; Roughley et al., 1980).

In many sites, desiccation effects will compound those of temperature. Al Rashidi et al. (1982) have shown strain differences in tolerance of soil desiccation; Kaul and Sekhon (1977) obtained much greater response to inoculation and mulch than to inoculation alone.

High soil temperatures also delay nodule formation, reduce nodule number and truncate the seasonal profile of N_2 fixation (Galletti et al., 1971; Dart et al., 1976; Munevar and Wollum, 1981b; 1982), with most host/strain combinations unable to form nodules at 35 C. It has not been reported with soybean, but with other legumes high surface temperatures can reduce crown nodulation, with most nodules formed deeper in the soil (Munns et al., 1977a; Graham and Rosas, 1978). Influence of cultivar on the effects of temperature have also been claimed (Galletti et al., 1971; Munevar and Wollum, 1982), but as yet are not well documented.

PLANT NUTRITION

Whereas the mineral nutrition of tropical legumes in general, and of soybean plants grown under temperate conditions, is well documented, the literature on the nutrition of soybean crops in the tropics is weak. Most fertilizer studies fail to distinguish between N-fertilized and N_2-dependent plants, while those where inoculation is involved deal mainly with N, P and K application (Singh and Saxena, 1973; Sachansky, 1977; Sobhan, 1978; Freire and Sarruge, 1979; Haque et al., 1980), or response to individual micronutrients (Freire and Sarruge, 1979; Haque and Bundu, 1980). There are no comprehensive fertility studies with tropical soybeans in which nodulation and N_2 fixation are considered. This is lamentable in a region where P, S, Ca and Mo deficiencies are common.

N_2-dependent soybeans require P for energy transfer and nodule development, and so have higher requirements for P than plants supplied fertilizer N (Cassman et al., 1981b). As a consequence P deficiency affects nodule development and function more than root or shoot growth (Cassman et al., 1980; 1981b), and will commonly restrict response to inoculation. There is little evidence in soybean of cultivar variation in P requirement (P. B. Vose, pers. commun.), but again this could reflect the narrow germplasm base of American cultivars. By contrast, strain differences in P requirement have been demonstrated (Cassman et al., 1981a; Beck and Munns, 1984), though at levels of P so low as to have limited practical application.

Several experiments report dual inoculation with *Rhizobium* and endomycorrhizal fungus. In each case yield and/or N_2 fixation was enhanced more by inoculation with both organisms, than with either microsymbiont alone. Thus Halos et al. (1982) in the Philippines reported yield increased from 1.90 g/plant without inoculation, to as much as 13.32 g/plant when both *Rhizobium* and endomycorrhiza were applied, and Bagyaraj et al. (1979) found yield raised from 44.49 g/plot in the absence of inoculation to 100.71 g/plot with both microsymbionts applied. In this trial, P uptake/plant was doubled. In Senegal, inoculation of soybean with *Glomus mosseae* increased N_2 fixation 31%, seed N by 17%, and the proportion of plant nitrogen due to fixation by 75% (Ganry et al., 1982). Skipper and Smith (1979) have reported cultivar specificity in response to mycorrhizal inoculation, with pH effects important in cultivar response.

Applied N affects response to inoculation in soybean. Commonly, in the tropics low levels of N fertilization will enhance N_2 fixation and yield (Kang, 1975; Eaglesham et al., 1982), whereas higher rates of application will reduce nodulation and the percentage of plant N due to fixation.

CONCLUSIONS

A majority of the inoculation trials undertaken with soybean in the tropics have been relatively simple experiments with one, or at most few, strains tested, and usually a single cultivar involved. While responses have generally been obtained, the yields have often been low, and affected by soil and environmental factors. More sophisticated studies on cultivar and strain variation, temperature tolerance, and resistance to acid soil conditions have been undertaken, but usually under simulated conditions in developed country or international institute laboratories. There is an urgent need for more field studies of inoculation as it pertains to soybean production in the tropics. This should detail cultivar and strain specificity in symbiosis, and evaluate promising new germplasm materials under a range of stress conditions. Consideration of alternate inoculation methodologies and of application rate would also be warranted, as are studies on the ecology and persistence of soybean rhizobia in tropical soils. As with many crops in the tropics, progress is likely to be limited by the availability of adequately trained researchers. Currently, there is little funding for collaborative research activities on soybean inoculation, restricting the scope and quality of research.

With pressure for increased soybean production, it is likely that lands previously considered unsuitable for the growth of soybean plants will be brought into use. This is already occurring in the "cerrado" of Brazil. In opening up new lands, the possibility of antagonism between soil organisms and inoculant rhizobia (Scotti et al., 1982) must be faced. In the case of the "cerrado", successful inoculant strains must tolerate high levels of antibiotics (Scotti et al., 1982). This is not to say that one should indulge in the wholesale testing of soil organisms for antagonistic effects on *Rhizobium*; it will, however, necessitate added care in the planning and interpretation of experiments.

NOTES

Peter H. Graham, Department of Soil Science, University of Minnesota, St. Paul, MN 55108.

956

REFERENCES

Albrecht, W. A. 1933. Inoculation of legumes as related to soil acidity. J. Am. Soc. Agron. 25:512-522.

Al-Rashidi, R. K., T. E. Loynachan and L. R. Frederick. 1982. Desiccation tolerance of four strains of *Rhizobium japonicum*. Soil Biol. Biochem. 14:489-493.

Awai, J. 1981. Inoculation of soya bean (*Glycine max* (L.) Merr. in Trinidad. Trop. Agric. (Trin.) 58:313-318.

Ayanaba, A. 1977. Toward better use of inoculants in the humid tropics. p. 181-187. *In* A. Ayanaba and P. J. Dart (eds.) Biological nitrogen fixation in farming systems of the tropics. John Wiley & Sons, Chichester.

Ayanaba, A. and A. L. Wong. 1982. Antibiotic resistant mutants identified from nodules of uninoculated soybeans grown in a strongly acid soil. Soil Biol. Biochem. 14:139-143.

Bagyaraj, D. J., A. Manjunath and R. B. Patil. 1979. Interaction between a vesicular-arbuscular mycorrhiza and *Rhizobium* and their effects on soybean in the field. New Phytol. 82:141-145.

Beck, D. P. and D. N. Munns. 1984. Phosphate nutrition of *Rhizobium* spp. Appl. Environ. Microbiol. 47:278-282.

Boonchee, S. and J. M. Schiller. 1979. Inoculation responses of soybean grown under rainfed conditions in Northern Thailand. Thai J. Agric. Sci. (1978) 87-104. Field Crops Abstract 32, No. 6907.

Bromfield, E. S. P. and A. Ayanaba. 1980. The efficacy of soybean inoculation on acid soil in tropical Africa. Plant Soil 54:95-106.

Bushby, H. V. A., R. A. Date and K. L. Butler. 1983. *Rhizobium* strain evaluation of *Glycine max* cv Davis, *Vigna mungo* cv Regur and *V. unguiculata* cv Caloona for three soils in glasshouse and field experiments. Aust. J. Exp. Agric. Anim. Husb. 23:43-53.

Carter, O. G., I. A. Rose and P. F. Reading. 1975. Variation in susceptibility to manganese toxicity in 30 soybean genotypes. Crop Sci. 15:730-732.

Cassman, K. G., D. N. Munns and D. P. Beck. 1981a. Growth of *Rhizobium* strains at low concentrations of phosphate. Soil Sci. Soc. Amer. J. 45:520-523.

Cassman, K. G., A. S. Whitney and R. L. Fox. 1981b. Phosphorus requirements of soybean and cowpea as affected by mode of N nutrition. Agron. J. 73:17-22.

Cassman, K. G., A. S. Whitney and K. R. Stockinger. 1980. Root growth and dry matter distribution of soybean as affected by phosphorus stress, nodulation and nitrogen source. Crop Sci. 20:239-244.

Chesney, H. A. D., M. A. Kahn and S. Bisessar. 1973. Performance of soybeans in Guyana as affected by inoculum (*Rhizobium japonicum*) and nitrogen. Turrialba 23:91-96.

Chowdhury, M. S. 1977. Response of soybean to *Rhizobium* inoculation at Morogoro, Tanzania. p. 245-253. *In* A. Ayanaba and P. J. Dart (eds.) Biological nitrogen fixation in farming systems of the tropics. John Wiley & Sons, Chichester.

Dart, P., J. Day, R. Islam and J. Döbereiner. 1976. Symbiosis in tropical grain legumes: Some effects of temperature and the composition of the rooting medium. p. 361-384. *In* P. S. Nutman (ed.) Symbiotic nitrogen fixation in plants. Cambridge Univ. Press, Cambridge.

Denaire, J. 1968. Inoculation de legumineuses a Madagascar: resultats experimentaux. Ann. Agron. 19:473-496.

Döbereiner, J., N. B. de Arruda and A. de Penteado. 1965. Problemas de inoculacao de soja em solos acidos. Proc. IX Int. Grassland Cong. Sao Paulo 2:1153-1157.

Dommergues, Y., J. L. Garcia and F. Ganry. 1980. Microbiological considerations of the nitrogen cycle in West African ecosystems. p. 55-72. *In* T. Rosswall (ed.) Nitrogen cycling in West African ecosystems. Royal Swedish Academy of Sciences, Stockholm.

Dube, J. N. 1976. Yield responses of soybean, chickpea, and pea and lentil to inoculation with legume inoculants. p. 203-207. *In* P. S. Nutman (ed.) Symbiotic nitrogen fixation in plants. Cambridge Univ. Press, Cambridge.

Dumont, R. 1981. L'experimentation sur le soja au Benin et au Togo jusqu' en 1976. Agron. Trop. 36:151-163.

Eaglesham, A. R. J., A. Ayanaba, V. Ranga Rao and D. L. Eskew. 1982. Mineral N effects on cowpea and soybean crops in a Nigerian soil. Development, nodulation, acetylene reduction and grain yield. Plant Soil 68:171-181.

Elkins, D. M., F. J. Olsen and E. Gower. 1976. Effects of lime and lime-pelleted seed on legume establishment and growth in South Brazil. Expl. Agric. 12:201-206.

Freire, F. M. and J. R. Sarruge. 1979. Producão de materia seca nodulacão e absorcão de nutrientes pela soja. (Glycine max (L.) Merrill, em funcão de niveis de fosforo e zinco em solos de Minas Gerais. Anais Escola Sup de Agric. "Luis de Queiros" 36:509-537.

Freire, J. R. J. 1978. Fixacão simbiotica do nitrogenio en soja. p. 120-169. In IX Reunion Latino Americana sobre Rhizobium, Mexico.

Freire, J. R. J. 1982. Research into the Rhizobium-Leguminosae symbiosis in Latin America. Plant Soil 67:227-239.

Galletti, P., A. A. Franco, H. Azevado and J. Döbereiner. 1972. Efeito da temperatura do solo na simbiose da soja anual. Pesq. Agropec. Bras. 6:1-8.

Ganry, F., H. G. Diem and Y. R. Dommergues. 1982. Effect of inoculation with Glomus mosseae on nitrogen fixation by field grown soybeans. Plant Soil 68:321-329.

Graham, P. H. and D. H. Hubbell. 1975. Interaccion suelo-planta-Rhizobium en la agricultura tropical. p. 217-233. In E. Bornemisza and A. Alvarado (eds.) Manejo de suelos en la America tropical. North Carolina State Univ., Raleigh.

Graham, P. H. and J. C. Rosas. 1978. Nodule development and nitrogen fixation in cultivars of Phaseolus vulgaris L. as influenced by planting density. J. Agric. Sci. (Camb.) 90:19-29.

Halos, P. M., E. M. Luis and M. S. Borja. 1982. Synergism between endomycorrhizas, Rhizobium japonicum CB1809 and soybean (Glycine max (L.) Merr.). Phil. Agric. 65:93-102.

Haque, I. and H. S. Bundu. 1980. Effects of inoculation, N, Mo and mulch on soybean in Sierra Leone. Comm. Soil Sci. Plant Anal. 11:477-483.

Haque, I., W. M. Walker and S. M. Funnah. 1980. Effects of phosphorus and zinc on soyabean in Sierra Leone. Comm. Soil Sci. Plant Anal. 11:1029-1040.

IITA. 1978. Annual report, International Institute of Tropical Agriculture. p. 130.

IITA. 1979. Annual report, International Institute of Tropical Agriculture. p. 152.

IITA. 1980. Annual report, International Institute of Tropical Agriculture. p. 185.

IITA. 1982. Annual report, International Institute of Tropical Agriculture. p. 217.

Jimenez, T. and E. Villalobos. 1980. Respuesta de frijol de soya a la inoculacion con Rhizobium japonicum y a la fertilizacion con nitrogeno y fosforo en Costa Rica. Agron Costarricense 4: 1-8.

Kang, B. T. 1975. Effect of inoculation and nitrogen fertilizer of soybean in Western Nigeria. Expl. Agric. 11:23-31.

Kang, B. T., D. Nangju and A. Ayanaba. 1977. Effects of fertilizer use on cowpea and soybean nodulation and nitrogen fixation in the lowland tropics. p. 205-216. In A. Ayanaba and P. J. Dart (eds.) Biological nitrogen fixation in farming systems of the tropics. John Wiley & Sons, Chichester.

Kaul, J. N. and H. S. Sekhon. 1977. Soybean yield improved by inoculation and mulching. World Crops Livestock 29:248-249.

Keyser, H. H., B. B. Bohlool, T. S. Hu and D. F. Weber. 1982. Fast-growing rhizobia isolated from nodules of soybeans. Science 215:1631-1632.

Kesyer, H. H. and D. N. Munns. 1979. Tolerance of rhizobia to acidity, aluminum and phosphate. Soil Sci. Soc. Amer. J. 43:519-523.

Mengel, D. B. and E. J. Kamprath. 1978. Effect of soil pH and liming on growth and nodulation of soybeans in histosols. Agron. J. 70:959-963.

Munevar, F. and A. G. Wollum. 1981a. Growth of Rhizobium japonicum strains at temperatures above 27°C. Appl. Environ. Microbiol. 42:272-276.

Munevar, F. and A. G. Wollum. 1981b. Effect of high root temperature and Rhizobium strain on nodulation, nitrogen fixation and growth of soybeans. Soil Sci. Soc. Amer. Proc. 45:1113-1120.

Munevar, F. and A. G. Wollum. 1982. Response of soybean plants to high root temperatures as affected by plant cultivars and Rhizobium strain. Agron. J. 74:138-142.

Munns, D. N., V. W. Fogle and B. G. Hallock. 1977a. Alfalfa root nodule distribution and inhibition of nitrogen fixation by heat. Agron. J. 69:377-380.

Munns, D. N., R. L. Fox and B. L. Koch. 1977b. Influence of lime on nitrogen fixation by tropical and temperate legumes. Plant Soil 46:591-601.

Munns, D. N., J. S. Hohenberg, T. L. Righetti and D. J. Lauter. 1981. Soil acidity tolerance of symbiotic and nitrogen fertilized soybeans. Agron. J. 73:407-410.

Nangju, D. 1980. Soybean response to indigenous rhizobia as influenced by cultivar origin. Agron. J. 72:403-406.

Ohki, K., D. O. Wilson and O. E. Anderson. 1980. Manganese deficiency and toxicity sensitivities of soybean cultivars. Agron. J. 72:713-716.

Pal, U. R. and M. C. Saxena. 1975. Response of soybean to symbiosis and nitrogen fertilization under humid sub-tropical conditions. Exp. Agric. 11:221-226.

Paterno, E. S., A. J. Alcantara, F. G. Gibe and J. S. Lales. 1979. Inoculation of soybean in two soils. Phil. Agric. 62:176-182.

Pulver, E. L., F. Brockman and H. C. Wien. 1982. Nodulation of soybean cultivars with *Rhizobium* spp. and their response to inoculation with *R. japonicum*. Crop Sci. 22:1065-1070.

Ramirez-Gama, R. M. 1982. Informe, programa de inoculacion en soya. 1978-1981. Faculdad Quimica, UNAM, Mexico.

Ranga Rao, V., G. Thotapilly and A. Ayanaba. 1982. Studies on the persistence of introduced strains of *Rhizobium japonicum* in soil during fallow and their effects on soybean growth and yield. p. 309-315. *In* P. H. Graham and S. C. Harris (eds.) Biological nitrogen fixation technology for tropical agriculture. CIAT, Cali.

Roughley, R. J., E. S. P. Bromfield, E. L. Pulver and J. Day. 1980. Competition between species of *Rhizobium* for nodulation of *Glycine max*. Soil Biol. Biochem. 12:467-470.

Sachansky, S. 1977. Effect of inoculation and NPK fertilizers on soybean (*Glycine max* (L.) Merr.). Trop. Grain Legume Bull. 7:15-17.

Salema, M. P. and M. S. Chowdhury. 1980. Rhizobial inoculation and fertilizer effect on soybean in the presence of native rhizobia at Morogoro, Tanzania. Beitr. Tropisch. Landwirt. Veterinarmed. 18:245-250.

Sapra, V. T., T. Mehrbrahtu and L. M. Mugwira. 1982. Soybean germplasm and cultivar aluminum tolerance in nutrient solution and Braden clay loam soil. Agron. J. 74:687-690.

Sartain, J. B. and E. J. Kamprath. 1978. Al tolerance of soybean cultivars based on root elongation in solution culture. Agron. J. 70:17-20.

Scotti, M. R. M. M. L., N. M. H. Sa, M. A. T. Vargas and J.Döbereiner. 1982. Susceptibility of *Rhizobium* strains to antibiotics: a possible reason for legume inoculation failure in cerrado soils. p. 195-200. *In* P. H. Graham and S. C. Harris (eds.) Biological nitrogen fixation technology for tropical agriculture. CIAT, Cali.

Scudder, W. T. 1974. *Rhizobium* inoculation of soybeans for subtropical and tropical soils. 1. Initial field studies. Soil Crop Sci. Soc. Fla. 34:79-82.

Singh, H. P. and K. V. B. R. Tilak. 1977. Response of soybean to inoculation with various commercial inoculants of *Rhizobium*. Indian J. Agron. 22:57-59.

Singh, N. P. and M. C. Saxena. 1973. Pattern of dry matter accumulation in soybean as affected by nitrogen, inoculation and phosphorus fertilization. Indian J. Agric. Sci. 43:62-66.

Sistachs, E. 1976. Inoculation and nitrogen fertilizer experiments on soybeans in Cuba. p. 281-288. *In* P. S. Nutman (ed.) Symbiotic nitrogen fixation in plants. Cambridge Univ. Press, Cambridge.

Skipper, H. D. and G. W. Smith. 1979. Influence of soil pH on the soybean-endomycorrhiza symbiosis. Plant Soil 53:559-563.

Smith, R. S., M. A. Ellis and R. E. Smith. 1981. Effect of *Rhizobium japonicum* inoculant rates on soybean nodulation in a tropical soil. Agron. J. 73:505-508.

Smith, R. S. and G. A. del Rio Escurra. 1982. Soybean inoculant types and rates evaluated under dry and irrigated field conditions. J. Agric. Univ. Puerto Rico 66:241-249.

Smith, R. S., W. H. Judy and M. Ramos. 1980. Effect of temperature, moisture and time on rhizobia inoculant quality mailed through the International Inoculant Shipping Evaluation (IISE). Agron. Abstr., 72 Annual Meetings American Society of Agronomy. ASA, Madison, Wisconsin.

Sobhan, A. 1978. Yield response of soybean to artificial inoculation of *Rhizobium* and NPK fertilization. Phil. Agric. 61:263-267.

Subba Rao, N. S. and V. R. Balasundaram. 1971. *Rhizobium* inoculants for soybean cultivation. Indian Farming 21(8):22-23.

USDA. 1979. *Rhizobium* culture collection catalogue. U.S. Department of Agriculture, Beltsville, p. 33.

Uratsu, S. L., H. H. Keyser, D. F. Weber and S. T. Lim. 1982. Hydrogen uptake (HUP) activity of *Rhizobium japonicum* from major US soybean production areas. Crop Sci. 22:600-602.

van Rensburg, H. J., B. W. Strijdom and C. J. Otto. 1983. Effective nodulation of soybeans by fast-growing strains of *Rhizobium japonicum*. South African J. Sci. 79:251-252.

Vargas, M. A. T. and A. R. Suhet. 1980. Efecto de tipos y niveis de inoculantes na soja cultivada em um solo de cerados. Pesq. Agropec. Bras. 15:343-347.

Weber, D. F. and V. L. Miller. 1972. Effect of soil temperature on *Rhizobium japonicum* serogroup distribution in soybean nodules. Agron. J. 64:796-798.

Wey, J. 1980. Premiers resultats concernant l'inoculation du soja au Senegal. p. 209-213. *In* T. Rosswall (ed.) Nitrogen cycling in West African ecosystems. Royal Swedish Academy of Sciences, Stockholm.

Wey, J. and H. Saint Macary. 1982. Inoculation du soja par le *Rhizobium japonicum* au Senegal. Agron. Trop. 37:24-29.

Zengbe, M. 1980. Presence et distribution de *Rhizobium japonicum* et de *Rhizobium* cowpea dans les sols de Côte d' Ivoire. p. 215-219. *In* T. Rosswall (ed.) Nitrogen cycling in West African ecosystems. Royal Swedish Academy of Sciences, Stockholm.

SOIL AND CROP MANAGEMENT

OPTIMIZING SEED PLACEMENT TO MAXIMIZE SEEDLING EMERGENCE UNDER NO TILLAGE

C. J. Baker and P. J. Desborough

The relationship between seeds, seedlings and their soil microenvironment during early crop establishment by direct drilling (in no-till seedbeds) has been studied extensively at Massey University, New Zealand and Grafton Research Station, Australia over a period of 15 years (Baker, 1983). This paper firstly summarizes data from seedling emergence experiments conducted under controlled conditions in temperate New Zealand, and secondly outlines the field performance of equipment developed therefrom in establishing no-till soybean crops in coastal New South Wales, Australia.

CONTROLLED CLIMATE STUDIES—NEW ZEALAND

Investigations centered on three, broad field situations: Where seeds were sown in (1) dry (or drying) soils, (2) optimal soils which subsequently became very wet, and (3) soils which remained optimal. Indoor controlled-environment, and field experiments were conducted with crop species ranging from soybean and cereals to pasture species and selected vegetables. Much of the detailed drill-slot microenvironmental work has involved cereal seeds, but the responses of direct-drilled seeds to their soil habitat appear applicable to a wide range of species, including soybean.

Controlled environment studies utilized portable steel bins, each containing 0.5 t of undisturbed soil, which were drilled on a special laboratory tillage-bin rig (Baker, 1969, 1972), and then relocated in climate controlled rooms (Choudhary and Baker, 1980).

Dry Soils

Evaporative moisture loss from residue-covered, no-till soils appears to occur preferentially from the disturbed, seed-slot area. This loss of vapor-phase slot-moisture was found to be strongly influenced by slot shape and the nature of the slot-covering medium, including any subsequent pressing or covering procedures, as well as by ambient humidity (Baker, 1970, 1971, 1976a; Choudhary and Baker, 1980, 1981a,b). The ability of a given drilled slot to retain moisture vapor for the few critical days between drilling and seedling emergence is referred to as 'moisture vapor potential captivity' (MVPC), which is equal to the reciprocal of the mean daily loss of relative humidity from the slot. Low MVPC values and seedling emergence counts were associated with failure of seeds to germinate (in U-shaped slots, which displayed intermediate MVPC values), as well as sub-surface seedling mortality (in V-shaped slots, which displayed the lowest MVPC of all slot shapes tested). Table 1 illustrates MVPC for various slot shapes.

Table 1. In-groove soil moisture drying rates of selected direct drilling opener designs and covering methods (Choudhary and Baker, 1981b).

Opener and slot shape	In-groove RH[a] loss rate	MVPC[b]
	% per day	
Winged opener/harrowed (inverted T-shape)	2.34	0.427
Hoe opener/harrowed (U-shape)	2.77	0.361
Hoe opener, 70 kPa pressure over seeds/harrowed (U-shape)	1.92	0.520
Triple disc opener/harrowed (V-shape)	4.23	0.236
Triple disc opener, 70 kPa pressure over seeds/harrowed (V-shape)	2.32	0.431
Triple disc opener with complete loose soil cover (applied manually), followed by 70 kPa pressure applied over row (V-shape)	2.43	0.413

[a]Relative humidity.
[b]MVPC = Moisture Vapor Potential Captivity.

An experimental inverted T-shaped slot consistently recorded the highest MVPC values and seedling emergence counts. Differences in both MVPC and seedling emergence (or at least seedling survival) increased with decreasing soil moisture levels, and were up to 14-fold in severe drought (Baker, 1979, 1983; Choudhary and Baker, 1982). Under these circumstances the alternative slot-covering media favored a residue-mulch/soil combination (a unique function of the inverted T-shaped slot), which was superior to loose soil (pressed or unpressed), which itself was superior to no cover at all.

Despite the apparent vigor of large seeds, increased seed size per se appeared not to compensate adequately for low MVPC values (Baker, 1976b).

Pressing directly on the seeds in V- and U-shaped slots before covering produced considerable improvement in numbers of emerged seedlings, but little or no response occurred when pressing was confined to the soil above the seeds after covering, or when pressing inverted T-shaped slots (as emergence had been high in the first place). Table 2 illustrates responses of wheat (*Triticum aestivum* L.) seedling emergence to pressing (Choudhary and Baker, 1980).

Wet Soils

Data for wet soils have so far involved saturated conditions at 15 to 25 C ambient temperature (A. D. Chaudhry, personal communication). In the absence of earthworms, opener design, and thus slot shape, had virtually no effect on oxygen diffusion rates within a 60 mm zone of the slot; nor did it influence seed or seedling survival. The single most important indirect factor that affected seed and seedling performance appeared to be earthworm activity. The number of earthworms and their channels showed a strong relationship with bulk density and oxygen diffusion rates around, and infiltration rates into, the slots. Numbers of earthworms were themselves influenced significantly by opener design and slot shape, which also interacted with the presence or absence of surface residues.

Table 2. Effects of intensity and position of applied pressure, opener type, and ambient relative humidity regime on fate of direct-drilled wheat seed in a dry soil (from Choudhary and Baker, 1980).

	Seedling emergence	Ungerminated seeds	Germinated but unemerged seeds
	%	%	%
Main Treatments			
Pressure applied over:			
covered seeds	25.9	36.2	38.2
uncovered seeds	41.6	18.8	39.8
Relative humidity regimes:			
high	38.6	22.6	38.6
low	28.2	32.1	39.3
Opener types and slot shapes:			
winged (inverted T-shape)	58.4	18.2	23.3
hoe (U-shape)	31.3	46.5	22.2
triple disc (V-shape)	10.5	17.8	71.7
Sub Treatments			
Pressure intensity (kPa):			
nil	28.8	33.4	38.7
35	34.6	24.0	41.1
70	37.9	24.7	37.3

The experimental winged opener (inverted T-shaped slot) retained maximum surface residue over the seed zone, and promoted the greatest earthworm activity, and thus seed and seedling survival, of a range of openers. This range included triple disc (V-shape), hoe (U-shape), power-till (100 mm wide U-shape), punch planter (discontinuous V-shaped holes), and surface broadcasting.

Where residue was pushed into the slot, seed and seedling performance was adversely affected, possibly because of fatty acid fermentation (Lynch, 1977, 1980; Lovett and Jessop, 1982), but this was confined to continuous V-shaped slots. The tendency of the triple disc opener to smear and produce localized compaction reduced earthworm activity (A. D. Chaudhary, personal communication) and restricted early root growth of a tap-rooted species, lupin (*Lupinus albus* L.) (Baker and Mai, 1982a,b). No smearing or compaction was evident in inverted T- or U-shaped slots.

The contrast between typical oxygen diffusion rate iso-lines drawn around V- and inverted T-shaped slots is illustrated in Figure 1. Table 3 lists typical earthworm and seedling responses to opener shape and surface residues.

Optimal Soils

Placing dry fertilizer in the slot, separated from the seed, had a marked beneficial effect on direct-drilled corn (*Zea mays* L.) yields, compared with surface broadcasting in a silt loam soil under otherwise favorable conditions (Sims and Baker, 1981; Baker and Afzal, 1981). The winged opener concept for creating inverted T-shaped slots was further developed to give it the capability of side banding fertilizer (by 10-20 mm) which created less seed damage than vertical separation by the same distance (Afzal, 1981).

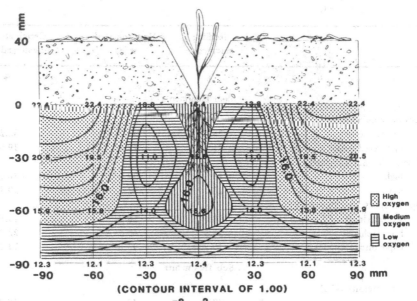

Oxygen diffusion rate (gm X 10^{-8}/cm^2/min): Tripple disc opener in residue.

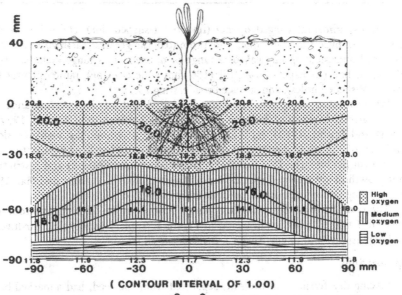

Oxygen diffusion rate (gm X 10^{-8}/cm^2/min): Winged opener in residue.

Figure 1. Oxygen diffusion rate iso-lines surrounding direct drilled slots (A. D. Chaudhry, pers. commun.).

Table 3. Effects of opener type and surface residue on wheat seedling emergence and earthworm numbers in a saturated, direct-drilled soil (A. D. Chaudhry, pers. commun.).

| Opener type | Seedling emergence | | | Earthworms | | |
	No residue	Residue	Mean	No residue	Residue	Mean
	– % –			– No.[a] –		
Winged	43.8	73.3	58.6	13.3	27.3	20.3
Triple disc	26.2	25.2	25.7	6.7	8.0	7.3
Hoe	32.9	63.8	48.3	11.3	26.7	19.0
Mean	34.3	54.1		10.4	20.7	

[a]Number within a 140 mm diameter core centered on a drilled row.

Accuracy of seed placement within the slot, relative to the opener, and the maintenance of a constant opener depth appears to influence both seedling survival patterns and evenness of emergence of otherwise healthy seedlings (Ritchie, 1981, 1982; Choudhary and Guo, 1984). An x-ray research technique for locating sown seeds precoated with lead oxide has enabled seed-placement studies to become more meaningful and practical (Barr, 1981; A. J. Campbell, unpublished data).

Opener Design

Development of a high-technology, winged, direct-drilling opener has occupied a design team for several years (Baker et al., 1979a,b; Brown, 1982). The opener is capable of creating an inverted T-shaped slot; retaining surface residue over the seed zone; maximizing MVPC; banding seed and fertilizer 10 to 20 mm apart; negotiating surface residue without blocking (even in 150 mm-spaced rows); preventing residue-seed contact; avoiding smearing or compaction; encouraging earthworm activity in and around the slot; accurately following contours; precisely ejecting seed within the slot relative to the opener; and minimizing costly wear. This opener, which is shown in Figure 2, is to be released commercially in 1984. It has the highest biological tolerance of sub-optimal soil and climatic conditions of any opener principle so far tested under controlled conditions at Massey University. It is expected, therefore, to contribute significantly to improving the seed habitat and microenvironment in the slot, and thereby, reduce substantially the biological risks currently associated with cropping by direct drilling in no-till soils.

FIELD PERFORMANCE—AUSTRALIA

The subtropical, coastal belt of eastern Australia, from latitudes 32°S to 26°S, has temperature and moisture regimes suitable for double-cropped soybean and winter cereals (Desborough, 1981). High rainfall (1,000 to 1,600 mm/year) and erodible soils have necessitated rapid adoption of no-till practices for sustained production from the predominantly undulating land.

In recent years a range of no-till equipment has been compared under field conditions over a range of soil types, soil moisture and surface residue levels. In each case, the equipment was of commercial size; i.e., with 8 to 10 rows in row-spacings of 25 to 35 cm. Soybeans sown in narrow rows are generally higher yielding (Boquet et al., 1983; Lawn et al. (1977) and smother late weed growth better than soybeans in wider row-spacings (>50 cm). Because winter cereals need to be sown in narrow rows, it is

**Massey University Winged opener
(inverted T shaped slot)**

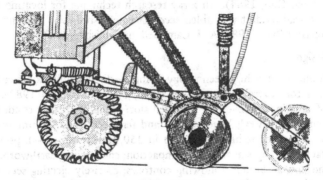

Triple disc opener (V shaped slot).

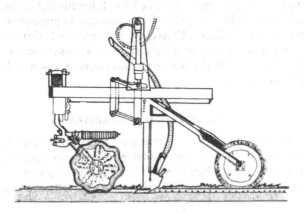

Chisel boot or hoe opener (U shaped slot)

Figure 2.   A typical range of direct drilling openers compared in no-till soils.

expedient to have one drill to sow both crops. Three machines were compared and are shown in Figure 2.
- a winged direct-drill opener, previously described, with integral, notched-coulter and press-wheels/depth wheels;
- triple-disc opener, with or without presswheels;
- fluted coulter with narrow, chisel-boot openers and presswheels in line.

Surface Residue

Two important considerations were blockages of ground engaging components and position of the residue after drilling. The frequency of blockages usually increased with quantity of residue and unevenness of residue distribution, but decreased with wider spacings between adjacent rows. Surface residue should be retained over the seeding slot to decrease temperature and moisture loss in the seed zone, which also aids nodulation.

On 30 cm row-spacings the winged opener was superior in residue handling, both in terms of lack of blockages and placement. This was evident in both winter cereal and soybean residues, wet or dry, standing or loose. Triple-disc configurations (in 25 or 30 cm spacings) were inferior in freedom from blockages only when the leading coulter was small (25-30 cm diameter), but equal to the winged opener with larger coulters (45-50 cm diameter). However, significant quantities of residue were pressed into the seeding slot by the double-disc openers. Chisle-boot-coulter combinations could handle significant quantities of surface residue in narrow row spacings (25-35 cm) only where adjacent rows were staggered on two gangs. After passage, residues did not cover the seeding slot. Minimal residue was found in the seed zone.

Seed Placement and Cover

Under no-till conditions optimum establishment of soybeans could be achieved only where depth of seed placement was accurate and, under drier soil conditions, where there was good seed-soil contact and soil cover. The winged opener was able to place soybean seed accurately and consistently at depths of approximately 12 mm, under high soil moisture conditions, and 60 mm, under dry conditions. Seedling emergence was reduced when drilling directly into herbicide-treated pasture when soil was wet, but only when sowing depth exceeded 12 to 15 mm. This depth restriction did not occur when direct drilling into no-till crop residues where soil disturbance was greater even with wet soils. Seed coverage was adequate under all conditions.

Depth of seed placement was usually acceptable where a press-wheel followed the triple-disc opener, but was sometimes too shallow where intermittent tight soil conditions were encountered. Seed coverage and seed-soil contact were adequate under drier conditions when sown into crop residue, but poor when sown into pasture or under wet soil conditions.

Due to wider spacing between the sowing boot and depth/presswheel in the chisel-boot-coulter combination, depth of seed placement was less consistent than with the winged opener, especially on uneven terrain. Seeds were covered adequately when drilled into moderate to drier soils, but often left uncovered when sown into wet soil or into pasture where loose soil and clods were thrown to the side of the wide seed slot. When drilling across slopes the amount of soil displaced to the downhill, inter-row area was increased with chisel openers.

Smearing

Smearing and compaction of the bottom and sides of drill-slots impaired downward root growth. This consideration became important when soil near the surface dried out in the first three to four weeks after sowing, often accompanied by root damage by insect larvae, such as white fringed weevil (*Graphognathus leucoloma*). The winged opener caused minimal smearing at or below seeding level, leading to less moisture stress under these conditions. Both the triple-disc and chisel-boot smeared both the bottom and sides of the slots under wetter conditions. Restrictions to downward root growth were observed in these situations, with higher seedling losses from moisture stress than with the winged opener.

Fertilizer Placement

Most of the soils of the soybean growing regions of coastal, eastern Australia are very deficient in phosphorus. When drilling soybeans or winter cereals, it is advantageous to place phosphatic fertilizer near to, but not with, the seed. Without major modifications to the drills examined, only the winged opener was capable of placing fertilizer in this way. The triple-disc left most of the fertilizer in contact with seeds in the slot.

NOTES

C. J. Baker, Director, Agricultural Machinery Research Centre, Massey University, Palmerston North, New Zealand.

P. J. Desborough, Research Agronomist, Agricultural Research Station, Grafton, NSW, Australia.

REFERENCES

Afzal, C. M. 1981. The requirements of dry fertilizer placement in direct drilled crops by an improved chisel coulter. M. Phil. Thesis. Massey University Library, Palmerston North, N.Z.

Baker, C. J. 1969. A tillage bin and tool testing apparatus for turf samples. J. Agric. Eng. Res. 14: 357-360.

Baker, C. J. 1970. A simple covering harrow for direct drilling. N.Z. Farm. 91(4):62-63.

Baker, C. J. 1971. Filling up spaces by surface seeding. Proc. 24th N.Z. weed & pest control conf. 24:75-79.

Baker, C. J. 1972. The mechanized collection of turf blocks and their utilization in field experimentation. p. 282-293. *In* Proc. 3rd int. conf. on mech. of field expts.

Baker, C. J. 1976a. Experiments relating to the techniques of direct drilling of seeds into untilled dead turf. J. Agric. Eng. Res. 21:133-145.

Baker, C. J. 1976b. Some effects of cover, seed size and soil moisture status on seedling establishment by direct drilling. N.Z. J. Exp. Agric. 5:47-53.

Baker, C. J. 1979. Equipment impact and recent developments. Proc. Conserv. Tillage Tech. Sem. Christchurch, N. Z.

Baker, C J. 1983. Direct drilling in N.Z.; machinery research and development. N.Z. Agric. Sci. 17(3): 288-292.

Baker, C. J. and C. M. Afzal. 1981. Some thoughts on fertilizer placement in direct drilling. Proc. Conserv. Tillage Tech. Sem. Christchurch, N.Z.

Baker, C. J. and T. V. Mai. 1982a. Physical effects of direct drilling equipment on undisturbed soils: 4. Techniques for measuring soil compaction in the vicinity of drilled grooves. N.Z. J. Agric. Res. 25:43-49.

Baker, C. J. and T. V. Mai. 1982b. Physical effects of direct drilling equipment on undisturbed soils: 5. Groove compaction and seedling root development. N.Z. J. Agric. Res. 25:51-60.

Baker, C. J., E. M. Badger, J. H. McDonald and C. S. Rix. 1979a. Developments with seed drill coulters for direct drilling: I. Trash handling properties of coulters. N.Z. J. Exp. Agric. 7:175-184.

Baker, C. J., J. H. McDonald, C. S. Rix, K. Seebeck and P. M. Griffiths. 1979b. Developments with seed drill coulters for direct drilling: 3. An improved chisel coulter with trash handling and fertilizer placement capabilities. N.Z. J. Exp. Agric. 7:189-196.

Barr, S. J. 1981. Evaluating performance of overdrilling machines. Proc. Conserv. Tillage Tech. Sem. Christchurch, N.Z.

Boguet, D. J., K. L. Koonce, and D. M. Walker. 1983. Row spacing and planting date effect on yield and growth responses of soybean. Louisiana Agric. Exp. Stn. Bull. No. 754.

Brown, S. W. 1982. An investigation into wear characteristics of a direct drilling coulter (opener). M.Agr.Sc. Thesis. Massey University Library, Palmerston North, N.Z.

Choudhary, M. A. and C. J. Baker. 1980. Physical effects of direct drilling equipment on undisturbed soils: 1. Wheat seedling emergence from a dry soil under controlled climates. N.Z. J. Agric. Res. 23:489-496.

Choudhary, M. A. and C. J. Baker. 1981a. Physical effects of direct drilling equipment on undisturbed soils: 2. Seed groove formation by a 'triple disc' coulter and seedling performance. N.Z. J. Agric. Res. 23:183-187.

Choudhary, M. A. and C. J. Baker. 1981b. Physical effects of direct drilling equipment on undisturbed soils: 3. Wheat seedling performance and in-groove micro-environment in a dry soil. N.Z. J. Agric. Res. 24:189-195.

Choudhary, M. A. and C. J. Baker. 1982. Effects of direct drill coulter design and soil moisture status on emergence of wheat seedlings. Soil & Tillage Res.:131-142.

Choudhary, M. A. and Guo Pei Yu. 1984. Measurement of seed placement and distribution, and effects on direct drilled seedling establishment. Soil & Tillage Res. (in press).

Desborough, P. J. 1981. A double-crop system for coastal NSW using direct drilled soybeans and winter cereals. Proc. national workshop on tillage systems for crop production. Roseworthy, S.A. Australia.

Lawn, R. J., D. E. Byth and V. E. Mungomery. 1977. Responses of soybeans to planting date in southeastern Queensland. III. Agronomic and physiological responses of cultivars to planting arrangements. Aust. J. Agric. Res. 28:63-79.

Lovett, J. V. and R. S. Jessop. 1982. Effects of residues of crop plants on germination and early growth of wheat. Aust. J. Agric. Res. 33:909-916.

Lynch, J. M. 1977. Phytotoxicity of acetic acid produced in the anaerobic decomposition of wheat straw. J. Appl. Bacteriol. 42:81-87.

Lynch, J. M. 1980. Production and phytotoxicity of acetic acid in anaerobic soils containing plant residues. Soil Biol. Biochem. 10:131-135.

Ritchie, W. R. 1981. Towards precision sowing in uncultivated ground. Proc. Conserv. Tillage Tech. Sem. Christchurch, N.Z.

Ritchie, W. R. 1982. Aspects of seed transfer with a direct drilling coulter (opener). M.Agr.Sc. Thesis. Massey University Library, Palmerston North, N.Z.

Sims, R. E. H. and C. J. Baker. 1981. Comparisons of four reduced time and energy seedbed preparation systems: Year one of a perennial experiment. N.Z. J. Exp. Agric. 9:299-305.

IRRIGATION WATER MANAGEMENT OF SOYBEANS

H. D. Scott

The soybean is grown in climatic and soil environments that often are unfavorable for growth. The water status of both environments is particularly dynamic and varies throughout the growing season and spatially over the field. Water stresses of field-grown soybean crops result from a deficient water supply in the root zone and from excessive atmospheric demand of water from the leaves. Therefore, some degree and duration of water stress is almost certain to occur during every growing season. As a result, soybean plants have developed numerous adaptations to deal with these variations in time and space. Examples include the ability of their root systems to extract water, nutrients, and oxygen from soil at considerable soil depths, of their stomata and leaf surface tissues to conserve water, and of their leaves to intercept and reflect solar radiation.

The major resources supplied to soybean plants by the soil and climatic environments are water, nutrients, carbon dioxide, oxygen and solar radiation. Of these, water is the most limiting on a world-wide scale. Water management is an important contributor to increased seed yields and maintenance of high yields in many parts of the world. In the United States alone about 83% of the water that is consumed is devoted to irrigation (U.S. Dept. Commerce, 1982). About 20 million hectares of the farmland are irrigated, and this area is increasing, particularly in the Mid-South (U.S. Dept. Commerce, 1982), southeast (Bruce et al., 1984), and in parts of the Great Plains. Soybean is a major component of the irrigation management schemes in these areas.

ENVIRONMENTAL FACTORS INFLUENCING IRRIGATION NEEDS

Soybean plants grow and develop with their roots in one environment, the soil, and shoots in another, the air. Their actual growth rates reflect the integration of temporal and spatial variations in each environment and the genetic potential of the crop. Knowledge of the effects of the soil and the meteorological environments upon crop growth and development is, therefore, required before soybean crops can be managed effectively.

For the most part, the soybean is grown in humid and semi-humid regions of the world. Experiments have shown that, when factors other than water are nonlimiting, soybean dry matter production and seed yield (Y) are proportional to transpiration (T):

$$Y = n T \qquad (1)$$

where n is a constant that is dependent upon crop and environmental conditions (Rosenberg et al., 1983), and transpiration, a plant characteristic, is a function of the soil and meteorological environments. Since transpiration and soil evaporation are similar physical processes, which occur simultaneously in the field, and are difficult to separate experimentally, they are usually combined into a term called evapotranspiration (ET). As equation (1) indicates, those soil and meteorological factors that affect ET will also affect Y.

Soil Factors

The soil is the storage reservoir of water for plant growth, and therefore, exerts a strong influence on the amount of water available to the roots. The amount of water in the reservoir at any one time depends upon the difference between the amount added and the amount lost. Additions of water occur primarily in the forms of rainfall and irrigation. From a practical view, it is the effective additions of water that are important to the crop, since often there may be considerable runoff when the rate of water added is greater than the rate of infiltration. Losses of water occur primarily in the forms of internal drainage and ET. Evaporation losses of water from some soil surfaces can be considerable, especially from those soils that can transmit water to the surface at a rate sufficient to keep the soil surface evaporating as rapidly as the meteorological conditions will allow.

In field-grown soybean crops evaporation losses are particularly important during early vegetative growth (Geddes et al., 1979). Transpirational losses of water are closely related to canopy development during vegetative growth, and to the water content of the soil surface (Ritchie, 1983a). Under non-water stressed conditions, transpiration is related to the leaf area index until a threshold value of approximately 3.0 is obtained (Ritchie, 1983b). Thereafter, transpiration occurs at a rate dependent upon the meteorological conditions. Under water-stress conditions, soybean plants control transpiration by changing stomatal aperture (Carlson et al., 1979; Jung and Scott, 1980), and leaf orientation (Meyer and Walker, 1981). These, in turn, affect other plant properties, such as leaf water potential, leaf temperature (Jung and Scott, 1980) and processes such as cell expansion, photosynthesis, and nitrogen fixation (Boyer, 1982). In general, with water stress, photosynthesis and transpiration continue but at limited levels, cell expansion is reduced, and carbohydrate storage increases.

The influence of the available water in a soil is important in determining when the crop will develop significant water stress. The concept of available water has recently taken on a new definition in that it depends upon both soil profile and crop characteristics (Ritchie, 1981). For soybean plants grown on a given soil, the available water is determined as the difference in the profile between the upper limit of available water and the lower limit of available water. The upper limit is determined as the water content distribution in the soil 2 or 3 days after near saturation of the profile. The lower limit is determined as the water content distribution of the profile after soybean plants have fully developed their root system without stress and are allowed to grow on stored soil water until severely distressed. Most studies have shown that transpiration is reduced after 70% of the available water in the profile has been depleted. However, this threshold percentage probably is soil and ET dependent.

Meteorological Factors

Weather plays a major role in determining rates of ET. In the absence of irrigation, water for ET is provided by rainfall, and water vapor is carried away from the crop canopy by the wind. Water lost by ET is a result of the great quantities of energy consumed in transforming water from the liquid to the vapor state. This energy comes primarily from solar radiation.

The relationship between energy sources and energy consumption is summarized in the energy balance equation.

$$R + H + S + LE + P + M = 0 \tag{2}$$

where R is the net radiation, S is the soil heat flux, H is the sensible heat flux, LE is the latent heat flux, and P and M represent photosynthesis and miscellaneous energy exchanges, respectively (Rosenberg et al., 1983). When water is present, LE is the major energy consumer and R is the major energy supplier for ET. Under well-watered conditions in humid regions, daily values of S in a soybean field are usually small.

The water status has been shown to affect the partitioning of available energy between LE and H in a soybean field in Nebraska (Baldocchi et al., 1983). When the soybeans were well-watered, all of R was consumed as LE, and H was directed toward the crop which caused LE to exceed R. When the crop was water stressed, only two-thirds of R was consumed as LE; the remainder was converted into H which was directed away from the crop. The reduction in LE under water-stressed conditions was attributed to stomatal closure.

UNDERSTANDING PLANT RESPONSES TO WATER STRESS

Mechanisms that enable the soybean to cope with drought have developed through genetic selection. Only in recent years have we begun to understand some of these mechanisms and their importance to soybean growth, development, and seed yield. Specific plant growth processes are affected by drought, and the soybean has evolved ways to change these processes, leading to adaptation. Plant growth is controlled directly by its water status, and only indirectly by soil water stresses. Recent research studies have been conducted on the effects of plant age on the response of soybean plants to water stress. These studies are useful in furthering our understanding of the mechanisms involved and the daily adjustments that occur in field-grown soybean plants in responding to water stress.

Germination and Emergence

Although any phase of the soybean life cycle can be affected by water stress, a particularly sensitive time occurs during germination and emergence. The germinating soybean has few roots and usually is located in relatively dry soil. Meyer and Boyer (1981) found that hypocotyl growth of soybean seedlings was rapidly decreased when the water supply around the roots was limited. Solutes accumulated in the elongating region of the hypocotyl and caused an osmotic adjustment. Root growth increased after solute accumulation in the hypocotyl was complete. The origin of the solute was thought to be the cotyledons. The rate of osmotic adjustment depended upon the growth rate of the hypocotyl.

In Mississippi, Heatherly and Russell (1979) evaluated the effects of four soil water potentials on the rate of soybean emergence on a clay and a silt loam in the greenhouse. They found that emergence from the clay was faster and more uniform over a wider range of soil water potentials than it was from the silt loam. Total emergence from the clay after 14 days was not significantly different over the range of soil water potentials used. When compared to the two intermediate soil water potentials, significantly fewer plants emerged from the silt loam at the wettest (-0.01 MPa) and driest (-0.07 MPa) soil water potentials. They concluded that the optimum range of water potentials for emergence was lower on the silt loam than on the clay.

Seedling Stages

Thomas and Andre' (1982) studied gross photosynthesis, O_2 uptake, CO_2, transpiration, root respiration, and ion uptake of soybean seedlings during 19 days, which included two periods of water stress. They found that short term water stress (a) decreased light interception by leaf wilting; (b) decreased photosynthesis; (c) decreased stomatal conductance, reducing both losses of water and the entry of CO_2 for assimilation; and (d) stimulated O_2 uptake. The ratio of O_2 uptake to CO_2 assimilation changed from 1.0 before water stress to 1.4 for several days after. Root respiration was less affected by the water stress than ion uptake and shoot gas exchange.

Vegetative Stages

Water stress during vegetative growth reduces the rate of dry weight accumulation, plant height, and leaf area index. As the leaf water potential decreases from less water uptake or more atmospheric demand, leaf enlargement and nitrogen fixation decrease. These effects of moisture stress on vegetative growth are reflected in smaller leaves, reduced stem diameter, and reduced plant height which collectively make less photosynthate available for later seed development.

Fukutoku and Yamada (1981) found that proline and asparagine accumulated in the leaves of 5-week old soybean plants grown in the greenhouse under severe water stress. Their results suggested that water stress altered protein metabolism.

Reproductive Stages

Water stress during reproductive growth varies, depending upon flower and seed development. Soybean plants stressed for water during flower production and development produce fewer flowers, pods and seeds because of a shorter flowering period and abortion of some flowers. Seed yield, as measured by total seed weight or total protein, is reduced most by stress when it occurs during early seed formation and pod filling. Water stress, however, apparently has no effect on the final protein content per seed.

Reicosky and Deaton (1979) found only small differences in the midday leaf water potential between two cultivars grown on a sandy loam in South Carolina. Evapotranspiration of the nonirrigated soybean plants was about two-thirds that of irrigated plants 25 and 32 days into the drought. Significant soil water extraction occurred at lower soil depth in the nonirrigated treatments near the end of the drought. The maximum within-canopy air temperature during drought of nonirrigated cultivar Davis was 6.7 C higher than in the irrigated treatment.

Reicosky et al. (1982) investigated the diurnal relationship between ET and leaf water potential (P) for field-grown soybean plants as affected by irrigation on a silt loam in Iowa. They found that the P-ET relationship of nonirrigated plants was hysteredic, but not with irrigated plants. They concluded that the simplified Ohm's law does not seem to describe adequately the relationship of plant water flux to resistance and potential differences under field conditions. This was attributed to the dynamic nature of the field environment, which reflects the general nonsteady-state flow conditions.

Carlson et al. (1979) studied the relationships between leaf water potential and leaf conductance by two indeterminate soybean cultivars grown on a silt loam in Iowa. The soybeans were subjected to four independent dry-down stress periods during reproductive growth. They found leaf water conductance to be significantly correlated with leaf water potential over most days; however, a unique nonlinear relationship between these two parameters was not established. Atmospheric demand, time of day, and some type of moisture-stress hardening contributed as significant sources of variation. The two cultivars differed significantly in their leaf water potential-leaf conductance relationships. In the same study, these same authors also found that moisture stress affected yield components, such as seed number or seed size. Moisture stress-cultivar interactions were difficult to identify. No comparison was made between stressed and nonstressed soybeans.

Snyder et al. (1982) evaluated the effects of moisture stress during reproductive growth on four cultivars grown in potometers in the field. Seed yield reductions for all soil water treatments varied from 2 to 27%. Preconditioning soybean plants resulted in less yield reduction than if the plants were unconditioned. The plants subjected to three periods of soil-water stress produced fewer but heavier seeds than plants stressed only during the third period. The response of the yield components varied extensively between cultivars and moisture treatments.

Brown and Scott (1984), in their review of the dependence of crop growth and yield on root development and activity, suggested that in-row root distributions of well-watered soybean plants increases to a maximum during early reproductive growth. Thereafter, root density declines until maturity. Root densities in the row middles increases throughout reproductive growth. During this same period roots immediately under the row and near the soil surface decay. Usually, severely water-stressed soybeans produce fewer roots but may have a greater proportion of their roots deeper in the profile. They also reported on a field experiment conducted by Brown in which soybean roots were grown under minimal soil stresses. Soybean roots were allowed to grow uninhibited within a porous membrane envelope 15 cm below the soil surface. All nutrients (except nitrogen) and oxygen were obtained from the soil. In this system Brown found excellent nodule development throughout the entire root system and significant increases in number of pods and seeds per pod. This work demonstrated the restrictive effects of soil on the potential genetic expression of soybean root systems.

Robertson et al. (1980) found no effects of irrigation on seed yield or soybean root distribution in a sandy soil in Florida. The determinations were made soon after full canopy development.

IRRIGATION RESEARCH STUDIES

Water is essential for plant growth, but too much or too little at any given stage of growth may be detrimental. Management practices, such as row spacing, mulching, planting density, and minimum tillage help to conserve water. Excess water may be more difficult to manage, but the use of land forming, drainage tiles, terraces, dams, and other water management practices help to reduce crop damage caused by too much water.

The general approach used to assess the genetic potential of a given soybean cultivar is to determine its productivity under conditions that are non-limiting. This method is known as the "abundant resource" approach, and has the advantage that the environment is reasonably predictable and that the genotypes can be adapted to it. Yield increases due to irrigation in humid regions generally have ranged up to 3000 kg/ha. Depending upon the climatic and soil environments, and with limited irrigation water available, early maturing cultivars may perform better than late maturing cultivars in dry regions.

The effects of water stress on soybean yields vary according to growth stage, adaptation to previous stresses, and growth habit. Assuming soybeans are planted in a profile of adequate water, the general ranges of yield reduction for determinate soybean crops in Arkansas differ with growth stage stressed. Approximate yield reductions for field-grown, determinate types as a function of growth stage can range from 0 to 25% at the late vegetative stage or 0 to 15% at early reproductive stages (R2 to R3). These reductions will be 0 to 20% with stress at R3 to R5. The reduction in seed yields is determined as a percentage of the remaining potential yield. For example, if the potential seed yield of a given cultivar grown on a given soil is 3300 kg/ha, a severe water stress during late vegetative growth would decrease actual seed yield to 2475 kg/ha, assuming no more water stresses occurred during the growing season. If a severe water stress also occurred between R3 and R4, the seed yield would be reduced to 1980 kg/ha.

Lawn (1982) found that dry matter production of field-grown soybean plants was a function of water use, and that water-use efficiency and seed yield depended upon the rainfall characteristics of the growing season. He compared the results of three cultural management schemes, irrigated, rainfed-fallowed and rainfed-double cropped and found that the irrigated soybean crops had the highest dry matter accumulation, seed yield, and seed weight. No significant differences were found in the number of seeds per pod, harvest index, and water-use efficiency. The relationship between aboveground dry matter at maturity and seasonal water use was linear. When soil water in the profile was limited, such as under double cropping situations, the soybean plants depleted most of the available soil water before podfill. As a result, they were in severe water stress during the sensitive late reproductive phases.

Ritchie (1984) applied the concept of risk analysis to the prediction of irrigation needs of soybean plants grown under conditions of varying soil water deficits at planting. Varying the soil water deficit at planting simulated the water depletion effects of a winter crop, such as wheat (*Triticum aestivum* L.), followed by a soybean crop in a double cropping management scheme. Risk analysis involves the development of cumulative probability curves for soil water deficits using long term weather data. It provides an opportunity to quantitatively compare alternative management situations. Using 50 years of simulated weather data near Stuttgart, Arkansas and with a soybean crop growing on a clay, Ritchie showed that (a) there is a 50% probability that a soil water deficit at maturity of wheat on or about June 1 will be 1.50 m of water, (b) cumulative soil water deficits increase during the growing season irrespective of the water deficit at planting, (c) there is a 50% probability that a water deficit between 1.20 and 2.25 m will occur by 13 August, if the initial soil water deficit is zero, and (d) the effects of the soil water deficit at planting are not linearly translated to higher deficits during the growing season. When starting with a full profile of water, about 1.20 m of irrigation water would be required on the average, and no irrigation would be required in 20% of the years if the soil could provide at least 1.2 m of water to the soybean plants without crop loss. If the soybean crop follows wheat, a full soil water supply at planting was shown to be unrealistic.

FUTURE RESEARCH NEEDS

Two distinct and complimentary pathways should be followed in soybean-water management research. One pathway is to develop cultivars that possess high yield potentials in water-limiting environments. Both well-watered and water stressed conditions should be used to determine the response of genotypes in the field. Methods of selection should be based upon characteristics that benefit yield.

The other research pathway is to develop environments that improve plant productivity in light of the conditions that exist in the field. Both soil and climatic environments need to be improved for crop growth. The objectives of these studies are to develop management schemes that will bring actual productivity closer to the potential. These management schemes should result from research involving a greater understanding of the mechanisms of adaptation of the crop to drought. In particular, it is imperative that there be a better understanding of the ways in which water management affects those physiological processes occurring on a time scale of minutes, and how they in turn, are integrated to affect the growth of soybean plants over longer periods, such as days, weeks, and even the entire growing season.

Greater use of risk analysis should be made to benefit irrigation management information transfer. This will enable the soybean grower to quantitatively determine the probabilities of the various production factors. These production factors, when combined with economic factors, are especially useful in evaluation of water management practices.

NOTES

H. D. Scott, University of Arkansas, Fayetteville, Arkansas.

REFERENCES

Baldocchi, D. D., S. B. Verma, N. J. Rosenberg, B. L. Blad, A. Garay and J. E. Specht. 1983. Influence of water stress on the diurnal exchange of mass and energy between the atmosphere and a soybean canopy. Agron. J. 75:543-548.

Boyer, J. S. 1982. Plant productivity and environment. Science 218:443-448.

Brown, D. A. and H. D. Scott. 1984. Dependence of crop growth and yield on root development and activity. In Root, nutrient and water influx and plant growth. American Society of Agronomy, Madison, Wisconsin.

Bruce, R. R., V. L. Quisenberry, H. D. Scott and A. W. Thomas. 1984. Irrigation in the Southeast United States. In D. Hillel (ed.) Vol. III: Advances in Irrigation. Academic Press, Inc., New York (in press).

Carlson, N. N. Momen, O. Arjmand and R. H. Shaw. 1979. Leaf conductance and leaf-water potential relationships for two soybean cultivars grown under controlled irrigation. Agron. J. 71:321-325.

Fukutoku, Y. and Y. Yamada. 1981. Sources of proline-nitrogen in water stressed soybean (Glycine max L.) I. Protein metabolism and proline accumulation. Plant Cell Physiol. 22:1397-1404.

Geddes, R. E., H. D. Scott and R. E. Oliver. 1979. Growth and water use by common cocklebur (Xanthium pensylvanicum) and soybeans (Glycine max) under field conditions. Weed Sci. 27: 206-212.

Heatherly, L. G. and W. J. Russel. 1979. Effect of soil water potential of two soils on soybean emergence. Agron. J. 71:980-982.

Jung, P. K. and H. D. Scott. 1980. Leaf water potential, stomatal resistance and temperature relations in field grown soybeans. Agron. J. 72:986-990.

Lawn, R. J. 1982. Responses of four grain legumes to water stress in south-eastern Queensland. III. Dry matter production, yield and water use efficiency. Aust. J. Agric. Res. 33:511-521.

Meyer, R. F. and J. S. Boyer. 1981. Osmoregulation, solute distribution, and growth in soybean seedlings having low water potentials. Planta 151:482-489.

Meyer, W. S. and S. Walker. 1981. Leaflet orientation in water stressed soybeans. Agron. J. 73:1071-1074.

Reicosky, D. C. and D. E. Deaton. 1979. Soybean water extraction, leaf water potential, and evapotranspiration during drought. Agron. J. 71:45-50.

Reicosky, D. C., T. C. Kaspar and H. M. Taylor. 1982. Diurnal relationship between evapotranspiration and leaf water potential of field-grown soybeans. Agron. J. 74:667-673.

Ritchie, J. T. 1981. Available Water. Plant Soil. 58:327-338.

Ritchie, J. T. 1983a. Efficient water use in crop production: Discussion on the generality of relations between biomass production and evapotranspiration. *In* Limitations to efficient water use in crop production. American Society of Agronomy, Madison, Wisconsin.

Ritchie, J. T. 1983b. Integrating weather, management, genetic and soil information for crop yield models. Ben J. Altheimer Lecture Series. Agric. Publ. Office, University of Arkansas, Fayetteville.

Robertson, W. K., L. C. Hammond, J. T. Johnson and K. J. Boote. 1980. Effects of plant-water stress on root distribution of corn, soybeans, and peanuts in sandy soil. Agron. J. 72:548-550.

Rosenberg, N. J., B. L. Blad and S. B. Verma. 1983. Microclimate: The Biological Environment. John Wiley and Sons, New York.

Snyder, R. L., R. E. Carlson and R. H. Shaw. 1982. Yield of indeterminate soybeans in response to multiple periods of soil-water stress during reproduction. Agron. J. 74:855-859.

Thomas, D. A. and M. Andre'. 1982. The respone of oxygen and carbon dioxide exchanges and root activity to short term water stress in soybean. J. Exp. Bot. 33:393-405.

United States Department of Commerce. 1982. 1978 Census of Agriculture. V. The 1979 Farm and Ranch Irrigation Survey. AC78-SR-8. U.S. Government Printing Office, Washington, D.C.

IRRIGATION MANAGEMENT FOR SOYBEAN YIELD ENHANCEMENT

Larry G. Heatherly

Irrigation is used to replenish soil water that has been extracted through evapotranspiration. In most instances, variability in both the occurrence and amount of rainfall during the growing season will probably result in at least one, and usually more, drought periods that result from evapotranspiration exceeding the soil water supply. These periods of drought can be of dire consequence if they occur during reproductive development of the soybean crop, since this is the time when the seed yield and economic return of the crop are being determined. The response of soybean seed yield to irrigation during this time is variable. In order to determine the proper irrigation strategy to maximize soybean seed yield, and increase net return, a knowledge of the patterns of response to various schemes of water application is necessary. This chapter will focus on a) the varied responses of soybeans to irrigation in relation to timing of application and stage of development of the plant; and b) irrigation for maximum soybeanseed yield (unlimited water supply) vs. irrigation for maximum response per unit of applied water (limited water supply).

IRRIGATION TIMING AND SOYBEAN YIELD

A key to proper management of any production input is timing. Planting, application of pesticides, and harvesting are examples of operations that produce maximum benefits when done at the optimum time. For maximum benefits to accrue from irrigation, proper timing of water application is necessary.

Numerous studies (Ashley and Ethridge, 1978; Brady et al., 1974; Constable and Hearn, 1978; Doss et al., 1974; Heatherly, 1983; 1984; Martin et al., 1979) have been conducted to investigate soybean yield response to both full-season irrigation (water applied as needed during both the vegetative and reproductive phases of development, designated VE) and irrigation during reproductive development only (water applied as needed from flowering to maturity, designated FL). In all cited cases, VE irrigation produced no appreciable yield advantage above that realized from FL irrigation (Table 1). Also, the irrigation efficiency, defined here as increase in seed yield per hectare due to irrigation (kg/ha/cm), was usually higher for the FL treatment than for the VE treatment. Thus, irrigation of a monoculture soybean crop before bloom appears to be of no benefit.

Delaying irrigation until the beginning of podset (PS) or beginning of podfill or seed development (SD), in years where rainfall was limited during early reproductive stages, resulted in significantly lower seed yields than those realized from either VE or FL irrigation, but did produce yield increases above the nonirrigated (NI) treatment

Table 1. Effect of timing of irrigation on seed yield and irrigation efficiency of soybean crops.

Year	Cultivar or maturity group [a]	Irrigation treatment [b]	Yield	Yield increase above nonirrigated	Irrigation efficiency [c]
			— kg/ha —		kg/ha/cm
GA (USA)—Ashley and Ethridge (1978)					
1972	Hampton 266A (VIII)	NI	390	–	–
		VE	2130	1740	74
		FL	2265	1875	102
		SD	1940	1550	124
1973	Ransom (VII)	NI	2045	–	–
		VE	3650	1605	82
		FL	3575	1530	116
		SD	3615	1570	159
	Coker 102 (VIII)	NI	1205	–	–
		VE	2670	1465	75
		FL	2520	1315	100
		SD	2150	945	95
1974	Hampton 266A	NI	1930	–	–
		VE	2010	80	5
		FL	1945	15	2
	Ransom	NI	2370	–	–
		VE	2925	555	34
		FL	2985	615	62
KS (USA)—Brady et al. (1974)					
1972	Calland (III)	NI	3080	–	–
		VE	3550	470	10
		FL	3315	235	5
		PS	3820	740	24
1973	Calland	NI	2730	–	–
		VE	3295	565	14
		FL	3335	605	18
		PS	3280	550	28
NSW (AUST)—Constable and Hearn (1978)					
1975	Ruse (V)	NI	1705	–	–
		VE	2735	1030	NA [d]
		FL	2530	825	
	Bragg (VII)	NI	1910	–	–
		VE	2945	1035	NA
		FL	3095	1185	
AL (USA)—Doss et al. (1974)					
1970	Bragg	NI	2610	–	–
		VE	2790	180	NA
		FL	2670	60	
1971	Bragg	NI	3060	–	–
		VE	3250	190	NA
		FL	3120	60	
1972	Bragg	NI	1880	–	–
		VE	2940	1060	28
		FL	2770	890	37
MS (USA)—Heatherly (1983)					
1979	Bedford (V)	NI	2750	–	–
		PS	2670	-80	0
		SD	2580	-170	0

Table 1. Continued.

Year	Cultivar or maturity group[a]	Irrigation treatment[b]	Yield	Yield increase above nonirrigated	Irrigation efficiency[c]
			— kg/ha —		kg/ha/cm
	Tracy (VI)	NI	3370	--	--
		PS	3380	10	1
		SD	3300	-70	0
	Bragg	NI	3170	--	--
		PS	3210	40	4
		SD	3630	460	51
1980	Bedford	NI	730	--	--
		FL	2000	1270	26
		PS	1750	1020	21
		SD	1110	380	11
	Tracy	NI	1150	--	--
		VE	2890	1740	29
		FL	2810	1660	26
		SD	1470	320	11
	Bragg	NI	1320	--	--
		VE	3560	2240	32
		FL	3220	1900	33
		SD	1790	470	16
		MS (USA)—Heatherly (1984)			
1979	Forrest (V)	NI	3510	--	--
		SD	3625	115	15
1980	Forrest	NI	1125	--	--
		FL	2980	1855	60
		PS	2480	1355	59
		SD	2450	1325	77
		NE (USA)—Korte et al. (1983a)			
1977	II, III, IV	NI	3000	--	
		F	3000	0	NA[d]
		P	3250	250	
		S	3030	30	
1978	II, III, IV	NI	2570	--	--
		F	2650	-80	NA
		P	3150	580	
		S	3540	970	
1979	II, III, IV	NI	2610	--	--
		F	2700	90	NA
		P	3020	410	
		S	2760	150	
		NC (USA)—Martin et al. (1979)			
1977	Ransom	NI	1730	--	--
		VE	2065	335	12
		FL	2125	395	26
		PS	1685	-45	0

[a]Where only maturity group or groups are given, tabular values are average of more than one cultivar in a group or groups.

[b]NI = no supplemental water (except at planting in some cases); VE, FL, PS, SD = irrigation started before flowering (VE), at beginning of flowering (FL), at beginning of podset (PS), or at beginning of podfill or seed development (SD) and continued as needed until near maturity; F, P, S = irrigation at beginning of flowering (F) only, at beginning of podset (P) only, or at beginning of podfill (S) only.

[c]Values are kg/ha yield increase (treatment yield minus NI yield) divided by total amount of irrigation water in cm added to that treatment.

[d]NA means that the amount of irrigation water was not given in the cited references.

in most cases (Ashley and Ethridge, 1978; Martin et al., 1979; Heatherly, 1983; Heatherly, 1984). In a year when rainfall was adequate during all reproductive phases except podfill (Heatherly, 1983), Bragg produced significantly more seed from one irrigation at this stage than was produced by either the PS or NI treatments. Earlier maturing cultivars, which were essentially through filling pods before this late-season stress occurred, did not show the same response. Obviously the stage of development, in relation to the occurrence of a period of limited rainfall, will have a dominant effect on a cultivar's response to irrigation.

In some cases irrigation at any stage of development did not increase yields above levels of the NI treatment (Table 1; Ashley and Ethridge, 1978; Doss et al., 1974; Heatherly, 1983; Heatherly, 1984; Martin et al., 1979). This usually was because rainfall was sufficient to produce yields of 2500 to 3500 kg/ha (Heatherly, 1983; 1984; Doss et al., 1974). However, Ashley and Ethridge (1978) reported that irrigated yields of Hampton 266A in 1974 were no greater than the NI yield of 1940 kg/ha, and this lack of response was attributed to greater lodging and a severe infestation of powdery mildew (*Erysiphe polygoni* DC) in the irrigated treatments. By comparison, yield from the FL treatment of Ransom was 2925 kg/ha, and this was 615 kg/ha greater than the NI treatment yield in the same year. The relatively low response to either VE or FL irrigation obtained by Martin et al. (1979) is difficult to understand, especially since the NI treatment yield was only 1730 kg/ha. It is possible that the approximately 15 cm of irrigation water applied to the FL treatment in their study was not sufficient to achieve maximum yield, especially in the loamy soil which was used.

Several research efforts have sought to establish a critical period for irrigation of soybeans. Sionit and Kramer (1977) found that stress during either the pod formation or podfill resulted in greater yield reductions than stress during flower induction or flowering. This finding is supported by the data of Doss et al. (1974), who obtained the lowest seed yield when they withheld irrigation during podfilling. In both cases, adequate water had been supplied to the plants up to the point of the induced stress. Stress during the podfill stage produced the smallest seed, but did not reduce the total number of seeds below the level produced by plants that were irrigated during all stages of growth (Sionit and Kramer, 1977). Korte et al. (1983a) found that one irrigation at the pod elongation stage (P) or the seed enlargement stage (S) significantly enhanced 3-year average yields above both the nonirrigated check and the treatment that received a single irrigation at flowering (F). Most of this effect, however, was the result of large differences in 1978 compared with those in both 1977 and 1979 (Table 1), and this resulted in a significant year by irrigation interaction. The smallest seed in this study were obtained from the treatment irrigated only at flowering (Korte et al., 1983b), but seed per plant were similar for all irrigation treatments.

Doss et al. (1974) concluded that the podfill stage of soybeans is the most critical period for irrigation to obtain maximum seed yield. This would appear to be supported by the studies of Sionit and Kramer (1977) and Korte et al. (1983a). However, in the former two cases, all of the results are from situations where water was available from either rainfall or irrigation during all other phases of growth leading up to the designated stress stage. This would allow the plant to set the maximum fruit load allowed by the available moisture. If that water was suddenly cut off (as was the case), then the plant would be unable to maintain the yield potential established earlier, and the result would be smaller seed, as found by Sionit and Kramer (1977) and Korte et al. (1983b). Conversely, plants stressed only during flowering produced significantly fewer seed than those stressed during podfill, but the seed were equal in size to those of the well-watered control plants (Sionit and Kramer, 1977). Yield from this treatment (stressed during flowering only) was between the yield of the other two.

In a field or producer situation, these artificial environments would not exist unless a very limited amount of water was available for irrigation. If irrigation is started at flowering in a dry year, or at the period of first stress in a relatively wet year, as supported by the data of Table 1, then it should continue until the seed are fully developed. If only a limited amount of water is available (not enough for full-season irrigation), the preponderance of data supports the practice of utilizing this limited water for irrigation during podfill. However, this latter practice will not produce maximum yields unless adequate rainfall has been received prior to this time. For maximum yield in dry years, irrigation of a soybean crop should be started at beginning of flowering and, in the absence of rainfall, continued until seed are fully developed.

OPTIMUM IRRIGATION AND SOYBEAN YIELD

The data in Table 2 are the results of numerous research experiments where NI and irrigated (I) treatments were employed. In the cited studies, row spacing was generally 75 to 100 cm, plant population was 225,000 to 250,000 plants per ha, and planting date was between 1 May and 15 June (or the equivalent). All of the NI treatments received only rainfall, and the I treatments generally received water only during reproductive development, although irrigation in some cases may have been started just before beginning of flowering. Watering was begun in most cases whenever 40 to 60% of the available water in the upper 30 to 60 cm of the soil profile has been depleted, and in most cases was continued on this same basis until near maturity.

Yield increases from irrigation exceeded 500 kg/ha in two-thirds of the situations presented in Table 2, and exceeded 1000 kg/ha in one-half of the situations. Yields of NI treatments of about 2500 to 3500 kg/ha were measured in 11 of the 20 cases where little or no response to irrigation was measured, and this is generally the same range of yields that were achieved with irrigation where responses were obtained. Therefore, based on the results of the majority of the available research, lack of a response to irrigation in these 11 cases is not surprising.

The nine situations in which NI yields were generally less than 2000 kg/ha, and little or no response to irrigation was obtained, deserve closer evaluation. Two of the cases (Ashley and Ethridge, 1978; Doss and Thurlow, 1974) involved similar cultivars (Hampton 266A and Hampton 266, respectively), and increased lodging was observed in both situations. In addition, powdery mildew infected the Hampton 266A which had been irrigated, and this also contributed to the lack of response to irrigation. The 15 cm of irrigation water applied to a loamy sand soil by Martin et al. (1979) probably fell short of the amount needed to significantly increase seed yield. Taylor et al. (1982) noted that seed from their irrigated treatment were poorly filled and were significantly smaller than those from the nonirrigated treatment. This strongly indicates that water was deficient during podfill, and the effects of drought stress at this stage on previously well-watered plants have been discussed earlier. Thus, they probably stopped irrigation too soon. Cutting off water too early was also partially blamed for the lack of measured response to irrigation of Hampton 266 (Doss and Thurlow, 1974). Karlen et al. (1982) attributed the lack of a yield response in their 1979 study to a lower plant density (20 plants/m^2) and "possible O_2 stress", assumed due to a large amount of rainfall during podfill. Obviously, some yield-limiting factor was present, because their 1978 study used the same cultivars and achieved yields of 3070 to 3510 kg/ha with yield increases of 1340 to 1490 kg/ha in the irrigated treatments. Reicosky and

Table 2. Yield response from irrigation of soybean plants during reproductive development.

Cultivar or maturity group[a]	Planting date	Yield[b] NI	Yield[b] I	Difference[c]
	Mo/Yr		– kg/ha –	
GA (USA)—Ashley and Ethridge (1978)				
Hampton 266A (VIII)	5/72	390	2265	1875
Ramson (VII)	5/73	2045	3575	1530
Coker 102 (VIII)	5/73	1205	2520	1315
Hampton 266A	5/74	1930	1945	15
Ransom	5/74	2370	2985	615
KS (USA)—Brady et al. (1974)				
Calland (III)	5/72	3080	3315	235
Calland	5/73	2730	3335	605
ND (USA)—Cassel et al. (1978)				
SRF100 (I)	6/72	1065	1875	810
I	5/73	210	3105	2895
I	5/74	350	2305	1955
NSW (AUST)—Constable and Hearn (1978)				
Ruse (V)	12/75	1705	2535	830
Bragg VII	12/75	1910	3100	1190
AL (USA)—Doss and Thurlow (1974)				
Bragg	6/68	1770	2570	800
Bragg	6/69	2120	3240	1120
Bragg	6/70	2360	2520	160
Hampton 266	6/68	1770	2070	300
Hampton 266	6/70	1960	2160	200
AL (USA)—Doss et al. (1974)				
Bragg	5-6/70	2610	2790	180
Bragg	5-6/71	2060	3250	1190
Bragg	5-6/72	1880	2940	1060
MS (USA)—Heatherly (1983)				
Bedford (V)	6/79	2750	2670	-80
Tracy (VI)	6/79	3370	3380	10
Bragg	6/79	3170	3630	460
Bedford	5/80	730	2000	1270
Tracy	5/80	1150	2810	1660
Bragg	5/80	1320	3220	1900
MS (USA)—Heatherly (1984)				
Forrest (V)	6/79	3510	3625	115
Bedford	5/79	3425	3260	-165
Bragg	5/79	3715	3910	195
Bedford	6/79	3190	2800	-390
Bragg	6/79	3650	3705	55
Forrest	5/80	1125	2980	1855
Bedford	5/80	990	2730	1740
Bragg	5/80	1330	3525	2195
Bedford	6/80	1150	3150	2000
Bragg	6/80	1520	2980	1460
Bedford	5/81	980	2780	1800
Braxton (VII)	5/81	1030	3275	2245
Bedford	6/81	1050	2375	1325
Braxton	6/81	1695	2940	1245
Bedford	5/82	975	2245	1270
Braxton	5/82	1010	2715	1705

Table 2. Continued.

Cultivar or maturity group[a]	Planting date	Yield[b]		Difference[c]
		NI	I	
	Mo/Yr	— kg/ha —		
Bedford	6/82	880	1670	790
Braxton	6/82	1185	2345	1160
	SC (USA)—Hunt et al. (1981)			
Bragg	6/78	1275	3090	1815
	AR (USA)—Jung and Scott (1980)			
Forrest	5/78	905	3455	2550
	SC (USA)—Karlen et al. (1982)			
Coker 338 (VIII)	6/78	1590	3070	1480
Bragg	6/78	1550	2890	1340
Ransom	6/78	2020	3510	1490
Coker 338	5/79	1440	1730	290
Bragg	5/79	1670	1960	290
Ransom	5/79	1740	2320	580
	NC (USA)—Mahler and Wollum (1981)			
Ransom	5/79	1295	2070	775
	NC (USA)—Martin et al. (1979)			
Ransom	5/77	1685	2135	450
	OH (USA)—Mederski and Jeffers (1973)			
I	5/-	2160	2925	765
II	5/-	2215	3100	885
III	5/-	2285	3015	730
	TX (USA)—Musick et al. (1976)			
IV, V	NA[d]	560	2640	2080
	SC (USA)—Reicosky and Deaton (1979)			
Davis (VI)	5/75	1990	2340	350
McNair 800 (VIII)	5/75	2040	2405	365
	IA (USA)—Taylor et al. (1982)			
Wayne (III)	5/79	2105	2110	5

[a]Where only maturity group or groups are given, tabular values are average of more than one cultivar in a group or groups.

[b]NI = nonirrigated; I = irrigated from beginning of flowering until seed are fully developed.

[c]Values are yield increase or decrease from irrigation (I yield minus NI yield).

[d]NA means that planting date was not given in the cited reference.

Deaton (1979) offered no explanation for the lack of a significant response to irrigation in their study.

CONCLUSIONS

Based upon data presented here, irrigation can be utilized to increase seed yield of soybeans significantly in years that have periods of low rainfall. Irrigation should begin at or near flowering, if needed, and continue until seed are fully developed in order to realize maximum yield potential. Soybeans are very sensitive to drought stress during the podfill stage if adequate water has been supplied up to that point. If a limited amount of water is available; i.e., not enough for irrigation during the entire

987

reproductive period, to obtain maximum benefit this water should be utilized during the podfill stage. In years that have adequate rainfall during most of the growing season, irrigation should be utilized to offset the potential yield-reducing effects of short periods of drought that may occur during reproductive development.

NOTES

Larry G. Heatherly, Research Agronomist, U.S. Department of Agriculture-ARS, Soybean Production Research Unit, P.O. Box 196, Stoneville, MS 38776.

REFERENCES

Ashley, D. A. and W. J. Ethridge. 1978. Irrigation effects on vegetative and reproductive development of three soybean cultivars. Agron. J. 70:467-471.

Brady, R. A., L. R. Stone, C. D. Nickell and W. L. Powers. 1974. Water conservation through proper timing of soybean irrigation. J. Soil Water Cons. 29:266-268.

Cassel, D. K., Armand Bauer and D. A. Whited. 1978. Management of irrigated soybeans on a moderately coarse-textured soil in the upper Midwest. Agron. J. 70:100-104.

Constable, G. A. and A. B. Hearn. 1978. Agronomic and physiological responses of soybean and sorghum crops to water deficits. I. Growth, development, and yield. Aust. J. Plant Physiol. 5:159-167.

Doss, B. D., R. W. Pearson and H. T. Rogers. 1974. Effect of soil water stress at various growth stages on soybean yield. Agron. J. 66:297-299.

Doss, B. D. and D. L. Thurlow. 1974. Irrigation, row width, and plant population in relation to growth characteristics of two soybean varieties. Agron. J. 66:620-623.

Heatherly, L. G. 1983. Response of soybean cultivars to irrigation of a clay soil. Agron. J. 75:859-864.

Heatherly, L. G. 1984. Soybean response to irrigation of Mississippi River Delta soils. U.S. Dept. Agric-Southern Series Res. Rep. (in press).

Hunt, P. G., A. G. Wollum, II and T. A. Matheny. 1981. Effects of soil water on Rhizobium japonicum infection, nitrogen accumulation, and yield in Bragg soybeans. Agron. J. 73:501-505.

Jung, P. K. and H. D. Scott. 1980. Leaf water potential, stomatal resistance, and temperature relations in field-grown soybeans. Agron. J. 72:986-990.

Karlen, D. L., P. G. Hunt and T. A. Matheny. 1982. Accumulation and distribution of P, Fe, Mn, and Zn by selected determinate soybean cultivars grown with and without irrigation. Agron. J. 74:297-303.

Korte, L. L., J. H. Williams, J. E. Specht and R. C. Sorensen. 1983a. Irrigation of soybean genotypes during reproductive ontogeny. I. Agronomic responses. Crop Sci. 23:521-527.

Korte, L. L., J. E. Specht, J. H. Williams and R. C. Sorensen. 1983b. Irrigation of soybean genotypes during reproductive ontogeny. II. Yield component responses. Crop Sci. 23:528-533.

Mahler, R. L. and A. G. Wollum, II. 1981. The influence of irrigation and Rhizobium japonicum strains on yields of soybeans grown in a Lakeland sand. Agron. J. 73:647-651.

Martin, C. K., D. K. Cassel and E. J. Kamprath. 1979. Irrigation and tillage effects on soybean yield in a coastal plain soil. Agron. J. 71:592-594.

Mederski, H. J. and D. L. Jeffers. 1973. Yield response of soybean varieties grown at two soil moisture stress levels. Agron. J. 65:410-412.

Musick, J. T., L. L. New and D. A. Dusek. 1976. Soil water depletion—yield relationships of irrigated sorghum, wheat, and soybeans. Trans. Am. Soc. Agric. Eng. 19:489-493.

Reicosky, D. C. and D. E. Deaton. 1979. Soybean water extraction, leaf water potential, and evapotranspiration during drought. Agron. J. 71:45-50.

Sionit, N. and P. J. Kramer. 1977. Effect of water stress during different stages of growth of soybean. Agron. J. 69:274-278.

Taylor, H. M., W. K. Mason, A. T. P. Bennie and H. R. Rowse. 1982. Responses of soybeans to two row spacings and two soil water levels. I. An analysis of biomass accumulation, canopy development, solar radiation interception, and components of seed yield. Field Crops Res. 5:1-14.

THE IMPACT OF SOYBEANS ON SOIL PHYSICAL PROPERTIES AND SOIL ERODIBILITY

D. V. McCracken, W. C. Moldenhauer and J. M. Laflen

Following soybean crops, a soil's physical condition differs in definite, observable ways from its condition after corn (*Zea mays* L.) or small grain (Moldenhauer and Duncan, 1968). On many soils not subject to erosion by wind or water, or severe surface sealing, the effect of a previous soybean crop is beneficial. Soybean plants enhance tilth—allowing high infiltration, good water transmission, a seedbed that is firm but not too firm or too loose, soil that is well granulated for good seed contact, easy root penetration, high water-holding capacity and good aeration. Unfortunately, few soils are not subject to surface sealing or to erosion by wind or water. The same properties that make soil following a soybean crop so desirable for a seedbed also make it much more subject to erosion.

POSITIVE EFFECTS OF SOYBEAN ON SOIL PHYSICAL PROPERTIES

The soybean, in rotation with other crops, leaves heavy, dark-colored soils in better tilth than corn or small grains (Browning, 1949; Calland, 1949). Soil in good tilth favors infiltration of rainfall, adequate air exchange and possesses low resistance to root growth (Baver et al., 1972). Greater infiltration of rainfall occurs following a soybean crop than following corn, if the soil is protected from raindrop impact and prevented from crusting (Kidder et al., 1943; Mannering and Johnson, 1969). Siemens and Oschwald (1978) measured less resistance to a penetrometer after soybean than after corn, which implies lower resistance to root penetration following soybean (Barley and Greacen, 1967).

Soil is loosened by soybean crops and requires minimal energy to prepare a favorable seedling environment. Armbrust et al. (1982) found that, after soybean, soil contained fewer dry aggregates larger than 0.84 mm than soil after either sorghum (*Sorghum bicolor* [L.] Moench) or wheat (*Triticum* spp.), a finding that has implications both for seedbeds and wind erodibility. The mechanical stability of dry aggregates was also less following soybean. Browning and Norton (1948) noted that, following soybean, satisfactory seedbeds could be prepared without plowing. Modern tillage recommendations continue to take advantage of soybean's loosening effect and encourage direct planting into soybean stubble (Griffith et al., 1977; Smith, 1980).

NEGATIVE EFFECTS OF SOYBEAN ON SOIL PHYSICAL PROPERTIES

Soybean can harm the tilth of silty, light-colored soils (Browning, 1949). Such soils are, by nature, structurally weak. Following soybean, a soil's aggregate strength and size are less than before the crop. Surface crusting and soil erosion by wind and water can be severe on weak soils following soybean if sensible management practices are abandoned.

Soil after a soybean crop is prone to crust under rain impact (Kidder et al., 1943). Rain crusts affect crop growth adversely soon after planting, because they resist seedling emergence (Hanks and Thorpe, 1957; Brossman et al., 1982). Yield reduction occurs if the surface seal prevents a good stand.

Rain crusts also greatly restrict a soil's ability to admit water (Duley, 1940). Bare soil following soybean seals more rapidly than it does following corn (Kidder et al., 1943). Soybean aggravates the erodibility of weak soils, in part, by promoting their tendency to crust. When the intensity of a storm exceeds the ability of water to infiltrate the soil and the capacity for surface storage, runoff begins, carrying soil with it. By increasing runoff, crusts reduce the water available for soil storage and may adversely affect crop growth during periods when precipitation is intense and infrequent.

The lack of resistance to crusting after soybean is related to the size distribution of water-stable aggregates which soybean promotes. Crusting is most complete on soils with low wet stability of aggregates (McIntyre, 1955). Stauffer (1946) compared the aggregation of plots in continuous corn and continuous soybean. Over a four-year period, corn resulted in more water-stable aggregates larger than 0.5 mm than did soybean. Conaway and Strickling (1962) reported similar results. Fahad et al. (1982) found the geometric mean diameter of water-stable aggregates to be smaller under continuous soybean than in soybean after corn, soybean after sorghum or soybean after fallow.

The ease with which soil is detached and transported following soybean crops is also related to the strength and size distribution of aggregates that they foster (Chepil, 1958; Bryan, 1968). Weak aggregates are more easily worn and broken by abrasion, wetting and raindrop impact. Small aggregates are more readily rolled, lifted and thrown by wind, channel flow and splash. The mean size of aggregates is smaller following soybean than following most other agronomic crops (Armbrust et al., 1982; Fahad et al., 1982), and in part due to this, the potential for soil loss following soybean can be unacceptably high.

EFFECT OF SOYBEAN ON SOIL LOSS AND SOIL ERODIBILITY

In the 1940s, agronomists rated soil loss under rowed soybean to be no more than that under other row crops appearing at the same place in the rotation (Walker, 1946; Browning, 1947; 1949). With little evidence, the same observers recognized that soybean increased the potential for erosion after harvest and that soil following soybeans may require special care to prevent excessive soil loss. Data gathered since then confirm their first impressions.

To give it meaning, soil loss due to soybean should be compared with that caused by another crop. Soybean competes directly with corn for land area in much of the USA. Data are presently sufficient to trace and compare the soil loss caused by both crops.

During the first half of the growing season, soil loss attributable to standing corn and soybean appears similar. With natural rainfall, Laflen and Moldenhauer (1979) found no difference between soil loss from soybean after corn and corn after corn (Wischmeier and Smith, 1978). Using a rainfall simulator to evaluate tillage systems and soybean-corn rotations, Laflen and Colvin (1981) reported similar results. Mannering and Johnson (1969) used simulated rainfall to evaluate different row spacings.

They measured no significant differences in soil loss from soybean and corn in the first 8 weeks of the growing season, when they compared the same row spacings. During this period the crop canopy cover remained similar (Colvin and Laflen, 1981).

Late in the summer soybean permits less soil loss than corn. Soybean leaf drop in this period greatly increases ground cover (Colvin and Laflen, 1981). Measured reduction in soil loss under soybean during this period, relative to that under corn (Laflen and Moldenhauer, 1979; Laflen and Colvin, 1981), may be attributable to the increased ground cover and to the lower, sometimes more continuous (Colvin and Laflen, 1981) canopy of a soybean crop.

Tillage and residue management strongly influence soil loss from corn and soybean ground after harvest and into the next growing season. With good management, corn residue is greater in coverage and reduces soil loss more effectively than soybean residue (Laflen and Colvin, 1981). But, even when the canopy and residue cover in the next crop is equivalent, soil loss is greater the season after soybean than it is following corn (Laflen and Moldenhauer, 1979; Laflen and Colvin, unpublished).

Measured soil loss confirms what simple, physical properties hint: soil is more erodible following soybean than following corn. But, this distinction is not always apparent. The effects of soil conditioning and residue cover on soil loss can be difficult to separate. Both affect erosion, and both vary together across tillages and, in the case of soybean and corn, across crop rotations.

With no tillage, soil loss is contained by an extensive residue cover and, very likely, well-consolidated soil (Van Doren et al., 1984). Mannering et al. (pers. commun.) measured the surface cover left by corn and soybean in the spring, after performing various fall tillages (Table 1). Siemens and Oschwald (1978) reported penetrometer data taken on corn and soybean plots in the spring, after providing different fall tillages (Table 2).

Table 1. Residue cover left by corn and soybeans in the spring after various fall tillages (Mannering, Griffith, Johnson and Wheaton, pers. commun.).

Prior crop	Tillage			
	None	Disk	Chisel	Plow
	— % cover —			
Soybean	21	11	10	2
Corn	64	65	37	4

Table 2. Cone Index Readings[a] from corn and soybean plots in the spring after various fall tillages (Siemens and Oschwald, 1978).

Prior crop	Tillage		
	None	Chisel	Plow
	— kPa —		
Soybean (1975)	510	421	296
Corn (1974)	676	627	386

[a]Average 0 to 15 cm readings taken with an ASAE standard penetrometer.

With conventional tillage (Griffith et al., 1977), the influence of the prior crop on the physical condition of the soil may determine soil loss (Siemens and Oschwald, 1978; Laflen and Moldenhauer, 1979). Crop residues are largely buried by moldboard plows. Soil is not well-covered between plowing and canopy establishment. Its surface is finely divided immediately after seedbed preparation and each cultivation. The ease with which clods and crusts are broken by disks and cultivators, and the size distribution of stable aggregates which result, is influenced by present and past crops. When viewed under conventional tillage, it is apparent that soybean conditions the soil to be more erodible than corn does in the next growing season.

Laflen and Moldenhauer (1979) measured soil and water losses from soybean-corn rotations and continuous corn. Tillage was simulated conventional: residues were spaded under and the seedbed prepared with a rototiller and harrow. Thirty-seven percent greater soil loss from corn after soybean than from corn after corn implied that a prior crop of soybean conditioned the soil to be more erodible than did a prior crop of corn. Laflen and Colvin (unpublished) studied soil loss from soybean-corn rotations under different tillages and row spacings. They reported that, on the average, when all residue cover was buried by conventional tillage, soil erosion after soybeans would be about 50% greater than after corn.

Soybean leaves the soil loose and erodible (Browning et al., 1943). But management systems that maximize surface cover and minimize the extent of soil disturbance help keep soil in place. On erodible landscapes, excessive tillage buries essential residue cover and compounds the effect of soybean on soil erodibility by stirring already loose soil.

REASONS FOR THE SOYBEAN INFLUENCE ON SOIL PHYSICAL PROPERTIES

We are not sure why soybeans affect physical properties the way they do. It is unlikely that one factor is solely responsible. Instead, an array of separate factors probably interact. Browning et al. (1943) suggested three influences that may work together to leave a ready supply of easily detachable aggregates following soybean: (1) the protective canopy of growing soybeans, (2) extreme fluctuation in soil moisture surrounding their roots, and (3) the decomposition of soybean roots and tops.

To the extent that crop canopies protect the soil surface from rainfall, they limit crusting and promote a loose, open surface. Browning et al. (1943) likened increased probe penetration under a soybean canopy to that under a burlap covering. Laflen and Moldenhauer (1979) reasoned that soybean's lower canopy and leaf drop helped to keep the soil surface open after cultivation, which allowed greater infiltration under soybean than under corn, late in the season.

Extreme fluctuations of soil moisture around living soybean roots may aid in loosening the soil. Soybean's root system is less extensive than the root systems of corn and sorghum (Teare et al., 1973; Allmaras et al., 1975b) and so extracts water from a smaller soil volume. Soybean also places a high demand on soil water (Rawson et al., 1977). As a result, soybean depletes soil water in the surface layers more completely than either corn or sorghum (Allmaras et al., 1975a,b; Burch et al., 1978). The relative abruptness of swings from dry to wet and back under soybean may contribute to perceived differences in soil structure. Rapid wetting of dry soil aggregates causes them to break apart. Dry aggregates break down more readily, when suddenly wet, than aggregates slightly more moist (Panabokke and Quirk, 1957).

The decomposition of soybean roots and residue promotes water-stable aggregation, though to a smaller degree than corn. After 4 months incubation under constant, optimal conditions, soil amended with 1% soybean roots and residue produced fewer water-stable aggregates larger than 2 mm than the 1% corn treatments (McCracken, 1984). With soybean roots and residue applied at half the rate of corn, more representative of field conditions, the difference was much wider. In addition to residue abundance, the quantity of nitrogen available to microbial decomposers can significantly affect aggregation. When a corn treatment (corn C:N = 49.5:1) was supplemented with nitrogen, the degree of aggregation it produced after 4 months incubation was indistinguishable from that of the soybean treatment (soybean C:N = 27.8:1) (McCracken, 1984). In field and lab studies, others reported similar results for nitrogen addition (Ram and Zwerman, 1960; Harris et al., 1966; Verma and Singh, 1974).

Earthworms may play a part in encouraging soil looseness after soybean. They appear to be more abundant under soybean than under corn. Preliminary results in Indiana show 6 times more worms in soybean plots than in corn plots that received the same tillage (Kladivko, pers. commun.).

METHODS OF EROSION CONTROL

The effectiveness of any tillage method or cropping practice in controlling erosion ultimately depends on how well the soil surface is protected. Soybean crops permit no more soil loss than other row crops; the hazard appears after harvest. Row-cropped soybeans leave a residue cover that is relatively small in amount and easily decomposed. Soybean residue on the soil surface is readily buried by any tillage and measurably diminishes each time an operation, such as planting or spraying, is carried out.

Minimizing tillage trips and other field operations keeps soybean residue on the soil surface, where it checks soil loss. Under natural rainfall, McGregor et al. (1975) compared soil loss from continuously row cropped soybean in no-till and conventional tillage. Their results showed that no-till soybean reduced annual soil loss to 15% of that from conventionally tilled soybeans. In a single, simulated rainstorm before planting in the spring, Siemens and Oschwald (1978) found a similar reduction in a comparison of soil loss from the same tillages.

Complete residue cover is also the best control against wind erosion. If fall tillage of soybean ground is considered essential, Moldenhauer and Duncan (1968) recommend forming ridges at right angles to the prevailing wind with a duck-foot cultivator or chisel. Ridges should be 5 to 12 cm high and appear on a 1:4 height-spacing ratio.

Surface roughness reduces wind and water erosion, but the protection may be short-lived because raindrops smooth the soil surface. Whereas tillage opens up the surface to allow greater intake of water, at least for a time, much soil is lost if runoff occurs, because tillage loosens the soil for easy detachment (Moldenhauer et al., 1983).

With no-till on moderate slopes (2.5%), row direction affects soil loss little (Moldenhauer et al., 1983). Where tillage is necessary for pest or disease control on these slopes, contouring successfully controls soil loss. With conventional tillage on a 2% slope, contour planting of soybean reduced annual soil loss to a quarter of that found with rows up-and-down-hill (Van Doren et al., 1951).

On slopes steeper than 4%, the residue cover left by soybean no longer sufficiently controls erosion by water on many soils. Cover crops, such as legumes, certain grasses, and wheat or rye (*Secale cereale* L.) seeded into the senescing soybean crop, provide a management alternative that allows continued, responsible soybean cropping. The cover crop is killed with a contact herbicide in the spring, and planting is done into the soybean and cover crop residue (Moldenhauer et al., 1983).

In the southern Corn Belt and southeastern USA, many growers take advantage of the longer growing season and adequate rainfall to double-crop soybeans and a winter grain. Planting soybean after the winter grain harvest keeps land under productive cover and effectively limits soil erosion (Moldenhauer et al., 1983).

CONCLUSION

Soybean improves many soils for crop production by modifying structure and enhancing tilth. Unfortunately, these improvements make soil more susceptible to surface sealing and erosion by wind and water. Good management is required to take advantage of the tilth enhancement made by growing soybean while controlling the hazards of surface sealing and accelerated erosion. Management alternatives include, but are not limited to, no tillage, fall-seeded cover crops, double-cropping with a winter grain, and planting on the contour. The most effective way to reduce erosion and surface sealing is to grow crops with a good canopy and leave the soil covered with residue from the previous crop.

NOTES

D. V. McCracken, Research Agronomist, Purdue University; W. C. Moldenhauer, Soil Scientist, USDA, ARS, National Soil Erosion Laboratory, West Lafayette, IN; and J. M. Laflen, Agricultural Engineer, USDA, ARS, Iowa State University, Ames, IA.

Cooperative project of the USDA Agricultural Research Service at Ames, Iowa and West Lafayette, Indiana in cooperation with Iowa State and Purdue Universities. Purdue Journal No. 10,089.

REFERENCES

Allmaras, R. R., W. W. Nelson and W. B. Voorhees. 1975a. Soybean and corn rooting in southwestern Minnesota: I. Water-uptake sink. Soil Sci. Soc. Am. Proc. 39:764-771.

Allmaras, R. R., W. W. Nelson and W. B. Voorhees. 1975b. Soybean and corn rooting in southwestern Minnesota: II. Root distributions and related water inflow. Soil Sci. Soc. Am. Proc. 39:771-777.

Armbrust, D. V., J. D. Dickerson, E. L. Skidmore and O. G. Russ. 1982. Dry soil aggregation as influenced by crop and tillage. Soil Sci. Soc. Am. J. 46:390-393.

Barley, K. P. and E. L. Greacen. 1967. Mechanical resistance as a soil factor influencing the growth of roots and underground shoots. Adv. Agron. 19:1-43.

Baver, L. D., W. H. Gardner and W. R. Gardner. 1972. Soil Physics. John Wiley and Sons, New York.

Brossman, G. D., J. J. Vorst and G. C. Steinhardt. 1982. A technique for measuring crust strengths. J. Soil Water Conserv. 37:225-226.

Browning, G. M. 1947. Erosion. Soybean Dig. 7(8):10-12.

Browning, G. M. 1949. Soybeans and the fertility level. Soybean Dig. 9(11):58, 61.

Browning, G. M. and R. A. Norton. 1948. Tillage, structure and irrigation: tillage practices with corn and soybeans in Iowa. Soil Sci. Soc. Am. Proc. 12:491-496.

Browning, G. M., M. B. Russell and J. R. Johnston. 1943. The relation of cultural treatment of corn and soybeans to moisture condition and soil structure. Soil Sci. Soc. Am. Proc. 7:108-113.

Bryan, R. B. 1968. The development, use and efficiency of indices of soil erodibility. Geoderma 2: 5-26.

Burch, G. J., R. C. G. Smith and W. K. Mason. 1978. Agronomic and physiological responses of soybean and sorghum crops to water deficits. II. Crop evaporation, soil water depletion and root distribution. Aust. J. Plant Physiol. 5:169-177.

Calland, J. W. 1949. What soybeans do to your land. Soybean Dig. 9(7):15-18.

Chepil, W. S. 1958. Soil conditions that influence wind erosion. U.S. Dept. Agr. Tech. Bul. 1185.

Colvin, T. S. and J. M. Laflen. 1981. Effect of corn and soybean row spacing on plant canopy, erosion, and runoff. Trans. Am. Soc. Agr. Eng. 24:1227-1229.

Conaway, A. W. and E. Strickling. 1962. A comparison of selected methods for expressing soil aggregate stability. Soil Sci. Soc. Am. Proc. 26:426-430.

Duley, F. L. 1940. Surface factors affecting the rate of intake of water by soils. Soil Sci. Soc. Am. Proc. 4:60-64.

Fahad, A. A., L. N. Mielke, A. D. Flowerday and D. Swartzendruber. 1982. Soil physical properties as affected by soybean and other cropping sequences. Soil Sci. Soc. Am. J. 46:377-381.

Griffith, D. R., J. V. Mannering and W. C. Moldenhauer. 1977. Conservation tillage in the eastern Corn Belt. J. Soil Water Conserv. 32:20-28.

Hanks, R. J. and F. C. Thorp. 1957. Seedling emergence of wheat, grain, sorghum, and soybeans as influenced by soil crust strength and moisture content. Soil Sci. Soc. Am. Proc. 21:357-359.

Harris, R. F., G. Chesters and O. N. Allen 1966. Dynamics of soil aggregation. Adv. Agron. 18:107-169.

Kidder, E. H., R. S. Stauffer and C. A. Van Doren. 1943. Effect on infiltration of surface mulches of soybean residues, corn stover, and wheat straw. Agr. Eng. 24:155-159.

Laflen, J. M. and T. S. Colvin. 1981. Effect of crop residue on soil loss from continuous row cropping. Trans. Am. Soc. Agr. Eng. 24:605-609.

Laflen, J. M. and W. C. Moldenhauer. 1979. Soil and water losses from corn-soybean rotations. Soil Sci. Soc. Am. J. 43:1213-1215.

Mannering, J. V. and C. B. Johnson. 1969. Effect of crop row spacing on erosion and infiltration. Agron. J. 61:902-905.

McCracken, D. V. 1984. Influence of corn and soybean residue decomposition on soil aggregate wet stability. M.S. Thesis. Purdue Unviersity, West Lafayette, IN.

McGregor, K. C., J. D. Greer and L. E. Gurley. 1975. Erosion control with no-till cropping practices. Trans. Am. Soc. Agr. Eng. 18:918-920.

McIntyre, D. S. 1955. Effect of soil structure on wheat germination in a red-brown earth. Aust. J. Agr. Res. 6:797-803.

Moldenhauer, W. C. and E. R. Duncan. 1968. Principles and methods of wind-erosion control in Iowa. Iowa State Agr. Exp. Stn. Spec. Rep. No. 62, Ames, IA.

Moldenhauer, W. C., G. W. Langdale, Wilbur Frye, D. K. McCool, R. I. Papendick, D. E. Smika and D. William Fryrear. 1983. Conservation tillage for erosion control. J. Soil Water Conserv. 38:144-151.

Panabokke, C. R. and J. P. Quirk. 1957. Effect of initial water content on stability of soil aggregates in water. Soil Sci. 83:185-195.

Ram, D. N. and P. J. Zwerman. 1960. Influence of management systems and cover crops on soil physical conditions. Agron. J. 52:473-476.

Rawson, H. M., J. E. Begg and R. G. Woodward. 1977. The effect of atmospheric humidity on photosynthesis, transpiration and water use efficiency of leaves of several plant species. Planta 134:5-10.

Siemens, J. C. and W. R. Oschwald. 1978. Corn-soybean tillage systems: erosion control, effects on crop production, costs. Trans. Am. Soc. Agr. Eng. 21:293-302.

Smith, D. 1980. Grow soybeans without soil loss? Farm J. 108(8):22-23.

Stauffer, E. 1946. Effect of corn, soybeans, their residues and a straw mulch on soil aggregation. Agron. J. 38:1010-1017.

Teare, I. D., E. T. Kanemasu, W. L. Powers and H. S. Jacobs. 1973. Water-use efficiency and its relation to crop canopy area, stomatal regulation, and root distribution. Agron. J. 65:207-211.

Van Doren, C. A., R. S. Stauffer and E. H. Kidder. 1951. Effect of contour farming on soil loss and runoff. Soil Sci. Soc. Am. Proc. 15:413-417.

Van Doren, D. M., W. C. Moldenhauer and G. B. Triplett. 1984. Influence of long-term tillage and crop rotation on water erosion. Soil Sci. Soc. Am. J. 48:636-640.

Verma, S. M. and N. T. Singh. 1974. Effect of some indigenous organic materials on soil aggregation. J. Indian Soc. Soil Sci. 22:220-225.

Walker, E. D. 1946. Soybeans and soil erosion. Soybean Dig. 6(5):16-17.

Wischmeier, W. H. and D. D. Smith. 1978. Predicting rainfall erosion losses: A guide to conservation planning. Agr. Handbk. 537. U.S. Dept. Agr., Washington, D.C.

SOYBEAN ROOT GROWTH IN RESPONSE TO SOIL ENVIRONMENTAL CONDITIONS

H. M. Taylor and T. C. Kaspar

Soybean roots anchor the plant, provide sites for nodules containing nitrogen-fixing bacteria, synthesize hormones that regulate growth and development, and absorb water and minerals required by the plant for growth. Growth and morphology of root systems affect each of these functions. Morphology of a root system is genetically controlled, but often is greatly modified by soil environment. Growth occurs because of cell division and expansion. Root cells expand if the hydrostatic pressure within them is sufficient to overcome the constraining influence of cell walls and any external media. Thus, hydrostatic-pressure-driven growth is limited by (a) permeability of cell walls and membranes to solutes and water, (b) quantity of solutes, which provide the osmotic component of the hydrostatic pressure, and (c) extensibility of cell walls and the surrounding soil matrix. Furthermore, for growth to continue, supplies of carbohydrates, of cell wall and membrane components, of growth-regulatory substances, of water, and of certain mineral elements must be available within the elongating region.

Many gaps exist in knowledge of the internal mechanisms that control root growth and morphology. This lack of knowledge is unfortunate because soil environmental effects are superimposed on these endogenous controls. Thus, interpretation of the observed form and growth of root systems in the field is difficult. Nonetheless, many reports are available in the literature that examine the effects of soil environment on root-system morphology and growth.

SOIL TEMPERATURE

Earley and Carter (1945) grew soybeans under uniform conditions for various lengths of time and then exposed the roots to different soil temperatures while maintaining the shoots at ambient greenhouse temperatures. In general, root mass increased with increasing soil temperature to 22 C and decreased above 27 C. Shoot dry weight followed the same trend. Response to soil temperature, however, varied depending on when the soil-temperature treatments were imposed.

Stone and Taylor (1983) also observed an interaction between soil temperature and time after emergence on root growth. They found that elongation of soybean taproots maintained at 29 C was about 60 mm/day during the first day after emergence, but decreased to about 40 mm/day by the 18th day. At 21 C, however, elongation rate of taproots decreased from about 40 mm/day soon after emergence to 30 mm/day on day 18. Additionally, during the period 8 to 19 days after emergence, lateral roots grown at 29 C maintained an elongation rate of about 50 mm/day, whereas lateral roots grown at 21 C increased their rate of elongation from 3 to 3.50 mm/day.

Soil temperature, not only affects root extension rates, but also influences the angle at which lateral roots begin to grow away from the taproot. Kaspar et al. (1981) found that this angle, averaged over 10 cultivars, was 15.5, 18.6, 20.6, and 19.7 degrees from the horizontal at soil temperatures of 10.0, 15.6, 21.1, and 26.7 C, respectively. Because lateral roots sometimes extend 0.30 to 0.50 m horizontally before turning downward, their growth angle determines how close to the soil surface lateral roots will be.

Stone et al. (1983) modelled rooting depth for four soybean cultivars as functions of cultivar, time, and soil temperature. The model was developed from the extension data of Stone and Taylor (1983) and the lateral inclination data of Kaspar et al. (1981), and required inputs of soil temperatures as functions of soil depth and time. Output from the model compared favorably with actual rooting depths of eight cultivars growing in a field of loess soil (Ida silt loam soil near Castana, Iowa, Table 1). Because this model accurately predicted rooting depth, Stone et al. (1983) concluded that soil temperature was a major factor controlling rooting depth at the site that year.

Nitrogen fixation by the soybean-rhizobia symbiosis is also influenced by soil temperature. Soil temperature increase from 28 to 40 C caused decreased number of nodules per plant, nodule fresh weight, and specific nitrogenase activity of nodules, and therefore, a decrease in the amount of nitrogen fixed. Furthermore, reductions in the level of nitrogenase activity can occur at temperatures even lower than 28 C, if the plant is water stressed. Pankhurst and Sprent (1976) found that nitrogenase activity decreased at soil temperatures above 30 C in unstressed plants, but in water-stressed soybean, nitrogenase activity decreased whenever soil temperatures increased above 15 C.

SOIL AERATION

Soybean roots require oxygen for growth and maintenance respiration. Additionally, unless ethylene, carbon dioxide, and other gases are transported from the rhizosphere, their concentrations will increase to inhibitory levels. Therefore, because diffusion of gases is restricted in waterlogged soils, soybean and most other agricultural crops usually grow better in a free-draining soil than in a completely waterlogged soil environment.

Table 1. Depth of soybean root penetration at 56 and 77 days after planting for predicted values of taproots and lateral roots from model of Stone et al. (1983) and for the observed average depths of eight cultivars in the field (Kaspar et al., 1984).

	Rooting depth				
	Predicted				
Days after planting	Beeson and Hawkeye		Wayne and Harosoy 63		Observed
	Taproots	Lateral	Taproots	Lateral	
	cm	cm	cm	cm	cm
56	162	131	152	134	131
77	205	173	198	179	172

Huck (1971) induced soil oxygen stress by maintaining a flowing stream of nitrogen gas around soybean roots. He found that a 4- to 5-h period without oxygen stopped root elongation and killed most of cells in the root tip and elongation zone. If a taproot was killed in this way, lateral root formation was enhanced after normal aeration was resumed, possibly because the taproot apex was no longer exerting apical dominance over the laterals.

Roots of soybean are able to penetrate into water tables only a short distance before a lack of oxygen, or an accumulation of ethylene, stops growth. In most instances, roots concentrate in the moist, aerated zones directly above the water table. Water uptake by the roots concentrated near a water table may account for a large percentage of a soybean plant's total water uptake (Reicosky et al., 1972). For example, Ogunremi et al. (1981) reported that water tables 0.15 m below the soil surface were optimum for vegetative growth and that seed yield was greatest with a water-table depth of 0.40 m. Similarly, researchers in Queensland Australia observed that static water tables near the surface increased shoot dry-matter accumulation and yield when compared with irrigated controls (Nathanson et al., 1984). Additionally, after an initial acclimation period of 2 to 3 weeks, nitrogen accumulation rates and nodule mass of plants grown in the presence of water tables were greater than those of the control (Troedson et al., 1982; Table 2). These researchers speculated that yield and nitrogen fixation increased because stomata of plants grown in irrigated plots partly closed for several hours each midday, whereas stomata of plants growing in plots with water tables remained open. High photosynthesis rates, thus, occurred for more hours per day in water-table plots.

Water tables sometimes rise for short periods in otherwise drained profiles. Stanley et al. (1980) investigated the effects of temporary water tables imposed for 7 days at three stages of growth on soybean root and shoot response. When water tables were imposed before flowering, roots below the water-table level stopped growing downward, but remained alive and resumed growth after the water table was lowered. When water tables were imposed either during flowering or rapid pod-filling, however, roots below the water table became discolored and began to decompose after the water was lowered, and no new roots grew into the soil volumes that had been below water, even though new root growth occurred above this depth. Obviously, lack of root growth below the previous water-table level implies that soil volumes below this level were inhibitory to root growth. Plant height was reduced compared with irrigated controls for temporary water tables imposed before or during flowering. Stem elongation had almost stopped by the time the pod-filling, water-table treatment was imposed; therefore, this treatment did not reduce plant height.

Table 2. Dry-matter and nitrogen accumulation rates before and during seed development for conventionally irrigated and for water-table culture treatments (Troedson et al., 1982).

	Preceding seed development		During seed development	
	Conventional	Water-table	Conventional	Water-table
Total DM, g	14.28	11.44	6.71	9.88
Grain DM, g	–	–	6.29	7.78
Total N, mg	400	315	125	312
Grain N, mg	–	–	406	514

SOIL COMPACTION

When soil is compacted, air-filled porosity is decreased, and the relative fractions of solid particles and water are increased. These changes, not only reduce oxygen diffusion to and ethylene diffusion from roots, but also alter transmission of nutrients and water to roots and increase mechanical resistance (soil strength) to penetration by roots.

Several workers (Rogers and Thurlow, 1973; Nash and Baligar, 1974; Baligar et al., 1980, 1981; Hallmark and Barber, 1981; Giles, 1983) have investigated the effects of soil compaction on root growth of soybean. In general, their results indicated (a) that increased soil strength reduced root elongation but increased root diameter, (b) that decreased water content at a specified bulk density increased root growth when the soil was saturated with water, (c) that decreased water content at or below field capacity usually increases soil strength, especially for sandy loam or coarser soils, and (d) that soil compaction decreased or increased mineral nutrient uptake, depending on whether root interception, root diameter, or the diffusion coefficient for the nutrient was the most important parameter controlling uptake.

Elongation of corn (*Zea mays* L.) roots into high-strength soil volumes was studied by Lachno et al. (1982). They found that the concentration of abscissic acid (ABA) in root tips was not increased when roots penetrated high-strength soil volumes, but that indole acetic acid (IAA) concentration increased 3.5 times that of unimpeded root tips. Whether the increased IAA concentration was the cause of the reduced root elongation or the result of it is not known. Furthermore, impeded soybean roots may not react the same as corn roots with respect to ABA and IAA concentrations. There is a need for research to determine the effect of exogenously applied plant-growth regulators on growth of impeded soybean root systems. It seems possible that certain plant-growth regulating substances may alter the ability of roots to grow into and through compact soil volumes.

CALCIUM DEFICIENCY AND ALUMINUM TOXICITY

In many areas of the world, soybean yields are reduced by calcium deficiency or aluminum toxicity, especially during drought. In these soils, the plow layer often is limed sufficiently to allow adequate root growth, but the subsoil may have too little calcium or too much aluminum in the soil solution to allow extensive root growth. This situation occurs most often in subhumid to humid areas on soils low in cation exchange capacity.

The calcium concentration required for soybean root elongation is quite low if other cation concentrations also are low and if no toxic ions are present. Lund (1970) observed that a 0.25 μg/L concentration of calcium ion was sufficient for rapid elongation at pH 5.6, but 2.50 μg/L of calcium was required at pH 4.5. Aluminum in solution, however, reduced root elongation of soybean whenever the ratio of activities of Al to Ca was greater than 0.02. In the Brazilian Cerrado, soybean yield was reduced when aluminum saturation on the cation exchange complex exceeded 40% (Goedert, 1983). Substantial progress has been made, both in developing practices that increase downward movement of calcium ions and in selecting cultivars tolerant to high aluminum concentrations.

NOTES

H. M. Taylor, Rockwell Professor of Soil Science, Department of Plant and Soil Science, Texas Tech University, Lubbock, TX 79409.

T. C. Kaspar, Agricultural Research Service, USDA, and Department of Agronomy, Iowa State University, Ames, IA 50011.

Joint contribution from School of Agriculture, Texas Tech University, Agriculture Research Service, USDA, and Journal Paper J-11468 of the Iowa Agriculture and Home Economics Experiment Station, Ames; Project 2659.

REFERENCES

Baligar, V. C., V. E. Nash, F. D. Whisler and D. L. Myhre. 1981. Sorghum and soybean growth as influenced by synthetic pans. Commun. Soil Sci. Plant Anal. 12:97-107.

Baligar, V. C., F. D. Whisler and V. E. Nash. 1980. Soybean seedling root growth as influenced by soil texture, matric suction and bulk density. Commun. Soil Sci. Plant Anal. 11:903-915.

Earley, E. B. and J. L. Cartter. 1945. Effect of the temperature of the root environment on growth of soybean plants. J. Am. Soc. Agron. 37:727-735.

Giles, J. F. 1983. Soil compaction and crop growth. N. D. Farm Res. 41:34-35.

Goedert, W. J. 1983. Management of the Cerrado soils of Brazil: A review. J. Soil Sci. 34:405-428.

Hallmark, W. B. and S. A. Barber. 1981. Root growth and morphology, nutrient uptake, and nutrient status of soybeans as affected by soil K and bulk density. Agron. J. 73:779-782.

Huck, M. G. 1971. Variation in taproot elongation rate as influenced by composition of the soil air. Agron. J. 62:815-818.

Kaspar, T. C., D. G. Woolley and H. M. Taylor. 1981. Temperature effect on the inclination of lateral roots of soybeans. Agron. J. 73:383-385.

Kaspar, T. C., H. M. Taylor and R. M. Shibles. 1984. Taproot-elongation rates of soybean cultivars in the glasshouse and their relation to field rooting depth. Crop Sci. 24:916-920.

Lachno, D. R., R. S. Harrison-Murray and L. J. Audus. 1982. The effects of mechanical impedance to growth on the levels of ABA and IAA in root tips of *Zea mays* L. J. Exp. Bot. 33:943-951.

Lund, Z. F. 1970. The effect of calcium and its relation to several cations in soybean root growth. Soil Sci. Soc. Am. Proc. 34:456-459.

Nash, V. E. and V. C. Baligar. 1974. The growth of soybean (*Glycine max*) roots in relation to soil micromorphology. Plant Soil 41:81-89.

Nathanson, K., R. J. Lawn, P. L. M. DeJabrun and D. E. Byth. 1984. Growth, nodulation, and nitrogen accumulation by soybean in saturated soil culture. Field Crops Res. 8:73-92.

Ogunremi, L. T., R. Lal and O. Babalola. 1981. Effects of water table depth and calcium peroxide application on cowpea (*Vigna unguiculata*) and soybean (*Glycine max*). Plant Soil 63:275-281.

Pankhurst, C. E. and Janet I. Sprent. 1976. Effects of temperature and oxygen tension on the nitrogenase and respiratory activities of turgid and water-stressed soybean and French bean root nodules. J. Exp. Bot. 27:1-9.

Reicosky, D. C., R. J. Millington, A. Klute and D. B. Peters. 1972. Patterns of water uptake and root distribution of soybeans in the presence of a water table. Agron. J. 64:292-297.

Rogers, H. T. and D. L. Thurlow. 1973. Soybeans restricted by soil compaction. Highlights Agric. Res. 20:10.

Stanley, C. D., T. C. Kaspar and H. M. Taylor. 1980. Soybean top and root response to temporary water tables imposed at three different stages of growth. Agron. J. 72:341-346.

Stone, J. A., T. C. Kaspar and H. M. Taylor. 1983. Predicting soybean rooting depth as a function of soil temperature. Agron. J. 75:1050-1054.

Stone, J. A. and H. M. Taylor. 1983. Temperature and development of the taproot and lateral roots of four indeterminate soybean cultivars. Agron. J. 75:613-618.

Troedson, R. J., D. E. Byth and R. J. Lawn. 1982. Wet soil culture of soybeans. Papers presented to Australian Soybean Research Workshop, Narrabari, N.S.W., March 1982.

SOYBEAN CROP RESPONSES TO SOIL ENVIRONMENTAL STRESSES

Alvin J. M. Smucker

The existence of a yield plateau of soybeans in the United States suggests that there is a large unrealized genetic potential for increased yields. Much of this yield potential could be realized if soybean plant types were better adapted to the physico-chemical environments in which they are grown (Boyer, 1982). Recent results from New Jersey indicate that nearly 7000 kg/h of soybeans can be produced when appropriate management systems are utilized (Flannery, 1984). Although genetic selection for adaptation to adverse environments may have contributed to greater crop production, the fundamental mechanisms of root tolerance to soil stresses have been rarely understood.

Compaction and the incipient air and water, mechanical impedance and nutrient stresses of most intensely tilled agricultural soils restrict the growth of soybean roots and have been reported to reduce the nutrient uptake and status of soybean plants (Hallmark and Barber, 1981; Silberbush et al., 1983). Previous research has resolved many of these soil stresses by manipulating the mineral substrate (e.g., greater tillage and fertilization rates, etc.). However, as our concerns for the environment, our natural resources, and the cost of energy increase, more consideration will be given to the efficiency of crop production. One of the least understood and perhaps most important components that must be investigated before an efficient crop production system can be developed is the root-soil interface.

SOYBEAN ROOT RESPONSES TO SOIL COMPACTION

Excessive secondary tillage and traffic significantly increases the bulk density of most soils. Figure 1 shows that this compaction process may continue to depths of at least 45 cm, and it may influence deeper horizons as axle loads increase (Voorheese, 1985). Therefore, it seems that most of the roots of field grown crops encounter the adverse effects of soil compaction for the duration of the growing season.

Root penetration of a soil is inversely related to the bulk density of that soil. The penetration capacity of roots to specific soil densities has been tested by planting seedlings into 7.6 by 17.8 cm cores and measuring, after 14 days, the root penetration ratios (RPR) of the middle 2.5 cm core. These cores may contain any soil type in which the middle core is compressed to a given bulk density and placed between the upper and lower cores having lower bulk densities. Figure 2 indicates an inverse relationship between the root penetration ratios (i.e., the number of roots that exit the lower portion of the central core divided by the number of roots that enter the upper portion of the central core) of soybean and drybean (*Phaseolus vulgaris* L.), and the

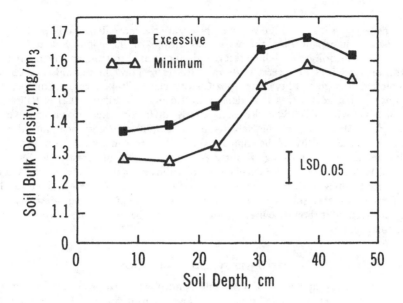

Figure 1. Bulk density profile responses to secondary tillage and traffic of a Conover loam soil. Samples were taken by the method of Srivastava et al. (1982).

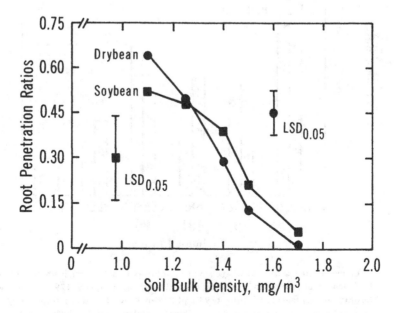

Figure 2. Root penetration responses of 14-day-old seedlings of soybean and drybean to increasing soil bulk densities of a Conover loam using the soil core seedling method.

bulk densities of a loam soil. Fewer roots penetrate the more dense soils, probably due to mechanical impedance and the anoxic conditions of the compacted soils (Asady et al., 1985). This soil core seedling test has also been used to compare the relative root penetration capacities of several soybean cultivars to both the natural (low fertility) and augmented (high fertility) rates of a Charity clay (illitic, calcareous, mesic, Aeric Haplaquept) compressed to a bulk density of 1.6 mg/m^3 (Figure 3). The RPR values of these six soybean cultivars are significantly different when grown at lower soil fertility rates. More roots of the Nebsoy and Beason-80 appear to penetrate the compacted soil as the fertility of the compacted middle core is increased. However, greater fertility levels appeared to significantly reduce the RPR of Corsoy-79, SRF 101, Wells and Weber. There are also excellent correlations (r^2 = 0.96) between the soil core seedling RPR values of Corsoy and Nebsoy and crop yield (Figure 4). Since the RPR measurements of this soil core seedling test can be made without destroying the entire root system, it should be useful for determining the inheritance of root tolerance to soil compaction.

RESPONSES TO SOIL COMPACTION

Soil compaction and many unfavorable conditions of the root and soil environments limit the development and function of soybean root systems. The limited supply of photoassimilates allocated to stressed roots are often wasted by higher respiration rates, increased exudation of soluble nitrogen and carbon compounds, greater root decomposition rates, etc. Compensatory root growth, one of many avoidance mechanisms, has been reported to consume more carbon per gram of root than nonstressed

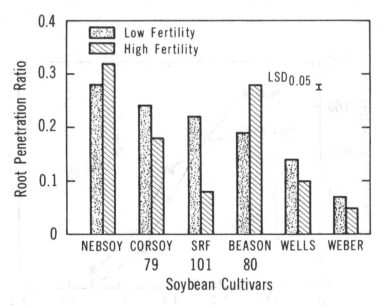

Figure 3. Root penetration capacities of six soybean cultivars to a Conover loam compressed to a bulk density of 1.80 mg/m^3. High fertility soil contained 623, 663, 1295 and 242, and low fertility soil contained 193, 206, 1393 and 242 of P, K, Ca and Mg, respectively. Manganese and zinc concentrations were 14.6 (high), 9.5 (low) and 4.5 (high), 3.4 (low) μg/g of soil, respectively.

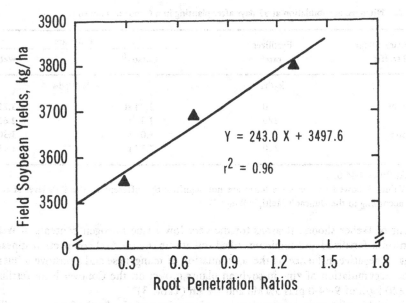

Figure 4. Correlation between the root penetration ratios of two soybean cultivars grown at three levels of compaction on a Charity clay in the soil core seedling test, and yield of soybeans grown in the field on the same soil having similar levels of compaction. Each value is the mean of eight replications.

controls (Schumacher and Smucker, 1984). This inefficient utilization of carbon by stressed root systems reduces the retention of carbon by the shoot portion of the soybean, resulting in a net loss of photosynthetically active leaf area (Table 1). Although more carbon appears to be allocated to the root systems of soybean plants growing in compacted soils, less was utilized for the biological fixation of nitrogen.

It has been reported that soil compaction adversely influences soybean growth and production by having a greater influence on root growth than on potassium movement in the soil (Silberbush et al., 1983). Therefore, it seems that the growth of soybean plants on compacted soils could be enhanced by the addition of greater quantities of fertilizer. The addition of approximately 20 kg/ha of ammonium nitrate fertilizer appeared to increase the tissue nitrogen content of Corsoy (Table 2), with a greater effect occurring where the soil was compacted by excessive secondary tillage and traffic (Figure 1). Neither tillage nor fertilization appeared to influence the nitrogen

Table 1. Influence of secondary tillage and traffic compaction on the bulk density of a clay soil and leaf area at R1.

Secondary tillage and traffic	Soil bulk density[a]		Total leaf area[a]	
	0-7.5 cm	7.5-15 cm	Nebsoy	Wells II
	— mg/m^3 —		— cm^2 —	
Minimum	1.09 r	1.16 rs	467 t	327 s
Conventional	1.23 rst	1.33 tuvw	446 t	326 s
Excessive	1.29 stuv	1.36 uvwx	258 s	169 r

[a]Values for each parameter followed by the same letter are not significantly different at the 95% level of confidence according to the Duncan's Multiple Range Test.

Table 2.   Nitrogen accumulation at 93 days after planting in a Conover loam soil.

Secondary tillage and traffic	Fertilizer rate[a]	Cultivars	
		Corsoy[b]	Weber[b]
	kg/ha	— % N/gdw —	
Minimum	0	3.24 st	3.22 st
	220	3.36 t	2.62 r
Excessive	0	3.01 s	3.30 t
	220	3.33 t	3.43 t

[a]Analyses 9-44-0.

[b]Values followed by the same letter are not significantly different at the 95% level of confidence according to the Duncan's Multiple Range Test.

content of Weber shoots. Reasons for the very low tissue nitrogen contents of Weber grown on a fertilized and minimum tilled soil are unknown. Soil compaction appeared to have a negative influence on the accumulation of manganese and a positive influence on the accumulation of zinc in soybean plants grown on the Conover loam fertilized with 220 kg/h of 9-44-0 plus 5% Mn and 2% Zn (Table 3).

The yields of Corsoy appeared to be independent of the level of secondary tillage and traffic when grown on fine and medium textured soils (Table 4). However, yields of irrigated SRF-101 grown in coarse textured sand were 21% lower (i.e., 660 kg/ha) in the absence of fertilizer. This yield deficit was reduced to 288 kg/ha by addition of

Table 3.   Influence of secondary tillage and traffic on soybean tissue concentrations of five elements 62 days after planting.

	Secondary tillage and traffic		Percentage change
	Minimum	Excessive	
Nitrogen, %	2.50	2.63	+5
Phosphorus, %	0.37	0.38	+3
Potassium, %	2.66	2.61	+2
Manganese, μg/g	95.2	79.4	-20
Zinc, μg/g	45.7	51.5	+13

Table 4.   Yield responses of Corsoy to secondary tillage, traffic, fertilization and type of soil.

Secondary tillage and traffic	Fertility level	Conover loam[a]	Charity clay[a]
		— kg/ha —	
Minimum	Low	3272 rst	3163 rs
	High[b]	2954 rst	3234 rs
Excessive	Low	2958 rst	3414 s
	High[b]	3230 rst	3292 rs

[a]Values of each column followed by the same letter are not significantly different at the 95% level of confidence according to the Duncan's Multiple Range Test.

[b]220 kg/ha of 9-44-0.

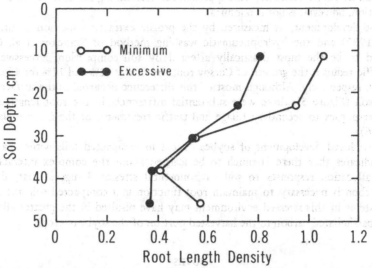

Figure 5. Root length density responses of Corsoy to secondary tillage and traffic with respect to soil depth. Each value is the mean of 12 replications.

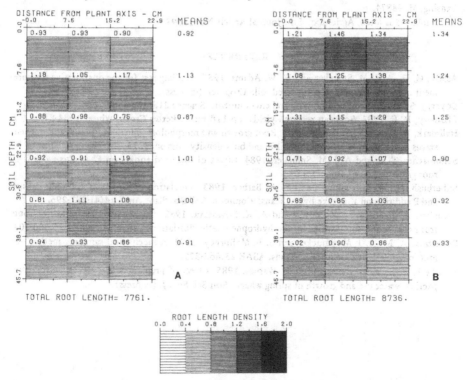

Figure 6. Root zone profiles of soybean root length density responses to secondary tillage and traffic of a Charity clay soil. Root length density profiles A and B represent minimum and excessive secondary tillage and traffic soil treatments, respectively. Each of the 18 subunits of the profile represent a mean of four replications.

220 kg/ha of 9-44-0 fertilizer. This approach for overcoming the adverse effects of soil compaction, however, is not economical.

Root development, as measured by the profile extraction method of Srivastava et al. (1982) and the hydropneumatic washing method of Smucker et al. (1982), appeared to be the most drastically affected by soil compaction. Excessive tillage and traffic reduced the growth of Corsoy roots by nearly 42% and 11% for a loam and clay soil, respectively. Although most of this difference occurred in the surface 30 cm of the soil (Figure 5), there were substantial differences in the root length density profile responses to secondary tillage and traffic treatments of the Charity clay soils (Figure 6).

The reduced development of soybean roots in compacted soils without a loss in yield indicates that there is much to be learned about the complex interaction of carbon allocation responses to soil environmental stresses. I suggest that, although more carbon is necessary to maintain root function in a compacted soil, the smaller root systems in this stressed environment may have resulted in the greater allocation of photoassimilated carbon to the harvested portion of the soybean crop.

NOTES

Alvin J. M. Smucker, Department of Crop and Soil Sciences, Michigan State University, East Lansing, MI 48824.
Published as Mich. Agric. Exp. Stn. Journal Article No. 11393.

REFERENCES

Asady, G. H., A. J. M. Smucker and M. W. Adams. 1985. Seedling test for the quantitative measurement of root tolerances to compacted soil. Crop Sci. (in press).

Boyer, J. S. 1982. Plant productivity and environment. Science 218:443-448.

Flannery, R. F. 1983. Soybean research yields top 118 bu/A. Better Crops. Winter 1983-84. p. 6-7.

Hallmark, W. B. and S. A. Barber. 1981. Root growth and morphology, nutrient uptake, and nutrient status of soybeans as affected by soil K and bulk density. Agron. J. 73:779-782.

Schumacher, T. E. and A. J. M. Smucker. 1984. Effect of localized anoxia on *Phaseolus vulgaris* L. root growth. J. Exp. Bot. 35:156.

Silberbush, M., W. B. Hallmark and S. A. Barber. 1983. Simulation of effects of soil bulk density and P addition on K uptake by soybeans. Commun. Soil Sci. Plant Anal. 14(4):287-296.

Smucker, A. J. M., S. L. McBurney and A. K. Srivastava. 1982. Quantitative separation of roots from compacted soil profiles by the hydropneumatic elutriation system. Agron. J. 74:500-503.

Srivastava, A. K., A. J. M. Smucker and S. L. McBurney. 1982. A mechanical sampling method for multiple soil-plant root studies. Trans. ASAE 25:868-871.

Voorhees, W. B., S. D. Evans and D. D. Warnes. 1985. Effect of preplant wheel traffic on soil compaction, water use and growth of spring wheat. Soil Sci. Soc. J. (in press).

FERTILIZER RATE AND PLACEMENT EFFECTS ON NUTRIENT UPTAKE BY SOYBEANS

Stanley A. Barber

Fertilizer is applied to soybean crops to compensate for a deficiency in the soil nutrient supply. The amount of fertilizer needed depends on both the nutrient supply level of the unfertilized soil and the efficiency of the use of applied fertilizer. Placement can affect fertilizer effectiveness. The method of fertilizer application has usually been either as a band placed near the row at planting or as a broadcast application, applied and mixed with the soil by tillage. Application as a band by the row at planting has not been as common for soybean as for wheat (*Triticum vulgare* L.) or corn (*Zea mays* L.), because early growth response from the row fertilizer usually is not observed. Because the seed is large, it supplies nutrients for early seedling growth, so that less response occurs from having a high soil nutrient level near the seed. In addition, the seed is sensitive to salt injury. Since the soybean does not show early growth responses from row fertilizer, it has sometimes been assumed that it does not give as large yield increases due to fertilization as are obtained with corn. However, in long term experiments at the Purdue Agronomy farm, I have found that soybeans respond to phosphate and potassium similarly to corn. Data in Table 1 give the relative yields of soybeans and corn as affected by the average annual application of P and K. Fertilizer applications were made in the row for corn and wheat in a corn-soybean-wheat rotation, or were broadcast and plowed under ahead of the corn crop. Values are means of yields for a 21-year period. Relative soybean yield response to applied phosphate and potassium was greater for P and slightly less for K than that obtained for corn.

Table 1. Mean relative yields of corn and soybean as affected by P and K applications to a corn, corn, soybean, wheat rotation; 1963-1983.

Average annual application		Relative yield	
P	K	Corn	Soybeans
−kg/ha −			
0	46	0.84	0.64
11	140	0.93	0.87
22	140	1.00	1.00
44	0	0.71	0.79

MECHANISMS OF NUTRIENT SUPPLY TO SOYBEANS

Fertilizer uptake depends both on the rate of nutrient supply by the soil and the uptake characteristics of the plant root system. A mechanistic mathematical model that describes the nutrient uptake process has been developed (Claassen and Barber, 1976). The model assumes roots absorb nutrients from the soil solution; the rate being determined by concentration in solution at the root surface. Uptake rate increases asymptotically to a maximum as nutrient level of the soil solution increases. The level in solution at the root surface is determined by the relation between the rate of flow of nutrients to the root by mass flow and the diffusion and uptake rate of the root. This gives a value for uptake rate by a unit of root as related to time. Total plant uptake is obtained by summing the uptake vs. time of each additional unit of root as root growth occurs. This model has been verified for predicting P and K uptake by soybean plants, using plants grown in pots in a controlled climate chamber (Silberbush and Barber, 1983) and plants grown in the field (Silberbush and Barber, 1984). The relation between observed P uptake by two soybean cultivars growing on three soils vs. predicted uptake is shown in Figure 1. The regression coefficient was not significantly different from a value of 1.0, indicating agreement between uptake predicted by the model and amount of P found in the plants. The relation between observed K uptake by five cultivars growing in the field and predicted K uptake is shown in Figure 2. The

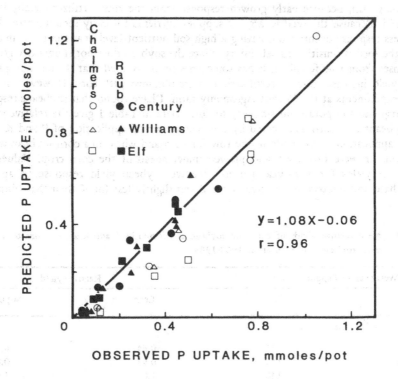

Figure 1. The relation between predicted P uptake and observed uptake of P by three soybean cultivars grown on two soils in a growth chamber (from Silberbush and Barber, 1983).

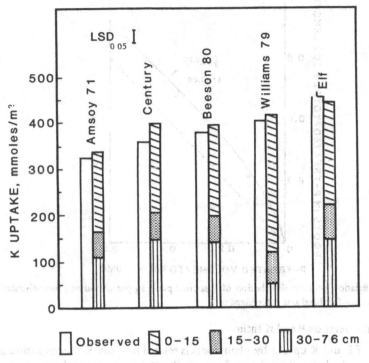

Figure 2. The relation between predicted K uptake and observed uptake of K by five soybean cultivars grown on Raub silt loam in the field (from Silberbush and Barber, 1984).

uptake from soil in the field was calculated separately for root growth in each soil depth, since K level varies with soil depth. The mechanistic model gave predicted uptake values that agreed closely with P and K uptake measured from plant analysis. The close agreement from both pot and field studies indicates that the model can be used to predict useful information on the effects of fertilizer placement on P and K uptake.

Placement Effects on Root Growth

When P fertilizer is mixed with only a portion of the soil, there is greater root growth in the fertilized portion of the soil than in the unfertilized portion, since the fertilized portion stimulates root growth. Borkert and Barber (unpublished) investigated this effect in a pot experiment for P application and obtained the relation shown in Figure 3. In this research the same amount of P per pot was mixed with portions of soil varying from 2 to 75% of the total soil volume. After 3 weeks, root lengths in the fertilized and unfertilized soil were measured separately and the data plotted in Figure 3. Application of P stimulated root growth in the fertilized portion. This relation is similar to that obtained for corn. Hence, it can be used to predict the effect of fertilizer placement on root distribution. Similar studies with K show that K distribution does not influence root distribution. Where P and K are added together in a fertilizer, the stimulation of root growth would be expected to be the same as where P was added alone, and hence K placement would benefit from combined application with P.

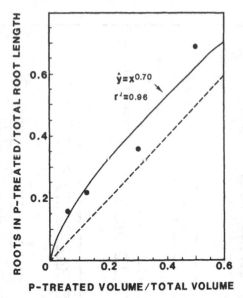

Figure 3. Relation between distribution of P applied per 3 kg pot of soil and root distribution between fertilized and unfertilized soil.

Placement Effects on P and K Influx

Rate of P and K uptake by plant roots is related to P and K concentration in solution. Figure 4 shows the relation between P uptake rate by soybean plants and P concentration in stirred solution. Since only part of the root system is fertilized when fertilizer is placed in the soil, it is important to know how placement affects the relation shown in Figure 4. When only part of the root system was exposed to solution containing P, so that plant demand for P per unit of root increased, influx by the

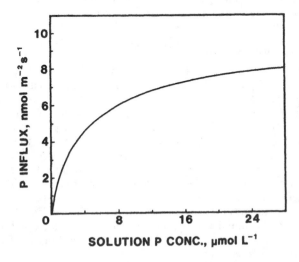

Figure 4. Relation between P influx by 25-day-old soybean plants and P concentration in stirred, aerated solution.

exposed roots did not change. Uptake rate was regulated by the uptake mechanisms in the plant root rather than by shoot demand. Uptake per plant depended on the proportion of the root system exposed to P. With a prolonged reduction in uptake rate per plant due to fertilizing only part of the roots, a reduction in P concentration of the shoot will occur. This reduction in the shoot P concentration does cause influx to increase somewhat, but the increase does not compensate for the reduction in root length supplied with P. Similar results have been obtained for K. Hence when P and K fertilizer are placed in the soil so only part of the roots are contacted, the amount of uptake is less than when all roots are in contact with soil fertilized at the same rate of P and K per unit of fertilized soil. From the plant's aspect, the most favorable situation is to have all roots in contact with the fertilized soil.

Placement Effects on P and K Soil Levels

When fertilizer is mixed with the soil, part of the P or K added is adsorbed rather tightly by the soil, so that its contribution to maintaining the P and K level of the soil solution is limited. The amount of P and K "tied-up" in this way varies with soil properties. As the rate of application is increased, the amount of P or K "tied up" from each increment added decreases. So, at very high rates only small amounts may be tied up by the soil. This is why fertilizer often is banded, to reduce the amount of fixation by the soil as compared to that occurring when the fertilizer is mixed with all the soil. The restriction of P or K to a small volume of the soil reduces fixation and leaves more of the nutrient in an available form for uptake by plant roots. However, this restriction also means that a small fraction of the root system will be in contact with fertilized soil. This will tend to reduce uptake. The most efficient placement would be somewhere between the extremes of complete mixing and mixing with 1 or 2% of the soil, which occurs with banding at planting. A balance between the reduction in availability as the fertilizer is mixed with more soil and the increase due to contacting more roots will determine the most effective placement.

CALCULATION OF THE MOST EFFECTIVE PLACEMENT

A mechanistic mathematical model that describes nutrient uptake by roots growing in soil can be used to determine most effective placement. The model assumes uptake by roots is determined by soil solution nutrient concentration, and supply through the soil to the root by mass flow and diffusion, which is determined by the soil nutrient level. When fertilizer is mixed with part of the soil, the model calculates uptake separately by the roots growing in the fertilized soil and the roots growing in unfertilized soil. The more roots in the highly fertilized portion, the greater uptake. As more soil is fertilized, nutrient level in the fertilized soil decreases and rate of supply to the roots in this soil decreases.

The mechanistic mathematical model was used to calculate uptake for various P or K placements in the soil. The soil parameters of the model that determine supply to the root by mass flow and diffusion were measured from data obtained after application of a wide range of P and K rates to the soil. The root distribution between fertilized and unfertilized soil was determined from the relation shown in Figure 3. The uptake kinetics for soybean roots, as related to shoot composition, also was determined. This gave a series of curves, similar to the curve shown in Figure 4, that could be used for calculating uptake. Then uptake by roots growing in the fertilized soil and in the unfertilized soil were calculated separately by running the model with its computer program for each. The two were then added together to get the total predicted uptake for each placement.

The results of calculating the effect of P placement on predicted P uptake by soy-bean plants grown on Raub silt loam soil are shown in Figure 5. This figure shows that increasing the volume of soil the fertilizer is mixed with increases predicted P uptake until a maximum is reached, and then uptake decreases as P is mixed with a greater soil volume. Where the rate was 30 mg P per kg, maximum P uptake was predicted with 6% of the soil fertilized. As the rate was increased to 60 mg P/kg, the value of fertilized soil for maximum predicted P uptake was 10%. With a rate of 120 mg P/kg, it was 20%.

The reason the curves have the shape shown in Figure 5 is that, with the P mixed with small volumes of soil, the P level is higher than that needed for maximum uptake by the root. Uptake is increased by contacting more roots. As the fertilizer P is mixed

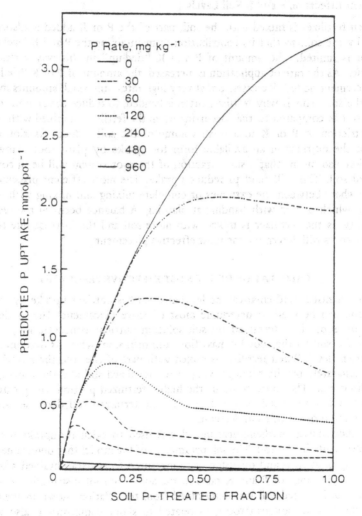

Figure 5. Relation between predicted P uptake by soybean plants and placement of several rates of P in Raub silt loam soil.

with more soil, it contacts more roots increasing uptake. Eventually, however, mixing the same amount of P with increasing amounts of soil causes greater P "tie-up" by the soil, so that predicted P uptake maximizes and then decreases as greater mixing occurs. The rate of P supply to the root becomes much less than that needed to maximize uptake rate. The increase in the slope of the curve is because more roots are contacted. The decrease is because even though more roots are contacted, the reduction in availability of P due to increased tie-up by the soil becomes the dominating factor, and predicted P uptake declines with further mixing of the applied P with the volume of soil present, even though more roots are contacted. The higher the rate of application, the greater the volume of soil that should be fertilized to maximize uptake.

Traditionally, fertilizer has been applied at planting. It is mixed with about 1% of the soil or broadcast and mixed with all the soil. The results from Figure 5 would indicate that uptake would be greater if the fertilizer P were mixed with 5 to 20% of the soil, the actual amount depending on rate applied, initial P level in the soil, and the degree of "tie-up" of P by the soil. In pot experiments where P was mixed with 1.5, 3.0, 6, 12.5, 25, 50, and 75% of the soil, predicted P uptake from these treatments agreed with observed uptake (Figure 6), verifying the predictions of the simulation model and indicating it is suitable for predicting the optimum placement. The uptake values of Figure 5 may be converted to crop yield using the relation shown in Figure 7.

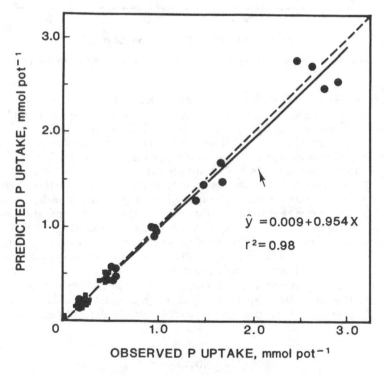

$$\hat{y} = 0.009 + 0.954X$$

$$r^2 = 0.98$$

Figure 6. Relation between predicted P uptake by soybean plants, from a wide range of P placements, and observed uptake from the same placements.

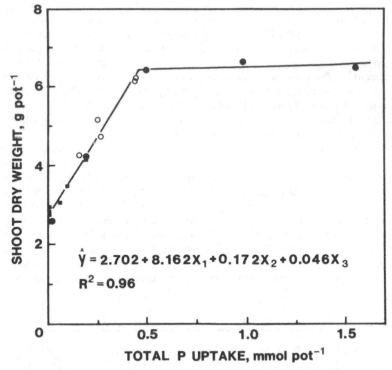

Figure 7. Relation between P uptake by soybean plants, from a range of P placements, and plant dry weight.

Few placement experiments have been conducted with soybean. Placement experiments have been conducted with corn, however. Treatments of P and K applied as a band near the row at planting, as a broadcast application, and as a strip application have been compared. The strip application is a procedure for fertilizing an intermediate portion of the soil volume and is done by applying the fertilizer in strips that cover 10 to 20% of the soil surface, then moldboard plowing to incorporate the fertilizer into the soil. The fraction of soil fertilized will be 10 to 20% of that occurring where a uniform broadcast application is made and plowed under. The yield of maize (average for 5 years) as a percentage of broadcast application was 0.96 for row application and 1.08 for strip application. The % K in the leaf was 1.45 for broadcast, 1.50 for row, and 1.80 for strip application. The strip application gave greater nutrient uptake, as predicted. Similar results could be expected for soybean, since its root system responds to fertilization in a manner similar to that of maize.

Significance of placement depends on the degree the soil fixes P or K, the initial soil test level, and the rate of application. At very high rates of application broadcasting the fertilizer to build-up the fertility level of all the soil will be most effective.

NOTES

Stanley A. Barber, Department of Agronomy, Purdue University, West Lafayette, Indiana 47907.

REFERENCES

Claassen, N. and S. A. Barber. 1976. Simulation model for nutrient uptake from soil by a growing plant root system. Agron. J. 68:961-964.

Silberbush, M. and S. A. Barber. 1983. Prediction of phosphorus and potassium uptake by soybeans with a mechanistic mathematical model. Soil Sci. Soc. Am. J. 47:262-265.

Silberbush, M. and S. A. Barber. 1984. Phosphorus and potassium uptake of field-grown soybean cultivars predicted by a simulation model. Soil Sci. Soc. Am. J. 48:592-596.

Cropping Systems

INTERCROPPING—COMPETITION AND YIELD ADVANTAGE

C. A. Francis

Multiple cropping systems are those that intensify the culture of two or more crop species in either time or space, or both. Long practiced by small farmers with a limited land resource, these cropping patterns are characterized by species diversity, intensive use of labor, low use of outside inputs by the farmer, and little attention from the research establishment (Francis et al., 1976; Navarro et al., this volume). As contrasted to the better known monoculture systems, where output is measured by total grain yield or net income, the farmer who practices intercropping often measures success in terms of one or more of the following criteria: food supply from diverse crops throughout the year; maximum use of labor over time; stability of production rather than maximum output under optimal conditions; maximum outputs with limited or no purchased inputs; and minimum risk from existing or new technology (Francis, 1981). Needless to say, the research scientist often is not prepared to face this complexity, or to decide how to measure the results of experiments. Further, the answers to whether intercropping is successful, or should be recommended, depends on the economic and cultural situation in which the farmer is operating.

Among the many and diverse systems practiced, intercropping is defined as the culture of two or more crops in the field with a significant overlap of their growth cycles, and consequently, competition for light, water, and nutrients during a major part of crop growth. Practically, this category includes any crops planted simultaneously, or with a second crop planted before flowering date of the first crop. A pattern with the second crop planted after flowering of the first crop would be considered a relay system (Whigham, this volume). When cropping cycles do not overlap, but two or more species are planted in the same field sequentially in the same year, this is termed double cropping, triple cropping, and so on (Caviness and Collins, this volume). These definitions are consistent with the agreed terminology published in the A.S.A. Symposium on Multiple Cropping (Andrews and Kassam, 1976). Other synonymous terms are listed in this report, and we are attempting to coordinate and standardize the use of these several descriptors in order to aid communication, and to bring better understanding by researchers to an already complex field.

INTERCROPPING

As defined, intercropping is an intensification in the space dimension with two crops in the field at the same time. Many variations are possible on this theme, as shown by the several intercrop examples in Figure 1. The maize/bush bean (*Zea mays* L./*Phaseolus vulgaris* L.) intercrop is common in Brazil and other Latin American

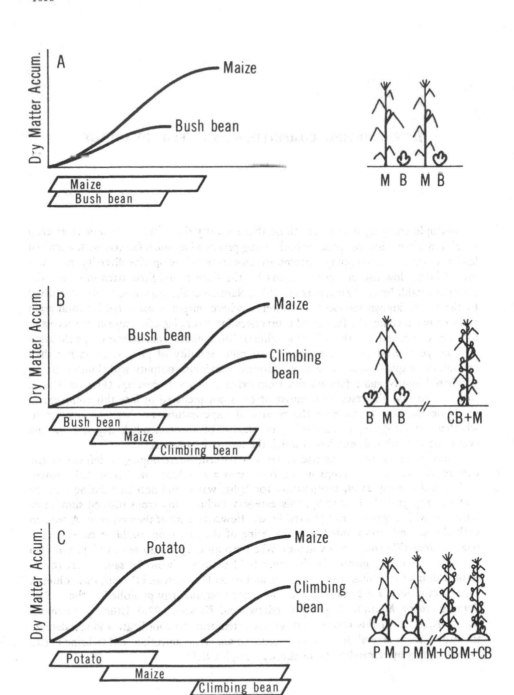

Figure 1. Three intercrop patterns found in Colombia at different altitudes.

countries, especially at lower altitudes. With simultaneous planting of the two crops (Figure 1A), there is an intense competition in the early stages until the bean crop matures, usually before the maize. In Colombia, the bean matures in about 70 to 90 days, depending on the temperature, while the maize may stay in the field from 100 to 120 days in the lowlands. The second example (Figure 1B) from the middle elevation in Colombia shows a bush bean planting, followed by maize planted in the middle of every other space between bean rows. After the maize is 20 to 30 cm tall, a climbing bean seed or two is planted at the base of every other maize plant. Finally, the potato (*Solanum tuberosum* L.)/maize/climbing bean pattern in the highlands of Colombia (Figure 1C) is an example of an intercrop-relay crop situation, where three crops are harvested in a period of about 13 to 14 months. If earlier cultivars of the three crops could be developed, it would be possible, with well-distributed rainfall, to harvest all three crops within one calendar year.

A yield advantage from intercropping, called "overyielding", occurs when the two or more crops in combination produce more than a monoculture of any single crop. For an excellent review of intercropping, the reader is invited to examine Willey's papers (Willey, 1979a,b). Several methods are used to measure whether overyielding has occurred. The Land Equivalent Ratio (LER) is one of the most common measures, and is the sum of the relative yields (fraction of monoculture) of the two or more components of the mixture (IRRI, 1974).

$$LER = \sum_i I_i/M_i$$

where I_i = intercrop yield of the i th species; and M_i = monoculture yield of the same species.

When LER is significantly more than 1, overyielding has occurred. When LER = 1, there has been an expected additive response of individual crop components, or if one has yielded poorly, another has compensated by producing relatively more. If the LER is less than 1, competition has reduced yields of both species, or has drastically reduced one of the components. In this case, intercropping is clearly undesirable if the objective is grain yield.

Other approaches have been suggested and used. One of the more exciting of these is the Total Equivalent Yield suggested by Federer (pers. commun.). In this approach, yields of component crops are converted to equivalent yields of a single crop, and the total yield expressed in terms of production of that single crop. This has the advantage of analysis of yields rather than ratios such as with the LER. This alternative seems to be gaining in popularity.

SOYBEAN INTERCROP RESULTS

With these definitions and measurement tools in mind, it is possible to compare a series of soybean results from monoculture and intercropping. The competition of a taller companion crop consistently reduces the seed yield of soybean. In a study by Chui and Shibles (1984) in Iowa, soybean yields, when associated with maize, were reduced 89% in 1980 and 85% in 1981. Maize yields were not reduced significantly, though production was 1 to 14% less for intercropped maize than in monoculture. The combined maize plus soybean yield generally was greater than monoculture maize at zero N levels, while there was no significant difference at the 135 kg/ha N level. Chui and Shibles concluded that soybeans provided a bonus yield at low N

In a similar study, Pavlish et al. (1983) studied soybean-sorghum intercrops in Nebraska in several row configurations: 4-row strips, 2-row strips, and intensive inter-cropping in the same row. All were compared to monocultures of the two component crops. Soybean yields on a per row basis were 8% higher in the 2-row alternating strips than in either 4-row strips or in monoculture, and this was consistent both in irri-gated culture at Clay Center and under dryland conditions at Mead. The irrigated soy-bean yields in the former location were about 19% higher than the dryland yields in the latter. Sorghum yields were reduced in the intercropping and strip cropping by 2% under irrigated culture at Clay Center, and by 10% in dryland conditions at Mead. When both crops were planted in the same row, this intensive intercrop resulted in 61% and 29% yield reductions of soybeans and sorghum, respectively, under irrigation at Clay Center. In this system the intercrop caused 38% and 57% yield reductions of soybeans and sorghum, respectively, under dryland conditions at Mead. When moisture is limiting, the soybean competes successfully for the available soil water, and yield reductions are relatively less than under irrigated conditions. When moisture is not limiting, sorghum does relatively better and soybean suffers more from the intense competition. Total equivalent yields calculated from these same experiments (Table 1) showed that the more intensive combinations of crops such as same-row planting and 2-row strips produced more than the monoculture and 4-row strips, both under irrigated and dryland conditions. Reasons for these differences are not clear.

In a study of several combinations of sorghum and soybean densities, Carter et al. (1983) showed consistent overyielding of about 10% for intercrops as compared with monocultures of the two species (Figure 2). With a constant sorghum density, higher levels of soybean density produced yields of about 2 t/ha of soybeans and 3 t/ha of sorghum (Figure 2A). Lower densities of soybean produced about 0.5 t/ha of soy-beans and 5 t/ha of sorghum. With a constant soybean density, different densities of sorghum produced from 2.4 t/ha soybeans and 2 t/ha sorghum to 0.5 t/ha soybeans and 5 t/ha sorghum (Figure 2B). As shown, the yields are consistently above the line connecting the two monoculture yields, a line which represents a series of soybean + sorghum monoculture total yields, if the farmer were to divide the field into two plots for the two species grown separately. The conclusion is that the farmer with limited resources, who wants to plant an intercrop of the two species, has a wide range of

Table 1. Total equivalent yields expressed in sorghum kg/ha for monocultures and intercrops of soybeans and sorghum (from Pavlish et al., 1983).

| Treatments[a] | Standardized system yields | |
	Clay Center[b] (irrigated)	Mead[b] (non-irrigated)
	kg/ha	
Intensive	6150 r	5010 r
2 Row	5770 s	4850 rs
Monoculture	5650 s	4770 rst
4 Row	5560 s	4650 st
Soybean monoculture	5610 s	4505 t
Sorghum monoculture	5700 s	5040 r

[a]See text.
[b]Yields followed by different letters are significantly different.

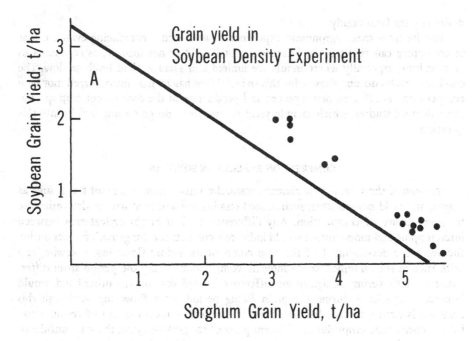

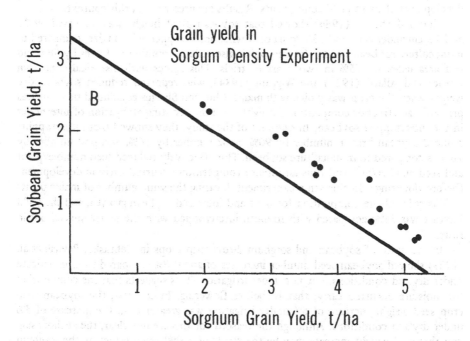

Figure 2. Soybean and sorghum yields in monoculture and several intercropping alternatives (Carter, 1984).

options, all of which will overyield about the same amount, and the choice of density combinations would depend on the relative prices of the two crops or the production desired by the farm family.

Results from these agronomic experiments show rather convincing evidence that intercropping can result in an increase in total yield or net income, as compared to monoculture, especially where inputs are limited and average yield levels are low. The empirical results do not show why this overyielding has or has not occurred, nor how competition has affected development and production of the component crop species. More detailed studies, which include yield components, do give some insight into this question.

COMPETITION EFFECTS ON SOYBEAN

In some of these same experiments, researchers have examined plant traits, such as height, and yield components, such as seed number/m^2 and seed size, to determine the temporal nature of competition. Any difference in plant height or leaf area between intercropping and monoculture would indicate a competition for growth factors during the vegetative development of the crop components, before flowering. Likewise, any difference in seed number would indicate competition during the period from differentiation to flowering. A significant difference in seed size, on the other hand, would indicate competition during the grain filling period, after flowering. Although this evidence is circumstantial, because there is not precise measurement of resource use by the component crops during different parts of the growth cycle, there is confidence in the growth patterns of cereals and legumes and established evidence of the temporal development of these yield components. Results are presented in this context.

Chui and Shibles (1984) showed that soybean plant height was increased by 13 and 22 cm under zero and 135 kg/ha of N when intercropped with maize, compared to monoculture soybean height. Intercropping also reduced number of leaves by 58% and leaf area index by 75% in two years of trials. This agrees with historical data from Garner and Allard (1920) and Wiggans (1934), who reported reduced soybean dry weight when the crop was grown with maize. Chui and Shibles concluded that this was primarily an effect of competition for light, with the resulting elongation of internodes in the intercropped soybean. In one year of the study, they showed that intercropping reduced soybean branch number by 90%, node number by 56%, and pod number by 76% as compared to monoculture soybean. This effectively reduced seed number/plant and seed number/m^2, and thus an intense competition occurred early in development (before flowering). In this same experiment, locating the same number of maize plants in hills reduced the competition for light and increased soybean growth, and the yield increase was 36%, compared with soybean intercropped with the maize spaced out in rows.

In the study of soybean and sorghum 2-row strip crops in Nebraska, Pavlish et al. (1983) showed soybean seed number increases of about 8% compared to monoculture under dryland conditions, but not under irrigation. This suggests that the competition for moisture occurred early; that is, before flowering. In contrast, the soybean strip crop seed weight increase of 6% under irrigation was greater than the increase of 3% under dryland conditions. Although the reasons for this are not clear, the authors suggest that seed weight increase may be the result of a sheltering effect by the sorghum in the strip crop, which reduces water loss from evapotranspiration and less dissipation of the CO_2 lost through respiration. This would explain the greater seed weight increase under dryland conditions, as compared with irrigation, where there is reduced or no moisture stress. In summary, under irrigation soybean showed a significant increase in seed size, but not in seed number, whereas under dryland soybean increased in seed number but not in size. This indicates a difference in temporal competition between the two conditions, although these were confounded with location in this experiment.

Reduced soybean yields in the experiments by Elmore and Jackobs (1984) were associated with fewer pods/plant in both years, with reduced number of seeds/pod in the first year, and with reduced seed weight in the second year. In previous experiments in Illinois, Wahua and Miller (1978a,b) reported that reduced soybean yields were associated with lower seed number/pod when the soybean was intercropped with both tall and short sorghum, and associated with fewer pods/plant when soybeans were intercropped with the short sorghum.

Results from these several studies with soybean intercropped with maize and with sorghum indicate the complexity of crop species interactions, and the types of yield component changes that occur as a result of competition. Further study is needed on the performance of the understory legume crops in these cereal-legume mixtures, and more detailed evaluation of the effects of both components during the course of crop growth.

CONCLUSIONS

Results presented from experiments where soybean was intercropped with maize and with sorghum show that overyielding is a common phenomenon with intercropping. This is especially frequent when the yield potential of an environment is low due to poor fertility or other limiting factors. The evidence shows why farmers with limited resources continue to plant intercrops rather than change their cropping patterns to monoculture. At the other end of the yield spectrum, under favorable conditions there is less advantage of intercropping, and results may be conflicting.

NOTES

C. A. Francis, Rodale Research Center, Box 323, RD 1, Kutztown, PA 19530.

REFERENCES

Ahmed, S. and M. R. Rao. 1982. Performance of maize-soybean intercrop combination in the tropics: results of a multi-location study. Field Crops Res. 5:147-161.

Andrews, D. J. and A. H. Kassam. 1976. The importance of multiple cropping in increasing world food supplies. p. 1-10. *In* R. I. Papendick et al. (eds.) Multiple cropping, ASA Special Publication No. 27. American Society of Agronomy, Madison, Wis.

Carter, D. C., C. A. Francis, L. A. Pavlish, G. M. Heinrich and R. V. Matthews. 1983. Sorghum and soybean density interactions in one intercrop pattern. Agron. Abstr. p. 43.

Chui, J. A. N. and R. Shibles. 1984. Maize spatial arrangements influence performance of an associated soybean intercrop. Field Crops Res. 8:187-198.

Dalal, R. C. 1977. Effect of intercropping maize with soya bean on grain yield. Tropical Agric. (Trinidad) 59:189-191.

Elmore, R. W. and J. A. Jackobs. 1984. Yield and yield components of sorghum and soybeans of varying plant heights when intercropped. Agron. J. 76:561-564.

Francis, C. A. 1981. Rationality of cropping systems practiced by small farmers in the tropics. Farming Systems Workshop. Kansas State Univ., Manhattan, Kansas.

Francis, C. A., C. A. Flor and S. R. Temple. 1976. Adapting varieties for intercropped systems in the tropics. p. 235-253. *In* R. I. Papendick et al. (eds.) Multiple cropping, ASA Special Publ. No. 27. American Society of Agronomy, Madison, Wis.

Garner, W. W. and H. S. Allard. 1920. Effect of the relative length of day and night and other factors of the environment on growth and reproduction in plants. J. Agric. Res. 18:553-606.

IRRI. 1974. Annual Report. International Rice Research Institute, Los Banos, Philippines.

Mohta, N. K. and R. De. 1980. Intercropping maize and sorghum with soya bean. J. Agric. Sci. (Cambr.) 95:117-122.

Pavlish, L. A., C. A. Francis, T. R. Zweifel and J. F. Rajewski. 1983. Intercropping versus monoculture of sorghum and soybeans in two contrasting environments in Nebraska. Agron. Abstr. p. 44.

Singh, J. N., P. S. Negi and S. K. Tripathi. 1973. Study on the intercropping of soybean with maize and jowar. Indian J. Agron. 18:75-78.

Wahua, T. A. T. and D. A. Miller. 1978a. Relative yield totals and yield components of intercropped sorghum and soybeans. Agron. J. 70:287-291.

Wahua, T. A. T. and D. A. Miller. 1978b. Effects of intercropping on soybean N_2-fixation and plant composition on associated sorghum and soybeans. Agron. J. 70:292-295.

Wiggans, R. G. 1934. The effect of growing corn and soybeans in combination on the percentage of dry matter in the two crops. J. Am. Soc. Agron. 26:59-65.

Willey, R. W. 1979a. Intercropping: its importance and research needs. Part 1. Competition and yield advantages. Field Crops Abstr. 32:1-10.

Willey, R. W. 1979b. Intercropping: its importance and research needs. Part 2. Agronomy and research applications. Field Crops Abstr. 32:73-65.

STRIP AND RELAY INTERCROPPING OF SOYBEANS

D. K. Whigham

Although there are many forms of multiple cropping, I will discuss only two types of intercropping. Strip intercropping is the production of more than one crop simultaneously in different strips that are narrow enough for the crops to interact and wide enough to permit independent cultivation of the different crops. Relay intercropping is the production of more than one crop simultaneously for a portion of each crop's growing season (Andrews and Kassam, 1976). An example of relay intercropping is when oat (*Avena sativa* L.) is planted first, and soybean is interseeded into the oat crop later. Soybean has increased in popularity during the past 25 years as a component of intercropping systems.

STRIP INTERCROPPING

Strip intercropping is practiced by farmers throughout the world. In areas of the world where labor is plentiful, the width of each strip can be narrow, since hand labor and small machines are used for planting, cultivating and harvesting. The contrary may be true if labor is scarce and large machines are utilized.

In general, soybean and maize (*Zea mays* L.) strip-intercropping systems have produced more maize and less soybeans than a comparable area of each grown as a sole crop. Pendleton et al. (1963) found that alternate strips of soybean and maize produced a 20% increase in maize yield and a 20% decrease in soybean yield, compared with sole crops. Crookston and Hill (1979) found similar results when they evaluated several different strip widths. Shen (1984) observed similar results in studies of 2.5 m strips of soybean alternated with 3 m strips of maize (Table 1). Soybean produced fewer pods/plant when strip intercropped with maize. This reduction was attributed to shading by the taller maize crop. The decrease in soybean yield was always accompanied by a decline in photosynthetically active radiation (PAR) at the top of the soybean canopy. An increase in maize plant density from 62,500 to 87,500 resulted in a significant soybean yield decrease (Table 1). Maize yield was increased as its plant density changed from 37,500 to 62,500 plants/ha. The intercrop maize yield did not change significantly when plant density increased from 62,500 to 87,500; however, the sole-crop maize yield decreased significantly with the same change in plant density. Strip intercropping also allowed more PAR into the maize canopy than did sole cropping. The inverse relationship was true for soybean; sole-crop soybean received significantly more PAR at the top of the canopy than did strip-intercropped soybean.

Land-use efficiency by strip intercropping varies considerably. LER, the land area required in sole crops to produce the same yield of the same crops on one unit of land

Table 1. Soybean and maize yields as affected by cropping system, and by maize plant density and row direction when intercropped (Shen, 1984).

| | Yield | |
	Soybean	Maize
	— kg/ha —	
Cropping system		
Sole crop	4080	6920
Strip intercrop	3120	7940
LSD (0.05)	240	370
Maize plant density, plants/ha		
37,500	3350	6570
62,500	3240	7920
87,500	2820	7810
LSD (0.05)	190	450
Row direction		
N-S	3360	7520
E-W	3960	6920
NE-SW	3430	7120
NW-SE	3650	8170
LSD (0.05)	300	520

of a multiple cropping system (Andrews and Kassam, 1976), values from soybean and maize strip intercropping in temperate environments have ranged from no advantage to a 10% advantage compared to sole cropping (Pendleton et al., 1963; Crookston and Hill, 1979; Shen, 1984). In tropical and subtropical environments, intercropping appears to be more efficient. LERs of 1.4 to 1.6 have been reported in the Philippines when different rates of nitrogen were applied to intercropped soybean and maize (IRRI, 1976).

Snaydon and Harris (1979) indicate both direct and indirect evidence that intercrop yield advantages occur because of below-ground interactions. Legumes and non-legumes are usually the most compatible because they have different types of roots, utilize different sources of nutrients, and require differing amounts of both nutrients and water. The demand varies during different periods of the growing season and for different root zones. Little research has been reported on the below-ground interaction of companion species in intercropping systems.

Cultivar Selection

Selection of the companion crops for an intercropping system is a critical management decision. Soybean is usually the dominated crop, and the selection of a compatible dominant crop determines the relative yield produced. Tall dominant crops increase the amount of shade on the dominated crop, and usually causes lower yield of the shorter crop. Elmore and Jackobs (1984) reported that soybean yields were reduced more by tall than short sorghum (*Sorghum bicolor* [L.] Moench) as a dominant companion crop. The yield reduction was primarily attributed to fewer pods/plant. Similar results were observed by Beets (1982) and Wahua and Miller (1978). Intercrop soybean and sorghum yields were less than sole crop yields, but the LERs favored intercropping. Therefore, cultivar selection within a crop species may also be a critical decision. Environmental factors may also affect the relative yield of cultivars or species from year-to-year.

In Iowa, cultivar selection was found to be important for maize when strip inter-cropped with soybean. Soybean grown as a companion crop with an upright-leaf maize cultivar produced higher yields at the low plant density (37,500 plants/ha) of maize. However, at medium and high plant densities the soybean yields were not different. Soybean cultivars have responded differently when intercropped (IRRI, 1983). Twelve soybean cultivars were evaluated when intercropped with maize and as a sole crop. The sole-crop yield ranged from 690 to 1820 kg/ha. The intercropped yield was re-duced, but was more stable and ranged from 340 to 830 kg/ha across all cultivars. Galal et al. (1974) evaluated 21 soybean cultivars when intercropped with maize. They concluded that partial shading by the maize reduced the PAR available to the soybean, which resulted in yield reductions for 19 of the cultivars. Two cultivars, however, increased their grain yield from 1 to 4% when intercropped. Gomez and Gomez (1983) reported a study that included 24 soybean cultivars that were evaluated under five environments, including two intercrop environments. Stability indices were determined for all cultivars and significant differences were found. The cultivar Wayne was more stable than cultivar Cobb. Wayne had a very narrow yield range of 1032 to 1033 kg/ha over the two intercrop environments, and Cobb had a range of 660 to 1179 kg/ha. During the dry season, intercrop soybean yields ranged from 538 to 1160 kg/ha across all 24 cultivars. The wet season yield ranged from 717 to 1244 kg/ha. Cultivars with stable indices are less risky if unforeseen stresses on the crop occur during the growing season. However, stable cultivars are less likely to produce very high yields when the environmental factors are favorable. These and other results suggest that cultivar and species selection are extremely important to the success of an intercropping system.

Row Direction

The effect of row direction on strip intercropping is dependent on the location of the field, crop species and environmental variables. Solar angle of incidence, direction and degree of slope all affect the amount of shade the dominant crop casts on the dominated crop of an intercropping system. Conflicting reports exist concerning the effect of row direction on the yield of sole or intercrop systems. Many reports have shown no effect on yield due to row direction, and some have identified a favorable row direction. Where favorable row directions have been identified the authors have attributed the benefit to longer duration of sunlight.

When comparing north-south (N-S) versus east-west (E-W) rows, Pendleton et al. (1963) found no row direction affect on strip intercropped soybean and maize. How-ever, they did report that the northern soybean rows in the E-W row direction yielded higher than the southern rows. In the N-S rows, the eastern rows yielded more than the western rows. The authors concluded that those rows received more sunlight which produced the yield differences.

Shen (1984) found that strip intercropped soybean and maize produced the highest soybean yields in E-W rows and the lowest yields in N-S rows (Table 1). In the same study the NW-SE row direction produced the highest maize yield. The E-W soybean rows produce significantly more pods and seeds per plant. However, when the LERs were calculated the N-S row direction proved to make the most efficient use of the land (Table 2). The relative yield of maize (L_m) was highly correlated with LER (r=0.984).

Windbreaks

Windbreaks benefit crops by providing shelter, which may reduce mechanical damage caused by high velocity wind, and by changing the microclimate of the crop

Table 2. Relative yields and Land Equivalent Ratios (LER) of row directions for strip intercropped soybean and maize (Shen, 1984).

Row direction	L_s[a]	L_m[b]	LER
N-S	0.33	0.77	1.10
E-W	0.36	0.55	0.91
NE-SW	0.37	0.67	1.04
NW-SE	0.34	0.66	1.00
LSD (0.05)	0.02	0.11	0.11

[a] Relative yield of soybeans.
[b] Relative yield of maize.

canopy. Wind speed is reduced on the lee side of the windbreak and reduces soil moisture evaporation. Radke and Burrows (1970) found that two rows of temporary maize windbreak in a soybean field produced taller soybean plants, more dry matter, a larger leaf area index and higher soybean yields than nonsheltered soybean plants. Water-use efficiency was also improved. Maize rows perpendicular to the prevailing wind were the most effective in sheltering soybean. Effective spacing between the windbreak strips were from 6H to 15H (H = height of windbreak). Soybean yields were increased five of the six years studied. The yield increase resulting from the temporary shelter was significant and ranged from 5.3 to 27.8%. The authors conclude that the soybean yield increase was not due to reduced mechanical damage caused by wind velocity. The results were indirectly due to changes in windspeed and air turbulence within and directly above the soybean canopy, which favored soybean production. Maize yields from the temporary windbreak strips were about 75% of those expected from sole cropping, but the land area occupied by maize was only 14% of the total strip intercropped area. Windspeed and potential evaporation were reduced significantly for 6 m on the lee side of maize temporary windbreaks (Radke and Hagstrom, 1973). Two maize rows 76 cm apart formed the temporary shelter.

RELAY INTERCROPPING

Relay intercropping is an established crop management practice in many areas of the world where productive land area is limited. Asian farmers routinely relay intercrop and produce many crops on the same field during the same year. A surplus of labor, irrigation facilities and year round temperatures favoring crop production make relay intercropping practical. In temperate environments where large scale mechanized crop production is practiced, relay intercropping is seldom utilized. When the growing season is sufficiently long to allow sequential cropping, double cropping is sometimes practiced. If the season is too short, or the rainfall is too irregular for sequential cropping an alternative is relay intercropping. To plant the second crop into the first crop mechanically without causing serious damage, correct timing and special equipment may be necessary.

Relay intercropping with soybean is most frequently practiced in the United States with a small grain cereal crop. Chan et al. (1980) evaluated winter wheat (*Triticum aestivum* L.) and spring oat (*Avena sativa* L.) as the first crop and soybean as the second crop (Table 3). They found that interseeding soybeans into wheat and oat crops did not affect the yield of the first crop. However, soybean yields were reduced significantly when compared to sole crop yields. Row spacings (21-81 cm) were also

Table 3. Effect of relay intercropping on the yield of oats, wheat and soybeans[a] in 41 cm rows (Chan et al., 1980).

| | Yield | | Relative |
	Sole crop	Intercrop	yield
		– kg/ha –	
1976			
Oats	4365	3960	0.91
Soybeans	2876	2050	0.71
Land Equivalent Ratio			1.62
Wheat	2472	2767	1.12
Soybeans	2785	978	0.35
Land Equivalent Ratio			1.47
1977			
Oats	3218	3234	1.01
Soybeans	3593	1835	0.51
Land Equivalent Ratio			1.52

[a]Soybean cultivar Williams.

evaluated, and small grain yields decreased as row width increased in both cropping systems. Sole crop soybean yields were not affected by changes in row spacings, but intercropped soybean yielded more in wide rows. Similar results were observed in Iowa for oat yields, but soybeans in 25 cm rows yielded significantly more than 76 cm rows when interseeded into oats. Oat yields were also decreased as row width was changed from 25 to 51 cm. Brown and Graffis (1976) found that, by harvesting the intercropped oat crop for silage, soybean produced higher yields (2688 kg/ha) than when oats were harvested for grain (1882 kg/ha). The selection of an early maturing species, cultivar or methods of early harvest, which will reduce the length of simultaneous competition, will likely result in higher yields from the second crop.

Time of Planting

The appropriate time of planting soybeans into a small grain crop can be evaluated by the affect on each crop. The earlier the soybeans are interplanted, the less the competition for light, since the small grain crop is less developed. Moisture may also be less limiting if rainfall patterns peak during the early part of the small grain growing season. Sufficient moisture at the time of planting the soybeans is critical for good stand establishment. Early soybean planting also allows for a longer growing season, which usually increases yield. A significant soybean yield increase occurred when soybeans were planted eight days before oat heading, compared to planting at the time of oat heading. Soybean yields were reduced even more by planting eight days after oat heading. However, oat yield was not affected by the different planting dates. Mechanically planting soybeans into an established oat field after heading appeared to cause some damage to the oat plants, but the damage was not reflected in reduced yield.

Interplanting into an established crop creates management problems unique to an intercropping system. Special equipment may be needed to plant into an undisturbed soil environment in order to properly place the seed and firm the soil for good soil-seed contact. Weed control may create the most challenging dilemma, especially if the system included a legume and a non-legume. Mechanical cultivation may not be an available option if the crops are intimately associated, such as a small grain and soybean. Therefore, hand weeding or chemical weed control may be required. If labor is plentiful, weeding can be completed without serious damage to the crops. If herbicides are to be used, careful selection and application are necessary to control the weeds without damaging either crop.

Careful management of nutrient application also is important to the success of an intercrop system. Non-legumes usually require nitrogen fertilization for optimum yield, but legumes require little or no applied nitrogen. Other nutrient elements are easier to manage and may be applied before planting of the first crop.

Cultivar Selection

Soybean cultivar selection for relay intercropping may also be critical to the success of the system. Chan et al. (1980) evaluated Elf, a short determinate cultivar, and Williams, a tall indeterminate cultivar, for their response to relay intercropping with an oat crop. Soybean yield and plant height were significantly reduced for the determinate cultivar when intercropped. Elf had reached its maximum height before the oats were harvested, but Williams continued to grow after the oats were harvested. Rapid growing soybeans may be tall enough to receive mechanical damage when the small grain crop is harvested. However, the selection of a short determinate cultivar does not seem to improve the soybean yield response to relay intercropping system when compared to the taller indeterminate cultivar.

McBroom et al. (1981) compared the performance of several soybean cultivars when relay intercropped with oat and wheat. Grain yield, plant height and lodging scores were significantly different among cultivars and were reduced significantly by relay intercropping. The genotype X cropping system interaction for yield was not significant for soybean. Mean and relative yields are shown in Table 4. The authors conclude that, based on their research, a separate breeding program to develop soybean cultivars for relay intercropping with small grain cereal crops is not warranted.

Soybean is relay intercropped with maize, cotton (*Gossypium hirsutum* Linn.) and cassava (*Manihot esculenta* Crantz) in Thailand (Benjasil and Lampang, 1984). Paired soybean rows between maize rows produced the highest LER (1.74) when soybeans were interseeded 80 days after the planting of maize. The LER was 1.10 when

Table 4. Mean and relative yields of seven soybean cultivars compared in relay intercropping with oat and wheat crops, and in sole cropping (McBroom et al., 1981).

Cultivar	Relay intercropping	Sole cropping	Relative yield
	— kg/ha —		
Harcor	2237	3630	0.62
Cumberland	2007	3463	0.58
Corsoy	1993	3300	0.60
Beeson	1990	3440	0.58
Wells	1767	3313	0.53
Oakland	1720	3343	0.52
Hark	1833	2903	0.63
LSD (0.05)	305	305	

both crops were planted at the same time. The maize growing season was 110 days in length. The optimum time of planting for cotton was 20 days after soybeans were planted. The LERs ranged from 1.34 to 1.39 for different row combinations of the two crops. Soybeans and cassava were planted at the same time. Soybean yield increased with an increase in plant density (50,000 to 800,000 plants/ha), but the cassava yield decreased. The LER range for the plant population range was from 1.32 to 1.61.

NOTES

D. K. Whigham, Agronomy Department, Iowa State University, Ames, Iowa 50011 U.S.A.

REFERENCES

Andrews, D. J. and A. H. Kassam. 1976. The importance of multiple cropping in increasing world food supplies. p. 1-10. *In* R. I. Papendick, R. A. Sanchez and G. B. Triplett (eds.) Multiple cropping. ASA Spec. Pub. No. 27. American Society of Agronomy, Madison, Wisconsin.

Beets, W. C. 1982. Multiple cropping and tropical farming systems. Westview Press Inc., Boulder, Colorado.

Benjasil, V. and A. N. Lampang. 1984. Soybean in cropping systems in Thailand: some technical and socio-economic aspects. *In* Proc. International symposium on soybean. Asian Vegetable Research and Development Center, Tainan, Taiwan. (in press)

Brown, C. M. and D. W. Graffis. 1976. Intercropping soybeans and sorghum in oats. Illinois Res. 18(2):3-4.

Chan, L. M., R. R. Johnson and C. M. Brown. 1980. Relay intercropping soybeans into winter wheat and spring oats. Agron. J. 72:35-39.

Crookston, R. K. and D. S. Hill. 1979. Grain yield and last equivalent ratios from intercropping corn and soybeans in Minnesota. Agron. J. 71:41-44.

Elmore, R. W. and J. A. Jackobs. 1984. Yield and yield components of sorghum and soybeans of varying plant heights when intercropped. Agron. J. 76:561-564.

Galal, S., Jr., L. H. Hinidi, A. F. Ibrahim and H. H. El-Hinnaway. 1974. Intercropping tolerance of soybean in different local corn stalk (*Zea mays* L.). Z. Acker Pflanzanbau. 139(2):135-145.

Gomez, A. A. and K. A. Gomez. 1983. Multiple cropping in the humid tropics of Asia. International Development Research Centre, Ottawa, Canada.

IRRI. 1976. Annual report for 1975. International Rice Research Institute, Los Banos, Philippines.

IRRI. 1983. Annual report for 1982. International Rice Research Institute, Los Banos, Philippines.

McBroom, R. L., H. H. Hadley, C. M. Brown and R. R. Johnson. 1981. Evaluation of soybean cultivars in monoculture and relay intercropping systems. Crop Sci. 21:673-676.

Pendleton, J. W., C. D. Bolen and R. D. Seif. 1963. Alternating strips of corn and soybean vs. solid plantings. Agron. J. 55:293-295.

Radke, J. K. and W. C. Burrows. 1970. Soybean plant response to temporary field windbreaks. Agron. J. 62:424-429.

Radke, J. K. and R. T. Hagstrom. 1973. Plant-water measurements on soybeans sheltered by temporary corn windbreak. Crop Sci. 13:543-548.

Shen, L. 1984. Yield response of corn and soybean strip intercropping in different row directions. M.S. Thesis, Iowa State University Library, Ames.

Snaydon, R. W. and P. M. Harris. 1979. Interactions below ground—the use of nutrients and water. p. 188-201. *In* R. W. Willey (ed.) International workshop on intercropping. International Crops Research Institute for the Semi Arid Tropics, Hyderabad, India.

Wahua, T. A. T. and D. A. Miller. 1978. Relative yield totals and yield components of intercropped sorghum and soybeans. Agron. J. 70:287-291.

DOUBLE CROPPING

C. E. Caviness and F. C. Collins

Double cropping usually is defined as growing two successive crops on the same land during one year. Double cropping soybean after winter small grains occupies the largest area given to this practice, but some area is double cropped after vegetables or other crops that are harvested in late spring or early summer. Some producers in the southeastern United States are interested in double cropping soybean behind soybean, actually producing two soybean crops per year, but this area is small.

In double cropping systems of small grains and soybean, the small grains may be utilized in various ways, such as for grain, silage, hay, grazing, or green chop. But the largest is the system where the small grains, such as wheat (*Triticum aestivum* L.), oat (*Avena sativa* L.) or barley (*Hordeum vulgare* L.), are harvested for grain and soybean is grown as a second crop for grain. This is the principal system that will be discussed.

The major advantages for double cropping are (a) increased profits resulting from better utilization of climate, land, and other resources; (b) reduced soil and water losses by having the soil covered during most of the year with a plant canopy; and (c) more intensive land use and better utilization of machinery, labor, and capital investments. It is well known that double cropping requires careful management for success, and this has been one of the major factors that has limited its use in certain areas.

An extensive review of double cropping soybean after small grains was presented by Lewis (1976). He pointed out that double cropping in the United States has proven effective under the longer growing season in the Southern and Coastal Plains States. In recent years, a substantial increase has occurred in double cropping in the southern part of the Corn Belt in states such as Kentucky, Illinois, Indiana, and Ohio. Soybean also is double cropped with small grains in many other areas of the world, with sizable acreages in Brazil and Argentina.

DOUBLE CROPPING SYSTEMS

Data show that conventionally planted soybean generally yields 15 to 20% more than double-crop soybean, but the total net income usually is higher from the double crop production system (Herbek, 1974; Hinkle, 1975; Swearingin, 1973). A large number of different systems have been used, such as interseeding soybeans in immature wheat, seeding in killed wheat, seeding no-till in wheat, or burning wheat straw and seeding directly in undisturbed soil or in a prepared seedbed. In general, the data indicate that the most desirable system is one where the wheat is harvested and soybeans are planted immediately after the wheat is harvested. The method of planting generally depends upon soil and climatic conditions at the time the small grain is harvested.

Data from many parts of the United States and other countries show that double cropping of small grains and soybean generally is a favorable practice, but the specific system varies greatly depending on soil series and climatic conditions (Crabtree and Rupp, 1980; Sanford, 1979; 1983; Touchton and Johnson, 1983). However, it is generally agreed that (a) soybean must be planted as soon as possible after small grain harvest to produce high yields; (b) planting practices should be used that result in a good soybean stand; and (c) straw should be managed so that weeds can be controlled and damage from diseases and insects is minimized.

Planting Date

Most research in areas where double cropping is practiced shows that yields of soybeans planted after 10 to 15 June usually are reduced (Table 1). Where soybean cannot be planted this early in a double crop system, it is imperative that it be planted as soon as possible after small grain harvest to obtain high yields. This may be accomplished by planting no-till directly into the stubble, or in a burned field after harvest. Some delay may occur when land is prepared after small grain harvest, but this may be required under weedy conditions or other situations.

The delay in planting may also be reduced by growing an early maturing cultivar of small grain. An early wheat cultivar, 'Doublecrop', developed in Arkansas permits soybeans to be planted about 5 days earlier than with other wheat cultivars. There appears to be a good potential for development of earlier small grain cultivars for double cropping as one way to increase soybean yields. Earliness in wheat also can be achieved by (a) proper fertilization (fall N plus early spring N); (b) early planting date; and (c) harvesting at high moisture.

Yield of double crop soybean also can be increased under most conditions by close row spacing (less than 60 cm) (Boquet and Walker, 1984; Lewis and Phillips, 1976). Data generally show that soybean must form a canopy by the time of flowering to produce maximum yields. Under most conditions soybean does not form a canopy in conventional row widths when planted after about 15 or 20 June; therefore, closer spacing generally will result in higher yields of double crop soybeans (Table 2). The actual width needed to obtain canopy closure is dependent upon many variables such as soil fertility, moisture, soil series, cultivar, etc.

Interseeding in Standing Wheat

Interseeding soybeans into immature wheat has been termed "relay intercropping." This practice is used to some extent in the Midwest where the system allows double

Table 1. Effect of planting date on soybean performance, Stuttgart, Arkansas (3-year average).[a]

Planting date	Yield	Plant height
	kg/ha	cm
15 May	2623 r	94 r
15 June	2757 r	97 r
15 July	2421 s	66 s

[a]Data are given as an average for Forrest, Davis, and Bragg. Rows are spaced 81 cm apart at the May and June dates of planting and 46 cm apart at the July date. Means followed by the same letter do not differ significantly at the 5% level of probability.

Table 2.   Influence of row spacing, plant population, and planting date on Bragg soybean, Stuttgart, AR (3-year average).

Row width	Population density	Yield[c]	Lodging score[b]
cm	plants/m	kg/ha	
Recommended planting date[a]			
91	21.6	2287 r	2.3
15	6.9	2005 r	3.6
15	9.2	2219 r	3.9
15	11.1	2085 r	4.2
15	17.7	2085 r	4.2
Late planting date[a]			
91	24.6	1480 r	1.1
15	6.9	1749 rs	1.1
15	12.1	2018 s	1.4
15	14.8	1950 s	1.8
15	18.0	1950 s	1.8

[a]Recommended planting date was last week in May each year and late date was during first week in July.

[b]1 = negligible lodging and 5 = severe lodging.

[c]Means followed by the same letter are not significantly different at the 5% level of probability.

cropping of soybean and wheat, which would otherwise be impractical because of the relatively short growing season. The length of the growing season in southern USA is ample for doublecropping soybean after wheat, but late planting generally reduces soybean yields. Interseeding soybean has been successful in areas of the midwest United States, and recent research indicates it may be desirable in southern USA under some conditions (Chan et al., 1980; Jeffers and Triplett, 1978).

Excellent stands and relatively high yields have been obtained by interseeding soybeans into wheat with ground equipment; however, the planting equipment also damages the wheat. Over the past four years in Arkansas, wheat yields have been severely reduced using a 3.66 m drill and tractor (Table 3). The amount of wheat loss was related to the date of interseeding; plants that were more mature recuperated less than younger plants. The use of skips or "tramlines" for running the tractor results in less damage to the wheat (Crabtree and Rupp, 1980). The use of narrow tractor tires and wider seeding equipment also reduces the wheat loss.

Even though substantial wheat losses may result from interseeding, the data indicate that, during most years, interseeding was more profitable than double cropping,

Table 3.   Yield loss in wheat from interseeding soybeans in Arkansas (3-year average).

Date soybeans seeded	Growth stage of wheat	Loss in wheat yield
		%
Mid-April	Early boot	28
Early May	Late flowering	30
Late May	Soft dough	40

where irrigation was not available. The advantage of interseeding is to obtain a soybean stand when soil temperature and moisture are optimum. If irrigation is available, there is little justification in southern USA for interseeding. However, in the Midwest where the growing season is short, interseeding appears to offer the most potential.

STRAW MANAGEMENT

Management of the straw after harvest of small grains is of prime consideration in most double cropping systems. A 4036 kg/ha small grain yield may produce from 4036 to 8071 kg/ha straw. Occasionally, this straw is removed from the field before planting soybeans, but in most cases it remains on the field. If not managed properly, it can be a major problem during development of the soybean crop (Boquet and Walker, 1984; Sanford, 1983).

No-till Planting

By leaving the straw chopped and evenly spread on the field and using one of the currently available no-till planters, one can plant soybeans immediately after wheat harvest. Conventional tillage frequently delays planting. This problem is overcome in the no-till system and provides a great advantage, particularly in the northern fringe of the double cropping areas. Another important advantage of the no-till system is soil and water conservation. No-till planting should have high priority in rolling hills where soil erosion may occur.

Stubble height has been shown to affect soybean yield and seedling emergence in no-till double cropping systems (Boquet and Walker, 1984). Seedling emergence at stubble heights of 15 and 23 cm was significantly greater than at stubble heights of 8 and 38 cm (Table 4). The 46-cm stubble height tended to reduce seedling emergence to less than that obtained in 15- and 23-cm stubble. Yield began to decline when the stubble height was 30 cm, but was not significantly affected until the stubble height was 38 cm. The reduced yields from the 38- and 46-cm stubble were not caused entirely by low plant population, because a similar plant population in the 8-cm stubble produced significantly higher yields. Observations indicated that reduced yields from planting in the tall stubble resulted primarily from etiolation of soybean seedlings and less effective weed control. The etiolated seedling in the tall stubble tended to

Table 4. Comparison of yield and seedling emergence of no-till double-crop soybeans planted in wheat stubble of different heights, averaged over 2 years, Northeast Research Station, St. Joseph, LA (Boquet and Walker, 1984).

Stubble height	Yield	Plant population[a]
cm	kg/ha	plants/ha
8	2959	89,944
15	2959	129,728
23	2959	129,480
30	2757	102,794
38	2555	86,979
46	2622	102,299
L.S.D.—0.05	269	35,830

[a]Planted 321,230 seeds/ha.

lodge more, and weed infestations were more severe because of the difficulty of effective herbicide application.

Straw Incorporation into the Seedbed

No-till double cropping has limitations, particularly where perennial weeds exist. Also, certain soils and weather conditions may dictate that a seedbed be prepared by working the straw into the soil. Generally, better stands can be assured in some soils by preparing a seedbed, and weed problems can be reduced through tillage (Sanford et al., 1973; Boquet and Walker, 1984). This alternative has several disadvantages:

1. It requires extra fuel and time; normally two additional trips over the field with a disk are necessary.
2. Cultivation is hindered by residue that often accumulates on the cultivator shank.
3. Soil-borne seedling diseases of soybean may be more severe. The primary pathogens are *Sclerotium rolfsii, Rhizoctonia solani*, and *Fusarium* spp. Severity is associated with the amount of residue and weather conditions at planting. High temperatures and humidity are conducive to development of these diseases.
4. Research at the University of Arkansas has shown that phytotoxins are present in wheat residue, which can slow soybean growth and potentially reduce soybean yields (Collins and Caviness, 1978). These substances appear to be phenolic acids and are present in the straw and in the root residue. Soybean growth during the first 4 to 6 weeks can be reduced 20 to 30% in soil containing 2% wheat straw (Table 5). The growth retardant appears to dissipate after about 4 to 6 weeks of decomposition. Results of recent studies indicate that soybean cultivars vary in their reaction to the toxic material in wheat straw with Davis and Centennial being much more tolerant than Dare and Mack.

Limited data collected in 1982 indicate that yields may be reduced slightly under some conditions, but this is really a minor factor in successful double cropping (Table 6).

Table 5. Effect of wheat straw on dry matter production of 6-week-old soybean plants in Arkansas, 1982.[a]

Cultivar	Soybean growth		Relative growth
	Wheat straw	Check	
	g/4 plants		%
Davis	3.17	3.83	82
Centennial	4.38	5.50	79
Lee 74	3.96	5.64	70
Bragg	3.76	5.60	67
Forrest	2.77	4.40	62
Hood 75	2.49	4.19	59

Treatment L.S.D. (.05) = .71 g.

[a]Wheat straw was added to the soil at the rate of 2% on a dry weight basis.

Table 6. Comparison of soybean cultivars double cropped with wheat and single cropped in Arkansas, 1982.[a]

Cultivar	After fallow	After wheat	Percentage of fallow
		– kg/ha –	
Forrest	2757	2152	78
Bedford	2556	1681	66
Davis	2623	2421	93
Centennial	2690	2757	103
Braxton	2219	2219	100
Wright	2287	2219	97

[a]Planted on 15 June in rows 97 cm apart.

Straw Burning Alternatives

In Arkansas and surrounding states, a sizable area of soybean is planted no-till or with minimum tillage on heavy clay soils after the straw is removed by burning (Hinkle, 1975; Boquet and Walker, 1984). On most other soils, a seedbed is usually prepared by disking and harrowing after burning. When the straw is dry, it burns quickly and leaves the field almost completely bare. As a consequence, fewer tillage operations are required to prepare a seedbed, heat from the fire destroys many weeds and kills seeds on the soil surface, seedling diseases are not as severe, ease of cultivation is enhanced, and the phytotoxic effect of the wheat residue is minimized.

Burning has many disadvantages and every effort should be made to double crop without resorting to this practice. Smoke from burning may cause excessive pollution, often is hazardous to traffic, and can be an irritant to persons suffering from asthma and allergies.

In addition to the health hazards, burning for several years may reduce soil productivity by organic matter depletion and increased soil compaction. It is well-recognized that burning is not a desirable agronomic practice and should be used only in those situations where erosion is not a problem and satisfactory production cannot be obtained where straw remains on the soil.

OTHER CONSIDERATIONS

Cultivar Selection

Research generally has shown that highest yields of soybeans in a double cropping program are from cultivars that take advantage of the remaining growing season, but mature before frost. Examples of maturity groups for different areas are given in Table 7. Boerma (1978) in Georgia has developed cultivars specifically adapted for double cropping and has shown that the Duocrop cultivar generally produces higher yields than other cultivars when planted extremely late. Research in other southern and southeastern states is designed to develop cultivars that are specifically adapted for double cropping.

Weed Control

The success of double cropping in the southern United States frequently is associated with the ability to control weeds (Graves et al., 1980). There are differences in

Table 7. Soybean maturity groups adapted to double cropping systems.

State	Soybean maturity groups		
Southern Illinois	II	III	IV
Kentucky	IV	V	VI
Arkansas	V	VI	VII
Georgia	VI	VII	VIII

philosophies and opinions among researchers, extension personnel, and growers regarding the cropping and weed control practices that should be used. Many weed control options are available and there is general agreement that a system can usually be found for the weed complex that is present in any particular field. A discussion of the possible weed control options, however, is beyond the scope of this paper.

NOTES

C. E. Caviness, Department of Agronomy, University of Arkansas, Fayetteville, AR 72701. F. C. Collins, Rohm and Haas Seed, Inc., P.O. Box 420, Marion, AR 72364.

REFERENCES

Boerma, H. R. 1978. Breeding soybeans for double cropping. p. 57-62. *In* Proc. Eighth Soybean Seed Research Conf. Amer. Seed Trade Assoc., Washington, D.C.

Boquet, D. J. and D. M. Walker. 1984. Wheat-soybean double cropping; stubble management, tillage, row spacing and irrigation. Louisiana Agric. Exp. Stn. Bull. No. 760.

Chan, L. M., R. R. Johnson and G. M. Brown. 1980. Relay intercropping soybeans into winter wheat and spring oats. Agron. J. 72:35-39.

Collins, F. C. and C. E. Caviness. 1978. Growth of soybeans on wheat straw residue. Agron. Abstr. 70:101.

Crabtree, J. J. and R. N. Rupp. 1980. Double and monocropped wheat and soybeans under different tillage and row spacing. Agron. J. 72:445-448.

Graves, C. R., Tom McCutchen, L. S. Jeffery, J. R. Overton and R. M. Hayes. 1980. Soybean-wheat cropping systems: Evaluation of planting methods, varieties, row spacings, and weed control. Tenn. Agric. Exp. Stn. Bull. 597.

Herbek, J. 1974. Double cropping. p. 70-75. *In* Proc. No-Tillage Research Conference. Univ. of Kentucky, Lexington.

Hinkle, D. A. 1975. Use of no-tillage in double cropping wheat with soybeans or grain sorghum. Arkansas Agric. Exp. Stn. Rep., Series 223.

Jeffers, D. L. and G. B. Triplett, Jr. 1978. Management for relay intercropping wheat and soybeans. p. 63-70. *In* Proc. Eighth Soybean Seed Research Conf. Amer. Seed Trade Assoc., Washington, D.C.

Lewis, W. M. and J. A. Phillips. 1976. Double cropping in the eastern United States. p. 41-50. *In* R. I. Pappendick et al. (eds.) Multiple Cropping. Am. Soc. Agron. Spec. Pub. No. 27. Am. Soc. of Agron., Madison, Wis.

Lewis, W. M. 1976. Double cropping soybeans after winter small grains. p. 44-52. *In* L. D. Hill (ed.) World Soybean Research. The Interstate Printers and Publishers, Inc., Danville, Ill.

Sanford, J. D. 1979. Establishing wheat after soybeans in double cropping. Agron J. 71:109-112.

Sanford, J. D. 1983. Straw and tillage management practices in soybean-wheat double cropping. Agron. J. 74:1032-1035.

Sanford, J. E., D. L. Myhre and Norman C. Merwine. 1973. Double cropping systems involving no-tillage and conventional tillage. Agron. J. 65:978-982.

Swearingin, M. L. 1973. Double cropping in the Corn Belt. Better Crops with Plant Food 2:20-23.

Touchton, J. T. and J. W. Johnson. 1983. Soybean tillage and planting method effects on yield of double cropped wheat and soybeans. Agron. J. 74:57-59.

ECONOMICS OF INTERCROPPING

Luis A. Navarro and Donald L. Kass

Intercropping is: "growing two or more crops *simultaneously* on the same field". This includes cases of crops arranged in a geometric pattern (row intercropping, strip intercropping), or in no noticeable arrangement (mixed cropping), and where only part of the life cycle of each species overlaps in time (relay intercropping). It is a case of 'multiple cropping', or the growing of more than one crop on the same land in one year (Andrews and Kassam, 1976).

To discuss the economics of intercropping requires the study of its evolution and dispersion in order to interpret its benefits and importance for society. It also implies interpreting its advantages and limitations as an enterprise and as a production system that permits a rational use of resources and provides other benefits for society in the present and future. This is partly attempted in this document. Particular attention will be given to soybeans in intercropping.

HISTORY AND WORLD PERSPECTIVE

General Background

Intercropping, as a form of multiple cropping, is of widespread use in the world (Beets, 1982). However, specific documentation of its early history is not readily available. Some authors believe it was one of the primitive cropping systems used by man at the dawn of agriculture, when people began selecting from nature some useful crops and followed the patterns in which they grew (Mandel, 1968).

Before urbanization became widespread, agriculture and economics were inseparable. Because of the limited development of markets for produce, agricultural enterprises were mostly for subsistence and were a family affair. Production systems were not as varied as at present, and farmers were slow to respond to technical, economic and environmental changes. The only sources of energy were human and animal power. Soil fertility could be maintained and pest problems reduced, although not always successfully, by the use of crop rotations, animal manures, composts and ash resulting from burning existing vegetation. Mixed cropping, including intercropping, was a suitable and rational production system under those conditions. It was common on most farms (Beets, 1982).

The first Europeans to arrive in the Americas found the Indians growing mixtures of maize (*Zea mays* L.), beans (*Phaseolus vulgaris* L.) and squash (*Cucurbita* sp.) from Argentina to the Great Lakes. Recent history of multiple cropping and intercropping is better documented for Asia than for Africa and America. In Asia, development

(Wang and Yu, 1975; Beets, 1982).

Even though the history of soybean as a crop goes back to ancient times in China, when it was mostly used for human consumption, no clear reference is found to its being intercropped previous to its introduction in the West. However, the tradition of intercropping in this region makes it likely that soybean has long been intercropped there. Furthermore, Bernard (1975) states that in the central Chinese province of Kirim, virtually all soybeans are intercropped with maize. Following their introduction to the United States in the late nineteenth century (Probst and Judd, 1973), soybeans were principally grown as a forage crop and often interplanted with maize to improve silage quality (Wiggans, 1937). Thatcher (1925) reported that 56% of the soybean hectarage in Ohio was grown in mixtures with maize in 1923, but only 40% in 1924. According to Hackleman et al. (1928), nearly 60% of the 260,000 ha soybean crop in Illinois during the 1923-1926 period was mixed with maize. It would seem that as soybean changed from a forage to a grain and oil crop in the period 1920-1940 intercropping was gradually replaced by monoculture (Probst and Judd, 1973; Crookston and Hill, 1979).

General Dispersion

The present dispersion and importance of intercropping in the world are mostly associated with the agriculture of less developed countries, particularly in tropical areas (Finlay, 1974; Beets, 1982; Gomez and Gomez, 1983). There, in turn, intercropping is associated with the diversified farming systems found on the numerous small farms and also with the production of food and feed crops. An important proportion of these crops is used for consumption at the farm level. The agricultural technology found on these farms is considered traditional and sometimes archaic in relation to what is known as "modern" in agriculture. This consideration is usually extended to the different forms of intercropping, and contributes to their dismissal as a form of production with economic potential for contemporary and future agriculture. In many respects, the production and economic conditions under which intercropping is widely used are similar to those of agriculture in the past. Under those restricted conditions, intercropping becomes a rational form of using agriculture resources to benefit society.

In macro terms, the most important determinants of the comparative importance and benefits of intercropping and monocropping for a given society are the relative availability of labor, land and capital inputs for agriculture and the relative importance of the latter for the economy at large. Intercropping, in most of its modern forms, can provide more production, employment and even income per unit of land, in relation to most cases of monocropping, when a given amount of capital is used in the same environment (Finlay, 1974). Thus, traditional intercropping tends to be associated with situations of relative abundance of labor with respect to land and capital. However, different forms of intercropping are found under other conditions, including relative abundance and development of capital with respect to land and labor.

The two most common situations in which intercropping occurs are: (1) where there is abundant labor and both land and capital are limiting, as in southeast Asia and parts of Africa and Latin America; and (2) where there is abundant land and both capital and labor are limiting, as in the slash-and-burn agriculture of Africa and Latin America. The latter situation also occurs in certain tree-pasture associations and where tree crops are grown within natural forest (Von Platen et al., 1982; Van Tienhoven et al., 1982). Some forms of intercropping associated with the production of vegetables, fruits and flowers are also appearing in the situation where capital is more abundant than both land and labor, as in western Europe, Japan and parts of the U.S. Examples include the case of olives (*Olea europea*) and grapes (*Vitis* sp.) associated with other crops practiced in Mediterranean countries. This situation has led to considerable innovativeness in the development of practices that are most efficient in the use of land, since a high degree of technological infrastructure is already available. Finally, there is the situation where only capital is limiting; thus intercropping may be the cheapest way of providing shade needed by certain plantation crops, as in the cases of cocoa (*Theobroma cacao* L.) and coffee (*Coffea* sp.) (Pinchinat et al., 1976). There are other situations where land is abundant but limited for agriculture due to water restrictions, as in the case of semi-arid tropical areas with very short, rainy seasons. Intercropping of maize and sorghum (*Sorghum* sp.) or of sorghum and millet (*Pennisteum* sp.) are ways of better exploiting soil water under those conditions. The interplanting of vegetable or field crops in the establishment stage of perennials is another way of improving resource use efficiency (Pinchinat et al., 1976). Intercropping is seldom found in situations where labor is the only limiting resource, as in most of the U.S. agriculture.

The soybean is commonly intercropped with maize and sorghum in Southeast Asia (Shanmugasundaran et al., 1980). In this region the soybean is principally cropped for human consumption. This is not the situation in other regions of the world where the soybean is not regularly grown by small holders who practice intercropping. However, other legumes are frequently intercropped with cereals. It is estimated that 98% of the cowpea (*Vigna unguiculata* [L.] Walp.) grown in Africa, where it is the most important food legume, is associated with other crops. Francis et al. (1976) report that the production of beans in Colombia, Brazil and Guatemala comes 90, 80 and 73%, respectively, from associated plantings. Other food legumes frequently intercropped are pigeon pea (*Cajanus cajan* [L.] Millsp.) (Dalal, 1974) and peanuts (*Arachis hypogaca* L.) (Evans, 1960). Soybean has been intercropped in the U.S., usually for the production of forage and not uncommonly as a green manure. Other forms of relay cropping with cereals have also been attempted in the U.S. and in Brazil (Beets, 1982).

ADVANTAGES OF INTERCROPPING

The economic rationale for intercropping stems from the possibility of growing two or more crops together, diminishing their competition and increasing their supplementarity and complementarity in the use of available resources, and of their production profiles in a given area and time. Most of the characteristics of intercropping have short-run as well as long-run economic implications. They include:

a) *Better utilization of environmental resources.* Plants of different growth habits and cycles often differ in their environmental requirements or in their use of resources; therefore, many intercrops exhibit complementarity in the structure of their canopies and rooting systems. Thus, they are able to exploit light, nutrients and water more fully than monocultures. The more complete ground cover by the intercrops can reduce erosion, weed competition and moisture loss through evaporation. One of the species in the mixture may benefit from the capacity to fix nitrogen or possess mycorrhiza, which brings more phosphorus into the system. In case of failure of one of the crops because of an environmental accident, the other can exploit the resources unused by the crop that failed. This would not be possible if the crops were planted in different fields. Spread of insects and diseases in one species may be slowed down by the presence of non-susceptible species within the planting.

b) *Input and output management flexibility.* Particularly when land is limited, crops could be selected and their relative sowing and planting dates arranged so as to fit the profiles and optimize the use of available labor, inputs, implements and money per hectare during cultivation and harvesting. Similarly, the crops and planting dates could be selected to obtain a production profile and composition which would fit farm consumption requirements or market availability for the different products. This would also contribute to the improvement of nutrition and cash flow, and diminish storage losses.

c) *Risk minimization and profit maximization.* Better utilization of resources and stabilization of production, which counteract environmental variations or the attack of insects and diseases, tend to reduce the risks of production. Flexibility in the use of labor and other farm resources, plus an extended and varied supply of produce, tend to minimize marketing risks in the face of price fluctuations, as well as the risk of storage losses. Generally, a higher yield and greater gross return per hectare can be obtained by intercropping two or more crops. Sometimes this higher output can be obtained from a less than proportional addition of inputs, usually labor or specialized labor-substituting-capital. When the opportunity cost for this extra input is sufficiently small, intercropping results in an increase of net income per hectare in relation to monocropping.

Observations which reflect these advantages of intercropping are reported by many authors (Darlrymple, 1971; Willey, 1975; Kass, 1978; Beets, 1982).

ECONOMIC MEASUREMENT AND INTERCROPPING

Economic measurements are seldom reported in the literature on intercropping of soybean. This reflects the inherent complexities of intercropping, but also the general bias towards agronomic evaluations and the lack of familiarity with appropriate tools in research.

In their literature reviews, Kass (1978) and Crookston and Hill (1979) found that, in the 1920-1940 period, most of the U.S. agricultural experimental stations in states where soybeans were grown carried on research on crop mixtures. Although the main interest of these studies was forage or silage production, grain yields were sometimes reported, but no related economic analysis was given. In most cases the objective was greater production of protein, or sometimes, total digestible nutrients per hectare by intercropping soybeans in different arrangements with maize or other crops. Brown (1935) found increased soil organic matter, N and P in plots following five years of a maize-soybean mixture in comparison with maize planted alone. Wiggans (1937) stated that maize-soybean mixtures produced considerable benefits for a slight increase in costs, but his economic analysis went no further.

The most common indicator used to evaluate the advantages of intercropping is the Land Equivalent Ratio (LER). This index relates the land area needed by monocrops of all component crops to give the same productivity as a unit land area of the intercrop. It has been modified to take into consideration the time period during which the ground would be used in each case (Hiebsch, 1980). Usually the LER obtained from intercropping is greater than one. However, Crookston and Hill (1979) report data from different sites in Minnesota over a three-year period where the LER for a maize-soybean intercrop under different plant populations was seldom greater than one. This was explained by the short growing season which prevented a better temporal arrangement of the component crops. This is not the case found in most tropical areas, even though short, rainy seasons in semi-arid areas may impose similar problems (Makena and Doto, 1980). Some reported values for the LER in maize-soybean intercrops are: 1.43 in Kentucky (Kinney and Roberts, 1921); 1.32 in Tennessee (Mooers, 1927); 1.48 in the Philippines (Sastrawinata, 1976) and 1.22 in Alabama (Allen and Obura, 1983).

Other evaluations consider the trade-off between the decrease in yields of the preferred crop in intercropping compared to yields of the companion crop. On some occasions, the associated monetary value is also considered, but on fewer occasions the cost side is explicitly accounted for.

One of the main research lines associated with intercropping of soybean is related to the development and evaluation of mechanical harvesting. Beste (1976) recommended planting maize and soybeans in the same rows, as damage by topping (Weber, 1955) would be less than the problem of running machinery over the soybean rows. Strip intercropping can provide some of the benefits of intercropping while allowing mechanization (Dolezal, 1983).

Just recently, economic analysis is being incorporated as a normal component of agricultural research. The basic tools include different forms of budget analysis. Some problems are related to price projection and labor measurement under research conditions. Possibly, the most beneficial in this trend is the motivation for researchers to observe and evaluate the use of inputs and resources, which is necessary to properly identify and weigh the economic benefits and outputs of intercropping (Johda, 1979).

In an economic evaluation of 20 trials to compare the association of maize and beans with their respective monocrops in Colombia, Francis and Sanders (1978) found that monocropping was more profitable over a wide range of relative prices, but the risk was also higher. However, the probability of obtaining a consistent income with relatively lower investment was higher for the association. Profit increase projections up to 40% were obtained from trials of cassava intercropped with maize or beans in relation to the monocrop in Costa Rica (Navarro, 1978; Meneses et al., 1983). Nevertheless, these results are highly dependent on price relations. The intercropping of soybean with castorbean (*Ricinus* sp.) was the only cropping pattern that failed to produce a profit in four years of experiments in India (Reddy et al., 1965). On the other hand, in the Philippines, the intercropping of ginger (*Zingiber officinalis* L.) with soybean and ginger with soybean and vegetables was more profitable than ginger planted alone (Paner, 1975).

Intercropping has also proved to be promising in relation to monocropping when the energy intake and production is budgeted (Kass, 1976).

Clearly, further work is needed to develop appropriate tools and procedures to fully determine and evaluate short-run as well as long-run economic costs and benefits of intercropping. This is one possible line of research. Others include the development of appropriate machinery that will determine the future use of intercropping under a capital intensive agriculture.

PERSPECTIVES FOR RESEARCH ON INTERCROPPING

Since intercropping has been considered a traditional practice, not suited to mechanization and other means of increasing efficiency, it did not receive much attention from investigators in the 1940-1970 period when rapid increases of productivity were the main research goals. However, with the realization of the limitation of this approach and the introduction of the "farming systems" approach to research, it became clear that present knowledge and research technology were often inadequate for dealing with problems of fertilization, weed control and cultivar improvement, and for improving efficiency in traditional systems.

A new interest in understanding traditional farming practices and investigation efforts for upgrading them by utilizing their basic advantages are part of the agendas in International Research Centers including IRRI, IITA, CIAT, ICRISAT, CIMMYT, AVRDC, and ICARDA, and also in some Regional Research Centers, including CATIE in Latin America and different national research and extension institutions in Asia, Africa and Latin America. Many U.S. universities also are making efforts of this kind as part of their own international programs, and also as part of different joint efforts developed under the Title XII programs. Furthermore, several international institutions supporting agricultural development are encouraging these efforts.

NOTES

L. A. Navarro and D. L. Kass, Senior Specialists in Agricultural Economics and Soil Management, respectively, Plant Production Department, Tropical Agricultural Research and Training Center (CATIE), Turrialba, Costa Rica.

REFERENCES

Allen, J. and R. Obura. 1983. Yield of corn, cowpea and soybean under different intercropping systems. Agron. J. 75:1005-1009.

Andrews, A. and A. Kassam. 1976. The importance of multiple cropping in increasing world food supplies. p. 1-10. In R. I. Pappendick, P. A. Sanchez and G. B. Triplett (eds.) Multiple cropping, Am. Soc. Agron. Spec. Publ. 27.

Beets, W. 1982. Multiple Cropping and Tropical Farming Systems. Westview Press, Boulder, Colorado.

Bernard, R. 1975. Soybeans in the People's Republic of China. Soybean News 26(2):1-4.

Beste, C. 1976. Co-cropping sweetcorn and soybean. Hort. Sci. 11:236-238.

Brown, H. 1935. Effect of soybean on corn yields. Louisiana Agric. Exp. Stn. Bull. 265.

Crookston, R. and D. Hill. 1979. Grain yields and land equivalent ratios from intercropping corn and soybeans in Minnesota. Agron. J. 71:41-44.

Dalal, R. 1974. Effects of intercropping maize with pigeon pea on grain yield and nutrient uptake. Exp. Agric. 10:219-224.

Darlrymple, D. 1971. Survey of multiple cropping in less developed nations. For. Econ. Dev. Bull. 12. USDA-AID, Washington, D.C.

Dolezal, J. 1983. Strip-intercropping corn and soybeans. Crops and Soils Magazine 36(2):18-20.

Evans, A. 1960. Studies of intercropping. Part I. Maize or sorghum with groundnuts. E. Afr. Agric. and For. J. 26:1-10.

Finlay, R. 1974. Intercropping soybean with cereals. INTSOY. Ser. No. 6:77-85. U. of Illinois, Urbana, Ill.

Francis, C., C. Flor and S. Temple. 1976. Adapting varieties of intercropped systems in the tropics. p. 235-253. In R. I. Pappendick, P. A. Sanchez and G. B. Triplett (eds.) Multiple cropping, Amer. Soc. Agron. Spec. Publ. 27.

Francis, C. and J. Sanders. 1978. Economic analysis of bean and maize systems: Monoculture versus associated cropping. Field Crop Research Rep. CIAT, Cali, Colombia.

Gomez, A. and K. Gomez. 1983. Multiple cropping in the Humid Tropics of Asia. IDRC-176 e. Ottawa, Ont., Canada.

Hackleman, J., O. Sears and W. Burlison. 1928. Soybean production in Illinois. Univ. of Ill. Agr. Exp. Stn. Bull. 310:465-531.

Hiebsch, C. 1980. Principles of intercropping: Effects of nitrogen fertilization, plant population and crop duration on equivalency ratios in intercrop versus monoculture comparisons. Ph.D. Dissertation. North Carolina State Univ. Library, Raleigh, N.C.

Jodha, N. 1979. Intercropping in traditional farming systems. *In* Proc. of the International Workshop in Intercropping. ICRISAT, Hyderabad, India.

Kass, D. 1976. Simultaneous polyculture of tropical food crops with special reference to the management of sandy soils of the Brazilian Amazon. Ph.D. Dissertation. Cornell Univ. Library, Ithaca, N.Y.

Kass, D. 1978. Polyculture cropping systems: review and analysis. Cornell Int. Agric. Bull. 32. Cornell University, Ithaca, N.Y.

Kinney, E. and G. Roberts. 1921. Soybeans. Kentucky Agr. Exp. Stn. Bull. 232.

Makena, M. and A. Doto. 1980. Soybean-cereal intercropping and its implications in soybean breeding. *In* Proc. Second Symposium on Intercropping in Semi-Arid Areas. IDRC, Canada.

Mandel, E. 1968. Marxist Economic Theory. Modern Reader, New York.

Meneses, R., L. Navarro and R. Moreno. 1983. Efecto de diferentes poblaciones de maíz (*Zea mays*) en la producción de raíces de yuca (*Manihot esculenta*) al cultivarlos en asocio. II Aspectos Económicos. Turrialba 33(3):291-296.

Mooers, C. 1927. Influence of cowpea crop on yield of corn. Univ. Tennessee Agr. Exp. Stn. Bull. 137.

Navarro, L. 1978. Evaluación económica de algunos sistemas incluidos en el experimento central durante el período 1976-1977 (mimeo). DPV-CATIE. Turrialba, Costa Rica.

Paner, V. 1975. Multiple cropping research in Philippines. p. 310-316. *In* Proc. of the Cropping Systems Workshop. IRRI, Los Baños, Philippines.

Pinchinat, A., J. Soria and R. Bazan. 1976. Multiple cropping in Tropical America. p. 51-62. *In* R. I. Pappendick, P. A. Sanchez and G. B. Triplett (eds.) Multiple Cropping. Am. Soc. Agr. Spec. Publ. 27.

Probst, A. and R. Judd. 1973. Origin, U.S. history and world distribution. p. 1-15. *In* B. E. Caldwell (ed.) Soybeans improvement, production and use. Am. Soc. Agron. Monograph 16.

Reddy, G., C. Rao and P. Reddy. 1965. Mixed cropping in castor. Indian Oilseeds J. 9:310-316.

Sastrawinata, S. 1976. Nutrient uptake, insect disease, labor use, and productivity characteristics of selected traditional intercropping patterns which together affect their continued use by farmers. Ph.D. Dissertation, Univ. of Philippines, Los Baños.

Shanmugasundaran, S., G. Kuo and A. Nalampang. 1980. Adaptation and utilisation of soybeans in different environments and agricultural systems. p. 265-277. *In* R. J. Summerfield and A. H. Bunting (eds.) Advances in legume science. Royal Botanic Gardens, Kew.

Thatcher, L. 1925. The soybean in Ohio. Ohio Agr. Exp. Stn. Bull. 384.

Van Tienhoven, N., J. Icaza and J. Lageman. 1982. Farming systems in Jinotega, Nicaragua. Tech. Rep. No. 31. CATIE, Costa Rica.

Von Platen, H., G. Rodriguez and J. Lageman. 1982. Farming systems in Acosta-Puriscal. Tech. Rep. No. 30. CATIE, Costa Rica.

Wang, Y. and T. Yu. 1975. Historical evolution and future prospect of multiple crop diversification in Taiwan. The Philippine Economic J. XIV:26-46.

Weber, L. 1955. Effects of defoliation and topping simulating hail injury to soybean. Agron. J. 47:262-266.

Wiggans, R. 1937. Soybeans in the Northeast. Am. Sci. Agron. J. 29:227-235.

Willey, R. 1975. Intercropping—Its importance and research needs. Part 1. Competition and yield advantages. Field Crop Abstr. 32:1-10.

MACHINERY FOR CROPPING SYSTEMS

Richard R. Johnson

THE PAST

The history of U.S. soybean production has been one of steady improvement. It is difficult to identify breakthroughs, yet from 1924 to 1983 area harvested for grain has increased from about 0.2 to 25 million hectares, average yields have increased from 0.7 to 1.7 t/ha, and draft horses and mules have been replaced by tractors and self-propelled combines. Viewed over a 60-year period, these are major changes, but no single change occurred in a 1-, 2-, or 3-year period, and no single discovery accounted for any one change.

Soybean producers perhaps most recognize the machinery industry for large gains that have been made in farm equipment productivity. USDA (1982) labor efficiency statistics show that labor efficiency has greatly improved throughout this century. Whether expressed on an area or crop yield basis, man-hours used to produce soybeans have decreased greatly (Figure 1). Although the decrease in man-hours per unit of production has slowed during the last two decades, the percentage increase in efficiency for each decade compared with the previous decade has remained about constant.

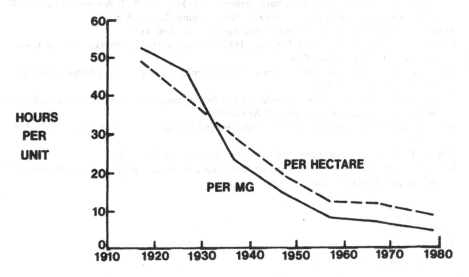

Figure 1. Man-hours used to produce 1 hectare or 1 Mg (megagram, or t) of U.S. soybeans (USDA, 1982).

Past presentations on farm equipment have emphasized advances in either planting or harvesting equipment. Many of these improvements resulted not only from the need to increase productivity, but also from the need to reduce harvest losses and to improve capabilities of planting in narrower rows. However, another significant change has been the trend toward greater use of conservation tillage, a trend that affects machine productivity, use and selection. Conservation tillage involves an entire cropping system, an important part of which is the machinery system used. My discussion will emphasize important equipment considerations. I emphasize at the outset that conservation tillage is part of an overall concept, not a specific system. The diverse range of soils, climate and cropping systems dictate the need for several different equipment types that can be used to meet the needs of the conservation tillage concept.

WHERE WE ARE

Before 1960 clean tillage was used because it assisted in economic control of weed, insect and disease pests. After World War II, technology in the chemical industry began to furnish pesticides that provided alternate methods of pest control. Chemical pesticides, along with improved seeding equipment, have allowed production agriculture to adopt what is now becoming known as conservation tillage. Conservation tillage systems are designed to provide a rough, residue-covered soil surface that is resistant to wind and water erosion. No-tillage represents the extreme in conservation tillage, since seed is planted in a previously undisturbed soil, and the only tillage used is that necessary to place seed in the soil. Less extreme forms of conservation tillage are usually referred to as reduced tillage, since the entire field is often tilled, but in such a way that crop residue is still present on the soil surface at planting time. In a review of soil erosion control with conservation tillage, Laflen et al. (1981) have emphasized that it is generally the percentage residue cover, rather than the particular tillage system, that reduces erosion.

In addition to soil conservation, a number of other reasons are advanced to promote conservation tillage. These include conservation of labor, moisture, energy, and money. Compared to clean tillage, the reduced number of tillage operations in conservation tillage systems generally do conserve soil, labor, and moisture. However, cost and energy reductions associated with reduced machinery operations are sometimes offset by an increased need for pesticides and fertilizers (Siemens and Oschwald, 1978; Lockeretz, 1983; Jolly et al., 1983). Soybean yields obtained with different tillage systems have differed (Siemens and Oschwald, 1978; Bauder et al., 1979; Nave et al., 1980; Colvin and Erbach, 1982; Touchton and Johnson, 1982). In general, drought-prone and well-drained soils yield more with conservation tillage because of increased moisture conservation. However, fine-textured and poorly drained soils often yield less with conservation than with clean tillage, largely due to wetter and cooler soils that delay planting and reduce early season crop growth. Rotating corn (*Zea mays* L.) and soybean has helped eliminate yield reductions experienced in continuous corn when conservation tillage has been used (Triplett and Van Doren, 1977; Erbach, 1982; Mulvaney, 1984).

Use of conservation tillage in soybean production has steadily grown. No-Till Farmer (1983) has surveyed state agronomists of the Soil Conservation Service each year since 1972. In 1972, about 2, 12 and 86% of the soybean area was reported to be in the no-, reduced and clean tillage categories, respectively (Table 1). By 1982, about 7, 38 and 55% of the soybean area was in these same categories. Of the 1982

Table 1. Use of conservation tillage in U.S. soybean production compared to all planted crops (No-till Farmer, 1983).

Crop	Year	No tillage	Reduced tillage	Clean tillage
		\- % of planted area \-		
All soybean crops	1972	2	12	86
	1977	4	20	76
	1982	7	38	55
Doublecrop soybeans	1982	22	39	39
All crops	1982	4	32	65

soybean area that was no-till planted, nearly 80% was double-cropped soybean. Current use of conservation tillage for soybean production exceeds the average use in all crops planted. A 1982 Office of Technology Assessment (OTA) report suggests that USDA estimates of conservation tillage use are somewhat lower than those of No-Till Farmer, but follow the same general trend. The OTA report projects that 75% of U.S. cropland will eventually be in some form of conservation tillage (Figure 2), but cites other estimates ranging from 50 to 84% adoption.

TILLAGE CONSIDERATIONS

An important concept of conservation tillage is the mulch of residue left on the soil surface. Table 2 is from a review of Colvin et al. (1981) and shows the amount of residue remaining on the surface after a single tillage pass in different cropping situations. The previous crop, time of tillage and sequence of tillage events all affect residue remaining. Speed, tillage depth, and ground engaging tools used also affect

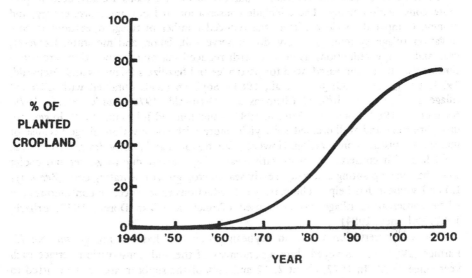

Figure 2. The proportion of planted cropland using conservation tillage in the past and projections for completion of adoption. Adapted from Office of Technology Assessment (1982).

Table 2. Percentage pretillage residue cover remaining after a single tillage pass (Colvin et al., 1981).

Tillage implement	Fall without previous tillage		Spring without previous tillage		Spring following previous tillage	
	Average	Range	Average	Range	Average	Range
— Percentage of pretillage surface cover after corn —						
Moldboard plow	4	0-10	7	5-10	–	–
Disk	84	–	50	42-73	80	46-100
Chisel	56	40-85	56	44-68	–	–
Field cult.	–	–	–	–	84	–
Till plant (sweep)	–	–	62	59-66	–	–
Plant double-disk opener)	–	–	90	82-100	80	–
— Percentage of pretillage surface cover after soybean —						
Moldboard plow	2	–	3	–	–	–
Disk	–	–	–	–	58	56-60
NH$_3$ knife on 762 mm centers	–	–	39	27-54	44	43-45
Chisel	14	–	28	25-31	115	106-130
Till plant (sweep)	–	–	–	–	74	73-76
Plant (double-disk opener)	–	–	81	70-94	100	76-113

amounts of residue left by a single pass of a particular tillage machine, and account for the ranges given for each implement in Table 2. Some companies currently market over 30 different sweeps, shovels or spikes for use on chisel plows (Johnson, 1982). Sweeps tend to incorporate small amounts of residue, whereas spikes and twisted shovels tend to incorporate intermediate and large amounts of residue, respectively. Some of the ranges for residue left on the surface in Table 2 appear to be conservative and will likely change as tillage machines are modified. For instance, moldboard plows are currently available that can be manually or hydraulically adjusted to vary width of cut from 36 to 61 cm per bottom. When operated at narrow widths, these plows can leave up to 25% residue cover in corn stubble, but will incorporate most residue if adjusted to the widest cut (Johnson, 1982). Thus, in conservation tillage, the manner in which a machine is equipped and operated can be as important as selection of the particular implement.

Compared with corn, soybean produces residue that is less in quantity and is subject to more rapid decomposition. Soil erosion following soybean is greater than following corn (Siemens and Oschwald, 1978; Laflen and Moldenhauer, 1979). Thus, where soybean is grown in rotation, tillage systems used after the soybean crop may be more important for erosion control, because of the small amount of residue present, than those used to prepare for the soybean crop. As shown in Table 2, a given tillage tool often incorporates a greater percentage of soybean than corn residue. Following the soybean crop, delaying all tillage until spring provides the most effective erosion control. In the corn-soybean rotation, anhydrous ammonia is often applied as the nitrogen fertilizer for corn. Anhydrous knife applicators may leave less surface residue cover than some other tillage tools and should be considered as a tillage tool when managing residue (Table 2).

There is a trend in the farm equipment industry to provide multicomponent tillage machines. For example, disk gangs have been combined with chisel shanks, field cultivator shanks and/or harrow attachments. These machines can often reduce the number of field passes. In addition, the strengths of each component can have a synergistic effect. For example, a disk gang in front of sweeps can size and orient crop residue so that it flows through the machine while the sweeps and harrow attachments both till soil and return residue to the soil surface. The net result can be a tillage machine that both maintains surface residue, yet thoroughly tills the soil.

Some soybean herbicides must be soil incorporated and others tend to provide more consistent weed control if incorporated. Thompson et al. (1981) have provided an extensive review of the incorporation capabilities of several different types of tillage tools. As in residue management, successful herbicide incorporation is often as dependent on how the machine is equipped and operated as on the general type of implement used. Several implement companies offer spray equipment as a part of the tillage system. For example, spray tanks and nozzle systems can be purchased as a component or an auxillary attachment to the tillage implement. Sufficient fore-aft spacing is also being provided within some multi-component tillage tools to allow alternate locations for spray nozzles. In fields with uneven soil surfaces, it may be desirable to locate nozzles after a section of ground engaging tools that have leveled the soil surface. In level field surfaces, optimum nozzle location may be in front of the machine.

PLANTING EQUIPMENT

Throckmorton (1980) reviewed machines available for solid seeded, intermediate, and wide row planting. He noted that cost restrictions have largely restricted solid seeded (15 to 25 cm widths) planting to various types of drills. Intermediate row widths (30 to 70 cm) can be seeded with either drills or row crop planters. Row widths greater than 70 cm are almost always seeded with row crop planters. Colvin and Erbach (1982) reported that any one of several different drills could be used to plant successfully in very narrow row widths under reduced tillage conditions.

No-till seedbeds often have an abundant quantity of surface residue and are firmer than reduced tillage seedbeds. No-till doublecrop seedbeds are generally the most adverse because the small grain crop depletes surface soil moisture and leaves an abundance of fresh residue. Machines differ in their capacity to plant in the wide range of no-till seedbeds. In general, row crop planters designed to plant in row widths wider than 30 cm have the capability of no-till planting in a wider range of seedbeds than grain drills designed to plant in row widths of 15 to 25 cm. Most planters capable of operating under no-till conditions use some type of coulter to open the soil for the seeding device. Each coulter requires several hundred kg of down-force to guarantee penetration in dry, firm seedbeds. For example, several no-till planters supply 200 to 300 kg per planter unit, but at least one manufacturer offers a drill that supplies up to 550 kg per planter unit. Heavy-duty drills with close row spacings are available, but require more planting units and weight—both requirements that increase machine cost. A second option to open a seed furrow is to use powered tillage blades, a factor that also increases machine cost. Thus, heavy-duty drills capable of no-till planting under adverse conditions are expensive per unit of width.

A number of lighter duty drills are also available with coulters. These machines have limited ability to no-till plant, but are effective in fields where some full-width tillage has been conducted. In looser soils or fields where irrigation can be used to moisten the topsoil, these machines can serve effectively as no-till planters. Several row crop planters capable of no-till planting in row widths of 36 to 50 cm have recently been introduced. Compared with no-till drills, these machines can often successfully operate in adverse seedbeds at a lower cost per unit width.

Selection of proper coulter type is important for no-till planting. Smooth coulters require the least down-pressure for penetration, but prepare a very narrow furrow that must be followed by an aggressive furrow opener. Rippled coulters have a straight sharp edge, but ripples located beyond the coulter edge do some limited seedbed preparation. Fluted coulters have a curved edge that prepares a seedbed 2 to 5 cm in width. Smooth or rippled coulters generally work better in surface residue and cover a wider range of soil conditions than do fluted coulters. Compared to fluted coulters, smooth or ripple coulters: a) require less weight to penetrate hard, dry soil, b) incorporate less crop residue into the seed zone, and c) operate at faster speeds and in wetter soils without removing soil from the seed zone. The wider seed zone prepared by fluted coulters helps decrease misalignment problems with the seed opener and is a factor that can be especially advantageous when planting contoured rows.

CULTIVATION AND HARVESTING

Cultivation remains an effective low-cost method of weed control. Several companies offer cultivators capable of operating in large amounts of surface residue, including fields that were no-till planted. Cultivators used for reduced tillage situations generally have more clearance between sweeps and use a shank that requires more force to cause tripping. No-till cultivators typically have a lead coulter followed by a single heavy-duty trip shovel. Disk hillers may be used on either reduced or no-till cultivators. Where skip-row planters are used to plant rows wider than 30 cm, row-crop cultivators are also used in narrow-row soybean production.

An important aspect of conservation tillage is uniform residue distribution behind the combine. Straw choppers are more effective in spreading residue than are straw spreaders. Where headers wider than 6 to 7 m are used, extended vanes on the straw chopper have assisted in providing uniform straw spreading.

COSTS AND RETURNS

Hanthorn and Duffy (1983) surveyed the 1980 cropping season costs and returns of clean, reduced and no-tillage soybean producers in the Midwest, Midsouth and Southeastern U.S. Herbicide use and cost differed by region, but additional herbicide applications were generally substituted for any reductions in tillage and mechanical cultivation. Insecticide use and costs were not significantly different among tillage strategies. The number of mechanical cultivations did not differ between reduced and clean tillage systems, but ranged from 1.12 to 1.79 cultivations per season among the three regions. No-tillage soybeans averaged 0, 0.19 and 0.47 cultivations in the Southeast, Midsouth and Midwest, respectively. Midwestern producers received the highest returns, but in this region no-tillage soybeans generated significantly lower returns than clean tillage soybeans—largely due to lower yields. Returns in the Midsouth and Southeast did not differ with tillage system. The authors concluded that no one tillage strategy shows a clear economic advantage over others.

SUMMARY

The relatively uniform set of tillage practices of the past have evolved into more complex management systems. Optimum tillage practices have become site specific, much like fertilizer and pesticide recommendations. Tillage systems differ not only from one region to another, but from field-to-field, and in some cases, practices within a field change from one year to the next. Use of conservation tillage has tripled during the past decade to the point that about 7, 38 and 55% of the 1982 U.S. soybean area was in the no-, reduced and clean tillage categories. When considering machine options for conservation tillage, the manner in which an implement is equipped and operated can be as important as selection of the particular implement. Capability to operate in and maintain surface residue, as well as integration of pesticide application, are all important aspects of machines used in conservation tillage. Planting devices need to be chosen based on seedbed conditions as well as row width and cost requirements. Uniform straw distribution behind the combine aids in successful conservation tillage. As long as crop yields are similar, no one tillage strategy has shown a clear economic advantage over others.

NOTES

Richard R. Johnson, Staff Agronomist, Deere & Company Technical Center, Moline, IL.

REFERENCES

Bauder, J. W., G. W. Randall, J. B. Swan, J. A. True and C. F. Halsey. 1979. Tillage practices in south-central Minnesota. Minn. Coop. Ext. Ser. Pamphlet PM-864.

Colvin, T. S. and D. C. Erbach. 1982. Soybean response to tillage and planting methods. Trans. Am. Soc. Agric. Eng. 25:1533-1535.

Colvin, T. S., J. M. Laflen and D. C. Erbach. 1981. A review of residue reduction by individual tillage implements. *In* Crop production with conservation in the 80's. Amer. Soc. Agric. Eng., St. Joseph, Michigan.

Erbach, D. C. 1982. Tillage for continuous corn and corn-soybean rotation. Trans. Am. Soc. Agric. Eng. 25:906-911.

Johnson, R. R. 1982. The impact of changing soybean cultural practices on farm equipment. p. 43-50. *In* Proc. 12th Soybean Seed Res. Conf., American Seed Trade Association, Washington, D.C.

Jolly, R. W., W. M. Edwards and D. C. Erbach. 1983. Economics of conservation tillage in Iowa. J. Soil Water Cons. 38:291-294.

Laflen, J. M. and W. C. Moldenhauer. 1979. Soil and water losses from corn-soybean rotations. Soil Sci. Soc. Am. J. 43:1213-1215.

Laflen, J. M., W. C. Moldenhauer, and T. S. Colvin. 1981. Conservation tillage and soil erosion on continuously row-cropped land. *In* Crop production with conservation in the 80's. American Society of Agricultural Engineering, St. Joseph, Michigan.

Lockeretz, W. 1983. Energy implications of conservation tillage. J. Soil Water Cons. 38:207-211.

Mulvaney, D. L. 1984. Conservation tillage in northern Illinois. p. 57-59. *In* 36th Illinois custom spray operator training school manual. Coop. Ext. Ser., Univ. of Illinois, Urbana.

Nave, W. R., L. M. Wax and J. W. Hummel. 1980. Tillage for corn and soybeans. ASAE Paper No. 80-1013, American Society of Agricultural Engineering, St. Joseph, Michigan.

No-Till Farmer. 1983. 1982-1983 tillage survey. February. No-Till Farmer, Inc., Brookfield, Wisconsin.

Office of Technology Assessment. 1982. Croplands. p. 91-133. *In* Impacts of technology on U.S. cropland and rangeland productivity. U.S. Govt. Printing Office, Washington, D.C.

Siemens, J. C. and W. R. Oschwald. 1979. Corn-soybean tillage systems, erosion control, effects on crop production, costs. Trans. Am. Soc. Agric. Eng. 21:293-302.

Thompson, L., W. A. Skroch and E. O. Beasley. 1981. Pesticide incorporation: Distribution of dye by tillage implements. North Carolina Agric. Ext. Ser. Publ. AG-250.

Throckmorton, R. I. 1980. Equipment for narrow-row and solid plant soybeans. p. 485-491. *In* F. T. Corbin (ed.) World soybean research conference II: Proceedings. Westview Press, Boulder, Colorado.

Touchton, J. T. and J. W. Johnson. 1982. Soybean tillage and planting method effects on yield of double-cropped wheat and soybeans. Agron. J. 74:57-59.

Triplett, G. B. and D. M. Van Doren. 1977. Agriculture without tillage. Sci. Amer. 236(1):28-33.

USDA. 1982. Agricultural Statistics 1981. U.S. Department of Agriculture, Washington, D.C.

ROTATIONAL BENEFITS TO SOYBEANS AND FOLLOWING CROPS

L. F. Welch

Extensive research has established that rotating crops almost always results in higher yields than from monoculture. The exact cause of this effect is not nearly so well established. The difficulty in establishing the causative beneficial effects of rotations, or the detrimental effects of monoculture, may be because the effects are due to several interacting factors.

A large part of the beneficial effect of rotations may be relatively easy to explain in certain situations. For example, with a soil low in nitrogen and without fertilizer nitrogen added, much of the benefit of a nonlegume following a legume may be due to nitrogen contributed by the legume.

Even though rotations may increase the yield of a particular crop species, as compared to monoculture, the economic value of other crops may be less than if only one species was grown in monoculture. The importance of determining the cause of the beneficial effect of rotations is that one can then decide if management can cause the same desirable effect on preceding crops, without growing the less economical crop in rotation.

It is also desirable to know the magnitude of any beneficial effect of a crop on following crops when considering the economics of various crops. Additional value should be assigned to a crop that has beneficial effects on following crops.

SOYBEAN EFFECT ON FOLLOWING CROPS

Corn (*Zea mays* L.) and soybean are the primary row crops in many soybean-producing states in the United States. Thus, much of the research on rotations has included corn following soybean.

Corn following soybean has yielded an average of 17% more than continuous corn on the Morrow Plots in Illinois (Table 1). This is equivalent to 1518 kg/ha more corn. The advantage of corn following soybean ranged from 6.5% in 1969 to 30.6% in 1979. These plots received 336 kg/ha of nitrogen fertilizer added each year when corn was grown, and soil phosphorus and potassium were maintained at high levels.

Data in Table 2 indicate that corn following either soybean or fallow yielded an average of about 10% more than continuous corn. Corn following 2 years of soybean yielded somewhat higher than when following only 1 year of soybean. Corn yielded about the same when no crop was grown the preceding year as when the preceding crop was soybean. All plots received 224 kg/ha of fertilizer nitrogen when corn was grown.

Table 1. Corn yields on the Morrow Plots at Urbana, Illinois when the preceding crop was either corn or soybean (Odell et al., 1982).

Year	Corn yield when following:		Increase for corn follow-ing soybean
	Corn	Soybean	
	– kg/ha –		%
1969	8,543	9,101	6.5
1971	9,158	10,594	15.7
1973	9,321	11,259	20.8
1975	10,117	12,005	18.7
1977	7,088	8,925	25.9
1979	7,270	9,496	30.6
1981	10,807	11,547	6.8
Average	8,900	10,418	17.0

Table 2. Corn yields following either corn, soybean, or fallow at West Lafayette, Indiana, 1972-80 (Barber, pers. commun.).

Corn following:	Corn yield	Increase compared with continu-ous corn
	kg/ha	%
Corn	8844	–
Soybean	9597	8.5
Soybean-Soybean	9848	11.4
Fallow	9785	10.6

OTHER CROPS EFFECT ON SOYBEAN

Soybean was grown in Minnesota on plots that differed in cropping history the three preceding years (Table 3). Continuous soybean produced the lowest yield. Corn

Table 3. Soybean yields as influenced by rotations in Minnesota (Crookston, 1984).

Cropping history[a]			1981 soybean yield	Increase from rotation
1978	1979	1980		
			kg/ha	%
S	S	S	3228	–
S	C	S	3631	12.5
S	S	C	3766	16.7
C	C	C	4102	27.1

[a]C = corn; S = soybean.

grown in only one of the two preceding years resulted in considerably higher soybean yields. Soybean grown the year after only 1 year of corn yielded 16.7% more than continuous soybeans. Corn grown 2 years earlier resulted in 12.5% higher soybean yields than continuous soybean. These data suggest that the effect of corn occurs quickly, but the effect persists for more than 1 year. Even though 1 year of corn increases yields of soybean that follow, the highest soybean yield was obtained following 3 years of corn.

It is generally accepted that soybean yields are higher when they follow some crop other than soybean. Data from Illinois, Indiana, and Nebraska provide more evidence (Table 4). Soybean following 2 years of corn yielded the same as when one year of corn was replaced by one year of wheat (*Triticum aestivum* L.) in the rotation. Soybean in both cropping systems yielded 13.6% more than continuous soybean in the Illinois study. The Indiana and Nebraska data again indicate that only 1 year of corn has an appreciable effect on the yield of soybeans that follow. The average increase in yields of soybeans grown in rotation was about 14% for the three states.

Mulvaney and Paul (1984) reported the results of soybean yields when grown continuously and when alternated with corn. Their study included three primary tillage systems (Table 5). Soybean yields were influenced more by the preceding crop than

Table 4. Soybean yields as influenced by preceding crop. The Illinois data are from Slife (pers. commun.). The Indiana and Nebraska data are cited from Christenson (1982).

Rotation[a]	Soybean yield	Increase from rotation
	kg/ha	%
Illinois, 10 years		
S-S-S	2959	–
C-C-S	3362	13.6
W-C-S	3362	13.6
Indiana, 6 years		
S-S	3161	–
C-S	3362	6.4
Nebraska, 8 years		
S-S	2421	–
C-S	2959	22.2

[a]C = corn; S = soybeans; W = wheat.

Table 5. Soybean yields at Elwood, Illinois when following either corn or soybean for tillage systems, 1979-82 (Mulvaney and Paul, 1984).

Primary tillage system	Soybean yields when following:		Increase for following corn
	Soybean	Corn	
	– kg/ha –		%
Moldboard plow	2555	3093	21.1
Chisel plow	2421	2959	22.2
Disk	2219	2892	30.3
Average	2398	2981	24.3

by the tillage system. When averaged over all tillage systems, soybean following corn yielded about 24% more than continuous soybean.

HOW ROTATIONS INFLUENCE CROP PERFORMANCE

Plant growth factors may be classified as physical, chemical, and biotic. Any favorable effect of rotations is apparently due to physical or chemical changes in soil brought about by rotations, or to biotic factors (weeds, diseases, and insects) in or on the soil. Furthermore, any favorable changes in plant growth factors caused by rotations must persist until a following crop is seeded. Otherwise, there would be no favorable effect of rotations on succeeding crops.

Chemical Changes

Crop rotations were once strongly advocated in many countries as a practice that helped maintain soil productivity. A forage legume was often included in the rotation. Prior to extensive use of inorganic nitrogen fertilizer, much of the benefit attributed to rotations was increased nitrogen available to nonlegumes when preceded by a legume.

The data in Tables 6-9 suggest that more than increased availability of nitrogen is responsible for the higher yields of corn and sorghum when preceded by soybean or alfalfa (*Medicago sativa* L.) crops. That is, even at rates of added fertilizer nitrogen (up to 360 kg/ha of N) considered adequate for nonlegumes, corn and sorghum [*Sorghum bicolor* (L.) Moench] yielded higher in rotations than in continuous monoculture. The studies in Minnesota, Illinois, and Kansas included only soybean in the rotations along with sorghum and corn (Tables 6, 7). Thus, the benefit of soybean preceding the nonlegumes can be attributed solely to soybean. The Iowa and Wisconsin

Table 6. Corn yields at various rates of added nitrogen when the preceding crop was either corn or soybean; Waseca, Minnesota (Randall and Langer, 1982); Elwood, Illinois (Mulvaney et al., 1980).

Fertilizer N added for corn	Corn yield when following:		Increase for corn following soybean
	Corn	Soybean	
	— kg/ha —		%
Waseca, Minn.			
0	4642	7025	51.3
45	5959	8279	38.9
90	6586	8844	34.3
135	6962	9032	29.7
180	7527	9534	26.7
225	7652	9408	22.9
Average	6555	8687	32.5
Elwood, Ill.			
0	3550	4795	35.1
90	6165	7286	18.2
180	7037	8095	15.0
270	7348	8407	14.4
360	7660	8531	11.4
Average	6352	7423	16.9

Table 7. Sorghum yields following sorghum or soybeans at Powhattan, Kansas, 1980-83 (Claassen and Kissel, 1984).

Fertilizer N	Sorghum yields when following:		Increase for following soybeans
	Sorghum	Soybeans	
kg/ha	kg/ha	kg/ha	%
0	2959	5514	86.3
45	3166	6187	64.3
90	5178	5851	13.0
135	5582	5851	4.8
225	5783	5985	3.5

Table 8. Corn yields at various rates of added nitrogen when following either corn, soybean, oat, or alfalfa at Clarion, Iowa, 1971-81 (Voss and Webb, 1983).

Fertilizer N added for corn	Corn yield	Increase compared with continu- ous corn[a]
kg/ha	kg/ha	%
Continuous corn		
0	3262	–
67	6021	–
134	7715	–
201	8279	–
Corn following soybean (C-S-$\underline{C}$-O_L)		
0	6962	113.4
67	8091	34.4
134	8718	13.0
201	8844	6.8
Corn following oat$_L$ ($\underline{C}$-S-C-O_L)		
0	6837	109.6
67	8342	38.5
134	8969	16.3
201	8844	6.8
Corn following alfalfa ($\underline{C}$-C-O-A)		
0	8969	175.0
67	8907	47.9
134	9032	17.1
201	8969	8.3

[a]Underlined crop in the rotation is the one being reported. C = corn, S = soybean, O_L = oat (*Avena sativa* L.) with a forage legume, A = alfalfa.

rotations included crops other than soybean and corn (Tables 8, 9). The increased yield of corn following soybean crops in these latter studies may be partly attributable to crops other than soybean, but this cannot be precisely determined.

The difference in corn or sorghum yields in monoculture versus rotation decreased as the rate of added fertilizer nitrogen increased. For example, sorghum yields following soybean was 86.3 and 3.5% greater than when following sorghum with 0 and 225

Table 9. Corn yields at various rates of added nitrogen when following either corn, soybean, or alfalfa at Lancaster, Wisconsin, 1967-74 (Higgs et al., 1976).

Fertilizer N added for corn	Corn yield	Increase compared with continu- ous corn[a]
kg/ha	kg/ha	%
Continuous corn		
0	4892	–
84	7150	–
168	7840	–
336	7276	–
Corn following soybean (C-S-C-OA)		
0	7840	60.3
84	8279	15.8
168	8405	7.2
336	7966	9.5
Corn following alfalfa (C-S-C-O-A)		
0	7778	59.0
84	8593	20.2
168	8405	7.2
336	8405	15.5

[a]Underlined crop in the rotation is the one being reported. C = corn, S = soybeans, O = oats, A = alfalfa.

kg/ha of fertilizer nitrogen, respectively (Table 7). The same general trend occurred when corn followed legumes other than soybean. With no fertilizer nitrogen added, the increase for corn following soybean was about the same as when following alfalfa in the Wisconsin study (Table 9), but corn yields were greater following alfalfa than when following soybean in the Iowa study (Table 8). There was no yield increase from fertilizer nitrogen when corn followed alfalfa in Iowa.

Even at the highest nitrogen rate, corn following soybean yielded about 11 and 23% higher than continuous corn in the Illinois and Minnesota studies, respectively. Again, the data suggest that the rotational benefits are due to something other than just increased nitrogen available for corn.

In many cases the favorable influence of rotations cannot be explained by increased availability of nitrogen. The increased yield of soybeans following corn is surely not due to the effect of corn on the status of soil nitrogen. Also, nonlegumes often increase the yields of nonlegumes that follow, and this would not be attributable to symbiotic nitrogen fixation.

The yield difference in monoculture versus rotation may be due to both beneficial effects of rotations and inhibitory effects of monoculture. Allelopathic effects of various crops are now receiving considerable attention from researchers. Certain chemical compounds produced during growth or decomposition may slow growth of following plants.

Physical Changes

The soil is usually more loose and friable following soybean than when following corn. The more loose soil would be expected to enhance aeration, which would enhance nutrient uptake and other growth processes. The more friable soil following soybean would also be conducive to good seedbed preparation. The preceding discussion might partly help explain the benefits of corn following soybean, but it would not help explain the benefits to soybean following corn.

The difference in susceptibility to soil erosion after growing corn and soybean suggests physical properties are affected differently by the two crops. Laflen and Moldenhauer (1982) reported about 13% more water runoff and about 40% more soil erosion with corn following soybean than with continuous corn. Corn produces about five times as much above-ground stover as soybean. Where tillage for following crops results in considerable residue remaining on the soil surface, one would expect greater soil protection from corn residue after harvest than from soybean residue.

Biotic Factors

Rotations help reduce the severity of adverse effects from weeds, diseases, and insects. Rotations allow a greater diversity of herbicide use, and this provides for better weed control than monoculture. Rotations are often a practical way to prevent severe infestations of diseases and insects. Even though the effect of rotations on biotic factors often may be visually evident, in many other cases the effect may not be evident even to the trained observer. However, the latter cases may still result in greater yields for rotations than for monoculture.

Other Factors

Differences in yields of rotations versus monoculture may be due to some growth factor other than those discussed. Indeed, the effect may be due to a combination of growth factors rather than to only a single factor. Readers may wish to add other factors to the list. The yield difference between rotations and monoculture is of such magnitude that it merits additional scientific thought and research.

NOTES

L. F. Welch, Professor of Soil Fertility, Department of Agronomy, University of Illinois at Urbana-Champaign, Urbana, Illinois 61801.

REFERENCES

Christenson, D. M. 1982. Cropping sequence effects on yields. Ninth Michigan seed, weed, and fertilizer school. Michigan State University, East Lansing, Michigan.

Claassen, M. M. and D. E. Kissel. 1984. Rotation with soybeans increases corn and grain sorghum yields. Better Crops. 68(Summer):28-30.

Crookston, R. K. 1984. The rotation effect. Crops and Soils 36(March):12-14.

Higgs, R. L., W. H. Paulson, J. W. Pendleton, A. E. Peterson, J. A. Jackobs and W. D. Shrader. 1976. Crop rotations and nitrogen. Wisconsin Coll. Agr. and Life Sciences. Res. Bull. R-2761.

Laflen, J. M. and W. C. Moldenhauer. 1982. Soil erosion in rotations. Crops and Soils 32(June-July): 9-10.

Mulvaney, D. L., R. R. Bell, D. E. Harshbarger and L. V. Boone. 1980. Northern Illinois and Northeastern Illinois agronomy research centers—report of research results. Illinois Agric. Exp. Stn. AG-2023.

Mulvaney, D. L. and L. Paul. 1984. Rotating crops and tillage. Crops and Soils 36(April-May):18-19.

Odell, R. T., W. M. Walker, L. V. Boone and M. G. Oldham. 1982. The Morrow plots—a century of learning. Illinois Agric. Exp. Stn. Bull. 775.

Randall, G. W. and D. K. Langer. 1982. A report of field research in soils. Minnesota Agric. Exp. Stn. Misc. Pub. 2.

Voss, R. D. and J. R. Webb. 1983. Does corn respond to residual P and legume N? Iowa Fertilizer and Ag Chemicals Conference Proceedings. Iowa State University, Ames.

RESIDUAL EFFECTS OF CORN AND SOYBEAN ON THE SUBSEQUENT CORN CROP

R. M. Cruse, I. C. Anderson, and F. B. Amos, Jr.

In the upper midwest of the United States, the grain yield of corn (*Zea mays* L.) is greater following soybean than following corn (*Zea mays* L.), even when the soil is clean tilled and the corn crop is adequately fertilized with nitrogen. This effect mainly has been documented from long-term rotation experiments with various levels of nitrogen fertilizer (Shrader and Voss, 1980; Sundquist et al., 1982; Welch, 1977; Voss and Shrader, 1979). The magnitude of the effect appears to be greater during dry years than during wet years (Benson, pers. commun.). Presently some of the most severe effects of corn after corn are being experienced in the irrigated areas of Nebraska (Stamp, pers. commun.).

The commonly accepted reasons for the greater corn grain yield when corn follows a legume have been the nitrogen provided by the legume or some aspect of disease or pest control. This explanation remained viable even though corn grain yields with corn grown in monoculture were lower than those grown in rotation when: 1) P and K fertility levels were adequate; 2) nitrogen fertilizer rates were high; and 3) insecticides were used on the corn crop. Recently Sperow (1983) indicated that corn grain yields were greater for corn following alfalfa (*Medicago sativa* L.) than for continuous corn, even when corn plants from both treatments had equal corn plant leaf nitrogen contents. Furthermore, the greatest rootworm damage found on her plots were in the third-year corn plots and not in the continuous corn plots as conventional wisdom would lead one to expect. While these findings do not disprove the theory that the legume effect can be explained by N fertility and insect pest control, they do suggest other factors may also be operative.

It was not until the late 1970s that a soybean effect, similar to that of meadow, was substantially documented. Table 1 illustrates the typical rotation effects. At the greatest rate of N, corn yielded 700 kg/ha more following a legume than did continuous corn. Also, at the greatest N application there was a "memory" effect of the legume during the following 2 to 3 years of corn. Second- and third-year corn yields were sequentially lower than first-year corn yields; however, they remained higher than the continuous corn yield. At zero N, the "memory" effect was even greater than at optimum N rates. For first-year corn after a legume, the commonly accepted explanation for the yield difference is residual N from the legume. However, it is difficult to attribute the "memory" effect for second- and third-year corn to residual N of the legume, particularly for soybean or for an oat (*Avena sativa* L.) with alfalfa catch crop. In fact, at zero N a part of the yield response of corn following soybean may not be due to residual N, but rather from some other effect of soybean. Instead of considering

Table 1.   Effect of fertilizer N on grain yield (t/ha) of corn in different crop rotations at the Clarion-Webster Research Center (Iowa); mean of 1971-79; C = corn, Sb = soybean, O_x = oats with alfalfa catch crop; M = alfalfa. The year in the rotation represented by the yields is underlined. (From Voss and Shrader, 1979).

N	Crop rotation							
kg/ha	Cont. C	$\underline{C}CCO_x$	$C\underline{C}CO_x$	$CC\underline{C}O_x$	$\underline{C}SbCO_x$	$CSb\underline{C}O_x$	$\underline{C}COM$	$CC\underline{O}M$
0	3.07	6.33	4.70	3.76	6.46	6.52	8.27	5.96
67	5.77	7.59	6.58	5.70	7.96	7.40	8.02	7.21
134	7.08	7.77	7.65	7.59	8.28	8.02	8.27	7.52
201	7.40	8.15	7.90	7.65	8.09	8.15	8.08	7.75

the response as a "memory" effect of the previous legume, one could argue that the toxic effect of corn after corn progressively increases with first-, second- and third-year corn. However, if this is true, the toxic effect is completely removed by one year of soybean or oat-alfalfa.

The main effect of the previous crop is not due to above-ground plant residue. Removing corn or soybean residue after harvest, or transferring the residues to the reciprocal corn or soybean areas, does not have much effect on the subsequent corn grain yield. The effect is in the soil where the crop was grown. Crookston (1982) reported that removing corn stover from corn plots tended to decrease grain yield of the subsequent corn crop, and adding corn stover to land previously cropped to soybean increased the yield of corn subsequently grown in that land. This is the opposite of the expected effect.

Break crops other than legumes may have some beneficial effects on corn yield (Hicks and Peterson, 1981), but the effect from legumes appears to be greater and more consistent. The magnitude of the legume effect is variable and ranges in the Midwest from 0 to 2000 kg/ha of corn yield when N is adequate. At latitudes below 40 degrees, the nonnitrogen effect of the legume decreases. Warmer winters would result in greater microbial activity, which could degrade compounds in the soil that may be responsible for the legume effect. There are rotation effects on soil structure and fertility that may be contributors, but these probably do not explain the major effect. For example, in a corn-soybean rotation, the soil K test is higher after soybean than after corn (Voss, pers. commun.). The previous crop may have effects on quantity of water stored in the rooting profile. There may be more moisture deep in the soil after soybean than after corn. Larger effects on depletion of soil moisture may occur with deep rooted legumes, such as alfalfa, particularly if plowed late in the fall, but this would have a depressing effect on the subsequent corn yield. Corn following a fallow treatment usually yields similar to corn following soybean. A portion of this yield improvement over corn following corn might be attributed to improved soil water conditions.

ALLELOPATHIC EFFECTS OF CORN

There are many allelopathic effects on plant growth reported in the literature (Rice, 1974). Jimenez et al. (1983) reported that corn pollen, at concentrations commonly present on soil, strongly inhibited growth of a number of plant species. In 1984, we planted a number of legumes with corn. Common crimson clover (*Trifolium incarnatum* L.), which had relatively sparce growth, was strongly allelopathic to corn.

Chou and Patrick (1976) identified 18 compounds from decomposing corn residue in soil that were phytotoxic to seed germination. Yakle and Cruse (1983) studied age and location of ground corn residue on growth of corn. Plants were most strongly inhibited by fresh residue and by residue placements where roots made direct contact with the residue. Growth of roots was much more strongly inhibited than growth of the corn shoot.

Garcia (1983) assayed soil from a field of second-year corn during the growing season. He used two types of bioassay. One was the growth of corn seedlings in sand-nutrient culture with some soil sample added to the sand. Another assay consisted of collecting circulating leachates from the soil sample on a partially hydrophobic resin, and measuring toxicity from the resin by cress seed (*Lepidium sativum* L.) germination. Both assays showed similar results. In the spring, there were large concentrations of inhibitors in the soil, during June the inhibitory effect disappeared and the soil was slightly stimulatory, during July inhibitors began to accumulate and remained large at the last sampling on 25 Sept. In a companion study he compared soil covered with plastic under the corn canopy to that of uncovered soil under the corn canopy. He found that about one-half of the inhibitory effect found in the soil was due to excretion from the living corn roots and the remaining from rain leachates of the shoot, or from pollen or other dropped plant parts. The total inhibition before harvest of the corn was less than that present in the spring, indicating that incorporated corn stover also made a contribution to the total inhibition present at planting in the spring.

Hicks and Peterson (1981), using five corn hybrids grown on plots with each of the hybrids the previous year, reported that previous hybrid affected grain yield. The results indicate that toxicity may vary with corn genotypes.

SIMULATORY EFFECTS OF SOYBEAN

In much of the world, beans (*Phaseolus* spp.) and corn are intercropped. Usually the bean has little effect on the yield of corn (Francis et al., 1978). If comparable amounts of weeds instead of beans were present, the yield of corn would be significantly depressed. Chui and Shibles (1984) reported that, at some densities of intercropped corn and soybean, there was a trend for an increase in yield of corn due to the soybean intercrop. These types of results would indicate production of some stimulatory allelochemical by beans and soybean, but there is no direct evidence of that effect.

Ries et al. (1977) reported that alfalfa meal used with N fertilizer resulted in greater crop yields than with adequate N alone. They proposed that the increased yield was due to the growth regulatory effect of triacontanol in the alfalfa meal. Since soybean leaves also have high levels of triacontanol, Assumpcao (1979) grew, in large pots, corn, soybean and alfalfa along with a fallow pot. After each two months of growth the plant tops were harvested, dried and incorporated into its own pot. The pots were replanted to give a continuous cropping effect. Soil samples were collected at various intervals during one year and frozen until bioassayed. The assay was the growth of corn seedlings in nutrient culture-sand with some of the soil sample added to the sand. Soil from the pots with soybeans and alfalfa usually increased height and weight of corn, whereas soil from the pot with corn decreased corn growth, particularly root growth, as compared with the fallow soil.

Kalantari (1981) extracted soil from a soybean field with chloroform after harvest. The extract was taken to dryness and isoproponol soluble material fractionated on a DEAE-sephadex column. The isopropanol insoluble material contained triacontanol.

Aliquots of the fractions were applied to germination paper, dried and used to germinate ten corn seedlings in the dark. Growth of roots and shoots were measured. There were two stimulatory peaks and two inhibitory peaks. Again, root growth was inhibited more than shoot growth.

FUTURE RESEARCH NEEDS

The most important need is to determine if the increased yield of corn following soybean is due to: 1) the adverse effects of corn on corn; 2) stimulatory effects of soybeans on subsequent corn; or 3) a combination of both effects. Knowing this answer would greatly improve chances of identifying the biology involved in the rotation effect.

The effect of interspecies allelopathic interaction has been well documented. However, the effect of genetic variability within a given species on allelopathic reactions is still a question. The effect of environmental variations on allelopathic responses has been noted, but not extensively addressed.

Nutrient budgets, particularly for nitrogen, in rotation studies would improve chances of drawing definite conclusions concerning fertility effects. The potential for substantial losses of fertilizer N to the environment through leaching and/or volitilization weaken the arguments of "sufficient nitrogen application for maximum corn yield" used in describing N application rates in many rotation x nitrogen rate studies.

We certainly do not know why corn yields greater after soybean than after corn. Stimulatory and inhibitory allelochemicals are well documented, particularly in other disciplines, and we believe that studying this approach to the problem is appropriate. We also realize that there are other possible causes for the soybean effect, such as the C:N ratio and other properties of recently added crop organic matter, or changes in the soil microflora and their direct or indirect allelochemical effects on plant growth.

NOTES

R. M. Cruse, I. C. Anderson, and F. B. Amos, Jr., Department of Agronomy, Iowa State University, Ames, IA 50011.

REFERENCES

Assumpcao, L. E. 1979. Allelopathic effects of four crop residues on corn seedlings. M.S. Thesis. Iowa State University Library, Ames.

Chou, C. and Z. A. Patrick. 1976. Identification and phytotoxic activity of compounds produced during decomposition of corn and rye residue in soil. J. Chem. Ecol. 2:369-387.

Chui, J. A. N. and R. Shibles. 1984. Influence of spatial arrangements of maize on performance of an associated soybean intercrop. Field Crops Res. 8:187-198.

Crookston, R. K. 1982. Field studies on the yield effect of corn and soybeans in rotation—are plant growth regulators involved? Plant Growth Regulatory Soc. Proc. 9:137-143.

Francis, C. A., C. A. Flor and M. Prager. 1978. Effects of bean association on yields and yield components of maize. Crop Sci. 18:760-764.

Garcia, A. G. 1983. Seasonal variation in allelopathic effects of corn residue on corn and cress seedlings. Ph.D. Dissertation. Iowa State University Library, Ames.

Hicks, D. R. and R. H. Peterson. 1981. Effects of corn variety and soybean rotation on corn yield. p. 89-93. *In* Proceedings 36th Annual Corn and Sorghum Research Conference. American Seed Trade Association, Washington, D.C.

Jimenez, J. J., A. Schultz, A. L. Anaya, J. Hernandez and O. Espejo. 1983. Allelopathic potential of corn pollen. J. Chem. Ecology 9:1011-1025.

Kalantari, I. 1981. Stimulation of corn seedling growth by allelochemicals from soybean residue. Ph.D. Dissertation, Iowa State University Library, Ames.

Rice, E. L. 1974. Allelopathy. Academic Press, New York.

Ries, S. K., V. Wert, C. C. Sweeley and R. A. Leavitt. 1977. Triacontanol: A new natural occurring plant growth regulator. Science 195:1339-1341.

Shrader, W. D. and R. D. Voss. 1980. Crop rotation vs. monoculture. Crops Soils 32:8-11.

Sperow, K. A. 1983. Effects of crop rotation and nitrogen fertilization on corn grain and stover production. M.S. Thesis. Iowa State University Library, Ames.

Sundquist, W. B., K. M. Menz and C. F. Neumeyer. 1982. A technology assessment of commercial corn production in the United States. Univ. of Minn. Agric. Exp. Stn. Bull. 546.

Voss, R. D. and W. D. Shrader. 1979. Crop rotations—effects on yield and response to nitrogen. Iowa State Univ. Ext. Ser. Pm 905.

Welch, L. F. 1977. Soybeans good for corn. Soybean News 28:1-4.

Yakle, G. A. and R. M. Cruse. 1983. Corn plant residue age and placement effects upon early corn growth. Can. J. Plant Sci. 63:871-897.

SOYBEAN CROP MODELING FOR PRODUCTION SYSTEM ANALYSIS

J. W. Jones, K. J. Boote, and J. W. Mishoe

The systematic integration of disciplinary knowledge into improved systems for crop production has lagged behind information development. At the University of Florida, an interdisciplinary research team began addressing this problem about 5 years ago, with a goal of determining optimal strategies for soybean irrigation and pest management practices. The approach was to develop a soybean production system model to integrate crop, soil, insect, weather, and management components, to test the model and then to use it.

MODEL OVERVIEW

In designing the structure of the model, several criteria were established. For management applications, the model would have to be integrated with other models such as soil, insect, and economic models. In addition, many runs would be required to determine optimal strategies or to evaluate the effects of variable weather patterns. Therefore, rapid computer execution was considered important. For integrating pest and weather-related stress effects, a process-oriented approach was needed. Finally, and perhaps most importantly, model usability was considered in several ways: (a) ability to transport the model to other computer systems, (b) ability to easily obtain inputs for the model, (c) ability to calibrate the model for various locations and cultivars using easily collected data, and (d) ability to include the effects of stresses.

One version of the soybean model has been completed and documented (SOYGRO V4.2, Wilkerson et al., 1983a,b; Jones et al., 1982). The model is programmed in FORTRAN and requires standard, class A weather station data. Growth analysis and phenology data can be used to calibrate the model, and the relationships between stresses and processes are measurable by combinations of growth analysis and canopy gas exchange techniques. A second version, SOYGRO V5.0, has been completed and is being tested for application over a broader range of cultivars, soil types, and locations. Version 4.2 will be described with applications under Florida soil and climate conditions.

Crop Phenology

The partitioning of dry matter into growth of different plant parts in soybean depends on stage of development. In order to predict soybean growth and yield, one must be able to predict accurately the timing and duration of various crop growth phases. The phases, delineated by the reproductive (R) stages described by Fehr et al. (1971), established the time basis for changing partitioning, setting pods, initiating protein remobilization, leaf senescence, and onset of maturity. In addition to R2, R4, R7, and R8, one additional stage was defined for the model, corresponding to the latest date that new pods can be formed. In SOYGRO V4.2, a table look-up function based on site and cultivar specific data was used to predict these reproductive stages. A soybean phenology model for Version 5.0, with cultivar specific parameters that depend on night length and temperature, has been developed (J. W. Mishoe, J. D. Hesketh, K. J. Boote, D. Herzog and Y. Tsai, unpublished).

Carbon Balance

Biomass growth of the crop is based on a carbohydrate balance, which includes photosynthesis, respiration, partitioning, remobilization of protein, and senescence (Wilkerson et al., 1983a). Carbohydrate supply is from gross photosynthesis, which is based on daily photosynthetically active radiation, temperature, leaf area index, water stress, and nitrogen concentration of leaves. Carbohydrate is used for biomass growth and respiration. Efficiencies for conversion of CH_2O to biomass, consisting of protein, fat and structural carbohydrate, are calculated from the work of Penning de Vries (1976). There is no differentiation in the calculated energy cost for protein synthesis whether the N source was from N-fixation or from nitrate reduction.

The overall carbon balance equation is written as

$$\frac{dW}{dt} = E(P_G - R_m) - \dot{S}_L - \dot{S}_S \tag{1}$$

where W = plant biomass, g/m^2; E = efficiency of converting CH_2O to plant biomass, g biomass/g CH_2O; P_G = gross photosynthetic rate g CH_2O/m^2 day; R_m = maintenance respiration, g CH_2O/m^2 day; $\dot{S}_L$ = leaf senescence rate, g biomass/m^2 day; and $\dot{S}_S$ = petiole senescence rate, g biomass/m^2 day.

Partitioning of biomass into various plant parts is based on coefficients that vary with phenology. For example, X_L, which changes with phenology, partitions new growth into leaves by:

$$\frac{dW_L}{dt} = X_L [E(P_G - R_m)] - \dot{S}_L - \dot{M}_L \tag{2}$$

where W_L = leaf weight, g/m^2; and $\dot{M}_L$ = biomass loss rate due to protein remobilization, g/m^2 day.

Similar equations describe growth rates of other plant parts (Wilkerson et al., 1983a). These equations were developed to meet the goal of using combinations of growth and gas analysis measurements to characterize growth processes and stress effects. Note that X_L can be solved in terms of dW_L/dt, dW/dt, $\dot{S}_L$, $\dot{S}_S$, and $\dot{M}_L$ which can be determined from field experiments without gas analysis techniques. Since P_G is in equation 1, gas analysis measurements can be used to measure changes caused by short-term stresses or by environmental conditions.

Pod setting and the initiation of seed growth are important processes in SOYGRO. Flowers are considered plentiful and their mass is ignored, whereas podwall growth depends on phenology as well as on carbohydrate supply. The number of seeds per pod, seed size, and growth rate per seed are cultivar characteristics required for running SOYGRO. Pods are initiated at a rate determined by temperature and by the average number of seed that the plant could have supported during the last 7 days, starting at R4 for determinate cultivars. Stresses that reduce carbohydrate supply also reduce pod initiation. After podwall expansion, the number of seed to start growth is based on the photosynthate supply less maintenance respiration. Seeds have first priority for CH_2O over other plant parts. Stresses that reduce photosynthate supply can delay seed initiation and growth. However, the soybean plant is flexible in that a full seed load may be set after a short stress is alleviated. Cultivars with longer pod addition phrases have more flexibility to recover from short periods of stress than those with short pod addition phases (Cure et al., 1983).

Stress Effects

Stresses were included by their effects on one or more of the processes in SOY-GRO (photosynthesis, respiration, light interception, water use, N remobilization, partitioning, pod addition rate, phenological development, leaf shedding), and/or by directly reducing the living plant material (leaves, stems, pods, seed) (Boote et al., 1983). Water stress, calculated from a soil water balance, reduces P_G and leaf area expansion. Insect defoliators reduce leaf area and, thus, indirectly may effect both P_G and transpiration. In addition, availability of N for remobilization to seed is reduced by leaf eaters. Pod feeding insects reduce seed number and potential growth rates of damaged seed. Foliar diseases reduce efficiency of leaves for fixing CH_2O and reduce light interception by unaffected leaves.

MODEL IMPLEMENTATION

Component models of the soybean production system (crop, soil, insect, disease, management, and economic) formed the basis for several specialized application models. By combining these component models with different user-friendly computer interfaces and other algorithms, applications were developed for decision making, education, research, forecasting, and policy evaluation (Mishoe et al., 1984a). For educational applications, we developed SOYGAME. This user-friendly version uses historical weather data and allows a user to practice growing a crop of soybeans. He must decide when to scout his field, when to spray for insects and when to irrigate. At the end of the season, outputs include final irrigation amount and costs, pesticide application types, amounts, and costs, crop yield and net profit.

Decision models for pest control (PESTDEC) and irrigation (IRDEC) were designed to assist growers in making optimal within-season decisions for a specific field. Procedurally, the crop is simulated from planting until "today" using actual weather data. From today, the model simulates various decisions using several years of historical weather data. The decisions with the highest expected profits, and/or the ones with the lowest risk of losses, are provided to the user. The Soybean Integrated Crop Management (SICM) and STRATEGY models are designed to evaluate management strategies, interactively and in a batch mode, respectively. These models include water and pest stresses in combination (Wilkerson et al., 1983b; Boggess et al., 1984). SOYVAL was designed for yield forecasting. The models are implemented on minicomputers at the University of Florida, and SOYGAME and SOYGRO V5.0 are also on microcomputers. Screen formatting and menu selectable options allow users without programming knowledge to run each model easily. An automatic weather data acquisition system has been implemented to supply current-season weather data for the models.

APPLICATIONS

Irrigation Decisions

Three levels of studies have been performed using SOYGRO V4.2 for analyzing the economics of soybean irrigation management in Florida. First, an analysis of the expected profit and risks associated with various irrigation strategies was conducted by Boggess et al. (1983) for soybeans growing on a Lake fine sand in Florida. Because of the erratic rainfall patterns and low water holding capacity of the soil, the optimal profit and lowest risk was achieved by frequent, light irrigations (1 to 2 cm at 30 to 40% water depletion from the root zone). A range of 14 to 21 cm of irrigation water was required on the average under these strategies for full season Bragg planted in June.

Swaney et al. (1983) developed a real-time irrigation decision model (IRDEC), which evaluates the expected benefit from an irrigation on a selected date during the year. Short-term rainfall forecast information is used to estimate this expected value and evaluate the risk of waiting a few days. This decision model includes the costs of irrigation, the expected price of the crop, and previous field irrigation history in making the evaluation. Because this decision model includes SOYGRO, it takes into account the differing sensitivities of the crop to water stress during different stages.

Boggess and Amerling (1983) applied SOYGRO to evaluate the risks associated with investing in various irrigation systems for soybeans in Florida on two soil types. They found that the expected increase in net profit for investing in a 53 ha, low pressure (40 psi) center pivot system for soybean was about $8000 over a 15-year period on sandy soil, but was negative on sandy loam soil. The analysis indicated that, on the average, it would pay to invest in the center pivot system for irrigating soybeans on sandy soil. However, the probability of losing money on the investment was about 40%. They found that investment in an irrigation system for corn or peanuts on sandy soil in Florida was a "sure-bet". Irrigation of soybeans in rotation with maize (*Zea mays* L.) and peanuts (*Arachis hypogaea* L.) also would be profitable with low risks. Figure 1 shows the simulated cumulative probabilities of net profit for irrigating soybean alone and in rotations with corn and peanuts on Lake fine sand in Florida (adapted from Boggess and Amerling, 1983). Although simulated results indicated that it was more profitable and less risky to irrigate corn and peanuts compared with

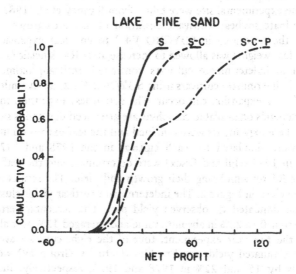

Figure 1. Simulated increase in net profit with a 53 ha center pivot irrigation system when irrigating soybean crops each year over a 15-year period. Soybean (S), soybean-corn rotation (S-C), and soybean-corn-peanut rotations (S-C-P), for Lake fine sand in North Florida. Adapted from Boggess and Amerling (1983).

soybeans, other factors such as pest management and land allotments may cause producers to rotate these crops. This example demonstrates the capability of the model to incorporate uncertainty and risk in evaluating crop production systems.

Double Crop Soybeans: Cultivar Selection and Cultural Operations

The purposes of this study were to evaluate the potential for double cropping soybean in North Florida and to determine combinations of cultivars to plant for maximum yield. Our approach was to use SOYGRO V4.2 to simulate crops planted in sequence for a two-year period (1978; 1979) in Gainesville, Florida. Boote (1981) planted a range of maturity groups in Gainesville during those two years to identify high yielding cultivars that matured by late June, so that a second crop could be planted. He found that cultivars of Maturity Group OO would mature as much as 3 weeks earlier than Maturity Group IV cultivars. Earliness of the first crop could be a determining factor in cultivar selection if delays in planting the second crop were expected. Also, timely planting of a full-season Maturity Group VII or VIII cultivar may result in enough yield advantage to offset a low yielding OO cultivar. In our study, we first evaluated the ability of SOYGRO V4.2 to simulate this range of cultivars. Then sequences of crops were simulated by following a March planted crop with either Bragg or Cobb. The effect of timeliness of harvesting the first crop and planting the second crop was investigated by delaying the second crop planting by 0, 1, 2, 3, or 4 weeks after the first crop reached R8.

Simulating the Early Planting Experiment. Two steps were necessary to simulate the Boote (1981) early planting experiment: (a) obtain necessary input data to run the model and (b) modify SOYGRO to simulate the early-planted, indeterminate cultivars. Boote (1981) grew five cultivars in 1978 and nine in 1979 with three of the cultivars grown both years. He recorded cultural practices, reproductive stages of growth, maturity dates, yield, and 100-seed weights. Daily weather data for the two years were obtained from the Agronomy Weather Station at the University of Florida. Soil characteristics at the experimental site were taken from Swaney et al. (1983).

For determinate soybean cultivars, cessation of leaf area growth occurs approximately at R4, the time used in SOYGRO V4.2 to end leaf expansion. Leaf weight (thus, specific leaf weight) was allowed to increase after R4. Sinclair (1984) found that leaf expansion in indeterminate cultivars continued until approximately R5. Thus, for simulating indeterminate cultivars using SOYGRO V4.2, we extended the time during which leaf area expansion can occur. The actual leaf expansion in the indeterminate cultivars depends on available carbohydrate after seed demand is satisfied.

After the phenology inputs were obtained and the leaf expansion modification was made, yields were simulated for each cultivar in the 1978 and 1979 experiments. Two cultivars in 1979 (Hill and Essex) were determinate and the leaf area expansion was stopped at R4 for simulating their growth and yield. The results of the simulated experiments are given in Figure 2. The indeterminate cultivar yields clustered about the 1:1 line on the simulated vs. observed yield plot. The percentage error in simulated yields ranged from 0 to 22% in absolute value and averaged 12% over all indeterminate cultivars. For the two-year experiment, four of the eight cultivars averaged less than 5% error. The simulated yields for Franklin (Maturity Group IV) were higher than observed yields by 15 and 22% in 1978 and 1979, respectively. In Boote's experiment, other MG IV Cultivars yielded 470 to 530 kg/ha higher than Franklin. Hesketh et al. (1981) found that Franklin had lower leaf photosynthetic rates than many other cultivars in Illinois. Although there may be other reasons why Franklin did not perform as well as other Maturity Group IV cultivars, a lower genetic potential for

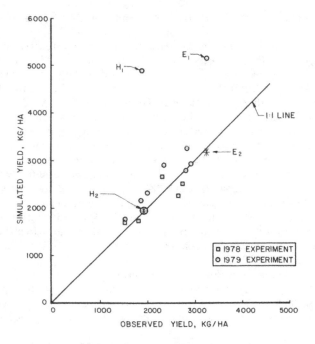

Figure 2. Plot of simulated yields vs. observed yields from Boote (1981) for 1978 and 1979 using
Gainesville weather. The H_2 and E_2 points were obtained by reducing pod addition rates
to demonstrate the effects of short night lengths on Maturity Group V cultivars, Hill and
Essex, respectively. The H_1 and E_1 points were the original simulation results.

photosynthesis could cause such a result. No attempt was made to adjust photosynthesis in SOYGRO V4.2 for simulating different cultivars; in fact, photosynthetic parameters were unchanged from the full season Bragg experiments.

The planting dates for the two determinate cultivars were outside their normal
range, and they responded by pod abortion and cessation of pod addition as night
lengths decreased. In addition, maturity was delayed until after August, the seeds
developed very slowly, and quality of seeds was poor. Because of the late maturity
and low quality of seeds, Boote (1981) concluded that Maturity Group V and above
were not suitable for early planting. Mann and Jaworski (1970) also reported that pod
addition and seed set was slowed by short night lengths. To demonstrate this effect
in the model, the maximum pod addition rates were reduced from 106 pods/m² day
to 40 and 17 for Essex and Hill, respectively. The original and modified simulated
yields were plotted against observed yields for these two cultivars in Figure 2. The
attempt to simulate Essex and Hill outside of their adapted area is a good example
where lack of model fit can point out areas of needed research or model modification. In SOYGRO V5.0 the pod addition rate has been related to night length.

Simulated Double Crop Experiment. One cultivar from each of Maturity Groups
OO, II, III, and IV were selected as candidates for the first soybean crop in a two-
year simulation study (Maple Arrow, Ansoy-71, Williams, and Cutler-71, respectively).
Two cultivars adapted to our location, Bragg (Maturity Group VII) and Cobb (Maturity
Group VIII), were selected for the second crop. The date of planting of the first crop
was March 15, whereas the second crop was delayed 0, 1, 2, 3, and 4 weeks after R8

occurred on the first crop. This resulted in 80 simulations, 40 cultivar sequences x 2 years each.

Combined yield results from the simulated double crop sequences are plotted in Figure 3, averaged over the two years. The low yield from MG OO was not offset by timely planting of the second crop, regardless of whether the second crop was Bragg or Cobb. The Maturity Group II cultivar was a better choice, but the double-crop yields were lower than those from Maturity Group III or IV. Regardless of the first crop, Bragg was the better choice for the second crop, if there was no delay in planting the second crop. As the delay increased, Cobb replaced Bragg as the better second crop cultivar. The reason for this replacement is that the simulated average yield for Cobb was higher than that for Bragg, if planted after the first week in July. Simulated two-crop yields were between 6000 and 7000 kg/ha when Bragg or Cobb followed Maturity Group III or IV cultivars. In 1979 and 1980 experiments in Gainesville, two-crop yields averaged 6110 and 6340 kg/ha, respectively, for the best early crop cultivar plus Cobb as the second crop (K. J. Boote, unpublished). The general conclusion of the study is that a Maturity Group IV or III cultivar planted in mid March should be followed with Bragg if the second crop could be planted before 8-10 July, and with Cobb if later.

This simulated double crop experiment demonstrates the ability of the crop model to evaluate yield benefits by double cropping and cultivar choices for optimal yield. The benefit of such an exercise is to gain confidence that the approach is valid at locations where experimental data are available. Then, within limits, the model can be used

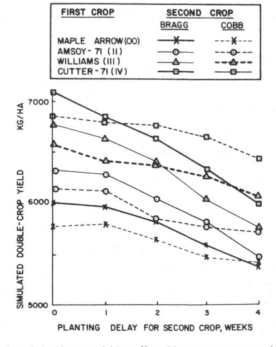

Figure 3. Simulated total double crop yield as affected by crop sequence and the delay in planting the second crop after the first crop matured. First crop planting date was 15 March for both years.

to simulate the potential for double cropping soybean, and select optimal cultivar sequences for other locations where experimental data are not available, and where soil types, rainfall, night lengths and other conditions are different.

NOTES

J. W. Jones, Professor, Agricultural Engineering Department; K. J. Boote, Associate Professor, Agronomy Department; and J. W. Mishoe, Professor, Agricultural Engineering Department, University of Florida, Gainesville, Florida.

This research was supported by the Institute of Food and Agricultural Sciences, the USDA and EPA supported Consortium for Integrated Pest Management (CIPM) Research Program, a USDA cooperative agreement, and the American Soybean Association.

REFERENCES

Boggess, W. G. and C. B. Amerling. 1983. A bioeconomics simulation of irrigation investments. S. J. Agric. Econ. 16:85-91.

Boggess, W. G., D. J. Cardelli and C. S. Barfield. 1984. A bioeconomic simulation approach to multi-species insect management. S. J. Agric. Econ. (In Press).

Boggess, W. G., G. D. Lynne, J. W. Jones and D. P. Swaney. 1983. Risk-return assessment of irrigation decisions in humid regions. S. J. Agric. Econ. 15:135-43.

Boote, K. J. 1981. Response of soybeans in different maturity groups to March plantings in Southern USA. Agron. J. 73:854-859.

Boote, K. J., J. W. Jones, J. W. Mishoe and R. D. Berger. 1983. Coupling pests to crop growth simulators to predict yield reductions. Phytopathology 73:1581-1587.

Cure, J. D., C. D. Raper, Jr., R. P. Patterson and W. A. Jackson. 1983. Water stress recovery in soybeans as affected by photoperiod during seed development. Crop Sci. 23:110-115.

Fehr, W. R., C. E. Caviness, D. T. Burmood and J. S. Pennington. 1971. Stage of development descriptions for soybeans, *Glycine max* (L.) Merrill. Crop Sci. 11:929-931.

Hesketh, J. D., W. L. Ogren, W. E. Hageman and D. B. Peters. 1981. Correlations among leaf CO_2-exchange rates, areas and enzyme activities among soybean cultivars. Photosyn. Res. 2:21-30.

Jones, J. W., L. C. Hammond and K. J. Boote. 1982. Predicting crop yield in response to irrigation practices. Florida Agric. Res. 82(1):20-22.

Mann, J. D. and E. G. Jaworski. 1970. Comparison of stresses which may limit soybean yield. Crop Sci. 10:620-624.

Mishoe, J. W., J. W. Jones, D. P. Swaney and G. G. Wilkerson. 1984. Using crop and pest models for management applications. Ag. Systems (In press).

Penning, de Vries, F. W. T. 1976. Use of assimilates in higher plants. p. 459-480. In J. P. Cooper (ed.) Photosynthesis and productivity in different environments. Cambridge University Press, Cambridge.

Sinclair, T. R. 1984. Cessation of leaf emergence in indeterminate soybeans. Crop Sci. 24:483-486.

Swaney, D. P., J. W. Mishoe, J. W. Jones and W. G. Boggess. 1983. Using crop models for management: Impact of weather probabilities on decisions. Trans. ASAE 26:1808-1814.

Wilkerson, G. G., J. W. Jones, K. J. Boote, K. T. Ingram and J. W. Mishoe. 1983a. Modeling soybean growth for crop management. Trans. ASAE 26:63-73.

Wilkerson, G. G., J. W. Mishoe, J. W. Jones, J. L. Stimac, D. P. Swaney and W. G. Boggess. 1983b. SICM: Florida soybean integrated crop management model. Report AGE 83-1. Agric. Engr. Dept., Univ. of Fla., Gainesville.

WEED SCIENCE

WEED SCIENCE

Herbicide Technology

SOYBEAN HERBICIDE DEVELOPMENT

D. L. Barnes

Early research reports on chemical weed control in soybean crops began to appear in the late 1940s and early 1950s in the United States. These reports describe using available herbicides that were generally not selective. Applications were made using directed or shielded sprayers in the crop to achieve selectivity during the growing season. Timing of application was also used as a means of gaining selectivity, by making applications early post-emergence or after the crop was mature to desiccate weeds and facilitate harvesting. Wax (1973) provided a good review of soybean herbicide development from the early 1950s through the mid-1970s. This review gives a historical account of the development and evolution of soybean herbicides and the subsequent changes in cultural and application techniques that resulted from the introduction of new chemicals.

There are 45 different herbicides registered in the U.S. for use on soybean crops, with 103 different companies holding registrations (based on information received from EPA through "freedom of information"). Many of the registrations are for off-patent products and are held by formulators who have their own private label. Based on soybean herbicide recommendations in the *Weed Control Manual and Herbicide Guide* (Anon., 1984), only 35 of the 45 registered herbicides are available in the market.

In Brazil there are 27 different herbicides registered for use on soybean (Von Hertwig, 1983), only one of which is not registered in the United States. In Argentina, there are approximately 25 different products registered for use on soybean crops, with only one of the products not being registered in the United States. This illustrates the importance put on the large U.S. soybean market by the major companies involved with developing new products for soybean, even though the weed problems, soil types and cultural practices are different in these two countries.

The four major soybean producing countries, U.S., Brazil, People's Republic of China and Argentina, account for over 90% of the planted area and over 94% of the world soybean production (ASA, 1983). The U.S. alone accounts for 55% of the world soybean hectarage and over 65% of total production. Based on market size, current economic conditions, uncertainty of government regulations, property rights and protection, it is not difficult to understand why products are developed for the U.S. market, and then the technology transferred to other countries.

CURRENT HERBICIDE DEVELOPMENT

The high cost of investment in terms of time, capital and human resources to discover, develop, register and commercialize new products dictate that a new product have an established or well defined potential commercial base in the United States before expanding commercialization to other producing countries. The support data required to clear regulatory hurdles, such as residue, metabolism, environmental impact and long term toxicology studies are essentially the same for the major soybean producing countries. This is costly and time consuming and makes it difficult to justify developing a product solely for soybean markets outside the United States. New product development for use outside the U.S. will continue to lag behind the U.S. because of uncertainties in protection of patents and property rights, regulatory requirements and long-term economic outlook.

Research and development expenditures by the U.S. chemical industry, expanded in real terms over a 12-year period (1967-1978), are reported by CAST (1981). The information used in my report was obtained from the NACA (National Agricultural Chemicals Association) and EPA (Environmental Protection Agency) and reflects an overall decline in productivity of research and development efforts if the introduction of new compounds or new formulations of existing compounds are used as measures.

The estimated research and development costs for commercializing a new product is in excess of $20 million, although exact information is difficult to obtain because of the complexibility of registering a new product. Gilbert (1978) highlighted the multiple expenditures for pesticide development, along with projected timelines for a company to realize a profit from the investment. These timelines are based on a modest rate of return and illustrate that a return on the investment would not be expected until well into or past the patent life of the compound. A shorter time interval to reach a breakeven point could be anticipated if the annual rate of return is increased to 12% as pointed out by Garbett (1984) in a recent review, but even then it requires 14 years. This leaves a limited amount of time to recover investment plus realize a return while still under the coverage of a patent, depending of course, on when the patent is issued during the research and development period.

The analysis made by Gilbert (1978) concluded that, based on the high risk and expense involved with developing new pesticides, there would be a gradual decline in pesticide research; fewer products would reach the market, borderline companies would merge or withdraw and the pesticide industry would become composed of a comparatively small number of companies able to afford the risk. This prophecy has been proven over the past 4-5 years, with a number of U.S. companies dropping out of the pesticide business or merging with other companies.

There are, however, currently a number of new products just introduced, or being developed, for the soybean market which will expand the flexibility and timeliness for the farmer over the products presently available. The development and commercialization of new products in countries outside the U.S. will continue to lag behind the U.S. as companies seek protection of industrial property rights and patents. Patents on proprietary technology are the life blood of multinational chemical companies investing millions of dollars on research for new products and in areas of new technology. The search for new and better products is a high risk business and without adequate protection, new technology will be restricted to countries providing protection of a company's industrial property rights.

In a more positive light, it would be expected that the time interval between the transfer and adoption of new herbicide technology can at times be shortened, because of the development work and experience gained in the U.S. However, the introduction of new technology is often delayed because of economic, political and regulatory hurdles rather than scentific, technical or marketing reasons.

HERBICIDE USE PATTERNS

Herbicide use patterns in the U.S. (USDA, 1983) show that the percentage of soybean hectares receiving a herbicide treatment increased from 68% in 1971 to 93% in 1982. The pattern for use during the 1976-1982 period did not change significantly for the established products. The major increase in product use has taken place with products introduced in the mid-70s, which are used as mixtures or as sequential treatments with established products.

The average yield for soybeans in the U.S. has gradually increased (17%) over the same time period (ASA, 1983). There is year-to-year variation due to the differences in growing season, but the long-term trend is toward increasing yields. The increasing use of herbicides has played a role in improving U.S. soybean yields over the long term.

There is a continued adoption of conservation tillage practices for soybean crops, both no-till and reduced tillage systems, according to figures comparing 1980 and 1982 tillage systems estimates (USDA, 1984). The long-term trend for conservation tillage adoption in the U.S. shows that reduced tillage crop production is being rapidly adopted across the United States (Lessiter, 1984).

The per-hectare weed control costs vary in the U.S., depending upon the region and the tillage practices employed by the farmers (USDA, 1983). Comparisons within regions show that the Midwest farmers who are no-tilling soybean crops are using significantly greater amounts of herbicide at a higher total weed control cost compared with the farmer that uses a reduced or conventional tillage system. These additional costs are partly offset by reduced costs of tillage, but the major advantage is likely a savings in time where soybeans are planted behind winter grains. In the Midsouth, herbicide costs are significantly higher for farmers using no-till and reduced tillage systems, compared with conventional tillage. The substitution of herbicides for tillage is resulting in higher herbicide cost, but total weed control costs are the same when figuring the cost of in-crop cultivations.

The growing trend of reduced tillage in the U.S. for soybean production does not appear to have a significant effect on the amount of herbicide used, the use pattern of established herbicides, or the total weed control cost (USDA, 1983; 1984).

A recent study sponsored by Pioneer Hi-Bred International, Inc. (Anonymous, 1982) provides insight on the attitudes and experiences of farmers regarding conservation tillage. The single most often mentioned factor involved with dissatisfaction of the conservation tillage was inadequate weed control. This, in turn, is also a major reason for farmers not adopting conservation tillage. The farmers perceive that there are inconsistencies associated with herbicide performance, and that herbicide costs will be too expensive if they shift to conservation tillage. One of the key factors listed by farmers in this survey, in terms of technical or new product developments that could influence the growth of conservation tillage, is better herbicides. The primary deterrent to faster adoption to conservation tillage is the fear of inadequate weed control.

One possible way to improve the consistency of preemergence herbicides in no-till soybean crops is to make early preplant applications and follow these with at-planting or postemergence applications. This type of herbicide program decreases the performance risks associated with dry weather following at-planting treatments. This program in Iowa has given superior weed control in dry seasons, and can eliminate the need for a preplant "knock-down" treatment to control weed growth at planting time.

There has not been a substantial change in the major herbicides used by soybean farmers using different tillage practices (USDA, 1984). The herbicide use figures are misleading, however, with a substantial percentage of the no-till planted acres being reported as treated with trifluralin, a preplant, incorporated herbicide. The figures, however, do point out increases in the percentage acres treated with preemergence grass herbicides and the apparent multiple herbicide treatments, either as preemergence tank mixes or as postemergence treatments following preemergence applications.

FUTURE PRODUCT DEVELOPMENT

The majority of new products under development or just recently introduced into the soybean market offer the grower alternatives to already established products. The longer term fit and acceptance of these new products is not known, and these products may be used as supplements to established products rather than as total replacements. A systems approach to weed management in soybean crops using a combination of preplant/preemergence herbicides, cultivation and postemergence herbicides has been demonstrated to be the most effective system for maximizing yields (Gebhradt, 1981; Robinson et al., 1984).

An economic deterrent for new product development and commercialization is the pending patent expiration of several of the widely used soybean herbicides. This will likely lead to multiple suppliers and subsequent price erosion, which could discourage companies from investing in research and development for new products. Expansion of product lines is expected to include package mixes and new formulations for products coming off patent. The farmer will be able to choose from a broader range of products and a variety of packaging and formulation options.

New formulation development of established products could lead to changes in biological activity, handling and storage requirements, and the delivery system to the farmer. There is already a growing trend toward bulk handling at the farm level for a number of herbicides. The combination of generic suppliers, plus product line expansion by major manufacturers, reduces the technical and economic opportunities for new product introduction, except for small segments of the soybean market or as supplements to established products.

The trend toward reduced tillage is expected to grow in popularity as farmers gain experience and confidence in managing their weed problems with less tillage and more dependence on herbicides. The use pattern for herbicides is more likely to change rather than the amount of herbicide used (USDA, 1982). The introduction of the new selective postemergence herbicides now provides the farmer with options over in-crop cultivations, if the preemergence products fail to perform. These new selective postemergence products should stimulate adoption to narrow-row soybean culture, where the farmer is totally dependent on chemical weed control after planting. If the preemergence products do not give adequate control, or if hard-to-control weeds emerge as a problem, these newer products provide the farmer with better options than in the past.

There is a need for additional research and development work to maximize the effectiveness of current products in reduced tillage systems, in addition to determining how to best utilize the new products. One of the key issues facing weed scientists is to develop and recommend sound weed management programs to the farmer using reduced tillage, or narrow-row soybean cultures with little or no dependence on tillage, either preplant or in-crop.

Continued research efforts are needed that are directed toward more effectively utilizing herbicides as part of an integrated weed management system to reduce weed control costs and maximize returns to the farmer. A better understanding of weed biology and economic threshold infestation levels is the key to providing soybean farmers sound weed management programs using the available herbicides and the new ones just coming onto the market. The use of integrated weed management systems will include the judicious use of selective herbicides (Shaw, 1982; Hill, 1982). The efforts of weed scientists will continue toward developing integrated weed management systems using cultural, mechanical and chemical means of weed control. These efforts will be aided by the introduction of new products to complement the current products. The goal is to maximize returns to the farmer while preserving the natural resource base and protecting the environment.

There are multiple research opportunities for developing new weed management technology beyond the conventional soybean herbicides now available or currently under development. A number of improved and expanded uses of available herbicides have recently been described by Wakaboyashi and Matsunka (1982) and could be used to develop new uses for herbicides not currently used on soybeans. Techniques to enhance selectivity, such as developing new application methods, modifying formulations, altering chemical structure to obtain selectivity, developing antidotes or safeners, and developing resistant cultivars using new plant breeding techniques are all possible. As an example, improved selectivity of soybeans to s-triazines, through development of antidotes or safeners would make available an effective class of herbicides for use on soybeans, plus reduce concern over use of these products on crops grown in rotation with soybean.

NOTES

D. L. Barnes, Monsanto Agricultural Products, St. Louis, Missouri.

REFERENCES

Anonymous. 1982. Soil conservation: Attitudes and practices. Pioneer Hi Bred International, Inc. Des Moines, Iowa.

Anonymous. 1984. Ag consultant and field man: Weed control manual and herbicide guide. A. Meister Publishing Company, Willoughby, Ohio.

ASA. 1983. Soya Bluebook. American Soybean Association, St. Louis, Missouri.

CAST. 1981. Impact of government regulation on the development of chemical pesticides for agriculture and forestry. Report 87. Council for Agricultural Science and Technology. Ames, Iowa.

Garbett, M. 1984. Expressions. Agrichemical Age 28(3):32.

Gebhradt, M. R. 1981. Cultural and chemical weed control systems in soybeans (*Glycine max*). Weed Sci. 29:133-138.

Gilbert, C. H. 1978. The incresing riskiness of the pesticide business. Farm Chemicals 141(4):20-27.

Hill, G. D. 1982. Herbicide technology for integrated weed management systems. Weed Sci. Suppl. 1, 30:35-39.

Lessiter, F. 1984. 1983-1984 No-till farmer acreage survey. No-Till Farmer 13(3):10-11.

Robinson, E. L., G. W. Langdale and J. H. Stuedemann. 1984. Effect of three weed control regimes on no-till and tilled soybeans (*Glycine max*). Weed Sci. 32:17-19.

Shaw, W. C. 1982. Integrated weed management systems technology for pest management. Weed Sci.: Suppl. 1, 30:2-12.

U.S. Department of Agriculture. 1982. 1980 pesticide use on soybeans in major producing states. ERS Staff Report No. AGES820106. U.S. Government Printing Office, Washington, D.C.

U.S. Department of Agriculture. 1983. Outlook and situation. Economic Research Service, Washington, D.C.

U.S. Department of Agriculture. 1984. Inputs outlook and situation. Economic Research Service, Washington, D.C.

Von Hertwig, K. C. 1983. Manual de Herbicidas, desfolhantes, dessecantes, fitorreguladores e bio-estimulantes. Editora Agronomica Ceres, Sao Paulo, Brazil.

Wakaboyashi, K. and S. Matsunaka. 1982. Crop safening in Japan. p. 439-450. *In* Proceedings 1982 British crop protection conference-weeds.

Wax, L. M. 1973. Weed Control. p. 417-457. *In* B. E. Caldwell (ed.) Soybeans: improvement, production and uses. American Society of Agronomy, Madison, Wisconsin.

RECENT DEVELOPMENTS IN POSTEMERGENCE SOYBEAN HERBICIDES

H. von Amsberg, B. Menck and W. McAvoy

BACKGROUND AND HISTORY

The history of postemergence herbicides goes back to the 1950s when the phenoxy-type herbicides were used mainly for broadleaf weed control in grass crops. The ability of legumes to metabolize 2,4-DB into nonlethal components made this herbicide the first selective postemergence herbicide for soybean.

Herbicides Controlling Broadleaf Weeds

2,4-DB. This herbicide was generally applied at early growth stages of soybean crops as a directed spray. An alternative use was at late soybean growth stages as an over-the-top spray. The main weeds controlled were *Xanthium* spp. and *Ipomoea* spp. Disadvantages of these uses included reduced yields from competition and/or crop injury.

Dinoseb. Dinoseb was used at-cracking or as a directed spray. It provided broad-spectrum weed control, but the application time was limited and crop injury was frequently observed.

Chloroxuron. A major contribution to the acceptance of postemergence herbicides was the discovery of chloroxuron in the 1960s. Chloroxuron at 1.1 kg/ha at four L/ha, gave good control of *Chenopodium album* (L.), *Amaranthus retroflexus* (L.), and *Brassica kaber* (DC.) in Minnesota (Andersen et al., 1969). Kapusta and Rouwenhorst (1969) reported satisfactory control of several broadleaf weeds with chloroxuron at 1.5 kg/ha plus oil (9.35 L/ha) or surfactants (1.17 L/ha) applied at the first to fifth trifoliolate stage of soybean. Crop injury was in the form of soybean leaf drop. Furthermore, the authors observed more late germinating weeds in the earlier treatments compared with the later treatments. The importance of crop selectivity and season-long weed control will be discussed later.

Bentazon. The major contribution of bentazon was its ability to control large-seeded broadleaf weeds including *Abutilon theophrasti* (Medic.), *Xanthium* spp. (L.), *Helianthus annus* (L.), *Ambrosia trifida* (L.), *Datura stramonium* (L.) and perennial weeds, such as *Cirsium arvense* (L.) Scop. and *Cyperus esculentus* (L.). Bentazon controls these weeds and others without serious leaf necrosis or stunting of soybean plants, thereby giving the crop a major competitive advantage over later-germinating weeds.

Acifluorfen. The more intensive use of metribuzin and bentazon during the mid-1970s brought about a shift in the weed spectrum, particularly in the southern USA. *Xanthium* spp. is a highly competitive weed (Barrentine, 1974) that out-competes certain other weeds. When *Xanthium* spp. is effectively controlled, this competition is removed and the population of *Ipomoea* spp. increases. The discovery of acifluorfen filled the gap by not only controlling some of these species, but also controlling *Sesbania exaltata* (Raf.) Cory, *Amaranthus* spp. (L.) and *Solanum nigrum* (L.).

Herbicides Controlling Grass Weeds

In retrospect, it is surprising that effective postemergence grass control lagged behind that of broadleaf weed control. This is even more amazing because the opposite was true for preemergence soybean herbicides where excellent grass control was obtained from propachlor, alachlor and trifluralin. Several compounds, such as chloroxuron, dinoseb and dalapon, were used postemergence for the control of annual grasses, but all had limitations under practical conditions. Therefore, the development of postemergence grass control really did not begin until diclofop was discovered.

Diclofop. This postemergence grass herbicide belongs to the general class of phenoxy-phenoxy type herbicides. Diclofop was effective in controlling *Avena fatua* (L.), *Setaria* spp. (Beauv.) and *Zea mays* (L.) (volunteer corn). The development of diclofop gave a soybean grower the tool to control some of the grasses escaping preemergence treatments, or to control grasses where he could not apply the preemergence treatments. One of the shortcomings of diclofop was that grasses beyond the 2 to 4 leaf stage were unsatisfactorily controlled.

Alloxydim. Although this graminicide was never developed, it nevertheless was the forefather of a new class of herbicide, the cyclohexanediones. The grass spectrum controlled was similar to diclofop; however, several species such as *Digitaria* spp. (Heist.), *Brachiaria* spp. (Ledeb.), *Eleusine indica* (L.) Gaertn., *Panicum* spp. (L.) and *Echinochloa* spp. (Beauv.) were controlled more effectively with alloxydim. Alloxydim was also the first graminicide providing control of *Sorghum halepense* (L.) Pers.

RECENT DEVELOPMENTS IN POSTEMERGENCE HERBICIDES

Graminicides

The Phenoxy-phenoxies. The original discovery of diclofop methyl by Hoechst scientists was followed by a rash of new compounds fitting into this class of chemistry. Hoechst, Maag, Ishihara, Nissan and Dow all discovered new compounds that were different from the original diclofop in herbicidal activity and spectrum.

Fluazifop butyl. The excellent crop tolerance observed earlier for diclofop was maintained when this graminicide was applied to soybean plants. Application rates varying from 0.2 to 0.4 kg/ha provided control of many annual grasses and such important perennial grasses as *Sorghum* spp., *Cynodon dactylon* (L.) Pers. and *Agropyron repens* (L.) Beauv. It is interesting that some annual grass species, such as *Setaria* spp., *Digitaria* spp., and *Panicum* spp., remain on the list of more difficult-to-control grasses.

Haloxyfop methyl. This graminicide is presently under development and is scheduled for registration in the U.S. in 1986. Data from university research show this compound to be roughly twice as active as fluazifop butyl. Another interesting improvement appears to be its faster destruction of plant tissue and a broader spectrum of grass control when compared with fluazifop butyl. DPX-Y-6202, a development compound from Nissan and DuPont, has similar unit activity to haloxyfop.

The phenoxy-phenoxy compounds all appear to have certain characteristics in common. Their physiological mode of action in plants and animals and their observed symptoms are very characteristic. There are also some interesting differences in soil degradation and activity.

The Cyclohexanediones. The original chemistry from Nippon Soda Company, Ltd. was first developed with the common name of alloxydim sodium. Although an effective herbicide for the control of some annual grasses, it provided erratic control under unfavorable environmental conditions and perennial grasses were only suppressed.

Sethoxydim. This herbicide was not only an improvement in unit activity (approx. 2-3 times) as compared to alloxydim, but it also provided broader spectrum as well as more reliable grass control. Sethoxydim also benefits from the addition of oil concentrate.

Broadleaf Herbicides

Diphenylethers. This herbicide family is an old one and has developed into an exceptionally effective one as exemplified in acifluorfen for selective weed control in soybean crops.

Fomesafen. This compound is under development by ICI and reflects excellent weed control of *Amaranthus* spp., *Ipomoea* spp., *Ambrosia* spp. (L.), *Polygonum pensylvanicum* (L.) and *Sesbania exaltata* (Raf.) Cory at application rates as low as 0.25 kg/ha. When compared with acifluorfen, fomesafen appears to be safer on soybean plants at rates that give good weed control. Soil residual activity has been observed, but can be considered of questionable value for soybean, which is particularly competitive in preventing late germination of weeds (Zarecor et al., 1977).

Lactofen. Another addition to the new arsenal of effective diphenylethers is this herbicide from Pittsburgh Plate Glass. Soybean tolerance varies from fair to acceptable and weed control rates start at 0.1 kg/ha. Good control of *Solanum nigrum* and *Sida spinosa* (L.) are special features of this herbicide.

There are several other new herbicides, under investigation and development by American Cyanamid and DuPont, that are highly active and can be used postemergence for control of many broadleaf weeds in soybean.

THE IMPACT OF POSTEMERGENCE HERBICIDES

Never before has there been such an excellent arsenal of postemergence herbicides for any single crop. With the exception of some annual weeds (*Cassia obtusifolia* (L.), *Euphorbia* spp. (L.)), and perennial weeds (*Asclepias syriaca* (L.), *Campsis radicans* (L.) Seem., *Cyperus rotundus* (L.)), growers can now control practically all important weeds in a soybean field using postemergence herbicides.

Weeds need to be controlled before they cause yield reductions. A wealth of research has established that, when even the most competitive weeds, such as cocklebur (Barrentine, 1974) and velvetleaf (Staniforth, 1965; Staniforth and Weber, 1956), were removed prior to the third trifoliolate growth stage of soybeans (or about 24 days after planting), no significant yield reductions could be observed. Annual grasses, especially *Setaria* spp., do not compete as vigorously as velvetleaf or cocklebur. However, populations of grasses tend to be higher than for broadleaf weeds. Grass weeds differ in their ability to compete, according to McCarty (1983). He showed clearly that soybean plants are excellent competitors through shading of potentially late germinating grasses. Thus, competition from *Panicum dichotomiflorum* Michx., found between soybean rows, was significant, while there were no significant yield reductions from *Panicum dichotomiflorum* within soybean rows. Ambrose and Coble (1975) reported that soybean yields were not significantly reduced if fields were weed-free for two or more weeks after emergence.

Johnsongrass, *Sorghum halepense* (L.) Pers., is one of the most vigorous grass weed competitors to soybeans, as reported by McWhorter and Hartwig (1972). Timely removal of *Sorghum* spp. Moech. using any of the new graminicides will maintain the maximum yield potential of the soybean crop. In summary (Dierker and Roof, 1982), if weeds are removed within four weeks after soybean planting, then full yields can be obtained even though there is no residual herbicide activity. Soybean is a very excellent competitor at this stage if it is not injured by the herbicide treatment. As such, growing soybean either in a solid seeding pattern or narrow rows will function as a season-long weed control mechanism. We now have these effective, selective postemergence soybean herbicides.

PROBLEMS AND CHALLENGES FOR POSTEMERGENCE HERBICIDE TECHNOLOGY

Herbicide Combinations

From the previous discussion, it can be concluded that a combination of any of the previously mentioned graminicides with bentazon and/or acifluorfen will provide broad spectrum control of both grass and broadleaf weeds. However, the question arises: How many of these are truly important and how often do all or some occur together in a particular soybean field? Generally, there are no more than six important broadleaf weeds and three important grasses. However, for each farmer, the problem weeds that occur in his fields are the most important. From several Experimental Use Permits (EUP) Programs conducted in soybean fields from 1974 to 1982, it has become evident that rarely are there more than four broadleaf and three grass weeds in a field. The choice of which broadleaf herbicide(s) to combine with which graminicide is then at the discretion of the grower. Naturally, this choice will depend on many individual factors, such as choice of fields, growing practice, seasonal variations, preference of preemergence versus postemergence programs, prices of herbicides and many more factors.

Herbicide Antagonism and Additive Effects

The new opportunities to tank mix various herbicides are not without problems. There is considerable evidence that there is some antagonism among all broadleaf herbicides and graminicides. For particular weed species there may even be a negative effect for a herbicide combination where one may expect additive results. For example, acifluorfen added to bentazon actually decreases the degree and consistency of *Abutilon theophrasti* control.

Nalewaja et al. (1983) and Dexter et al. (1984) did comprehensive studies with various broadleaf herbicides and graminicides. The latter herbicides differed not only in their ability to control various grass species but also were affected differently by various broadleaf herbicides when applied as tank mixtures. It appears as if the phenoxy type broadleaf herbicides are more additive to the graminicides, in terms of grass control, than are herbicides like bentazon or desmediphan. It has been found that the antagonism from the broadleaf herbicides can be overcome by increasing the rate of the graminicide, but again this may differ among herbicide combinations and grass species found in the field. In one field test conducted by BASF field personnel, it was observed that the addition of 2,4-D low volatile ester to sethoxydim increased the control of *Dactylis glomerata* above that of sethoxydim applied alone.

Another challenge is how to optimize the activity of the graminicides. The addition of a non-phytotoxic oil concentrate has been discussed previously and appears to enhance the activity of all graminicides. Research personnel at BASF have conducted various trials evaluating the enhanced effect of reduced water carrier volume on grass control.

OPPORTUNITIES FOR TOTAL POSTEMERGENCE WEED CONTROL

Assuming that during the next 3 to 5 years we understand how to apply the previously discussed herbicide combinations optimally, then 26 combinations are possible using one, two or three particular herbicides at two use rates. A grower could rely entirely on postemergence weed control practices in soybean fields if he knows his weeds.

During 1983, BASF technical personnel conducted 13 trials in Iowa, Nebraska and Minnesota in which the present grower practice (mostly preplant incorporated or preemergence herbicides) was compared with a total postemergence herbicide program. These tests were conducted in side-by-side comparisons using minimum blocks of 4 ha each for conventionally tilled soybeans, minimum-till soybeans and no-till soybeans.

The total postemergence herbicide system contained sethoxydim + bentazon + acifluorfen. The preemergence system varied depending upon whether alachlor + metribuzin or linuron were the major products used. In the no-tillage system, glyphosate, paraquat and 2,4-D were used as burn-down treatments. There was a basic superiority of the efficacy of postemergence herbicides as compared to preemergence herbicides where refuse on the soil surface and environmental conditions modified preemergence herbicide performance (Table 1).

It is our belief that with the excellent, selective postemergence herbicides available now and in the future, this practice will grow substantially, especially for soybean farmers using reduced or no-tillage practices. The benefits derived from reduced cultivation have been discussed previously, and are complimented by the postemergence herbicides available to the soybean grower.

Table 1. Weed control of a total postemergence herbicide program (Post) versus preemergence (Pre) and preplant incorporated (PPI) programs under various tillage systems (unpublished data of J. A. Antognini, BASF Wyandotte Corp., Parsippany, N.J.).

Herbicide system	Weed species					
	Setaria spp.	*Zea mays*	*Abutilon theophrasti*	*Amaranthus* spp.	*Chenopodim album*	*Polygonum pensylvanicum*
			— % control —			
No tillage						
Pre	84	28	89	84	85	84
Post	91	97	97	90	99	90
Minimum tillage						
Pre	94	10	75	86	80	97
Post	98	99	97	96	95	88
Conventional tillage						
Pre	97	--				
PPI	98	62	88			
Post	96	98	97			

NOTES

H. von Amsberg, BASF Wyandotte Corp., Parsippany, N.J.; B. Menck, BASF AG, Ludwigshafen, W. Germany; and W. McAvoy, BASF Wyandotte Corp., Parsippany, N.J.

REFERENCES

Ambrose, L. G. and H. D. Coble. 1975. Fall panicum competition in soybeans. Proc. Southern Weed Sci. Soc. 28:36.

Anderson, R. N. and D. J. Ziehart 1969. Evaluation of herbicides in a soybean weed nursery. Res. Rep. North Central Weed Control Conf. 26:152-155.

Barrentine, W. L. 1974. Common cocklebur competition in soybeans. Weed Sci. 22:600-603.

Dexter, A. G., J. D. Nalewaja and S. D. Miller. 1984. Antagonism between broadleaf and grass control herbicides. Abstr. Weed Sci. Soc. of Amer. No. 13.

Dierker, W. W. and M. E. Roof. 1982. Review of weed competition in soybeans and corn. Publ. MP 540. Univ. of Missouri Extension Division, University of Missouri, Columbia.

Kapusta, G. and D. L. Rouwenhorst. 1969. Weed control efficacy as influenced by chloroxuron applications at several growth stages. Res. Rept. North Central Weed Control Conf. 26:133.

McCarty, M. T. 1983. Economic thresholds of annual grasses in agronomic crops. Ph.D. Dissertation, North Carolina State Univ. Library, Raleigh.

McWhorter, C. G. and E. E. Hartwig. 1972. Competition of johnsongrass and cocklebur with six soybean varieties. Weed Sci. 20:56-59.

Nalewaja, J. D., S. D. Miller and A. G. Dexter. 1983. Grass and broadleaf control herbicides mixtures. Proc. North Central Weed Control Conf. 38:29.

Staniforth, D. W. 1965. Competitive effects of three foxtail species on soybeans. Weeds 13:191-193.

Staniforth, D. W. and C. R. Weber. 1956. Effects of annual weeds on the growth and yield of soybeans. Agron. J. 48:467-471.

Zarecor, D., E. Ellison, C. Cole and J. Lunsford. 1977. Fluchloralin and bentazon in a systems approach to soybean weed control. Proc. South. Weed Sci. Soc. 30:91-92.

LOW VOLUME APPLICATIONS OF POSTEMERGENCE HERBICIDES
USING ELECTROSTATICS, CDA'S, AND SOYBEAN OIL

L. E. Bode and L. M. Wax

Applications of pesticides with ground sprayers typically range from 100 L/ha for some soil-applied herbicides to 500 L/ha for contact pesticides that require complete coverage on the foliage. A major factor limiting the reduction of spray volumes has been that the nozzles required for low-volume application have very small orifices that easily become clogged. Another reason for using higher-volume spray nozzles is that high-volume nozzles produce larger droplets, which are more likely to reach the target and less likely to drift out of the field. The development of practical controlled droplet atomizers (CDAs) and electrostatic sprayers have created a renewed interest in reducing the amount of spray solution required for applying pesticides (Bals, 1975; Law, 1978).

Equipment for low-volume application has also made it economically feasible to use crop oils as carriers for pesticides. Using a crop oil in place of water as a carrier has the potential for improving application efficiency. Spray droplets of oil are relatively nonevaporative, may penetrate into plant foliage more easily than water, and should drift less than water droplets of the same size (ASA, 1984).

Developing equipment and techniques to apply pesticides at 20 L/ha or less is difficult, but the advantages make such developments desirable. Tests were conducted at the University of Illinios to measure the physical properties of soybean oil over a range of temperatures, and to determine the flow characteristics of soybean oil through several types of spray nozzles. Several field experiments were also conducted to evaluate the potential of low-volume application of pesticides using two CDAs, two electrostatic sprayers, and experimental air-assist flat fan nozzles. Water and soybean oil were compared as pesticide carriers.

PROCEDURE

The Micromax spinning nozzle, the Spraying Systems prototype CDA, and the Deere Electrostatic sprayer were used to apply postemergence grass and broadleaf weed herbicides for pest control in soybean plots. Water and soybean oil as carriers were applied at 15 to 30 L/ha. CDA units were operated at disk speeds of 2,000 and 3,500 rpm, and the electrostatic unit was operated with and without a charge on the spray cloud. An experimental, low-volume, flat fan nozzle having an external air assist to maintain a full fan angle when applying viscous soybean oil solutions was compared to conventional flat fan nozzles. The ICI electrostatic atomizer (the Electrodyn) was used to apply fluazifop-butyl (Fusilade) at 0.6, 0.3 and 0.15 kg/ha in a total spray volume of 1 L/ha (Coffee, 1979). Comparisons of weed control were made with 187 L/ha applications using conventional flat fan nozzles.

RESULTS

Viscosity of Crop Oil Mixtures

The viscosity of several crop oils and pesticide mixtures was measured over a temperature range of 5 to 10 C. Results for some of the measurements are given in Table 1. The viscosity of the crop-oil-based mixtures is much higher than that of water, especially at low temperatures. The viscosity of water at 40 C was 0.6 centistoke (cSt), while the viscosity of soybean and cottonseed oil at the same temperature was over 30 cSt. The viscosity of pesticide formulations varied greatly. For example, Pydrin without the active ingredient (blank Pydrin) had a viscosity of 1.1 cSt at 40 C, whereas the viscosity of blank Pounce increased to 2.7 cSt. Adding the active ingredient to Pounce did not change the viscosity, but the active ingredient increased the viscosity of blank Pydrin from 1.1 to 2.3 cSt. Concentrations of pesticides mixed in soybean oil had different viscosities from those of pure soybean oil, but the changes were small in comparison with the difference between the viscosity of water and that of soybean oil. For example, 1% Pydrin 2.4 EC concentration in fully refined soybean oil decreased the viscosity from 30.5 cSt to 29.9 cSt at 40 C. An 8.3% Pydrin concentration further reduced the viscosity to 24.4 cSt (Table 1).

Table 1. Viscosity of several crop oil mixtures over a temperature range of 5 to 40 C.

Fluid	Temperature (C)							
	5	10	15	20	25	30	35	40
	— Viscosity (centistokes) —							
Water	1.5	1.3	1.1	1.0	0.9	0.8	0.7	0.6
Pydrin blank (diluent only)	1.9	1.7	1.5	1.4	1.3	1.2	1.2	1.1
Pydrin 2.4 EC	5.3	4.4	4.0	3.6	3.1	2.8	2.5	2.3
Pounce blank (diluent only)	5.9	5.1	4.7	4.0	3.6	3.3	2.9	2.7
Pounce 3.2 EC	6.6	5.6	4.9	4.2	3.7	3.4	2.9	2.7
Fully refined (FR) soybean oil	138.0	101.0	86.5	67.3	54.6	42.6	36.5	30.5
Once refined (OR) soybean oil	145.7	101.1	81.2	64.0	52.7	43.2	36.4	30.6
Cottonseed oil	154.5	109.1	89.3	70.7	58.6	46.5	39.1	30.8
Trifluralin, oil formulation	163.3	118.0	94.2	74.6	59.9	44.7	38.6	31.8
1% Pydrin 2.4 EC in FR soyoil	132.2	97.7	79.7	62.8	52.3	42.9	35.1	29.9
1% Pydrin 2.4 EC in OR soyoil	130.3	95.5	73.7	63.2	52.5	42.9	35.1	29.9
8.3% Pydrin 2.4 EC in FR soyoil	94.1	75.4	62.2	49.1	40.9	34.2	28.1	24.2
8.3% Pydrin 2.4 EC in OR soyoil	101.0	73.6	58.4	48.9	42.4	33.9	28.0	24.4
1.6% Pounce 3.2 EC in FR soyoil	124.2	93.9	73.8	62.1	51.5	43.9	34.9	29.5
1.6% Pounce 3.2 EC in OR soyoil	118.9	94.2	74.9	62.3	52.0	43.1	34.5	29.3
3.1% Pounce 3.2 EC in FR soyoil	119.5	89.2	69.7	58.6	49.4	42.9	33.0	28.1
3.1% Pounce 3.2 EC in OR soyoil	120.1	89.1	71.9	59.5	48.5	41.1	33.0	28.0

Flow Rate of Soybean Oil through Spray Nozzles

High-viscosity soybean oil-pesticide mixtures affect flow rates and nozzle spray patterns. Flow rates and patterns were measured for several nozzle types at temperatures of 10, 25, and 40 C. Results for three of the nozzle types tested are given in Table 2. Flow rates of soybean oil through small, regular flat fan nozzles are similar to those of water. At 10 C, the flow rate of soybean oil is slightly less than that of water. At a temperature of 25 C, the flow rate of oil is slightly greater than that of water and increases further at a temperature of 40 C. The fan angle is reduced when soybean oil instead of water is sprayed. Table 3 shows the fan angle obtained over a range of pressures and temperatures. Much higher spray pressure or high temperature is required to develop a full fan angle from regular flat fan nozzles.

Flow rates through flooding nozzles decreased dramatically when soybean oil was sprayed as compared with when water was sprayed. As the temperature increased,

Table 2. Flow rate of soybean oil through nozzle tips at various pressures and temperatures.

Nozzle	Pressure	Water flow rate	Temperature (C)		
			10	25	40
	kPa	mL/min	− Flow rate of soyoil (mL/min) −		
Flat fan 8001	138	261	246	265	271
	276	371	357	385	397
	345	413	400	432	442
Flat fan 8002	138	537	474	538	574
	276	761	716	792	820
	345	852	802	894	918
Flat fan 8001 LP	138	458	413	464	479
	276	647	622	684	712
	345	723	682	758	784
Flat fan 8002 LP	138	833	768	861	902
	276	1177	1152	1263	1281
	345	1317	1305	1422	1401
Flooding TK.5	138	265	124	142	144
	276	375	203	229	272
	345	420	234	264	312
Flooding Tk1	138	526	274	339	304
	276	746	429	530	556
	345	836	482	598	630
Hollow cone TX1	138	45	34	44	54
	276	64	63	72	85
	345	72	75	84	98
Hollow cone TX4	138	178	304	378	367
	276	254	468	536	502
	345	284	522	602	534
Hollow cone TX6	138	269	405	450	432
	276	379	592	636	552
	345	424	666	682	608

Table 3. Spray patterns from flat fan and flooding nozzles with soybean oil.

Nozzle	Pressure	Temperature (C)		
		10	25	40
	kPa	— Spray pattern angle, degrees —		
Flat fan 8001	138	28	30	50
	276	44	45	70
	345	50	50	75
Flat fan 8002	138	28	35	60
	276	45	60	75
	345	55	75	80
Flat fan 8001 LP	138	50	45	65
	276	60	65	80
	345	75	80	80
Flat fan 8002 LP	138	45	55	60
	276	70	75	80
	345	75	80	80
Flooding TK.5	138	0	0	0
	276	0	0	0
	345	0	0	0
Flooding TK1	138	0	0	0
	276	0	0	90
	345	0	0	90

the flow rate was closer to water, but at 40 C the flow rate was still about 25% less than that for water (Table 2). It is almost impossible to obtain a spray pattern from flooding nozzles, unless unrealistic spray pressures are used (Table 3).

Flow rates through hollow cone nozzles increased when soybean oil was sprayed compared with when water was sprayed. As temperatures increased from 10 to 25 C, the flow rate for the TX4 hollow cone nozzle increased, but at a temperature of 40 C the flow rate decreased (Table 2). For the TX6 hollow cone nozzle, the flow rate at 40 C was generally less than the flow rate at 10 C. Spray patterns were very erratic when applied through hollow cone nozzles.

Field Study Using Soybean Oil as a Carrier

Tests have shown that properties of pesticide mixtures rapidly approach the properties of water as water is added to the mixture. Conventional nozzles can be used to apply soybean oil if the oil can be mixed with water and applied uniformly. Table 4 summarizes the results of a study where two rates of a postemergence grass herbicide were mixed at varying concentrations of water and soybean oil and applied at a total volume of 37 L/ha.

In this study, foxtail control was about the same when the herbicide was applied with water at 37 L/ha as it was when applied at 187 L/ha. However, as oil was added to the water, grass control actually decreased, especially with the 50% concentration. The decreased control was not caused by the activity of the oil. Rather the cause was the poor spray pattern and change in flow rate that occurred when oil was used. For example, when a 50% oil-water mix was applied through the flooding nozzle,

Table 4. Foxtail control from two rates of a postemergence herbicide applied at 37 L/ha.

Equipment	0[b]	12.5	25	50	Average
		Percent soybean oil[a]			
			– % control[c] –		
Flat fan 80067 at 234 kPa	74.8 rs	55.8 s	70.8 rs	55.0 s	64.1 r
Flooding TK.5 at 248 kPa	79.8 r	60.8 rs	59.2 rs	25.0 t	56.2 rs
Micromax, 3,500 rpm	64.8 rs	58.3 rs	56.7 s	21.7 t	50.4 s
Electrostatic charged	64.2 rs	72.5 rs	61.6 rs	59.2 rs	64.4 r
Electrostatic uncharged	66.7 rs	64.2 rs	53.3 s	53.3 s	59.4 rs
(Flat fan at 187 L/ha)	(78)				
(Flooding at 187 L/ha)	(82)				
Rate			– % control[d] –		
High	87.1 r	71.0 s	75.0 s	52.3 t	
Low	53.0 t	53.7 t	45.7 t	33.3 u	
Average	70.1 r	62.3 rs	60.3 s	42.8 t	

[a]85% soybean oil, 15% emulsifier.

[b]1% crop oil concentrate added.

[c]Data followed by the same letters in these five columns are not significantly different at the 10% level.

[d]Data followed by the same letters in these rows are not significantly different at the 10% level.

the spray pattern was not acceptable. The result was inadequate spray coverage on the weeds.

Applications with the Electrodyn

Fluazifop-butyl was applied to soybean plots with the Electrodyn and flat fan nozzles when the grass was 5 to 10 cm tall, and again when it was 15 to 20 cm tall. Table 5 summarizes the foxtail (*Setaria* spp.) control ratings taken 17 days after the first treatment and 10 days following the second treatment. Fluazifop-butyl was more effective when applied to 5- to 10-cm-tall grass than when applied to 15- to 20-cm grass. The Electrodyn resulted in excellent control when applying the 0.6 kg/ha rate on short grass, but control was reduced for the 15- to 20-cm grass. Flat fan nozzle applications

Table 5. Percentage foxtail control[a] in soybean plots from fusilade.

Equipment	Foxtail height	Fusilade rate (kg/ha)		
		0.6	0.3	0.15
	cm		– % control[b] –	
Electrodyne at 1 L/ha	5 to 10	93.8 r	47.5 tu	37.5 uv
Flat fan at 187 L/ha		95.8 r	86.3 rs	52.5 t
Electrodyne at 1 L/ha	15 to 20	75.0 s	47.5 tu	30.0 v
Flat fan at 187 L/ha		97.0 r	91.3 r	32.5 v

[a]Rating was taken 17 days after first treatment and 10 days after second treatment. 0 = no control; 100 = complete control.

[b]Numbers followed by the same letters are not significantly different at the 5% level.

at 187 L/ha out-performed the Electrodyn when Fluazifop-butyl was applied at reduced rates, and when the grass was taller and under stress. The flow rate (0.15 kg/ha)
gave ineffective control regardless of the method of application.

Foliar Herbicide Studies

Tables 6 and 7 summarize the results of studies in which low-volume applications
of postemergence herbicides were compared with conventional applications. Table 6
shows results using electrostatic spray units and the Micromax CDA to apply the herbi
cides in a total spray solution of 28 L/ha. Regular flat fan and flooding flat fan nozzles
were used for applications of 140 and 280 L/ha. A postemergence grass and a postemergence broadleaf herbicide were each applied at two rates. The most consistent
foxtail control was obtained from 280 L/ha applications using either the flooding or
flat fan nozzles, but there was no significant difference in control between the Micromax operating at 3,500 rpm and the conventional nozzles at either 140 or 280 L/ha.
The high herbicide rate resulted in significantly better grass control than the low rate.

Velvetleaf (*Abutilon theophrasti* Medic.) and pigweed (*Amaranthus* spp.) control
from the postemergence broadleaf herbicide was essentially the same regardless of the
method of application. Soybean injury from the broadleaf herbicides generally relates

Table 6. Results[a] from foliar applications of herbicides to soybean plots.

	Foxtail control	Velvetleaf control	Pigweed control	Soybean injury	Yield
	%	%	%	%	kg/ha
Equipment					
Micromax (28 L/ha)					
2,000 rpm	75.8 tu	80.0 r	68.3 rs	0.0 s	2068 r
3,500 rpm	87.5 rs	83.3 r	71.7 rs	0.0 s	2351 r
Electrostatic (28 L/ha)					
Charged	68.3 u	85.8 r	73.3 rs	0.0 s	2317 r
Uncharged	83.3 st	77.5 r	77.5 r	0.0 s	2384 r
Flat fan					
8002 (140 L/ha)	88.3 rs	75.8 r	65.0 rs	10.8 r	2074 r
8003 (280 L/ha)	93.3 r	85.8 r	70.8 rs	10.8 r	2088 r
Flooding					
TK2 (140 L/ha)	91.7 rs	83.3 r	63.3 rs	12.5 r	2398 r
TK3 (280 L/ha)	95.0 r	78.3 r	51.7 s	13.3 r	2018 r
Herbicide					
Grass herbicide					
High rate	88.3 r	--	--	0.0 s	2088 s
Low rate	82.5 s	--	--	0.0 s	2142 s
Broadleaf herbicide					
High rate	--	78.1 r	72.3 r	13.3 r	2425 r
Low rate	--	84.4 r	63.1 r	10.8 r	2243 rs

[a]Numbers followed by the same letters in a column are not significantly different at the 5%
level.

Table 7. Foxtail and velvetleaf control[d] and soybean injury from postemergence herbicides applied at 19 L/ha[a].

Equipment	Foxtail control			Velvetleaf control			Soybean injury		
	Water[b]	Soyoil[c]	Average	Water	Soyoil	Average	Water	Soyoil	Average
	— % —			— % —			— % —		
Micromax									
2,000 rpm	76.6 s	85.0 rs	80.8 s	93.0 r	86.7 rs	89.3 r	13.3 r	11.7 rs	12.5 r
3,500 rpm	76.7 s	81.7 rs	79.2 s	92.7 r	60.0 tuv	76.3 rs	11.7 rs	6.7 stu	9.2 rs
S.S. CDA									
2,000 rpm	83.3 rs	63.3 t	73.3 s	89.3 rs	40.0 vw	64.7 s	11.7 rs	5.0 tuv	8.3 rs
3,500 rpm	75.0 s	80.0 rs	77.5 s	92.7 r	80.0 rst	86.3 r	13.3 r	6.7 stu	10.0 rs
Electrostatic									
Charged	43.3 u	33.3 uv	38.3 t	63.3 tu	76.6 rst	70.0 s	0.0 v	5.0 tuv	2.5 t
Uncharged	23.3 vw	20.0 w	21.7 u	30.0 w	43.3 uvw	36.7 t	0.0 v	1.7 uv	0.8 t
Air-assist flat fan	91.7 r	88.0 r	89.9 r	70.0 rst	68.3 st	69.2 s	10.0 rst	6.7 stu	8.3 s
(Flat fan at 187 L/ha)	(65.0)			(87.0)			(5.0)		
Average	67.1 r	64.5 r		75.9 r	65.0 s		8.6 r	6.2 s	

[a]Data are for the high rate of application only.
[b]1% crop oil concentrate added.
[c]85% soybean oil and 15% emulsifier.
[d]Data followed by the same letters in these columns are not significantly different at the 10% level.

to the coverage obtained on the foliage. Injury occurred only from high-volume applications, with flooding nozzles giving slightly more injury than flat fan nozzles. There was no difference in soybean yield from any type of application (Table 6).

Table 7 summarizes two studies comparing low-volume application of postemergence herbicides using a total volume of 19 L/ha. Both water and soybean oil were used as carriers. Applications of water included a 1% concentration of a commercially available crop oil concentrate, and the soybean oil carrier consisted of 85% soybean oil and 15% emulsifier. Two rates of each herbicide were applied, but the data in Table 7 are only for the high rate.

Foxtail control was best from applications with the air-assist flat fan nozzle. There was no difference between the two CDA units at either disk speed. The electrostatic unit resulted in the poorest control. Foxtail control was 65% when the herbicide was applied with flat fan nozzles at 187 L/ha. Averaged over all types of application, there was no significant difference in foxtail control between water and soybean oil as a carrier. However, except for the 2,000-rpm speed with the SS unit, CDA units indicate a trend for improved control with soybean oil.

Velvetleaf control was generally best from water applications with the CDA units (Table 7). Water was significantly better than soybean oil as a carrier, especially for the applications with the CDA units. Soybean injury from applications of the broadleaf herbicide followed the same trends as the velvetleaf control data (Table 7). Soybean yields were collected and analyzed, but there were no significant differences in yield among the treatments.

NOTES

L. E. Bode, Agricultural Engineering Department, University of Illinois, Urbana, IL 61801. L. M. Wax, U.S. Department of Agriculture, ARS, Urbana,IL 61801.

Trade names are used for the purpose of providing specific information. Mention of specific equipment does not constitute an endorsement by the U.S.D.A. or the University of Illinois, and does not imply approval or the exclusion of other products that may be suitable.

REFERENCES

ASA. 1984. Ag-Chem uses of soybean oil. Proceedings of a workshop. American Soybean Association, St. Louis, Missouri.

Bals, E. J. 1975. The importance of controlled droplet application (CDA) in pesticide applications. p. 153-160. *In* Proceedings 8th British insecticide and fungicide conference.

Coffee, R. A. 1979. Electrodynamic energy: a new approach to pesticide application. p. 777-789. *In* Proceedings 1979 crop protection conference.

Law, S. R. 1978. Embedded electrode electrostatic induction spray charging nozzle: theoretical and engineering design. Trans. ASAE 21:1096-1104.

THE ROLE OF SEED DORMANCY AND GERMINATION IN
DEVISING WEED CONTROL METHODS

R. B. Taylorson

It is widely recognized that seed dormancy and germination[1] strategies contribute to the perpetuation of weeds as agricultural pests. In fact, without dormancy few annual plant species would be serious weeds, particularly in temperate climates. Also, the status of weed and crop germination contributes to the safe use, effectiveness, and to some extent, selectivity of herbicides.

The activity of preemergence herbicides is based on preventing the emergence of susceptible germinating weeds (or destroying them once they emerge), but leaving the crop seeds relatively unaffected. Under proper conditions, herbicides are remarkably effective in preventing the establishment of weeds in crops. It is commonly recognized, however, that preemergence herbicides fail to destroy dormant weed seeds. Thus, seeds that fail to germinate (remain dormant) during the period when effective herbicide concentrations are present, escape chemical control measures. The proportion of seeds that escape may be large compared to the total buried seed population. This suggests that, by understanding seed dormancy in weeds, we might advance weed control technology and better understand the impact of currently used cropping practices on weed behavior.

BIOLOGY OF SEEDS IN SOILS

It is necessary to consider some facts about the biology of weed seeds in soils. Firstly, the numbers of weed seeds in agricultural soils are usually very large and also quite variable; often about 100 million seeds/ha within the tilled layer (Chancellor, 1981). Secondly, the number of seedlings that emerge in any one year is quite low relative to the total number of seeds present; perhaps only 5 to 10% of the total seed populations; however, studies on rates of depletion of buried seed populations indicate annual losses of 20% or more (Roberts, 1970). Differences between the number that emerge and the total losses in one year may be ascribed to predation, failure to emerge after germination, etc. These numbers also indicate that a very high percentage of the total buried seed population must survive from one year to the next. In fact, seeds of many weed species are capable of prolonged survival in soils (Sagar, 1970). Persistence for several years is very common and survival for decades is not uncommon. Some species may survive for centuries (Odum, 1965). It is therefore axiomatic that seed dormancy contributes greatly to the weed trait of persistence.

Whether a preemergence or postemergence type, most herbicides act on the weed seeds that germinate, but their activity on dormant seeds is low (Mitchell and Brown,

1947). In other words, they act on the 10% or less of the total weed seed population that germinates, but do not destroy the much larger portion that remains dormant. In reality, then, herbicides give a rather cosmetic effect in terms of reduction of total seed populations, destroying only those that would have germinated anyway and leaving the remaining dormant seeds to infest subsequent crops.

Recognition of these facts by weed scientists has engendered several new approaches to the weed problem, based on influencing seed dormancy. One of these is to develop techniques whereby dormancy can be overcome (i.e., germination stimulated) in a large portion of the soil-borne population, followed by conventional destruction of the resultant seedlings. Barring major inputs of fresh seeds, such techniques could minimize requirements for future herbicide use and other weed control techniques, assuming no vegetatively propagated perennial species were present. Another approach is to attack the dormant seed directly and destroy its viability by either chemical, biological, or physical agents.

MECHANISMS OF SEED DORMANCY

In order to devise weed control technologies based on influencing seed dormancies, one must understand the actual dormancy mechanisms occurring under field and laboratory conditions. A detailed discussion of seed dormancy is far beyond the scope of this report (for a more complete analysis see Bewley and Black, 1982). However, the major points that relate particularly to the development of techniques leading to stimulation of germination of dormant seeds can be presented.

Seeds display various mechanisms for enforcing dormancy upon dehiscence from the parent plant. Such mechanisms have been termed primary, or innate. In many cases, it is not easy to categorize the type of primary dormancy within a species. One reason for this is that many species have multiple types of primary dormancy. That is, more than one requirement must be met before germination can be evoked. It is important to recognize, however, that one type is usually less intense than another, and that the most intense may block the entire system until its requirement has been met. Many weed species display this sequential behavior and pass through stages of dormancy intensity. An example is fall panicum (*Panicum dichotomiflorum* Michx.) (Taylorson, 1980). The initial intense dormancy, during which the seeds are virtually insensitive to most stimuli, may be overcome by a period of after-ripening. Subsequently, the seeds become sensitive to stimulation by light. A transition from intensely dormant to non-dormant proceeds in stages, which are shown graphically in Figure 1.

The scheme depicts the transition from a relatively intense dormancy to a relatively non-dormant condition as a continuous process, marked at certain points by responsiveness to particular stimuli. In the most intense stage, few if any, short-term stimuli can overcome dormancy. In the least intense stage, only an appropriate manipulation of temperature may be required. It should be evident that individual seeds within a population of a species may be distributed over a substantial portion of the dormancy stages. Thus, providing appropriate temperature may bring about only a small amount of germination, a brief irradiance may add another increment, and a prolonged irradiance still more, etc. Similarly, separate collections of a given species often behave differently. Part of the differences can be accounted for by variations during maturation, which bring the seeds to various stages of the transition. The scheme also suggests the transitions of dormancy are reversible; e.g., induction of secondary dormancy by unfavorable environments. Also, the transitions, whether proceeding from intense to

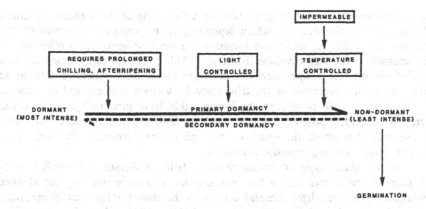

Figure 1. Changes in dormancy intensity as stages in a sequential process.

non-dormant or vice versa, do not necessarily pass through each depicted stage. For example, seeds that require prolonged chilling do not necessarily pass through a light-requiring stage. The scheme provides a conceptual framework that depicts dormancy as a sequential series of events. The term sequential dormancy is suggested as an alternative to separate kinds of dormancy.

So far as our interest in trying to manipulate dormancy in buried seed populations is concerned, it is the author's opinion that analysis of the intense type of primary dormancy, e.g. comparable to the intense stage of fall panicum, is not warranted. Such mechanisms are viewed as temporary blocks to prevent germination, and while of academic interest, are of no great concern in attempting to stimulate germination of weed seeds in soils. The less intense types of primary dormancy and the so-called secondary types of dormancy (or acquired) are of interest, however. Secondary dormancies arise chiefly under unfavorable environmental conditions in seeds that were relatively non-dormant initially or in a less intense type of primary dormancy. Thus, freshly matured seeds of many weeds have an intense primary dormancy that serves mainly to prevent their immediate germination. For many species overwintering in the soil, other, less intense, types of dormancy exist. It is these types that have to be dealt with in devising methods to stimulate germination.

Several types of dormancy seem to be prevalent in seeds that have been over-wintered. There are two well-known types: Seeds that have impermeable seedcoats and seeds in which germination is light controlled. A third type is related to temperature control (requiring a certain minimum or maximum, or an amplitude of alternation), but it may be so interrelated with requirements for light, so as to be difficult to consider separately. Seeds requiring only a temperature related stimulus possess the least intense form of dormancy. Also, it is evident that a temperature stimulus apparently can act independently to initiate some germination in the field. A fourth general type may be those seeds that have entered a secondary dormancy, which prevents response to most stimuli.

Seeds with covering structures that are impermeable to water, and possibly gases, are common within certain plant families; e.g., the Convolvulaceae, Geraniaceae, Malvaceae, and especially the Leguminosae. This condition is often termed hardseededness. Some species with impermeable seeds are among the most persistent known. Dormancy is imposed so long as water is prevented from reaching the embryo. Other than the restriction in water uptake, there is usually no other form of dormancy in many of these seeds. Hardseededness is thought to be overcome naturally by microorganisms, weathering, abrasion or a fracturing of the seedcoat. A structural discontinuity in the seedcoat, which depending upon species etc. is termed the hilum, chalaza, or strophiole, sometimes furnishes the route of water entry other than by a disturbance of the seed envelope (Rolston, 1978). In certain cases these may be controlled via discrete temperature and moisture conditions. Impermeability to water may be readily overcome in the laboratory by various physical and chemical techniques that degrade or disrupt the impermeable layer surrounding the embryo. So far, however, little progress has been made in finding chemicals, other than the corrosive ones that attack the seedcoat, that relieve the dormancy. Heating, especially by fire, often induces germination in nature.

Another major type of dormancy under field conditions is controlled by light. In seeds buried deeper than a few mm, dormancy is maintained by the absence of light. Perception of a light stimulus is through the almost ubiquitous, photoreversible plant pigment, phytochrome (Shropshire and Mohr, 1983). Phytochrome participates in many plant processes including growth and development, flowering, etc. In simple terms, the low energy requiring responses are initiated by red irradiations of about 660 nm, which convert inactive phytochrome (P_r) to the active form (P_{fr}). Subsequent far-red irradiations (730 nm) return the pigment to its inactive form, P_r. Germination is usually dependent upon the formation of an adequate amount of P_{fr}. Germination of many kinds of seeds is controlled by changes in phytochrome, either as a primary dormancy or as an acquired type once the seed is incorporated in the soil environment. In the latter case, I refer to seeds that were initially unresponsive to light, either because a more intense primary dormancy prevented the response or because relatively non-dormant seeds assumed a secondary dormancy because of unfavorable germination conditions, which ultimately led to light-requiring seeds (Figure 1). High temperature, water stress and prolonged far-red irradiation commonly bring about such changes in non-dormant seeds. Good evidence exists for phytochrome control of weed germination under field conditions. Phenomena such as flushes of weed germination following soil disturbances (Wesson and Wareing, 1969), and effects of crop canopy shading on reducing weed germination appear to act through phytochrome. In contrast to hardseeds, stimulation of germination in light-requiring seeds is readily accomplished in the laboratory by a variety of chemicals, from simple inorganics such as nitrate ion to substances such as gibberellins, ethylene, etc. (Bewley and Black, 1982).

OTHER ASPECTS OF SEED DORMANCY

Under field conditions, changes in seed dormancies (especially in light-requiring seeds) may be quite dynamic. The changes are manifested by profound differences in response to light and temperature, and possibly other stimuli, depending upon the season. It has long been recognized that emergence behavior in many weeds is characterized by season of the year. For example, some are recognized as being spring or early summer weeds, others as summer and fall emerging weeds. Less well recognized is that these observations bear not only on environmentally favorable germination conditions but also on the relative dormancy of the seed. Thus, seeds held under field conditions and exhumed at intervals often reveal a cyclical dormancy behavior (Karssen, 1980). Therefore, at some times of the year, particularly corresponding with normal emergence times, seed dormancy is easily overcome when the appropriate temperature and irradiance requirements are satisfied. However, at other times, seeds may completely fail to respond to the same (and perhaps other) stimuli.

There is some evidence to suggest that depth of placement of seeds within the soil profile affects dormancy (Taylorson, 1972). Generally, seeds at shallow depths tend to undergo more drastic changes in relative dormancy than deeper ones. Thus, persistence may increase as depth increases. Such changes may be expected owing to the extremes of temperature and moisture change that occur near the surface, as opposed to more moderate changes as depth increases. Also, light as a factor in germination is limited to within a few mm of the soil surface unless cracks, action of raindrops, and movement of soil by tillage allows light to penetrate more deeply.

Most weed seed germination seems to occur within a centimeter or so of the soil surface, especially among the small-seeded species. Since alternating temperatures are more favorable for germination in many species than constant ones, the more frequent and drastic oscillations near the surface play a significant role in both changes in relative dormancy and initiation of germination. In the latter case, temperature may function either as an independent stimulus or to render light-requiring seeds maximally sensitive to very brief or low energy irradiations. Conversely, unfavorable temperatures may induce secondary dormancy. Shading by foliage also minimizes amplitudes of temperature change. Species requiring alternating temperatures may use this mechanism to recognize an open site. The extremes in temperature and water stress encountered near the surface must interact strongly with levels of relative dormancy.

Unfavorable gaseous composition of the soil atmosphere may deter weed germination as depth increases. However, recent evidence indicates that under most conditions, oxygen penetration into soils is not limiting. Similarly, accumulations of CO_2 to about 5%, or depletion of O_2 levels to 5% or less, may be required before germination of many species is impaired (Koller, 1972). Movement of gases is restricted in wet soils. Also, where vigorous microbial growth occurs, microsites of reduced O_2 availability may arise, particularly where diffusion rates are limited. I question, therefore, the idea that poor germination at depths below a few centimeters is associated with a lack of O_2 or accumulation of CO_2, except under stress conditions. Much less is known about the impact of soil gaseous environment on changes in relative dormancies in seeds.

LONGEVITY

Generally, persistence of seeds in soils can be related to the sum of positive and negative impetuses received by the seed that effect its ability to overcome dormancy. Thus, seeds held in an environment most unfavorable for germination will generally persist longest. Obviously, this neglects other major considerations—e.g., attack by predators and microorganisms, adequacy of storage materials for preservation of life processes, etc. Most of the early "classic" studies on persistence may be misinterpreted, owing chiefly to excessive depth of placement and inadequate sampling. More recently, studies by Roberts (1970) have provided data more applicable to agricultural situations. Typically, Roberts finds that maximum rates of decay (ca. 45%/year averaged over the entire seed population) occur where tillage is performed several times a year. Minimum rates of decay (ca. 22%/yr) occur with no soil disturbance (e.g., no-till). Rates of 50%/yr indicate that after 7 years, seed populations will be ca. 1% of that originally present. Work by Lueschen and Anderson (1980) points out that, when starting from very high numbers, it may be difficult to reduce populations adequately by natural attrition.

Not all weed species are capable of prolonged persistence. In fact, under agricultural conditions the average viability period for many species is less than 5 years. However, replenishment of seeds by maturing weeds can quickly offset any reductions of the soil populations. Using current technologies, it is difficult technically and economically to maintain the level of weed control necessary to achieve substantial diminution of seed populations. Accordingly, it is desirable that new techniques to control soil-borne weed populations be devised.

MANIPULATION OF DORMANCY AS A WEED CONTROL STRATEGY

It has long been recognized that weed seed populations in soils could be greatly reduced if dormancy could be relieved all at once and the resultant seedling flush destroyed. Similarly, induction of some long lasting form of dormancy, or destruction of seed viability, would reduce the severity of weed infestations. These approaches identify the soil-borne population of seeds as the basis of the annual weed problem, and control of dormancy or viability as a rational means of attacking it.

Prevention of Germination

Induction of long-lasting dormancies in soil seed populations, as a weed control approach, has never received significant research attention. To simply induce a dormancy would mean that the seed might ultimately be capable of germination except where viability was lost while the dormancy effector was still active.

Methods of directly attacking seed viability represent potentially ideal means of reducing soil-seed populations. Both biological and chemical approaches may be taken. Until now, there has been little interest in using biologicals for this purpose. Use of chemicals to destroy viability of seeds in soils is a well established practice. The intent is usually to "sterilize"; i.e., to destroy pathogenic organisms, insects, and weed seeds all at once. For these purposes, the chemicals used are referred to as fumigants, and are gaseous, require confinement, act over short times, and permit planting soon after treatment. Other "sterilants" are persistent herbicides that prevent plant growth for long periods and are dissimilar to fumigants. Use of fumigants is too expensive for large areas, and often severely disrupts the soil's beneficial organisms. Further, among the soil inhabitants, weed seeds are generally the least sensitive to chemicals. Accordingly, some seeds escape the toxic effects, notably those with impermeable seed coats, and probably, some with an intense dormancy. Such escapes become evident as weed populations reestablish in treated soils. Clearly, much progress could be made in reducing weed seed populations in soils with more specific "seedicides." Interestingly, the techniques used to evaluate potential herbicides may have delayed progress in the development of seedicides. This is because most evaluate for preemergence activity, and since a criterion is whether or not some planted target species emerge, non-dormant seeds are used as test objects. In such tests, dormant seeds would fail to emerge and, thus, not permit an assessment of activity. It is the germination process itself that must be removed from evaluation of potential seedicides, because for effectiveness, these materials must act on dormant seeds as well as those that germinate freely. This reduces to a question of how to determine whether a non-germinating seed is viable. Technically, this is a difficult problem, the solution to which would allow rapid progress in seedicide development, much like that which occurred for preemergence herbicides following development of adequate methodology. Existing tests for viability are often equivocal, difficult to apply to small seeds, and very time consuming.

Stimulation of Germination

Currently, the search for chemicals to overcome dormancy in weed seeds is underway. Details of the rationale and some problems involved have been elaborated in previous sections. Here also, lack of appropriate methodology for testing chemical activity limits commercial interest. From a practical point of view, there are a few biological and physical properties that a candidate stimulant might possess. One of these, obviously, is that the substance should act on the dormancy encountered in the target species under natural conditions. For many species, this means the stimulant should cause high germination of light-requiring seeds in the absence of irradiation, which mimics the situation of seeds beneath the soil surface. Other forms of dormancy might require a different kind of chemical. Secondly, it is commonly assumed that substances with a relatively low molecular weight penetrate easier. Over a broad range this is probably so. Uptake of gibberellin, for example, is limited and often requires steep external concentration gradients to achieve active levels internally. A third consideration is whether the substance has some volatility, which is desirable to facilitate distribution in the soil. Non-volatile compounds must either contact the seed directly or be mixed well enough so that diffusion distances to target seeds are not excessive.

Assuming dormancy control in buried weed populations is predominantly through phytochrome, the probability of finding substances to influence this dormancy is good. Many substances overcome this dormancy under laboratory conditions (Taylorson, 1984). The few attempts to stimulate weed seed germination under field conditions have not given very promising results, however. Partly, this is due to the nature of the substances employed, each having particular limitations with respect to the "ideal" criteria outlined above. Chemical control of other forms of dormancy, especially hardseededness, has not been successfully achieved even in the laboratory. Even with the discovery of new highly active chemicals with the desirable criteria, there are still limitations on what to expect under field conditions. For example, it is more feasible to try to stimulate germination of either a single species or a closely related small group of species than the entire complex of species typically present in agricultural soil. Further, owing to cyclical dormancy among species, it is probable that use will require proper timing and periodic applications over the entire season, rather than just a single application.

To conclude, seed dormancy and its control is an extremely important consideration in attempting a diminution of the annual weed problem. Technology can furnish tools to assist humans in combating weeds, but is unlikely to provide a panacea to simply end the problem.

NOTES

R. B. Taylorson, USDA-ARS, Beltsville Agricultural Research Center-West, Beltsville, MD 20705.

[1] Because of dissimilarities of usage some definition of terms is needed. A "dormant" seed fails to germinate when held under adequate temperature, moisture and gaseous composition. Germination refers to protrusion of the first structure, usually the radicle, through the seedcoat, whereas emergence is appearance of a seedling above the soil surface.

REFERENCES

Bewley, J. D. and M. Black. 1982. Physiology and biochemistry of seeds. Vol. II. Viability, dormancy and environmental control. Springer-Verlag, New York.

Chancellor, R. J. 1981. The manipulation of weed behavior for control purposes. Phil. Trans. R. Soc. Lond. B. 295:103-110.

Karssen, C. M. 1980. Patterns of change in dormancy during burial of seeds in soil. Israel J. Bot. 29: 65-73.

Koller, D. 1972. Environmental control of seed germination. p. 1-101. *In* T. T. Kozlowski (ed.) Seed biology. Vol. II. Academic Press, New York.

Lueschen, W. L. and R. N. Anderson. 1980. Longevity of velvetleaf (*Abutilon theophrasti*) seeds in soil under agricultural practices. Weed Sci. 28:341-346.

Mitchell, J. W. and J. W. Brown. 1947. Relative sensitivity of dormant and germinating seeds to 2,4-D. Science 106:266-267.

Odum, S. 1965. Germination of ancient seeds. Dansk Botanisk Arkiv 24:1-70.

Roberts, H. A. 1970. Viable weed seeds in cultivated soils. Rep. Nat. Veg. Res. Stn. 1969 (U.K.). p. 25-38.

Rolston, M. P. 1978. Water Impermeable seed dormancy. Bot. Rev. 44:365-396.

Sagar, G. R. 1970. Factors controlling the size of plant populations. p. 965-979. *In* Proc. 10th British Weed Control Conf.

Shropshire, W., Jr. and H. Mohr. 1983. Photomorphogenesis. Springer-Verlag, New York.

Taylorson, R. B. 1972. Phytochrome controlled changes in dormancy and germination of buried weed seeds. Weed Sci. 20:417-422.

Taylorson, R. B. 1980. Aspects of seed dormancy in fall panicum (*Panicum dichotomiflorum*). Weed Sci. 28:64-67.

Taylorson, R. B. 1984. Chemical control of germination. *In* J. L. Hilton (ed.) Agricultural chemicals of the future. Rowman & Allenheld, New Jersey. (In press.)

Wesson, G. and P. F. Wareing. 1969. The role of light in the germination of naturally occurring populations of buried weed seeds. J. Exp. Bot. 20:402-413.

TILLAGE EFFECTS OF ANNUAL WEED GERMINATION

R. J. Chancellor

Ever since man began to plant crops he has needed to cultivate the soil. Basically there are three objectives: to bury the trash remaining from the previous crop, to loosen the soil in order to make a seedbed for the next crop and lastly to control weeds.

Mechanical control of weeds can be achieved in various ways: (1) by cutting up and burying weeds so that they die, (2) by bringing roots and rhizomes to the surface so they dry out, (3) by repeated stimulation of buds of perennials into growth so that they exhaust their food reserves and (4) by stimulating dormant weed seeds in the soil to germinate so that they become vulnerable to further tillage or to herbicides. This review paper is concerned only with the effects of tillage on the germination of weed seeds in the soil.

TILLAGE EFFECTS ON FACTORS THAT STIMULATE GERMINATION

Weed seeds in the soil that are ungerminated, either through innate dormancy or through dormancy enforced by burial, need a stimulus to initiate germination. Seeds of some weeds require one particular stimulus, such as a period of chilling, whereas others appear to be stimulated by any of a number of stimuli either alone or in combination. Many of these stimuli arise either through seasonal changes in climate or through artificial changes in environmental factors caused by man's activities.

Seasonal changes in climate that stimulate seeds to germinate recur inevitably. Arable soils usually contain a moderate level of nitrate, which also can be stimulatory. The three main factors that can be changed by tillage and have a great influence on seed germination are light, soil atmosphere and the amplitude of temperature fluctuations.

Light and Seed Germination

Light can be of great importance in triggering seed germination. Seeds that are stimulated by light are positively photoblastic, but as shown by Wesson and Wareing (1969a), stimulation is not invariable. Although seeds of many species are insensitive to light when fresh, they can acquire a need for light after a period of burial. These authors examined the response of seeds of arable weeds to light after a period of burial in the field. Fields in undisturbed grass for 2.5 and 6 years were selected, as this enabled tests to be made on buried seeds showing enforced dormancy rather than short-term innate dormancy.

Pits 75 cm square and 5, 17.5 and 30 cm deep were dug in the field. Pits of each depth were covered with a sheet of light-excluding asbestos, transparent glass or left uncovered. No seedlings appeared in the pits from which light was excluded, whereas both treatments exposed to light produced many seedlings. Freshly collected seeds of the 22 species occurring in the fields were then tested for their light response. They found that seeds of 12 of the species were either light insensitive or were inhibited by it. These results show clearly that whereas seeds are not always stimulated by light, it is of over-riding importance to seeds that have been buried in the soil. Tillage will expose some of the seeds in the soil to light and so stimulate weed seedling emergence.

Oxygen Level and Seed Germination

The composition of the soil atmosphere, and especially oxygen concentration, has been suggested to be important in regulating the germination of weed seeds (Bibbey, 1948). However, it has been recognized more recently that other factors interact, and the system is more complicated than originally supposed.

Very little oxygen is required for germination to occur. However, when imbibition occurs, oxygen starvation may result, for the gas can then only reach the embryo by dissolving in the imbibed water. Most weed seeds are in a state of imbibition. Furthermore, phenolic substances if present in the seed coat can fix oxygen, and so diminish the supply. Temperature can also play a part, for the higher the temperature the smaller the quantity of oxygen available, although the need of the embryo is greater. Thus, at high temperatures, a critical level of oxygen may be reached, which can result in a secondary dormancy known as thermodormancy (Come and Tissaoui, 1973).

A further interrelationship of oxygen in seed dormancy has been elucidated for *Sinapis arvensis* L. (Edwards, 1969). She found that the resistance of the seed coat to diffusion of oxygen was very high, and although the rate of diffusion failed to keep pace with consumption, the dormancy of the seeds was not due directly to the lack of oxygen. Dormancy resulted from growth inhibiting substances, but the concentration of these increased with a decreasing oxygen supply.

Investigations by Wesson and Wareing (1969b) indicated the presence of a gaseous inhibitor in the soil atmosphere which prevented weed seed germination. This inhibitor appeared to arise from the seeds themselves and was not carbon dioxide.

Carbon dioxide is another factor in the soil atmosphere. At high levels it inhibits germination of most species, but in the laboratory at levels of 2-12% it has been found to stimulate germination of seeds of *Setaria* spp. (James and Staniforth, 1969). The range found in soils is about 0.03-12%. Similarly, ethylene can occur in waterlogged soils at concentrations that will stimulate weed seed germination (Smith and Russell, 1969).

It is impossible to generalize from the data available on various species, but it seems likely that any form of tillage will affect the composition of the soil atmosphere and consequently could influence weed seed germination.

Amplitude of Temperature Fluctuations

Although light is frequently an essential requirement for seed germination to occur, there are occasions when seedlings have been found to originate from seeds buried at a depth to which light cannot penetrate. It seems that these seeds germinate as a result of increased diurnal fluctuations in temperature, due to the changing seasons. However, the amplitude is mitigated by the depth of burial, as has been shown by Stoller and Wax (1973). They measured soil temperatures continuously at six depths between 1.3 and 20.3 cm beneath the uncropped surface of a silty clay loam at Urbana, Illinois. Daily fluctuations in soil temperature were greater in summer than in winter, and down to 5.1 cm depth they were greater than the fluctuations in air temperature. But, at 10.2 cm and deeper they were less. There was a pattern of amplitude reduction and phase retardation with increasing depth. Seeds are, therefore, even less likely to germinate at lower depths.

Vegetation also exerts a considerable influence on the amplitude of diurnal temperature fluctuations of the soil. Thompson (quoted by Grime, 1981) found in grassland in the north of England that the amplitude beneath a complete grass sward was about 5 C, but in gaps in the sward the amplitude increased depending on the size of the gap: the amplitude was about 9 C daily in gaps 15 cm in diameter and over 13 C in 25 cm gaps. Trash from a previous crop on the soil surface can similarly restrict temperature fluctuations and also light penetration.

This is a remarkable system for regulating seed germination, so that it occurs only at depths from which the seedling can emerge and in areas where the resultant seedling is relatively free from competition. This system presumably can operate equally with seeds of arable weeds, so that seeds brought nearer the surface by tillage will then be stimulated by the greater temperature fluctuations even if light cannot penetrate to that depth.

Interactions

The three factors considered above are influential on seed germination and all can be provided by tillage. These factors and others have been shown to interact by Roberts (1973). He investigated the interactions between light, nitrate and fluctuating temperatures on the germination of seeds of various species.

Seeds of *Rumex crispus* L. incubated without nitrate in light at a constant temperature of 25 C failed to germinate, as did seeds given nitrate and incubated in the dark at that temperature, and those at that temperature without nitrate. However, seeds given temperatures alternating between 15 and 25 C for 16/8 h, respectively, in the dark without nitrate gave 1% germination, and at these alternating temperatures with light gave 68% germination. When given nitrate, light and alternating temperatures in combination, 100% germination occurred.

Thus, although some species require only a single factor to trigger germination, others require a combination of two or more factors and some of these can be provided by tillage operations. These results indicate the importance of tillage for weed seed germination.

TILLAGE AT DIFFERENT TIMES

One of the more remarkable aspects of annual weed germination is that each species has its own particular period or periods during which seedlings emerge, although many overlap to some extent. Emergence may occur in only 1 or 2 weeks of the year, or at the opposite extreme, it can occur over several months in both spring and autumn. As a general rule, in Europe the longer the emergence period of a weed the greater its importance in agriculture, for it is better able to withstand the vagaries of crop planting time in rotations. Although the timing of a weed's emergence period may vary by a week or two from one year to the next depending on the climate, this does not normally affect the sequence of emergence of different species.

As each weed will only germinate within its set period, and as tillage provides various stimuli for seed germination, then one would expect that the date the soil is tilled will determine to a large extent which weeds occur in the crop that is planted at that time.

The influence of date of tillage and additionally the effects of climate on seedling emergence have been investigated by Roberts and Potter (1980). They cultivated sets of plots carefully to a depth of 10 cm, each plot once only, at intervals of 2 weeks over a period of 4 years and monitored the subsequent emergence of seedlings. In the first year, the numbers of seedlings recorded per m^2 ranged from 38 to 601 for individual cultivations during the growing season. As expected, contributions made by a species to the batch of seedlings emerging after each cultivation were influenced by its periodicity of emergence. For example, *Polygonum aviculare* L. produced the greatest number of seedlings after cultivations in March or early April, but its contributions to the weed populations decreased until the end of May when its emergence virtually ceased. Other species were less definite and germinated throughout the growing season, but even these had a consistent seasonal pattern. For example, *Poa annua* L., which was a major weed component, tended to occur in relatively low numbers following cultivations up to the end of May. Thereafter, emergence increased during the summer and declined once more after cultivations made later than mid-September.

A second factor of importance was rainfall. In the spring there was an initial flush of seedlings, which was probably associated with increasing temperatures, but subsequent to this seedling emergence was related to the pattern of rainfall. In each year there were periods when lack of soil moisture restricted emergence. This appeared to be an over-riding factor determining seedling numbers. Interestingly, after a flush of seedlings had emerged following cultivation, the numbers appearing subsequently fell to the level prevailing on plots that had not been disturbed. However, if soil moisture was deficient over a long period, as it was in Britain in 1976, the flush of seedlings resulting from tillage did not appear until sufficient rain fell; i.e., the effects of tillage in June were delayed by about 3 months until September when the soil was sufficiently moist again.

These results illustrate well that the season when the crop is planted determines which weeds will occur in the crop, and rainfall can regulate the timing of the subsequent seedling emergence.

FREQUENCY OF TILLAGE AND SEED GERMINATION

As seed germination is encouraged by disturbing the soil, it would be expected that the more frequently the soil is tilled, the greater the number of seedlings that appear. By tilling the soil at intervals the greatest possible number of seeds would receive the stimulus of light, higher oxygen levels, greater amplitude of temperature fluctuations, etc., and so germinate. The effects of three frequencies of cultivation upon weed seedling emergence, and the relationship of the number emerging from the viable seed population in the soil, has been investigated over 6 years by Roberts and Dawkins (1967).

Plots of soil were either dug-over each year in March, June, September and December, or dug in March and September only, or not dug at all. The soil, a sandy loam, was dug to a depth of 23 cm. Seedlings were assessed in thirty random 15 x 15 cm quadrants on the average of 6 to 7 times a year and then killed chemically. No seed shedding occurred. More seedlings emerged on plots dug four times a year than on those dug twice a year during the first 3 years, whereas the reverse was true in the last 3 years because, presumably, the seeds had declined more in the four-times-a-year plots. The differences were small, however, and statistically significant only in the first and fifth years. On the plots that were dug-over, the numbers of seedlings that emerged each year decreased exponentially once the regimes had become established. The rates of decrease were 28% on those dug four times and 23% per year on those dug twice.

The numbers of seedlings of individual species on these plots also decreased exponentially. For *Chenopodium album* L., one of the most frequent species, the rates of decline were almost the same on the two series of tilled plots, but other species tended to decline more rapidly on plots dug four times a year. On both sets of plots *Poa annua* L. and *Stellaria media* (L.) Vill. decreased more rapidly than other species.

On undisturbed plots, the results were very different. They were tilled only once at the start of the experiment, and in the first year the number of seedlings that emerged was only half of those on plots dug twice. The numbers of seedlings emerging thereafter declined rapidly at about 70% per year until the fourth year when very few emerged.

The numbers of seedlings emerging, when expressed as a percentage of the viable seeds present at the beginning of each year, show that on plots dug four times a year the seedlings represented between 6 and 11% of the seeds, on those dug twice, 5 to 9%, and on the undisturbed plots, 0.2 to 5%. The seeds decreased in the soil of undisturbed plots at a rate of 22% per year, largely as a result of natural decay. If this slow decline continued unchanged then 1% of the original seed population would still be viable after 18 years.

These data represent what can occur in short-term vegetable cropping, but nonetheless illustrate the basic principles that (a) the more the soil is tilled the more seedlings emerge and the faster the seed population in the soil declines, and (b) that when soil remains undisturbed, once seed in the surface soil is exhausted, seedling numbers decline, but seeds at lower depth can persist for very much longer. These principles, of course, only apply provided that no further seeds are shed onto the soil at any time and when soil moisture is not limiting.

PLANTING CROPS WITHOUT ANY PRELIMINARY TILLAGE

In England, monitoring of the weed flora on experiments with different tillage systems in the same field has shown what the effects of various cultural practices are on weed flora composition (Froud-Williams et al., 1983). Five experiments were continued over several years in which no tillage was compared with plowing to 20 cm, and two sites also included a comparison with shallow tillage to 5 cm. At two sites, the cropping sequence was continuous winter wheat (*Triticum aestivum* L.) and at the other three, a rotation of winter rape (*Brassica napus* L.), winter wheat, winter oats (*Avena sativa* L.) and winter wheat. Because the primary objective was to compare crop yields under these tillage regimes, straw was burned after harvest and the best available herbicides were used for weed control.

The effects of the cultivation treatments on the total weed numbers depended upon which weeds were dominant. At two sites where grass weeds were not numerous, the total numbers of seedlings recorded were greater on plowed plots than on undisturbed ones. This confirms the results of Roberts and Dawkins (1967) detailed above. However, the reverse was true at the other sites where the weed populations had become dominated by grass weeds. These grass weeds, mainly *Poa annua* L., *P. trivialis* L. and *Alopecurus myosuroides* Hudson, were much more frequent on untilled plots. There were two possible reasons for this reversal. First, lack of soil disturbance would probably favor the establishment of these small-seeded grasses, for many of them have a light requirement for germination, and their seeds would be left at or near the surface when there was no tillage. The other factor is that they were not always controlled very satisfactorily, probably because ash resulting from the straw burning reduced the effectiveness of the soil-acting herbicides used. Plowing, of course, would tend to dilute this effect by mixing the ash-containing surface layers with deeper soil.

Although these annual grasses were favored by untilled conditions, most dicotyledonous annuals were much more frequent on the plowed plots. These are what could be termed the traditional weeds, which are very frequently found in the arable areas of England, e.g. *Anagallis arvensis* L., *Papaver rhoeas* L., *Polygonum* spp. and *Viola arvensis* Murray. This apparent preference for plowed plots may have been due to previously buried seeds that had lost their dormancy being brought up to near the surface where they could germinate. On the undisturbed plots, however, the seeds near the surface gradually germinated and their seedlings were effectively controlled each year by the herbicides. As no more seeds were brought up to the surface, this led to exhaustion of the seeds in the uppermost layers of the soil, and so, in comparison to the plowed plots fewer seedlings appeared.

Other species, such as *Matricaria* species, although common in arable land, showed generally an inconsistent response to the tillage regime. These species, although dicotyledonous plants, differ from the other traditional arable weeds in having, like the grasses, a much greater dependence upon light for seed germination.

CONCLUSIONS

Innate dormancy tends to prevent germination of weed seeds until they have been buried by tillage, and then, even if they lose this dormancy, a majority will still remain ungerminated because dormancy is enforced by burial. Weed seeds in tilled soils can, therefore, build-up to very high densities. A weedy field may contain as many as 25,000/m^2 and very weedy ones may contain many more (Roberts, 1981).

As stated at the beginning, one of the main purposes of tillage is to control weeds and one of the main methods of doing this is to stimulate dormant weed seeds to germinate, so that the potential weed population can be reduced. Tillage acts by bringing weed seeds nearer to the surface and providing the necessary stimuli to encourage them to germinate. Tillage can also ensure that none of the resultant seedlings survive and set seed, although this last is now usually accomplished by herbicides. However, even with frequent tillage the rate of decline of seeds in the soil is slow and tillage needs to be continued over several years to have an appreciable effect. Furthermore, this will only succeed if no seeds at all are shed during that period, which is unlikely to be achieved for any length of time.

Thus, although tillage is an excellent method of reducing seeds in the soil, it is very slow. As an alternative, one of the more promising ways of reducing seed numbers would be to find a chemical that will break seed dormancy and encourage germination even of buried seeds. Various substances such as nitrates and thiourea will help stimulate some species to germinate, but ethylene is probably one of the most effective known. Ethylene can be injected directly into the soil or applied as an ethylene-generating chemical such as ethephon (2-chloroethylphosphonic acid). It has been found to be an effective stimulant on various weeds (Egley and Dale, 1970; Chancellor et al., 1971). Further investigations are needed to establish the range of species that respond to its use. The search for other chemicals to break dormancy should be increased, for such treatments combined with efficient control of the resulting seedlings could prove to be the next major step forward in weed control, and consequently, also lead to a reduction in the amount of tillage needed.

NOTES

R. J. Chancellor, Agricultural and Food Research Council, Weed Research Organization, Begbroke Hill, Yarnton, Oxford, England.

REFERENCES

Bibbey, R. O. 1948. Physiological studies of weed seed germination. Plant Physiol. 23:467-484.

Chancellor, R. J., C. Parker and T. Teferedegn. 1971. Stimulation of weed seed germination by 2-chloroethylphosphonic acid. Pestic. Sci. 2:35-37.

Come, D. and T. Tissaoui. 1973. Interrelated effects of imbibition, temperature and oxygen on seed germination. p. 157-167. *In* W. Heydecker (ed.) Seed ecology. Butterworths, London.

Edwards, M. M. 1969. Dormancy in seeds of charlock. IV. Interrelationships of growth, oxygen supply and concentration of inhibitor. J. Exp. Bot. 20:876-894.

Egley, G. H. and J. E. Dale. 1970. Ethylene, 2-chloroethylphosphonic acid, and Witchweed germination. Weed Sci. 18:586-589.

Froud-Williams, R. J., D. S. H. Drennan and R. J. Chancellor. 1983. Influence of cultivation regime on weed floras of arable cropping systems. J. Appl. Ecol. 20:187-197.

Grime, J. P. 1981. Plant strategies and vegetation processes. John Wiley and Sons, Chichester.

James, A. L. and D. W. Staniforth. 1969. Some influence of soil atmosphere on germination of annual weeds. Weed Sci. Soc. Am. Abst. 148.

Roberts, E. H. 1973. Oxidative processes and the control of seed germination. p. 189-216. *In* W. Heydecker (ed.) Seed ecology. Butterworths, London.

Roberts, H. A. 1981. Seed banks in soils. Adv. Appl. Biol. 6:1-55.

Roberts, H. A. and P. A. Dawkins. 1967. Effect of cultivation on the numbers of viable weed seeds in soil. Weed Res. 7:290-301.

Roberts, H. A. and M. E. Potter. 1980. Emergence patterns of weed seedlings in relation to cultivation and rainfall. Weed Res. 20:377-386.

Smith, K. A. and R. S. Russell. 1969. Occurrence of ethylene and its significance in anaerobic soil. Nature 222:769.

Stoller, E. W. and L. M. Wax. 1973. Temperature variations in the surface layers of an agricultural soil. Weed Res. 13:273-282.

Wesson, G. and P. F. Wareing. 1969a. The role of light in the germination of naturally occurring populations of buried weed seeds. J. Exp. Bot. 20:402-413.

Wesson, G. and P. F. Wareing. 1969b. The induction of light sensitivity in weed seeds by burial. J. Exp. Bot. 20:414-425.

METHODS FOR CROP-WEED COMPETITION RESEARCH

M. M. Schreiber

The primary objective of crop-weed competition research is to develop new knowledge that will be adaptable to control strategies needed for integrated weed management systems. This applies to all cropping systems and most weeds associated with those crops. The specific methods employed are designed to answer specific questions relative to the interaction of a crop and associated weed(s). It is generally accepted that, when two or more plants attempt to grow in the same place, interference occurs between them. This interference can be divided into two possibilities, competition and allelopathy. Allelopathy is defined as an effect dependent on the addition to the environment of a chemical substance by an agent (plant). Although this is a legitimate concern, it is not the subject of this paper. Competition, as used, involves the removal or reduction of a resource(s) in the environment that is required by both plants sharing the habitat.

There have been no major changes in the general methods for evaluating crop-weed competition in the last two decades. Weed scientists are still determining when competition begins for specific species at specific densities and the converse, the determination of the required weed-free-period for maximum yield. Some of the methods used to answer these questions are more important today because of the rapid changes in weed control technology. Within the last several years many new postemergence herbicides have been developed for grass and broadleaf weed control in soybean fields. This does not mean that preemergence treatments will reduce significantly, but that more options are now available to growers than ever before. One of the major questions constantly faced is how much weed control is needed, or in more practical terms, what is the cost/benefit ratio of reduced weed competition. Cossans (1980) concluded that in most situations eradication of weeds is not a practicable goal for most weed species, but seed production is one of the critical factors.

SEED PRODUCTION AND SURVIVAL

Many researchers do not place weed seed production and survival in the area of weed competition research. In reality, it is *the* basis of the problem. Harper (1977) described the "soil seed bank" as a record of past and present vegetation and as representing the potential vegetation to come.

Two general methods have been used to determine the seed bank in the soil: seedling emergence and washing flotation methods. The general procedure for soil sampling is to use a large number of small samples, usually soil cores. These may be stratified for soil depth and the total numbers may be a factor that determines the method used.

The seedling emergence method, which is a method for determining viable non-dormant seeds, is both time and space consuming. The bulk of the sample is usually reduced by sieving the soil. The soil is then spread on flats and placed in the greenhouse. Seedlings are identified, counted, and removed as they emerge. Samples are kept for observation for periods ranging from months to years. Jensen (1969) provides a summary of the methodology as used by several authors.

In the washing-flotation method the samples are washed in water and in solutions of various densities. Most commonly, chemical dispersion of soil can be accomplished with a mixture of sodium hexametaphosphate, sodium bicarbonate and magnesium sulfate. Some agitation is needed before decanting debris and sieving. The procedures used have to be standardized and may be repeated several times depending on the sample. Once the organic matter fraction is separated from the mineral fraction, the seeds can be separated by hand, identified by species, and counted. This method accounts for all seeds in the soil: broken seeds, partially decayed seeds, viable and nonviable intact seeds. With experience, the broken and decayed seeds can be discounted. The viability of the intact seed is usually not determined, but tests for viability can be conducted by the use of the tetrazolium chloride method.

These types of determinations for weed seed populations in the soil should be utilized in all weed competition studies, particularly if they are conducted for long periods of time. Much data are needed to determine the potential seed bank in the soil following various density studies on weed competition. Although low densities may not affect soybean yield in the year the study was conducted, a significant seed population potential for a subsequent year could alter conclusions drawn from the first year.

ADDITIVE METHOD

One of the most common methods of evaluating weed competition is the additive method. In this method two species are grown together where one species, usually the crop, is kept at a constant density, and with the other species, usually a weed, the density is varied. Many variations of this method have been used depending on the objective of the researcher. The weed density factor itself is not as simple as it seems.

If one is attempting to determine what density of a weed causes significant yield reduction in soybean fields, one has to be aware that the time of incidence of the weed in the crop, and the duration of the density of the weed, must be considered. Peters (1972) points out that onset of competition is difficult to determine when weed species have prolonged emergence characteristics. Thus, one has a population composed of different age groups and, hence, competitive ability. If the later emerging weeds are removed by hand or hoeing, some soil disturbances occur that could influence the growth of the crop.

Studies of densities using perennial weeds become even more difficult, not only because of emergence over long periods of time, but because there may be plants from seeds and from propagules such as rhizomes. Color coding with tags (Holroyd, 1972) can be helpful in timing incidence, but populations are still different.

There is some mention of threshold concepts in weed science, but these have not been clearly defined. Certainly, how long weed-free conditions exist, or how long weeds can be allowed to grow in a soybean crop, become part of the threshold concept. The relationship of weed growth to crop growth cycle is critical (Nieto et al., 1968). As an example of application of the concept in terms of weed management strategy, the rate of herbicidal action can become important. Some of the newer postemergence grass herbicides may take a week or more before herbicidal action is completed.

One major consideration in additive studies may not necessarily be related to weed density per se, but to the total dry matter accumulation as well. Topham and Lawson (1982) point this out in the development of diversity indices for crop-weed competition.

REPLACEMENT SERIES METHOD

The substitutive design, or more commonly called the replacement series (de Wit, 1960), involves sowing a crop and a weed species in varying proportions, but maintaining a constant total density. This procedure works well under controlled conditions in a greenhouse or growth room, but can also be used in the field. Dekker and Meggitt (1983a,b) recently showed how effective this method is in a field study of velvetleaf (*Abutilon theophrasti* Medic) competition with soybean.

A range of mixtures of two species is grown, starting with a pure monoculture of species one (crop), then gradually substituting plants of species one (crop) with those of species two (weed) until a pure monoculture of species two is obtained. This provides very precise measures of the effect of the two species on each other and allows predictions of yield responses. It can point out the degree of adaptability of each species and various growth parameters can be determined with time.

MATRIX MODELS

The matrix model was first developed by Mortimer et al. (1978) for application to weed competition. This mathematical model can be used for simulating populations of weeds to predict the size and growth under different weed management practices.

The matrix analysis consists of a transition matrix, which represents "age states" of a weed, and a column vector, which represents the number of individuals in each "age state" at a set time. By multiplication of the transition matrix and the column vector, one can obtain a set of numbers of individuals in each age class in the next generation.

ENVIRONMENTAL FACTORS

Regardless of the specific objective and method used in weed competition research, one cannot fully explain the responses recorded unles one is able to describe in detail changes in the environment within the systems studied. Sagar (1968) put it as well as anyone when he said: "It (environment) is a matter which is too often ignored in experimental studies of competition."

Much research is now being conducted on how soybean cultural systems of row-spacing and tillage influence weed management. Likewise, similar studies are underway to determine how weeds modify the microenvironment of the crop and how the crop responds to the changed environment. The monitoring of the microenvironment in weed competition studies is essential, if one is to truly understand why the responses occurred and develop principles that may be transferred from one location to another.

New and more versatile instruments are available to measure most of the parameters of the microenvironment. Line quantum sensors from L1-COR can be used to measure photosynthetic radiation in a soybean canopy across the row, or in the row, and at various positions in the canopy stratum. Portable spectroradiometers are available to measure the energy within bandwidths of the same canopy. What better way to indicate the actual existence and measurement of light competition due to weed growth. Similarly, there are available instruments that will measure nondestructively the leaf area and water status of the components of that same canopy.

Microloggers can record and process these environmental data quickly, accurately, and remotely so that growth and development can be evaluated in a continually changing environment.

Weed scientists are interested in methods that will give us predictive capabilities. However, these will not be forthcoming unless the environment within the system being researched can be accurately described. These same data are essential for model development of weed competition, and for interfacing with soybean crop models.

NOTES

M. M. Schreiber, Research Agronomist, USDA-ARS and Departments of Botany and Plant Pathology, Purdue University, West Lafayette, Indiana 49707.

Journal Paper No. 9989 of Purdue University Agricultural Experiment Station.

Trade names are used solely for the purpose of providing specific information. Mention of the trade name, proprietary product, or specific equipment does not constitute a guarantee or warranty by the USDA or Purdue University, and does not imply approval or the exclusion of other products that may be suitable.

REFERENCES

Cossans, G. W. 1980. Strategic planning for weed control—a researcher's view. Proc. 15th British Weed Control Conf. 3:823-831.

Dekker, J. and W. F. Meggitt. 1983a. Interference between velvetleaf (*Abutilon theophrasti* Medic.) and soybean [*Glycine max* (L.) Merr.]. I. Growth. Weed Res. 23:91-101.

Dekker, J. and W. F. Meggitt. 1983b. Interference between velvetleaf (*Abutilon theophrasti* Medic.) and soybean [*Glycine max* (L.) Merr.]. II. Population dynamics. Weed Res. 23:103-107.

deWit, C. T. 1960. On competition. Versl. Landbouwk. Onderz. 66:1-82.

Harper, J. L. 1977. Population Biology of Plants. Academic Press, London.

Holroyd, J. 1972. Techniques for the assessment of *Avena* spp. in the field. Proc. 11th British Weed Control Conf. 1:119-122.

Jensen, H. A. 1969. Content of buried seeds in arable soil in Denmark and its relation to the weed population. Dansk Botanisk Arkiv 27:1-56.

Mortimer, A. M., P. D. Putwain and D. J. McMahon. 1978. A theoretical approach to the prediction of weed population sizes. Proc. 14th British Weed Control Conf. 2:467-474.

Nieto, J., M. A. Brondo and J. T. Gonzalez. 1968. Critical periods of the crop growth cycle for competition from weeds. PANS(C) 14:159-166.

Peters, N. C. B. 1972. Methods for evaluating weed competition using systems of handweeding or hoeing. Proc. 11th British Weed Control Conf. 1:116-118.

Sagar, G. R. 1968. Factors affecting the outcome of competition between crops and weeds. Proc. 9th British Weed Control Conf. 3:1157-1162.

Topham, P. B. and H. M. Lawson. 1982. Measurement of weed species diversity in crop/weed competition studies. Weed Res. 22:285-293.

ASPECTS OF WEED-CROP INTERFERENCE RELATED TO
WEED CONTROL PRACTICES

L. M. Wax and E. W. Stoller

Losses from weed interference in soybean production account for more losses than from all other pests combined (McWhorter and Patterson, 1979). Losses from weed interference have been reduced over the past decade in the main areas of production, primarily due to increased use and effectiveness of herbicides. In some situations, however, losses can be substantial when herbicides are not used or are not effective because of environmental factors. Also, in other situations, certain weeds that remain uncontrolled are now becoming dominant and causing losses. In addition, changes in tillage practices are causing a change in the weed spectrum, making control more difficult.

Thus, weed interference continues to be a significant problem in soybean production. Researchers are continuing their studies to assess the extent and nature of the weed interference problem. It is the objective of this paper to review some of the results of these studies on weed-crop interference, with special emphasis on soybean-weed interference, and to attempt to relate the results to weed control practices. Some references will be cited relating to recent work. For a great deal of the earlier work, the reader is referred to excellent summaries by Aldrich (1984), McWhorter and Patterson (1979) and Zimdahl (1980).

TYPES OF INTERFERENCE

Interference is the net effect of one plant on another and consists of the combination of allelopathy, competition and other factors (Harper, 1977). Allelopathy involves the harmful effect of one plant or another through the production of chemicals released into the environment. While allelopathy has been demonstrated in weed-crop relationships in laboratory, greenhouse, and in some field situations, the evidence is insufficient to ascribe a significant portion of the total weed interference to allelopathic factors in most soybean production areas where conventional tillage and fairly good weed control practices are utilized. Allelopathy could play a role in very heavy, dense infestations of weeds and/or in reduced tillage situations where considerable plant residues are at the soil surface.

Competition among plants growing in association occurs when the supply of an essential growth factor does not meet their combined demands (Harper, 1977). Weeds compete with soybean plants for limited resources such as light, moisture, nutrients, and perhaps growing space. From the numerous reports of studies on competition for available resources among weeds and crops, we conclude that light is the resource most often in limited supply in most soybean cropping situations in the Corn Belt. Competition for light is especially important where dicotyledonous weeds with large leaves grow substantially taller than the soybean plants and shade them. Competition for moisture occurs less often than competition for light. We consider that competition for nutrients is the least important of these three factors. Competition for moisture and nutrients is important when soybean plants compete with weeds that do not overtop them, and in droughty soils of relatively low fertility.

Other factors that contribute to interference, such as weeds that harbor pests that attack soybean plants or weeds that are parasitic, are known, but research on the importance of such factors in soybean production are lacking. We believe that the competitive properties of weeds are the most severe of the interfering factors affecting soybean performance.

FACTORS AFFECTING COMPETITION

Some of the factors that affect the amount of weed-soybean competition are (a) the crop cultivar, (b) the cultural practices, (c) the weed and (d) the environment (weather).

Soybean cultivars can vary greatly in maturity and growth habit, and have been shown to have substantial differences in ability to withstand weed competition (McWhorter and Hartwig, 1972). Cultivars may also vary in response to herbicides, and there may be an interaction among response to herbicides and the ability to compete with weeds for the available growth factors (Aldrich, 1984; McWhorter and Patterson, 1979).

Cultural factors, such as row spacing, plant spacing within the row, planting date, and tillage before and after planting markedly affect competitive relationships (Aldrich, 1984; Felton, 1976; McWhorter and Patterson, 1979). Several researchers have reported on the interaction of weed competition and row spacing and plant spacing within the row, with mixed results. In general, soybean plants gain a competitive advantage over weeds as soybean populations increase within rows spaced 75 to 100 cm apart. At some point, however, the advantage gained in improved soybean competition is not worth the losses caused by increased soybean lodging. As row spacing goes from 100 down to 20 cm, the soybean canopy closes rapidly and suppresses the weeds. Narrow rows are usually more effective in suppressing weed growth by competition than are wide rows, assuming weeds are not controlled by cultivation or herbicides in either row spacing. However, wide rows are preferred over narrow rows in situations where weeds need to be controlled by cultivation.

Tillage practices influence the amount and kind of weeds that affect the soybean-weed competitive relationship. Results (Aldrich, 1984; McWhorter and Patterson, 1979; Wax and Schoper, 1983) indicate that primary and secondary tillage regimes affect the number and kind of weeds that occur in subsequent soybean fields. This effect is cumulative and becomes apparent after a few years of the same tillage regime. A move toward reduced tillage tends to result in higher populations of certain annual weeds as well as a trend toward increased populations of perennial weeds. Reduced tillage may also necessitate increasingly higher levels of weed management to achieve the desired level of control. For many soybean production situations, delaying planting gives a competitive advantage to the soybean plants over the weeds (Aldrich, 1984; Oliver, 1979). In some instances, in a natural infestation of weeds, delaying soybean planting from mid-May to mid-June can reduce the losses from weed competition considerably. However, the gain may not be worth the potential loss in soybean yields resulting from delayed planting. The best procedure may be to combine a moderate planting date with good cultural practices and a sound weed management program.

The species of weed involved, and its population in the field, is a very important factor in determining the extent of the competitive effects. Often, too, there is a mixture of weeds present, which further complicates the matter. Of equal importance is the time at which weeds emerge, which affects the duration of competition. These aspects of weed competition concerning species, density and duration will be discussed below.

Environmental factors often override all of the other factors and contribute greatly to the season-to-season variability in losses. Temperature and available moisture during the season affect the weed species present and their growth rate. Timing of precipitation throughout the season affects the relative growth rates of the soybean plants as well as the weeds, and the resulting losses caused by the weeds. Specialized situations, where soybean plants are grown with irrigation or in very definite dry season-wet season climates, may alter the competitive relationships somewhat from that associated with the majority of soybean production areas.

Compared with some other crops, soybean is a good competitor with weeds when grown under optimum management conditions. Soybean is efficient in the capture and use of available light. When soybean plants rapidly establish a canopy, they occupy the space and utilize all the available light, thereby gaining the competitive advantage. When soybean gets the 'head start' on weeds, the weeds have difficulty growing through the canopy and shading the crop.

TYPES OF COMPETITION STUDIES

Several types of competition studies have been conducted to assess the nature and extent of losses caused by weed competition in soybean fields. Early studies on mixed, high populations were helpful in establishing the degree of loss sustained in uncontrolled, severe weed populations, and were essential in showing the value of the use of good control methods (McWhorter and Patterson, 1979; Zimdahl, 1980). More recently, as most soybean fields are treated with herbicides, weed infestations often exist at low populations of single or multiple species. Thus, studies have been designed to provide more precise ways to improve methods of control and to determine the weed density at which control measures are justified.

Past and present studies on weed competition in soybean crops involve: (a) single weed species, (b) various weed densities of a single species, and (c) duration of competition. In general, the duration studies can be grouped into two types: (a) studies to determine the length of early control required from soybean emergence until soybean plants form a canopy and are capable of effectively competing with weeds for the remainder of the season, and (b) studies to determine the optimum time of removal of weeds from the crop after they have been allowed to compete with the soybean plants from emergence through the first part of the growing season.

A number of experiments from U.S. production areas have involved studies on various densities of a weed species, where the weed was allowed to compete with the soybean plants for the entire growing season. In spite of climatic, cultural, and specific differences in the various experiments, certain similarities remain, allowing us to summarize the data and draw some general conclusions about comparison of various species and density of weeds in terms of competitive relationships with soybean.

Some of the most predominant weeds throughout the Corn Belt, if not controlled, are the foxtails (*Setaria* spp.). Giant foxtail (*Setaria faberi* Herrm.) is especially prevalent and is a vigorous competitor. Because of its prolific seed production, giant foxtail often emerges in very high populations and competes with other weeds and with soybeans. Much of the early work was conducted by Knake and Slife (1962) in Illinois and Staniforth (1965) in Iowa. These studies showed that, whereas low populations of giant foxtail did not affect yield of soybeans greatly, higher natural populations confined to the row by cultivation, were capable of reducing soybean yields 10 to 20%. Other, more recent studies with this species and other annual grass species confirm that heavy stands can reduce soybean yields 10 to 20%, or more in some instances, but that losses attributed to individual grass plants were small (Harrison et al., 1984; Rathmann and Miller, 1981).

Two common broadleaf weeds that infest soybean fields, especially in the Corn Belt of the U.S., are velvetleaf (*Abutilon theophrasti* Medic.) and jimsonweed (*Datura stramonium* L.). Experiments with these species demonstrate greater losses per plant and higher potential soybean yield losses from either one than from annual grasses, such as the foxtails (Kirkpatrick et al., 1983; McWhorter and Patterson, 1979; Oliver, 1979). These broadleaf plants are similar in that they often escape early herbicide treatments, and by mid-season grow tall and shade the soybean plants, and compete for moisture and nutrients. A density of one velvetleaf or jimsonweed per m^2 may reduce soybean yields significantly, while a heavy, natural infestation may reduce yields 30 to 40%.

Other broadleaf weed species, smartweeds (*Polygonum* spp.) and pigweeds (*Amaranthus* spp.), while somewhat different in growth habits, are highly competitive and capable of causing substantial yield losses in soybean fields (Coble and Ritter, 1978; Moolani et al., 1964; McWhorter and Patterson, 1979). The data indicate that smooth pigweed (*Amaranthus hybridus* L.) is present in most soybean production areas in the U.S. and consistently causes greater yield reduction per plant than does either velvetleaf or jimsonweed. Pigweeds grow much taller than soybean plants, and produce substantial above-ground forage. A natural infestation in the soybean row may reduce soybean yields by 50% or more. Smartweed (primarily *Polygonum pensylvanicum* L.) is also very competitive, like pigweed, and reduces soybean yield more than either velvetleaf or jimsonweed.

One of the most competitive weed species in northern and southern U.S. soybean production areas is common cocklebur (*Xanthium strumarium* L.) (Bloomberg et al., 1982; McWhorter and Patterson, 1979). A tall-growing broadleaf weed that produces massive amounts of forage and heavy shading, common cocklebur, at very low populations causes more damage per plant than any of the other weeds discussed above. A heavy population may reduce soybean yields 70 to 90%, if not controlled.

Recent studies on the losses caused by volunteer corn (*Zea mays* L.) show that this weed may be the most competitive of all weeds found in soybean fields in the Corn Belt (Andersen et al., 1982; Beckett and Stoller, unpublished data). In recent years, volunteer corn has become a more prevalent problem in situations where corn-soybean rotations are combined with conservation tillage. Only a light to moderate stand of volunteer corn has been shown to cause up to a 30 to 40% yield reduction of soybean crops, indicating the economic feasibility of controlling only sparse stands of volunteer corn.

Several other weed species have been investigated with respect to competitive ability. Quackgrass (*Agropyron repens* L. Beauv.), johnsongrass (*Sorghum halepense* L. Pers.), sicklepod (*Cassia obtusifolia* L.), annual morningglories (*Ipomoea* spp.), common ragweed (*Ambrosia artemiisifolia* L.), common sunflower (*Helianthus annuus* L.), and wild mustard (*Sinapis arvensis* L.) are substantial weed problems in one production area or other, and cause significant soybean yield reductions (Coble et al., 1981; Cordes and Bauman, 1984; Irons and Burnside, 1982; McWhorter and Anderson, 1979; Thurlow and Buchanan, 1972; Williams and Hayes, 1984). These species vary in their competitive ability but, on a per plant basis, lie somewhere between the extremes of the smaller annual grasses like foxtails and the larger species, such as common cocklebur and volunteer corn.

A summary of the density and species competition work shows that species vary greatly in their ability to compete on an individual plant basis. In general, the amount of yield reduction caused by these weeds decreases with increasing weed population. This is partly due to the increase in competition among the weeds themselves as their density increases. Density data may be helpful in evaluating threshold levels for planning control procedures. However, the data points from most all the density studies are at such high densities that the curves must be extrapolated to apply to the weed densities that exist in most growers' fields.

Duration studies have been conducted to determine the length of weed-free conditions needed after soybean emergence to ensure that the soybean plants could suppress any weeds emerging thereafter. These studies have their utility in determining how long after planting weeds need to be controlled. Some early work on this concept was conducted in Illinois by Knake and Slife (1965) with giant foxtail. Other researchers have subsequently conducted similar studies, at various locations in the soybean production areas, using a variety of species under various environmental conditions (Aldrich, 1984; Bloomberg et al., 1982; Coble et al., 1981; Kirkpatrick et al., 1983; McWhorter and Anderson, 1979; McWhorter and Patterson, 1979).

The results varied somewhat over species and locations, but some general conclusions can be drawn. When weeds emerge at the same time as do soybean seedlings, they produce vigorous growth and reduce soybean yields from 20 to 80%, depending on species and populations present. In contrast, only a 2-week delay in weed emergence often reduces losses due to weeds in half. Further, when weeds emerge 4 to 6 weeks after the soybean seedlings, losses are very low, and in most instances statistically insignificant.

These data provide guidelines for length of control needed under most conditions in soybean fields grown under optimum conditions in the temperate soybean production areas. These lengths of needed control are usually shorter as row spacings decrease and soybean populations increase. The intervals would also be the same or slightly shorter when weeds fail to germinate and grow due to early and mid-season drought.

The shorter intervals would also likely occur if the weeds in question were annuals that tend to germinate only early in the season, such as foxtails, common lambsquarters (*Chenopodium album* L.), smartweeds, and velvetleaf. The lengths of time of weed control that are required would likely be longer than average when perennials, or annuals with mid-to-late season, sporadic germination were involved; i.e., johnsongrass, annual morningglories, and common cocklebur. Production areas with higher rainfall and longer growing seasons, where weeds are both shade tolerant and have late, sporadic germination, or an abundance of perennial weeds, may also require longer periods of weed-free conditions early in the season than the average of 4 to 6 weeks that seems to apply to many weed-soybean competitive relationships. With the increasing trend toward conservation tillage, and reduced cultivation after planting, a longer period of weed control may be required if cultivation is omitted in wide-row soybean production.

Another kind of duration-of-competition study has been conducted where soybeans and weeds are given an 'equal start' by being allowed to emerge and grow together, then "removed" at various times to determine when weeds need to be "removed" to avoid any significant reduction in quantity or quality of the harvested soybean crop (Aldrich, 1984; Bloomberg et al., 1982; Coble et al., 1981; Harrison et al., 1984; Kirkpatrick et al., 1983; McWhorter and Patterson, 1979; Williams and Hayes, 1984). Certain conclusions can be drawn from the data. In general, with most cropping practices now in use for soybean production, weeds can be allowed to emerge and grow with the soybeans for about 4 weeks after emergence, and then be removed, without causing any significant reduction in soybean yield or quality.

In many of the early studies the weeds were carefully removed by hand weeding, since effective and selective, herbicidal treatments were not available to kill the weeds under study. In recent years, interest in these types of studies, and applicability of the data have increased greatly, since the introduction of selective postemergence herbicides such as bentazon, acifluorfen, sethoxydim, fluazifop, and others.

We believe that losses from weeds are a continuous function of the time they are present in the field throughout the season, with losses increasing as the time present increases. Losses of less than 10% usually are not significant, utilizing conventional statistical methods. It is important in conducting these kinds of experiments to have well-designed studies that have several removal times, are well-replicated, and repeated over time and space.

OTHER CONSIDERATIONS

While the development of guidelines from these kinds of studies is possible and valuable, numerous factors may affect the outcome of such studies and should be considered when interpreting and applying the resulting data. For example, ecotypes of weeds occur, and they may differ markedly in growth habit and, very likely, in competitive ability. Weather and location also play important roles in affecting the delicate balance in the weed-crop relationship, making repeated experiments over various environments essential.

There may be differences in results gained from handremoval of weeds versus those gained from actual treatment of the weeds with herbicides, because herbicides sometimes are only partly effective and the weeds grow for some time after application. This causes some difficulty in relating studies where weeds were removed by hand to those removed by herbicides. This factor may come into play in explaining why the yield reduction from a "natural" stand of giant foxtail removed with postemergence applications of fluazifop at various times were greater than the reductions reported from studies where foxtail was removed by hand (Harrison et al., 1984).

There is a need to combine the results of weed density and duration of competition into a single analysis. This occurs in situations where the crop previously has been treated with herbicides, and a decision is needed as to whether the remaining, escaped, weeds need to be controlled with postemergence herbicides, and at what time, to preclude losses. Oliver and his colleagues in Arkansas have reported excellent work in the area of weed-soybean competition, and have demonstrated clearly the effects of population density on losses sustained by various periods of morningglory competition in soybean fields. They also recently reported on a new way of presenting these kinds of data involving both times and densities, using response surface analyses (Keisling et al., 1984).

Much of the times-of-removal and density studies have involved individual species, whereas in actual practice, many fields contain a mixture of several weedy species.

More work is needed to predict losses from mixed stands of weeds, using data collected from studies on individual species.

MANAGEMENT THRESHOLDS IN WEED COMPETITION

We would hope that some practical use can come from the application of all these competition data discussed. To that end, we will discuss the possible use of the management threshold concept. As used here, management threshold is defined as the minimum weed population for which control becomes justified. The situation is complicated in that there are many different reasons that justify control. Presently, it seems that the control or management of pests is normally thought of in terms of economic thresholds. However, there are several kinds of thresholds, largely dependent on the individual situation and the attitudes of the grower and his associates. For example, while the most obvious threshold may be based on crop yield and the losses sustained, other thresholds might be based on esthetic considerations. Often, control of escaped weeds is motivated by a grower's sense of pride in his fields' appearance (not to mention the motivation effected by the attitudes of friends, neighbors, and landlords!). Crop quality may justifiably be another valid threshold, as in the case where Eastern black nightshade (*Solanum ptycanthum* Dun.) may not substantially reduce yields at low densities, it will likely cause harvest problems and reduce seed quality, justifying control measures.

Another threshold could well be weed seed production. One prolific weed per 10 m^2 might never reduce yields, but could produce enough seed to cause substantial problems for years to come. These management thresholds (esthetic, crop quality, and weed seed production), and perhaps others, must be considered by the grower, but are variable and subjective, and would be very difficult to determine. We are not aware of reported work in this regard.

An economic threshold, based on crop yield, should be easier to determine and apply than the other management thresholds discussed above. Definitions of economic thresholds differ: some would consider any loss due to weeds to be an economic threshold that justified control measures. In general, however, considering the cost of modern-day treatments and application, the economic threshold is roughly equal to about 5 to 10% of the value of the crop. Thus control measures which cost about 5 to 10% of the crop value would be justified by a population of weeds sufficient to reduce yields 5 to 10%.

Some investigators have calculated economic thresholds relating to postemergence control of certain broadleaf weeds in soybean fields (Marra and Carlson, 1983). These authors presented economic thresholds for a number of weeds; however, they also demonstrated the complexity of such procedures. In their analyses, the economic threshold was shown to be affected by species, treatment cost and effectiveness, as well as by the yield of the crop and the price received for it.

These studies, and the increasing storehouse of knowledge on competition, increase the likelihood of our being able to develop and use economic thresholds for weeds in soybean fields. At the present time, the development and use of such thresholds will likely apply to postemergence control of weeds that are present and whose populations (or some other measure of intensity of potential severity) can be determined by the grower, scout, or custom applicator. The accurate application of economic models for soil-applied herbicides may have to await the development of methods to evaluate and predict weed poulations under a variety of production systems.

SUMMARY

Interference of weeds with soybean growth, development and yield consists of both competition, allelopathy and other factors. However, in most cropping situations, competition likely plays the principal role in reducing soybean yield. The weed species present, the weed density, and the length of time the weed is present affect the degree of competitiveness expressed. Weed competition is a continuous function of the amount of time the weed grows with the soybeans. In areas of reasonably well-timed rainfall and fertile soils, light may be the principal resource in limited supply for which weeds and soybean plants compete during the season.

Management thresholds such as economic, esthetic, weed seed production, and crop quality are important, but difficult to evaluate. With the appropriate inputs, economic thresholds can be prepared. However, variables such as species, treatment cost and effectiveness, as well as crop price and yield affect economic thresholds.

The widespread use of preplanting and preemergence herbicide treatments, combined with good cultural practices, and followed with postemergence treatments, if necessary, based on economic thresholds has much potential for reducing losses to a minimum in growers' fields.

NOTES

Lloyd M. Wax, USDA-ARS, W-321 Turner Hall, University of Illinois; and E. W. Stoller, Department of Agronomy, University of Illinois, Urbana, Illinois.

REFERENCES

Aldrich, R. J. 1984. Weed-Crop Ecology. Breton Publishers, North Scituate, Massachusetts.

Andersen, R. N., J. H. Ford and W. E. Lueschen. 1982. Controlling volunteer corn (*Zea mays*) in soybeans (*Glycine max*) with diclofop and glyphosate. Weed Sci. 30:132-136.

Bloomberg, J. R., B. L. Kirkpatrick and L. M. Wax. 1982. Competition of common cocklebur (*Xanthium pensylvanicum*) with soybean (*Glycine max*). Weed Sci. 30:507-513.

Coble, H. D. and R. L. Ritter. 1978. Pennsylvania smartweed (*Polygonum pensylvanicum*) interference in soybeans (*Glycine max*). Weed Sci. 26:556-559.

Coble, H. D., F. M. Williams and R. L. Ritter. 1981. Common ragweed (*Ambrosia artemisiifolia*) interference in soybeans (*Glycine max*). Weed Sci. 29:339-342.

Cordes, R. C. and T. T. Bauman. 1984. Field competition between ivyleaf morningglory (*Ipomoea hederacea*) and soybean (*Glycine max*). Weed Sci. 32:364-370.

Felton, W. L. 1976. The influence of row spacing and plant population on the effect of weed competition in soybeans (*Glycine max*). Aust. J. Exp. Agric. and Anim. Husb. 16:926-931.

Harper, J. L. 1977. Population Biology of Plants. Academic Press, New York.

Harrison, S. K., C. S. Williams and L. M. Wax. 1984. Interference and control of giant foxtail (*Setaria faberi*) in soybeans (*Glycine max*). Weed Sci. 32:In press.

Irons, S. M. and O. C. Burnside. 1982. Competitive and allelopathic effects of sunflower (*Helianthus annuus*). Weed Sci. 30:372-377.

Keisling, T. C., L. R. Oliver, R. H. Crowley and F. L. Baldwin. 1984. Potential use of response surface analyses for weed management in soybean (*Glycine max*). Weed Sci. 32:In press.

Knake, E. L. and F. W. Slife. 1962. Competition of *Setaria faberii* with corn and soybeans. Weed Sci. 10:26-29.

Knake, E. L. and F. W. Slife. 1965. Giant foxtail seeded at various times in corn and soybeans. Weeds 13:331-334.

Kirkpatrick, B. L., L. M. Wax and E. W. Stoller. 1983. Competition of jimsonweed with soybean. Agron. J. 75:833-836.

Marra, M. C. and G. A. Carlson. 1983. An economic threshold model for weeds in soybeans (*Glycine max*). Weed Sci. 31:604-609.

Moolani, M. K., E. L. Knake and F. W. Slife. 1964. Competition of smooth pigweed with corn and soybeans. Weeds 12:126-128.

McWhorter, C. G. and J. M. Anderson. 1979. Hemp sesbania (*Sesbania exaltata*) competition in soybeans (*Glycine max*). Weed Sci. 27:58-63.

McWhorter, C. G. and E. E. Hartwig. 1972. Competition of johnsongrass and cocklebur with six soybean varieties. Weed Sci. 20:56-59.

McWhorter, C. G. and D. T. Patterson. 1979. Ecological factors affecting weed competition in soybeans. p. 371-392. In F. T. Corbin (ed.) World Soybean Res. Conf. II: Proceedings. Westview Press, Boulder, Colorado.

Oliver, L. R. 1979. Influence of soybean (*Glycine max*) planting date on velvetleaf (*Abutilon theophrasti*) competition. Weed Sci. 27:183-188.

Rathmann, D. P. and S. D. Miller. 1981. Wild oat (*Avena fatua*) competition in soybean (*Glycine max*). Weed Sci. 29:410-414.

Staniforth, D. W. 1965. Competitive effects of three foxtail species on soybeans. Weeds 13:191-193.

Thurlow, D. L. and G. A. Buchanan. 1972. Competition of sicklepod with soybeans. Weed Sci. 20:379-384.

Wax, L. M. and J. B. Schoper. 1983. Weed control and crop yields as affected by crop rotation, tillage and levels of pest management. Weed Sci. Soc. Am. Abstr. No. 12, p. 5-6.

Williams, C. S. and R. M. Hayes. 1984. Johnsongrass (*Sorghum halepense*) competition in soybeans (*Glycine max*). Weed Sci. 32:In press.

Zimdahl, R. L. 1980. Weed-Crop Competition. International Plant Protection Center, Oregon State University, Corvallis.

DIMENSIONS OF WORLD PRODUCTION, UTILIZATION, AND RESEARCH

WORLD SOIL EROSION: PROBLEMS AND SOLUTIONS

W. E. Larson and F. J. Pierce

Erosion is the reshaping of the earth's land surface by the forces of nature. It is both a cause and an effect, sharing responsibility for shaping of the land surfaces and for their deterioration. Whereas eroded sediment was a source of soil fertility to the ancient farmers along the Nile River, accelerated erosion is a major problem to those who use the land surface for production of food, clothing, and shelter.

Soil erosion, equated by Bennett (1947) with accelerated erosion, frequently occurs at rates that are orders of magnitude greater than 'natural' erosion, and it is directly related to human activities. The primary effects of human activity have been to expose the soil surface by removal of natural vegetation, thereby reducing the soil's resistance to erosive forces and allowing for the full effect of weather's erosive energy. Accelerated soil erosion phenomena are not new, extending to early civilizations (Stallings, 1957). However, two facts clearly demonstrate the sense of urgency for solutions to the world soil erosion problem. First, the soil resource in the world is finite, and is deteriorating in quality (FAO, 1978; Brink, 1977; Eckholm, 1976; Brown, 1978). Secondly, the world population is increasing, placing substantial stresses on the soil resource base (The Global 2000 Report to the President, 1980). Dudal (1980) estimates that agricultural production must increase 60% by the year 2000 to meet the food requirements of the growing world population. Demand for food and other agricultural products could be three times present levels by the time population stability is reached in the year 2110 (Dudal, 1982). The overriding problem is how the world is going to feed its growing population in the context of a finite soil resource and degradation of these resources by erosion and other causes.

A number of books (Bennett, 1939; El-Swaify et al., 1982; Holy, 1980; Hudson, 1981; Stallings, 1957; Zachar, 1982) document the forms of erosion, the processes involved, erosion effects, and measures for its control. It is not our focus to repeat the familiar warning that erosion reduces the soil's productive potential and deteriorates the quality of our environment. Rather, our thesis is that determination of the adequacy of the world's soil resources to meet the present and future food demand requires: (1) estimates of the inherent productivity of the world's soils, their distribution, and areal extent, (2) knowledge of the rate of change in productivity due to the many processes that threaten to degrade soil, and (3) the rate of occurrence of the degradation process; i.e., erosion. The extent to which these are documented in the world and information gaps are the topics explored here.

INHERENT AND POTENTIAL PRODUCTIVITY

Knowledge of the quality, areal extent and distribution of the world soils is prerequisite to understanding world soil erosion. Land resources and population are unevenly distributed in the world, and the quality of the soil varies considerably, as does the suitability of the climate for growing crops. The land use and population in major regions of the world are given in Table 1 (Dudal, 1982; FAO, 1983). Dudal (1982) estimates that about 11% of the land area, or 1,461 million ha, of the world is presently cultivated and an additional 12%, or 1,570 million ha, is potentially cultivable. About 77% of the potential cultivable land in the developed countries is presently cultivated, whereas only 36% is presently cultivated in the developing nations. The United States includes 916 million ha of land of which 18%, or 167 million ha, is cultivated and an additional 6%, or 51 million ha, is suitable for cropland (U.S. Dept. Agr. and Council of Env. Quality, 1981). This corresponds to over 11% of the cultivated land and a little over 7% of the cultivable land in the world. Africa and South America account for 11 and 8%, respectively, of the cultivated land, but 26 and 27%, respectively, of the potentially cultivable land in the world. Asia, on the other hand, accounts for 29%, or 456 million ha, of the cultivated land, but less than 16% of the cultivable land in the world. Asia, however, accounts for nearly 56% of the world population. With projected population increases to the year 2110, and all cultivable lands cultivated, the persons per hectare of cultivated land increases from 5.4 to 12.4 in Asia while the remainder of the world population average rises from 1.4 to 1.8. Africa would remain at 2.8 if all cultivable lands were cultivated. The developed countries would maintain at a ratio of 1.6. The ratio for the United States is presently at 1.4. These statistics illustrate the uneven distribution of land suitable for agriculture and population. It, however, says nothing of the inherent potential or quality of the soil in each area.

The agricultural potential of cropland can be determined inductively or deductively.

Table 1.   Land use and population in major regions of the world.

Region	Land area	Arable land	Irrigated land[a]	Population 1981[a]	Potentially cultivable hectares
	— Mha —			millions	Mha
World	13392	1368	211	4513	ND[c]
Africa	2886	163	8	484	789
North and Central America	2136	265	28	381	ND[c]
United States[b]	913	189	21	230	540
South America	1770	124	7	246	819
Asia	2677	456	134	2625	472
Europe	473	127	15	486	ND[c]
Oceania	843	45	2	23	ND[c]
USSR	2227	227	18	268	ND[c]
Developed	5773	677	59	1178	877
Developing	7619	784	152	3335	2154

[a]Data for Africa, South America, Asia, and Developed countries taken from Dudal (1982). Other data taken from FAO (1982).

[b]U.S. Dept. Agr. and Council of Env. Qual. (1981).

[c]ND means not determined.

In the United States, soil potentials have at times been deduced from long-term crop yield records. This is especially true for forest soils where an acceptable measure of quality is site index (Carmean, 1975). The Crop Equivalent Rating (Rust and Hanson, 1975) of Minnesota soils is another example of a deductive scheme to rate soil potentials. Inductive schemes for rating soil potentials base their evaluation on attributes of the soils, landscapes, and climates and their interaction with each other; and plants, animals, and human management practices. The Storie index (Storie, 1978), a multiplicative index of soil productivity, is an often-cited inductive rating scheme (Huddleston, 1984). Land evaluation methods have recently been reviewed by FAO (1974). The Agroecological zones project of the FAO (1978-1981) was a global attempt to match soil and climatic inventories with the soil and climatic requirements of 11 crops at two levels of inputs (Higgins and Kassam, as cited by Dudal, 1982). Attempts at assessing production potential have resulted in summaries of the potential over large areas in the forms of maps or summaries of land suitabilities, such as the agro-ecological project (FAO, 1978-1981) or the World Soil Atlas (FAO, 1971-1981).

One problem common to all schemes that estimate the potential of the land is the lack of available data on soil as it occurs in the landscape, the context in which soil erosion can be analyzed (Larson et al., 1983). Even the most simple indices or evaluation schemes require inputs unavailable even on a gross scale in most areas of the world. As is the case in the humid tropics, comprising nearly 5 billion ha of area (Dudal, 1980), inadequate knowledge of soils and of their food-crop potential is the greatest source of uncertainty in reaching quantitative decisions about the ability of the world to feed its growing population (Greenland, 1981). A second problem is the definition of soil productivity, which is compounded by the climate, crop, level of technology, and management system. This dependence of crop production on non-soil factors complicates the isolation of the inherent productivity of soil and limits worldwide comparisons of productivity on a soils base alone.

In the dynamic view of erosion, some numerical measure of productivity is necessary to comply with the second requirement expressed earlier, that the rate of change in productivity due to erosion be known or be estimated. One approach has been to evaluate the distribution of world soils in the context of the climate in which they occur and the crops produced in the region (FAO, 1978-81). Such studies can provide some measure of world productive potential, although at a higher level of abstraction than desirable. Another approach has been to identify the soil-related constraints to production of various crops as an inverse index to their inherent and potential productivity.

Buringh (1979) summarized the major soil units of the world and their potential for use as cropland (Table 2) from the Soil Map of the World (FAO, 1971-1981). Alfisols, Oxisols, and Inceptisols account for nearly 50% of the potential cropland in the world. About one-fourth of all tropical land is potentially suitable for cultivation, accounting for two-fifths of the world's potential cropland (Buringh, 1979). Ultisols are among the best soils of the tropics, allowing deep rooting. The Oxisols are typical of the humid tropics and are deep soils with good physical, but poor chemical, conditions. The Alfisols and Inceptisols occur throughout the world, although they are not typical of the tropics.

Soil-related constraints can provide some measure of the adequacy of the world soil resource for food production (IRRI, 1980; Munn and Franco, 1983). Usually soil-related constraints must be viewed in the context of the climate in which the soils occur. Constraints to crop production are more widely reported, but the potential of soil is more informative, as evidenced by the recent emphasis in soil survey interpretations

Table 2. Major soil units of the world.[a]

Major soil unit[b]	U.S. taxonomy	Total area	Proportion	Potential cropland Area	Proportion
		Mha	%	Mha	%
Acrisols, Nitosols	Alfisols, Ultisols	1050	8.0	300	9
Andosols	Inceptisols	101	0.8	80	2
Cambisols	Inceptisols	925	7.0	500	15
Chernozems, Greyzems, Phaeozems	Mollisols	108	1.1	200	6
Ferralsols	Oxisols	1068	8.1	450	14
Fluvisols	Entisols	316	2.4	250	8
Gleysols	Inceptisols	623	4.7	250	8
Histosols	Histosols	240	1.8	10	0
Lithosols, Rendzinas, Rankers	Mollisols	2264	17.2	0	0
Luvisols	Alfisols	922	7.0	650	20
Planosols	Alfisols	120	0.9	20	1
Podzols	Spodisols	478	3.6	130	4
Podzuluvisols	Alfisols	264	2.0	100	3
Regosols, Arenosols	Entisols	1330	10.1	30	1
Solonchaks, Solonetz	Aridisols	268	2.0	50	2
Vertisols	Vertisols	311	2.4	150	5
Xerosols, Kastanozems	Aridisols	896	6.8	100	3
Yermosols	Aridisols	1176	8.9	0	0
Miscellaneous land units		420	3.2	0	0
Total		13180		3270	

[a]From Buringh (1979).
[b]FAO, 1971-1981.

on soil potentials rather than on limitations (Olson, 1981). Both constraints and potentials are limited in that both are crop or land-use specific. The major soil-related constraints of the world's soil resources are given in Table 3 for major regions of the world (Dent, 1980). Severe limitations exist in all areas. However, Europe and the United States, in terms of percentage of land area, fare the best, especially if nutritional deficiencies can be alleviated economically. Munn and Franco (1983) list a set of soil properties likely to constrain legume production (Table 4) and the percentage of tropical land expressing that constraint. The first four, droughtiness, excess strength, erodibility and salinity, are unlikely to be altered easily or economically and are "capability limiters", whereas acidity and nutrient deficiencies are often economically remedied (Munn and Franco, 1983). In areas of traditional farming in the tropics, soil nutrient deficiencies are the main soil constraint (Buringh, 1979). Whereas most importance is attached to nutrient deficiencies, Dudal (1980) suggests that more active consideration be given to constraints due to physical characteristics, especially erodibility, workability, and water storage capacity. The key in this approach is to determine the constraints to food production, develop mechanisms and approaches for further identifying these constraints, and technologies to remove them (Brady, 1980). Erosion intensifies the constraints to crop production through deterioration of soil characteristics or the removal of favorable soil material.

The level of data available for assessment of the potential of soils in many regions of the world is inadequate (The Global 2000 Report to the President, 1980). Since

Table 3. World soil resources and their major limitations for agriculture (Dent, 1980).

Region	Proportion of total land area with limitation[a]					
	Drought	Mineral stress[b]	Shallow depth	Water excess	Perma-frost	No serious limitation
			– % –			
North America	20	22	10	10	16	22
Central America	32	16	17	10	–	25
South America	17	47	11	10	–	15
Europe	8	33	12	8	3	36
Africa	44	18	13	9	–	16
South Asia	43	5	23	11	–	18
North and Central Asia	17	9	38	13	13	10
Southeast Asia	2	59	6	19	–	14
Australia	55	6	8	16	–	15
World	28	23	22	10	6	11

[a]Data compiled from FAO/UNESCO soil map of the world.
[b]Nutritional deficiencies or toxicities related to chemical composition or mode of origin.

Table 4. Soil properties likely to constrain legumes and their distribution in the tropics (Munn and Franco, 1983).

	Soil constraints
%	
45	Droughtiness, due to shallowness, impedances, and pore characteristics
25-30	Excessive strength, cohesiveness, and ease of compaction
75	Excessive erodibility
7	Salinity, alkalinity, sudicity
35	Acidity and related problems
80	Phosphorus deficiency
10 S,K,Zn	Nutrient deficiencies

quantification of the inherent and potential productivity is possible only on broad scale bases, the difficulty in assessing its rate of change with degradation processes such as erosion is great.

RATE OF CHANGE IN PRODUCTIVITY

Soil degradation processes are dynamic. Soil erosion is one of a number of such degradation processes. Because it is dynamic, the effects of erosion on soil productivity are best viewed in terms of its rate of change. While it may be argued that soil is a finite resource that is, at best, slowly renewable, and that it will, in time, degrade completely if erosion rates exceed renewal rates, it is the dynamic rather than a static view of erosion that provides a proper framework for analysis. A static view is one, for example, whereby erosion is assessed by comparing the depth of topsoil of a specified soil with its thickness in an ideal soil type, or relative to some reference position (Riquier, 1978). In the dynamic view, the soil erosion problem can be viewed as follows: There is a natural rate of erosion, which for many soils is in equilibrium with or less than the soil renewal rates. Human activity has accelerated this natural rate through management practices resulting in the chemical, physical, and biological degradation of soil. Erosion is not entirely separable from other degradation processes.

The sensitivity of soil to changes in its properties due to erosion and the rate at which these changes occur are documented for only a few soils. This is, in part, due to the static rather than dynamic view of changes in productivity due to erosion that were taken in the past. In that view, the fact that erosion caused serious damage was established and the rate of change in productivity was assumed to be proportional to the amount of erosion. Evidence as to the detrimental effects of erosion was deemed sufficient and, consequently, research was focused primarily on the erosion process and the extent to which it occurs in the world (U.S. Congress, OTA, 1982).

This static view of erosion and associated assumptions about the rate of change of productivity with erosion rate have not served us well. Lal (1976) has shown that very small changes in soil properties that result from erosion can be devastating to soil productivity in Nigeria (Figure 1). The data also illustrate that the rate of change in productivity with soil removed is variable. Shrader et al. (1963) indicate that yield

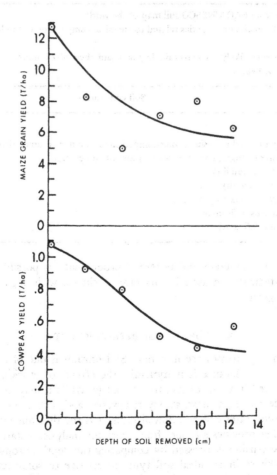

Figure 1. Effect of depth of soil removed on maize (*Zea mays* L.) and cowpea (*Vigna unguiculata* [L.] Walp.) yield in Nigeria (from Lal, 1976).

declines on certain deep soils can be offset with addition of fertilizers.

The use of technological inputs to offset erosion induced deficiencies, especially fertilizers, is often described as "masking" the effects of erosion on soil productivity (Brown, 1984; National Soil Erosion-Soil Productivity Research Planning Committee, 1981). Erosion research in the United States before intensive fertilizer use is often heavily relied on to document the deleterious effects of erosion (Pimentel et al., 1976), an approach described as inadequate for the U.S. by Langdale and Shrader (1982). In the tropics, however, crop yields are a small proportion, 10 to 20%, of the maximum, with the main constraint being soil fertility (Buringh, 1979). El-Swaify et al. (1982) caution against the use of soil characteristics alone to establish erosion loss tolerance levels in underdeveloped countries, as limited resources are available to the typical small farmer.

The dynamic and three dimensional nature of the soil provides the key to its vulnerability to erosion (Larson et al., 1983). All too frequently studies are cited documenting reductions in crop yields due to erosion with little regard for the soil, the environment in which it occurs, or the technology available at the time of the study, suggesting that such yield reductions are common on all soils under a variety of conditions (Pimentel et al., 1976; Brown, 1984). Of the nearly 14,000 soil series presently recognized in the United States, few have been the subject of studies where soil productivity, erosion rates, and changes in productivity have been measured on the same site.

The evidence for the detrimental effects of erosion on productivity is based upon scattered but alarming examples of soil deterioration (Brown, 1978, 1984) sufficient to warrant documentation of soil productivity and its change from erosion. A methodology for assessing soil degradation has been proposed by FAO (1978, 1979) for predicting "the possible decrease in productivity in the future and to evaluate the input of conservation measures or cultural practices necessary to avoid such loss" (Riquier, 1978). In the United States, a significant effort to model the processes involved in erosion and crop production has been made (Williams et al., 1983). A model developed by Kiniry et al. (1983) was applied by Pierce et al. (1984) to soils in the Corn Belt Region of the United States to estimate long-term effects of erosion on soil productivity. For rainfed corn under a high level of management, changes in long-term soil productivity,weighted by areal extent, for Major Land Resource Regions (USDA-SCS, 1981) were projected to be less than 8% over the next 100 years (Table 5), assuming USDA-SCS estimated erosion rates remained constant. This estimate is based on a model that relates long-term nonreplaceable productivity to the soil's sufficiency as a rooting medium for crop growth and the soil's ability to store plant-available water. It assumes a high level of management and sufficient added nutrients. Deep fertile soils and a preponderance of cropland in nearly level to gently rolling terrain (Table 6) buffer the Corn Belt Region from severe losses over the next 100 years under a high level of management. However, significant losses in soil productivity, up to 100%, were projected for soils with unfavorable characteristics in the subsoil horizons. "The fact that most of the country's erosion occurs on a relatively small amount of land," states a recent OTA report (U.S. Congress, OTA, 1982), "has only recently been widely recognized by national policy makers."

The study by Pierce et al. (1984) provided estimates of the effects of erosion on soil productivity over a broad geographic region. It was made possible because significant sources of data on soil properties and spatial estimates of erosion rates were compiled by the Soil Conservation Service. Inputs even to simple methods such as that of FAO (1978, 1979) or Pierce et al. (1983, 1984) are difficult to obtain and exist for

Table 5. Estimated reduction in soil productive potential from erosion in 100 years (Pierce et al., 1984).

Major land resource area	Estimated erosion	Estimated reduction in soil productive potential from erosion in 100 years[a]
	t/ha·year	%
102A	6	1.7
102B	19	2.3
103	10	3.1
104	13	2.7
105	24	6.4
106	21	5.9
107	38	2.7
108	19	2.4
109	38	7.1
110	10	4.9
111	10	5.7
112	16	7.8
113	19	4.4
114	14	5.5
115	22	3.9

[a]Assumes erosion continues at rates given in column 2.

Table 6. Areal extent of cropland in the Corn Belt of the United States by slope class.

Slope class	Area	Proportion of total
	kha	%
0-2	18030	45.3
2-6	15142	38.0
6-12	5375	13.5
12-20	1212	3.0
20-45	72	0.2
45	6	0.1

only a few soils worldwide. The exact form of the model and the assumptions used by Pierce et al. (1983, 1984) are likely unique to the Corn Belt Region of the United States and do not necessarily apply elsewhere. Their approach, however, is dynamic and lends itself to definition of the erosion problem, a first step in its solution.

The rate of change in soil productivity due to erosion is not known on a global level, and a satisfactory estimate will probably not be obtainable in the near future. However, critical decisions need to be made now. The challenge to the world's scientists is apparent.

EXTENT OF EROSION

If both the inherent productivity of soil and its rate of change with erosion are known, the final information necessary to assess world soil erosion is the rate at which erosion occurs. Soil erosion is difficult to measure. Static estimates of erosion are those that define erosion relative to some standard reference ideal soil type. These include estimation of the quantity of soil removed from measurable changes in soil level throughout the field (Dunne, as cited by El-Swaify et al., 1982) or as qualitative reconnaissance estimates, such as that used in the United States where soils are assigned to descriptive soil classes (USDA, 1957). Dynamic estimates are based on rate measures of the process, including trapping and measuring the quantity of soil removed with time, relating sediment loads of streams or rivers to erosion rates within a watershed, and evaluation of potential erosion rates through physically modeling the erosion process (El-Swaify et al., 1982). Actual measurements are too difficult, costly, and time consuming for many applications. Sediment delivery ratios necessary for relating stream sediment loads to soil erosion are fraught with problems (Walling, 1983) and can, at best, give average rates for the watershed. Erosion is most commonly estimated by simple empirical equations. Most widely used are the Universal Soil Loss Equation, USLE, (Wischmeier and Smith, 1978) and the Wind Erosion Equation (Woodruff and Siddoway, 1965). While subject to misuse (Wischmeier, 1976), the USLE has provided a useful tool for estimation of water erosion in many parts of the world. Quantitative values for the parameters in the equations must be derived from on-site studies (El-Swaify et al., 1982). Quantitative values for these parameters do not exist for most of the world areas.

The extent to which erosion by water occurs on arable land is determined by (i) the erosivity of the rainfall, (ii) the erodibility of the soil, (iii) management practices, and (iv) the land form, which includes the scale and geometry of the slopes (Hudson, 1977). This is shown diagramatically in Figure 2.

Erosivity

The importance of falling raindrops in the erosion process was first recognized by Ellison in 1944 (Stallings, 1957). Erosivity is generally related to intensity of the rainfall and described in terms of the kinetic energy (KE) of the storm. The seasonal distribution of the energy load of rainstorms often follows the annual distribution of rains (Kowal and Kassam, 1977) such that mean annual rainfall gives a gross estimate of potential soil erosion (Hudson, 1981). Zachar (1982) cautions that the erosive influence of raindrops, expressed in terms of rainfall intensity, is often overestimated and the erosive effect of flowing water is underestimated. Raindrop erosion is more

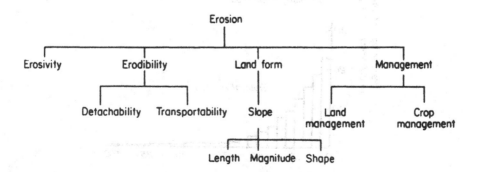

Figure 2. Diagram showing that erosion is determined by (a) the erosivity, (b) the erodibility, (c) land form, and d) management (from Hudson, 1977).

important on the upper parts of slopes. Flowing water is more important in the lower parts, where the splashing effect of raindrops is strongly inhibited by the water layer.

The percentage of rainstorms with potential to produce erosion varies. Lal (1977), for example, reported that a majority of tropical rainstorms fall in the erosive category, whereas only 5 to 10% of the temperate storms are erosive. Lal attributed the high erosivity of tropical storms to the higher intensity rainfall (associated with large drop sizes and total kinetic energy) and increased wind velocities of tropical storms. Comparison of the frequency distribution of rainfall intensity for temperate and tropical rainfall is illustrated in Figure 3 (Hudson,1981). Neither rainfall amount nor rainfall intensity are highly correlated with soil erosion (Lal, 1977). As a result, empirical formulae for estimating erosivity have been widely used (Hudson, 1981; Lal, 1980; El-Swaify et al., 1982). The erosivity index (EI) of Wischmeier and Smith (1958) has found wide use, but should be used with caution and not extrapolated beyond the situations for which it was intended (Wischmeier, 1976). Hudson (1981) discusses the folly of such extrapolation and warns against the deduction of EI values from other rainfall parameters, as those reported in Atheshian (1974, as cited by Hudson, 1981).

Erodibility

The erodibility of a soil refers to its susceptibility to erosion due to the forces causing detachment and transport of soil particles (Lal, 1977). Detachment refers to

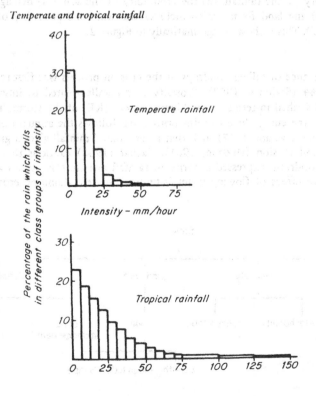

Figure 3. Frequency distribution of rainfall intensity for temperate and tropical rainfall (from Hudson, 1981).

the breaking away of soil particles by forces generated to a large degree by raindrop impact and is negatively related to the degree of soil cover by vegetation or residue. The resistance of the soil to detachment is largely determined by soil structure, which also largely determines the rate that water can enter soil (Greenland, 1977). The degree of aggregation and its stability are important characteristics of soil structure that are related to soil properties including texture, organic matter content, nature of the clays, presence of cementing agents such as iron and aluminum oxides, and balance of cations on the exchange complex (Lal, 1977). The exact relationship between erodibility and soil structure is not known (Greenland, 1981). Erodibility is a dynamic soil factor, although it is often treated as a soil constant. Erodibility is often deduced from other more easily measurable properties (Hudson, 1981), such as the nomograph for determining the K factor of the Universal Soil Loss Equation (Wischmeier and Smith, 1978). Such empirical relationships are not always applicable (El-Swaify et al., 1982).

Transport of sediment is a function of the amount of water runoff and energy associated with the runoff. The amount of runoff is a function of the hydraulic properties of the soil, its initial wetness, the rainfall rate, and capacity of the surface for storage. The energy associated with the runoff is a function of the amount of runoff and the slope characteristics, including the gradient and geometry and the resistance of the surface to flow. Management and vegetation affect the hydraulic properties of the soil and the resistance of the surface to flow. Thus, land form and land management influence the degree to which erosion will occur.

Erosion estimates for many countries are available and have been summarized in tables and maps. El-Swaify et al. (1982) recently summarized available data on erosion for the tropics. Global estimates on erosion of continents have been made (Table 7). In the United States, the best available estimates of erosion were compiled in the National Resource Inventories (NRI) of 1977 and 1982 by the Soil Conservation Service. A graphic summary of the 1977 erosion estimates for the United States is given in Figure 4.

Erosion by wind depends on the wind force, the structure of the soil, the water content of the soil, and the density of the vegetation cover (Zachar, 1982). Wind erosion may occur, according to Lyles (1977), "whenever the surface soil is finely divided, loose, and dry; the surface is smooth and bare; and the field is unsheltered, wide and improperly oriented with respect to the prevailing wind direction." The main

Table 7. Rates of erosion of the continents[a] (El-Swaify et al., 1982).

	Area	Denudation rates	
		Mechanical[b]	Chemical[c]
	Mha	— t/ha·year —	
Africa	2981	0.47	0.25
Asia	4489	1.66	0.42
Australia	796	0.32	0.11
Europe	967	0.43	0.32
North and Central America	2044	0.73	0.40
South America	1798	0.93	0.55

[a]From river data.
[b]Suspended sediment.
[c]Dissolved solids.

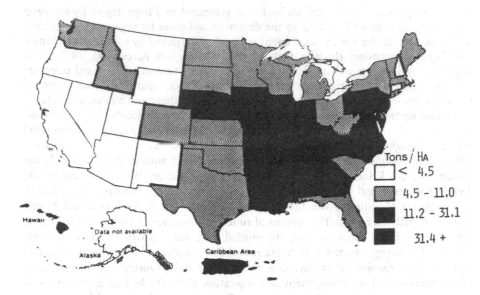

Figure 4. Graphic summary of the erosion estimates for 1977 for the United States (USDA-SCS, 1981).

cause for accelerated erosion is the removal of vegetation and the drainage of land (Zachar, 1982). Zachar (1982) estimates the total area of deserts, semideserts, and wind threatened regions of the world to be 3 billion hectares, or approximately 20% of land area. Wind erosion estimates are not available, to any extent, on a global level. Estimates of wind erosion for the Great Plains of the United State are given in Table 8.

Table 8. Wind erosion on cropland and rangeland in the Great Plains States of the United States, 1977.

	Cropland			
	Erosion (t/ha·year)			
	4.5	4.5-11.0	11.1-31.4	41.4
	— kha —			
Colorado	1963	724	825	979
Kansas	8023	1598	1533	509
Montana	3310	1517	1076	313
Nebraska	7165	658	411	146
New Mexico	292	140	267	226
North Dakota	7578	2266	999	44
Oklahoma	3333	558	625	254
South Dakota	3982	2275	954	139
Texas	5256	794	2530	3743
Wyoming	855	110	213	24
Total	41757	10639	9433	6377

Movement of soil particles by wind can have several degrading effects on the soil. In one case, material is removed from the surface, exposing less favorable soil materials for plant growth. In a contrasting case, coarse material from desert or desert-like areas is deposited on the surface of more favorable soil materials, making the entire medium less favorable for plant growth. This process is usually called desertification. In a third case, because of the sorting action of wind, the finer soil fraction (silt, clay, organic matter) may be removed, leaving the coarse material (sand) on the soil surface. All of these effects can be devastating to soil productivity if allowed to proceed beyond safe limits.

The value of erosion estimates lies in determination of how much erosion occurs and where it is occurring. Erosion must be viewed in the context of the soil as it occurs in the landscape (Larson et al., 1983) and so erosion estimates must be site-specific and combined with knowledge of the productivity of soil and its rate of change with soil removed. This, along with knowledge of the fate of eroded sediments, will provide a good perspective of the problem of soil erosion.

EROSION FROM SOIL CROPPED TO SOYBEANS

The world's need for high protein food and vegetable oil has resulted in dramatic increase in soybean production. In many of the countries, the areas with intense soybean production are also areas with alarming rates of erosion. Soybean production increased dramatically in some of the South and North American countries between 1972 and 1982. For example, production increased by 4300% in Argentina, 400% in Brazil, and 200% in Mexico. All of these countries have excessively high erosion rates. Erosion rates are great in areas where soybean is a major crop. Soybean production increased by 79% in the United States in the decade ending in 1982, and has, no doubt, contributed to high erosion rates in some areas.

Soybean has long been considered a crop that accentuates erosion. However, in most areas of the developing countries where erosion has increased with a corresponding increase in soybean area, it is not possible to separate the effects of growing soybeans on erosion from the effects of land clearing, and consequently, increasing area under cultivation.

If other practices are the same, soil erosion is frequently greater after the soil has been cropped to soybean than after the soil has been cropped to corn (Siemens and Oschwald, 1976; 1978; Oschwald and Siemens, 1976). Erosion and water runoff may be less, however, during the year of cropping to soybean as compared to corn (Van Doren and Stauffer, 1943; Browning et al., 1942; Smith and Whitt, 1947).

Several studies in Illinois and Iowa (Siemens and Oschwald, 1976; 1978; Oschwald and Siemens, 1976) found a higher percentage of simulated rain ran off the year after a soybean crop than after corn. Table 9 from data obtained in Iowa shows that runoff and erosion were usually greater when the prior crop was soybean than when it was corn. Fine- or medium-textured soil is often more mellow, less cloddy, and contains more small aggregates following soybean, in contrast to corn. The small aggregates may be susceptible to dispersion and sealing of the soil surface. Fahad et al. (1982) found that the rank order of the geometric mean diameter of soil aggregates for several soybean cropping sequences was: continuous soybean ⟨ soybean after corn ⟨ soybean after sorghum ⟨ soybean after fallow. Surface sealing enhances runoff and the dispersed soil material may be readily transported in the runoff water.

Soybean does not produce as much residue as corn or cereal crops, and the residue rapidly decomposes. Thus, residue cover is small for protection of the soil surface

Table 9. Runoff and soil erosion, averaged over three tillage systems for rainfall simulation studies in 1977 and 1978 (Laflen and Colvin, pers. commun.).

Location	Prior crop	Residue cover	Runoff	Erosion
		%	cm	t/ha
June 1977, Canopy = 10%				
WRC	Corn	14	5.1	37.4
	Soybean	13	4.6	77.8
Ames	Corn	27	1.5	1.0
	Soybean	15	2.5	3.7
Early July 1978, Canopy = 60%				
WRC	Corn	27	3.0	8.7
	Soybean	13	3.6	25.3
Ames	Corn	15	2.3	3.2
	Soybean	9	4.1	9.1
Early August 1978, Canopy = 90%				
WRC	Corn	35	3.8	3.5
	Soybean	29	7.3	6.6
Ames	Corn	27	1.5	1.3
	Soybean	22	2.8	2.4

from raindrop impact. In contrast to soybean, corn and many cereal crops produce large quantities of residues that decompose less rapidly. If the residues from corn are left on the surface by appropriate management, they are an effective deterrent to runoff and erosion. During tillage the soil surface is often left more rough and cloddy following corn and cereals. All of these factors contribute to the usually observed greater runoff and erosion following soybean as contrasted to corn or cereals.

In most soybean growing areas of the world it is probably feasible to plant a fall or winter cover crop of cereals, grasses or legumes immediately after soybean harvest to reduce damage from erosion. Because of the binding action of the roots and crop cover, water or wind erosion would be reduced on most soils. The cover crop could be killed by tillage or herbicide before planting of the succeeding crop. Whereas feasible, planting of a cover crop is not commonly practiced, probably because of cost.

SOCIO-ECONOMIC ASPECTS

The greater challenge in soil conservation is in coming to terms with the nontechnical constraints (Hudson, 1981). In writing about deteriorating mountain environments, Eckholm (1976) states "there is no major mountain problem for which technological solutions are not already known. . . it is in general easy to recommend technological answers to ecological problems. Political and cultural factors are invariably the real bottlenecks holding up progress." Rieger (1978/79) suggests that determinants of human behavior must be understood before any concerted attack on humanity's destruction of the ecosystem can be made with any hope of success. Even after 50 years of conservation efforts in the United States, soil erosion remains the main conservation problem on 50% of the nation's cultivated cropland (USDA, 1981). This is in spite of the fact that the technology to control erosion is readily available. It must be concluded, therefore, that the erosion problem transcends the technology of its control.

Rieger (1978/79) sketches four behavioral phenomena that contribute to the ineffectiveness of soil conservation programs. These derive from the divergence of individual rationality from collective rationality. They are Future Ignoring Behavior (FIB), Social Cost Ignoring Practices (SCIP), Lack of Collective Organization (LOCO), and Lack of Identification with Public Property (LIPP). Under the FIB phenomenon there is a disregard for the future effects of one's actions, since future repercussions are felt by the next generations only, and a priority exists for present over future actions. This type of behavior is a typical response to the often low benefit-cost ratio of erosion control. For example, in comparing the costs of depletion of soil resources by erosion with the costs of preventing depletion, Rosenberry et al. (1980) concluded that the cost of erosion control to farmers is greater than the economic return from controlling it. Under the SCIP phenomenon, social costs are ignored in production or technological decisions. Examples of SCIP are off-site costs of erosion and non-point source pollution. The LOCO phenomenon occurs when the perception centers on individual gain versus collective gain. Either one does collectively undesirable actions for fear someone else will do it and gain from it, or conversely, one fails to do collectively desirable actions for fear nobody else will. In either case, suggested Rieger (1978/79), the system is stable at a non-optimal position. The "sod-buster" problem in the Great Plains Region of the United States, and failure to implement erosion control practices, exemplify the two aspects of this phenomenon. Under the LIPP phenomenon, the higher levels of social organization are viewed as unwarranted restrictions on traditional rights, rather than an expression of the individual's efforts to further the welfare of society. Federal and state conservation legislation is often undermined by the LIPP phenomenon. Soil conservation programs can only be effective when moved from below (Dudal, 1981) and will not succeed when imposed from above, even if they are technically correct (Hudson, 1981).

Other determinants of human behavior exist and are important. Rieger (1978/79) chose to describe these four to demonstrate that individual rationality can be at odds with collective rationality, a concept important to bear in mind when reaching for solutions to the erosion problem.

General educational levels and training in soil conservation technology are important factors in determining human behavior. The importance of understanding the host of interactions among practices, soils, available resources, and local culture cannot be overestimated (Kellogg, 1975). Dudal warns that the potential of soils will not be fully developed unless the physical resource base needs are matched to prevailing social, economic, and health conditions that may considerably alter production perspectives. This is especially true as technology produced in the developed countries, especially the United States, is transferred to the developing countries (El-Swaify et al., 1982).

CONCLUSIONS

Soil erosion exists to varying degrees throughout the world. The data necessary to quantify erosion, as to its extent and effects on the world, are far from adequate. The available data, however, provide sufficient evidence to warrant a high priority be given to the erosion problem worldwide.

To determine the adequacy of the soil resource base in light of world population growth and degradation processes, such as soil erosion, it is necessary to know: the inherent and potential productivity of soils, the rate of change in productivity due to processes that degrade it, and the rate at which the degradation process is occurring. Great strides must be made in each of these areas. To sustain soil resources, three factors must be considered: a reduction in the rate at which degradation processes occur, technological developments that enhance crop production, and an understanding of the determinants of human behavior, so that best management practices will be used on land suitable for food production.

NOTES

W. E. Larson, Professor and Head, Department of Soil Science, University of Minnesota, St. Paul, MN 55108. F. J. Pierce, Assistant Professor, Department of Crop and Soil Sciences, Michigan State University, East Lansing, MI 48824.

REFERENCES

Bennett, H. H. 1939. Soil Conservation. McGraw-Hill, New York.

Bennett, H. H. 1947. Elements of Soil Conservation. McGraw-Hill, New York.

Brady, N. C. 1980. Welcome Address, p. 1-2. *In* Priorities for alleviating soil-related constraints to food production in the tropics. International Rice Research Institute and New York State College of Agriculture and Life Sciences, Cornell University, Ithaca, N.Y.

Brink, R. A. 1977. Soil deterioration and the growing world demand for food. Science 197:625-630.

Brown, L. R. 1978. The worldwide loss of cropland. Worldwatch Institute, Washington.

Brown, L. R. 1984. Conserving Soils. p. 54-73. *In* State of the World—1984. Worldwatch Institute, Washington, D.C.

Browning, G. M., M. B. Russell and J. R. Johnston. 1942. The relation of cultural treatment of corn and soybean to moisture condition and soil structure. Soil Sci. Soc. Am. Proc. 7:108-113.

Buringh, P. 1979. Introduction to the study of soils in tropical and subtropical regions. 3rd edition. Centre for Agricultural Publishing and Documentation. Wageningen, The Netherlands.

Carmean, W. H. 1975. Forest site quality evaluation in the United States. Adv. Agron. 27:209-269.

Dent, F. J. 1980. Major production systems and soil-related constraints in southeast Asia. p. 79-106. *In* Priorities for alleviating soil-related constraints to food production in the tropics. International Rice Research Institute and New York State College of Agriculture and Life Sciences, Cornell University, Ithaca, N.Y.

Dudal, R. 1980. Soil-related constraints to agricultural development in the tropics. p. 23-37. *In* Priorities for alleviating soil-related constraints to food production in the tropics. International Rice Research Institute and New York State College of Agriculture and Life Sciences, Cornell University, Ithaca, N.Y.

Dudal, R. 1981. An evaluation of conservation needs. p. 3-12. *In* R. P. C. Morgan (ed.) Soil conservation: problems and prospects. Wiley, Chichester, England.

Dudal, R. 1982. Land degradation in a world perspective. J. Soil Water Conserv. 37:245-249.

Eckholm, E. P. 1976. Losing Ground: Environmental Stress and World Food Prospects. Norton, New York.

El-Swaify, S. A., E. W. Dangler and C. L. Armstrong. 1982. Soil erosion by water in the tropics. Research Extension Series, HITAHR, College of Tropical Agriculture and Human Resources, University of Hawaii.

Fahad, A. A., L. N. Mielke, A. D. Flowerday and D. Swartzendruber. 1982. Soil physical properties as affected by soybean and other cropping sequences. Soil Sci. Soc. Am. J. 46:377-381.

FAO. 1971-1981. FAO/UNESCO. Soil map of the world (1:5,000,000) Vol 1-10. UNESCO, Paris, France.

FAO. 1974. A framework for land evaluation. Soil Bull. No. 32. Rome, Italy.

FAO. 1978. FAO/UNEP Methodology for assessing soil degradation. FAO, Rome.

FAO. 1978-1981. Reports of the agro-ecological zones project. Vol. 48/1 Africa; 48/2 Southwest Asia; Vol. 48/3 South and Central America; Vol. 48/4 Southeast Asia. Rome, Italy.

FAO. 1979. FAO/UNEP/UNESCO. Provisional Methodology for Assessing Soil Degradation. Rome, Italy.

FAO. 1983. FAO production yearbook. FAO Statistic Series No. 40, Vol. 36.

Greenland, D. J. 1981. Soils and crop production in the lowland humid tropics. p. 1-7. In D. J. Greenland (ed.) Characterization of soils in relation to their classification and management for crop production: examples from some areas of the humid tropics. Clarendon Press, Oxford.

Greenland, D. J. 1977. Soil structure and the erosion hazard. p. 17-24. In D. J. Greenland and R. Lal (eds.) Soil conservation and management in the humid tropics. John Wiley and Sons, New York.

Holy, M. 1980. Erosion and Environment. Environmental Sciences and Applications, Volume 9. Pergamon Press, New York.

Huddleston, J. H. 1984. Development and use of soil productivity ratings in the United States. Geoderma. 3:297-317.

Hudson, N. 1977. The factors determining the extent of erosion. p. 11-16. In D. J. Greenland and R. Lal (eds.) Soil conservation and management in the humid tropics. John Wiley and Sons, New York.

Hudson, N. 1981. Soil Conservation, second edition. Cornell University Press, Ithaca, N.Y.

IRRI, International Rice Research Institute and New York State College of Agriculture and Life Sciences, Cornell University. 1980. Priorities for alleviating soil-related constraints to food production in the Tropics. IRRI, Manila, Philippines.

Kellogg, C. E. 1975. Agricultural Development: Soil, food, people, work. Soil Sci. Soc. Am., Inc., Madison, Wisconsin.

Kiniry, L. N., C. L. Scrivner and M. E. Keener. 1983. A soil productivity index based upon predicted water depletion and root growth. Missouri Agric. Exp. Stn. Res. Bull. 1051.

Kowal, J. M. and A. H. Kassam. 1977. Energy load and instantaneous intensity of rainstorms at Samaru, Northern Nigeria. p. 57-70. In D. J. Greenland and R. Lal (eds.) Soil conservation and management in the humid tropics. John Wiley and Sons, New York.

Lal, R. 1976. Soil erosion problems on an Alfisol in western Nigeria and their control. Monograph No. 1. Ibadan, IITA.

Lal, R. 1977. Analysis of factors affecting rainfall erosivity and soil erodibility. p. 49-56. In D. J. Greenland and R. Lal (eds.) Soil conservation and management in the humid tropics. John Wiley and Sons, New York.

Lal, R. 1980. Soil erosion as a constraint to crop production. p. 405-424. In Priorities for alleviating soil-related constraints to food production in the tropics. International Rice Research Institute and New York State College of Agriculture and Life Sciences, Cornell University, Ithaca, N.Y.

Langdale, G. W. and W. D. Shrader. 1982. Soil erosion effects on soil productivity of cultivated cropland. p. 41-51. In B. L. Schmidt, R. R. Allmaras, J. V. Mannering and R. I. Papendick (eds.) Determinants of soil loss tolerance. ASA Special Pub.: No. 45. American Society of Agronomy, Madison, Wisconsin.

Larson, W. E., F. J. Pierce and R. H. Dowdy. 1983. The threat of soil erosion to long-term crop production. Science 219:458-465.

Lyles, L. 1977. Wind Erosion: Processes and effect on soil productivity. Trans. Am. Soc. Agric. Eng. 20:880-884.

Munn, D. N. and A. A. Franco. 1983. Soil constraints to legume production. p. 133-152. In P. Graham and S. C. Harris (eds.) Biological nitrogen fixation technology for tropical agriculture: papers presented at Workshop held at CIAT. Centro Internacional de Agricultura Tropical. Cali, Colombia.

National Soil Erosion-Soil Productivity Research Planning Committee, NSE-SPRPC. 1981. Soil erosion effects on soil productivity: A research perspective. J. Soil Water Cons. 36:82-90.

Olson, G. W. 1981. Soils and the environment—a guide to soil surveys and their application. Chapman and Hall, New York.

Oschwald, W. R. and J. C. Siemens. 1976. Soil erosion after soybeans. p. 79-81. *In* L. D. Hill (ed.) World Soybean Research. Proc. of the World Soybean Research Conference. Interstate Printers, Danville, IL.

Pierce, F. J., W. E. Larson, R. H. Dowdy and W. A. P. Graham. 1983. Productivity of soils: Assessing long-term changes due to erosion. J. Soil Water Conserv. 38:39-44.

Pierce, F. J., R. H. Dowdy, W. E. Larson and W. A. P. Graham. 1984. Soil Productivity in the Corn Belt: An assessment of erosion's long-term effects. J. Soil Water Conserv. 39:131-136.

Pimentel, D., E. C. Terhune, R. Dyson-Hudson, S. Rocherau, R. Sames, E. A. Smith, D. Denman, D. Reifschneider and M. Shepard. 1976. Land Degradation: Effects on food and energy resources. Science 194:149-155.

Rieger, H. C. 1978/79. Socio economic aspects of environmental degradation in the Himalayas. J. Nepal Res. Centre. No. 2/3 (Sciences):177-184.

Riquier, J. 1978. A methodology for assessing soil degradation. *In* FAO/UNEP Methodology for assessing soil degradation.

Rosenberry, P., R. Knutson and L. Harmon. 1980. Predicting the effects of soil depletion from erosion. J. Soil Water Conserv. 35:131-134.

Rust, R. H. and L. D. Hanson. 1975. Crop equivalent rating guide for soils of Minnesota. Minnesota Agric. Exp. Stn. Misc. Rep. 132.

Shrader, W. D., H. P. Johnson and J. F. Timmons. 1963. Applying erosion control principles. J. Soil Water Conserv. 18:195-199.

Siemens, J. C. and W. R. Oschwald. 1976. Erosion for corn tillage systems. Trans. Am. Soc. Agric. Eng. 19:69-72.

Siemens, J. C. and W. R. Oschwald. 1978. Corn-soybean tillage systems; erosion control, effects on crop production, costs. Trans. Am. Soc. Agric. Eng. 21:293-302.

Smith, D. D. and D. M. Whitt. 1947. Estimating soil losses from field areas of claypan soils. Soil Sci. Soc. Am. Proc. 12:485-490.

Stallings, J H. 1957. Soil Conservation. Prentice-Hall, Englewood Cliffs, N.J.

Storie, R. E. 1978. Storie index rating. Univ. of California-Davis, Division of Agriculture, Sci. Spec. Pub. 3203.

The Global 2000 Report to the President: The Technical Report Vol. II. 1980. Council of Environmental Quality and U.S. Dept. of State.

U.S. Congress, Office of Technical Assessment. 1982. Washington, D.C.

U.S. Department of Agriculture and Council of Environmental Quality. 1981. National Agriculture Lands-Study, Final Report. Government Printing Office, Washington, D.C.

USDA-SCS Soil Survey Staff. 1981. Land resource regions and major land resource areas of the United States. Agric. Handbook No. 296. USDA, Soil Conservation Service, Washington, D.C.

USDA. 1957. Soil Survey Manual. Agric. Handbook No. 18. Washington, D.C.

USDA. 1981. Soil and Water Resources Conservation Act Appraisal. Part I. Washington, D.C.

Van Doren, C. A. and R. S. Stauffer. 1943. Effect of crop and surface mulches on runoff, soil losses and soil aggregation. Soil Sci. Soc. Am. Proc. 8:97-101.

Walling, D. E. 1983. The sediment delivery problem. Scale problems in hydrology. J. Hydrol. 65:209-237.

Williams, J. R., K. G. Renard and P. T. Dybe. 1983. EPIC. A new method for assessing erosion's effect on soil productivity. J. Soil Water Conserv. 38:381-383.

Wischmeier, W. H. and D. D. Smith. 1958. Rainfall energy and its relationship to soil loss. Trans. Amer. Geophys. Union 39(2):285-291.

Wischmeier, W. H. 1976. Use and misuse of the Universal Soil Loss Equation. J. Soil Water Conserv. 31:5-9.

Wischmeier, W. H. and D. D. Smith. 1978. Predicting rainfall-erosion losses—a guide to conservation planning. Agric. Handbook No. 537. USDA, Washington, D.C.

Woodruff, N. P. and F. H. Siddoway. 1965. A wind erosion equation. Soil Sci. Soc. Am. Proc. 29:602-608.

Zachar, D. 1982. Soil Erosion. Developments in Soil Science 10, Elsevier Scientific Publishing Company, New York.

World Soil Erosion:
Problems and Solutions

PRODUCTION CONSTRAINTS AND SOIL EROSION IN THE HUMID
TROPICS OF DENSELY POPULATED JAVA

Enrique M. Barrau

Java's ideal climatic conditions for plant growth and lack of natural catastrophies such as hurricanes and typhoons have made this island an ideal place to live. Improvements in medicine, without parallel gains in family planning programs, have resulted in tremendous population increases. Population growth in Indonesia was 66% from 1961 to 1980.

The island of Java has the highest population density in the world. High population density and high food demand are aggravated by the fact that 21% of the land in Java is under government forest and over 5% is in government-run estate crops. Approximately 65% of the total population are farmers, and average farm size in irrigated land is 0.26 ha and in non-irrigated land about 0.6 ha.

Rice (*Oryza sativa* L.) production increases in Indonesia from 1974 to 1983 have been nearly 100%. After importing more than 2 million tons in 1980, the country obtained self-sufficiency by 1982. Although the government has been very successful in its effort to increase irrigated rice production, other crops such as groundnut (*Arachis hypogaea* L.), maize (*Zea mays* L.) and especially soybean have been practically ignored until recent years. While soybean poduction has remained almost the same for the last 10 years (Table 1), consumption has increased by about 60%, reflecting the 65% population increase during 1961 to 1980. The demand for soybeans as a source of concentrated feed for livestock, and especially for the poultry industry, has also increased significantly. As a result of the demand for soybeans and lack of local production, imports have increased from 150 to 330,000 t in the last 10 years at a cost of approximately $110,000,000 in 1983.

The majority of the soybean production in Indonesia comes from Java, which has 78% of the total production and about 77% of the total area harvested (Handy and Mardanus, 1978).

MAJOR CONSTRAINTS TO SOYBEAN PRODUCTION

The majority of the soils in Java are strongly acid with a pH of about 5, thus Al levels are usually high. Work underway by the Food Crops Research Institute at West Sumatra (Oct-Dec 1983 Quarterly Report) shows increases in soybean production of up to 300% by liming acid soils. Similar results have been reported by Sanchez et al. (1983) in Yurimaguas, Peru on acid Ultisols, which gave impressive yield increases by modest applications of fertilizers and lime. Table 2 shows the levels of fertilizers used in these 10-year, continuously cultivated, plots and Figure 1 shows the yields.

Table 1. Changes in soybean production in Indonesia, four-year averages. (Statistical Information on Indonesian Agriculture, Second Edition 1968/1980, Department of Agriculture, R.I.)

Regions	Average Periods	
	1968/1971	1977/1980
Net Area Harvested (1,000 ha)		
Java and Madura	549	579
Sumatra	28	60
Kalimantan	1	1
Sulawesi	7	23
Maluku and Irian Jaya	1	2
Bali and Nusa Tenggara	65	60
Indonesia	651	728
Production (1,000 t Dry, Shelled)		
Java and Madura	395	500
Sumatra	19	49
Kalimantan	1	3
Sulawesi	6	16
Maluku and Irian Jaya	0	1
Bali and Nusa Tenggara	35	47
Indonesia	456	616
Average Yield (kg/ha Dry, Shelled)		
Java and Madura	720	850
Sumatra	670	820
Kalimantan	800	750
Sulawesi	760	680
Maluku and Irian Jaya	500	500
Bali and Nusa Tenggara	540	780
Indonesia	700	850

Table 2. Fertilizer requirements for continuous cropping of annual rotations of rice-corn-soybean or rice-peanut-soybean on an acid Ultisol in Yurimagas, Peru (Sanchez et al., 1983).

Amendment[a]	Elemental rate (kg/ha)	Frequency
Lime	3 t CaCO3-equivalent	Per 3 years
N	80-100	Only rice and corn
P	25	Per crop
K	100	Per crop, split application
Mg	25	Per crop, unless dolomite used
Cu	1	Per year or two[b]
Zn	1	Per year or two[b]
B	1	Per year or two[b]
Mo	20 (g/ha)	Only with legume seed

[a]Plant Ca and S needs as supplied by lime, triple superphosphate and Mg, Cu and Zn carriers.
[b]According to soil test recommendations.

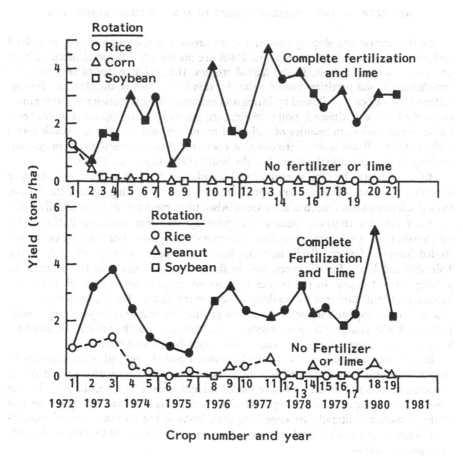

Figure 1. Yield record for two continuously cultivated plots at Yurimaguas with and without fertilization and lime (Sanchez et al., 1983).

Pest control is the second greatest constraint to greater yields in Indonesia, followed by the lack of quality germplasm. Although there are several very promising cultivars, they are not available to farmers and are seldom produced in quantities greater than that needed for research trials.

The greatest source of potential expansion for soybean production in Java is the rainfed uplands where production can easily double. The irrigated lands are used mainly to produce rice, the staple food of Indonesia, and soybeans are planted only as a third crop during the dry season. The third crop of soybean, if irrigation water is not available, effectively uses the residual moisture in the paddy field, where soybean is planted with no tillage or additional fertilizer application. In the rainfed upland, soybeans are planted in February during the second cropping season either as a monoculture or intercropped with corn (*Zea mays* L.). This is especially the case in East Java Province. Soybean, generally called a secondary food crop on irrigated lands, is a major crop in the rainfed uplands. Thus, rainfed land is the major potential source in Java for soybean production.

SOIL EROSION, THE GREATEST THREAT TO AGRICULTURE PRODUCTION

Under rainfed and sloping conditions soil erosion is the major threat to upland agriculture in Java, where slopes up to 100% are intensively cultivated, often without any conservation practices. In the humid tropics, the major source of erosion is the detachment of soil particles, caused either by raindrop splash or the abrasive effect of surface flow of water, followed by lifting and transport of soil particles as water moves down the hillside. "Although policy makers are generally more impressed by the spectacular forms and stark realities of gully and tunnel erosion, it is the less visible forms such as sheet, rill and inter rill erosion that cause the greatest cumulative damage—particularly to the agricultural potential of the land" (El-Swaify et al., 1982).

Table 3 shows the rate of erosion and population densities in some of the major watersheds of Java. These erosion figures are, however, misleading because the bulk of eroded soil comes from much smaller areas. Also, these measurements are usually based on water samples collected from major rivers by the Department of Public Works, which uses them for irrigation or dam construction purposes. The majority of the soil eroded from the farmers' fields, however, has been trapped before reaching the rivers. Tolerable soil losses for cultivated soils in the U.S. range from 2 to 11 t/ha year, depending on soil depth. In the tropics the rate of soil weathering is much faster; the depletion of soil nutrients by leaching is also much faster. Thus, productivity is restricted to the topmost segment of the soil profile. Precipitation in Java ranges from 1,500 to 3,500 mm/year for most rainfed agriculture areas, with above 2,000 mm/year being the most common pattern, mainly distributed during nine months.

In Java soil erosion is caused mainly by water, especially in lands with slopes above 20% that are intensely cultivated to annual food crops. Population density, lack of job opportunities, and food demand have forced marginal, steep lands to be placed under annual food crop production. People farming these lands often are among the poorest of the farmers and literally are exploiting their lands to produce subsistence crops. As a rule most of the critical lands are away from major roads and not visible to the public or policy-makers.

PROGRAMS IN CONSERVATION

Politically, development investments are usually made on highly visible and quickly responsive activities. With conservation programs the returns are usually too far in the future to be considered attractive to the politically oriented policy-maker as well as to the poor farmers.

Land degradation is not the result of cultivation per se, but rather the improper management of the soil. Throughout Java, lands frequently with 100% slope that are

Table 3. Sediment loads of selected Indonesian river basins (El-Swaify et al., 1983; P3PRDAS, Ministry of Forestry, Indonesia).

River Basin	t/ha year	People/km^2	Area (ha)
Cimanuk	78	682	325,000
Citanduy	37	654	446,000
Bengawan Solo	23	700	1,614,300

under irrigated rice, have nearly perfect terraces, and at elevations above 1000 m, as in Tawangmangu, Central Java, lands under vegetable production have perfect contour ridges and grassed waterways. In both of these cases, the land is highly productive and the additional effort put into conservation is obviously worth it. It is also not uncommon to find land under permanent vegetation in mature forests with gullies one and a half meters deep. On sloping land, it is not the trees that stop soil erosion but the grass, legumes and shrubs in the understorage. When they are not present, or in great enough density to provide sufficient ground cover, soil erosion will take place. If sloping land is to be planted in annual crops, the topography of the land needs to be modified by building back-slope, bench terraces.

The terraces should have a drainage ditch against the hillside to divert excess water into waterways to discharge runoff. The terrace risers should be revegetated to protect the terrace lip and prevent water from cascading down from terrace to terrace, which causes erosion and clogs the drainage ditch. If the drainage ditches are not functioning, the water will not flow into the waterway but cascade from terrace to terrace causing rill and gully erosion. Hudson (1983) wrote, "There is little chance of someone inventing a magic solution to soil erosion when the land is used to produce clean-weeded maize on 45% slopes.... The other technical problem is that we really do not have a practical solution to the problem of surface runoff on steep slopes requiring bench terracing". T. C. Sheng (1984), Watershed Management Officer, FAO, states that in humid countries where steep lands are under cultivation, bench terraces with reverse slope toward the hillside are the best suited areas for conservation. Under the agroclimatic and socio-economic conditions of Java, bench terracing is a practical solution, since land holdings are so small (0.6 ha/family) and the land is intensively cultivated.

One rule should be allowed to stand: No one should be allowed to use the land in such a way that soil erosion exceeds permissible soil losses. Recommendations on land use are seldom followed by farmers and are extremely difficult to enforce. Still, the public should be aware of the effects of soil erosion on production. Man should not exploit the land but be caretakers of a non-renewable resource that belongs to humankind.

PRODUCTION AS AN INCENTIVE TO ACHIEVE CONSERVATION

Farming income under rainfed conditions is, in most cases, less than subsistence level. However, over 42% of the farmers in Java work rainfed lands and have holdings of approximately 0.6 ha per family. Their present farm income is usually between $15 to $20 per month, and usually they have to walk from 5 to 10 km to purchase their farm inputs and sell the products. Still, the potential for tripling the present farm output is real and easily attained by using lime, better ratios of N-P-K and slightly increased fertilizer applications, better pest control management, plant cultivars more responsive to fertilizer inputs and resistant to diseases, revegetation of bench terrace risers with forage grasses and legumes, and by employing small ruminants to utilize the farm by-products and forage.

Another constraint to increased production is the availability of inputs, especially good quality seeds and the capital to purchase the required inputs to fully utilize the land resources. The lowest commercial interest rate, which is seldom available to farmers, is 24% per year.

Population increases and consequent demands for food in Java will cause even greater pressures on the non-irrigated uplands. Production of food crops high in protein and cash value, such as soybean and groundnut, will continue mainly in the uplands.

However, under the present soil management and conservation practices, production is very low while soil erosion is extremely high. Programs to increase production will only aggravate soil erosion, and soil conservation programs by themselves will have little acceptance by farmers with a barely subsistence level of living. The farmers can, on their own, modify their land topography by building back-slope, bench terraces with terrace risers stabilized with grasses and legumes. Under such conditions the soil is very stable and capable of producing any crop adapted to that agroclimate, with minimum danger of erosion. Subsidies to increase production can be given by the government to those farmers who have followed a soil conservation program and have built bench terraces to specifications. This exchange of government inputs for farmers' efforts in conservation is working well in the upland conservation and agriculture development programs going on in the Citanduy Watershed of Java (Barrau and Djati, 1983).

Soil conservation programs should be conceived as investment programs to improve conservation and production, as well as to provide inputs and basic resources to increase the capacity for sustainable production for the poor farmers.

There is a real need on the part of agricultural scientists to create public awareness about agriculture production and the seriousness of soil erosion problems to both citizens and policy-makers. This is especially true if it is taken into consideration that the overwhelming majority of the decisions on technical issues are never made by technical persons.

NOTES

Enrique M. Barrau, USAID/AGR, Jakarta, Indonesia.

The thoughts expressed in this paper are those of the author, and not necessarily those of USAID.

REFERENCES

Barrau, E. M. and Ketut Djati. 1983. Soil and Water Conservation Techniques: An Approach to Watershed Stabilization on Java, The Citanduy Project. (In press.) Paper presented at the "Second International Conference on Erosion and Conservation", January 16-22, 1983. Honolulu, Hawaii.

El-Swaify, S. A., S. Arsyad and P. Krishnarajah. 1983. Soil Erosion by Water. p. 99-161. *In* Richard A. Carpenter (ed.) Natural Systems for Development. Macmillan, New York.

El-Swaify, S. A., E. W. Dangler and C. L. Armstrong. 1982. Soil Erosion by Water in the Tropics. College of Tropical Agriculture and Human Resources, University of Hawaii. Research Extension Series 024.

Handy, R. and D. Mardanus. 1984. Menjadi Bangsa Tempe Yang Baik, Majalah Promosi. March:17-19.

Hudson, N. W. 1983. Soil Conservation Strategies in the Third World. J. Soil Water Cons. 38:446-450.

Sanchez, P. A., D. E. Bandy, J. H. Villachica and J. J. Nicholasides, III. 1983. Continuous cultivation and nutrient dynamics. p. 11-15. *In* J. J. Nicholaides, II, W. Conto and M. K. Wade (eds.) Agronomic-Economic Research on Soils of the Tropics, 1980-1981 Technical Report. Contract AID/ta-C-1236. Soil Science Dept., North Carolina State University, Raleigh, NC.

Sheng, T. C. 1974. Protection of Cultivated Slopes (Terracing Steep Slopes in Humid Regions). FAO, Rome.

SOYBEAN PRODUCTION AND SOIL EROSION PROBLEMS
AND SOLUTIONS—AFRICA

R. Lal

Africa cultivates only 0.76% of the world's area under soybean and contributes only 0.47% to its annual production (FAO, 1982). The mean soybean yield in Africa for 1982 was 1094 kg/ha in comparison with the world's average of 1772 kg/ha. It was only due to Zimbabwe that the soybean yield was near the world's average, while the mean yields for Nigeria (385 kg/ha) and Tanzania (240 kg/ha) were the lowest in Africa. Nigeria has the most land area and produces the largest share of the total soybean production in West Africa, although neither its total land area under soybean nor total production has changed over the past decade (Figure 1).

Low soybean yields in Africa are due to poor soil fertility (Shannon, 1983), the unavailability of improved cultivars, poor seed quality due to inadequate storage facilities (Ndimande et al., 1981) and problems of crop establishment due to high soil temperatures and frequent drought stress experienced during the seedling stage (Nduguru, 1974; Lal, 1979). Crop establishment and yield also depend on the severity of runoff and soil erosion before and during soybean growth.

A land suitability map for rainfed production of soybean in Africa has been produced by FAO. The middle regions of the Soudanian and Zambesian zones have the most suitable ecologies for soybean production (FAO, 1978). The rainfall in these zones is generally monomodal with a mean annual range of 800 to 1300 mm. This region extends from approximately 7° to 12° north and south of the equator and is dominated by ferruginous soils. Soils in the northern or southern fringes of this area are affected by aeolian materials. These soils cap or crust due to raindrop impact. Most soils are dominated by low activity clays with CEC ranging from 1 to 10 me/100 g and have low levels of soil organic matter. These soils are structurally unstable to raindrop impact, have a shallow effective rooting depth, have low available water and nutrient holding capacities, are easily compacted when cultivated by tractor-driven farm equipment, and crops grown on them are prone to drought stress even in the regions of high annual rainfall. Risk of erosion, both by wind and water, is high. The Guinean, Sahelian and the Eastern ecological zones are either marginal or unsuitable for soybean production.

SOIL EROSION

The potential denudation rate, as estimated from the climatic coefficient, for the subhumid and semiarid regions of Africa is 2000 t/km^2 year (Fournier, 1962). For the continent as a whole, however, Fournier (1962) estimated the mean annual soil loss to

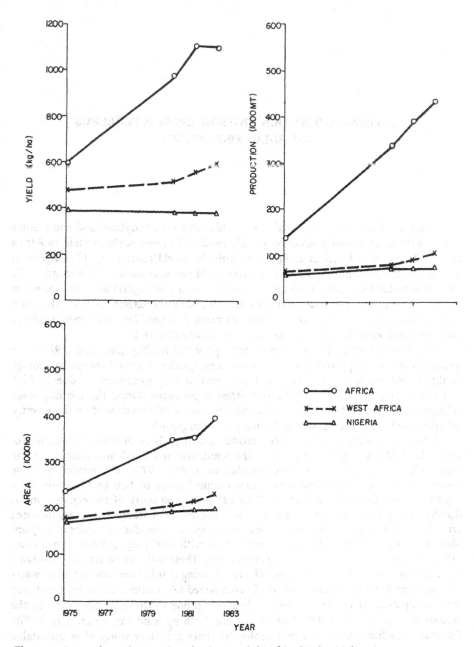

Figure 1. Area under soybeans and production trends in Africa (FAO, 1982).

be 715 t/km² year, a denudation rate that is about nine times higher than that of Continental Europe. Since these estimates were made two decades ago, the sediment loads in African rivers have increased substantially (Oyebande, 1981), because of the rapid expansion of arable land area.

The soils of the Soudanian and Zambesian zones are particularly vulnerable to accelerated soil erosion. High rainfall erosivity (Roose, 1977) and the lack of protective vegetative cover at the onset of the monsoons are responsible for high risk of soil erosion. The annual rainfall erosivity index for this region ranges from 800 to 2000 t/year. The soil erodibility for the Soudanian zone ranges from 0.16 to 0.23 t/ha (Roose, 1977). Considering the mean soil erodibility to be 0.20 t/ha (Lal, 1976a; Roose, 1977), the potential erosion hazard for the most suitable ecology for soybean production in Africa ranges from 160 t/ha year to 400 t/ha year. Field measurements made at Ibadan, Nigeria corroborate these estimates (Lal, 1976a).

In addition to water erosion, semiarid regions are also vulnerable to wind erosion during the dry season (Riquier, 1980). Long distance transportation of Sub-Saharan soil is observed throughout west Africa, and the intensity of wind erosion has reportedly increased over the past two decades (Michon, 1973; Delwaulle, 1973a,b), probably due to increase in cropping intensity.

The acceptable rates of soil loss for most soils in west Africa are low. Consequently, the loss of fertile top soil often leads to drastic yield reductions. Although little research data relating soybean yield to erosion is available for tropical Africa, many researchers have observed severe yield reductions by erosion in other grain crops. Fournier (1963) observed about a 50% decrease (from 727 to 352 kg/ha) in millet (*Pennisetum americanum*) yield in the Niangoloko region of Upper Volta when soil erosion increased from 143 to 1318 t/km^2. Lal (1983) reported the rate of decline in maize (*Zea mays* L.) grain yield caused by natural erosion to be 0.26 t/ha mm of soil depth eroded. Mbagwu et al. (1984) observed that the loss of topsoil cannot be compensated for by the addition of chemical fertilizers alone.

SOYBEAN CULTIVATION AND SOIL EROSION

Soybean, especially with a good stand and when adequately managed, is a soil-conserving crop. For example, Aina et al. (1977) observed the least water runoff and soil erosion from plots sown to a soybean-soybean rotation, in comparison with that from cassava (*Manihot esculenta* Crantz), pigeon pea (*Cajanus cajan* [L.] Millsp.), and maize and cassava grown as a mixed crop (Table 1). The low level of soil erosion in soybean is attributed to its ability to establish an early canopy cover. The soybean canopy intercepts more rain than other crops, and being shorter than maize or cassava, has a cover close to the soil surface. The soybean canopy effectively intercepts rains, even though its LAI is generally low in comparison with maize, cassava or pigeon pea. Consequently, the C factor of the Universal Soil Loss Equation for soybean is low (Table 1).

Table 1. Soil erosion and water runoff from soybean compared with other tropical crops[a] (Aina et al., 1977).

Cropping systems	Time to 50% canopy cover	Rainfall interception at 90% canopy cover	Soil erosion	Runoff	C-factor	
					First[b] season	Second[b] season
	d	%	t/ha year	%		
Soybean-soybean	38	58	39	15.1	0.19	0.02
Maize + cassava mixed	45	40	69	20.9	0.43	0.05
Cassava	63	28	109	27.9	0.72	0.39
Pigeonpea-pigeonpea	50	--	96	31.2	--	--

[a] Each value is the average of 4 plots.
[b] First season is from April to July, and the second is from August to November.

In spite of its favorable canopy, soil erosion in soybean exceeds the tolerance level. It is necessary, therefore, to adopt appropriate methods of seedbed preparation and other conservation measures to control runoff and reduce soil erosion. If unchecked, the soil loss of 39 t/ha year (Table 1) can rapidly deplete the fertile top soil. Soybean leaves little or no crop residue to protect the soil. Soybean residue, which has a low C:N ratio, decomposes rapidly and renders the soil bare and vulnerable to erosion. Consequently, soil erosion following soybean can be more severe than after maize. Lal (1976a) observed that soil erosion following cowpea was also substantially more than after maize.

Soil structure, on the other hand, is generally better after soybean than maize. The high infiltration rate generally observed after a soybean crop decreases water run-off and soil erosion. For example, Aina et al. (1977) observed that the cumulative infiltration in 3 h was 150, 60, 50, and 30 cm for soybean, cassava, cassava + maize and plowed fallow treatments, respectively. Soybean also is grown at a closer plant spacing and narrow row width than maize. The closer spacing of soybean grown in contour rows offers more resistance to runoff flow than widely spaced maize or cassava. Decrease in runoff velocity decreases its shearing and sediment carrying capacity, and prolongs the time for water to infiltrate into the soil.

EROSION CONTROL MEASURES FOR SOYBEAN

Runoff Management

Terraces. Terraces and other mechanical devices for water diversion are often recommended for management of steep and undulating terrains. If properly constructed and regularly maintained, a combination of graded channel terraces and grassed waterways can decrease runoff velocity and reduce soil losses. Terraces do not reduce runoff amount, but decrease soil loss by decreasing runoff velocity. Inter-terrace soil management, however, is very important for effective erosion control. Even with the use of local materials, terraces can be prohibitively expensive.

Contour Ridges. The ridge-furrow system is used to allow more time for water to infiltrate into the soil. Ridges can be appropriately spaced to allow seeding with tractor-driven equipment. Two adjacent ridges are sometimes tied together to develop a series of small basins that permit rainwater to infiltrate the soil where it falls. The ridge-furrow system is, however, effective only for soils with stable structure and for gentle slopes. Moreover, the effects of furrowing and tied ridges are short-lived.

Whatever the advantages in soil conservation, ridges rapidly lose soil moisture reserves and experience wide fluctuations in soil temperature. The maximum soil temperature on ridges is far greater than on flat and mulched seedbeds. The data in Table 2 show soil temperatures measured under soybean seedlings one week after sowing during the first (April) and second season (August) crops at IITA. It is apparent that supraoptimal soil temperatures prevailed on ridges during the first growing season. Crop establishment, seedling growth and grain yields during the first season were least on the ridged seedbed. Leaf area development of soybeans was negatively correlated with soil temperature.

Soil Surface Management

No-till Farming. Soil erosion is drastically reduced by growing soybean with a no-till system. With the no-tillage system of seedbed preparation, it is often unnecessary to use other erosion control measures, such as terraces and diversion channels, as long

Table 2. Effects of soil surface conditions on soil temperature at 5-cm under soybean seedlings one
week after sowing (Lal, 1979).

Treatment	First season		Second season	
	Grain yield	Soil temper- ature at 1500 h	Grain yield	Soil temper- ature at 1500 h
	t/ha	C	t/ha	C
Black polythene mulch	1.9	39.5	1.3	28.4
Clear polythene mulch	1.2	45.0	1.3	30.7
Straw mulch	1.7	35.1	1.6	26.0
Ridges	0.2	45.4	1.4	28.0
Flat unmulched seedbed	1.6	41.9	1.5	25.5
Aluminum soil mulch	2.1	32.5	1.5	26.3
LSD (.05)	0.7	–	0.4	–

as there is an adequate quantity of crop residue mulch. The data in Table 3 indicate
that the no-till method of growing soybean effectively prevented soil erosion on slopes
of up to 15%. Similar results were reported by Aina et al. (1977), who observed no
soil loss in a no-till soybean crop, in comparison with an annual soil erosion of 0.4,
21.9, 66.1 and 68.0 t/ha on 1, 5, 10 and 15% conventionally tilled slopes and sown
to soybean. The annual water runoff was 0.7, 1.1, 1.7, and 1.7% of the rainfall of 1150
mm in no-till soybean, in comparison with 3.7, 18.2, 22.9, and 15.0% for the conventionally plowed plots of soybean for the 1, 5, 10 and 15% slopes.

Management requirements for the no-till system include effective and timely weed
control, appropriate seeding equipment, adequate quantity of crop residue mulch,
and other inputs such as a balanced fertilizer and pest control. With no-till mulch,
soil properties under soyean are generally more favorably maintained than with conventional tillage consisting of plowing and harrowing (Table 4).

Cover Crops. Frequent use of cover crops in rotation is recommended to provide
residue mulch, restore soil physical and chemical properties, and keep weeds and pests
under control. Suitable legumes that can be grown in sequence are Centrosema, Pueraria, Stylosanthes, Mucuna, Psophocarpus and Calapogonium. These cover crops can
be suppressed by chemical or mechanical means to grow soybean through them. These

Table 3. Soil and water losses from plowed and no-till soybean plots. Rainfall = 42 mm. Maximum
intensity = 10 cm/h. I_3 = 5.8 cm/h. From Lal (1976b).

Slope	Runoff		Soil erosion	
	No-till	Plowed	No-till	Plowed
%	– % –		– kg/ha –	
1	0.9	8.8	0.2	40
5	2.1	24.5	.5	620
10	2.4	13.6	1	7,620
15	2.4	21.4	1	6,910
Mean	1.9	17.1	2	3,800

Table 4. Effects of no-tillage and conventional plowing methods of seedbed preparation for a soybean crop on soil properties (0-10 cm layer) (modified from Lal, 1976b).

Soil property	No-tillage	Conventional tillage	LSD (.05)
Organic carbon (%)	1.24	0.76	0.19
Total nitrogen (%)	0.11	0.07	0.03
Bray-1 P (ppm)	55	23	29
Cation exchange capacity (me/100 g)	6.6	5.6	0.6
Exchangeable Ca^{++} (me/100 g)	4.8	4.5	0.9
Earthworm activity (casts/m^2)	42	3	--
Soil bulk density (g/cm^3)	1.37	1.48	--
Degree-hours of soil temperature exceeding 30 C at 5 cm depth	364	1943	--

covers are known to improve soil structure and water infiltration rate, but this improvement is easily eliminated by mechanical methods of seedbed preparation (Pereira et al., 1954; Wilkinson, 1975). It is important, therefore, to adopt a no-tillage method of soybean production following the growth of legumes or grass covers.

Deeply rooted woody perennials also are used to improve water acceptance of soils in the tropics. In East Africa, Pereira et al. (1954) observed significant improvements in the water infiltration rate of soil by fallowing with *Leucaena glauca*. Similar observations were reported for western Nigeria by fallowing with *Cajanus cajan* and *Leucaena leucocephala* (Juo and Lal, 1977). Growing soybean in association with woody perennials by a system of alley cropping may also be a soil conserving technique.

Some grasses (e.g. *Paspalm notatum* or *Setaria* sp.) are specifically grown on contour bunds and in graded channels to stabilize them and to decrease runoff velocity. These contour furrows are sometimes levelled (rather than graded) to improve water seepage into the soil. These practices are widely used for soil and water conservation in Rwanda and in other East African countries (Musema Uwimana, 1979; Hindson, 1981). The use of biological measures including grass and shrub hedges is also recommended for erosion control on arable lands in Ivory Coast (Roose, 1977).

CONCLUSIONS

Soil erosion, high soil temperature, low soil moisture availability, crusting and compaction are soil physical constraints that affect soybean production in tropical Africa. Soil erosion is particularly severe when soybean is grown with the conventional plowing and harrowing system of seedbed preparation. Graded channel terraces reduce soil erosion if used in conjunction with appropriate soil management techniques, but are expensive. Contour ridges, especially those with cross ties, are effective on soils of gentle slopes and stable structure. The ridges, however, have the disadvantages of having a supra-optimal soil temperature. Soybean grown on ridges is liable to suffer from drought stress more than those grown on a flat seed bed. Soil erosion is effectively reduced by growing soybean with a no-till system of seedbed preparation.

NOTES

R. Lal, International Institute of Tropical Agriculture, Oyo Road, Ibadan, Nigeria.

REFERENCES

Aina, P. O., R. Lal and G. S. Taylor. 1977. Soil and crop management in relation to soil erosion in the rainforest of western Nigeria. p. 75-84. *In* G. Foster (ed.) Soil erosion: Prediction and control. SCSA Special Publication 21. Ankeny, Iowa, USA.

Delwaulle, J. C. 1973a. Desertification de l'Afrique au sud du Sahara. Bois et Forêts des Tropiques 149:3-20.

Delwaulle, J. C. 1973b. Résultats de six ans d'observations sur l'erosion au Niger. Bois Forêts Trop. 150:15-36.

FAO. 1978. Report on the agroecological zones project. Vol. I. Methodology and results for Africa. World Soil Resources Report No. 48. FAO, Rome.

FAO. 1982. Production Yearbook. FAO, Rome.

Fournier, F. 1962. Carte du Danger d'Erosion en Afrique au Sud du Sahara. CEE-CCTA Presses, Univ. Paris.

Fournier, F. 1963. The soils of Africa. p. 221-248. *In* A review of the natural resources of the African continent. UNESCO, Paris.

Hindson, J. 1981. Proposed remedy for erosion: contour seepage furrows. Appropriate Technol. 8:10.

Juo, A. S. R. and R. Lal. 1977. The effect of fallow and continuous cultivation on the chemical and physical properties of an Alfisol in Western Nigeria. Plant Soil 47:567-584.

Lal, R. 1976a. Soil erosion problems on Alfisols in Nigeria and their control. IITA Monograph 1. IITA, Ibadan, Nigeria.

Lal, R. 1976b. No-tillage effects on soil properties under different crops in western Nigeria. Soil Sci. Soc. Am. J. 40:762-768.

Lal, R. 1979. Soil and micro-climatic consideration for developing tillage systems in the tropics. p. 48-63. *In* R. Lal (ed.) Soil Tillage and Crop Production. IITA, Ibadan, Nigeria.

Lal, R. 1983. No-till farming. IITA Monograph 2. IITA, Ibadan, Nigeria.

Michon, P. 1973. Le Sahara avance-t-il vers le sud? Bois et Forêts des Tropiques 150:69-80.

Mbagwu, J. S. C., R. Lal and T. W. Scott. 1984. Effects of desurfacing of Alfisols and Ultisols in Southern Nigeria. I. Crop Performance. Soil Sci. Soc. Am. J. 48:828-833.

Ndimande, B. N., H. C. Wien and E. A. Kueneman. 1981. Soybean seed deterioration in the tropics. Field Crops Res. 4:113-122.

Nduguru, B. J. 1974. Comparative germination studies of cowpea and soybean under tropical temperature conditions. M.Sc. Thesis. Univ. Reading, U.K.

Musema Uwimana, A. 1979. The problem of soil conservation in Rwanda. Bull. Agric. Rwanda 124: 206-211.

Oyebande, L. 1981. Sediment transport and river basin management in Nigeria. p. 201-226. *In* R. Lal and E. W. Russell (eds.) Tropical agricultural hydrology. J. Wiley & Sons, New York.

Pereira, H. C., E. M. Chenery and W. R. Mills. 1954. The transient effects of grasses on the structure of tropical soils. Emp. J. Exp. Agric. 22:148-160.

Riquier, J. 1980. Small-scale mapping of present and potential erosion. p. 23-28. *In* M. De Boodt and D. Gabriels (eds.) Assessment of erosion. J. Wiley & Sons, New York.

Roose, E. J. 1977. Erosion et ruissellement en Afrique de l' Ouest. Travaux et Document De l' ORSTOM, Paris.

Shannon, D. A. 1983. A study of factors responsible for variable growth of soybean in the southern Guinea Savanna of Nigeria. Ph.D. Thesis. Cornell University Library, Ithaca, New York.

Wilkinson, G. E. 1975. Effect of grass fallow rotations on the infiltration of water into a savanna zone soil of northern Nigeria. Trop. Agric. 52:97-103.

SOYBEAN PRODUCTION AND THE SOIL EROSION PROBLEMS AND SOLUTIONS—SOUTH AMERICA

N. P. Cogo

Soil erosion studies directly related to soybean production in South America have been limited. Most of the information on this subject comes from Brazil, the largest South American soybean producer and the country that has the greatest erosion problems. Brazil accounts for approximately 75% of both the total cropped area (about 11,000,000 ha) and the total grain production (about 20 million metric tons) of soybeans in this part of the world. For this reason, most of the data used were taken from Brazilian references.

Rainfall erosion has been serious and is greatly increasing in the major soybean production areas of South America. Though a conservation needs inventory is still unavailable, it is believed that at least 80% of the total soybean hectareage suffers significantly from soil erosion problems. These problems are mainly represented by soil loss rates exceeding tolerable limits, as well as large runoff water, seed, lime, and fertilizer losses. As a consequence, the productivity levels have been considerably affected and production costs have increased. Other erosion-associated problems, such as sedimentation and water pollution also have been costly. Gianluppi et al. (1979) accomplished a field survey to estimate some of the erosion damages caused by intensive rains that occurred over a period of 15 consecutive days in November, 1978 in Rio Grande do Sul State, Brazil. Land area considered in this survey was 2,600,000 ha of conventionally tilled recently planted soybean fields, representing at that time 70% of the total soybean hectareage in Rio Grande do Sul. The estimated soil losses over the 15-day rainfall period averaged the spectacular rate of 90 t/ha, besides the large losses of soybean seed, lime, and fertilizers in the runoff water. The total cost for these losses averaged U.S. $33,425,000 (costs related to land degradation, yield reduction, sedimentation, and water pollution were not taken into account). Situations like that are not infrequent, clearly showing the magnitude of the erosion problem in these major soybean production areas.

FACTORS CONTRIBUTING TO EROSION

Many factors have directly or indirectly contributed to erosion in these regions. The main direct factors include (a) erosive rains, (b) deteriorated soil physical conditions, (c) steep and long slopes, (d) inadequate tillage and management practices, and (e) limited use of complementary soil conservation practices. Indirect factors are represented by the (a) limited conservation consciousness of the people, (b) limited teaching by schools about the preservation of natural resources, (c) unsatisfactory teaching of soil and water conservation in universities, (d) low farmer educational level, (e) unsatisfactory knowledge of professionals on the subject, (f) limited erosion research under local conditions, (g) ineffective soil conservation programs, (h) very limited financial support, (i) predominance of an agricultural policy that emphasizes cash crops for export, and (j) a large disordered increase in the soybean hectareage without an accompanying policy on soil conservation. Discussion of erosion causes in this paper will be limited to those factors influencing it directly.

Rainfall Erosivity

The potential for rainfall erosion in these soybean production areas, as expressed by the R factor of the Universal Soil Loss Equation (Wischmeier and Smith, 1978), can be regarded to be great. Reported rainfall erosivity values, R, have ranged from 500 to 1,200 MJ cm/ha h (Cogo et al., 1983; Lombardi Neto et al., 1981; Biscaia et al., 1982; Irutia et al., 1983). Most of these values are above 650 MJ cm/ha h. In addition, in some places approximately 40 to 60% of the annual erosivity usually is concentrated over a short period of time, from soil tillage to about 6 to 8 weeks after soybean planting (Eltz et al., 1977; Cogo et al., 1978; Lombardi Neto and Moldenhuer, 1981; Saraiva et al., 1981). This is the most critical period for erosion, especially under conventional tillage systems, since the soil is not yet sufficiently protected by the plant.

Soil Erodibility

The susceptibility of soils to erosion in these areas, as expressed by the K factor of the Universal Soil Loss Equation, varies greatly. Reported soil erodibility values, K, range from 0.06 to 0.49 t ha h/ha MJ cm (Encontro Nacional de Pesquisa Sobre Conservação do Solo, 1978; Irutia et al., 1983). The higher values are for clayey soils classified as Oxisols and Ultisols, the first one being the dominant soil used for soybean production in Brazil (EMBRAPA, 1980). The higher values are for silty soils. Based on K values, most of the soils cropped to soybean in these regions could be regarded as having moderate to low resistance to erosion. However, these soils erode more easily than one would expect from examination of their K values. This evidence is probably related to the method of erodibility computation, which may not be accounting for the effects of soil properties other than those indicated by the parameters in the Wischmeier's nomograph.

Soil Topography

This factor has influenced erosion strongly (Vernetti, 1974; EMBRAPA, 1980; Irurtia et al., 1983). Except for some places in Argentina, where topography is relatively flat, the steep and/or long slopes on most of the soils cropped to soybean in South America favor large runoff rates and amounts, causing prominent rills and spectacular gullies on agricultural fields. It is common to grow soybean on soils having slopes steeper than 10% and slope lengths several hundred meters long, many times in the absence of any soil erosion control practice. Under these conditions, erosion rates have been increased greatly.

Tillage Methods

Doubtless, tillage methods presently used for soybean production in South America represent one of the principal causes for erosion exceeding tolerable limits. In the majority of the production areas, soybean traditionally has been planted using clean-tillage methods. The generalized use of conservation tillage methods is still limited, though in some areas they are gradually being adopted, especially the no-tillage system.

However, conventional tillage is still the most popular method. This system is usually characterized by one disk-plowing operation for primary tillage (sometimes a heavy disk is substituted for the disk-plow) and two to four disking operations for secondary tillage. This large amount of secondary tillage, many times under inadequate soil moisture conditions, excessively pulverizes the soil surface and creates subsurface compacted layers on many of the agricultural fields. As a consequence, the soil water infiltration capacity has been drastically reduced, resulting in large runoff rates and amounts, and consequently, erosion. Field experiments (Eltz et al., 1977; Vieira et al., 1977; Saraiva et al, 1981; Eltz et al., 1983) have shown consistently that soil losses under conventional tillage exceed by two or three times the tolerable limits (estimations of the maximum permissible soil loss rates for the more tolerable soils are in the order of 12 to 14 t/ha year). Fortunately, farmers today are gradually changing to less intensive forms of soil tillage.

Cover and Management Practices

Besides clean-tillage systems, inadequate cover and management practices presently used represent another important factor influencing erosion in these regions. In many places, soybean has traditionally been planted following wheat (*Triticum aestivum* L.), in a continuous double-cropping sequence. This intensive use of the soil, with the same tillage, the same crops, and the same land year after year has severely degraded the soil physical conditions, increasing erosion. Effective crop rotations are almost nonexistent. It was once a common practice, especially in Brazil, to burn the wheat residue before preparing the seedbed for soybean and to plow-incorporate the soybean residue before preparing the seedbed for wheat, in both cases using conventional tillage methods. Under this residue management and cropping system, the soil surface was essentially bare and excessively pulverized right at the time rainfall erosion potential was greatest. (About 40 to 60% of the total rainfall erosivity accumulated over the year usually occurs during the initial stages of soybean growth.) Under most conditions, about 80 to 90% of the total soil loss observed during the entire soybean crop period occurs from planting time to approximately 6 to 8 weeks later, regardless of the tillage method (Vieira et al., 1977). This situation becomes more critical as the land is continuously subjected to the same tillage and cropping system. It was once a common practice in southern Brazil to perform up to seven field operations for growing wheat (four passes for tillage and three passes for cultural practices and harvesting) and nine field operations for growing soybean (five passes for tillage and four passes for cultural practices and harvesting), in both cases using conventional tillage. This large number of field operations, after fifteen years of cropping, was shown almost to double soil bulk density values and to decrease by half both macropores and organic matter content, as compared to values for the same soil under forest. As a consequence, soil compaction indexes have reached values as high as 20 kg/cm^2, or even greater in some situations, and the final water infiltration capacity of the soils has been reduced to values as low as 5.0 mm/h (Cintra et al., 1983). Under these adverse soil physical conditions, erosion rates have been high. In a field experiment under natural rainfall conditions, on a 12% slope Ultisol, Eltz et al. (1983) showed the 5-year average soil loss rate for traditional double-cropped wheat and soybeans under conventional tillage to be about 30 t/ha year. This represents an erosion rate about five times greater than the maximum permissible rate for this soil.

Complementary Soil Conservation Practices

Contouring, strip-cropping, and terracing should be used when erosion cannot be controlled to tolerable values solely by means of tillage and management practices. The limited use of these practices has been another factor responsible for the high erosion levels observed in soybean production areas. It is still common in many places to plant soybeans parallel with the slope. Strip-cropping is generally not used. The most popular of these complementary soil conservation practices, but still limited in use, has been terraces. Besides the high cost for terracing the land, many other problems are usually associated with such an erosion control structure. The most important of these problems are (a) inaccurate design due to the lack of basic information, such as rainfall duration-intensity-recurrence interval relationships, (b) unsatisfactory construction due to the inappropriate machinery used and the experience of the operator, (c) lack of proper maintenance, (d) channel sedimentation, and (e) lack of well-planned vegetated waterways. All these imperfections have led to frequent failures of terraces. It is also important to point out that, where terraces are used, they are usually used as an isolated practice; i.e., the interterraced areas continue to be badly managed. Under these conditions, erosion is not effectively controlled. Sometimes, erosion rates are even greater on lands where the terrace system fails because of increased concentration of runoff on the vulnerable soil areas. Also common is the presence of prominent rills on the interterraced areas, showing that terraces were not appropriately spaced.

SOLVING EROSION PROBLEMS

The soil erosion problems in the major soybean production areas of South America have not yet been effectively faced, at least to the degree to which they should be. This is obvious from the gradually increasing erosion rates. Integrated actions between government, researchers, extension educators, and farmers are still to be taken. Some isolated programs have been developed, but they have not succeeded well because of limitations in financial support, research information, specialists, extension personnel, and other requirements needed to satisfactorily carry them out. Much more is being done in terms of research than in terms of extension education, simply because financial support and other facilities have been much more limited for the latter. A general recognition of the seriousness of erosion is still to be achieved.

Alternative solutions to the erosion problems in major soybean production areas of South America are several. The main ones can be grouped into (a) increased application of erosion control practices, which are already known, (b) adapting erosion research approaches to local conditions, and (c) changes in the present agricultural policy.

Application of Known Erosion Control Practices

The immediate action that can be taken to solve partly the erosion problems in these regions is to lower the excessive soil loss rates to values closer to tolerable limits. The simplest and quickest way of accomplishing this is by increasing application of those conservation techniques that are already known. Among these are use of (a) conservation tillage systems, (b) mulches of crop residues, and (c) crop rotations.

Conservation Tillage Systems. With conservation tillage, soil losses can be reduced to 10 to 30% of that observed under conventional tillage (Vieira et al., 1977; Eltz et al., 1977, 1983; Saraiva et al., 1981). This level of erosion reduction was observed for disking-in chopped residues and for a no-till system. Another promising tillage system for these areas would be the chisel-plow. Conservation tillage represents one of the most effective and economical ways to control erosion on agricultural lands. For most of the major soybean production areas, soils and climate are favorable to conservation tillage. Problems eventually associated with these systems, such as weeds, insects, and diseases, as well as effects on yields and production costs, need to be better investigated, however.

Mulches of Crop Residues. The maintenance of plant residues from the previous crops on the soil surface represents another effective and economical way to substantially reduce erosion. In a field study with simulated rainfall, on a 7.5% slope sandy loam soil, Lopes and Cogo (unpublished data) compared soil losses under residue cover with that observed with no cover. Mulches of corn, wheat, and soybean residues, in amounts from 0 to 10 t/ha (0 to 100% soil cover) evenly spread on a freshly and conventionally tilled soil surface, were used in the experiment. The derived mulch factor (MF) residue cover (RC, in percent) equations were MF = $e^{-0.045\ RC}$ for corn, MF = $e^{-0.046\ RC}$ for wheat, and MF = $e^{-0.030\ RC}$ for soybean. Based on these findings, a 20% soil cover with soybean residue reduces erosion by 45%, compared to zero cover, whereas a 50% soil cover reduces erosion by 78%. Somewhat larger erosion reductions are obtained with corn and wheat residues. Cogo et al. (1983, 1984) also have pointed out the importance of soil surface conditions, such as random roughness index and mulch cover for controlling erosion on sloping lands. The equations relating residue cover (RC, in percent) to mass of residue (RM, in kg/ha) were RC = $100(1 - e^{-0.00031\ RM})$ for corn, RC = $100(1 - e^{-0.000326\ RM})$ for wheat and RC = $100(1 - e^{-0.000252\ RM})$ for soybean. Erosion reductions that are a little larger can be obtained with both corn and wheat residues. These results clearly show the importance of mulches of crop residues for controlling erosion on sloping soils.

Crop Rotations. These represent an alternative for slowing erosion on the more severely degraded soils, or on lands not continuously suited for annual row crops. However, they should be effective crop rotations, like those including pastures or any other close-growing crops (either for harvesting or as cover crops). In a 5-year study under natural rainfall conditions on a 12% slope sandy clay loam soil, Eltz et al. (1983) obtained an average soil loss rate of 7.0 t/ha year for a 4-year clover + 1-year wheat-soybean rotation, as opposed to 30 t/ha year for continuous double-cropped wheat and soybean using conventional tillage. About 80% of the total soil loss measured over the 5-year crop rotation period occurred in the year having wheat-soybean with conventional tillage. These findings show the benefits of a crop rotation for controlling erosion. A good way of establishing a crop rotation is to combine it with strip-cropping on the contour.

Contouring can also be used, but in many situations it is impractical and not sufficiently effective for controlling erosion. Terraces are expensive and not well adapted to mechanization. Therefore, it is advisable not to encourage the widespread use of terraces before implementing conservation tillage, mulches of crop residues, and crop rotations. It is important to remember that, if terraces are going to be used, they should not be used as an isolated practice. Sound tillage, cover, and crop management practices are much more important than terraces alone.

Strengthening Erosion Research Under Local Conditions

Parallel to the increased application of known conservation techniques, local erosion research should be strengthened to obtain on-site parameter values. The main research needs for these regions include studies on the (a) rainfall characteristics, (b) soil erodibility, (c) tillage methods, (d) management practices, and (e) soil loss tolerances.

Rainfall Characteristics. A better knowledge of rainfall characteristics would be valuable for soil conservation study. Local determinations of the rainfall duration-intensity-recurrence interval relationships are necessary for accurately designing erosion control structures. Determinations of the rainfall erosivity values and their temporal distribution are also needed for efficiently selecting tillage methods and management practices. Techniques for accomplishing these studies need to be better investigated, however.

Soil Erodibility. Better information on the relative susceptibility of the soils to erosion is needed in order to decide on which soils more emphasis on erosion control should be placed. Efforts should be concentrated on direct measurements of soil erodibility, however, since indirect methods like Wischmeier's nomograph have not produced satisfactory results.

Tillage Methods. The effects of tillage methods on soil erosion need to be understood better. The erosion control effectiveness of both conventional tillage and conservation tillage has varied greatly from region to region. Therefore, it is advisable to obtain as many on-site tillage-parameter values as possible to more precisely assess the erosion control effectiveness of different tillage systems. It is important to quantitatively describe the soil surface conditions induced by tillage, especially the percentage of soil area covered by crop residue and the random roughness index. This would permit derivation of valuable quantitative relationships between soil losses and the several soil physical parameters induced by tillage.

A great variety of tillage tools exist in these soybean production areas. However, they have not yet been sufficiently tested with respect to producing favorable soil physical conditions or reducing erosion. Therefore, the different plow types, disks, harrows, field cultivators, scarifiers, subsoilers, and other tillage tools that are being used for soybean production, must be tested locally before their widespread application.

Management Practices. More information is also needed on management practices. Crop, residue, and land management all should be better investigated not only with respect to their effects on erosion control, but also on how and to what extent they affect the physical, chemical, and biological properties of the soil that relate to erosion. Crop rotations, crop sequences, cultural practices, crop residues, cover crops, contouring, strip-cropping, and terracing all are aspects of management which deserve much more attention for soybean production in these regions.

Soil Loss Tolerances. In conservation planning, it is necessary to know the maximum permissible soil loss rate that a given soil will tolerate without losing significantly its productivity potential. Therefore, determination of the soil loss tolerance (T value of the Universal Soil Loss Equation) for the different soils is important in order to compare erosion rates under different tillage and management practices. This will permit better assessment of the effectiveness of the several erosion control practices.

Changes in Agricultural Policy

Usually, technical problems should be resolved by technical solutions. However, many technical solutions can only be efficiently implemented through sound agricultural policy. Because this subject is rather complex, I can only outline policy aspects in need of improvement. Most importantly, there is need for (a) gradual change from monoculture systems to multiple cropping systems, especially on small farms, (b) development of a stable agriculture based on sound principles of soil and water conservation, (c) establishment of feasible land use policies, (d) financial support for soil conservation works, (e) a mutual sharing of the responsibility for soil conservation between government and land owners, (f) development of effective, nationwide soil conservation programs, and (g) development of educational programs for the preservation of natural resources.

CHALLENGES

To farmers: The soil that is serving you today is the same soil that will serve your children tomorrow. Therefore, preserve it as if it were a part of yourself.

To soil conservationists: The excuse that, because research data are limited or even nonexistent one can stand aside and do nothing, can no longer be justified. It certainly is much more worthwhile to make mistakes trying to accomplish something than not to make mistakes by doing nothing. Therefore, use what is available to move forward now.

To governments: The most valuable heritage of a nation is the surface soil, and its most powerful enemy is soil erosion. It would be more dignified and worthwhile, and less costly, to fight erosion with a bunch of straw than to fight ideologies with weapons.

NOTES

N. P. Cogo, Department of Soils, Faculty of Agronomy, Federal University of Rio Grande do Sul, P.O. Box 776, 90.000 - Porto Alegre, RS, Brazil.

REFERENCES

Biscaia, R. C. M., F° Castro, and F. C. Henklain. 1982. Mapa de erosivida de do Estado do Paraná; 1a. Aproximação do Solo, 4., e Encontro Nacional de Pesquisa Sobre Conservação do Solo, 4., Campinas, SP, Brasil, 1982. Programa de Resumos. SBCS, Campinas.

Cintra, F. L. D., J. Mielniczuk, and I. Scopel. 1983. Caracterização do impedimento mecânico em um Latossolo Roxo do Rio Grande do Sul. R. Bras. Ci. Solo 7:323-327.

Cogo, N. P., C. R. Drews, and C. Gianello. 1978. Índice de erosividade das chuvas dos municípios de Guaíba, Ijuíe Passo Fundo. pp. 145-152. *In* Encontro Nacional de Pesquisa Sobre Conservação do Solo, s°, Passo Fundo, RS, Brasil, 1978. Anais. EMBRAPA/CNPTrigo, Passo Fundo.

Cogo, N. P., W. C. Moldenhauer and G. R. Foster. 1983. Effect of crop residue, tillage-induced roughness, and runoff velocity on size distribution of eroded soil aggregates. Soil Sci. Soc. Am. J. 47: 1005-1008.

Cogo, N. P., W. C. Moldenhauer and G. R. Foster. 1984. Soil loss reductions from conservation tillage practices. Soil Sci. Soc. Am. J. 48:368-373.

Eltz, F. L. F., N. P. Cogo, and J. Mielniczuk. 1977. Perdas por erosão em diferentes manejos de solo e coberturas vegetais em solo Laterítico Bruno Avermelhado Distrófico (São Jerônimo). I. Resultados do primeiro ano. R. Bras. Ci. Solo 1:123-127.

Eltz, F. L. F., E. A. Cassol, I. Scopel, and M. Guerra. 1983. Perdas de solo e água em diferentes sistemas de manejo e coberturas vegetais em solo Lateritico Bruno Avermelhado Distrófico (São Jerônimo) sob chuva natural. Resultados dos primeiros cinco anos. R. Bras. Ci. Solo (in press).

EMBRAPA. Centro Nacional de Pesquisa de Soja. 1980. Subsídios para a Revisão do Programa Nacional de Pesquisa de Soja. Londrina, PR, Brasil.

Encontro Nacional de Pesquisa Sobre Conservação do Solo, 2° 1978. Passo Fundo, RS, Brasil, 1978. Anais. EMBRAPA/CNPTrigo, Passo Fundo, 481 p.

Gianluppi, D., I. Scopel, and J. Mielniczuk. 1979. Alguns prejuízos da ersão do solo no Rio Grande do Sul. p. 92. *In* Congresso Brasileiro de Ciência do Solo, 17°, Manaus, AM, Brasil, 1979. Resumos. SBCS, Manaus.

Irurtia, C. B., J. C. Musto, and P. Cullot. 1983. Evaluacion a nivel nacional del potencial de degrada-cion y riesgo de erosion de los suelos. *In* Reunion Del Grupo de Especialistas in Conservacion de Suelos de los Paises de la Cuenca del Plata, Montevideo, Uruguay, 1983. FAO-Comite Inter-gubernamental Coordinador de Los Paises de La Cuenca Del Plata, Montevideo.

Lombardi Neto, F. and W. C. Moldenhauer. 1981. Erosividade da chuva sua distribuição e relação com as perdas de solo em Campinas, SP. p. 158. *In* Encontro Nacional de Pesquisa Sobre Conservação do Solo, 3°, Recife, PE, Brasil, 1981. Anais. UFRPe, Recife.

Lombardi Neto, F., I. R. da Silva, and O. M. de Castro. 1981. Potencial de erasão das chuvas no Estado de São Paulo. p. 135. *In* Encontro Nacional de Pesquisa Sobre Conservação do Solo, 3°, Recife, PE, Brasil, 1981. Anais. UFRPe, Recife.

Saraiva, O. F., N. P. Cogo, and J. Mielniczuk. 1981. Erosividade das chuvas e perdas por erosão em diferentes manejos de solo e coberturas vegetais em solo Laterítico Bruno Avermelhado Distró-fico. I. Rsultados do segundo ano. Pesq. Agropec. Bras. 16(1):121-128.

Vernetti, F. J. 1974. A cultura da soja no Paraguay; programa de pesquisa. IICA Zona Sur, Monte-video. 1v.

Vieira, M. J., N. P. Cogo, and E. A. Cassol. 1977. Perdas por erosão em diferentes sistemas de preparo do solo para a cultura da soja (*Glycine max* (L.) Merr.) em condições de chuva simulada. R. Bras. Ci. Solo 2:209-214.

Wischmeier, W. H. and D. H. Smith. 1978. Predicting rainfall-erosion losses—a guide to conservation planning. SEA-USDA Agricultural Handbook No. 537. U.S. Government Printing Office, Wash-ington, D.C.

SOYBEAN PRODUCTION AND SOIL EROSION PROBLEMS—NORTH AMERICA

J. M. Laflen and W. C. Moldenhauer

About 98% of soybean production in North America is concentrated in the United States. The United States also produces more than 60% of the world's soybeans. The impact of soybean production on North American or world soil erosion problems should be evident in the United States. One objective of this paper is to improve understanding of the current soil erosion problems in the United States and how crop production changes, particularly those involving soybean production, are related to these problems. Another objective is to discuss briefly some solutions to these soil erosion problems.

Recent surveys give current estimates of the soil erosion problems in the United States (Soil Conservation Service, 1982). However, there were no good scientific surveys preceding the 1960s; hence, there is no good quantitative evidence of how cropping changes over the long term have impacted the current U.S. soil erosion problem. The approach taken here will be to show how land use in the United States has changed and how these changes have affected soil erosion.

The methods of analysis and the data bases used in this paper give results that can be used for broad, regional generalizations. The data bases available required assumptions about crop rotations in order that the analysis could be conducted, but these assumptions are subject to question. Hence, we caution that results should be viewed from a regional or national perspective.

Even though wind erosion is a major concern in certain regions of the United States, water erosion is of the most concern where soybeans are produced. Wind erosion will not be discussed in this paper.

SOYBEAN PRODUCTION AND CROPPING CHANGES IN THE U.S.

In the last 60 years, total planted area in the United States has been relatively constant, ranging from about 120 to 150 million ha/yr. The area in various crops responds to government programs and economic factors. The most important crops in soybean-producing regions are corn (*Zea mays* L.), cotton (*Gossypium* spp.), soybean, oat (*Avena sativa* L.), and wheat (*Triticum aestivum* L.). Although other crops, such as rice (*Oryza sativa* L.), sugarcane (*Saccharum officinarum* L.), sorghum (*Sorghum bicolor* [L.] Moench) and barley (*Hordeum vulgare* L.), are produced in these regions, they have relatively little impact on soybean production and soil erosion by water.

As shown in Figure 1, there have been major shifts in corn, cotton, wheat, oat, and soybean plantings. The most obvious is that the area planted to soybean has

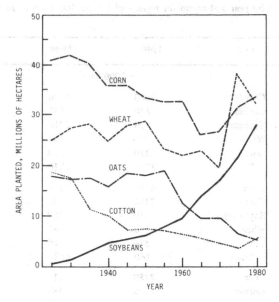

Figure 1. Area of major crops grown in the United States, 1925-1980.

increased continuously from about 1 million ha in 1925 to about 28 million ha in 1980.

Another major change has been the reduction in oat and cotton planted. The area in cotton decreased about 50% in the 1925-45 period, with a gradual reduction from then to the mid-1970s. Total cotton area is presently about 30% of that of 1925. The early decrease in cotton area greatly exceeded the increase in soybean area, leading to the conclusion that the early reduction in cotton during that period was little related to soybean cropping.

The reduction in oat plantings also has implications for several specific regions of the United States. Oats usually are produced in sod-based rotations that include a row crop, small grain, and meadow. The reduction in oat area implies a reduction in these sod-based rotations.

Cropping by regions of the United States for 1940 and 1980 is given in Table 1 for regions where significant amounts of soybeans are produced. These regions are shown in Figure 2. Also given in Table 1 is the average annual area change by crop over the 1940-1980 period for a particular region, computed as the slope of a linear regression of area versus year, based on areas planted at 5-year intervals from 1940 to 1980. As shown in Table 1, soybean has replaced mostly corn and cotton in the Appalachian, Southeast, and Delta regions. There has also been a major decrease in cropland area in the Southeast. In the Corn Belt and Lake State regions, there has been a major increase in soybean and corn area, partly because of a reduction in oat area and partly because of the replacement of rotation hay and meadow with corn and soybean. The Northern Plains area has had an increase in soybean area, with an equal decrease in oat area. Wheat area has also increased in the Northern Plains, but only in recent years.

In Table 2 are given estimates that we have made of the areal extent of crop rotations in 1940 and 1980 for important soybean producing regions in the United States. Generally, our assumptions were that oat crops have nearly always been produced in a row-crop, small-grain, meadow rotation, and that some wheat has also been produced

Table 1. Cropping area by year and region for regions of United States where soybeans are grown in quantity.

Region	Crop	1940	1980	Area change[a]
		\- 1000 ha \-		1000 ha/yr
Appalachian	Corn	3690	2100	-46
	Cotton	640	140	-13
	Soybeans	520	2810	54
	Wheat	760	660	-4
	Oats	220	120	-4
	Total	5830	5840	-13
Southeast	Corn	4170	1270	-78
	Cotton	2100	250	-45
	Soybeans	210	2760	59
	Wheat	170	480	6
	Oats	510	70	-14
	Total	7160	4830	-72
Delta	Corn	2780	110	-69
	Cotton	2300	980	-34
	Soybeans	550	4960	117
	Wheat	15	580	14
	Oats	240	20	-5
	Total	5880	6650	22
Corn Belt	Corn	11290	15740	113
	Cotton	170	100	-2
	Soybeans	2710	12750	248
	Wheat	2280	2610	-7
	Oats	5440	860	-132
	Total	21890	32060	221
Lake	Corn	3320	5830	52
	Cotton	–	–	–
	Soybeans	260	2470	54
	Wheat	1010	1840	14
	Oats	3540	1140	-64
	Total	8130	11270	56
Northern Plains	Corn	5310	5540	-11
	Cotton	–	–	–
	Soybeans	40	1770	42
	Wheat	10990	12860	-2
	Oats	2960	1600	-41
	Total	19300	21770	-13

[a]Slope of linear regression of areas vs. year.

in such a rotation. In 1980, our estimate of the area in sod-based rotation was based on the area of rotation hay and pasture reported in the 1977 National Resource Inventory (Soil Conservation Service, 1982). In a sod-based rotation, we assumed that corn was usually the preferred row crop, with cotton and soybeans produced in a sod-based rotation to only a limited extent. Also, soybeans were assumed never to be produced continuously in 1940, and continuously only in the southern states

Table 2. Estimates of area for crop rotations for 1940 and 1980.[a]

| | 1940 | | | 1980 | | | |
| | Sod-based rotation | Continuous cropping | | Sod-based rotation | Continuous cropping | | Double-crop wheat-soybean |
		Without soybean	With soybean		Without soybean	With soybean	
	– 1000 ha –			– 1000 ha –			
Appalachian	2920	2840	1030	1390	--	4590	160
Southeast	2030	5380	420	210	--	4210	240
Delta	760	4280	1090	570	--	5860	210
Corn Belt	23180	1010	5430	7280	1710	25500	--
Lake	10740	970	--	8930	380	4930	--
Northern Plains	16060	8590	--	7370	13310	3540	--
Total	55690	23100	7970	25770	15400	48620	600

[a]Assumptions: Sod-based rotation = 3 x smaller of row crop (corn, cotton, soybeans) or small grain (wheat, oats) area for 1940. Smaller of 3 x area in rotation hay and pasture taken from NRI (SCS, 1982) or small grain area for 1980. Double-crop wheat-soybeans = ½ of wheat area, after allowing for sod-based rotation. Double cropping allowed only in Appalachian, Southeast and Delta regions, and only for 1980. Continuous cropping with soybean = 2 x area of soybean after allowing for sod-based rotation and double cropping. Continuous cropping without soybean = area not included in other categories.

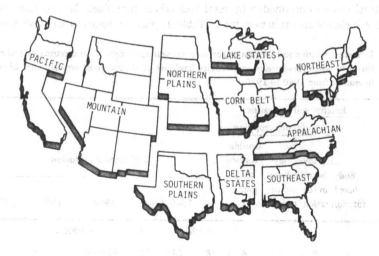

Figure 2. Cropping regions of the United States.

(Appalachian, Southeast, and Delta) in 1980. We also assumed that all sod-based rotations were 1/3 row crop, 1/3 small grain, and 1/3 meadow. More detailed assumptions are given in Table 2.

The major shift in crop rotations has been the replacement of nearly 30 million ha in sod-based rotations with a slightly greater area in continuous cropping. Most area in continuous cropping now includes soybean, and nearly all area that includes soybean in continuous cropping would be planted to a corn-soybean rotation because soybean, except in the south, is seldom grown continuously. Other continuous cropping would also be row crop, except for a major area of continuous wheat in the Northern Plains. Our estimate was that there is presently only a limited amount of double-cropped wheat and soybeans, which, because of climatic limitations, is limited to southern states. Although there are other regions of the United States that have some crop rotations given in Table 2, those regions produce very few soybeans.

EFFECT OF SOYBEAN ON CURRENT U.S. SOIL EROSION PROBLEMS

In Table 3, we give soil erosion estimates for 1940 based on the area in the various crop rotations in 1940 and in 1980, and on erosion estimates made in 1977. Relative soil loss ratios were determined for the various crop rotations by use of the USLE, Universal Soil Loss Equation, (Wischmeier and Smith, 1978) and assuming conventional tillage (Table 3).

Relative soil loss ratios for the rotations given in Table 3 are generalized averages over the regions given. There was little difference in most ratios for regions having common crops in the rotation. However, in the Northern Plains, the use of wheat in rotation gave a lower relative soil loss ratio when soybean was not in the rotation. Based on yields in 1940 and 1980, and on soil loss ratios used in the USLE, relative soil-loss was estimated to be 12% greater in 1940 than in 1980 for similar rotations.

Relative total soil erosion in 1940 and in 1980 was computed as the product of the area in these rotations (given in Table 2) and the relative soil loss ratios of the rotations, then summed over all rotations. Relative erosion per unit cropped area was relative total soil erosion divided by total area taken from Table 2. Note that the area in Table 2 includes all the area given in Table 1, plus the rotation hay and meadow.

Table 3. Estimated relative soil loss ratios for crop sequences of Table 2, and estimated soil erosion for 1940 when present-day productivity levels are used. Also given are estimates of changes in erosion from 1940 to 1980.

| | Relative soil loss ratios | | | | Estimated soil erosion | | | | | |
| | Continuous cropping | | | | | | | | | |
	Sod-based rotation	No soy-bean in rotation	Soy-bean in rotation	Double-cropped small grain-soybean	1940	1980	Change	1940	1980	Change
					— t/ha —		%	— 1000 t —		%
Appalachia	.3	.9	1	.4	18	20	+12	121800	123600	+1
Southeast	.3	.9	1	.4	13	14	+12	99800	66300	-34
Delta	.3	.9	1	.4	17	16	-3	103100	108700	+5
Corn Belt	.3	.9	1	–	10	17	+69	301400	591000	+96
Lake	.3	.9	1	–	4	6	+42	49100	85000	+73
Northern Plains	.3	.6	1	–	6	8	+26	150300	185100	+23
Weighted average	–	–	–	–	10	13	+35	–	–	+40

Erosion per unit cropped area for each region for 1980 was assumed to be that given in the 1977 National Resource Inventory (Soil Conservation Service, 1982). Erosion per unit cropped area for 1940 was then computed as the product of the ratio of the relative erosion per unit cropped area for 1940 to 1980 and the erosion per unit cropped area for 1980. Total erosion from cropped area in 1940 and 1980 is the product of the area given in Table 2 and erosion per unit cropped area. We recognize that the cropped area, as used herein, does not include all area in crops. However, for 1980, it does include most of the area classified as land used for cultivated crops in the 1977 National Resource Inventory (Soil Conservation Service, 1982). It includes over 90% of cultivated area for the Appalachian, Corn Belt, and Lake States regions; over 80% for the Southeast and Delta regions, and nearly 70% for the Northern Plains regions. Lower percentages in the Delta and Southeast are likely due to our not having included rice and sugarcane, while in the Northern Plains, not having included barley and sorghum reduced the percentages of cultivated area in our analysis.

An additional factor in soil loss estimates was the changes that are occurring in the tillage systems used in U.S. agriculture. Our estimate was that, in 1980, conservation tillage systems were not used effectively enough to greatly affect the erosion estimates. This assumption is questionable today, but for 1980 conditions the assumption may have been quite acceptable. This analysis also assumes that the land base for crop production in 1940 and in 1980 has the same inherent susceptibility to soil erosion and that, where there have been changes in crop production, these changes have not resulted in land shifts that concentrate crops on a particular class of land in terms of its susceptibility to soil erosion.

As shown in Table 3, erosion rates in Appalachia have increased about 12% over the last 40 years. Our estimate, shown in Table 2, is that sod-based rotations have greatly decreased, while rotations with soybean have greatly increased. As shown in Table 1, area in corn production has been reduced by about 40% of 1940 levels, cotton area has decreased by about 80%, and oat area, although never very significant in Appalachia, has been reduced by about 50%. The major change has been more than a 5-fold increase in soybean hectarage, mostly at the expense of corn. Much of the effect of soybean has been to substitute one row crop for another. Although soil erosion is greater after soybean than after corn (Laflen and Moldenhauer, 1979), the effect of soybean on soil erosion problems in Appalachia has not been great, because of the reduced erosion due to improved yields. The effect of cropping changes apparently are quite small when viewed from the regional perspective.

For the southeastern United States, there has been a major reduction in cropped area. Area planted to corn and cotton has been drastically reduced. Area planted to corn has been reduced more than the increase in planted area of soybean. Cotton area has decreased nearly as much as soybean area has increased. Oat area has been reduced, but wheat area has increased considerably. The net result is a 12% increase in soil erosion per unit of cropped area, with a 34% reduction in total soil erosion. The effect of soybean on soil erosion in the Southeast has been relatively minor, inasmuch as it has not resulted in an increase in row-crop area.

The Delta region is similar to the southeastern region in that the area planted to corn, cotton, and oats has decreased, while wheat acreage has increased considerably. However, the area in soybean, and total cropland area, has increased substantially. The net result is a minor decrease in erosion per unit of cropped area and about a 5% increase in total erosion in the Delta. As shown in Table 2, sod-based rotations evidently were never very important in the Delta.

The Corn Belt region has had about a 70% increase in soil erosion, expressed on a unit-area basis, with nearly a doubling of total soil erosion since 1940. Since 1940, there has been a doubling of row-cropped area, about half of it at the expense of rotation hay and meadow, the other half likely from previously uncropped areas. Oat has virtually disappeared as an important crop (relative to the area of corn and soybean). As shown in Table 2, our estimate is that the area in sod-based rotations has decreased by about 2/3 and that about 80% of Corn Belt land produces soybeans at least once every 2 years. If the 50% increase in cropped area has occurred on land less suited to cropping, our estimates of soil erosion increases in the Corn Belt could be lower than the actual increases.

Erosion in the Lake States also is estimated to have increased greatly, by slightly more than 40% per unit of cropped area and by about 75% on a total erosion basis. This is because of about a 20% increase in cropped area, with increases being quite similar for corn and soybean, but with a large decrease in oat area. This has resulted in a relatively small decrease in sod-based rotations, but about a 5-fold increase in area in continuous row cropping.

For the Northern Plains, there has been an increase in soil erosion of nearly 25% for each unit of cropped area and about the same increase in total erosion. Soybean is a relatively minor crop in the Northern Plains; hence, it has likely had a minor effect on the soil erosion increase in the Northern Plains. Seemingly, there has been a decrease in sod-based rotations in the Northern Plains, with a major increase in the area planted continuously to wheat. The relative soil losses for crop rotations in the Northern Plains are adjusted to account for the use of fallow in these areas—based on the proportions of fallow to wheat computed from the 1977 National Resource Inventory (Soil Conservation Service, 1982).

In the major soybean-producing regions of the United States, the analysis clearly shows that erosion has increased greatly in the last 40 years, with erosion per unit of cropped area increasing about 35% and total erosion increasing by about 40%. However, changes have been variable among regions. In the southern regions (Appalachian, Southeast, and Delta) of the United States, total erosion has likely decreased in the last 40 years. Although soybean has been a crop that has greatly increased in area, in these regions there has been a bigger decrease in the planted area of corn and cotton, and increases in yields have decreased erosion rates somewhat. The net result has been essentially a maintenance of the status quo in the southern areas when viewed from a regional perspective. A soil loss of over 11 t/ha is usually not considered acceptable, in terms of the long-term maintenance of the soil resource. Hence, as shown in Table 3, the status quo in the southern regions is hardly acceptable.

The large increases in erosion have occurred in the northern regions (Corn Belt, Lake and Northern Plains) of the United States, although erosion rates per unit of cropped area average somewhat less than those in the southern United States. The increased area in cropping and in row crop in the northern regions, principally the Corn Belt, has caused large erosion increases in the last 40 years. The switch from sod-based rotations to continuous row cropping has been a major cause of the severity of the soil erosion problem in the northern states. Much of the increase in row cropping is a direct result of the increase in soybean area, although the increase in corn area also has been substantial. Much of the increase in row-cropped area has occurred on land previously in small grain and rotation hay and meadow, more than tripling erosion rates on those lands.

SOLUTIONS TO EROSION PROBLEMS UNDER SOYBEAN PRODUCTION

Soil erosion on cropland can be controlled in a number of ways. Usually, for a specific field, there are several conservation practice alternatives. For some sites, one practice will adequately control erosion; for others, adequate erosion control requires more than one practice. Each site, and its proposed use, must be evaluated to determine the specific erosion control practice or sets of practices to use. Most practices can be used when soybean is in the crop rotation even though erosion when soybeans are produced is greater than when the rotation does not include soybean (Laflen and Moldenhauer, 1979).

The array of conservation practices available include conservation tillage, sod-based rotations, contour farming, strip-cropping, terracing, grassed waterways, water and sediment control basins and vegetative cover. The performance of most of these practices is well known, and they are regularly implemented by practicing conservationists. The production of soybeans does present some special problems for conservation tillage where crop residue is used for erosion control.

Conservation tillage is "any tillage sequence that reduces loss of soil or water relative to conventional tillage" (Soil Conservation Society of America, 1982). Conservation tillage is the major method for controlling soil erosion in the United States. However, conservation tillage is effective only when adequate amounts of crop residue are left on the soil surface.

Several studies have shown that conservation tillage can be effective in reducing soil erosion when soybeans are produced in a continuous row-crop rotation. Laflen and Colvin (1981) have shown that soybean residue reduces soil erosion as effectively as corn residue when both cover the same proportion of the soil surface. Soybean residue is very effective in covering a large proportion of the soil surface after harvest. However, subsequent tillage and traffic destroy much of the soybean residue, making it very difficult to maintain an effective surface cover during the high erosion periods occurring after planting of the following crop. Although there are few data for soybean residue, most indicate a maximum residue cover after soybeans of about 40% for no-till after planting the following crop. If any tillage is performed, residue cover usually is less than 15%. Estimates of the residue cover and relative erosion expected for different tillage systems are given in Table 4 for three tillage systems and corn-soybean rotations. These estimates are made with the assumption that tillage after soybean is minimized as much as possible, so that maximum residue is maintained for erosion control.

As shown in Table 4, tillage systems properly used can greatly reduce soil erosion after corn, but the only conservation tillage system that is effective in reducing erosion after soybean is no-till. When the increased erosion after soybeans due to the soil's condition is combined with the effect of low residue covers, it is obvious that soil erosion after soybean is difficult to control with the use of any conservation tillage system, unless no-till is used.

Soil erosion can be further reduced when a good conservation tillage system is combined with other conservation practices. The use of conservation tillage with terraces should extend the useful life of terraces, reducing maintenance costs. Conservation tillage above grassed waterways can also extend the life of waterways. The effectiveness of most conservation practices is increased when conservation tillage is used.

SUMMARY

Soil erosion problems in the United States have greatly increased in the last 40 years, principally because of the increase in row crop area. Continuous row crop production has nearly eliminated the less erosive, but generally less profitable, sod-based

Table 4. Residue cover and relative soil erosion rates for corn-soybean rotations.

	Tillage system		
	Moldboard plow	Chisel plow	No-till
Residue cover—%			
Soybean or corn after corn	4	25	75
Corn after soybean	2	15	40
Soil erosion—relative rates			
Soybean or corn after corn	1.00	0.48	0.08
Corn after soybean	1.40	0.92	0.35

rotations.

The advent of soybean production and the increase in row crops has had a major impact on soil erosion in the northern states; the impact has been relatively minor in the southern states. Total erosion in the Corn Belt on cropland is double what it likely was in 1940; on a unit area basis, it has increased about 70% in the last 40 years.

Soil erosion can be greatly reduced with wise selection and careful use of conservation tillage systems. No-till in a corn-soybean rotation will reduce soil erosion by about 75%, as compared with a clean-tilled field. The technology for such systems is available now. Such systems, when combined with other needed conservation practices, can effectively control soil erosion on most lands.

NOTES

J. M. Laflen, Agricultural Engineer, USDA, ARS, Ames, IA; and W. C. Moldenhauer, Soil Scientist, USDA, ARS, National Soil Erosion Laboratory, West Lafayette, IN.

This is a contribution from the USDA-ARS and is Journal Paper No. J-11614 of the Iowa Agriculture and Home Economics Experiment Station, Ames, Iowa, Project No. 2684.

REFERENCES

Laflen, J. M. and T. S. Colvin. 1981. Effect of crop residue on soil loss from continuous row cropping. Trans. Am. Soc. Agric. Eng. 24:605-609.

Laflen, J. M. and W. C. Moldenhauer. 1979. Soil and water losses from corn-soybean rotation. Soil Sci. Soc. Am. J. 43:1213-1215.

Soil Conservation Service. 1982. Basic statistics: 1977 national resources inventory. U.S. Dep. Agric. Stat. Bull. 686.

Soil Conservation Society of America. 1982. Resource conservation glossary. Soil Conservation Society of America, Ankeny, Iowa.

U.S. Department of Agriculture. 1943. Agricultural Statistics. U.S. Government Printing Office.

U.S. Department of Agriculture. 1963. Agricultural Statistics. U.S. Government Printing Office.

U.S. Department of Agriculture. 1983. Agricultural Statistics. U.S. Government Printing Office.

Wischmeier, W. H. and D. D. Smith. 1978. Predicting rainfall erosion losses. Agric. Handbook 537. U.S. Dep. Agric., Washington, D.C.

Seed Programs in the Tropics

SOYBEAN SEED PROGRAMS IN LATIN AMERICA

Flavio Popinigis

A seed program is defined as "the measures to be implemented and activities being carried out in a country to achieve the timely production and supply of (seed of) described quality in the quantities needed" (Douglas, 1980). Being part of the state, regional or national agricultural development strategy, programs must be efficiently coordinated if they are to become successful.

Seed programs vary among countries, states or regions according to various factors, the most important being the stage of agricultural development and the available rural infrastructure. According to Douglas (1980) four broad stages of seed activities can be delineated. In Stage 1, nearly all farmers save their own seed and most cultivars and production practices are traditional. In Stage 2, a few farmers and seed production units are multiplying seed; improved cultivars are being developed and are beginning to replace the traditional ones. In Stage 3, there are many components of a seed program and the supply of seed ranges from fair to adequate. Improved cultivars are rapidly replacing the traditional ones. Some seed enterprises are being formed, but farmers still use much less seed than that available for distribution. At Stage 4, a seed law is in force; there is a clear seed policy aimed at strengthening commercial seed production and marketing. The agricultural sector is well advanced and links are established among the various components of the program. All these stages are present in seed programs of Latin American countries.

In this paper, I report on seed legislation, cultivar development and promotion, crop area, grain and seed production, and seed needs and supply with respect to the seed programs in the main soybean-producing countries of Latin America.

CULTIVAR DEVELOPMENT, EVALUATION AND PROMOTION

In Latin America, most soybean breeding is carried out by government institutions, although some private companies are initiating cultivar development research in Argentina, Brazil and Colombia. This situation is mostly a result of the fact that soybean is a self-pollinated crop and breeders rights legislation didn't exist until recently. Although Argentina, Bolivia, Mexico and Uruguay now have passed plant variety protection laws, these have not yet produced the expected results. In Argentina, where such legislation has been in effect longer, since 1973, there is some soybean crop improvement being done by private companies, especially transnationals. This, however, consists mostly of testing cultivars and breeding lines introduced from abroad, where these companies maintain their main breeding efforts.

Requirements for releasing improved soybean cultivars to seed producers, and ultimately to farmers, vary among the various Latin American countries. In most of them, especially in those that do not have breeders rights laws, there is a cultivar trial system run directly by, or under the supervision of, a governmental institution. A final list of recommended cultivars is approved by a legally established authority. This is the case for Bolivia, Brazil, Colombia, Ecuador, Mexico, Peru and Uruguay. In Argentina and Paraguay, the sponsor of the cultivar must submit field data showing its agronomic value; the governmental office may, eventually, subject the cultivar to further laboratory and field tests.

The main soybean cultivars in use in Latin America are shown in Table 1. Some have wide adaptation and have been cropped for many years in many countries, like Bossier and Bragg, which originated in the U.S. However, as the breeding program progresses, locally developed, adapted cultivars tend to perform better and replace the early introductions.

In Brazil, for example, productivity was 1,060 kg/ha in the period 1960 to 1968, when the most frequently planted cultivars were Hill, Hood, Majos, Bienville and Hampton, most introduced from the U.S. It increased progressively, reaching 1,740 kg/ha in 1980, when most of the predominantly used cultivars were locally developed— BR-2, BR-3, BR-4, Ivai, Vila Rica, União, Cobb, Lancer, IAC-5, IAC-6, IAC-7, UFV-2, UFV-3, Cristalina and Doko (Kaster and Bonato, 1981). Increased yields were due not only to new cultivars, but resulted also from the use of improved production technology developed by agricultural research and adopted by farmers. However, improved cultivars made, no doubt, an important contribution to that increase.

BREEDER'S AND FOUNDATION SEED

In order to effectively express their superior traits on farms, improved cultivars must be maintained as released and described through seed multiplication, processing and distribution. Plant breeders are responsible for maintaining cultivar characteristics and a continuous supply of small amounts of genetically pure seed to the seed program. This initial seed supply is identified as Breeders and Foundation (or Basic) seed. The governmental research institution is, generally, responsible for the multiplication

Table 1. Main soybean cultivars currently in use in Latin America.

Argentina	Hood, Bragg, Ogden, Bossier, Stuart, Forrest, Hardee, Williams, Crawford, Cutler, Hutton, Dowling, IAS-1, IAC-4 (Zelarayan, 1983).
Bolivia	UFV-1, Bossier, Cristalina (Morales, personal communication).
Brazil	Paraná, Bragg, Bossier, BR-1, Davis, Cristalina, IAS-4, IAS-5, UFV-1, IAC-7, IAC-8, Doko, FT-1, FT-2, Santa Rosa, Viçoja, BR-5, Tropical (Kiihl, personal communication).
Colombia	ICA-Tunia, Victoria, Pelican SM, ICA Lili, Mandarin S4 ICA, ICA Taroa, ICA Pance, ICA Caribe, SV-77, Soy ICA P31, SOY ICA P32, SOY ICA N21 (Bastidas and Camacho, personal communication).
Ecuador	INIAP 301, INIAP 302, INIAP Jupiter (Escudero and Andrade, personal communication).
Mexico	Cajeme, Tetabiate, Mayo 80, Yaqui 80, Suprema, Jupiter, Davis, Bragg (Menchaca, personal communication).
Paraguay	Paraná, Bragg, Bossier, Viçoja, UFV-1, Santa Rosa (Moreno, personal communication).
Peru	Mandarin S4-ICA, Jupiter I, Jupiter II, Improved Pelican, Tulumayo, Nacional (Apolitano S., 1983; Lopez, personal communication).
Uruguay	Ransom, Davis, Paraná, Bragg, Forrest, Prata, Pérola, Lancer, Lee, Hutton, IAS-2, IAS-4, IAS-5 (de la Rosa, personal communication).

of these initial quantities of seed, supplying Foundation seed to commercial seed producers. Private companies that have developed their own cultivars also produce Foundation seed. This seed, however, generally is not sold, but is used for further multiplication under the company's control; seed available for sale, then, is of the Certified class.

SEED PRODUCTION SYSTEMS

In Latin American soybean-producing countries, there are two seed production systems: Certification and Inspection. Seed certification, as internationally practiced, aims essentially to guarantee trueness to cultivar characteristics. In Latin America, however, as in many other parts of the world, other seed quality aspects are considered including the lot's physical, physiological and sanitary conditions.

Under the Seed Certification system there is strict generation control, starting from seed produced under the direct supervision of the plant breeder, originator, or owner of the cultivar. Cultivar purity is maintained through the Foundation or Basic, Registered and Certified seed classes by the seed certification agency. In Latin America, seed certification programs are run by governmental agencies. Except in Argentina and Paraguay, only those cultivars recommended by an official office, after tested for their agronomic characteristics and performance, are eligible for seed certification.

In Brazil and Paraguay, legislation provides for an "inspection" seed production system. This differs from certification in that there is no generation control, and quality control is the seed producer's responsibility, under the supervision of a governmental authority.

Any seed produced under a quality-guarantee system is generally referred to as "improved" seed. Based on crop area, theoretical seed needs and improved seed distributed in 1982, the rate of improved soybean seed use in Latin America has been estimated (Table 2). All soybean seed used in Colombia and Brazil is either certified or inspected. Use of improved seed decreases progressively in Ecuador, Argentina, Uruguay, Mexico, Paraguay and Bolivia, and is negligible in Peru where the area cropped to soybean is very small and the seed program is just being started.

Table 2. Cropped area, grain production, productivity, estimated seed needs, seed distributed, and soybean seed use in Latin America for 1982 (FAO, 1983; CIAT, 1983).

	Area cropped	Grain produced	Productivity	Estimated seed needs [a]	Distributed seed	Improved seed use rate
	1000 ha	1000 t	kg/ha	t	t	%
Argentina	1,988	4,000	2,012	159,040	91,032	57
Bolivia	30	51	1,700	2,400	400	17
Brazil	8,202	12,835	1,565	656,160	655,298	100
Colombia	47	94	2,000	3,760	3,960	100
Ecuador	29	44	1,500	2,320	1,510	65
Mexico	391	672	1,719	31,280	11,200	36
Paraguay	400	600	1,500	32,000	7,500	23
Peru	4	8	1,862	320	_b	_b
Uruguay	22	25	1,136	1,760	901	51

[a]Using an 80 kg/ha seeding rate average.
[b]No data available. Negligible.

Although in many countries the use of common or one's own saved seed is not forbidden, incentives are given for the use of improved seed. These incentives generally take the form of granting subsidized credit for crop operations only to farmers who use seed produced under one of these quality-guarantee systems.

MARKETING QUALITY CONTROL

All countries in Latin America have marketing control legislation that specifies that any seed lot offered for sale must be labeled and is subject to sampling by seed law enforcement inspectors. The label must include data that specifies the actual quality or the minimum quality standards that the seed meets.

As an illustration, Table 3 shows the results from marketing control on soybean seed offered for sale in the State of Rio Grande do Sul, Brazil, from 1976 through 1982. Sampling ranged from 4% to 12% during this period. Seed lots within standards varied from a minimum of 65% in 1978 to 89% in 1981. Germination percentage, weed seed and cultivar mixtures were factors contributing to low seed quality in some years. Cowpeas (*Vigna ungiuculata* [L.] Walp.), considered a noxious weed seed in soybeans in Brazil, consistently were an important factor in low soybean seed quality in Rio Grande do Sul during the 1977-82 period.

Seed quality problems, relating to cultivar identity and purity, inert matter, weed seed and other crop seed, result from the production system in Rio Grande do Sul, in which the seed company itself is responsible for quality control (sampling, testing and labeling) of the seed it produces. Certainly these problems would be eliminated or greatly reduced if this seed had been produced under a certification system. Low germination, however, is associated with climatic conditions, and becomes a major problem in years of frequent rainfall at harvest time. Under these conditions, low vigor seed is obtained, and consequently, germination losses occur in storage.

Marketing control aims to protect seed buyers so they get seed of good quality that meets minimum legally established standards. To be effective, it must act promptly at the appropriate time. As a public activity, its effectiveness varies among states, provinces, and countries according to the available infrastructure (inspectors, laboratories, vehicles, etc.), enforcement policies, and the degree and severity of penalties.

In the example given, the low sampling rate is a consequence of the shortage of

Table 3. Quality of soybean seed offered for sale in the state of Rio Grande do Sul, Brazil, as determined by seed law enforcement agents and official seed testing laboratories (Ministério da Agricultura, 1983).

| | Seed distributed | Inspection sampling rate | Seed meeting minimum standards | Factors contributing to the nonconformity of seed lots to legal minimum standards | | | | | |
				Germination	Inert matter	Other crop seed	Weed seed	Other cultivars	Cowpeas
	1000 t	%	%						
1976	-	-	80	10	24	30	36	-	-
1977	330	6	76	40	29	2	15	-	33
1978	603	7	65	15	18	0	23	30	40
1979	366	11	78	15	8	0	31	12	51
1980	488	10	85	35	5	11	14	-	46
1981	444	12	89	21	3	12	16	12	53
1982	402	4	81	42	10	-	16	9	47

inspectors and of physical and financial resources. Although Rio Grande do Sul has the laboratories, trained inspectors and vehicles that allowed, from 1979 to 1981, a sampling rate of 10-12% of the soybean seed being marketed, only a 4% rate was accomplished in 1982. This was due mainly to a shortage of funds for travel, which prevented inspectors from inspecting seed being sold at any distance from their offices.

Although I don't have data on sampling rate and quality of seed being marketed in other countries, it is almost certain that this activity offers great opportunity for improving the quality of seed offered for sale to Latin American farmers.

INFRASTRUCTURE

Besides the need for legislation, improved cultivars, foundation seed supply and quality control, many other components are needed for a seed program to fulfill its objectives. Among these are trained personnel, good seed producers, processing and storage facilities, and seed testing laboratories. Table 4 shows the number of registered seed producers, processing plants and seed testing laboratories involved in soybean seed production programs in Latin America.

As a general rule, the public sector is the first to make physical structure investments when a program is getting started (Bolivia, Peru). More seed processing plants and seed testing laboratories are built and operated by seed producers as the program develops and the private sector gets more involved (Argentina, Brazil).

CONCLUSIONS AND PERSPECTIVES

In the Latin American soybean producing countries, the great majority of cultivars currently in use by farmers are public cultivars, developed or introduced by governmental institutions. Cultivar description and breeder's seed supply hasn't been satisfactory in all cases. Where plant variety protection is not involved, it is very common for the breeder, or institution which developed or sponsors the cultivar, to supply breeder's seed only at the time of release. In many cases this is not perfectly pure, containing off-types. Registration of cultivars has been established in most countries. This probably will reduce or eliminate progressively problems relating to imprecise or incomplete cultivar descriptions. Foundation seed has received considerable attention in

Table 4. Seed producers and facilities available for soybean seed production and testing in Latin America (CIAT, 1983).

	Registered seed producers		Seed processing plants		Seed testing laboratories	
	Public	Private	Public	Private	Public	Private
			— number —			
Argentina	2	381	25	287	10	27
Bolívia	0	4	2	0	2	0
Brazil	4	1000	23	300	40	110
Colombia	1	5	1	5	2	5
Ecuador[a]	1	2	-	-	-	-
Mexico[a]	-	600	-	-	38	-
Paraguay	2	12	1	10	3	4
Peru	1	1	1	1	2	1
Uruguay[a]	2	-	1	-	3	12

[a]Data only partially available.

the last few years, especially in Brazil where a specific nationwide Basic Seed Production Service was set up by EMBRAPA to supply commercial seed producers with Basic seed of major food crops.

Seed certification programs vary in effectiveness according to various factors, among which one of the most important is governmental support of the certification agency's activities. Considering the increasing difficulty of public institutions to operate these systems in Brazil, due to the reduced availability of financial resources and personnel, Popinigis (1983) suggested that this activity be transferred to seed producers associations under permanent supervision of the government. This would allow the government to concentrate its action on seed marketing inspection, using the additional labor force and financial resources now used in seed certification. Seed marketing inspection can also be substantially improved, if present limitations in personnel and financial resources are removed. Clear policies must be established regarding seed programs, especially concerning the role of government, and what is expected from and is allowed the private sector to do. Incentives must be provided in order to get the private sector involved in the seed program; the possibility of making a profit must be clear. Credit to seed producers requires longer pay-back periods than those allowed farmers.

Personnel training must be done intensively by the government at early stages of the program, and should continue as the program goes forth. This activity must be intensified in Latin America in general. Considerable training has been done in Brazil in the last twenty years, and more recently on a regional basis by CIAT. However, in most countries not enough training is done, especially of technicians from the private seed companies.

Research in seed technology is an area that has been greatly neglected in many countries. Most technical information is adapted from the literature, or from observations made in the U.S. or other countries. However, many problems are local and must be solved within the country. Extreme difficulties are being encountered in Brazil, for example, in producing, conditioning, storing and distributing soybean seed under the hot and humid conditions below 10° latitude, even though cropping for grain has proven very successful using locally developed cultivars. Under these conditions, soybean seed germination is generally low at harvest. Even when germination is high, it decreases rapidly during storage and has poor planting value by the next sowing season. Refrigerated and dehumidified storage is not feasible, due to the high cost of the long storage period (7-8 months). Breeding and selection for high vigor seed or impermeable seed coat may become mandatory for this region. Maybe new solutions must be sought to overcome these problems. There is a need for a simple and reasonably low-cost storage technology capable of preventing seed deterioration.

NOTES

Flavio Popinigis, Assistant, Quality Control, Basic Seed Production Service, EMBRAPA, Ed. Palácio Desenvolvimento, 9°andar, 70.057-Brasília, DF.

REFERENCES

Apolitano, S. 1983. El cultivo de soja en el Peru. Primera Reunión Latino-Americana de Mejoramiento de Soja, Palmira, Colombia.

Centro Internacional de Agricultura Tropical—CIAT. 1983. Survey on seed legislation and seed programs in Latin America. Workshop on "Development and Projection of the Seed Sector in Latin America". Cali, Colombia.

Douglas, J. E. 1980. Successful seed programs: a planning and management guide. Westview Press, Boulder, Colorado.

Food and Agriculture Organization of the United Nations—FAO. 1983. Soybean production. *FAO Monthly Bulletin of Statistics* 6(7-8):43.

Kaster, M. and E. R. Bonato. 1981. Evolução da cultura da soja no Brasil. p. 58-64. *In* S. Miyasaka and J. C. Medina (eds.) A Soja no Brasil.

Ministéro da Agricultura Brasil. 1983. Fiscalização do comércio de sementes fiscalizadas-soja. Delegacia Federal de Agricultura, Porto Alegre, BRASIL.

Popinigis, F. 1983. Seed production under Seedsmen Association responsibility. Braz. Seed J. 5(3): 133-145.

Zelarayan, E. 1983. Actividades de investigación y producción de soja en la Republica Argentina. Primera Reunión Latino-Americana de Mejoramiento de Soja, Palmira, Colombia.

SOYBEAN SEED PROGRAMS IN SOUTHERN AFRICA

D. M. Naik

Soybean is increasingly becoming an important crop in several African countries, not only because of its multiple uses but also because of its potential as a crop in diverse agricultural systems in Africa. The slogan, "seeds are the bullets of the green revolution", on the back of a vehicle belonging to a seed officer in Zambia epitomizes an essential ingredient required to promote production of soybeans. The success of a soybean seed and production program is determined by the scope of utilization of the soybean. The completeness of a system, including the extension of production, marketing and processing of soybeans, will be the ultimate determinant of the success of the crop.

Soybean seed programs *per se* encompass several activities that complement each other to form a whole system. The various activities incumbent upon a successful soybean seed program should include (1) cultivar development, release and maintenance, (2) seed multiplication under a regulatory system and (3) seed marketing and distribution. While not attempting to present an exhaustive overivew of all the soybean seed programs in Africa, it is my intention to concentrate primarily on those countries comprising the Southern Africa Development Coordinating Conference (SADCC): Tanzania, Mozambique, Malawi, Zambia, Angola, Zimbabwe, Botswana, Swaziland and Lesotho.

UTILIZATION

Utilization of soybeans in Southern Africa takes various forms. The major ways in which soybeans are used include: (1) baby foods, (2) cookies, (3) blending with flour and (4) stockfeeds. Use of soybeans in the manufacture of cooking oil is particularly prevalent in countries of Southern Africa. The movement of soybeans from the growers through marketing channels to processors and finally to the consumer is illustrated in Figure 1.

CULTIVAR DEVELOPMENT

The major criteria for development of soybean cultivars suitable for production in various areas of Africa have been identified (International Institute of Tropical Agriculture, 1982) as:

(1) suitable oil and protein content
(2) adaptation to high and low fertility and having uniform maturity within different maturity groups of 80 to 130 days

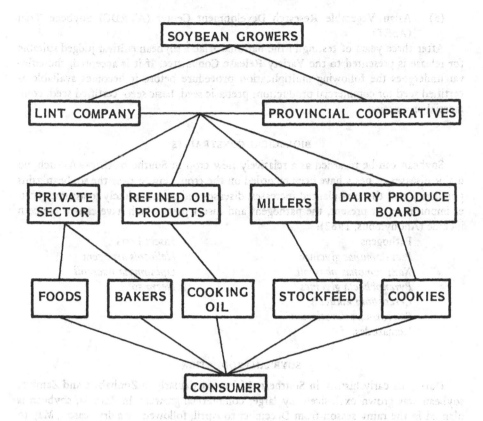

Figure 1. A flow-chart for soybean from growers to the consumer in Zambia.

(3)   capacity for efficient nodulation with indigenous *Rhizobium*
(4)   resistance to prevalent diseases
(5)   resistance to shattering
(6)   resistance to lodging and having suitable pod height and synchronous matur-
       ity of pods and stems
(7)   good seed longevity
(8)   tolerance to acid soil conditions.

Faced with a situation in which breeders are scarce and breeding programs minimal or nonexistent, a comprehensive cultivar development program which could be responsive to the previously enumerated needs is often unattainable. However, limited objectives can be achieved with a minimal program in cooperation with international and regional research centers.

In the case of Zambia, limited objectives are being achieved through an intensive cultivar testing program in collaboration with international research centers. The cultivar trials program includes (Anonymous, 1983):

(1)   National Soybean Variety Trial
(2)   Zimbabwe/Zambia Soybean Variety Trial
(3)   International Soybean Variety Evaluation Experiment (ISVEX)–INTSOY
(4)   Soybean Preliminary Observation Trial (SPOT)–INTSOY

(5)   Asian Vegetable Research Development Center (AVRDC) Soybean Trial (ASET)

After three years of testing in the national trial, a soybean cultivar judged suitable for release is presented to the Variety Release Committee. If it is accepted, the cultivar undergoes the following multiplication procedure before it becomes available as certified seed for commercial production: prebasic seed, basic seed, certified seed, commercial production.

BIOLOGICAL CONSTRAINTS

Soybean can be regarded as a relatively new crop in Southern Africa. As such, no major diseases or pests have been recorded on the crop. However, as the soybean gains popularity and the area planted increases, diseases and pests are likely to assume greater importance. At present, the pathogens and insect pests which have caused concern include (Anonymous, 1982):

Pathogens	Insect Pests
Pseudomonas glycinea	*Heliothis armigera*
Xanthomonas phaseoli	*Ophiomyia phaseoli*
Pyrenochaeta glycines	*Nezra* sp.
Cercospora kikuchii	
Sclerotinia sclerotiorum	
Nematodes.	

SOYBEAN PRODUCTION

During its early history in Southern Africa, particularly in Zimbabwe and Zambia, soybean was grown exclusively by large, commercial growers. In Zambia, soybean is planted in the rainy season from December to April, followed by a dry-season, May to September, irrigated crop of wheat (*Triticum aestivum* L.).

However, during the last two years, there has been an increase in interest in the soybean among small farmers. Cultivar characteristics required for the two types of production systems are shown below:

	Larger Grower	Small Grower
(1)	high podding	
(2)	resistance to lodging	
(3)	shatter resistant	shatter resistant
(4)	medium to late maturity	early to medium maturity
(5)	100 to 200 ha	less than one ha

These characteristics relate to planting and harvesting procedures which differ between large growers, who plant and harvest by machine, and small growers, who do the work by hand.

Small growers who are unfamiliar with biological technicalities and who lack refrigeration facilities are unable to exploit the advantages accruing from application of *Rhizobium* inoculum. It behoves breeders to investigate the phenomenon of promiscuous nodulation in order to identify soybean lines with a capacity to nodulate efficiently with indigenous *Rhizobium*.

SOYBEAN SEED PROGRAMS

The foregoing cursory overview of the importance, cultivar characteristics and production practices of soybeans in Southern Africa provides a background against which

Table 1.   Soybean production in relation to marketed seed, and use of improved cultivars in 12 African countries (Food and Agriculture Organization of the United Nations, 1981)

	Area	Marketed seed	Improved cultivars	Quality control
	1000 ha	t	% area	
Egypt	39.7	5572	100	present
Ethiopia	6.6	396	100	absent
Rwanda	5.1	20	13	absent
Malawi	- a	24	-	present
Tanzania	-	2	-	present
Gabon	-	-	-	present
Mozambique	-	-	-	absent
Malagasy	-	-	-	present
Ivory Coast	-	-	-	present
Congo	-	-	-	absent
Cameroon	-	-	-	absent
Angola	-	-	-	absent

aStatistic not provided.

the current seed programs in the region can be considered. The FAO Seed Review of 1979-1980 (Food and Agriculture Organization of the United Nations, 1981) while not being exhaustive, highlights the situation in Africa, where there is a great diversity in the completeness of seed programs. Out of the 25 countries which responded to questionnaires, 12 reported some activity in soybean seed multiplication. Five of the latter reported data for marketed soybean seed, and six indicated the existence of seed quality control operations (Table 1).

As noted in Table 1, there is a dearth of soybean seed programs in Africa, in general, and in Southern Africa in particular. However, regulations and standards for the multiplication of soybean seed have been declared in Zambia and Zimbabwe. Similar regulations and standards for seed certification are also present in Malawi (H. Mwandemere, personal communication) and Tanzania (Cooney, 1982).

STANDARDS FOR CERTIFIED SOYBEAN SEED

The standards for the certification of soybean seed in Zambia and Zimbabwe are summarized in Table 2.

THE PRODUCTION OF SOYBEAN SEED

Soybean seed in Zambia, Zimbabwe and Malawi is produced by contract growers under largely similar certification rules applicable in each country. In Tanzania, soybean and other seed are produced on state-owned seed farms (Cooney, personal communication). The regulatory procedures applicable to a grower of certified seed in Zambia include (Republic of Zambia, 1975):

(1)   Application for registration as a certified seed producer of a specific crop.
(2)   When registered, a grower receives a Certificate of Registration which is valid for only one cropping season.

Table 2. Standards for certified soybean seed in Zambia and Zimbabwe (Republic of Zambia, 1975; Republic of Zimbabwe, 1971).

Requirement	Zambia	Zimbabwe
Parent seed	Basic or Certified Seed	Breeder, basic or certified seed
Minimum inspections	2 fields, 1 post harvest	1 field
Isolation (m):		
basic seed	10^a-100^b	3^a-100^b-100^c
certified seed	5^a-100^b	3^a-100^b- 3^c
Crop rotation.	No soybean in:	
basic seed	previous 2 seasons	one season
certified seed	previous season	one season
Field standards	0.1% off-types^d 0.1% plants with seed-borne diseases	0.2% off-types^d --
Seed standards (%):		
purity	98	98
maximum defect^e	2	8
weed seeds	0.1	0.2
other crop seeds	0.5	0.5
germination	70	75

[a]Minimum distance from any basic crop of the same cultivar.

[b]Minimum distance from any crop of soybean.

[c]Minimum distance from any other soybean crop where there is a difference of more than 4 weeks in the time of flowering.

[d]"Off-type" means a plant which does not exhibit the recognized and accepted habits and characteristic growth, botanical formation either in leaf or flower, shape or color, or is in any way obviously different from accepted characteristics of the cultivar being grown.

[e]"Defective seed" means seed which is discolored by disease, insect damage, broken or chipped.

(3)   After harvest and when the seed satisfies the minimum standards of purity and germination, the seed is certified by a Certificate of Seed Certification. Certified seed is then sold to the Zambia Seed Company which is responsible for the marketing and distribution of seed through private, retail outlets, provincial cooperative unions or directly through its own outlets.

Seed growers in Zambia are members of the Zambia Seed Producers Association. The association in collaboration with the seed company allocates the area of a seed crop to be grown by a farmer in a given cropping season. Collaboration between the association and the company facilitates proper planning for the production of sufficient seed to meet the country's requirements.

NOTES

D. M. Naik, Chief Agricultural Research Officer, Mount Makulu Central Research Station, Chilanga, Zambia.

REFERENCES

Anonymous. 1983. Annual Report, Soybean Breeding Program in Zambia. Mount Makulu Research Station, Chilanga, Zambia.

Anonymous. 1982. Report of the Planning and Evaluation Committee Meeting for Soybean, Eastern and Southern Regions. High Yielding Varieties Technology Project. International Institute of Tropical Agriculture, Ibadan, Nigeria.

Food and Agriculture Organization of the United Nations. 1981. FAO Seed Review 1979-80. FAO, Rome.

International Institute of Tropical Agriculture. 1982. Final Report. High Yielding Varieties Technology Project. IITA, Ibadan, Nigeria.

Republic of Zambia. 1975. p. 239-293. *In* Agriculture (Seeds) (General) Regulations, 1975. Lusaka, Zambia.

Republic of Zimbabwe. 1971. Seeds (Certification Scheme) Notice, 1971. Harare, Zimbabwe.

SOYBEAN SEED PRODUCTION IN ASIA

P. Wannapee, D. Gregg and S. Jongvanich

Soybean is a major crop in Asia, but production is short of that needed for oil and protein. The soybean faces many production constraints; a major factor limiting increases in production and yield is simply the lack of good quality seed of improved cultivars. In most Asian countries, a seed industry capable of supplying quality soybean seed has not yet developed. As a result, soybean production during 1971 to 1981 in the Asia-Pacific region increased at an annual rate of only 0.4%, while the rest of the world increased at 7.6%, and developed countries at 6.2% (FAO, 1983). Due to the costs, risks and technical problems of producing seed in the humid tropics and subtropics, this seed shortage is likely to remain a limitation until special emphasis is directed toward creating the necessary organizational support and implementing adequate levels of operational technology.

SEED REQUIREMENTS AND SUPPLY

Complete statistics on seed production, quality, and supply are not readily available, but present seed use can be estimated (Table 1). Most soybean "seed" used is not truly "seed", nor a yield-increasing production input; it is "grain beans" used as seed. Too often, little emphasis is placed on supplying seed of yield-supporting quality to farmers, and establishing organized seed programs to assure a dependable, recurring seed supply. Production from presently cropped areas could probably be increased significantly by adequate supplies of good quality, genetically pure seed of improved cultivars.

There has been considerable Asian interest in "security seedstocks", to supply farmers stricken by all too common disasters, such as drought and flood. However, the susceptibility to deterioration and poor storability of soybean seed under present organizational and technological conditions do not permit supply of improved seed to plant the usual cropped areas, much less maintain reserve stocks. Soybean security stocks can be economically maintained only by producing additional seed beyond present needs, and then discarding it as grain after a few months if it is not needed for planting, or by collecting the best quality grain for seed use.

ORGANIZATION TO SUPPLY SEED

Some Asian countries have organized seed supply programs, but not all supply soybean seed; those that do, usually supply only a small part of the needs (INTSOY, 1981; Fiestritzer, 1981).

Table 1. Estimated soybean seed requirements in Asia in 1982 (from FAO, 1983; INTSOY, 1981).

Country	Crop area	Average planting rate	Seed needs	Annual growth rate	Seed needed for 10% increase
	1000 ha	kg/ha	t	%	t
Bangladesh	6	80	480	-	48
Burma	27	50	1,350	1.9	135
China, PR	8,012	80	640,960	-0.5	64,096
India	680	50	34,000	63.6	3,400
Indonesia	770	50	38,500	1.5	3,850
Iran	50	50	2,500	-	250
Iraq	1	50	50	-	5
Japan	147	50	7,350	3.3	735
Kampuchea	3	50	150	-8.1	15
Korea, DPR	310	50	15,500	0.8	1,550
Korea, Rep	183	50	9,150	-3.8	915
Laos	6	50	300	4.1	30
Pakistan	4	75	300	4.1	30
Philippines	11	45	495	24.6	50
Sri Lanka	2	68	136	-	14
Thailand	128	45	5,760	9.9	576
Vietnam	100	50	5,000	0.2	500
Total	10,465	-	763,356	-	76,337
Average	-	55	-	0.4	-

Bangladesh: Small-scale production of Bragg and Davis seed was undertaken by the Agricultural Development Corporation, the main producer of other crop seed.

Burma: The main seed producers are the Applied Research Division and the Agricultural Research Institute of the Agriculture Corporation. In 1979-80, an estimated 11% of the 23,000 ha of soybeans were planted with seed of improved cultivars.

People's Republic of China: Each commune has a specialized team to produce its seed requirements. Experienced communes are contracted to produce seed for national, provincial and regional government seed companies, which market seed for the whole country. Ninety-five percent of the soybean area is planted to improved cultivars.

India: Although soybeans were not reported, seed are produced by state seed farms, state seed corporations, the National Seeds Corporation, the State Farm Corporation, and private-sector seed companies.

Indonesia: Although soybeans were not mentioned, the major seed organization is the National Seed Corporation.

Korea, Republic of: In 1979, an estimated 16% of the 207,000 ha of soybeans was planted with improved cultivars. The National Office of Seed Production and Distribution produced 2,116 tons of seed. Provincial governments now produce soybean seed, distributed by the OSPD.

Malaysia: Large plantations supply seed of some crops. The Federal Department of Agriculture produces and distributes seed to small farmers. Only small amounts of soybean seed have been produced.

Nepal: Some government farms and stations produce soybean seed, distributed to farmers by the Nepal Grain Legume Improvement Programme.

Pakistan: Soybean seed is multiplied by the Punjab Corporation, provincial agriculture departments, and progressive farmers. Provincial agriculture departments and the Ghee Corporation supply seed in different soybean-growing areas.

Philippines: In 1979-80, 11.2 tons of soybean seed were reported from the major seed producers.

Sri Lanka: Soybean seed is produced by the Division of Seed and Planting Materials, Ministry of Agricultural Development and Research.

Thailand: The DOAE Seed Division supplies soybean seed, 1,100 tons in 1983. Targets are 6.4% of soybean seed needs in 1984, rising to 12.8% in 1986. Soybean seed is not profitable for private sector seed companies (major source of seed), but is being considered as a secondary, supporting seed crop as technical capability grows (Anonymous, 1983; Seed Division records).

Vietnam: Provincial seed production centers re-multiply and distribute seed to district centers, which produce seed for farmers. All seed distribution is by the Central Agency for Seed Production. Soybean seed in 1979 totalled 400 tons.

PROBLEMS OF SEED SUPPLY

Soybean seed production and supply in Asia, produced by small farmers under hot humid tropical or semi-tropical climates, is plagued with low seed quality, rapid loss of germination, poor storability, loss of lots, and high costs.

Organizational Problems

Soybean seed must be produced, harvested, threshed, dried, processed, stored and distributed with special care, employing high technology and special attention to timely performance. Low-technology farmer methods are not successful for seed. Special organizations are required to focus resources on, and carry out, the special procedures necessary to supply seed adequate in both quantity and quality. Such organizations generally do not exist, or have inadequate staff, funds, facilities, or technical procedures to be effective on a scale that would influence total soybean production.

Technical Problems

Technical requiremens for seed in hot humid areas are very restrictive; minor delays or oversights result in serious losses.

Production. Other crops (e.g., rice) are often of primary importance, with soybean production adjusted to fit. Seed is harvested during periods of rain or high humidity. Less attention to production techniques—cultivation, pest and weed control, etc.—lowers yields. Low seed prices and inadequate technical supervision influence the use of practices such as roguing.

Post Harvest. An estimated 90% of soybean seed quality problems occur between the time seed reach physiological maturity, and they are planted. To clear fields for a following crop, or due to lack of irrigation water, seed are often harvested before they reach proper maturity. Improper or delayed threshing and drying, excessive exposure to unfavorable environments, improper handling, storage at excessively high temperatures or seed moisture, or over-long storage cause losses.

POTENTIALS AND RECOMMENDATIONS FOR RESOLVING PROBLEMS

Much of the technology required to produce and supply soybean seed of acceptable quality is known. Where transferred technology is not completely adequate, local research can devise specific procedures for local needs. Successful organizational structures and technology for other crop seed can be adapted to supply soybean seed.

Organization and Implementation Requirements

1. Adequate organizational structures must be established to implement the necessary actions. A carefully organized program must be created to produce soybean seed of the desired cultivars and quality in the required amounts.
2. It must have the authorizations required to carry out timely operations at the necessary technology levels.
3. It must have highly trained staff capable of analyzing local conditions, identifying the causes of problems, and developing and implementing technical procedures to resolve or prevent problems.
4. It must have the operating funds and facilities necessary to carry out timely seed production, harvest, drying, conditioning, packaging, storage and distribution at adequate technology levels.
5. It must establish a distribution system which can get seed to farmers in good condition at the proper time. A "seed-for-grain" exchange may be established to serve small, low-income farmers. Adequate ongoing promotion must be maintained.
6. It must establish a special body to coordinate research, basic seed supply, seed production, and seed distribution.
7. It should offer special incentives to encourage private-sector seed production.
8. It will need to determine seed production targets from actual surveys of farmer needs to avoid excessive production that must be discarded or carried over at great expense.

Technical and Operational Procedures

1. Select crop rotations, and cultivars of suitable maturity periods, so soybean seed can mature properly and be harvested during favorable weather.
2. All seed fields should be sown with stock seed of adequate genetic purity.
3. Use planting methods which ensure good stands without excessive seed rates, especially in wet and clay soils, such as rice paddies. Use plant spacings and populations which give high yields and facilitate operations such as roguing.
4. Use improved crop (insect control, etc.) and seed production (roguing, etc.) practices, for high yields and good seed quality.
5. Provide adequate water for seed to mature properly and be harvested only when fully field-mature.
6. Dry harvested plants immediately; thresh them and deliver seed immediately to the conditioning plant for safe drying and handling.
7. Condition seed immediately, using facilities designed and operated to minimize mechanical injury to seed.
8. Package seed and move it into safe storage without delay. Soybean seed often will not survive for the required storage period under ambient conditions; conditioned (cooled and dehumidified) storage is required. When seed in vapor-porous bags are removed from conditioned storage, keep them in a warm dehumidified room for 1-2 weeks, so they will not gain moisture rapidly from the

hot humid ambient atmosphere.

9. Where quality problems during storage and distribution are severe, dry seed to low moisture (6%) and hermetically seal them in vaporproof packages. These packages permit longer storage under ambient conditions, and protect seed during distribution to farmers.

10. Limit seed storage to the shortest possible periods; produce seed in one season for planting in the immediately following season.

11. Handle seed gently. During distribution, keep seed dry, shaded, and as cool as possible. Establish a network of suitable storages so seed will be exposed to unfavorable ambient conditions for the shortest possible time before planting.

12. During storage and handling, test seed at monthly intervals. Seed beginning to decline in germination should be planted as soon as possible. Lots with below standard germination, or which will probably be below standard by the time they reach farmers, should not be used for seed. Regardless of cost, low-quality seed should be discarded—never supplied to small farmers.

NOTES

Petcharat Wannapee, Bill Gregg and Sommart Jongvanich, Director, Senior Seed Specialist (MSU), and Quality Control Specialist, respectively, Seed Division, Department of Agricultural Extension, Ministry of Agriculture and Cooperatives, Bangkok, Thailand.

REFERENCES

Anonymous. 1983. Directory of the Thailand Seed Industry. Seed Division, DOAE. Bangkok.

FAO. 1983. Selected Indicators of Agricultural Development in Asia-Pacific Region. FAO-RAPA Monograph No. 2. Bangkok.

Feistritzer, W. P. (editor). 1981. FAO Seed Review, 1979-80. FAO, Rome.

INTSOY. 1981. Country Reports. p. 143-148. *In* Soybean seed quality and stand establishment. INTSOY Series No. 22.

SOYBEAN PRODUCTION AND UTILIZATION IN SUB-SAHARAN AFRICA

L. E. N. Jackai, K. E. Dashiell, D. A. Shannon and W. R. Root

About a quarter of a century ago, the soybean was cultivated only in a few locations in Africa, mainly in areas where early missionaries and explorers had introduced the crop. Today, many more countries are either producing soybeans as an export crop, or for use in local industries and diets. During the past decade, soybean production has shown a steady increase for several reasons. The nutritional value of the crop is one of the most important. In a continent that imports vegetable oil, which always seems to be in short supply, introduction of another indigenous source is most welcome. Traditionally, Africa has been an importer of soybean meal for annual feed. In years gone by when cash flow problems were of little concern, importation of both oil and meal was met with little difficulty. Today, the foreign reserves of most African nations are dwindling at an alarming rate, and this situation has forced many countries to encourage local production of oil seed crops, including soybeans. As a result, practically every country on the African continent is today involved in some level of production research on the soybean, and several countries are already well-established producers. Unfortunately, this enthusiasm has met with a series of production and utilization problems created by the environment and local dietary habits.

These problems and others, such as poor seed viability of high yielding cultivars, slowed the initial rate of soybean production on the continent. However, collaborative research efforts by national, international and bilateral agencies have, by and large, alleviated most of these initial problems. Soybean producing countries on the continent can now look forward to a future when less foreign reserves will be spent on the importation of vegetable oil and soybean meal for local industries and consumption, and to when a sizable percentage of the population will obtain more nutritional value from their diets.

PRODUCTION

According to FAO (1982) figures, soybean production in Africa increased by over 200% between 1976 and 1982. The area under soybean cultivation also increased in the same period by about 100%, while yields appear to have stabilized around 1500 kg/ha since 1980. The same trends are generally true for almost every producing country in Africa.

Ecological Requirements and Geographical Spread

The principal regions in Africa where the soybean is most likely to be grown successfully are the savanna areas where rainfall is not seriously limiting (see Kuenemann et al., Figure 1, this volume). In West Africa, this corresponds to the Guinea Savanna. In the more arid Sudan Savanna, it may be possible to produce a rainfed crop, but the short duration of the cropping season is likely to lead to conflict in the timing of planting of cereals and soybeans. The chances of drought during seedling establishment and seed-filling are high. Areas with a growing season shorter than that of the Sudan Savanna are wholly unsuitable to rainfed soybean production.

The high rainfall forest zones meet the moisture requirements of the soybean, but this zone has high humidity throughout most of the year. This condition is not conducive to maintaining seed viability. Dehumidified chambers are needed to store seed from one season to the next. Secondly, soils of large areas of this zone consist of potentially acid ultisols and oxisols. Some of these soils may require liming to eliminate Al or Mn toxicity before soybeans can be grown. Oxisols and ultisols cover large areas in central and eastern Africa and are also important along much of the coast of west Africa from The Gambia to the Ivory Coast and in southeastern Nigeria (Sanchez, 1976). Though soybeans are grown in high rainfall regions, the wetter parts of the savanna, with a growing season of 120 to 180 days and a dry season characterized by low humidity, are likely to become the most important soybean growing areas in Africa because of their more favorable climate and soils. Soybeans are also likely to be grown in the mid-elevation regions in Africa when adapted cultivars become available.

Producing Countries

The largest area under soybean cultivation in Africa is in Nigeria. Estimates range from 70,000 ha (Oyekan, 1984) to 195,000 ha (FAO, 1982). However, the largest seed producer is Egypt, which produced 170,000 metric tons in 1982 and also gave highest overall yields. Figure 1 shows overall production trends of some of the major producing countries during 1980 to 1982.

West Africa. Several African countries are at different stages of developing soybean production. There has been renewed interest in soybean production mainly due to availability of high yielding cultivars, some of which can nodulate without inoculants. In Nigeria, a boom in the poultry industry, followed shortly afterwards by a shortfall in foreign currency for importation of feed, has helped boost soybean research in the past few years. In West Africa, the Ivory Coast and Nigeria are the largest producers. In the Ivory Coast, production has increased steadily since the beginning of the decade. However, problems of loss of seed viability in storage have necessitated importing seed from Brazil, which also supplies technical support for soybean production to the Ivory Coast. The Institut des Recherches Agronomiques Tropicales (IRAT), working with the Institut Senegalais de Recherche Agricole (ISRA), has a soybean program in Senegal supported by agronomists and microbiologists. Through this program, adapted cultivars and inoculant have been introduced to west and central African Francophone countries. Other countries in the region have embarked on large projects aimed at expanding soybean production. Many of these are in the form of bilateral projects aimed at encouraging small farmers to incorporate soybean in their present farming systems. Others are government supported seed farms, such as in Togo, where the West German Agency for Technical Cooperation (GTZ) provides technical aid. In the Republic of Benin, the Catholic Relief Services (CRS) has initiated an ambitious project with plans to reach 50,000 rural farm families. If this succeeds, it is expected that other Francophone countries will follow the example. In several other west African countries research activities on soybean production are receiving high attention. It is likely that there will be large-scale soybean production in most of these countries before the turn

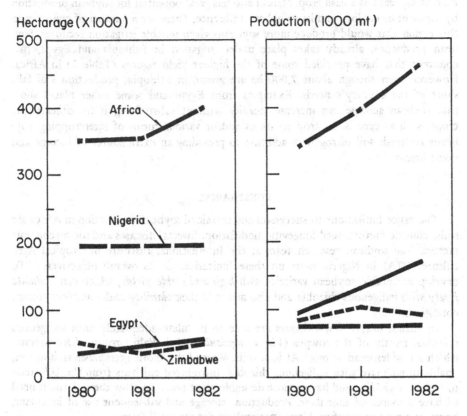

Figure 1. Soybean hectarage and production in Egypt, Nigeria and Zimbabwe in 1980-82 (FAO, 1982).

of the century.

Central Africa. Many countries in this region have large mid-altitude areas of cool climate and adequate rainfall where soybeans can be produced. Zaire, Rwanda and Cameroon have the greatest potential in this region. In Cameroon production has been restricted to the higher altitude region of the Western Province. Production has been on a rather small scale mainly with the cooperation of IRAT. As in Cameroon, Rwanda and Zaire have oriented production towards the small farmer. In Zaire, where 75% of the estimated 29 million people are dependent on agriculture and malnutrition is ever increasing, a bilateral program with USAID has intensified efforts to grow more soybeans. Soybeans are regarded an excellent remedy for kwashiokor, and several baby and adult foods and snacks are made either wholly or partially from soybeans.

Eastern and Southern Africa. The best soybean production program on the continent is in Zimbabwe, which is one of the major producers. Soybean production in this country is essentially large scale, commercial and highly mechanized. However, indigenous small-holders are now being encouraged to cultivate the soybean. This is expected to gather momentum with the availability of promiscuous, high-yielding cultivars (Jacob Tichagwa, pers. comm.). Zambia is currently expanding its area under soybean production. Some sources put the cultivated area in the country at 8,000 ha (FAO, 1982), whereas others put it at 30,000 ha (Anon., 1983). So much interest has been generated in soybeans that several hundred small farmers grow up to 2 hectares each of soybeans as a cash crop. Malawi also has great potential for soybean production by virtue of its ecology and its neighbors' influence. There are a number of countries in this region that would produce more soybeans given reliable irrigation facilities. Soybean production already takes place under irrigation in Ethiopia and Egypt, two countries that have provided some of the highest yield figures (Table 1) in Africa. However, even though about 7,000 ha are grown in Ethiopia, production still falls short of the country's needs. Examples from Egypt and some other places show that soybean acreage can increase steadily without substituting it for other staple crops, such as cereals. In crop rotations and/or various forms of intercropping, soybeans replenish soil nitrogen in addition to providing an extra source of income and nourishment.

CONSTRAINTS

The major limitations to successful and sustained soybean production in Africa are soils, climatic factors, seed longevity, nodulation, insects, diseases and socio-economic factors. The soybean research team at the International Institute of Tropical Agriculture (IITA) in Nigeria sums up these limitations in its overall objectives: "To develop acceptable soybean varieties with *high and stable yields*, which can *nodulate freely* with indigenous rhizobia and also maintain their *viability* under ambient storage conditions."

In Africa, only a few cultivars are able to nodulate adequately with indigenous rhizobia, mostly of the cowpea (*Vigna unguiculata* [L.] Walp.) group of *Rhizobium*, which are widespread in most African soils. Most of the exotic germplasm is, however, unable to nodulate with indigenous rhizobia. Introduced cultivars from the U.S. tend to be higher yielding and have a more desirable plant type. To grow these, farmers need to have a source of inoculant. Production, storage and subsequent use of inoculant, however, pose a series of problems, particularly for the small farmer.

In countries such as Zimbabwe, where inoculants are produced locally, supply is steady and abundant, albeit at an extra cost. This situation, however, does not exist in most other African countries, and there is need for cultivars that do not require inoculation with rhizobia.

Seed deterioration; i.e., loss of viability before planting, is a widespread phenomenon in the tropics. Up to 80% of seed is lost in this way. The principal causes of this

Table 1. Soybean yields from selected countries in Africa (FAO, 1982).

	1980	Yield 1981	1982
		—kg/ha—	
Egypt	2671	2837	3036
Ethiopia	2923	2889	2857
Ivory Coast	787	900	1000
Rwanda	821	825	842
Uganda	750	1000	1000
Zaire	1169	1169	1169
Zambia	954	1015	964
Zimbabwe	1727	2746	1756
Mean	1475	1673	1578

phenomenon seem to be high temperatures and humidity (Kueneman, 1982), but fungi play a secondary role (Ndimande et al., 1981). As will be discussed later, these, and the problem of nonpromiscuity, have been successfully tackled through systematic breeding (Anon., 1983), such that high yielding cultivars with increased longevity and ability to nodulate with indigenous rhizobia have been produced.

Insect and disease pests are presently highly localized and of less significance on a region-wide basis. In some areas in Nigeria, for example, soybeans can hardly be grown without weekly applications of insecticides during podding (Ayoade, 1982). The major insect problem is the stink bug complex, mainly *Nezara viridula*. This pest complex has also been reported as the major problem in the Ivory Coast, Zambia and several other countries. In one location in Nigeria, the stink bug population greatly exceeds the economic threshold established anywhere (Figure 2), and the associated loss in yield is as high as 60% (Jackai and Kueneman, 1982; Jackai and Raheja, 1984). Other insect pests are reported from various countries, but mostly as minor pests. These include pod, flower and foliage pests. Recently, a nodule-feeding fly (initial identification *Rivelia* sp.) has been found in Maradi, Niger Republic. Even though Maradi is not a

Figure 2.   Population fluctuation of stink bugs on soybean (cv. TGx 17-2GE) at Mokwa Cattle Ranch, Nigeria. (Trial planted July 2, 1983.)

typical soybean growing area, the possibility of this pest spreading into other areas cannot be overlooked.

Ezueh and Dina (1979) give a long list of insect pests associated with soybeans in Nigeria. This list is comprehensive enough to cover species reported from many countries. Aphids are important vectors of the soybean mosaic virus, and an unpublished report from Zaire puts the number of aphid vector species at thirty-one. Beetles (mainly *Chrysolagria* sp. and *Lagria villosa*), lepidopterous larvae of *Plusia* sp., *Heliothis armigera*, *Spodoptera* sp., and *Maruca testulalis* among others have been reported from Zimbabwe, Zaire, Nigeria, Egypt and Ethiopia. Termites and leafhoppers can also become a problem in certain areas.

In general, most soybean growers do not spray their crops but where they do the number of sprays never exceeds three, even then only as a preventive measure. The present status of insect pests, however, does not suggest that research efforts should be relaxed. Studies of the biology, behavior and bionomics of these insects must be conducted to gather basic information required to address future pest problems, which are almost certain to come with increase in soybean acreage (Kogan, 1981). No cultivars resistant to any of the major insect pests in Africa are presently available, but resistant sources for several of the minor pests may already be present in existing germplasm.

The problem of diseases is not fully recognized by the farmer as a major constraint to production. Increase in knowledge and availability of experts have made the farmer more aware of the potential importance of diseases in soybean production. A few diseases have made their mark in almost every producing country on the continent. Commonly occurring diseases are listed in Table 2. Among these, the most important are bacterial pustule, Pyrenochaeta leaf spot and soybean mosaic virus (SMV). SMV is seed-borne and is, therefore, a potentially serious disease in many soybean growing areas of Africa, particularly with increasing movement of soybeans across borders. However, where effective quarantine regulations exist, this problem is adequately controlled. Other diseases include frog-eye leaf spot (*Cercospora sojina*), anthracnose, charcoal rot (*Macrophomina phaseolina*), which is mainly found in hot dry areas, fusarium wilt and root rot, found in hot humid areas, and bacterial blight.

OVERCOMING PRODUCTION CONSTRAINTS

International, national and private agencies have played a major role in overcoming some of the obstacles to soybean production and in promoting its use.

Cultivar Introduction and Breeding

The International Soybean Program (INTSOY) has been instrumental in making available to researchers throughout Africa cultivars with high yield potential and wide adaptability. Through their ISVEX trials, they have demonstrated that high soybean yields can be obtained in many parts of Africa. The French Institut de Recherches Agronomiques Tropicales et des Cultures Vivrières (IRAT), for many years conducted a breeding program in Senegal, and selections from this program have been distributed throughout much of French-speaking Africa. Cultivars from this breeding program are currently being grown in Cameroon and Zaire.

The soybean breeding program at IITA, which began in 1976, now has cultivars of varying maturity adapted to west and central Africa. New high-yielding cultivars with improved seed longevity and promiscuous nodulation are being evaluated in 29 countries in Africa through an international testing scheme. These should soon be available to farmers.

Table 2. Most common soybean diseases in Africa.

Disease	Classification	Countries reported
Soybean mosaic virus (SMV)	Viral	Most
Bacterial pustule (*Xanthomonas campestris* pv. *phaseoli* (Smith) Dye	Bacterial	Nigeria, Zaire
Bacterial blight (*Pseudomonas syringe* pv. *glycinea*)	Bacterial	Nigeria, Zaire, Madagascar, Cameroon.
Pyrenochaeta leaf spot (*Pyrenochaeta glycines*)	Fungal	Nigeria, Zambia, Cameroon, Zaire, Ethiopia, Zimbabwe.
Cercospora leaf spot (*Cercospora sojina*)	Fungal	Zaire, Nigeria, Zambia.
Anthracnose (*Colletotrichum dematium* var. *truncatum*)	Fungal	Nigeria, Zaire
Alternaria leaf spot (*Altanaria* sp.)	Fungal	Nigeria
Target leaf spot (*Coryonespora cassicola*)	Fungal	Nigeria, Cameroon, Zaire.
Seedling damping-off (*Macrophomina phaseolina*)	Fungal	Zaire, Zambia, Cameroon, Nigeria
Web blight (*Rhizoctoria solani*)	Fungal	Nigeria, Zaire, Zambia, Cameroon.
Fusarium rot (*Fusarium* sp.)	Fungal	Zaire, Nigeria, Cameroon.
Purple stain (*Cercospora kikuchii*)	Fungal	Nigeria, Zaire, Ivory Coast.
Nematodes (*Meloidogyne* sp.)	Nematode	Several

Two cultivars developed in Nigeria by the Institute of Agricultural Research were released in 1983 as Samsoy-1 and Samsoy-2. The Soybean Breeding Program of the Crop Research Institute of Zimbabwe has had the largest impact to date. It is one of the oldest soybean breeding programs in Africa, and its cultivars have enabled Zimbabwe to become one of the largest soybean producers in subsaharan Africa, as well as becoming an exporter of soybean oil and meal.

Production Research

Research on production methods and constraints has been conducted by IRAT, IITA, several national research institutes, and through bilateral aid projects. Though considerably more work needs to be done, some information is already available on topics such as fertilization, spacing and insect control. Research on inoculants has been conducted by IRAT, IITA and some national research institutes, such as that in

Zimbabwe. Because of lack of inoculants, soybean breeders at IITA have developed cultivars capable of nodulating effectively with indigenous rhizobia. This approach will help to overcome problems associated with producing and distributing inoculants, but it will not rule out using inoculants where they are available. Yields above 3 t/ha have been obtained with these cultivars without inoculation (IITA, 1982).

Insect pests are easily overcome using appropriate chemicals. However, other control methods are being investigated. In Zimbabwe, scientists have developed methods for biological control of the looper (Rex Tattersfield, pers. comm.), and at IITA the breeders and entomologists are working towards resistance to stink bugs. Diseases, Al toxicity, shattering, lodging and field weathering are also being tackled through breeding.

The problem of seed longevity has caused a lot of anxiety. However, physiologists at IITA identified cultivars with superior seed longevity in near-wild types of soybeans from Indonesia. Lines with unacceptable plant type were crossed with cultivars with superior agronomic characteristics. Cultivars are now available that are useful in local diets. Introducing processing plants in towns and villages will lead to greater interest in production for food and cash.

SOYBEAN UTILIZATION

During the Soybean Utilization Workshop for Africa recently held in Nigeria, and jointly sponsored by IITA and INTSOY, delegates expressed the need for more information on commercial and home preparation of soybeans for the African consumer. All over Africa, one can find examples where soybeans have been well accepted by the people and are being consumed as part of the daily diet. For instance, in Nigeria the Federal Institute of Industrial Research developed a product called "soy-oge" which is used as a baby food. This product is being produced commercially. It is 20% protein and is made with soybean, maize (*Zea mays* L.) or sorghum (*Sorghum bicolor* [L.] Moench) plus minerals and vitamins. Also, soybeans are used in place of melon ("Egusi") seed to thicken soup. Most of the soybeans produced are used to make a fermented product called "daddawa". This is traditionally made with locust bean (*Parkia filicoiddes*), and is used to flavor soup and stew. Some private investors in Nigeria have plans to open soybean processing factories to make soy-milk and other soybean products.

In Rwanda, soybean is used as a full-fat flour to prepare "bean sauce", as oil or as soycake. The "bean sauce" is prepared by soaking, grinding and then boiling the soybean into a sauce. There is at least one small oil press plant in Rwanda, and installation of a soy-milk plant is being considered by a German company. The demand for soybeans is already high and is likely to increase beyond production capacity in the near future.

Almost all the soybeans produced in Zaire are processed into full-fat flour. Some of the flour is packaged and sold mainly for mixing with cassava (*Manihot esculenta* Crantz) or maize flour. There are also small scale industries that make weaning food and cookies sold throughout the country. In one area, a medical doctor required mothers of malnourished children to purchase soybean flour as a prerequisite to medical treatment. In addition, soybeans are already established in many areas for treating malnutrition. There is also keen interest among health and rural development planners to popularize soybean more with the population.

In Cameroon where soybean is a relatively new crop, the primary aim is to develop it as an industrial crop, but if this is not achieved, they are promoting local consumption of soybeans. In one area, soybeans are being mixed with sorghum in preparing weaning food for infants. Courses have also been conducted in preparing soybeans, and recipe books have been published.

The Catholic Relief Services (CRS) is in collaboration with five Ministries in the Republic of Benin promoting the use of soybeans in local diets. The CRS has six teams conducting both production and utilization demonstrations at schools, in villages and at Mother-Child Health Centers. The demand for soy-enriched flour is high at many centers, and there are plans to install flour mills for processing soybeans. A locally produced enriched flour used as a weaning food is made of 2 parts cereal, 1 part field beans (*Phaseolus vulgaris* L.), and 1 part peanuts. Comparable soybean enriched flour also is produced locally.

Efforts are being made in Zambia to introduce intermediate level technology to process soybeans. One example is a percolator used to cook soybeans as a component in hog feed. The percolator is constructed using a 100 L drum with an open top, a perforated disk from a disk harrow, and a 2 cm pipe about as long as the oil drum. The percolator can cook 75 kg of soybeans in 2 h.

There is need for training in soybean utilization and for stepping up research in soybean utilization with particular emphasis on incorporating soybean into local foods. For Africa, high level technology is bound to come, but before this the emphasis must be on intermediate technology, which is the link to increased production and utilization in the years to come.

FUTURE OF SOYBEANS IN AFRICA

The future of soybeans in Africa looks bright. The major constraints to expanding soybean production in Africa are being overcome through breeding. Further breeding will permit expansion of soybean production into areas with shorter growing seasons, into higher elevations and onto acid soils where present cultivars are not well suited. Efforts need to be intensified to multiply and distribute improved cultivars to farmers and to provide them with proper management information.

With the widespread deficit in foreign exchange being experienced by many countries in Africa, increased soybean production provides an opportunity for saving this valuable asset by reducing imports of oil and meal. With increasing demand and soaring costs of protein foods, soybeans offer one of the best solutions for improving human nutrition in Africa.

NOTES

L. E. N. Jackai, K. E. Dashiell, D. A. Shannon and W. R. Root, International Institute of Tropical Agriculture, Oyo Road, PMB 5320, Ibadan, Nigeria.

The authors wish to thank Dr. Peter Oyekan for providing information on diseases.

REFERENCES

Anonymous. 1983. Soybean improvement at the International Institute of Tropical Agriculture—A position paper. IITA, Ibadan.

Ayoade, K. A. 1982. Preliminary investigation on the insect pests of soyabeans and their control. p. 46-49. *In* Proc. 2nd national meeting, Nigerian soybean scientists. Publ. No. II, IAR Ahmadu Bello Univ., Zaria, Nigeria.

Ezueh, M. I. and S. O. Dina. 1979. Pest problems of soybeans and their control in Nigeria. p. 275-283. *In* F. T. Corbin (ed.) World Soybean Conference II: Proceedings, Westview Press. Boulder, Colorado.

FAO. 1982. Production Yearbook, Vol. 36. FAO, Rome.

IITA. 1982. Annual Report for 1981. International Institute of Tropical Agriculture, Ibadan.

Jackai, L. E. N. and E. A. Kueneman. 1983. Survey of the major pests of soybeans at two locations in Nigeria. Paper presented at the 3rd national meetings of Nigerian soybean scientists, 7-9 February, Makurdi, Benue State.

Jackai, L. E. N. and A. K. Raheja. 1984. Soybean Entomology: Pest survey and yield loss assessment. *In* Annual Report for 1983. International Institute of Tropical Agriculture, Ibadan.

Kogan, M. 1981. Dynamics of insect adaptations to soybean: Impact of integrated pest management. Environ. Entomol. 10:363-371.

Kueneman, E. A. 1982. Soybean improvement at IITA. p, 77-95. *In* Proc 2nd national meeting of Nigerian soybean scientists. Publ. No. 2, IAR Ahmadu Bello University. Zaria, Nigeria.

Ndimande, B. N., H. C. Wien and E. A. Kueneman. 1981. Soybean seed deterioration in the tropics. The role of physiological factors and fungal pathogens. Field Crops Research 4:113-121.

Oyekan, P. O. 1984. Nationally co-ordinated research project on soybeans in Nigeria. Paper presented at the Joint Planning and Evaluation Meeting of the EEC/SAFGRAD Project. IITA, Ibadan, 5-9 March.

Sanchez, P. A. 1976. Properties and management of soils of the tropics. John Wiley and Sons. New York.

SOYBEAN PRODUCTION IN ASIA

Ricardo M. Lantican

The soybean has been a traditional crop for millenia in China and adjoining areas of East and Southeast Asia. The crop has been grown for centuries in the foothills of the Himalayas, but only recently has it assumed prominence in India. Aside from China and India, the other dominant soybean producing areas in Asia are Korea, Indonesia, Thailand and Japan (see FAO Production Yearbook for latest data).

In view of the emergence of soybeans as a commercial crop and its importance as a source of edible oil and protein for food and feed, many countries have become interested in establishing their own production industries. Iran and the Philippines have a nascent production industry. Other countries (Turkey, Pakistan, Bangladesh, Sri Lanka, Nepal and Malaysia) are undertaking similar efforts, although production has not yet made headway.

PRODUCTION STATISTICS

The region, although the ancestral home of the soybean, has not equalled the production nor the hectarage of the USA. Statistics from 1977 to 1981 showed that Asia accounted for 24% of world hectarage and 14% of world production, while the U.S. accounted for 52% of hectarage and 61% of world production. As a result of low hectarage and productivity, soybean production has not met the requirements of the region. Thus, the whole of Asia has been a net importer of soybeans, requiring an average per year of 2.67 million tons of raw beans, 1.26 million tons of oil and 0.94 million tons of cake and meal (see FAO Trade Yearbook for specific country data).

PRODUCTION SYSTEMS IN ASIA

The production systems and cultivars used across the region are highly variable as a consequence of the diversity of ecological conditions. In Asia, the existing and potential production areas for soybean range from the equatorial belt to middle and high latitudes.

Humid Equatorial Belt

This production area in Asia, which extends to the lower latitudes north and south of the equator, includes Thailand, Burma, Vietnam, Philippines, Indonesia and central and south India. In this region, rainfall is the chief limiting factor to yields. The yield potential of the area is 2 t/ha.

Early-maturing soybean cultivars (75 to 120 days) are grown in this area during the monsoonal wet and dry seasons in upland and lowland areas. In the upland areas, the crop is rotated or intercropped with cereal and other upland species, such as sugarcane (*Saccharum officinarum* L.). For example, in the southern parts of India, soybean is intercropped with sorghum (*Sorghum bicolor* [L.] Moench), cotton (*Gossypium* spp.) and pigeon pea (*Cajanus cajan* [L.] Millsp.). In the lowland rice (*Oryza sativa* L.) fields, it is common to plant soybeans after the rice harvest without the benefit of soil tillage. In Indonesia, seed is broadcast before the rice crop is harvested, while in Thailand and Taiwan, the seeds are dibbled into the soil at the base of the rice stubble. In these systems, the crop subsists on residual moisture or is irrigated two to three times. Soil moisture is conserved by mulching with rice straw.

Middle to High Latitudes (25° to 50°N)

Prevailing production systems and types of cultivars in this zone have been influenced by dominant factors such as temperature, rainfall, duration of frost-free periods and daylength. Countries in this zone; i.e. China, Korea, Japan and the northern part of India, have a yield potential of 3 t/ha.

The cultivars used in this region range in maturity from Group 00 to Group X, each with specificity in adaptation to latitude. The early-maturing cultivars are generally grown in the northern latitudes, with the late-maturing ones in the southern areas. In the northern provinces in China, indeterminate and daylength-sensitive cultivars are grown with many of them being green- and black-seeded, whereas in the lower latitudes the determinate, yellow- and green-seeded types predominate. The situation in Korea is more like in China where both indeterminate and determinate types are grown. In the hilly regions of northern India, the indigenous cultivars which are grown are long-maturing, viny, shattering and black-seeded. However, in the newly developed production areas in the northern plain, the introduced early-maturing and determinate types are cultivated. In Japan, most cultivars are determinate and large-seeded with excellent grain quality.

Cropping systems vary considerably. For example, in the northern provinces in China, soybean is spring-sown and grown as a full-season crop, in rotation with maize (*Zea mays* L.) and spring wheat (*Triticum aestivum* L.) In the northern plain, a major wheat producing area, soybean is grown in summer after winter wheat. In the southern region, where rainfall is abundant and temperature is not limiting, soybean crops can be grown in spring, summer or autumn in rotation with rice, winter wheat and barley (*Hordeum vulgare* L.), or interplanted with maize or sweet potato (*Ipomoea batatas* [L.] Lam.). The systems in use are similar to those employed in Taiwan. In Japan, where production is largely mechanized, soybean is usually intercropped or rotated with wheat or planted in rice paddies. In the northern plains of India, soybean has replaced maize and sorghum and is grown as a sole crop or intercropped with pigeon pea, cotton, sugarcane, sunflower (*Helianthus annus* L.) and maize.

In Korea, the cropping systems also vary. In the northern mountainous region, soybean is usually planted as a full-season crop, whereas in the northern to the central plain where much of the soybean is produced, it is intercropped with wheat and barley. In the southern plains the crop is cultivated following barley or intercropped with other upland crops like sorghum, corn, sweet potato and sesame (*Sesamum indicum* L.).

PRESENT AND FUTURE RESEARCH NEEDS

In view of the wide range of latitudes and cropping systems in which soybeans are grown in Asia, breeding and production research has to be intensified and systematized in accordance with the interrelated effects of variety, plant population density, soil fertility level, season of planting, rainfall, daylength and cropping systems used in a given location. In Japan and China, where soybean is important, research on cultivar development and production has been active over a long time. The richness in genetic diversity of germplasm in these countries, including that evolved through breeding, has become an invaluable source of material to all breeding programs around the world.

In other Asian countries organized research was begun only in the 1960s and consisted mainly in adapting introduced cultivars from the U.S. and high latitudes. Some of these American-bred cultivars have served as the starting point for commercial production of soybeans in India, Iran and the Philippines. Production research subsequently developed based on the culture of these introduced cultivars. In Indonesia and Thailand, where soybeans are traditionally grown, culture has relied on the use of Asian types, including those developed through breeding. Other countries like Pakistan, Bangladesh, Sri Lanka, Nepal and Malaysia are at a cultivar screening stage and production research is just getting organized.

Investigators and extension specialists in Asia have developed familiarity with the crop to enable them to identify production constraints, such as poor crop stands, pests, diseases, and nonsuitability of cultivars. A critical problem in the tropics relates to the rapid loss of seed viability in storage, which makes it difficult for farmers to establish a good stand of the crop. While improved packaging techniques of seed can mitigate the problem, the technology becomes inconsequential in countries where no commercial seed production is in operation. Thus, this problem remains one of the chief causes of low productivity in the region.

Another biologic constraint of regional significance is the seriousness of occurrence of certain pests and diseases, particularly pod borer, beanfly, yellow mosaic virus, rust and noxious upland and paddy weeds. Development of cultivar resistance has been achieved, but sustainment of resistance needs constant vigilance in research, owing to its transitory nature. Strain differences of the rust organism are also known to occur among countries, adding complications to the problem of identification of gene sources for resistance.

Photoperiodic response to short days in the tropics and the nonnodulating habits of introduced cultivars from high latitudes are problems that need to be addressed in cultivar development. Nodulation and rhizobial activities in the tropics are not well understood, especially under adverse conditions of soil acidity, element toxicity, low organic matter content of soil, high temperatures, puddled conditions of paddy fields and shading by perennial crop canopies. It has been observed that natural populations of *Rhizobia* thrive under extreme ecologic conditions in the tropics. Isolation of effective tropical strains and their commercialization need to be worked out. In the tropics, commercial inocula are not readily available, and artificial inoculation to improve yields is not widely practiced.

The unique practice in tropical Asia of growing soybean in cropping schemes, such as the rice-based system and intercropping with annual and perennial crops is beset with several problems. In the rice-based system of soybean production, several factors come into play: puddling of soil that is associated with rice culture, nontillage, reliance on residual moisture and fertility, high temperatures, and shifts in daylength during production. On the other hand, in intercropping, shading, nutrient competition, and allelopathy are important factors. Genetic advance through breeding can be realized if the mechanisms responsible for cultivar adaptation were identified. In the rice paddy system, the important selection indices for soybean seem to be fast rate of vegetative development during the seedling stage, less sensitivity to photoperiod, heavy seed weight and fast rate of grain filling. Larger yields can be realized with increased plant density and application of two or three flush irrigations. The important selection criteria for adaptation to intercropping are high pod number, high harvest index and resistance to foliar diseases. The effects of diseases are usually accentuated under reduced sunlight. Since yields in both the rice-based and intercropping systems are usually low, the economics of soybean production for the different cropping systems needs to be established in order to determine their relative competitiveness.

With the overall importance of soybean in Asia, there is a need to organize the available expertise into an inter-Asian network that can address common problems besetting soybean production in the region.

NOTES

Ricardo M. Lantican, Professor of Agronomy and Director, Institute of Plant Breeding, University of the Philippines at Los Baños, College, Laguna, Philippines.

PRODUCTION, CULTURAL PRACTICES AND UTILIZATION OF SOYBEAN IN EUROPE

R. Blanchet

In this paper, I will briefly examine some ecological and agricultural characteristics of Europe, present production and its evolution, the main cultural practices, some aspects of utilization, and finally the research problems which seem to be the principal keys of further developments of soybean production. I will not discuss problems of international economy and politics, although they certainly have an influence on soybean production and utilization in Europe.

ECOLOGICAL CHARACTERISTICS OF EUROPE

Temperature

Europe is situated mainly between 40N and the arctic polar circle. In spite of the beneficial influence of the oceanic Gulf Stream in the western part, it is not a warm continent. Large areas where average daily summer temperatures exceed 20 C for a month or more are only situated in the Soviet Union, in the eastern basin of the Danube (Bulgaria, Romania, Hungary), in Italy, Spain, and some parts of France. Lack of warm conditions almost precludes soybean culture in all of northwestern Europe, at least with present cultivars. Some cultivars have been bred for adaptation to low temperature and long day conditions, for instance the well-known Swedish cultivar Fiskeby, but their production is hardly competitive with other better adapted crops.

Water Availability

Aside from the question of rainfall distribution during the year, in large areas of the south of the Soviet Union, and in central and east Spain, where temperature is adequate for soybean production, average annual precipitation is less than 500 mm. In many other regions, it is less than 750 mm, and exceeds 1000 mm mainly on mountains and in the coastal areas of the northwest that are not suited for soybean production. Where temperature allows it, annual potential evapotranspiration is about 800 to 1200 mm, and maximum evapotranspiration of a soybean crop is about 500 to 800 mm, depending on situations and years. Available water, then, is generally a limiting factor. Only small areas, in the north of Italy (Po Valley), southwestern France, northwestern Spain and Portugal, and in some parts of Central Europe, have both adequate temperature and water. These are already well known for their fitness for maize production. In other regions with favorable temperatures, high soybean yields generally require irrigation. Otherwise yields are variable among years, from 1.5 to 2.5 t/ha for instance, and soybean frequently cannot compete with other crops. In the warmer regions, double-cropping is possible after winter crops, but generally requires irrigation.

Soils

Though deep chernozems cover large areas in the Ukraine and in some plains of the Danube, in southwestern Europe soils are frequently superficial, often calcareous, and even stony. The water-holding capacity is often between 50 and 100 mm; rainfall occurs mainly in winter, autumn or spring according to regions, but not in summer.

A few soils are strongly acid, and they have often been limed. Some chlorosis effects are possible on calcareous soils. Generally, the cropped soils are correctly fertilized in areas of intensive production. Mineral elements other than nitrogen seem rarely a limiting factor to soybean.

SOME GENERAL AGRICULTURAL CHARACTERISTICS OF SOUTHERN EUROPE

Where soybean can reasonably be grown, farming is generally intensive or semi-intensive, well mechanized, and chemicals (fertilizers, herbicides, pesticides) are used. Both plant and animal production are practiced, with specialization of some regions either on plants or on animals. Family farms, sometimes small, are dominant in the west, and large collective farms dominate in the east. The prices of inputs and products depend upon government policies. During the past decade cereals have been favored in the west, perhaps more than in the east, but everywhere soybean has to compete with other summer crops, mainly maize (*Zea mays* L.), sunflower (*Helianthus annuus* L.), and sometimes grain sorghum (*Sorghum bicolor* [L.] Moench), sugar beet (*Beta vulgaris* L.), etc., both in lands equipped or not for irrigation. Sprinkle irrigation is often dominant. Figure 1 gives an indication of the yields which can be reached for competitive crops with different water availabilities, and of their relative gross margins given

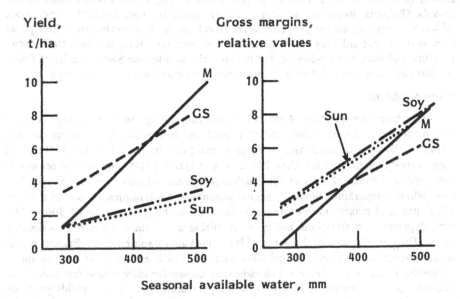

Figure 1. Yields and relative gross margins expected for summer field crops according to available water (E.E.C. prices). M = maize; GS = grain sorghum; Soy = soybean; Sun = sunflower.

present European Economic Community prices. Until recently, the ratio price of soybean:price of maize was about 2.0; it increased only to about 2.5 during recent years, but is greater in the east, from 2.5 to 4.0.

PRODUCTION

The Soviet Union has for a long time been the main producer, followed by Romania (Table 1). In 1973, only Bulgaria, Yugoslavia and Spain were significant soybean producers. Production increased greatly during the past decade in Romania, Bulgaria and Yugoslavia. In the Yugoslavian plain of Voïvodina high yields are obtained (about 2.5 t/ha in 1982). These three countries intend to extend and increase production.

Hungary, where the important work of mapping environmental zones and maturity groups has been completed, has started production. More recently, it also started in Italy (mainly in the Po Valley, here good yields are obtained), and to a lesser extent in France, Czechoslovakia, and Greece. Some small production occurs in other countries: Poland, East and West Germany, Austria, Switzerland, Portugal. All these countries have research programs. So, in general, production is not great in Europe, but it has increased quite significantly during the past ten years (Cejudo Fernandez, 1983).

The maturity groups used depend, of course, on latitude, temperature and cropping system, with dominance of the 0, I, and II groups. Both American and national cultivars are grown; Merit, Chippewa, Amsoy and Amsoy 71 are well known and have been used in the past, as is true of Maple Arrow and Evans in the North and Williams and Calland in the South. Indeterminate types are quite generally preferred.

MAIN CULTURAL PRACTICES

Of course, practices differ according to climatic, agronomic and farming conditions, but some general characteristics can be outlined: Planting is generally made in April to May, with a tendency to plant before maize, in row spacings varying from about 35 to 60 cm. Plant density depends on climate and irrigation: 50-60 plants/m^2 for Maturity Group 00 or in double-cropping, about 40 plants/m^2 under rainfed conditions for Maturity Groups I and II, and a little less with irrigation in order to prevent lodging.

Table 1. Estimates of soybean production and its change in 10 years

	Approximate area in production (1000 ha)		Approximate production 1983 (1000 tons)
	1973	1983	
Soviet Union	840	800	600
Romania	180	450	550
Bulgaria	20	100	150
Yugoslavia	10	110	250
Italy	-	30	80
Hungary	-	20	30
France	-	12	30
Spain	10	10	15
Czechoslovakia	-	5	5
Greece	-	3	5

When a choice of maturity group is possible, this choice is often made according to expected date of harvest, allowing for good conditions for planting winter wheat (*Triticum aestivum* L.) after soybean. Optimum harvesting date varies from September in the east to October in the west, where winter comes later.

Inoculation is practiced. Originally, *Rhizobium japonicum* was not present in soils. Various countries have developed inoculum manufacturing, often with peat as a base material. Strains are selected for their efficiency and competitiveness, with a preference for efficient and not too competitive strains, in order to retain the option of changing them in the future.

Weed control is generally ensured by chemicals, no or little cultivation being done. Specific chemicals and application methods depend on weed species, which are quite different from east to west, and even from north to south. Trifluralin is becoming more popular.

Till now, due to the relatively small areas cropped, diseases have not been an important factor. Those occurring in similar climates of North America are also known in Europe, except *Phytophtora megasperma*, but the attacks seem not too damaging. The more common are the various viruses. In the South, red and yellow mites (*Tetranychus atlanticus*) do some damage in warm summers; generally no treatment is applied.

In areas equipped for irrigation this is an important consideration, as already mentioned. There have been studies involving irrigation for a long time. Table 2 illustrates typical responses. Practical approaches are now known: except in case of very severe drought, it is not necessary to irrigate until the beginning of flowering, but it is important to prevent water stress during pod setting and up until the beginning of grain filling. Equipment used differs greatly among countries and farming systems.

Harvest is made by combine. Potential losses are one of the reasons for using higher plant densities for early cultivars, so that pods do not set too low on the stems. Under irrigation, lodging can be a problem. High plant densities are more likely to lodge, but on the contrary too low a density can also result in losses from branch breakage. Optimum plant density depends on conditions, and even on cultivar used. As far as possible, harvest is made when the seeds are at about 16% moisture. Crop residues are either plowed under or left on the surface, as for example when winter wheat is planted soon after with a no-tillage procedure.

UTILIZATION

Meat consumption by Europeans has increased, whereas direct consumption of plant proteins has decreased during the last decade (Table 3). This modification in food

Table 2. Influence on yield of irrigation amounts and periods at Szarvas, Hungary, Maturity Groups 0 and I (from Posgay, 1976).

Treatments	Water applied[a]				Total	Yield
	May	June	July	August	mm	t/ha
a	60	60	60	70	277	1.80
b	60	60	70	70	298	2.00
c	60	60	80	70	316	2.52
d	60	70	80	70	325	2.68
e	60	80	80	70	330	2.72
f	Potential evapotranspiration				480	2.43
	Significant difference, 5%					0.186

[a]As percent of evaporation from a "Class A" pan for the respective months.

Table 3. Evolution of protein and lipid consumption in several European countries, g/person d.

Countries	Total proteins, plants and animals	Plant proteins	Animal proteins	Lipids
West Germany				
1955-59	83.1	38.1	45.0	143.0
1969-70	86.0	32.0	54.0	153.0
1976-77	88.0	30.0	58.0	155.0
United Kingdom				
1955-59	89.8	37.7	52.1	132.0
1969-70	93.9	36.7	57.2	148.0
1976-77	91.7	35.7	56.0	139.0
Italy				
1955-59	81.2	53.2	28.0	
1969-71	96.0	56.0	40.0	120.0
1976-78	98.1	53.4	44.7	123.0
Spain				
1955-59	74.4	52.2	22.2	73.4
1969-70	82.0	43.0	40.0	100.0
1976-77	93.5	44.9	48.6	127.0
Europe				
1955-59	90.0	52.0	38.0	
1969-70	92.5	44.4	48.1	126.0
1976-77	96.0	43.2	52.8	134.0

habits has stimulated a great need for protein for animal feeding, and there have been large imports of soybean seeds and meal from the USA and South America (Knipscheer, 1980). This way of consumption can be considered as a sort of wastage, but it is more or less a general fact in Europe. Some countries, for instance Yugoslavia, have developed an industry for soybean protein extraction for direct human utilization. This also occurs in many other countries, but only covers a small part of the human protein consumption.

Lipid consumption also has increased, but not to so great an extent, and the oil sources are more diversified (Table 4). Though soybean oil is one of the most used, sunflower, rapeseed, cottonseed, palm, olive, and groundnut oils are also of common usage, depending on local production and industries. Soybean oil and meal uses are not equivalent, requiring large imports of meal, especially by the E.E.C. where soybean meal represents about 75% of the total meal consumption (about 65% for all of Europe). There have been attempts to develop other plant protein sources, for example *Pisum sativum, Vicia faba* (see Table 5 for data for France), and to increase the oil and protein production of rapeseed, sunflower, etc. But until now these various oil-protein crops have provided only a rather small part of the protein requirement for intensive husbandry. Where significant soybean production exists in southeastern Europe, it is often locally processed by large Kombinats[1] into oil and meal, for pig, poultry and milk production. In the future, perhaps the extruding process will allow direct utilization of seeds by animals, even in smaller units.

Table 4. Estimates of annual European consumption of various oils and meals in 1000 metric tons (Oil World Sources, 1983).

Products		E.E.C.	Other W. Europe	East Europe	U.S.S.R.
Soybean	- oil	1,510	350	490	370
	- meal	15,700	3,500	4,200	4,000
Sunflower	- oil	610	430	550	1,800
	- meal	1,500	560	1,100	1,600
Rapeseed	oil	780	110	400	40
	- meal	1,700	300	600	50
Groundnut	- oil	340	40	-	-
	- meal	370	60	270	30
Cottonseed	- oil	30	20	-	700
	- meal	700	70	50	1,850
Palm	- oil	640	70	-	300
	- meal	650	100	-	-
Olive oil		950	470	-	-

MAIN RESEARCH PROBLEMS

Some efforts, concerning general plant productivity, are common to about all regions: canopy structure and light penetration, photosynthesis, translocation, relationships to pod and grain setting and filling, and all the general processes of growth, development and yield elaboration, with their physiological and genetic implications (Ecochard et al., 1979). But, some aspects have peculiar significance to Europe, taking into account its ecological and agricultural characteristics.

Adaptation to Cool Temperatures and Long Days

This problem concerns primarily all the large areas situated at about 50°N (USSR, Poland, Czechoslovakia, E. and W. Germany, parts of France), and also countries whose climate is influenced by mountains (Switzerland, Austria, etc.) (Soldati et al., 1983). In such areas, there is a question of whether there could be sufficient and regular productivity from 00 or 000 Maturity Group types. Those countries have breeding programs, sometimes using mutagenesis, trying to adapt the soybean to those rather severe conditions.

Table 5. Oil protein crop hectarage in France, 1982.

Crop		Hectares
Pea	- Winter	2,500
	- Spring	93,000
Field bean (*Vivia faba minor*)	- Winter	21,000
	- Spring	12,000
Rapeseed	- Winter	460,000
	- Spring	15,000
Sunflower		280,000
Soybean		10,000

In southern areas, from north of Spain to Romania, there is more a question of flexibility to variations of temperature for photosynthesis, translocation and pod and grain setting. Some physiological studies are on-going on these topics, and breeders are trying to improve flexibility.

Drought Tolerance and Water Use Efficiency

As mentioned previously, this is a general problem in Europe, even under irrigated conditions. Many aspects are involved: root penetration in subsoil (Taylor, 1979), moderate transpiration and leaf area index, the process of water transport in the plant and control of gas exchanges, fructification and ability to recover after stress. Agronomic and physiological researches are on-going on these subjects, and plant breeders are giving special attention to drought tolerance. Screening tests are under development, though these are not easy to apply because water relationships are complex (Sammons et al., 1978; 1979; Vidal et al., 1981).

Nitrogen Nutrition and Interactions with Water Status

If the *Rhizobium*-soybean symbiosis and nitrogen nutrition are general problems, their importance is magnified in sub-humid areas; the sensitivity of root nodules to water stress results in a large decrease in nitrogen fixation when water becomes limiting in the upper soil layers. A double stress then arises: water stress and nitrogen stress, with important consequences to photosynthesis and fructification (Streeter et al., 1979; Bouniols et al., 1982; Obaton et al., 1982). Practical ways to improve nitrogen nutrition under such conditions are not yet clear.

Other Problems

Research on cropping systems for the various ecological and agricultural conditions leading to specific cultural practices is important. Until now plant protection against pests has not been a problem, but it could become so if larger areas are cropped. Weed control by herbicides is not yet satisfactory, and obviously differs according to soil, climate and weed species.

CONCLUSIONS

Despite a large soybean requirement, especially for meal, European soybean production is low although increasing significantly. The difficulties in production are quite real, and are mainly due to ecological conditions, temperature and rainfall. Progress in adaptation, through both short- and long-term research, without neglecting other significant plant characters, is needed. Then Europe can take advantage of the general progress on soybean in the whole world (Kauffman, 1983; Leffel, 1983), and certainly other countries and regions will benefit from the knowledge and experience gained in Europe, often under difficult conditions.

NOTES

R. Blanchet, Institut National de la Recherche Agronomique (INRA), Centre de Recherches de Toulouse, Station d'Agronomie, B.P. 12, 31320—Castanet-Tolosan—France.

I thank Messrs A. Vidal and A. Merrien for their help in the preparation and review of this paper.

[1]"Kombinats" are large collective institutions of eastern Europe, sometimes agro-industrial, that process the main agricultural products.

REFERENCES

Bouniols, A., J. Puech and M. Mondies. 1982. Influence d'un déficit hydrique appliqué durant la fructification sur la production du soja. Bull. Gr. Fr. Humid. Neutron. 11:39-56.

Cejudo-Fernandez, J. 1983. Futuro de la soja en Europa. Eurosoya 1:7-8.

Ecochard, R., M. Gallais, M. H. Paul and C. Planchon. 1979. Héritabilité et réponse à la sélection de caractéres physiologiques liés au rendement chez le soja. Ann. Am Pl. 29:493-514.

Kauffman, H. E. 1983. Potential for global collaboration in soybean research and development. Eurosoya 1:35-40.

Knipscheer, H. C. 1980. Demand for soybeans and soybean meal in the European Common Market. p, 807-822. In F. T. Corbin (ed.) World soybean research conference II: Proceedings. Westview Press, Boulder, CO.

Leffel, R. C. 1983. The U.S. soybean improvement program in transition. Eurosoya 1:17-20.

Obaton, M., M. Miquel, P. Robin, G. Conejero, A. M. Domenach, and R. Bardin. 1982. Influence du déficit hydrique sur l'activité nitrate réductase et nitrogénase chez le soja (Glycine max L. Merr. cv Hodgson). C.R. Acad. Sc. (Paris) 294:1007-1012.

Posgay, E. 1976. Results of irrigated soybean growing trials. Növeny-termeles 25:359-368.

Sammons, D. J., D. B. Peters and T. Hymowitz. 1978. Screening soybeans for drought resistance, I. Growth chamber procedure. Crop Sci. 18:1050-1055.

Sammons, D. J., D. B. Peters and T. Hymowitz. 1979. Screening soybeans for drought resistance. II. Drought box procedure. Crop Sci. 19:719-722.

Soldati, A., E. R. Keller, H. Brenner and J. Schmid. 1983. Adaptation of soybeans to northern regions: a review of research work conducted at the Swiss Federal Institute of Technology. Eurosoya 1:27-34.

Streeter, J. G., H. J. Mederski and R. A Ahmad. 1979. Coupling between photosynthesis and nitrogen fixation. p. 129-138. In F. T. Corbin (ed.) World soybean research conference II: Proceedings. Westview Press, Boulder, CO.

Taylor, H. M. 1979. Postponement of severe water stress in soybeans by rooting modifications: a progress report. p. 161-178. In F. T. Corbin (ed.) World soybean research conference II: Proceedings. Westview Press, Boulder, CO.

Vidal, A., D. Arnaudo and M. Arnoux. 1981. La résistance à la sécheresse du soja. Agronomie 1: 295-313.

HISTORY AND DEVELOPMENT OF SOYBEAN PRODUCTION
IN SOUTH AMERICA

Warney M. C. Val

Only during the past 20 years has soybean production gained importance in South America. Brazil, Argentina and Paraguay are responsible for the greatest increase (Table 1). In the beginning, the increased production did not result from replacement of any other crop, since soybean was grown in new areas or was double-cropped with wheat (*Triticum aestivum* L.). Nevertheless, in recent years soybean is competing for space with traditional crops.

The major soybean production area in South America is located between 15° and 33° south latitude, and the major producing countries are Brazil, Argentina, Paraguay, Uruguay, Bolivia and Chile, in that order. In all these countries there exists the possibility of increasing the hectarage of soybean, since new land is available. However, the infrastructure needs some improvement.

Most cultivars that were introduced into South America originated in the United States. Only some of them were adapted to tropical conditions. New cultural practices had to be developed by research institutions in order to obtain better yields. Brazil is leading in soybean research in South America. The cultivars developed there are well adapted to different regions, and are successfully utilized by neighboring countries.

The IICA-CONE SUL/BID agreement has played an important role in exchange of germplasm and technology among the signatory countries. Several approaches, such as technical meetings, courses, seminars and technical consultations, have been organized to discuss and solve practical and technical problems.

Table 1. Area, yield and production of soybeans, 1982-83 (USDA, 1984).

	Area	Yield	Prod
	1000 ha	kg/ha	1000 t
United States	28,256	2,147	60,677
China	8,414	1,073	9,030
Brazil	8,227	1,793	14,750
Argentina	2,116	1,687	3,570
Paraguay	350	1,429	500
Others	4,983	1,100	5,480
Total	52,346	1,796 (ave.)	94,007

REGIONAL PRODUCTION

Argentina

During the last three years soybean production has been about four million metric tons produced on two million hectares. Ninety percent of this production is concentrated in the "Pampeana" region, in the central part of the country, including the provinces of Buenos Aires, Cordoba, Santa Fé and Entre Rios. The remaining 10% is produced in the North. As soybeans already occupy the majority of the best area for the crop, any further expansion will take place in drier regions. The expected lower yield in these areas, along with the uncertainty of future demand and the world price, possibly will slow down the rate of soybean expansion in Argentina.

Most of the production is for export. In 1982 there was a significant increase in the domestic use of soybeans and an 18% reduction in exports. The production of soybean meal increased 26% and exports 80%, at the same time oil production increased 12% and exports declined 24%. There is a substantial increase in the domestic consumption of soybean oil (Programa Cooperativo de Investigacion Agricola, 1983).

The future of this crop in Argentina is promising mainly because of its ability to adapt to distinct soil and climatic conditions, its ability to fix nitrogen and certain advantages in relation to other crops. A limiting factor is the price, because after 1981 in the "Pampeana" region corn started competing with soybean.

Soybean research in Argentina is coordinated by the Instituto Nacional de Tecnologia Agropecuaria (INTA), centered at the Experiment Station at Marcos Juarez. Research also is being conducted by 73 research scientists in 22 experimental units throughout the production regions, which are situated mainly within 30° and 35° south latitude. The research program is directed to four major fields: plant breeding, crop protection, crop management, and economics. Also, some work is carried out in soil science, genetics, entomology, and plant pathology. Technology transfer is accomplished by the INTA Extension Service and other private and official institutions.

The Japanese government, through a cooperative agreement established in 1978, has been providing technical assistance as well as field and laboratory equipment.

Other official institutions, such as universities, state institutions and private seed companies, are active in research. Development of new cultivars showing greater yield potential, resistance to lodging, better seed quality, and resistance to insects and diseases such as viruses and *Sclerotinia sclerotiorum*, are the main objectives of the breeding program.

Bolivia

The soybean was introduced into Bolivia in the region of Santa Cruz de la Sierra about 1960. Initially, however, research on this crop was insignificant, mainly because of low prices and marketing difficulties. Later, the Experiment Station of "los Llanos Saavedra" initiated a research program on soybean, and production areas developed in the states of Tarija and Santa Cruz.

Lack of a national policy of good prices, and marketing conditions, contributed to the slow expansion of the crop. Today, with the establishment of a new processing plant for vegetable oil in Tarija, there is a better chance for opening up new soybean producing areas.

Up to 1977, the cultivars used were Acadian, Pelicano and Halesoy 71, with an average yield of 1,500 kg/ha. The introduction of new cultivars as Rillito, Davis, Bragg, Williams and Bossier, from the U.S., Colombia, and Argentina has improved yields. In 1978, the Brazilian cultivar UFV-1 was introduced in the "Gran Chaco" region with good results. However, Bossier from the U.S. gave the best results, yielding around 2,500 kg/ha in the southern part of "Gran Chaco".

The Instituto Boliviano de Tecnologia Agropecuaria (IBTA) and Centro de Investigaciones en Agricultura Tropical (CIA) are in charge of the research work, and are located in the states of Tarija and Santa Cruz, respectively, the major soybean producing areas in the country. Other national institutes, universities, and private enterprises, as well as foreign organizations, such as the Chinese Agricultural Mission and USAID (Chemonics), are involved. IICA-CONE SUL/BID cooperates in training, consulting and technical exchange (Programa Cooperativo de Investigacion Agricola, 1984). Only eight full-time research scientists are engaged in the soybean program. A few others are involved part time.

Bolivia still faces serious difficulties in establishing an efficient seed program. Fifty percent of the needed seed is imported from Brazil, despite great effort by the local seed industry. Around 150 cultivars and 80 advanced lines were introduced from Brazil, Puerto Rico, Argentina and Asia.

Brazil

The crop was introduced in 1914 in the southernmost state, Rio Grande do Sul, but the first commercial fields were not established until 1931. It was only during the sixties that the crop became economically feasible and expanded through the states of Paraná and São Paulo. During the seventies production increased from 1.5 to 15 million tons. This increase was mainly due to expansion of cultivated area, which increased to about 8.7 million hectares. At the same time, yield improved by 571 kg/ha (Empresa Brasileira de Pesquisa Agropecuária, 1981).

Rapid expansion of the crop was associated with its high profitability, along with the possibility of double-cropping with wheat. Other factors that promoted soybean production in Brazil were subsidized credit, mechanization, active participation of cooperatives, and the necessary support from research and extension services.

Since 1978, the cultivated area and production have stabilized at around 8.5 million hectares and 15 million tons. For the 1983-84 season, the estimated area planted to soybean was around 9 million hectares. The increase was mainly due to the reduction in U.S. production, with the subsequent price reaction in international markets (Fundação Instituto Brasileiro de Geografia e Estatistica, 1983).

According to Bonato (1981), soybean in Brazil is grown in three regions, which are distinguished as traditional, expanding and potential regions, according to the size of the cultivated area and the level of technology applied.

Traditional Region. This region comprises the southern states: Rio Grande do Sul, Santa Catarina, Paraná and São Paulo. The initial development of the crop was accomplished by importing technology from the U.S. south. Research played a major role in the improvement of yield, as well as in area expansion.

In 1975 the traditional region was responsible for 90% of the national production (9.5 million tons); today it represents 72.8% (11.65 million tons). Average yield of 1,718 kg/ha is considered low and could be increased substantially if available technology were fully applied.

Expanding Region. This area is located in the central part of the country ("Cerrado") including the states of Mato Grosso do Sul, Mato Grosso, Goiás, Maranhão, Minas Gerais and Bahia. With 2.5 million hectares and a production of 4.5 million tons, this region is responsible for 27.2% of the national production. The cultivated area has increased 10.4% since 1975, and yield has been raised by 439 kg/ha.

All the technology utilized in this part of the country has been developed in Brazil, based on experience with the crop in the south. This technology has resulted in an average yield of 1,818 kg/ha; however, if better soil management is applied, yields may be improved.

   Potential Region. Development of new cultivars adapted to low latitudes (below 10°) allowed the introduction and expansion of the crop into the northern and northeastern states. However, the economic, social and climatic conditions may limit the expansion of the crop in this region.

   Research in Brazil is coordinated by the National Soybean Research Center (CNPS), belonging to the Brazilian Agricultural Research Enterprise (EMBRAPA). This program is carried out in 15 states, two territories and the Federal District, and experiments are conducted by state institutes and enterprises, research centers, universities, private companies and cooperatives. Research areas considered by the National Research Program to be important are: (a) the development of cultivars adapted to low latitudes; (b) soil management involving conservation, tillage practices and organic matter in the soil; (c) plant nutrition; (d) crop management: weed control, soil and water relationship, sequential cropping and management in different cropping systems; (e) diseases and pests: insects, fungi, bacteria, viruses and nematodes; (f) seed production and technology: germination and vigor, cultivar purity, drying systems; (g) mechanization: improvement of machinery for planting, cultural practices and harvest, as well as adaptation of research equipment (Bonato, 1981).

Chile

   Soybean was first introduced in the '30s; however, the area cultivated was insignificant (Compradora de Maravilla, 1968). Traditionally, the country is not selfsufficient in the production of edible vegetable oil, even though the government and private industries are attempting to stimulate growing of the crop by establishing better oil prices than the international market. Nevertheless, soybean is not competitive with other oil crops, such as sunflower, which has an oil price 30% above that of soybean oil.

   Even though soybean has not been economically attractive to the domestic oil industries, "COMARSA—Compradora de girassol" (an organization of several oil industries) has attempted to stimulate soybean production since 1966. Thus, during the period of 1970-75 about 1,150 ha were grown annually with an average yield of 1,200 kg/ha. Later, COMARSA was dissolved and the individual oil industries took charge of promotion, and The Sugar Refining Company of Viña del Mar S.A. (GRAVAL) started a program for soybean utilization in diets of children in 1974.

   Up to 1981, COMARSA was the only promoter of this crop, providing technical and other necessary support to the producers. Soybean was grown in rotation with corn in a small irrigated area situated at 34° south latitude. The major cultivar used was Amsoy, and the level of mechanization was very high. This program was not successful at all, and today soybean oil is imported at lower prices from the international market.

   In spite of this situation, some research work was carried out by the Instituto de Investigaciones Agropecuarias (INIA) during 1965-80. The last research projects conducted in Chile were those done under international agreements, as the one with the University of Minnesota. There are very fine research scientists working on soybean part time.

   According to some experimental results and the experience of commercial producers, soybean could be profitably grown in the vast irrigated area in the central part of the country (Valdivia, 1981). Almost 100% of the edible oil is imported, and the possibility of success depends upon competition between national and international prices. Any new attempt to promote this crop will depend on the establishment of a safe and solid national market, as well as on the interest of the processing industry.

Paraguay

Soybean was introduced into the country in 1921, but real expansion occurred only in the 1960s. After 1968, demand and government incentives greatly stimulated soybean expansion and commercialization (Programa Cooperativo de Investigacion Agricola, 1983). All the production initially was processed by the local oil industries. After 1967, the country started to export grain due to the growing demand in the international market.

There was a large expansion in cultivated area between 1961 and 1981, mainly after 1965. By 1982 soybean was already the number one crop in the country with an area of 600,000 ha and a production of 800,000 tons. Over the last 4 years, cultivated area has stabilized at around 600,000 ha, and yield has remained around 1,330 kg/ha.

The major soybean area in Paraguay is located close to the Paraná river, in the localities of Itapuã, Alto Paraná, Canendiyu and Amambay, which are responsible for 48%, 19%, 9% and 6% of the total production, respectively. These areas have great potential for future expansion. The remaining 18% is distributed in the north and central parts of the country, where the need for basic inputs, and competition from traditional crops, are limiting factors.

Seventy percent of the production is exported as grain and the remaining 30% is processed internally, but some of the by-products are also exported. Estimates for this year indicate that approximately 650,000 ha will be planted to soybeans, and the expected production is around 900,000 tons. There is a good potential for the expansion of this crop in Paraguay, since the soil and climatic conditions are adequate and manpower is not a limiting factor. The region near the Paraná river is best suited for soybean. There is no competition with traditional crops, such as cotton (*Gossypium* spp.) and tobacco (*Nicotiona tobacum* L.); besides, there is the possibility of crop rotation with wheat and oats (*Avena sativa* L.).

Research on the soybean was initiated in 1970, and since that time work has been done to solve soybean production problems throughout the country. The major objectives of the program were to develop higher yielding cultivars and to study the different yield components for the establishment of better crop management systems. The Ministry of Agriculture and Dairy (MAG), through the Direccion de Investigaciones y Extension Agropecuaria y Florestal (DIEAF), is responsible for this work in Paraguay.

Experiments were conducted in the Instituto Agronomico Nacional (IAN) located in Caacupé and in the Centro Regional de Investigacion Agricola (CRIA) located in Capitão Miranda, as well as in some production areas in the country. By 1980, research was strengthened through specific projects, as the soybean project belonging to the Cooperative Project of Agricultural Research, IICA-CONE SUL/BID. This project, along with the Ministry of Agriculture, organized specialized courses on soybean production and technology transfer for training agronomists (Programa Cooperativo de Investigacion Agricola, 1984).

Also in 1980, the Integrated Program for Paraguay (PIDAP II) was initiated, which aided in technology transfer. This was considered of high priority because of its social and economic importance. This program provides funds for hiring and training of personnel, purchase of equipment and scientific and technical literature, and for building facilities at the experimental centers.

Uruguay

In 1973, there were only 6,000 ha sown to soybean in the country. This area increased rapidly, reaching 50,000 ha, where it remains today (Programa Cooperativo de Investigacion Agricola, 1983). As regards soybean production, the country could be divided into two regions: the coastal and north-northwestern regions. The coastal region is the traditional agricultural zone, producing wheat, sunflower (*Helianthus annuus* L.) and sorgum (*Sorghum bicolor* [L.] Moench). Soils are fertile but weeds are a problem. Double-cropping is common after wheat, and the growers have a choice of many crops. The farmers do not accept new technologies with ease, and low yields are mainly due to lack of adequate technology, weeds, late sowing and soils with a low water-holding capacity.

In the north-northwestern region, a relatively new agricultural area, rice (*Oryza sativa* L.) is the major crop. Soybean was introduced without replacing any other crop. Improvements in soil conservation and management, as well as weed control, are necessary for increasing soybean production in this region.

Soybean yield in Uruguay averages 1,500 kg/ha and varies from 1,000 to 2,000 kg/ha, depending upon the technology applied. Commonly grown cultivars are Bragg and Paraná. Research work on soybean in Uruguay started in 1970 with the evaluation of the newly introduced cultivars. In 1973 a soybean research program was established by the Agricultural Research Center "Alberto Boerger" (CIAAB), at the East, North and Estanzuela Experiment Stations. In addition to CIAAB, universities, cooperatives and official institutes are involved in soybean research. Ten research scientists are involved, but only two are full-time with soybean.

Some work on the sequential cropping of soybean and rice on heavy soils has been conducted at the East experimental station. Most of the experimental results are very similar to those obtained in Brazil, and due to this fact, the majority of soybean cultivars grown in Uruguay between the latitudes 15° and 20° south are imported from Brazil: UFV-1, Paraná, Planalto and IAS-5. Consequently, all the research results available in the south and southeastern regions of Brazil are used in Uruguay. In addition to this, Uruguay utilizes the Port of Paranagua, Brazil to export its products.

NOTES

Warney M. C. Val, EMBRAPA/CNPS, C.P. 1061, 86 100 Londrina, PR, Brazil.

REFERENCES

Bonato, E. R. 1981. Programa Nacional de Pesquisa de Soja. p. 765-793. *In* Seminário Nacional de Pesquisa de Soja, Vol. 1. EMBRAPA-CNPS, Londrina.

Compradora de Maravilla S.A. (COMARSA). 1968. Reseña historica de las oleaginosas en Chile. (Boletin Tecnico, 22). COMARSA, Santiago.

Empresa Brasileira de Pesquisa Agropecuária. 1981. Programa Nacional de Pesquisa de Soja. EMBRAPA-DID, Brasília.

Fundação Instituto Brasileiro de Geografia e Estatistica. 1983. Anuário estatístico do Brasil. FIBGE, Rio de Janeiro.

Programa Cooperativo de Investigacion Agricola. 1983. Cono Sur/BID. Plan indicativo; anexos (proyectos). PCIA, Montevideo.

Programa Cooperativo de Investigacion Agricola. 1984. Convenio IICA-Cono Sur/BID. Informe final. PCIA, Montevideo.

U.S. Department of Agriculture. 1984. World oilseed situation and U.S. export opportunities. Foreign Agriculture Circular. U.S.D.A., Washington.

Valdivia, V. A. 1981. Investigacion en soya en el periodo 1970-1980: Perspectivas del cultivo en Chile. INIA, Santiago.

International Programs and Networks

THE FOOD AND AGRICULTURE ORGANIZATION'S SOYBEAN PROGRAM

H. A. Al-Jibouri

Soybean is grown in all continents including the mediterranean region, and its cultivation has extended to the sub-tropical and tropical zones. Cultivated areas and production have doubled during the last decade, and yields per unit area have also increased significantly as shown in Table 1. Over 52 million hectares were cultivated with soybean in 1982 producing about 93 million tons. About 28 million tons are traded annually with a value of about 7 billion U.S. dollars (Table 2).

COUNTRY PROJECTS

Soybean is being introduced and tested in several hundreds of FAO-operated field projects on food crops in Africa, Asia, Europe, and Latin America. Under the auspices of the United Nations Development Program (UNDP) and recently under FAO's Technical Cooperation Program (TCP), the organization has assisted governments of several countries through small, medium and large scale projects for the development of soybean. (In 1000$ as follows: China, 500; Mozambique, 750; Sri Lanka, 2,500; Vietnam, 260; Zambia, 300.)

CONSULTANCIES

Under FAO's Regular Program about 30 short-term consultancies have been commissioned to China, Guinea, Indonesia, Iraq, Malaysia, Mozambique, Pakistan, Poland,

Table 1. World soybean area, yield and production (FAO, 1981, 1982a).

| | 1969-1971 | | | 1982 | | |
	Area	Yield	Production	Area	Yield	Production
	1000 ha	kg/ha	1000 mt	1000 ha	kg/ha	1000 mt
World	29,247	1,487	43,487	52,463	1,772	92,982
Africa	193	413	80	400	1,094	437
North America	17,308	1,831	31,684	29,402	2,160	63,502
South America	1,438	1,218	1,751	10,736	1,643	17,635
Asia	9,334	999	9,329	10,450	964	10,080
Europe	108	1,082	117	546	1,444	788
Oceania	5	1,111	5	46	1,717	79
USSR	860	606	521	876	525	460

Table 2. World soybean imports and exports (FAO, 1982b).

	Imports	Value	Exports	Value
	mt	1000$	mt	1000$
World	28,238,501	7,306,016	28,916,013	7,025,615
Africa	95,712	30,970	727	319
North America	1,091,245	290,559	25,652,256	6,275,172
South America	1,437,352	358,143	2,878,358	643,055
Asia	7,239,459	1,909,228	1,755,103	51,661
Europe	16,858,105	4,251,730	210,296	55,727
Oceania	10,621	3,317	50	22
USSR	1,506,097	382,069	--	--

Philippines, Sudan, Thailand, Vietnam, Romania, Bolivia, Colombia, Ecuador, and Peru, where specialist advice was urgently required to resolve technical problems, development of programs, appraisal of potential and/or formulation of suitable projects for soybean production and improvement.

REGIONAL NETWORKS AND PROJECTS

One of FAO's aims is to foster regional and sub-regional cooperation in research on the development of basic food crops, which includes soybean, with the aim of eventually working under the auspices of an institution or an agency to be estbalished for this purpose.

In 1976 FAO established a European Cooperative Network on Soybean (see paper by Arnoux, this symposium) in which 14 countries participate in a program to develop regional cooperation among European national scientific and technical institutes and universities. The purpose is to promote voluntary exchange of information and data concerning experiments on various aspects of the production of soybean.

Five sub-regional coordination programs on food legumes, including soybean, have been established in Latin America recently. One covers the Andean countries, the second is concerned with Rio de La Plata Basin countries, the third with Central America, the fourth with the Caribbean Spanish-speaking countries, and the fifth with the Caribbean English-speaking countries.

A regional project entitled "Technical Cooperation Among Developing Countries for Research and Development of Food Legumes in the Tropics and Sub-tropics of Asia", costing about 1 million dollars and involving 22 national institutes in 14 Asian countries, is presently in operation. The soybean is one of the crops included in that project whose headquarters is situated in Bogor, Indonesia.

SEED EXCHANGE

Soybean materials have been obtained from the International Soybean Program (INTSOY), The International Institute for Tropical Agriculture (IITA), The Asian Vegetable Research and Development Centre (AVRDC), The People's Republic of China and Thailand, and have been distributed to research technicians through the FAO Seed Exchange Laboratory. In 1982, 4424 seed samples were distributed to 14 countries.

GENETIC RESOURCE ACTIVITIES

The International Board for Plant Genetic Resources (IBPGR), the secretariat of which is located at FAO headquarters, is supporting national genetic resource programs in several countries, and encourages as far as possible regional and inter-regional cooperation in these activities. Missions to collect cultivated and annual and perennial wild species of soybean have been undertaken and additional missions are planned for the future. The IBPGR provided a small grant to CSIRO, Australia for multiplication and distribution of perennial *Glycine* species. With financial support from the IBPGR a total of 61 land races were collected in Indonesia. A few samples were also collected in Thailand, Zambia and Zimbabwe during multicrop collecting missions in 1982. A list of soybean descriptors was agreed upon in 1982. In collaboration with INTSOY, the IBPGR secretariat initiated the inventory and documentation of world soybean germplasm collections in 1983.

BIOLOGICAL NITROGEN FIXATION

The present aim of the FAO program on Biological Nitrogen Fixation (BNF) is to coordinate ongoing field projects with current research programs in order to promote efficient and widespread use of nitrogen fixation in current farming practices. These objectives include:
- Practical demonstrations of inoculation of legumes in the field.
- Improvement of traditional legume crops by the use of selected strains of Rhizobia.
- Introduction of new legume crops and then specific inoculants into developing countries where the lack of Rhizobium in the soil has previously precluded legume cultivation.
- Improvement of soil fertility by the use of suitable legume crops.

FAO assisted in the setting up of microbiology laboratories to facilitate work on BNF with respect to soybean in Burma, Mozambique, Sri Lanka and Vietnam. Three recent technical publications have been devoted to BNF.
1. Technical Handbook on Symbiotic Nitrogen Fixation Legume/Rhizobium. FAO, 1983.
2. World Catalogue on Rhizobium Collections. FAO/UNESCO, 1983.
3. Legume Inoculants and Their Use. FAO, 1984.

A number of national and international institutions are cooperating with FAO in the program on BNF, such as the Instituto Nacional de Technologia Agropecuaria (INTA), Argentina; Institut National de la Recherche Agronomique (INRA), France; Commonwealth Scientific and Industrial Research Organization (CSIRO), Australia; University of Queensland, Brisbane, Australia; the project on Nitrogen Fixation in Tropical Agricultural Legumes (NIFTAL), University of Hawaii, USA; Microbiological Resource Centres (MIRCENS) in Porto Alegre, Brazil; and the following international Centres: CIAT, ICRISAT, and IITA.

TRAINING

Another FAO aim is to aid in manpower development through study tours, visiting scholar programs and fellowships for training at the graduate level, in order to build the theoretical base necessary for top quality research. Through the organization of suitable training courses for agronomists and extensionists FAO helps develop national competences in solving problems at the farm level. FAO is collaborating with several institutes in organizing national and regional training courses on soybean production and improvement. Two courses in Cuba and one in Panama were organized in 1981, 1982 and 1983. A national soybean production course for Turkey is planned in 1984. A regional training course for Latin American countries was organized in Palmira, Colombia in November, 1982.

WORKSHOPS

The objective of workshops is to promote and intensify regional cooperation, to exploit the genetic potential of the soybean, and explore the way in which international organizations and programs can help countries accelerate soybean production and improvement. FAO has collaborated with INTSOY and ICA (Instituto Colombiano Agropécuario) in the organizing of a workshop for soybean breeders/agronomists in Latin America, which was held in Palmira, Colombia in 1983. It is now collaborating with INTSOY for the organization of another workshop for Asian soybean breeders/agronomists that will be held in Indonesia in 1984.

MEETINGS

FAO has collaborated with INTSOY and USAID in organizing two International Conferences, one on Irrigated Soybean Production in Arid and Semi-Arid Regions, which was held in Cairo, Egypt in 1979, and the other on Soybean Seed Quality and Stand Establishment, held in Colombo, Sri Lanka in 1981. FAO is also associated with the organization of World Soybean Research Conference III. The Second Conference on Irrigated Soybean is planned for 1985 or 1986.

INFORMATION SERVICES

Collection and dissemination of information concerning new developments in crop production and improvement of soybean is one very important part of FAO's central assistance role. This is achieved through the preparation of technical publications, reports, and other information material. Some recent publications and reports devoted to soybean are:

1. Genetic Resources of Soybean, 1983
2. Soybean Agronomy in the People's Republic of China, 1983
3. Soybean Genetics and Plant Breeding in the People's Republic of China, 1983
4. Soybean Pathology/Nematology in the People's Republic of China, 1983
5. Soybean Production in the Tropics, 1982
6. Potential for Soybean Production in the Sudan, 1982
7. Soybean Production Development in Vietnam, 1982
8. Soybean Development in Mozambique, 1982
9. Soybean Assessment in the People's Republic of China, 1981
10. Soybean Breeding for Selected Tropical Asian Countries (Indonesia, Malaysia, Philippines, Thailand), 1979

ASSISTANCE AVAILABLE

FAO will cooperate with and help governments implement recommendations to the extent that resources permit. The services of FAO's Seed Exchange Laboratory are available to interested researchers, and our training program for the development of manpower at all levels is open to research technicians. If desired, FAO might also help identify and formulate projects suited to international funding, apart from providing short-term consultants, holding meetings, seminars, symposia, workshops and conferences, and providing technical information through our documentation and dissemination system.

NOTES

H. A. Al-Jibouri, Senior Officer, Plant Production and Protection Division, Food and Agriculture Organization of the United Nations, Via delle Terme di Caracalla, Rome 00100, Italy.

REFEREBCES

FAO. 1981. Production Yearbook, V 35. FAO, Rome.
FAO. 1982a. Production Yearbook, V 36. FAO, Rome.
FAO. 1982b. Trade Yearbook, V 36. FAO, Rome.

INTSOY—PARTICIPANT IN THE INTERNATIONAL SOYBEAN RESEARCH NETWORK

H. E. Kauffman

Production of edible oil and protein in most less-developed countries (LDCs) in the tropics and subtropics has not kept pace with the expanding needs of the rapidly increasing population (Thompson, 1981; ASA, 1983). This shortage exists in many countries in spite of the dramatic gains in global production of crops such as soybeans and oil palm (*Eleise guiniensis* L.) and modest increases in rapeseed (*Brassica napus* L.) and sunflower (*Helianthus annuus* L.) in several developing countries.

Domestic shortages in most LDCs are made worse by the wide fluctuation in prices of edible oil and protein meal on the international market. LDCs do not have the financial means to import unlimited amounts of soybeans or soybean products (World Bank, 1983). They, therefore, must establish sound production, marketing, processing and utilization programs if they are to benefit from soybeans.

This paper looks at prospects for soybean development in the tropics, reviews activities of the International Soybean Program (INTSOY), and outlines the need for a bold collaborative international soybean research effort between LDCs and developed countries to better exploit the potential of soybeans.

PROBLEMS OF SOYBEAN DEVELOPMENT IN TROPICAL REGIONS

The dramatic gain in soybean production in the temperate regions of the western hemisphere in the 20th century is unparalleled in the history of global agriculture. First, the U.S. and then Brazil surpassed China as the leading soybean producer (Howell, 1983) (de Miranda et al., 1983). Argentina may do so by the end of the decade. The rapid rise to prominence of soybeans has been founded on strong research programs developing improved high-yielding cultivars, effective herbicides, increased fertility, improved machinery and the increased markets and uses for soybean products. The high transferability of technology from North America to South America allowed for the quick establishment of successful soybean industries in temperate regions of South America. However, soybean technology does not transfer directly from temperate regions to tropical regions.

Some have raised the question whether soybeans can profitably be grown in the tropics, since tropical conditions pose different problems than temperate conditions, and therefore, require a range of new production and processing techniques. Others have argued that scarce land in most parts of the tropics should be devoted to food crops such as cereals and edible legumes to meet burgeoning population needs, and that cash and oilseed crops, such as soybean, have a comparative advantage in the temperate regions.

Several international soybean programs such as INTSOY, the Asian Vegetable Research and Development Center (AVRDC), and the International Institute of Tropical Agriculture (IITA), have had active research and development programs during the past decade that have focused on overcoming production and processing constraints unique to the tropics (IITA, 1983; Shanmugasundaram, 1983). A brief examination of some questions that have been raised about soybean production in the tropics and insights that have been gained from these and other programs, give a good background as to why INTSOY was established and to the future direction of INTSOY programs.

Can Tropical Soybean Yields Equal Those in Temperate Regions?

Average yields in major soybean producing countries like the U.S. and Brazil are nearly double those of tropical countries, such as Indonesia, Thailand, and India. Since soybeans originated and evolved in the temperate regions of the Orient, and since major cultivar improvement efforts have concentrated in the temperate regions, it is natural that soybean yields are higher in the temperate regions than in the tropics. On the research stations of many tropical countries, however, yields of new tropical soybean cultivars can equal those of their temperate counterparts (Jackobs et al., 1984). Breeding efforts in Brazil have successfully moved high-yielding soybean production technology farther and farther into the tropics where good farmers are now obtaining up to 3 t/ha (de Miranda et al., 1983). The low yields in tropical Asia can be explained partly because soybeans are grown as a part of a complex cropping system and short duration cultivars are required (Suzuki and Konno, 1982). Conditions are more favorable for the primary crops grown in the rotation than soybeans, which are usually grown during the off-season under less than favorable conditions.

There is plenty of evidence to show that soybeans do have potential for high production in tropical environments, and as tropical breeding programs continue to mature and more traditional tropical and subtropical germplasm is made available from the center of origin, significant gains in production of tropical soybean cultivars can be expected (Ma and Zhang, 1983). Pests and diseases are expected to become more serious in tropical and subtropical areas compared to temperate areas, but if managed properly, as they are in the southern U.S. and northern Brazil, they can be controlled effectively.

Can Soybeans Be Produced Efficiently on Small Farms?

Since soybeans in the United States and South America are primarily grown on large mechanized farms, and a good share of the soybean production in northeast China is on large state farms, soybeans generally have the image of a crop adapted better to large mechanized farm conditions than to small farm conditions. On close examination, vast hectarages of soybean in central and south China and southeast Asian countries, especially Indonesia, Vietnam, and Thailand, are on small land holdings. Yields historically have been low in tropical China and southeast Asia, in part due to the government and farmers treating the soybean as a secondary crop (Suzuki and Konno, 1982).

More recently, the rapidly developing soybean production in India and Sri Lanka has been exclusively on small farms. Yields have been very good in many locations in these countries, although weather related problems have frequently reduced production. The question is not the size of the farm but whether soil, moisture and weather conditions are conducive to soybean production, and whether government and private industry will provide the infrastructure and inputs to make soybeans economically competitive for farmers living on small land holdings.

Can Soybeans Be Introduced into the Tropics Without Replacing Important Food Crops?

The soybean has been criticized for its replacement of important food crops in several countries. When soybean was spreading rapidly in Brazil, for example, in some areas it replaced the field bean (*Phaseolus vulgaris* L.), which is a major part of the Brazilian diet.

Contrary to the general image in the western world, soybean is a food crop. Fortunately, it is also a good cash crop which, in some countries, leads to production for export markets. In countries like Brazil, where soybean first replaced the field bean, government policy was adjusted to encourage increased production of the field bean. In a number of other countries, such as India and Sri Lanka, the soybean has become an integral part of cropping systems without replacing food crops (Anonymous, 1982).

Can the Complex Processing Associated with Soybeans Be Established in Less Developed Countries?

Unlike some grain legume seeds, soybeans require considerable processing before consumption for animal or human food is possible. The need to eliminate several anti-nutritional factors necessitates a heating process that requires considerable energy, which is scarce in LDCs. Solvent extraction, the most efficient processing technique, is complex and requires a high volume, which makes it impractical for a number of developing countries.

In the Orient, soybeans have historically been processed and marketed by small entrepreneurs and by family businesses (Shurtleff and Aoyagi, 1984). In recent years, a range of intermediate processing methods for whole soybeans have been developed at the commercial, village, and home level. Research and development results in Sri Lanka are especially notable, as a number of soybean products are becoming widely used. The same has taken place in India and a number of African countries, and in Guatemala where several relief organizations have established successful village-wide soybean production and processing programs.

It is no longer a question as to whether appropriate soybean processing techniques can be established for the developing countries, but a question of establishing the proper mix of old and new processes to meet each country's need. If LDCs are to consume soybeans widely as a human food, everyone agrees that soybeans must be grown in areas where they are consumed.

INTSOY'S PAST INVOLVEMENT IN TROPICAL SOYBEAN RESEARCH AND DEVELOPMENT

INTSOY was formally organized in 1973 as a collaborative effort of the University of Illinois at Urbana-Champaign and the University of Puerto Rico, Mayaguez. The antecedents of INTSOY domestically date to the late 1800s when the University of Illinois pioneered soybean research in the United States. Internationally, the antecedents date to the mid-1960s when the University of Illinois development teams initiated soybean work in India.

The United States Agency for International Development (USAID) has provided the basic program support for INTSOY, augmented by support from a number of other donors. INTSOY programs have focused on germplasm improvement, plant protection, nitrogen fixation, and soybean food utilization.

INTSOY's three interrelated cultivar testing programs—the Soybean Initial Evaluation Variety Experiment (SIEVE), the Soybean Preliminary Observation Trial (SPOT),
and the International Soybean Variety Experiment (ISVEX)—compose the major germplasm activity. Trials have been conducted in 115 countris since 1953, with more than
200 scientists cooperating. The modest tropical breeding program of INTSOY, first
headquartered in Puerto Rico and now located in Cali, Colombia, complement regional
and national efforts in breeding and testing. The trials have stimulated breeding programs in various countries, which now contribute more than half of the entries evaluated. The ISVEX results give many leads on breeding strategies for tropical and subtropical environments. The results indicate that existing soybean cultivars perform
reasonably well in many environments on experimental stations, but to obtain high
stable yields in a wide range of cropping system environments, a vastly increased range
of soybean cultivars will be needed. Countries such as India, Sri Lanka, Peru, Egypt,
Turkey, Ecuador, Costa Rica, Nepal, and Guatemala are among the more than 20
countries now commercially growing soybean cultivars that initially were introduced
into the INTSOY trials. Many more breeding materials were used in hybridization.

INTSOY entomologists and pathologists have worked on component technology in
the development of integrated pest management practices collaboratively with scientists in a number of tropical countries. Insect problems in the tropics have been minimized in a number of countries as a result of research and development activities.
Major diseases have been characterized and in some cases strains, such as soybean
mosaic virus, have been identified. A range of research, from the applied screening activities for disease and insect resistance to more basic research on DNA activities in
viruses, has been undertaken. Research on the distribution and survival of *Rhizobium*
in different types of soil as well as studies on the enhancement of nodulation has significant implications for establishing highly productive soybean technology for various
cropping systems. INTSOY has worked collaboratively with IITA in the evaluation of
cultivars that perform well with indigenous strains of *Rhizobium*.

A number of INTSOY conferences and workshops have provided forums for
scientists from over 50 countries to contribute important knowledge relating to soybean production and utilization practices. INTSOY maintains several types of services
and publications to assist soybean scientists. The Soybean Insect Research Information
Center (SIRIC) maintains computerized files on the scientific literature of soybean
arthropods and supplies biographic listings and copies of documents. The INTSOY
Newsletter contains information on current INTSOY activities and research, and
features news items of interest to soybean workers around the world. The INTSOY
training course in technical aspects of soybean production has been host to 101 participants. The course in soybean processing for food uses has been host to 79 trainees
from 36 countries. In addition, regional short courses have been held in several countries.

INTSOY'S INVOLVEMENT IN AN INTERNATIONAL SOYBEAN RESEARCH NETWORK

In the absence of an international agriculture research center focusing exclusively
on soybeans, there is a need to establish a more formal cooperative arrangement
between international, regional and national soybean programs. An "International
Soybean Research Network" has been proposed (Figure 1) which will serve as a mechanism to intensify soybean research.

As a part of an International Soybean Research Network, INTSOY plans to focus
its activities in areas where INTSOY has unique expertise or a comparative advantage.

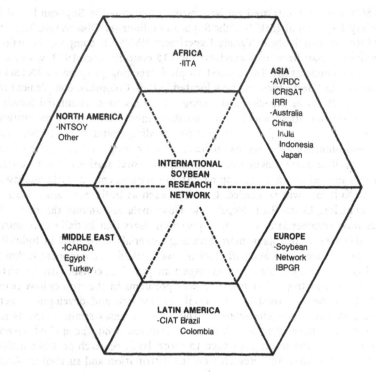

Figure 1. Proposed International Soybean Research Network.

We envision these areas to be as follows:

Soybean Processing and Utilization

Major emphasis will be placed on simple techniques to produce soymilk and whole soybean extruded products and to extract soybean oil. A modest research program will be undertaken at the University of Illinois with strong extension and training linkages to scientists in Asia, Africa, and Latin America through the regional networks that are being established.

Germplasm Preservation, Evaluation and Enhancement

INTSOY will work with the International Board of Plant Genetic Resources (IBPGR) to help national programs collect, evaluate and preserve soybean germplasm. The international soybean testing program (ISVEX) will serve as a link between regional testing programs. INTSOY will work with others to develop disease, insect and problem-soil trials.

INTSOY germplasm enhancement work will be centered at ICA/CIAT in Colombia. Major objectives of the breeding program will be to develop improved cultivars for the tropical regions of South America and the Caribbean Basin. Biotechnology research at the University of Illinois will focus on wide crosses for pest and disease resistance and tolerance to several stresses.

Improved Production Technology for Various Cropping Systems

INTSOY scientists will concentrate research in areas with broad application, such as integrated pest management, and link up with those of several international centers and national programs to work on location-specific problems, such as control of major diseases like rust.

Education and Communication

INTSOY will continue to co-sponsor conferences and workshops and conduct training programs. The training programs will focus more on regional needs, and therefore, an increasing number will be held in collaboration with national and other international organizations. Significant emphasis will be placed on degree training of soybean scientists, who will give leadership to soybean industries being established in an increasing number of countries. An international soybean research newsletter will be initiated collaboratively with AVRDC, IITA, and others to provide a communication mechanism for soybean scientists, producers, processors, and policy makers to facilitate rapid exchange of information about soybeans.

CONCLUSION

Soybeans have brought benefits to several Asian countries for many centuries. During the past few decades, a number of temperate countries in the western hemisphere have begun to share in their benefits. There are, however, still tremendous needs for soybean products in most LDCs in the tropics. An "International Soybean Research Network" can help develop and extend appropriate production, processing, and utilization technology to all countries; thus, bringing additional benefits from soybeans to all participating countries.

NOTES

H. E. Kauffman, Director, International Soybean Program (INTSOY), University of Illinois, 113 Mumford Hall, 1301 W. Gregory Drive, Urbana, Illinois 61801.

REFERENCES

American Soybean Association. 1983. 2002: First the questions, now the answers. ASA, St. Louis.

Anonymous. 1982. Soybean Success. p. 127. *In* India Today.

Howell, R. W. 1983. Historical development of the United States soybean industry. p. 11-15. *In* Soybean research in China and the United States. INTSOY Publication Series number 25. INTSOY, Champaign, Illinois.

IITA. 1983. Soybean improvement at the International Institute of Tropical Agriculture. IITA, Ibadan, Nigeria.

Jackobs, J. A., C. A. Smyth, and D. R. Erickson. 1984. International soybean variety experiment—8th report of results, 1980-81. INTSOY Publication Series number 26. INTSOY, Champaign, Illinois.

Kauffman, H. E., J. A. Jackobs, and S. Kolivalli. 1983. The challenge of strengthening soybean research and development activities in the tropics and subtropics. *In* Proceedings of the international symposium on soybean in tropical and subtropical cropping systems (in press).

Ma, Rhu-Hwa and Zhang Kan. 1983. Historical Development of Soybean Production in China. p. 16-18. *In* Soybean Research in China and the U.S. INTSOY Publication Series number 25. INTSOY, Champaign, Illinois.

de Miranda, M. A. D., E. A. Bullisana, and H. A. A. Mascarenhas. 1983. Soybeans in Brazil. Proceedings of the international symposium on soybean in tropical and subtropical cropping systems (in press).

Shanmugasundaram, S. 1983. Role of AVRDC in soybean and mungbean improvement. p. 137-166. *In* Grain legume production in Asia. Asian Production Organization, Nordica International Limited, Hong Kong.

Shurtleff, W., and A. Aoyagi. 1984. The soyfoods industry and market directory and databook. The Soyfoods Center, Lafayette, California.

Suzuki, F., and S. Konno. 1982. Regional report on grain legume production in Asia. p. 15-93. *In* Grain legume production in Asia. Asian Productivity Organization, Nordica International Limited, Hong Kong.

Thompson, W. N. 1981. Increasing the supply of soybean. *In* World conference on soya processing and utilization. American Oil Chemists Society.

World Bank. 1983. World development report. Oxford University Press, New York.

THE ASIAN VEGETABLE RESEARCH AND DEVELOPMENT CENTER'S SOYBEAN PROGRAM

S. Shanmugasundaram

The Asian Vegetable Research and Development Center (AVRDC) is an international agricultural research center dedicated to improving the agronomic potential and nutritive value of legumes and vegetables in the lowland tropics and subtropics. AVRDC has conducted basic and applied research on a wide range of problems related to tropical soybean since 1973. The major problems associated with low soybean yields in the tropics and subtropics include: Low yield potential, lack of improved cultivars for specific purposes, sensitivity to short photoperiods and high temperatures, susceptibility to diseases and insects, lack of good quality seeds, and lack of adoption of required management practices. The average yield of soybean in the tropics is about 800 kg/ha. Given improved cultivars and appropriate, economical management practices, soybean could become an important protein crop in developing countries.

THE AVRDC SOYBEAN PROGRAM

The goal of AVRDC's soybean program is the development of stable tropical and subtropical cultivars with wide adaptability, high yield, disease and pest resistance, and good seed quality. Specific objectives include:

1. Collection, documentation, utilization, and maintenance of soybean germplasm.
2. The development of stable, higher yielding selections.
3. The development of cultivars with good seed quality and resistance to weathering.
4. The development of cultivars with wide adaptability to different photoperiods and high temperatures.
5. Disease research and management.
    i) Emphasis: (According to priority) resistance/tolerance, cultural control, and chemical control.
    ii) Diseases: (According to priority) soybean rust, bacterial diseases, viruses, downy mildew, seed quality and storability diseases, and anthracnose.
6. Insect management.
    i) Priority: Resistance/tolerance, cultural control, and chemical control.
    ii) Insects: (According to priority) beanflies, pod borers, stink bugs, and defoliators.
7. The development of appropriate management technologies for maximum economic yield.

8. Studies on *Rhizobium* and Vescicular Arbuscular Mycorrhizae ecology and potential.
9. The development of green vegetable- and beansprout-type soybeans.
10. The development of soybeans suitable for intercropping or mixed cropping.
11. Expansion of international cooperative network trials.
12. Transfer of improved technologies through training, communication, and bilateral projects.

PROGRESS

By the end of 1983, AVRDC's soybean germplasm collection totalled 10,523 accessions, one of the largest collections in the world. AVRDC plant pathologists, entomologists, and plant breeders have screened the majority of the germplasm, and have identified resistance or tolerance to soybean rust and beanflies, and insensitivity to photoperiod (Chiang and Talekar, 1980; Shanmugasundaram, 1981; Tschanz and Wang, 1984). Thus far, the Center's hybridization program has made 3,554 crosses to combine desirable characteristics from new selections and local cultivars. Soybean can be planted at AVRDC (23°07'N latitude and 120°17'E longitude) three times a year: In January-February (spring season), June-July (summer season), and September (autumn season). Because photoperiod and temperature conditions vary according to season "disruptive seasonal selection" is used to identify breeding lines adapted to diverse agro-climatic conditions. Soybean rust (*Phakopsora pachyrhizi* Sydow) and downy mildew (*Peronospora manshurica* (Naum.) syd. ex Gaum.) are serious in the spring and autumn, while bacterial diseases such as bacterial pustule (*Xanthomonas campestris* pv. phaseoli (Smith) Dye), are severe in the summer rainy season. Similarly, beanfly (*Melanagromyza* sp.) is an important pest in the autumn, whereas defoliators are a problem in the spring and summer. The seasonal diversity of these pests allows AVRDC scientists to effectively screen for resistance and tolerance.

Selected breeding lines undergo replicated, preliminary yield trials (PYT) across all three seasons during a single year. Selections from the PYTs undergo another year of three season intermediate, replicated yield trials (IYT). Promising entries from the IYTs are coded as AVRDC *Glycine* Selections (AGS), and are tested for an additional year in advanced yield trials. The photoperiodic response of AGS entries is evaluated in pot culture studies using artificial and natural lighting and dark rooms. Selections with high yield potential across all three seasons have been identified, some of which yield up to 10,000 kg/ha/year. Some high-yielding AGS entries combine resistance or tolerance to diseases and reduced sensitivity to photoperiod.

Beanfly, *Melanagromyza sojae* Gehntner, resistance has been identified in Korean *G. soja* accessions G 3089, G 3091, G 3104, and G 3122. Efforts are currently underway to transfer this resistance to cultivated soybean. However, each of the Korean cultivars is susceptible to another beanfly species, *Ophiomyia phaseoli* Tryon (Chiang and Talekar, 1980).

The epidemiology of soybean rust has also been studied. Yield losses due to rust can reach 90%. Vertical resistance genes available in the soybean germplasm are being incorporated into improved breeding lines. Rate reducing resistance has also been identified at AVRDC. However, the concept of tolerance to soybean rust, defined as relative yielding ability under severe soybean rust epidemics, appears to have more promise than vertical resistance, which tends to break down because of the pathogen's multiplicity of races (Tschanz and Wang, 1984).

Soybeans are planted continuously on the same land in countries such as Indonesia.

Results from two years of continuous cropping of two AGS entries (three times a year) on the same land showed no obvious adverse effects. Both entries yielded 8,000 kg/ha/year.

The results of a three-year study of genotypes continuously planted with either maximum or minimum inputs in different seasons revealed that the yield gap between minimum and maximum input treatments varied according to season (Figure 1). Genotypic differences also were observed. The same study also showed that cultivars suited to low input cultivation can be selected from promising cultivars in maximum input, advanced yield trials. Furthermore, genotypes with optimum yield potential that are indifferent to inputs in specific seasons may also be selected (Shanmugasundaram et al., 1982; Shanmugasundaram and Toung, 1984).

AVRDC economists have surveyed soybean production problems in Taiwan, Thailand, and the Philippines (AVRDC, 1976). The principal reasons for the decline in soybean area and production in these countries were attributed to poor economic return from soybean compared with other crops such as vegetables, adzuki bean (*Phaseolus angularis* [Willd.] W. F. Wright), and vegetable-type soybean (Calkins and Huang, 1978; Liu and Shanmugasundaram, 1984). Difficulties in procuring good quality seeds, low average yields, high production costs, and lack of adoption of appropriate management technology also were listed as problems. Appropriate location- and season-specific technologies need to be identified to help farmers obtain maximum economic yields.

Mulching or plowing combined with nitrogen top dressing was found to increase

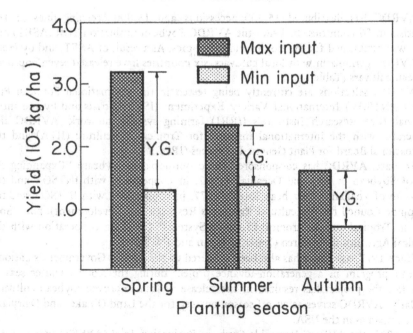

Figure 1. Yield gap (YG) between maximun inputs and minimum inputs over three seasons and three years (1980, 1981, 1982). Maximum input includes herbicide application, irrigation, and insecticide and fungicide application, while minimum input consisted of planting and harvesting.

soybean yields. Deep plowing and compost application significantly increased both nodule formation and yield (AVRDC, 1984). Traditionally, soybean is planted by broadcasting seed with or without tillage followed by rice (*Oryza sativa* L.) straw mulching to cover the seeds. An experiment to compare the broadcast method with row planting and space planting in different seasons showed that under certain conditions the broadcast method might be successful. However, there is considerable risk in using this technique (Shanmugasundaram, 1982).

The number of days to maturity are less in the tropics and subtropics than in the temperate zone because of shorter photoperiods. Early maturity is desirable in the tropics because soybean is frequently used in multiple cropping systems. However, to obtain large yields with a shorter growing period, it is necessary to have denser plant populations. For example, results from a population density experiment with narrow and broad leaflet AGS genotypes planted in three consecutive seasons, suggested that population density and within- and between-row spacing is cultivar- and season-specific. Nevertheless, a 400,000 to 600,000 plants/ha range appears to be optimum for maximum yield under AVRDC conditions (AVRDC, 1981).

Vegetable-type soybeans are an excellent source of protein and their popularity is increasing (P. Shurtleff, pers. comm.). AVRDC vegetable-type soybean cultivars suited to tropical and subtropical environments are currently being evaluated and distributed to national programs.

INTERNATIONAL COOPERATION

AVRDC has distributed 15,500 accessions and 15,800 breeding lines to 164 scientists in 56 countries. In 1980, the AVRDC Soybean Evaluation Trial (ASET) network was established for the tropics and subtropics. As a result of ASET, and by crossing AVRDC germplasm with local cultivars, six countries have released seven improved soybean cultivars (Table 1).

AVRDC selections are currently being tested in the International Soybean Program's (INTSOY) International Variety Experiment (ISVEX) trials and by the International Rice Research Institute's (IRRI) farming systems network. AVRDC also cooperates with the International Institute for Tropical Agriculture (IITA) and the International Board for Plant Genetic Resources (IBPGR).

To date, AVRDC has co-sponsored three symposia on soybean: "Expanding the Use of Soybean in Asia and Oceania" (1977, in cooperation with INTSOY and the Kingdom of Thailand); "Soybean Rust" (1977, in cooperation with INTSOY and the Philippine Council for Agricultural Resources Research and Development); and "Soybean in Tropical and Subtropical Cropping Systems" (1983, in cooperation with the Tropical Agricultural Research Center of Japan and INTSOY).

Since 1977, assistance has also been rendered to the Korean Government's national soybean program in a generation-advance project during the Korean winter season. Thus far, the project has resulted in the release of four improved soybean cultivars. Similarly, AVRDC serves as an off-season nursery for the Land O'Lakes and Dairyland Seed companies in the USA.

In 1983, the AVRDC Vegetable Soybean Evaluation Trial (AVSET) network was established in Taiwan. AVSET will be extended to cooperators overseas in 1985. AVRDC was one of the key organizers of the International Working Group on Soybean Rust (IWGSR), and has published the Group's "Soybean Rust Newsletter" since 1977. To date, AVRDC has trained 102 specialists from ten countries in soybean production, research and extension.

Table 1. New soybean cultivars released to farmers from AVRDC breeding materials.

Country	Cultivar name	Year released	Remarks
India	KM-1	1980	Adapted to rice-fallows
Indonesia	Taiwan G 2120	1980	High yield and adapted to crude cultivation
	Wilis	1983	High yield, earlier maturing than Taiwan G 2120
Honduras	Darco-1	1980	High yielding, early maturing, and resistant to alternaria leaf spot
U.S.A.	Dowling[a]	1978	AVRDC assisted in screening for soybean rust resistance
Malaysia	T 30050	1982	Higher yielding
Taiwan	Kaohsiung No. 9	1982	Adapted to the spring and summer season

[a]This cultivar was not developed from AVRDC breeding material.

THE ASIAN SOYBEAN IMPROVEMENT NETWORK (ASIN)

INTSOY, IITA, the United Nation's Food and Agricultural Organization (FAO), the United Nations Development Program (UNDP), the International Atomic Energy Agency (IAEA), and the European Cooperative Soybean Network have active international soybean research and development programs. These projects include the development of improved breeding lines and appropriate technology, as well as crop assistance programs in a number of developing countries. AVRDC is planning to initiate a global network for soybean improvement in association with these organizations.

To improve soybean production in developing countries, it is necessary to consolidate national, regional, and international research efforts. The proposed Asian Soybean Improvement Network (ASIN) is expected to bring available expertise into focus and improve research coordination. The following objectives are suggested for ASIN:

1. Provide a mechanism for cooperative project planning and review by national programs and AVRDC.
2. Group soybean growing countries according to specific needs to facilitate research and development.
3. Develop improved cultivars for the soybean growing areas of the region.
4. Develop appropriate management technology for specific regions.
5. Assist AVRDC in the transfer of suitable methodologies and appropriate technologies to national programs.
6. Promote long-term upgrading of national programs.

Initially the soybean growing countries in the region will be divided into two major groups. The first group will be comprised of countries with strong national research programs and sufficient numbers of subject matter specialists; the second group will be characterized by countries interested in soybean but who lack resources.

India, China, Japan, Taiwan, Indonesia, Thailand, and the Philippines will be tentatively placed in the first group, while Sri Lanka, Nepal, Bangladesh, Pakistan, Burma,

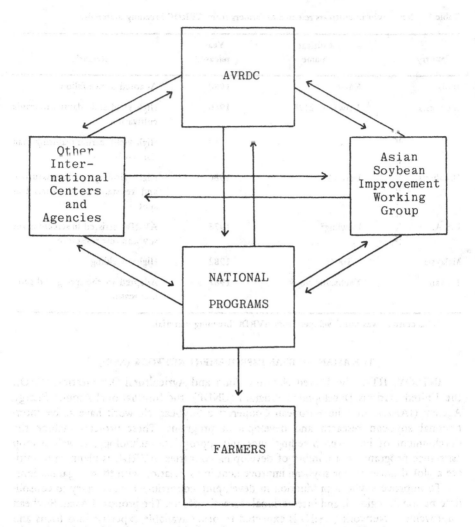

Figure 2. Relationship of AVRDC with the Asian Soybean Improvement Working Group.

Malaysia, and Vietnam could be placed in the second. It is essential that a committee be formed with representatives from both groups to implement collaborative research. One of the committee's functions will be to select testing sites in individual countries in consultation with national program members.

The link between ASIN and national programs will vary from country to country, depending on the capabilities of the existing research program. ASIN will cooperate with national programs in the region, and with international and national programs in other regions.

AVRDC will (1) assume responsibility for coordinating the network and will provide secretariat services; (2) offer appropriate short-term research and extension training programs; (3) organize working group meetings to promote the exchange of information and planning, and conduct monitoring tours to study problems and progress; (4) offer improved germplasm, cultivars, and technology packages for testing; (5) provide consultancy services for program development; and (6) offer national program scientists the opportunity to work with AVRDC researchers. The relationship between the Asian Soybean Improvement Working Group, AVRDC, other international agencies, and the national programs is shown in Figure 2.

The following major topics are suggested for consideration: (1) cultivar improvement and coordinated cultivar trials; (2) research on maximum yield and maximum economic yield; (3) component technology research for disease and insect resistance/tolerance; (4) research on cultivar and *Rhizobium* and/or Mycorrhizae interaction; (5) vegetable-type soybean improvement, coordinated trials and utilization.

NOTES

S. Shanmugasundaram, Plant Breeder and Legume Program Leader, the Asian Vegetable Research and Development Center (AVRDC), P.O. Box 42, Shanhua, Tainan 741, Taiwan, Republic of China.

REFERENCES

Asian Vegetable Research and Development Center (AVRDC). 1976. Soybean Report '75. Shanhua, Taiwan, Republic of China.

Asian Vegetable Research and Development Center. 1981. Progress Report for 1980 (AVRDC 80-155). Shanhua, Taiwan, Republic of China.

Asian Vegetable Research and Development Center. 1984. Annual Report for 1983. Shanhua, Taiwan, Republic of China (in press).

Calkins, P. H. and K. R. Huang. 1978. Soybean production in Taiwan: A farm survey. AVRDC Tech. Bull. No. 11 (78-89). Shanhua, Taiwan, Republic of China.

Chiang, H. S. and N. S. Talekar. 1980. Identification of sources of resistance to the beanfly and two other agromyzid flies in soybean and mungbean. J. Econ. Entomol. 73(2):197-199.

Liu, C. P. and S. Shanmugasundaram. 1984. Frozen vegetable soybean industry in Taiwan. In Vegetables and Ornamentals in the Tropics. University Pertanian Malaysia (in press).

Shanmugasundaram, S. 1981. Varietal differences and genetic behavior for the photoperiodic responses in soybeans. Inst. Trop. Agric. Kyushu Univ. (Japan) Bull. 4:1-61.

Shanmugasundaram, S. 1982. Role of AVRDC in Soybean and Mungbean Improvement. p. 137-166. In Grain Legumes Production in Asia. Asian Productivity Organization, Tokyo, Japan.

Shanmugasundaram, S., Chung-Ruey Yen and T. S. Toung. 1982. Genotypic response to maximum and minimum input in soybean. p. 157-164. In Sung-Ching Hsieh and Dah-Jiang Liu (eds.) Proc. Symp. Plant Breeding. Agricultural Association of China and the Regional Society of SABRAO, Taiwan, Republic of China.

Shanmugasundaram, S. and T. S. Toung. 1984. Soybean genotypic responses for minimum and maximum input in different seasons. In S. Shanmugasundaram and E. W. Sulzberger (eds.) Soybean in Tropical and Subtropical Cropping Systems, AVRDC, Shanhua, Taiwan, Republic of China (in press).

Tschanz, A. T. and T. C. Wang. 1984. Recent advances in soybean rust research. In S. Shanmugasundaram and E. W. Sulzberger (eds.) Soybean in Tropical and Subtropical Cropping Systems, AVRDC, Shanhua, Taiwan, Republic of China (in press).

THE EUROPEAN COOPERATIVE RESEARCH NETWORK ON SOYBEAN

Maurice Arnoux

ORGANIZATION AND STRUCTURE

Over the past 10 years, the FAO Regional Office for Europe, on the recommendation of the European Commission on Agriculture, and in agreement with interested European Governments, has encouraged the organization of research networks on topics of particular concern to Europe. Thus, ten networks, of which soybean is one, have been successively established since 1974. Today these networks make up the European System of Cooperative Research Networks in Agriculutre: ESCORENA.

ESCORENA aims firstly at forging a link between the research institutions, universities, and the scientific and technical organizations present in different countries, so as to establish international scientific and technical contacts. Another objective is to stimulate cooperation that could improve a given research activity and speed up both the achievement and dissemination of results. This form of cooperation essentially comprises voluntary exchanges of information, techniques and materials, as well as concerted studies undertaken jointly by several countries with a view to seeking more rapid results to certain problems of specific concern to Europe.

The network structure is simple and flexible. Members who join voluntarily select a limited number of research topics, if possible of common interest. A sub-network is then formed for each topic and draws up its programs and organizes its activities. On attainment of its objectives, the sub-network usually ceases its activity. At any given moment, it is possible to establish a new sub-network if another research topic is considered of high priority. For specific studies, working groups can be organized. These generally are interdisciplinary and open to any world specialist not yet cooperating. Each sub-network and working group is headed by a Liaison Officer while the overall activities of the network are placed under the responsibility of a Coordinator, elected for 4 years.

The operation of the Network is also very flexible. Each institute or researcher continues his own personal research program, but can join in and contribute to any joint research being undertaken by the network and participate in the analysis and publication of results. The research programs, the responsibilities of each participant, and the modalities for collaboration are determined during the various encounters of the researchers organized by the network. Individual missions, informal meetings, workshops and the biennial plenary consultations all serve to bring researchers together.

Expenses incurred by the joint research are borne by the participants. FAO assistance, which is very important and often indispensable, is limited to organizing or facilitating the contacts and meetings between researchers. In addition, FAO also guarantees the expenses for diffusion of meeting reports as well as the printing of individual network newsletters which disseminate general information with special emphasis on research results. It is within this general framework that the European Cooperative Research Network on Soybean, set up in 1976, carries out its activities.

This Network presently groups 30 Institutions belonging to 20 countries in Europe, North Africa and the Near East. Institutes in USA and Canada and several international organizations, like EUCARPIA (European Association for Research on Plant Breeding) and INTSOY, also collaborate with the Network. The Network started its life with six sub-networks and recently, as a result of increasing knowledge and evolving programs, three supplementary working groups have been organized (Table 1).

ACTIVITIES

At the beginning, because European experience with soybean was limited, the Network decided as a first step to undertake actions of general interest which would not only widen existing knowledge, but also promote contacts and encourage a form of collaboration which would become progressively closer and take on a specific orientation. In this spirit, six joint research topics were selected and still constitute the six sub-networks currently in operation (Table 1).

Collection, Exchange and Study of Breeding Material

The objective is to establish, maintain, and disseminate an inventory of soybean cultivars existing in European collections and available for exchange between countries. This sub-network is also now in charge of registering and regularly publishing a list of breeding populations (F_2 to F_5) which have been discarded by various breeders but which might be interesting for other breeders working in other environments.

Comparative Cultivar Trials

The aim is to organize cultivar trials in Europe to test the agronomic value and adaptability of new cultivars. This research generally comprises 30 to 40 trials spread

Table 1.   The European Cooperative Research Network on Soybean

Coordination - M. ARNOUX - France		
Sub-networks	**Liaison Officers**	
Exchange of breeding material	B. BELIC	- Yugoslavia
Comparative cultivar trials	S. DENCESCU	- Romania
Map of regions suitable for soybean	E. KURNIK	- Hungary
Weed control	G. ZANIN	- Italy
Diseases and pests	P. SIGNORET	- France
Rhizobium inoculation	M. A. CHAMBER-	Spain
Working Groups		
Soybean in northern Europe and Canada	A. SOLDATI	- Switzerland
Cooperative intermating program	A. VIDAL	- France
Breeding methods	R. ECOCHARD	- France

over 15 countries each year, compares an average of 20 to 30 cultivars, whose maturity, in view of European climatic conditions, usually falls within groups 00 to IV.

Map of European Regions Suitable for Soybean

This sub-network aims at studying the main European regions where soybean might be cultivated and at drawing-up maps detailing the respective quality and potentiality of these regions. This important study, which up to now has laid particular emphasis on Central Europe, has resulted in maps which, taking into account the pedo-climatic conditions and types of crops (principal or catch, irrigated or non-irrigated), give precise indications about the potentialities of each region, as well as the types of cultivars suitable for them. These maps will soon be published and this sub-network will then have completed its task.

Chemical Weed Control

This important activity, which was not very advanced at the beginning of the sub-network's activity, has successfully generated information that now allows growers to keep weeds under control.

Continuation of joint work, however, proved increasingly difficult because of differences in species, soil, climate and agricultural practices existing in the various European countries. Thus, this sub-network now only acts as an information center, gathering and diffusing any new results obtained, particularly that from new commercial or experimental chemicals.

Diseases and Pests

Despite the presence of various fungi and viruses, no diseases are currently important in Europe. While it is true that the areas under cultivation are limited, they are also constantly expanding, and in view of the potential importance of diseases, it seemed important to draw up a list of the main organisms existing in Europe and to analyze their respective importance as well as the risk of their possible development. In addition, this sub-network evaluates cultivar resistance and compares treatment methods and sanitary tests of seeds.

Rhizobium Inoculation

Rhizobium japonicum does not exist in European soils and seeds have to be inoculated at the time of sowing. The development of suitable inoculation techniques and production of inoculants, and the exchange and comparison of races of *Rhizobium* are essentially the bases of the joint work of this sub-network. Research is presently being undertaken to study factors limiting nitrogen fixation, as well as the selection of more efficient and less competitive strains of *Rhizobium*.

ORIENTATION OF RESEARCH

As a result of research and the experience progressively acquired with increased cultivation, European agriculture today is well able to manage its soybean production and, in the absence of serious diseases, to obtain yields generally between 1 to 3 metric tons/ha. These yields are, however, irregular and often at a level insufficient to enable soybean to be competitive with other summer crops. Improvement in the level and regularity of yields is sought through improved cutural techniques and a better knowledge of the physiology of the plant, particularly of its symbiotic nitrogen fixation and nitrogen metabolism. Improvement is also sought, above all, through the breeding of cultivars better adapted to European environments, ones that would be more regularly productive than the American types presently being grown. The cultivar problem, which is an overriding and general objective, must be attacked differently in the two large regions of Europe.

In the North; i.e., in the countries situated approximately above 50°N latitude, as well as in the central mountainous areas, rainfall is generally sufficient but temperatures are low and days are long. There are some very early cultivars adapted to these climates, like Fiskeby for example, but their productivity is too weak and irregular to enable them to be exploited. In this large Northern area, acquisition of more knowledge about photoperiod-temperature relationships and their influence on earliness and yield must precede breeding of new cultivars. Need for information on this fundamental research topic motivated the establishment, in collaboration with EUCARPIA, of an international and multi-disciplinary working group on the promotion of soybean in northern Europe and Canada. This group, comprising 24 institutes in 10 countries in Europe and North America, has undertaken a joint study of growth and development phenomena of eight early cultivars (Groups 000, 00 and 0) in 24 trials, spread out above 50°N latitude.

In the South summer rainfall is generally insufficient and irregular, the reserve of groundwater is variable, temperatures are normally sufficient but with frequent daily and periodic variations, while the role of photoperiod seems less important. In these environments with very different pedo-climatic characteristics, soybean provides yields which can attain 2.5 tons/ha; but yields are irregular, and breeders are seeking better adapted, better performing and more stable cultivars.

In attempt to enlarge the genetic base of the breeding material existing in Europe, a cooperative intermating program has been set up. In this second working group, the breeders of eight countries divide up among themselves the various crosses to be undertaken in a "circular" system design, so as to increase genetic recombinations, and thus, obtain significant quantities of crossed seeds.

Finally, a third working group of interest to all breeders has been created to undertake a study of breeding methods. This group comprises 11 institutes from 9 European countries. Its activity is particularly oriented toward the single seed descent technique, and there are plans to compare, after 4 years of activity, the results obtained in different countries.

An information bulletin, *EUROSOYA*, has been published by the Network since 1983.

NOTES

Maurice Arnoux, Coordinator, I.N.R.A., 9 place Viala, 34060 Montpellier, France.

THE SOYBEAN IMPROVEMENT PROGRAM AT THE INTERNATIONAL INSTITUTE OF TROPICAL AGRICULTURE

E. A. Kueneman, W. R. Root, K. E. Dashiell and S. R. Singh

Within the network of international agricultural research centers the International Institute of Tropical Agriculture (IITA) has been given responsibility for research on soybeans for the tropics with major emphasis on Africa. For the past five years IITA's program had focused primarily on development of cultivars for the lowland, humid and semi-humid tropics with emphasis on the West African Savanna, which is characterized by sandy soils with low fertility and a growing season ranging from 100 to 150 days. Recently work has been initiated to develop material adapted to higher elevations and to potential growing areas in southern and eastern Africa. Many breeding lines developed at IITA in Africa will be useful in tropical environments in other regions of the world. IITA, in collaboration with the International Soybean Program (INTSOY) based at the University of Illinois, USA and the Asian Vegetable Research and Development Center (AVRDC) based in Taiwan, is planning an integrated trial network to make improved soybean lines available to cooperators around the world. In this network IITA will take a major coordinating role for trials in Africa.

Soybean production in Africa is presently only a very small percentage of what is possible. There were fewer than 10,000 tons produced in developing countries south of the Sahara in 1979. An FAO study (1978) estimated that 145 million hectares are suitable to very suitable for growing soybeans under low input farming, and about 179 million hectares could be used for soybean production if an intensive input farming system was used (Figure 1). The area where soybeans can be grown is similar to the areas where maize (*Zea mays* L.) is adapted, and soybean production is likely to expand in Africa, primarily as a rotation crop with maize.

Food production per capita in Africa has declined over the past ten years. Soybeans, having approximately 40% protein, could play a major role in increasing the protein intake for the people of Africa. Locally processed soybean oil could replace the large quantities of vegetable oil which are now imported.

In 1972, research was initiated on soybeans at IITA at Ibadan, Nigeria. Researchers identified cultivars, primarily from the southern USA, with superior yield potential for the humid tropics. They also identified two major difficulties that have prevented or slowed the expansion of soybean as a crop in Africa and other tropical areas.

The first problem is that soybeans lose their viability rapidly when stored in warm, humid environments. The second problem is that cultivars introduced from the USA require inoculation with *Rhizobium japonicum*. Most developing countries lack inoculant industries to produce and distribute inoculant, and most African farmers lack facilities to store it and the expertise to apply it. Consequently, researchers at IITA

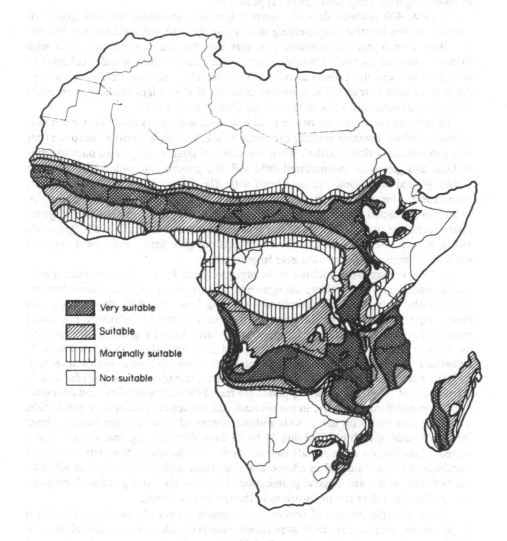

Figure 1. Soybean generalized agro-climatic suitability assessment for rainfed production (FAO, 1978).

decided to develop 'promiscuous' soybean cultivars capable of nodulating effectively with the indigenous cowpea-type rhizobia which are ubiquitous in Africa. In 1976-1977 soybean germplasm was screened for seed storability and for ability to nodulate with rhizobia indigenous to Africa. In 1978 major efforts were initiated at IITA to develop cultivars with superior seed storability and promiscuous nodulation. The methods used at IITA to breed for improved seed longevity are presented in a companion paper in this volume: "Breeding for Management of Soybean Seed Diseases." This paper will focus primarily on breeding promiscuous soybean lines capable of nodulating with indigenous cowpea-type rhizobia.

In 1978, 400 soybean lines of diverse origin were evaluated for their ability to nodulate at five locations representing diverse climatic and soil conditions in Nigeria. Ten lines were highly 'promiscuous', in that they formed effective symbiosis with rhizobia native to all sites. Rhizobia strains were isolated from nodules collected at each location, and the promiscuous lines were found to be compatible with a large majority of native strains. The improved cultivars of U.S. origin established an effective symbiosis with only a few of the isolates (Pulver et al., 1984).

To evaluate the efficiency of the symbiosis with indigenous rhizobia, a study was conducted which involved grafting the shoot of a U.S. cultivar 'Bossier', onto the root of a promiscuous cultivar 'Orba'. When the shoot of Bossier was grafted onto the root of Orba and grown in uninoculated field soil, the growth and yields were equal to Bossier grafted onto itself and inoculated with *Rhizobium japonicum*. Similar results were observed for grafting experiments involving the U.S. cultivar 'Jupiter' and Orba. In studies conducted to compare responses of promiscuous and non-promiscuous germplasm accessions to inoculation and nitrogen fertilizer (Pulver et al., 1982; Rao et al., 1981), it was observed that promiscuous accessions gave little or no yield response while non-promiscuous accessions gave large responses.

In 1978 crosses were initiated to incorporate genes for promiscuity and superior seed longevity into high-yielding backgrounds. The frequency of promiscuous progeny in segregating generations is relatively high (ranging from 13 to 34%, depending on the cross), suggesting inheritance of promiscuity is conditioned by relatively few major genes (IITA, 1980). Selections for promiscuity are made by growing the breeding nurseries in fields that are low in nitrogen and have never been inoculated with *R. japonicum*. In early generations individual plants that give vigorous growth and are well nodulated are selected. Plants are dug at physiological maturity and assessed for nodule mass. Seeds of selected F_5 and F_6 plants are tested for seed storability, and are multiplied in progeny rows for use in multi-locational, replicated, preliminary yield trials, which are also grown on low N soils without history of inoculant application. Plants from both ends of each plot are dug 65 to 75 days after planting and nodule mass is scored. Lines that do not nodulate well across sites are discarded. Nodulation and seed storability are also evaluated in advanced and uniform trials. This method of selection has been very successful. Several promiscuous lines that also have good seed longevity have performed well at several locations in the savanna of Nigeria.

To test the effectiveness of symbiotic nitrogen fixation with rhizobium other than *R. japonicum*, promiscuous lines were grown with and without supplemental nitrogen fertilizer at the rate of 50 kg/ha applied 30, 60 and 90 days after sowing. For lines identified as promiscuous, no consistent differences have been observed between plots grown with ample nitrogen fertilizer and plots using only fixed N (Table 1). IITA microbiologists also have tested the most promising breeding lines with and without *R. japonicum*, and preliminary results are similar to trials mentioned previously. Large responses to inoculants are commonly observed for standard non-promiscuous cultivars, whereas inoculation generally results in only small yield increases when applied to promiscuous lines.

An international testing program was initiated in 1982 by IITA with a focus on Africa. The entries in these trials are elite IITA breeding lines. The objectives are to assist the breeders at IITA in determining how their breeding lines are adapted to various tropical environments, and to distribute germplasm to national programs. Lines

Table 1. Yields of some soybean lines grown with and without supplemental N applied in 50 kg/ha doses at 30, 60 and 90 days after sowing at two sites in West Africa in 1982.

Cultivar	N level	Odienne, Ivory Coast	Sotouboua, Togo
		— kg/ha —	
TGx 326-034D	+N	1792	2303
	-N	2750	2212
TGx 442-01C	+N	1650	1667
	-N	1817	1727
TGx 304-059D	+N	1400	1333
	-N	1608	1212
TGx 340-1C-Y	+N	1542	1333
	-N	1417	1697
Non-Promiscuous Check[a]	+N	1725	1606
	-N	1392	1364
Trial Mean	+N	1337	1479
	-N	1474	1652
LSD Nitrogen		NS	151
LSD Cultivar		531	476

[a]The check variety at Odienne was 'Bossier'; the check at Sotouboua was 'Jupiter'.

that perform well in these trials will then be entered in the new global testing program involving IITA, INTSOY and AVRDC. IITA will coordinate the African trial network.

While exciting progress has been made, there is still scope for improvement. Future breeding efforts will continue to focus on seed longevity and promiscuous nodulation. Emphasis will also be placed on improving stem strength, virus resistance, and shattering resistance. Most of IITA's improved cultivars mature from 115 to 140 days. Cultivars with 100-day maturation would be useful for some cropping systems, and selections are currently being made with this in mind.

NOTES

E. A. Kueneman, W. R. Root, K. E. Dashiell and S. R. Singh, Grain Legume Improvement Program, International Institute of Tropical Agriculture, Oyo Road, Ibadan, Nigeria.

REFERENCES

FAO. 1978. Report on the agro-ecological zones project. Vol. 1. Methodology and results for Africa. International Institute of Tropical Agriculture. 1980. Annual Report for 1979. Ibadan, Nigeria.

Pulver, E. L., F. Brockman and C. Wien. 1982. Nodulation of soybean cultivars with Rhizobium spp. and their response to inoculation with *R. japonicum*. Crop Sci. 22:1065-1070.

Pulver, E. L., E. A. Kueneman and V. Ranga-Rao. 1984. Identification of promiscuous nodulating soybean efficient in N_2 fixation. (Submitted to Crop Sci.).

Rao, V. R., A. Ayanaba, A. J. R. Eaglesham, and E. A. Kueneman. 1981. Exploiting symbiotic nitrogen fixation for increasing soybean yields in Africa. p. 153-167. *In* Emejuiwe, O. Ogunbi, and S. O. Sanni (eds.) GIAM VI. Global Academic Press, London.

SOYBEAN IN BRAZIL—PRODUCTION AND RESEARCH

Emídio Rizzo Bonato and Amélio Dall'Agnol

SOYBEAN PRODUCTION IN BRAZIL

Development of Soybean Production

The first reference to soybean in Brazil was made in 1882 (D'utra) in the Northeastern state of Bahia. In São Paulo the crop was introduced in 1892 (Daffert, 1892), and in the southernmost state of Rio Grande do Sul in 1901 (Minssen, 1901). Commercial production was started in the 1940s in Rio Grande do Sul. In 1941, soybean was mentioned for the first time in the state statistics (Vernetti, 1977), and soybeans were processed for the first time (Vernetti, date unknown). As a soybean producer, Brazil was mentioned for the first time in the international statistics in 1949 (Miyasaka, 1961).

The crop became economically important in the 1960s. In 1960 total Brazilian production was 202,969 metric tons. In 1969, production exceeded one million metric tons (1,056,584 t), and was almost entirely concentrated in the state of Rio Grande do Sul.

The greatest increase in soybean production occurred in the 1970s (Table 1). From 1970 to 1977 the rate of annual increase was 32%. From 1960 to 1980 soybean was the farm product with greatest increase in area (Figure 1). The area jumped from 169,800 ha in 1960 to 8,774,000 ha in 1980, and production went from 202,969 t to 15,155,800 t. In the same period, yield increased from 1,195 kg/ha to 1,727 kg/ha. According to Bonato (1978) the following factors were responsible for such a phenomenal increase:

Table 1. Evolution of soybean area (1,000 ha) and production (1,000 t) in the traditional and expanding regions of Brazil from 1941 to 1983.

Year	Area		Production	
	Traditional region	Expanding region	Traditional region	Expanding region
1941	0.7	-	0.5	-
1950	24.3	-	33.7	-
1960	169.3	0.5	202.7	0.3
1970	1,303.5	15.3	1,487.9	20.6
1980	7,479.5	1,294.5	12,955.3	2,200.5
1983	6,254.0	1,883.0	10,955.0	3,638.0

Source: IBGE (Instituto Brasileiro de Geografia e Estatística 1941 to 1983).

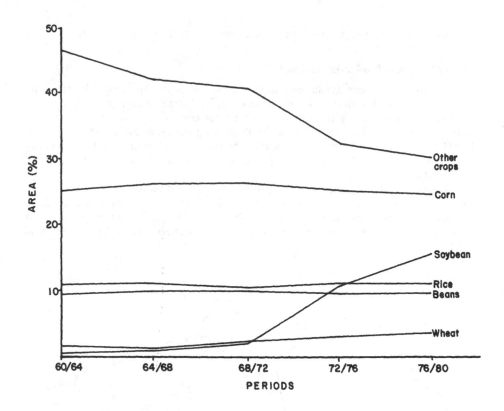

Figure 1. Percent change in area occupied by the main crops and other crops in Brazil, 1960 to 1980.

a. favorable marketing conditions (especially for exporting);
b. double cropping with wheat (*Triticum aestivum* L.), allowing the use of the same land, equipment and labor for both crops in the same year;
c. a government policy aimed at sufficiency in wheat which gave incentives favoring this crop, which also favored the soybean;
d. the possibility of total mechanization of the crop;
e. government incentives for production and marketing;
f. good participation of cooperatives in the production and marketing stages;
g. development of technology adapted to Brazilian conditions, resulting in increased yield and expansion of the crop to areas of lower latitudes and;
h. good cooperation between research and extension, resulting in faster release and efficient flow of information to farmers.

Although there was a tremendous expansion of soybean growing area in the 1960s and 1970s, total soybean area in Brazil decreased by 667,000 ha from 1980 to 1983 (Table 1). A government policy favoring other summer crops, such as corn (*Zea mays* L.) and beans (*Phaseolus vulgaris* L.) and lower market prices for soybeans were factors responsible for such a reduction. Increased production costs were also important in this reduction. Considering the period of 1977-78 to 1980-81, the cost of production in Rio Grande do Sul, the largest producing state, increased by 20.7%, while the average soybean price received by the farmer decreased 24.6% (Bonato, 1983).

In 1983-84, the soybean area increased anew, mainly as a result of high prices for soybeans and products in the international market, particularly during August and September of 1983.

Areas of Soybean Production in Brazil

Traditional Soybean Production Region. This region is represented by the four southern states of Rio Grande do Sul, Santa Catarina, Paraná and São Paulo (Figure 2). Until 1970, this region was responsible for almost the entire soybean production in Brazil. Currently, it has about 75% of the Brazilian production and 76.8% of the area. Rio Grande do Sul and Paraná are, respectively, the first and second producers. Paraná has the highest yield (2.1 t/ha) and Rio Grande do Sul one of the lowest yields (1.55 t/ha) of the entire country.

The initial development of the crop was based on technologies brought from the United States, with cultivars of groups VI, VII and VIII being used. These cultivars were the basis of soybean expansion. A few of them are still being grown (Bragg, Davis and Hardee), along with new ones developed by local researchers. The soybean area in this

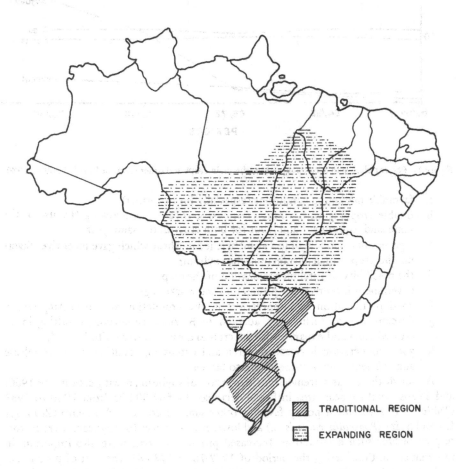

TRADITIONAL REGION

EXPANDING REGION

Figure 2. Areas traditionally grown to soybean in Brazil and areas of present expansion.

region increased up to 1980, and then started to decline. From 1980 to 1983 this region experienced a significant reduction (1,255,000 ha) in soybean area, which was partially compensated for by increases in the expanding region. Storage conditions are good but transportation is mainly by roads, which contributes greatly to increasing the final cost of the soybeans. Soybean farms in the region are of variable size, but most are small (less than 30 ha) to medium (less than 100 ha).

Until 1972, almost 100% of the soybean area was double-cropped with wheat, which was the main crop at that time. However, successive failures of wheat crops and reduced prices turned soybean into the main crop. Currently, wheat area represents around 30% of that planted to soybean.

Expanding Region. This region comprises the central-western part of Brazil. It includes the states of Mato Grosso, Mato Grosso do Sul, Goiás, the south of Maranhão, the west of Bahia and Minas Gerais, and the Federal District. It is traditionally a livestock region, but annual crops are gradually replacing the pastures. The vegetation is mostly "Cerrado" (Savana), where natural grasses predominate. Trees grow short, twisted and sparcely. Soils are usually chemically poor with low base saturation, low pH, low organic matter, medium to low potassium and traces of phosphorus. Nevertheless, they are physically good and topographically favorable for mechanization. Land is still cheap compared to prices in the traditional soybean region. The rain pattern is characterized by a rainy season (October to April) and a dry season (May to September). With irrigation during the dry season, it is possible to grow wheat or even a second soybean crop.

After clearing the land, upland rice has been the most common crop for one or two years, followed by improved pasture or soybean. Presently, soybean is being used as a first crop, after clearing the "cerrado", with excellent results. Mato Grosso do Sul has already become the third soybean producing state, and if market prices remain good, this region is expected to become the most important soybean area in the country. In 1970, the region was responsible for only 1.2% (15,288 ha) of the total soybean area, and in 1983 this percentage increased to 23.2% (1,833,000 ha), larger than the total hectarage in Brazil in 1971. Presently, soybean is the best option for farmers located in the region.

Basically, this region uses technologies developed in Brazil, particularly cultivars suited for the tropical environment. Research carried out in the "cerrado" area has shown that the poor soils can be made highly productive by applying 3 to 4 t/ha of lime and 5 kg/ha of zinc every 4 to 5 years and by the application of 35 to 40 kg/ha of phosphorus (P) and 40 to 50 kg/ha of potassium (K) at planting.

The region is characterized by large farms, commonly more than 500 ha. Usually they are organized into agricultural enterprises. Most of the soybean farmers come from the traditional soybean region, where land is scarce and highly priced. As a result many small farm owners from the South sell their properties and buy large farms in the "cerrado". With the excellent agronomic performance of soybean in the area, these farmers are earning good profits, which are usually invested in more land. A few farmers grow over 3,000 ha of soybean.

Transportation and storage are still problems in the region. Road transportation is responsible for moving almost the entire production, but roads are still deficient. Harbors are hundreds of miles away from this production area, and only recently have some crushing industries been established in the region. Silos for grain storage also are being built by farmers, cooperatives or the government.

Potential Soybean Region. Recently released cultivars adapted to low latitudes made soybean production possible all over Brazil, and millions more of hectares can be turned into soybean production. Areas potentially fitted for soybean can still be found within the traditional region and the region of present expansion, as well as in new regions of north and northeast Brazil. Also, semi-arid land in northeastern Brazil can be turned into soybean production by irrigation, and the wet lowlands of Rio Grande do Sul can produce soybeans if properly drained. Other limitations can be overcome by cultural practices, development of cultivars with particular characteristics, and/or by government investments providing less developed regions with the basic infrastructure for transportation, storage and industry.

Soybean in the Brazilian Economy

The participation of soybean in the total Brazilian agricultural production was insignificant until the end of the 1960s. This participation increased significantly during the 1970s, from 2.5% in 1970 to 14.5% in 1980.

In 1980, the soybean complex was responsible for 11.5% of total Brazilian exports and for 27% of the total exports of raw products, representing an income of $2.24 billion. For 1984, more than $3 billion are expected from soybean exports. Since 1981, the soybean complex has been the most important item in the export of Brazilian raw products.

SOYBEAN RESEARCH IN BRAZIL

Soybean research began in Brazil when Gustavo D'utra (1882) introduced and tested the first plants in the state of Bahia. Isolated research efforts continued until the end of the 1940s when some programs became well established and were important for the early development of the crop. Special reference should be made to research organizations such as IPEAS (Instituto de Pesquisa Agropecuária do Sul) and IPAGRO (Instituto de Pesquisas Agronômicas) in Rio Grande do Sul; IPEAS and IPEAME (Instituto de Pesquisa Agropecuária Meridional) in Paraná; IAC (Instituto Agronômico, Campinas) in São Paulo and UFV (Universidade Federal de Viçosa) and IPEACO (Instituto de Pesquisa Agropecuária do Centro-Oeste) in Minas Gerais. Each program worked independently of the other.

The first effort to integrate the isolated research programs was made in 1964 when the Ministry of Agriculture created the National Soybean Commission, which organized annual national soybean meetings to evaluate cultivars and inoculants. In 1969, the state (IPAGRO) and federal (IPEAS) research organizations in Rio Grande do Sul joined efforts to evaluate cultivars and dates of planting. In Minas Gerais, in 1970, a similar effort was made by the Secretary of Agriculture, Ministry of Agriculture and the University of Viçosa.

With the establishment in 1972 of the National Soybean Project, whose objectives were to assist programs throughout the country, soybean research stepped closer to the organization of a National Soybean Research Program (NSRP), established in 1976 after the creation of the National Soybean Research Center (NSRC) in 1975.

National Soybean Research Program

Research Planning. Soybean research planning is organized into two distinct events: first, a national meeting is held every three years to define the research focus and priorities of the NSRP and second, regional meetings are held annually to evaluate the previous year's research and to plan for the following season at a regional level.

The first national meeting was held in 1977. The main problems were discussed by representatives of all social and economic sectors related to soybeans (private and public research organizations, extension service, farmers, cooperatives, industries, etc.) and research guidelines and priorities were established for the NSRP. In 1980 and 1984 similar meetings were held to reevaluate the NSRP. In order to facilitate research planning the country was divided into three regions. Thus, three regional meetings are held annually, one for each region, and each time at a different location.

The three regions have different requirements in terms of soybean development. Region I is represented by the states of Rio Grande do Sul and Santa Catarina; Region II is represented by central-south Brazil, and Region III comprises north and northeast Brazil. Research projects are proposed according to regional priorities, and their results presented for discussion. Technical information is usually released at the end of each meeting. The NSRC organizes and coordinates the meetings. EMBRAPA (Brazilian Research Organization for Agriculture), through the NSRC, provides financial support for the NSRP.

Research Execution. Following discussion and approval by regional research agronomists, along with NSRC representatives, research projects are conducted by the staff of local research organizations with the supervision and assistance of NSRC scientists. Frequent assistance has been needed for central-west, north and northeast Brazil where researchers are few and poorly trained.

Research Evaluation. Evaluation of the quality of the research carried out by the NSRP is accomplished by follow-up field visits, annual reports prepared for each project by the project coordinator, annual regional meetings, and the National Soybean Research Seminar held every two years.

Transfer of Technology. New technologies are most commonly transferred to soybean growers through the extension service. Eventually, it can be done directly by the research organization responsible for the development of the technology. Research and extension in Brazil are carried out by separate organizations, but cooperation between the two is very close. Most research units have staff that promote the interaction of research and extension personnel.

Scope of the Program. The NSRP involves 29 research organizations throughout the country (Figure 3). Seventeen of the 25 federal units of Brazil are participating in the program, totalling over 180 research projects.

General Objectives of the Program.

   a. Production increase through better use of land and labor in the farms, mainly by the development of rotation and double-cropping systems; expansion of the soybean frontier by the development of cultivars and production systems suited for the new areas; yield increase by a more rational use of fertilizers, pesticides, and soil management practices, and by training farmers to use new soybean technologies more efficiently.
   b. Efficient utilization and conservation of natural resources through proper soil tillage and conservation practices; adequate cultural practices with emphasis on crop rotation and use of biological control, instead of pesticides, to control insects and weeds.
   c. Increase farmer's net income, mainly by reducing production costs.
   d. Increase the use of soybean in human consumption by developing cultivars with high protein content, good taste, and suitability for use in the food industry. Rejection barriers of consumers for soybean food have to be overcome.

Research Capability of the Program. Approximately 300 scientists are full- or part-time involved with soybean research in Brazil. Unfortunately, they are not distributed according to the needs of the research program. Near two-thirds of them are working in the traditional soybean region where satisfactory information is already available.

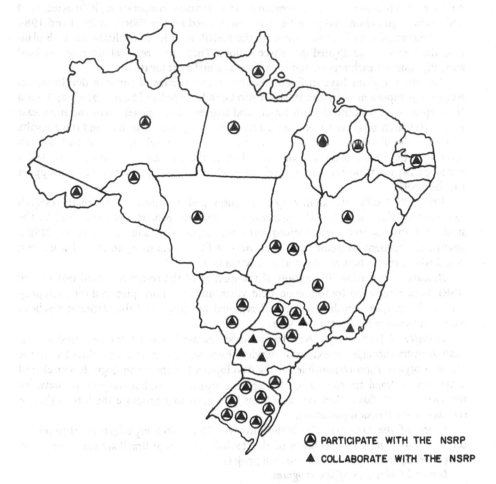

Figure 3. Number and location of soybean research organizations in Brazil that participate or collaborate with the National Soybean Research Program (NSRP).

Reallocation of these scientists is very difficult because many of them belong to state organizations and current government wage policy does not provide incentives for working in less developed regions. Needless to say, however, most of the technologies developed in the traditional soybean region are also fitted, or can be adapted, to other parts of the country.

Another difficulty faced by the NSRP, even in the traditional region, is to adjust scientist training to the research priorities. For example, there are many soybean breeders but none or very few physiologists, soil physicists, seed technologists, etc. As coordinator of the NSRP, the NSRC is attempting to overcome research deficiencies in less developed regions, by encouraging its full-time research staff (16 Ph.D.s and 30 M.Sc.s), to dedicate part of their time in assisting local research agronomists in central-west, north and northeast Brazil.

Results Obtained by the Program. Due to lack of space, it will not be possible to list all the accomplishments of NSRP, but some results that were responsible for economic and social impacts are worth commenting upon.

Many cultivars were developed for the traditional soybean region, but the greatest contribution the breeders gave to Brazilian soybean production was the development of cultivars adapted to the tropical environment. These new cultivars were responsible for the soybean boom in central-west Brazil, where average yield is higher (1,932 kg/ha) than in the traditional region (1,750 kg/ha). The main cultivars for this region are IAC-2, IAC-7, IAC-8, Doko and Cristalina, but recently released cultivars will soon become established and, possibly, replace those presently grown. The development of tropical cultivars made soybean production profitable at latitudes below 15°. Currently, around 1,500 ha of soybean are cultivated in South Maranhão (5-10° latitudes), where yields are close to those of central-west Brazil. Preliminary studies indicate that over 20,000,000 ha are suitable for soybean growing in this region. New cultivars have been released recently (Timbira, Terezina and Carajás), giving more options to farmers near the equator. It is expected that soybean will expand rapidly in this area due to the excellent results obtained so far. Farmers are looking at the soybean as a better option than cattle and rice (*Oryza sativa* L.), the basic local economic activity.

Assisted by foreign consultants in the early stages, Brazilian scientists have implemented an integrated pest management (IPM) approach that is now routinely applied by the farmers. As a result, thousands of farmers who used to apply insecticides 4 to 5 times during the growing season have cut-back to one or two applications. This had a tremendous social and economic impact by reducing pollution and production costs. Conservative estimates indicate that 15% of the Brazilian soybean growers use the IPM technology. Insect management was further advanced by the use of *Baculovirus anticarsia* to control the velvetbean caterpillar (*Anticarsia gemmatalis*). Very simple techniques were developed for farmers to use and produce their own biological insecticide. Each farmer, interested in using the technology, is supplied with enough inoculum to treat one ha. From this treated area the farmer can collect enough inoculum to treat his whole farm.

The cost of herbicides is a significant component of the total cost of soybean production. A simple technique was developed to integrate the use of herbicide with mechanical control. The use of herbicide was reduced by half, limiting its area of application to the soybean row. Planting and herbicide applications are simultaneous operations. Weeds growing between soybean rows are controlled by cheaper mechanical operations.

R. japonicum strains commonly used in the traditional soybean area proved inefficient on soybean crops grown in the "cerrado". A successful research project was set up to overcome the problem and specific *R. japonicum* strains were selected. From these selected strains new inoculants were developed and presently are successfully nodulating soybean plants in the "cerrado".

Research observations indicated that thousands of hectares grown to soybean, mostly in the state of Paraná, were improperly fertilized. Repeated applications of fertilizers high in phosphorus but low in potassium were commonly used by soybean farmers. This practice was based on the information that Brazilian soils are very poor in phosphorus but rich in potassium. That is generally true, but continuous soybean crops that removed, respectively, less phosphorus and more potassium than the amounts applied as fertilizers, required that recommendations be changed from 33 kg/ha of P and 20 kg of K to 15 kg/ha of P and 37 kg/ha of K. As a result, productivity in those areas was increased and the total cost of fertilizer reduced.

The most appropriate time for planting soybean in the traditional region is from 15 October to the end of November. Large soybean farms usually are unable to make all the plantings within this period. Efforts were made to develop cultivars fitted for early or late planting. Paranagoiana and Cristalina are two cultivars that were recently released for early and late planting, respectively. Paranagoiana is a late-maturing selection from the early cultivar Paraná. It was released in Paraná for planting from 15 September through 15 October. If sown in late September its maturation coincides with early cultivars planted up to November. Cristalina is also a late-maturing cultivar and is recommended in Paraná for late plantings (December to January). Its time from sowing to harvest is drastically reduced when planted in December or January. Sowing time can be almost doubled when Paranagoiana and Cristalina are used along with normal cultivars. As a result, large farm operations can greatly increase their profit by increasing production and reducing the cost of machinery and labor.

Field surveys (Oliveira et al., 1980) indicated that around 10% of the production is lost during harvesting. About 60% of these losses can be eliminated by proper adjustment of the combine. Proper seed bed preparation and adequate cultural practices can also reduce losses during harvesting. In order to make soybean farmers aware of the amount of soybeans they were losing, 70,000 plastic cups, especially manufactured for easy assessment of grain losses, were distributed directly to farmers or through the extension service.

Other important research findings could be listed as accomplishments of NSRP. The implementation of a no-till system was initiated in the early 1970s. Besides greatly reducing soil erosion, the system allows soybean planting at a more appropriate time when double-cropped with wheat in the south of Paraná and in the state of Rio Grande do Sul. Currently, over 500,000 ha of soybean is grown under no-till, and its adoption is rapidly increasing.

NOTES

E. Bonato and A. Dall'Agnol, EMBRAPA/CNPS—C.P. 1061, 86100 Londrina, PR, BRAZIL.

REFERENCES

Bonato, E. R. 1978. Realidade da soja brasileira. Ciclo de palestras e debates sobre atualidades agronômicas. Universidade de Passo Fundo. Passo Fundo, RS, Brasil.

Bonato, E. R. 1983. Aumento da produtividade como meio de minimizar custo de produção. p. 110-118. *In* Anais do 3º Seminário sobre Soja. Fundação Centro de Estudos do Comércio Exterior. Sao Paulo, SP, Brasil.

Daffert, F. W. 1892. Relatório anual do instituto agronômico do estado de São Paulo. Campinas, SP, Brasil.

D'utra, G. 1882. A soja. J. Agric. 4(7):185-188.

Minssen, G. 1901. A soja. Revista agrícola do Rio Grande do Sul 5(1):2-4.

Miyasaka, S. 1961. Instruções para a cultura da soja. Bull. 124, Secretaria da Agricultura do Estado de São Paulo. Campinas, SP, Brasil.

Oliveira, F. T. G., A. C. Roessing, C. M. Mesquita, J. B. Silva, E. F. Queiróz, N. P. Costa and J. B. França Neto. 1980. Retorno dos investimentos em pesquisa feitos pela EMBRAPA: redução de perdas na colheita da soja. Empresa Brasileira de Pesquisa Agropecuária Doc. DDT. 3. Brasília, DF, Brasil.

Vernetti, F. de J. and R. E. Kalekmann. Date unknown. Cultura e adubação da soja. p. 3-16. *In* Instituto Agronômico do Sul. Pelotas, RS, Brasil.

Vernetti, F. de J. 1977. História e importância da soja no Brasil. A Lavoura 81:21-24.

CONTINUING COMMITTEE FOR WORLD SOYBEAN
RESEARCH CONFERENCE–IV

Soybean specialists in attendance at World Soybean Research Conference-III approved a constitution for the orderly continuation of similar conferences in the future. The constitution provides for a Continuing Committee composed of persons from four regions. The members of the Continuing Committee that were elected, and the conference at which their duties will end, are provided below. W. Fehr will serve as a member of the committee by virtue of his position as chair of the immediate past conference. The Continuing Committee elected S. Shanmugasundaram to serve as its Chair.

Region	Name and Country	Term
1	J. Sinclair, United States	WSRC-IV
1	K. Smith, United States	WSRC-IV
1	D. Hume, Canada	WSRC-V
1	J. Schillinger, United States	WSRC-V
2	J. Hatkin, Mexico	WSRC-IV
2	F. Moscardi, Brazil	WSRC-IV
2	A. Lattanzi, Argentina	WSRC-V
2	F. Vernetti, Brazil	WSRC-V
3	V. Benjasil, Thailand	WSRC-IV
3	R. Lantican, Philippines	WSRC-IV
3	P. Bhatnagar, India	WSRC-V
3	J. Gai, People's Republic of China	WSRC-V
4	S. Shanmugasundaram, Taiwan	WSRC-IV
4	A. Tanaka, Japan	WSRC-IV
4	R. Lawn, Australia	WSRC-V
4	R. Summerfield, U.K.	WSRC-V
ex off.	W. Fehr, United States	WSRC-IV

CONSTITUTION FOR WORLD SOYBEAN RESEARCH CONFERENCES

1. **ORGANIZATION NAME: WORLD SOYBEAN RESEARCH CONFERENCE**

2. **PURPOSE AND OBJECTIVES:**

The purpose of the World Soybean Research Conference is to promote interchange of scientific information on all aspects of soybean research. The conference shall be held about every five years to consider research progress since the previous conference, the current status of research, and opportunities for progress in the future.

3. **MEMBERSHIP:**

Attendance at the Conference is open to any person interested in soybean research in any country of the world. The membership (registration) fee shall be set by the host institution in consultation with the Continuing Committee. There shall be two types of membership.

A. Full members (with voting rights at Conference Business Meetings).

Payment of the full registration fee gives the individual voting rights at the Business Meetings held during the Conference. The host institution has the right to waive the membership fee of persons invited to address the Conference. Persons whose membership fee is waived will have full voting privileges.

B. Associate members (without voting rights).

Associate members are persons who are charged a fee that is less than the full registration. This includes persons who register on a daily basis, students, and family members.

4. **VOTING PROCEDURES:**

For the purpose of conducting conference business, only full members of the Conference are entitled to cast one vote for each issue in which an election is held. Voting will be by a show of hands, unless a ballot is specified by the Chair of the Continuing Committee. Evidence of voting eligibility may be required by the Continuing Committee. The Continuing Committee shall maintain a permanent record of all elections that are held.

5. **AMENDMENT TO THE CONSTITUTION:**

Proposed amendments to the Constitution may be made by any individual who was a full member at the immediate past conference. They must be sent in writing to the Chair of the Continuing Committee at least six months before the Conference. The Continuing Committee will decide by a simple majority if the amendment should be brought to vote at the next conference. Details of the proposed amendment will be announced by the Chair of the Continuing Committee at the First Business Meeting and the final decision for adoption will be made by voting at the Final Business Meeting of the same Conference. Two-thirds of the total valid votes which are cast must be affirmative for the amendment to be adopted. Any amendment that is passed will come in force at the end of the Conference in which it has been adopted.

6. **THE CONTINUING COMMITTEE:**

A.   The members of the Continuing Committee shall include the President of the immediate past Conference and four soybean specialists from each of the following regions.

Region 1.   Canada and United States of America

Region 2.   Mexico, Central America, South America, and the Caribbean Islands

Region 3.   Continental Asia (including Bangladesh, Burma, Cambodia, India, Nepal, Laos, Malaysia, Pakistan, Thailand, Vietnam, People's Republic of China, North Korea, and South Korea)

Region 4.   Africa and adjacent island countries, Indonesia, Europe, Middle East, Oceania, Philippines, Sri Lanka, Taiwan, Australia, New Zealand, Japan, and USSR

B.   Two persons from each region will be elected at each Conference. The procedure of election shall be as follows:

(i)   The Chair of the Continuing Committee shall appoint a Nominating Committee of four members, one from each region, who shall nominate at least two candidates from each region. The nominations shall be submitted to the Chair of the Continuing Committee at least one month before the First Business Meeting. Nominations also will be solicited from full members of the Conference at the First Business Meeting.

(ii)   The full members of the Conference shall vote for the candidates after the First Business Meeting. Individuals shall vote for only those candidates from their region. The Chair shall announce the new members of the Continuing Committee at the Final Business Meeting.

(iii)   The term of office for each member of the Continuing Committee shall not exceed two terms. A term is the period between the close of the Final Business Meeting of two consecutive conferences. The terms of office shall be arranged so that eight members will be elected at each Conference. To establish this rotation, 16 new members will be elected at Conference-III, of which eight will serve until Conference-IV and eight until Conference-V.

(iv)   If a member of the Continuing Committee cannot serve the full term, the Chair shall solicit names of at least two candidates from the same region. The Continuing Committee shall vote on the candidates and the individual with the majority shall be elected to complete the term.

C.   Organization and Responsibilities of the Continuing Committee.

(i)   The Continuing Committee shall elect from among its members a Chair. For the purpose of electing a new Chair, the retiring Chair will convene the Continuing Committee after the Final Business Meeting and hold the election either by raising hands or written ballot. The newly-elected Chair of the Continuing Committee will assume her/his responsibilities immediately after being elected.

(ii)   The outgoing Chair shall submit to the new Chair a review of the activities of the Continuing Committee during her/his term of office.

(iii)   The Continuing Committee shall (a) ensure that action is taken on all the resolutions adopted at the past Conference, (b) conduct any other necessary affairs that require attention between Conferences, (c) select a substitute if the original host is unable to proceed with a Conference, and (d) serve as an advisory committee to the host for planning the forthcoming Conference.

(iv) The Chair shall receive invitations from potential hosts of the next Conference at least three months before the forthcoming Conference. The Chair shall put the names of the potential hosts to vote by the members of the Continuing Committee before or during the Conference. The potential hosts will provide details of the personnel and facilities that they can offer. If one potential host receives more than half of the votes, they will be elected to hold the next Conference. If no potential host receives over half of the votes, the Chair shall put the names of the two potential hosts receiving the highest number of votes before the Continuing Committee. The potential host with the majority in the second election will be declared the host for the next Conference. The Chair of the Continuing Committee shall declare the result at the Final Business Meeting of the Conference.

(v) The Chair shall maintain a list of members of at least the past two Conferences and of other interested persons to whom the host should send the first circular regarding the forthcoming Conference.

(vi) The Chair shall preside over the business meetings at the Conference and convene the Continuing Committee after the Final Business Meeting for the purpose of electing the new Chair.

7. HOST AND CONFERENCE PRESIDENT:

A. The host shall have the authority to name the President of the Conference. The Conference President should be a soybean specialist.

B. The host and Conference President will be responsible for planning and implementing the Conference. The Continuing Committee will provide guidelines to the host.

C. The collection, disbursement, and accounting of money to support the Conference will be the responsibility of the host. Any surplus funds available at the completion of the Conference will be dispersed at the discretion of the host.

D. At the completion of the Conference, the host and President should provide as much information as possible to the organizers of the next Conference.

E. After completion of the Conference, the President will become a member of the Continuing Committee for one term or shall appoint a representative.

8. CONFERENCE SESSIONS:

At the beginning of the Conference, the First Business Meeting shall be called by the Chair of the Continuing Committee. Details of this meeting will be determined by the Chair in consultation with the Conference President. The Final Business Meeting will be held near the end of the Conference.

9. CONFERENCE PUBLICATIONS:

The host will be responsible for printing abstracts of invited and contributed papers for distribution at the Conference. The proceedings of the Conference will be published and distributed by the host within one year after the Conference is completed.

AUTHOR INDEX